Principles of
Physical Science

Principles of Physical Science

Editor

Donald R. Franceschetti, PhD

The University of Memphis

SALEM PRESS

A Division of EBSCO Information Services, Inc.
Ipswich, Massachusetts

GREY HOUSE PUBLISHING

Publisher's Cataloging-In-Publication Data
(Prepared by The Donohue Group, Inc.)

Names: Franceschetti, Donald R., 1947- editor.
Title: Principles of physical science / editor, Donald R. Franceschetti, PhD, the University of Memphis.
Description: [First edition]. | Ipswich, Massachusetts : Salem Press, a division of EBSCO Information Services, Inc. ; [Amenia, New York] : Grey House Publishing, [2017] | Series: Principles of | Includes bibliographical references and index.
Identifiers: ISBN 978-1-68217-326-8 (hardcover)
Subjects: LCSH: Physical sciences.
Classification: LCC Q158.5 .P75 2017 | DDC 500.2--dc23

PRINTED IN THE UNITED STATES OF AMERICA

CONTENTS

Publisher's Note vii
Editor's Introduction ix
Contributors xiv

Acoustics 1
Aeronautics and aviation 10
Air-quality monitoring 17
Algebra 23
Antiballistic missile defense systems 31
Applied mathematics 38
Applied Physics 45
Artificial intelligence 52
Atmospheric sciences 58
Audiology and hearing aids 66
Automated processes and servomechanisms 72
Avionics and aircraft instrumentation 78

Barometry 85
Bioengineering 91
Biofuels and synthetic fuels 99
Biomathematics 105
Biomechanical Engineering 110
Biomechanics 115
Bionics and biomedical engineering 121
Biophysics 127
Bose condensates 133
Bridge design and barodynamics 135

Calculus 140
Ceramics 145
Chaotic systems 150
Charge-coupled devices 155
Civil engineering 159
Climate engineering 166
Climate modeling 172
Climatology 178
Communication 184
Communications satellite technology 191
Computer engineering 197
Computer graphics 203
Computer languages, compilers, and tools 208
Computer science 214
Cracking 221
Cryogenics 227
Cryptology and cryptography 233

Detergents 240
Digital logic 246
Diode technology 252
Distillation 257

Earthquake engineering 264
Electrical engineering 271
Electrical measurement 278
Electric automobile technology 284
Electrochemistry 292
Electromagnet technologies 300
Electrometallurgy 306
Electronic materials production 311
Electronics and electronic engineering 316
Electron spectroscopy analysis 323
Engineering 328
Engineering mathematics 334
Engineering seismology 339
Environmental chemistry 345
Environmental Engineering 351

Fiber-optic communications 358
Fiber technologies 363
Fluid dynamics 368
Fuel cell technologies 373

Geometry 379
Glass and glassmaking 385
Graphene 390
Gravitational radiation 392

Heat-exchanger technologies 394
Hybrid vehicle technologies 399
Hydraulic engineering 404

Information technology 410
Integrated-circuit design 417

Laser interferometry 423
Laser technologies 428
Light-emitting diodes 434
Liquid crystal displays 439
Liquid crystal technology 441
Lithography 446

Magnetic resonance Imaging 452
Magnetic storage 457
Mechanical engineering 463
Metallurgy. 472
Meteorology. 480
Mineralogy 486
Mirrors and lenses 493

Nanotechnology. 498
Nuclear technology 502
Numerical analysis 508

Ocean and tidal energy technologies. 513
Optics 517

Parallel computing. 523
Pattern recognition (science) 528
Petroleum extraction and processing. 532
Photography science 537
Photonics 542
Planetology and astrogeology 546
Polymer science 551
Propulsion technologies 556

Radio. 562
Radio astronomy 567

Scanning probe microscopy 573
Space environments for humans 577
Space science 581
Spectroscopy. 587
Surface and interface science 592

Telephone technology and networks 597
Telescopy 602
Temperature measurement. 609
Terrestrial magnetism 614
Time measurement 617
Transistor technologies 622
Transuranic elements 626

Weight and mass measurement. 629

Timeline 635
Biographical Dictionary of Scientists 701
General Bibliography. 717
Index. 745

Publisher's Note

Salem Press is pleased to add *Principles of Physical Science* as the fifth title in the *Principles of* series that includes Chemistry, Physics, Astronomy, Computer Science, and Physical Science. This new resource introduces students and researchers to the fundamentals of physical science using easy-to-understand language, giving readers a solid start and deeper understanding and appreciation of this complex subject.

The 112 entries range from Audiology to Weight and Mass Measurement and are arranged in an A to Z order, making it easy to find the topic of interest. Entries include the following:

- Related fields of study to illustrate the connections between the various branches of physical science including algebra, biomechanical engineering, computer engineering, cryogenics, laser technologies, and nanotechnology;
- A brief, concrete summary of the topic and how the entry is organized;
- Key terms and concepts that are fundamental to the discussion and to understanding the concepts presented;
- Text that gives a definition and basic principles; describes the topic's impact on industry, including government and academic research; an explanation of how it works; career options and course work information;
- Illustrations that clarify difficult concepts via models, diagrams, and charts of such key topics as automated processes, photography, ocean and tidal energy technologies, geometry, and earthquake engineering;
- Photographs of significant contributors to the field of physical science;
- Fascinating facts about the topic or field;
- Further reading lists that relate to the entry.

This reference work begins with a comprehensive introduction to the field, written by editor Donald R. Franceschetti, PhD, starting with an explanation of the distinction between the physical and the life sciences, an overview of the three significant periods of the history of science, and concluding with a look into the future significance of physical science as we grapple with problems related to finding safe sources of energy and space travel.

The book's backmatter is another valuable resource and includes:

- Timeline, listing important historical events in science in chronological order;
- Biographical Dictionary of Scientists;
- General Bibliography; and
- Index.

Salem Press and Grey House Publishing extend their appreciation to all involved in the development and production of this work. The entries have been written by experts in the field. Their names and affiliations follow the Editor's Introduction.

Principles of Physical Science, as well as all Salem Press reference books, is available in print and as an e-book. Please visit www.salempress.com for more information.

Editor's Introduction

Physical science is the complement to the life sciences. Taken together the physical and life sciences constitute the whole of empirical (experimental) science. The origin of the term physical science dates back to the time when living (or once-living) matter was assumed to possess special properties that could only be found in living matter—that is to say, organic matter. This assumption was made according to the theory of vitalism—a now discredited theory that posited living organisms to be fundamentally different from non-living entities on the basis that living (animate) organisms contain some non-physical element (sometimes referred to as élan or soul) or are governed by different principles than are inanimate things. Once Friedrich Wöhler (1800–1882) demonstrated that urea, an organic compound, could be formed by the isomerization of the inorganic compound ammonium cyanate, there appeared to be no clear boundary that could be drawn between organic and inorganic matter. Widespread belief in vitalism as a scientific theory started to wane; however, the distinction between life sciences and physical sciences continues to be useful.

Living processes are now regarded as something that has evolved over many millennia as the living organisms these processes support have become increasingly complex. On the other hand, the physical sciences tend to study phenomena that are less dependent on evolution and the passage of time. Physical science includes astronomy, physics, chemistry, and planetary science—which includes the science of climate and the structure of astrophysical bodies (planets and stars). Were one to write a curriculum for any of these areas a century ago, there would have been clear distinctions between them, keeping them essentially independent of one another. Today, the separation between the distinct areas of physical science is not so clear-cut. Basic nuclear physics along with the electron theory of matter unites physics and chemistry. The behavior of materials is largely a reflection of the motions of electrons. The abundance of isotopes is now understood as a consequence of stellar evolution. The myriad biochemicals produced by living beings can all be made in the laboratory.

A BRIEF HISTORY OF THE PHYSICAL SCIENCES

We can divide the history of science into three broad epochs: First came a period of practical discoveries but competing doctrines; next came the triumph of empiricism (experiment); and finally, the modern period. The period of practical discoveries coupled with competing doctrines extends from the earliest days of man's scientific discoveries up through the time of Newton and Galileo. These practical discoveries included, for example, the use of straw to strengthen brick as described in the Bible, without there being a theory of material properties to guide the process of discovery. The earth, it seemed obvious to these early scientists, was the center of the universe and, equally obvious, once could distinguish between natural motions that restored objects to their rightful places and violent motions like the launching of a projectile by a catapult. Views of the chemical elements in this first period of science were various. Among the alchemists, it was generally assumed that elements, at least the metals, could transform into each other and that mineral ores within the earth gradually ripened over time to become more perfect, as if the planet were a gigantic womb. Astronomical arguments in this era assumed that planets moved in perfect circles around the earth, because celestial objects were deemed to be perfect. This sort of circular reasoning is typical of the early scientific literature.

The scientific revolution began with the work of Newton (1642–1727) and Galileo (1564–1642) in Physics and the enshrinement of Inductive Logic as found in the writings of Sir Francis Bacon (1561–1626). Newton's great book, *Philosophiæ Naturalis Principia Mathematica* (first published in 1687 and often referred to simply at *Principia*; The Mathematical Principles of Natural Philosophy), took Galileo's Principle of Inertia as its foundation. He then added two more laws of motion and a law of universal gravitation. He further established the hypothetico-deductive method, in which one begins with a reasonable hypothesis and then determines its truth value based on the predictions that follow. Newton's followers produced a mechanics that served admirably to describe the motion of celestial and terrestrial bodies.

Chemistry developed somewhat more slowly than physics partly because of religious dogma and partly because the mechanism of chemical reactions could not be seen, even under the most powerful of microscopes. In the twelfth century, Saint Thomas Aquinas (1225-1274) adopted the Aristotelian distinction between matter and form in order to explain the Eucharist and the doctrine of transubstantiation. This explanation become Catholic dogma and was used with a heavy hand to discourage scientists working in Catholic countries (in other words, Europe) from theorizing about the nature of chemical transformation.

Modern students of chemistry are often surprised to learn that there was any debate about the theory that chemical bonds were responsible for holding together atoms and combinations, but there continued to be doubts until about 1905, when Albert Einstein (1871-1955) showed that the Brownian motion of particles suspended in water could be explained as the result of collisions between the particles and the water molecules. While chemists could point to many compounds as examples of Dalton's Law of Definite proportions, most solids exhibited a finite range of stoichiometry.

The very notion of a chemical *element* was far from certain throughout this era. The four basic elements of the ancient Greeks—air, earth, fire and water—gave way to the three basic elements (the *tria prima*) of the alchemists—sulfur, salt and mercury. Robert Boyle (1527-1691), a contemporary of Newton, wrote *The Skeptical Chymist* (1661) to show the defects of these philosophical systems. The first modern list of elements came from Antoine-Laurent de Lavoisier (1743-1794), a French aristocrat who was ultimately beheaded during the French Revolution. By 1865, enough was known of the elements that Dmitri Mendeleev (1834-1907) could organize a periodic table in which the elements appear in order of increasing atomic mass (with minor exceptions), broken up into periods so that elements with similar chemical properties were located above each other.

While, the emergence of chemical theory had to wait until the late eighteenth century, a practical understanding of the properties of materials was not so delayed. For example, the Romans developed an insight into stress and strain in composite materials that allowed them to span great distances with aqueducts. This understanding laid the foundations for the gothic style of architecture, in which flying buttresses made it possible for stained glass windows to be placed in walls that would otherwise be too heavy.

At the beginning of the twentieth century, there was a belief that science had advanced as far as it was going to go. There was talk of closing the patent office because everything useful had already been discovered. Physicists tried to discourage their offspring from following in their footsteps. They could not have been more incorrect. In fact, the modern period begins in earnest in the first decades of the twentieth century with the discovery of sub-atomic particles; relativity and the quantum theory; and an understanding of a new property—intrinsic angular momentum or spin—that a great many particles possess. Physicist George Gamow (1904–1968) christened this period "the thirty years that shook physics."

The modern understanding of science follows these lines: electromagnetic radiation and other physically meaningful quantities are often quantized. An electromagnetic wave can only interact with matter by exchanging energy in units of hf, where f is the frequency of the wave and h is Planck's constant $(6.626070040(81) \times 10^{-34} \times \text{J} \cdot \text{s})$. Angular momentum is also quantized, in units of $h/2\varpi$. All elementary particles have a spin equal to an integer or half odd integer multiple of $h/2\varpi$. The spin can only change by h/ϖ, which means that the spin particles may be separated into two classes: *fermions*, with half-integer spin, and *bosons* with integer spin. The quantum mechanical rules require that the wave function for any system of identical fermions change sign when any two identical fermions are interchanged. This requirement is the basis of Wolfgang Pauli's (1900–1958) Exclusion Principle: Two or more identical fermions (particles with half-integer spin) cannot occupy the same quantum state within a quantum system simultaneously.

Further there is no necessity to believe that two observers in relative motion will obtain the same results when they measure time intervals or lengths. Instead it seems that nature is in a grand conspiracy such that the velocity of light, no matter what the source or how it is moving, will always be measured to be a constant 3.0×10^8 m/s.

The interaction of elementary particles may be long-range or short-range. The weakest interaction is gravitational. The idea that the gravitational field was capable of sustaining waves was predicted by Albert Einstein (1879–1955) about a hundred years ago, but was only detected definitively in 2015 by laser measurements over a 2000-km baseline. Next in strength is the nuclear weak force, which has been termed the "cosmic alchemist," because it is involved in the transformation of protons into neutrons. Without that transformation, the bottom two thirds of the periodic table would not exist.

The weak force involves that mysterious particle first predicted by Wolfgang Pauli and named *neutrino* (from the Italian for the little neutral one) in 1934 by Enrico Fermi (1901–1954), but not actually discovered for 30 years. Neutrinos interact very weakly with normal matter. The average solar neutrino passes through the earth with no interaction, and the chances are that a human body will interact with a neutrino is about 1 in 10^{25}. Nonetheless neutrino telescopes have been built that can detect a few dozen argon atoms produced by neutrinos in a tank of carbon tetrachloride over a week's time, and which show a difference in neutrino flux when a star in a nearby galaxy explodes as a supernova.

The electromagnetic force is one between moving charges. It involves an electrostatic component dependent on the size of each charge and a magnetic component dependent on the magnitude of their relative velocity. While a few features of the electromagnetic force were understood as early as 1800, the invention of the voltaic cell accelerated experimentation so that by 1865, James C. Maxwell (1831–1879) could write down the basic equations of the electric and magnetic field and demonstrate that in empty space electromagnetic waves could travel indefinitely at 3.0×10^8 m/s and that the index of refraction of a transparent medium could be determined from its dielectric properties. A complete theory of the electromagnetic wave would follow.

Albert Einstein and several others were fascinated by the possibility of unifying electromagnetic theory and gravitation, but this never worked out in detail. Instead Stephen Weinberg (1933–) produced a unification of the weak force and electromagnetism, and a number of theories that try to treat the weak and strong forces on a common footing.

The electromagnetic force is about 100,000 times stronger than the weak interaction, although only the weak interaction can convert protons into neutrons and positrons or into protons plus electrons. Electromagnetic force features prominently in both chemical and electrical phenomena. Thus the energy released when a substance burns is basically electrical

in nature, as is the energy radiated from an antenna.

Electricity and magnetism are the basis of current technology, though they were not understood until the nineteenth century. In fact, the earth itself is an electromagnetic system. One of the earliest practical uses of the earth's magnetism is the magnetic compass, which has been used as an aid to navigation since about 200–300 BCE. At first, it was believed that the magnetic needle was attracted by the North Star (Polaris); this does appear to make sense, as the North Star is roughly on the axis of the celestial sphere. William Gilbert (1544–1603), physician to Queen Elizabeth I of England, built a large sphere (*terrella*) and uniformly magnetized it. Moving his compass along the surface, he found that his sphere reproduced aspects of the earth's magnetic field, including magnetic dip.

The earth's magnetic field is found to change with time. There is a slow change from year to year and then sudden reversals. Though made of ferromagnetic materials, the interior of the Earth is above the Curie temperature of any likely material. The currently held belief is that the Earth's magnetic field is generated by electrical currents associated with the flow of conducting materials. While the pattern of current flow is generally slow to change, changes can occur rapidly when the Earth collides with another astronomical body. These collisions are also associated with the extinction of biological species. Iron ores tend to retain the magnetization that they have as they pass through the Curie temperature from above, so that they retain a record of the magnetic field changes.

The Earth's electrical field is also significant, enough so that metal airplanes have to be grounded when they land. The average electrical field of the Earth is about 100 V/m, reflecting an excess surface charge corresponding to a couple of extra electrons per square meter. The compensating positive charge is to be found in the ionosphere about 50 km above the ground. A quick calculation shows that the charge separation would disappear within a half-hour or so. Therefore, there must be some mechanism to restore it. Current thinking is that it is thunderstorms that are responsible for the surface charge of the earth.

The Earth's electric and magnetic field are of great interest to the geosciences, especially at the current time, when most scientists believe that the globe is becoming warmer due to continuing generation of greenhouse gases. Today it is known that the Earth's surface temperature represents a balance of several factors. One factor is the decay of radioisotopes in our planet's interior. Another is the greenhouse effect: If the earth's atmosphere had no carbon dioxide, most of the Earth's surface would be below the freezing point of water. We are currently facing the opposite problem, with too much carbon dioxide as a result of the burning of fossil fuels. This means that we must cope with rising temperatures, making the search for alternative sources of power increasingly important.

The biggest controversy pertains to role of nuclear reactors concerning both their inherent safety and the problem of radioactive waste. Nuclear power plants could be either nuclear fission reactors, which harness the energy stored in oversized (enriched) uranium nuclei when they are split or decay into smaller particles (now used for power generation) and nuclear fusion reactors that would be powered by the fusion of small nuclei to make larger ones. Fission reactors generate waste products that are themselves radioactive and remain so for a very long time. Fusion reactors would have nonradioactive waste products that could be returned to the environment safely. But fusion requires very high temperatures and so we are confronted, like the sorcerer's apprentice with the universal solvent. How would we store it?

Monitoring celestial bodies has led to the acquisition of much new knowledge, starting with the first Earth satellites launched by the USSR in 1957 and the US in 1958. The van Allen belts were discovered in 1957–58. Space-based telescopes are not restricted to wavelength regions to which the Earth's atmosphere is transparent, so we are gradually able to study objects that adsorb anywhere from the x-ray to microwave regions. Neutrino telescopes and the search for proton decay continue to grow our understanding and knowledge of the basic materials and processes of life itself.

THE FUTURE

Barring unforeseen catastrophes, we can expect many new developments in the years to come. While lecturing to Caltech undergraduates in 1963 following the discovery that the weak nuclear force, and the understanding that unlike the strong force,

electromagnetic and gravitational forces did not conserve parity, the eminent physicist, Richard P. Feynman (1918–188) made the following statement: "Outside the nucleus, we seem to know all." Discoveries that were yet to come included the rest mass of the neutrino, the fact that there are not one but three kinds of neutrino (and three kinds of antineutrino, also) and the fact that solar neutrinos transform into each other during their seven-minute journey from the sun, meaning that the number of neutrinos that are detected is only one-third of the theoretically expected number. Surely other discoveries await.

—Donald R. Franceschetti, PhD

FOR FURTHER READING

Atkins, P W. *The Periodic Kingdom: A Journey into the Land of the Chemical Elements.* New York: Basic-Books, 1995. Print.

Bacon, Francis, Elizabeth S. Haldane, G R. T. Ross, David E. Smith, Marcia L. Latham, W H. White, Amelia H. Stirling, Francis Bacon, Francis Bacon, René Descartes, René Descartes, René Descartes, René Descartes, and Benedictus . Spinoza. *Advancement of Learning: Novum Organum ; New Atlantis.* Chicago: Encyclopaedia Britannica, 1990. Print.

Eiseley, Loren C. *The Man Who Saw Through Time.* New York: C. Scribner's Sons, 1983. Print.

Emsley, John. *Nature's Building Blocks: An A-Z Guide to the Elements.* Oxford: Oxford University Press, 2001. Print.

Feynman, Richard P, Robert B. Leighton, and Matthew L. Sands. The Feynman Lectures on Physics. Reading, Mass: Addison-Wesley Pub. Co, 1963. Print.

Gamow, George. *Thirty Years That Shook Physics: The Story of Quantum Theory.* New York: Dover Publications, 1985. Print.

Hewitt, Paul G. *Conceptual Physics.* Boston: Pearson, 2015. Print.

Pais, Abraham. *Inward Bound: Of Matter and Forces in the Physical World.* Oxford Oxfordshire: Clarendon Press, 1986. Print.

Contributors

Jeongmin Ahn, PhD
Syracuse University

Ezinne Amaonwu, MAPW
Rockville, MD

Michael P. Auerbach, MA
Marblehead, MA

Craig Belanger, MST
Journal of Advancing Technology

Raymond D. Benge, Jr., MS
Tarrant County College

Harlan H. Bengston,
Southern Illinois University, Edwardsville

Joseph I. Brownstein, MS
Atlanta, GA

Byron D. Cannon, PhD
University of Utah

Richard P. Capriccioso, MD
University of Phoenix

Christina Capriccioso,
University of Michigan College of Engineering

Christine M. Carroll, RN, MBA
American Medical Writers Association

Michael J. Caulfield, PhD
Gannon University

Martin Chetlen, MCS
Moorpark College

Edward N. Clark, PhD
Worcester Polytechnic Institute

Robert L. Cullers, PhD
Kansas State University

Joseph Di Rienzi, PhD
College of Notre Dame of Maryland

Jack Ewing
Boise, ID

June Gastón, EdD
City University of New York

Gina Hagler, MBA
Washington, DC

Robert M. Hordon, PhD
Rutgers University

Carol L. Huth, JD
Publications Services Inc.

April D. Ingram,
Kelowna, British Columbia

Micah L. Issitt, MA
Philadelphia, PA

Vincent Jorgensen,
Sunnyvale, CA

Bassam Kassab, MSc
Santa Clara Valley Water District

Marylane Wade Koch, MSN, RN
University of Memphis

Narayanan M. Komerath, PhD
Georgia Institute of Technology

Jeanne L. Kuhler, PhD
Bendictine University

Lisa LaGoo, MS
Medtronic

Dawn A. Laney, CCRC
Atlanta, GA

M. Lee, MA
Independent Scholar

Donald W. Lovejoy, PhD
Palm Beach Atlantic University

R.C. Lutz, PhD
Cll Group

Marianne M. Madsen, MS
University of Utah

Mary E. Markland, MA
Argosy University

Sergei A. Markov, PhD
Austin Peay State University

Juila M. Meyers, PhD
Duquesne University

Randall L. Milstein, PhD
Oregon State University

Terrence R. Nathan, PhD
University of California, Davis

David Olle, MS
Eastshire Communications

Robert J. Paradowski, PhD
Rochester Institute of Technology

Ellen E. Anderson Penno, MS, MD, FRCSC, Dip. ABO
The Mayo Clinic

John R. Phillips, PhD
Purdue University Calumet

George M. Plitnik, PhD
Frostburg State University

Corie Ralston, PhD
Lawrence Berkeley National Laboratory

Richard M.J. Renneboog, MSc
Independent Scholar

Charles W. Rogers, PhD
Southwestern Oklahoma State University

Lars Rose, PhD
Author: *The Nature of Matter*

Julia A. Rosenthal, MS
Chicago, IL

Joseph R. Rudolph, Jr., PhD
Towson University

Sibani Sengupta, PhD
American Medical Writers Association

Martha A. Sherwood, PhD
Kent Anderson Law Office

Paul P. Sipiera, PhD
Harper College

Polly D. Steenhagen, MS
Delaware State University

Judith L. Steininger, MA
Milwaukee School of Engineering

Martin V. Stewart, PhD
Middle Tennessee State University

Robert E. Stoffels, MBA
St. Petersburg, FL

Bethany Thivierge, MPH
Technicality Resources

Anh Tran, PhD
Indiana University - Bloomington

Christine Watts, PhD
University of Sydney

George M. Whiston III, PhD
University of Texas at Tyler

Edwin G. Wiggins, PhD
Webb Institute

Barbara Woldin,
American Medical Writers Association

Robin L Wulffson, MD, FACOG
American College of Obstetrics and Gynecology

Principles of
Physical Science

A

ACOUSTICS

FIELDS OF STUDY

Electrical, chemical, and mechanical engineering; architecture; music; speech; psychology; physiology; medicine; atmospheric physics; geology; oceanography.

SUMMARY

Acoustics is the science dealing with the production, transmission, and effects of vibration in material media. If the medium is air and the vibration frequency is between 18 and 18,000 hertz (Hz), the vibration is termed "sound." Sound is used in a broader context to describe sounds in solids and underwater and structure-borne sounds. Because mechanical vibrations, whether natural or human induced, have accompanied humans through the long course of human evolution, acoustics is the most interdisciplinary science. For humans, hearing is a very important sense, and the ability to vocalize greatly facilitates communication and social interaction. Sound can have profound psychological effects; music may soothe or relax a troubled mind, and noise can induce anxiety and hypertension.

KEY TERMS AND CONCEPTS

- **cochlea:** Inner ear, which converts pressure waves of sound into electric impulses that are transmitted to the brain via the auditory nerves.
- **decibel (db):** Unit of sound intensity used to quantify the loudness of a vibration.
- **destructive interference:** Interference that occurs when two waves having the same amplitude in opposite directions come together and cancel each other.
- **Doppler effect:** Apparent change in frequency of a wave because of the relative motions of the source and an observer. Wavelengths of approaching objects are shortened, and those of receding objects are lengthened.
- **hertz (Hz):** Unit of frequency; the number of vibrations per second of an oscillation.
- **infrasound:** Air vibration below 20 hertz; perceived as vibration.
- **physical acoustics:** Theoretical area concerned with the fundamental physics of wave propagation and the use of acoustics to probe the physical properties of matter.
- **pesonance:** Large amplitude of vibration that occurs when an oscillator is driven at its natural frequency.
- **sound:** Vibrations in air having frequencies between 20 and 20,000 hertz and intensities between 0 and 135 decibels and therefore perceptible to humans.
- **sound spectrum:** Representation of a sound in terms of the amount of vibration at a each individual frequency. Usually presented as a graph of amplitude (plotted vertically) versus frequency (plotted horizontally).
- **spectrogram:** Graph used in speech research that plots frequency (vertical axis) versus the time of the utterance (horizontal axis). The amplitude of each frequency component is represented by its darkness.
- **transducer:** Device that transmutes one form of energy into another. Acoustic examples include microphones and loudspeakers.
- **ultrasound:** Frequencies above 20,000 hertz used by bats for navigation and by humans for industrial applications and nonradiative ultrasonic imaging.

DEFINITION AND BASIC PRINCIPLES

The words "acoustics," and "phonics" evolved from ancient Greek roots for hearing and speaking, respectively. Thus, acoustics began with human communication, making it one of the oldest if not the most basic of sciences. Because acoustics is ubiquitous in human endeavors, it is the broadest and most interdisciplinary of sciences; its most profound contributions have occurred when it is commingled with an independent field. The interdisciplinary nature of acoustics has often consigned it to a subsidiary role as an minor subdivision of mechanics, hydrodynamics, or electrical engineering. Certainly, the various technical aspects of acoustics could be parceled out to larger and better established divisions of science, but then acoustics would lose its unique strengths and its source of dynamic creativity. The main difference between acoustics and more self-sufficient branches of science is that acoustics depends on physical laws developed in and borrowed from other fields. Therefore, the primary task of acoustics is to take these divergent principles and integrate them into a coherent whole in order to understand, measure, and control vibration phenomena.

The Acoustical Society of America subdivides acoustics into fifteen main areas, the most important of which are ultrasonics, which examines high-frequency waves not audible to humans; psychological acoustics, which studies how sound is perceived in the brain; physiological acoustics, which looks at human and animal hearing mechanisms; speech acoustics, which focuses on the human vocal apparatus and oral communication; musical acoustics, which involves the physics of musical instruments; underwater sound, which examines the production and propagation of sound in liquids; and noise, which concentrates on the control and suppression of unwanted sound. Two other important areas of applied acoustics are architectural acoustics (the acoustical design of concert halls and sound reinforcement systems) and audio engineering (recording and reproducing sound).

BACKGROUND AND HISTORY

Acoustics arguably originated with human communication and music. The caves in which the prehistoric Cro-Magnons displayed their most elaborate paintings have resonances easily excited by the human voice, and stalactites emit musical tones when struck or rubbed with a stick. Paleolithic societies constructed flutes of bird bone, used animal horns to produce drones, and employed rattles and scrapers to provide rhythm.

In the sixth century BCE, Pythagoras was the first to correlate musical sounds and mathematics by relating consonant musical intervals to simple ratios of integers. In the fourth century BCE, Aristotle

deduced that the medium that carries a sound must be compressed by the sounding body, and the third century BCE philosopher Chrysippus correctly depicted the propagation of sound waves with an expanding spherical pattern. In the first century BCE, the Roman architect and engineer Marcus Vitruvius Pollio explained the acoustical characteristics of Greek theaters, but when the Roman civilization declined in the fourth century, scientific inquiry in the West ceased for the next millennium.

In the seventeenth century, modern experimental acoustics originated when the Italian mathematician Galileo explained resonance as well as musical consonance and dissonance, and theoretical acoustics got its start with Sir Isaac Newton's derivation of an expression for the velocity of sound. Although this yielded a value considerably lower than the experimental result, a more rigorous derivation by Pierre-Simon Laplace in 1816 obtained an equation yielding values in complete agreement with experimental results.

During the eighteenth century, many famous mathematicians studied vibration. In 1700, French mathematician Joseph Sauveur observed that strings vibrate in sections consisting of stationary nodes located between aggressively vibrating antinodes and that these vibrations have integer multiple frequencies, or harmonics, of the lowest frequency. He also noted that a vibrating string could simultaneously produce the sounds of several harmonics. In 1755, Daniel Bernoulli proved that this resultant vibration was the independent algebraic sum of the various harmonics. In 1750, Jean le Rond d'Alembert used calculus to obtain the wave equation for a vibrating string. By the end of the eighteenth century, the basic experimental results and theoretical underpinnings of acoustics were extant and in reasonable agreement, but it was not until the following century that theory and a concomitant advance of technology led to the evolution of the major divisions of acoustics.

Although mathematical theory is central to all acoustics, the two major divisions, physical and applied acoustics, evolved from the central theoretical core. In the late nineteenth century, Hermann von Helmholtz and Lord Rayleigh (John William Strutt), two polymaths, developed the theoretical aspects. Helmholtz's contributions to acoustics were primarily in explaining the physiological aspects of the ear. Rayleigh, a well-educated wealthy English baron, synthesized virtually all previous knowledge of acoustics and also formulated an appreciable corpus of experiment and theory.

Experiments by Georg Simon Ohm indicated that all musical tones arise from simple harmonic vibrations of definite frequency, with the constituent components determining the sound quality. This gave birth to the field of musical acoustics. Helmholtz's studies of instruments and Rayleigh's work contributed to the nascent area of musical acoustics. Helmholtz's knowledge of ear physiology shaped the field that was to become physiological acoustics.

Underwater acoustics commenced with theories developed by the nineteenth-century mathematician Siméon-Denis Poisson, but further development had to await the invention of underwater transducers in the next century.

Two important nineteenth-century inventions, the telephone (patented 1876) and the mechanical phonograph (invented 1877), commingled and evolved into twentieth-century audio acoustics when united with electronics. Some products in which sound production and reception are combined are microphones, loudspeakers, radios, talking motion pictures, high-fidelity stereo systems, and public sound-reinforcement systems. Improved instrumentation for the study of speech and hearing has stimulated the areas of physiological and psychology acoustics, and ultrasonic devices are routinely used for medical diagnosis and therapy, as well as for burglar alarms and rodent repellants. Underwater transducers are employed to detect and measure moving objects in the water, while audio engineering technology has transformed music performance as well as sound reproduction. Virtually no area of human activity has remained unaffected by continually evolving technology based on acoustics.

How it Works

Ultrasonics. Dog whistles, which can be heard by dogs but not by humans, can generate ultrasonic frequencies of about 25 kilohertz (kHz). Two types of transducers, magnetostrictive and piezoelectric, are used to generate higher frequencies and greater power. Magnetostrictive devices convert magnetic energy into ultrasound by subjecting ferric material (iron or nickel) to a strong oscillating magnetic field. The field causes the material to alternately expand and contract, thus creating sound waves of the same frequency as that of the field. The resulting sound waves

have frequencies between 20 Hz and 50 kHz and several thousand watts of power. Such transducers operate at the mechanical resonance frequency where the energy transfer is most efficient.

Piezoelectric transducers convert electric energy into ultrasound by applying an oscillating electric field to a piezoelectric crystal (such as quartz). These transducers, which work in liquids or air, can generate frequencies in the megahertz region with considerable power. In addition to natural crystals, ceramic piezoelectric materials, which can be fabricated into any desired shape, have been developed.

Physiological and psychological acoustics. Physiological acoustics studies auditory responses of the ear and its associated neural pathways, and psychological acoustics is the subjective perception of sounds through human auditory physiology. Mechanical, electrical, optical, radiological, or biochemical techniques are used to study neural responses to various aural stimuli. Because these techniques are typically invasive, experiments are performed on animals with auditory systems that are similar to the human system. In contrast, psychological acoustic studies are noninvasive and typically use human subjects.

A primary objective of psychological acoustics is

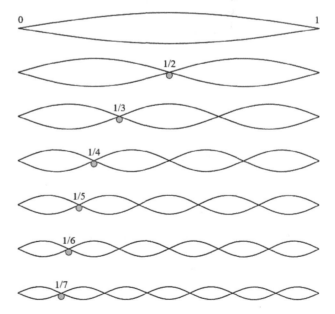

The fundamental and the first 6 overtones of a vibrating string. The earliest records of the study of this phenomenon are attributed to the philosopher Pythagoras in the 6th century BCE to define the psychological correlates to the physical parameters of sound waves. Sound waves in air may be characterized by three physical parameters: frequency, intensity, and their spectrum. When a sound wave impinges on the ear, the pressure variations in the air are transformed by the middle ear to mechanical vibrations in the inner ear. The cochlea then decomposes the sound into its constituent frequencies and transforms these into neural action potentials, which travel to the brain where the sound is evidenced. Frequency is perceived as pitch, the intensity level as loudness, and the spectrum determines the timbre, or tone quality, of a note.

Another psychoacoustic effect is masking. When a person listens to a noisy version of recorded music, the noise virtually disappears if the music is being enjoyed. This ability of the brain to selectively listen has had important applications in digitally recorded music. When the sounds are digitally compressed, such as in MP3 (MPEG-1 audio layer 3) systems, the brain compensates for the loss of information; thus one experiences higher fidelity sound than the stored content would imply. Also, the brain creates information when the incoming signal is masked or nonexistent, producing a psychoacoustic phantom effect. This phantom effect is particularly prevalent when heightened perceptions are imperative, as when danger is lurking.

Psychoacoustic studies have determined that the frequency range of hearing is from 20 to about 20,000 Hz for young people, and the upper limit progressively decreases with age. The rate at which hearing acuity declines depends on several factors, not the least of which is lifetime exposure to loud sounds, which progressively deteriorate the hair cells of the cochlea. Moderate hearing loss can be compensated for by a hearing aid; severe loss requires a cochlear implant.

Speech acoustics. Also known as acoustic phonetics, speech acoustics deals with speech production and recognition. The scientific study of speech began with Thomas Alva Edison's phonograph, which allowed a speech signal to be recorded and stored for later analysis. Replaying the same short speech segment several times using consecutive filters passing through a limited range of frequencies creates a spectrogram, which visualizes the spectral properties of vowels and consonants. During the first half of the twentieth century, Bell Telephone Laboratories

invested considerable time and resources to the systematic understanding of all aspects of speech, including vocal tract resonances, voice quality, and prosodic features of speech. For the first time, electric circuit theory was applied to speech acoustics, and analogue electric circuits were used to investigate synthetic speech.

Musical acoustics. A conjunction of music, craftsmanship, auditory science, and vibration physics, musical acoustics analyzes musical instruments to better understand how the instruments are crafted, the physical principles of their tone production, and why each instrument has a unique timbre. Musical instruments are studied by analyzing their tones and then creating computer models to synthesize these sounds. When the sounds can be recreated with minimal software complications, a synthesizer featuring realistic orchestral tones may be constructed. The second method of study is to assemble an instrument or modify an existing instrument to perform nondestructive (or on occasion destructive) testing so that the effects of various modifications may be gauged.

Underwater sound. Also know as hydroacoustics, this field uses frequencies between 10 Hz and 1 megahertz (MHz). Although the origin of hydroacoustics can be traced back to Rayleigh, the deployment of submarines in World War I provided the impetus for the rapid development of underwater listening devices (hydrophones) and sonar (sound navigation ranging), the acoustic equivalent of radar. Pulses of sound are emitted and the echoes are processed to extract information about submerged objects. When the speed of underwater sound is known, the reflection time for a pulse determines the distance to an object. If the object is moving, its speed of approach or recession is deduced from the frequency shift of the reflection, or the Doppler effect. Returning pulses have a higher frequency when the object approaches and lower frequency when it moves away.

Noise. Physically, noise may be defined as an intermittent or random oscillation with multiple frequency components, but psychologically, noise is any unwanted sound. Noise can adversely affect human health and well-being by inducing stress, interfering with sleep, increasing heart rate, raising blood pressure, modifying hormone secretion, and even inducing depression. The physical effects of noise are no less severe. The vibrations in irregular road surfaces caused by large rapid vehicles can cause adjacent buildings to vibrate to an extent that is intolerable to the buildings' inhabitants, even without structural damage. Machinery noise in industry is a serious problem because continuous exposure to loud sounds will induce hearing loss. In apartment buildings, noise transmitted through walls is always problematic; the goal is to obtain adequate sound insulation using lightweight construction materials.

Traffic noise, both external and internal, is ubiquitous in modern life. The first line of defense is to reduce noise at its source by improving engine enclosures, mufflers, and tires. The next method, used primarily when interstate highways are adjacent to residential areas, is to block the noise by the construction of concrete barriers or the planting of sound-absorbing vegetation. Internal automobile noise has been greatly abated by designing more aerodynamically efficient vehicles to reduce air turbulence, using better sound isolation materials, and improving vibration isolation.

Aircraft noise, particularly in the vicinity of airports, is a serious problem exacerbated by the fact that as modern airplanes have become more powerful, the noise they generate has risen concomitantly. The noise radiated by jet engines is reduced by two structural modifications. Acoustic linings are placed around the moving parts to absorb the high frequencies caused by jet whine and turbulence, but this modification is limited by size and weight constraints. The second modification is to reduce the number of rotor blades and stator vanes, but this is somewhat inhibited by the desired power output. Special noise problems occur when aircraft travel at supersonic speeds (faster than the speed of sound), as this propagates a large pressure wave toward the ground that is experienced as an explosion. The unexpected sonic boom startles people, breaks windows, and damages houses. Sonic booms have been known to destroy rock structures in national parks. Because of these concerns, commercial aircraft are prohibited from flying at supersonic speeds over land areas.

Construction equipment (such as earthmoving machines) creates high noise levels both internally and externally. When the cabs of these machines are not closed, the only feasible manner of protecting operators' hearing is by using ear plugs. By carefully designing an enclosed cabin, structural vibration can be reduced and sound leaks made less significant, thus quieting the operator's environment. Although manufacturers are attempting to reduce the external noise, it is a daunting task because the rubber tractor treads occasionally used to replace metal are not as durable.

Applications and Products

Ultrasonics. High-intensity ultrasonic applications include ultrasonic cleaning, mixing, welding, drilling, and various chemical processes. Ultrasonic cleaners use waves in the 150 to 400 kHz range on items (such as jewelry, watches, lenses, and surgical instruments) placed in an appropriate solution. Ultrasonic cleaners have proven to be particularly effective in cleaning surgical devices because they loosen contaminants by aggressive agitation irrespective of an instrument's size or shape, and disassembly is not required. Ultrasonic waves are effective in cleaning most metals and alloys, as well as wood, plastic, rubber, and cloth.

Ultrasonic waves are used to emulsify two nonmiscible liquids, such as oil and water, by forming the liquids into finely dispersed particles that then remain in homogeneous suspension. Many paints, cosmetics, and foods are emulsions formed by this process.

Although aluminum cannot be soldered by conventional means, two surfaces subjected to intense ultrasonic vibration will bond—without the application of heat—in a strong and precise weld. Ultrasonic drilling is effective where conventional drilling is problematic, for instance, drilling square holes in glass. The drill bit, a transducer having the required shape and size, is used with an abrasive slurry that chips away the material when the suspended powder oscillates. Some of the chemical applications of ultrasonics are in the atomization of liquids, in electroplating, and as a catalyst in chemical reactions.

Low-intensity ultrasonic waves are used for nondestructive probing to locate flaws in materials for which complete reliability is mandatory, such as those used in spacecraft components and nuclear reactor vessels. When an ultrasonic transducer emits a pulse of energy into the test object, flaws reflect the wave and are detected. Because objects subjected to stress emit ultrasonic waves, these signals may be used to interpret the condition of the material as it is increasingly stressed. Another application is ultrasonic emission testing, which records the ultrasound emitted by porous rock when natural gas is pumped into cavities formed by the rock to determine the maximum pressure these natural holding tanks can withstand.

Low-intensity ultrasonics is used for medical diagnostics in two different applications. First, ultrasonic waves penetrate body tissues but are reflected by moving internal organs, such as the heart. The frequency of waves reflected from a moving structure is Doppler-shifted, thus causing beats with the original wave, which can be heard. This procedure is particularly useful for performing fetal examinations on a pregnant woman; because sound waves are not electromagnetic, they will not harm the fetus. The second application is to create a sonogram image of the body's interior. A complete cross-sectional image may be produced by superimposing the images scanned by successive ultrasonic waves passing through different regions. This procedure, unlike an X ray, displays all the tissues in the cross section and also avoids any danger posed by the radiation involved in X-ray imaging.

Physiological and psychological acoustics. Because the ear is a nonlinear system, it produces beat tones that are the sum and difference of two frequencies. For example, if two sinusoidal frequencies of 100 and 150 Hz simultaneously arrive at the ear, the brain will, in addition to these two tones, create tones of 250 and 50 Hz (sum and difference, respectively). Thus, although a small speaker cannot reproduce the fundamental frequencies of bass tones, the difference between the harmonics of that pitch will re-create the missing fundamental in the listener's brain.

Another psychoacoustic effect is masking. When a person listens to a noisy version of recorded music, the noise virtually disappears if the individual is enjoying the music. This ability of the brain to selectively listen has had important applications in digitally recorded music. When sounds are digitally compressed, as in MP3 systems, the brain compensates for the loss of information, thus creating a higher fidelity sound than that conveyed by the stored content alone.

As twentieth-century technology evolved, environmental noise increased concomitantly; lifetime exposure to loud sounds, commercial and recreational, has created an epidemic of hearing loss, most noticeable in the elderly because the effects are cumulative. Wearing a hearing aid, fitted adjacent to or inside the ear canal, is an effectual means of counteracting this handicap. The device consists of one or several microphones, which create electric signals that are amplified and transduced into sound waves redirected back into the ear. More sophisticated hearing aids incorporate an integrated circuit to control volume, either manually or automatically, or to switch to volume contours designed for various listening environments, such conversations on the telephone or where excessive background noise is present.

Speech acoustics. With the advent of the computer age, speech synthesis moved to digital processing, either by bandwidth compression of stored speech or by using a speech synthesizer. The synthesizer reads a text and then produces the appropriate phonemes on demand from their basic acoustic parameters, such as the vibration frequency of the vocal cords and the frequencies and amplitudes of the vowel formants. This method of generating speech is considerably more efficient in terms of data storage than archiving a dictionary of prerecorded phrases.

Another important, and probably the most difficult, area of speech acoustics is the machine recognition of spoken language. When machine recognition programs are sufficiently advanced, the computer will be able to listen to a sentence in any reasonable dialect and produce a printed text of the utterance. Two basic recognition strategies exist, one dealing with words spoken in isolation and the other with continuous speech. In both cases, it is desirable to teach the computer to recognize the speech of different people through a training program. Because recognition of continuous speech is considerably more difficult than the identification of isolated words, very sophisticated pattern-matching models must be employed. One example of a machine recognition system is a word-driven dictation system that uses sophisticated software to process input speech. This system is somewhat adaptable to different voices and is able to recognize 30,000 words at a rate of 30 words per minute. The ideal machine recognition system would translate a spoken input language into another language in real time with correct grammar. Although some progress is being made, such a device has remained in the realm of speculative fantasy.

Musical acoustics. The importance of musical acoustics to manufacturers of quality instruments is apparent. During the last decades of the twentieth century, fundamental research led, for example, to vastly improved French horns, organ pipes, orchestral strings, and the creation of an entirely new family of violins.

Underwater sound. Applications for underwater

acoustics include devices for underwater communication by acoustic means, remote control devices, underwater navigation and positioning systems, acoustic thermometers to measure ocean temperature, and echo sounders to locate schools of fish or other biota. Low-frequency devices can be used to explore the seabed for seismic research.

Although primitive measuring devices were developed in the 1920's, it was during the 1930's that sonar systems began incorporating piezoelectric transducers to increase their accuracy. These improved systems and their increasingly more sophisticated progeny became essential for the submarine warfare of World War II. After the war, theoretical advances in underwater acoustics coupled with computer technology have raised sonar systems to ever more sophisticated levels.

Noise. One system for abating unwanted sound is active noise control. The first successful application of active noise control was noise-canceling headphones, which reduce unwanted sound by using microphones placed in proximity to the ear to record the incoming noise. Electronic circuitry then generates a signal, exactly opposite to the incoming sound, which is reproduced in the earphones, thus canceling the noise by destructive interference. This system enables listeners to enjoy music without having to use excessive volume levels to mask outside noise and allows people to sleep in noisy vehicles such as airplanes. Because active noise suppression is more effective with low frequencies, most commercial systems rely on soundproofing the earphone to attenuate high frequencies. To effectively cancel high frequencies, the microphone and emitter would have to be situated adjacent to the user's eardrum, but this is not technically feasible. Active noise control is also being considered as a means of controlling low-frequency airport noise, but because of its complexity and expense, this is not yet commercially feasible.

IMPACT ON INDUSTRY

Acoustics is the focus of research at numerous governmental agencies and academic institutions, as well as some private industries. Acoustics also plays an important role in many industries, often as part of product design (hearing aids and musical instruments) or as an element in a service (noise control consulting).

Government research. Acoustics is studied in many government laboratories in the United States, including the U.S. Naval Research Laboratory (NRL), the Air Force Research Laboratory (AFRL), the Los Alamos National Laboratory, and the Lawrence Livermore National Laboratory. Research at the NRL and the AFRL is primarily in the applied acoustics area, and Los Alamos and Lawrence Livermore are oriented toward physical acoustics. The NRL emphasizes fundamental multidisciplinary research focused on creating and applying new materials and technologies to maritime applications. In particular, the applied acoustics division, using ongoing basic scientific research, develops improved signal processing systems for detecting and tracking underwater targets. The AFRL is heavily invested in research on auditory localization (spatial hearing), virtual auditory display technologies, and speech intelligibility in noisy environments. The effects of high-intensity noise on humans, as well as methods of attenuation, constitute a significant area of investigation at this facility. Another important area of research is the problem of providing intelligible voice communication in extremely noisy situations, such as those encountered by military or emergency personnel using low data rate narrowband radios, which compromise signal quality.

Academic research. Research in acoustics is conducted at many colleges and universities in the United States, usually through physics or engineering departments, but, in the case of physiological and psychological acoustics, in groups that draw from multiple departments, including psychology, neurology, and linguistics. The Speech Research Laboratory at Indiana University investigates speech perception and processing through a broad interdisciplinary research program. The Speech Research Lab, a collaboration between the University of Delaware and the A. I. duPont Hospital for Children, creates speech synthesizers for the vocally impaired. A human speaker records a data bank of words and phrases that can be concatenated on demand to produce natural-sounding speech.

Academic research in acoustics is also being conducted in laboratories in Europe and other parts of the world. The Laboratoire d'Acoustique at the Université de Maine in Le Mans, France, specializes in research in vibration in materials, transducers, and musical instruments. The Andreyev Acoustics Institute of the Russian Acoustical Society brings together researchers from Russian universities, agencies, and businesses to conduct fundamental and applied research in ocean acoustics, ultrasonics, signal processing, noise and vibration, electroacoustics, and bioacoustics. The Speech and Acoustics Laboratory at the Nara Institute of Science and Technology in Nara, Japan, studies diverse aspects of human-machine communication through speech-oriented multimodal interaction. The Acoustics Research Centre, part of the National Institute of Creative Arts and Industries in New Zealand, is concerned with the impact of noise on humans. A section of this group, Acoustic Testing Service, provides commercial testing of building materials for their noise attenuation properties.

Industry and business. Many businesses (such as the manufacturers of hearing aids, ultrasound medical devices, and musical instruments) use acoustics in their products or services and therefore employ experts in acoustics. Businesses also are involved in many aspects of acoustic research, particularly controlling noise and facilitating communication. Raytheon BBN technologies (Cambridge, Massachusetts) has developed low data rate Noise Robust Vocoders (electronic speech synthesizers) that generate comprehensible speech at data rates considerably below other state-of-the-art devices. Acoustic Research Laboratories in Sydney, Australia, designs and manufactures specialized equipment for measuring environmental noise and vibration, in addition to providing contract research and development services.

CAREERS AND COURSE WORK

Career opportunities occur in academia (teaching and research), industry, and national laboratories. Academic positions dedicated to acoustics are few, as are the numbers of qualified applicants. Most graduates of acoustics programs find employment in research-based industries in which acoustical aspects of products are important, and others work for government laboratories.

Although the subfields of acoustics are integrated into multiple disciplines, most aspects of acoustics can be learned by obtaining a broad background in a scientific or technological field, such as physics, engineering, meteorology, geology, or oceanography. Physics probably provides the best training for almost any area of acoustics. An electrical engineering major is useful for signal processing and synthetic speech research, and a mechanical engineering background

---

FASCINATING FACTS ABOUT ACOUSTICS

- Scientists have created an acoustic refrigerator, which uses a standing sound wave in a resonator to provide the motive power for operation. Oscillating gas particles increase the local temperature, causing heat to be transferred to the container walls, where it is expelled to the environment, cooling the interior.
- A cochlear implant, an electronic device surgically implanted in the inner ear, provides some hearing ability to those with damaged cochlea or those with congenital deafness. Because the implants use only about two dozen electrodes to replace 16,000 hair cells, speech sounds, although intelligible, have a robotic quality.
- MP3 files contain audio that is digitally encoded using an algorithm that compresses the data by a factor of about eleven but yields a reasonably faithful reproduction. The quality of sound reproduced depends on the data sampling rate, the quality of the encoder, and the complexity of the signal.
- Sound cannot travel through a vacuum, but it can travel four times faster through water than through air.
- The cocktail party effect refers to a person's ability to direct attention to one conversation at a time despite the many conversations taking place in the room.
- Continued exposure to noise over 85 decibels will gradually cause hearing loss. The noise level on a quiet residential street is 40 decibels, a vacuum cleaner 60-85, a leafblower 110, an ambulance siren 120, a rifle 163, and a rocket launching from its pad 180.

---

is requisite for comprehending vibration. Training in biology is expedient for physiological acoustic research, and psychology course work provides essential background for psychological acoustics. Architects often employ acoustical consultants to advise on the proper acoustical design of concert halls, auditoriums, or conference rooms. Acoustical consultants also assist with noise reduction problems and help design soundproofing structures for rooms. Although background in architecture is not a prerequisite for becoming this type of acoustical consultant, engineering or physics is.

Acoustics is not a university major; therefore, specialized knowledge is best acquired at the graduate level. Many electrical engineering departments have at least one undergraduate course in acoustics, but most physics departments do not. Nevertheless, a firm foundation in classical mechanics (through physics programs) or a mechanical engineering vibration course will provide, along with numerous courses in mathematics, sufficient underpinning for successful graduate study in acoustics.

SOCIAL CONTEXT AND FUTURE PROSPECTS

Acoustics affects virtually every aspect of modern life; its contributions to societal needs are incalculable. Ultrasonic waves clean objects, are routinely employed to probe matter, and are used in medical diagnosis. Cochlear implants restore people's ability to hear, and active noise control helps provide quieter listening environments. New concert halls are routinely designed with excellent acoustical properties, and vastly improved or entirely new musical instruments have made their debut. Infrasound from earthquakes is used to study the composition of Earth's mantle, and sonar is essential to locate submarines and aquatic life. Sound waves are used to explore the effects of structural vibrations. Automatic speech recognition devices and hearing aid technology are constantly improving.

Many societal problems related to acoustics remain to be tackled. The technological advances that made modern life possible have also resulted in more people with hearing loss. Environmental noise is ubiquitous and increasing despite efforts to design quieter machinery and pains taken to contain unwanted sound or to isolate it from people. Also, although medical technology has been able to help many hearing- and speech-impaired people, other individuals still lack appropriate treatments. For example, although voice generators exist, there is considerable room for improvement.

—George R. Plitnik, MA, PhD

FURTHER READING

Bass, Henry E., and William J. Cavanaugh, eds. *ASA at Seventy-five.* Melville, N.Y.: Acoustical Society of America, 2004. An overview of the history, progress, and future possibilities for each of the fifteen major subdivisions of acoustics as defined by the Acoustical Society of America.

Beyer, Robert T. *Sounds of Our Times: Two Hundred Years of Acoustics.* New York: Springer-Verlag, 1999. A history of the development of all areas of acoustics. Organized into chapters covering twenty-five to fifty years. Virtually all subfields of acoustics are covered.

Crocker, Malcolm J., ed. *The Encyclopedia of Acoustics.* 4 vols. New York: Wiley, 1997. A comprehensive work detailing virtually all aspects of acoustics.

Everest, F. Alton, and Ken C. Pohlmann. *Master Handbook of Acoustics.* 5th ed., New York: McGraw-Hill, 2009. A revision of a classic reference work designed for those who desire accurate information on a level accessible to the layperson with limited technical ability.

Rossing, Thomas, and Neville Fletcher. *Principles of Vibration and Sound.* 2d ed. New York: Springer-Verlag, 2004. A basic introduction to the physics of sound and vibration.

Rumsey, Francis, and Tim McCormick. *Sound and Recording: An Introduction.* 5th ed. Boston: Elsevier/ Focal Press, 2004. Presents basic information on the principles of sound, sound perception, and audio technology and systems.

Strong, William J., and George R. Plitnik. *Music, Speech, Audio.* 3d ed. Provo, Utah: Brigham Young University Academic Publishing, 2007. A comprehensive text, written for the layperson, which covers vibration, the ear and hearing, noise, architectural acoustics, speech, musical instruments, and sound recording and reproduction.

Swift, Gregory. "Thermoacoustic Engines and Refrigerators." *Physics Today* (July, 1995): 22-28. Explains how sound waves may be used to create more efficient refrigerators with no moving parts.

WEB SITES

Acoustical Society of America
http://asa.aip.org

Institute of Noise Control Engineering
http://www.inceusa.org

International Commission for Acoustics
http://www.icacommission.org

National Council of Acoustical Consultants
http://www.ncac.com

See also: Applied Physics; Communication; Noise Control; Pattern Recognition.

AERONAUTICS AND AVIATION

FIELDS OF STUDY

Algebra; calculus; inorganic chemistry; organic chemistry; physical chemistry; optics; modern physics; statics; aerodynamics; thermodynamics; strength of materials; propulsion; propeller and rotor theory; vehicle performance; aircraft design; avionics; orbital mechanics; spacecraft design.

SUMMARY

Aeronautics is the science of atmospheric flight. Aviation is the design, development, production, and operation of flight vehicles. Aerospace engineering extends these fields to space vehicles. Transonic airliners, airships, space launch vehicles, satellites, helicopters, interplanetary probes, and fighter planes are all applications of aerospace engineering.

KEY TERMS AND CONCEPTS

- **airfoil:** Structure, such as a wing or a propeller, designed to interact, in motion, with the surrounding airflow in a manner that optimizes the desired reaction, whether that be to minimize air resistance or to maximize lift.
- **boundary layer:** Thin region near a surface of an aircraft where the flow slows down because of viscous friction.
- **bypass ratio:** Ratio of turbofan engine mass flow rate bypassing the hot core, to that through the core.
- **Delta V:** Speed difference corresponding to the difference in energies between two orbital states.
- **fuselage:** Body of an aircraft, other than engines, wings, tails, or control surfaces.
- **lift to drag ratio:** Ratio of the lift to drag in cruise; the aerodynamic efficiency metric for transport aircraft and gliders.
- **oblique shock:** Thin wave in a supersonic flow through which flow turns and decelerates sharply.
- **Prandtl-Meyer expansion:** Ideal model of a supersonic flow accelerating through a turn.
- **stall:** Condition in which flow separates from most of a lifting surface, sharply lowering lift and raising drag.

- **takeoff gross weight:** Mass or weight of an aircraft at takeoff with full payload and fuel load; the highest design weight for liftoff.
- **wind tunnel:** Facility where a smooth, uniform flow helps simulate flow around an object in flight.
- **wing:** Object that generates lift with low drag and supports the weight of the aircraft in flight.

DEFINITION AND BASIC PRINCIPLES

Aeronautics is the science of atmospheric flight. The term ("aero" referring to flight and "nautics" referring to ships or sailing) originated from the activities of pioneers who aspired to navigate the sky. These early engineers designed, tested, and flew their own creations, many of which were lighter-than-air balloons. Modern aeronautics encompasses the science and engineering of designing and analyzing all areas associated with flying machines.

Aviation (based on the Latin word for "bird") originated with the idea of flying like the birds using heavier-than-air vehicles. "Aviation" refers to the field of operating aircraft, while the term "aeronautics" has been superseded by "aerospace engineering," which specifically includes the science and engineering of spacecraft in the design, development, production, and operation of flight vehicles.

A fundamental tenet of aerospace engineering is to deal with uncertainty by tying analyses closely to what is definitely known, for example, the laws of physics and mathematical proofs. Lighter-than-air airships are based on the principle of buoyancy, which derives from the law of gravitation. An object that weighs less than the equivalent volume of air experiences a net upward force as the air sinks around it.

Two basic principles that enable the design of heavier-than-air flight vehicles are those of aerodynamic lift and propulsion. Both arise from Sir Isaac Newton's second and third laws of motion. Aerodynamic lift is a force perpendicular to the direction of motion, generated from the turning of flowing air around an object. In propulsion, the reaction to the acceleration of a fluid generates a force that propels an object, whether in air or in the vacuum of space. Understanding these principles allowed aeronauts to design vehicles that could fly steadily despite being much heavier than the air they

Lighter than air geostationary airship telecommunications satellite

displaced and allowed rocket scientists to develop vehicles that could accelerate in space. Spaceflight occurs at speeds so high that the vehicle's kinetic energy is comparable to the potential energy due to gravitation. Here the principles of orbital mechanics derive from the laws of dynamics and gravitation and extend to the regime of relativistic phenomena. The engineering sciences of building vehicles that can fly, keeping them stable, controlling their flight, navigating, communicating, and ensuring the survival, health, and comfort of occupants, draw on every field of science.

BACKGROUND AND HISTORY

The intrepid balloonists of the nineteenth century were followed by aeronauts who used the principles of aerodynamics to fly unpowered gliders. The Wright brothers demonstrated sustained, controlled, powered aerodynamic flight of a heavier-than-air aircraft in 1903. The increasing altitude, payload, and speed capabilities of airplanes made them powerful weapons in World War I. Such advances improved flying skills, designs, and performance, though at a terrible cost in lives.

The monoplane design superseded the fabric-and-wire biplane and triplane designs of World War I. The helicopter was developed during World War II and quickly became an indispensable tool for medical evacuation and search and rescue. The jet engine, developed in the 1940's and used on the Messerschmitt 262 and Junkers aircraft by the Luftwaffe and the Gloster Meteor by the British, quickly enabled flight

in the stratosphere at speeds sufficient to generate enough lift to climb in the thin air. Such innovations led to smooth, long-range flights in pressurized cabins and shirtsleeve comfort. Fatal crashes of the de Havilland Comet airliner in 1953 and 1954 focused attention on the science of metal fatigue.

The Boeing 707 opened up intercontinental air travel, followed by the Boeing 747, the supersonic Concorde, and the EADS Airbus A380. A series of manned research aircraft designated X-planes since the 1930's investigated various flight regimes and also drove the development of better wind tunnels and high-altitude simulation chambers. German ballistic missiles led to U.S. and Soviet missile programs that grew into a space race, culminating in the first humans landing on the Moon in 1969. Combat-aircraft development enabled advances that resulted in safer and more efficient airliners.

HOW IT WORKS

Force balance in flight. Five basic forces acting on a flight vehicle are aerodynamic lift, gravity, thrust, drag, and centrifugal force. For a vehicle in steady level flight in the atmosphere, lift and thrust balance gravity (weight) and aerodynamic drag. Centrifugal force due to moving steadily around the Earth is too weak at most airplane flight speeds but is strong for a maneuvering aircraft. Aircraft turn by rolling the lift vector toward the center of curvature of the desired flight path, balancing the centrifugal reaction due to inertia. In the case of a vehicle in space beyond the atmosphere, centrifugal force and thrust counter gravitational force.

Aerodynamic lift. Aerodynamics deals with the forces due to the motion of air and other gaseous fluids relative to bodies. Aerodynamic lift is generated perpendicular to the direction of the free stream as the reaction to the rate of change of momentum of air turning around an object, and, at high speeds, to compression of air by the object. Flow turning is accomplished by changing the angle of attack of the surface, by using the camber of the surface in subsonic flight, or by generating vortices along the leading edges of swept wings.

Propulsion. Propulsive force is generated as a reaction to the rate of change of momentum of a fluid moving through and out of the vehicle. Rockets carry all of the propellant onboard and accelerate it out through a nozzle using chemical heat release, other heat sources, or electromagnetic fields. Jet engines "breathe" air and accelerate it after reaction with fuel. Rotors, propellers, and fans exert lift force on the air and generate thrust from the reaction to this force. Solar sails use the pressure of solar radiation to push large, ultralight surfaces.

Static stability. An aircraft is statically stable if a small perturbation in its attitude causes a restoring aerodynamic moment that erases the perturbation. Typically, the aircraft center of gravity must be ahead of the center of pressure for longitudinal stability. The tails or canards help provide stability about the different axes. Rocket engines are said to be stable if the rate of generation of gases in the combustion chamber does not depend on pressure stronger than by a direct proportionality, such as a pressure exponent of 1.

Flight dynamics and controls. Static stability is not the whole story, as every pilot discovers when the airplane drifts periodically up and down instead of holding a steady altitude and speed. Flight dynamics studies the phenomena associated with aerodynamic loads and the response of the vehicle to control surface deflections and engine-thrust changes. The study begins with writing the equations of motion of the aircraft resolved along the six degrees of freedom: linear movement along the longitudinal, vertical and sideways axes, and roll, yaw, and pitch rotations about them. Maneuvering aircraft must deal with coupling between the different degrees of freedom, so that roll accompanies yaw, and so on.

The autopilot system was an early flight-control achievement. Terrain-following systems combine information about the terrain with rapid updates, enabling military aircraft to fly close to the ground, much faster than a human pilot could do safely. Modern flight-control systems achieve such feats as reconfiguring control surfaces and fuel to compensate for damage and engine failures; or enabling autonomous helicopters to detect, hover over, and pick up small objects and return; or sending a space probe at thousands of kilometers per hour close to a planetary moon or landing it on an asteroid and returning it to Earth. This field makes heavy use of ordinary differential equations and transform techniques, along with simulation software.

Orbital missions. The rocket equation attributed to Russian scientist Konstantin Tsiolkovsky related the speed that a rocket-powered vehicle gains to the amount and speed of the mass that it ejects. A vehicle launched from Earth's surface goes into a trajectory where its kinetic energy is exchanged for gravitational potential energy. At low speeds, the resulting trajectory intersects the Earth, so that the vehicle falls to the surface. At high enough speeds, the vehicle goes so far so fast that its trajectory remains in space and takes the shape of a continuous ellipse around Earth. At even higher kinetic energy levels, the vehicle goes into a hyperbolic trajectory, escaping Earth's orbit into the solar system. The key is thus to achieve enough tangential speed relative to Earth. Most rockets rise rapidly through the atmosphere so that the acceleration to high tangential speed occurs well above the atmosphere, thus minimizing air-drag losses.

Hohmann transfer. Theoretically, the most efficient way to impart kinetic energy to a vehicle is impulsive launch, expending all the propellant instantly so that no energy is wasted lifting or accelerating propellant with the vehicle. Of course, this would destroy any vehicle other than a cannonball, so large rockets use gentle accelerations of no more than 1.4 to 3 times the acceleration due to gravity. The advantage of impulsive thrust is used in the Hohmann transfer maneuver between different orbits in space. A rocket is launched into a highly eccentric elliptical trajectory. At its highest point, more thrust is added quickly. This sends the vehicle into a circular orbit at the desired height or into a new orbit that takes it close to another heavenly body. Reaching the same final orbit using continuous, gradual thrust would require roughly twice as much expenditure of energy. However, continuous thrust is still an attractive option for long missions in space, because a small amount of thrust can be generated using electric propulsion engines that accelerate propellant to extremely high speeds compared with the chemical engines used for the initial ascent from Earth.

APPLICATIONS AND PRODUCTS

Aerospace structures. Aerospace engineers always seek to minimize the mass required to build the vehicle but still ensure its safety and durability. Unlike buildings, bridges, or even (to some degree) automobiles, aircraft cannot be made safer merely by making them more massive, because they must also be able to overcome Earth's gravity. This exigency has driven development of new materials and detailed, accurate methods of analysis, measurement, and construction. The first aircraft were built mostly from wood frames and fabric skins. These were superseded by all-metal craft, constructed using the monocoque concept (in which the outer skin bears most of the stresses). The Mosquito high-speed bomber in World War II reverted to wood construction for better performance. Woodworkers learned to align the grain (fiber direction) along the principal stress axes. Metal offers the same strength in all directions for the same thickness. Composite structures allow fibers with high tensile strength to be placed along the directions where strength is needed, bonding different layers together.

Aeroelasticity. Aeroelasticity is the study of the response of structurally elastic bodies to aerodynamic loads. Early in the history of aviation, several mysterious and fatal accidents occurred wherein pieces of wings or tails failed in flight, under conditions where the steady loads should have been well below the strength limits of the structure. The intense research to address these disasters showed that beyond some flight speed, small perturbations in lift, such as those due to a gust or a maneuver, would cause the structure to respond in a resonant bending-twisting oscillation, the perturbation amplitude rapidly rising in a "flutter" mode until structural failure occurred. Predicting such aeroelastic instabilities demanded a highly mathematical approach to understand and apply the theories of unsteady aerodynamics and structural dynamics. Modern aircraft are designed so that the flutter speed is well above any possible speed achieved. In the case of helicopter rotor blades and gas turbine engine blades, the problems of ensuring aeroelastic stability are still the focus of leading-edge research. Related advances in structural dynamics have enabled development of composite structures and of highly efficient turbo machines that use counter-rotating stages, such as those in the F135 engines used in the F-35 Joint Strike Fighter. Such advances also made it possible for earthquake-surviving high-rise buildings to be built in cities such as San Francisco, Tokyo, and Los Angeles, where a number of sensors, structural-dynamics-analysis software, and actuators allow the correct response to dampen the effects of earth movements even on the upper floors.

Smart materials. Various composite materials such as carbon fiber and metal matrix composites have come to find application even in primary aircraft structures. The Boeing 787 is the first to use a composite main spar in its wings. Research on nano materials promises the development of materials with hundreds of times as much strength per unit mass as steel. Another leading edge of research in materials is in developing high-temperature or very low-temperature (cryogenic) materials for use inside jet and rocket engines, the spinning blades of turbines, and the impeller blades of liquid hydrogen pumps in rocket engines. Single crystal turbine blades enabled the development of jet engines with very high turbine inlet temperatures and, thus, high thermodynamic efficiency. Ceramic designs that are not brittle are pushing turbine inlet temperatures even higher. Other materials are "smart," meaning they respond actively in some way to inputs. Examples include piezoelectric materials.

Wind tunnels and other physical test facilities. Wind tunnels, used by the Wright brothers to develop airfoil shapes with desirable characteristics, are still used heavily in developing concepts and proving the performance of new designs, investigating causes of problems, and developing solutions and data to validate computational prediction techniques. Generally, a wind tunnel has a fan or a high-pressure reservoir to add work to the air and raise its stagnation pressure. The air then flows through means of reducing turbulence and is accelerated to the maximum speed in the test section, where models and measurement systems operate.

The power required to operate a wind tunnel is proportional to the mass flow rate through the tunnel and to the cube of the flow speed achieved. Low-speed wind tunnels have relatively large test sections and can operate continuously for several minutes at a time. Supersonic tunnels generally operate with air blown from a high-pressure reservoir for short durations. Transonic tunnels are designed with ventilating slots to operate in the difficult regime where there may be both supersonic waves and subsonic flow over the test configuration. Hypersonic tunnels require heaters to avoid liquefying the air and to simulate the high stagnation temperatures of hypersonic flight and operate for millisecond durations. Shock tubes generate a shock from the rupture of a diaphragm, allowing high-energy air to expand into stationary air in the tube. They are used to simulate the extreme conditions across shocks in hypersonic flight. Many other specialized test facilities are used in structural and materials testing, developing jet and rocket engines, and designing control systems.

Avionics and navigation. Condensed from the term "aviation electronics," the term "avionics" has come to include the generation of intelligent software systems and sensors to control unmanned aerial vehicles (UAVs), which may operate autonomously. Avionics also deals with various subsystems such as radar and communications, as well as navigation equipment, and is closely linked to the disciplines of flight dynamics, controls, and navigation.

During World War II, pilots on long-range night missions would navigate celestially. The gyroscopes in their aircrafts would spin at high speed so that their inertia allowed them to maintain a reference position as the aircraft changed altitude or accelerated. Most modern aircraft use the Global Positioning System (GPS), Galileo, or GLONASS satellite constellations to obtain accurate updates of position, altitude, and velocity. The ordinary GPS signal determines position and speed with fair accuracy. Much greater precision and higher rates of updates are available to authorized vehicle systems through the differential GPS signal and military frequencies.

Gravity assist maneuver. Yuri Kondratyuk, the Ukrainian scientist whose work paved the way for the first manned mission to the moon, suggested in 1918 that a spacecraft could use the gravitational attraction of the moons of planets to accelerate and decelerate at the two ends of a journey between planets. The Soviet Luna 3 probe used the gravity of the Moon when photographing the far side of it in 1959. American mathematician Michael Minovitch pointed out that the gravitational pull of planets along the trajectory of a spacecraft could be used to accelerate the craft toward other planets. The Mariner 10 probe used this "gravitational slingshot" maneuver around Venus to reach Mercury at a speed small enough to go into orbit around Mercury. The Voyager missions used the rare alignment of the outer planets to receive gravitational assists from Jupiter and Saturn to go on to Uranus and Neptune, before doing another slingshot around Jupiter and Saturn to escape the solar system. Gravity assist has become part of the mission planning for all exploration missions and even for missions near Earth, where the gravity of the Moon is used.

IMPACT ON INDUSTRY

Aeronautics and aviation have had an immeasurable impact on industry and society. Millions of people fly long distances on aircraft every day, going about their business and visiting friends and relatives, at a cost that is far lower in relative terms than the cost of travel a century ago.

Every technical innovation developed for aeronautics and aviation finds its way into improved industrial products. Composite structural materials are found in everything from tennis rackets to industrial machinery. Bridges, stadium domes, and skyscrapers are designed with aerospace structural-element technology and structural-dynamics instrumentation and testing techniques. Electric power is generated in utility power plants using steam generators sharing jet engine turbo machine origins.

Satellite antennae are found everywhere. Much digital signal processing, central to digital music and cell phone communications, came from research projects driven by the need to extract low-level signatures buried in noise. Similarly, image-processing algorithms that enable computed tomography (CT) scans of the human body, eye and cardiac diagnostics, image and video compression, and laser printing came from aerospace image-processing projects. The field of geoinformatics has advanced immensely, with most mapping, navigation, and remote-sensing enterprises assuming the use of space satellites. The GPS has spawned numerous products for terrestrial drivers on land and navigators on the ocean. Aerospace medicine research has developed advances in diagnosing and monitoring the human body and its responses to acceleration, bone marrow loss, muscular degeneration, and their prevention through exercise, hypoxia, radiation protection, heart monitoring, isolation from microorganisms, and drug delivery. Teflon coatings developed for aerospace products are also used in cookware.

CAREERS AND COURSE WORK

Aerospace engineers work on problems that push the frontiers of technology. Typical employers in this Aeronautics and Aviation industry are manufacturers of aircraft or their parts and subsystems, airlines, government agencies and laboratories, and the defense services. Many aerospace engineers are also sought by financial services and other industries seeking those with excellent quantitative (mathematical and scientific) skills and talents.

University curriculum generally starts with a year of mathematics, physics, chemistry, computer graphics, computer science, language courses, and an introduction to aerospace engineering, followed by sophomore-year courses in basic statics, dynamics, materials, and electrical engineering. Core courses include low-speed and high-speed aerodynamics, linear systems analysis, thermodynamics, propulsion, structural analysis, composite materials, vehicle performance, stability, control theory, avionics, orbital mechanics, aeroelasticity and structural dynamics, and a two-semester sequence on capstone design of flight vehicles. High school students aiming for such careers should take courses in mathematics, physics, chemistry and natural sciences, and computer graphics. Aerospace engineers are frequently required to write clear reports and present complex issues to skeptical audiences, which demands excellent communication skills. Taking flying lessons or getting a private pilot license is less important to aerospace engineering, as exhilarating as it is, and should be considered only if one desires a career as a pilot or astronaut.

The defense industry is moving toward using aircraft that do not need a human crew and can perform beyond the limits of what a human can survive, so the glamorous occupation of combat jet pilot may be heading for extinction. Airline pilot salaries are also coming down from levels that compared with surgeons toward those more comparable to bus drivers. Aircraft approach, landing, traffic management, emergency response, and collision avoidance systems may soon become fully automated and will require maneuvering responses that are beyond what a human pilot can provide in time and accuracy.

Opportunities for spaceflight may also be minimal unless commercial and military spaceflight picks up to fill the void left by the end of civilian programs discussed below. This is not a unique situation in aviation history. Early pilots, even much later than the intrepid "aeronauts," also worked much more for love of the unparalleled experience of flying, rather than for looming prospects of high-profile careers or the salaries paid by struggling startup airline companies. The only reliable prediction that can be made about aerospace careers is that they hold many surprises.

SOCIAL CONTEXT AND FUTURE PROSPECTS

Airline travel is under severe stress in the first part of the twenty-first century. This is variously attributed to airport congestion, security issues, rising fuel prices, predatory competition, reduction of route monopolies, and leadership that appears to offer little vision beyond cost cutting. Meanwhile, the demand for air travel is rising all over the world. Global demand for commercial airliners is estimated at nearly 30,000 aircraft through 2030 and is valued at more than $3.2 trillion—in addition to 17,000 business jets valued at more than $300 billion.

Detailed design and manufacturing of individual aircraft are distributed between suppliers worldwide, with the wings, tails, and engines of a given aircraft often designed and built in different parts of the world. Japan and China are expected to increase their aircraft manufacturing, while major U.S. companies appear to be moving more toward becoming system integrators and away from manufacturing.

The human venture in space is also under stress as the U.S. space shuttle program ends without another human-carrying vehicle to replace it. The future of the one remaining space station is in doubt, and there are no plans to build another.

On the other hand, just over one century into powered flight, the human venture into the air and beyond is just beginning. Aircraft still depend on long runways and can fly only in a very limited range of conditions. Weather delays are still common because of uncertainty about how to deal with fluctuating winds or icing conditions. Most airplanes still consist of long tubes attached to thin wings, because designing blended wing bodies is difficult with the uncertainties in modeling composite structures. The aerospace and aviation industry is a major generator of atmospheric carbon releases. This will change only when the industry switches to renewable hydrogen fuel, which may occur faster than most people anticipate.

The human ability to access, live, and work in space or on extraterrestrial locations is extremely limited, and this prevents development of a large space-based economy. This situation may be expected to change over time, with the advent of commercial space launches. New infrastructure will encourage commercial enterprises beyond Earth.

The advancements in the past century are truly breathtaking and bode well for the breakthroughs that one may hope to see. Hurricanes and cyclonic storms are no longer surprise killers; they are tracked from formation in the far reaches of the oceans, and their paths are accurately predicted, giving people plenty of warning. Crop yields and other resources are accurately tracked by spacecraft, and ground-penetrating radar from Earth-sensing satellites has discovered much about humankind's buried ancient heritage and origins. Even in featureless oceans and deserts, GPS satellites provide accurate, reliable navigation information. The discovery of ever-smaller distant planets by orbiting space telescopes, and of unexpected forms of life on Earth, hint at the possible discovery of life beyond Earth.

—*Narayanan M. Komerath, PhD*

FURTHER READING

Anderson, John D., Jr. *Introduction to Flight.* 5th ed. New York: McGraw-Hill, 2005. This popular textbook, setting developments in a historical context, is derived from the author's tenure at the Smithsonian Air and Space Museum.

Bekey, Ivan. *Advanced Space System Concepts and Technologies, 2010-2030+.* Reston, Va.: American Institute of Aeronautics and Astronautics, 2003. Summaries of various advanced concepts and logical arguments used to explore their feasibility.

Design Engineering Technical Committee. *AIAA Aerospace Design Engineers Guide.* 5th ed. Reston, Va.: American Institute of Aeronautics and Astronautics, 2003. A concise book of formulae and numbers that aerospace engineers use frequently or need for reference.

Gann, Ernest K. *Fate Is the Hunter.* 1961. Reprint. New York: Simon and Schuster, 1986. Describes an incident that was the basis for a 1964 film of the same name. Autobiography of a pilot, describing the early days of commercial aviation and coming close to the age of jet travel.

Hill, Philip, and Carl Peterson. *Mechanics and Thermodynamics of Propulsion.* 2d ed. Upper Saddle River, N.J.: Prentice Hall, 1991. A classic textbook on propulsion that covers the basic science and engineering of jet and rocket engines and their components. Also gives excellent sets of problems with answers.

Jenkins, Dennis R. *X-15: Extending the Frontiers of Flight.* NASA SP-2007-562. Washington, D.C.: U.S. Government Printing Office, 2007. Contains various copies of original data sheets, memos, and pictures from the days when the X-15 research vehicle was developed and flown.

Lewis, John S. *Mining the Sky: Untold Riches from the Asteroids, Comets and Planets.* New York: Basic Books, 1997. The most readable answer to the question "What resources are there beyond Earth to make exploration worthwhile?" Written from a strong scientific background, it sets out the reasoning to estimate the presence and accessibility of extraterrestrial water, gases, minerals, and other resources that would enable an immense space-based economy.

Liepmann, H. W., and A. Roshko. *Elements of Gas Dynamics.* Mineola, N.Y.: Dover Publications, 2001. A textbook on the discipline of gas dynamics as applied to high-speed flow phenomena. Contains several photographs of shocks, expansions, and boundary layer phenomena.

O'Neill, Gerard K. *The High Frontier: Human Colonies in Space.* 3d ed. New York: William Morrow & Company, 1977. Reprint. Burlington, Ontario, Canada: Apogee Books, 2000. Sets out the logic, motivations, and general parameters for human settlements in space. This formed the basis for NASA/ASEE (American Society for Engineering Education) studies in 1977-1978 and beyond, to investigate the design of space stations for permanent habitation. Fascinating exposition of how ambitious concepts are systematically analyzed and engineering decisions are made on how to achieve them, or why they cannot yet be achieved.

Peebles, Curtis. *Road to Mach 10: Lessons Learned from the X-43A Flight Research Program.* Reston, Va.: American Institute of Aeronautics and Astronautics, 2008. A contemporary experimental flight-test program description.

WEB SITES

Aerospace Digital Library
http://www.adl.gatech.edu

American Institute of Aeronautics and Astronautics
http://www.aiaa.org

Jet Propulsion Laboratory A Gravity Assist Primer
http://www2.jpl.nasa.gov/basics/grav/primer.php

National Aeronautics and Space Administration
Born of Dreams, Inspired by Freedom. U.S. Centennial of Flight Commission
http://www.centennialofflight.gov

See also: Applied Mathematics; Applied Physics; Atmospheric Sciences; Avionics and Aircraft Instrumentation; Communications; Computer Science; Propulsion Technologies.

AIR-QUALITY MONITORING

FIELDS OF STUDY

Meteorology; electronics; engineering; environmental planning; environmental studies; statistics; physics; chemistry; mathematics.

SUMMARY

Air-quality monitoring involves the systematic sampling of ambient air, analysis of samples for pollutants injurious to human health or ecosystem function, and integration of the data to inform public policy decision making. Air-quality monitoring in the United States is governed by the federal Clean Air Act of 1963 and subsequent amendments and by individual state implementation plans. Data from air monitoring are used to propose and track remediation strategies that have been credited with dramatically lowering levels of some pollutants in recent years.

KEY TERMS AND CONCEPTS

- **aerosol:** Suspension of solid particles or very fine droplets of liquid, one of the major components of smog.
- **air pollutants:** Foreign or natural substances occurring in the atmosphere that may result in adverse effects to humans, animals, vegetation, or materials.
- **air quality index (AQI):** Numerical index used for reporting severity of air pollution levels to the public. The AQI incorporates five criteria pollutants — ozone, particulate matter, carbon monoxide, sulfur dioxide, and nitrogen dioxide —into a single index, which is used for weather reporting and public advisories.
- **carbon monoxide (CO):** Colorless, odorless gas resulting from the incomplete combustion of hydrocarbon fuels. CO interferes with the blood's ability

to carry oxygen to the body's tissues and results in numerous adverse health effects. More than 80 percent of the CO emitted in urban areas is contributed by motor vehicles.

- **continuous emission monitoring (CEM):** Used for determining compliance of stationary sources with their emission limitations on a continuous basis by installing a system to operate continuously inside of a smokestack or other emission source.
- **criteria air pollutant:** Air pollutant for which acceptable levels of exposure can be determined and for which an ambient air-quality standard has been set. Examples include: ozone, carbon monoxide, nitrogen dioxide, sulfur dioxide, PM10, and PM2.5.
- **ozone:** Reactive toxic chemical gas consisting of three oxygen atoms, a product of the photochemical process involving the Sun's energy and ozone precursors, such as hydrocarbons and oxides of nitrogen. Ozone in the troposphere causes numerous adverse health effects and is a major component of smog.
- **PM 10 and PM 2.5:** Particulate matter with maximum diameters of 10 and 2.5 micrometers (μm), respectively.
- **regional haze:** Haze produced by a multitude of sources and activities that emit fine particles and their precursors across a broad geographic area. National regulations require states to develop plans to reduce the regional haze that impairs visibility in national parks and wilderness areas.
- **volatile organic compounds (VOCs):** Carbon-containing compounds that evaporate into the air. VOCs contribute to the formation of smog and may themselves be toxic.

DEFINITION AND BASIC PRINCIPLES

Air-quality monitoring aims to track atmospheric levels of chemical compounds and particulate matter that are injurious to human health or cause some form of environmental degradation. The scope varies, from single rooms to the entire globe. There has been a tendency since the 1990's to shift focus away from local exterior air pollution to air quality inside buildings, both industrial and residential, to regional patterns including air-quality issues that cross international boundaries, and to worldwide trends including global warming and depletion of the ozone layer.

Data from air-quality monitoring are used to alert the public to hazardous conditions, track remediation efforts, and suggest legislative approaches to environmental policy. Agencies conducting local and regional exterior monitoring include national and state environmental protection agencies, the National Oceanic and Atmospheric Administration (NOAA, popularly known as the Weather Bureau), and the National Aeronautics and Space Administration (NASA). Industries conduct their own interior monitoring with input from state agencies and the Occupational Safety and Health Administration (OSHA). There is an increasing

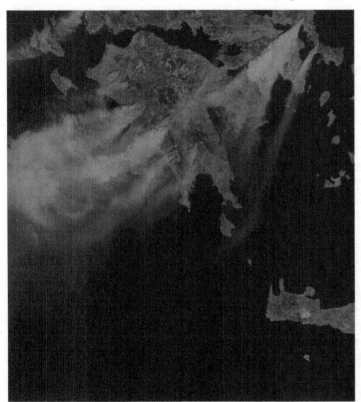

Wildfires give rise to an elevated AQI in parts of Greece

market for inexpensive devices homeowners can install themselves to detect household health hazards.

The Clean Air Act of 1963 and amendments of 1970 and 1990 mandate monitoring of six criteria air pollutants—carbon monoxide, nitrogen dioxide, sulfur dioxide, ozone, particulate matter, and lead—with the aim of reducing release into the environment and minimizing human exposure. With the exception of particulate matter, atmospheric concentrations of these pollutants have decreased dramatically since the late 1980's.

A key concept in implementing monitoring programs is that of probable real-time exposure, with monitoring protocols matched to real human experience. Increasingly compact and automated equipment that can be left permanently at a site and collect data at intervals over a period of weeks and months has been a real boon to establishing realistic tolerance levels.

BACKGROUND AND HISTORY

The adverse effects of air pollution from burning coal were first noticed in England in the Middle Ages and were the reason for the ban on coal in London in 1306. By the mid-seventeenth century, when John Evelyn wrote *Fumifugium or, the Inconvenience of the Aer and the Smoake of London Dissipated* (1661) the problem of air pollution in London had become acute, and various measures, including banning certain industries from the city, were proposed to combat it. Quantifying or even identifying the chemicals responsible exceeded contemporary scientific knowledge, but English statistician John Graunt, correlating deaths from lung disease recorded in the London bills of mortality with the incidence of "great stinking fogs," obtained objective evidence of health risks, and scientist Robert Boyle proposed using the rate of fading of strips of dyed cloth to monitor air quality.

Chemical methods for detecting nitrogen and sulfur oxides existed in the mid-eighteenth century. By the late nineteenth century, scientists in Great Britain were undertaking chemical analyses of rainwater and measuring rates of soot deposition with the aim of encouraging industries either to adopt cleaner technologies or to move to less populated areas. Early examples of environmental legislation based on scientific evidence are the Alkali Acts, the first of which was enacted in 1862, requiring industry to mitigate dramatic environmental and human health effects by reducing hydrogen chloride (HCl) emissions by 95 percent.

Until the passage of the first national Clean Air Act in 1963, monitoring and abatement of air pollution in the United States was mainly a local matter, and areas in which the residents were financially dependent on a single polluting industry were reluctant to take any steps toward abatement. Consequently, although air quality in most large metropolitan areas improved in the first half of the twentieth century, grave health hazards remained in some industries. This discrepancy was highlighted by the 1948 tragedy in Donora, Pennsylvania, when twenty people died and almost 6,000 (about half the population) became acutely ill during a prolonged temperature inversion that trapped effluents from a zinc smelter, including highly toxic fluorides. Although autopsies and blood tests revealed a pattern of chronic and acute fluoride poisoning, no specific legislation regulating the industry followed the investigation. Blood tests and autopsies have also been used to track incidences of lead and mercury poisoning, some of it from atmospheric pollution.

HOW IT WORKS

Structure of air-monitoring programs. In the United States, state environmental protection agencies, overseen by the federal government, conduct the largest share of air-quality monitoring out of doors, while indoor monitoring is usually the responsibility of the owner or operator.

The Clean Air Act and its amendments require each state to file a State Implementation Plan (SIP) for air-pollution monitoring, prevention, and remediation. States are responsible for most of the costs of implementing regulations, which may be more stringent than federal guidelines but cannot fall below them. Detailed regulations specify which pollutants must be monitored, frequency and procedures for monitoring, and acceptable equipment. The regulations change constantly in response to developing technology, shifting patterns of pollution, and political considerations. There is a tendency to respond rapidly and excessively to new threats while grandfathering in older programs, such as those aimed at

asbestos and lead, which address hazards that are far less acute than they once were.

The United States Environmental Protection Agency (EPA) collects and analyzes data from state monitoring stations to track national and long-term trends, make recommendations for expanded programs, and ensure compliance. Some pollutants can be tracked using remote sensing from satellites, a function performed by NASA. The World Health Organization (WHO), a branch of the United Nations, integrates data from national programs and Air-Quality monitoring conducts monitoring of its own. Air pollution is an international problem, and the lack of controls in developing nations spills over into the entire biosphere.

Workplace air monitoring falls under the auspices of state occupational safety and health administrations. In addition to protecting workers against the by-products of manufacturing processes, monitoring also identifies allergens, ventilation problems, and secondhand tobacco smoke.

Monitoring methods and instrumentation. Methods of monitoring are specific to the pollutant. A generalized monitoring device consists of an air pump capable of collecting samples of defined volume, a means of concentrating and fixing the pollutants of interest, and either an internal sensor that registers and records the level of the pollutant or a removable collector.

An ozone monitor is an example of a continuous emission sampler based on absorption spectroscopy. A drop in beam intensity is proportional to ozone concentration in the chamber. Absorption spectrometers exist for sulfur and nitrogen oxides. This type of technology is portable and relatively inexpensive to run and can be used under field conditions, for example monitoring in-use emissions of motor vehicles. Absorption spectroscopy is also used in satellite remote sensing and has been adapted to remote sensing devices deployed on the ground to measure vehicular emissions.

There are a number of methods for measuring particulate matter, the simplest of which, found in some home smoke detectors, involves a photoelectric cell sensitive to the amount that a light beam is obscured. More sophisticated mass monitors measure scattering of a laser light beam, with the degree of scattering proportional to particle size and density. Forcing air through a filter traps particles for further analysis. X-ray fluorescence, in which a sample is bombarded with X rays and emitted light is measured, will detect lead, mercury, and cadmium at very low concentrations. Asbestos fibers present will turn up either by X-ray fluorescence or visual inspection of filters. Pollen and mold spores, important as allergens, are detected by visual inspection. A drawback of filters is cost and the skilled labor required to process them.

A total hydrocarbon analyzer used by the auto industry uses the flame ionization detection principle to identify specific hydrocarbons in auto exhaust.

Volatile organic compounds (VOCs) present a challenge because total concentration is low outside of enclosed spaces and certain industrial sites, and rapid efficient methodology is not available for distinguishing between different classes of organic compounds.

APPLICATIONS AND PRODUCTS

Reporting and predicting air quality. Media report air-quality indices along with other weather data, and the general public has become accustomed to using this information to plan activities such as outdoor recreation. Projections are also used to schedule unavoidable industrial-emissions release to coincide with favorable weather patterns and minimize public inconvenience. Predicting air quality is an evolving and inexact science involving predicting and tracking weather patterns but also integrating myriad human activities.

Vehicular emissions. Federal law mandates that urban areas with unhealthy levels of vehicular pollution require testing of automobile emissions. Laws concerning testing vary considerably from state to state and jurisdictions within states. Additionally, new vehicles manufactured or sold in the United States must undergo factory testing, both of the model and of the individual units, to ensure the vehicle meets federal standards.

Typical vehicle-inspection protocol requires motorists to bring the vehicle to a garage where automatic equipment samples exhaust and analyzes it for CO, aggregate hydrocarbons, nitrogen and sulfur oxides, and particulate matter. Some state departments of motor vehicles operate their own inspection stations, while others license private garages. There are a number of compact units on the market that provide the required information with little operator input.

With improving air quality and a decreasing proportion of older cars that lack pollution-control equipment, some jurisdictions are withdrawing from vehicle testing.

Public-health effects of air-pollution monitoring and mitigation on public health deserve mentioning. Adverse effects on human health were the principal rationale for instituting laws curbing air pollution, and the decline in certain health problems associated with atmospheric pollution since those laws were enacted is testimony to their effectiveness.

Rates of lung cancer and chronic obstructive pulmonary disease (COPD) have declined, and ages of onset have increased substantially since the mid-twentieth century. Although the bulk of this is due to declining tobacco use, some of it is related to pollution control.

IMPACT ON INDUSTRY

Impact on industry is twofold, including the effects, both positive and negative, of complying with emissions regulations, and the creation and marketing of new products to meet the needs of monitoring.

Costs of installing pollution-mitigation equipment and monitoring emissions are nontrivial, and the benefits to a community of cleaning up an industry do not necessarily compensate the manufacturer's bottom line. Earlier fears that stringent environmental regulations would force plants to close or relocate elsewhere have to some extent been realized, although the relative contributions of environmental legislation, labor costs, tax structure, and other government policies to the exodus of heavy industry from the United States are debated. When existing facilities are exempted from regulations but new facilities must comply, it discourages growth and investment. Few people doubt that the benefits of clean air outweigh the costs.

Air-pollution monitoring and abatement is a major industry in its own right, employing people in state and federal environmental agencies, university- and industry-based research laboratories, and in the manufacture, sale, and servicing of pollution-control devices and monitoring equipment. Because devices and their deployment must comply with constantly shifting state and federal regulations, this is an industry in which the manufacturing end for industrial and research equipment remains based in the

United States.

The consumer market for pollution-detection devices and services is growing rapidly, particularly for interior air. Smoke detectors have become a standard household fixture and CO monitors are recommended for fuel-oil- and gas-heated homes. Testing for formaldehyde, mold, and in some localities radon is often part of building inspections during real estate sales.

Although the field is not expanding at the rate it once was, identification of additional areas of concern, the rapid expansion of industry in other parts of the world, and the continuing need for trained technicians make this one of the better employment prospects for would-be research scientists.

CAREERS AND COURSE WORK
Air-pollution monitoring is a field with solid career prospects, in government and private industry. The majority of openings call for field technicians who supervise monitoring facilities, conduct a varying amount of chemical and physical analyses (now mostly automated), and collect and analyze data. For this type of position a bachelor's degree in a field that includes substantial grounding in chemistry, mathematics, and data management, and training on the types of equipment used in monitoring, is usually required. A number of state colleges offer undergraduate degrees in environmental engineering that provide a solid background. Online programs offered by for-profit institutions lack the rigor and hands-on experience necessary for this demanding occupation.

For research positions in government laboratories and educational institutions, an advanced degree in meteorology, environmental science, or environmental engineering is generally required.

This is a rapidly evolving field for which knowledge of the latest techniques and regulations is essential. Degree programs that offer a solid internship program, integrating students into working government or industrial laboratories, offer a tremendous advantage in a job market where actual experience is essential.

The development, sale, and servicing of monitoring equipment offers other employment options. While course work can provide the general level of knowledge necessary to sell, adjust, and repair sophisticated automated electronic equipment, such a career objective will also require extensive on-the-job training. Some manufacturers offer factory training for service people.

SOCIAL CONTEXT AND FUTURE PROSPECTS
The regulations and remediation efforts that monitoring informs and supports have clearly had a positive impact on the health and well-being of Americans in the nearly half-century since the passage of the

---

FASCINATING FACTS ABOUT AIR-QUALITY MONITORING

- In 1987, the Environmental Protection Agency (EPA) ranked indoor air pollution from cigarette smoke, poorly functioning heating systems, and formaldehyde from building materials as fourth in thirteen environmental health problems analyzed.

- Ninety percent of Californians were exposed to unhealthful air conditions at some point in 2008.

- The earliest known complaint about noxious air pollution is one registered in Nottingham, England, in 1257 by Queen Eleanor of Provence.

- Medieval and early modern physicians were quick to associate the smell of sulfur in coal smoke as a health hazard because of the prevailing belief that disease was caused by miasmas, or vaporous exhalations from swamps and rotting material including sewage and garbage.

- In 1955 the city of Los Angeles instituted the first systematic air-monitoring program to track levels of the city's notorious smog, caused mainly by motor vehicles. The first monitors used the rate of degradation of thin rubber strips to determine when smog levels were unhealthful and vulnerable individuals should stay indoors.

- Tetraethyl lead, a gasoline additive and significant pollutant, was removed from gasoline not because of human health issues but because it poisoned catalytic converters, needed for fuel efficiency.

- Haze from coal-fired electric generating plants is adversely affecting tourism in Great Smoky Mountains National Park and other Southeastern vacation destinations noted for splendid vistas.

- Ironically, children who engage in vigorous exercise out of doors are vulnerable to respiratory damage from smog, making real-time monitoring of sulfur, nitrogen oxides, and ozone and prompt reporting to the public an important public-health measure.

Clean Air Act of 1963. Heavy-metal exposure from atmospheric sources has dropped dramatically. Mandatory pollution-control devices on new passenger vehicles have curbed emissions in states where annual Air-Quality monitoring vehicle emission tests are not even required. Older diesel trucks, farm vehicles, and stationary engines remain a concern in the United States.

Air pollution remains a significant problem in the developing world, especially in China, where the rapid growth of coal-fired industries has created conditions in urban areas reminiscent of Europe in the nineteenth century. Addressing these problems is a matter of international concern, because air pollution is no respecter of national boundaries.

Low-end environmental monitoring devices for the consumer market are a growth industry. With respect to genuine hazards such as smoke and carbon monoxide, this is a positive development, but there is concern that overzealous salespeople and environmental-consulting firms will exaggerate risks and push for costly solutions in order to enhance their bottom line, as occurred with asbestos abatement in the 1970's and 1980's.

Although often criticized, an integrated approach to environmental policy that includes "cap and trade"—allowing industries to use credits for exceeding standards in one area to offset lagging performance in other areas, or to sell these credits to other industries so long as an industry-wide target is met—helps ease the nontrivial burdens of complying with constantly evolving environmental standards.

Much of the information-gathering and tracking system developed to address emissions of criteria pollutants is being integrated into the effort to slow global warming due to carbon dioxide (CO_2) emissions from fossil fuel burning. While elevated CO_2 does not directly affect human health, and is actually beneficial to plants, the overall projected effects of global warming are sufficiently dire that efforts to reduce CO_2 emissions deserve a high priority in environmental planning.

—*Martha A. Sherwood, PhD*

FURTHER READING

Brimblecombe, Peter. *The Big Smoke: A History of Air Pollution in London Since Medieval Times.* 1987. New York: Routledge, 2011. A readable and fact-filled account of the history of air pollution.

Collin, Robert W. *The Environmental Protection Agency: Cleaning Up America's Act.* Westport, Conn.: Greenwood Press, 2006. Detailed history of environmental regulation, aimed at the general college-educated reader.

Committee on Air Quality Management in the United States, Board on Environmental Studies and Toxicology, Board on Atmospheric Sciences and Climate, Division on Earth and Life Studies. *Air Quality Management in the United States.* Washington, D.C.: National Academies Press, 2004. A reference for public agencies and policymakers, with much specific information.

Magoc, Chris J. *Environmental Issues in American History: A Reference Guide with Primary Documents.* Westport, Conn.: Greenwood Press, 2006. The chapters on tetraethyl lead and the Donora disaster deal with air pollution.

World Health Organization. Text edited by David Breuer. *Monitoring Ambient Air Quality for Health Impact Assessment.* Copenhagen: Author, 1999. Good coverage of efforts outside of the United States.

WEB SITES

Air and Waste Management Association
http://www.awma.org/Public

Ambient Monitoring Technology Information Center
http://www.epa.gov/ttn/amtic

American Academy of Environmental Engineers
http://www.aaee.net

National Association of Environmental Professionals
http://www.naep.org
United States Environmental Protection Agency

See also: Environmental Engineering; Meteorology.

ALGEBRA

FIELDS OF STUDY

College algebra; precalculus; calculus; linear algebra; discrete mathematics; finite mathematics; computer science; science; engineering; finance.

SUMMARY

Algebra is a branch of applied mathematics that goes beyond the practical and theoretical applications of the numbers of arithmetic. Algebra has a definitive structure with specified elements, defined operations, and basic postulates. Such abstractions identify algebra as a system, so there are algebras of different types, such as the algebra of sets, the algebra of propositions, and Boolean algebra. Algebra has connections not only to other areas of mathematics but also to the sciences, engineering, technology, and other applied sciences. For example, Boolean algebra is used in electronic circuit design, programming, database relational structures, and complexity theory.

KEY TERMS AND CONCEPTS

- **complement of a set:** Group of all elements that are not in a particular group but are in a larger group; the complement of a set A is the set of all elements that are not in A but are in the universal set and is written A'.
- **conjunction:** Sentence formed by connecting two statements with "and"; the conjunction of statements p and q is written $p \wedge q$.
- **cisjunction:** Sentence formed by connecting two statements with "or"; the disjunction of statements p or q is written $p \vee q$.
- **cqual sets:** Sets whose members are identical; if every element in set A is in set B, and if every element in set B is in set A, A and B are equal.
- **cquivalent sets:** Sets that have the same cardinality.
- **negation:** Statement that changes the truth value of a given statement to its opposite truth value; negation of statement p, or not p, is written as $\sim p$.
- **set:** Well-defined collection of objects or elements.
- **set intersection:** Elements that two sets have in common; the intersection of set A and set B is the set of all elements in A and in B and is written $A \cap B$.
- **set union:** Set of elements that are in either or both of two sets; the union of set A and set B is the set of elements that are in either set A or set B or in both sets and is written $A \cup B$.
- **subset:** Group of elements all of which are contained within a larger group of elements; if every element of set A is in set B, A is a subset of B.
- **universal set:** Universe of discourse or fixed set from which subsets are formed; written U.

DEFINITION AND BASIC PRINCIPLES

Algebra is a branch of mathematics. The word "algebra" is derived from an Arabic word that links the content of classical algebra to the theory of equations. Modern algebra includes a focus on laws of operations on symbolic forms and also provides a systematic way to examine relationships between such forms. The concept of a basic algebraic structure arises from understanding an important idea. That is, with the traditional definition of addition and multiplication, the identity, associative, commutative, and distributive properties characterize these operations with not only real numbers and complex numbers but also polynomials, certain functions, and other sets of elements. Even with modifications in the definitions of operations on other sets of elements, these properties continue to apply. Thus, the concept of algebra is extended beyond a mere symbolization of arithmetic. It becomes a definitive structure with specified elements, defined operations, and basic postulates. Such abstractions identify algebra as a system, and therefore, there are algebras of many different types, such as the algebra of sets, the algebra of propositions, and Boolean algebra.

The algebra of sets, or set theory, includes such fundamental mathematical concepts as set cardinality and subsets, which are a part of the study of various levels of mathematics from arithmetic to calculus and beyond. The algebra of propositions (logic or propositional calculus) was developed to facilitate the reasoning process by providing a way to symbolically represent statements and to perform calculations based on defined operations, properties, and truth tables. Logic is studied in philosophy, as well as various areas of mathematics such as finite mathematics. Boolean algebra is the system of symbolic logic used primarily in computer science applications; it is studied in areas of applied mathematics such as discrete mathematics.

Boolean algebra can be considered a generalization of the algebra of sets and the algebra of propositions. Boolean algebra can be defined as a nonempty set B together with two binary operations, sum (symbol +) and product (symbol ×). There is also a unary operation, complement (symbol ¢). In set B, there are two distinct elements, a zero element (symbol 0) and a unit element (symbol 1), and certain laws or properties hold. The laws and properties table shows how laws and properties used in the algebra of sets and the algebra of propositions relate to those of Boolean algebra.

BACKGROUND AND HISTORY

The algebra of sets. In 1638, Italian scientist Galileo published Discorsi e dimostrazioni matematiche: Intorno à due nuove scienze attenenti alla mecanica e i movimenti locali (Dialogues Concerning Two New Sciences, 1900). In this work, Galileo recognized the basic concept of equivalent sets and distinguishing characteristics of infinite sets. During the nineteenth century, Bohemian mathematician Bernhard Bolzano studied infinite sets and their unique properties; English mathematician George Boole took an algebraic approach to the study of set theory. However, it was German mathematician Georg Cantor who developed a structure for set theory that later led to the modernization of the study of mathematical analysis.

Cantor had a strong interest in the arguments of medieval theologians concerning continuity and the infinite. With respect to mathematics, Cantor realized that not all infinite sets were the same. In 1874, his controversial work on infinite sets was published. After additional research, he established set theory as a mathematical discipline known as *Mengenlehre* (theory of assemblages) or *Mannigfaltigkeitslehre* (theory of manifolds).

The algebra of propositions and Boolean algebra. During the nineteenth century, Boole, English mathematician Charles Babbage, German mathematician Gottlob Frege, and Italian mathematician Giuseppe Peano tried to formalize mathematical reasoning by an "algebraization" of logic.

Boole, who had clerical aspirations, regarded the human mind as God's greatest accomplishment. He wanted to mathematically represent how the brain processes information. In 1847, his first book, *The Mathematical Analysis of Logic: Being an Essay Towards a Calculus of Deductive Reasoning*, was published with limited circulation. He rewrote and expanded his ideas in an 1854 publication, *An Investigation of the Laws of Thought: On Which Are Founded the Mathematical Theories of Logic and Probabilities*. Boole introduced the algebra of logic and is considered the father of symbolic logic.

Boole's algebra was further developed between 1864 and 1895 through the contributions of British mathematician Augustus De Morgan, British economist William S. Jevons, American logician Charles Sanders Peirce, and German mathematician Ernst Schröder. In 1904, American mathematician Edward V. Huntington's *Sets of Independent Postulates for the Algebra of Logic* developed Boolean algebra into an abstract algebraic discipline with different interpretations. With the additional work of American mathematician Marshall Stone and Polish American logician Alfred Tarski in the 1930's, Boolean algebra became a modern mathematical discipline, with connections to several other branches of mathematics, including topology, probability, and statistics.

In his 1940 Massachusetts Institute of Technology master's thesis, Claude Elwood Shannon used symbolic Boolean algebra as a way to analyze relay and switching circuits. Boole's work thus became the foundation for the development of modern electronics and digital computer technology.

Outside the realm of mathematics and philosophy, Boolean algebra has found applications in such diverse areas as anthropology, biology, chemistry, ecology, economics, sociology, and especially computer science. For example, in computer science, Boolean algebra is used in electronic circuit design, programming, database relational structures, and complexity theory.

HOW IT WORKS

Boolean algebra achieved a central role in computer science and information theory that began with its connection to set theory and logic. Set theory, propositional logic, and Boolean algebra all share a common mathematical structure that becomes apparent in the properties or laws that hold.

Set theory. The language of set theory is used in the definitions of nearly all mathematical elements, and set theory concepts are integrated throughout the mathematics curriculum from the elementary to the college level. In primary school, basic set concepts may be introduced in sorting, combining, or classifying objects even before the counting process is introduced. Operations such as set complement, union, and intersection can be easily understood in this context.

For example, let the universal set U consist of six blocks, each of which is a different color. A block may be red, orange, yellow, violet, blue, or green. Using set notation, U = {red, orange, yellow, violet, blue, green}. Let four of the six blocks be sorted into two subsets, A and B, such that A = {red, yellow} and B = {blue, green}. The complement of set A is the set of blocks that are neither red nor yellow, $A¢$ = {orange, violet, blue, green}. The union of sets A and B is the set that contains all of the blocks in set A or set B or both, if there were any colors in common: A È B = {red, yellow, blue, green}. The intersection of sets A and B is the set of blocks that are in set A and in set B, any color that both sets have in common. Because the two sets of blocks have no color in common, A Ç B = Æ.

Above the primary level, the concepts of logic are introduced. Daily life often requires that one construct valid arguments, apply persuasion, and make meaningful decisions. Thus, the development of the ability to organize thoughts and explain ideas in clear, precise terms makes the study of reasoning and the analysis of statements most appropriate.

Logic. In propositional algebra, statements are either true or false. A statement may be negated by using "not." Statements can be combined in a variety of ways by using connectives such as "and" and "or." The resulting compound statements are either true or false, based on given truth tables.

A compound statement such as "The First International Conference on Numerical Algebra and Scientific Computing was held in 2006 and took place at the Institute of Computational Mathematics of the Chinese Academy of Sciences in New York" can thus be easily analyzed, especially when written symbolically. The "and" connective indicates that the compound statement is a conjunction. Let p be "The First International Conference on Numerical Algebra and Scientific Computing was held in 2006," a true statement; let q be "(it) took place at the Institute of Computational Mathematics of the Chinese Academy of Sciences in New York," a false statement

A page from Al-Khw rizm 's al-Kit b al-mu ta ar f is b al- abr wa-l-muq bala

because the institute is in Beijing. The truth table for the conjunction indicates that the given compound statement is false: T Ù F º F.

Compound symbolic statements may require multistep analyses, but established properties and truth tables are still used in the process. For example, it is possible to analyze the two symbolic compound statements ~(p Ú q) and ~p Ù ~q and also to verify that they are logically equivalent. The truth tables for each compound statement can be combined in one large table to facilitate the process. The first two columns of the table show all possibilities for the truth values of two statements, p and q. The next three columns show the analysis of each of the parts of the two given compound statements, using the truth tables for negation, disjunction, and conjunction. The last two columns of the table have exactly the same corresponding T and F entries, showing that the truth value will be the same in all cases. This verifies that the two compound statements are logically equivalent. Note that the equivalence of these two propositions is one of De Morgan's laws: ~(p Ú q) º ~p Ù ~q.

Computer circuits. Shannon showed how logic could be used to design and simplify electric circuits. For example, consider a circuit with switches p and q that can be open or closed, corresponding to the Boolean binary elements, 0 and 1. A series circuit corresponds to a conjunction because both switches must be closed for electric current to flow. A circuit where electricity flows whenever at least one of the switches is closed is a parallel circuit; this corresponds to a disjunction. Because the complement for a given switch is a switch in the opposite position, this corresponds to a negation table. When a circuit is represented in symbolic notation, its simplification may use the laws of logic, such as De Morgan's laws. The simplification may also use tables in the same way as the analysis of the equivalence of propositions, with 1 replacing T and 0 replacing F. Other methods may use Karnaugh maps, the Quine-McCluskey method, or appropriate software. Computer logic circuits are used to make decisions based on the presence of multiple input signals. The signals may be generated by mechanical switches or by solid-state transducers. The various families of digital logic devices, usually integrated circuits, perform a variety of logic functions through logic gates. Logic gates are the basic building blocks for constructing digital systems. The gates implement the hardware logic function based on Boolean algebra. Two or more logic gates may be combined to provide the same function as a different type of logic gate. This process reduces the total number of integrated circuit packages used in a product. Boolean expressions can direct computer hardware and also be used in software development by programmers managing loops, procedures, and blocks of statements.

Boolean searches. Boolean algebra is used in information theory. Online queries are input in the form of logical expressions. The operator "and" is used to narrow a query and "or" is used to broaden it. The operator "not" is used to exclude specific words from a query. For example, a search for information about "algebra freeware" may be input as "algebra or freeware," "algebra and freeware," or perhaps "algebra and freeware not games." The amount of information received from each query will be different. The first query will retrieve many documents because it will select those that contain "algebra," those that contain "freeware," and those that contain both terms. The second query will retrieve fewer documents because it will select only those documents that contain both terms. The last query will retrieve documents that contain both "algebra" and "freeware" but will exclude items containing the term "games."

APPLICATIONS AND PRODUCTS

Logic Machines, Calculating Machines, and Computers. The "algebraization" of logic, primarily the work of De Morgan and Boole, was important to the transformation of Aristotelian logic into modern logic, and to the automation of logical reasoning. Several machines were built to solve logic problems, including the Stanhope demonstrator, Jevons's logic machine, and the Marquand machine. In the mid-nineteenth century, Jevons's logic machine, or logic piano, was among the most popular; it used Boolean algebra concepts. Harvard undergraduates William Burkhardt and Theodore Kalin built an electric version of the logic piano in 1947.

In the 1930's, Boolean algebra was used in wartime calculating machines. It was also used in the design of the first digital computer by John Atanasoff and his graduate student Clifford Berry. During 1944-1945, John von Neumann suggested using the binary mathematical system to store programs in computer memory. In the 1930's and 1940's, British mathematician Alan Turing and American mathematician Shannon recognized that binary logic was well suited to the development of digital computers. Just as Shannon's work served as the basis for the theory of switching and relay circuits, Turing's work became the basis for the field of automata theory, the theoretical study of information processing and computer design.

By the end of World War II, it was apparent that computers would soon replace logic machines. Later computer software and hardware developments confirmed that the logic process could be mechanized. Although research work continues to provide theoretical guidelines, automated reasoning programs such as those used in robotics development, are in demand by researchers to resolve questions in mathematics, science, engineering, and technology.

Integrated circuit design. Boolean algebra became indispensable in the design of computer microchips and integrated circuits. It is among the fundamental concepts of digital electronics that are essential to understanding the design and function of different types of equipment.

Many integrated circuit manufacturers produce complex logic systems that can be programmed to perform a variety of logical functions within a single integrated circuit. These integrated circuits include gate array logic (GAL), programmable array logic (PAL), the programmable logic device (PLD), and the complex programmable logic device (CPLD).

Engineering approaches to the design and analysis of digital logic circuits involves applications of advanced Boolean algebra concepts, including algorithmic state and machine design of sequential circuits, as well as digital logic simulation. The actual design and implementation of sizeable digital design problems involves the use of computer-aided design (CAD).

Computer algebra systems. During the 1960's and 1970's, the first computer algebra systems (CASs) emerged and evolved from the needs of researchers. Computer algebra systems are software that enable users to do tedious and sometimes difficult algebraic tasks, such as simplifying rational functions, factoring polynomials, finding solutions to a system of equations, and representing information graphically in two or three dimensions. The systems offer a programming language for user-defined procedures. Computer algebra systems have not only changed how algebra is taught but also provided a convenient tool for mathematicians, scientists, engineers, and technicians worldwide.

Among the first popular computer algebra systems were Reduce, Scratchpad, Macsyma (later Maxima), and Mu-Math. Later popular systems include MATLAB, Mathematica, Maple, and MathCAD. In 1987, Hewlett-Packard introduced HP-28, the first handheld calculator series with the power of a computer algebra system. In 1995, Texas Instruments released the TI-92 calculator with advanced CAS capabilities based on Derive software. Manufacturers continue to offer devices such as these with increasingly powerful functions; such devices tend to decrease in size and cost with advancements in technology.

IMPACT ON INDUSTRY

Government and university research. Boolean algebra has roots and applications in many areas, including topology, measure theory, functional analysis, and ring theory. Research and study of Boolean algebras therefore includes structure theory and model theory, as well as connections to other logics. Some of the techniques for analyzing Boolean functions have been used in such areas as computational learning theory, combinatorics, and game theory.

Computer algebra, originally known as algebraic computing, is concerned with the development, implementation, and application of algorithms that manipulate and analyze mathematical expressions. Practical and theoretical research includes the development of effective and efficient algorithms for use in computer algebra systems. Research includes engineering, scientific, and educational applications.

Linear algebra begins with the study of linear equations, matrices, determinants, function spaces, eigenvalues, and orthogonality. Research and development in applied linear algebra includes theoretical studies, algorithmic designs, and implementation of advanced computer architectures. Such research involves scientific, engineering, and industrial applications.

Engineering and technology. The study of associative digital network theory comprises computer science, electrical engineering digital circuit design, and number theory. Such theory is of interest to researchers at industrial laboratories and instructors and students at technical institutions. The focus is on new research and developments in modeling and designing digital networks with respect to both mathematics and engineering disciplines. The unifying associative algebra of function composition (semigroup theory) is used in the study of the three main computer functions: sequential logic (state machines), arithmetic, and combinational (Boolean) logic.

Applied science. There has been a dramatic rise in the power of computation and information technology. With it have come vast amounts of data in fields such as business, engineering, and science. The challenge of understanding the data has led to new tools and approaches, such as data mining. Data mining involves the use of algorithms to identify and verify structure from data analysis. Developments in the field of data mining have brought about increased focus on higher level mathematics. Such areas as topology, combinatorics, and algebraic structures (lattices and Boolean algebras) are often included in research.

CAREERS AND COURSE WORK

The applications of algebra are numerous, which means that those interested in algebra can pursue jobs and careers in a wide range of fields, including business, engineering, and science, particularly computer science.

Data analyst or data miner. Data mining is a broad mathematical area that involves the discovery of patterns and hidden information in large databases, using algorithms. In applications of data mining, career opportunities emerge in e-commerce, security, forensics, medicine, bioinformatics and genomics, astrophysics, and chemical and electric power engineering. Course work should include a focus on

LAWS AND PROPERTIES

Law or Property	Algebra of Sets (set theory)	Algebra of Propositions (logic or proposition calculus)	Boolean Algebra
	For nonempty sets A, B, and C that are subsets of a universal set U ($\varnothing$ designates the empty set)	For propositions p, q, and r (T is a true proposition. F is a false proposition.)	For any elements x, y, and z in set B (The operation symbol $\times$ may be omitted.)
Identity property	$A \cup \varnothing = A$ $A \cap U = A$	$p \vee F \equiv p$ $p \wedge T \equiv p$	$x + 0 = x$ $x \times 1 = x$
Complement law	$A \cup A' = U$ $A \cap A' = \varnothing$	$p \vee \sim p \equiv T$ $p \wedge \sim p \equiv F$	$x + x' = 1$ $x \times x' = 0$
Involution law	$(A')' = A$	$\sim (\sim p) \equiv p$	$(x')' = x$
Commutative property	$A \cup B = B \cup A$ $A \cap B = B \cap A$	$p \vee q \equiv q \vee p$ $p \wedge q \equiv q \wedge p$	$x + y = y + x$ $x \times y = y \times x$
Associative property	$(A \cup B) \cup C = A \cup (B \cup C)$ $(A \cap B) \cap C = A \cap (B \cap C)$	$(p \vee q) \vee r \equiv p \vee (q \vee r)$ $(p \wedge q) \wedge r \equiv p \wedge (q \wedge r)$	$(x + y) + z = x + (y + z)$ $(x \times y) \times z = x \times (y \times z)$
Distributive property	$A \cup (B \cap C) = (A \cup B) \cap (A \cup C)$ $A \cap (B \cup C) = (A \cap B) \cup (A \cap C)$	$(p \vee (q \vee r) \equiv (p \vee q) \wedge (p \vee r)$ $(p \wedge (q \wedge r) \equiv (p \wedge q) \vee (p \wedge r)$	$x + (y \times z) = (x + y) \times (x + z)$ $x \times (y + z) = (x \times y) + (x \times z)$
De Morgan's laws	$(A \cup B)' = A' \cap B'$ $(A \cap B)' = A' \cup B'$	$\sim (p \vee q) \equiv \sim p \wedge \sim q$ $\sim (p \wedge q) \equiv \sim p \vee \sim q$	$(x + y)' = x' \times y'$ $(x \times y)' = x' + y'$
Idempotent law	$(A \cup A) = A$ $(A \cap A) = A$	$p \vee p \equiv p$ $p \wedge p \equiv p$	$x + x = x$ $x \times x = x$
Absorption law			$x + (x \times y) = x$ $x \times (x + y) = x$
Domination law	$(A \cup U) = U$ $(A \cap \varnothing) = \varnothing$	$p \vee T \equiv T$ $p \wedge F \equiv F$	$x + 1 = 1$ $x \times 0 = 0$

higher level mathematics in such areas as combinatorics, topology, and algebraic structures.

Materials engineer. Materials science is the study of the properties, processing, and production of such items as metallic alloys, liquid crystals, and biological materials. There are many career opportunities in research, manufacturing, and development in aerospace, electronics, biology, and nanotechnology. The design and analysis of materials depends on mathematical models and computational tools. Course work should include a focus on applied mathematics, including differential equations, linear algebra, numerical analysis, operations research, discrete mathematics, optimization, and probability.

Computer animator or digital artist. Computer animation encompasses many areas, including mathematics, computer science, physics, biomechanics, and anatomy. Career opportunities arise in medical diagnostics, multimedia, entertainment, and fine arts. The algorithms for computer animation come from scientific relationships, statistics, signal processing, linear algebra, control theory, and computational geometry. Recommended mathematics course work includes statistics, discrete mathematics, linear algebra, geometry, and topology.

Financial analyst. As quantitative methods transform the financial industry, banking, insurance, investment, and government regulatory institutions are among those relying on mathematical tools and computational models. Such tools and models are used to support investment decisions, to develop and price new securities, to manage risk, and to guide portfolio selection, management, and optimization. Course work should include a focus on the mathematics of finance, linear algebra, linear programming, probability, and descriptive statistics.

SOCIAL CONTEXT AND FUTURE PROSPECTS

Algebra is part of two broad, rapidly growing fields, applied mathematics and computational science. Applied mathematics is the branch of mathematics that develops and provides mathematical methods to meet scientific, engineering, and technological needs. Applied mathematics includes not only discrete mathematics and linear algebra but also numerical analysis, operations research, and probability. Computational science integrates applied mathematics, science, engineering, and technology to create a multidisciplinary field developing and using

FASCINATING FACTS ABOUT ALGEBRA

- Algebra has been studied since 2000 BCE., making it the oldest branch of written mathematics. Babylonian, Chinese, and Egyptian mathematicians proposed and solved problems in words, that is, using "rhetorical algebra."

- In 1869, British logician William S. Jevons, a student of the mathematician Augustus De Morgan, created a logic machine that used Boolean algebra. The popular machine was known as the logic piano because it had ivory keys and resembled a piano.

- English logician John Venn was heavily influenced by English mathematician George Boole, and his Venn diagrams, developed around 1880, facilitated conceptual and procedural understanding of Boolean algebra.

- In his 1936 paper, "On Computable Numbers, with an Application to the Entscheidungs problem," British mathematician Alan Turing characterized which numbers and functions in mathematics are effectively computable. His paper was an early contribution to recursive function theory, which was a topic of interest in several areas, including logic.

- With the publication of A Symbolic Analysis of Relay and Switching Circuits (1940) and "A Mathematical Theory of Communication" (1948), American mathematician Claude Elwood Shannon introduced a new area for the application of Boolean algebra. He showed that the basic properties of series and parallel combinations of electric devices such as relays could be adequately represented by this symbolic algebra. Since then, Boolean algebra has played a significant role in computer science and technology.

- In applied algebra, properties of groups can be used to analyze transformations and symmetry. The transformations include translating, rotating, reflecting, and dilating a pattern such as one in an M. C. Escher painting, or parts of an object such as a Rubik's cube.

- Applied algebra is used in cryptography to study codes and ciphers in problems involving data security and data integrity.

- Applied algebra is used in chemistry to study symmetry in molecular structure.

TRUTH TABLES FOR ALGEBRA OF PROPOSITIONS & BOOLEAN ALGEBRA

Algebra of Propositions			Boolean Algebra		
For propositions p and q (T is a true proposition; F is a false proposition)			For any elements x and y in set B		
Negation			Negation		
p	~p		x	x'	
T	F		1	0	
F	T	0	1		
Disjunction	Sum				
p	q	p $\vee$ q	x	y	x + y
T	T	T	1	1	1
T	F	T	1	0	1
F	T	T	0	1	1
F	F	F	0	0	0
Conjunction			Product		
p	q	p $\wedge$ q	x	y	x × y
T	T	T	1	1	1
T	F	F	1	0	0
F	T	F	0	1	0
F	F	F	0	0	0

innovative problem-solving strategies and methodologies. Applied mathematics and computational science are used in almost every area of science, engineering, and technology. Business also relies on applied mathematics and computational science for research, design, and manufacture of products that include aircraft, automobiles, computers, communication systems, and pharmaceuticals. Research in applied mathematics therefore often leads to the development of new mathematical models, theories, and applications that contribute to diverse fields.

—*June Gastón, MSEd, MEd, EdD*

FURTHER READING

Barnett, Raymond A., Michael R. Ziegler, and Karl E. Byleen. *Finite Mathematics for Business, Economics, Life Sciences, and Social Sciences.* 12th ed. Boston: Prentice Hall, 2011. Covers the mathematics of finance, linear algebra, linear programming, probability, and descriptive statistics, with an emphasis on cross-discipline principles and practices. Helps develop a functional understanding of mathematical concepts in preparation for application in other areas.

Cohen, Joel S. *Computer Algebra and Symbolic Computation: Elementary Algorithms.* Natick, Mass.: A. K. Peters, 2002. Examines mathematical fundamentals, practical challenges, formulaic solutions, suggested implementations, and examples in a few programming languages appropriate for building a computer algebra system. Further reading recommendations provided.

Cooke, Roger. *Classical Algebra: Its Nature, Origins, and Uses.* Hoboken, N.J.: Wiley-Interscience, 2008. Broad coverage of classical algebra that includes

TRUTH TABLE VERIFYING

$\sim(p \vee q) \equiv \sim p \wedge \sim q$

p	q	$\sim p$	$\sim q$	$p \vee q$	$\sim(p \vee q)$	$\sim p \wedge \sim q$
T	T	F	F	T	F	F
T	F	F	T	T	F	F
F	T	T	F	T	F	F
F	F	T	T	F	T	T

its history, pedagogy, and popularization. Each chapter contains thought-provoking problems and stimulating questions; answers are provided in the appendix.

Dunham, William. *Journey Through Genius: The Great Theorems of Mathematics.* New York: Penguin, 1991. Provides historical and technical information. Each chapter is devoted to a mathematical idea and the people behind it. Includes proofs.

Givant, Steven, and Paul Halmos. *Introduction to Boolean Algebras.* New York: Springer, 2009. An informal presentation of lectures given by the authors on Boolean algebras, intended for advanced undergraduates and beginning graduate students.

Van der Waerden, B. L. *Algebra.* New York: Springer, 2003. Reprint of the first volume of the 1970 translation of van der Waerden's *Moderne Algebra* (1930), designated one of the most influential mathematics textbooks of the twentieth century. Based in part on lectures by Emmy Noether and Emil Artin.

WEB SITES

American Mathematical Society
http://www.ams.org
Mathematical Association of America
http://www.maa.org
Society for Industrial and Applied Mathematics
http://www.siam.org

See also: Applied Mathematics; Calculus; Computer Languages, Compilers, and Tools; Electronics and Electronic Engineering; Engineering Mathematics; Geometry; Numerical Analysis; Pattern Recognition.

ANTIBALLISTIC MISSILE DEFENSE SYSTEMS

FIELDS OF STUDY

Astronautical engineering; physics; electronics engineering; electrical engineering; mechanical engineering; optical engineering; software engineering; systems engineering.

SUMMARY

To protect people and their possessions from harm by incoming missiles launched by a potential enemy, antiballistic missile defense systems have been designed to detect, track, and destroy incoming missiles. These systems are designed to fire guided defensive missiles to hit the incoming missiles before they strike their targets.

KEY TERMS AND CONCEPTS

- **ballistic missile:** Missile that is powered only during the first part of its flight; includes missiles powered and steered during the end of their flight.
- **boost phase:** Initial part of a missile's flight, when it is under power.
- **electromagnetic pulse (EMP):** Burst of electromagnetic radiation given off by a nuclear explosion that sends a burst of energy through electronic devices and may severely damage them.
- **exoatmospheric:** Outside or above the atmosphere.

- **hit-to-kill:** Destroying a warhead by hitting it precisely, often compared to hitting a bullet with a bullet.
- **kill vehicle:** Speeding mass that smashes into a warhead using its energy of motion (kinetic energy) to destroy the warhead.
- **midcourse phase:** Middle portion of a missile's flight, as it coasts through space.
- **terminal phase:** End of a missile's flight, as it reenters the atmosphere and nears its target.
- **warhead:** Payload of the missile that does the desired damage; it can consist of high explosives, a nuclear bomb, or a kill vehicle.

DEFINITION AND BASIC PRINCIPLES

An intercontinental ballistic missile (ICBM) can deliver enough nuclear explosives to devastate a city. Because ICBMs plunge from the sky at very high speeds, unseen by the eye until they hit, they are difficult to defend against. That is the purpose of an antiballistic missile (ABM) system. Such systems have several essential tasks: to detect a missile launch or ascent, to track the missile during its midcourse and its terminal flight, and to calculate an intercept point for the defensive missile, fire a defensive missile, track the defensive missile, guide it to its target, and verify that the target has been destroyed.

Using multiple independently targeted reentry vehicle (MIRV) technology, ten to twelve independently targeted warheads and decoys can be delivered by a single ICBM (which is against the signed but unratified 1993 Strategic Arms Limitations Talks II treaty). Each deployed warhead requires a defensive missile to destroy it. Countries with MIRV technology could attack with enough warheads and decoys to saturate any ABM system, allowing at least some nuclear warheads to reach their targets. If the targeted country retaliates, the situation is likely to escalate into full-scale nuclear war.

Although ABM systems cannot prevent nuclear war, they do have some uses. For example, a limited system could defend against a few accidental launches of ICBMs or against a few missiles from rogue nations.

BACKGROUND AND HISTORY

The evolution of ABM systems has been driven by politics and perceived need. In the 1960's, the strategy of "mutual assured destruction," or MAD,

A Ground-Based Interceptor of the United States' Ground-Based Midcourse Defense system, loaded into a silo at Fort Greely, Alaska, in July 2004

was articulated by U.S. secretary of defense Robert McNamara. If either side launched a first strike against the other, the nonaggressor would still have enough warheads left to devastate the aggressor. An effective ABM system would have upset this balance. The ABM treaty of 1972 was signed by U.S. president Richard Nixon and the Soviet Union's general secretary, Leonid Brezhnev, and was amended in 1974 to allow each side to have only one ABM site. The Soviet Union elected to place its single ABM system around Moscow, a system still maintained by Russia.

The United States built the Safeguard ABM system at the Grand Forks Air Force Base in North Dakota, where it guarded Minuteman III missiles. Safeguard operated for only a few months before it was closed down. The U.S. Missile Defense Agency was formed in

1983 (under a different name) and given the charge to develop, test, and prepare missile defense systems. Also in 1983, President Ronald Reagan announced a change: The offensive MAD would become the defensive building of an impenetrable shield under the Strategic Defense Initiative, popularly dubbed "Star Wars." Although Edward Teller, chief architect of the hydrogen bomb, assured President Reagan that the Strategic Defense Initiative could be implemented, it proved to be unfeasible.

After the Cold War ended with the dissolution of the Soviet Union in 1991, both the United States and Russia eventually concluded that the most likely use of their ABM systems would be against a limited strike by a country such as North Korea or Iran. To guard against Iranian and North Korean missiles, the United States has twenty-six ground-based interceptor (GBI) missiles at Fort Greeley, Alaska, and four at Vandenberg Air Force Base in California. (Two sites would not have been allowed under the ABM treaty, but the United States withdrew from the ABM treaty in June, 2002.) Warships equipped with the Aegis Combat System (a system that uses radar and computers to guide weapons aimed at enemy targets), with Standard Missile-3's (SM-3's), are stationed in the Mediterranean and the Black Sea to protect Europe.

How it Works

An ABM system must successfully perform several different functions to work properly. Initial detection may be done by a remote ground radar, by an airborne radar plane, or even by infrared sensors in space. Infrared sensors are particularly effective at spotting the hot rocket plume of a launching ICBM. Normal radar is line of sight and cannot detect targets beyond the curve of the Earth. Therefore, airborne radar is used; being higher, it can see farther.

To guard the United States from attack by ICBMs or submarine-launched missiles, three PAVE PAWS (Precision Acquisition Vehicle Entry Phased-Array Warning System) radars look outward from American borders. PAVE is a U.S. Air Force program, and the three radar systems are located on Air Force bases in Massachusetts, Alaska, and California. The systems track satellites and can spot a car-sized object in space 5,500 kilometers away. The initial detection of a long-range missile would probably come from a PAVE PAWS radar system. The radar and associated computers must also classify the objects they detect and determine if the objects are threatening.

If PAVE PAWS spots a suspicious object, a defensive missile site will be notified and will begin tracking the object with its on-site phased-array radar. If the object is still deemed a threat, permission to fire on it must be given either by direct command or by standing orders. Aided by a fire-control computer, the operator selects the target and fires one or two missiles at it. The missiles are guided using ground radar until they approach the target, when the missile's own radar or infrared sensors assume the tracking duties.

Modern missiles are usually hit-to-kill missiles, although some carry conventional explosives to ensure the kill. Next, the radar and computer must see if the target has been destroyed, and if not, defenses closer to the ICBM's target must be activated. These actions are all coordinated through the "command, control, and communications" resources of the on-site unit.

Initially, antiballistic missiles were designed to approach their targets and detonate a nuclear warhead, so great accuracy was not required. Close was good enough. The Safeguard ABM system had a long-range Spartan missile with a 5-megaton yield and the short-range Sprint missile, which carried a neutron bomb (an "enhanced radiation bomb"). The Spartan was to engage targets in space, and any surviving targets would be destroyed in the atmosphere by the Sprint. Unfortunately, the nuclear explosions would produce electromagnetic pulses, which would blind the Spartan/Sprint guiding radar. This problem encouraged designers to work toward a hit-to-kill technology. Russia has an ABM system around Moscow and has removed the nuclear warheads from these missiles and replaced them with conventional explosives.

Applications and Products

The hardware of an ABM system includes several key parts. Radars and infrared sensors are the eyes of an ABM system. Defensive missiles destroy the invading missiles, and computers calculate trajectories and direct the defensive missiles.

Radar. A phased-array radar antenna is a key component of any modern ABM system. It consists of an array of hundreds or thousands of small antennas mounted in a regular array of rows and columns on a wall. The radar can project a beam in a certain direction or receive a return echo from a particular direction by activating the small antennas in certain patterns. Because this is all done electronically without the radar antenna moving, many scan patterns can be run simultaneously. The Aegis SN/SPY-1 radar can simultaneously track more than one-hundred targets at a distance of up to 190 kilometers. Some multi-mission Navy destroyers have the AN/SPY-3 radar. It combines the functions of several radars into one, requires fewer operators, and is less visible to other radars. An advanced radar can also interrogate an IFF (identify friend or foe) device. Incoming missiles can also be identified by their flight paths and radar signatures.

Missiles. The Patriot system began as an antiaircraft system but was upgraded to defend against tactical missiles. It seemed to do well against scud missiles during the 1990-1991 Gulf War, but later analysis showed that most of the claimed kills were actually ripped apart by air resistance. The Patriot system is highly mobile because all its modules are truck or trailer mounted. One hour after the unit arrives on site, it can be up and running. The system uses the Patriot Advanced Capability-2 (PAC-2) missile with a range of 160 kilometers, a ceiling of 24 kilometers, a speed of Mach 5, and an explosive warhead. Its phased-array radar is difficult to jam.

The PAC-3 missile is smaller; four of them will fit in the space taken by one PAC-2 missile. This gives a missile battery more firepower. The missile travels at Mach 5, has a ceiling of 10 to 15 kilometers, and a maximum range of 45 kilometers. The warhead is hit-to-kill, backed up with an exploding fragmentation bomb on a proximity fuse. Two missiles are fired at a target, with the second missile firing a few seconds after the first. The second missile targets whatever might be a warhead left in the debris from the first warhead's impact. Under test conditions, the PAC-3 missile has scored twenty-one intercepts out of thirty-nine attempts.

The Theater High Altitude Area Defense missile system (THAAD) is designed to shoot down short, medium, and intermediate range missiles during their terminal phases. It uses the AN/TPY-2 radar. THAAD missiles also have some limited capability against ICBMs. Their effective range is about 200 kilometers with a peak altitude of 150 kilometers. Their warhead is a kinetic kill vehicle (KKV). When the missile nears its target, the KKV is explosively separated from its spent rocket. Guided by an advanced

infrared sensor, steering rockets adjust the KKV's course so that it will hit dead on. In the six tests since 2006, the THAAD missile hit its target all six times.

The Ground-Based Midcourse Defense (GMD) system is designed to defend against a limited attack by intermediate- and long-ranged missiles. Its missile is the Ground-Based Interceptor (GBI), a three-stage missile with an exoatmospheric kill vehicle (EKV). The GBI is not mobile but is fired from an underground silo. Although there is on-site tracking radar, the GMD can receive early warnings from radars hundreds of kilometers away. The GBI travels at about 10 kilometers per second and has a ceiling of about 2,000 kilometers. Out of fourteen tests, GBIs have hit

---

FASCINATING FACTS ABOUT ANTIBALLISTIC MISSILE DEFENSE SYSTEMS

- The first ABM system deployed by the United States, the Safeguard system, had a fatal flaw that was known before it was built. Exploding warheads on the first salvo of Sprint missiles would blind Sprint's tracking radar.

- In 2008, the United States used a Standard Missile-3 to shoot down an orbiting satellite.

- American ABM systems can defend against an accidental launch or a limited attack by a rogue nation but not a full-scale attack by Russia or China.

- The United States can field a layered missile defense, where each successive layer deals with missiles that survived the previous layer. It would start with the long-range Ground-Based Interceptor, then the Aegis standard missiles, the THAAD missile, and finally the short-range Aegis missiles.

- The "sizzler" is a Russian-built cruise missile with a 300-kilometer maximum range. It flies at Mach 0.8 until it nears its target, when it accelerates to nearly Mach 3 and takes a zigzag path to its target—making it difficult to shoot it down. China, India, Algeria, and Vietnam have all purchased it.

- It may not be possible to defend against a full-scale attack by a determined aggressor who could place ten warheads and decoys in a single missile (banned by treaty). Each warhead and possibly each decoy would need to be targeted by a separate missile, unless several kill vehicles could fly on a single ABM (also banned by treaty).

their targets eight times.

Aegis. Ticonderoga-class cruisers and Arleigh Burke-class destroyers all have Aegis Combat Systems (ACSs). Aegis was built to counter short- and medium-range ballistic missiles, aircraft, and other ships. Aegis combines several key parts: the AN/SPY-1 phased-array radar, the MK 99 Fire Control System, the Weapons Control System, the Command and Decision Suite, and the Standard Missile-2 (SM-2).

The SM-2's speed is Mach 3.5, and its range is up to 170 kilometers. The missile has radar and an infrared seeker for terminal guidance, and it has a blast fragmentation warhead. The SM-2 is being replaced by the SM-6, which has twice the range and better radar and is more agile so that it can better deal with the Russian "sizzler" cruise missile.

The third Aegis missile is the Standard Missile-3 (SM-3). It has four stages, a range of more than 500 kilometers, a ceiling of more than 160 kilometers, and a kinetic kill vehicle (KKV). It is guided by ground radar and by onboard infrared sensors. On February 21, 2008, an SM-3 missile was used to shoot down a failed U.S. satellite. The satellite was 240 kilometers above the ground, and the missile approached it at 36,667 kilometers per hour. The satellite had never reached its proper orbit and was coming down because of air resistance. The reason given for shooting it down was the large amount of toxic hydrazine fuel still aboard. Many viewed it as an excuse to test the anti-satellite capability of the SM-3 because it was likely that the hydrazine would have been dispersed and destroyed when the satellite reentered the atmosphere. When equipped with the SM-3, Aegis can serve as an ABM system for assets within range. As has been noted, Aegis-equipped warships with the Standard Missile-3 (SM-3) are stationed in the Mediterranean and the Black Sea to protect Europe from Iranian missiles.

Lasers. The Airborne Laser Test Bed (ALTB) is mounted in a modified Boeing 747 designated the Boeing YAL-1. It has a megawatt-class chemical laser that gets its energy from the chemical reaction between oxygen and iodine compounds. It has successfully destroyed target missiles in flight, but they were not far away. It is unlikely that the laser's range will ever exceed 300 kilometers, and if the aircraft must loiter that close to the launch site, it is in danger of being shot down by the enemy nation's air defense. In 2009, Secretary of Defense Robert Gates Antiballistic Missile Defense Systems recommended that the ALTB project be cut back to limited research.

Another laser project that showed promise was the Tactical High Energy Laser (THEL). The THEL is a deuterium fluoride chemical laser with a theoretical power of 100 kilowatts. It was a joint project with the United States and Israel and was able to shoot down Katyusha rockets but nothing larger. Although lasers show promise, it seems unlikely that they will be used in an ABM system anytime soon.

IMPACT ON GOVERNMENTS AND INDUSTRY

ABM systems have encouraged governments to cooperate. The Patriot system is to be replaced by the Medium Extended Air Defense System (MEADS), a joint project of the United States, Germany, and Italy. It is designed for quick setup so that it will be ready almost as quickly as it is unloaded. Although it is more capable than the Patriot system, MEADS has been streamlined so that it requires only one-fifth of the cargo flights to deliver it to its operation site. Its purpose is to protect against tactical ballistic missiles, unmanned aerial drones, cruise missiles, and aircraft. The ceiling of the PAC-3 MSE (missile segment enhancement) is 50 percent higher and its range is twice the range of the PAC-3. The United States together with Israel developed the Arrow missile to protect Israel from Iranian missiles. The United States, Russia, Israel, Japan, China, the Republic of China (Taiwan), India, North Korea, and Iran all have vigorous ABM programs and active defense industries. Other countries have smaller programs and use hardware manufactured elsewhere. For example, Patriot missile systems manufactured by Raytheon are in thirteen countries.

In the following list of the top defense companies involved in ballistic missile defense, no distinction has been made between companies that are the prime contractor or a subcontractor, and employee numbers and revenues (as opposed to profits) are for 2009. The numbers in parentheses are the ranks, by revenue, of these U.S. defense contractors.

Boeing Company is involved with the Airborne Warning and Control System (AWACS radar), Aegis SM-3, Arrow Interceptor, Ground-based Midcourse Defense (GMD) system, Patriot Advanced Capability-3 (PAC-3), and Strategic Missile and Defense Systems. Boeing has about 160,000 employees and a revenue of $68.3 billion.

Lockheed Martin has 140,000 employees and revenue of $45 billion. It produces the Terminal High Altitude Area Defense (THHAD) weapon system, the Medium Extended Air Defense System (MEADS), Airborne Laser Test Bed, Space-Based Infrared System (SBIRS), and various other missiles and satellites.

Northrop Grumman has developed a Kinetic Energy Interceptor (KEI), is developing a satellite that will spot and track ICBMs, and is modernizing the Minuteman III missiles. The company has about 120,000 employees and revenue of about $32 billion.

General Dynamics is involved with Aegis and several satellite systems with infrared sensors that can track ballistic missiles. Its revenue is about $32 billion, and it has 91,200 employees.

Raytheon builds exoatmospheric kill vehicles, Standard Missiles-2, -3, and -6, airborne radar, the Patriot missile system, and other defense-related equipment. The company had more than 72,000 employees and earned $27 billion.

L-3 Communications is involved with the Ground-Based Midcourse Defense system, the Standard Missile-3, and the Aegis system. It has a revenue of $14 billion and a workforce of 64,000.

Orbital Sciences manufactures the GMD three-stage boost vehicle. Orbital has 3,100 employees and a revenue of $1.2 billion per year.

Altogether, these companies have around 650,000 employees and about $220 billion in revenue. Research shows that for every job in the defense industry, one to four more jobs are created in the community to meet the needs of the defense workers and their families. Reasonable guesses are that 20 percent of the employees work on antiballistic missile projects, and that each job draws one other job to the area. The effect on the economy of antiballistic missile programs is then 260,000 jobs and $88 billion each year.

CAREERS AND COURSE WORK

Many defense industry jobs in the United States require a security clearance; this means the applicant must be a U.S. citizen. A strong background in the physical sciences is necessary for the aerospace industry. High school students should take all the courses in physics, chemistry, computer science, and mathematics that they can. At least a bachelor's degree in science or engineering is required. Employees will need to write reports and make presentations, so students should take some classes in writing and speech. They may eventually become a team or unit leader; if so, they may wish they had taken a simple business management course. Those who are involved with research and development need a feel for how things work. It helps if they like to build or repair things. They should be creative and be able to think of new ways to do things.

Bachelor's degrees are sufficient for a number of aerospace positions: astronautical, computer, electrical, mechanical, optical engineer or physicist. Employees generally start as junior members of a team, but as time passes, they can become senior members and then perhaps team leaders with more responsibility. Astronautical engineers design, test, and supervise the construction of rockets, missiles, and satellites. Computer engineers interface hardware with computers, writing and debugging programs that instruct the hardware to do what is wished. Electrical engineers design, develop, and produce radio frequency data links for missile applications. Mechanical engineers design, analyze, and integrate cryogenic components and assemblies. Optical engineers develop solutions to routine technical problems and work with signal processing analysis and design as well as sensor modeling and simulation. Systems engineers design systems for missile guidance and control, computational fluid mechanics analysis, and wind tunnel testing. Physicists who work with ABM systems test electrical and mechanical components, measure radiation effects, and mitigate them if necessary.

SOCIAL CONTEXT AND FUTURE PROSPECTS

People have always wondered whether money spent on an ABM system could be better spent elsewhere. Some maintain that even an excellent system would be unlikely to protect against some city destroyers. ICBMs with a dozen warheads and decoys could overwhelm any ABM system. Furthermore, if nation A thought that nation B was installing an effective ABM system, nation A might launch a preemptive strike before the ABM system became operational. At the very least, nation A would probably build more ICBMs and escalate the international arms race. Because of such considerations, it is generally conceded that a limited ABM system to deal with accidental launches or a few missiles from rogue nations makes sense.

The United States proposed defending Europe against Iranian missiles by placing defensive missiles and radar in Poland and the Czech Republic. Russia saw this move as a means to blunt a Russian attack on the United States. It threatened to respond with nuclear weapons if Poland or the Czech Republic allowed the installations. In 2009, President Barack Obama scrapped that plan and announced that Aegis-equipped warships would be stationed in the Mediterranean and Black Seas, where they could defend Europe. Russia welcomed this change, which makes it plain that simply building ABM installations may have serious political consequences.

Another consideration is that sooner or later another asteroid will hit Earth, and humans might want to do something about it. For example, asteroid (29075) 1950 DA has a 0.0033 percent (one-third of 1 percent) chance of hitting Earth on March 16, 2880. Experience and technology developed from the various ABM programs will most likely be of some use in dealing with errant asteroids, as their technology of guiding missiles toward a target may be adapted to deflect an incoming asteroid by just enough to prevent it from destroying the planet.

—*Charles W. Rogers, MS, PhD*

FURTHER READING

Burns, Richard Dean. *The Missile Defense Systems of George W. Bush: A Critical Assessment.* Santa Barbara, Calif.: Praeger, 2010. A critical look at the fiscal and political costs to deploy a ground-based ABM system and the effects of trying to extend it to Europe.

Denoon, David. *Ballistic Missile Defense in the Post-Cold War Era.* Boulder, Colo.: Westview Press, 1995. An overview of various proposed ABM systems along with ways to judge if they would be worth the expense of constructing them.

Hey, Nigel. *The Star Wars Enigma: Behind the Scenes of the Cold War Race for Missile Defense.* Dulles, Va.: Potomac Books, 2006. An interesting story of who pushed for the Strategic Defense Initiative (SDI) and other ABM systems.

O'Rourke, Ronald. *Navy Aegis Ballistic Missile Defense (BMD) Program: Background and Issues for Congress.* Washington, D.C.: Congressional Research Service, 2010. O'Rourke, a specialist in naval affairs, discusses the past, present, and future of Aegis and especially the politics behind decisions affecting it.

Payne, Keith B. *Strategic Defense: "Star Wars" in Perspective.* Mansfield, Tex.: Hamilton Press, 1986. A wide-ranging examination of the various issues of SDI.

Sloan, Elinor C. "Space and Ballistic Missile Defense." In *Security and Defense in the Terrorist Era: Canada and the United States Homeland.* Montreal: McGillQueen's University Press, 2010. How Canada's response to the threat of ballistic missiles is, and should be, different from that of the United States.

WEB SITES

Boeing Defense, Space, and Security
http://www.boeing.com/bds

Federation of American Scientists Military Analysis Network
http://www.fas.org/programs/ssp/man/index.html

Lockheed Martin Missiles and Fire Control
http://www.lockheedmartin.com/mfc/products.html

Northrop Grumman Missile Defense
http://www.northropgrumman.com/missiledefense/index.html

Raytheon Missile Systems
http://www.raytheon.com/businesses/rms

Union of Concerned Scientists Nuclear Weapons and Global Security
http://www.ucsusa.org/nuclear_weapons_and_global_security

U.S. Department of State Arms Control and International Security
http://www.state.gov

See also: Aeronautics and Aviation.

APPLIED MATHEMATICS

FIELDS OF STUDY

Mechanical design; mechanical engineering; fluid dynamics; hydraulics; pneumatics; electronic engineering; physics; process modeling; physical chemistry; chemical kinetics; geologic engineering; geographic information systems; computer science; statistics; actuarial science; particle physics; epidemiology; investments; game theory; game design.

SUMMARY

Applied mathematics is the application of mathematical principles and theory in the real world. The practice of applied mathematics has two principal objectives: One is to find solutions to challenging problems by identifying the mathematical rules that describe the observed behavior or characteristic involved, and the other is to reduce real-world behaviors to a level of precise and accurate predictability. Mathematical rules and operations are devised to describe a behavior or property that may not yet have been observed, with the goal of being able to predict with certainty what the outcome of the behavior would be.

KEY TERMS AND CONCEPTS

- **algorithm:** Effective method for solving problems that uses a finite series of specific instructions.
- **boundary conditions:** Values and properties that are required at the boundary limits of a mathematically defined behavior.
- **markov chain:** Probability model used in the study of processes that are considered to move through a finite sequence of steps, with repeats allowed.
- **modeling:** Representation of real-world events and behaviors by mathematical means.
- **particle-in-a-box model:** Classic mathematical analogy of the motion of an electron in an atomic orbital, subject to certain conditions, such as the requirement to have specific values at the limiting boundaries of the box.
- **sparse system:** Any system whose parameter coefficients can be represented in a matrix that contains mostly zeros. Such systems are described as loosely coupled and typically represent members that are linked together linearly rather than as a network.
- **string theory:** Mathematical representation of all subatomic particles as having the structures of strings rather than discrete spheres.
- **wavelet minimization:** Mathematical process whereby small random variations in frequency-dependent measurements (background noise) are removed to allow a clearer representation of the measurements.

DEFINITION AND BASIC PRINCIPLES

Applied mathematics focuses on the development and study of mathematical and computational tools. These tools are used to solve challenging problems primarily in science and engineering applications and in other fields that are amenable to mathematical procedures. The principal mathematical tool is calculus, often referred to as the mathematics of change. Calculus provides a means of quantitatively understanding how variables that cannot be controlled directly behave in response to changes in variables that can be controlled directly. Thus, applied mathematics makes it possible to make predictions about the behavior of an environment and thus gain some mastery over that environment.

For example, suppose a specific characteristic of the behavior of individuals within a society is determined by the combination of a large number of influencing forces, many of which are unknown and perhaps unknowable, and therefore not directly controllable. Study of the occurrence of that characteristic in a population, however, allows it to be described in mathematical terms. This, in turn, provides a valid means of predicting the future occurrence and behavior of that characteristic in other situations. Applied mathematics, therefore, uses mathematical techniques and the results of those techniques in the investigation or solving of problems that originate outside of the realm of mathematics.

The applications of mathematics to real-world phenomena rely on four essential structures: data structures, algorithms, theories and models, and computers and software. Data structures are ways of organizing information or data. Algorithms are specific methods of dealing with the data. Theories and models are used in the analysis of both data and ideas and represent the rules that describe either the way the data were formed or the behavior of the data. Computers and software are the physical devices that are used to manipulate the data for analysis and application. Algorithms are central to the development of software, which is computer specific, for the manipulation and analysis of data.

BACKGROUND AND HISTORY

Applied mathematics, as a field of study, is newer than the practices of engineering and building. The mathematical principles that are the central focus of applied mathematics were developed and devised from observation of physical constructs and behaviors and are therefore subsequent to the development of those activities. The foundations of applied mathematics can be found in the works of early Egyptian and Greek philosophers and engineers. Plane geometry is thought to have developed during the reign of the pharaoh Sesostris, as a result of agricultural land measurements necessitated by the annual inundation of the Nile River. The Greek engineer Thales of Miletus is credited with some of the earliest and most profound applications of mathematical and physical principles in the construction of some of his devices, although there is no evidence that he left a written record of those principles. The primary historical figures in the development of applied mathematics are Euclid and Archimedes. It is perhaps unfortunate that the Greek method of philosophy lacked physical experimentation and the testing of hypotheses but was instead a pure thought process. For this reason, there is a distinction between the fields of pure mathematics and applied mathematics, although the latter depends strictly on the former.

During the Middle Ages, significant mathematical development took place in Islamic nations, where *al geber*, which has come to be known as algebra, was developed, but the field of mathematics showed little progress in Europe. Even during the Renaissance period, mathematics was the realm almost exclusively of astronomers and astrologers. It is not certain that even Leonardo da Vinci, foremost of the Renaissance engineers and artists, was adept at mathematics despite the mathematical brilliance of his designs. The major historical development of applied mathematics began with the development of calculus by Sir Isaac Newton and Gottfried Wilhelm Leibniz in

the seventeenth century. The applicability of mathematical principles in the development of scientific pursuits during the Industrial Revolution brought applied mathematics to the point where it has become essential for understanding the physical universe.

HOW IT WORKS

Applied mathematics is the creation and study of mathematical and computational tools that can be broadly applied in science and engineering. Those tools are used to solve challenging problems in these and related fields of practice. In its simplest form, applied mathematics refers to the use of measurement and simple calculations to describe a physical condition or behavior.

A simple example might be the layout or design of a field or other area of land. Consider a need to lay out a rectangular field having an area of 2,000 square meters (m^2) with the shortest possible perimeter. The area (A) of any rectangular area is determined as the product of the length (l) and the width (w) of the area in question. The perimeter (P) of any rectangular area is determined as the sum of the lengths of all four sides, and in a rectangular area, the opposite sides are of equal length. Thus, $P = 2l + 2w$, and $A = l \times w = 2000 \ m^2$. By trial and error, pairs of lengths and widths whose product is 2,000 may be tried out, and their corresponding perimeters determined. A field that is 2,000 meters long and 1 meter wide has an area of 2,000 square meters and a perimeter of 4,002 meters. Similarly, a field that is 200 meters long and 10 meters wide has the required area, and a perimeter of only 420 meters. It becomes apparent that the perimeter is minimized when the area is represented as a square, having four equal sides. Thus, the length of each side must be equal to the square root of 2,000 in magnitude. Having determined this, the same principles may be applied to the design of any rectangular area of any size.

The same essential procedures as those demonstrated in this simple example apply with equal validity to other physical situations and are, in fact, the very essence of scientific experimentation and research. The progression of the development of mathematical models and procedures in many different areas of application is remarkably similar. Development of a mathematical model begins with a simple expression to which refinements are made, and the results of the calculation are compared with the actual behavior of the system under investigation. The changes in the difference between the real and calculated behaviors are the key to further refinements that, ideally, work to bring the two into ever closer agreement. When mathematical expressions have been developed that adequately describe the behavior of a system, those expressions can be used to describe the behaviors of other systems.

A key component to the successful application of mathematical descriptions or models is an understanding of the different variables that affect the behavior of the system being studied. In fluid dynamics, for example, obvious variables that affect a fluid are the temperature and density of the fluid. Less obvious perhaps are such variables as the viscosity of the fluid, the dipolar interactions of the fluid atoms or molecules, the adhesion between the fluid and the surface of the container through which it is flowing, whether the fluid is flowing smoothly (laminar flow) or turbulently (nonlaminar flow), and a number of other more obscure variables. A precise mathematical description of the behavior of such a system would include corrective terms for each and every variable affecting the system. However, a number of these corrective terms may be considered together in an approximation term and still produce an accurate mathematical description of the behavior.

An example of such an approximation may be found in the applied mathematical field of quantum mechanics, by which the behavior of electrons in molecules is modeled. The classic quantum mechanical model of the behavior of an electron bound to an atomic nucleus is the so-called particle-in-a-box model. In this model, the particle (the electron) can exist only within the confines of the box (the atomic orbital), and because the electron has the properties of an electromagnetic wave as well as those of a physical particle, there are certain restrictions placed on the behavior of the particle. For example, the value of the wave function describing the motion of the electron must be zero at the boundaries of the box. This requires that the motion of the particle can be described only by certain wave functions that, in turn, depend on the dimensions of the box. The problem can be solved mathematically with precision only for the case involving a single electron and a single nuclear proton that defines the box in which the electron is found. The calculated results agree extremely well with observed measurements of electron energy.

For systems involving more particles (more electrons and more nuclear protons and neutrons), the number of variables and other factors immediately exceeds any ability to be calculated precisely. A solution is found, however, in a method that uses an approximation of the orbital description, known as a Slater-type orbital approximation, rather than a precise mathematical description. A third-level Gaussian treatment of the Slater-type orbitals, or STO-3G analysis, yields calculated results for complex molecular structures that are in excellent agreement with the observed values measured in physical experiments. Although the level of mathematical technique is vastly more complex than in the simple area example, the basic method of finding an applicable method is almost exactly identical.

APPLICATIONS AND PRODUCTS

Applied mathematics is essentially the application of mathematical principles and theories toward the resolution of physical problems and the description of behaviors. The range of disciplines in which applied mathematics is relevant is therefore very broad. The intangible nature of mathematics and mathematical theory tends to restrict active research and development to the academic environment and applied research departments of industry. In these environs, applied mathematics research tends to be focused rather than general in nature. Applied mathematics is generally divided into the major areas of computational mathematics: combinatorics and optimization, computer science, pure mathematics, and statistical and actuarial science. The breadth of the research field has grown dramatically, and the diversity of subject area applications is indicated by the applied mathematics research being conducted in atmospheric and biological systems applications, climate and weather, complexity theory, computational finance, control systems, cryptography, pattern recognition and data mining, multivariate data analysis and visualization, differential equation modeling, fluid dynamics, linear programming, medical imaging, and a host of other areas.

Computational mathematics. Simply stated, computational mathematics is the process of modeling systems quantitatively on computers. This is often referred to in the literature as *in silico*, indicating that the operation or procedure being examined is carried out as a series of calculations within the silicon-based electronic circuitry of a computer chip and not in any tangible, physical manner. Research in computational mathematics is carried out in a wide range of subject areas.

The essence of computational mathematics is the development of algorithms and computer programs that produce accurate and reliable models or depictions of specific behaviors. In atmospheric systems, for example, one goal would be to produce mathematical programs that precisely depict the behavior of the ozone layer surrounding the planet. The objective of such a program would be to predict how the ozone layer would change as a result of alterations in atmospheric composition. It is not feasible to observe the effects directly and would ultimately be counterproductive if manifesting an atmospheric change resulted in the destruction of the ozone layer. Modeling the system *in silico* allows researchers to institute virtual changes and determine what the effect of each change would be. The reliability of the calculated effect depends directly on how accurately the model describes the existing behavior of the system.

Medical imaging. An area in which applied mathematics has become fundamentally important is the field of medical imaging, especially as it applies to magnetic resonance imaging (MRI). The MRI technique developed from nuclear magnetic resonance (NMR) analysis commonly used in analytical chemistry to determine molecular structures. In NMR, measurements are obtained of the absorption of specific radio frequencies by molecules held within a magnetic field. The strength of each absorption and specific patterns of absorptions are characteristic of the structure of the particular molecule and so can be used to determine unequivocally the exact molecular structure of a material.

One aspect of NMR that has been greatly improved by applied mathematics is the elimination of background noise. A typical NMR spectrum consists of an essentially infinite series of small random signals that often hide the detailed patterns of actual absorption peaks and sometimes even the peaks themselves. The Fourier analysis methodology, in which such random signals can be treated as a combination of sine and cosine waves, also known as wavelet theory, eliminates a significant number of such signals from electromagnetic spectra. The result is a much more clear and precise record of the actual absorptions. Such basic NMR spectra are only

one dimensional, however. The second generation modification of NMR systems was developed to produce a two-dimensional representation of the NMR absorption spectrum, and from this was developed the three-dimensional NMR imaging system that is known as MRI. Improvements that make the Fourier analysis technique ever more effective in accord with advances in the computational abilities of computer hardware are the focus of one area of ongoing applied mathematics research.

Population dynamics and epidemiology. Population dynamics and epidemiology are closely related fields of study. The former studies the growth and movements of populations, and the latter studies the growth and movements of diseases and medical conditions within populations. Both rely heavily for their mathematical descriptions on many areas of applied mathematics, including statistics, fluid dynamics, complexity theory, pattern recognition, data visualization, differential equation modeling, chaos theory, risk management, numerical algorithms and techniques, and statistical learning.

In a practical model, the movements of groups of individuals within a population are described by many of the same mathematical models of fluid dynamics that apply to moving streams of particles. The flow of traffic on a multilane highway or the movement of people along a busy sidewalk, for example, can be seen to exhibit the same gross behavior as that of a fluid flowing through a system of pipes. In fact, any population that can be described in terms of a flow of discrete particles, whether molecules of water or vehicles, can be described by the same mathematical principles, at least to the extent that the variables affecting their motion are known. Thus, the forces of friction and adhesion that affect the flow of a fluid within a tube are closely mimicked by the natural tendencies of drivers to drive at varying speeds in different lanes of a multilane highway. Window-shoppers and other slow-moving individuals tend to stay to the part of the sidewalk closest to the buildings, while those who walk faster or more purposefully tend to use the part of the sidewalk farthest away from the buildings, and this also follows the behavior of fluid flow.

The spread or movement of diseases through a population can also be described by many of the same mathematical principles that describe the movements of individuals within a population. This is especially true for diseases that are transmitted directly from person to person. For other disease vectors, such as animals, birds, and insects, a mathematical description must describe the movements of those particular populations, while at the same time reflecting the relationship between those populations and the human population of interest.

Statistical analysis and actuarial science. Perhaps the simplest or most obvious use of applied mathematics can be found in statistical analysis. In this application, the common properties of a collection of data points, themselves measurements of some physical property, are enumerated and compared for consistency. The effectiveness of statistical methods depends on the appropriately random collection of representative data points and on the appropriate definition of a property to be analyzed.

Statistical analysis is used to assess the consistency of a common property and to identify patterns of occurrence of characteristics. This forms the basis of the practice of statistical process control (SPC) that has become the primary method of quality control in industry and other fields of practice. In statistical process control, random samples of an output stream are selected and compared to their design standard. Variations from the desired value are determined, and the data accumulated over time are analyzed to determine patterns of variation. In an injection-molding process, for example, a variation that occurs consistently in one location of the object being molded may indicate that a modification to the overall process must be made, such as adjusting the temperature of the liquid material being injected or an alteration to the die itself to improve the plastic flow pattern.

In another context, one that is tied to epidemiology, the insurance and investment industries make very detailed use of statistical analysis in the assessment of risk. Massive amounts of data describing various aspects of human existence in modern society are meticulously analyzed to identify patterns of effects that may indicate a causal relationship. An obvious example is the statistical relationship pattern between healthy lifestyle and mortality rates, in which obese people of all age groups have a higher mortality rate than their counterparts who maintain a leaner body mass. Similarly, automobile insurance rates are set much higher for male drivers between the ages of sixteen and twenty-five than for female drivers in

that age group and older drivers because statistical analysis of data from accidents demonstrates that this particular group has the highest risk of involvement in a traffic accident. This type of data mining is a continual process as relationships are sought to describe every factor that plays a role in human society.

IMPACT ON INDUSTRY

Statistical process control. The practice of statistical process control has had an unprecedented effect on modern society, especially in the manufacturing industry. Before the development of mass-production methods, parts and products were produced by skilled and semiskilled craftspeople. The higher the level of precision required for a production piece, the more skilled and experienced an artisan or craftsperson had to be. Manpower and cost of production quickly became the determining factors in the availability of goods.

The advent of World War II, however, ushered in an unprecedented demand for materials and products. The production of war materials required the development of methods to produce large numbers of products in a short period of time. Various accidents and mistakes resulting from the rapid pace of production also indicated the need to develop methods of quality control to ensure that goods were produced to the expected level of quality and dependability. The methods of quality assurance that were developed at that time were adopted by industries in postwar Japan and developed into a rigorous protocol of quality management based on the statistical analysis of key features of the goods being produced. The versatility of the system, when applied to mass-production methods, almost instantly eliminated the dependency on skilled tradespeople for the production of precision goods.

In the intervening half century, statistical process control became a required component of production techniques and an industry unto itself, spawning such rigorous programs as the International Organization for Standardization's ISO 9000, ISO 9001, and ISO 14000; the Motorola Corporation's Six Sigma; and Lean Manufacturing (derived from the Toyota Production System), which have become global standards.

Economics. The analysis of data by mathematical means, especially data mining and pattern recognition in population dynamics, has provided much of the basis of the economic theory that directs the conduct of business on a global scale. This has been especially applicable in regard to the operation of the stock market. In large part, the identification of patterns and trends in the historical data of trade and business provides the basic information for speculation on projected or future consumer trends. Although this is important for the operation of publicly traded businesses on the stock market, it is more important in the context of government. The same trends and patterns play a determining role in the design and establishment of government policies and regulations for both domestic and international affairs.

Medicine and pharmaceuticals. Applied mathematics, particularly the use of Fourier analysis and wavelet theory, has sparked explosive growth in the fields of medicine and pharmaceutical development. The enhanced analytical methods available to chemists and bioresearchers through NMR and Fourier transform infrared (FTIR) spectroscopy facilitate the identification and investigation of new compounds that have potential pharmaceutical applications. In addition, advanced statistical methods, computer modeling, and epidemiological studies provide the foundation for unprecedented levels of research.

The science of medical diagnostic imaging has grown dramatically because of the enhancements made available through applied mathematics. Computer science, based on the capabilities of modern digital computer technology, has replaced the physical photographic method of X-ray diagnostics with a digital version that is faster and more efficient and has the additional capability of zooming in on any particular area of interest in an X-ray image. This allows for more accurate diagnostics and is especially important for situations in which a short response time is essential.

The methodology has enabled entirely new procedures, including MRI. Instead of relatively plain, two-dimensional images, real-time three-dimensional images can be produced. These images help surgeons plan and rehearse delicate surgical maneuvers to perfect a surgical technique before ever touching a patient with a scalpel. MRI data and modifications of MRI programming can be used to control the sculpting or fabrication of custom prosthetic devices. All depend extensively on the mathematical manipulation of data to achieve the desired outcome.

CAREERS AND COURSE WORK

The study of applied mathematics builds on a solid and in-depth comprehension of pure mathematics. The student begins by taking courses in mathematics throughout secondary school to acquire a solid foundational knowledge of basic mathematical principles before entering college or university studies. A specialization in mathematics at the college and university level is essential to practically all areas of study. As so many fields have been affected by applied mathematics, the methodologies being taught

---

FASCINATING FACTS ABOUT APPLIED MATHEMATICS

- Applied mathematics research toward the development of the quantum computer, which would operate on the subatomic scale, has led some researchers to postulate the simultaneous existence of multiple universes.
- Applied mathematics may have begun as geometry in ancient Egypt during the reign of the pharaoh Sesostris, who instituted the measurement of changes in land area for the fair assessment of taxes in response to the annual flooding of the Nile River.
- Ancient engineers apparently used certain esoteric principles of mathematics in the construction of the pyramids in Egypt and elsewhere, as the dimensions of these structures reflect the golden ratio (φ).
- The flow of traffic along a multilane highway can be described by the same mathematics that describes the flow of water in a system of pipes.
- The behavior of an electron bound to a single proton is described mathematically as a particle in a box, because it must conform to specific conditions determined by the confines of the box.
- The seemingly uniform movement of individual birds in flocks of flying birds demonstrates the chaos theory, in which discrete large-scale patterns arise from unique small-scale actions. This is also known as the butterfly effect.
- The growth rates of bananas in the tropics and other natural behaviors can be described exactly by differential equations.
- Three-dimensional MRI can be used to practice and perfect surgical techniques without ever touching an actual patient.

in undergraduate courses are a reflection of the accepted concepts of applied mathematics on which they are constructed. As the depth of the field of applied mathematics indicates, career options are, for all intents and purposes, unlimited. Every field of endeavor, from anthropology to zoology, has a component of applied mathematics, and the undergraduate must learn the mathematical methods corresponding to the chosen field.

Students who specialize in the study of applied mathematics as a career choice and proceed to postgraduate studies take courses in advanced mathematics and carry out research aimed at developing mathematics theory and advancing the application of those developments in other fields.

Particular programs of study in applied mathematics are included as part of the curriculum of other disciplines. The focus of such courses is on specific mathematical operations and techniques that are relevant to that particular field of study. All include a significant component of differential calculus, as appropriate to the dynamic nature of the subject matter.

SOCIAL CONTEXT AND FUTURE PROSPECTS

Human behavior is well suited and highly amenable to description and analysis through mathematical principles. Modeling of population dynamics has become increasingly important in the contexts of service and regulation. As new diseases appear, accurate models to predict the manner in which they will spread play an ever more important role in determining the response to possible outbreaks. Similarly, accurate modeling of geological activities is increasingly valuable in determining the best ways to respond to such natural disasters as a tsunami or earthquake. The Earth itself is a dynamic system that is only marginally predictable. Applied mathematics is essential to developing models and theories that will lead to accurate prediction of the occurrence and ramifications of seismic events. It will also be absolutely necessary for understanding the effects that human activity may be having on the atmosphere and oceans, particularly in regard to the issues of global warming and the emission of greenhouse gases. Climate models that can accurately and precisely predict the effects of such activities will continue to be the object of a great deal of research in applied mathematics.

The development of new materials and engineering new applications with those materials is an ongoing human endeavor. The design of extremely small nanostructures that employ those materials is based on mathematical principles that are unique to the realm of the very small. Of particular interest is research toward the development of the quantum computer, a device that operates on the subatomic scale rather than on the existing scale of computer construction.

—*Richard M. J. Renneboog, MSc*

FURTHER READING

Anton, Howard. Calculus: A New Horizon. 6th ed. New York: John Wiley & Sons, 1999. Offers a clear and progressive introduction to many of the basic principles of applied mathematics.

De Haan, Lex, and Toon Koppelaars. Applied Mathematics for Database Professionals. New York: Springer, 2007. Focuses on the use of set theory and logic in the design and use of databases in business operations and communications.

Howison, Sam. Practical Applied Mathematics: Modelling, Analysis, Approximation. New York: Cambridge University Press, 2005. Provides a basic introduction to several practical aspects of applied mathematics, supported by in-depth case studies that demonstrate the applications of mathematics to actual physical phenomena and operations.

Kurzweil, Ray. The Age of Spiritual Machines: When Computers Exceed Human Intelligence. New York: Penguin Books, 2000. Presents a thought-provoking view of the nature of artificial intelligence that may arise as a direct result of applied mathematics.

Moore, David S. The Basic Practice of Statistics. 5th ed. New York: Freeman, 2010. Introduces statistical analysis at its most basic level and progresses through more complex concepts to an understanding of statistical process control.

Naps, Thomas L., and Douglas W. Nance. Introduction to Computer Science: Programming, Problem Solving, and Data Structures. 3d ed. St. Paul, Minn.: West Publishing, 1995. Presents a basic introduction to the principles of computer science and the development of algorithms and mathematical processes.

Rubin, Jean E. Mathematical Logic: Applications

and Theory. Philadelphia: Saunders College Publishing, 1990. An introduction to the principles of mathematical logic that requires the reader to work closely through the material to obtain a good understanding of mathematical logic.

Washington, Allyn J. Basic Technical Mathematics with Calculus. 9th ed. Upper Saddle River, N.J.: Pearson Prentice Hall, 2009. Provides thorough and well-explained instruction in applied mathematics for technical pursuits in a manner that allows the reader to grasp many different mathematical concepts.

WEB SITES

Mathematical Association of America
http://www.maa.org

Society for Industrial and Applied Mathematics
http://www.siam.org

See also: Algebra; Calculus; Computer Science; Engineering Mathematics; Game Theory; Numerical Analysis; Pattern Recognition.

APPLIED PHYSICS

FIELDS OF STUDY

Physics; mathematics; applied mathematics; mechanical engineering; civil engineering; metrics; forensics; biological physics; photonics; microscopy and microanalysis; biomechanics; materials science; surface science; aeronautics; hydraulics; nanotechnology; robotics; medical imaging; radiology.

SUMMARY

Applied physics is the study and application of the behavior of condensed matter, such as solids, liquids, and gases, in bulk quantities. Applied physics is the basis of all engineering and design that requires the interaction of individual components and materials. The study of applied physics now extends into numerous other fields, including physics, chemistry, biology, engineering, medicine, geology, meteorology, and oceanography.

KEY TERMS AND CONCEPTS

- **biomechanics:** The science of physical movement in living systems.
- **condensed matter:** Matter is a physical quantity that is normally recognized as having the form of a solid, a liquid, or a gas, also known as a bulk quantity.
- **forensics:** The science of analysis to determine cause and effect after an event has occurred, highly dependent upon the application of laws of physics in the case of physical events.
- **magnetic levitation:** A technology that utilizes the mutual repulsion of like magnetic fields to maintain a physical separation between an object and a track (such as a high-speed train and its track bed), while keeping the two in virtual contact.
- **magnetic resonance imaging (MRI):** A noninvasive diagnostic imaging technology based on nuclear magnetic resonance that provides detailed images of internal organs and the structures of living systems.
- **nanoprobe:** A device designed to perform a specific function, constructed of nanometer (10^{-9} meter) scale components.
- **newtonian mechanics:** The behavior of condensed matter as described by Newton's laws of motion and inertia.
- **nuclear magnetic resonance (NMR):** A property of certain atomic nuclei having unpaired electrons to absorb energy of a specific electromagnetic wavelength within a constant magnetic field.
- **prosthetics:** The branch of applied physics that deals with the design of artificial components that replace or augment natural biological components; the artificial components that are designed and constructed to replace normal biological components.
- **quantum mechanics:** The physical laws describing the behavior of matter in quantities below the nanometer (10^{-9} meter) scale, especially of individual atoms and molecules.
- **spectrometry:** An analytical technology that utilizes the specific electromagnetic energy of specific wavelengths to identify or quantify specific components of a material or solution.

DEFINITION AND BASIC PRINCIPLES

Applied physics is the study of the behavior and interaction of condensed matter. Condensed matter is commonly recognized as matter in the phase forms of solids, liquids, and gases. Each phase has its own unique physical characteristics, and each material has its own unique physical suite of properties that derive from its chemical identity. The combination of these traits determines how something consisting of condensed matter interacts with something else that consists of condensed matter.

The interactions of constructs of condensed matter are governed by the interaction of forces and other vector properties that are applied through each construct. A lever, for example, functions as the medium to transmit a force applied in one direction so that it operates in the opposite direction at another location. Material properties and strengths are also an intimate part of the interaction of condensed matter. A lever that cannot withstand the force applied to it laterally will bend or break rather than function as a lever. Similarly, two equally hard objects, such as a train car's wheels and the steel tracks upon which they roll, rebound elastically with little or no generation of friction. If, however, one of the materials is less hard than the other, as in the case of a steel rotor and a composition brake pad, a great deal of friction characterizes their interaction.

BACKGROUND AND HISTORY

Applied physics is of necessity the oldest of all practical sciences, dating back to the first artificial use of an object by an early hominid. The basic practices have been in use by builders and designers for many thousands of years. With the development of mathematics and measurement, the practice of applied physics has grown apace, relying as it still does upon the application of basic concepts of vector properties (force, momentum, velocity, weight, moment of inertia) and the principles of simple machines (lever, ramp, pulley).

The modern concept of applied physics can be traced back to the Greek philosopher Aristotle (c. 350 BCE), who identified several separate fields for systematic study. The basic mathematics of physics was formulated by Pythagoras about two hundred years before, but was certainly known to the Babylonians as early as 1600 BCE. Perhaps the greatest single impetus for the advance of applied physics was the development of calculus by Sir Isaac Newton and the fundamental principles of Newtonian mechanics in the seventeenth century. Those principles describe well the behaviors of objects composed of condensed matter, only failing as the scale involved becomes very small (as below the scale of nanotechnology). Since the Industrial Revolution and into the twenty-first century, the advancing capabilities of technology and materials combine to enable even more advances and applications, yet all continue to follow the same basic rules of physics as the crudest of constructs.

HOW IT WORKS

As the name implies, applied physics means nothing more or less than the application of the principles of physics to material objects. The most basic principles of applied physics are those that describe and quantify matter in motion (speed, velocity, acceleration, momentum, inertia, mass, force). These principles lead to the design and implementation of devices and structures, all with the purpose of performing a physical function or action, either individually or in concert.

A second, and equally important, aspect of applied physics is the knowledge of the physical properties and characteristics of the materials being used. This includes such physical properties as melting and boiling points, malleability, thermal conductivity, electrical resistivity, magnetic susceptibility, density, hardness, sheer strength, tensile strength, compressibility, granularity, absorptivity, and a great many other physical factors that determine the suitability of materials for given tasks. Understanding these factors also enables the identification of new applications for those materials.

Design. Applied physics is the essential basis for the design of each and every artificial object and construct, from the tiniest of nanoprobes and the simplest of levers to the largest and most complex of machines and devices. Given the idea of a task to be performed and a device to perform that task, the design process begins with an assessment of the physical environment in which the device must function, the forces that will be exerted against it, and the forces that it will be required to exert in order to accomplish the set task. The appropriate materials may then be selected, based on their physical properties and characteristics. Dimensional analyses determine the necessary size of the device, and also play a significant role in determining the materials that will be utilized.

All of these factors affect the cost of any device, an important aspect of design. Cost is an especially important consideration, as is the feasibility of replacing a component of the designed structure. For example, a device operating in a local environment in which replacement of a component, such as an actuating lever, is easily carried out may be constructed using a simple steel rod. If, however, the device is something like the Mars Rover, for which replacement of worn parts is not an option, it is far more reasonable and effective to use a less-corrosive, stronger, but more costly material, such as titanium, to make movable parts. The old design tenet that form follows function is an appropriate rule of thumb in the field of applied physics.

APPLICATIONS AND PRODUCTS
In many ways, applied physics leads to the idea of creating devices that build other devices. At some point in time, an individual first used a handy rock to break up another rock or to pound a stick into the ground. This is a simple example of applied physics, when that individual then began to use a specific kind of rock for a specific purpose; this action advanced again when someone realized that attaching the rock to a stick provided a more effective tool. Applied physics has advanced far beyond the crude rock-on-a-stick hammer, yet the exact same physical principles apply now with the most elegant of impact devices. It would not be possible to itemize even a small percentage of the applications and products that have resulted from applied physics, and new physical devices are developed each day.

Civil society. Applied physics underpins all physical aspects of human society. It is the basis for the design and construction of the human environment and its supporting infrastructure, the most obvious of which are roads and buildings, designed and engineered with consideration of the forces that they must withstand, both human-made and natural. In this, the highest consideration is given to the science and engineering of materials in regard to the desired end result, rather than to the machines and devices that are employed in the actual construction process. The principles of physics involved in the design and construction of supporting structures, such as bridges and high-rise towers, are of primary importance. No less important, however, are the physical systems

that support the function of the structure, including such things as electrical systems, internal movement systems, environmental control systems, monitoring systems, and emergency response systems, to name a few. All are relevant to a specific aspect of the overall physical construct, and all must function in coordination with the other systems. Nevertheless, they all also are based on specific applications of the same applied physics principles.

Transportation. Transportation is perhaps the single greatest expenditure of human effort, and it has produced the modern automotive, air, and railway transportation industries. Applied physics in these areas focuses on the development of more effective and efficient means of controlling the movement of the machine while enhancing the safety of human and nonhuman occupants. While the physical principles by which the various forms of the internal combustion engine function are the same now as when the devices were first invented, the physical processes used in their operation have undergone a great deal of mechanical refinement.

A particular area of refinement is the manner in which fuel is delivered for combustion. In earlier gasoline-fueled piston engines, fuel was delivered via a mechanical fuel pump and carburetor system that functions on the Venturi principle. This has long since been replaced by constant-pressure fuel pumps and injector systems, and this continues to be enhanced as developers make more refinements based on the physical aspects of delivering a liquid fuel to a specific point at a specific time. Similarly, commercial jet engines, first developed for aircraft during World War II, now utilize the same basic physical principles as did their earliest counterparts. Alterations and developments that have been achieved in the interim have focused on improvement of the operational efficiency of the engines and on enhancement of the physical and combustion properties of the fuels themselves.

In rail transport, the basic structure of a railway train has not changed; it is still a heavy, massive tractor engine that tows a number of containers on very low friction wheel-systems. The engines have changed from steam-powered behemoths made up of as many as one-quarter-million parts to the modern diesel-electric traction engines of much simpler design.

Driven by the ever-increasing demand for limited resources, this process of refinement for physical field of applied physics.

Medical. No other area of human endeavor demonstrates the effects of applied physics better than the medical field. The medical applications of applied physics are numerous, touching every aspect of medical diagnostics and treatment and crossing over into many other fields of physical science. The most obvious of these is medical imaging, in which X-ray diagnostics, magnetic resonance imaging (MRI), and other forms of spectroscopic analysis have become primary tools of medical diagnostics. Essentially, all of the devices that have become invaluable as medical tools began as devices designed to probe the physical nature of materials.

Spectrographs and spectrometers that are now routinely used to analyze the specific content of various human sera were developed initially to examine the specific wavelengths of light being absorbed or emitted by various materials. MRI developed from the standard technique of physical analytical chemistry called nuclear magnetic resonance spectrometry, or NMR. In this methodology, magnetic signals are recorded as they pass through a material sample, their patterns revealing intimate details about the three-dimensional molecular structure of a given compound. The process of MRI that has developed from this simpler application now permits diagnosticians to view the internal structures and functioning of living systems in real time. The diagnostic X ray that has become the single most common method of examining internal biological structures was first used as a way to physically probe the structure of crystals, even though the first practical demonstration of the existence of X rays by German physicist Wilhelm Röntgen in 1895 was as an image of the bones in his wife's hand.

Military. It is not possible to separate the field of applied physics from military applications and martial practice, and no other area of human endeavor so clearly illustrates the double-edged sword that is the nature of applied physics. The field of applied physics at once facilitates the most humanitarian of endeavors and the most violent of human aggressions. Ballistics, which gets its name from the Roman Latin word for "war," is the area of physics that applies to the motion of bodies moving through a gravitational field. The mathematical equations that describe the motion of a baseball being thrown from center field equally describe the motion of an arrow or a bullet

---

FASCINATING FACTS ABOUT APPLIED PHYSICS

- When the semiconductor junction transistor was invented in 1951, it initiated the digital electronics revolution and formed the functional basis of the personal computer. In 1954, scientists at RAND envisioned a home computer in the year 2000, complete with a teletype Fortran-language programming station, a large CRT television monitor, rows of blinking lights and dials, and twin steering-wheels.

- Nuclear magnetic resonance (NMR) spectrometry is an analytical method that is sensitive enough to differentiate and measure the magnetic environments of individual atoms in molecules.

- Creating a Stone Age hammer by fastening a rock to a stick is as much a product of applied physics as is the creation of the International Space Station.

- A rock thrown through the air obeys exactly the same laws of ballistics as does a bullet fired from a high-powered rifle or a satellite orbiting a planet.

- The B2 Spirit stealth bomber, a U.S. military aircraft representing sophisticated applied physics, has a unit cost of more than US$1.15 billion (as of 2003).

- Spectrography, initially invented to identify specific wavelengths of emitted or absorbed light, is the basis for all analytical methods that rely on electromagnetic radiation.

- The formal identification of the science known as applied physics can be traced back to Aristotle, to circa 350 BCE.

- All human technology is the product of applied physics. Spin-off from such advanced activities as space research translates into a great number of everyday products, including cell phones, home computers, and digital egg timers, with an overall economic value of trillions of U.S. dollars annually.

---

efficiency progresses in every aspect of transportation on land, sea, and air. Paradoxically, it is in the area of railway transport that applied physics offers the greatest possibility of advancement with the development of nearly frictionless, magnetic levitation systems upon which modern high-speed bullet trains travel. Physicists continue to work toward the development of materials and systems that will be superconducting at ambient temperatures. It is believed that such materials will completely revolutionize not only the transportation sector but also the entire in flight, of the motion of the International Space Station in its orbit, and of the trajectory of a warhead-equipped rocket. The physical principles that have permitted the use of nuclear fission as a source of reliable energy are the same principles that define the functioning of a nuclear warhead.

In the modern age it is desired that warfare be carried out as quickly and efficiently as possible as the means to restore order and peace to an embattled area, with as little loss of life and destruction of property as possible. To that end, military research relies heavily on the application of physics in the development of weapons, communications, and surveillance systems.

Digital electronics. Applied physics has also led to the development of the transistor, commonly attributed to William Shockley, in 1951. Since that time, the development of ever-smaller and more efficient systems based on the semiconductor junction transistor has been rapid and continuous. This has inexorably produced the modern technology of digital electronics and the digital computer. These technologies combine remarkably well to permit the real-time capture and storage of data from the many and various devices employed in medical research and diagnostics, improve the fine control of transportation systems and infrastructure, and advance a veritable host of products that are more efficient in their use of energy than are their nondigital counterparts.

IMPACT ON INDUSTRY

There is a common misconception that persons with degrees and expertise within the field of physics are not presented with sufficient opportunities for employment after graduating. This misconception reflects a difference in academic terminology versus business terminology only, and not reality. It is true that few physics graduates actually work as traditional physicists. Given the training and ability in mathematics and physical problem-solving that physicists (anyone with a degree in physics) have developed, they tend to acquire positions as industrial physicists instead, holding positions with industrial titles as managers, engineers, computer scientists, and technical staff.

Data from the American Institute of Physics indicate that physicists represent a significantly higher-than-average income group. About one-fifth of all physicists enter military careers, while about one-half of those with a doctorate remain in academia and teaching and about one-half of those with master's degrees make their careers in industry.

Given that applied physics is the fundamental basis of all human technology upon which modern society has been built, it is essentially impossible to assign a monetary value to the overall field. The direct economic value that derives from any individual aspect of applied physics is often measured in billions of U.S. dollars, and in some cases the effects are so far-reaching as to be immeasurable. The amount spent on space research, for example, by the U.S. government through the National Aeronautics and Space Administration (NASA), is less than 1 percent of the annual national budget of approximately $2.5 trillion. The return on this investment in terms of corporate and personal taxes, job creation, and economic growth, however, is conservatively estimated to be no less than seven-to-one. That is to say, every $1 million invested in space research generates a minimum of $7 million in tax revenues and economic growth.

Not the least significant of effects is the spin-off that enters the general economy as a direct result of applied physics research and development. Space research has provided a long list of consumer products and technical enhancements that touch everything from golf balls and enriched baby foods to industrial process modeling, virtual reality training, and education aids. The list of spin-off technologies and products from this source alone is very long, and it continues to grow as research advances. It is interesting to note that many of the products of space research that were developed with the intent of bettering a human presence on other planets have proved eminently applicable to bettering the human presence on Earth.

The wireless communications industry is one of the most obvious spin-offs from space research, and it now represents a significant segment of the overall economy of nations around the world. The value of this industry in Canada, for example, provides a good indication of the value of wireless communications to other nations. The wireless communications industry in Canada serves a country that has four persons per square kilometer, as compared with thirty persons per square kilometer in the United States. In 2010, the economic value of the wireless communications industry in Canada was approximately US$39 billion, of which $16.9 billion was a direct contribution to the gross domestic product (GDP) through the sale of goods and services. This contribution to the GDP compared favorably with those of the automotive manufacturing industry, the food manufacturing industry, and agricultural crop production.

Wireless communications produced up to 410,000 Canadian jobs, or about 2.5 percent of all jobs in Canada; this marks a significantly higher average salary than the national average. The value-added per employee in the wireless telecommunications industry was also deemed to be more than 2.5 times higher than the national average. While a direct extrapolation to the effect of this industry on the economy of the United States, and other nations, is not appropriate because of the differences in population distribution and geography, it is nevertheless easy to see that the wireless communications industry now represents a significant economic segment and has far-reaching effects.

Furthermore, wireless communication can greatly reduce unproductive travel time and the associated economic costs, while significantly improving the logistics of business operations. Wireless communications empowers small businesses, especially in rural areas. Medical devices represent another large economic segment that has grown out of applied physics research and development. Recent surveys (2008) suggest that it is one segment that is relatively unaffected by economic upswings and downswings. Given that medical methodology continues to advance on the strength of diagnostic devices, while the development of medical technology continues to advance in step with advancing medical methodology, it is logical that the effects of economic changes are mitigated in this field.

As of 1999, medical research in the United States represented an investment of more than $35 billion. The economic value of medical research was estimated in 1999 to be approximately $48 trillion for the elimination of death from heart disease and $47 trillion for a cure for cancer. Simply reducing cancer mortality by a mere 1 percent would generate an economic return of $500 billion. These estimates are all based on the assumed economic "value of life" as benefits to persons deriving from an extended ability to participate in society. The research and development of medical knowledge that would generate these economic returns cannot be called applied physics.

Rather, the various technologies by which the overall medical research program is supported are the direct result of applied physics.

In the realm of military expenditure, it is estimated that approximately 10 percent of the U.S. GDP is directly related to military purposes. This includes research and development in applied physics technologies.

CAREERS AND COURSEWORK

Programs of study for careers in applied physics or in fields that are based on applied physics include advanced mathematics, physics, and materials science. Depending on the area of career specialization, coursework may also include electronics engineering, mechanical engineering, chemistry and chemical engineering, biology and biomechanics, instrumentation, and more advanced courses in mathematics and physics. The range of careers is broad, and the coursework required for each will vary accordingly.

Credentials and qualifications required for such careers range from a basic associate's degree from a community college to one or more postgraduate degrees (master's or doctorate) from a top-tier specialist university. An advanced degree is not a requirement for a career in an applied physics discipline, and the level of education needed for a successful career will depend on the nature of the discipline chosen.

Each discipline of applied physics has a core knowledge base of mathematics and physics that is common to all disciplines. Students should therefore expect to take several courses in these areas of study, including advanced calculus and a comprehensive program of physics. Other courses of study will depend on the particular discipline that a person chooses to pursue. For disciplines that focus on medical applications, students will also take courses in chemistry and biology, because understanding the principles upon which biological systems operate is fundamental to the effective application of physics and physical devices within those systems. Engineering disciplines, the most directly related fields of applied physics, require courses in materials science, mechanics, and electronics. As may be expected, there is a great deal of overlap of the study requirements of the different disciplines in the field of applied physics.

It is also worth noting that all programs require familiarity with the use of modern computers and general software applications, and that each discipline will make extensive use of specialized applications, which may be developed as proprietary software rather than being commercially available.

SOCIAL CONTEXT AND FUTURE PROSPECTS

Applied physics has become such a fundamental underpinning of modern society that most people are unaware of its relevance. A case in point is the spin-off of technologies from the NASA space program into the realm of everyday life. Such common household devices as microwave ovens and cellular telephones, medical devices such as pacemakers and monitors, and the digital technology that permeates modern society were developed through the efforts of applied physics in exploring space. For most laypersons, the term "applied physics" conjures images of high-energy particle accelerators or interplanetary telescopes and probes, without the realization that the physical aspect of society is based entirely on principles of applied physics. As these cutting-edge fields of physical research continue, they will also continue to spawn new applications that are adapted continually into the overall fabric of modern society.

The areas of society in which applied physics has historically played a role, particularly those involving modern electronics, will continue to develop. Personal computers and embedded microcontrollers, for example, will continue to grow in power and application while decreasing in size, as research reveals new ways of constructing digital logic circuits through new materials and methods. The incorporation of the technology into consumer products, transportation, and other areas of the infrastructure of society is becoming more commonplace. As a result, the control of many human skills can be relegated to the autonomous functioning of the corresponding device. Consider, for example, the now-built-in ability of some automobiles to parallel park without human intervention.

The technologies that have been developed through applied physics permit the automation of many human actions, which, in turn, drives social change. Similarly, the ability for nearly instantaneous communication between persons in any part of the world, which has been provided through applied physics, has the potential to facilitate the peaceful and productive cooperation of large numbers of people toward resolving significant problems.

—Richard M. Renneboog, MSc

FURTHER READING

Askeland, Donald R. *The Science and Engineering of Materials.* 3d ed. London: Chapman & Hall, 1998. Includes chapters on the atomic nature of matter, the microstructure and mechanical properties of materials, the engineering of materials, the physical properties of materials, and the failure modes of materials.

The Britannica Guide to the One Hundred Most Influential Scientists. London: Constable & Robinson, 2008. Provides brief biographical and historical information about renowned people of science, from Thales of Miletus to Tim Berners-Lee. Includes chapters on Bacon, da Vinci, Kepler, Descartes, Watt, Galvani, Volta, Faraday, Kelvin, Maxwell, Tesla, Planck, Rutherford, Schrödinger, Fermi, Turing, Feynman, Hawking, and many other physical scientists.

Clegg, Brian. *Light Years: An Exploration of Mankind's Enduring Fascination with Light.* London: Judy Piatkus, 2001. Examines the history and relevance of the various concepts of philosophy and science in attempting to understand the physical nature of light.

Giancoli, Douglas C. *Physics for Scientists and Engineers with Modern Physics* Toronto, Ont.: Pearson Prentice Hall, 2008. Provides fundamental instruction in the basic physical laws, including Newtonian mechanics, thermodynamic properties, magnetism, wave mechanics, fluidics, electrical properties, optics, and quantum mechanics.

Kaku, Michio. *Visions: How Science Will Revolutionize the Twenty-first Century.* New York: Anchor Books, Doubleday, 1997. Examines the author's view of three great influences of the twenty-first century: computer science, biomolecular science, and quantum science.

Kean, Sam. *The Disappearing Spoon.* New York: Little, Brown, 2010. Provides a thoughtful and entertaining account of the persons and histories associated with each element on the periodic table, and how they have variously affected the course of society.

Sen, P. C. *Principles of Electric Machines and Power Electronics.* 2d ed. New York: John Wiley & Sons, 1997. Includes chapters on magnetic circuits, transformers, DC machines, asynchronous and synchronous machines, and special machines such as servomotors and stepper motors.

Valencia, Raymond P., ed. *Applied Physics in the Twenty-first Century*. New York: Nova Science, 2010. Includes chapters on nano metal oxide thin films, spectroscopic analysis, computational studies, magnetically modified biological materials, and plasma technology.

WEB SITES

American Institute of Physics
www.aip.org

See also: Aeronautics and Aviation; Applied Mathematics; Bionics and Biomedical Engineering; Biophysics; Civil Engineering; Communication; Computer Science; Digital Logic; Space Science.

ARTIFICIAL INTELLIGENCE

FIELDS OF STUDY

Expert systems; knowledge engineering; intelligent systems; computer vision; robotics; computer-aided design and manufacturing; computer programming; computer science; cybernetics; parallel computing; electronic health record; information systems; mobile computing; networking; business; physics; mathematics; neural networks; software engineering.

SUMMARY

Artificial intelligence is the design, implementation, and use of programs, machines, and systems that exhibit human intelligence, with its most important activities being knowledge representation, reasoning, and learning. Artificial intelligence encompasses a number of important subareas, including voice recognition, image identification, natural language processing, expert systems, neural networks, planning, robotics, and intelligent agents. Several important programming techniques have been enhanced by artificial intelligence researchers, including classical search, probabilistic search, and logic programming.

KEY TERMS AND CONCEPTS

- **automatic theorem proving:** Proving a theorem from axioms, using a mechanistic procedure, represented as well-formed formulas.
- **computer vision:** Technology that allows machines to recognize objects by characteristics, such as color, texture, and edges.
- **first-order predicate calculus:** System of formal logic, including Boolean expressions and quantification, that is rich enough to be a language for mathematics and science.

- **game theory:** Technology that supports the development of computer programs or devices that simulate one or more players of a game.
- **intelligent agent:** System, often a computer program or Web application, that collects and processes information, using reasoning much like a human.
- **logic programming:** Programming methodology that uses logical expressions for data, axioms, and theorems and an inference engine to derive results.
- **natural language processing:** How humans use language to represent ideas and to reason.
- **neural network:** Artificial intelligence system modeled after the human neural system.
- **planning:** Set of processes, generally implemented as a program in artificial intelligence, that allows an organization to accomplish its objectives.
- **robotics:** Science and technology used to design, manufacture, and maintain intelligent machines.

DEFINITION AND BASIC PRINCIPLES

Artificial intelligence is a broad field of study, and definitions of the field vary by discipline. For computer scientists, artificial intelligence refers to the development of programs that exhibit intelligent behavior. The programs can engage in intelligent planning (timing traffic lights), translate natural languages (converting a Chinese Web site into English), act like an expert (selecting the best wine for dinner), or perform many other tasks. For engineers, artificial intelligence refers to building machines that perform actions often done by humans. The machines can be simple, like a computer vision system embedded in an ATM (automated teller machine); more complex, like a robotic rover sent to Mars; or very complex, like an automated factory that builds an exercise machine with little human intervention. For cognitive scientists, artificial intelligence refers to building models of human intelligence to better understand human behavior. In the early days of artificial intelligence, most models of human intelligence were symbolic and closely related to cognitive psychology and philosophy, the basic idea being that regions of the brain perform complex reasoning by processing symbols. Later, many models of human cognition were developed to mirror the operation of the brain as an electrochemical computer, starting with the simple Perceptron, an artificial neural network described by Marvin Minsky in 1969, graduating to the backpropagation algorithm described by David E. Rumelhart and James L. McClelland in 1986, and culminating in a large number of supervised and nonsupervised learning algorithms.

When defining artificial intelligence, it is important to remember that the programs, machines, and models developed by computer scientists, engineers, and cognitive scientists do not actually have human intelligence; they only exhibit intelligent behavior. This can be difficult to remember because artificially intelligent systems often contain large numbers of facts, such as weather information for New York City; complex reasoning patterns, such as the reasoning needed to prove a geometric theorem from axioms; complex knowledge, such as an understanding of all the rules required to build an automobile; and the ability to learn, such as a neural network learning to recognize cancer cells. Scientists continue to look for better models of the brain and human intelligence.

BACKGROUND AND HISTORY

Although the concept of artificial intelligence probably has existed since antiquity, the term was first used by American scientist John McCarthy at a conference held at Dartmouth College in 1956. In 1955-1956, the first artificial intelligence program, Logic Theorist, had been written in IPL, a programming language, and in 1958, McCarthy invented Lisp, a programming language that improved on IPL. *Syntactic Structures* (1957), a book about the structure of natural language by American linguist Noam Chomsky, made natural language processing into an area of study within artificial intelligence. In the next few years, numerous researchers began to study artificial intelligence, laying the foundation for many later applications, such as general problem solvers,

intelligent machines, and expert systems.

In the 1960's, Edward Feigenbaum and other scientists at Stanford University built two early expert systems: DENDRAL, which classified chemicals, and MYCIN, which identified diseases. These early expert systems were cumbersome to modify because they had hard-coded rules. By 1970, the OPS expert system shell, with variable rule sets, had been released by Digital Equipment Corporation as the first commercial expert system shell. In addition to expert systems, neural networks became an important area of artificial intelligence in the 1970's and 1980's. Frank Rosenblatt introduced the Perceptron in 1957, but it was *Perceptrons: An Introduction to Computational Geometry* (1969), by Minsky and Seymour Papert, and the two-volume *Parallel Distributed Processing: Explorations in the Microstructure of Cognition* (1986), by Rumelhart, McClelland, and the PDP Research Group, that really defined the field of neural networks. Development of artificial intelligence has continued, with game theory, speech recognition, robotics, and autonomous agents being some of the best-known examples.

HOW IT WORKS

The first activity of artificial intelligence is to understand how multiple facts interconnect to form knowledge and to represent that knowledge in a machine-understandable form. The next task is to understand and document a reasoning process for arriving at a conclusion. The final component of artificial intelligence is to add, whenever possible, a learning process that enhances the knowledge of a system.

Knowledge representation. Facts are simple pieces of information that can be seen as either true or false, although in fuzzy logic, there are levels of truth. When facts are organized, they become information, and when information is well understood, over time, it becomes knowledge. To use knowledge in artificial intelligence, especially when writing programs, it has to be represented in some concrete fashion. Initially, most of those developing artificial intelligence programs saw knowledge as represented symbolically, and their early knowledge representations were symbolic. Semantic nets, directed graphs of facts with added semantic content, were highly successful representations used in many of the early artificial intelligence programs. Later, the nodes of the semantic nets were expanded to contain more information, and the resulting knowledge representation was referred to as frames. Frame representation of knowledge was very similar to object-oriented data representation, including a theory of inheritance.

Another popular way to represent knowledge in artificial intelligence is as logical expressions. English mathematician George Boole represented knowledge as a Boolean expression in the 1800's. English mathematicians Bertrand Russell and Alfred Whitehead expanded this to quantified expressions in 1910, and French computer scientist Alain Colmerauer incorporated it into logic programming, with the programming language Prolog, in the 1970's. The knowledge of a rule-based expert system is embedded in the if-then rules of the system, and because each if-then rule has a Boolean representation, it can be seen as a form of relational knowledge representation.

Neural networks model the human neural system and use this model to represent knowledge. The brain is an electrochemical system that stores its knowledge in synapses. As electrochemical signals pass through a synapse, they modify it, resulting in the acquisition of knowledge. In the neural network model, synapses are represented by the weights of a weight matrix, and knowledge is added to the system by modifying the weights.

Reasoning. Reasoning is the process of determining new information from known information. Artificial intelligence systems add reasoning soon after they have developed a method of knowledge representation. If knowledge is represented in semantic nets, then most reasoning involves some type of tree search. One popular reasoning technique is to traverse a decision tree, in which the reasoning is represented by a path taken through the tree. Tree searches of general semantic nets can be very time-consuming and have led to many advancements in tree-search algorithms, such as placing bounds on the depth of search and backtracking.

Reasoning in logic programming usually follows an inference technique embodied in first-order predicate calculus. Some inference engines, such as that of Prolog, use a back-chaining technique to reason from a result, such as a geometry theorem, to its antecedents, the axioms, and also show how the reasoning process led to the conclusion. Other inference engines, such as that of the expert system shell CLIPS, use a forward-chaining inference engine to see what

facts can be derived from a set of known facts.

Neural networks, such as backpropagation, have an especially simple reasoning algorithm. The knowledge of the neural network is represented as a matrix of synaptic connections, possibly quite sparse. The information to be evaluated by the neural network is represented as an input vector of the appropriate size, and the reasoning process is to multiply the connection matrix by the input vector to obtain the conclusion as an output vector.

Learning. Learning in an artificial intelligence system involves modifying or adding to its knowledge. For both semantic net and logic programming systems, learning is accomplished by adding or modifying the semantic nets or logic rules, respectively. Although much effort has gone into developing learning algorithms for these systems, all of them, to date, have used ad hoc methods and experienced limited success. Neural networks, on the other hand, have been very successful at developing learning algorithms. Backpropagation has a robust supervised learning algorithm in which the system learns from a set of training pairs, using gradient-descent optimization, and numerous unsupervised learning algorithms learn by studying the clustering of the input vectors.

APPLICATIONS AND PRODUCTS

There are many important applications of artificial intelligence, ranging from computer games to programs designed to prove theorems in mathematics. This section contains a sample of both theoretical and practical applications.

Expert systems. One of the most successful areas of artificial intelligence is expert systems. Literally thousands of expert systems are being used to help both experts and novices make decisions. For example, in the 1970's, Dell developed a simple expert system that allowed shoppers to configure a computer as they wished. In the 2010's, a visit to the Dell Web site offers a customer much more than a simple configuration program. Based on the customer's answers to some rather general questions, dozens of small expert systems suggest what computer to buy. The Dell site is not unique in its use of expert systems to guide customer's choices. Insurance companies, automobile companies, and many others use expert systems to assist customers in making decisions.

There are several categories of expert systems, but

by far the most popular are the rule-based expert systems. Most rule-based expert systems are created with an expert system shell. The first successful rule-based expert system shell was the OPS 5 of Digital Equipment Corporation (DEC), and the most popular modern systems are CLIPS, developed by the National Aeronautics and Space Administration (NASA) in 1985, and its Java clone, Jess, developed at Sandia National Laboratories in 1995. All rule-based expert systems have a similar architecture, and the shells make it fairly easy to create an expert system as soon as a knowledge engineer gathers the knowledge from a domain expert. The most important component of a rule-based expert system is its knowledge base of rules. Each rule consists of an if-then statement with multiple antecedents, multiple consequences, and possibly a rule certainty factor. The antecedents of a rule are statements that can be true or false and that depend on facts that are either introduced into the system by a user or derived as the result of a rule being fired. For example, a fact could be red-wine and a simple rule could be if (redwine) then (it-tastes-good). The expert system also has an inference engine that can apply multiple rules in an orderly fashion so that the expert system can draw conclusions by applying its rules to a set of facts introduced by a user. Although it is not absolutely required, most rule-based expert systems have a user-friendly interface and an explanation facility to justify its reasoning.

Theorem provers. Most theorems in mathematics can be expressed in first-order predicate calculus. For any particular area, such as synthetic geometry or group theory, all provable theorems can be derived from a set of axioms. Mathematicians have written programs to automatically prove theorems since the 1950's. These theorem provers either start with the axioms and apply an inference technique, or start with the theorem and work backward to see how it can be derived from axioms. Resolution, developed in Prolog, is a well-known automated technique that can be used to prove theorems, but there are many others. For Resolution, the user starts with the theorem, converts it to a normal form, and then mechanically builds reverse decision trees to prove the theorem. If a reverse decision tree whose leaf nodes are all axioms is found, then a proof of the theorem has been discovered. Gödel's incompleteness theorem (proved by Austrian-born American mathematician Kurt Gödel) shows that it may not be possible to automatically prove an arbitrary theorem in systems as complex as the natural numbers. For simpler systems, such as group theory, automated theorem proving works if the user's computer can generate all reverse trees or a suitable subset of trees that can yield a proof in a reasonable amount of time. Efforts have been made to develop theorem provers for higher order logics than first-order predicate calculus, but these have not been very successful.

Computer scientists have spent considerable time trying to develop an automated technique for proving the correctness of programs, that is showing that any valid input to a program produces a valid output. This is generally done by producing a consistent model and mapping the program to the model. The first example of this was given by English mathematician Alan Turing in 1931, by using a simple model now called a Turing machine. A formal system that is rich enough to serve as a model for a typical programming language, such as C++, must support higher order logic to capture the arguments and parameters of subprograms. Lambda calculus, denotational semantics, von Neuman geometries, finite state machines, and other systems have been proposed to provide a model onto which all programs of a language can be mapped. Some of these do capture many programs, but devising a practical automated method of verifying the correctness of programs has proven difficult.

Intelligent tutor systems. Almost every field of study has many intelligent tutor systems available to assist students in learning. Sometimes the tutor system is integrated into a package. For example, in Microsoft Office, an embedded intelligent helper provides popup help boxes to a user when it detects the need for assistance and full-length tutorials if it detects more help is needed. In addition to the intelligent tutors embedded in programs as part of a context-sensitive help system, there are a vast number of stand-alone tutoring systems in use.

The first stand-alone intelligent tutor was SCHOLAR, developed by J. R. Carbonell in 1970. It used semantic nets to represent knowledge about South American geography, provided a user interface to support asking questions, and was successful enough to demonstrate that it was possible for a computer program to tutor students. At about the same time, the University of Illinois developed its PLATO computer-aided instruction system, which provided a general language for developing intelligent tutors with touch-sensitive screens, one of the most famous of which was a biology tutorial on evolution. Of the thousands of modern intelligent tutors, SHERLOCK, a training environment for electronic Artificial Intelligence troubleshooting, and PUMP, a system designed to help learn algebra, are typical.

Electronic games. Electronic games have been played since the invention of the cathode-ray tube for television. In the 1980's, games such as Solitaire, Pac-Man, and Pong for personal computers became almost as popular as the stand-alone game platforms. In the 2010's, multiuser Internet games are enjoyed by young and old alike, and game playing on mobile devices is poised to become an important application. In all of these electronic games, the user competes with one or more intelligent agents embedded in the game, and the creation of these intelligent agents uses considerable artificial intelligence. When creating an intelligent agent that will compete with a user or, as in Solitaire, just react to the user, a programmer has to embed the game knowledge into the program. For example, in chess, the programmer would need to capture all possible configurations of a chess board. The programmer also would need to add reasoning procedures to the game; for example, there would have to be procedures to move each individual chess piece on the board. Finally, and most important for game programming, the programmer would need to add one or more strategic decision modules to the program to provide the intelligent agent with a strategy for winning. In many cases, the strategy for winning a game would be driven by probability; for example, the next move might be a pawn, one space forward, because that yields the best probability of winning, but a heuristic strategy is also possible; for example, the next move is a rook because it may trick the opponent into a bad series of moves.

IMPACT ON INDUSTRY

United States government support has been essential in artificial intelligence research, including funding for the 1956 conference at which McCarthy introduced the term "artificial intelligence." The Defense Advanced Research Projects Agency (DARPA) was a strong early supporter of artificial intelligence, then reduced support for a number of years before again providing major support for research in basic and applied artificial intelligence. Industry support for development of artificial intelligence has generally emphasized short-range projects, and university research has developed both theory and applications. Although estimates of the total value of the goods and services produced by artificial intelligence technology in a year are impossible to determine, it is clear that it is in the range of billions of dollars a year.

Government, industry, and university research. Many government agencies have provided support for basic and applied research in artificial intelligence. In 1985, NASA released the CLIPS expert system shell, and it remains the most popular shell. The National Science Foundation (NSF) supports a wide range of basic research in artificial intelligence, and the National Institutes of Health (NIH) concentrates on applying artificial intelligence to health systems. Important examples of government support for artificial intelligence are the many NSF and NIH grants for developing intelligent agents that can be embedded in health software to help doctors identify at-risk patients, suggest best practices for these patients, manage their health care, and identify cost savings. DARPA also has a very active program in artificial intelligence, including the Deep Learning project, which supports basic research into "hierarchical machine perception and analysis, and applications in visual, acoustic and somatic sensor processing for detection and classification of objects and activities." The goal is to develop a better biological model of human intelligence than neural networks and to use this to provide machine support for high-level decisions made from sensory and learned information.

Industry and business sectors. Many, if not all, software companies use artificial intelligence in their software. For example, the Microsoft Office help system is based on artificial intelligence. Microsoft's Visual Studio programming environment, uses IntelliSense, an intelligent code completion system, and Microsoft's Xbox, like all game systems, uses many artificial intelligence techniques.

Medical technology uses many applications from artificial intelligence. For example, MEDai (an Elsevier company) offers business intelligence solutions that improve health care delivery for clinics and hospitals. Other businesses that make significant use of artificial intelligence are optical character recognition companies, such as Kurzweil Technologies; speech recognition companies, such as Dragon

---

FASCINATING FACTS ABOUT ARTIFICIAL INTELLIGENCE

- In 1847, George Boole developed his algebra for reasoning that was the foundation for first-order predicate calculus, a logic rich enough to be a language for mathematics.
- In 1950, Alan Turing gave an operational definition of artificial intelligence. He said a machine exhibited artificial intelligence if its operational output was indistinguishable from that of a human.
- In 1956, John McCarthy and Marvin Minsky organized a two-month summer conference on intelligent machines at Dartmouth College. To advertise the conference, McCarthy coined the term "artificial intelligence."
- Digital Equipment Corporation's XCON, short for eXpert CONfigurer, was used in house in 1980 to configure VAX computers and later became the first commercial expert system.
- In 1989, international chess master David Levy was defeated by a computer program, Deep Thought, developed by IBM. Only ten years earlier, Levy had predicted that no computer program would ever beat a chess master.
- In 2010, the Haystack group at the Computer Science and Artificial Intelligence Laboratory at the Massachusetts Institute of Technology developed Soylent, a word-processing interface that lets users edit, proof, and shorten their documents using Mechanical Turk workers.

Speaking Naturally; and companies involved in robotics, such as iRobot.

CAREERS AND COURSE WORK

A major in computer science is the most common way to prepare for a career in artificial intelligence. One needs substantial course work in mathematics, philosophy, and psychology as a background for this degree. For many of the more interesting jobs in artificial intelligence, one needs a master's or doctoral degree. Most universities teach courses in artificial intelligence, neural networks, or expert systems, and many have courses in all three. Although artificial intelligence is usually taught in computer science, it is also taught in mathematics, philosophy, psychology, and electrical engineering. Taking a strong minor in any field is advisable for someone seeking a career in artificial intelligence because the discipline is often applied to another field.

Those seeking careers in artificial intelligence generally take a position as a systems analyst or programmer. They work for a wide range of companies, including those developing business, mathematics, medical, and voice recognition applications. Those obtaining an advanced degree often take jobs in industrial, government, or university laboratories developing new areas of artificial intelligence.

SOCIAL CONTEXT AND FUTURE PROSPECTS

After artificial intelligence was defined by McCarthy in 1956, it has had a number of ups and downs as a discipline, but the future of artificial intelligence looks good. Almost every commercial program has a help system, and increasingly these help systems have a major artificial intelligence component. Health care is another area that is poised to make major use of artificial intelligence to improve the quality and reliability of the care provided, as well as to reduce its cost by providing expert advice on best practices in health care.

Ethical questions have been raised about trying to build a machine that exhibits human intelligence. Many of the early researchers in artificial intelligence were interested in cognitive psychology and built symbolic models of intelligence that were considered unethical by some. Later, many artificial intelligence researchers developed neural models of intelligence that were not always deemed ethical. The social and ethical issues of artificial intelligence are nicely represented by HAL, the Heuristically programmed ALgorithmic computer, in Stanley Kubrick's 1968 film *2001: A Space Odyssey*, which first works well with humans, then acts violently toward them, and is in the end deactivated.

Another important ethical question posed by artificial intelligence is the appropriateness of developing programs to collect information about users of a program. Intelligent agents are often embedded in Web sites to collect information about those using the site, generally without the permission of those using the Web site, and many question whether this should be done.

—*George M. Whitson III, MS, PhD*

FURTHER READING

Giarratano, Joseph, and Peter Riley. *Expert Sys*

tems: Principles and Programming. 4th ed. Boston: Thomson Course Technology, 2005. Provides an excellent overview of expert systems, including the CLIPS expert system shell. Minsky, Marvin, and Seymour Papert. *Perceptrons: An Introduction to Computational Geometry.* Rev. ed. Boston: MIT Press, 1990. Originally printed in 1969, this work introduced many to neural networks and artificial intelligence.

Rumelhart, David E., James L. McClelland, and the PDP Research Group. *Parallel Distributed Processing: Explorations in the Microstructure of Cognition.* 1986. Reprint. 2 vols. Boston: MIT Press, 1989. Volume 1 gives an excellent introduction to neural networks, especially backpropagation. Volume 2 shows many biological and psychological relationships to neural nets.

Russell, Stuart, and Peter Norvig. *Artificial Intelligence: A Modern Approach.* 3d ed. Upper Saddle River, N.J.: Prentice Hall, 2010. The standard textbook on artificial intelligence, it provides a complete overview of the subject, integrating material from expert systems and neural networks.

Shapiro, Stewart, ed. *Encyclopedia of Artificial Intelligence.* 2d ed. New York: John Wiley & Sons, 1992. Contains articles covering the entire field of artificial intelligence.

WEB SITES
Association for the Advancement of Artificial Intelligence
http://www.aaai.org/home.html

Computer Society
http://www.computer.org

Defense Advanced Research Projects Agency Deep Learning
http://www.darpa.mil/i2o/programs/deep/deep.asp

Institute of Electrical and Electronics Engineers
http://www.ieee.org

Massachusetts Institute of Technology Computer Science and Artificial Intelligence Laboratory
http://www.csail.mit.edu

Society for the Study of Artificial Intelligence and Simulation of Behaviour
http://www.aisb.org.uk

See also: Computer Engineering; Computer Languages, Compilers, and Tools; Computer Science; Parallel Computing; Pattern Recognition.

ATMOSPHERIC SCIENCES

FIELDS OF STUDY

Climatology; meteorology; climate change; climate modeling; hydroclimatology; hydrometeorology; physical geography; chemistry, physics.

SUMMARY

Atmospheric sciences includes the fields of physics and chemistry and the study of the composition and dynamics of the layers of air that constitute the atmosphere. Related topics include climatic processes, circulation patterns, chemical and particulate deposition, greenhouse gases, oceanic temperatures, interaction between the atmosphere and the ocean, the ozone layer, precipitation patterns and amounts, climate change, air pollution, aerosol composition, atmospheric chemistry, modeling of pollutants both indoors and outdoors, and anthropogenic alteration of land surfaces that in turn affect conditions within the ever-changing atmosphere.

KEY TERMS AND CONCEPTS

- **atmosphere:** Gaseous layer that surrounds the Earth and is held in place by gravity.
- **atmospheric pressure:** Gravitational force caused by the weight of overlying layers of air.
- **atmospheric water:** Water in the atmosphere; it can be in a gaseous, liquid, or frozen form.
- **Coriolis effect:** Rightward (Northern Hemisphere) and leftward (Southern Hemisphere) deflection of air caused by the Earth's rotation.
- **energy balance:** Return to space of the Sun's energy that was received by the Earth.
- **front:** Boundary between airmasses that differ in

temperature, moisture, and pressure.

- **greenhouse effect:** Process by which longwave radiation is trapped in the atmosphere and then returned to the Earth's surface by counterradiation.
- **occluded gront:** Front produced when a cold front overtakes a warm front and forces the air upward.
- **troposphere:** Lowest level of the Earth's atmosphere that contains water vapor that can condense into clouds.
- **urban heat island:** Urbanized area that experiences higher temperatures than surrounding rural lands.

DEFINITION AND BASIC PRINCIPLES

Atmospheric sciences is the study of various aspects of the nature of the atmosphere, including its origin, layered structure, density, and temperature variation with height; natural variations and alterations associated with anthropogenic impacts; and how it is similar to or different from other atmospheres within the solar system. The present-day atmosphere is in all likelihood quite dissimilar from the original atmosphere. The form and composition of the present-day atmosphere is believed to have developed about 400 million years ago in the late Devonian period of the Paleozoic era, when plant life developed on land. This vegetative cover allowed plants to take in carbon dioxide and release oxygen as part of the photosynthesis process.

The atmosphere consists of a mixture of gases that remain in place because of the gravitational attraction of the Earth. Although the atmosphere extends about 6,000 miles above the Earth's surface, the vast proportion of its gases (97 percent) are located in the lower 19 miles. The bulk of the atmosphere consists of nitrogen (78 percent) and oxygen (21 percent). The last 1 percent of the atmosphere contains all the remaining gases, including an inert gas (argon), which accounts for 0.93 percent of the 1 percent, and carbon dioxide (CO_2), which makes up a little less than 0.04 percent. Carbon dioxide has the ability to absorb longwave radiation leaving the Earth and shortwave radiation from the Sun; therefore, any increase in carbon dioxide in the atmosphere has profound implications for global warming.

BACKGROUND AND HISTORY

Evangelista Torricelli, an Italian physicist, mathematician, and secretary to Galileo, invented the barometer, which measures barometric pressure, in 1643. The first attempt to explain the circulation of the global atmosphere was made in 1686 by Edmond Halley, an English astronomer and mathematician. In 1735, George Hadley, an English optician, described a pattern of air circulation that became known as a Hadley cell. In 1835, Gustave-Gaspard Coriolis, a French engineer and mathematician, analyzed the movement of air on a rotating Earth, a pattern that became known as the Coriolis effect. In 1856, William Ferrel, an American meteorologist, developed a model of hemispheric circulation of the atmosphere that became known as a Ferrel cell. Christophorus Buys Ballot, a Dutch meteorologist, explained the relationship between the distribution of pressure, wind speed, and direction in 1860.

Manned hot-air balloon flights beginning in the mid-nineteenth century facilitated high-level observations of the atmosphere. For example, in 1862, English meteorologist James Glaisher and English pilot Henry Coxwell reached 29,000 feet, at which point Glaisher became unconscious and Coxwell was partially paralyzed so that he had to move the control valve with his teeth. In 1902, Léon Teisserenc de Bort of France was able to determine that air temperatures begin to level out at 39,000 feet and actually increase at higher elevations. In the twentieth century, additional information about the upper atmosphere became available through radio waves, rocket flights, and satellites.

HOW IT WORKS

A knowledge of the basic structure and dynamics of the atmosphere are a necessary foundation for understanding applications and practical uses based on atmospheric science.

Layers of the atmosphere. The heterosphere and the homosphere form the two major subdivisions of the Earth's atmosphere. The uppermost subdivision (heterosphere) extends from about 50 miles above the Earth's surface to the outer limits of the atmosphere at about 6,000 miles. Nitrogen and oxygen, the heavier elements, are found in the lower layers of the heterosphere, and lighter elements such as hydrogen and helium are found at the uppermost layers of the atmosphere. The homosphere (or lowest layer) contains gases that are more uniformly mixed, although their density decreases with height. Some exceptions to this statement occur with the existence of an ozone layer at an altitude of 12 to 31 miles and with variations in concentrations of carbon dioxide, water vapor, and air pollutants closer to the Earth's surface.

The atmosphere can be divided into several zones based on decreasing or increasing temperatures as elevation increases. The lowest zone is the troposphere, where temperatures decrease from sea level up to an altitude of 10 miles in equatorial and tropical regions and up to an altitude of 4 miles at the poles. This lowermost zone holds substantial amounts of water vapor, aerosols that are very small, and light particles that originate from volcanic eruptions, desert surfaces, soot from forest and brush fires, and industrial emissions. Clouds, storms, and weather systems occur in the troposphere.

The tropopause marks the boundary between the troposphere and the next higher layer, the stratosphere, which reaches an altitude of 30 miles above the Earth's surface. Circulation in this layer occurs with strong winds that move from west to east. There is limited circulation between the troposphere and the stratosphere. However, manned balloons, certain types of aircraft (Concorde and the U-2), volcanic eruptions, and nuclear bomb tests are able to break through the tropopause and enter the stratosphere.

The gases in the stratosphere are generally uniformly mixed, with the major exception of the ozone layer, which is found at an altitude range of 12 to 31 miles above the Earth. This layer is extremely important because it shields life on Earth from the intense and harmful ultraviolet radiation from the Sun. The ozone layer has been diminishing because of the release of chlorofluorocarbons (CFCs), organic compounds containing chlorine, fluorine, and carbon, used as propellants in aerosol sprays and in synthetic chemical compounds used for refrigeration purposes. In 1978, the use of CFCs in aerosol sprays was banned in the United States, but they are still used in some refrigeration systems. Other countries continue to use CFCs, which eventually get into the ozone layer and result in ozone holes of considerable size. In 1987, members of the international community took steps to reduce CFC production through the Montreal protocol, and by 2003, the rate of ozone depletion had began to slow down. Although the manufacture and use of CFCs can be controlled, natural events that are detrimental to the ozone layer cannot be prevented. For example, the 1991 eruption of Mount

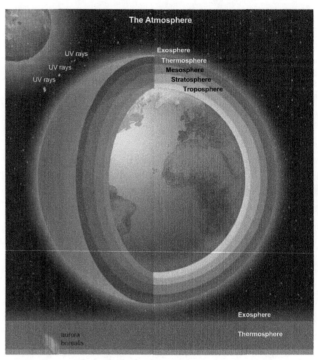

The Earth's atmosphere is divided into five layers.

Pinatubo in the Philippines reduced the ozone layer in the midlatitudes nearly 9 percent.

Temperatures decrease with elevation at the stratopause, where the mesosphere layer begins at about 30 miles and continues to an altitude of about 50 miles. The mesopause at about 50 miles marks the beginning of the thermosphere, where the density of the air is very low and holds minimal amounts of heat. However, even though the atmospheric density is minimal at altitudes above 155 miles, there is enough atmosphere to have a drag effect on spaceships.

Atmospheric pressure. The gas molecules in the atmosphere exert a pressure due to gravity that amounts to about 15 pounds per square inch on all surfaces at sea level. As the distance from the Earth gets larger, in contrast to the various increases and decreases in atmospheric temperature, atmospheric pressure decreases at an exponential rate. For example, air pressure at sea level varies from about 28.35 to 31.01 inches of mercury, averaging 29.92 inches. The pressure at the top of Mount Everest at 20,029 feet can get as low as 8.86 inches. This means that each inhalation of air at this altitude is about one-third of the pressure at sea level, producing severe shortness of breath.

Earth's global energy balance. The Earth's elliptical orbit about the Sun ranges from 91.5 million miles at perihelion (closest point to the Sun) on January 3 to 94.5 million miles at aphelion (furthest from the Sun) on July 4, averaging 93 million miles. The Earth intercepts only a tiny fraction of the total energy output of the Sun. Upon reaching the Earth, part of the incoming radiation is reflected back into space, and part is absorbed by the atmosphere, land, or oceans. Over time, the incoming shortwave solar radiation is balanced by a return to outer space of longwave radiation.

Earth-Moon differences. Scientists believe that the moon's surface has a large number of craters formed by the impact of meteorites. In contrast, there are relatively few meteorite craters on the Earth, even though, based simply on its size, the Earth is likely to have been hit by as many or even more meteorites than the Moon. This notable difference is attributed to the Earth's atmosphere, which burns up incoming meteorites, particularly small ones (the Moon does not have an atmosphere). Larger meteorites can pass through the Earth's atmosphere, but their impact craters may have been filled in or washed away over millions of years. Only the more recent ones, such as Meteor Crater in northern Arizona, with a diameter of 4,000 feet and a depth of 600 feet, remain easily recognizable.

Air masses. Different types of air masses within the troposphere, the lowest layer of the atmosphere, can be delineated on the basis of their similarity in temperature, moisture, and to a certain extent, air pressure. These air masses develop over continental and maritime locations that strongly determine their physical characteristics. For example, an air mass starting in the cold, dry interior portion of a continent develops thermal, moisture, and pressure differences that can be substantially different from an air mass that develops over water. Atmospheric dynamics also allow air masses to modify their characteristics as they move from land to water and vice versa.

Air mass and weather front terminology were developed in Norway during World War I. Norwegian meteorologists were unable to get weather reports from the Atlantic theater of operations; consequently, they developed a dense network of weather stations that led to impressive advances in atmospheric modeling that are still being used.

The radiation budget. The incoming solar energy that reaches the Earth is primarily in the shortwave (or visible) portion of the electromagnetic spectrum. The Earth's energy balance is attained by about one-third of this incoming energy being reflected back to space and the other two-thirds leaving the Earth as outgoing longwave radiation. This balance between incoming and outgoing energy is known as the Earth's radiation budget. The decades-long, ongoing National Aeronautics and Space Administration (NASA) program known as Clouds and the Earth's Radiant Energy System (CERES) is designed to measure how much shortwave and longwave radiation leaves the Earth from the top of the atmosphere.

Clouds play a very important role in the global radiation balance. For one thing, they constantly change over time and in type. Some clouds, such as high cirrus clouds found near the top of the troposphere at 40,000 feet, can have a substantial impact on atmospheric warming. Accordingly, the value of CERES is based on its ability to observe if human or natural changes in the atmosphere can be measured even if they are smaller than large-scale energy variations.

Greenhouse effect. Selected gases in the lower parts of the atmosphere trap heat and then radiate some of that heat back to Earth. If there was no natural greenhouse effect, the Earth's overall average temperature would be close to 0 degrees Fahrenheit, rather than the existing 57 degrees Fahrenheit.

The burning of coal, oil, and gas makes carbon dioxide (CO) the major greenhouse gas, accounting for nearly half of the total amount of heat-producing gases in the atmosphere. Before the Industrial Revolution in Great Britain in the mid-eighteenth century, the estimated level of carbon dioxide in the atmosphere was about 280 parts per million by volume (ppmv). Estimates for the natural range of carbon dioxide for the past 650,000 years range from 180 to 300 ppmv. All these values are less than the 391 ppmv recorded in January, 2011. Carbon dioxide levels have been increasing since 2000 at a rate of 1.9 ppmv each year. The radiative effect of carbon dioxide accounts for about one-half of all the factors that affect global warming. Estimates of carbon dioxide levels at the end of the twenty-first century range from 490 to 1,260 ppmv.

natural and human induced)—results in a doubling of the amount of this gas over what would be produced solely by wetland decay.

Chlorofluorocarbons (CFCs) absorb longwave energy (warming effect), but they also have the ability to destroy stratospheric ozone (cooling effect). The warming radiative effect is three times greater than the cooling effect. CFCs account for about 10 percent of all global warming factors. Tropospheric ozone from air pollution and nitrous oxide (NO) from motor vehicle exhaust and bacterial emissions from nitrogen fertilizers account for about 10 percent and 5 percent, respectively, of all global warming factors.

Several kinds of human actions lead to a cooling of the Earth's climate. For example, the burning of fossil fuels results in the release of tropospheric aerosols, which acts to scatter incoming solar radiation back into space, thereby lowering the amount of solar energy that can reach the Earth's surface. These aerosols also lead to the development of low and bright clouds that are quite effective in reflecting solar radiation back into space.

APPLICATIONS AND PRODUCTS

Atmospheric science is applied in many ways. It is used to help people better understand their global and interplanetary environment and to make it possible for them to live safely and comfortably within that environment. By using the principles of this field, researchers, engineers, and space scientists have developed a vast number of applications. Among the most important are those used to track and predict weather cycles and climate.

Remote sensing techniques. Oceans cover about 71 percent of the Earth's surface, which means that large portions of the world do not have weather stations or places where precipitation can be measured with standard rain gauges. To provide more information about precipitation in the equatorial and tropical parts of the world, NASA and the Japan Aerospace Exploration Agency initiated the Tropical Rainfall Monitoring Mission (TRMM) in 1997. The orbit of the TRMM satellite monitors the Earth between 35 degrees north and 35 degrees south latitude. The goal of the study is to obtain information about the extent of precipitation, along with its intensity and length of occurrence. The major instruments on the satellite are radar to detect rainfall, a passive microwave imager that can acquire data

The second most important greenhouse gas is methane (CH), which accounts for about 14 percent of all global warming factors. The origin of this gas is attributed to the natural decay of organic matter in wetlands, but anthropogenic activity—rice paddies, manure from farm animals, the decay of bacteria in sewage and landfills, and biomass burning (both about precipitation intensity and the extent of water vapor, and a scanner that can examine objects in the visible and infrared portions of the electromagnetic spectrum. The goal of data collection is to obtain the necessary climatological information about atmospheric circulation in this portion of the Earth to develop better mathematical models for determining large-scale energy movement and precipitation.

Geostationary satellites. Geostationary operational environmental satellites (GOES) enable researchers to view images of the planet from what appears to be a fixed position above the Earth. The satellites are actually circling the globe at a speed that is in step with the Earth's rotation. This means that a satellite at an altitude of 22,200 miles will make one complete revolution in the same twenty-four hours and direction that the Earth is turning above the equator. At this height, the satellite is in a position to view nearly one-half of the planet at any time. Onboard instruments can be activated to look for special weather conditions such as hurricanes, flash floods, and tornadoes. On-board instruments are also used to make precipitation estimates during storm events.

Doppler radar. Doppler radar was first used in England in 1953 to pick up the movement of small storms. The basic principle guiding this type of radar is that back-scattered radiation frequency detected at a certain location changes over time as the target, such as a storm, moves. A transmitter is used to send short but powerful microwave pulses. When a foreign object (or target) is intercepted, some of the outgoing energy is returned to the transmitter, where a receiver can pick up the signal. An image (or echo) from the target can then be enlarged and shown on a screen. The target's distance is revealed by the time that elapses between transmission and return. The radar screen cannot only indicate where the precipitation is taking place but also reveal the intensity of the rain by the amount of the echo's brightness. In short, Doppler radar has become a very useful device for determining the location of a storm and the intensity of its precipitation and for obtaining good estimates of the total amount of precipitation.

Responses to climate change. Since the 1970's, many scientists have pointed out the possibility that human activity is having more than a short-term impact on the atmosphere and therefore on weather and climate. Although much debate continues on the full impact of human activities and greenhouse gas emissions, the atmospheric sciences have led to conferences, United Nations conventions, and agreements among nations on ways that human beings can alter their behavior to halt or at least mitigate the possibility of global climate change. The impact of these agreements, still in their infancy, remains unknown—as does the overall effect of human activity on weather and climate (the models for which are highly complex). However, the insights contributed by the atmospheric sciences to the overall debate on whether climate change is primarily anthropogenic (human caused)—and whether global warming is actually taking place—have caused many nations and individuals to modify their attitudes toward human relationships with the global environment, resulting in national and intergovernmental changes in policies concerning carbon emissions, as well as personal decisions ranging from the consumption of "green" building materials to the purchase of vehicles fueled by noncarbon sources of energy.

IMPACT ON INDUSTRY

Global perspective. The World Meteorological Organization, headquartered in Geneva, Switzerland, was established to encourage weather station networks to acquire many types of atmospheric data. Accordingly, in 2007, its members decided to expand the Global Observing System (GOS) and other related observing systems, including the Global Ocean Observing System (GOOS), Global Terrestrial Observing System (GTOS), and the Global Climate Observing System (GCOS). Data are being collected from some 10,000 manned and automatic surface weather stations, 1,000 upper-air stations, more than 7,000 ships, 100 moored and 1,000 floating buoys that can drift with the currents, several hundred radars, and more than 3,000 commercial airplanes that can acquire key data on aspects of the atmosphere, land, and ocean surfaces on a daily basis.

Government research. About 180 countries maintain meteorological departments. Although many of these departments are small, the larger countries tend to have well-established governmental organizations. The major U.S. agency involved in the atmospheric sciences is the National Oceanic and Atmospheric Administration (NOAA). The agency's National Climatic Data Center (NCDC) in Asheville, North Carolina, has meteorological records going back to 1880 for both the world and the United States. These records provide invaluable historical information. For example, sea ice in the Arctic Ocean typically reaches its maximum extent in March. The coverage at the end of March, 2010, was 5.8 million square miles, which marked the seventeenth consecutive March with below-average areal extent. The National Climatic Data Center issues monthly temperature and precipitation summaries for all of the states in addition to many specialized climate data publications. It also publishes monthly mean climatic data for temperature, precipitation, barometric pressure, sunshine, and vapor pressure (that portion of atmospheric pressure that is attributed to water vapor at the time of measurement) on a global scale for about 2,000 surface sites. In addition, monthly mean upper air temperatures, dew point depressions, and wind velocities are collected and published for about 500 locations scattered around the world.

University research. Forty-eight states (all but Rhode Island and Tennessee) have either a state climatologist or someone with comparable responsibility. Most of the state climatologists are connected with state universities, in particular, the land grant institutions. The number of cooperative weather stations established since the late nineteenth century to take daily readings of temperature and precipitation in each of the forty-eight states has varied over time. These cooperative weather stations include public and private water supply facilities, colleges and universities, airports, and interested citizens.

Industry and business sectors. The number, size, and capability of private consulting firms has increased since the latter part of the twentieth century. Perhaps among the earliest entrants into this market were frost-warning service providers for citrus and vegetable growers in Arizona, Florida, and California. These private companies expanded as better forecasting and warning techniques were developed. One example of a private enterprise using atmospheric sciences is AccuWeather.com, which has seven global and fourteen regional forecast models for the United States and North America and prepares a daily weather report for *The New York Times*.

CAREERS AND COURSE WORK

The study of the physical characteristics of the atmosphere falls within the purview of atmospheric scientists. Those interested in a career in this technical area should recognize that there are several categories of specialization. The major group of specialists are operational meteorologists, who are responsible for weather forecasts. They have to carefully study the temperature, humidity, wind speed, and barometric pressure from a number of weather stations to make daily and long-range forecasts. They use data from weather satellites, radar, special sensors, and observation stations in other locations to make forecasts.

In contrast to meteorologists, who focus on short-term weather forecasts, the study of changes in weather over longer periods of time such as months, years, and in some cases centuries, is handled by climatologists. Other atmospheric scientists concentrate on research. For example, physical meteorologists are concerned with various aspects of the atmosphere, such as its chemical and physical properties, energy transfer, severe storm mechanics, and the spread of air pollutants over urbanized areas. The growing interest in air pollution and water shortages has led to another group of research scientists known as environmental meteorologists.

Given the importance of weather forecasting on a daily basis, operational meteorologists who work in weather stations may work on evenings, weekends, and holidays. Research scientists who are not engaged in weather forecasts may work regular hours.

In 2009, the American Meteorological Society estimated that there are about one hundred undergraduate and graduate atmospheric science programs in the United States that offer courses is such departments as physics, earth science, environmental science, geography, and geophysics. Entry-level positions usually require a bachelor's degree with at least twenty-four semester hours in courses covering atmospheric science and meteorology. The acquisition of a master's degree enhances the chances of employment and usually means a higher salary and more opportunities for advancement. A doctorate is required only for those who want a research position at a university.

In 2008, excluding research positions in colleges and universities, about 9,400 atmospheric scientists were working, and about one-third were employed by the federal government. Above-average employment growth is projected until 2018, representing a 15 percent increase from 2008. The median annual average salary for atmospheric scientists in May, 2008, was $81,290. The middle 50 percent had earnings between $55,140 and $101,340.

SOCIAL CONTEXT AND FUTURE PROSPECTS

Climate change may be caused by both natural internal/external processes in the Earth-Sun system or by human-induced changes in land use and the atmosphere. Article 1 of the United Nations Framework Convention on Climate Change (entered into force March, 1994) states that the term "climate change" should refer to anthropogenic changes that affect the composition of the atmosphere rather than natural causes, which should be referred to "climate variability." An example of natural climate variability is the global cooling of about 0.5 degrees Fahrenheit in 1992-1993 that was caused by the 1991 eruption of Mount Pinatubo in the Philippines. The 15 million to 20 million tons of sulfuric acid aerosols ejected into the stratosphere reflected incoming radiation from the sun, thereby creating a cooling effect. Many suggest that the above-normal temperatures experienced in the first decade of the twenty-first century provide evidence of climate change caused by human activity. Based on a variety of techniques that allow scientists to estimate the temperature in previous centuries, the year 2005 was the warmest in the last thousand years. A 2009 article published by the American Geophysical Union suggests that human intervention in Earth systems has reached a point where the Holocene epoch of the past 12,000 years is becoming a new Anthropocene epoch in which human systems have become primary Earth systems rather than simply influencing natural systems.

Numerous observations strongly suggest a continuing warming trend. Snow and ice have retreated from areas such as Mount Kilimanjaro in Tanzania, which at 19,340 feet is the highest mountain in Africa. Glaciated areas in Switzerland also provide evidence of this warming trend. The Special Report on Emission Scenarios issued in 2001 by the Intergovernmental Panel on Climate Change (IPCC) examined the broad spectrum of possible concentrations of greenhouse gases by considering the growth of population and industry along with the efficiency of energy use. The IPCC computer climate models were used to estimate future trends. For example, a global temperature increase of 35.2 to 39.2 degrees Fahrenheit by the year 2100 is a IPCC standard estimate.

—*Robert M. Hordon, MA, PhD*

FURTHER READING

Christopherson, Robert W. *Geosystems: An Introduction to Physical Geography.* 8th ed. Upper Saddle River, N.J.: Pearson Prentice Hall, 2012. Covers many topics in atmospheric sciences. Color illustrations.

Coley, David A. *Energy and Climate Change: Creating a Sustainable Future.* Hoboken, N.J.: John Wiley & Sons, 2008. A detailed review of energy topics and their relationship to climate change and energy technologies.

Ellis, Erle C., and Peter K. Haff. "Earth Science in the Anthropocene: New Epoch, New Paradigm, New Responsibilities." *EOS, Transactions, American Geophysical Union* 90, no. 49 (2009): 473. Makes the point that human systems are no longer simply influencing the natural world but are becoming part of it.

Gautier, Catherine. *Oil, Water, and Climate: An Introduction.* New York: Cambridge University Press, 2008. A good discussion of the impact of fossil fuel burning, particularly on climate change.

Lutgens, Frederick K., and Edward J. Tarbuck. *The Atmosphere: An Introduction to Meteorology.* 11th ed. Upper Saddle River, N.J.: Prentice Hall, 2010. A very good textbook written with considerable clarity.

Strahler, Alan. *Introducing Physical Geography.* 5th ed. Hoboken, N.J.: John Wiley, & Sons 2011. Covers weather and climate; contains superlative illustrations, clear maps, and lucid discussions of the material.

Wolfson, Richard. *Energy, Environment, and Climate.* New York: W. W. Norton, 2008. Provides an extensive discussion of the relationship between energy and climate change.

WEB SITES

American Geophysical Union
http://www.agu.org

American Meteorological Society
http://www.ametsoc.org

Intergovernmental Panel on Climate Change
http://www.ipcc.ch

International Association of Meteorology and Atmospheric Sciences
http://www.iamas.org

National Oceanic and Atmospheric Administration
National Climatic Data Center
http://www.Ncdc.noaa.gov

National Weather Association
http://www.nwas.org

U.S. Environmental Protection Agency Air Science
http://www.epa.gov/airscience

U.S. Geological Survey Climate and Land Use

Change
http://www.usgs.gov/climate_landuse

World Meteorological Organization Climate
http://www.wmo.int/pages/themes/climate/
index_en.php

See also: Barometry; Climate Engineering; Climate Modeling; Climatology; Meteorology; Temperature Measurement.

AUDIOLOGY AND HEARING AIDS

FIELDS OF STUDY

Biology; chemistry; physics; mathematics; anatomy; physiology; genetics; pharmacology; neurology; head and neck anatomy; acoustics.

SUMMARY

Audiology is the study of hearing, balance, and related ear disorders. Hearing disorders may be the result of congenital abnormalities, trauma, infections, exposure to loud noise, some medications, and aging. Some of these disorders may be corrected by hearing aids and cochlear implants. Hearing aids amplify sounds so that the damaged ears can discern them. Some fit over the ear with the receiver behind the ear, and some fit partially or completely within the ear canal. Cochlear implants directly stimulate the auditory nerve, and the brain interprets the stimulation as sound.

KEY TERMS AND CONCEPTS

- **analogue signal processing:** Process in which sound is amplified without additional changes.
- **audiologist:** Licensed professional who assesses hearing loss and oversees treatment.
- **auditory nerve:** Nerve that carries stimuli from the cochlea to the brain.
- **cochlea:** Coiled cavity within the ear that contains nerve endings necessary for hearing.
- **deafness:** Full or partial inability to detect or interpret sounds.
- **digital signal processing:** Process in which sound

is received and mathematically altered to produce clearer, sharper sound.
- **hearing:** Ability to detect and process sounds.
- **ototoxicity:** Damage to the hair cells of the inner ear, resulting in hearing loss.
- **sound:** Waves of pressure that a vibrating object emits through air or water.

DEFINITION AND BASIC PRINCIPLES
Audiology is the study of hearing, balance, and related ear disorders. Audiologists are licensed professionals who assess hearing loss and related sensory input and neural conduction problems and oversee treatment of patients.

Hearing is the ability to receive, sense, and decipher sounds. To hear, the ear must direct the sound waves inside, sense the sound vibrations, and translate the sensations into neurological impulses that the brain can recognize.

The outer ear funnels sound into the ear canal. It also helps the brain determine the direction from which the sound is coming. When the sound waves reach the ear canal, they vibrate the eardrum. These vibrations are amplified by the eardrum's movement against three tiny bones behind the eardrum. The third bone rests against the cochlea; when it transmits the sound, it creates waves in the fluid of the cochlea.

The cochlea is a coiled, fluid-filled organ that contains 30,000 hairs of different lengths that resonate at different frequencies. Vibrations of these hairs trigger complex electrical patterns that are transmitted along the auditory nerve to the brain, where they are interpreted.

BACKGROUND AND HISTORY

The first hearing aids, popularized in the sixteenth century, were large ear trumpets. In the nineteenth century, small trumpets or ear cups were placed in acoustic headbands that could be concealed in hats and hairstyles. Small ear trumpets were also built into parasols, fans, and walking sticks.

Electrical hearing devices emerged at the beginning of the twentieth century. These devices, which had an external power source, could provide greater amplification than mechanical devices. The batteries were large and difficult to carry; the carrying cases were often disguised as purses or camera cases. Zenith introduced smaller hearing aids with vacuum tubes and batteries in the 1940's.

In 1954, the first hearing aid with a transistor was introduced. This led to hearing aids that were made to fit behind the ear. Components became smaller and more complex, leading to the marketing of devices that could fit partially into the ear canal in the mid-1960's and ones that could fit completely into the ear canal in the 1980's.

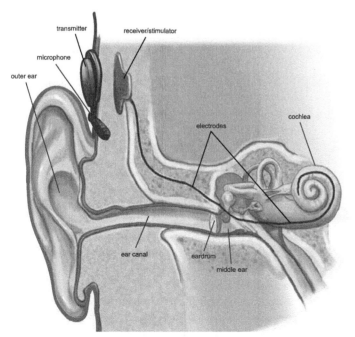

Cochlear Implant Hearing Aid

A cochlear implant differs from a hearing aid, delivering a utilitarian representation of sound.

HOW IT WORKS

Audiologists are concerned with three kinds of hearing loss: conductive hearing loss, in which sound waves are not properly transmitted to the inner ear; sensorineural hearing loss, in which the cochlea or the auditory nerve is damaged; and mixed hearing loss, which is a combination of these.

Conductive hearing loss. Otosclerosis, inefficient movement of the three bones in the middle ear, results in hearing loss from poor conduction. This disease is treatable with surgery to replace the malformed, misaligned bones with prosthetic pieces to restore conductance of sound waves to the cochlea.

Meniere's disease is thought to result from an abnormal accumulation of fluid in the inner ear in response to allergies, blocked drainage, trauma, or infection. Its symptoms include vertigo with nausea and vomiting and hearing loss. The first line of treatment is motion sickness and antinausea medications. If the vertigo persists, treatment with a Meniett pulse generator may result in improvement. This device safely applies pulses of low pressure to the middle ear to improve fluid exchange.

Hearing loss may result from physical trauma, such as a fracture of the temporal bone that lies just behind the ear or a puncture of the eardrum. These injuries typically heal on their own, and in most cases, the hearing loss is temporary.

A gradual buildup of earwax may block sound from entering the ear. Earwax should not be removed with a cotton swab or other object inserted in the ear canal; that may result in infection or further impaction. Earwax should be softened with a few drops of baby oil or mineral oil placed in the ear canal twice a day for a week and then removed with warm water squirted gently from a small bulb syringe. Once the wax is removed, the ear should be dried with rubbing alcohol. In stubborn cases, a physician or audiologist may have to perform the removal.

Foreign bodies in the ear canal, most commonly toys or insects, may block sound. If they can be seen clearly, they may be carefully removed with tweezers. If they cannot be seen clearly or moved, they may be floated out. For toys, the ear canal is flooded with warm water squirted gently from a small bulb syringe. For insects, the ear canal should first be filled with mineral oil to kill the bug. If the bug still cannot be seen clearly, the ear canal may then be flooded with warm water squirted gently from a small bulb syringe. Once the object is removed, the ear canal should be dried with rubbing alcohol.

Sensorineural hearing loss. Some medications have adverse effects on the auditory system and may cause hearing loss. These medications include large doses of aspirin, certain antibiotics, and some chemotherapy agents. Doses of antioxidants such as vitamin C, vitamin E, and ginkgo biloba may ameliorate these ototoxic effects.

Exposure to harmful levels of noise, either over long periods or in a single acute event, may result in hearing loss. If the hair cells in the cochlea are destroyed, the hearing loss is permanent. Hearing aids or cochlear implants may be necessary to compensate.

An acoustic neuroma is a noncancerous tumor that grows on the auditory nerve. It is generally surgically removed, although patients who are unable to undergo surgery because of age or illness may undergo stereotactic radiation therapy instead.

Presbycusis is the progressive loss of hearing with age as a result of the gradual degeneration of the cochlea. It may be first noticed as an inability to hear high-pitched sounds. Hearing aids are an appropriate remedy.

Mixed hearing loss. Infections of the inner ear by the viruses that cause mumps, measles, and chickenpox and infections of the auditory nerve by the viruses that cause mumps and rubella may cause permanent hearing loss. Because fluid builds up and viruses do not respond to antibiotics, viral ear infections may require surgical treatment. In a surgical procedure called myringotomy, a small hole is created in the eardrum to allow the drainage of fluid. A small tube may be inserted to keep the hole open long enough for drainage to finish.

Some children are born with hearing loss as the result of congenital abnormalities. Screening to determine the nature and severity of the hearing loss is difficult. Children are not eligible for cochlear implants before the age of twelve months. Young children often do well with behind-the-ear hearing aids, which are durable and can grow with them.

APPLICATIONS AND PRODUCTS

Audiological products consist mainly of hearing aids and cochlear implants.

Hearing aids. Although hearing aids take many forms, basically, all of them consist of a microphone that collects sound, an amplifier that magnifies the sound, and a speaker that sends it into the ear canal. They require daily placement and removal, and must be removed for showering, swimming, and battery replacement. Most people do not wear them when sleeping.

Hearing aids that sit behind the ear consist of a plastic case, a tube, and an earmold. The case contains components that collect and amplify the sound, which is then sent to the earmold through the tube. The earmold is custom-made to fit comfortably. This type of hearing aid is durable and well suited for children, although their earmolds must be replaced as they grow.

Hearing aids that sit partially within the ear canal are self-contained and custom-molded to the ear canal, so they are not recommended for growing children. Because of the short acoustic tube that channels the amplified sound, they are prone to feedback, which makes them less than ideal for people with profound hearing loss. Newer models offer feedback cancellation features.

Hearing aids that sit completely within the ear canal are self-contained and custom-molded. However, thin electrical wires replace the acoustic tube, so this type of hearing aid is free of feedback and sound distortion. They are not well suited for elderly people because their minimal size limits their volume capabilities.

The first hearing aids were analogue, a process that amplified sound without changing its properties. Although amplification was adjustable, the sound was not sorted, so background noise was also amplified. New hearing aids are digital, a process in which sound waves are converted into binary data that can be cleaned up to deliver clearer, sharper sounds.

The choice of hearing aid depends on the type and severity of hearing loss. Hearing aids will not restore natural hearing. However, they increase a person's awareness of sounds and the direction from which they are coming and heighten discernment of words.

Cochlear implants. The U.S. Food and Drug Administration approved cochlear implants in the mid1980's. A cochlear implant differs from a hearing aid in its structure and its function. It consists of a microphone that collects sound, a speech processor that sorts the incoming sound, a transmitter that relays the digital data, a receiver/stimulator that converts the sound into electrical impulses, and an electrode array that sends the impulses from the stimulator to the auditory nerve. Hearing aids compensate for damage within the ear; cochlear implants avoid the damage within the ear altogether and directly stimulate the auditory nerve.

Whereas hearing aids are completely removable, cochlear implants are not. The internal receiver/stimulator and electrode array are surgically implanted. Three to six weeks after implantation, when the surgical incision has healed, the external portions are fitted. The transmitter and receiver are held together through the skin by magnets. Thus, the external portions may be removed for sleep, showering, and swimming.

IMPACT ON INDUSTRY

Audiology is practiced in many clinics and hospitals throughout the United States, and research in the field is conducted at governmental agencies and facilities associated with universities and medical institutions. Hearing aids, cochlear implants, and other devices designed to assist people with hearing loss are the focus of many medical equipment manufacturers.

University research. A Massachusetts Institute of Technology research team, led by electrical engineering professor Dennis M. Freeman, is studying how the ear sorts sounds in an effort to create the next generation of hearing aids. This team is studying how the tectorial membrane of the cochlea carries sound waves that move from side to side, while the basilar membrane of the cochlea carries sound waves that move up and down. Together, these two membranes detect individual sounds. This team has genetically engineered mice that lack a crucial gene and protein of the tectorial membrane to study the compromised hearing.

Second-year audiology students at the University of Western Australia in Perth are conducting research projects in the development of new hearing assessment tests for children of all ages. These tests include electrophysiological studies for infants, efficient screening for children beginning school, and audiovisual games for grade school students.

Government research. Researchers funded

FASCINATING FACTS ABOUT AUDIOLOGY AND HEARING AIDS

- More than 31.5 million Americans have some degree of hearing loss.
- Three out of every 1,000 babies in the United States are born with hearing loss.
- Only one in five people who would benefit from a hearing aid actually wears one.
- Approximately 188,000 people around the world have cochlear implants. In the United States, 41,500 adults and 25,500 children have had the device implanted.
- Noise exposure is the most common cause of hearing loss. One study has shown that people who eat substantial quantities of salt are more susceptible to hearing damage from noise.
- Addiction to Vicodin (hydrocodone with acetaminophen) can result in complete deafness.
- After its discovery in 1944, streptomycin was used to treat tuberculosis with great success; however, many patients experienced irreversible cochlear dysfunction as a result of the drug's ototoxicity.
- Medications that can cause hearing loss in adults are aspirin in large quantities, antibiotics such as streptomycin and neomycin, and chemotherapy agents such as cisplatin and carboplatin. Taking antioxidants such as vitamin C, vitamin E, zinc, and ginkgo biloba helps combat ototoxicity.
- The first nationally broadcast television program to show open captioning was The French Chef with Julia Child, which appeared on station WGBH from Boston on August 5, 1972.
- The National Captioning Institute broadcast the first closed-captioned television series on March 16, 1980. Real-time closed captioning was developed in 1982.
- The Television Decoder Circuitry Act of 1990 mandated that by mid-1993 all television sets 13 inches or larger must have caption-decoding technology.

---

by the National Institute on Deafness and Other Communication Disorders (NIDCD) are looking at ways that hearing aids can selectively enhance sounds to improve the comprehension of speech. A related research team is studying ears on animals to fine-tune directional microphones that may make conversations more easily understood. Another related research The U.S. Department of Defense oversees the

team is learning to use computer-aided design programs to create and fabricate better fitting and performing hearing aids. research to American fighting
forces. Its Committee for Audiology Research is investigating alternative helmet designs to protect hearing
without blocking auditory cues and also developing a
bone conduction microphone for insertion into hel-
mets to facilitate hands-free radio operation.

Medical clinics. In addition to providing standard
audiological care, medical clinics put into practice
the knowledge obtained by research. William H.
Lippy, founder of the Lippy Group for Ear, Nose, and
Throat in Warren, Ohio, is one of the few surgeons
in the world who has perfected the stapedectomy, an
operation in which the stapes (the bone closest to
the cochlea) is replaced with a prosthesis to restore
hearing. He created the artificial stapes and many
surgical instruments for this procedure. He is sharing
his knowledge through peer-reviewed articles, book
chapters, and an online video library.

Researchers at the Mayo Clinic are working to improve hearing aid use in younger persons, who are
less likely to wear amplification devices than older
people. They are developing remote programming
of digital hearing aids over the Internet, disposable hearing aids (because children break or lose
things easily), and devices that wearers can adjust
themselves.

Assistive device manufacturers. Companies are
continuing to design, manufacture, and market assistive devices for individuals with hearing loss. These
devices can be divided into two categories: listening
devices and alerting devices. Listening devices include amplified telephones and cell phones, wireless headsets for listening to television, and FM or
infrared receivers for use in theaters and churches.
Alerting devices signal users through flashing lights,
vibration, or increased sound. They work in combination with smoke and carbon monoxide detectors,
baby monitors, alarm clocks, and doorbells.

Some companies offer patch cords that connect
the speech processors of cochlear implants to assistive listening devices for improved effect. Such patch
cords make it easier to use cell phones and enjoy
music on MP3 players. They are available with and
without volume controls.

CAREERS AND COURSE WORK

Audiologists are licensed professionals but not medical doctors. Physicians who specialize in ears, nose,
and throat disorders are called otorhinolaryngolo-
gists. Audiologists must have a minimum of a master's degree in audiology; a doctoral (Au.D.) degree
is becoming increasingly desirable in the workplace
and is required for licensure in eighteen states. State
licensure is required to practice in all fifty states. To
obtain a license, an applicant must graduate from
an accredited audiology program, accumulate 300
to 375 hours of supervised clinical experience, and
pass a national licensing exam. To get into a graduate
program in audiology, applicants must have had undergraduate courses in biology, chemistry, physics,
anatomy, physiology, math, psychology, and communication. Forty-one states have continuing education
requirements to renew a license. An audiologist must
pass a separate exam to mold and place hearing aids.
The American Board of Audiology and the American
Speech-Language-Hearing Association (AHSA) offer
professional certification programs for licensed
audiologists.

Audiologists can go into private practice as a sole
proprietor or as an associate or partner in a larger
practice. They may also work in hospitals, outpa-
tient clinics, and rehabilitation centers. Some are
employed in state and local health departments and
school districts. Some teach at universities and conduct academic research, and others work for medical
device manufacturers. Audiologists may specialize
in working with specific age groups, conducting
hearing protection programs, or developing therapy
programs for patients who are newly deaf or newly
hearing.

SOCIAL CONTEXT AND FUTURE PROSPECTS

In 2008, there were 12,800 audiologists working in
the United States. About 64 percent were employed
in health care settings and 14 percent in educational
settings. The number of audiologists was expected
to grow by 25 percent until 2018. The projected increased need can be traced to several factors.

As the population of older people continues to
grow, so will the incidence of hearing loss from aging.
In addition, the market demand for hearing aids is
expected to increase as devices become less noticeable and existing wearers switch from analogue to
digital models. At the same time, advances in medical

treatment are increasing the survival rates of premature infants, trauma patients, and stroke patients, populations who may experience hearing loss.

Hearing aid manufacturers are on their fourth generation of products, and digital devices are becoming smaller and providing increasingly better sound processing. Neurosurgical techniques also are improving. Public health programs are promoting hearing protection. Excessive noise is the most common cause of hearing loss, and one-third of noise-related hearing loss is preventable with proper protective equipment and practices. Researchers are continuing to study the genes and proteins related to specialized structures of the inner ear. They hope to discover the biological mechanisms behind hearing loss in order to interrupt them or compensate for them on the molecular level.

—Bethany Thivierge, MPH

FURTHER READING

Dalebout, Susan. *The Praeger Guide to Hearing and Hearing Loss: Assessment, Treatment, and Prevention.* Westport, Conn.: Praeger, 2009. Guides those who are experiencing hearing loss through the process of assessment and describes possible treatments and assistive devices.

DeBonis, David A., and Constance L. Donohue. *Survey of Audiology: Fundamentals for Audiologists and Health Professionals.* 2d ed. Boston: Pearson/ Allyn and Bacon, 2008. Provides excellent coverage of audiology, focusing on assessment and covering pediatric audiology.

Gelfand, Stanley A. *Essentials of Audiology.* 3d ed. New York: Thieme Medical Publishers, 2009. A comprehensive introductory textbook with abundant figures and study questions at the end of each chapter.

Kramer, Steven J. *Audiology: Science to Practice.* San Diego, Calif.: Plural, 2008. Examines the basics of audiology and how they relate to practice.

Contains a chapter on hearing aids by H. Gustav Mueller and Earl E. Johnson and a chapter on the history of audiology by James Jerger.

Lass, Norman J., and Charles M. Woodford. *Hearing Science Fundamentals.* Philadelphia: Mosby, 2007. Covers the anatomy, physiology, and physics of hearing. Contains figures, learning objectives, and chapter questions.

Moore, Brian C. J. *Cochlear Hearing Loss: Physiological, Psychological and Technical Issues.* 2d ed. Hoboken, N.J.: John Wiley & Sons, 2007. Comprehensive coverage of issues associated with cochlear hearing loss, including advances in pitch and speech perception.

Roeser, Ross J., Holly Hosford-Dunn, and Michael Valente, eds. *Audiology.* 2d ed. 3 vols. New York: Thieme, 2008. Volumes in this set cover diagnosis, treatment, and practice management. Focus is on clinical practice.

WEB SITES

American Academy of Audiology
http://www.audiology.org

American Speech-Language-Hearing Association
http://www.asha.org

Audiological Resource Association
http://www.audresources.org

Educational Audiology Assocation
http://www.edaud.org

Military Audiology Association
http://www.militaryaudiology.org

National Institute on Deafness and Other Communication Disorders
http://www.nidcd.nih.gov

AUTOMATED PROCESSES AND SERVOMECHANISMS

FIELDS OF STUDY

Electronics; hydraulics and pneumatics; mechanical engineering; computer programming; machining and manufacturing; millwright; quality assurance and quality control; avionics; aeronautics.

SUMMARY

An automated process is a series of sequential steps to be carried out automatically. Servomechanisms are systems, devices, and subassemblies that control the mechanical actions of robots by the use of feedback information from the overall system in operation.

KEY TERMS AND CONCEPTS

- **CNC (Computer Numeric Control):** An operating method in which a series of programmed logic steps in a computer controls the repetitive mechanical function of a machine.
- **control loop:** The sequence of steps and devices in a process that regulates a particular aspect of the overall process.
- **feedback:** Information from the output of a process that is fed back as an input for the purpose of automatically regulating the operation.
- **proximity sensor:** An electromagnetic device that senses the presence or absence of a component through its effect on a magnetic field in the sensing unit.
- **weld cell:** An automated fabrication center in which programmed robotic welding units perform a specified series of welds on successive sets of components.

DEFINITION AND BASIC PRINCIPLES

An automated process is any set of tasks that has been combined to be carried out in a sequential order automatically and on command. The tasks are not necessarily physical in nature, although this is the most common circumstance. The execution of the instructions in a computer program represents an automated process, as does the repeated execution of a series of specific welds in a robotic weld cell. The two are often inextricably linked, as the control of

the physical process has been given to such digital devices as programmable logic controllers (PLCs) and computers in modern facilities.

Physical regulation and monitoring of mechanical devices such as industrial robots is normally achieved through the incorporation of servomechanisms. A servomechanism is a device that accepts information from the system itself and then uses that information to adjust the system to maintain specific operating conditions. A servomechanism that controls the opening and closing of a valve in a process stream, for example, may use the pressure of the process stream to regulate the degree to which the valve is opened.

The stepper motor is another example of a servomechanism. Given a specific voltage input, the stepper motor turns to an angular position that exactly corresponds to that voltage. Stepper motors are essential components of disk drives in computers, moving the read and write heads to precise data locations on the disk surface.

Another essential component in the functioning of automated processes and servomechanisms is the feedback control systems that provide self-regulation and auto-adjustment of the overall system. Feedback control systems may be pneumatic, hydraulic, mechanical, or electrical in nature. Electrical feedback may be analogue in form, although digital electronic feedback methods provide the most versatile method of output sensing for input feedback to digital electronic control systems.

BACKGROUND AND HISTORY

Automation begins with the first artificial construct made to carry out a repetitive task in the place of a person. Early clock mechanisms, such as the water clock, used the automatic and repetitive dropping of a specific amount of water to accurately measure the passage of time. Water-, animal- or, wind-driven mills and threshing floors automated the repetitive action of processes that had been accomplished by humans. In many underdeveloped areas of the world, this repetitive human work is still a common practice.

With the mechanization that accompanied the Industrial Revolution, other means of automatically controlling machinery were developed, including self-regulating pressure valves on steam engines.

Modern automation processes began in North America with the establishment of the assembly line as a standard industrial method by Henry Ford. In this method, each worker in his or her position along the assembly line performs a limited set of functions, using only the parts and tools appropriate to that task.

Servomechanism theory was further developed during World War II. The development of the transistor in 1951, and hence, digital electronics, enabled the development of electronic control and feedback devices. The field grew rapidly, especially following the development of the microcomputer in 1969. Digital logic and machine control can now be interfaced in an effective manner, such that today's automated systems function with an unprecedented degree of precision and dependability.

How it Works

An automated process is a series of repeated, identical operations under the control of a master operation or program. While simple in concept, it is complex in practice and difficult in implementation and execution. The process control operation must be designed in a logical, step-by-step manner that will provide the desired outcome each time the process is cycled. The sequential order of operations must be set so that the outcome of any one step does not prevent or interfere with the successful outcome of any other step in the process. In addition, the physical parameters of the desired outcome must be established and made subject to a monitoring protocol that can then act to correct any variation in the outcome of the process.

A plain analogy is found in the writing and structuring of a simple computer programming function. The definition of the steps involved in the function must be exact and logical, because the computer, like any other machine, can do only exactly what it is instructed to do. Once the order of instructions and the statement of variables and parameters have been finalized, they will be carried out in exactly the same manner each time the function is called in a program. The function is thus an automated process.

The same holds true for any physical process that has been automated. In a typical weld cell, for example, a set of individual parts are placed in a fixture that holds them in their proper relative orientations. Robotic welding machines may then act upon the setup to carry out a series of programmed welds to join the individual pieces into a single assembly. The series of welds is carried out in exactly the same manner each time the weld cell cycles. The robots that carry out the welds are guided under the control of a master program that defines the position of the welding tips, the motion that it must follow, and the duration of current flow in the welding process for each movement, along with many other variables that describe the overall action that will be followed. Any variation from this programmed pattern of movements and functions will result in an incorrect output.

The control of automated processes is carried out through various intermediate servomechanisms. A servomechanism uses input information from both the controlling program and the output of the process to carry out its function. Direct instruction from the controller defines the basic operation of the servomechanism. The output of the process generally includes monitoring functions that are compared to the desired output. They then provide an input signal to the servomechanism that informs how the operation must be adjusted to maintain the desired output. In the example of a robotic welder, the movement of the welding tip is performed through the action of an angular positioning device. The device may turn through a specific angle according to the voltage that is supplied to the mechanism. An input signal may be provided from a proximity sensor such that when the necessary part is not detected, the welding operation is interrupted and the movement of the mechanism ceases.

The variety of processes that may be automated is practically limitless given the interface of digital electronic control units. Similarly, servomechanisms may be designed to fit any needed parameter or to carry out any desired function.

Applications and Products

The applications of process automation and servomechanisms are as varied as modern industry and its products. It is perhaps more productive to think of process automation as a method that can be applied to the performance of repetitive tasks than to dwell on specific applications and products. The commonality of the automation process can be illustrated by examining a number of individual applications, and the products that support them.

"Repetitive tasks" are those tasks that are to be carried out in the same way, in the same circumstances, and for the same purpose a great number of times. The ideal goal of automating such a process is to ensure that the results are consistent each time the process cycle is carried out. In the case of the robotic weld cell described above, the central tasks to be repeated are the formation of welded joints of specified dimensions at the same specific locations over many hundreds or thousands of times. This is a typical operation in the manufacturing of subassemblies

---

FASCINATING FACTS ABOUT AUTOMATED PROCESSES

- The ancient Egyptians used a remote hydraulic system to monitor the level of the water in the Nile River to determine the beginning and end of annual religious festivals.
- The water clock designed to measure the passage of a specific amount of time was perhaps the first practical, artificial automated process.
- Grain threshing mills powered by animals, wind, or water to automate the separation of grain kernels from harvested plants are still in use today.
- In relatively recent times, the turning of roasting spits in roadhouse kitchens was automated through the use of dogs called turnspits, which ran inside a treadmill wheel. A morsel of food suspended just beyond the reach of the dog was the "servomechanism" that kept the dog running.
- Control mechanisms of automated processes can use either analogue signal processing or digital signal processing.
- Computer control of physical operations is much more effective than human control, but human intervention is far superior for the subjective functions of management and quality control.
- Servomechanisms that perform some kind of automated control function can be found in unexpected places, including the spring-loaded switch that turns on the light in a refrigerator when its door is opened.
- Modern digital electronics permits extremely fine control of many physical actions and movements through servomechanisms.
- Process automation is as applicable to the fields of business management and accounting as it is to mass production in factories.

---

in the automobile industry and in other industries in which large numbers of identical fabricated units are produced.

Automation of the process, as described above, requires the identification of a set series of actions to be carried out by industrial robots. In turn, this requires the appropriate industrial robots be designed and constructed in such a way that the actual physical movements necessary for the task can be carried out. Each robot will incorporate a number of servomechanisms that drive the specific movements of parts of the robot according to the control instruction set. They will also incorporate any number of sensors and transducers that will provide input signal information for the self-regulation of the automated process. This input data may be delivered to the control program and compared to specified standards before it is fed back into the process, or it may be delivered directly into the process for immediate use.

Programmable logic controllers (PLCs), first specified by the General Motors Corporation in 1968, have become the standard devices for controlling automated machinery. The PLC is essentially a dedicated computer system that employs a limited-instruction-set programming language. The program of instructions for the automated process is stored in the PLC memory. Execution of the program sends the specified operating parameters to the corresponding machine in such a way that it carries out a set of operations that must otherwise be carried out under the control of a human operator.

A typical use of such methodology is in the various forms of CNC machining. CNC (computer numeric control) refers to the use of reduced-instruction-set computers to control the mechanical operation of machines. CNC lathes and mills are two common applications of the technology. In the traditional use of a lathe, a human operator adjusts all of the working parameters such as spindle rotation speed, feed rate, and depth of cut, through an order of operations that is designed to produce a finished piece to blueprint dimensions. The consistency of pieces produced over time in this manner tends to vary as operator fatigue and distractions affect human performance. In a CNC lathe, however, the order of operations and all of the operating parameters are specified in the control program, and are thus carried out in exactly the same manner for each piece that is produced. Operator error and fatigue do not affect production, and the machinery produces the desired pieces at the same rate throughout the entire working period. Human intervention is required only to maintain the machinery and is not involved in the actual machining process.

Servomechanisms used in automated systems check and monitor system parameters and adjust operating conditions to maintain the desired system output. The principles upon which they operate can range from crude mechanical levers to sophisticated and highly accurate digital electronic-measurement devices. All employ the principle of feedback to control or regulate the corresponding process that is in operation.

In a simple example of a rudimentary application, units of a specific component moving along a production line may in turn move a lever as they pass by. The movement of the lever activates a switch that prevents a warning light from turning on. If the switch is not triggered, the warning light tells an operator that the component has been missed. The lever, switch, and warning light system constitute a crude servomechanism that carries out a specific function in maintaining the proper operation of the system.

In more advanced applications, the dimensions of the product from a machining operation may be tested by accurately calibrated measuring devices before releasing the object from the lathe, mill, or other device. The measurements taken are then compared to the desired measurements, as stored in the PLC memory. Oversize measurements may trigger an action of the machinery to refine the dimensions of the piece to bring it into specified tolerances, while undersize measurements may trigger the rejection of the piece and a warning to maintenance personnel to adjust the working parameters of the device before continued production.

Two of the most important applications of servomechanisms in industrial operations are control of position and control of rotational speed. Both commonly employ digital measurement. Positional control is generally achieved through the use of servomotors, also known as stepper motors. In these devices, the rotor turns to a specific angular position according to the voltage that is supplied to the motor. Modern electronics, using digital devices constructed with integrated circuits, allows extremely fine and precise control of electrical and electronic factors, such as voltage, amperage, and resistance. This, in

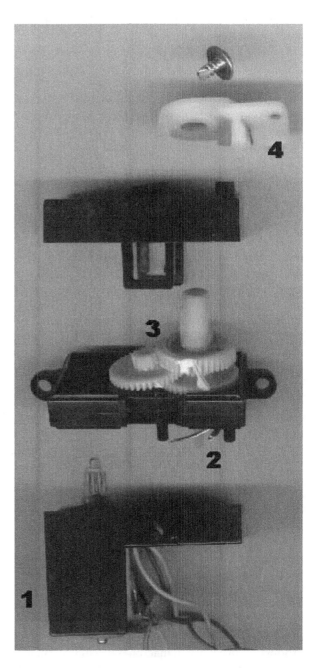

Small R/C servo mechanism1. 1. electric motor 2. position feedback potentiometer 3. reduction gear 4. actuator arm

turn, facilitates extremely precise positional control. Sequential positional control of different servomotors in a machine, such as an industrial robot, permits precise positioning of operating features. In other robotic applications, the same operating principle allows for extremely delicate microsurgery that would not be possible otherwise.

The control of rotational speed is achieved through the same basic principle as the stroboscope. A strobe light flashing on and off at a fixed rate can be used to measure the rate of rotation of an object. When the strobe rate and the rate of rotation are equal, a specific point on the rotating object will always appear at the same location. If the speeds are not matched, that point will appear to move in one direction or the other according to which rate is the faster rate. By attaching a rotating component to a representation of a digital scale, such as the Gray code, sensors can detect both the rate of rotation of the component and its position when it is functioning as part of a servomechanism. Comparison with a digital statement of the desired parameter can then be used by the controlling device to adjust the speed or position, or both, of the component accordingly.

IMPACT ON INDUSTRY

Automated processes, and the servomechanisms that apply to them, have had an immeasurable impact on industry. While the term "mass production" does not necessarily imply automation, the presence of automated systems in an operation does indicate a significant enhancement of both production and precision. Mass production revolutionized the intrinsic nature of industry in North America, beginning with the assembly-line methods made standard by Ford. This innovation enabled Ford's industry to manufacture automobiles at a rate measured in units per day rather than days per unit.

While Ford's system represents a great improvement in productions efficiency, that system can only be described as an automatic system rather than an automated process. The automation of automobile production began when machines began to carry out some of the assembly functions in the place of humans. Today, it is entirely possible for the complete assembly process of automobiles, and of other goods, to be fully automated. This is not the most effective means of production, however, as even the most sophisticated of computers pales in comparison to the

human brain in regard to intuition and intelligence. While the computer excels at controlling the mechanical function of processes, the human values of aesthetics and quality control and management are still far beyond computer calculation. Thus, production facilities today utilize the cooperative efforts of both machines and humans.

As technology changes and newer methods and materials are developed, and as machines wear out or become obsolete, constant upgrading of production facilities and methods is necessary. Process automation is essential for minimizing costs and for maximizing performance in the delivery of goods and services, with the goal of continuous improvement. It is in this area that process automation has had another profound effect on modern industry, identified by the various total quality management (TQM) programs that have been developed and adopted in many different fields throughout the world.

TQM can be traced to the Toyota method, developed by the Toyota Motor Corporation in Japan following World War II. To raise Japan's economy from the devastating effects of its defeat in that conflict, Toyota executives closely examined and analyzed the assembly-line methods of manufacturers in the United States, then enhanced those methods by stressing the importance of increasing efficiency and minimizing waste. The result was the method that not only turned Toyota into the world's largest and most profitable manufacturer of automobiles and other consumer goods, but revolutionized the manner in which manufacturers around the world conducted their businesses. Today, TQM programs such as Lean, 6Sigma, and other International Organization for Standardization (ISO) designations use process automation as a central feature of their operations; indeed, TQM represents an entirely new field of study and practice.

CAREERS AND COURSEWORK

Students looking to pursue a career that involves automated processes and servomechanisms can expect to take foundational courses in applied mathematics, physics, mechanics, electronics, and engineering. Specialization in feedback and control systems technology will be an option for those in a community college or pursuing an associate's, degree. Industrial electronics, digital technology, and machining and millwrighting also are optional routes at this level.

More advanced levels of studies can be pursued through college or university programs in computer programming, mechanical engineering, electrical engineering, and some applied sciences in the biomedical field.

As may be imagined, robotics represents a significant aspect of work in this field, and students should expect that a considerable amount of their coursework will be related to the theory and practice of robotics.

As discussed, process automation is applicable in a variety of fields in which tasks of any kind must be repeated any number of times. The repetition of tasks is particularly appropriate in fields for which computer programming and control have become integral. It is therefore appropriate to speak of automated accounting practices, automated blood-sample testing, and other biomedical testing both in research and in treatment contexts, automated traffic control, and so on. One may note that even the procedure of parallel parking has become an automated process in some modern automobiles. The variety of careers that will accept specialized training geared to automated processes is therefore much broader than might at first be expected through a discussion that focuses on robotics alone.

SOCIAL CONTEXT AND FUTURE PROSPECTS

While the vision of a utopian society in which all menial labor is automated, leaving humans free to create new ideas in relative leisure, is still far from reality, the vision does become more real each time another process is automated. Paradoxically, since the mid-twentieth century, knowledge and technology have changed so rapidly that what is new becomes obsolete almost as quickly as it is developed, seeming to increase rather than decrease the need for human labor.

New products and methods are continually being developed because of automated control. Similarly, existing automated processes can be re-automated using newer technology, newer materials, and modernized capabilities.

Particular areas of growth in automated processes and servomechanisms are found in the biomedical fields. Automated processes greatly increase the number of tests and analyses that can be performed for genetic research and new drug development. Robotic devices become more essential to the success

of delicate surgical procedures each day, partly because of the ability of integrated circuits to amplify or reduce electrical signals by factors of hundreds of thousands. Someday, surgeons will be able to perform the most delicate of operations remotely, as normal actions by the surgeon are translated into the miniscule movements of microscopic surgical equipment manipulated through robotics.

Concerns that automated processes will eliminate the role of human workers are unfounded. The nature of work has repeatedly changed to reflect the capabilities of the technology of the time. The introduction of electric street lights, for example, did eliminate the job of lighting gas-fueled streetlamps, but it also created the need for workers to produce the electric lights and to ensure that they were functioning properly. The same sort of reasoning applies to the automation of processes today. Some traditional jobs will disappear, but new types of jobs will be created in their place through automation.

—*Richard M. Renneboog, MSc*

FURTHER READING

Bryan, L. A., and E. A. Bryan. *Programmable Controllers: Theory and Implementation*. Atlanta: Industrial Text, 1988. This textbook provides a sound discussion of digital logic principles and the digital electronic devices of PLCs. Proceeds through a detailed discussion of the operation of PLCs, data measurement, and the incorporation of the systems into more complex and centralized computer networks.

James, Hubert M. *Theory of Servomechanisms*. New York: McGraw-Hill, 1947. Discusses the theory and application of the principles of servomechanisms before the development of transistors and digital logic.

Automated Processes and Servomechanisms

Kirchmer, Mathias. *High Performance Through Process Excellence*. Berlin: Springer, 2009. Examines the processes of business management as related to automation and continuous improvement in all areas of business operations.

Seal, A. M. *Practical Process Control*. Oxford, England: Butterworth-Heinemann/Elsevier, 1998. An advanced technical book that includes a good description of analogue and digital process control mechanisms and a discussion of several specific industrial applications.

Seames, Warren S. *Computer Numerical Control Concepts and Programming*. 4ª ed. Albany, N.Y.: Delmar, Thomson Learning, 2002. Provides an overview of numerical control systems and servomechanisms, with an extended discussion of the specific functions of the technology.

Smith, Carlos A. *Automated Continuous Process Control*. New York: John Wiley & Sons, 2002. A comprehensive discussion of process control systems in

chemical engineering applications.

WEB SITES

Control-Systems-Principles
http://www.control-systems-principles.co.uk

See also: Applied Mathematics; Applied Physics; Computer Engineering; Computer Science; Digital Logic; Electronics and Electrical Engineering; Hydraulic Engineering; Mechanical Engineering.

AVIONICS AND AIRCRAFT INSTRUMENTATION

FIELDS OF STUDY

Aerodynamics; aeronautical engineering; computer science; electrical engineering; electronics; hydraulics; mechanical engineering; meteorology; physics; pneumatics.

SUMMARY

Flight instrumentation refers to the indicators and instruments that inform a pilot of the position of the aircraft and give navigational information. Avionics comprises all the devices that allow a pilot to give and receive communications, such as air traffic control directions and navigational radio and satellite signals. Early in the history of flight, instrumentation and avionics were separate systems, but these systems have been vastly improved and integrated. These systems allow commercial airliners to fly efficiently and safely all around the world. Additionally, the integrated systems are being used in practically all types of vehicles—ships, trains, spacecraft, guided missiles, and unmanned aircraft—both civilian and military.

KEY TERMS AND CONCEPTS

- **airspeed indicator:** Aircraft's speedometer, giving speed based on the difference between ram and static air pressure.
- **artificial horizon:** Gyroscopic instrument that displays the airplane's attitude relative to the horizon.
- **directional gyro (bertical compass):** Gyroscopically controlled compass rose.
- **global positioning system (GPs):** Satellite-based navigational system.

- **horizontal situation indicator (HSI):** Indicator that combines an artificial horizon with a directional gyro.
- **instrument flight:** Flight by reference to the flight instruments when the pilot cannot see outside the cockpit because of bad weather.
- **instrument landing system (ILS):** Method to guide an airplane to the runway using a sensitive localizer to align the aircraft with the runway and a glide slope to provide a descent path to the runway.
- **radar altimeter:** Instrument that uses radar to calculate the aircraft's height above the ground.
- **ram:** Airflow that the aircraft generates as it moves through the air.
- **turn and bank indicator:** Gyroscopic instrument that provides information on the angle of bank of the aircraft.

DEFINITION AND BASIC PRINCIPLES

Flight instrumentation refers to the instruments that provide information to a pilot about the position of the aircraft in relation to the Earth's horizon. The term "avionics" is a contraction of "aviation" and "electronics" and has come to refer to the combination of communication and navigational devices in an aircraft. This term was coined in the 1970's after the systems were becoming one integral system.

The components of basic flight instrumentation are the magnetic compass, the instruments that rely on air-pressure differentials, and those that are driven by gyroscopes and instruments. Air pressure decreases with an increase in altitude. The altimeter and vertical-speed indicator use this change in pressure to provide information about the height of the aircraft above sea level and the rate that the aircraft is climbing or descending. The airspeed indicator uses ram air pressure to give the speed that the aircraft is traveling through the air.

Other instruments use gyroscopes to detect changes in the position of the aircraft relative to the Earth's surface and horizon. An airplane can move around the three axes of flight. The first is pitch, or the upward and downward position of the nose of the airplane. The second is roll, the position of the wings. They can be level to the horizon or be in a bank position, where one wing is above horizon and the other below the horizon as the aircraft turns. Yaw is the third. When an airplane yaws, the nose of the airplane moves to the right or left while the airplane is in level flight. The instruments that use gyroscopes to show movement along the axes of flight are the turn and bank indicator, which shows the angle of the airplane's wings in a turn and the rate of turn in degrees per second; the artificial horizon, which indicates the airplane's pitch and bank; and the directional gyro, which is a compass card connected to a gyroscope. Output from the flight instruments can be used to operate autopilots. Modern inertial navigation systems (INS's) use gyroscopes, sometimes in conjunction with a Global Positioning System (GPS), as integrated flight instrumentation and avionics systems.

The radios that comprise the avionics of an aircraft include communications radios that pilots use to talk to air traffic control (ATC) and other aircraft and navigation radios. Early navigation radios relied on ground-based radio signals, but many aircraft have come to use GPS receivers that receive their information from satellites. Other components of an aircraft's avionics include a transponder, which sends a discrete code to ATC to identify the aircraft and is used in the military to discern friendly and enemy aircraft, and radar, which is used to locate rain and thunderstorms and to determine the aircraft's height above the ground.

BACKGROUND AND HISTORY

Flight instruments originally were separated from the avionics of an aircraft. The compass, perhaps the most basic of the flight instruments, was developed by the Chinese in the second century BCE. Chinese navy commander Zheng He's voyages from 1405 to 1433 included the first recorded use of a magnetic compass for navigation.

The gyroscope, a major component of many flight instruments, was named by French physicist Jean-Bernard-Léon Foucault in the nineteenth century. In 1909, American businessman Elmer Sperry invented the gyroscopic compass that was first used on

U.S. naval ships in 1911. In 1916, the first artificial horizon using a gyroscope was invented. Gyroscopic flight instruments along with radio navigation signals guided American pilot Jimmy Doolittle to the first successful all-instrument flight and landing of an airplane in 1929. Robert Goddard, the father of rocketry, experimented with using gyroscopes in guidance systems for rockets. During World War II, German rocket scientist Wernher von Braun further developed Goddard's work to build a basic guidance system for Germany's V-2 rockets. After the war, von Braun and 118 of his engineers immigrated to the United States, where they worked for the U.S. Army on gyroscopic inertial navigation systems (INSs) for rockets. Massachusetts Institute of Technology engineers continued the development of the INS to use in Atlas rockets and eventually the space shuttle. Boeing was the first aircraft manufacturer to install INSs into its 747 jumbo jets. Later, the Air Force introduced the system to their C-141 aircraft.

Radios form the basis of modern avionics. Although there is some dispute over who actually invented the radio, Italian physicist Guglielmo Marconi first applied the technology to communication. During World War I, in 1916, the Naval Research Laboratory developed the first aircraft radio. In 1920, the first ground-based system for communication with aircraft was developed by General Electric. The earliest navigational system was a series of lights on the ground, and the pilot would fly from beacon to beacon. In the 1930's, the nondirectional radio beacon (NDB) became the major radio navigation system. This was replaced by the very high frequency omnidirectional range (VOR) system in the 1960's. In 1994, the GPS became operational and was quickly adapted to aircraft navigation. The great accuracy that GPS can supply for both location and time was adapted for use in INS.

HOW IT WORKS

Flight instruments. Flight instruments operate using either gyroscopes or air pressure. The instruments that use air pressure are the altimeter, the vertical speed indicator, and the airspeed indicator.

The Airbus A380 glass cockpit featuring pull-out keyboards and two wide computer screens on the sides for pilots

Airplanes are fitted with two pressure sensors: the pitot tube, which is mounted under a wing or the front fuselage, its opening facing the oncoming air; and the static port, which is usually mounted on the side of the airplane out of the slipstream of air flowing past the plane. The pitot tube measures ram air; the faster the aircraft is moving through the air, the more air molecules enter the pitot tube. The static port measures the ambient air pressure, which decreases with increasing altitude. The airspeed indicator is driven by the force of the ram air calibrated to the ambient air pressure to give the speed that the airplane is moving through the sky. The static port's ambient pressure is translated into altitude above sea level by the altimeter. As air pressure can vary from location to location, the altimeter must be set to the local barometric setting in order to receive a correct altimeter reading. The vertical speed indicator also uses the ambient pressure from the static port. This instrument can sense changes in altitude and indicates feet per minute that the airplane is climbing or descending.

Other flight instruments operate with gyroscopes. These instruments are the gyroscopic compass, the turn and bank indicator, and the artificial horizon. The gyroscopic compass is a vertical compass card connected to a gyroscope. It is either set by the pilot or slaved to the heading indicated on the magnetic compass. The magnetic compass floats in a liquid that allows it to rotate freely but also causes it to jiggle in turbulence; the directional gyro is stabilized by its gyroscope. The magnetic compass will also show errors while turning or accelerating, which are eliminated by the gyroscope. The turn and bank indicator is connected to a gyroscope that remains stable when the plane is banking. The indicator shows the angle of bank of the airplane. The artificial horizon has a card attached to it that shows a horizon, sky, and ground and a small indicator in the center that is connected to the gyroscope. When the airplane pitches up or down or rolls, the card moves with the airplane, but the indicator is stable and shows the position of the aircraft relative to the horizon. Pilots use the artificial horizon to fly when they cannot see the natural horizon. The artificial horizon and directional gyro can be combined into one instrument, the horizontal situation indicator (HSI). These instruments can be used to supply information to an autopilot, which can be mechanically connected to the flight surfaces of the aircraft to fly it automatically.

Ground-based avionics. Ground-based avionics provide communications, navigational information, and collision avoidance. Communication radios operate on frequencies between 118 and 136.975 megahertz (MHz). Communication uses line of sight. Navigation uses VOR systems. The VOR gives a signal to the aircraft receiver that indicates the direction to or from the VOR station. A more sensitive type of VOR, a localizer, is combined with a glide slope indicator to provide runway direction and a glide path for the aircraft to follow when it is landing in poor weather conditions and the pilot does not have visual contact with the runway. Collision avoidance is provided by ATC using signals from each aircraft's transponders and radar. ATC can identify the aircrafts' positions and advise pilots of traffic in their vicinity.

Satellite-based systems. The limitation of line of sight for ground-based avionic transmitters is a major problem for navigation over large oceans or in areas of the world that have large mountain ranges or few transmitters. The U.S. military was very concerned about these limitations and the Department of

Defense spearheaded the research and implementation of a system that addresses these problems. GPS is the United States's satellite system that provides navigational information. GPS can give location, movement, and time information. The system uses a minimum of twenty-four satellites orbiting the Earth that send signals to monitoring receivers on Earth. The receiver must be able to get signals from a minimum of four satellites in order to calculate an aircraft's position correctly. Although originally designed solely for military use, GPS is widely used by civilians.

Inertial navigation systems (INS). The INS is a self-contained system that does not rely on outside radio or satellite signals. INS is driven by accelerometers and gyroscopes. The accelerometer houses a small pendulum that will swing in relation to the aircraft's

acceleration or deceleration and so can measure the aircraft's speed. The gyroscope provides information about the aircraft's movement about the three axes of flight. Instead of the gimbaled gyros, more precise strap-down laser gyroscopes have come to be used. The strap-down system is attached to the frame of the aircraft. Instead of the rotating wheel in the gimbaled gyroscopes, this system uses light beams that travel in opposite directions around a small, triangular path. When the aircraft rotates, the path traveled by the beam of light moving in the direction of rotation appears shorter than the path of the other beam of light moving in the opposite direction. The length of the path causes a frequency shift that is detected and interpreted as aircraft rotation. INS must be initialized: The system has to be able to detect its initial position or it must be programmed with its initial position before it is used or it will not have a reference point from which to work.

APPLICATIONS AND PRODUCTS

Military INS and GPS uses. Flight instrumentation and avionics are used by military aircraft as well as civilian aircraft, but the military have many other applications. INS is used in guided missiles and submarines. It can also be used as a stand-alone navigational system in vehicles that do not want to communicate with outside sources for security purposes. INS and GPS are used in bombs, rockets, and, with great success, unmanned aerial vehicles (UAVs) that are used for reconnaissance as well as delivering ordnance without placing a pilot in harm's way. GPS is used in almost all military vehicles such as tanks, ships, armored vehicles, and cars, but not in submarines as the satellite signals will not penetrate deep water. GPS is also used by the United States Nuclear Detonation Detection System as the satellites carry nuclear detonation detectors.

Navigation. Besides the use of flight instrumentation and avionics for aircraft navigation, the systems can also be used for almost all forms of navigation. The aerospace industry has used INS for guidance of spacecraft that cannot use earthbound navigation systems, including satellites that orbit the planet. INS systems can be initialized by manually inputting the craft's position using GPS or using celestial fixes to direct rockets, space shuttles, and long-distance satellites and space probes through the reaches of the solar system and beyond. These systems can be

---

FASCINATING FACTS ABOUT AVIONICS AND AIRCRAFT INSTRUMENTATION

- During the Civil War, surveillance balloons communicated with ground crews using telegraph. The telegraph wires from the balloon could be connected with ground telegraph wires so that the observations could be relayed directly to President Lincoln in the White House.
- Before radios were installed in airplanes, inventors tried to communicate with people on the ground by dropping notes tied to rocks and by smoke signals.
- The technology that tells airplanes how high in the sky they are is used by skydivers to know when to open their parachutes and by scuba divers to know how far under the water they are.
- Some researchers are investigating using robotic dirigibles flying in the stratosphere to replace expensive communication satellites and unsightly cell phone towers.
- GPS, which is commonly used in the family car and is an "app" in smart phones, was originally developed by the military for defense.
- GPS can be used to help lost children and pets get home safely.
- Aircraft flight instrumentation is used to construct flight simulators, including flight simulators that can be used on a home computer.
- Someday unmanned airplanes may be used in commercial transportation, carrying freight and passengers.

---

synchronized with computers and sensors to control the vehicles by moving flight controls or firing rockets. GPS can be used on Earth by cars, trucks, trains, ships, and handheld units for commercial, personal, and recreational uses. One limitation of GPS is that it cannot work where the signals could be blocked, such as under water or in caves.

Cellular phones. GPS technology is critical for operating cellular phones. GPS provide accurate time that is used in synchronizing signals with base stations. If the phone has GPS capability built into it, as many smart phones do, it can be used to locate a mobile cell phone making an emergency call. The GPS system in cell phones can be used in cars for navigation as well for recreation such as guidance while hiking, biking, boating, or geo-caching.

Tracking systems. In the same manner that GPS can be used to locate a cell phone, GPS can be used to find downed aircraft or pilots. GPS can be used by biologists to track wildlife by placing collars on the animals, a major improvement over radio tracking that was line-of-sight and worked only over short ranges. Animals that migrate over great distances can be tracked by using only GPS. Lost pets can be tracked through GPS devices in their collars. Military and law enforcement use GPS to track vehicles.

Other civilian applications. Surveyors and mapmakers use GPS to mark boundaries and identify locations. GPS units installed at specific locations can detect movements of the Earth to study earthquakes, volcanoes, and plate tectonics.

Next Generation air transportation system (NextGen). While ground-based navigational systems such as the VOR are still used by pilots and radar is used by ATC to locate airplanes, the Federal Aviation Administration (FAA) is researching and designing NextGen, a new system for navigation and tracking aircraft that will be based on GPS in the National Airspace System (NAS). Using NextGen GPS navigation, aircraft will be able to fly shorter and more direct routes to their destinations, saving time and fuel. ATC will change to a satellite-based system of managing air traffic.

IMPACT ON INDUSTRY

Throughout the history of aviation, the united states has been a leader in research and development of flight instrumentation and avionics. During the Cold War, the Soviet Union and China worked to develop their own systems but lagged behind the United States. China has produced its own avionics but has not found an international market for its products. China has partnered with U.S. companies such as Honeywell and Rockwell Collins to produce avionics.

Government and university research. Most government research is spearheaded by the FAA and conducted in government facilities or in universities and businesses supported by grants. The FAA is concentrating on applying GPS technology to aviation. Ohio University's Avionics Engineering Center is researching differential GPS applications for precision runway approaches and landings, as GPS is not sensitive enough to provide the very precise positioning required. The University of North Dakota is researching GPS applications for NextGen. Embry-Riddle Aeronautical University is working with the FAA's Office of Commercial Space Transportation to develop a space-transportation information system that will tie in with air traffic control. Embry-Riddle is also partnering with Lockheed Martin, Boeing, and other companies on the NextGen system. The FAA has agreements with Honeywell and Aviation Communication & Surveillance Systems to develop and test airfield avionics systems to be used in NextGen. The military also does research on new products, but their efforts are usually classified.

Commercial and industry sectors. The major consumers of flight instrumentation and avionics are aircraft manufacturers and airlines. In the United States, Boeing is the foremost commercial manufacturer for civilian as well as military aircraft, while Lockheed Martin, Bell Helicopter, and Northrop Grumman produce military aircraft. In Europe, Airbus is the major aircraft manufacturer.

The GPS technology developed for aircraft has also been used by companies such as Verizon and AT&T for cellular phones, Garmin for GPS navigation devices, and by power plants to synchronize power grids.

Spacecraft and space travel sectors have several companies actively manufacturing rocket ships, spacecraft, and navigational systems. The British company Virgin Galactic and the American company Space Adventures operate spacecraft, and private citizens can book flights, although it is extremely expensive.

Flight instrumentation and avionics are being used to construct UAVs for the military. Northrop

Grumman is working with the military to develop this technology further, and it has been used very successfully in Iraq and Afghanistan. Other companies and universities are researching the possibility of having unmanned commercial aircraft in the United States, although the problems with coordinating unmanned aircraft with traditional aircraft and the safety concerns of operating an aircraft without a pilot are daunting.

Major corporations. Several companies manufacture avionics, most of which are in the United States. Some of these companies have joint ventures with other countries and have established factories in those countries. Major avionics and flight instrument manufacturers are Honeywell, Rockwell Collins, Bendix/King, ARINC, Kollsman, Narco Avionics, Sigtronics, and Thales.

Many companies continue to develop new flight and avionics systems for aircraft. These products often emphasize integrating existing systems and applying declassified military advances to civilian uses. For example, Rockwell Collins's Pro Line Fusion integrated avionics system uses the military head-up display (HUD) technology for civilian business jets.

CAREERS AND COURSE WORK

The possible careers associated with flight instruments and avionics include both civilian and military positions ranging from mechanics and technicians to designers and researchers. The education required for these occupations usually requires at least two years of college or technical training, but research and design may require a doctorate.

Maintenance and avionics technicians install and repair flight instruments and avionics. They may work on general aviation airplanes, commercial airliners, or military aircraft. With more and more modes of transportation using INS and GPS, mechanics and technicians may also be employed to install and repair these systems on other types of vehicles—ships, trains, guided missiles, tanks, or UAVs. NASA and private companies employ technicians to work with spacecraft. Most of these positions require an associate's degree with specialization as an aircraft or avionics technician, or the training may be acquired in the military.

As computers are becoming more and more important in these fields, the demand for computer technicians, designers, and programmers will increase. Jobs in these fields range from positions in government agencies such as the FAA or National Aeronautics and Space Administration (NASA) or the military to private-sector research and development. The education required for these occupations varies from high school or vocational computer training to doctorates in computer science or related fields.

Flight instrument and avionics systems are being designed and researched by persons who have been educated in mechanical, electrical, and aeronautical engineering, computer science, and related fields. Some of these occupations require the minimum of a bachelor's degree, but most require a master's or doctorate.

SOCIAL CONTEXT AND FUTURE PROSPECTS

Aviation, made possible by flight instrumentation and avionics, has revolutionized how people travel and how freight is moved throughout the world. It has also dramatically changed how wars are fought and how countries defend themselves. In the future, flight instrumentation and avionics will continue to affect society not only through aviation but also through applications of the technology in daily life.

Military use of UAVs controlled by advances in flight instrumentation and avionics will continue to change how wars are fought. However, in the not-toodistant future this technology may be used in civilian aviation. UAVs could be used to inspect pipelines and perform surveys in unpopulated areas or rough terrain, but it is unsure whether they will be used for passenger flights. Many people will certainly be fearful of traveling in airplanes with no human operators. The FAA would have to develop systems that would incorporate unmanned aircraft into the airspace. However, the use of unmanned vehicles may be an important part of future space exploration.

Perhaps the avionics system that has had the most impact on society is GPS. As GPS devices are being made more compact and more inexpensively, they are being used more and more in daily life. GPS can permit underdeveloped countries to improve their own air-navigation systems more rapidly without the expensive of buying and installing expensive ground-based navigational equipment or radar systems used by air traffic control facilities.

—Polly D. Steenhagen, MS

FURTHER READING

Collinson, R. P. G. *Introduction to Avionics Systems.* 2d ed. Boston: Kluwer Academic, 2003. A comprehensive and well-illustrated review of both civilian and military flight and avionics systems.

Dailey, Franklyn E., Jr. *The Triumph of Instrument Flight: A Retrospective in the Century of U.S. Aviation.* Wilbraham, Mass.: Dailey International, 2004. A history of instrument flight with factual information as well as the author's personal flying experiences.

El-Rabbany, Ahmed. *Introduction to GPS: The Global Positioning System.* 2d ed. Norwood, Mass.: Artech House, 2006. A thorough overview of GPS and a discussion of future applications for the technology.

Federal Aviation Administration. *Instrument Flying Handbook.* New York: Skyhorse, 2008. The FAA's official manual on instrument flying includes discussions of flight instruments and avionics and the two allow a pilot to fly by reference to the instruments.

Johnston, Joe. *Avionics for the Pilot: An Introduction to Navigational and Radio Systems for Aircraft.* Wiltshire, England: Airlife, 2007. A straightforward basic explanation of aircraft avionics, what they do, and how they operate.

Tooley, Mike. *Aircraft Digital Electronic and Computer Systems: Principles, Operation and Maintenance.* Burlington, Mass.: Butterworth-Heinemann, 2007. A more technical look at aircraft avionics and computer systems and how they work.

WEB SITES

Aircraft Electronics Association
http://www.aea.net

Aviation Instrument Association
http://www.aia.net

Federal Aviation Administration
http://www.faa.gov

Official Government Information About the GPS
http://www.gps.gov

See also: Aeronautics and Aviation; Computer Science; Electrical Engineering; Mechanical Engineering; Meteorology.

B

BAROMETRY

FIELDS OF STUDY

Atmospheric science; physics; chemistry; fluid mechanics; electromagnetics; signal processing, meteorology;

SUMMARY

The science and engineering of pressure measurement in gases take its practitioners far beyond its original realm of weather prediction. The sensors used in barometry range from those for the near vacuum of space and the small amplitudes of soft music to those for ocean depths and the shock waves of nuclear-fusion explosions.

KEY TERMS AND CONCEPTS

- **aneroid barometer:** Instrument that measures atmospheric pressure without using liquid.
- **bar:** Unit of pressure equal to 100,000 newtons per square meter.
- **pascal:** Unit of pressure equal to 1 newton per square meter.
- **piezoelectric material:** Any material that generates a voltage when the pressure acting on it changes.
- **pounds per square inch absolute (psia):** Unit of pressure in pounds per square in absolute.
- **pounds per square inch gauge (psig):** Unit of pressure in pounds per square inch relative to some reference pressure.
- **pressure-sensitive paint:** Liquid mixture that, when painted and dried on a surface and illuminated with ultraviolet light, emits radiation, usually infrared, the intensity of which changes with the pressure of air acting on the surface.
- **torr:** Unit of pressure equal to one part in 760 of a standard atmosphere, also equal to the pressure at the bottom of a column of mercury 1 millimeter high with vacuum above it.
- **torricelli barometer:** Instrument to measure absolute atmospheric pressure, consisting of a graduated tube closed at the top end, containing a vacuum at the top above a column of mercury, and standing in a reservoir of mercury.
- **u-tube manometer:** Instrument consisting of a graduated tube containing liquid and shaped like the letter U with each arm connected to a source of pressure, so that the difference in liquid levels between the two tubes indicates the differential pressure.

DEFINITION AND BASIC PRINCIPLES

Barometry is the science of measuring the pressure of the atmosphere. Derived from the Greek words for "heavy" or "weight" (*baros*) and "measure" (*metron*), it refers generally to the measurement of gas pressure. In gases, pressure is fundamentally ascribed to the momentum flowing across a given surface per unit time, per unit area of the surface. Pressure is expressed in units of force per unit area. Although pressure is a scalar quantity, the direction of the force due to pressure exerted on a surface is taken to be perpendicular and directed onto the surface. Therefore, methods to measure pressure often measure the force acting per unit area of a sensor or the effects of that force. Pressure is expressed in newtons per square meter (pascals), in pounds per square foot (psf), or in pounds per square inch (psi). The pressure of the atmosphere at standard sea level at a temperature of 288.15 Kelvin (K) is 101,325 pascals, or 14.7 psi. This is called 1 atmosphere. Mercury and water barometers have become such familiar devices that pressure is also expressed in inches of water, inches of mercury, or in torrs (1 torr equals about 133.3 pascals).

The initial weather-forecasting barometer, the Torricelli barometer, measured the height of a liquid column that the pressure of air would support, with a vacuum at the closed top end of a vertical tube. This barometer is an absolute pressure instrument. Atmospheric pressure is obtained as the product of the height, the density of the barometric liquid, and the acceleration because of gravity at the Earth's surface. The aneroid barometer uses a partially evacuated box the spring-loaded sides of which expand or contract depending on the atmospheric pressure, driving a clocklike mechanism to show the pressure on a circular dial. This portable instrument was convenient to carry on mountaineering, ballooning, and mining expeditions to measure altitude by the change in atmospheric pressure. A barograph is an aneroid barometer mechanism adapted to plot a graph of the variation of pressure with time, using a stylus moving on a continuous roll of paper. The rate of change of pressure helps weather forecasters to predict the strength of approaching storms.

The term "manometer" derives from the Greek word *manos*, meaning "sparse," and denotes an instrument used to measure the pressure relative to a known pressure. A U-tube manometer measures the pressure difference from a reference pressure by the difference between the height of a liquid column in the leg of the U-tube that is connected to a known pressure source and the height of the liquid in the other leg, exposed to the pressure of interest. Manometers of various types have been used extensively in aerospace engineering experimental-test facilities such as wind tunnels. The pitot-static tubes used to measure flow velocity in wind tunnels were initially connected to water or mercury manometers. Later, electronic equivalents were developed. Inclined tube manometers were used to increase the sensitivity of the instrument in measuring small pressure differences amounting to fractions of an inch of water.

BACKGROUND AND HISTORY

In 1643, Italian physicist and mathematician Evangelista Torricelli proved that atmospheric pressure would support the weight of a thirty-five-foot water column leaving a vacuum above that in a closed tube and that this height would change with the weather. Later Torricelli barometers used liquid mercury to reduce the size of the column and make such instruments more practical.

The technology of pressure measurement has evolved gradually since then, with the aneroid barometer demonstrating the reliability of deflecting a diaphragm. This led to electrical means of measuring the amount of deflection. The most obvious

method was to place strain gauges on the diaphragm and directly measure the strain. Later electrical methods used the change in capacitance caused by the changing gap between two charged plates. Such sensors dominated the market until the 1990's at the low end of the measurement range. Piezoresistive materials expanded the ability of miniaturized strain gauge sensors to measure high pressures changing at high frequency. Microelectromechanical system (MEMS) technology enabled miniaturized solid-state sensors to challenge the market dominance of the diaphragm sensors. In the early twenty-first century, pressure-sensitive paints allowed increasingly sensitive and faster-responding measurements of varying pressure with very fine spatial resolution.

HOW IT WORKS

Barometry measures a broad variety of pressures using an equally broad variety of measurement techniques, including liquid column methods, elastic element methods, and electrical sensors. Electrical sensors include resistance strain gauges, capacitances, piezoresistive instruments, and piezoelectric devices. The technologies range from those developed by French mathematician Blaise Pascal, Greek mathematician Archimedes, and Torricelli to early twenty-first century MEMS sensors and those used to conduct nanoscale materials science.

Pressures can be measured in environments from the near vacuum of space to more than 1,400 megapascals (MPa) and from steady state to frequencies greater than 100,000 cycles per second. Sensors that measure with respect to zero pressure or absolute vacuum are called absolute pressure sensors, whereas those that measure with respect to some other reference pressure are called gauge pressure sensors. Vented gauge sensors have the reference side open to the atmosphere so that the pressure reading is with respect to atmospheric pressure. Sealed gauges report pressure with respect to a constant reference pressure.

Where rapid changes in pressure must be measured, errors due to the variation of sensitivity with the rate of change must be considered. A good sensor is one whose frequency response is constant over the entire range of frequency of fluctuations that might occur. Condenser microphones with electromechanical diaphragms have long been used to measure acoustic pressure in demanding applications such as music recording, with flat frequency response from 0.1 cycles per second (hertz) to well over 20,000 hertz, covering the range of human hearing. Pressure-sensitive paint in certain special formulations has been shown to achieve excellent frequency response to more than 1,600 hertz but only when

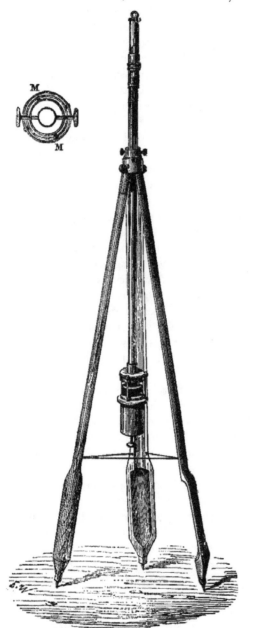

Fig. 279. — Baromètre de Fortin et son pied. — MM, suspension à la Cardan du baromètre de Fortin.

Fortin barometer

the fluctuation amplitude is quite large, near the upper limit of human tolerance. Using digital signal processing, inexpensive sensors can be corrected to produce signals with frequency response quality approaching that of much more expensive sensors.

In the 1970's, devices based on the aneroid barometer principle were developed, in which the deflection of a diaphragm caused changes in electrical capacitance that then were indicated as voltage changes in a circuit. In the 1980's, piezoelectric materials were developed, enabling electrical voltages to be created from changes in pressure. Micro devices based on these largely replaced the more expensive but accurate diaphragm-based electromechanical sensors. Digital signal processing enabled engineers using the new small, inexpensive devices to recover most of the accuracy possessed by the more expensive devices.

APPLICATIONS AND PRODUCTS

Barometry has ubiquitous applications, measured by a broad variety of sensors. It is key to weather prediction and measuring the altitude of aircraft as well as to measuring blood pressure to monitor health. Pressure-sensitive paints enable measurement of surface pressure as it changes in space and time. The accuracy of measuring and controlling gas pressure is fundamental to manufacturing processes.

Weather forecasting. Scientists learned to relate the rate of change of atmospheric pressure to the possibility of strong winds, usually bringing rain or snow. For example, if the pressure drops by more than three millibar per hour, winds of up to fifty kilometers per hour are likely to follow. Powerful storms may be preceded by drops of more than twenty-four millibar in twenty-four hours. If the pressure starts rising, clear calm weather is expected. However, these rules change with regional conditions. For instance in the Great Lakes region of the United States, rising pressure may indicate an Arctic cold front moving in, causing heavy snowfall. In other regions, a sharply dropping pressure indicates a cold front moving in, followed by a quick rise in pressure as the colder weather is established. As a warm front approaches, the pressure may level out and rise slowly after the front passes. Modern forecasters construct detailed maps showing contours of pressure from sensors distributed over the countryside and use these to predict weather patterns. Aircraft pilots use such maps to identify safe routes and areas to avoid. Using Doppler radar wind measurements, infrared temperature maps and cloud images from satellites, and computational fluid dynamics, modern weather bureaus are able to issue warnings about severe weather several hours in advance for smaller local weather fronts and storms and several days ahead for major storms moving across continents or oceans. However, the number of weather-monitoring pressure sensors available to forecasters is quite inadequate to issue accurate predictions for minor weather changes, particularly when predicting rain or snow.

Electrical gauges. Gauges operating on the electrical changes induced by deflection of a diaphragm are used in industrial process monitoring and control where computer interfacing is required. Unsteady pressure transducers come in many ranges of amplitude and frequency. Piezoresistive sensors are integrated into an electrical-resistance bridge and constructed as miniature self-contained, button-like sensors. These are suitable for high amplitudes and frequencies, such as those encountered in shock waves and explosions, and transonic or supersonic wind-tunnel tests. Condenser microphones are used in acoustic measurements. As computerized data-acquisition systems became common, but pressure sensors remained expensive, pressure switches enabled dozens of pressure-sensing lines connected through the switches to each sensor to be measured one at a time. This required a long time to collect data from all the sensors, spending enough time at each to capture all the fluctuations and construct stable averages, making it unsuitable for rapidly changing conditions. Inexpensive, miniaturized, and highly sensitive solid-state piezoelectric sensors and fast, multichannel analogue-digital converters have made it possible to connect each pressure port to an individual sensor, vastly reducing the time between individual measurements at each sensor.

Aircraft testing. Water and mercury manometers were used extensively in aerospace test facilities such as wind tunnels, where banks of manometers indicated the distribution of pressure around the surfaces of models from pressure-sensing holes in the models. Pressure switches connecting numerous pressure-sensing ports to a single sensor became common in the 1970's. In the 1990's, inexpensive sensors based on microelectromechanical systems technology enabled numerous independent sensing channels to be monitored simultaneously.

Aphygmomanometers for blood pressure. Other than weather forecasting, the major common application of pressure measurement is in measuring blood pressure. The device used is called a sphygmomanometer. The high and low points of pressure reached in the heartbeat cycle are noted on a mercury manometer tube synchronized with the heartbeat sounds detected through a stethoscope.

Bourdon tubes for household barometry. The Bourdon tube is a pressure-measuring device in which a coiled tube stretches and uncoils depending on the difference between pressures inside and outside the tube and drives a levered mechanism connected to an indicator dial. Diaphragm-type pressure gauges and Bourdon-tube gauges are still used in the vast majority of household and urban plumbing. These instruments are highly reliable and robust, but they operate over fairly narrow ranges of pressure.

Pressure-sensitive paints to map pressure over surfaces. So-called pressure-sensitive paints (PSPs) offer an indirect technique to measure pressure variations over an entire surface, using the fact that the amount of oxygen felt at a surface is proportional to the density and thus to the pressure if the temperature does not change. These paints are luminescent dyes dispersed in an oxygen-permeable binder. When illuminated at certain ultraviolet wavelengths, the dye molecules absorb the light and move up into higher energy levels. The molecules then release energy in the infrared wavelengths as they relax to equilibrium. If the molecule collides with an oxygen molecule, the energy gets transferred without emission of radiation. Therefore, the emission from a surface becomes less intense if the number of oxygen molecules being encountered increases. This occurs when the pressure of air increases. The observed intensity from a painted surface is inversely proportional to the pressure of oxygen-containing air. Light-intensity values at individual picture element (pixel) are converted to numbers, compared with values at some known reference pressure, and presented graphically as colors. Typically, an accurate pressure sensor using either piezoelectric or other technology is used for reference. As of 2011, pressure-sensitive paints had reached the sensitivity required to quantify pressure distributions over passenger automobiles at moderate highway speeds, given expert signal processing and averaging a large number of images.

Smart pressure transmitters for automatic control systems. Wireless pressure sensors are used in remote applications such as weather sensing. Modern automobiles incorporate tire pressure transmitters. Manifold pressure sensors send instantaneous readings of the pressure inside automobile engine manifolds so that a control computer can calculate the best rate of fuel flow to achieve the most efficient combustion. Smart pressure transmitters incorporating capacitance-type diaphragm pressure sensors and microprocessors can be configured to adjust their settings remotely, perform automatic temperature compensation of data, and transmit pressure data and self-diagnosis data in digital streams.

Nuclear explosion sensors. Piezoresistive transducers have been developed to report the extremely high overpressure, as high as sixty-nine megapascals, of an air blast of a nuclear weapon, with the microsecond rise time required to measure the blast wave accurately. One design uses a silicon disk with integral diffused strain-sensitive regions and thermal barriers. Another design uses the principle of Fabry-Perot interferometry, in which laser light reflecting in a cavity changes intensity depending on the shape of the cavity when the diaphragm bounding the cavity flexes because of pressure changes. This sensor has the response speed and ruggedness required to operate in a hostile environment, where there may be very large electromagnetic pulses. In such environments a capacitance-based sensor or piezoelectric sensor may not survive.

Extreme applications of barometry. The basic origins of pressure can be used to explain the pressure due to radiation as the momentum flux of photons. At Earth's orbit around the Sun, the solar intensity of 1.38 kilowatts per square meter causes a radiation pressure of roughly 4.56 micropascals. Solar sails have been proposed for long-duration missions in space, driven by this pressure. Close to the center of the Earth, the pressure reaches 3.2 to 3.4 million bars. Inside the Sun, pressure as high as 250 billion bars is expected, while the explosion of a nuclear-fusion weapon may produce a quarter of that. Metallic solid hydrogen is projected to form at pressures of 250,000 to 500,000 bars.

IMPACT ON INDUSTRY

Government and university research. Barometers enabled rapid development of scientific weather forecasting. Weather forecasting, in turn, has had a tremendous effect on emergency preparedness. Research sponsored by the defense research offices provides fertile opportunities and challenges for new pressure-measurement techniques. Any experiment that uses fluids, either flowing or in containers, requires monitoring and often rapid measurement of pressure. The development of pressure-sensitive paint technology is a frontier in research in the early twenty-first century. Both the sensitivity and the frequency response of such paints need substantial improvement before they can be routinely used in laboratory measurements and transitioned to industrial measurements.

Industry and business. Quantitative knowledge of the detailed surface pressure distribution on wind tunnel models of flight vehicles enables engineers to develop modifications to improve the performance of the vehicle and reduce fuel consumption. The ability to monitor pressure is critical in the nuclear and petroleum industries as well. Submarine and oil-rig crews depend on pressure measurements for their lives. Oil exploration involves several steps in which the pressure must be tracked with extreme care, especially when there is a danger of gas rising through drilling tubes from subterranean reservoirs. Pressure buildup in steam or other gas circuits is critical in the nuclear industry and in most of the chemical industry wherever leaks of gas into the atmosphere must be strictly controlled.

CAREERS AND COURSE WORK

Because barometry is so important to so many industries and so many branches of scientific research, most students who are planning on a career in engineering, other technological jobs, and the sciences must understand it.

Modern pressure-measurement technology integrates ideas from many branches of science and engineering derived from physics and chemistry. The pressure-measurement industry includes experts in weather forecasting, plumbing, atmospheric sciences, aerospace wind-tunnel experimentation, automobile-engine development, the chemical industry, chemists developing paint formulations, electrical and electronics engineers developing microelectromechanical sensors, software engineers developing smart sensor logic, and the medical community interested in using barometry to monitor patients' health and vital signs.

Pressure measurement therefore comes up as a subject in courses offered in schools of mechanical, chemical, civil, and aerospace engineering. The numerous other related issues come up in specialized courses in materials science, electronics, atmospheric sciences, and computer science.

SOCIAL CONTEXT AND FUTURE PROSPECTS

Instrumentation for measuring pressure, normal and shear stresses, and flow rate from numerous sensors are becoming integrated into computerized measurement systems. In many applications, such sensors are mass-produced using facilities similar to those for making chips for computers. Very few ideas exist for directly measuring pressure, as it changes rapidly at a point in a flowing fluid, without intrusive probes of some kind. Such nonintrusive measurements, if they become possible, could help us to understand the nature of turbulence and assist in a major breakthrough in fluid dynamics.

Measurement of pressure is difficult to make inside flame environments, where density and temperature fluctuate rapidly. Better methods of measuring pressure in biological systems, such as inside blood vessels, would have major benefits in diagnosing heart disease and improving health.

As of 2011, pressure-measurement systems are still too expensive to allow sufficient numbers to be deployed to report pressure with enough spatial and time resolution to permit development of a real-time three-dimensional representation. Research in this area will doubtless improve the resolution and response and hopefully bring down the cost. With more pressure sensors distributed over the world, weather prediction will become more accurate and reliable.

—*Narayanan M. Komerath, PhD*

FURTHER READING

American Society of Mechanical Engineers. *Pressure Measurement.* New York: American Society of Mechanical Engineers, 2010. Authoritative document with guidance on determining pressure values, according to the American Society of Mechanical Engineers performance test codes. Discusses how to choose and use the best methods, instrumentation, and corrections, as well as the allowable uncertainty.

Avallone, Eugene A., Theodore Baumeister III, and Ali M. Sadegh. *Marks' Standard Handbook for Mechanical Engineers.* 11th ed. New York: McGraw-Hill, 2006. Best reference for solving mechanical engineering problems. Discusses pressure sensors and measurement techniques and their applications in various parts of mechanical engineering.

Benedict, Robert P. *Fundamentals of Temperature, Pressure, and Flow Measurements.* 3d ed. New York: John Wiley & Sons, 1984. Suited for practicing engineers in the process control industry.

Burch, David. *The Barometer Handbook: A Modern Look at Barometers and Applications of Barometric Pressure.*

Seattle: Starpath Publications, 2009. Written to assist the practicing weather forecaster, with chapters on weather forecasting on land and sea. Contains an excellent history of the field as well as methods for instrument calibration and maintenance.

Gillum, Donald R. *Industrial Pressure, Level, and Density Measurement.* 2d ed. Research Triangle Park, N.C.: International Society for Automation, 2009. Teaching and learning resource on the issues and methods of pressure measurement, especially related to industrial control systems. Contains assessment questions at the end of each section.

Green, Don W., and Robert H. Perry. *Perry's Chemical Engineers' Handbook.* 8th ed. New York: McGraw-Hill, 2008. Still considered the best source for bringing together knowledge from various parts of the field of chemical engineering, where the student can find the different applications of pressure measurement and many other things. Contains a succinct, illustrated explanation of pressure-measurement techniques.

Ryans, J. L. "Pressure Measurement." In *Kirk-Othmer Encyclopedia of Chemical Technology.* 5th ed. Hoboken, N.J.: John Wiley & Sons, 2000.

Describes mechanical and electronic sensors and instrumentation for pressure measurements from 1,380 megapascals to near vacuum.

Taylor, George Frederic. *Elementary Meteorology.* New York: Prentice Hall, 1954. This classic remains an excellent resource for the basic methods of weather prediction, including the methods for measuring temperature, pressure, and humidity, as well as the methods for using these measurements in predicting the weather.

WEB SITES

American Meteorological Society
http://www.ametsoc.org

American Society of Mechanical Engineers
http://www.asme.org

National Oceanic and Atmospheric Administration
Office of Oceanic and Atmospheric Research
http://www.oar.noaa.gov

See also: Atmospheric Sciences; Civil Engineering; Mechanical Engineering; Meteorology.

BIOENGINEERING

FIELDS OF STUDY

Cell biology; molecular biology; biochemistry; physiology; ecology; microbiology; pharmacology; genetics; medicine; immunology; neurobiology; biotechnology; biomechanics; bioinformatics; physics; mechanical engineering; electrical engineering; materials science; buildings science; architecture; chemical engineering; genetic engineering; thermodynamics; robotics; mathematics; computer science; biomedical engineering; tissue engineering; bioinstrumentation; bionics; agricultural engineering; human factors engineering; environmental health engineering; biodefense; nanotechnology; nanoengineering.

SUMMARY

Bioengineering is the field in which techniques drawn from engineering are used to tackle biological problems. For example, bioengineers may use mechanics principles—knowledge about how to design and construct mechanical objects using the most ideal materials—to create drug delivery systems. They may work on developing efficient ways to irrigate and drain land for growing crops, or they may be involved in building artificial environments that can support life even in the harsh climate of outer space. A highly interdisciplinary, collaborative field that synthesizes expertise from multiple research areas, bioengineering has had a significant impact on many fields of study, including the health sciences, technology, and agriculture.

KEY TERMS AND CONCEPTS

- **biocompatible material:** Material used to replace or repair tissues in the body, or to perform a biological function in a living organism.
- **bioinformatics:** Application of data processing, retrieval, and storage techniques to biological research, especially in genomics.

- **biomechanics:** Application of mechanical principles to questions of motor control in biological systems.
- **bioreactor:** Tool or device that generates chemical reactions to create a product.
- **bioremediation:** Use of bacteria and other microorganisms to solve environmental problems, such as neutralizing hazardous waste.
- **geoengineering:** Use of engineering techniques to modify environmental or geological processes, such as the weather, on a global scale.
- **prosthetic pevice:** Artificial part or implant designed to replace the function of a lost or damaged part of the body.
- **regenerative medicine:** Therapies that aim to restore the function of tissues that have been damaged or lost through injury or disease by using tissue or cells grown in laboratories or compounds created in laboratories.
- **systems biology:** Theoretical branch of bioengineering that creates models of complex biological processes or systems, using them to predict future behavior.
- **transgenic organism:** Plant or animal containing genetic information taken from another species.

DEFINITION AND BASIC PRINCIPLES

Bioengineering is an interdisciplinary field of applied science that deals with the application of engineering methods, techniques, design approaches, and fundamental knowledge to solve practical problems in the life sciences, including biology, geology, environmental studies, and agriculture. In many contexts, the term bioengineering is used to refer solely to biomedical engineering. This is the application of engineering principles to medicine, such as in the development of artificial limbs or organs. However, the field of bioengineering has many applications beyond the field of health care. For example, genetically modified crops that are resistant to pests, suits that protect astronauts from the ultra-low pressures in space, and brain-computer interfaces that may allow soldiers to exercise remote control over military vehicles all fall under the wide umbrella of bioengineering.

Each of the subdisciplines within bioengineering relies on different sets of basic engineering principles, but a few fundamental approaches can be said to apply broadly across the entire field. From an engineering perspective, three basic steps are involved in solving any problem: an analysis of how the system in question works, an attempt to synthesize the information gathered from this analysis and generate potential solutions, and finally an attempt to design and test a useful product. Bioengineers apply this three-stage problem-solving process to problems in the life sciences. What is somewhat novel about this approach is that it is a holistic one. In other words, it treats biological entities as systems—sets of parts that work together and form an integrated whole—rather than looking at individual parts in isolation. For example, to develop an artificial heart, bioengineers need to consider not just the structure of the heart itself on a cellular or tissue level but also the complex dynamics of the organ's interactions with the rest of the body through the circulatory system and the immune system. They must build a device whose parts can mimic the functionality of a healthy heart and whose materials can be easily integrated into the body without triggering a harmful immune response.

BACKGROUND AND HISTORY

Principles of chemical and mechanical engineering have been applied to problems in specific biological systems for centuries. For example, bioengineering applications include the fermentation of alcoholic beverages, the use of artificial limbs (which are documented as far back as 500 BCE), and the building of heating and cooling systems that regulate human environments.

Bioengineering did not emerge as a formal scientific discipline, however, until the middle of the twentieth century. During this period, more and more scientists began to be interested in applying new technologies from electronic and mechanical engineering to the life sciences. As the United States, Japan, and Europe began to enter a period of economic recovery and growth following World War II, governments increased funding for bioengineering efforts. The cardiac pacemaker and the defibrillator, both developed during this postwar period, were two of the earliest and most significant inventions to come out of the quickly developing field. In 1966, the Engineers Joint Council Committee on Engineering Interaction with Biology and Medicine first used the term "bioengineering." At about the same time, academic institutions began to form specialized departments and programs of study to train professionals in the application of engineering principles to biological problems. In the twenty-first century, rapid technological advances continue to produce growth in the field of bioengineering.

HOW IT WORKS

Because bioengineering is such a large and diverse field, it would be impossible to enumerate all the processes involved in creating the totality of its applications. The following are a few of the most significant examples of the types of technological tools used in bioengineering.

Materials science. One of the most important areas of bioengineering is the intersection of materials science and biology. Scientists working in this field are charged with developing materials that, although synthetic, are able to successfully interact with living tissues or other natural biological systems without impeding them. (For example, it is vital that biocompatible materials not allow blood platelets to adhere to them and form clots, which can be fatal.) Depending on the specific application in question, other properties, such as tensile strength, resistance to wear, and permeability to water, gases, and small biological molecules, are also important. To manipulate these properties to achieve a desired end, engineers must carefully control both the chemical structure and the molecular organization of the materials. For this reason, biocompatible materials are generally made out of some kind of synthetic polymer—substances with simple and extremely regular molecular structures that repeat again and again. In addition, additives may be incorporated into the materials, such as inorganic fillers that allow for greater mechanical flexibility or stabilizers and antioxidants that keep the material from becoming degraded over time.

Biochemical engineering. Since living cells are essentially chemical systems, the tools of chemical engineering are especially applicable to biology. Biochemical engineers study and manipulate the behavior of living cells. Their basic tool for doing this is a fermenter, a large reactor within which chemical processes can be carried out under carefully controlled conditions. For example, the modern production of virtually all antibiotics, such as penicillin and tetracycline, takes place inside a fermenter. A central vessel, sealed tight to prevent contamination and surrounded by jackets filled with coolants to control its temperature, contains propellers that stir around the

nutrients, culture ingredients, and catalysts that are associated with the reaction at hand.

Genetic engineering is a subfield of biochemical engineering that is growing increasingly significant. Scientists alter the genetic information in one cell by inserting into it a gene from another organism. To do this, a vector such as a virus or a plasmid (a small strand of DNA) is placed into the cell nucleus and combines with the existing genes to form a new genetic code. The technology that enables scientists to alter the genetic information of an organism is called gene splicing. The new genetic information created by this process is known as recombinant DNA. Genetic engineering can be divided into two types: somatic and germ line. Somatic genetic engineering is a process by which gene splicing is carried out within specific organs or tissues of a fully formed organism; germ-line genetic engineering is a process by which gene splicing is carried out within sex cells or embryos, causing the recombinant DNA to exist in every cell of the organism as it grows.

Electrical engineering. Electrical engineering technologies are an essential part of the bioengineering tool kit. In many cases, what is required is for the bioengineer to find some way to convert sensory data into electric signals, and then to produce these electric signals in such a way as to enable them to have a physiological effect on a living organism.

The cochlear implant is an example of one such development. The cochlea is the part of the brain that interprets sounds, and a cochlear implant is designed for people who are profoundly deaf. A cochlear implant uses electronic devices that capture sounds and relay them to the cochlea. The implant has four parts: a microphone, a tiny computer processor, a radio transmitter, and a receiver, which surgeons implant in the user's skull. The microphone picks up nearby sounds, such as human speech or music emerging from a pair of stereo speakers. Then the processor converts the sounds into digital information that can be sent through a wire to the radio transmitter. The software used by the processor separates sounds into different channels, each representing a range of frequencies. In turn, the radio transmitter translates the Scientists alter the genetic information in one cell digital information into radio signals, which it relays by inserting into it a gene from another organism. through the skull to the receiver. The receiver then To do this, a vector such as a virus or a plasmid (a turns the radio signals into electric impulses, which small strand of DNA) is placed into the cell nucleus directly stimulate the nerve endings in the cochlea. and combines with the existing genes to form a new It is these electric signals that the brain is able to interpret as sounds, allowing even profoundly deaf people to hear.

Another example of how electric signals can be used to direct biological systems can be found in brain-computer interfaces (BCIs). BCIs are direct channels of communication between a computer and the neurons in the human brain. They work because activity in the brain, such as that produced by thoughts or sensory processing, can be detected by bioinstruments designed to record electrophysiological signals. These signals can then be transmitted to a computer and used to generate commands. For example, BCIs allow stroke victims who have lost the use of a limb to regain mobility; a patient's thoughts about movement are transmitted to an external machine, which in turn transmits electric signals that precisely control the movements of a cradle holding his or her paralyzed arm.

APPLICATIONS AND PRODUCTS

Biomedical applications. Biomedical engineering is a vast subdiscipline of bioengineering, which itself encompasses multiple fields of interest. The many clinical areas in which applications are being developed by biomedical engineers include medical imaging, cell and tissue engineering, bioinstrumentation, the development of biocompatible materials and devices, biomechanics, and the emerging field of bionanotechnology.

Medical imaging applications collect data about patients' bodies and turn that data into useful images that physicians can interpret for diagnostic purposes. For example, ultrasound scans, which map the reflection and reduction in force of sounds as they bounce off an object, are used to monitor the development of fetuses in the wombs of pregnant women. Magnetic resonance imaging (MRI), which measures the response of body tissues to high-frequency radio waves, is often used to detect structural abnormalities in the brain or other body parts.

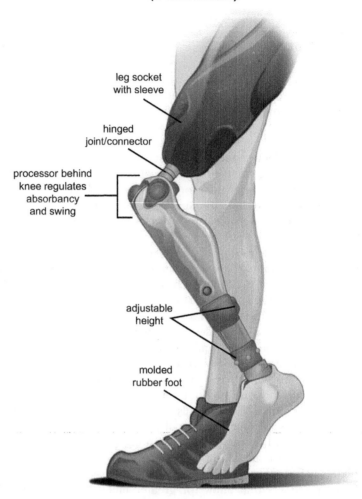

PROSTHETIC LEG
(C-LEG model)

leg socket with sleeve

hinged joint/connector

processor behind knee regulates absorbancy and swing

adjustable height

molded rubber foot

Bioengineers have made significant advances in the field of prosthetics; the C-Leg prosthesis was introduced in 1997.

Cell and tissue engineering is the attempt to exploit the natural characteristics of living cells to regenerate lost or damaged tissue. For example, bioengineers are working on creating viable replacement heart cells for people who have suffered cardiac arrests, as well as trying to discover ways to regenerate brain cells lost by patients with neurodegenerative disorders such as Alzheimers disease. Genetic engineering is a closely related area of biomedicine in which DNA from a foreign organism is introduced into a cell so as to create a new genetic code with desired characteristics.

Bioinstrumentation is the application of electrical engineering principles to develop machines that can sense and respond to biological or physiological signals, such as portable devices for diabetics that measure and report the level of glucose in their blood. Other common examples of bioinstrumentation include electroencephalogram (EEG) machines that continuously monitor brain waves in real time, and electrocardiograph (ECG) machines that perform the same task with heartbeats.

Many biomedical engineers work on developing materials and devices that are biocompatible, meaning that they can replace or come into direct contact with living tissues, perform a biological function, and refrain from triggering an immune system response. Pacemakers, small artificial devices that are implanted within the body and used to stimulate heart muscles to produce steady, reliable contractions, are a good example of a biocompatible device that has emerged from the collaboration of engineers and clinicians.

Biomechanics is the study of how the muscles and skeletal structure of living organisms are affected by and exert mechanical forces. Biomechanics applications include the development of orthotics (braces or supports), such as spinal, leg, and foot braces for patients with disabling disorders such as cerebral palsy, multiple sclerosis, or stroke. Prostheses (artificial limbs) also fall under the field of biomechanics; the sockets, joints, brakes, and pneumatic or hydraulic controls of an artificial leg, for example, are manufactured and then combined in a modular fashion, in much the same way as are the parts of an automobile in a factory.

Bionanotechnology. Nanotechnology is a fairly young field of applied science concerned with the manipulation of objects at the nanoscale (about 1-100 nanometers, or about one-thousandth the width of a strand of human hair) to produce machinery. Bionanotechnological applications within medicine include microscopic biosensors installed on small chips; these can be specialized to recognize and flag specific proteins or antibodies, helping physicians conduct extremely fast and inexpensive diagnostic tests. Bioengineers are also developing microelectrodes on a nanoscale; these arrays of tiny electrodes can be implanted into the brain and used to stimulate specific nerve cells to treat movement

disorders and other diseases.

Military applications. Bioengineering applications are making themselves felt as a powerful presence on the front lines of the military. For example, bioengineering students at the University of Virginia designed lighter, more flexible, and stronger bulletproof body armor using specially created ceramic tiles that are inserted into protective vests. The armor is able to withstand multiple impacts and distributes shock more evenly across the wearer's body, preventing damaging compression to the chest. Others working in the field are creating sophisticated biosensors that soldiers can use to detect the presence of potential pathogens or biological weapons that have been released into the air.

One of the most significant contributions of bioengineering to the military is in the development of treatments for severe traumas sustained during warfare. For example, stem cell research may one day enable military physicians to regenerate functional tissues such as nerves, bone, cartilage, skin, and muscle—an invaluable tool for helping those who have lost limbs or other body parts as a result of explosives. The United States military was responsible for much of the early research done in creating safe, effective artificial blood substitutes that could be easily stored and relied on to be free of contamination on the battlefield.

Agriculture. Agricultural engineering involves the application of both engineering technologies and knowledge from animal and plant biology to problems in agriculture, such as soil and water conservation, food processing, and animal husbandry. For example, agricultural engineers can help farmers maximize crop yields from a defined area of land. This technique, known as precision farming, involves analyzing the properties of the soil (factors such as drainage, electrical conductivity, pH [acidity] level, and levels of chemicals such as nitrogen) and carefully calibrating the type and amount of seeds, insecticides, and fertilizers to be used.

Farm machinery and implements represent another area of agriculture in which engineering principles have made a big impact. Tractors, harvesters, combines, and grain-processing equipment, for example, have to be designed with mechanical and electrical principles in mind and also must take into account the characteristics of the land, the needs of the human operators, and the demands of working with particular agricultural products. For example, many crops require specialized equipment to be successfully mechanically harvested. Thus a pea harvester may have several components—one that lifts the vines and cuts them from the plant, one that strips pea pods from the stalk, and one that threshes the pods, causing them to open and release the peas inside them. Another example of an agricultural engineering application is the development of automatic milking machines that attach to the udders of a cow and enable dairy farmers to dispense with the arduous task of milking each animal by hand.

The management of soil and water is also an important priority for bioengineers working in agricultural settings. They may design structures to control the flow of water, such as dams or reservoirs. They may develop water-treatment systems to purify wastewater coming out of industrial agricultural production centers. Alternatively, they may use soil walls or cover crops to reduce the amount of pesticides and nutrients that run off from the soil, as well as the amount of erosion that takes place as a result of watering or rainfall.

Environmental and ecological applications. Environmental and ecological engineers study the impact of human activity on the environment, as well as the ways in which humans respond to different features of their environments. They use engineering principles to clean, control, and improve the quality of natural spaces, and find ways to make human interactions with environmental resources more sustainable. For example, the reduction and remediation of pollution is an important area of concern. Therefore, an environmental engineer may study the pathways and rates at which volatile organic compounds (such as those found in many paints, adhesives, tiles, wall coverings, and furniture) react with other gases in the air, causing smog and other forms of air pollution. They may design and build sound walls in residential areas to cut down on the amount of noise pollution caused by airplanes taking off and landing or cars racing up and down highways.

The life-support systems designed by bioengineers to enable astronauts to survive in the harsh conditions of outer space are also a form of environmental engineering. For example, temperatures around a space shuttle can vary wildly, depending on which side of the vehicle is facing the Sun at any given moment. A complex system of heating, insulation, and

ventilation helps regulate the temperature inside the cabin. Because space is a vacuum, the shuttle itself must be filled with pressurized gas. In addition, levels of oxygen, carbon dioxide, and nitrogen within the cabin must be controlled so that they resemble the atmosphere on Earth. Oxygen is stored on board in tanks, and additional supplies of the essential gas are

---

FASCINATING FACTS ABOUT BIOENGINEERING

- Bioengineering has enabled scientists to grow replacement human skin, tracheas, bladders, cartilage, and other tissues and organs in the laboratory.

- Materials scientists and clinical researchers are working together to develop contact lenses that can deliver precise doses of drugs directly into the eye.

- By genetically engineering crops that are naturally resistant to insects, bioengineers have helped reduce the need to use harmful pesticides in industrial farming.

- Bacteria whose genetic information has been carefully re-engineered may eventually provide an endless supply of crude oil, helping meet the world's energy needs without engaging in damaging drilling.

- In 2009, an MIT bioengineer invented a new way to pressurize space suits that does not use gas, making them far sleeker and less bulky than conventional astronaut gear.

- One military application of bioengineering is a robotic system that seeks out and identifies tiny pieces of shrapnel lodged within tissue, then guides a needle to those precise spots so that the shrapnel can be removed.

- Bionic men and women are not just the stuff of television and motion-picture fantasy. In fact, anyone who has an artificial body part, such as a prosthetic leg, a pacemaker, or an implanted hearing aid, can be considered bionic.

- Some bioengineers are working on developing artificial noses that can detect and diagnose disease by smell--literally sniffing out infections and cancer, for example.

- One day, it may be possible to "print out" artificial organs using a three-dimensional printer. Layer by layer, cells would be deposited onto a glass slide, building up specialized tissues that could be used to replace damaged kidneys, livers, and other organs.

- Each year, more women choose to enter the field of biomedical engineering than any other specialty within engineering.

---

produced from electrolyzed water; in turn, carbon dioxide is channeled out of the shuttle through vents.

Geoengineering. Geoengineering is an emerging subfield of bioengineering that is still largely theoretical. It would involve the large-scale modification of environmental processes in an attempt to counteract the effects of human activity leading to climate change. One proposed geoengineering project involves depositing a fine dust of iron particles into the ocean in an attempt to increase the rate at which algae grows in the water. Since algae absorbs carbon dioxide as it photosynthesizes, essentially trapping and containing it, this would be a means of reducing the amount of this greenhouse gas in the atmosphere. Other geoengineering proposals include the suggestion that it might be possible to spray sulfur dust into the high atmosphere to reflect some of the Sun's light and heat back into space, or to spray drops of seawater high up into the air so that the salt particles they contain would be absorbed into the clouds, making them thicker and more able to reflect sunlight.

IMPACT ON INDUSTRY
Bioengineering is a global industry that boasts consistent revenues. With an aging public, biomedical engineering looks poised for significant growth. The global focus on reversing climate change, ensuring an adequate supply of food and clean water, and improving and maintaining health means that bioengineering is likely to be an ever-expanding field. Those nations that embrace its possibilities will find themselves reaping the rewards in a better quality of life for their citizens.

Government and university research. The United States is generally considered to be the world leader in bioengineering research, especially within the field of biomedical engineering. However, significant strides are being made in many European countries, including France and Germany, as well as by many growing economies in Asia, such as China, Singapore, and Taiwan. In the United States, the main governmental organization funding studies in this area is the National Institute of Biomedical Imaging and Bioengineering, a branch of the National Institutes of Health. Among the United States universities whose faculty and students are recognized as conducting the most leading-edge research in bioengineering are The Johns Hopkins School of Medicine, the

Massachusetts Institute of Technology (MIT), and the University of California, San Diego—all of which were ranked at the top of the 2009 *U.S. News and World Report*'s list of the best biomedical/bioengineering schools in the country.

Major corporations. Major biomedical engineering corporations include Medtronic, Abbot, Merck, and Glaxo-Smith Kline, all international producers of products such as pharmaceuticals and medical devices. Medtronic, for example, manufactures items such as defibrillators, pacemakers, and heart valves, while Merck produces drugs to treat cancer, heart disease, diabetes, and infections. Within the field of agricultural biotechnology, Monsanto and DuPont are industry leaders. Both corporations produce, patent, and market seeds for transgenic crops. The plants grown from these seeds possess traits attractive to industrial farmers, such as resistance to pesticides, longer ripening times, and higher yield.

Industry and business. Bioengineering is considered a strong growth industry with a great deal of potential for expansion in the twenty-first century. The production of biomedical devices and biocompatible materials alone, for example, is a market worth an estimated $170 billion per year. In the United States, Department of Labor statistics indicate that in 2006, engineers working in the biomedical, agricultural, health and safety, and environmental engineering fields—all of which fall under the bioengineering umbrella—held a total of nearly 100,000 jobs nationwide. In the following years, the department estimated that each of these sectors would add jobs at a rate that either keeps pace with or far exceeds the national average for all occupations.

CAREERS AND COURSE WORK
Although bioengineering is a field that exists at the intersection between biology and engineering, the most common path for professionals in the field is to first become trained as engineers and later apply their technical knowledge to problems in the life sciences. (A less common path is to pursue a medical degree and become a clinical researcher.) At the high school level, it is important to cover a broad range of mathematical topics, including geometry, calculus, trigonometry, and algebra. Biology, chemistry, and physics should also be among an aspiring bioengineer's course work. At the college level, a student should pursue a bachelor of science in engineering.

At many institutions, it is possible to further concentrate in a subfield of engineering: Appropriate subfields include biomedical engineering, electrical engineering, mechanical engineering, and chemical engineering. Students should continue to take electives in biology, geology, and other life sciences wherever possible. In addition, English and humanities courses, especially writing classes, can provide the aspiring bioengineer with strong communication skills—important for working collaboratively with colleagues from many different disciplines.

Many, though not all, choose to pursue graduate-level degrees in biomedical engineering, agricultural engineering, environmental engineering, or another subfields of bioengineering. Others go through master's of business administration programs and combine this training with their engineering background to become entrepreneurs in the bioengineering industry. Additional academic training beyond the undergraduate level is required for careers in academia and higher-level positions in private research and development laboratories, but entry-level technical jobs in bioengineering may require only a bachelor's degree. Internships (such as at biomedical companies) or evidence of experience conducting original research will be helpful in obtaining one's first job.

A variety of career options exist for bioengineers; many work as researchers in academic settings, private industry, government institutions, or research hospitals. Some are faculty members, and some are administrators, managers, supervisors, or marketing consultants for these same organizations. Others are engaged in designing, developing, and conducting safety and performance testing for bioengineering instruments and devices.

SOCIAL CONTEXT AND FUTURE PROSPECTS

Bioengineering is a field with the capacity to exert a powerful impact on many aspects of social life. Perhaps most profound are the transformations it has made in health care and medicine. By treating the body as a complex system—looking at it almost as if it were a machine—bioengineers and physicians working together have enabled countless patients to overcome what once might have seemed to be insurmountable damage. After all, if the body is a machine, its parts might be reengineered or replaced entirely with new ones—as when the damaged cilia of individuals with hearing impairments are replaced

with electro-mechanical devices. Some aspects of bioengineering, however, have drawn concern from observers who worry that there may be no limit to the scientific ability to interfere with biological processes. Transgenic foods are one area in which a contentious debate has sprung up. Some are convinced that the ecological and health ramifications of growing and ingesting crops that contain genetic information from more than one species have not yet been fully explored. Stem cell research is another area of controversy; some critics are uncomfortable with the fact that human embryonic stem cells are being obtained from aborted fetuses or fertilized eggs that are left over from assisted reproductive technology procedures.

One aspect of bioengineering that has been the subject of both fear and hope in the twenty-first century is the question of whether it might be possible to stop or even reverse the harmful effects of climate change by carefully and deliberately interfering with certain geological processes. Some believe that geo-engineering could help the international community avoid the devastating effects of global warming predicted by scientists, such as widespread flooding, droughts, and crop failure. Others, however, warn that any attempt to interfere with complex environmental systems on a global scale could have wildly unpredictable results. Geoengineering is especially controversial because such projects could potentially be carried out unilaterally by countries acting without international agreement and yet have repercussions that could be felt all across the world.

—*M. Lee, MA*

FURTHER READING

Artmann, Gerhard M., and Shu Chien, eds. Bioengineering in Cell and Tissue Research. New York: Springer, 2008. Examines bioengineering's role in cell research. Heavily illustrated with diagrams and figures; includes a comprehensive index and references after each section.

Enderle, John D., Susan M. Blanchard, and Joseph D. Bronzino, eds. Introduction to Biomedical Engineering. 2d ed. Boston: Elsevier Academic Press, 2005. A broad introductory textbook designed for undergraduates. Each chapter contains an outline, objectives, exercises, and suggested reading.

Huffman, Wallace E., and Robert E. Evenson.

Science for Agriculture: A Long-Term Perspective. 2d ed. Ames, Iowa: Blackwell, 2006. A history of agricultural engineering research within the United States. Includes a glossary and list of relevant acronyms.

Madhavan, Guruprasad, Barbara Oakley, and Luis G. Kun, eds. Career Development in Bioengineering and Biotechnology. New York: Springer, 2008. An extensive guide to careers in bioengineering, biotechnology, and related fields, written by active practitioners. Covers both traditional and alternative job opportunities.

Nemerow, Nelson Leonard, et al., eds. Environmental Engineering. 3 vols. 6th ed. Hoboken, N.J.: John Wiley & Sons, 2009. Discusses topics such as food protection, soil management, waste management, water supply, and disease control. Each section includes references and a bibliography.

WEB SITES
Biomedical Engineering Society
http://www.bmes.org
National Institutes of Health
National Institute of Biomedical Imaging and Bioengineering http://www.nibib.nih.gov
Society for Biological Engineering
http://www.aiche.org/sbe

See also: Audiology and Hearing Aids; Biochemical Engineering; Biomathematics; Bionics and Biomedical Engineering; Bioprocess Engineering; Climate Engineering; Electrical Engineering.

BIOFUELS AND SYNTHETIC FUELS

FIELDS OF STUDY

Biology; microbiology; plant biology; chemistry; organic chemistry; biochemistry; agriculture; biotechnology; bioprocess engineering; chemical engineering.

SUMMARY

The study of biofuels and synthetic fuels is an interdisciplinary science that focuses on development of clean, renewable fuels that can be used as alternatives to fossil fuels. Biofuels include ethanol, biodiesel, methane, biogas, and hydrogen; synthetic fuels include syngas and synfuel. These fuels can be used as gasoline and diesel substitutes for transportation, as fuels for electric generators to produce electricity, and as fuels to heat houses (their traditional use). Both governmental agencies and private companies have invested heavily in research in this area of applied science.

KEY TERMS AND CONCEPTS

- **biodiesel:** Biofuel with the chemical structure of fatty acid alkyl esters.
- **biogas:** Biofuel that contains a mixture of methane (50-75 percent), carbon dioxide, hydrogen, and carbon monoxide.
- **biomass:** Mass of organisms that can be used as an energy source; plants and algae convert the energy of the sun and carbon dioxide into energy that is stored in their biomass.
- **ethanol:** Colorless liquid with the chemical formula C_2H_5OH that is used as a biofuel; also known as ethyl alcohol, grain alcohol, or just alcohol.
- **Fischer-Tropsch process:** Process that indirectly converts coal, natural gas, or biomass through syngas into synthetic oil or synfuel (liquid hydrocarbons).
- **fuel:** Any substance that is burned to provide heat or energy.
- **gasification:** Conversion of coal, petroleum, or biomass into syngas.
- **methane:** Colorless, odorless, nontoxic gas, with the molecular formula CH_4, that is the main chemical component of natural gas (70-90 percent) and is used as a biofuel.
- **molecular hydrogen:** Also known by its chemical symbol H_2, a flammable, colorless, odorless gas; hydrogen produced by microorganisms is used as a biofuel and is called a biohydrogen.
- **synfuel:** A synthetic liquid fuel (synthetic oil) obtained via the Fischer-Tropsch process or methanol-to-gasoline conversion.
- **synthesis gas:** Synthetic fuel that is a mixture of carbon monoxide and H_2; also known as syngas.

DEFINITION AND BASIC PRINCIPLES

The science of biofuels and synthetic fuels deals with the development of renewable energy sources, alternatives to nonrenewable fossil fuels such as petroleum. Biofuels are fuels generated from organisms or by organisms. Living organisms can be used to generate a number of biofuels, including ethanol (bioethanol), biodiesel, biomass, butanol, biohydrogen, methane, and biogas. Synthetic fuels (synfuel and syngas) are a class of fuels derived from coal or biomass. Synthetic fuels are produced by a combination of chemical and physical means that convert carbon from coal or biomass into liquid or gaseous fuels.

Around the world, concerns about climate change and possible global warming due to the emission of greenhouse gases from human use of fossil fuels, as well as concerns over energy security, have ignited interest in biofuels and synthetic fuels. A large-scale biofuel and synthetic fuel industry has developed in many countries, including the United States. A number of companies in the United States have conducted research and development projects on synthetic fuels with the intent to begin commercial production of synthetic fuels. Although biofuels and synthetic fuels still require long-term scientific, economic, and political investments, investment in these alternatives to fossil fuels is expected to mitigate global warming, to help protect the global climate, and to reduce U.S. reliance on foreign oil.

BACKGROUND AND HISTORY

People have been using biofuels such as wood or dried manure to heat their houses for thousands of years. The use of biogas was mentioned in Chinese literature more than 2,000 years ago. The first biogas plant was built in a leper colony in Bombay, India, in the middle of the nineteenth century. In Europe, the first apparatus for biogas production was built in Exeter, England, in 1895. Biogas from this digester was used to fuel street lamps. Rudolf Diesel, the inventor of the diesel engine, used biofuel (peanut oil) for his engine during the World Exhibition in Paris in 1900. The version of the Model T Ford built by Henry Ford in 1908 ran on pure ethanol. In the 1920's, 25 percent of the fuels used for automobiles in the United States were biofuels rather than petroleum-based fuels. In the 1940's, biofuels were replaced by inexpensive petroleum-based fuels.

Gasification of wood and coal for production of syngas has been done since the nineteenth century. Syngas was used mainly for lighting purposes. During World War II, because of shortages of petroleum, internal combustion engines were modified to run on syngas and automobiles in the United States and the United Kingdom were powered by syngas. The United Kingdom continued to use syngas until the discovery in the 1960's of oil and natural gas in the North Sea.

The process of converting coal into synthetic liquid fuel, known as the Fischer-Tropsch process, was developed in Germany at the Kaiser Wilhelm Institute by Franz Fischer and Hans Tropsch in 1923. This process was used by Nazi Germany during World War II to produce synthetic fuels for aviation.

During the 1970's oil embargo, research on biofuels and synthetic fuels resumed in the United States and Europe. However, as petroleum prices fell in the 1980's, interest in alternative fuels diminished. In the twenty-first century, concerns about global warming and increasing oil prices reignited interest in biofuels and synthetic fuels.

HOW IT WORKS

Biofuels and synthetic fuels are energy sources. People have been using firewood to heat houses since prehistoric time. During the Industrial Revolution, firewood was used in steam engines. In a steam engine, heat from burning wood is used to boil water; the steam produced pushes pistons, which turn the wheels of the machinery.

Biofuels and synthetic fuels such as ethanol, biodiesel, butanol, biohydrogen, and synthetic oil can be used in internal combustion engines, in which the combustion of fuel expands gases that move pistons or turbine blades. Other biofuels such as methane, biogas, or syngas are used in electric generators. Burning of these fuels in electric generators rotates a coil of wire in a magnetic field, which induces electric current (electricity) in the wire.

Hydrogen is used in fuel cells. Fuel cells generate electricity through a chemical reaction between molecular hydrogen (H) and oxygen (O). Ethanol, the most common biofuel, is produced by yeast fermentation of sugars derived from sugarcane, corn starch, or grain. Ethanol is separated from its fermentation broth by distillation. In the United States, most ethanol is produced from corn starch. Biodiesel,

A bus powered by biofuel.

another commonly used biofuel, is made mainly by transesterification of plant vegetative oils such as soybean, canola, or rapeseed oil. Biodiesel may also be produced from waste cooking oils, restaurant grease, soap stocks, animal fats, and even from algae. Methane and biogas are produced by metabolism of microorganisms. Methane is produced by microorganisms called *Archaea* and is an integral part of their metabolism. Biogas produces by a mixture of bacteria and archaea.

Industrial production of biofuels is achieved mainly in bioreactors or fermenters of some hundreds gallons in volume. Bioreactors or fermenters are closed systems that are made of an array of tanks or tubes in which biofuel-producing microorganisms are cultivated and monitored under controlled conditions.

Syngas is produced by the process of gasification in gasifiers, which burn wood, coal, or charcoal. Syngas can be used in modified internal combustion engines. Synfuel can be generated from syngas through Fischer-Tropsch conversion or through methanol to gasoline conversion process.

APPLICATIONS AND PRODUCTS

Transportation. Biofuels are mainly used in transportation as gasoline and diesel substitutes. As of the early twenty-first century, two biofuels—ethanol and

biodiesel—were being used in vehicles. In 2005, the

U.S. Congress passed an energy bill that required that ethanol sold in the United States for transportation be mixed with gasoline. By 2010, almost every fuel station in the United States was selling gasoline with a 10 percent ethanol content. The U.S. ethanol industry has lobbied the federal government to raise the ethanol content in gasoline from 10 to 15 percent. Most cars in Brazil can use an 85 percent/15 percent ethanol-gasoline mix (E85 blend). These cars must have a modified engine known as a flex engine. In the United States, only a small fraction of all cars have a flex engine.

Biodiesel performs similarly to diesel and is used in unmodified diesel engines of trucks, tractors, and other vehicles and is better for the environment. Biodiesel is often blended with petroleum diesel in ratios of 2, 5, or 20 percent. The most common blend is B20, or 20 percent biodiesel to 80 percent diesel fuel. Biodiesel can be used as a pure fuel (100 percent or B100), but pure fuel is a solvent that degrades the rubber hoses and gaskets of engines and cannot be used in winter because it thickens in cold temperatures. The energy content of biodiesel is less than that of diesel. In general, biodiesel is not used as widely as ethanol, and its users are mainly governmental and state bodies such as the U.S. Postal Service; the U.S. Departments of Defense, Energy, and Agriculture; national parks; school districts; transit authorities; public utilities; and waste-management facilities. Several companies across the United States (such as recycling companies) use biodiesel because of tax incentives.

Hydrogen power ran the rockets of the National Aeronautics and Space Administration for many years. A growing number of automobile manufactures around the world are making prototype hydrogen-powered vehicles. These vehicles emit only water, no greenhouse gases, from their tailpipes. These automobiles are powered by electricity generated in the fuel cell through a chemical reaction between H and O. Hydrogen vehicles offer quiet operation, rapid acceleration, and low maintenance costs because of fewer moving parts. During peak time, when electricity is expensive, fuel-cell hydrogen automobiles could provide power for homes and offices. Hydrogen for these applications is obtained mainly from natural gas (methane and propane), through steam reforming, or by water electrolysis. As of 2010, hydrogen was used only in experimental applications. Many problems need to be overcome before hydrogen becomes widely used and readily available. The slow acceptance of biohydrogen is partly caused by the difficulty in producing it on a cost-effective basis. For hydrogen power to become a reality, a great deal of research and investment must take place.

Methane was used as a fuel for vehicles for a number of years. Several Volvo automobile models with Bi-Fuel engines were made to run on compressed methane with gasoline as a backup. Biogas can also be used, like methane, to power motor vehicles.

Electricity generation. Biogas and methane are mainly used to generate electricity in electric generators. In the 1985 film *Mad Max Beyond Thunderdome*, starring Mel Gibson, a futuristic city ran on methane generated by pig manure. While the use of methane has not reached this stage, methane is a very good alternative fuel that has a number of advantages over biofuels produced by microorganisms. First, it is easy to make and can be generated locally, eliminating the need for an extensive distribution channel. Second, the use of methane as a fuel is a very attractive way to reduce wastes such as manure, wastewater, or municipal and industrial wastes. In farms, manure is fed into digesters (bioreactors), where microorganisms metabolize it into methane. There are several landfill gas facilities in the United States that generate electricity using methane. San Francisco has extended its recycling program to include conversion of dog waste into methane to produce electricity and to heat homes. With a dog population of 120,000, this initiative promises to generate a significant amount of fuel and reduce waste at the same time.

Heat generation. Some examples of biomass being used as an alternative energy source include the burning of wood or agricultural residues to heat homes. This is a very inefficient use of energy, because typically only 5 to 15 percent of the biomass energy is actually used. Burning biomass also produces harmful indoor air pollutants such as carbon monoxide. On the positive side, biomass is an inexpensive resource whose costs are only the labor to collect it. Biomass supplies more than 15 percent of the energy consumed worldwide. Biomass is the number-one source of energy in developing countries; in some countries, it provides more than 90 percent of the energy used.

In many countries, millions of small farmers maintain a simple digester for biogas production to generate heat energy. More than 5 million household digesters are being used in China, mainly for cooking and lighting, and India has more than 1 million biogas plants of various capacities.

IMPACT ON INDUSTRY
In 2009, the annual revenue of the global biofuels industry was $46.5 billion, with revenue in the United States alone reaching $20 billion. The United States

is leading the world in research on biofuels and synthetic fuels. Significant biofuels and synthetic fuels research has also been taking place in many European countries, Russia, Japan, Israel, Canada, Australia, and China.

Government and university research. Many governmental agencies such as the U.S. Department of Energy (DOE), the National Science Foundation (NSF), and the U.S. Department of Agriculture provide funding for research in biofuels and synthetic fuels. The DOE has several national laboratories (such as the National Renewable Energy Laboratory in Golden, Colorado) where cutting-edge research on biofuels and synthetic fuels is performed. In addition, three DOE research centers are concentrated entirely on biofuels. These centers are the BioEnergy Science Center, led by Oak Ridge National Laboratory; the Great Lakes Bioenergy Research Center, led by the University of Wisconsin, Madison; and the Joint BioEnergy Institute, led by Lawrence Berkeley National Laboratory.

In 2007, DOE established the Advanced Research Projects Agency-Energy (ARPA-E) to fund the development and deployment of transformational energy technologies in the United States. Several projects funded by this agency are related to biofuels, such as the development of advanced or second-generation biofuels. Traditional biofuels such as biomass (wood material) and ethanol and biodiesel from crops are sometimes called first-generation biofuels. Second-generation biofuels such as cellulosic ethanol are produced from agricultural and forestry residues and do not take away from food production. Another second-generation biofuel is biohydrogen.

Biofuels such as butanol are referred to as third-generation biofuels. Butanol ($CH OH$) is an alcohol fuel, but compared with ethanol, it has a higher energy content (roughly 80 percent of gasoline energy content). It does not absorb water as ethanol does, is not as corrosive as ethanol is, and is more suitable for distribution through existing gasoline pipelines.

Scientists are trying to create "super-bugs" for superior biofuel yields and studying chemical processes and enzymes to improve existing bioprocesses for biofuels production. They also are working to improve the efficiency of the existing production process and to make it more environmentally friendly. Engineers and scientists are designing and developing new apparatuses (bioreactors or fermenters) for fuel generation and new applications for by-products of fuel production.

Industry and business sectors. The major products of the biofuel industry are ethanol, biodiesel, and biogas; therefore, research in industry has concentrated mainly on these biofuels. Some small businesses in the biofuel industry include startup research and development companies that study feedstocks (such as cellulose) and approaches for production of biofuels at competitive prices. These companies, many of which are funded by investment firms or government agencies, analyze biofuel feedstocks, looking for new feedstocks or modifying existing ones (corn, sugarcane, or rapeseed).

Big corporations such as Poet Energy, ExxonMobil, and BP are spending a significant part of their revenues on biofuel or synthetic fuels research. One area of biofuel research examines using algae to generate biofuels, especially biodiesel. More than fifty research companies worldwide, including GreenFuel Technologies, Solazyme, and Solix Biofuels, are conducting research in this area. Research conducted by the U.S. Department of Energy Aquatic Species Program from the 1970's to the 1990's demonstrated that many species of algae produce sufficient quantities of oil to become economical feedstock for biodiesel production. The oil productivity of many algae greatly exceeds the productivity of the best-producing oil crops. Algal oil content can exceed 80 percent per cell dry weight, with oil levels commonly at about 20 to 50 percent. In addition, crop land and potable water are not required to cultivate algae, because algae can grow in wastewater. Although development of biodiesel from algae is a very promising approach, the technology needs further research before it can be implemented commercially.

CAREERS AND COURSE WORK

The alternative fuels industry is growing, and research in the area of biofuels and synthetic fuels is increasing. Growth in these areas is likely to produce many jobs. The basic courses for students interested in a career in biofuels and synthetic fuels are microbiology, plant biology, organic chemistry, biochemistry, agriculture, bioprocess engineering, and chemical engineering. Many educational institutions are offering courses in biofuels and synthetic fuels, although actual degrees or concentrations in these disciplines are still rare. Several community colleges offer associate degrees and certificate programs that prepare students to work in the biofuel and synthetic fuel industry. Some universities offer undergraduate courses in biofuels and synthetic fuels or concentrations in these areas. Almost all these programs are interdisciplinary. Graduates of these programs will have the knowledge and internship experience to enter directly into the biofuel and synthetic fuel workforce. Advanced degrees such as a master's degree or doctorate are necessary to obtain top positions in academia and industry related to biofuels and synthetic fuels. Some universities such as Colorado State University offer graduate programs in biofuels.

Careers in the fields of biofuels and synthetic fuels can take different paths. Ethanol, biodiesel, or biogas industries are the biggest employers. The available jobs are in sales, consulting, research, engineering, and installation and maintenance. People who are interested in research in biofuels and synthetic fuels can find jobs in governmental laboratories and in universities. In academic settings, fuel professionals may share their time between research and teaching.

SOCIAL CONTEXT AND FUTURE PROSPECTS
The field of biofuels and synthetic fuels is undergoing expansion. Demands for biofuels and synthetic fuels are driven by environmental, social, and economic factors and governmental support for alternative fuels.

The use of biofuels and synthetic fuels reduces the U.S. dependence on foreign oil and helps mitigate the devastating impact of increases in the price of oil, which reached a record $140 per barrel in 2008. The production and use of biofuels and synthetic fuels reduces the need for oil and has helped hold world oil prices 15 percent lower than they would have been otherwise. Many experts believe that biofuels and synthetic fuels will replace oil in the future.

Pollution from oil use affects public health and causes global climate change because of the release of carbon dioxide. Using biofuels and synthetic fuels as an energy source generates fewer pollutants and little or no carbon dioxide.

The biofuel and synthetic fuel industry in the United States was affected by the economic crisis in 2008 and 2009. Several ethanol plants were closed, some plants were forced to work below capacity, and other companies filed for Chapter 11 bankruptcy protection. Such events led to layoffs and hiring freezes.

Nevertheless, overall, the industry was growing and saw a return to profitability in the second half of 2009. Worldwide production of ethanol and biodiesel is expected to grow to $113 billion by 2019; this is more than 60 percent growth of annual earnings. One segment of the biofuel and synthetic fuel industry, the biogas industry, was not affected by recession at all. More than 8,900 new biogas plants were built worldwide in 2009. According to market analysts, the biogas industry has reached a turning point and may grow at a rate of 24 percent from 2010 to 2016. Research and development efforts in biofuels and synthetic fuels actually increased during the economic crisis. In general, the future of biofuels and synthetic fuels is bright and optimistic.

—*Sergei A. Markov, PhD*

FURTHER READING
Bart, Jan C. J., and Natale Palmeri. *Biodiesel Science and Technology: From Soil to Oil.* Cambridge, England: Woodhead, 2010. A comprehensive book on biodiesel fuels.

Bourne, Joel K. "Green Dreams." *National Geographic* 212, no. 4 (October, 2007): 38-59. An interesting discussion about ethanol and biodiesel fuels and their future.

Glazer, Alexander N., and Hiroshi Nikaido. *Microbial Biotechnology: Fundamentals of Applied Microbiology.* New York: Cambridge University Press, 2007. Provides an in-depth analysis of ethanol and biomass for fuel applications.

Mikityuk, Andrey. "Mr. Ethanol Fights Back." *Forbes,* November 24, 2008, 52-57. Excellent discussion about the problems and the hopes of the ethanol industry. Examines how the ethanol industry fought the economic crisis in 2008 and returned to profitability in 2009.

Probstein, Ronald F., and Edwin R. Hicks. *Synthetic Fuels.* Mineola, N.Y.: Dover Publications, 2006. A comprehensive work on synthetic fuels. Contains references and sources for further information.

Service, Robert F. "The Hydrogen Backlash." *Science* 305, no. 5686 (August 13, 2004): 958-961. Discusses the future of hydrogen power, including its maturity. Written in an easy-to-understand manner.

Wall, Judy, ed. *Bioenergy.* Washington, D.C.: ASM Press, 2008. Provides the information on generation of biofuels by microorganisms and points out future areas for research. Ten chapters focus on ethanol production from cellulosic material.

WEB SITES
Advanced BioFuels USA
http://advancedbiofuelsusa.info

International Energy Agency Bioenergy
http://www.ieabioenergy.com

U.S. Department of Agriculture, Economic Research

Service Featuring Bioenergy
http://www.ers.usda.gov/features/bioenergy

U.S. Department of Energy Energy Efficiency and Renewable Energy
http://www.energy.gov/energysources/index.htm

U.S. Energy Information Administration Renewable and Alternative Fuels
http://www.eia.doe.gov/fuelrenewable.html

See also: Hybrid Vehicle Technologies.

BIOMATHEMATICS

FIELDS OF STUDY

Algebra; geometry; calculus; probability; statistics; cellular biology; genetics; differential equations; molecular biology; oncology; immunology; epidemiology; prokaryotic biology; eukaryotic biology

SUMMARY

Biomathematics is a field that applies mathematical techniques to analyze and model biological phenomena. Often a collaborative effort, mathematicians and biologists work together using mathematical tools such as algorithms and differential equations in order to understand and illustrate a specific biological function. Biomathematics is used in a wide variety of applications from medicine to agriculture. As new technologies lead to a rise in the amount of biological data available, biomathematics will become a discipline that is increasingly in demand to help analyze and effectively utilize the data.

KEY TERMS AND CONCEPTS

- **algorithm:** Use of symbols and a set of rules for solving problems.
- **biology:** Study of living things.
- **cell:** Unit that is the basis of an organism and encompasses genetic material, as well as other molecules, and is defined by a cell membrane.
- **deoxyribonucleic acid (DNA):** Nucleic acid that forms the molecular basis for heredity.
- **differential equation:** Mathematic expression that

uses variables to express changes over time. Differential equations can be linear or nonlinear.
- **genetics:** Study of an organism's traits, including how they are passed down through generations.
- **matrix:** Mathematical structure for arranging numbers or symbols that have particular mathematical rules for use.
- **oncology:** Field of science that studies the cause and treatment of cancer.

DEFINITION AND BASIC PRINCIPLES
Biomathematics is a discipline that quantifies were developed to describe a cellular function known biological occurrences using mathematical tools. Biomathematics is related to and may be a part of other disciplines including bioinformatics, biophysics, bioengineering, and computational biology, as these disciplines include the use of mathematical tools in the study of biology.

Biologists have used different ways to explain biological functions, often employing words or pictures. Biomathematics allows biologists to illustrate these functions using techniques such as algorithms and differential equations. Biological phenomena vary in both scale and complexity, encompassing everything from molecules to ecosystems. Therefore, the creation of a model requires the scientist to make some assumptions in order to simplify the process. Biomathematical models vary in length and complexity and several different models may be tested.

The use of biomathematics is not limited to modeling a biological function and includes other techniques, such as structuring and analyzing data.

Scientists may use biomathematics to organize data or analyze data sets, and statistics are often considered an integral tool.

BACKGROUND AND HISTORY

As early as the 1600's, mathematics was used to explain biological phenomena, although the mathematical tools used date back even farther. In 1628, British physician William Harvey used mathematics to prove that blood circulates in the body. His model changed the belief at that time that there were two kinds of blood. In the mid-1800's, Gregor Mendel, an Augustinian monk, used mathematics to analyze the data he obtained from his experiments with pea plants. His experiments would become the basis for genetics. In the early 1900's, British mathematician R. A. Fisher applied statistical methods to population biology, providing a better framework for studying the field. In 1947, theoretical physicist Nicolas Rashevsky argued that mathematical tools should be applied to biological processes and created a group dedicated to mathematical biology. Despite the fact that some dismiss Rashevsky's work as being too theoretical, many view him as one of the founders of mathematical biology. In the 1950's, the Hodgkin-Huxley equations Biomathematics is a discipline that quantifies were developed to describe a cellular function known biological occurrences using mathematical tools. as ion channels. These equations are still used. In the 1980's, the Smith-Waterman algorithm was created to aid scientists in comparing DNA sequences. While the algorithm was not particularly efficient, it paved the way for the BLAST (Basic Local Alignment Search Tool) software, a program that has allowed scientists to compare DNA sequences since 1990. Despite the fact that mathematical tools have been applied to some biological problems during the second half of the twentieth century, the practice has not been all-inclusive. In the twenty-first century, there has been a renewed interest in biology becoming more quantitative, due in part to an increase in new data.

HOW IT WORKS

Basic mathematical tools. Biomathematicians may use mathematical tools at different points during the investigation of a biological function. Mathematical tools may be used to organize data, analyze data, or even to generate data. Algorithms, which use symbols and procedures for solving problems, are employed in biomathematics in several ways. They may be used to analyze data, as in sequence analysis. Sequence analysis uses specifically developed algorithms to detect similarity in pieces of DNA. Specifically developed algorithms are also used to predict the structure of different biological molecules, such as proteins. Algorithms have led to the development of more useful biological instruments such as specific types of microscopy. Statistics are another common way of analyzing biological data. Statistics may be used to analyze data, and this data may help create an equation to describe a theory: Statistics was used to analyze the movement of single cells. The data taken from the analysis was then used to create partial equations describing cell movement.

Differential equations, which use variables to express changes over time, are another common technique in biomathematics. There are two kinds of differential equations: linear and nonlinear. Nonlinear equations are commonly used in biomathematics. Differential equations, along with other tools, have been used to model the functions of intercellular processes. Differential equations are utilized in several of the important systems used in biomathematics for modeling, including mean field approaches. Other modeling systems include: patch models, reaction-diffusion equations, stochastic models, and interacting particle systems. Each modeling system provides a different approach based on different assumptions. Computers have helped in this area by providing an easier way to apply and solve complex equations. Computer modeling of dynamic systems, such as the motion of proteins, is also a work in progress.

New methods and technology have increased the amount of data being obtained from biological experimentation. The data gained through experimentation and analysis may be structured in different ways. Mathematics may be used to determine the structure. For example, phylogenetic trees (treelike graphs that illustrate how pieces of data relate to one another) use different mathematical tools, including matrices, to determine their structure. Phlyogenic trees also provide a model for how a particular piece of data evolved. Another way to organize data is a site graph, or hidden Markov model, which uses probability to illustrate relationships between the data.

Modeling a biological function. The scientist may be at different starting points when considering a mathematical model. He or she may be starting with data already analyzed or organized by a mathematical technique or already described by a visual depiction or written theory. However, there are several considerations that scientists must take into account when creating a mathematical model. As biology covers a wide range of matter, from molecules to ecosystems, when creating a model the scale of phenomena must be considered. The time scale and complexity must also be considered, as many biological systems are dynamic or interact with their environment. The scientist must make assumptions about the biological phenomena in order to reduce the parameters used in the model. The scientist may then define important variables and the relationships between them. Often, more than one model may be created and tested.

APPLICATIONS AND PRODUCTS

The field of biomathematics is applicable to every area of biology. For example, biomathematics has been used to study population growth, evolution, and genetic variation and inheritance. Mathematical models have also been created for communities, modeling competition or predators, often using differential equations. Whether the scale is large or small, biomathematics allows scientists a greater understanding of biological phenomena.

Molecules and cells. Biomathematics has been applied to various biological molecules, including DNA, ribonucleic acid (RNA), and proteins. Biomathematics may be used to help predict the structure of these molecules or help determine how certain molecules are related to one another. Scientists have used biomathematics to model how bacteria can obtain new, important traits by transferring genetic material between different strains. This information is important because bacteria may, through sharing genetic material, acquire a trait such as a resistance to an antibiotic. To model the sharing of a trait, scientists have combined two of the ways to structure data: the phylogenetic tree and the site graph. The phylogenetic tree illustrates how the types of bacteria are related to one another. The site graph illustrates how pieces of genetic material interact. Then, scientists use a particular algorithm to determine the parameters of the model. By using such tools, scientists can predict which areas of genetic material are most likely to transfer between the bacterium.

Biomathematics has been used in cellular biology to model various cellular functions, including cellular division. The models can then be used to help scientists organize information and gain a deeper understanding about cellular functions. Cellular movement is one example of an application of biomathematics to cellular biology. Cellular movements can be seen as a set of steps. The scientists first considered certain cellular steps or functions, including how a cell senses a signal and how this signal is used within the cell to start movement. Scientists also considered the environment surrounding the cell, how the signal was provided, and the processes that occurred within the cell to read the signal and start movement. The scientists were then able to build a mathematical function that takes these steps into account. Depending on the particular question, the scientists may chose to focus on any of these steps. Therefore, more than one model may be used.

Organisms and agriculture. Biomathematics has been used to create mathematical models for different functions of organisms. One popular area has been organism movement, where models have been created for bacteria movement and insect flight. A more complete understanding of organisms through mathematical models supports new technologies in agriculture. Biomathematics may also be used to help protect harvests. For example, biomathematics has been used to model a type of algae bloom known as brown tide. In the late 1980's, brown tide appeared in the waters near Long Island, New York, badly affecting the shellfish population by blocking sunlight and depleting oxygen. Four years later, the algae blooms receded. Both mathematicians and scientists collaborated in order to create a model of the brown tide in order to understand why it bloomed and whether it will bloom in the future. To create a model, the collaborators used differential equations. They focused on the population density, which included factors such as temperature and nutrients. The collaborators had to consider many variables and remove the ones they did not consider important. For instance, they hypothesized that a period of drought followed by rain may have affected growth. They also considered fertilizers and pesticides that were used in the area. A better understanding of the brown tide may help protect the shellfish harvest in future years.

Medical uses. Biomathematical models have been developed to illustrate various functions within the human body, including the heart, kidneys, and cardiac and neural tissue. Biomathematics is useful in modeling cancer, enabling scientists to learn more about the type of cancer, thereby allowing them to study the efficacy of different types of treatment. One project has focused on modeling colon cancer on a genetic and molecular level. Not only did scientists gain information about the genetic mutations that are present during colon cancer, but they also developed a model that predicted when tumor cells would be sensitive to radiation, which is the most common way to treat colon cancer. Studies such as this can be built on in future experimentations, the results of which may someday be used by doctors to create more effective cancer treatments.

Biomathematics has also been used to organize and analyze data from experiments dealing with drug efficacy and gene expression in cancer cells. Using matrices, statistics, and algorithms, scientists have been able to understand if a particular drug is more likely to work based on the patient's cancer cell's gene expression. Biomathematics has also been integral in epidemiology, the field that studies diseases within a population. Biomathematics may be used to model various aspects of a disease such as human immunodeficiency virus (HIV), allowing for more comprehensive planning and treatment.

Impact on Industry

Government and university research. Biomathematics is often developed through a collaborative effort between biologists and mathematicians. In the United States, the National Science Foundation has a mathematical biology program that provides grants for research to develop mathematical applications related to biology. In addition, many U.S. universities offer biomathematics programs. Some universities have biomathematics departments, such as Ohio State University's Mathematical Biosciences Institute, which focuses on creating mathematical applications to help solve biological problems. This institute provides research and education opportunities, including workshops and public lectures.

Each biomathematics department or program may emphasize a different aspect of biomathematics: Some focus on medical applications, others focus on the need to quantify biological phenomena using mathematical tools. UCLA's department of biomathematics conducts research in areas such as statistical genetics, evolutionary biology, molecular imaging, and neuroscience.

Biomathematics is also being developed internationally. Many universities in the United Kingdom have biomathematics programs. The University of Oxford has a Centre for Mathematical Biology. There are also independent research institutions and organizations that focus on biomathematics. The Institute for Medical BioMathematics, in Israel, works on developing analytical approaches to treating cancer and infectious diseases. The Society for Mathematical Biology, in Boulder, Colorado, has provided an international forum for biomathematics for more than twenty years. The International Biometric Society in Washington, D.C., also addresses the application of

---

FASCINATING FACTS ABOUT BIOMATHEMATICS

- Scientists are using biomathematics to create a virtual patient. First, biomathematics is used to model the human body. Using the virtual patient, scientists can test cancer-prevention drugs. The result is a quicker and cheaper way to develop drugs.

- Biomathematics has applications in nature. By using biomathematics, the pigment patterns of a leopard's spots or the patterns of seashells can be modeled.

- Scientists and mathematicians have found that by using biomathematics and computers they can simulate kidney functions. This kidney simulation helps doctors understand kidney disease better and can help them provide more effective treatments.

- Biomathematics is being used to model the workings of a heart in order to improve artificial heart valves. With biomathematics models, designs can be tested more quickly and efficiently.

- Mathematical models of biofilms systems, which are layers of usually nonresistant microorganisms such as bacteria that attach to a surface, are critical to the medical and technical industries. Biofilm systems can cause infections in humans and corrosion and deterioration in technical systems. Biomathematics can be used to model biofilms systems to help understand and prevent their formation.

- Biomathematics was used to sort and analyze data from the Human Genome Project. Completed in 2003, the thirteen-year-long project identified the entire human genome.

mathematical tools to biological phenomena. The European Society for Mathematical and Theoretical Biology, founded in 1991, promotes biomathematics, and the Society for Industrial and Applied Mathematics in Philadelphia has an activity group on the life sciences that provides a platform for researchers working in the area of biomathematics.

Industry. Biomathematics may lead to developments in medical treatments or technology, which may then be marketed. The pharmaceutical industry is a good example of this development. Biomathematics is often used to create models of diseases that can lead to a deeper understanding of the disease and new medicines. In addition, biomathematics provides tools that may be used throughout the drug-creation process and can be used to predict how well a drug will work or how safe a drug will be for a group of patients with a particular genetic makeup. The engineering of microorganisms has also benefited from biomathematics. Biomathematics is being used to create models to understand the fundamentals of microorganisms better. The end result of this understanding may be the changed metabolism or structure of an organism to produce more milk or a sweeter wine. Finally, some companies are targeting software to aid with biological modeling, and others provide consulting in the field of biomathematics. BioMath Solutions in Austin, Texas, is a company that provides analytical software in the area of molecular biology.

CAREERS AND COURSE WORK

Degree programs in biomathematics are gaining popularity in universities. Some schools have biomathematics departments, and others have biomathematics programs within the mathematics or biology departments. Undergraduate course work for a B.S. in biomathematics encompasses classes in mathematics, biology, and computer science, including: calculus, chemistry, genetics, physics, software development, probability, statistics, organic chemistry, epidemiology, population biology, molecular biology, and physiology. A student may also choose to receive a B.S. in biology or mathematics. In addition, students may seek additional opportunities outside their program. Ohio State's Mathematical Biosciences Institute offers summer programs for undergraduate and graduate students.

In the field of biomathematics, a doctorate is required for many careers. Doctoral programs in biomathematics include course work in statistics, biology, probability, differential equations, linear algebra, cellular modeling, genetics modeling, computer programming, pharmacology, and clinical research methods. In addition, doctoral candidates often will perform biomathematics research with support from departmental faculty. As with the undergraduate degrees, a student may also choose to pursue a doctorate in biology or mathematics.

With the influx of biological data from new technologies and tools, a degree in biomathematics is imperative. Those who receive a Ph.D. may choose to enter a postdoctoral program, such as the one at Ohio State's Mathematical Biosciences Institute, which offers postdoctoral fellowships as well as mentorship and research opportunities. Other career paths include research in medicine, biology, and mathematics with universities and private research institutions; work with software development and computer modeling; teaching; or collaborating with other professionals and consulting in an industry such as pharmaceuticals or bioengineering.

SOCIAL CONTEXT AND FUTURE PROSPECTS

While mathematical tools have been applied to biology for some time, many scientists believe there is still a need for increased quantitative analysis of biology. Some call for more emphasis on mathematics in high school and undergraduate biology classes. They believe that this will advance biomathematics. As more universities develop biomathematics departments and degrees, more mathematics classes will be added to the curriculum. A concern has been raised in the biomathematics field about the assumptions used to create simplified mathematical models. More complex and accurate models will likely be developed.

Important future applications for biomathematics will be in the bioengineering and medical industries. The development of mathematical models for complex biological phenomena will aid scientists in a deeper understanding that can lead to more effective treatments in such areas such as tumor therapy. As new tools and methods continue to develop, biomathematics will be a field that expands to sort and analyze the large influx of data.

—Carly L. Huth, JD

FURTHER READING

Hochberg, Robert, and Kathleen Gabric. "A Provably Necessary Symbiosis." *The American Biology Teacher* 72, No. 5 (2010): 296-300. This article describes some mathematics that can be taught in biology classrooms.

Misra, J. C., ed. *Biomathematics: Modelling and Simulation.* Hackensack, N.J.: World Scientific, 2006. This book provides an in-depth guide to several modern applications of biomathematics and includes many helpful illustrations.

Schnell, Santiago, Ramon Grima, and Philip Maini. "Multiscale Modeling in Biology." *American Scientist* 95 (March-April, 2007): 134-142. This article gives an overview of how biological models are created and provides several modern examples of biomathical applications.

WEB SITES
International Biometric Society

http://www.tibs.org

National Science Foundation
http://www.nsf.gov

Ohio State University Mathematical Biosciences Institute
http://mbi.osu.edu

Society for Industrial and Applied Mathematics
http://www.siam.org

Society for Mathematical Biology
http://www.smb.org

See also: Bioengineering; Biomechanical Engineering; Bionics and Biomedical Engineering; Biophysics; Calculus; Computer Science; Geometry.

BIOMECHANICAL ENGINEERING

FIELDS OF STUDY

Biomedical engineering; biomechanics; physiology; nanotechnology; implanted devices; modeling; bioengineering; bioinstrumentation; computational biomechanics; cellular and molecular biomechanics; forensic biomechanics; tissue engineering; mechanobiology; micromechanics; anthropometics; imaging; biofluidics.

SUMMARY

Biomechanical engineering is a branch of science that applies mechanical engineering principles such as physics and mathematics to biology and medicine. It can be described as the connection between structure and function in living things. Researchers in this field investigate the mechanics and mechanobiology of cells and tissues, tissue engineering, and the physiological systems they comprise. The work also examines the pathogenesis and treatment of diseases using cells and cultures, tissue mechanics, imaging, microscale biosensor fabrication, biofluidics, human motion capture, and computational methods. Real-world applications include the design and evaluation of medical implants, instrumentation, devices, products, and procedures. Biomechanical engineering is a multidisciplinary science, often fostering collaborations and interactions with medical research, surgery, radiology, physics, computer modeling, and other areas of engineering.

KEY TERMS AND CONCEPTS

- **angular motion:** Motion involving rotation around a central line or point known as the axis of rotation.
- **biofluidics:** Field of study that combines the characterization of fluids focused on flows in the body as well as environmental flows involved in disease process.
- **dynamics:** Branch of mechanics that studies systems that are in motion, subject to acceleration or deceleration.
- **kinematics:** Study of movement of segments of a body without regard for the forces causing the movement.
- **kinesiology:** Study of human movement.
- **kinetics:** Study of forces associated with motion.
- **linear motion:** Motion involving all the parts of a

body or system moving in the same direction, at the same speed, following a straight (rectilinear) or curved (curvilinear) line.
- **mechanics:** Branch of physics analyzing the resulting actions of forces on particles or systems.
- **mechanobiology:** Emerging scientific field that studies the effect of physical force on tissue development, physiology, and disease.
- **modeling:** Computerized analytical representation of a structure or process.
- **statics:** Branch of mechanics that studies systems that are in a constant state of motion or constant state of rest.

DEFINITION AND BASIC PRINCIPLES
Biomechanical engineering applies mechanical engineering principles to biology and medicine. Elements from biology, physiology, chemistry, physics, anatomy, and mathematics are used to describe the impact of physical forces on living organisms. The forces studied can originate from the outside environment or generate within a body or single structure. Forces on a body or structure can influence how it grows, develops, or moves. Better understanding of how a biological organism copes with forces and stresses can lead to improved treatment, advanced diagnosis, and prevention of disease. This integration of multidisciplinary philosophies has lead to significant advances in clinical medicine and device design. Improved understanding guides the creation of artificial organs, joints, implants, and tissues. Biomechanical engineering also has a tremendous influence on the retail industry, as the results of laboratory research guide product design toward more comfortable and efficient merchandise.

BACKGROUND AND HISTORY
The history of biomechanical engineering, as a distinct and defined field of study, is relatively short. However, applying the principles of physics and engineering to biological systems has been developed over centuries. Many overlaps and parallels to complementary areas of biomedical engineering and biomechanics exist, and the terms are often used interchangeably with biomechanical engineering. The mechanical analysis of living organisms was not internationally accepted and recognized until the definition provided by Austrian mathematician Herbert Hatze in 1974: "Biomechanics is the study of the structure and function of biological systems by means of the methods of mechanics." Aristotle introduced the term "mechanics" and discussed the movement of living beings around 322 BCE in the first book about biomechanics, *On the Motion of Animals.* Leonardo da Vinci proposed that the human body is subject to the law of mechanics in the 1500's. Italian physicist and mathematician Giovanni Alfonso Borelli, a student of Galileo's, is considered the "father of biomechanics" and developed mathematical models to describe anatomy and human movement mechanically. In the 1890's German zoologist Wilhelm Roux and German surgeon Julius Wolff determined the effects of loading and stress on stem cells in the development of bone architecture and healing. British physiologist Archibald V. Hill and German physiologist Otto Fritz Meyerhof shared the 1922 Nobel Prize for Physiology or Medicine. The prize was divided between them: Hill won "for his discovery relating to the production of heat in the muscle"; Meyerhof won "for his discovery of the fixed relationship between the consumption of oxygen and the metabolism of lactic acid in the muscle."

The first joint replacement was performed on a hip in 1960 and a knee in 1968. The development of imaging, modeling, and computer simulation in the latter half of the 1900's provided insight into the smallest structures of the body. The relationships between these structures, functions, and the impact of internal and external forces accelerated new research opportunities into diagnostic procedures and effective solutions to disease. In the 1990's, biomechanical engineering programs began to emerge in academic and research institutions around the world.

HOW IT WORKS
Biomechanical engineering science is extremely diverse. However, the basic principle of studying the relationship between biological structures and forces, as well as the important associated reactions of biological structures to technological and environmental materials, exists throughout all disciplines. The biological structures described include all life forms and may include an entire body or organism or even the microstructures of specific tissues or systems. Characterization and quantification of the response of these structures to forces can provide insight into disease process, resulting in better treatments and diagnoses. Research in this field extends

beyond the laboratory and can involve observations of mechanics in nature, such as the aerodynamics of bird flight, hydrodynamics of fish, or strength of plant root systems, and how these findings can be modified and applied to human performance and interaction with external forces.

As in biomechanics, biomechanical engineering has basic principles. Equilibrium, as defined by British physicist Sir Isaac Newton, results when the sum of all forces is zero and no change occurs and energy cannot be created or destroyed, only converted from one form to another.

The seven basic principles of biomechanics can be applied or modified to describe the reaction of forces to any living organism.

The lower the center of mass, the larger the base of support; the closer the center of mass to the base of support, and the greater the mass, the more stability increases.

The production of maximum force requires the use of all possible joint movements that contribute to the task's objective.

The production of maximum velocity requires the use of joints in order—from largest to smallest.

The greater the applied impulse, the greater increase in velocity.

Movement usually occurs in the direction opposite that of the applied force.

Angular motion is produced by the application of force acting at some distance from an axis, that is, by torque.

Angular momentum is constant when a body or object is free in the air.

The forces studied can be combinations of internal, external, static, or dynamic, and all are important in the analysis of complex biochemical and biophysical processes. Even the mechanics of a single cell, including growth, cell division, active motion, and contractile mechanisms, can provide insight into mechanisms of stress, damage of structures, and disease processes at the microscopic level. Imaging and computer simulation allow precise measurements and observations to be made of the forces impacting the smallest cells.

APPLICATIONS AND PRODUCTS
Biomechanical engineering advances in modeling and simulation have tremendous potential research and application uses across many health care

disciplines. Modeling has resulted in the development of designs for implantable devices to assist with organs or areas of the body that are malfunctioning. The biomechanical relationships between organs and supporting structures allow for improved device design and can assist with planning of surgical and treatment interventions. The materials used for medical and surgical procedures in humans and animals are being evaluated and some redesigned, as biomechanical science is showing that different materials, procedures, and techniques may be better for reducing complications and improving long-term patient health. Evaluating the physical relationship between the cells and structures of the body and foreign implements and interventions can quantify the stresses and forces on the system, which provides more accurate prediction of patient outcomes.

Biomechanical engineering professionals apply their knowledge to develop implantable medical devices that can diagnose, treat, or monitor disease and health conditions and improve the daily living of patients. Devices that are used within the human body are highly regulated by the U.S. Food and Drug Administration (FDA) and other agencies internationally. Pacemakers and defibrillators, also called cardiac resynchronization therapy (CRT) devices, can constantly evaluate a patient's heart and respond to changes in heart rate with electrical stimulation. These devices greatly improve therapeutic outcomes in patients afflicted with congestive heart failure. Patients with arrhythmias experience greater advantages with implantable devices than with pharmaceutical options. Cochlear implants have been designed to be attached to a patient's auditory nerve and can detect sound waves and process them in order to be interpreted by the brain as sound for deaf or hard-of-hearing patients. Patients who have had cataract surgery used to have to wear thick corrective lenses to restore any standard of vision but with the development of intraocular lenses that can be implanted into the eye, their vision can be restored, often to a better degree than before the cataract developed.

Artificial replacement joints comprise a large portion of medical-implant technology. Patients receive joint replacement when their existing joints no longer function properly or cause significant pain because of arthritis or degeneration. More than 220,000 total hip replacements were performed in the United States in 2003, and this number is expected to grow significantly as the baby boomer portion of the population ages. Artificial joints are normally fastened to the existing bone by cement, but advances in biomechanical engineering have lead to a new process called "bone ingrowth," in which the natural bone grows into the porous surface of the replacement joint. Biomechanical engineering contributes considerable knowledge to the design of the artificial joints, the materials from which they are made, the surgical procedure used, fixation techniques, failure mechanisms, and prediction of the lifetime of the replacement joints.

Computer-aided (CAD) design has allowed biomechanical engineers to create complex models of organs and systems that can provide advanced analysis and instant feedback. This information provides insight into the development of designs for artificial organs that align with or improve on the mechanical properties of biological organs.

Biomechanical engineering can provide predictive values to medical professionals, which can help them develop a profile that better forecasts patient outcomes and complications. An example of this is using finite element analysis in the evaluation of aortic-wall stress, which can remove some of the unpredictability of expansion and rupture of an abdominal aortic aneurysm. Biomechanical computational methodology and advances in imaging and processing technology have provided increased predictability for life-threatening events.

Nonmedical applications of biomechanical engineering also exist in any facet of industry that impacts human life. Corporations employ individuals or teams to use engineering principles to translate the scientifically proven principles into commercially viable products or new technological platforms. Biomechanical engineers also design and build experimental testing devices to evaluate a product's performance and safety before it reaches the marketplace, or they suggest more economically efficient design options. Biomechanical engineers also use ergonomic principles to develop new ideas and create new products, such as car seats, backpacks, or even equipment and clothing for elite athletes, military personnel, or astronauts.

Impact on Industry

Biomechanical engineering is a dynamic scientific field, and its vast range of applications is having

---

FASCINATING FACTS ABOUT BIOMECHANICAL ENGINEERING

- Biomechanical engineers design and develop many items used and worn by astronauts.
- Of all the engineering specialties, biomechanical engineering has one of the highest percentages of female students.
- Synovial fluid in joints has such low friction that engineers are trying to duplicate it synthetically to lubricate machines.
- Many biomechanical engineering graduates continue on to medical school.
- On October 8, 1958, in Sweden, forty-three-year-old Arne Larsson became the first person to receive an implanted cardiac pacemaker. He lived to the age of eighty-six.
- Many biomechanical engineers have conducted weightlessness experiments aboard the National Aeronautics and Space Administration's (NASA) C-9 microgravity aircraft.
- Cardiac muscles rest only between beats. Based on 72 beats a minute, they will pump an average of
- 3.0 trillion times during an eighty-year life span.

---

a significant impact and influence on industry. Corporations realize the value of having their designs and products evaluated by biomechanical engineers to optimize the comfort and safety of consumers. Small modifications in the design of a product can influence consumers to select one product over several others, be it clothing, furniture, sporting equipment, or beverage and food containers. With so many products to choose from, it is important that one stand out as safer, more comfortable, or more efficient than another.

Biotechnology and health care are highly competitive, multibillion dollar industries. Biotechnology is an extremely research-intensive industry, so biotech companies employ teams of biomechanical engineers to research and develop devices, treatments, and diagnostic and monitoring devices to be used in health care. Implantable devices are used in the treatment and management of cardiovascular, orthopedic, neurological, ophthalmic, and various other chronic disorders. Orthopedic implants are the most common and include reconstructive joint

replacements, spinal implants, orthobio-
logics, and trauma implants. In the United
States, the demand for implantable medical
devices is expected to reach a market value
of $48 billion by 2014.

Surgeons, medical personnel, and health
care administrators are always looking for
new options that will provide their patients
with optimal care, earlier diagnosis, pain reduction, and a decreased risk of complications. Aging and disabled persons have more
effective options than ever before that allow
them to continue vital, independent, and
active lives. Industry is aligning with these
developments, demands, and discoveries,
and there is great competition among companies to be the first to capitalize. The FDA
and other international regulatory bodies
have strict standards and high levels of control for any device or implement that will be
used on patients. The application and approval process can take many years. It is not
uncommon for a biotechnology company to
invest millions of dollars in the research and
development of a medical device before it
can be presented to the consumer.

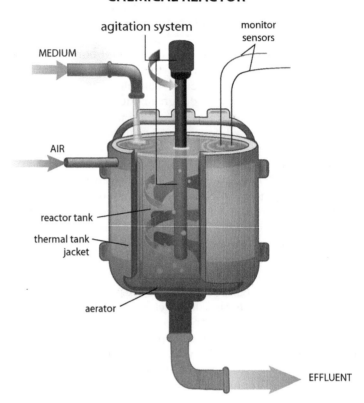

CHEMICAL REACTOR

Biomedical instrumentation amplifier schematic used in monitoring low voltage biological signals, an example of a biomedical engineering application of electronic engineering to electrophysiology.

CAREERS AND COURSE WORK
There are a variety of career choices in biomechanical engineering, and study in this
field often evolves into specialized work in
related areas. Students who earn a bachelor's degree
from an accredited biomechanical engineering program may begin working in areas such as medical device, implant, or product design. Most teaching positions require a master's or doctoral degree. Some
students continue to medical school.

Biomechanical engineering programs are composed of a cross section of course work from many disciplines. Students should have a strong aptitude for
mathematics as well as biological sciences. Elements
from engineering, physics, chemistry, anatomy, biology, and computer science provide core knowledge that is applied to mathematical modeling and
computer simulation. Experimental work involving
biological, mechanical, and clinical studies are performed to illustrate theoretical models and solve
important research problems. The principles of biomechanical engineering can have vast applications,

ranging from building artificial organs and tissues
to designing products that are more comfortable for
consumers.

Biomechanical engineering programs often are
included as a subdiscipline of engineering or biomedicine. However, some schools, such as Stanford
University, are creating interdisciplinary programs
that offer undergraduate and graduate degrees in
biomechanical engineering.

SOCIAL CONTEXT AND FUTURE PROSPECTS
The diversity of studying the relationship between
living structure and function has opened up vast opportunities in science, health care, and industry. In
addition to conventional implant and replacement
devices, the demand is growing for implantable tissues for cosmetic surgery, such as breast and tissue
implants, as well as implantable devices to aid in
weight loss, such as gastric banding.

Reports of biomechanical engineering triumphs and discoveries are appearing in the mainstream media, making the general public more aware of the scientific work being done and how it impacts daily life. Sports fans learn about the equipment, training, and rehabilitation techniques designed by biomechanical engineers that allow their favorite athletes to break performance records and return to work sooner after being injured or having surgery. The public is accessing more information about their own health options than ever before, and they are becoming knowledgeable about the range of treatments available to them and the pros and cons of each.

Biomechanical engineering and biotechnology is an area that is experiencing accelerated growth, and billions of dollars are being funneled into research and development annually. This growth is expected to continue.

—*April D. Ingram*

FURTHER READING

Ethier, C. Ross, and Craig A. Simmons. *Introductory Biomechanics: From Cells to Organisms.* Cambridge, England: Cambridge University Press, 2007. Provides an introduction to biomechanics and also discusses clinical specialties, such as cardiovascular, musculoskeletal, and ophthalmology.

Hall, Susan J. *Basic Biomechanics.* 5th ed. New York: McGraw-Hill, 2006. A good introduction to biomechanics, regardless of one's math skills.

Hamill, Joseph, and Kathleen M. Knutzen. *Biomechanical Basis of Human Movement.* 3d ed. Philadelphia: Lippincott, 2009. Integrates anatomy, physiology, calculus, and physics and provides the fundamental concepts of biomechanics.

Hay, James G., and J. Gavin Reid. *Anatomy, Mechanics, and Human Motion.* 2d ed. Englewood Cliffs, N.J.: Prentice Hall, 1988. A good resource for upper high school students, this text covers basic kinesiology.

Peterson, Donald R., and Joseph D. Bronzino, eds. *Biomechanics: Principles and Applications.* 2d ed. Boca Raton, Fla.: CRC Press, 2008. A collection of twenty articles on various aspects of research in biomechanics.

Prendergast, Patrick, ed. *Biomechanical Engineering: From Biosystems to Implant Technology.* London: Elsevier, 2007. One of the first comprehensive books for biomechanical engineers, written with the student in mind.

WEB SITES
American Society of Biomechanics
http://www.asbweb.org

Biomedical Engineering Society
http://www.bmes.org

European Society of Biomechanics
http://www.esbiomech.org

International Society of Biomechanics
http://www.isbweb.org

World Commission of Science and Sports
http://www.wcss.org.uk

See also: Bioengineering; Biomechanics; Bionics and Biomedical Engineering; Biophysics; Calculus; Computer Science; Nanotechnology.

BIOMECHANICS

FIELDS OF STUDY

Kinesiology; physiology; kinetics; kinematics; sports medicine; technique/performance analysis; injury rehabilitation; modeling; orthopedics; prosthetics; bioengineering; bioinstrumentation; computational biomechanics; cellular/molecular biomechanics; veterinary (equine) biomechanics; forensic biomechanics; ergonomics.

SUMMARY

Biomechanics is the study of the application of mechanical forces to a living organism. It investigates the effects of the relationship between the body and forces applied either from outside or within. In humans, biomechanists study the movements made by the body, how they are performed, and whether the forces produced by the muscles are optimal for the intended result or purpose. Biomechanics integrates the study of anatomy and physiology with physics, mathematics, and engineering principles. It may be considered a subdiscipline of kinesiology as well as a scientific branch of sports medicine.

KEY TERMS AND CONCEPTS

- **angular motion:** Motion involving rotation around a central line or point known as the axis of rotation.
- **dynamics:** Branch of mechanics that studies systems in motion, subject to acceleration or deceleration.
- **kinematics:** Study of movement of segments of a body without regard for the forces causing the movement.
- **kinesiology:** Study of human movement.
- **kinetics:** Study of forces associated with motion.
- **lever:** Rigid bars (in the body, bones) that move around an axis of rotation (joint) and have the ability to magnify or alter the direction of a force.
- **linear motion:** Motion involving all the parts of a body or system moving in the same direction, at the same speed, following a straight (rectilinear) or curved (curvilinear) line.
- **mechanics:** Branch of physics analyzing the resulting actions of forces on particles or systems.
- **qualitative movement:** Description of the quality of movement without the use of numbers.
- **quantitative movement:** Description or analysis of movement using numbers or measurement.
- **sports medicine:** Branch of medicine studying the clinical and scientific characteristics of exercise and sport, as well as any resulting injuries.
- **statics:** Branch of mechanics that studies systems that are in a constant state of motion or constant state of rest.
- **torque:** Turning effect of a force applied in a direction not in line with the center of rotation of a nonmoving axis (eccentric).

DEFINITION AND BASIC PRINCIPLES

Biomechanics is a science that closely examines the forces acting on a living system, such as a body, and the effects that are produced by these forces. External forces can be quantified using sophisticated measuring tools and devices. Internal forces can be measured using implanted devices or from model calculations. Forces on a body can result in

movement or biological changes to the anatomical tissue. Biomechanical research quantifies the movement of different body parts and the factors that may influence the movement, such as equipment, body alignment, or weight distribution. Research also studies the biological effects of the forces that may affect growth and development or lead to injury. Two distinct branches of mechanics are statics and dynamics. Statics studies systems that are in a constant state of motion or constant state of rest, and dynamics studies systems that are in motion, subject to acceleration or deceleration. A moving body may be described using kinematics or kinetics. Kinematics studies and describes the motion of a body with respect to a specific pattern and speed, which translate into coordination of a display. Kinetics studies the forces associated with a motion, those causing it and resulting from it. Biomechanics combines kinetics and kinematics as they apply to the theory of mechanics and physiology to study the structure and function of living organisms.

BACKGROUND AND HISTORY

Biomechanics has a long history even though the actual term and field of study concerned with mechanical analysis of living organisms was not internationally accepted and recognized until the early 1970's. Definitions provided by early biomechanics specialists James G. Hay in 1971 and Herbert Hatze in 1974 are still accepted. Hatze stated, "Biomechanics is the science which studies structures and functions of biological systems using the knowledge and methods of mechanics."

Highlights throughout history have provided insight into the development of this scientific discipline. The ancient Greek philosopher Aristotle was the first to introduce the term "mechanics," writing about the movement of living beings around 322 BCE. He developed a theory of running techniques and suggested that people could run faster by swinging their arms. In the 1500's, Leonardo da Vinci proposed that the human body is subject to the law of mechanics, and he contributed significantly to the development of anatomy as a modern science. Italian scientist Giovanni Alfonso Borelli, a student of Galileo, is often considered the father of biomechanics. In the mid-1600's, he developed mathematical models to describe anatomy and human movement mechanically. In the late 1600's, English physician and mathematician Sir Isaac Newton formulated mechanical principles and Newtonian laws of motion (inertia, acceleration, and reaction) that became the foundation of biomechanics.

British physiologist A. V. Hill, the 1923 winner of the Nobel Prize in Physiology or Medicine, conducted research to formulate mechanical and structural theories for muscle action. In the 1930's, American anatomy professor Herbert Elftman was able to quantify the internal forces in muscles and joints and developed the force plate to quantify ground reaction. A significant breakthrough in the understanding of muscle action was made by British physiologist Andrew F. Huxley in 1953, when he described his filament theory to explain muscle shortening. Russian physiologist Nicolas Bernstein published a paper in 1967 describing theories for motor coordination and control following his work studying locomotion patterns of children and adults in the Soviet Union.

HOW IT WORKS

The study of human movement is multifaceted, and biomechanics applies mechanical principles to the study of the structure and function of living things. Biomechanics is considered a relatively new field of applied science, and the research being done is of considerable interest to many other disciplines, including zoology, orthopedics, dentistry, physical education, forensics, cardiology, and a host of other medical specialties. Biomechanical analysis for each particular application is very specific; however, the basic principles are the same.

Newton's laws of motion. The development of scientific models reduces all things to their basic level to provide an understanding of how things work. This also allows scientists to predict how things will behave in response to forces and stimuli and ultimately to influence this behavior.

Newton's laws describe the conservation of energy and the state of equilibrium. Equilibrium results when the sum of forces is zero and no change occurs, and conservation of energy explains that energy cannot be created or destroyed, only converted from one form to another. Motion occurs in

two ways, linear motion in a particular direction or rotational movement around an axis. Biomechanics explores and quantifies the movement and production of force used or required to produce a desired objective.

Seven principles. Seven basic principles of biomechanics serve as the building blocks for analysis. These can be applied or modified to describe the reaction of forces to any living organism.

The lower the center of mass, the larger the base of support; the closer the center of mass to the base of support and the greater the mass, the more stability increases.

The production of maximum force requires the use of all possible joint movements that contribute to the task's objective.

The production of maximum velocity requires the

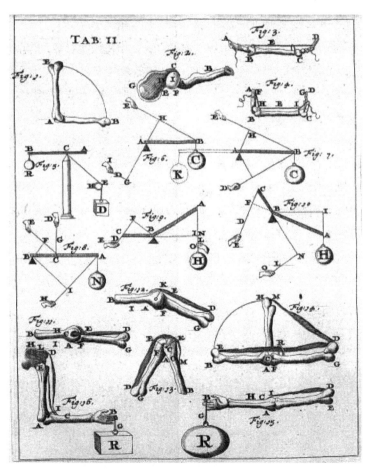

Page of one of the first works of Biomechanics (De Motu Animalium of Giovanni Alfonso Borelli) in the 17th century

---

use of joints in order, from largest to smallest.

The greater the applied impulse, the greater increase in velocity.

Movement usually occurs in the direction opposite that of the applied force.

Angular motion is produced by the application of force acting at some distance from an axis, that is, by torque.

Angular momentum is constant when an athlete or object is free in the air.

Static and dynamic forces play key roles in the complex biochemical and biophysical processes that underlie cell function. The mechanical behavior of individual cells is of interest for many different biologic processes. Single-cell mechanics, including growth, cell division, active motion, and contractile mechanisms, can be quite dynamic and provide insight into mechanisms of stress and damage of structures. Cell mechanics can be involved in processes that lie at the root of many diseases and may provide opportunities

as focal points for therapeutic interventions.

APPLICATIONS AND PRODUCTS

Biomechanics studies and quantifies the movement of all living things, from the cellular level to body systems and entire bodies, human and animal. There are many scientific and health disciplines, as well as industries that have applications developed from this knowledge. Research is ongoing in many areas to effectively develop treatment options for clinicians and better products and applications for industry.

Dentistry. Biomechanical principles are relevant in orthodontic and dental science to provide solutions to restore dental health, resolve jaw pain, and manage cosmetic and orthodontic issues. The design of dental implants must incorporate an analysis of load bearing and stress transfer while maintaining the integrity of surrounding tissue and comfortable function for the patient. This work has lead to the development of new materials in dental practices such as reinforced composites rather than metal frameworks.

Forensics. The field of forensic biomechanical analysis has been used to determine mechanisms of injury after traumatic events such as explosions in military situations. This understanding of how parts of the body behave in these events can be used to develop mitigation strategies that will reduce injuries. Accident and injury reconstruction using biomechanics is an emerging field with industrial and legal applications.

Biomechanical modeling. Biomechanical modeling is a tremendous research field, and it has potential uses across many health care applications. Modeling has resulted in recommendations for prosthetic design and modifications of existing devices. Deformable breast models have demonstrated capabilities for breast cancer diagnosis and treatment. Tremendous growth is occurring in many medical fields that are exploring the biomechanical relationships between organs and supporting structures. These models can assist with planning surgical and treatment interventions and reconstruction and determining optimal loading and boundary constraints during clinical procedures.

Materials. Materials used for medical and surgical procedures in humans and animals are being evaluated and some are being changed as biomechanical science is demonstrating that different materials, procedures, and techniques may be better for reducing complications and improving long-term patient health. Evaluation of the physical relationship between the body and foreign implements can quantify the stresses and forces on the body, allowing for more accurate prediction of patient outcomes and determination of which treatments should be redesigned.

Predictability. Medical professionals are particularly interested in the predictive value that biomechanical profiling can provide for their patients. An example is the unpredictability of expansion and rupture of an abdominal aortic aneurysm. Major progress has been made in determining aortic wall stress using finite element analysis. Improvements in biomechanical computational methodology and advances in imaging and processing technology have provided increased predictive ability for this life-threatening event.

As the need for accurate and efficient evaluation grows, so does the research and development of effective biomechanical tools. Capturing real-time, real-world data, such as with gait analysis and range of motion features, provides immediate opportunities for applications. This real-time data can quantify an injury and over time provide information about the extent that the injury has improved. High-tech devices can translate real-world situations and two-dimensional images into a three-dimensional framework for analysis. Devices, imaging, and modeling tools and software are making tremendous strides and becoming the heart of a highly competitive industry aimed at simplifying the process of analysis and making it less invasive.

IMPACT ON INDUSTRY

Many companies have discovered the benefit of biomechanics in various facets of their operations, including the development of products and of workplace procedures and practices. Most products that are made to assist or interact with people or any living being have probably been designed with the input of a biomechanical professional. Corporations want to protect their investment and profits by ensuring that their products will effectively meet the needs of the consumer and comply with strict safety standards. Biomechanics personnel work with product development engineers and designers to create new products and improve existing ones. Athletic equipment has

been redesigned to produce better results with the same exertion of force. Two major sports products that have received international attention and led to world-record-breaking performances have been clap skates (used in speed skating) and the LZR Racer swimsuit, designed to reduce drag on swimmers.

Biomechanics. Sporting equipment for athletes and the general public is constantly being redesigned to enhance performance and reduce the chance of injury. Sport-specific footwear is designed to maximize comfort and make it easier for athletes to perform the movements necessary for their sport. Equipment such as bicycles and golf clubs are designed using lighter and stronger materials, using optimal angles and maximizing strength to provide athletes with the best experience possible. Sports equipment goes a step further by analyzing the individual athlete and adjusting a piece of equipment specifically to his or her body. A small change to an angle or lever can produce dramatic results for an individual. This customization goes beyond sporting equipment to rehabilitation implements, wheelchairs, and prosthetic devices.

Biomechanics has a profound influence on many industries that are outside of sports. Most products used or handled on a daily basis have undergone biomechanical evaluation. Medicine bottle tops have been redesigned for easy opening by those with arthritic hands, and products from kitchen gadgets to automobiles have all been altered to improve comfort, safety, and effectiveness and to reduce the need for physical exertion.

Corporations are facing stricter regulations regarding the workplace environment. Safety and injury prevention are key to keeping productivity optimal. In a process called ergonomic assessment, equipment and workstations are biomechanically assessed, and adjustments are made that will limit acute or chronic injuries. Employees also receive instruction on the proper procedures to follow, such as lifting techniques. Industry procedures need to be in place and diligently followed to protect both employees and the company.

Accident litigation is becoming more common, and biomechanical science is playing a large role in accident re-creation and law-enforcement training. Investigations at accident or crime scenes can reveal more evidence, with greater accuracy than ever before, leaving less room for speculation when reconstructing the event.

CAREERS AND COURSE WORK

Careers in biomechanics can be dynamic and can take many paths. Graduates with accredited degrees may pursue careers in laboratories in universities or in private corporations researching and developing ways of improving and maximizing human performance. Beyond research, careers in biomechanics can involve working in a medical capacity in sport medicine and rehabilitation. Biomechanics experts may also seek careers in coaching, athlete development, and education.

Consulting and legal practices are increasingly seeking individuals with biomechanics expertise who are able to analyze injuries and reconstruct accidents involving vehicles, consumer products, and the environment.

Biomechanical engineers commonly work in industry, developing new products and prototypes and evaluating their performance. Positions normally require a biomechanics degree in addition to mechanical or biomedical engineering degrees.

Private corporations are employing individuals with biomechanical knowledge to perform employee fitness evaluations and to provide analyses of work environments and positions. Using these assessments, the biomechanics experts advise employers of any ergonomic changes or job modifications that will reduce the risk of workplace injury.

Individuals with a biomechanics background may chose to work in rehabilitation and prosthetic design. This is very challenging work, devising and modifying existing implements to maximize people's abilities and mobility. Most prosthetic devices are customized to meet the needs of the patient and to maximize the recipient's abilities. This is an ongoing process because over time the body and needs of a patient may change. This is particularly challenging in pediatrics, where adjustments become necessary as a child grows and develops.

SOCIAL CONTEXT AND FUTURE PROSPECTS

Biomechanics has gone from a narrow focus on athletic performance to become a broad-based science, driving multibillion dollar industries to satisfy the needs of consumers who have become more knowledgeable about the relationship between science, health, and athletic performance. Funding

for biomechanical research is increasingly available from national health promotion and injury prevention programs, governing bodies for sport, and business and industry. National athletic programs want to ensure that their athletes have the most advanced training methods, performance analysis methods, and equipment to maximize their athletes' performance at global competitions.

Much of the existing and developing technology is focused on increasingly automated and digitized systems to monitor and analyze movement and force. The physiological aspect of movement can be examined at a microscopic level, and instrumented athletic implements such as paddles or bicycle cranks allow real-time data to be collected during an event or performance. Force platforms are being reconfigured as starting blocks and diving platforms to measure reaction forces. These techniques for biomechanical performance analysis have led to revolutionary technique changes in many sports programs and rehabilitation methods.

Advances in biomechanical engineering have led to the development of innovations in equipment, playing surfaces, footwear, and clothing, allowing people to reduce injury and perform beyond previous expectations and records.

Computer modeling and virtual simulation training can provide athletes with realistic training opportunities, while their performance is analyzed and measured for improvement and injury prevention.

—April D. Ingram

FURTHER READING

Hamill, Joseph, and Kathleen Knutzen. *Biomechanical Basis of Human Movement.* 3d ed. Philadelphia: Lippincott, Williams & Wilkins, 2009. This introductory text integrates basic anatomy, physics, and physiology as it relates to human movement. It also includes real-life examples and clinically relevant material.

Hatze, H. "The Meaning of the Term 'Biomechanics.'" *Journal of Biomechanics* 7, no. 2 (March, 1974): 89-90. Contains Hatze's definition of biomechanics, then an emerging field.

Hay, James G. *The Biomechanics of Sports Techniques.* 4th ed. Englewood Cliffs, N.J.: Prentice Hall, 1993. A seminal work by an early biomechanics expert, first published in 1973.

Kerr, Andrew. *Introductory Biomechanics.* London: El-sevier, 2010. Provides a clear, basic understanding of major biomechanical principles in a workbook style interactive text.

Peterson, Donald, and Joseph Bronzino. *Biomechanics: Principles and Applications.* Boca Raton, Fla.: CRC Press, 2008. A broad collection of twenty articles on various aspects of research in biomechanics.

Watkins, James. *Introduction to Biomechanics of Sport and Exercise.* London: Elsevier, 2007. An introduction to the fundamental concepts of biomechanics that develops knowledge from the basics. Many applied examples, illustrations, and solutions are included.

WEB SITES

American Society of Biomechanics
http://www.asbweb.org

European Society of Biomechanics
http://www.esbiomech.org

International Society of Biomechanics
http://www.isbweb.org

World Commission of Science and Sports
http://www.wcss.org.uk

See also: Bioengineering; Biophysics.

BIONICS AND BIOMEDICAL ENGINEERING

FIELDS OF STUDY DEFINITION AND BASIC PRINCIPLES

Biology; physiology; biochemistry; engineering; orthopedic bioengineering; physics; bionanotechnology; biomechanics; biomaterials; neural engineering; genetic engineering; tissue engineering; prosthetics.

SUMMARY

Bionics combines natural biologic systems with engineered devices and electrical mechanisms. An example of bionics is an artificial arm controlled by impulses from the human mind. Construction of bionic arms or similar devices requires the integrative use of medical equipment such as electroencephalograms (EEGs) and magnetic resonance imaging (MRI) machines with mechanically engineered prosthetic arms and legs. Biomedical engineering further melds biomedical and engineering sciences by producing medical equipment, tissue growth, and new pharmaceuticals. An example of biomedical engineering is human insulin production through genetic engineering to treat diabetes.

KEY TERMS AND CONCEPTS

- **biologics:** Medicines produced from genes by manipulating genes and using genetic technology.

- **biomaterials:** Substances, including metal alloys, plastic polymers, and living tissues, used to replace body tissues or as implants.

- **bionanotechnology:** Construction of materials on a very small scale, enabling the use of microscopic machinery in living tissues.

- **bionic:** Integrating biological function and mechanical devices.

- **clone:** Genetically engineered organism with genetic composition identical to the original organism.

- **human genetic engineering:** Genetic engineering focused on altering or changing visible human characteristics through gene manipulations.

- **prosthesis:** Artificial or biomechanically engineered body part.

- **recombinant DNA:** DNA created by the combination of two or more DNA sequences that do not normally occur together.

The fields of biomedical engineering and bionics focus on improving health, particularly after injury or illness, with better rehabilitation, medications, innovative treatments, enhanced diagnostic tools, and preventive medicine.

Bionics has moved nineteenth-century prostheses, such as the wooden leg, into the twenty-first century by using plastic polymers and levers. Bionics integrates circuit boards and wires connecting the nervous system to the modular prosthetic limb.

Controlling artificial limb movements with thoughts provides more lifelike function and ability. This mind and prosthetic limb integration is the "bio" portion of bionics; the "nic" portion, taken from the word "electronic," concerns the mechanical engineering that makes it possible for the person using a bionic limb to increase the number and range of limb activity, approaching the function of a real limb.

Biomedical engineering encompasses many medical fields. The principle of adapting engineering techniques and knowledge to human structure and function is a key unifying concept of biomedical engineering. Advances in genetic engineering have produced remarkable bioengineered medications. Recombinant DNA techniques (genetic engineering) have produced synthetic hormones, such as insulin. Bacteria are used as a host for this process; once human-insulin-producing genes are implanted in the bacteria, the bacteria's DNA produce human insulin, and the human insulin is harvested to treat diabetics. Before this genetic technique was developed in 1982 to produce human insulin, insulin-dependent diabetics relied on insulin from pigs or cows. Although this insulin was life saving for diabetics, diabetics often developed problems from the pig or cow insulin because they would produce antibodies against the foreign insulin. This problem disappeared with the ability to engineer human insulin using recombinant DNA technology.

BACKGROUND AND HISTORY

In the broad sense, biomedical engineering has existed for millennia. Human beings have always envisioned the integration of humans and technology to increase and enhance human abilities. Prosthetic devices go back many thousands of years: A three-thousand-year-old Egyptian mummy, for example, was found with a wooden big toe tied to its foot. In the fifteenth century, during the Italian Renaissance, Leonardo da Vinci's elegant drawings demonstrated some early ideas on bioengineering, including his helicopter and flying machines, which melded human and machine into one functional unit capable of flight. Other early examples of biomedical engineering include wooden teeth, crutches, and medical equipment, such as stethoscopes.

Electrophysiological studies in the early 1800's produced biomedical engineering information used to better understand human physiology. Engineering principles related to electricity combined with human physiology resulted in better knowledge of the electrical properties of nerves and muscles.

X rays, discovered by Wilhelm Conrad Röntgen in 1895, were an unknown type of radiation (thus the "X" name). When it was accidentally discovered that they could penetrate and destroy tissue, experiments were developed that led to a range of imaging technologies that evolved over the next century. The first formal biomedical engineering training program, established in 1921 at Germany's Oswalt Institute for Physics in Medicine, focused on three main areas: the effects of ionizing radiation, tissue electrical characteristics, and X-ray properties.

In 1948, the Institute of Radio Engineers (later the Institute of Electrical and Electronics Engineers), the American Institute for Electrical Engineering, and the Instrument Society of America held a conference on engineering in biology and medicine. The 1940's and 1950's saw the formation of professional societies related to biomedical engineering, such as the Biophysics Society, and of interest groups within engineering societies. However, research at the time focused on the study of radiation. Electronics and the budding computer era broadened

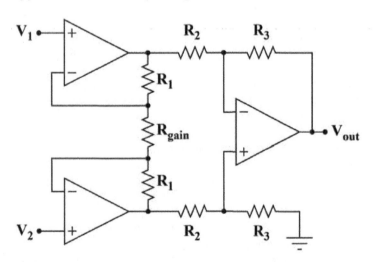

Biomedical instrumentation amplifier schematic used in monitoring low voltage biological signals, an example of a biomedical engineering application of electronic engineering to electrophysiology.

interest and activities toward the end of the 1950's.

James D. Watson and Francis Crick identified the DNA double-helix structure in 1953. This important discovery fostered subsequent experimentation in molecular biology that yielded important information about how DNA and genes code for the expression of traits in all living organisms. The genetic code in DNA was deciphered in 1968, arming researchers with enough information to discover ways that DNA could be recombined to introduce genes from one organism into a different organism, thereby allowing the host to produce a variety of useful products. DNA

recombination became one of the most important tools in the field of biomedical engineering, leading to tissue growth as well as new pharmaceuticals.

In 1962, the National Institutes of Health created the National Institute of General Medical Sciences, fostering the development of biomedical engineering programs. This institute funds research in the diagnosis, treatment, and prevention of disease.

Bionics and biomedical engineering span a wide variety of beneficial health-related fields. The common thread is the combination of technology with human applications. Dolly the sheep was cloned in 1996. Cloning produces a genetically identical copy of an existing life-form. Human embryonic cloning presents the potential of therapeutic reproduction of needed organs and tissues, such as kidney replacement for patients with renal failure.

In the twenty-first century, the linking of machines with the mind and sensory perception has provided hearing for deaf people, some sight for the blind, and willful control of prostheses for amputees.

How it Works

Restorative bionics integrates prosthetic limbs with electrical connections to neurons, allowing an individual's thoughts to control the artificial limb. Tiny arrays of electrodes attached to the eye's retina connect to the optic nerve, enabling some visual perception for previously blind people. Deaf people hear with electric devices that send signals to auditory nerves, using antennas, magnets, receivers, and electrodes. Researchers are considering bionic skin development using nanotechnology to connect with nerves, enabling skin sensations for burn victims requiring extensive grafting.

Many biomedical devices work inside the human body. Pacemakers, artificial heart valves, stents, and even artificial hearts are some of the bionic devices correcting problems with the cardiovascular system. Pacemakers generate electric signals that improve abnormal heart rates and abnormal heart rhythms. When pulse generators located in the pacemakers sense an abnormal heart rate or rhythm, they produce shocks to restore the normal rate. Stents are inserted into an artery to widen it and open clogged blood vessels. Stents and pacemakers are examples of specialized bionic devices made up of bionic materials compatible with human structure and function.

Cloning. Cloning is a significant area of genetic

FASCINATING FACTS ABOUT BIONICS AND BIOMEDICAL ENGINEERING

- In 2010, scientists at the University of California, San Diego, developed biosensor cells that can be implanted in the brain to help monitor receptors and chemical signals that allow cells in the brain to communicate with one another. These cells may help scientists understand drug addiction.
- Vanderbilt engineers in 2010 began testing a knowledge repository and interactive software that will help surgeons more accurately and rapidly place electrodes in the brains of people with Parkinson's disease in a procedure called deep brain stimulation. The data allow for faster surgery and the implementation of best practices.
- In 2010, scientists at Vanderbilt University developed a robotic prosthesis for the lower leg that has powered knee and ankle joints. Intent recognizer software takes information from sensors, determines what the user wants to do, and provides power to the leg.
- The Wadsworth Center at the New York State Department of Health in 2009 developed a brain-computer interface that translates brain waves into action. It allowed a patient with amyotrophic lateral sclerosis who was no longer able to communicate with others because of failing muscles to write e-mails and convey his thoughts to others.
- In 2008, a research team at the Johns Hopkins University modified chondroitin sulfate, a natural sugar, so it could glue a hydrogel (like the material used in soft contact lenses) to cartilage tissue. It is hoped that this technique may help those experiencing joint pain from oseoarthritis, in which the natural cartilage in a joint disappears.

engineering that allows the replication of a complete living organism by manipulating genes. Dolly the sheep, an all-white Finn Dorset ewe, was cloned from a surrogate mother blackface ewe, which was used as an egg donor and carried the cloned Dolly during gestation (pregnancy). An egg cell from the surrogate was removed and its nucleus (which contains DNA) was replaced with one from a Finn Dorset ewe; the resulting new egg was placed in the blackface ewe's uterus after stimulation with an electric pulse. The electrical pulse stimulated growth and cell duplication. The blackface ewe subsequently gave birth to the all-white Dolly. The newborn all-white Finn Dorset ewe was an identical genetic twin of the Finn Dorset that contributed the new nucleus.

Recombinant DNA. Another significant genetic engineering technique involves recombinant DNA. Human genes transferred to host organisms, such as bacteria, produce products coded for by the transferred genes. Human insulin and human growth hormone can be produced using this technique. Desired genes are removed from human cells and placed in circular bacterial DNA strips called plasmids. Scientists use enzymes to prepare these DNA formulations, ultimately splicing human genes into bacterial plasmids. These plasmids are used as vectors, taken up and reproduced by bacteria. This type of genetic adaptation results in insulin production if the spliced genes were taken from the part of the human genome producing insulin; other cells and substances, coded for by different human genes, can be produced this way. Many biologic medicines are produced using recombinant DNA technology.

APPLICATIONS AND PRODUCTS

Medical devices. Biomedical engineers produce life-saving medical equipment, including pacemakers, kidney dialysis machines, and artificial hearts. Synthetic limbs, artificial cochleas, and bionic sight chips are among the prosthetic devices that biomedical engineers have developed to enhance mobility, hearing, and vision. Medical monitoring devices, developed by biomedical engineers for use in intensive care units and surgery or by space and deep-sea explorers, monitor vital signs such as heart rate and rhythm, body temperature, and breathing rate.

Equipment and machinery. Biomedical engineers produce a wide variety of other medical machinery, including laboratory equipment and therapeutic equipment. Therapeutic equipment includes laser devices for eye surgery and insulin pumps (sometimes called artificial pancreases) that both monitor blood sugar levels and deliver the appropriate amount of insulin when it is needed.

Imaging systems. Medical imaging provides important machinery devised by biomedical engineers. This specialty incorporates sophisticated computers and imaging systems to produce computed tomography (CT), magnetic resonance imaging (MRI), and positron emission tomography (PET) scans. In naming its National Institute of Biomedical Imaging and Bioengineering (NIBIB), the U.S. Department of Health and Human Services emphasized the equal importance and close relatedness of these subspecialties by using both terms in the department's name.

Computer programming provides important circuitry for many biomedical engineering applications, including systems for differential disease diagnosis. Advances in bionics, moreover, rely heavily on computer systems to enhance vision, hearing, and body movements.

Biomaterials. Biomaterials, such as artificial skin and other genetically engineered body tissues, are areas promising dramatic improvements in the treatment of burn victims and individuals needing organ transplants. Bionanotechnology, another subfield of biomedical engineering, promises to enhance the surface of artificial skin by creating microscopic messengers that can create the sensations of touch and pain. Bioengineers interface with the fields of physical therapy, orthopedic surgery, and rehabilitative medicine in the fields of splint development, biomechanics, and wound healing.

Medications. Medicines have long been synthesized artificially in laboratories, but chemically synthesized medicines do not use human genes in their production. Medicines produced by using human genes in recombinant DNA procedures are called biologics and include antibodies, hormones, and cell receptor proteins. Some of these products include human insulin, the hepatitis B vaccine, and human growth hormone.

Bacteria and viruses invading a body are attacked and sometimes neutralized by antibodies produced by the immune system. Diseases such as Crohn's disease, an inflammatory bowel condition, and psoriatic arthritis are conditions exacerbated by inflammatory antibody responses mounted by the affected person's immune system. Genetic antibody production in the form of biologic medications interferes with or attacks mediators associated with Crohn's and arthritis and improves these illnesses by decreasing the severity of attacks or decreasing the frequency of flare-ups.

Cloning and stem cells. Cloned human embryos could provide embryonic stem cells. Embryonic stem cells have the potential to grow into a variety of cells, tissues, and organs, such as skin, kidneys, livers, or heart cells. Organ transplantation from genetically identical clones would not encounter the recipient's natural rejection process, which transplantations must overcome. As a result, recipients of genetically identical cells, tissues, and organs would enjoy more successful replacements of key organs and a better quality of life. Human cloning is subject to future research and development, but the promise of genetically identical replacement organs for people with failed hearts, kidneys, livers, or other organs provides hope for enhanced future treatments.

IMPACT ON INDUSTRY

Government and university research. The National Institute of Biomedical Imaging and Bioengineering was established in 2001. It is dedicated to health improvement by developing and applying biomedical technologies. The institute supports research with grants, contracts, and career development awards. Many major universities around the world actively research and develop bionics and biomedical engineering. Top engineering schools, such as the University of Michigan, have undergraduate and graduate degree programs for biomedical engineering. Biomedical engineers work in industries producing medical equipment and pharmaceutical companies. They also conduct research and develop genetic therapies useful for treating a wide variety of illnesses.

Industry and business sectors. The pharmaceutical industry, along with the medical equipment industry, employs biomedical engineers. The medical industry, including the suppliers of laboratory equipment, imaging technology, bionics, and pharmaceuticals, drives much of the employment demand for biomedical engineers. Medical hardware, such as artificial hearts, pacemakers, and renal dialysis machines, remains an important part of biomedical engineering employment.

CAREERS AND COURSE WORK

According to the 2010-2011 edition of the *Occupational*

Outlook Handbook, issued by the U.S. Department of Labor's Bureau of Labor Statistics, biomedical engineers are projected to have a 72 percent employment growth from 2010 to 2020, a rate of increase much greater than the average for all occupations. Biomedical engineers are employed as researchers and scientists, interfacing with a wide variety of disciplines and specialties.

Many career paths exist in biomedical engineering and bionics. Jobs exist for those holding bachelor's degrees through doctorates. Research scientists, usually with Ph.D. or M.D. degrees, can work in a variety of environments, from private companies to universities to government agencies, including the National Institutes of Health.

A typical undergraduate biomedical engineering curriculum for University of Michigan students, for example, includes three divisions of course study: subjects required by all engineering programs, advanced science and engineering mathematics, and biomedical engineering courses. All engineering students take courses in calculus, basic engineering concepts, computing, chemistry, physics, and humanities or social sciences. Required courses in advanced sciences in the second division include biology, chemistry, and biological chemistry, along with engineering mathematics courses. Biomedical engineering courses in the third division cover circuits and electrical systems, biomechanics, engineering materials, cell biology, physiology, and biomedical design. Students can concentrate in areas such as biomechanical, biochemical, and bioelectric engineering, with some modifications in course selection. The breadth and depth of the course work emphasize the link between engineering and biologic sciences—the essence of bionics and biomedical engineering.

A biomedical engineer is an engineer developing and advancing biomedical products and systems. This type of activity spans many specialties and presents many career opportunities. Biomedical engineering interfaces with almost every engineering discipline because bioengineered products require the breadth and depth of engineering knowledge. Major areas include the production of biomaterials, such as the type of plastic used for prosthetic devices. Bioinstrumentation involves computer integration in diagnostic machines and the control of devices. Genetic engineering presents many opportunities for life modifications. Clinical engineers help integrate new technologies, such as electronic medical records, into existing hospital systems. Medical imaging relieves the need for exploratory surgery and greatly enhances diagnostic capabilities. Orthopedic bioengineering plays important roles in prosthesis development and use, along with assisting rehabilitative medicine. Bionanotechnology offers the hope of using microscopic messengers to treat illness and advanced capabilities for artificial bionic devices. Systems physiology organizes the multidisciplinary approach necessary to complete complex bionic projects such as a functioning artificial eye.

SOCIAL CONTEXT AND FUTURE PROSPECTS

Bionics technologies include artificial hearing, sight, and limbs that respond to nerve impulses. Bionics offers partial vision to the blind and prototype prosthetic arm devices that offer several movements through nerve impulses. The goal of bionics is to better integrate the materials in these artificial devices with human physiology to improve the lives of those with limb loss, blindness, or decreased hearing.

Cloned animals exist but cloning is not a yet a routine process. Technological advances offer rapid DNA analysis along with significantly lower cost genetic analysis. Genetic databases are filled with information on many life-forms, and new DNA sequencing information is added frequently. This basic information that has been collected is like a dictionary, full of words that can be used to form sentences, paragraphs, articles, and books, in that it can be used to create new or modified life-forms.

Biomedical engineering enables human genetic engineering. The stuff of life, genes, can be modified or manipulated with existing genetic techniques. The power to change life raises significant societal concerns and ethical issues. Beneficial results such as optimal organ transplantations and effective medications are the potential of human genetic engineering.

—*Richard P. Capriccioso, MD, and Christina Capriccioso*

FURTHER READING

Braga, Newton C. *Bionics for the Evil Genius: Twenty-five Build-It-Yourself Projects.* New York: McGraw-Hill, 2006. Step-by-step projects that introduce basic concepts in bionics.

Fischman, Josh. "Merging Man and Machine: The Bionic Age." *National Geographic* 217, no. 1 (January, 2010): 34-53. A well-illustrated consideration of the latest advances in bionics, with specific examples of people aided by the most modern prosthetic technologies.

Hung, George K. *Biomedical Engineering: Principles of the Bionic Man.* Hackensack, N.J.: World Scientific, 2010. Examines scientific bioengineering principles as they apply to humans.

Bionics and Biomedical Engineering

Richards-Kortum, Rebecca. *Biomedical Engineering for Global Health.* New York: Cambridge University Press, 2010. Examines the potential of biomedical engineering to treat diseases and conditions throughout the world. Examines health care systems and social issues.

Smith, Marquard, and Joanne Morra, eds. *The Prosthetic Impulse: From a Posthuman Present to a Biocultural Future.* Cambridge, Mass.: MIT Press, 2007. Examines the developments in prosthetic devices and addresses the social aspects, including what it means to be human.

WEB SITES

American Society for Artificial Internal Organs
http://www.asaio.com

Biomedical Engineering Society
http://www.bmes.org

National Institute of Biomedical Imaging and Bioengineering
http://www.nibib.nih.gov/HomePage

Rehabilitation Engineering and Assistive Technology Association of North America
http://resna.org

Society for Biomaterials
http://www.biomaterials.org

BIOPHYSICS

FIELDS OF STUDY

Physics; physical sciences; chemistry; mathematics; biology; molecular biology; chemical biology; engineering; biochemistry; classical genetics; molecular genetics; cell biology.

SUMMARY

Biophysics is the branch of science that uses the principles of physics to study biological concepts. It examines how life systems function, especially at the cellular and molecular level. It plays an important role in understanding the structure and function of proteins and membranes and in developing new pharmaceuticals. Biophysics is the foundation for molecular biology, a field that combines physics, biology, and chemistry.

KEY TERMS AND CONCEPTS

- **circular dichroism (CD):** Differential absorption of left- and right-handed circularly polarized light.
- **electromagnetic waves:** Waves that can transmit their energy through a vacuum.
- **molecular genetics:** Branch of genetics that analyzes the structure and function of genes at the molecular level.
- **polarized light:** Light waves that vibrate in a single plane.
- **quantum mechanics:** Physical analysis at the level of atoms or subatomic fundamental particles.
- **thermodynamics:** Branch of physics that studies energy conversions.
- **vector:** Quantity that has both magnitude and direction.

DEFINITION AND BASIC PRINCIPLES

The word "biophysics" means the physics of life. Biophysics studies the functioning of life systems, especially at the cellular and molecular level, using the principles of physics. It is known that atoms make up molecules, molecules make up cells, and cells in turn make up tissues and organs that are part of an organism, or a living machine. Biophysicists use this knowledge to understand how the living machine works.

In photosynthesis, for instance, the absorption of sunlight by green plants initiates a process that culminates with synthesis of high-energy sugars such as glucose. To fully understand this process, one needs to look at how it begins—light absorption by the photo-systems. Photosystems are groups of energy-absorbing pigments such as chlorophyll and carotenoids that are located on the thylakoid membranes inside the chloroplast, the photosynthetic organelle in the plant cell. Biophysical studies have shown that once a chlorophyll molecule captures solar energy, it gets excited and transfers the energy to a neighboring unexcited chlorophyll molecule. The process repeats itself, and thus, packets of energy jump from one chlorophyll molecule to the next. The energy eventually reaches the reaction center, where it begins a chain of high-energy electron-transfer reactions that lead to the storage of the light energy in the form of adenosine triphosphate (ATP) and nicotinamide adenine dinucleotide phosphate (NADPH). In the second half of photosynthesis, ATP and NADPH provide the energy to make glucose from carbon dioxide.

Biophysics is often confused with medical physics. Medical physics is the science devoted to studying the relationship between human health and radiation exposure. For example, a medical physicist often works closely with a radiation oncologist to set up radiotherapy treatment plans for cancer patients.

BACKGROUND AND HISTORY

In comparison with other branches of biology and physics, biophysics is relatively new and, therefore, still evolving. Even though the use of physical concepts and instrumentation to explain the workings of life systems had begun as early as the 1840's, biophysics did not emerge as an independent field until the 1920's. Some of the earliest studies in biophysics were conducted in the 1840's by a group known as the Berlin school of physiologists. Among its members were pioneers such as Hermann von Helmholtz, Ernst Heinrich Weber, Carl F. W. Ludwig, and Johannes Peter Müller. This group used well-known physical methods to investigate physiological issues, such as the mechanics of muscular contraction and the electrical changes in a nerve cell during impulse transmission. The first biophysics textbook was written in 1856 by Adolf Fick, a student of Ludwig. Although these early biophysicists made significant advances, subsequent research focused on other areas.

In the 1920's, the first biophysical institutes were established in Germany and the first textbook with the word "biophysics" in its title was published. However, through the 1940's, biophysics research was primarily aimed at understanding the biophysical impact of ionizing radiation. In 1944, Austrian physicist Erwin Schrödinger published *What Is Life ? The Physical Aspect of the Living Cell*, based on a series of lectures that addressed biology from the viewpoint of a classical physicist. This cross-disciplinary work motivated several physicists to become interested in biology and thus laid the foundation for the field of molecular biology. From 1950 to 1970, the field of biophysics experienced rapid growth, tremendously accelerated by the discovery in 1953 of the double helix structure of DNA by James D. Watson and Francis Crick. Both Watson and Crick have stated that they were inspired by Schrödinger's work.

How it Works

Biophysicists study life at all levels, from atoms and molecules to cells, organisms, and environments. They attempt to describe complex living systems with the simple laws of physics. Often, biophysicists work at the molecular level to understand cells and their processes.

The work of Gregor Mendel in the late nineteenth century laid the foundation for genetics, the science of heredity. His studies, rediscovered in the twentieth century, led to the understanding that the inheritance of certain traits is governed by genes and that the alleles of the genes are separated during gamete formation. Experiments in the 1940's revealed that genes are made of DNA, but the mechanisms by which genes function remained a mystery. Watson and Crick's discovery of the double helix structure of DNA in 1953 revealed how genes could be translated into proteins.

Biophysicists use a number of physical tools and techniques to understand how cellular processes work, especially at the molecular level. Some of the important tools are electron microscopy, nuclear magnetic resonance (NMR) spectroscopy, circular

dichroism (CD) spectroscopy, and X-ray crystallography. For example, the discovery of Watson and Crick's double helix model was possible in part because of the X-ray images of DNA that were taken by Rosalind Franklin and Maurice H. F. Wilkins. Franklin and Wilkins, both biophysicists, made DNA crystals and then used X-ray crystallography to analyze the structure of DNA. The array of black dots arranged in an X-shaped pattern on the X-ray photograph of wet DNA suggested to Franklin that DNA was helical.

Electron microscopy. Electron microscopes use beams of electrons to study objects in detail. Electron microscopy can be used to analyze an object's surface texture (topography) and constituent elements and compounds (composition), as well as the shape and size (morphology) and atomic arrangements (crystallographic details) of those elements and compounds. Electron microscopes were invented to overcome the limitations posed by light microscopes, which have maximum magnifications of 500x or 1000x and a maximum resolution of 0.2 millimeter. To see and study subcellular structures and processes required magnification capabilities of greater than 10,000x. The first electron microscope was a transmission electron microscope (TEM) built by Max Knoll and Ernst Ruska in 1931. The invention of the scanning electron microscope (SEM) was somewhat delayed (the first was built in 1937 by Manfred von Ardenne) because the field had to figure out how to make the electron beam scan the sample.

NMR spectroscopy. Nuclear magnetic resonance (NMR) spectroscopy is an extremely useful tool for the biophysicist to study the molecular structure of organic compounds. The underlying principle of NMR spectroscopy is identical to that of magnetic resonance imaging (MRI), a common tool in medical diagnostics. The nuclei of several elements, including the isotopes carbon-12 and oxygen-16, have a characteristic spin when placed in an external magnetic field. NMR focuses on studying the transitions between these spin states. In comparison with mass spectroscopy, NMR requires a larger amount of sample, but it does not destroy the sample.

CD spectroscopy. Circular dichroism (CD) spectroscopy measures differences in how left-handed and right-handed polarized light is absorbed. These differences are caused by structural asymmetry. CD spectroscopy can determine the secondary and tertiary structure of proteins as well as their thermal stability. It is usually used to study proteins in solution.

X-ray crystallography. X rays are electromagnetic waves with wavelengths ranging from 0.02 to 100 angstroms (Å). Even before X rays were discovered in 1895 by Wilhelm Conrad Röntgen, scientists knew that atoms in crystals were arranged in definite patterns and that a study of the angles therein could provide clues to the crystal structure. As is true of all forms of radiation, the wavelength of X rays is inversely proportional to its energy. Because the wavelength of X rays is smaller than that of visible light, X rays are powerful enough to penetrate most matter. As X rays travel through an object, they are diffracted by the atomic arrangements inside and thus provide a guideline for the electron densities inside the object. Analysis of this electron density data offers a glimpse into the internal structure of the crystal. As of 2010, about 90 percent of the structures in the Worldwide Protein Data Bank had been elucidated through X-ray crystallography.

APPLICATIONS AND PRODUCTS

Biophysical tools and techniques have become extremely useful in many areas and fields. They have furthered research in protein crystallography, synthetic biology, and nanobiology, and allowed scientists to discover new pharmaceuticals and to study biomolecular structures and interactions and membrane structure and transport. Biophysics and its related fields, molecular biology and genetics, are rapidly developing and are at the center of biomedical research.

Biomolecular structures. Because structure dictates function in the world of biomolecules, understanding the structure of the biomolecule (with tools such as X-ray crystallography and NMR and CD spectroscopy), whether it is a protein or a nucleic acid, is critical to understanding its individual function in the cell. Proteins function as catalysts and bind to and regulate other downstream biomolecules. Their functional basis lies in their tertiary structure, or their three-dimensional form, and this function cannot be predicted from the gene sequence. The sequence of nucleotides in a gene can be used only to predict the primary structure, which is the amino acid sequence in the polypeptide. Once the structure-function relationship has been analyzed, the next step is to make mutants or knock out the gene via techniques such

as ribonucleic acid interference (RNAi) and confirm loss of function. Subsequently, a literature search is performed to see if there are any known genetic disorders that are caused by a defect in the gene being studied. If so, the structure-function relationship can be examined for a possible cure.

Membrane structure and transport. In 1972, biologist S. J. Singer and his student Garth Nicolson conceived the fluid mosaic model of the plasma membrane. According to this model, the plasma membrane is a fluid lipid bilayer largely made up of phospholipids arranged in an amphipathic pattern, with the hydrophobic lipid tails buried inside and the hydrophilic phosphate groups on the exterior. The bilayer is interspersed with proteins, which help in cross-membrane transport. Because membranes control the import and export of materials into the cell, understanding membrane structure is key to coming up with ways to block transport of potentially harmful pathogens across the membrane.

Electron microscopes were used in the early days of membrane biology, but fluorescence and confocal microscopes have come to be used more frequently. The development of organelle-specific vital stains has rejuvenated interest in evanescent field (EF) microscopy because it permits the study of even the smallest of vesicles and the tracking of the movements of individual protein molecules.

Synthetic biology. In the 2000's, the term "systems biology" became part of the field of life science, followed by the term "synthetic biology." To many people, these terms appear to refer to the same thing, but they do not, even though they are indeed closely related. While systems biology focuses on using a quantitative approach to study existing biological systems, synthetic biology concentrates on applying engineering principles to biology and constructing novel systems heretofore unseen in nature. Clearly, synthetic biology benefits immensely from research in systems and molecular biology. In essence, synthetic biology could be described as an engineering discipline that uses known, tested functional components (parts) such as genes, proteins, and various regulatory circuits in conjunction with modeling software to design new functional biological systems, such as bacteria that make ethanol from water, carbon dioxide, and light. The biggest challenge to synthetic biologists is the complexity of lifeforms, especially higher eukaryotes such as humans,

FASCINATING FACTS ABOUT BIOPHYSICS

- The concept of biophysics was first developed by ancient Greeks and Romans who were trying to analyze the basis of consciousness and perception.
- Human sight begins when the protein rhodopsin absorbs a unit of light called the quanta. This energy absorption triggers an enzymatic cascade that culminates in an amplified electric signal to the brain and enables vision.
- Crystallin, the lens protein, is made only in the lens of the human eye, and melanin, the skin pigment, is made only in skin cells, or melanocytes, even though all the cells in the human body have the genes to make crystallin and melanin.
- A technique called footprinting allows scientists to determine exactly where a protein binds on DNA and how much of the protein actually binds.
- Genes in human cells are selectively turned on and off by proteins called regulators. A defect in this regulatory mechanism can cause diseases such as cancer.

and the possible existence of unknown processes that can affect the synthetic biological systems.

Drug discovery. In the pharmaceutical world, the initial task is to identify the aberrant protein, the one responsible for generating the symptoms in any disease or disorder. Once that is done, a series of biophysical tools are used to ensure that the target is the correct one. First, the identity of the protein is confirmed using techniques such as N-terminal sequencing and tandem mass spectroscopy (MS-MS). Second, the protein sample is tested for purity (which typically should be more than 95 percent) using methods such as denaturing sodium dodecyl sulfate polyacrylamide gel electrophoresis (SDS-PAGE). Third, the concentration of the protein sample is determined by chromogenic assays such as the Bradford or Lowry assay. The fourth and probably the most important test is that of protein functionality. This is typically carried out by either checking the ligand binding capacity of the protein (with biacore ligand binding assays) or by testing the ability of the protein to carry out its biological function. All these thermodynamic parameters need to be tested to develop a putative drug, one that could somehow correct or restrain the ramifications of the

protein's malfunction.

Nanobiology. With the aid of biophysical tools and techniques, the field of biology has moved from organismic biology to molecular biology to nanobiology. To get a feel for the size of a nanometer, picture a strand of hair, then visualize a width that is 100,000 times thinner. Typically nanoparticles are about the size of either a protein molecule or a short DNA segment. Nanomedicine, or the application of technology that relies on nanoparticles to medicine, has become a popular area for research. In particular, the search for appropriate vectors to deliver drugs into the cells is an endless pursuit, especially in emerging therapeutic approaches such as RNA interference. Because lipid and polymer-based nanoparticles are extremely small, they are easily taken up by cells instead of being cleared by the body.

IMPACT ON INDUSTRY

As one would expect, biophysics research worldwide has progressed faster in countries that traditionally have had a large base of physicists. The Max Planck Institute of Biophysics, one of the earliest pioneers in this field, was established in 1921 in Frankfurt, Germany, as the Institut für Physikalische Grundlagen der Medizin. The aim of the first director director, Friedrich Dessauer, was to look for ways to apply the knowledge of radiation physics to medicine. He was followed by Boris Rajewsky, who coined the term "biophysics." In 1937, Rajewsky established the Kaiser Wilhelm Institute for Biophysics, which incorporated the institute led by Dessauer. The Max Planck Institute of Biophysics has become one of the world's foremost biophysics research institutes, with scientists and students analyzing a wide array of topics in biophysics such as membrane biology, molecular neurogenetics, and structural biology. In addition to Germany, countries active in biophysics research include the United States, Japan, France, Great Britain, Russia, China, India, and Sweden.

The International Union of Pure and Applied Biophysics (IUPAB) was created to provide a platform for biophysicists worldwide to exchange ideas and set up collaborations. The IUPAB in turn is a part of the International Council for Science (ICSU). The primary goal of IUPAB is to encourage and support students and researchers so that the field continues to grow and flourish. By 2010, the national biophysics societies of about fifty countries had become affiliated with the IUPAB. In the United States, the Biophysical Society was created in 1957 to facilitate propagation of biophysics concepts and ideas.

Government and university research. To further broaden its mission, the Biophysical Society includes members from universities as well those in government research agencies such as the National Institutes for Health (NIH)and National Institute of Standards and Technology (NIST), many of whom are also part of the American Institute for Advancement of Sciences (AAAS) and the National Science Foundation (NSF). These members provide useful feedback to federal agencies such as the National Science and Technology Council, National Science Board, and the White House's Office of Science and Technology Policy, which are responsible for formulating national policies and initiatives.

Industry and business. Most countries at the forefront of pharmaceutical breakthroughs have industries that are heavily invested in biophysical and biochemical research. The pharmaceutical industry has been spending billions of dollars to find treatments and cures for diseases and disorders that affect millions of people, including stroke, arthritis, cancer, heart disease, and neurological disorders. Biophysics is at the forefront of the field of drug discovery because it provides the tools for conducting research in proteomics and genomics and allows the scientific community to identify opportunities for drug design.

Karl Pearson, who first coined the term biophysics, at work in 1910.

The next step after drug design is to plan the method of drug delivery, and biophysical research can help provide suitable vectors, including nanovectors. The pharmaceutical industry in the United States— companies such as Novartis, Eli Lilly, Bristol-Myers Squibb, and Pfizer—and the National Institutes of Health have a combined budget of about $60 billion per year. However, neither the industry nor the government support basic research at the interface of life sciences with physics and mathematics. Without this support, biophysics is unlikely to produce new tools to revolutionize or accelerate the ten-to-twelve-year drug development cycle. If this impediment can be overcome, the number of new drugs added to the market every year is likely to grow.

CAREERS AND COURSE WORK
Few universities offer undergraduate majors in biophysics, but several universities offer graduate programs in biophysics. Students interested in pursuing a career in biophysics can major in molecular biology, physics, mathematics, or chemistry and supplement that with courses outside their major but relevant to biophysics. For example, a mathematics major would take supplementary courses in biology and vice versa. An ideal undergraduate curriculum for the future biophysicist would include courses in biology (genetics, cell biology, molecular biology), physics (thermodynamics, radiation physics, quantum mechanics), chemistry (organic, inorganic, and analytical), and mathematics (calculus, differential equations, computer programming, and statistics). The student should also have hands-on research experience, preferably as a research intern in a laboratory. To become an independent biophysicist, a graduate degree is required, usually a doctorate in biophysics, although some combine that with a medical degree. While in graduate school, the student should determine an area of research to pursue and engage in postdoctoral research for several years. Typically, this is the last step before one becomes an independent biophysicist running his or her own laboratory.

SOCIAL CONTEXT AND FUTURE PROSPECTS
The discovery of the structure of DNA set off a revolution in molecular biology that has continued into the twenty-first century.

131

In addition, modern scientific equipment has made study at the molecular level possible and productive. Many biophysicists, especially those who have also had course work in genetics and biochemistry, are working in molecular biology, which promises to be an active and exciting area for the foreseeable future.

Organisms are believed to be complex machines made of many simpler machines, such as proteins and nucleic acids. To understand why an organism behaves or reacts a certain way, one must determine how proteins and nucleic acids function. Biophysicists examine the structure of proteins and nucleic acids, seeking a correlation between structure and function. Once proper function is understood, scientists can prevent or treat diseases or disorders that result from malfunctions. This understanding of how proteins function enables scientists to develop pharmaceuticals and to find better means of delivering drugs to patients, and someday, this knowledge may allow scientists to design drugs specifically for a patient, thus avoiding many side effects. In addition, the scientific equipment developed by biophysicists in their research has been adapted for use in medical imaging for diagnosis and treatment. This transformation of laboratory equipment to medical equipment is likely to continue.

Biophysics applications have played and will continue to play a large role in medicine and health care, but future biophysicists may be environmental scientists. Biophysics is providing ways to improve the environment. For example, scientists are modifying microorganisms so that they produce electricity and biofuels that may lessen the need for fossil fuels. They are also using microorganisms to clean polluted water. As biophysics research continues, its applications are likely to cover an even broader range.

—*Sibani Sengupta, MS, PhD*

FURTHER READING

Bischof, Marco. "Some Remarks on the History of Biophysics and Its Future." In Current Development of Biophysics, edited by Changlin Zhang, Fritz Albert Popp, and Marco Bischof. Hangzhou, China: Hangzhou University Press, 1996. This paper delivered at a 1995 symposium on biophysics in Neuss, Germany, examines how the field of biophysics got its start and predicts future developments.

Claycomb, James R., and Jonathan Quoc P. Tran. Introductory Biophysics: Perspectives on the Living State. Sudbury, Mass.: Jones and Bartlett, 2011. This textbook considers life in relation to the universe. Contains a compact disc that allows computer simulation of biophysical phenomena. Relates biophysics to many other fields and subjects, including fractal geometry, chaos systems, biomagnetism, bioenergetics, and nerve conduction.

Glaser, Roland. Biophysics. 5th ed. New York: Springer, 2005. Contains numerous chapters on the molecular structure, kinetics, energetics, and dynamics of biological systems. Also looks at the physical environment, with chapters on the biophysics of hearing and on the biological effects of electromagnetic fields.

Goldfarb, Daniel. Biophysics Demystified. Maidenhead, England: McGraw-Hill, 2010. Examines anatomical, cellular, and subcellular biophysics as well as tools and techniques used in the field. Designed as a self-teaching tool, this work contains ample examples, illustrations, and quizzes.

Herman, Irving P. Physics of the Human Body. New York: Springer, 2007. Analyzes how physical concepts apply to human body functions.

Kaneko, K. Life: An Introduction to Complex Systems Biology. New York: Springer, 2006. Provides an introduction to the field of systems biology, focusing on complex systems.

WEB SITES

Biophysical Society
http://www.biophysics.org

International Union of Pure and Applied Biophysics
http://iupab.org

Worldwide Protein Data Bank
http://www.wwpdb.org

See also: Applied Physics; Biofuels and Synthetic Fuels; Magnetic Resonance Imaging.

BOSE CONDENSATES

FIELDS OF STUDY

Quantum physics; Particle physics

SUMMARY

Bose condensates are a unique form of matter that is formed when individual particles exist in the same quantum state and 'condense' into a macroscopic structure having the same quantum state. First theorized by Satyendra Nath Bose in 1924, they have since been observed in many materials at temperatures near absolute zero. They are believed to be responsible for superfluidity and superconductivity.

PRINCIPAL TERMS

- **atomic model:** a theoretical representation of the structure and behavior of an atom based on the nature and behavior of its component particles.
- **electron:** a fundamental subatomic particle with a single negative electrical charge, found in a large, diffuse cloud around the nucleus.
- **electron configuration:** the order and arrangement of electrons within the orbitals of an atom or molecule.
- **electron shell:** a region surrounding the nucleus of an atom that contains one or more orbitals capable of holding a specific maximum number of electrons.
- **laminar flow:** the even and stable flow of a fluid; opposite of turbulent flow.
- **quantum mechanics:** the branch of physics that deals with matter interactions on a subatomic scale, based on the concepts that energy is quantized, not continuous, and that elementary particles exhibit wavelike behavior.
- **quantum state:** the condition of a physical system as defined by its associated quantum attributes
- **standard model:** a generally accepted unified framework of particle physics that explains electromagnetism, the weak interaction, and the strong interaction as products of interactions between different types of elementary particles.

QUANTUM STATES

The current atomic model is described by the standard model, in which atoms consist of a very small, dense nucleus composed of protons and neutrons that is surrounded by a very large, and very diffuse, cloud of electrons. The various forces at work within the nucleus of an atom ultimately define the specific energies that the surrounding electrons are 'allowed' to have in the space about the nucleus. A particular electron cannot have just any energy within an atom. Instead, it can possess only the energy that is defined by the mathematics of quantum mechanics and that is associated with a specific 'quantum number'. This restricts an electron to occupy only a specific region about the nucleus that is generally called an 'orbital'. Various properties of subatomic particles such as the electron include 'spin' and 'angular momentum'. These are ascribed quantum attributes, and do not have the same meaning as their counterparts in classical physics; electrons don't actually spin like a top, and angular momentum is just the name given to indicate a particular vector quantity in quantum mechanics. Accordingly, the quantum energy levels allowed within an atom define an electron shell, and the ordering of electrons within the electron shells of an atom define its electron configuration. The vector properties associated with each electron shell split that shell into a corresponding number of subunits, usually termed 'orbitals', having very slightly different energies. The sum of the specific quantum properties of the electrons as they are ordered in the atom define the quantum state of the atom. Depending on the number of electrons in a particular atom, a corresponding number of quantum 'microstates' are possible.

BOSE-EINSTEIN CONDENSATION

In 1924, Satyendra Nath Bose described a quantum mechanical rationalization for a condition at absolute zero of temperature in which bosonic particles, or certain atoms, that had acquired the same quantum state individually could 'condense' into a macroscopic structure defined by the same quantum state. His paper describing this "Bose condensation" was translated into German and submitted for publication by Albert Einstein on Bose's behalf. The effect has been generally called Bose-Einstein condensation because of this, even though the primacy of the concept is correctly attributed only to Bose. The effect has been observed experimentally in gases many times since it was first described theoretically. In a typical experiment, a diffuse gas is cooled to a temperature very close to absolute zero, and the formation of macroscopic structures consisting of a number of atoms or molecules of the gas confirms the formation of a Bose condensate. The formation of a Bose condensate is classed as a phase change, in the same sense that the change from solid to liquid or liquid to gas is a phase change. However, the phase change in the formation of a Bose condensate is based on the indistinguishable character and quantum wave functions of the individual particles, rather than on any specific interactions between particles.

POTENTIAL OF BOSE CONDENSATES

While Bose condensation and Bose condensates may seem to be extremely exotic, and they are, they are perhaps not unusual. The formation of the condensate can be likened to the change from normal light from a light bulb to the coherent light of a laser, or from turbulent flow to laminar flow in a fluid. The movement of electrons from atom to atom in individual atoms has a corresponding stepwise nature due to the barrier of space between the atoms and their individual quantum states. However, in a Bose condensate that barrier is eliminated, and the electrons can move freely from atom to atom because they are described by the exact same quantum state. This homogeneity is thought to be the cause of superfluidity in liquid helium and other fluids that exhibit superfluidity. It is also believed to be the basis of superconductivity, a feature that is being found in materials at temperatures well above absolute zero.

Further Reading

Ueda, Masahito *Fundamentals and New Frontiers of Bose-Einstein Condensation* Hackensack, NJ: World Scientific, 2010.

Martelucci, Sergio, Chester, Arthur N., Aspect, Alain and Inguscio, Massimo, eds. *Bose-Einstein Condensates and Atom Lasers* New York, NY: Kluwer, 2002.

Pethick, C.J. And Smith, H. *Bose-Einstein Condensation in Dilute Gases* New York, NY: Cambridge University Press, 2008.

PRACTICE AND APPLICATON

The number of quantum microstates, N, that a particular atom can have in a particular electron shell is determined by the number of electrons that the shell is allowed to accommodate, n, and the number of electrons that are actually present in that shell, x. It is calculated as

$$N = (n!)/(n-x)!(x!)$$

where ! indicates the factorial calculation of the number before it. Calculate the number of microstates possible for the *f* shell of an atom in which there are two electrons in the *f* shell, given that the *f* shell can accommodate a total of 14 electrons.

Answer:

$$n = 14 \text{ and } x = 2$$
$$\text{therefore}$$
$$N = (14!)/(14-2)!(2!)$$

$$= (14 \times 13 \times 12!)/(12!)(2!)$$

$$= (14 \times 13)/(2)$$

$$= 91$$

Therefore there are 91 microstates for this quantum condition.

Volovik, Grigory E. *The Universe in a Helium Droplet* New York, NY: Oxford University Press, 2009.

Annett, James F. *Superconductivity, Superfluids and Condensates* New York, NY: Oxford University Press, 2004.

—*Richard M. Renneboog MSc*

BRIDGE DESIGN AND BARODYNAMICS

FIELDS OF STUDY

Chemistry, civil engineering, construction, material engineering, mechanics, physics, structural engineering.

SUMMARY

Barodynamics is the study of the mechanics of heavy structures that may collapse under their own weight. In bridge building, barodynamics is the science of the support and mechanics of the methods and types of materials used in bridge design to ensure the stability of the structure. Concepts to consider in avoiding the collapse of a bridge are the materials available for use, what type of terrain will hold the bridge, the obstacle to be crossed (such as river or chasm), how long the bridge needs to be to cross the obstacle, what types of natural obstacles or disasters are likely to occur in the area (high winds, earthquakes), the purpose of the bridge (foot traffic, cars, railway), and what type of vehicles will need to cross the bridge.

KEY TERMS AND CONCEPTS

- **arch bridge:** Type of bridge where weight is distributed outward along two paths that curve toward the ground in the shape of an arch.
- **beam bridge:** Simplest type of bridge consisting of two or more supports holding up a beam; ranges from a complex structure to a plank of wood.
- **cantilever bridge:** Bridge type where two beams, well anchored, support another beam.
- **cofferdam:** A temporary watertight structure that is pumped dry to enclose an area underwater and allow construction work on a bridge in the underwater area.
- **keystone:** The wedge-shape stone at the highest point of an arch; function is to keep the other stones locked into place through gravity.
- **pontoon bridge:** A floating bridge supported by floating objects that contain buoyancy sufficient to support the bridge and any load it must carry; often a temporary structure.
- **suspension bridge:** A bridge hung by cables, which, in turn, hang from towers; weight is transferred from the cables to the towers to the ground.
- **truss bridge:** A type of bridge supported by a network of beams in triangular sections.

DEFINITION AND BASIC PRINCIPLES

Barodynamics is a key component of any bridge design. Bridges are made of heavy materials, and many concepts, such as tension and compression of building materials, and other factors, such as wind shear, torsion, and water pressure, come into play in bridge building.

Bridge designers and constructors must keep in mind the efficiency (the least amount of material for the highest-level performance) and economy (lowest possible costs while still retaining efficiency) of bridge building. In addition, some aesthetic principles must be followed; public outcry can occur when a bridge is thought to be "ugly." Conversely, a beautiful bridge can become a landmark symbol for an area, such as the Golden Gate Bridge has become for San Francisco.

The four main construction materials for bridges are wood, stone, concrete (including prestressed concrete), and iron (from which steel is made). Wood is nearly always available and inexpensive but is comparatively weak in compression and tension. Stone, another often available material, is strong in compression but weak in tension. Concrete, or "artificial stone," is, like stone, strong in tension and weak in compression.

The first type of iron used in bridges, cast iron, is strong in compression but weak in tension. The use of wrought iron helped in bridge building, as it is strong in compression but still has tensile strength. Steel, a further refinement of iron, is superior in compression and tensile strength, making it a preferred material for bridge building. Reinforced concrete or prestressed concrete (types of concrete with steel bars running through concrete beams) are also popular bridge-building materials because of their strength and lighter-weight design.

BACKGROUND AND HISTORY

From the beginning of their existence, humans have constructed bridges to cross obstacles such as rivers and chasms using the materials at hand, such as trees,

Bridge Types

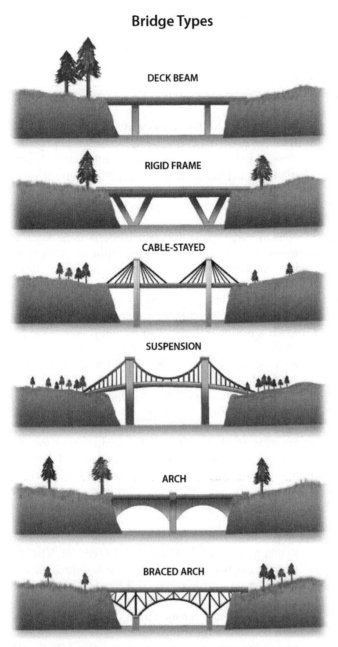

In describing the design of a bridge, there are four primary factors: span, placement, material, and form.

stones, or vines. In China during the third century BCE, the emperor of the Qin Dynasty built canals to transport goods, but when these canals interfered with existing roads, his engineers built bridges of

stone or wood over these canals.

However, the history of barodynamics in bridge building truly begins in Roman times, when Roman engineers perfected the keystone arch. In addition to the keystone and arch concepts, the Romans improved bridge-building materials such as cement and concrete and invented the cofferdam so that underwater pilings could be made for bridges. These engineers built a network of bridges throughout the Roman empire to keep communication with and transportation to and from Rome intact. The Romans made bridges of stone because of its durability, and many of these bridges are still intact today.

In the Middle Ages, bridges were an important part of travel and transportation of goods, and many bridges were constructed during this period to support heavy traffic. This is also the period when people began to live in houses built on bridges, in part because in walled cities, places to build homes were limited. Possibly the most famous inhabited bridge was London Bridge, the world's first stone bridge to be built over a tidal waterway where the water rose and fell considerably every twelve hours. However, in Paris in the sixteenth century, there were at least five inhabited bridges over the Seine to the Île de la Cité.

The first iron bridge was built in 1779. Using this material changed the entire bridge-building industry because of the size and strength of structure that became possible. In the 1870's, a fall in the price of steel made bridges made of this material even more popular, and in 1884, Alexandre Gustave Eiffel, of Eiffel Tower fame, designed a steel arch bridge that let wind pass through it, overcoming many of the structural problems with iron and steel that had previously existed. Iron and steel are still the most common materials to use in bridge building.

Suspension bridges began to be quite popular as they are the most inexpensive way to span a longer distance. In the early 1800's, American engineer John Roebling designed a new method of placing cables on suspension bridges. Famous examples of suspension bridges include the Golden Gate Bridge (completed in 1937) and Roebling's Brooklyn Bridge (completed in 1883).

Girder bridges were often built to carry trains in the early twentieth century. Though capable of carrying heavyweight railroad cars, this type of bridge is usually only built for short distances as is typical with beam-type bridges. In the 1950's, the box girder was designed, allowing air to pass through this type of bridge and making longer girder bridges possible.

HOW IT WORKS

The engineering principles that must be used to construct even a simple beam bridge are staggering. Supports must be engineered to hold the weight of the entire structure correctly as well as any traffic that will cross the bridge. The bridge itself, or "span," must be strong enough to bear the weight of traffic and stable enough to keep that traffic safe. Spans must be kept as short as reasonably possible but must sometimes be built across long distances, for example, over deep water.

Arch. The Roman arch concept uses the pressure of gravity on the material forming the arch to hold the bridge together with the outward thrust contained by buttresses. It carries loads by compressing and exerting pressure on the foundation, which must be prevented from settling and sliding. This concept allowed bridges to be built that were longer than ever before. For example, a surviving bridge over the Tagus River in Spain has two central arches that are 110 feet wide and 210 feet above the water level. These arches are made of uncemented granite and each keystone weighs eight tons. This type of bridge is constructed by building a huge timber structure to support the bridge during the building phase, then winching blocks into place with a pulley system. After the keystone to the arch is put into place, the scaffolding is removed, leaving the bridge to stand alone.

Beam. This is the most common form of bridge and may be as simple as a log across a stream. This type of bridge carries a load by bending, which horizontally compresses the top of the beam and simultaneously causes horizontal tension on the bottom of the beam.

Truss. A truss bridge is popular because it requires a relatively small amount of construction material to carry a heavy load. It works like a beam bridge, carrying loads by bending and causing compression and tension in the vertical and diagonal supports.

Suspension. Suspension bridges are essentially steel-rope bridges: Thick steel cables are entwined like ropes into a larger and stronger steel cable or rope. These thick, strong cables then suspend the

bridge itself between pylons that support the weight of the bridge. A suspension bridge can be thought of as an upside-down arch, as the curved cables use tension and compression to support the load.

Cantilever. Cantilevered means something that projects outward and is supported at only one end (similar to a diving board). This type of bridge is generally made with three spans with the outside spans supported on the shore and the middle span supported by the outside spans. This is a type of beam bridge that uses tension in the lower spans and compression in the upper span to carry a load.

Pontoon. A pontoon bridge is built across water with materials that float. Each pontoon, or floating object, can support a maximum load equal to the amount of water it displaces. If the load placed on one pontoon-supported section exceeds the water displaced, the pontoon will submerge and cause the entire bridge to sink.

APPLICATIONS AND PRODUCTS

Bridges are continuously being built to cross physical obstacles, and as the nature of materials changes, the ability to cross even larger obstacles becomes reality. Nature is the defining force on a bridge; most bridges fail because of flooding or other natural disasters.

Improvements in building materials are ongoing. For example, the Jakway Park Bridge in Buchanan County, Iowa, was the first bridge in North America to be built with ultrahigh performance concrete (UHPC) with pi-girders. This moldable material combines high compressive strength and flexibility and offers a wide range of design possibilities. It is very durable and has low permeability.

Bridge-building products may even be developed that help the environment. For example, the rebuilt I-35W bridge in Minnesota uses a concrete that is said to "eat smog." The concrete contains photo-catalytic titanium dioxide, which accelerates the decomposition of organic material. Other materials like this may change the future of bridge building.

IMPACT ON INDUSTRY

Bridges have a significant impact on travel and transportation of goods. They are, generally, made for the public, for transportation or travel, and are often built with public funds. Therefore, bridge building has its greatest impact in the governmental or public-transportation areas. Bridge engineers often find jobs with

FASCINATING FACTS ABOUT BRIDGE DESIGN AND BARODYNAMICS

- The Tacoma Narrows Bridge across Puget Sound tore apart and fell into the water below in 1940 because of engineering miscalculations concerning vertical and torsional motion. Viewing motion pictures of the disaster helped engineers rethink the aerodynamics of building bridges.
- The I-35W bridge in Minneapolis collapsed in 2007. The deaths and injuries led to a renewed interest in maintenance and retrofitting of public bridges across the United States.
- London Bridge is possibly the most famous bridge in the world. It was the only bridge across the River Thames from the tenth century until the eighteenth century, when the original stone bridge was torn down and rebuilt with stone. It was rebuilt again in 1973 of steel and concrete to accommodate the heavy traffic using the bridge. Robert McCulloch bought the old stone bridge, numbering each stone, and had it rebuilt in Lake Havasu City, Arizona, where it remains a tourist attraction.
- The origin of the word "bridge" goes back to an Old English word "brycg," which is believed to be derived from the German word root "brugj." The name of the Belgian city Brugge can be translated as "a bridge or a place of bridges."
- The Ponte Vecchio in Florence has an added covered walk forming a top story above the shops that line the bridge. This addition was constructed in 1565 so members of the Medici family could walk across the river from the Uffizi to the Palazzo Pitti without descending to the street level and mingling with commoners.
- The world's first iron bridge was built in 1779 over the River Severn in an English industrial area called Coalbrookdale.
- In *Ramayana*, a mythological Indian epic, tales are told of bridges constructed by the army of Sri Rama, a mythological king of Ayodhya, from India to the island of Sri Lanka, a feat thought to have been impossible in that time period. However, space images taken by the National Aeronautics and Space Administration reveal an ancient bridge, called Adam's Bridge, in the Palk Strait with a unique curvature and composition that indicate that it is man-made.
- The Chinese made suspension bridges as early as 200 BCE
- The oldest surviving stone bridge in China, the Zhaozhou Bridge, is thought to have been built during the Sui Dynasty, around the year 600.

government agencies, such as the U.S. Department of Transportation, or with private companies that are subcontractors on government projects.

CAREERS AND COURSE WORK

Those who engineer and design bridges may have backgrounds in a variety of fields, including architecture and design. However, those who are involved in the barodynamic aspects of bridge building are engineers, usually either civil, materials, or mechanical engineers. Earning a degree in one of these fields is required to get the training needed in geology, math, and physics to learn about the physical limitations and considerations in bridge building. Many bridge engineers have advanced degrees in a specific related field. After earning a degree, a candidate for this type of job usually works for a few years as an assistant engineer in a sort of apprenticeship, learning the specifics of bridge building such as drafting, blueprint reading, surveying, and stabilization of materials. To become a professional engineer (PE), one must then take a series of written exams to get his or her license.

SOCIAL CONTEXT AND FUTURE PROSPECTS

Barodynamics is a rapidly changing field. As new materials are created and existing materials change, the possibilities for future improvements in this field increase. Development of future lightweight materials may change the way bridges are engineered and designed to make the structure stable and avoid collapse. Just as innovations in iron refinement changed the face of bridge building in the late 1800's, new materials may refine and improve bridge building even further, bringing more efficient and economical bridges.

Materials engineers are usually the people who provide the technological innovation to create these kinds of new materials. They examine materials on the molecular level to understand how materials can be improved and strengthened in order to provide better building materials for structures such as bridges. Possible future bridge-building materials include ceramics, polymers, and other composites. Two other rapidly growing materials fields that may affect bridge barodynamics are biomaterials and nanomaterials.

—*Marianne M. Madsen, MS*

FURTHER READING

Blockley, David. *Bridges: The Science and Art of the World's Most Inspiring Structures.* New York: Oxford University Press, 2010. Written by a professor of engineering with a lay reader in mind; discusses basic forces such as tension, compression, and shear, and bridge failures. Includes a comprehensive history of bridge building with fifty illustrations.

Chen, Wai-Fah, and Lian Duan, eds. *Bridge Engineering Handbook.* Boca Raton, Fla.: CRC Press, 1999. Contains more than 1,600 tables, charts, and illustrations with step-by-step design procedures. Covers fundamentals, superstructure design, substructure design, seismic design, construction and maintenance, and worldwide practice; includes a special topics section.

Haw, Richard. *Art of the Brooklyn Bridge: A Visual History.* New York: Routledge, 2007. A visually interesting compilation of artists' renderings of the Brooklyn Bridge, contributing to the idea that bridges are artful as well as functional.

Tonias, Demetrios E., and Jim J. Zhao. *Bridge Engineering: Design, Rehabilitation, and Maintenance of Modern Highway Bridges.* 2d ed. New York: McGraw-Hill, 2007. Details the entire highway-bridge design process; includes information on design codes.

Unsworth, John F. *Design of Modern Steel Railway Bridges.* Boca Raton, Fla.: CRC Press, 2010. Focuses on new steel superstructures for railway bridges but also contains information on maintenance and rehabilitation and a history of existing steel railway bridges.

Van Uffelen, Chris. *Masterpieces: Bridge Architecture and Design.* Salenstein, Switzerland: Braun Publishing, 2010. Includes photos of sixty-nine bridges from around the world, displaying a variety of structures and materials.

Yanev, Bojidar. *Bridge Management.* Hoboken, N.J.: John Wiley & Sons, 2007. Contains case studies of bridge building and design; discusses bridge design, maintenance, and construction with topics such as objectives, tools, practices, and vulnerabilities.

WEB SITES

American Society of Civil Engineers
http://www.asce.org

Design-Build Institute of America
http://www.dbia.org

National Society of Professional Engineers
http://www.nspe.org/index.html

See also: Civil Engineering; Earthquake Engineering.

C

CALCULUS

FIELDS OF STUDY

Algebra; geometry; trigonometry.

SUMMARY

Calculus is the study of functions and change. It is the bridge between the elementary mathematics of algebra, geometry, and trigonometry, and advanced mathematics. Knowledge of calculus is essential for those pursuing study in fields such as chemistry, engineering, medicine, and physics. Calculus is employed to solve a large variety of optimization problems; one example is the so-called least squares solution method commonly used in statistics and elsewhere. The least squares function best fits a set of data points and can then be used to generalize or predict results based on that set.

KEY TERMS AND CONCEPTS

- **antiderivative:** Function whose derivative is equal to that of a given function.
- **continuity:** Characteristic manifested by a function when its output values are equal to the values of its limits.
- **converge:** Action of an improper integral or series with a finite value.
- **definite integral:** Limit of a Riemann sum as the number of terms approaches infinity.
- **derivative:** Function derived from a given function by means of the limit process, whose output equals the instantaneous rates of change of the given function.
- **diverge:** Action of an improper integral or series with no finite value.
- **gradient:** Vector whose components are each of the partial derivatives of a function of several variables.
- **indefinite integral:** Antiderivative of a given function.
- **limit:** Number that the output values of a function approach the closer the input values of the function get to a specified target.
- **Riemann sum:** Sum of the products of functional values and the lengths of the subintervals over which the function is defined.
- **series:** Formal sum of an infinite number of terms; it may be convergent or divergent.
- **Taylor series:** Series whose sums equal the output values of a given function, at least along some interval of input values.

DEFINITION AND BASIC PRINCIPLES

Calculus is the study of functions and their properties. Calculus takes a function and investigates it according to two essential ideas: rate of change and total change. These concepts are linked by their common use of calculus's most important tool, the limit. It is the use of this tool that distinguishes calculus from the branches of elementary mathematics: algebra, geometry, and trigonometry. In elementary mathematics, one studies problems such as "What is the slope of a line?" or "What is the area of a parallelogram?" or "What is the average speed of a trip that covers three hundred miles in five and a half hours?" Elementary mathematics provides methods or formulas that can be applied to find the answer to these and many other problems. However, if the line becomes a curve, how is the slope calculated? What if the parallelogram becomes a shape with an irregularly curved boundary? What if one needs to know the speed at an instant, and not as an average over a longer time period?

Calculus answers these harder questions by using the limit. The limit is found by making an approximation to the answer and then refining that approximation by improving it more and more. If there is a pattern leading to a single value in those improved approximations, the result of that pattern is called the limit. Note that the limit may not exist in some cases. The limit process is used throughout calculus to provide answers to questions that elementary mathematics cannot handle.

The derivative of a function is the limit of average slope values within an interval as the length of the interval approaches zero. The integral calculates the total change in a function based on its rate of change function.

BACKGROUND AND HISTORY

Calculus is usually considered to have come into being in the seventeenth century, but its roots were formed much earlier. In the sixteenth century, Pierre de Fermat did work that was very closely related to calculus's differentiation (the taking of derivatives) and integration. In the seventeenth century, René Descartes founded analytic geometry, a key tool for developing calculus.

However, it is Sir Isaac Newton and Gottfried Wilhelm Leibniz who share the credit as the (independent) creators of calculus. Newton's work came first but was not published until 1736, nine years after his death. Leibniz's work came second but was published first, in 1684. Some accused him of plagiarizing Newton's work, although Leibniz arrived at his results by using different, more formal methods than Newton employed.

Both men found common rules for differentiation, but Leibniz's notation for both the derivative and the integral are still in use. In the eighteenth century, the work of Jean le Rond d'Alembert and Leonhard Euler on functions and limits helped place the methods of Newton and Leibniz on a firm foundation. In the nineteenth century, Augustin-Louis Cauchy used a definition of limit to express calculus concepts in a form still familiar more than two hundred years later. German mathematician Georg Riemann defined the integral as a limit of a sum, the same definition learned by calculus students in the twenty-first century. At this point, calculus as it is taught in the first two years of college reached its finished form.

HOW IT WORKS

Calculus is used to solve a wide variety of problems using a common approach. First, one recognizes that the problem at hand is one that cannot be solved using elementary mathematics alone. This recognition is followed by an acknowledgment: There are some things known about this situation, even if they do not provide a complete basis for solution. Those known properties are then used to approximate a solution to the problem. This approximation may not be very good, so it is refined by taking a succession of better and better approximations. Finally, the limit is taken, and if the limit exists, it provides the exact answer to the original problem.

One speaks of taking a limit of a function $f(x)$ as x approaches a particular value, for example, $x = a$. This means that the function is examined on an interval around, but not including $x = a$. Values of $f(x)$ are taken on that interval as the varying x values get closer and closer to the target value of $x = a$. There is no requirement that $f(a)$ exists, and many times it does not. Instead the pattern of functional values is examined as x approaches a. If those values continue to approach a single target value, it is that value that is said to be equal to the limit of $f(x)$ as x approaches a. Otherwise, the limit is said not to exist. This method is used in both differential calculus and integral calculus.

Differential calculus. Differentiation is a term used to mean the process of finding the derivative of a function $f(x)$. This new function, denoted $f'(x)$, is said to be "derived" from $f(x)$. If it exists, $f'(x)$ provides the instantaneous rate of change of $f(x)$ at x. For curves (any line other than a straight line), the calculation of this rate of change is not possible with elementary mathematics. Algebra is used to calculate that rate between two points on the graph, then those two points are brought closer and closer together until the limit determines the final value.

Shortcut methods were discovered that could speed up this limit process for functions of certain types, including products, quotients, powers, and trigonometric functions. Many of these methods go back as far as Newton and Leibniz. Using these formulas allows one to avoid the more tedious limit calculations. For example, the derivative function of sine x is proven to be cosine x. If the slope of sine x is needed at $x = 4$, the answer is known to be cosine 4, and much time is saved.

Integral calculus. A natural question arises: If $f'(x)$ can be derived from $f(x)$, can this process be reversed? In other words, suppose an $f(x)$ is given. Can an $F(x)$ be determined whose derivative is equal to $f(x)$? If so, the $F(x)$ is called an antiderivative of $f(x)$; the process of finding $F(x)$ is called integration. In general, finding antiderivatives is a harder task than finding derivatives. One difficulty is that constant functions all have derivatives equal to zero, which means that without further information, it is impossible to determine which constant is the correct one. A bigger problem is that there are functions, such as sine (x^2), whose derivatives are reasonably easy to calculate but for which no elementary function serves as an antiderivative.

The definite integral can be thought of as an attempt to determine the amount of area between the graph of $f(x)$ and the x-axis, usually between a left and right endpoint. This cannot typically be answered using elementary mathematics because the shape of the graph can vary widely. Riemann proposed approximating the area with rectangles and then improving the approximation by having the width of the rectangles used in the approximation get smaller and smaller. The limit of the total area of all rectangles would equal the area being sought. It is this notion that gives integral calculus its name: By summing the areas of many rectangles, the many small areas are integrated into one whole area.

As with derivatives, these limit calculations can be quite tedious. Methods have been discovered and proven that allow the limit process to be bypassed. The crowning achievement of the development of calculus is its fundamental theorem: The derivative of a definite integral with respect to its upper limit is the integrand evaluated at the upper limit; the value of a definite integral is the difference between the values of an antiderivative evaluated at the limits. If one is looking for the definite integral of a continuous $f(x)$ between $x = a$ and $x = b$, one need only find any antiderivative $F(x)$ and calculate $F(b)$ -$F(a)$.

APPLICATIONS AND PRODUCTS

Optimization. A prominent application of the field of differential calculus is in the area of optimization, either maximization or minimization. Examples of optimization problems include What is the surface area of a can that minimizes cost while containing a specified volume? What is the closest that a passing asteroid will come to Earth? What is the optimal height at which paintings should be hung in an art gallery? (This corresponds to maximizing the viewing angle of the patrons.) How shall a business minimize its costs or maximize its profits?

All of these can be answered by means of the derivative of the function in question. Fermat proved

that if $f(x)$ has a maximum or minimum value within some interval, and if the derivative function exists on that interval, then the derivative value must be zero. This is because the graph must be hitting either a peak or the bottom of a valley and has a slope of zero at its highest or lowest points. The search for optimal values then becomes the process of finding the correct function modeling the situation in question, finding its derivative, setting that derivative equal to zero, and solving. Those solutions are the only candidates for optimal values. However, they are only candidates because derivatives can sometimes equal zero even if no optimal value exists. What is certain is that if the derivative value is not zero, the value is not optimal.

The procedure discussed here can be applied in two dimensions (where there is one input variable) or three dimensions (where there are two input variables).

Surface area and volume. If a three-dimensional object can be expressed as a curve that has been rotated about an axis, then the surface area and volume of the object can be calculated using integrals. For example, both Newton and Johannes Kepler studied the problem of calculating the volume of a wine barrel. If a function can be found that represents the curvature of the outside of the barrel, that curve can be rotated about an axis and pi (p) times the function squared can be integrated over the length of the barrel to find its volume.

Hydrostatic pressure and force. The pressure exerted on, for example, the bottom of a swimming pool of uniform depth is easily calculated. The force on a dam due to hydrostatic pressure is not so easily computed because the water pushes against it at varying depths. Calculus discovers the answer by integrating a function found as a product of the pressure at any depth of the water and the area of the dam at that depth. Because the depth varies, this function involves a variable representing that depth.

Arc length. Algebra is able to determine the length of a line segment. If that path is curved, whether in two or three dimensions, calculus is applied to determine its length. This is typically done by expressing the path in parametric form and integrating the function representing the length of the vector that is tangent to the path. The length of a path winding through three-dimensional space, for example, can be determined by first expressing the path in the

Isaac Newton developed the use of calculus in his laws of motion and gravitation.

parametric form $x = f(t)$, $y = g(t)$, and $z = h(t)$, in which f, g, and h are continuous functions defined for some interval of values of t. Then the square root of the sum of the squares of the three derivatives is integrated to find the length.

Kepler's laws. In the early seventeenth century, Kepler formulated his three laws of planetary motion based on his analysis of the observations kept by Tycho Brahe. Later, calculus was used to prove that these laws are correct. Kepler's laws state that any planet's orbit around the Sun is elliptical, with the Sun at one focus of the ellipse; that the line joining the Sun to the planet sweeps out equal areas in equal times; and that the square of the period of revolution is proportional to the cube of the length of the major axis of the orbit.

Probability. Accurate counting methods can be sufficient to determine many probabilities of a discrete random variable. This would be a variable whose values could be, for example, counting numbers such as 1, 2, 3, and so on, but not numbers inbetween, such as 2.4571. If the random variable is continuous, so that it can take on any real number within an interval, then its probability density function must

be integrated over the relevant interval to determine the probability. This can occur in two or three dimensions.

One common example is determining the likelihood that a customer's wait time is longer than a specified target, such as ten minutes. If the manager knows the average wait time that a customer experiences at an establishment is, for example, six minutes, then this time can be used to determine a probability density function. This function is integrated to determine the probability that a person's wait time will be longer than ten minutes, less than three minutes, between five and thirteen minutes, or within any range of times that is desired.

IMPACT ON INDUSTRY
The study of calculus is the foundation of more advanced work in mathematics, engineering, economics, and many areas of science. In some cases, it is these other disciplines that affect industry, but much of the time, the effect of calculus can be seen directly. Businesses, government, and industry throughout the developed world continue to apply calculus in a wide variety of settings.

Government and university research. Knowing the importance of calculus in many fields, the National Science Foundation funded many projects in the 1990's that were designed to renew and refresh calculus education. The foundation's Division of Mathematical Sciences continues to fund projects related to both education and research. Projects funded in 2009 included research on the suspension of aerosol particles in the atmosphere, including relating these concepts to the teaching of calculus, and research in the area of partial differential equations and their applications.

Industry and business. Many branches of engineering use calculus methods and results. In chemical engineering, knowledge of vector calculus and Taylor polynomials is particularly important. In electrical engineering, these same topics, together with an understanding of integration techniques, are emphasized. Mechanical engineering makes significant use of vector calculus and the solving of differential equations. The latter are often solved by numerical procedures such as Euler's method or the Runge-Kutta method when exact solutions are either impossible or impractical to obtain.

In finance, series are used to find the present value

FASCINATING FACTS ABOUT CALCULUS

- Archimedes, who lived in the third century BCE, derived the formula for the volume of a sphere by using a method that foreshadowed integration. This was understood only when Archimedes's explanation of his method was discovered on a palimpsest in 1906.

- The controversies as to whether Sir Isaac Newton or Gottfried Wilhelm Leibniz (or both) should be credited as the founders of calculus and whether Leibniz stole Newton's ideas led to Leibniz's disgrace. In fact, his secretary was the only mourner to attend Leibniz's funeral.

- The geometric figure known as Gabriel's horn or Toricelli's trumpet is found by revolving the curve $1/x$ about the x-axis, beginning at $x = 1$. Calculus shows that this object, an infinitely long horn shape, has an infinite surface area but only finite volume.

- The Bernoulli brothers, Jakob and Johann, proved that a chain hanging from two points has the shape of a catenary, not a parabola, and that of all possible shapes, the catenary has the lowest center of gravity and thus the minimal potential energy.

- The logical foundations of calculus were not established until well after the methods themselves had been employed. One of the critics who spurred this development was George Berkeley, a bishop of the Church of Ireland, who accused mathematicians of accepting calculus as a matter of faith, not of science.

- Calculus, combined with probability, is used in the pricing, construction, and hedging of derivative securities for the financial market.

of Actuaries or the Casualty Actuarial Society is also expected, which requires a thorough understanding of calculus.

In statistics, a master's degree is typically preferred, and to work as a physicist or mathematician, a doctorate is the standard. In terms of calculus-related course work, in addition to the calculus sequence, students will almost always take a course in differential equations and perhaps one or two in advanced calculus or mathematical analysis.

In 2008, about 428,000 people were working as either chemical, electrical, or mechanical engineers in the United States, most in either the manufacturing or service industries. Private industry and the government employed 23,000 statisticians. Insurance carriers, brokers, agents, or other offices provided jobs for 20,000 actuaries. About 15,000 physicists worked as researchers for private industry or the government. The Bureau of Labor Statistics counted only 3,000 individuals as simply mathematicians but tallied 55,000 mathematicians who teach at the college or university level. As of 2010, all these careers were projected to have a job growth rate at or above the national average.

SOCIAL CONTEXT AND FUTURE PROSPECTS

Calculus itself is not an industry, but it forms the foundation of other industries. In this role, it continues to power research and development in diverse fields, including those that depend on physics. Physics derives its results by way of calculus techniques. These results in turn enable developments in small- and large-scale areas. An example of a small-scale application is the ongoing development of semiconductor chips in the field of electronics. Large-scale applications are in the solar and space physics critical for ongoing efforts to explore the solar system. These are just two examples of calculus-based fields that will continue to have significant impact in the twenty-first century.

—Michael J. Caulfield, MS, PhD

FURTHER READING

Bardi, Jason Socrates. *The Calculus Wars: Newton, Leibniz, and the Greatest Mathematical Clash of All Time.* New York: Thunder's Mouth Press, 2006. Examines the controversy over who should be considered the originator of calculus.

Dunham, William. *The Calculus Gallery: Masterpieces*

of revenue streams with an unlimited number of perpetuities. Calculus is also used to model and calculate levels of risk and benefit for investment schemes. In economics, calculus methods find optimal levels of production based on cost and revenue functions.

CAREERS AND COURSE WORK

A person preparing for a career involving the use of calculus will most likely graduate from a university with a degree in mathematics, physics, actuarial science, statistics, or engineering. In most cases, engineers and actuaries are able to join the profession after earning their bachelor's degree. For actuaries, passing one or more of the exams given by the Society *from Newton to Lebesgue.* Princeton, N.J.: Princeton University Press, 2005. Focuses on thirteen individuals and their notable contributions to the development of calculus.

Kelley, W. Michael. *The Complete Idiot's Guide to Calculus.* 2d ed. Indianapolis: Alpha, 2006. Begins by examining what calculus is, then leads the reader through calculus basics.

Simmons, George F. *Calculus Gems: Brief Lives and Memorable Mathematics.* 1992. Reprint. Washington, D.C.: The Mathematical Association of America, 2007. Includes dozens of biographies from ancient times to the nineteenth century, together with twenty-six examples of the remarkable achievements of these people.

Stewart, James. *Essential Calculus.* Belmont, Calif.: Thomson Brooks/Cole, 2007. A standard text that relates all of the concepts and methods of calculus, including examples and applications.

WEB SITES
American Mathematical Society
http://www.ams.org

Mathematical Association of America
http://www.maa.org

Society for Industrial and Applied Mathematics
http://www.siam.org

See also: Applied Mathematics; Engineering Mathematics; Numerical.

CERAMICS

FIELDS OF STUDY

Calculus; chemistry; glass engineering; materials engineering; physics; statistics; thermodynamics.

SUMMARY

Ceramics is a specialty field of materials engineering that includes traditional and advanced ceramics, which are inorganic, nonmetallic solids typically created at high temperatures. Ceramics form components of various products used in multiple industries, and new applications are constantly being developed. Examples of these components are rotors in jet engines, containers for storing nuclear and chemical waste, and telescope lenses.

KEY TERMS AND CONCEPTS

- **ceramic:** Inorganic, nonmetallic solid processed or used at high temperatures; made by combining metallic and nonmetallic elements.
- **coke:** Solid, carbon-rich material derived from the destructive distillation of low-ash, low-sulfur bituminous coal.
- **glazing:** Process of applying a layer or coating of a glassy substance to a ceramic object before firing it, thus fusing the coating to the object.
- **kiln:** Thermally insulated chamber, or oven, in which a controlled temperature is produced and used to fire clay and other raw materials to form ceramics.

DEFINITION AND BASIC PRINCIPLES
Ceramic engineering is the science and technology of creating objects from inorganic, nonmetallic materials. A specialty field of materials engineering, ceramic engineering involves the research and development of products such as space shuttle tiles and rocket nozzles, building materials, as well as ball bearings, glass, spark plugs, and fiber optics.

Ceramics can be crystalline in nature; however, in the broader definition of ceramics (which includes glass, enamel, glass ceramics, cement, and optical fibers) they can also be noncrystalline. The most distinguishing feature of ceramics is their ability to resist extremely high temperatures. This makes ceramics very useful for tasks where materials such as metals and polymers alone are unsuitable. For example, ceramics are used in the manufacture of disk brakes for high-performance cars (such as race cars) and for heavy vehicles (such as trucks, trains, and aircraft). These brakes are lighter and more durable and can withstand greater heat and speed than the conventional metal-disk brakes. Ceramics can also be used to increase the efficiency of turbine engines used to operate helicopters. These aircraft have a limited travel range and cannot carry a great deal of weight because of the stress these activities place on engines made of metallic alloys. However, turbine engines using ceramic parts and thermal barrier coatings are currently in development—they already show superior performance when compared with existing engines. The ceramic engine parts are from 30 to 50 percent lighter than their metallic counterparts, and the thermal coatings increase the engine operating temperatures to 1,650 degrees Celsius (C). These qualities will enable future helicopters to travel farther and carry more weight.

BACKGROUND AND HISTORY

One of the oldest industries on Earth, ceramics date back to prehistoric times. The earliest known examples of ceramics, animal and clay figures that were fired in kilns, date back to 24,000 BCE. These ceramics were earthenware that had no glaze. Glazing was discovered by accident and the earliest known glazed items date back to 5000 BCE. Chinese potters studied glazing and first developed a consistent glazing technique. Glass was first produced around 1500 BCE. The development of synthetic materials with better resistance to very high temperatures in the 1500's enabled the creation of glass, cement, and ceramics on an industrial scale. The ceramics industry has grown in leaps and bounds since then.

Many notable innovators have contributed to the growth of advanced ceramics. In 1709, Abraham Darby, a British brass worker and key player in the Industrial Revolution, first developed a smelting process for producing pig iron using coke instead of wood as fuel. Coke is now widely used in the production of carbide ceramics. In 1888, Austrian chemist Karl Bayer first separated alumina from bauxite ore. This method, known as the Bayer process, is still used to purify alumina. In 1893, Edward Goodrich Acheson, an American chemist, electronically fused carbon and clay to create carborundum, also known as synthetic silicon carbide, a highly effective abrasive.

Other innovators include brothers Pierre and Jacques Curie, French physicists who discovered piezoelectricity around 1880; French chemist Henri Moissan, who combined silicon carbide with tungsten carbide around the same time as Acheson; and German mathematician Karl Schröter, who in 1923 developed a liquid-phase sintering method to bond cobalt with the tungsten-carbide particles created by Moissan.

The need for high-performance materials during World War II helped accelerate ceramic science and engineering technologies. Development continued throughout the 1960's and 1970's, when new types of ceramics were created to facilitate advances in atomic energy, electronics, and space travel. This growth continues as new uses for ceramics are researched and developed.

HOW IT WORKS

There are two main types of ceramics: traditional and advanced. Traditional ceramics are so called because the methods for producing them have been in existence for many years. The familiar methods of creating these ceramics—digging clay, molding the clay by hand, or using a potter's wheel, firing, and then decorating the object—have been around for centuries, and have only been improved and mechanized to meet increasing demand. Advanced ceramics, which cover the more recent developments in the field, focus on products that make full use of specific properties of ceramic or glass materials. For example, ferrites, a type of advanced ceramics, are very good conductors of electricity and are typically used in electrical transformers and superconductors. Zirconia, another type of advanced ceramics, is strong, tough, very resistant to wear and tear, and does not cause an adverse reaction when introduced to biological tissue. This makes it ideal for use in creating joint replacements in humans. It works particularly well in hip replacements, but it is also useful for knee, shoulder, and finger-joint replacements. The unique qualities of these, and other advanced ceramics, and the research into the variety of ways they can be applied, are what differentiate them from traditional ceramics.

There are seven basic steps to creating traditional ceramics. They are described in detail below.

Raw materials. In this first stage, the raw materials are chosen for creating a ceramic product. The type of ceramic product to be created determines the type of raw materials required. Traditional ceramics use natural raw materials, such as clay, sand, quartz, and flint. Advanced ceramics require the use of chemically synthesized powders.

Beneficiation. Here, the raw materials are treated chemically or physically to make them easier to process.

Batching and mixing. In this step, the parts of the

ceramic product are weighed and combined to create a more chemically and physically uniform material to use in forming, the next step.

Forming. The mixed material is consolidated and molded to create a cohesive body of the determined shape and size. Forming produces a "green" part, which is soft, pliable and, if left at this stage, will lose its shape over time.

Green machining. This step eliminates rough surfaces, smooths seams, and modifies the size and shape of the green part to prepare for sintering.

Drying. Here, the water or other binding agent is removed from the formed material. Drying is a carefully controlled process that should be done as quickly as possible. After drying, the product will be smaller than the green part. It is also very brittle and must be handled carefully.

Firing or sintering. The dried parts now undergo a controlled heating process. The ceramic becomes denser during firing, as the spaces between the individual particles of the ceramic are reduced as they heat. It is during this stage that the ceramic product acquires its heat-resistant properties.

Assembly. This step occurs only when ceramic parts need to be combined with other parts to form a complete product. It does not apply to all ceramic products.

This is not a comprehensive list. More steps may be required depending on the type of ceramic product being made. For advanced ceramics production, this list of steps will either vary or expand. For example, an advanced ceramic product may need to have forming or additives processes completed in addition to the standard forming processes. It may also require a post-sintering process such as machining or annealing.

APPLICATIONS AND PRODUCTS

Traditional ceramics. Applications and products include whiteware, glass, structural clay items, cement, refractories, and abrasives. Whiteware, so named because of its white or off-white color, includes dinnerware (plates, mugs, and bowls), sanitary ware (bathroom sinks and toilets), floor and wall tiles, dental implants, and decorative ceramics (vases,

figurines, and planters). Glass products include containers (bottles and jars), pressed and blown glass (wineglasses and crystal), flat glass (windows and mirrors), and glass fibers (home insulation). Structural clay products include bricks, sewer pipes, flooring, and wall and roofing tiles. Cement is used in the construction of concrete roads, buildings, dams, bridges, and sidewalks. Refractories are materials that retain their strength at high temperatures. They are used to line furnaces, kilns, incinerators, crucibles, and reactors. Abrasives include natural materials such as

A potter at work, 1605

diamonds and garnets, and synthetic materials such as fused alumina and silicon carbide, which are used for precision cutting as well.

Advanced ceramics. Advanced ceramics focus on specific chemical, biomedical, mechanical, or optical uses of ceramic or glass materials. Advanced ceramics fully came into being within the last few decades (beginning in the 1960's), and research and development is ongoing. The field has produced a wide range of applications and products. In aerospace, ceramics are used in space shuttle tiles, aircraft instrumentation and control systems, missile nose cones, rocket nozzles, and thermal insulation. Automotive applications include spark plugs, brakes, clutches, filters, heaters, fuel pump rollers, and emission control devices. Biomedical uses for ceramics include replacement joints and teeth, artificial bones and heart valves, hearing aids, pacemakers, dental veneers, and orthodontics. Electronic devices that use ceramics include insulators, magnets, cathodes, antennae, capacitors, integrated circuit packages, and superconductors. Ceramics are used in the chemical and petrochemical industry for ceramic catalysts, catalyst supports, rotary seals, thermocouple protection tubes, and pumping tubes. Laser and fiber-optics applications for ceramics include glass optical fibers (used for very fast data transmission), laser materials, laser and fiber amplifiers, lenses, and switches. Environmental uses of ceramics include solar cells, nuclear fuel storage, solid oxide fuel cells, hot gas filters (used in coal plants), and gas turbine components. Ceramic coatings include self-cleaning coatings for building materials, coatings for engine components, cutting tools, industrial wear parts, optical materials (such as lenses), and antireflection coatings.

Other products in the advanced ceramics segment include water-purification devices, particulate or gas filters, and glass-ceramic stove tops.

IMPACT ON INDUSTRY

Ceramic engineering has come a long way from its humble beginnings in the 1800's. Various government agencies, universities, businesses, and the general public have contributed to, or benefited from, the innovations made in advanced ceramics. The North American market for advanced structural ceramics was worth $2.7 million in 2007 and is predicted to grow to $3.7 million by 2012. This

represented a compound annual growth of 6 percent. In 2010, the global advanced ceramics market was predicted to grow to $ 56.4 million by 2015. These predictions are evidence that the field of advanced ceramics continues to grow even in the face of economic difficulties.

Japan is the largest regional market for advanced ceramics, and the market is expected to grow as it is driven by rapid developments in the information technology and electronics industries. The United States remains the largest market for military applications of advanced ceramics. The United Kingdom along with India, China, and Germany are making significant progress in research and development.

Government and university research. Fraunhofer-Gesellschaft, a German research organization that is partially funded by the German government, has been making great strides in developing new advanced ceramic products. Recently, the company developed pump components that use a diamond-ceramic composite it invented. Scientists at the company also created digital projector lenses using flat arrays of glass microlenses as well as a credit card-size platform that uses magnetic nanoparticles to detect sepsis. Each of these products was a result of the joint efforts of the various institutes that fall under Fraunhofer's substantial umbrella.

The United States Department of Energy is a major investor in advanced ceramics research and development. The agency recently awarded funding for three U.S.-China Clean Energy Research Centers (CERCs). These CERCs, which will focus on clean vehicle and clean coal technologies, will receive $50 million in funding (governmental and nongovernmental) over a period of five years.

A joint Chinese research team from Tsinghua University and Peking University has developed a lightweight, durable floating sponge for use in cleaning up oil spills. The sponge is made of randomly oriented carbon nanotubes and attracts only oil. It expands to hold nearly 200 times its weight and 800 times the volume of the oil. It automatically moves toward higher concentrations of the oil and can be squeezed clean and reused dozens of times.

Industry and business. Nearly all the major industries and business sectors rely to some degree on advanced ceramics. Corning, Alcoa, Boeing, and Motorola are just a few of the major corporations that use advanced ceramics in their products. DuPont

FASCINATING FACTS ABOUT CERAMICS

- Advanced ceramics are the basis of the lightest, most durable body armor used by soldiers for small- to medium-caliber protection.
- Snowboards have become stronger and tougher, thanks to special composite materials that combine innovative glass laminates and carbon fiber materials.
- Prior to the 1700's, potters were criticized for digging holes in the roads to obtain more clay. The name of this offense is still used: pothole.
- Ceramics tend to be hard but brittle. To reduce the incidence of cracks, they are often applied as coatings on other materials that are resistant to cracking.
- Clay bricks are the only building product that will not burn, melt, dent, peel, warp, rot, rust, or succumb to termites. Buildings built with bricks are also more resistant to hurricanes.
- Radioactive glass microspheres are being used to treat patients with liver cancer. The microspheres, which are about one-third the diameter of a human hair, are made radioactive in a nuclear reactor. They are inserted into the artery supplying blood to the tumor using a catheter, and the radiation destroys the tumor and causes only minimal damage to the healthy tissue.

recently opened a photovoltaic applications lab at its Chestnut Run facility in Wilmington, Delaware. The lab will support materials development for the fast-growing photovoltaic energy market. PolyPlus in Berkeley, California, is developing lightweight, high-energy, single-use batteries that can use the surrounding air as a cathode. The company is currently developing these batteries for the government, but it expects them to be on the market within a few years. The drive toward efficient, inexpensive, clean technology ensures continued advances in research and development in the field.

CAREERS AND COURSE WORK

Ceramic scientists and engineers work in a variety of industries, as their skills are applicable in various contexts. These fields include aerospace, medicine, mining, refining, electronics, nuclear technology, telecommunications, transportation, and construction. A bachelor's degree is required for entrance into the above-mentioned fields. A master's degree in ceramic engineering qualifies the holder for managerial, consulting, sales, research, development, and administrative positions.

Pursuing a career in this field requires one have an aptitude for the sciences. Most colleges require that high school course work include four years of English, four years of math (at least one of which should be an advanced math course), at least three years of science (one of which should be a laboratory science), and at least two years of a foreign language.

Typical course work for a bachelor's degree in ceramics engineering includes calculus, physics, chemistry, statistics, materials engineering, and glass engineering. It also includes biology, mechanics, English composition, process design, and ceramic processing. There are fewer than ten universities in the United States that offer bachelor's degrees in ceramic engineering.

These include the Inamori School of Engineering at Alfred University and Missouri University of Science and Technology. Other universities, such as Iowa State University and Ohio State University, offer a bachelor's degree in materials engineering with a specialization in ceramics engineering. Additionally, some schools offer a combined-degree option: Undergraduate students can combine undergraduate and graduate course work to earn a bachelor's and a master's or doctorate in materials engineering with a concentration in ceramics simultaneously.

However, many American universities offer a bachelor's degree in materials engineering with a specialization in ceramics engineering.

An alternative to acquiring a bachelor's degree in ceramics engineering is to acquire a degree in a related field and then pursue a master's degree in ceramics engineering. Examples of related fields include biomedical engineering, chemical engineering, materials engineering, chemistry, physics, and mechanical engineering.

About ten universities in the United States offer master's degrees in ceramic engineering. As a result, admission into these programs is extremely competitive. Graduate students focus primarily on research and development, though they are required to take classes. Doctoral candidates also focus on research and development and after they have been awarded their degree, they can choose to teach or continue working in research.

SOCIAL CONTEXT AND FUTURE PROSPECTS

As mentioned in preceding sections, the advanced ceramics segment of the field, which is the primary focus of ceramic engineering, has plenty of room for growth. In 2008, there were about 24,350 materials engineers (including ceramic engineers) employed nationally; and a small number of ceramic engineers taught in colleges. The number of job openings is expected to exceed the number of engineers available.

Many of the industries in which ceramic engineers work—stone, clay, and glass products; primary metals; fabricated metal products; and transportation equipment industries—are expected to experience little employment growth through the year 2018. However, employment opportunities are expected to grow in service industries (research and testing, engineering and architectural). This is primarily because more firms, and by extension, more ceramic engineers, will be hired to develop improved materials for industrial customers.

—*Ezinne Amaonwu, LLB, MAPW*

FURTHER READING

Barsoum, M. W. *Fundamentals of Ceramics.* London: Institute of Physics, 2003. This text provides a detailed yet easy to understand overview of ceramics engineering. It is a good introductory and reference text, especially for readers with experience in other fields of science such as physics and chemistry, as it approaches ceramics engineering from these viewpoints.

Callister, William D., Jr., and David G. Rethwisch. *Materials Science and Engineering: An Introduction.* 8th ed. Hoboken, N.J.: John Wiley & Sons, 2010. This text provides an overview of materials engineering, with useful content on ceramics.

King, Alan G. *Ceramic Technology and Processing.* Norwich, N.Y.: Noyes, 2002. Published posthumously by the author's son, this text provides a technical, detailed description of every step in the advanced ceramics-production process. It focuses on implementing the production process in a laboratory, describes common problems associated with each step, and offers solutions for these problems.

Kingery, W. D., H. K. Bowen, and D. R. Uhlmann. *Introduction to Ceramics.* 2d ed. New York: John Wiley & Sons, 1976. A well-written guide to the basics of ceramics engineering, with an exhaustive index and questions to test the reader's retention.

Rahaman, M. N. *Ceramic Processing and Sintering.* 2d ed. New York: Marcel Dekker, 2003. Provides a clear description of all the steps in ceramics processing.

WEB SITES

American Ceramic Society
http://ceramics.org

American Society for Testing and Materials
http://www.astm.org

ASM International
http://www.asminternational.org

See also: Calculus.

CHAOTIC SYSTEMS

FIELDS OF STUDY

Mathematics; chemistry; physics; ecology; systems biology; engineering; theoretical physics; theoretical biology; quantum mechanics; fluid mechanics; astronomy; sociology; psychology; behavioral science.

SUMMARY

Chaotic systems theory is a scientific field blending mathematics, statistics, and philosophy that was developed to study systems that are highly complex, unstable, and resistant to exact prediction. Chaotic systems include weather patterns, neural networks, some social systems, and a variety of chemical and quantum phenomena. The study of chaotic systems began in the nineteenth century and developed into a distinct field during the 1980's.

Chaotic systems analysis has allowed scientists to develop better prediction tools to evaluate evolutionary systems, weather patterns, neural function and development, and economic systems. Applications from the field include a variety of highly complex evaluation tools, new models from computer and electrical engineering, and a range of consumer products.

KEY TERMS AND CONCEPTS

- **aperiodic:** Of irregular occurrence, not following a regular cycle of behavior.
- **attractor:** State to which a dynamic system will eventually gravitate—a state that is not dependent on the starting qualities of the system.
- **butterfly effect:** Theory that small changes in the initial state of a system can lead to pronounced changes that affect the entire system.
- **complex system:** System in which the behavior of the entire system exhibits one or more properties that are not predictable, given a knowledge of the individual components and their behavior.
- **dynamical system:** Complex system that changes over time according to a set of rules governing the transformation from one state to the next.
- **nonlinear:** Of a system or phenomenon that does not follow an ordered, linear set of cause-and-effect relationships.
- **randomness:** State marked by no specific or predictable pattern or having a quality in which all possible behaviors are equally likely to occur.
- **static system:** System that is without movement or activity and in which all parts are organized in an ordered series of relationships.
- **system:** Group interrelated by independent units that together make up a whole.
- **unstable:** State of a system marked by sensitivity to minor changes in a variety of variables that cause major changes in system behavior.

DEFINITION AND BASIC PRINCIPLES

Chaotic systems analysis is a way of evaluating the behavior of complex systems. A system can be anything from a hurricane or a computer network to a single atom with its associated particles. Chaotic systems are systems that display complex nonlinear relationships between components and whose ultimate behavior is aperiodic and unstable.

Chaotic systems are not random but rather deterministic, meaning that they are governed by some overall equation or principle that determines the behavior of the system. Because chaotic systems are determined, it is theoretically possible to predict their behavior. However, because chaotic systems are unstable and have so many contributing factors, it is nearly impossible to predict the system's behavior. The ability to predict the long-term behavior of complex systems, such as the human body, the stock market, or weather patterns, has many potential applications and benefits for society.

The butterfly effect is a metaphor describing a system in which a minor change in the starting conditions can lead to major changes across the whole system. The beat of a butterfly's wing could therefore set in motion a chain of events that could change the universe in ways that have no seeming connection to a flying insect. This kind of sensitivity to initial conditions is one of the basic characteristics of chaotic systems.

BACKGROUND AND HISTORY

French mathematician Henri Poincaré and Scottish theoretical physicist James Clerk Maxwell are considered two of the founders of chaos theory and complexity studies. Maxwell and Poincaré both worked on problems that illustrated the sensitivity of complex systems in the late nineteenth century. In 1960, meteorologist Edward Lorenz coined the term "butterfly effect" to describe the unstable nature of weather patterns.

By the late 1960's, theoreticians in other disciplines began to investigate the potential of chaotic dynamics. Ecologist Robert May was among the first to apply chaotic dynamics to population ecology models, to considerable success. Mathematician Benoit Mandelbrot also made a significant contribution in the mid1970's with his discovery and investigation of fractal geometry.

In 1975, University of Maryland scientist James A. Yorke coined the term "chaos" for this new branch of systems theory. The first conference on chaos theory was held in 1977 in Italy, by which time the potential of chaotic dynamics had gained adherents and followers from many different areas of science.

Over the next two decades, chaos theory continued to gain respect within the scientific community and the first practical applications of chaotic systems analysis began to appear. In the twenty-first century, chaotic systems have become a respected and popular branch of mathematics and system analysis.

HOW IT WORKS

There are many kinds of chaotic systems. However, all chaotic systems share certain qualities: They are unstable, dynamic, and complex. Scientists studying complex systems generally focus on one of these qualities in detail. There are two ultimate goals for complex systems research: first, to predict the evolution of chaotic systems and second, to learn how to manipulate complex systems to achieve some desired result.

Instability. Chaotic systems are extremely sensitive to changes in their environment. Like the example of the butterfly, a seemingly insignificant change can become magnified into system-wide transformations. Because chaotic systems are unstable, they do not settle to equilibrium and can therefore continue developing and leading to unexpected changes. In the chaotic system of evolution, minor changes can lead to novel mutations within a species, which may eventually give rise to a new species. This kind of innovative transformation is what gives chaotic systems a reputation for being creative. Scientists often study complexity and chaos by creating simulations of chaotic systems and studying the way the systems react when perturbed by minor stimuli. For example, scientists may create a computer model of a hurricane and alter small variables, such as temperature, wind speed and other factors, to study the ultimate effect on the entire storm system.

Strange attractors. An attractor is a state toward which a system moves. The attractor is a point of equilibrium, where all the forces acting on a system have reached a balance and the system is in a state of rest.

Because of their instability, chaotic systems are less likely to reach a stable equilibrium and instead proceed through a series of states, which scientists sometimes call a dynamic equilibrium. In a dynamic equilibrium, the system is constantly moving toward some attractor, but as it moves toward the first attractor, forces begin building that create a second attractor. The system then begins to shift from the first to the second attractor, which in turn leads to the formation of a new attractor, and the system shifts again. The forces pulling the system from one attractor to the next never balance each other completely, so the system never reaches absolute rest but continues changing.

Visual models of complex systems and their strange attractors display what mathematicians call fractal geometry. Fractals are patterns that, like chaotic systems,

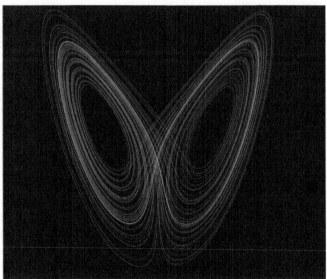

A plot of the Lorenz attractor for values $r = 28$, $= 10$, $b = 8/3$

are nonlinear and dynamic. Fractal geometry occurs throughout nature, including in the formation of ice crystals and the branching of the circulatory system in the human body. Scientists study the mathematics behind fractals and strange attractors to find patterns that can be applied to complex systems in nature. The study of fractals has yielded applications for medicine, economics, and psychology.

Emergent properties. Chaotic systems also display emergent properties, which are system-wide behaviors that are not predictable from knowledge of the individual components. This occurs when simple behaviors among the components combine to create more complex behaviors. Common examples in nature include the behavior of ant colonies and other social insects.

Mathematic models of complex systems also yield emergent properties, indicating that this property is driven by the underlying principles that lead to the creation of complexity. Using these principals, scientists can create systems, such as computer networks, that also display emergent properties.

APPLICATIONS AND PRODUCTS

Medical technology. The human heart is a chaotic system, and the heartbeat is controlled by a set of nonlinear, complex variables. The heartbeat appears periodic only when examined through an imprecise measure such as a stethoscope. The actual signals that compose a heartbeat occur within a dynamic equilibrium.

Applying chaos dynamics to the study of the heart is allowing physicians to gain more accuracy in determining when a heart attack is imminent, by detecting minute changes in rhythm that signify the potential for the heart to veer away from its relative rhythm. When the rhythm fluctuates too far, physicians use a defibrillator to shock the heart back to its rhythm. A new defibrillator model developed by scientists in 2006 and 2007 uses a chaotic electric signal to more effectively force the heart back to its normal rhythmic pattern.

Consumer products. In the mid-1990's, the Korean company Goldstar manufactured a chaotic dishwasher that used two spinning arms operated with an element of randomness to their pattern. The chaotic jet patterns were intended to provide greater cleaning power while using less energy.

Chaotic engineering has also been used in the manufacture of washing machines, kitchen mixers, and a variety of other simple machines. Although most chaotic appliances have been more novelty than innovation, the principles behind chaotic mechanics have become common in the engineering of computers and other electrical networks.

Business and industry. The global financial market is a complex, chaotic system. When examined mathematically, fluctuations in the market appear aperiodic and have nonlinear qualities. Although the market seems to behave in random ways, many believe that applying methods created for the study of chaotic systems will allow economists to elucidate hidden developmental patterns. Knowledge of these patterns might help predict and control recessions and other major changes well before they occur.

The application of chaos theory to market analysis produced a subfield of economics known as fractal market analysis, wherein researchers conduct economic analyses by using models with fractal geometry. By looking for fractal patterns and assuming that the market, like other chaotic systems, is highly sensitive to small changes, economists have been able to build more accurate models of market evolution.

IMPACT ON INDUSTRY

The fact that chaos theory applies to so many fields means that there are numerous funding opportunities available. Many countries offer some government funding for studies into chaos theory and related fields.

In the United States, the National Science Foundation offers a grant for researchers studying theoretical biology, which has been awarded to several research teams studying chaotic systems. The National Science Foundation, National Institute of Standards and Technology, National Institutes of Health, and the Department of Defense have also provided some funding for research into chaotic dynamics.

Any product created using chaotic principles may be subject to regulation governing the industry in question. For instance, the chaotic defibrillator is subject to regulation from the Food and Drug Administration, which has oversight over any new medical technology. Other equipment, such as the chaotic dishwasher, will be subject to consumer safety regulations at the federal and regional levels.

CAREERS AND COURSE WORK

Those seeking careers in chaotic systems analysis might begin by studying mathematics or physics at a university. Those with backgrounds in other fields such as biology, ecology, economics, sociology, and computer science might also choose to focus on chaotic systems during their graduate training.

There are a number of graduate programs offering training in chaotic dynamics. For instance, the Center for Interdisciplinary Research on Complex Systems at Northeastern University in Boston, Massachusetts, offers programs training students in many types of complex system analyses. The Center for Nonlinear Phenomena and Complex Systems at the Université Libre de Bruxelles offers programs in thermodynamics and statistical mechanics.

Careers in chaotic systems span a range of fields from engineering and computer network design to theoretical physics. Trained researchers can choose to contribute to academic analyses and other pertinent laboratory work or focus on creating applications for immediate consumer use.

SOCIAL CONTEXT AND FUTURE PROSPECTS

As chaotic systems analysis spread in the 1980's and 1990's, some scientists began theorizing that chaotic dynamics might be an essential part of the search for a grand unifying theory or theory of everything. The grand unifying theory is a concept that emerged in the early nineteenth century, when scientists began

FASCINATING FACTS ABOUT CHAOTIC SYSTEMS

- Chaos theory gained popular attention after the 1993 film adaptation of Michael Crichton's novel *Jurassic Park* (1990). In the film, a scientist played by Jeff Goldblum evokes chaos theory to warn that complex systems are impossible to control.

- Fractal image compression uses the fractal property of self similarity to compress images.

- Chaotic systems can be found at both the largest and the smallest levels of the universe, from the evolution of galaxies to molecular activity.

- Fractal geometry is commonly found in the development of organisms. Among plants with fractal organization are broccoli, ferns, and many kinds of flowers.

- Some philosophers and neuroscientists have suggested that the creative properties of complex systems are what allow for free will in the human mind.

- The Korean company Daewoo marketed a washing machine known as the "bubble machine" in 1990, which the company claimed was the first consumer device to use chaos theory in its operation.

theorizing that there might be a single set of rules and patterns underpinning all phenomena in the universe.

The idea of a unified theory is controversial, but the search has attracted numerous theoreticians from mathematics, physics, theoretical biology, and philosophy. As research began to show that the basic principles of chaos theory could apply to a vast array of fields, some began theorizing that chaos theory was part of the emerging unifying theory. Because many systems meet the basic requirements to be considered complex chaotic systems, the study of chaos theory and complexity has room to expand. Scientists and engineers have only begun to explore the practical applications of chaotic systems, and theoreticians are still attempting to evaluate and study the basic principles behind chaotic system behavior.

—*Micah L. Issitt, MA*

FURTHER READING

Ford, Kenneth W. *The Quantum World: Quantum Physics for Everyone.* Cambridge, Mass.: Harvard University Press, 2005. An exploration of quantum theory and astrophysics written for the general reader. Contains coverage of complexity and chaotic systems applied to quantum phenomena.

Gleick, James L. *Chaos: Making a New Science.* Rev. ed. New York: Penguin Books, 2008. Technical popular account of the development of chaos theory and complexity studies. Topics covered include applications of chaos systems analysis to ecology, mathematics, physics, and psychology.

Gribbin, John. *Deep Simplicity: Bringing Order to Chaos and Complexity.* New York: Random House, 2005. An examination of how chaos theory and related fields have changed scientific understanding of the universe. Provides many examples of complex systems found in nature and human culture.

Mandelbrot, Benoit, and Richard L. Hudson. *The Misbehavior of Markets: A Fractal View of Financial Turbulence.* New York: Basic Books, 2006. Provides a detailed examination of fractal patterns and chaotic systems analysis in the theory of financial markets. Provides examples of how chaotic analysis can be used in economics and other areas of human social behavior.

Stewart, Ian. *Does God Play Dice? The New Mathematics of Chaos.* 2d ed. 1997. Reprint. New York: Hyperion, 2005. An evaluation of the role that order and chaos play in the universe through popular explanations of mathematic problems. Includes accessible descriptions of complex mathematical ideas underpinning chaos theory.

Strogatz, Peter H. *Sync: How Order Emerges from Chaos in the Universe, Nature and Daily Life.* 2003. Reprint. New York: Hyperion Books, 2008. Strogatz provides an accessible account of many aspects of complex systems, including chaos theory, fractal organization, and strange attractors.

CHARGE-COUPLED DEVICES

FIELDS OF STUDY

Physics; electrical engineering; electronics; optics; mathematics; semiconductor manufacturing; solid-state physics; chemistry; mechanical engineering; photography; computer science; computer programming; signal processing.

SUMMARY

Originally conceptualized as a form of optical volatile memory for computers, charge-coupled devices were also seen to be useful for capturing light and images. The ability to capture images that can be read by a computer in digital form was profoundly useful, and charge-coupled devices quickly become known for their imaging abilities. Besides their utility in producing digital images that could be analyzed and manipulated by computers, charge-coupled devices also proved to be more sensitive to light than film. As technology advanced, charge-coupled devices became less expensive and more capable, and eventually they replaced film as the primary means of taking photographs.

KEY TERMS AND CONCEPTS

- **bias frame:** Image frame of zero length exposure, designed to measure pixel-to-pixel variation across the charge-coupled device array.
- **blooming:** Bleeding of charge from a full charge, often appearing as lines extending from overexposed pixels.
- **dark frame:** Image taken with the shutter closed, designed to measure noise and defects affecting images in the charge-coupled device.
- **flat field:** Image taken of uniform illumination across the charge-coupled device chip, designed to measure defects in the chip and optical system affecting images.
- **photoelectric effect:** Physical process whereby photons of light knock electrons off of atoms.
- **pixel:** Individual charge-collecting region and ultimately a single element of a final image.
- **quantum efficiency:** Measure of the percentage of photons hitting a device that will produce electrons.
- **radiation noise:** Spurious charges produced in an image due to ionizing radiation, frequently produced by cosmic rays.
- **readout noise:** Electronic noise affecting image quality due to shifting and reading the charge in the pixels.

DEFINITION AND BASIC PRINCIPLES

A charge-coupled device (CCD) is an electronic array of devices fabricated at the surface of a semiconductor chip. The typical operation of a CCD is to collect electrical charge in specific locations laid out as an array on the surface of the chip. This electrical charge is normally produced by light shining onto the chip, producing an electrical charge. Each collecting area is called a pixel and consists of an electrical potential well that traps charge carriers. The charge in one area of the array is read, and then the charges on other parts of the chip are shifted from potential well to potential well until all have been read. The brighter the light shining on a pixel, the more charge it will have. Thus, an image can be constructed from the data collected by each pixel. When the CCD is placed at the focal plane of an optical system, such as a lens, then the image on the chip will be the same as the image seen by the lens; thus CCD chips can be used as the heart of a camera system. Because the data is collected electronically in digital form for each pixel, the image from a CCD is inherently a digital image and can be processed by a computer. For this reason, CCDs (and digital cameras) have become more popular than film as a way to take photographs.

BACKGROUND AND HISTORY

The charge-coupled device was invented based on an idea by Willard S. Boyle and George E. Smith of AT&T Bell Laboratories for volatile computer memory. They were seeking to develop an electrical analogue to magnetic bubble memory. Boyle and Smith postulated that electric charge could be used to store data in a matrix on a silicon chip. The charge could be moved from location to location within that array, shifting from one holding cell to another by applying appropriate voltages. The name "charge-coupled device" stemmed from this shifting of electrical charge around on the device. Initial research aimed to perfect the CCD as computer memory, but the inventors soon saw that it may be even more useful in converting light values into electrical charge that could be read. This property led to the development of the CCD as an imaging array. Subsequent developments made pixel sizes smaller, the arrays larger, and the price lower. The CCD became the detector at the heart of digital cameras, and their ease of use quickly made digital cameras popular. By the early 2000's, digital cameras and their CCDs had become more popular than film cameras.

How it Works

At the heart of a charge-coupled device is a semiconductor chip with electrical potential wells created by doping select areas in an array on the surface of the device. Doping of a semiconductor involves fabricating it with a select impurity that would tend to have either an extra electron or one too few electrons for the normal lattice structure of the semiconductor. Fabricating the CCD with select areas so doped would tend to make any electrical charge created at the surface of the semiconductor chip want to stay in the region of the potential well. By applying the proper electrical voltage between adjacent wells, however, charge is permitted to flow from one well to another. In this way, charge anywhere in the array can be moved from well to well to a location where it can be read.

The photoelectric effect. The key to using a charge-coupled device is to put charge on it. This is done by shining light onto the surface of the semiconductor chip. When photons of sufficient energy shine on a material, the absorbed light can knock electrons from atoms. This is known as the photoelectric effect, a physical phenomenon first observed by German physicist Heinrich Hertz in 1887 and explained by German physicist Albert Einstein in 1905. While the photoelectric electric effect works in many materials, in a semiconductor it produces a free electron and a missing electron, called a hole. The hole is free to move just like an electron. Normally, the electron and hole recombine (the electron going back to the atom it came from in the semiconductor lattice) after the electron-hole pair are created. However, if an electric field is present, the electron may go one way in the semiconductor and the hole

another. Such an electric field is present in an operating CCD, and so the electrons will congregate in regions of lower electrical potential (electrical potential wells), building up an electrical charge proportional to the intensity of the light shining on the chip. The potential wells are arranged in an array across the surface of the CCD. The individual wells serve as the individual picture elements in a digital image and are called pixels. A nearly identical technology to the CCD uses different materials and is called the complementary metal oxide semiconductor (CMOS).

Shifting charges. Once charge is produced and captured on the CCD chip, it must be measured in order for the device to be of any use. Typically, charge is read only from one location on the CCD. After that charge is read, all of the charges in all the pixels on a row are shifted over to the adjacent pixel to allow the next charge to be read from the readout position. This process repeats until all charges in the row have been read. All columns of pixels are then shifted down one row, shifting the charges in the second row into the row that has just been read. The process of reading that row continues until all charges are read and all the columns shift down again, repopulating the empty row of charge with new charges. All the charges are shifted around until they have been read. If a pixel has too much charge (being overexposed), then the charge can bleed between adjacent pixels. This creates spurious surplus charge on charges throughout a column and appears as vertical lines running through the image. This image defect is known as blooming.

Color images. Charge-coupled devices respond to light intensity, not color. Therefore, all images are inherently black and white. There are several techniques to get color images from CCDs. The simplest and most cost-effective method is to take pictures using different color filters and then to create a final image by adding those images in the colors of the filters. This is the technique usually used in astronomy. The disadvantage of this technique, though, is that it requires at least three images taken in succession. This does not permit "live" or action photography. A separate technique is to have filters constructed over each pixel on the CCD array. This permits one chip to take multiple images simultaneously. The disadvantage of this technique, besides the cost of constructing such an array of filters, is that each image uses only about one-third of the pixels on the chip,

The charge-coupled device was invented in 1969 at AT&T Bell Labs by Willard Boyle and George E. Smith. Shown here in 2009

thus losing image quality and detail. Most color digital cameras use this technique. Though other strategies exist for making color images with CCDs, these two are the most common.

APPLICATIONS AND PRODUCTS

CCDs and CMOSs are the detectors in virtually every digital imaging system. CCDs have also displaced the imaging tubes in television cameras.

Cameras. Because CCDs produce digital images, they can be directly viewed using computers, sent by e-mail, coded into Web pages, and stored in electronic media. Digital images can also be viewed almost immediately after taking them, unlike film, which must first be developed. This has made digital cameras very popular. In addition to providing nearly instant pictures, CCDs can be manufactured quite small, permitting cameras that are much smaller than used to exist. This has allowed cameras to be placed in cell phones, computers, tablet computers, and many other devices. Traditionally, the pixels in CCDs have been larger than the grains in film, so digital image quality has suffered. However, technological advantages have permitted CCD pixels to be made much smaller, and by 2010, most commercial CCD camera systems compared very favorably with film systems in terms of image quality, with high-end CCD cameras often performing better than most film cameras.

Scientific applications. Some of the first applications of CCDs were in the scientific community. CCDs permit very accurate images, but these images also contain precise data about light intensity. This is particularly important in chemistry, where spectral lines

can be studied using CCD detectors. It is also important in astronomy, where digital images can be studied and measured using computers. Astronomical observatories were among the first to adopt CCD imaging systems, and they are still leaders in working with companies to develop new more powerful and larger CCD arrays.

Satellite images. Early satellite cameras used film that had to be dropped from orbit to be collected below. Soon, television cameras displaced film, but as soon as the first reliable charge-coupled devices capable of imaging became available, that technology was far better than any other. As of 2011, nearly all satellite imaging systems, both civilian and military, rely on a type of CCD technology.

IMPACT ON INDUSTRY

Government and university research. The digital nature of CCDs makes them of great use in scientific images, and since they are more sensitive than film, they can record dimmer light sources. This makes CCDs particularly useful for astronomy. By the 1990's, practically all professional astronomical research was done using CCDs. By 2010, nearly all amateur astronomers were using CCDs in astrophotography.

Industry and business. CCDs are used in fax machines, scanners, and even many small copiers. The ubiquity of these devices changes business communications. Web cameras attached to computers make videos easy to produce and have led to an explosion of individual amateur video clips being posted online in such places as YouTube.

Major corporations. The widespread use of CCDs in digital cameras has displaced film in most photographic activities. Many companies that used to make film have had to adjust their business model to accommodate the change in customer activities. Companies that built their reputations on fine film products have had to shift their product lines to stay competitive. Most of these companies now sell more digital-related products than film-related products.

CAREERS AND COURSE WORK

Charge-coupled devices have become the most common imaging system in use. Thus, any career that uses images will come into contact with CCDs. For most people using CCDs, there is no different training or course work than would be needed to use any camera.

Manufacturing CCDs or developing new CCD technology, however, requires specialized training. CCDs are semiconductor devices, so their manufacture requires a background in semiconductor-manufacturing technology. Both two-year and four-year degrees in semiconductor manufacturing technology exist, and these degrees require courses in electronics, semiconductor physics, semiconductor-manufacturing technology, mathematics, and related disciplines. Manufacturing CCDs is not much different from manufacturing any other semiconductor device.

Construction of equipment, such as cameras, to hold CCDs involves both optics and electronics. Manufacture of such equipment requires little detailed knowledge of these areas other than what is required for any other camera or electronic device. Most of the components are manufactured ready to be assembled. Some specialized electronics knowledge is needed in order to design the circuit boards to operate the CCDs, however.

Developing improved CCD technology or research into better CCDs or technologies to replace CCDs requires much more advanced training, generally a postgraduate degree in physics or electrical engineering. These degrees require extensive physics and mathematics courses, along with electrical engineering course work and courses in chemistry, materials science, electronics, and semiconductor technology. Most jobs related to research and development are in university or corporate laboratory settings.

SOCIAL CONTEXT AND FUTURE PROSPECTS

CCDs were once very esoteric devices. Early digital cameras had CCDs with large pixels, producing pictures of inferior quality to even modest film cameras. However, CCDs quickly became less expensive and of higher quality and soon became ubiquitous. Most cameras sold by 2005 were digital cameras. Most cell phones now come equipped with cameras having CCD technology. Since most people own a cell phone, this puts a camera in the hands of almost everyone all the time. The number of photographs and videos being made has skyrocketed far beyond what has ever existed since the invention of the camera. Since these images are digital, they can easily be shared by e-mail

---

FASCINATING FACTS ABOUT CHARGE-COUPLED DEVICES

- A charge-coupled device (CCD) is the imaging system at the heart of every digital camera, Web cam, and cell phone camera. CCDs are also found in scanners, fax machines, and many other places where digital images are needed.
- By 2005, digital-related sales revenue exceeded film-related sales revenue for Kodak, a company known for its high-quality film and film photography products.
- Astronomical and military applications of CCDs led the way before widespread commercial applications became available.
- In order to reduce thermal noise, astronomical charge-coupled devices are often cooled with cryogenic fluids, such as liquid nitrogen.
- The first experimental CCD imaging system, constructed in 1970 at AT&T Bell Laboratories, had only eight pixels.
- The first commercially produced CCD imaging system, made available by Fairchild Electronics in 1974, had only 10,000 pixels, arranged in a 100-by-100 array.
- Because charge-coupled devices are sensitive to infrared light, they can be used in some solid-state night-vision devices—and some commercial cameras and video cameras have night-vision capability.

---

or on Web pages. Social networking on the Internet has permitted these pictures to be more widely distributed than ever. There is no indication that this will change in the near future. The possibility exists that in the near future two-way video phone calls may become commonplace.

Though charge-coupled devices are far superior to film, other competing technologies that can do the same thing are being developed. A very similar technology (so similar that it is often grouped with CCD technology) is the complementary metal oxide semiconductor (CMOS). CMOS technology works the same as CCD technology for the end user, but the details of the chip operation and manufacture are different. However, with CCD technology becoming so commonplace in society, it is likely that other imaging technological developments will follow.

—Raymond D. Benge, Jr., MS

FURTHER READING

Berry, Richard, and James Burnell. *The Handbook of Astronomical Image Processing*. 2d ed. Richmond, Va.: Willmann-Bell, 2005. Though the book discusses charge-coupled devices somewhat, it mainly focuses on how to use the images taken by such devices to produce pictures and do scientific research. Comes with a CD-ROM.

Holst, Gerald C., and Terrence S. Lomheim. *CMOS/ CCD Sensors and Camera Systems*. 2d ed. Bellingham, Wash.: SPIE, 2011. A technical review of CCD and CMOS imaging systems and cameras with specifics useful for engineering applications and for professionals needing to know the specific capabilities of different systems.

Howell, Steve B. *Handbook of CCD Astronomy*. 2d ed. New York: Cambridge University Press, 2006. This excellent and easy to understand handbook covers how to use charge-coupled device cameras in astronomy. The book is written for the amateur astronomer but contains good information on how CCDs work.

Janesick, James R. *Scientific Charge-Coupled Devices*. Bellingham, Wash.: SPIE, 2001. A very technical, thorough overview of how charge-coupled devices work.

Jorden, P. R. "Review of CCD Technologies." *EAS Publications Series* 37 (2009): 239-245. A good overview of charge-coupled devices, particularly as they relate to astronomy and satellite imaging systems.

Nakamura, Junichi, ed. *Image Sensors and Signal Processing for Digital Still Cameras*. Boca Raton, Fla.: CRC Press, 2006. A thorough and somewhat technical overview of CCD and CMOS technologies. Includes a good history of digital photography and an outlook for future technological developments.

Williamson, Mark. "The Latent Imager." *Engineering & Technology* 4, no. 14 (August 8, 2009): 36-39. A very good history of the development of charge-coupled device imaging systems and some of their uses.

WEB SITES

American Astronomical Society
http://aas.org

Institute of Electrical and Electronics Engineers
http://www.ieee.org

Professional Photographers of America
http://www.ppa.com

See also: Computer Engineering; Computer Science; Electrical Engineering; Mechanical Engineering; Optics; Photography.

CIVIL ENGINEERING

FIELDS OF STUDY

Mathematics; chemistry; physics; engineering mechanics; strength of materials; fluid mechanics; soil mechanics; hydrology; surveying; engineering graphics; environmental engineering; structural engineering; transportation engineering.

SUMMARY

Civil engineering is the branch of engineering concerned with the design, construction, and maintenance of fixed structures and systems, such as large buildings, bridges, roads, and other transportation systems, and water supply and wastewater-treatment systems. Civil engineering is the second oldest field of engineering, with the term "civil" initially used to differentiate it from the oldest field of engineering, military engineering. The major subdisciplines within civil engineering are structural, transportation, and environmental engineering. Other possible areas of specialization within civil engineering are geotechnical, hydraulic, construction, and coastal engineering.

KEY TERMS AND CONCEPTS

- **abutment:** Part of a structure designed to withstand thrust, such as the end supports of a bridge or an arch.
- **aqueduct:** Large pipe or conduit used to transport water a long distance.
- **backfill:** Material used in refilling an excavated area.

- **cofferdam:** Temporary structure built to keep water out of a construction zone in a river.
- **design storm:** Storm of specified return period and duration at a specified location, typically used for storm water management design.
- **foundation:** Ground that is used to support a structure.
- **freeboard:** Difference in height between the water level and the top of a tank, dam, or channel.
- **girder:** Large horizontal structural member, supporting vertical loads.
- **invert:** Curved, inside, bottom surface of a pipe.
- **percolation test:** Test to determine the drainage capability of soil.
- **rapid sand filter:** System for water treatment by gravity filtration through a sand bed.
- **reinforced concrete:** Concrete that contains wire mesh or steel reinforcing rods to give it greater strength.
- **sharp-crested weir:** Obstruction with a thin, sharp upper edge, used to measure flow rate in an open channel.
- **tension member:** Structural member that is subject to tensile stress.
- **ultimate bearing capacity:** Theoretical maximum pressure that a soil can support from a load without failure

Definition and Basic Principles

Civil engineering is a very broad field of engineering, encompassing subdisciplines ranging from structural engineering to environmental engineering, some of which have also become recognized as separate fields of engineering. For example, environmental engineering is included as an area of specialization within most civil engineering programs, many colleges offer separate environmental engineering degree programs.

Civil engineering, like engineering in general, is a profession with a practical orientation, having an emphasis on building things and making things work. Civil engineers use their knowledge of the physical sciences, mathematics, and engineering sciences, along with empirical engineering correlations to design, construct, manage, and maintain structures, transportation infrastructure, and environmental treatment equipment and facilities.

Empirical engineering correlations are important in civil engineering because useable theoretical

equations are not available for all the necessary engineering calculations. These empirical correlations are equations, graphs, or nomographs, based on experimental measurements, that give relationships among variables of interest for a particular engineering application. For example the Manning equation gives an experimental relationship among the flow rate in an open channel, the slope of the channel, the depth of water, and the size, shape, and material of the bottom and sides of the channel. Rivers, irrigation ditches, and concrete channels used to transport wastewater in a treatment plant are examples of open channels. Similar empirical relationships are used in transportation, structural, and other specialties within civil engineering.

Background and History

Civil engineering is the second oldest field of engineering. The term "civil engineering" came into use in the mid-eighteenth century and initially referred to any practice of engineering by civilians for nonmilitary purposes. Before this time, most large-scale construction projects, such as roads and bridges, were done by military engineers. Early civil engineering projects were in areas such as water supply, roads, bridges, and other large structures, the same type of engineering work that exemplifies civil engineering in modern times.

Although the terminology did not yet exist, civil engineering projects were carried out in early times. Examples include the Egyptian pyramids (about 2700-2500 BCE), well-known Greek structures such as the Parthenon (447-438 BCE), the Great Wall of China (220 BCE), and the many roads, bridges, dams, and aqueducts built throughout the Roman Empire.

Most of the existing fields of engineering split off from civil engineering or one of its offshoots, as new fields emerged. For example, with increased use of machines and mechanisms, the field of mechanical engineering emerged in the early nineteenth century.

How it Works

In addition to mathematics, chemistry, and physics, civil engineering makes extensive use of principles from several engineering science subjects: engineering mechanics (statics and strength of materials), soil mechanics, and fluid mechanics.

Engineering mechanics—statics. As implied by

the term "statics," this area of engineering concerns objects that are not moving. The fundamental principle of statics is that any stationary object must be in static equilibrium. That is, any force on the object must be cancelled out by another force that is equal in magnitude and acting in the opposite direction. There can be no net force in any direction on a stationary object, because if there were, it would be moving in that direction. The object considered to be in static equilibrium could be an entire structure or it could be any part of a structure down to an individual member in a truss. Calculations for an object in static equilibrium are often done through the use of a free body diagram, that is a sketch of the object, showing all the forces external to that object that are acting on it. The principle then used for calculations is that the sum of all the horizontal forces acting on the object must be zero and the sum of all the vertical forces acting on the object must be zero. Working with the forces as vectors helps to find the horizontal and vertical components of forces that are acting on the object from some direction other than horizontal or vertical.

Engineering mechanics—strength of materials. This subject is sometimes called mechanics of materials. Whereas statics works only with forces external to the body that is in equilibrium, strength of materials uses the same principles and also considers internal forces in a structural member. This is done to determine the required material properties to ensure that the member can withstand the internal stresses that will be placed on it.

Soil mechanics. Knowledge of soil mechanics is needed to design the foundations for structures. Any structure resting on the Earth will be supported in some way by the soil beneath it. A properly designed foundation will provide adequate long-term support for the structure above it. Inadequate knowledge of soil mechanics or inadequate foundation design may lead to something such as the Leaning Tower of Pisa. Soil mechanics topics include physical properties of soil, compaction, distribution of stress within soil, and flow of water through soil.

Fluid mechanics. Fundamental principles of physics are used for some fluid mechanics calculations. Examples are conservation of mass (called the continuity equation in fluid mechanics) and conservation of energy (also called the energy equation or the first law of thermodynamics). Some fluid mechanics applications, however, make use of empirical (experimental) equations or relationships. Calculations for flow through pipes or flow in open channels, for example, use empirical constants and equations.

Knowledge from engineering fields of practice. In addition to these engineering sciences, a civil engineer uses accumulated knowledge from the civil engineering areas of specialization. Some of the important fields of practice are hydrology, geotechnical engineering, structural engineering, transportation engineering, and environmental engineering. In each of these fields of practice, there are theoretical equations, empirical equations, graphs or nomographs, guidelines, and rules of thumb that civil engineers use for design and construction of projects related to structures, roads, storm water management, or wastewater-treatment projects, for example.

Civil engineering tools. Several tools available for civil engineers to use in practice are engineering graphics, computer-aided drafting (CAD), surveying, and geographic information systems (GIS). Engineering graphics (engineering drawing) has

John Smeaton, the "father of civil engineering"

been a mainstay in civil engineering since its inception, for preparation of and interpretation of plans and drawings. Most of this work has come to be done using computer-aided drafting. Surveying is a tool that has also long been a part of civil engineering. From laying out roads to laying out a building foundation or measuring the slope of a river or of a sewer line, surveying is a useful tool for many of the civil engineering fields. Civil engineers often work with maps, and geographic information systems, a much newer tool than engineering graphics or surveying, make this type of work more efficient.

Codes and design criteria. Much of the work done by civil engineers is either directly or indirectly for the public. Therefore, in most of civil engineering fields, work is governed by codes or design criteria specified by some state, local, or federal agency. For example, federal, state, and local governments have building codes, state departments of transportation specify design criteria for roads and highways, and wastewater-treatment processes and sewers must meet federal, state, and local design criteria.

APPLICATIONS AND PRODUCTS

Structural engineering. Civil engineers design, build, and maintain many and varied structures. These include bridges, towers, large buildings (skyscrapers), tunnels, and sports arenas. Some of the civil engineering areas of knowledge needed for structural engineering are soil mechanics/geotechnical engineering, foundation engineering, engineering mechanics (statics and dynamics), and strength of materials.

When the Brooklyn Bridge was built over the East River in New York City (1870-1883), its suspension span of 1,595 feet was the longest in the world. It remained the longest suspension bridge in North America until the Williamsburg Bridge was completed in New York City in 1903. The Brooklyn Bridge joined Brooklyn and Manhattan, and helped establish the New York City Metropolitan Area.

The Golden Gate Bridge, which crosses the mouth of San Francisco Bay with a main span of 4,200 feet, had nearly triple the central span of the Brooklyn Bridge. It was the world's longest suspension bridge from its date of completion in 1937 until 1964, when the Verrazano-Narrows Bridge opened in New York City with a central span that was 60 feet longer than that of the Golden Gate Bridge. The Humber Bridge,

which crosses the Humber estuary in England and was completed in 1981, has a single suspended span of 4,625 feet and is the longest suspension bridge in the world.

One of the most well-known early towers illustrates the importance of good geotechnical engineering and foundation design. The Tower of Pisa, commonly known as the Leaning Tower of Pisa, in Italy started to lean to one side very noticeably, even during its construction (1173-1399). Its height of about 185 feet is not extremely tall in comparison with towers built later, but it was impressive when it was built. The reason for its extreme tilt (more than 5 meters off perpendicular) is that it was built on rather soft, sandy soil with a foundation that was not deep enough or spread out enough to support the structure. In spite of this, the Tower of Pisa has remained standing for more than six hundred years.

Another well-known tower, the Washington Monument, was completed in 1884. At 555 feet in height, it was the world's tallest tower until the Eiffel Tower, nearly 1,000 feet tall, was completed in 1889. The Washington Monument remains the world's tallest masonry structure. The Gateway Arch in St. Louis, Missouri, is the tallest monument in the United States, at 630 feet.

The 21-story Flatiron Building, which opened in New York City in 1903, was one of the first skyscrapers. It is 285 feet tall and its most unusual feature is its triangular shape, which was well suited to the wedge-shaped piece of land on which it was built. The 102floor Empire State Building, completed in 1931 in New York City with a height of 1,250 feet, outdid the Chrysler Building that was under construction at the same time by 204 feet, to earn the title of the world's tallest building at that time. The Sears Tower (now the Willis Tower) in Chicago is 1,450 feet tall and was the tallest building in the world when it was completed in 1974. Several taller buildings have been constructed since that time in Asia.

Some of the more interesting examples of tunnels go through mountains and under the sea. The Hoosac Tunnel, built from 1851 to 1874, connected New York State to New England with a 4.75-mile railway tunnel through the Hoosac Mountain in northwestern Massachusetts. It was the longest railroad tunnel in the United States for more than fifty years. Mount Blanc Tunnel, built from 1957 to 1965, is a 7.25-mile long highway tunnel under Mount

Blanc in the Alps to connect Italy and France. The Channel Tunnel, one of the most publicized modern tunnel projects, is a rather dramatic and symbolic tunnel. It goes a distance of 31 miles beneath the English Channel to connect Dover, England, and Calais, France.

Transportation engineering. Civil engineers also design, build, and maintain a wide variety of projects related to transportation, such as roads, railroads, and pipelines.

Many long, dramatic roads and highways have been built by civil engineers, ever since the Romans became the first builders of an extensive network of roads. The Appian Way is the most well known of the many long, straight roads built by the Romans. The Appian Way project was started in 312 BCE by the Roman censor, Appius Claudius. By 244 BCE, it extended about 360 miles from Rome to the port of Brundisium in southeastern Italy. The Pan-American Highway, often billed as the world's longest road, connects North America and South America. The original Pan-American Highway ran from Texas to Argentina with a length of more than 15,500 miles. It has since been extended to go from Prudhoe Bay, Alaska, to the southern tip of South America, with a total length of nearly 30,000 miles. The U.S. Interstate Highway system has been the world's biggest earthmoving project. Started in 1956 by the Federal Highway Act, it contains sixty-two highways covering a total distance of 42,795 miles. This massive highway construction project transformed the American system of highways and had major cultural impacts.

The building of the U.S. Transcontinental Railroad was a major engineering feat when the western portion of the 2,000-mile railroad across the United States was built in the 1860's. Logistics was a major part of the project, with the need to transport steel rails and wooden ties great distances. An even more formidable task was construction of the Trans-Siberian Railway, the world's longest railway. It was built from 1891 to 1904 and covers 5,900 miles across Russia, from Moscow in the west to Vladivostok in the east.

The Denver International Airport, which opened in 1993, was a very large civil engineering project. This airport covers more than double the area of all of Manhattan Island.

The first oil pipeline in the United States was a 5-mile-long, 2-inch-diameter pipe that carried 800 barrels of petroleum per day. Pipelines have become much larger and longer since then. The Trans-Alaska Pipeline, with 800 miles of 48-inch diameter pipe, can carry 2.14 million barrels per day. At the peak of construction, 20,000 people worked 12-hour days, seven days a week.

Water resources engineering. Another area of civil engineering practice is water resources engineering, with projects like canals, dams, dikes, and seawater barriers.

The oldest known canal, one that is still in operation, is the Grand Canal in China, which was constructed between 485 BCE and 283 CE. The length of the Grand Canal is more than 1,000 miles, although its route has varied because of several instances of rerouting, remodeling, and rebuilding over the years. The 363-mile-long Erie Canal was built from 1817 to 1825, across the state of New York from Albany to Buffalo, thus overcoming the Appalachian Mountains as a barrier to trade between the eastern United States and the newly opened western United States. The economic impact of the Erie Canal was tremendous. It reduced the cost of shipping a ton of cargo between Buffalo and New York City from about $100 per ton (over the Appalachians) to $4 per ton (through the canal).

The Panama Canal, constructed from 1881 to 1914 to connect the Atlantic and Pacific oceans through the Isthmus of Panama, is only about 50 miles long, but its construction presented tremendous challenges because of the soil, the terrain, and the tropical illnesses that killed many workers. Upon its completion, however, the Panama Canal reduced the travel distance from New York City to San Francisco by about 9,000 miles.

When the Hoover Dam was built from 1931 to 1936 on the Colorado River at the Colorado-Arizona border, it was the world's largest dam, at a height of 726 feet and crest length of 1,224 feet. The technique of passing chilled water through pipes enclosed in the concrete to cool the newly poured concrete and speed its curing was developed for the construction of the Hoover Dam and is still in use. The Grand Coulee Dam, in the state of Washington, was the largest hydrolectric project in the world when it was built in the 1930's. It has an output of 10,080 megawatts. The Itaipu Dam, on the Parana River, along the border of Brazil and Paraguay, is also one of the largest hydroelectric dams in the world. It began operation in 1984 and is capable of producing 13,320 megawatts.

Dikes, dams and similar structures have been used for centuries around the world for protection against flooding. The largest sea barrier in the world is a 2-mile-long surge barrier in the Oosterschelde estuary of the Netherlands, constructed from 1958 to 1986. Called the Dutch Delta Plan, the purpose of this project was to reduce the danger of catastrophic flooding. The impetus that brought this project to fruition was a catastrophic flood in the area in 1953. A major part of the barrier design consists of sixty-five huge concrete piers, weighing in at 18,000 tons each. These piers support tremendous 400-ton steel gates to create the sea barrier. The lifting and placement of these huge concrete piers exceeded the capabilities of any existing cranes, so a special U-shaped ship was built and equipped with gantry cranes. The project used computers to help in guidance and placement of the piers. A stabilizing foundation used for the concrete piers consists of foundation mattresses made up of layers of sand, fine gravel, and coarse gravel. Each foundation mattress is more than 1 foot thick and more than 650 feet by 140 feet, with a smaller mattress placed on top.

IMPACT ON INDUSTRY

In view of its status as the second oldest engineering discipline and the essential nature of the type of work done by civil engineers, it seems reasonable that civil engineering is well established as an important field of engineering around the world. Civil engineering is the largest field of engineering in the United States. The U.S. Bureau of Labor Statistics estimates that 278,400 civil engineers were employed in the United States in 2008. The bureau also projected that civil engineering employment would grow at the rate of 24 percent rate until 2018, which is much faster than average for all occupations. Civil engineers are employed by a wide variety of government agencies, by universities for research and teaching, by consulting engineering firms, and by industry.

Consulting engineering firms. This is the largest sector of employment for civil engineers. There are many consulting engineering firms around the world, ranging in size from small firms with a few employees to very large firms with thousands of employees. In 2010, the *Engineering News Record*

identified the top six U.S. design firms: AECOM Technology, Los Angeles; URS, San Francisco; Jacobs, Pasadena, California; Fluor, Irving, Texas; CH2M Hill, Englewood, Colorado; and Bechtel, San Francisco. Many consulting engineering firms have some electrical engineers and mechanical engineers, and some even specialize in those areas; however, a large proportion of engineering consulting firms are made up predominantly of civil engineers. About 60 percent of American civil engineers are employed by consulting engineering firms.

Construction firms. Although some consulting engineering firms design and construct their own projects, some companies specialize in constructing projects designed by another firm. These companies also use civil engineers. About 8 percent of American civil engineers are employed in the nonresident building construction sector.

Other industries. Some civil engineers are employed in industry, but less than 1 percent of American civil engineers are employed in an industry other than consulting firms and construction firms. The industry sectors that hire the most civil engineers are oil and gas extraction and pipeline companies.

Government agencies. Civil engineers work for many federal, state, and local government agencies. For example, the U.S. Department of Transportation uses civil engineers to handle its many highway and other transportation projects. Many road or highway projects are handled at the state level, and each state department of transportation employs many civil engineers. The U.S. Corps of Engineers and the Department of the Interior's Bureau of Reclamation employ many civil engineers for their many water resources projects. Many cities and counties have one or more civil engineers as city or county engineers, and many have civil engineers in their public works departments. About 15 percent of American civil engineers are employed by state governments, about 13 percent by local government, and about 4 percent by the federal government.

University research and teaching. Because civil engineering is the largest field of engineering, almost every college of engineering has a civil engineering department, leading to a continuing demand for civil engineering faculty members to teach the next generation of civil engineers and to conduct sponsored research projects. This applies to universities not only in the United States but also around the world.

CAREERS AND COURSE WORK

A bachelor's degree in civil engineering is the requirement for entry into this field. Registration as a professional engineer is required for many civil engineering positions. In the United States, a graduate from a bachelor's degree program accredited by the Accreditation Board for Engineering and Technology is eligible to take the Fundamentals of Engineering exam to become an engineer in training. After four years of professional experience under the supervision of a professional engineer, one is eligible to take the Professional Engineer exam to become a registered professional engineer.

A typical program of study for a bachelor's degree in civil engineering includes chemistry, calculus and differential equations, calculus-based physics, engineering graphics/AutoCAD, surveying, engineering mechanics, strength of materials, and perhaps engineering geology, as well as general education courses during the first two years. This is followed by fluid mechanics, hydrology or water resources, soil mechanics, engineering economics, and introductory courses for transportation engineering, structural engineering, and environmental engineering, as well as civil engineering electives to allow specialization in one of the areas of civil engineering during the last two years.

A master's degree in civil engineering that provides additional advanced courses in one of the areas of specialization, an M.B.A., or engineering management master's degree complement a bachelor's of science degree and enable their holder to advance more rapidly. A master's of science degree would typically lead to more advanced technical positions, while an M.B.A. or engineering management degree would typically lead to management positions.

Anyone aspiring to a civil engineering faculty or research position must obtain a doctoral degree. In that case, to provide proper preparation for doctoral level study, any master's-level study should be in pursuit of a research-oriented master of science degree rather than a master's degree in engineering or a practice-oriented master of science degree.

SOCIAL CONTEXT AND FUTURE PROSPECTS

Civil engineering projects typically involve basic infrastructure needs such as roads and highways, water supply, wastewater treatment, bridges, and public buildings. These projects may be new construction

---

FASCINATING FACTS ABOUT CIVIL ENGINEERING

- The Johnstown flood in Pennsylvania, on May 31, 1889, which killed more than 2,200 people, was the result of the catastrophic failure of the South Fork Dam. The dam, built in 1852, held back Lake Conemaugh and was made of clay, boulders, and dirt. An improperly maintained spillway combined with heavy rains caused the collapse.

- The I-35W bridge over the Mississippi River in Minneapolis, Minnesota, collapsed during rush-hour traffic on August 1, 2007, causing 13 deaths and 145 injuries. The collapse was blamed on undersized gusset plates, an increase in the concrete surface load, and the weight of construction supplies and equipment on the bridge.

- On November 7, 1940, 42-mile-per-hour winds twisted the Tacoma Narrows Bridge and caused its collapse. The bridge, with a suspension span of 2,800 feet, had been completed just four months earlier. Steel girders meant to support the bridge were blocking the wind, causing it to sway and eventually collapse.

- Low-quality concrete and incorrectly placed rebar led to shear failure, collapsing the Highway 19 overpass in Laval, Quebec, on September 30, 2006.

- The first design for the Gateway Arch in St. Louis, Missouri, had a fatal flaw that made it unstable at the required height. The final design used 886 tons of stainless steel, making it a very expensive structure.

- On January 9, 1999, just three years after it was built, the Rainbow Bridge, a pedestrian bridge across the Qi River in Sichuan Province, collapsed, killing 40 people and injuring 14. Concrete used in the bridge was weak, parts of it were rusty, and parts had been improperly welded.

- On December 7, 1982, an antenna tower in Missouri City, Texas, collapsed, killing 2 riggers on the tower and 3 who were in an antenna section that was being lifted. U-bolts holding the antenna failed, and as it fell, it hit a guy wire on the tower, collapsing the tower. The engineers on the project declined to evaluate the rigger's plans.

- On February 26, 1972, coal slurry impoundment dam 3 of the Pittston Coal Company in Logan County, West Virginia, failed, four days after it passed inspection by a federal mine inspector. In the Buffalo Creek flood that resulted, 125 people were killed. The dam, which was above dams 1 and 2, had been built on coal slurry sediment rather than on bedrock.

or repair, maintenance or upgrading of existing highways, structures, and treatment facilities. The buildup of such infrastructure since the beginning of the twentieth century has been extensive, leading to a continuing need for the repair, maintenance, and upgrading of existing structures. Also, governments tend to devote funding to infrastructure improvements to generate jobs and create economic activity during economic downturns. All of this leads to the projection for a continuing strong need for civil engineers.

—*Harlan H. Bengtson, MS, PhD*

FURTHER READING

Arteaga, Robert R. *The Building of the Arch*. 10th ed. St. Louis, Mo.: Jefferson National Parks Association, 2002. Describes how the Gateway Arch in St. Louis, Missouri, was built up from both sides and came together at the top. Contains excellent illustrations. Davidson, Frank Paul, and Kathleen Lusk-Brooke, comps. *Building the World: An Encyclopedia of the Great Engineering Projects in History*. Westport, Conn.: Greenwood Press, 2006. Examines more than forty major engineering projects from the Roman aqueducts to the tunnel under the English Channel.

Hawkes, Nigel. *Structures: The Way Things Are Built*. 1990. Reprint. New York: Macmillan, 1993. Discusses many well-known civil engineering projects. Chapter 4 contains information about seventeen projects, including the Great Wall of China, the Panama Canal, and the Dutch Delta Plan. Contains illustrations and discussion of the effect of the projects.

National Geographic Society. *The Builders: Marvels of Engineering*. Washington, D.C.: National Geographic Society, 1992. Documents some of the most ambitious civil engineering projects, including roads, canals, bridges, railroads, skyscrapers, sports arenas, and exposition halls. Discussion and excellent illustrations are included for each project.

Weingardt, Richard G. *Engineering Legends: Great American Civil Engineers—Thirty-two Profiles of Inspiration and Achievement*. Reston, Va.: American Society of Civil Engineers, 2005. Looks at the lives of civil engineers who were environmental experts, transportation trendsetters, builders of bridges, structural trailblazers, and daring innovators.

WEB SITES

American Institute of Architects
http://www.aia.org

American Society of Civil Engineers
http://www.asce.org

Institution of Civil Engineers
http://www.ice.org.uk

WBGH Educational Foundation and PBS
http://www.pbs.org/wgbh/buildingbig/index.html

See also: Bridge Design and Barodynamics; Hydraulic Engineering.

CLIMATE ENGINEERING

FIELDS OF STUDY

Atmospheric chemistry; ecology; meteorology; plant biology; ecosystem management; marine systems and chemistry; geoengineering; environmental engineering; aeronautical engineering; naval architecture; engineering geology; applied geophysics; Earth system modeling.

SUMMARY

Climate engineering (more commonly known as geoengineering) is a field of science that aims to deliberately control both micro (local) and macro (global) climates—by actions such as seeding clouds or shooting pollution particles into the upper atmosphere to reflect the Sun's rays—with the intention of reversing the effects of global warming.

KEY TERMS AND CONCEPTS

- **albedo:** Measure of the reflectivity of the Earth surface (that is, the proportion of solar radiation or energy that is reflected back into space).

- **carbon dioxide temoval (CDR):** Geoengineering techniques and technologies that remove carbon dioxide from the atmosphere in an attempt to combat climate change and global warming.
- **climate change:** Alterations over time in global temperatures and rainfall because of human-caused increases in greenhouse gases, such as carbon dioxide and methane.
- **fossil fuel:** Deposit of either solid (coal), liquid (oil), or gaseous (natural gas) hydrocarbon derived from the decomposition of organic plant and animal matter over many million of years within the Earth's crust.
- **ocean acidification:** Process in which the pH (acidity) level of the Earth's oceans becomes more acidic because of the increase in atmospheric carbon dioxide.
- **phytoplankton:** Microscopic photosynthesizing aquatic plants that are responsible for absorbing carbon dioxide and producing more than half of the world's oxygen.
- **radiative forcing:** Measure of the effect that climatic factors have in modifying the energy balance (incoming and outgoing energy) of the Earth's atmosphere.
- **solar radiation management (SRM):** Geoengineering techniques and technologies that reflect solar radiation from the Earth's atmosphere (or surface) back into space in an attempt to combat climate change and global warming.
- **stratosphere:** The region of the Earth's atmosphere (approximate altitude of 10 to 50 kilometers) below the mesosphere and above the troposphere.

DEFINITION AND BASIC PRINCIPLES

The term "geoengineering" comes from the Greek word "geo," meaning earth, and the word "engineering," a field of applied science that incorporates and uses data from scientific, technical, and mathematical sources in the invention and execution of specific structures, apparatuses, and systems.

Not to be confused with geotechnical engineering, which is related to the engineering behavior of earth materials, geoengineering is a field of science that aims to manipulate both micro (local) and macro (global) climates with the intention of reversing the effects of climate change and global warming. Although geoengineering, or climate engineering, can be undertaken on the local level, such as cloud seeding, most of the proposed technologies are based on a worldwide scale for the wholesale mediation of the global climate and the effects of climate change.

BACKGROUND AND HISTORY

Human society and ancient cultures had long believed power over the weather to be the province of gods, but the advent of human-made climate control began with the concept of rainmaking (pluviculture) in the mid-nineteenth century. James Pollard Espy, an American meteorologist, was instrumental in developing the thermal theory of storms. Although such a discovery placed him in the annals of scientific history, he also became known (and disparaged) for his ideas regarding artificial rainmaking. According to Espy, the burning of huge areas of forest would create sufficient hot air and updraft to create clouds and precipitation.

In 1946, Vincent Schaefer, a laboratory technician at General Electric Research Laboratory in New York, generated a cloud of ice from water droplets that had been supercooled by dry ice. That same year, Bernard Vonnegut discovered that silver iodide smoke produced the same result. These processes came to be known as cloud seeding. Cloud seeding attempts to encourage precipitation to fall in arid or drought-stricken agricultural areas by scattering silver iodide in rain clouds. Traditional climate modification has generally been limited to cloud seeding programs on a regional or local level.

In the following years, particularly during the Cold War, researchers in countries including the United States and the Soviet Union investigated climate control for its potential as a weapon. The Soviets also looked at climate control to warm the frozen Siberian tundra. However, in the 1970's, as concern regarding the greenhouse effect began to be expressed within conventional scientific circles, the concept of climate engineering took on new global significance.

Cesare Marchetti, an Italian physicist, first coined the term "geoengineering" in 1977. The word initially described the specific process of carbon dioxide capture and storage in the ocean depths as a means of climate change abatement. Since then, however, the term "geoengineering" has been used to cover all engineering work performed to manipulate the global and local climate. In later years, many researchers have begun using the term "climate engineering,"

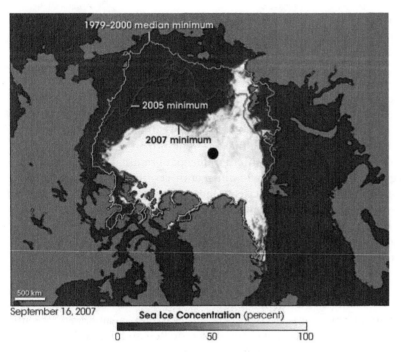

September 16, 2007

Sea Ice Concentration (percent)

0   50   100

Significant reduction in ice volume in the Arctic Ocean in the range between 1979 and 2007 years

which is a more accurate way to describe this type of applied science.

HOW IT WORKS

Significant scientific evidence points to human activities, particularly the burning of fossil fuels, as playing a role in global climate change. An increase in the level of greenhouse gases, in particular carbon dioxide and methane, has been identified as the main culprit in relation to global climate change. The majority of scientists agree that the safest and best way to tackle climate change is through a reduction in fossil fuel consumption via the implementation of green technology, societal change, and industry regulation. However, although carbon reduction technology is already available and affordable, many scientists are increasingly concerned that carbon reduction schemes will not be introduced in time to stop the possible and probable effects of climate change. As a result, interest in geoengineering technology as a means to provide rapid solutions to climate change has increased.

There are two main areas of geoengineering research—carbon dioxide removal (CDR) and solar radiation management (SRM). Techniques based on

CDR propose to remove carbon dioxide from the atmosphere and store it, while SRM techniques seek to reflect solar radiation from the Earth's atmosphere (or surface) back into space. Despite having the same objective of combating climate change and reducing global temperatures, these two approaches differ significantly in regard to their implementations, time scales, temperature effects, and possible consequences. According to the Royal Society of London, CDR techniques "address the root cause of climate change by removing greenhouse gases from the atmosphere," whereas SRM techniques "attempt to offset effects of increased greenhouse gas concentrations by causing the Earth to absorb less solar radiation."

The CDR approach removes carbon dioxide from the atmosphere and sequesters it underground or in the ocean. Many consider this approach to be the more attractive of the two as it not only helps reduce global temperatures but also works to combat issues such as ocean acidification caused by escalating carbon dioxide levels. Conversely, SRM techniques have no effect on atmospheric carbon dioxide levels, instead using reflected sunlight to reduce global temperatures.

APPLICATIONS AND PRODUCTS

Traditionally, climate modification took the form of regional or local cloud seeding programs. In cloud seeding, a substance, usually silver iodide, is scattered in rain clouds to cause precipitation in arid or drought-stricken agricultural areas. By the start of the twenty-first century, however, deleterious global climate change and increasing levels of atmospheric carbon dioxide had pushed climate engineering to the forefront of science. Geoengineering technology, once considered as more fringe then functional, is being investigated as a possible weapon in the fight against global warming.

Despite the surge in research and interest, no longitudinal or large-scale geoengineering projects have been conducted. Geoengineering applications and products are generally regarded as highly speculative and environmentally untested, with significant ambiguity in regard to global and institutional regulation. Although climate engineering theories abound, only a limited few have captured global attention for their feasibility and applicability.

Iron fertilization of the oceans. The intentional introduction of iron into the upper layers of certain areas of the ocean to encourage phytoplankton blooms is a form of CDR. The concept relies on the fact that increasing certain nutrients—such as iron— in nutrient-poor areas stimulates phytoplankton growth. Carbon dioxide is absorbed from the surface of the ocean during the processes of photosynthesis; when the phytoplankton, marine animals, and plankton die and sink in the natural cycle, that carbon is removed from the atmosphere and sequestered in the ocean's depths.

Scrubbers and artificial trees. Both scrubbers and artificial trees aim to remove and store carbon dioxide from the Earth's atmosphere and assist in reducing the effect of climate change. Scrubbing towers involve the use of large wind turbines funneling air into specially designed structures, where the air reacts with a number of chemicals to form water and carbonate precipitates, essentially capturing the carbon. These carbon precipitates can then be stored. The use of artificial trees seeks the same result but by a different method. Large artificial trees or structures act as filters to capture and convert atmospheric carbon dioxide into a carbonate, which is then removed and stored.

Biochar. Biochar is a form of charcoal created from the pyrolysis (chemical decomposition by heating) of plant and animal waste. It captures and stores carbon by sequestering it in biomass. The biochar can be returned to the soil as fertilizer, as it helps the soil retain water and necessary nutrients, as well as to store carbon.

Stratospheric sulfur aerosols. Stratospheric sulfur aerosols are minute sulfur-rich particles that are found in the Earth's stratosphere and are often observed following significant volcanic activity (such as after the 1991 Mount Pinatubo eruption). The presence of these aerosols in the stratosphere results in a cooling effect. The SRM geoengineering technique of intentionally releasing sulfur aerosols into the stratosphere is based on the concept that they would produce a cooling or dimming effect by reflecting solar radiation. A workable delivery system has not yet been developed, but proposals include using high-altitude aircraft, balloons, and rockets.

Orbital mirrors and space sunshades. The SRM technique of orbital mirrors and space sunshades entails the release of many billions (possibly trillions) of small reflective objects at a Lanrangian point in space to partially reflect solar radiation or impede it from entering the Earth's atmosphere. The theory is that the decrease in sunlight hitting the Earth's surface would help decrease average global temperatures.

Marine cloud whitening. Marine cloud whitening involves increasing the reflective properties of cloud cover so that solar radiation entering the Earth's atmosphere is reflected back into space. Proposed methods for achieving this include mounting large-scale mist-producing structures on seafaring vessels. The theory is that the spray of minute water droplets released by these structures would increase cloud cover and whitening, which would in turn increase sunlight reflection.

Reflective roofs. Reflective or white roofs are often considered to be the most cost effective and easily implemented SRM method of reducing global temperatures. The concept relies on reflecting solar radiation back into space by using white materials (or paint) on the surface of building roofs.

IMPACT ON INDUSTRY

Geoengineering has been experiencing a revolution. For the most part, many scientists and governments have not been interested in climate engineering because of questions about its feasibility or have been opposed to research and implementation of such technology. Because of the global consequences of climate change, however, many government agencies, universities, industries, corporations, and businesses worldwide are becoming interested in researching and working in this field of applied science.

Government and university research. The governments of the United States and the United Kingdom have increasingly expressed interest in researching geoengineering technology and determining its applicability. Although the U.S. government has traditionally shied away from the climate engineering controversy, the Congress heard testimony related to geoengineering in November, 2009. This follows on the July, 2009, joint seminar, "Geoengineering: Challenges and Global Impacts," held by the United Kingdom's House of Commons, the Institute of Physics, the Royal Society of Chemistry, and the Royal

Academy of Engineering. The seminar explored and discussed the possibility of applying geoengineering technology to help mitigate the effects of climate change.

Although a growing number of governments are studying the possible use of geoengineering technology to help alleviate the effects of climate change, many scientists and other experts hesitate to embrace and apply the technologies. They cite the unanticipated consequences of these yet untested theories and technologies.

An increasing number of universities are exploring the potential of climate engineering. Some of the more well-known institutions that conduct geoengineering research include the Department of Atmospheric Chemistry at the Max Planck Institute for Chemistry, the Carnegie Institution for Science's Department of Global Ecology at Stanford University, the Earth Institute at Columbia University, the Energy and Environmental Systems Group at the University of Calgary, and the Oxford Geoengineering Institute.

Major organizations. Many world-renowned organizations have entered into the geoengineering debate. Although there is growing support for increasing research into such technology, many organizations are hesitant to fully endorse its actual implementation and use. Such organizations include the Intergovernmental Panel on Climate Change, which published information about climate engineering in its fourth assessment report, "Climate Change 2007," and the American Meteorological Society, which released a guarded geoengineering policy statement in 2009. One of the most comprehensive reports, *Geoengineering the Climate: Science, Governance, and Uncertainty* (2009), was issued by the Royal Society of London, the world's oldest scientific academy. The society detailed the potential applications and costs of geoengineering and stated that although some climate engineering schemes are utterly implausible or are being endorsed without due consideration, some are more realistic. Its report, the first by any national science academy, concludes that much greater research is required before the implementation of any such technology is undertaken.

CAREERS AND COURSE WORK
Most commonly, students who wish to pursue careers in climate engineering begin by majoring in scientific fields such as atmospheric science, marine systems,

and civil engineering. However, given the multitude of fields covered in geoengineering, a career in this applied science could follow many different paths, and students should have a solid understanding of subjects such as atmospheric chemistry, ecology, meteorology, plant biology, ecosystem management, marine systems and chemistry, and engineering. The majority of graduate programs in this area are open to students who have backgrounds in engineering, applied sciences, or closely related disciplines. University research covers many areas, and students who obtain a doctorate or master's degree in climate engineering can expect to have careers in geoengineering research and design, atmospheric sciences, aeronautical and nautical engineering, and environmental management consulting.

SOCIAL CONTEXT AND FUTURE PROSPECTS
In the past, global geoengineering was regarded as more science fiction than fact. With greenhouse gas emissions continuing to increase from the burning of fossil fuel, however, the concept of deliberately engineering the Earth's climate is garnering interest and gaining credibility.

Engineering the climate could assist in lowering atmospheric carbon dioxide and reducing the impact of climate change. Although the majority of geoengineering scientists stress that such technology should be used only for emergency quick fixes or as a last resort, many are also stressing that research into such technology is imperative. The 2009 Royal Society report strongly advocated increased research and recommended the world's governments allocate some £100 million ($165 million), collectively, per year to examine geoengineering options. Given the probable economic and environmental costs associated with climate change, many researchers have claimed that implementing geoengineering technology may be cheaper and more viable than doing nothing. Many researchers are also quick to state, however, that while dollar costs may be affordable, the possible environmental and economic costs if the technology fails to work or has unexpected deleterious effects may be immeasurable.

Despite the move into more mainstream science, climate engineering is still controversial and full of significant technological, social, ethical, legal, diplomatic, and safety challenges. The concept of purposely modifying the climate to correct climate

FASCINATING FACTS ABOUT CLIMATE ENGINEERING

- Human activity, particularly the burning of fossil fuels such as oil and coal, is responsible for the release of some 30 billion metric tons of carbon dioxide into the Earth's atmosphere every year.
- Most climate engineering proposals fall into one of two fundamentally different approaches--the removal of carbon dioxide from the atmosphere or the reflection of solar radiation from Earth's atmosphere or surface back into space.
- Climate engineers claim they can cool the planet and reverse ice-sheet melting by mimicking a natural volcanic eruption through the release of sulfate aerosols into the Earth's stratosphere, where they will reflect sunlight back into space.
- Giant artificial trees, ocean fertilization, and trillions of space mirrors are some of the most popular proposals being considered and researched in the fight against global warming.
- Reflecting the Sun's rays back into the atmosphere could quickly affect global temperatures but would not have an impact on the levels of carbon dioxide in the atmosphere.
- Biochar, the end result of the burning of animal and agricultural waste, can help clean the air by storing carbon dioxide that would have been released into the atmosphere during decomposition of the waste, and it assists plants by aiding in the storage of carbon from photosynthesis.

change caused by humans has been labeled as, at best, ironic and, at worst, catastrophic. Significant concern has been raised about encouraging geoengineering research and technology. For example, many conservation organizations are concerned that access to such technology not only will promote the "if we build it, they will use it" mentality but also will lessen the resolve of governments and people to tackle climate change by reducing people's ecological footprint and consumption of fossil fuels.

In addition, geoengineering technology may ignite tensions between nations The ethical ramifications of climate engineering are unclear, and significant confusion and uncertainty exists regarding who should implement and control the global thermostat. If a country implements technology to fix one area and

inadvertently adversely alters the climatic patterns in another country, who pays for the mishap? The consequences of climate manipulation on a global scale will almost certainly be unequal across nations. Such concerns stress the importance of conducting further research and of using caution regarding any technological advance in geoengineering.

—*Christine Watts, PhD*

FURTHER READING

Flannery, Tim F. *The Weather Makers: How Man Is Changing the Climate and What It Means for Life on Earth.* New York: Grove Press, 2006. A comprehensive look at how human activity has influenced the global climate and its impact on life on the Earth.

Keith, David. "Geoengineering the Climate: History and Prospect." *Annual Review of Energy and the Environment* 25 (November, 2000): 245-284. Provides an interesting review of the history of climate control technology and future directions.

Launder, Brian, and J. Michael T. Thompson, eds. *Geoengineering Climate Change: Environmental Necessity or Pandora's Box?* New York: Cambridge University Press, 2010. Presents a comprehensive examination of the problems of climate change and the potential of different geoengineering technologies.

The Royal Society. *Geoengineering the Climate: Science, Governance, and Uncertainty.* London: The Royal Society, 2009. Twelve experts in the fields of science, economics, law, and social science conducted this study of the main techniques of climate engineering, focusing on how well they might work and examining possible consequences.

WEB SITES

American Geophysical Union Geoengineering the Climate
http://www.agu.org/sci_pol positions/geoengineering.shtml

Intergovernmental Panel on Climate Change
http://www.ipcc.ch

See also: Air-Quality Monitoring; Atmospheric Sciences; Biofuels and Synthetic Fuels; Climate Modeling; Climatology; Environmental Engineering; Meteorology.

CLIMATE MODELING

FIELDS OF STUDY

Physics; environmental science; oceanography; meteorology; climatology; atmospheric physics; earth sciences; computer science; computer programming; advanced mathematics; biology; sociology; chemistry; game theory; statistics; agriculture; forestry; anthropology.

SUMMARY

The goal of climate models is to provide insight into the Earth's climate system and the interactions among the atmosphere, oceans, and landmasses. Computer climate modeling provides the most effective means of predicting possible future changes and the potential effect such changes might have on societies and ecosystems. Shifting multiple variables and the unpredictability of climate model components have produced findings that fuel intense scientific and political debate. Climate models will continue to be refined as computing power increases, allowing models to be run with more integrative components and using wider, more refined space and time scales.

KEY TERMS AND CONCEPTS

- **climate:** Regional weather conditions averaged over an extended time period.
- **climate modification:** Changes to the climate due to either natural processes or human activities.
- **climate system:** System consisting of the atmosphere, hydrosphere, cryospher, land surface, and biosphere and their interactions.
- **cryosphere:** Portion of Earth covered by sea ice, mountain glaciers, snow cover, ice sheets, or permafrost.
- **feedback:** Process changing the relationship between a forcing agent acting on climate and the climate's response to that agent.
- **general circulation model:** Model simulating the state of the entire atmosphere, ocean, or both at chosen locations over a discrete, selected incremental time scale.
- **global warming:** Event characterized by measurable increases in annual mean surface temperature of the Earth, most noticeably occurring since about 1860.
- **greenhouse gases:** Atmospheric gases that absorb infrared radiation; the most important are carbon dioxide, methane, water vapor, and chlorofluorocarbons.
- **infrared absorbers:** Molecules such as carbon dioxide and water vapor that absorb electromagnetic radiation at infrared wavelengths.
- **infrared radiation:** Electromagnetic radiation with wavelengths between microwaves and visible red light.
- **microclimate:** Local climate regime distinguished by physical characteristics and processes associated with local geography.
- **radiative cooling:** Process by which the Earth's surface cools by emitting longwave radiation.
- **regime:** Preferred or dominant state of the climate system.

DEFINITION AND BASIC PRINCIPLES

All climate models are mathematical models derived from differential equations defining the principles of physics, chemistry, and fluid dynamics driving the observable processes of the Earth's climate. Climate models can be highly complex three-dimensional computer simulations using multiple variables and tens of thousands of differential equations requiring trillions of computations or fairly simple two-dimensional projections with a single equation defining a sole observable process. In all climate models, each additional physical process incorporated into the model increases the level of complexity and escalates the mathematical parameters needed to define the process's potential effects.

BACKGROUND AND HISTORY

The ability to predict weather and climate patterns could have a major impact on the health and well-being of societies. The possibility to warn of impending harm, prepare for changes, estimate outcomes, and chose more opportune times for essential human activities has been a dream of cultures for thousands of years. Although observational records of climate phenomena have been kept and interpreted for generations, the advent of computer models and advanced procedures for gathering and assimilating global data allows for quantitative assessments of the complex interactions between climate processes.

Climate models are governed by the laws of physics, which produce climate system components affecting the atmosphere, cryosphere, oceans, and land. No climate model would exist without an understanding of Newtonian mechanics and the fundamental laws of thermodynamics. By the late eighteenth century, scientific understanding of these physical laws allowed for numerical calculations to define observable climate processes. By the early twentieth century, scientists understood atmospheric phenomena as the product of preceding phenomena defined by physical laws. If one has accurate observable data of the atmosphere at a particular time and understands the physical laws under which atmospheric phenomena take place, it is possible to predict the outcome of future atmospheric changes. Mathematical equations were subsequently developed to define atmospheric motions and simulate observable processes and features of climate system components.

By the late 1940's, electronic computing provided a means by which greater numbers of equations could be assigned to data and calculated to define observable climate phenomena. Throughout the 1950's, predictive climate models were refined through changes to equations, resulting in modifications to models of atmospheric circulation. During the 1960's, similar equations were used to define ocean, land, and ice processes and their role in climate. Once this was done, the development of coupled models began. Coupled models involve combining observational processes to achieve more realistic outcomes. For each new process, however, multiple observations must be added and defined by additional equations. As the quality of observational data increases, the complexity of the climate model grows. Each observation of sea ice movement, cloud density, atmospheric chemistry, land cover, water

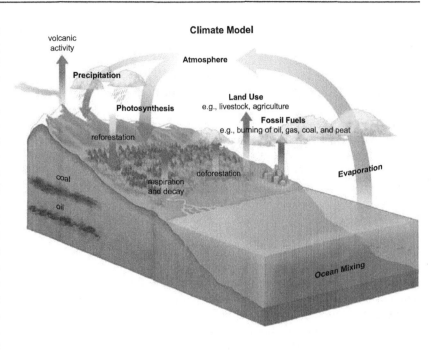

Climate Model

temperature, precipitation, and human activities can be defined mathematically and added to a climate model's program. By defining the multiple variables of climate processes and adding these to model programs, scientists became able to produce predictive simulations of climate ranging from small regional models to global general circulation models.

HOW IT WORKS

The purpose of all climate models is to simulate, over a given time period, regional or global patterns of climatic processes: wind, temperature, precipitation, ocean currents, sea levels, and sea ice. By imposing changes to the physical process within a model simulation—such as altering the amount of solar input, carbon dioxide, or ice cover—predictions can be made concerning possible future climate scenarios. Although models constructed from purely observational evidence result in less accurate predictions, the use of computers to process numerical models representing multiple levels of climate system feedbacks improves the accuracy of predictions for multiple components of the climate model.

When formulating a truly practical climate model, all components of Earth's climate system must be taken into consideration. The model must represent the atmosphere, the oceans, ice and snow cover on both land and sea, and landmasses, including their biomass. The interactions between these climate components occur in multiple ways, at differing intensities, and over varied time scales: Practical climate models must be able to reflect these dynamics.

The key to understanding climate and climatic change is energy balance. Multiple feedback processes occur within Earth's climate system, and these interactions constantly fluctuate to maintain an efficient energy balance. Positive feedback processes amplify variations to the climate system; negative feedbacks dampen them. Examples of feedbacks include radiation feedback from clouds and water vapor; changes to reflective power from ice, snow, and deserts; and changes in ocean temperatures. These and many other forms of feedback must be considered when preparing a climate model. Because the physics of energy transfer between climate system processes is so varied, resolution scale becomes a major problem in preparing data for modeling. Flux compensations must be made in models to account for differences in energy transfers among atmosphere, oceans, and land. Heat, momentum, water densities, size of phenomena such as currents or eddies, energy radiation, rainfall, wind, humidity, barometric pressure, atmospheric chemistry, and natural phenomena such as forest fires and volcanic eruptions all must be considered.

The purpose of preparing climate models is to answer the fundamental questions of what can be reliably predicted about climate and at what time scales these predictions can be made. The difficulty with all climate modeling is that the computers used must be powerful enough to deal with the complexity of observable climate system data; that the data used must be as free as possible of errors in interpretation; and that all observational data have limitations and these limit interpretive predictability. These factors are the limitations confining all existing climate-modeling capabilities and set the boundary ranges for predicting climate phenomena in space and time.

Variability of climate over space and time scales is a key component of Earth's history. Throughout Earth's history, climate has naturally fluctuated, often to extremes in comparison to present-day climatic conditions. Being able to distinguish between naturally occurring and human-induced climate changes is an important aspect of climate modeling.

The implications of climate models suggesting that human behaviors are altering the atmosphere—and in turn disrupting the planet's energy balance—drives vigorous debates and conflict.

APPLICATIONS AND PRODUCTS

All climate models are simulations that make predictions about climate processes. Climate change skeptics and deniers make that case that model predictions are merely simulations and not "real" data produced by "real" science. Without climate models, however, there would be no climate data. No climate observation, no satellite imagery or observation, no meteorological data, no atmospheric sampling, no remote sensing, and no chemical analysis exists without passing through a series of computer data models. Almost everything known about Earth's climatic processes exists because of models. At present, almost all knowledge and understanding of climate change comes from three kinds of computer models: simulation models of weather and climate; reanalysis models re-creating climate history from historical weather data, including worldwide weather stations and remote data platforms; and data models combining and adjusting measurements from many different sources, including ice cores, gas analysis, tree rings, observation data, and archaeological and geological data. Because the amount of weather and climate data available are so vast and diverse and from so many different sources and at differing time scales, it can be managed and organized only by computer analysis. The result of modern computer-simulated climate models has been to create a stable and reliable basis for making predictions about climate change and the byproducts of that change.

IMPACT ON INDUSTRY

Understanding how climate works and having the ability to predict possible changes to climate over certain regions and at specific time scales can have substantial impacts on the social, economic, and cultural well-being of any nation. More accurate prediction of cycles of increased precipitation or drought, rises in temperature, or shifts in oceanic currents allows people to determine more opportune times to plant and harvest crops, to change patterns of land use, to predict marine harvests, and to better plan other human activities. The ability to provide general predictions about these economically important aspects of society help control product costs and plan for social needs. High-quality climate models can also provide warnings that specific human behaviors may be affecting the environment and, as a result, people's quality of life. Climate models reflecting processes of deforestation, desertification, industrial insertion of greenhouse gases into the atmosphere, and ozone depletion all help establish grounds to alter human behaviors to increase survivability. Because of these factors and how they may affect the well-being and security of nations, governments throughout the world have created programs, earmarked research funds, and mandated agencies to study and model climate. A number of high-profile international treaties, most notably the Kyoto Protocol, have been initiated to address the observable and predicted effects of climate change.

Most major universities have climate change programs or research centers. A number of host government climate research centers and many universities have formed research consortiums to engage in regional climate studies and to share resources and supercomputing time. The result of high-profile interest in climate modeling related to issues of climate change provides large sums of research monies to qualified researchers and research facilities.

Although climate models are certainly helpful in predicting marine harvests, crop futures, energy needs, potential droughts, food shortages, and water resources, the biggest effect on industry from climate modeling is the concern they have generated regarding certain human behaviors. The use of environmentally detrimental technologies and the release of large amounts of greenhouse gases such as carbon dioxide, methane, sulfur dioxide, and chlorofluorocarbons into the atmosphere has begun to significantly alter Earth's future livability. Individuals do not wish to hear that driving their automobiles or heating their homes is damaging the planet. Companies do not want to be told that their methods of doing business are producing more harm than good and that changing these methods will require large financial investments. Additionally, governments do not wish to alter their strategies for national growth because their attempts to modernize are environmentally detrimental.

Since the mid-nineteenth century, many of humankind's economic, technological, and cultural advancements have been achieved through the exploitation of fossil fuels and other resources. Certain methods of forest harvesting have been shown to promote desertification, disrupt water resources, change the amount of radiation from the sun reflected in a region, and alter regional climate patterns. Because trees absorb carbon dioxide, the loss of large tracts of forest lessens Earth's ability to remove and recycle carbon dioxide from the atmosphere. Burning petroleum and coal results in the discharge of greenhouse gases and the release of harmful particulates into the atmosphere. Climate models examining historical and observational climate data and human behaviors have demonstrated a direct link between the use of fossil fuels and changes in the atmosphere. Climate observations and models show that increased greenhouse gas levels have produced real, noticeable, historically unprecedented, human-induced changes in the atmosphere and to climate patterns. In addition, the insertion of industrial chemicals—such as chlorofluorocarbons—into the atmosphere is known to deplete ozone and have a serious effect on both the climate and human health.

Laws and regulations. In attempts to control industrial and commercial release of greenhouse gases and other toxic chemicals into the atmosphere, governments have imposed regulations on consumers and companies. Laws and regulations have been enacted that mandate limits on automobile exhaust, industrial air pollution, aerosol propellants, power plant emissions, field burning, biofuels, and certain agricultural practices. The difficulty with these attempts to alter industrial, personal, and governmental behaviors is that their reach is limited. Although one nation, state, or individual seeks to reduce harmful emissions, not all nations, states, or individuals share the same goals. Earth's atmosphere is an open system of circulation without human-imposed boundaries; therefore, pollutants created in one region affect all others.

Financial impact. The impact of climate modeling on industry is often financial, decreasing corporations' profits. As climate models become more accurate in their predictions, certain industrial practices are likely to be found clearly responsible for some negative atmospheric effects. Altering business practices often requires large financial investments in new technologies; these investments cut into short-term profits. Although such investments are likely to result in long-term growth and allow the business to survive, unquestionably, they negatively affect short-term profits. For many nations and all nonprofit industries, the bottom line is of utmost importance. Changing the way one does business, retooling, reequipping, changing priorities, changing business models, and altering national and institutional

identities is unthinkable for many. Some nations and many businesses have reacted to regulations and attempts to formulate new laws restricting atmospheric pollution by denying the existence of climate change. They seek to maintain the status quo for as long as possible.

CAREERS AND COURSE WORK

Careers in climate modeling are limited. Most positions involving climate modeling are in academic institutions, government agencies, and private research organizations. The majority of climate modelers are employed within university settings, which usually means that their time must be divided between active research and teaching responsibilities. In some university consortiums, such as the University Corporation for Atmospheric Research, climate modelers work in teams. Climate modelers working for government agencies focus on predictive applications to help fulfill their governmental branch mandates. At the international level, a number of climate modelers work under contract or are funded by the United Nations and may do research for any of a number of United Nations agencies or advisory boards. Climate modelers at the federal level may find work with the U.S. Geological Survey, the Department of Agriculture, all branches of the military and intelligence communities, the National Oceanic and Atmospheric Agency, the National Aeronautics and Space Administration, the National Center for Atmospheric Research, and the Department of Energy. A limited number of climate modeling positions are available in private research organizations that offer services to businesses, lobbying organizations, agricultural interests, commodities traders, maritime shipping and fishing concerns, and politicians.

Students interested in careers involving climate modeling need to take classes in chemistry, physics, atmospheric sciences, geosciences, meteorology, oceanography, biology, mathematics, statistics, computer science, and environmental science, and obtain a bachelor of science degree. Almost all careers in climate modeling require a graduate-level education. Although obtaining a master's degree may allow for an entry-level position or journeyman status within certain agencies, for nearly all climate-modeling opportunities in academia, civil service, or private research, a doctorate is a necessity. In graduate school,

---

FASCINATING FACTS ABOUT CLIMATE MODELING

- Some of the earliest known long-term observations leading to predictive climate models are those of ancient native peoples of South and Central America who noted regional climate changes related to what are now known as El Niño and La Niña ocean currents.

- In the early 1960's, Warren M. Washington and Akira Kasahara, scientists with the National Center for Atmospheric Research, developed a computer model of atmospheric circulation. Data were input to a CDC computer using punch cards and seven-channel digital magnetic tape, and data were output through two line printers, a card punch, a photographic plotter, and standard magnetic tape.

- In December, 2009, the Copenhagen United Nations Climate Conference was the most-searched topic on the Google Internet search engine. In December, 2010, "climate change" was one of the top phrases searched daily on the Internet.

- In 2010, the National Center for Atmospheric Research released the Community Earth System Model, which creates computer simulations of Earth's past, present, and future climates. Experiments using this model will be part of the Intergovernmental Panel on Climate Change's 20131014 assessments.

- The complexity of climate models requires that they be run on supercomputers. The fastest supercomputer in the United States as of November, 2010, was the Cray XT5, at the U.S. Department of Energy's Oak Ridge Leadership Computing Facility. It can perform at 1.75 petaflops (quadrillions of calculations per second).

- Pioneering climate modeler Warren M. Washington was awarded the National Medal of Science by President Barack Obama on November 17, 2010. Washington was also a member of the Intergovernmental Panel on Climate Change, which received the 2007 Nobel Peace Prize.

studies in advanced mathematics and computer programming combined with intensely focused studies in climatology, oceanography, and physics are the norm. Individual areas of research interest require students to narrow their courses to reflect the direction of their research and make course and seminar selections accordingly.

As interest in climate change issues continues to grow, the need for qualified climate modeling scientists will increase. The rate at which observable changes to the climate begin to reflect predictions made by climate models will most likely raise the value and significance of trained climatologists and their predictive modeling skills.

SOCIAL CONTEXT AND FUTURE PROSPECTS

Using climate models to accurately predict future climate trends is the ultimate goal. If climate models can help define the difference between natural climate fluctuations and human-induced climate change, positive human activities can be developed to counter the impact of environmentally unsustainable behaviors. Accurate predictive climate models can also increase economic outcomes by indicating more opportune times for planting, harvesting, and fishing, and by predicting droughts and temperature shifts. Climate shifts are known to be associated with pandemic outbreaks of disease: Climate models may allow people to prepare in advance for disease-formulating conditions.

Existing climate modeling is limited by computational power and the nonlinear nature of climate system phenomena. As computing technology advances and observational data of climate processes over longer time scales becomes available, flux compensations will be more accurately defined and the accuracy of future climate models will increase. Public acceptance of climate model predictions will remain complicated as long as politics and economics, rather than observational facts, are allowed to drive the climate change debate.

—Randall L. Milstein, MS, PhD

FURTHER READING

Edwards, Paul N. *A Vast Machine: Computer Models, Climate Data, and the Politics of Global Warming.* Cambridge, Mass.: MIT Press, 2010. Tells the history of how scientists learned to understand the atmosphere.

Kiehl, J. T., and V. Ramanathan. *Frontiers of Climate Modeling.* New York: Cambridge University Press, 2006. A good general overview of climate modeling, with an emphasis on how greenhouse gases are altering the climate system.

McGuffie, Kendall, and Ann Henderson-Sellers. *A Climate Modelling Primer.* West Sussex, England: John Wiley & Sons, 2005. Explains the basis and mechanisms of existing physical-observation-based climate models.

Mote, Philip, and Alan O'Neill, eds. *Numerical Modeling of the Global Atmosphere in the Climate System.* Boston: Kluwer Academic, 2000. Meant for those actively creating climate models, this work explains the uses of numerical constraints to define climate processes.

Robinson, Walter A. *Modeling Dynamic Climate Systems.* New York: Springer, 2001. A basic book for understanding climate modeling that includes a good description of how climate systems function and interact with each other and vary over time and space.

Trenberth, Kevin E., ed. *Climate System Modeling.* New York: Cambridge University Press, 2010. A comprehensive textbook covering the most important topics for developing a climate model.

Washington, Warren M., and Claire L. Parkinson. *An Introduction to Three-Dimensional Climate Modeling.* Sausalito, Calif.: University Science Books, 2005. An introduction to the use of three-dimensional climate models. Includes a history of climate modeling.

WEB SITES

American Meteorological Society
http://www.ametsoc.org

Intergovernmental Panel on Climate Change
http://www.ipcc.ch

National Center for Atmospheric Research
http://ncar.ucar.edu

National Oceanic and Atmospheric Administration
National Climatic Data Center
http://www.Ncdc.noaa.gov

National Weather Association

http://www.nwas.org

University Corporation for Atmospheric Research
http://www2.ucar.edu

U.S. Geological Survey Climate and Land Use
    Change
http://www.usgs.gov/climate_landuse

World Meteorological Organization Climate
    http://www.wmo.int/pages/themes/climate/
    index_en.php

See also: Air-Quality Monitoring; Atmospheric
Sciences; Climate Engineering; Climatology;
Meteorology.

CLIMATOLOGY

FIELDS OF STUDY

Atmospheric sciences; meteorology; physical geography; climate change; climate classification; climate zones; tree-ring analysis; climate modeling; bioclimatology; climate comfort indices; hydroclimatology.

SUMMARY

Climatology deals with the science of climate, which includes the huge variety of weather events. These events change at periods of time that range from months to millennia. Climate has such a profound influence on all forms of life, including human life, that people have made numerous attempts to predict future climatic conditions. These attempts resulted in research efforts to try to understand future changes in the climate as a consequence of anthropogenic and naturally caused activity.

KEY TERMS AND CONCEPTS

- **climate:** Average weather for a particular area.
- **climograph:** Line graph that shows monthly mean temperatures and precipitation.
- **coriolis effect:** Rghtward (Northern Hemisphere) and leftward (Southern Hemisphere) deflection of air due to the Earth's rotation.
- **energy balance:** Balance between energy received from the Sun (shortwave electromagnetic radiation) and energy returned from the Earth (longwave electromagnetic radiation).
- **front:** Boundary between air masses that differ in temperature, moisture, and pressure.
- **greenhouse effect:** Process in which longwave radiation is trapped in the atmosphere and then radiated back to the Earth's surface.
- **occluded front:** Front that occurs when a cold front overtakes a warm front and forces the air upward.
- **semipermanent low-pressure centers:** Semipermanent patterns of low pressure that occur in northern waters off the Alaskan Aleutian Islands and Iceland.
- **sublimation:** Change from ice in a frozen solid state to water vapor in a gaseous state without going through a liquid state.
- **troposphere:** Lowest level of the atmosphere that contains water vapor that can condense into clouds.

DEFINITION AND BASIC PRINCIPLES
"Weather" pertains to atmospheric conditions that constantly change, hourly and daily. In contrast, "climate" refers to the long-term composite of weather conditions at a particular location, such as a city or a state. Climate at a location is based on daily mean conditions that have been aggregated over periods of time that range from months and years to decades and centuries. Both weather and climate involve measurements of the same conditions: air temperature, water vapor in the air (humidity), atmospheric pressure, wind direction and speed, cloud types and extent, and the amount and kind of precipitation.

Estimates of ancient climates, going back several thousand years or more, are produced in various ways. For example, the vast amount of groundwater discovered in southern Libya indicates that during some period in the past, that part of the Sahara Desert was much wetter. In ancient Egypt, Nilometers, stone markers built along the banks of the Nile, were used to gauge the height of the river from year to year. They are similar to the staff gauges that are used by the U.S. Geological Survey to indicate stream or canal elevation. The height of the Nilometer reflects the extent of precipitation and associated runoff in the headwaters of the Nile in east central Africa.

BACKGROUND AND HISTORY

Measurements of precipitation were being made and recorded in India during the fourth century BCE. Precipitation records were kept in Palestine about 100 CE, Korea in the 1440's, and in England during the late seventeenth century. Galileo invented the thermometer in the early 1600's. Physicist Daniel Fahrenheit created a measuring scale for a liquid-inglass thermometer in 1714, and Swedish astronomer Anders Celsius developed the centigrade scale in 1742. Italian physicist Evangelista Torricelli, who worked with Galileo, invented the barometer in 1643.

The first attempt to explain the circulation of the atmosphere around the Earth was made by English astronomer Edmond Halley, who published a paper charting the trade winds in 1686. In 1735, English meteorologist George Hadley further explained the movement of the trade winds, describing what became known as a Hadley cell, and in 1831, Gustave-Gaspard Coriolis developed equations to describe the movement of air on a rotating Earth. In 1856, American meteorologist William Ferrel developed a theory that described the mid-latitude atmospheric circulation cell (Ferrel cell). In 1860, Dutch meteorologist Christophorus Buys Ballot demonstrated the relationship between pressure, wind speed, and direction (which became known as Buys Ballot's law).

The first map of average annual isotherms (lines connecting points having the same temperature) for the northern hemisphere was created by German naturalist Alexander von Humboldt in 1817. In 1848, German meteorologist Heinrich Wilhelm Dove created a world map of monthly mean temperatures. In 1845, German geographer Heinrich Berghaus prepared a global map of precipitation. In 1882, the first world map of precipitation using mean annual isohyets (lines connecting points having the same precipitation) appeared.

HOW IT WORKS

Earth's global energy balance. The Earth's elliptical orbit about the Sun ranges from 91.5 million miles at perihelion (closest to the Sun) on January 3, to 94.5 million miles at aphelion (furthest from the Sun) on July 4, averaging 93 million miles. The Earth intercepts about two-billionth of the total energy output of the Sun. Upon reaching the Earth, a portion of the incoming radiation is reflected back into space, while another portion is absorbed by the atmosphere, land, or oceans. Over time, the incoming shortwave solar radiation is balanced by a return to outer space of longwave radiation.

The Earth's atmosphere extends to an estimated height of about 6,000 miles. Most of it is made up of nitrogen (78 percent by volume) and oxygen (about 21 percent). Of the remaining 1 percent, carbon dioxide (CO_2) accounts for about 0.0385 percent of the atmosphere. This is a minute amount, but carbon dioxide can absorb both incoming shortwave radiation from the sun and outgoing longwave radiation from the Earth. The measured increase in carbon dioxide since the early 1900's is a major cause for concern as it is a very good absorber of heat radiation, which adds to the greenhouse effect.

Air temperature. Air temperature is a fundamental constituent of climatic variation on the Earth. The amount of solar energy that the Earth receives is governed by the latitude (from the equator to the poles) and the season. The amount of solar energy reaching low-latitude locations is greater than that reaching higher-latitude sites closer to the poles. Another factor pertaining to air temperature is the fivefold difference between the specific heat of water (1.0) and dry land (0.2). Accordingly, areas near the water have more moderate temperatures on an annual basis than inland continental locations, which have much greater seasonal differences.

Anthropogenic (human-induced) changes in land cover in addition to aerosols and cloud changes can result in some degree of global cooling, but this is much less than the combined effect of greenhouse gases in global warming. The gases include carbon dioxide from the burning of fossil fuels (coal, oil, and natural gas), which has been increasing since the second half of the twentieth century. Other gases such as methane (CH_4), chlorofluorocarbons (CFCs), ozone (O_3), and nitrous oxide (NO_2) also create additional warming effects.

Air temperature is measured at 5 feet above the ground surface and generally includes the maximum and minimum observation for a twenty-four-hour period. The average of the maximum and minimum temperature is the mean daily temperature for that particular location.

Earth's available water. Water is a tasteless, transparent, and odorless compound that is essential to all biological, chemical, and physical processes. Almost all the water on the Earth is in the oceans, seas, and bays (96.5 percent) and another 1.74 percent is frozen in ice caps and glaciers. Accordingly, 98.24 percent of the total amount of water on this planet is either frozen or too salty and must be thawed or desalinated. About 0.76 percent of the world's water is fresh (not saline) groundwater, but a large portion of this is found at depths too great to be reached by drilling. Freshwater lakes make up 0.007 percent, and atmospheric water is about 0.001 percent of the total. The combined average flows of all the streams on Earth—from tiny brooks to the mighty Amazon River—account for 0.0002 percent of the total.

Air masses. The lowest layer of the atmosphere is the troposphere, which varies in height from 10 miles at the equator and lower latitudes to 4 miles at the poles. Different types of air masses within the troposphere can be delineated on the basis of their similarity in temperature, moisture, and to a certain extent, air pressure. Air masses develop over continental and maritime locations that strongly determine their physical characteristics. For example, an air mass starting in a cold, dry interior portion of a continent would develop thermal, moisture, and pressure characteristics that would be substantially different from those of an air mass that developed over water. Atmospheric dynamics also allow air masses to modify their characteristics as they move from land to water and vice versa.

Air mass and weather front terminology were developed in Norway during World War I. The Norwegian meteorologists were unable to get weather reports from the Atlantic theater of operations; consequently, they developed a dense network of weather stations that led to impressive advances in atmospheric modeling.

Greenhouse effect. Selected gases in the lower parts of the atmosphere trap heat and radiate some of that heat back to Earth. If there was no natural greenhouse effect, the Earth's overall average temperature would be close to 0 degrees Fahrenheit rather than 57 degrees Fahrenheit.

The burning of coal, oil, and gas makes carbon dioxide the major greenhouse gas. Carbon dioxide accounts for nearly half of the total amount of heat-producing gases in the atmosphere. In mid-eighteenth century Great Britain, before the Industrial Revolution, the estimated level of carbon dioxide was about 280 parts per million (ppm). Estimates for the natural range of carbon dioxide for the past 650,000 years are 180-300 ppm. All of these values are less than the October, 2010, estimate of 387 ppm. Since 2000, atmospheric carbon dioxide has been increasing at a rate of 1.9 ppm per year. The radiative effect of carbon dioxide accounts for about one-half of all the factors that affect global warming. Estimates of carbon dioxide values at the end of the twenty-first century range from 490 to 1,260 ppm.

The second most important greenhouse gas is methane (CH_4), which accounts for about 14 percent of all of the global warming factors. This gas originates from the natural decay of organic matter in wetlands, but anthropogenic activity in the form of rice paddies, manure from farm animals, the decay of bacteria in sewage and landfills, and biomass burning (both natural and human-induced) doubles the amount produced.

Chlorofluorocarbons (CFCs) absorb longwave energy (warming effect) but also have the ability to destroy stratospheric ozone (cooling effect). The warming radiative effect is three times greater than the cooling effect. CFCs account for about 10 percent of all of the global warming factors. Tropospheric ozone (O_3) from air pollution and nitrous oxide (N_2O) from motor vehicle exhaust and bacterial emissions from nitrogen fertilizers account for about 10 percent and 5 percent, respectively, of all the global warming factors.

Several human actions lead to a cooling of the Earth's climate. For example, the burning of fossil fuels results in the release of tropospheric aerosols, which acts to scatter incoming solar radiation back into space, thereby lowering the amount of solar energy that can reach the Earth's surface. These aerosols also lead to the development of low and bright clouds that are quite effective in reflecting solar radiation back into space.

APPLICATIONS AND PRODUCTS

Climatology involves the measurement and recording of many physical characteristics of the Earth. Therefore, numerous instruments and methods have been devised to perform these tasks and obtain accurate measurements.

Measuring temperature. At first glance, it would appear that obtaining air temperatures would be relatively simple. After all, thermometers have been around since 1714 (Fahrenheit scale) and 1742 (Celsius scale). However, accurate temperature measurements require a white (high-reflectivity) instrument shelter with louvered sides for ventilation, placed where it will not receive direct sunlight. The standard height for the thermometer is 5 feet above the ground.

Remote-sensing techniques. Oceans cover about 71 percent of the Earth's surface, which means that large portions of the world do not have weather stations and places where precipitation can be measured with standard rain gauges. To provide more information about precipitation in the equatorial and tropical parts of the world, the National Aeronautics and Space Administration (NASA) and the Japanese Aerospace Exploration Agency began a program called the Tropical Rainfall Monitoring Mission (TRMM) in 1997. The TRMM satellite monitors the area of the world between 35 degrees north and 35 degrees south latitude. The goal of the study is to obtain information about the extent of precipitation, its intensity, and length of occurrence. The major instruments on the satellite include radar to detect rainfall,

a passive microwave imager that can acquire data about precipitation intensity and the extent of water vapor, and a scanner that can examine objects in the visible and infrared portions of the electromagnetic spectrum. The goal of collecting this data is to obtain the necessary climatological information about atmospheric circulation in this portion of the Earth so as to develop better mathematical models for determining large-scale energy movement and precipitation.

Geostationary satellites. Geostationary orbiting earth satellites (GOES) enable researchers to view images of the planet from what appears to be a fixed position. To achieve this, these satellites circle the globe at a speed that is in step with the Earth's rotation. This means that the satellite, at an altitude of 22,200 miles, will make one complete revolution in the same twenty-four hours and direction that the Earth is turning above the Equator. At this height, the satellite is in a position to view nearly half the planet at any time. On-board instruments can be activated to look for special weather conditions such as hurricanes, flash floods, and tornadoes. The instruments can also be used to make estimates of precipitation during storm events.

Rain gauges. The accurate measurement of precipitation is not as simple as it may seem. Collecting rainfall and measuring it is complicated by the possibility of debris, dead insects, leaves, and animal intrusions occurring. Standards were established, although the various national climatological offices use more than fifty types of rain gauges. The location of the gauge, its height above the ground, the possibility for splash and evaporation, its distance from trees, and turbulence all affect the results. Accordingly, all gauge records are really estimates. Precipitation estimates are also affected by the number of gauges per unit area. The number of gauges in a sample area of 3,860 square miles for Britain, the United States, and Canada is 245, 10, and 3, respectively. Although the records are reported to the nearest 0.01 inch, discrepancies occur in the official

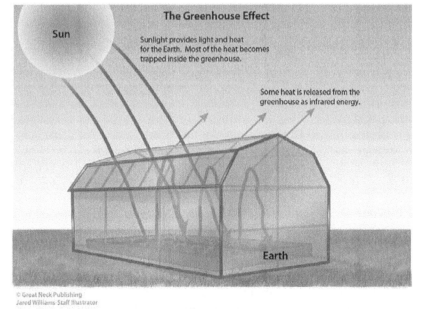

The Greenhouse Effect

Sunlight provides light and heat for the Earth. Most of the heat becomes trapped inside the greenhouse.

Some heat is released from the greenhouse as infrared energy.

Sun

Earth

© Great Neck Publishing
Jared Williams-Staff Illustrator

The greenhouse effect refers to the process in which longwave radiation is trapped in the atmosphere and then radiated back to the Earth's surface.

records. It is important to have a sufficiently dense network of rain gauges in urban areas. Some experts think that 5 to 10 gauges per 100 square miles is necessary to obtain an accurate measure of rainfall.

Doppler radar. Doppler radar was first used in England in 1953 to pick up the movement of small storms. The basic principle behind Doppler radar is that the back-scattered radiation frequency detected at a certain location changes over time as the target, such as a storm, moves. The mode of operation requires a transmitter that is used to send short but powerful microwave pulses. When a foreign object (or target) is intercepted, some of the outgoing energy is returned to the transmitter, where a receiver can pick up the signal. An image (or echo) from the target can then be enlarged and shown on a screen. The target's distance is revealed by the time between transmission and return. The radar screen can indicate not only where the precipitation is taking place but also its intensity by the amount of the echo's brightness. Doppler radar has developed into a very useful device for determining the location of storms and the intensity of the precipitation and for obtaining good estimates of the total amount of precipitation.

IMPACT ON INDUSTRY

Global perspective. The World Meteorological Organization, headquartered in Geneva, Switzerland, was established to encourage weather station networks that would facilitate the acquisition of climatic data. In 2007, the organization decided to expand the global observing system and other related observing systems, including the global ocean observing system, global terrestrial observing system, and the global climate observing system. Data are being collected by 10,000 manned and automatic surface weather stations, 1,000 upper-air stations, more than 7,000 ships, 100 moored and 1,000 floating buoys that can drift with the currents, several hundred radars, and more than 3,000 commercial airplanes, which record key aspects of the atmosphere, land, and ocean surfaces on a daily basis.

Government research. About 183 countries have meteorological departments. Although many are small, the larger countries have well-established organizations. In the United States, meteorology and climatology are handled by the National Oceanic and Atmospheric Administration. The agency's National Climatic Data Center has records that date back to 1880 and provide invaluable information about previous periods for the United States and the rest of the world. For example, sea ice in the Arctic Ocean typically reaches its maximum extent in March. The coverage at the end of March, 2010, was 5.8 million square miles. This was the seventeenth consecutive March with below-average coverage. The center issues many specialized climate data publications as well as monthly temperature and precipitation summaries for all fifty states.

University research. In the United States, forty-eight states (except Rhode Island and Tennessee) have either a state climatologist or someone who has comparable responsibility. Most of the state climatologists are connected with state universities, particularly the colleges that started as land-grant institutions.

The number of cooperative weather stations established to take daily readings of temperature and precipitation in each of these states has varied in number since the late nineteenth century.

Industry and business sectors. The number, size, and capability of private consulting firms has increased over the years. Some of the first of these firms were frost-warning services that served the citrus and vegetable growers in Arizona, Florida, and California. These private companies expanded considerably as better forecasting and warning techniques were developed. For example, AccuWeather.com has seven global forecast models and fourteen regional forecast models for the United States and North America and also prepares a daily weather report for *The New York Times*.

CAREERS AND COURSE WORK

Although many consider meteorologists to be people who forecast weather, the better title for such a person is atmospheric scientist. For example, climatologists focus on climate change, and environmental meteorologists are interested in air quality. Broadcast meteorologists work for television stations. The largest number of jobs in the field are with the National Weather Service, which employs about one-third of the atmospheric scientists who work for the federal government.

Meteorologists are predicted to have above-average employment growth in the 2010's. Employment at the National Weather Station generally requires a bachelor's degree in meteorology, or at least twenty-four credits in meteorology courses along with college physics and physical science classes. Anyone who wants to work in applied research and development needs a master's degree. Research positions require a doctorate. The median annual average salary in 2008 was $81,290. Entry-level positions at the National Weather Service earn about $35,000, and those in the highest 10 percent bracket earn more than $127,000.

SOCIAL CONTEXT AND FUTURE PROSPECTS

Climate change may be caused by both natural internal and external processes in the Earth-Sun system and human-induced changes in land use and the atmosphere. The United Nations Framework Convention on Climate Change states that the term "climate change" should refer to anthropogenic changes that affect the composition of the atmosphere as distinguished from natural causes, which should be referred to "climate variability." An example of natural climate variability is the global cooling of about 0.5 degrees Fahrenheit in 1992-1993 that was related to the 1991 Mount Pinatubo volcanic eruption in the Philippines. The 15 million to 20 million tons of sulfuric acid aerosols that were released into the stratosphere reflected incoming radiation from the sun and created a cooling effect. Many experts suggest that climate change is caused by human activity, as evidenced by the above-normal temperatures in the 2000's. Based on a variety of techniques that estimate temperatures in previous centuries, the year 2005 was the warmest in the last thousand years.

Numerous observations strongly suggest a continuing warming trend. Snow and ice have retreated from areas such as Mount Kilimanjaro, which at 19,340 feet is the highest mountain in Africa, and glaciated areas in Switzerland. In the Special Report on Emission Scenarios (2000), the Intergovernmental Panel on Climate Change examined the broad spectrum of possible concentrations of greenhouse gases by examining the growth of population and industry along with the efficiency of energy use. The panel estimated future trends using computer climate models. For example, it estimated that the global temperature would increase 35.2-39.2 degrees Fahrenheit by the year 2100.

Given the effect that climate change will have on humanity, many agencies and organizations will be doing research in the area, and climatologists are

---

FASCINATING FACTS ABOUT CLIMATOLOGY

- Mean sea levels for the world have increased an average of 0.07 inches per year from 1904 to 2003, an amount that is much larger than the average rate in the past.

- Serbian astronomer Milutin Milankovitch developed the astronomical hypothesis in 1938 by observing that glacial and interglacial periods were related to insolation variations caused by small cycles in the Earth's axial rotation and orbit about the Sun.

- The larger input of solar energy received at and near the equator creates a very large circulation path known as the Hadley cell. The air converges at a narrow zone known as the intertropical convergence zone, which seasonally varies from 15 degrees south in northern Australia to 25 degrees north in northern India, representing a latitudinal shift of 40 degrees.

- The coldest temperatures in the world were recorded at the Russian weather station at Vostok, Antarctica, at 78 degrees south: −127 degrees Fahrenheit in 1958 and −128.5 degrees Fahrenheit on July 21, 1983, at an ice sheet elevation of 11,220 feet.

- There is a fivefold difference between the specific heat of water (1.0) and dry land (0.2). This results in a one- to two-month lag in average maximum and minimum temperatures after the summer and winter solstices.

- Atlantic storm names can be repeated after six years unless the hurricane was particularly severe. The names of noteworthy storms--such as Camille (1969), Hugo (1989), Andrew (1992), Floyd (1999), and Katrina (2005)--are not used again, so as to avoid confusion in later years.

---

likely to be needed by a variety of governmental and private entities.

—*Robert M. Hordon, MA, PhD*

FURTHER READING

Coley, David A. Energy and Climate Change: Creating a Sustainable Future. Hoboken, N.J.: John Wiley & Sons, 2008. A detailed review of energy topics and their relationship to climate change and energy technologies.

Gautier, Catherine. Oil, Water, and Climate: An Introduction. New York: Cambridge University Press,

2008. A good discussion of the impact of fossil fuel burning on climate change.

Lutgens, Frederick K., and Edward J. Tarbuck. The Atmosphere: An Introduction to Meteorology. 9th ed. Upper Saddle River, N.J.: Prentice Hall, 2004. A useful and standard text that is written with considerable clarity.

Strahler, Alan. Introducing Physical Geography. 5th ed. Hoboken, N.J.: John Wiley & Sons, 2011. An excellent text covering weather and climate with superlative illustrations, clear maps, and lucid discussions.

Wolfson, Richard. Energy, Environment, and Climate. New York: W. W. Norton, 2008. Provides an extensive discussion of the relationship between energy and climate change.

WEB SITES

American Association of State Climatologists
http://www.stateclimate.org

American Meteorological Society
http://www.ametsoc.org
Intergovernmental Panel on Climate Change
http://www.ipcc.ch

International Association of Meteorology and Atmospheric Sciences
http://www.iamas.org

National Oceanic and Atmospheric Administration National Climatic Data Center
http://www.Ncdc.noaa.gov

National Weather Association
http://www.nwas.org

U.S. Geological Survey Climate and Land Use Change
http://www.usgs.gov/climate_landuse

World Climate Research Programme
http://www.wcrp-climate.org

World Meteorological Organization
http://www.wmo.int/pages/themes/climate/index_en.php

See also: Atmospheric Sciences; Barometry; Climate Engineering; Climate Modeling; Meteorology; Temperature Measurement.

COMMUNICATION

FIELDS OF STUDY

Interpersonal communication; conflict management; group dynamics; intercultural communication; linguistics; nonverbal communication; semiotics; communication technology; negotiation and mediation; organizational behavior; public speaking; speech pathology; rhetorical theory; diction; mass communication; advertising; broadcasting and telecommunications; journalism; media ethics; public relations.

SUMMARY

Communication is the complex, continuous, two-way process of sending and receiving information in the form of messages. Communication engages all the senses and involves speech, writing, and myriad nonverbal methods of data exchange. It is a vital component in the everyday existence of all living creatures. An interdisciplinary, multidimensional field that incorporates language, linguistic structure, symbols, interpretation, and meaning in both personal and professional life, communication is an essential part of any division of scientific study.

KEY TERMS AND CONCEPTS

- **etymology:** Study of the origins of words and their changes in meaning and form over time.
- **feedback:** Response of a recipient to a message received.
- **haptics:** Study of how touch and bodily contact affects communication.
- **kinesics:** Study of how nonverbal forms of communication such as posture, gestures, and facial expressions affect meaning.
- **paralinguistics:** Study of the nonverbal elements

that individualize speech to help convey meaning, including intonation, pitch, word stress, volume, and tempo.

- **phonetics:** Study of how speech is physically produced and received.
- **proxemics:** Study of the use of personal space in face-to-face communication.
- **semantics:** Interpretation of meaning in language.
- **semiotics:** Study of how signs, symbols, and signifiers create meaning.
- **syntax:** Study of the grammatical rules of a language that aid effective communication.

DEFINITION AND BASIC PRINCIPLES

Communication is the science and art of transmitting information. Communication science is objective, involving research, data, methodologies, and technological approaches to procedures of information exchange. Communication art is subjective, concentrating on the aesthetics and effectiveness with which messages are composed, sent, received, interpreted, understood, and acted on. Art and science are equally important in understanding how communication is supposed to work, why it succeeds or fails, and how best to use that knowledge to improve the dissemination of information. Communication is the oldest, broadest, most complicated, and most versatile of all scientific disciplines, with attitude-changing, life-influencing applications in every field of human endeavor.

The process of communication requires three basic components: a sender, a message, and a receiver. The sender encodes information into a message to send to the recipient for decoding. The message usually has an overt purpose, based on the sender's desire: to inform, educate, persuade, or entertain. Some messages can also be covert: to attract attention, make connections, gain support, or sell something.

In theory, the communication ideal is to match sender intent to receiver interpretation of a message closely enough to generate a favorable response. In practice, communication is often unsuccessful in achieving such objectives, considering the quirky, complicated nature of human beings and the variety and intricacy of languages used to transmit messages.

There are two fundamental forms of communication: verbal and nonverbal. Verbal communication includes interpersonal or electronically transmitted conversations and chats, lectures, audiovisual mass media such as television and radio, and similar forms of oral speech. Verbal communication also includes written communication: books, letters, magazines, signs, Web sites, and e-mails. Nonverbal communication incorporates an infinite range of facial expressions, gestures and sign language, body language, and paralanguage. In face-to-face interactions, nonverbal messages frequently convey more meaning than verbal content.

BACKGROUND AND HISTORY

Human communication has been a vital part of evolution. Protohumans communicated through gestures before developing the ability to speak, perhaps as far back as 150,000 years ago. An opposable thumb allowed early humans to make tools and to leave long-lasting marks, such as cave paintings. As communities formed, spoken language evolved. Verbal language spawned written language; pictographs appeared 10,000 years ago. Cuneiform originated several thousand years later in Mesopotamia about the time hieroglyphics were born in Egypt. By 3000 BCE, writing had developed independently in China, India, Crete, and Central America. Written symbols became alphabets, which were organized into words and structured linguistic systems.

From the dawn of civilization, humans have reached out to one another, devising ingenious solutions to expand range, enhance exactitude, and ensure the permanence of messages. A library was begun in Greece in the sixth century BCE, and paper was invented soon afterward. By 1455, Johannes Gutenberg had devised a printing press with movable type and printed his first communication: the Gutenberg Bible. Each innovation represented a leap forward in the spread of information.

Communication has greatly accelerated and expanded during the last two hundred years as it became a scientific study. The nineteenth century witnessed the development of inventions such as railroads, the telegraph, postal systems, the typewriter and phonograph, automobiles, and telephones. The twentieth century ushered in movies, airplanes, radio and television, audio and video recorders, communication satellites, photocopiers, and facsimile machines. Since the 1980's personal computers, cellular phones, fiber optics, the Internet, and a variety of mobile, handheld devices have widened and sped human integration into the information age.

HOW IT WORKS

Though necessary to interaction throughout human history, communication is still an imperfect science. As in the past, modern senders have vastly different abilities in composing comprehensible messages. Contemporary receivers, for a variety of reasons, may have an equally difficult time interpreting messages.

Channeling. How a message is sent—the medium of transmission, called a channel in communications science—is key to the communication process. An oral message too faint to be heard, a written message too garbled to be understood, or a missed nonverbal signal can interrupt, divert, delay, or derail the passage of information from one source to another. In the modern world, there are dozens of methods of sending messages: telephone, text messaging, ground mail, mass media, e-mail or social networking via the Internet, and face-to-face contact are most common. Deciding the best means of information transmission depends on a number of factors, including the purpose of the message, the amount and profundity of data to be sent, the audience for the message, and the desired outcome. All media have distinct advantages and disadvantages. Each type of communication requires a particular skill set from both sender and receiver.

A vital consideration is the relationship between sender and receiver: They must have something in common. A message written in Chinese and sent to someone who reads only English is ineffective. Likewise, shouting at the hearing impaired or sending text messages to newborn infants are counterproductive activities. Even among individuals with a great degree of similarity in language and culture there can be gaping discrepancies between sender intent and receiver interpretation. Ultimately, communication is a result-oriented discipline. Whatever the purpose of a message, a response of some kind from the recipient (feedback) is expected. If there is no reply, or an inappropriate response is generated, the process is incomplete and communication fails.

Context. Communication does not occur in a vacuum. Many factors determine and define how the process of information transfer will unfold and the likelihood of successful transmission, reception, and response to a message.

Psychological context, for example, entails the emotional makeup of individuals who originate and receive information: the desires, needs, values, and personalities of the participants, which may be in harmony or at variance. Environmental context deals with the physical setting of an interaction: weather, surroundings, noise level, or other elements with the potential to impair communication. Situational context is specific to the relationship between the participants in communication: Senders tailor messages differently to receivers who are friends, relatives, coworkers, or strangers. Linguistic context concerns the relationship among words, sentences, and paragraphs used throughout a speech or written work that help to clarify meaning. Interpretation of any part of a message is relative to the preceding and following parts of the message.

Particularly relevant to effective communication in the ever-widening global community is cultural context. Every culture has particular, ingrained rules governing verbal and nonverbal behavior among its members. What might be acceptable within one cultural group—such as extended eye contact, the use of profanity, or frequent touching during conversation—might be offensive to a different cultural group. High-context cultures are homogeneous communities with long-established, well-defined traditions that help preserve cultural values; such cultures are found throughout Asia, Africa, the Middle East, and South America. Low-context cultures, like the United States, are heterogeneous, a blend of many traditions that has produced a less rigid, broader-based, more open-ended set of behavioral rules. Though adhering to certain patterns learned from diverse national and ethnic heritages, Americans as a group are more geared toward individual values. When high- and low-context cultures collide in communication, such differences in attitude can wreak havoc on the implied intent of a message and the interpretation of meaning.

Communication barriers. There are universal emotions (such as fear, surprise, happiness, sadness, and anger), common to all peoples in all cultures at all times, that can serve as building blocks to understanding. However, numerous obstacles interfere with the clear, unambiguous transmission of messages at one end, and the full grasp of meaning at the other end. Some of the more prevalent physical, social, and psychological impediments to be overcome include racial or ethnic prejudice, human ego, noise (in transmission equipment or surroundings), and distractions. Gender issues are also a primary concern. For a number of reasons, men and women think, speak, and act quite differently in interpersonal relations. There are generational issues, too: Children, parents, and grandparents can speak different languages. It is the task of communication to identify and find methods of circumventing or accommodating such information blockers.

APPLICATIONS AND PRODUCTS

Linguistics. The scientific study of the structure and function of spoken, written, and nonverbal languages, linguistics has many academic, educational, social, and professional applications. A foundation in linguistics is essential in understanding how words are formed and fitted together to create meaning and establish connections among people who are simultaneously senders and receivers of billions of messages over a lifetime. The stronger a linguistic base, the easier it is to discriminate among the plethora of messages received daily from disparate sources, to interpret meaning and respond accordingly, to prioritize, and to bring personal order to a disordered world of information overload.

Linguistics incorporates numerous threads, each worthy of close examination in its own right—word origins, vocabulary, grammar, phonetics, symbols, gestures, dialects, colloquialisms, or slang—that contribute to the rich tapestry of communication. Linguistics can be approached from several basic, broadly overlapping directions.

Structural linguistics deals with how languages are built and ordered into rules for communication. This subdiscipline, a basis for scientific research, clinical studies, sociological explorations, or scholarly pursuits, focuses on how speech is physically produced. Structural linguistics involves acoustic features, the comparison of different languages, grammatical concepts, sentence construction, vocabulary function, and other analytical aspects of language.

Historical linguistics concentrates on the development of written, oral, and nonverbal languages over time. An appropriate field for historians, educators, and comparative linguists, the discipline is concerned with why, where, when, and how words change in pronunciation, spelling, usage, and meaning.

Geolinguistics, a newer discipline with sociological, political, and professional implications in the modern global marketplace, focuses on living languages. The study is primarily concerned with historical and contemporary causes and effects of language distribution and interaction as a means to improve international and intercultural communication.

Interpersonal communication. Interpersonal communication is that which takes place between two or more people. An understanding of the process is useful in grasping how message transfer, receipt, interpretation, and response succeed or fail in the real world.

A broad discipline relevant to all applied science—since it involves everyday relationships and interactions among people—interpersonal communication serves as the basis for a considerable amount of scientific and popular study. Thousands of books and articles are published annually on various aspects of the subject: how to talk to a spouse, how to deal with a child, how to succeed in business, how to communicate with pets.

Interpersonal communication can be subdivided by specific fields of study or professional emphasis.

Conflict management and problem solving are communication-based specialties invaluable for students of family therapy, sociology, psychology, psychiatry, law, criminology, business management, and political science.

Gender, racial, sexual, intergenerational, and cultural communication issues are the concerns of many disciplines, particularly sociologists, psychologists, intercultural specialists, and international business students.

Health communication, a subdiscipline that involves translation of jargon to plain language and the development of empathy, is of particular relevance to medical and psychology students, and to intercultural therapists and clinicians.

Organizational communication, including departmental interaction, group dynamics, and behavior, is indispensable to the successful function of businesses, professional associations, government, and the military.

Mass communication. This is the study of the dissemination of information through various media (newspapers, magazines, billboards, books, radio, television, Web sites, blogs, and a host of other methods) with the potential to influence large audiences. Modern mass media is the culmination of centuries of technology enhancing the power and glory of communication. For better or worse, messages sent and received are the manifestation of qualities and possibilities that have elevated humans to a predominant position on earth. As languages have been defined and refined, as methods of transmission have improved, the range of communication has expanded. The distribution of data, once confined to the limitations of the human voice, through a series of quantum leaps is now a worldwide phenomenon. There are multiple outlets (Internet, satellite television, international phone lines) available to send messages, solicit responses, and record aftereffects on a global scale. Mass communication offers numerous opportunities for specialization.

Advertising, marketing, and public relations represent the persuasive power of communication. The modern world is a global marketplace. There are billions of potential customers for every conceivable product, service, or cause, and there are thousands of businesses and agencies whose task it is to create consumer demand. Marketing is sales strategy: planning, researching, testing, budgeting, setting objectives, and measuring results. Advertising is message strategy: what to say; how, when, where, and how often to say it to achieve desired aesthetic and marketing goals. Public relations is image strategy: the manipulation of words and pictures that establish and preserve public perception of corporations, institutions, organizations, governmental entities, or individuals.

Performance or public communication often involves a particular motivation: to generate immediate response from an audience. Motivational speakers, lecturers, debaters, politicians, actors, and other public performers all have the common goal of eliciting instant emotion in listeners or viewers.

Telecommunications refers to electronic means of sending information, primarily via radio or television broadcast, but the field also incorporates broadband, mobile wireless, information technology, networks, cable, satellite, unified communication, and emergency communication. A burgeoning, far-ranging industry, telecommunications offers global possibilities in business management, on-air and on-screen performance, research, marketing, journalism, programming, advertising, editing, sales, information science, and technology.

Creative communication concerns the composition of messages for a variety of informational purposes and is a component found to some degree throughout all facets of mass media. Writers of all kinds, graphic artists, photographers, designers, illustrators, and critics have the ability to influence behavior through the use of words or pictures.

Related fields of study fall under the umbrella of mass communication. Demographics, the collection of data that quantify an intended audience according to economic, political, ethnic, religious, professional, or educational factors, is an important consideration for many segments of mass media. There are also numerous legal, ethical, environmental, political, and regulatory issues to be dealt with that require specialized training.

IMPACT ON INDUSTRY

A few facts demonstrate the scope and trend of contemporary communication, a field that continues to grow exponentially in all directions. More than a million new books were released in 2009, three-quarters of which were self-published; digital book sales are expected to increase 400 percent by 2015. In 2010, the leading one hundred worldwide corporations—led by Proctor & Gamble's $8.68 billion expenditures— spent an average of $1 billion apiece in a $900 billion advertising industry. In 2011, the telecommunications industry will comprise $4 trillion, about 3 percent of the entire global economy. Internet advertising, expected to top $105 billion, will soon surpass newspaper advertising. Though the United States leads the way in online spending, with Japan in second place, China has supplanted Germany for third place, with Central and Eastern Europe, the Middle East, and Africa all fast-growing regions for such spending. Some 1.2 billion people, more than 20 percent of Earth's inhabitants, were connected to the Internet in 2010, and the penetration of the medium (about 75 percent in the United States) to the farthest corners of the world is expected to continue apace throughout the century. There is a constant influx of new individuals and groups establishing a presence through Web sites, blogs, and social networks. Mind-boggling numbers of messages about a bewildering array of subjects are sent and received daily, making effective communication more important than ever.

Academics. Though rhetoric and oratory have been studied since the days of Aristotle, communication science as a formal discipline dates from the

1940's, when research facilities were established at Columbia University, the University of Chicago, and the University of Illinois, Urbana-Champaign. In the twenty-first century, virtually all universities and colleges worldwide offer communication studies, most have degree programs, and many offer postgraduate work in various areas of concentration. The Annenberg School for Communication at the University of Pennsylvania, for example, has a long tradition in examining the effects of mass media on culture. Rensselaer Polytechnic Institute specializes in technical communications. New York University has particular interest in exploring media ecology, the study of media environments. At many institutions, original research in a wide variety of communication-related studies is encouraged and supported via grants and scholarships.

Government. The U.S. Government Printing Office (GPO) was established 150 years ago to keep track of mountains of information generated in the course of governing behavior across a sprawling, complicated society. The agency physically or electronically prints and disseminates—to officials, libraries, and the public— Supreme Court documents, the text of all congressional proceedings and legislation, presidential and executive department communications, and the records or bulletins of many independent agencies. The Federal Communications Commission (FCC), founded in 1934 and segmented to deal with various issues, regulates wired and wireless interstate and international transmissions via radio, television, satellite, cable, and fiber optics. Likewise, such bureaus as the Federal Trade Commission (FTC) oversee the commerce generated from correspondence between businesses and citizens. Millions of dollars in research funding is available at federal and state government levels across a spectrum of concerns related to communication. Finally, many government agencies, bureaus, and administrations—from the Census Bureau to the National Aeronautics and Space Administration to the U.S. Geological Survey and the White House— disperse aggregated data, images, and other information to the tax-paying public.

Corporations. While universities study causes and effects and governments regulate legal form and use, the business world presents numerous opportunities for practical application of communication principles. A broad-based, interdisciplinary field, communication allows many points of corporate entry, such as technology, creative disciplines, public relations, management, or sales. A computer maven can find a niche in many modern industries; prowess in journalism can be adapted to other forms of writing; performance skills can translate from television to film to stage; artistic talent can be suited to graphic design or commercial illustration; an individual who can successfully sell one product can easily learn to sell another product. Corporations reward innovation.

Many professional organizations exist to provide detailed, updated data on issues that affect both for-profit and not-for-profit industries. The National Communication Association, the International Communication Association, and a United Nations agency, the International Telecommunications Union, are leaders in the exchange of ideas, knowledge, research, technology, and regulatory information pertinent to communication.

CAREERS AND COURSE WORK

Communication is an all-purpose discipline that offers a multitude of career paths to personally and professionally rewarding occupations in a booming, always relevant field. Core courses—English language and grammar, sociology, psychology, history and popular culture, speech and business—provide a firm foundation on which a successful specialization can be built in any of several broad, compatible areas always in demand.

In academics, communication offers careers in teaching, counseling, research, and technology. Students should plan to pursue degrees past the bachelor's level, adding courses in education, communication theory, media ethics and history, information technology, and telecommunications in the undergraduate program.

Many creative niches are available in corporate communications, journalism, consumer advertising, public relations, freelance writing and art, and the media. Verbal artists need course work in rhetoric, literature, composition, linguistics, persuasion, and technical writing. Visual artists require graphic design, illustration, semiotics, and computer-aided design. On-screen or on-air performers would benefit from courses in organization, diction, public speaking, nonverbal communication, and mass media.

There are dozens of jobs and professions (positions including editor, agent, producer, director, publicist, and journalist) in media, government, politics, corporations, nonprofits, and other organizations for those with skills and credentials in communication and its various subfields. Courses in interpersonal and intercultural communication, business management, group dynamics, interviewing, problem solving, negotiation, and motivation are particularly useful.

Communication also has many applications in the social and health sciences. These include family therapy, psychiatry, marriage counseling, speech pathology, gender and sexuality services, legal providers, ethical concerns, international relations, and intercultural, intergenerational, and interspecies specialties that often require advanced degrees and original research. Course work can include concentrated studies in sociology and psychology, cultural history and geolinguistics, law, and foreign languages.

SOCIAL CONTEXT AND FUTURE PROSPECTS

From the beginning of civilization, communication has been the glue that binds together all elements of human society. In the modern global community— with billions of information exchanges passing rapidly among members of a vast, diverse, largely receptive audience eager to connect—competence in oral and written skills is mandatory for personal and professional success. More than ever, messages must be concise, unambiguous, accurate, and compelling to cut through clutter streaming from dozens of sources.

Several issues are of particular concern in contemporary communication studies. Publishing is changing from print to electronic. There is high interest across the multitrillion-dollar telecommunications industry in green technology that reduces air and noise pollution. A new field, unified communications, which promotes system interoperability, is expected to reach worldwide revenues of $14.5 billion by 2015. With fresh markets of information exchangers emerging around the world, there is great demand for intercultural expertise.

All disciplines are subject to economic climate. If employment slumps in one field, such as journalism, skill sets can often be easily and profitably applied to a related field, such as advertising or public relations. People and management and verbal skills translate well across industries, geography, and time.

One of the greatest challenges facing communication in the age of the Internet is the educated assessment and consumption of the vast array of information available to receivers at all levels. The need for courses and instruction in critical thinking, always important, has risen exponentially as various electronic sources of information have proliferated—especially when those sources are not well understood, their authoritativeness is open to question, and their existence is ephemeral.

For the foreseeable future, there will always be news to report, information to supply, products and services to sell, causes to promote, politicians to elect, legislation to be enacted, and government actions to document and disseminate. There will always be ethical debates about what can be done versus what should be done. There will always be a place for those who advance the art and science of communication through performance, education, therapy, research, or innovation. And there will always be a need for people who can consistently connect successfully with fellow humans through the evocative use of words and images.

—Jack Ewing

FURTHER READING

Baran, Stanley J., and Dennis K. Davis. *Mass Communication Theory: Foundations, Ferment, and Future.* 6th ed. Boston: Wadsworth, 2011. This textbook examines the field of mass communication, exploring theories and providing examples that aid in understanding the roles and ethics of various media.

Belch, George, and Michael Belch. *Advertising and Promotion: An Integrated Marketing Communications Perspective.* 9th ed. New York: McGraw-Hill/Irwin, 2008. This work spotlights the variety of strategies and methods used to build relationships between advertisers and customers in highly competitive, in-demand communication specialties.

Knapp, Mark L., and Judith A. Hall. *Nonverbal Communication in Human Interaction.* 7th ed. Boston: Wadsworth, 2009. A thorough examination of the theory, research, and psychology behind one of the most significant aspects of interpersonal communication.

Seiler, William J., and Melissa L. Beall. *Communication: Making Connections.* 8th ed. Needham Heights, Mass.: Allyn & Bacon, 2010. This book concentrates on the theory and practice of speech communication as applied to public discourse, interpersonal relationships, and group dynamics in a variety of settings.

Tomasello, Michael. *Origins of Human Communication.* Cambridge, Mass.: MIT Press, 2008. An award-winning study of how human communication evolved from gestures, founded on a fundamental need to cooperate for survival.

Varner, Iris, and Linda Beamer. *Intercultural Communication in the Global Workplace.* 5th ed. New York: McGraw-Hill/Irwin, 2010. This work addresses the impact of the Internet in particular on communication, especially as it affects relationships between businesses and governments in a shrinking world.

WEB SITES

International Communication Association
http://www.icahdq.org

International Telecommunications Union
http://www.itu.int/en/pages/default.aspx

Linguistic Society of America
http://www.lsadc.org

National Communication Association
http://www.natcom.org

See also: Information Technology; Telephone Technology and Networks.

COMMUNICATIONS SATELLITE TECHNOLOGY

FIELDS OF STUDY

Avionics; computer engineering; computer information systems; computer network technology; computer numeric control technology; computer science; electrical engineering; electronics engineering; mechatronics; physics; telecommunications technology; television broadcast technology.

SUMMARY

Communications satellite technology has evolved from its first applications in the 1950's to become a part of most people's daily lives and thereby producing billions of dollars in yearly sales. Communications satellites were initially used to help relay television and radio signals to remote areas of the world and to aid navigation. Weather forecasts routinely make use of images transmitted from communications satellites. Telephone transmissions over long distances, including fax, cellular phones, pagers, and wireless technology, are all examples of the increasingly large impact that communications satellite technology continues to have on daily, routine communications.

KEY TERMS AND CONCEPTS

- **baseband:** Transmission method for communications signals that uses the entire range of frequencies (bandwidth) available. It differs from broadband transmission, which is divided into different frequency ranges to allow multiple signals to travel simultaneously.

- **bit:** Binary digit, of which there are only two, the numbers zero and one. Computer technology is based on the binary number system because early computers contained switches that could be only on or off.
- **browser:** Software program that is used to view Web pages on the Internet.
- **downlinking:** Transmitting data from a communications satellite or spacecraft to a receiver on Earth.
- **downloading:** Process of accessing information on the Internet and then allowing a browser to make a copy to save on a personal computer.
- **gravity:** Force of attraction between two objects, expressed as a function of their masses and the distance between them. Typically, the Earth's gravity is the most important consideration for satellites, and it has the constant value of 9.8 meters per second squared (m/s^2).
- **hyperlinks:** Clickable pointers to online content in Web pages other than the page one is reading.
- **internet service provider (ISP):** Organization that provides access to the Internet for a fee.
- **ionosphere:** Part of the upper atmosphere that is ionized because of radiation from the sun, and therefore it affects the propagation of the radio waves within the electromagnetic spectrum.
- **kilobit:** Quantity equal to 1,000 bits.
- **orbit:** Curved path that an object travels in space because of gravity.
- **period:** Time required to complete one revolution around the Earth.
- **satellite:** Object that travels around the Earth.
- **transponder:** The electronic component of a communications satellite that automatically amplifies and retransmits signals that are received.
- **uplinking:** Transmitting data from a station on Earth up to a communications satellite or spacecraft.

DEFINITION AND BASIC PRINCIPLES

Sputnik 1, launched on October 4, 1957, by the Soviet Union, was the first artificial satellite. It used radio transmission to collect data regarding the distribution of radio signals within the ionosphere in order to measure density in the atmosphere. In addition to space satellites, the most common artificial satellites are the satellites used for communication, weather, navigation, and research. These artificial satellites travel around the Earth because of human action, and they depend on computer systems to function. A rocket is used to launch these artificial satellites so that they will have enough speed to be accelerated into the most common types of circular orbits, which require speeds of about 27,000 kilometers per hour. Some satellites, especially those that are to be used at locations far removed from the Earth's equator, require elliptical-shaped orbits instead, and their acceleration speeds are 30,000 kilometers per hour. If a launching rocket applies too much energy to an artificial satellite, the satellite may acquire enough energy to reach its escape velocity of 40,000 kilometers per hour and break free from the Earth's gravity. It is important that the satellite be able to maintain a constant high speed. If the speed is too low, gravity may cause the satellite to fall back down to the Earth's surface. There are also natural satellites that travel without human intervention, such as the Moon.

BACKGROUND AND HISTORY

In 1945, science fiction writer Arthur C. Clarke first described the concept of satellites being used for the mass distribution of television programs in his article "Extra-Terrestrial Relays," published in *Wireless World.* John Pierce, who worked at Bell Telephone Laboratories, further expanded on the idea of using satellites to repeat and relay television channels, radio signals, and telephone calls in his article "Orbital Radio Relays," published in the April, 1955, issue of *Jet Propulsion.* The first transatlantic telephone cable was opened by AT&T in 1956. The first transatlantic call was made in 1927, but it traveled via radio waves. The cable vastly improved the signal quality. The Soviet Union launched Sputnik 1, the first satellite, in 1957, which began the Space Race between the Soviet Union and the United States. The Communications Satellite Act of 1962 was passed by the United States Congress to regulate and assist the developing communications satellite industry. The first American television satellite transmission was made on July 10, 1962, five years into the Space Race, with the National Space and Aeronautics

Administration's (NASA) launch of the world's first communications satellite, AT&T's Telstar.

The many new communications satellites followed, with names such as Relay, Syncom, Early Bird, Anik F2, Westar, Satcom, and Marisat. Since the 1970's, communications satellites have allowed remote parts of the world to receive television and radio, primarily for entertainment purposes. Technology advances have continued to evolve and now use these satellites to facilitate mobile phone communication and high-speed Internet applications.

HOW IT WORKS

Communications satellites orbit the Earth and use microwave radio relay technology to facilitate communication for television, radio, mobile phones, weather forecasting, and navigation applications by receiving signals within the six-gigahertz (GHz) frequency range and then relaying these signals at frequencies within the four-GHz range. Generally there are two components required for a communications satellite. One is the satellite itself, sometimes called the space segment, which consists of the satellite and its telemetry controls, the fuel system, and the transponder. The other key component is the ground station, which transmits baseband signals to the satellite via uplinking and receives signals from the satellite via downlinking.

These communications satellites are suspended around the Earth in different types of orbits, depending on the communication requirements.

Geostationary orbits. Geostationary orbits are most often used for communications and weather satellites because this type of orbit has permanent latitude at zero degrees, which is above the Earth's equator, and only longitudinal values vary. The result is that satellites within this type of orbital can use a fixed antenna that is pointed toward one location in the sky. Observers on the ground view these types of satellites as motionless because their orbit exactly matches the Earth's rotational period. Numerically, this movement equates to an orbital velocity of 1.91 miles per second, or a period of 23.9 hours. Because this type of orbit was first publicized by the science fiction writer Arthur C. Clarke in the 1940's, it is sometimes called a Clarke orbit. Systems that use geostationary satellites to provide images for meteorological applications include the Indian National Satellite System (INSAT), the European

Organisation for the Exploitation of Meteorological Satellites' (EUMETSAT) Meteosat, and the United States' Geostationary Operational Environmental Satellites (GOES). These geostationary meteorological satellites provide the images for daily weather forecasts.

Molinya orbits. Molniya orbits have been important primarily in the Soviet Union because they require less energy to maintain in the area's high latitudes. These high latitudes cause low grazing angles, which indicate angles of incidence for a beam of electromagnetic energy as it approaches the surface of the Earth. The angle of incidence specifically measures the deviation of this approach of energy from a straight line. As a result, geostationary satellites would orbit too low to the Earth's surface and their signals would have significant interference. Because of Russia's high latitudes, Molniya orbits are more energy efficient than geostationary orbits. The word Molniya comes from the Russian word for "lightning," and these orbits have a period of twelve hours, instead of the twenty-four hours characteristic of geostationary orbits. Molniya orbits have a large amount of incline, with an angle of incidence of about 63 degrees.

Low earth orbits. Low earth orbit (LEO) refers to a satellite orbiting between 140 and 970 kilometers above the Earth's surface. The periods are short, only about ninety minutes, which means that several of them are necessary to provide the type of uninterrupted communication characteristic of geostationary orbits, which have twenty-four-hour periods. Although a larger number of low earth orbits are needed, they have lower launching costs and require less energy for signal transmission because of how close to the Earth they orbit.

APPLICATIONS AND PRODUCTS

Dish network and direct broadcast satellites. DISH Network is a type of direct broadcast satellite (DBS) network that communicates using small dishes that have a diameter of only 18 to 24 inches to provide access to television channels. This DBS service is available in several countries through many commercial direct-to-home providers, including DIRECTV in the United States, Freesat and Sky Digital in the United Kingdom, and Bell TV in Canada. These satellites transmit using the upper portion of the microwave Kμ band, which has a range of between 10.95 and 14.5

GHz. This range is divided based on the geographic regions requiring transmissions. Law enforcement also uses the frequencies of the electromagnetic spectrum to detect traffic-speed violators.

Fixed service satellites. Besides the DBS services, the other type of communication satellite is called a fixed service satellite (FSS), which is useful for cable television channel reception, distance learning applications for universities, videoconferencing applications for businesses, and local television stations for live shots during the news broadcasts. Fixed service satellites use the lower frequencies of the Kμ bands and the C band for transmission. All of these frequencies are within the microwave region of the electromagnetic spectrum. The frequency range for the C band is about 4 to 8 GHz, and generally the C band functions better when moisture is present, making it especially useful for weather communication.

Intercontinental telephone service. Traditional landline telephone calls are relayed to an Earth station via the public switched telephone network (PSTN). Calls are then forwarded to a geostationary satellite to allow intercontinental phone communication. Fiber-optic technology is decreasing the dependence on satellites for this type of communication.

Iridium satellite phones. Iridium is the world's largest mobile satellite communications company. Satellites are useful for mobile phones when regular mobile phones have poor reception. These phones depend only on open sky for access to an orbiting satellite, making them very useful for ships on the open ocean for navigational purposes. Iridium manufactures several types of satellite phones, including the Iridium 9555 and 9505A, and models that have water resistance, e-mail, and USB data ports. Although it has the largest market share for satellite phones in Australia, Iridium does face competition from two other satellite phone companies: Immarsat and Globalstar.

Satellite trucks and portable satellites. Trucks equipped with electrical generators to provide the power for an attached satellite have found applications for mobile transmission of news, especially after natural disasters. Some of these portable satellites use the C-band frequency for the transmission of information via the uplink process, which requires rather large antennas, whereas other portable satellites were developed in the 1980's to use the Ku band for transmission of information.

Global positioning system (GPS). GPS makes use of communications satellite technology for navigational purposes. The GPS was first developed by the government for military applications but has become widely used in civilian applications in products such as cars and mobile phones.

IMPACT ON INDUSTRY

Cable television and satellite radio. Ted Turner led the way in the 1980's with the application of communications satellite technology for distributing cable television news and entertainment channels (CNN, TBS, and TNT, to name a few). Since the 1980's, access to additional cable television stations has continued to grow and has evolved to include access to radio stations as well, such as SiriusXM radio. Communications satellite technology is transforming the radio industry by allowing listeners to continue to listen to the radio stations of their choice no matter where they are. Specifically, many new cars

---

FASCINATING FACTS ABOUT COMMUNICATIONS SATELLITE TECHNOLOGY

- In 2011, there are more than 2,000 communications satellites orbiting the Earth.
- There are at least nineteen Orbiting Satellite Carrying Amateur Radio (OSCAR). These artificial communication satellites have been launched by individuals.
- The Communications Satellite Act was passed in 1962 with the intention of making the United States a worldwide leader in communications satellite technology so that it could encourage peaceful relations with other nations. This act also established the Communications Satellite Corporation (Comsat), with headquarters in Washington, D.C.
- In 2003, Optus and Defence C1 carried sixteen antennas to provide eighteen satellite beams across Australia and New Zealand, making it one of the most advanced communications satellites ever launched.
- Sputnik, the first artificial satellite, was only 58 centimeters in diameter.
- The first known ideas for communications satellites were actually made public not by scientists or engineers but by the writers Arthur C. Clarke and Edward Everett Hale.

incorporate the Sirius S50 Satellite Radio Receiver and MP3/WMA player. However, the overall effect of communications satellites on the radio industry has not been as significant as it has on the television industry.

Internet access and Cisco. In addition to providing access to additional television stations, the satellite dish can be used for Internet access as well. Downloading data occurs at speeds faster than 513 kilobits per second and uploading is faster than 33 kilobits, which is more than ten times faster than the dial-up modems that use the plain old telephone service. Cisco has been a leader in developing the tools to connect the world via the Internet for more than twenty years. High-speed connections to the Internet have been replacing the slower dial-up modems that connect via telephone lines. High-speed connections use fiber distributed data interface (FDDI), which is composed of two concentric rings of fiber-optic cable that allow it to transmit over a longer distance. This is expensive, so Cisco developed internet routing in space (IRIS), which will have a huge impact on the industry as it places routers on the communications satellites that are in orbit. These routers will provide more effective transmission of data, video, and voice without requiring any transmission hubs on the ground at all. IRIS technology will radically transform Internet connections in remote, rural areas. An Internet server connects to the antenna of a small satellite dish about the same size of the dishes used to enhance television reception. This connection speed can be as fast as 1.5 megabits per second and is a viable alternative to wireless networks. Wireless networks have been transforming the way that people communicate on a daily basis, by using radio and infrared signals for a variety of handheld devices (iPhones, iPods, iPads). Huge economic growth potential exists for Cisco and other companies that can effectively incorporate more reliable satellite technology, such as IRIS.

EchoStar and high-speed internet access. Further indication of the large economic effect of communications satellites is from the February, 2011, report that EchoStar Corporation is going to pay more than $1 billion to purchase Hughes Communications. Hughes Communications uses satellites to provide high-speed, broadband access to the Internet. EchoStar, located in Englewood, Colorado, is the major provider of satellites and fiber networking technology to Dish Network Corporation (the second largest television provider in the United States), as well as additional military, commercial, and government customers that involve various data, video, and voice communications. The combined impact of merging Hughes Communications and Echo-Star will be to continue the expansion of communications satellite technology.

CAREERS AND COURSE WORK

Careers working with communications satellite technology can be found primarily in the radio, television, and mobile phone industries. Specifically, these careers involve working with the wired telecommunications services that often include direct-to-home satellite television distributors as well as the newer wireless telecommunications carriers that provide mobile telephone, Internet, satellite radio, and navigational services. Government organizations also need employees who are trained in working with communications satellite technology for weather forecasting and other environmental applications as well as communication of data between public-safety officials.

The highest salaries are earned by those with a bachelor's degree in avionics technology, computer engineering, computer science, computer information systems, electrical engineering, electronics engineering, physics, or telecommunications technology, although a degree in television broadcast technology can also lead to lucrative career after obtaining several years of on-the-job-training. The work environments for those with these types of degrees are primarily office and technology, with more than 14 percent of the workers working more than 40 hours per week. Although the telecommunications and communications technology industries are expected to continue to grow faster than many other industries, the actual job growth is expected to be less than other high-growth industries because of computer optimization. Those without a bachelor's degree are the most at-risk, as their jobs also can involve lifting and climbing around electrical wires outdoors in a variety of weather and locations. The National Coalition for Telecommunications Education and Learning (NACTEL), the Communications Workers of America (CWA), and the Society of Cable Telecommunications Engineers (SCTE) are sources of detailed career information for anyone interested in communications satellite technology.

SOCIAL CONTEXT AND FUTURE PROSPECTS

Advances in satellite technology have accompanied the rapid evolution of computer technology to such an extent that some experts describe this media revolution as an actual convergence of all media (television, motion pictures, printed news, Internet, and mobile phone communications). In 1979, Nicholas Negroponte of the Massachusetts Institute of Technology began giving lectures describing this future convergence of all forms of media. As of the twenty-first century, this convergence seems to be nearly complete. Television shows can be viewed on the Internet, as can news from cable television news stations such as CNN, Fox, and MSNBC. Hyperlinks provide digital connections between information that can be accessed from almost anywhere in the world instantly because of communication satellite technology. The result is that there is a twenty-four-hour news cycle, and the effects are sometimes positive but can also be negative if the wrong information is broadcast. The instantaneous transmission of political and social unrest by communications satellite technology can lead to further actions, as shown by the 2011 protests in Egypt, Iran, Yemen, Libya, and Bahrain.

—Jeanne L. Kuhler, MS, PhD

FURTHER READING

Baran, Stanley J., and Dennis K. Davis. *Mass Communication Theory: Foundations, Ferment, and Future.* 5th ed. Boston: Wadsworth Cengage Learning, 2008. Provides a detailed historical discussion of the communications technologies and their impacts on society.

Bucy, Erik P. *Living in the Information Age: A New Media Reader.* 2d ed. Belmont, Calif.: Wadsworth Thomson, 2005. Describes the societal implications of the evolution of media technology and the convergence of communications media made possible by new technologies.

Giancoli, Douglas C. *Physics for Scientists and Engineers with Modern Physics.* 4th ed. Upper Saddle River, N.J.: Pearson Education, 2008. This introductory physics textbook provides mathematical information regarding satellites and their orbits.

Grant, August E., and Jennifer Meadows. *Communication Technology Update and Fundamentals.* 12th ed. Burlington, Mass.: Focal Press, 2010. This introductory textbook provides technical information regarding communications satellites in terms of historical development, detailed applications, and existing uses.

Hesmondhalgh, David. *The Cultural Industries.* 2d ed. Thousand Oaks, Calif.: Sage, 2007. Describes the worldwide cultural effects of communications technologies in the past and present.

Mattelart, Armand. *Networking the World: 1794-2000.* Translated by Liz Carey-Libbrecht and James A. Cohen. Minneapolis: University of Minnesota Press, 2000. Describes the impact of communications satellites and other technologies on political, economic, and cultural phenomena, including nationalism, liberalism, and universalism.

Parks, Lisa, and Shanti Kumar, eds. *Planet TV:, A Global Television Reader.* New York: New York University Press, 2003. This textbook focuses on the application of communication satellites to television programming and discusses its historical evolution as well as its impact on the societies of various nations, especially India.

WEB SITES

Communications Workers of America
http://www.cwa-union.org

National Coalition for Telecommunications Education and Learning
http://www.nactel.org

Society of Cable Telecommunications Engineers
http://www.scte.org

See also: Avionics and Aircraft Instrumentation; Computer Engineering.

COMPUTER ENGINEERING

FIELDS OF STUDY

Computer engineering; computer science; computer programming; computer information systems; electrical engineering; information systems; computer information technology; software engineering.

SUMMARY

Computer engineering refers to the field of designing hardware and software components that interact to maximize the speed and processing capabilities of the central processing unit (CPU), memory, and the peripheral devices, which include the keyboard, monitor, disk drives, mouse, and printer. Because the first computers were based on the use of on-and-off mechanical switches to control electrical circuits, computer hardware is still based on the binary number system. Computer engineering involves the development of operating systems that are able to interact with compilers that translate the software programs written by humans into the machine instructions that depend on the binary number system to control electrical logic circuits and communication ports to access the Internet.

KEY TERMS AND CONCEPTS

- **basic input-output system (BIOS):** Computer program that allows the central processing unit (CPU) of a computer to communicate with other computer hardware.
- **browser:** Software program that is used to view Web pages on the Internet.
- **icon:** Small picture that represents a file, program, disk, menu, or option.
- **internet service provider (ISP):** Organization that provides paid access to the Internet.
- **logical topology:** Pathway within the physical network devices that directs the flow of data. The bus and ring are the only two types of logical topology.
- **mainframe:** Large, stand-alone computer that completes batch processing (groups of computer instructions completed at once).
- **protocol:** Set of rules to be followed that allows for communication.
- **server:** Computer that is dedicated to managing resources shared by users (clients).

DEFINITION AND BASIC PRINCIPLES

Much of the work within the field of computer engineering focuses on the optimization of computer hardware, which is the general term that describes the electronic and mechanical devices that make it possible for a computer user (client) to utilize the power of a computer. These physical devices are based on binary logic. Humans use the decimal system for numbers, instead of the base two-number system of binary logic, and naturally humans communicate with words. A great deal of interface activity is necessary to bridge this communication gap, and computer engineering involves additional types of software (programs) that function as intermediate interfaces to translate human instructions into hardware activity. Examples of these types of software include operating systems, drivers, browsers, compilers, and linkers.

Computer hardware and software generally can be arranged in a series of hierarchical levels, with the lowest level of software being the machine language, consisting of numbers and operands that the processor executes. Assembly language is the next level, and it uses instruction mnemonics, which are machine-specific instructions used to communicate with the operating system and hardware. Each instruction written in assembly language corresponds to one instruction written in machine code, and these instructions are used directly by the processor. Assembly language is also used to optimize the runtime execution of application programs. At the next level is the operating system, which is a computer program written so that it can manage resources, such as disk drives and printers, and can also function as an interface between a computer user and the various pieces of hardware. The highest level includes applications that humans use on a daily basis. These are considered the highest level because they consist of statements written in English and are very close to human language.

BACKGROUND AND HISTORY

The first computers used vacuum tubes and

The motherboard used in a HD DVD player, the result of computer engineering efforts.

mechanical relays to indicate the switch positions of *on* or *off* as the logic units corresponding to the binary digits of 0 or 1, and it was necessary to reconfigure them each time a new task was approached. They were large enough to occupy entire rooms and required huge amounts of electricity and cooling. In the 1930's the Atanasoff-Berry Computer (ABC) was created at Iowa State University to solve simultaneous numerical equations, and it was followed by the electronic numerical integrator and computer (ENIAC), developed by the military for mathematical operations.

The transistor was invented in 1947 by John Bardeen, Walter Brattain, and William Shockley, which led to the use of large transistors as the logic units in the 1950's. The integrated circuit chip was invented by Jack St. Clair Kilby and Robert Norton Noyce in 1958 and caused integrated circuits to come into usage in the 1960's. These early integrated circuits were still quite large and included transistors, diodes, capacitors, and transistors. Modern silicon chips can hold these components and as many as 55

million transistors. Silicon chips are called microprocessors, because each microprocessor can hold these logic units within just over a square inch of space.

How it Works

Hardware. The hardware, or physical components, of a computer can be classified according to their general uses of input, output, processing, and storage. Typical input devices include the mouse, keyboard, scanner, and microphone that facilitate communication of information between the human user and the computer. The operation of each of these peripheral devices requires special software called a driver, which is a type of controller, that is able to translate the input data into a form that can be communicated to the operating system and controls input and output peripheral devices.

A read-only memory (ROM) chip contains instructions for the basic input-output system (BIOS) that all the peripheral devices use to interact with the CPU. This process is especially important when a user first turns on a computer for the boot process.

When a computer is turned on, it first activates the BIOS, which is software that facilitates the interactions between the operating system, hardware, and peripherals. The BIOS accomplishes this interaction by first running the power-on self test (POST), which is a set of routines that are always available at a specific memory address in the read-only memory. These routines communicate with the keyboard, monitor, disk drives, printer, and communication ports to access the Internet. The BIOS also controls the time-of-day clock. These tasks completed by the BIOS are sometimes referred to as booting up (from the old expression "lift itself up by its own bootstraps"). The last instruction within the BIOS is to start reading the operating system from either a boot disk in a diskette drive or the hard drive. When shutting down a computer there are also steps that are followed to allow for settings to be stored and network connections to be terminated.

The CPU allows instructions and data to be stored in memory locations, called registers, which facilitate the processing of information as it is exchanged between the control unit, arithmetic-logic unit, and any peripheral devices. The processor interacts continuously with storage locations, which can be classified as either volatile or nonvolatile types of memory. Volatile memory is erased when the computer is turned off and consists of main memory, called random-access memory (RAM) and cache memory. The fundamental unit of volatile memory is the flip-flop, which can store a value of 0 or 1 when the computer is on. This value can be flipped when the computer needs to change it. If a series of 8 flip-flops is hooked together, an 8-bit number can be stored in a register. Registers can store only a small amount of data on a temporary basis while the computer is actually on. Therefore, the RAM is needed for larger amounts of information. However, it takes longer to access data stored in the RAM because it is outside the processor and needs to be retrieved, causing a lag time. Another type of memory, called cache, is located in the processor and can be considered an intermediate type of memory between registers and main memory.

Nonvolatile memory is not erased when a computer is turned off. It consists of hard disks that make up the hard drive or flash memory. Although additional nonvolatile memory can be purchased for less money than volatile memory, it is slower. The hard drive consists of several circular discs called platters that are made from aluminum, glass, or ceramics and covered by a magnetic material so that they can develop a magnetic charge. There are read and write heads made of copper so that a magnetic field develops that is able to read or write data when interacting with the platters. A spindle motor causes the platters to spin at a constant rate, and either a stepper motor or voice coil is used as the head actuator to initiate interaction with the platters.

The control unit of the CPU manages the circuits for completing operations of the arithmetic-logic unit. The arithmetic-logic unit of the CPU also contains circuits for completing the logical operations, in addition to data operations, causing the CPU essentially to function as the brain of the computer. The CPU is located physically on the motherboard. The motherboard is a flat board that contains all the chips needed to run a computer, including the CPU, BIOS, and RAM, as well as expansion slots and power-supply connectors. A set of wires, called a bus, etched into the motherboard connects these components. Expansion slots are empty places on the motherboard that allow upgrades or the insertion of expansion cards for various video and voice controllers, memory expansion cards, fax boards, and modems without having to reconfigure the entire computer.

The motherboard is the main circuit board for the entire computer.

Circuit design and connectivity. Most makers of processor chips use the transistor-transistor logic (TTL) because this type of logic gate allows for the output from one gate to be used directly as the input for another gate without additional electronic input, which maximizes possible data transmission while minimizing electronic complications. The TTL makes this possible because any value less than 0.5 volt is recognized as the logic value of 0, while any value that exceeds 2.7 volts indicates the logic value of 1. The processor chips interact with external computer devices via connectivity locations called ports. One of the most important of these ports is called the Universal Serial Bus (USB) port, which is a high-speed, serial, daisy-chainable port in newer computers used to connect keyboards, printers, mice, external disk drives, and additional input and output devices.

Software. The operating system consists of software programs that function as an interface between the user and the hardware components. The operating system also assists the output devices of printers and monitors. Most of the operating systems being used also have an application programming interface (API), which includes graphics and facilitates use. APIs are written in high-level languages (using statements approximating human language), such as C++, Java, and Visual Basic.

APPLICATIONS AND PRODUCTS

Stand-alone computers. Most computer users rely on relatively small computers such as laptops and personal computers (microcomputers). Companies that manufacture these relatively inexpensive computers have come into existence only since the early 1980's and have transformed the lives of average Americans by making computer usage a part of everyday life. Before microcomputers came into such wide usage, the workstation was the most accessible smaller-size computer. It is still used primarily by small and medium-size businesses that need the additional memory and speed capabilities. Larger organizations such as universities use mainframe computers to handle their larger power requirements. Mainframes generally occupy an entire room. The most powerful computers are referred to as supercomputers, and they are so expensive that often several universities will share them for scientific and computational activities. The military uses them as well. They often require the space of several rooms.

Inter-network service architecture, interfaces, and inter-network interfaces. A network consists of two or more computers connected together in order to share resources. The first networks used coaxial cable, but now wireless technologies allow computer devices to communicate without the need to be physically connected by a coaxial cable. The Internet has been a computer-engineering application that has transformed the way people live. Connecting to the Internet first involved the same analogue transmission used by the plain old telephone service (POTS), but connections have evolved to the use of fiber-optic technology and wireless connections. Laptop computers, personal digital assistants (PDAs), cell phones, smart phones, RFID (radio frequency identification), iPods, iPads, and Global Positioning Systems (GPS) are able to communicate, and their developments have been made possible by the implementation of the fundamental architectural model for inter-network service connections called the Open Systems Interconnection (OSI) model. OSI is the layered architectural model for connecting networks. It was developed in 1977 and is used to make troubleshooting easier so that if a component fails on one computer, a new, similar component can be used to fix the problem, even if the component was manufactured by a different company.

OSI's seven layers are the application, presentation, session, transport, network, data link, and physical layers. The application and presentation layers work together. The application layer synchronizes applications and services in use by a person on an individual computer with the applications and services shared with others via a server. The services include e-mail, the World Wide Web, and financial transactions. One of the primary functions of the presentation layer is the conversion of data in a native format such as extended binary coded decimal interchange code (EBCDIC) into a standard format such as American Standard Code for Information Interchange (ASCII). As its name implies, the primary purpose of the session layer is to control the dialog sessions between two devices. The network file system (NFS) and structured query language (SQL) are examples of tools used in this layer.

The transport layer controls the connection-oriented flow of data by sending acknowledgments to data senders once the recipient has received data and also makes sure that segments of data are retransmitted if they do not reach the intended recipient. A router is one of the fundamental devices that works in this layer, and it is used to connect two or more networks together physically by providing a high degree of security and traffic control.

The data link layer translates the data transmitted by the components of the physical layer, and it is within the physical layer that the most dramatic technological advances made possible by computer engineering have had the greatest impact. The coaxial cable originally used has been replaced by fiber-optic technology and wireless connections.

Fiber Distributed Data Interface (FDDI) is the fiber-optic technology that is a high-speed method of networking, composed of two concentric rings of fiber-optic cable, allowing it to transmit over a longer distance but at greater expense. Fiber-optic cable uses glass threads or plastic fibers to transmit data. Each cable contains a bundle of threads that work by reflecting light waves. They have much greater bandwidth than the traditional coaxial cables and can carry data at a much faster rate of 100 gigabits per second because light travels faster than electrical signals. Fiber-optic cables are also not susceptible to electromagnetic interference and weigh less than coaxial cables.

Wireless networks use radio or infrared signals for cell phones and a rapidly growing variety of hand-held devices, including iPhones, iPods, and tablets. New technologies using Bluetooth and Wi-Fi for mobile connections are leading to the next phase of inter-network communications with the Internet called cloud computing, which is the direction for the most economic growth. Cloud computing is basically a wireless Internet application where servers supply resources, software, and other information to users on demand and for a fee.

Smart phones that use Google's new Android operating system for mobile phones have a PlayStation emulator called PSX4Droid that allows PlayStation games to be played on these phones. Besides making games easily accessible, cloud computing is also making it easier and cheaper to do business all around the world with applications such as Go To Meeting, which is one example of video conferencing technology used by businesses.

IMPACT ON INDUSTRY

The first developments in technology that made personal computers, the Internet, and cell phones more easily accessible to the average consumer have primarily been made by American companies. Computer manufacturers include IBM and Microsoft, which were followed by Dell, Compaq, Gateway, Intel, Hewlett-Packard, Sun Microsystems, and Cisco. Cisco continues to be the primary supplier of the various hardware devices necessary for connecting networks, such as routers, switches, hubs, and bridges. Mobile computing is being led by American companies Apple, with its iPhone, and Research in Motion (RIM), with its Blackberry. These mobile devices require collaboration with phone companies such as AT&T and Verizon, causing them to grow in worldwide dominance as well. Most U.S. technology companies are international in scope, with expansion still expected in the less-developed countries of Southeast Asia and Eastern Europe.

Microsoft was founded in 1975 by Bill Gates and Paul Allen. All the personal computers (PCs) that became more accessible to consumers depended on Microsoft's operating system, Windows. Over the years, Microsoft has made numerous improvements with each new version of its operating system, and it has continued to dominate the PC market to such an extent that the U.S. government sued Microsoft, charging it with violating antitrust laws. As of 2011, Microsoft remains the largest software company in the world, although it has been losing some of its dominance as an innovative technology company because of the growth of Google, with its Internet search engine software, and Apple, with its new, cutting-edge consumer electronics, including the iPod, iPad, and iPhone. In 2010, Microsoft started selling the Kinect for Xbox 360, which is a motion-capture device used for games that easily connects to a television and the Internet. More than 2.5 million of the Kinect for Xbox 360 units have been sold in the United States since November, 2010. To try to compete with Apple's dominance in the mobile-technology market, Microsoft announced in January, 2011, that it would collaborate with chip manufacturer Micron, to incorporate Micron's flash memory into mobile devices. A new version of Microsoft's Windows operating system was to use processor chips manufactured by ARM Holdings rather than those by Intel, because the ARM chips contain lower power, which facilitates mobile communication. Microsoft continues to dominate the PC operating-system market worldwide.

Apple is one of the most successful U.S. companies because of its Macintosh computer and iPod. Introduced in 2001, the iPod uses various audio file formats, such as MP3 and WAV, to function as a portable media player. To further expand on the popularity of the iPod, Apple opened its own online media and applications stores to allow consumers to access music and video applications via the Internet. In 2007, Apple expanded its market share by introducing the iPhone, and adding visual text messaging, emailing of digital photos using its own camera capabilities, and a multi-touch screen for enhanced access to the Internet. The Apple iPhone is essentially a miniature handheld computer that also functions as a phone. The apps that are be downloaded from the Apple Store online are computer programs that can be designed to perform just about any specific task. There are more than 70,000 games available for purchase for the iPhone, and more than 7 million apps have been sold as of January, 2011.

Google was started by Larry Page and Sergey Brin, who met as graduate students in Stanford University's computer science department, in 1996. Ten years later, the Oxford dictionary added the term "google,"

FASCINATING FACTS ABOUT COMPUTER ENGINEERING

- It is possible to warm up the grill before arriving home by connecting to the Internet and attaching an iGrill to a USB port. Recipes can also be downloaded directly from the Internet for the iGrill.
- The average laptop computer has more processing power than the ENIAC, which had more than 20,000 vacuum tubes and was large enough to fill entire rooms.
- By about 2015, batteries in digital devices will consume oxygen from the air in order to maintain their power.
- By 2015, it is expected that some mobile devices will be powered by kinetic energy and will no longer depend on batteries.
- A job growth rate of 31 percent is projected for computer engineers by 2018.

a verb meaning to do an Internet search on a topic using the Google search engine. Despite attempts by several competitors, including Microsoft, to develop search engine software, Google has dominated the market. In addition to its search engine, Google has developed many new innovative software applications, including Gmail, Google Maps, and its own operating system called Android, designed specifically to assist Google with its move into the highly lucrative mobile-technology market.

CAREERS AND COURSE WORK

Knowledge of the design of electrical devices and logic circuits is important to computer engineers, so a strong background in mathematics and physics is helpful. Since there is a great deal of overlap between computer engineering and electrical engineering, any graduate with a degree in electrical engineering may also find work within the computer field.

One of the most rapidly expanding occupations within the computer engineering field is that of network systems and data communication analyst. According to the Bureau of Labor Statistics, this occupation is projected to grow by 53.4 percent to almost 448,000 workers between by 2018, making it the second-fastest growing occupation in the United States. The related occupation of computer systems analyst is also expected to grow faster than the average for all jobs through 2014. Related jobs include network security specialists and software engineers. All of these occupations require at least a bachelor's degree, typically in computer engineering, electrical engineering, software engineering, computer science, information systems, or network security.

SOCIAL CONTEXT AND FUTURE PROSPECTS

The use of cookies, a file stored on an Internet user's computer that contains information about that user, an identification code or customized preferences, that is recognized by the servers of the Web sites the user visits, as a tool to increase online sales by allowing e-businesses to monitor the preferences of customers as they access different Web pages is becoming more prevalent. However, this constant online monitoring also raises privacy issues. In the future there is the chance that one's privacy could be invaded with regard to medical or criminal records and all manner of financial information. In addition, wireless technologies are projected to increase the usage of smart phones, the iPad, the iPod, and many new consumer gadgets, which can access countless e-business and social networking Web sites. Thus, the rapid growth of Internet applications that facilitate communication and financial transactions will continue to be accompanied by increasing rates of identity theft and other cyber crimes, ensuring that network security will continue to be an important application of computer engineering.

—Jeanne L. Kuhler, MS, PhD

FURTHER READING

Cassel, Lillian N., and Richard H. Austing. *Computer Networks and Open Systems: An Application Development Perspective.* Sudbury, Mass.: Jones & Bartlett Learning, 2000. This textbook describes the OSI architecture model.

Das, Sumitabha. *Your UNIX: The Ultimate Guide.* New York: McGraw-Hill, 2005. This introductory textbook provides background and instructions for using the UNIX operating system.

Dhillon, Gurphreet. "Dimensions of Power and IS Implementation." *Information and Management* 41, no. 5 (2004): 635-644. Describes some of the work done by managers when choosing which computer languages and tools to implement.

Irvine, Kip R. *Assembly Language for Intel-Based Computers.* 5th ed. Upper Saddle River, N.J.: Prentice Hall, 2006. This introductory textbook provides instruction for using assembly language for Intel processors.

Kerns, David V., Jr., and J. David Irwin. *Essentials of Electrical and Computer Engineering* 2d ed. Upper Saddle River, N.J.: Prentice Hall, 2004. A solid introductory text that integrates conceptual discussions with modern, relevant technological applications.

Magee, Jeff, and Jeff Kramer. *Concurrency: State Models and Java Programming.* Hoboken, N.J.: John Wiley & Sons, 2006. This intermediate-level textbook describes the software engineering techniques involving control of the timing of different processes.

Silvester, P. P., and D. A. Lowther. *Computer Engineering Circuits, Programs, and Data.* New York: Oxford University Press, 1989. This is an introductory text for engineering students.

Sommerville, Ian. *Software Engineering.* 9th ed. Boston: Addison-Wesley, 2010. Text presents a broad, upto-date perspective of software engineering that focuses on fundamental processes and techniques for creating software systems. Includes case studies and extensive Web resources.

WEB SITES
Computer Society
http://www.computer.org

Institute of Electrical and Electronics Engineers
http://www.ieee.org

Software Engineering Institute
http://www.sei.cmu.edu

See also: Computer Graphics; Computer Languages, Compilers, and Tools; Computer Science; Electrical Engineering; Engineering; Pattern Recognition.

COMPUTER GRAPHICS

FIELDS OF STUDY

Three-dimensional (3-D) design; calculus; computer programming; computer animation; digital modeling; graphic design; multimedia applications; software engineering; Web development; vector graphics and design; drawing; animation.

SUMMARY

Computer graphics involves the creation, display, and storage of images on a computer with the use of specialized software. Computer graphics fills an essential role in many everyday applications. 3-D animation has revolutionized video games and has resulted in box-office hits in theaters. Virtual images of people's bodies and tissues are used in medicine for teaching, surgery simulations, and diagnoses. Educators and scientists are able to develop 3-D models that illustrate principles in a more comprehensible manner than a two-dimensional (2-D) image can. Through the use of such imagery, architects and engineers can prepare virtual buildings and models to test options prior to construction. Finally, businesses use computer graphics to prepare charts and graphs for better comprehension during presentations.

KEY TERMS AND CONCEPTS

- **application programming interface (API):** Set of functions built into a software program that allows communication with another software program.
- **computer-aided design (CAD):** Software used by architects, engineers, and artists to create drawings and plans.
- **computer animation:** Art of creating moving images by means of computer technology.
- **graphical user interface (GUI):** Program by which a user interacts with a computer by controlling visual symbols on the screen.
- **graphics processing unit (GPU):** Specialized microprocessor typically found on a video card that accelerates graphics processing.
- **pixels:** Abbreviation for "picture elements," the smallest discrete components of a graphic image appearing on a computer screen or other graphics output device.
- **raster graphics:** Digital images that use pixels arranged in a grid formation to represent an image.
- **rendering:** Process of generating an image from a model by means of computer programs.
- **three-dimensional (3-D) model:** Representation of any 3-D surface of an object using graphics software.
- **vector graphics:** Field of computer graphics that uses mathematical relationships between points and the paths connecting them to describe an image.

DEFINITION AND BASIC PRINCIPLES
The field of computer graphics uses computers to create digital images or to modify and use images obtained from the real world. The images are created from internal models by means of computer programs. Two types of graphics data can be stored in a computer. Vector graphics are based on mathematical formulas that generate geometric images by joining straight lines. Raster graphics are based on a grid of dots known as pixels, or picture elements. Computer graphics can be expressed as 2-D, 3-D, or animated images.

The graphic data must be processed in order to

render the image and display on a computer movie screen. The work of computer-graphics programmers has been facilitated by the development of application programming interfaces (APIs), notably the Open Graphics Library (OpenGL). OpenGL provides a set of standard operations to render images across a wide variety of platforms (operating systems). The graphics processing unit (GPU) found in a video card facilitates presentation of the image.

There are subtle differences between the responsibilities of the computer-graphics specialist and graphic or Web designers. Computer-graphics specialists develop programs to display visual images or models, while designers are creative artists who use programs to communicate a specific message effectively. The end products of graphic designers are seen in various print media, while Web designers produce digital media.

BACKGROUND AND HISTORY

The beginning of computer graphics has been largely attributed to Ivan Sutherland, who developed a computer drawing program in 1961 for his dissertation work at the Massachusetts Institute of Technology. This program, Sketchpad, was a seminal event in the area of human-computer interaction, as it was one of the first to use graphical user interfaces (GUIs). Sutherland used a light pen containing a photoelectric cell that interacted with elements on the monitor. His method was based on vector graphics. Sketchpad provided vastly greater possibilities for the designer or engineer over previous methods based on pen and paper.

Other researchers further developed vector-graphics capabilities. Raster-based graphics using pixels was later developed and is the primary technology being used. The mouse was invented and proved more convenient than a light pen for selecting icons and other elements on a computer screen. By the early 1980's, Microsoft's personal computer (PC) and Apple's Macintosh were marketed using operating systems that incorporated GUIs and input devices that included the mouse as well as the standard keyboard.

Major corporations developed an early interest in computer graphics. Engineers at Bell Telephone Laboratories, Lawrence Berkeley National Laboratory, and Boeing developed films to illustrate satellite orbits, aircraft vibrations, and other physics principles. Flight simulators were developed by Evans & Sutherland and General Electric.

The invention of video graphics cards in the late 1980's followed by continual improvements gave rise to advances in animation. Video games and full-length animated motion pictures have become large components of popular culture.

HOW IT WORKS

Types of images. Vector graphics uses mathematical formulas to generate lines or paths that are connected at points called vertices to form geometric shapes (usually triangles). These shapes are joined in a meshwork on the surfaces of figures. Surfaces on one plane are two-dimensional, while connecting vertices in three dimensions will produce 3-D images. Two-dimensional images are more useful for applications such as advertising, technical drawing, and cartography. Raster images, on the other hand, develop images based on pixels, or picture elements. Pixels can be thought of as tiny dots or cells that contain minute portions of the image and together compose the image on the computer screen. The bits of information that the pixels are able to process determine the resolution or sharpness of the image. Raster images are much more commonly used in computer graphics and are essential for 3-D and animation work.

Graphics pipeline. The process of creating an image from data is known as the graphics pipeline. The pipeline consists of three main stages: modeling, rendering, and display.

Modeling begins with a specification of the objects, or components of a scene, in terms of shape, size, color, texture, and other parameters. These objects then undergo a transformation involving their correct placement in a scene.

Rendering is the process of creating the actual image or animation from the scene. Rendering is analogous to a photograph or an artist's drawing of a scene. Aspects such as proper illumination and the visibility of objects are important at this stage.

The final image is displayed on a computer monitor with the use of advanced software, as well as computer hardware that includes the motherboard and graphics card. The designer must keep in mind that the image may appear differently on different computers or different printers.

OpenGL. OpenGL is an API that facilitates

writing programs across a wide variety of computer languages and hardware and software platforms. OpenGL consists of graphics libraries that provide the programmer with a basic set of commands to render images. OpenGL is implemented through a rendering pipeline (graphics pipeline). Both vector and raster data are accepted for processing but follow different steps. At the rasterization stage, all data are converted to pixels. At the final step, the pixels are written into a 2-D grid known as a framebuffer. A framebuffer is the memory portion of the computer allocated to hold the graphics information for a single frame or picture.

Maya. Maya (trade name Autodesk Maya) is computer-graphics software that has become the industry standard for generating 3-D models for game development and film. Maya is particularly effective in producing dazzling animation effects. Maya is imported into OpenGL or another API to display the models on the screen.

Video games. Video-game development takes a specialized direction. The term "game engine" was coined to refer to software used to create and render video games on a computer screen. Many features need to work together to create a successful game, including 2-D and 3-D graphics, a "physics engine" to prepare realistic collision effects, sound, animation, and many other functions. Because of the highly competitive nature of the video-game industry, it is necessary to develop new games rapidly. This has led to reusing or adapting the same game engine to create new games. Some companies specialize in developing so-called middleware software suites that are conceived to contain basic elements on which the game programmer can build to create the complete game.

Television. To create 3-D images for television, individual objects are created in individual layers in computer memory. This way, the individual objects can move independently without affecting the other objects. TV graphics are normally produced a screen at a time and can be layered with different images, text, backgrounds, and other elements to produce rich graphic images. Editing of digital graphics is much faster and efficient than the traditional method of cutting and pasting film strips.

Film. The film *Avatar* (2009) illustrated how far 3-D animation had been developed. The production used a technique called performance or motion tracking. Video cameras were attached to computers and focused on the faces of human actors as they performed their parts. In this manner, subtle facial expressions could be transferred to their animated avatars. Backgrounds, props, and associated scenery moved in relation to the actors.

APPLICATIONS AND PRODUCTS

Game development. Game development has become a major consumer industry, grossing $20 billion in 2010. Ninety-two percent of juveniles play video or computer games. Major players in the field include Sony PlayStation, Nintendo, and Microsoft Xbox. The demographics of video-game players are changing, resulting in a

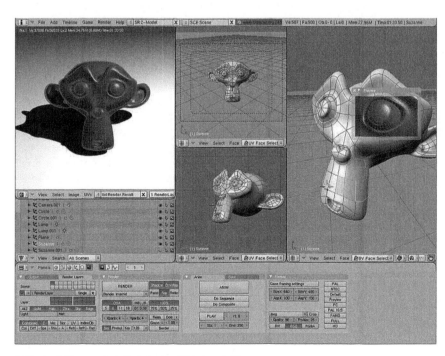

A Blender 2.45 screenshot, displaying the 3D test model Suzanne.

FASCINATING FACTS ABOUT COMPUTER GRAPHICS

- *Avatar* (2009) became the highest-grossing film of all time, earning $2.7 billion worldwide and demonstrating how compelling stories told using computer graphics had become.
- Creation of virtual persons allows surgeons to perform a trial run of their procedures before the actual surgery.
- Real-time visualization models prepared by architects allow buyers to walk through virtual homes and change the views they see.
- Scientists can study chemical reactions with molecules in real time to gain insight into the structural changes taking place among the molecules.
- Engineers can prepare a 3-D model of an engine under development and study the effects of various changes.
- The features visible in 3-D through Google Earth—topography, cities, street views, trees, and other details--dramatically demonstrate the uses of computer graphics when wedded to satellite technologies.
- 3-D visualization of atmosphere and terrain has resulted in improved weather forecasting and stunning presentations by news meteorologists.
- 3-D augmented reality is being studied to animate traffic-simulation models to aid in the planning and design of highway construction projects. In the 3-D simulation tests, the user has the opportunity to drive a virtual or real car.

greater number of female and adult players. Whereas previously video games focused on a niche market (juvenile boys), in the future game developers will be increasingly directing their attention to the mass market. These market changes will result in games that are easier to play and broader in subject matter.

Film. The influence of computer graphics in the film industry is largely related to animation. Of course, animation predates computer graphics by many years, but it is realism that gives animation its force. Entire films can be animated, or animation can play a supplemental role. The fantasy of Hollywood was previously based on constructing models and miniatures, but now computer-generated imagery can be integrated into live action.

Television. The conversion of the broadcast signal for television from analogue to digital, and later to high definition, has made the role of the computer-graphics designer even more important. It is common for many shows to have a computer-graphics background instead of a natural background, such as when a weatherperson stands in front of a weather map. This development results in more economical productions, since there are no labor costs involved in preparing sets or a need to store them.

Computer graphics was used in television advertising before film, since it is more economical to produce. A combination of dazzling, animated graphics with a product or brand-name can leave a lasting impression on the viewer.

Medicine. Computer graphics typically works in concert with other advanced technologies, such as computed tomography, magnetic resonance, and positron emission tomography, to aid in diagnosis and treatment. The images obtained by these technologies are reconstructed by computer-graphics techniques to form 3-D models, which can then be visualized. The development of virtual human bodies has proven invaluable to illustrate anatomical structures in healthy and diseased subjects. These virtual images have found application in surgery simulations and medical education and training. The use of patient-specific imaging data guides surgeons to conduct minimally invasive interventions that are less traumatic for the patient and lead to faster healing. Augmented reality provides a larger view of the surgical field and allows the surgeon to view structures that are below the observed surface.

Science. Computer graphics has proven valuable to illustrate scientific principles and concepts that are not easily visible in the natural environment. By preparing virtual 3-D models of molecular structures or viruses moving through tissues, a student or scientist who is a visual learner is able to grasp these concepts.

Architecture and engineering. Computer-aided design has greatly helped the fields of architecture and engineering. Computer graphics was initially used only as a replacement for drafting with pencil and paper. However, the profession has come to recognize its value in the early stages of a project to help designers check and reevaluate their proposed designs. Multimedia designs such as animations and panoramas are very useful in client presentations. The designs allow clients to walk through a building virtually and interactively look around interior spaces. Engineers can also test the effect of various inputs to a system, model, or circuit.

Business. Presentation of numeric data in graphs and charts is an important application of computer graphics in business. Market trends, production data, and other business information presented in graphic form are often more understandable to an audience or reader.

Education. Computer graphics has proven very useful in education because of the power of visualization in the learning process. There are many benefits to using computer graphics in education: Students learn at their own pace at their own time, the instruction is interactive rather than passive, the student is engaged in the learning process, and textual and graphic objects can be shared among applications via tools such as cutting and pasting, clipboards, and scrapbooks.

IMPACT ON INDUSTRY

The worldwide market for hardware, software, and services for 3-D visualization is expected to grow rapidly. Spending in the defense and government markets reached $16.5 billion in 2010 and is expected to increase to $20 billion by 2015. The popularity of games and the film industry has resulted in huge investments in developing 3-D technology. With the technology already in place, world governments can acquire 3-D visualization at a much more reasonable price than previously. The military sector is the largest market, followed by civil-aviation and civil-service sectors, including law-enforcement and emergency-response agencies.

Government and university research. Although the U.S. government played a leading role in the early development of computer graphics, it has come to play a more collaborative role with universities. A typical example is the Graphics and Visualization Center, which was founded by the National Science Foundation to pursue interdisciplinary research at five universities (Brown, California Institute of Technology, Cornell, University of North Carolina, and University of Utah). The center pursues research in four main areas of computer graphics: modeling, rendering, user interfaces, and high-performance architectures. The Department of Defense also conducts research involving computer graphics.

On the local level, computer graphics has proven useful in preparing presentation material, such as visually appealing graphs, charts, and diagrams.

Complex projects and issues can be presented in a manner that is more understandable than traditional speeches or handouts. County and community planners can explore "what if" scenarios for land-use projects and investigate the potential for change. Computers can enhance information used in the planning process or explain the scope of a project. The presentation material can be modified to incorporate community suggestions over a series of meetings.

Universities tend to offer courses and degree programs in the types of research they specialize in. Brown, Penn State, and several University of California campuses are working on scientific visualization that has applications in designing virtual humans and medical illustrations. Universities in Canada and Europe are also active in these areas. The University of Utah, a longtime leader in the field, conducts research in geometric design and scientific computing.

Industry and business. The computer-graphics industry will continue to grow, resulting in an increased demand for programmers, artists, scientists, and designers. Pixar, DreamWorks, Disney, Warner Bros., Square Enix, Sony, and Nintendo are considered top animation producers in the film and video-game industries. Architectural and engineering consulting firms have been contracted to use virtual images in planning public buildings such as stadiums. Virtual people help to determine traffic flow through buildings. If congestion points are observed as virtual crowds travel to concession stands or restrooms, for example, changes can be made to provide more efficient flow. Virtual models were also used to study traffic flow in Hong Kong harbor.

Computer graphics are a big improvement over architectural scale models. They can readily portray a variety of alternative plans without incurring large costs.

CAREERS AND COURSE WORK

Computer-graphics specialists must have a unique combination of artistic and computer skills. They must have good math, programming, and design skills, and be able to visualize 3-D objects. The specialist must be creative, detail oriented, and able to work well individually and as part of a team.

Course offerings can vary considerably among universities, so the student must consider the specialties of the prospective schools in relation to the field in which he or she is most interested. For example, the student may want to focus more on software design than on graphics design. Essential courses can include advanced mathematics, programming, computer animation, geometric design and modeling, multimedia systems and applications, and software engineering.

Computer graphics can be applied to a vast number of fields, and its influence can only increase. In addition to animation in video games and film, computer graphics has proven valuable in architectural and engineering design, education, medicine, business, and cartography. Typical positions for a computer-graphics specialist include 3-D animator or modeler, special effects designer, video game designer, and Web designer.

Computer-graphics specialists can work in the public or private sector; they can also work independently. Although companies prefer to hire candidates with a bachelor's degree, many workers in the field have only an associate's degree or vocational certificate.

In the private sector, computer-graphics specialists can be employed by architectural, construction, and engineering consulting firms, electronics and software manufacturing companies, and petrochemical, food processing, and energy industries. In the public sector, they can work at all levels of government and in hospitals and universities.

SOCIAL CONTEXT AND FUTURE PROSPECTS
Computer graphics will continue to have a profound effect on the visual arts, freeing the artist from the need to master technical skills in order to focus on creativity and imagination in his or her work. The artist can experiment with unlimited variations in structures and designs in a single work to see which produces the desired effect. Continuing advances in producing virtual images and 3-D animation will enhance understanding of scientific principles and processes in education, medicine, and science.

—David Olle, MS

FURTHER READING
McConnell, Jeffrey. *Computer Graphics: Theory into Practice.* Boston: Jones & Bartlett, 2006. The basic principles of graphic design are amply presented with reference to the human visual system. OpenGL is integrated with the material, and examples of 3-D graphics are illustrated. Shirley, Peter, and Steve Marschner. *Fundamentals of Computer Graphics.* 3d ed. Natick, Mass.: A. K. Peters, 2009. This book emphasizes the underlying mathematical fundamentals of computer graphics rather than learning particular graphics programs.
Vidal, F. P., et al. "Principles and Applications of Computer Graphics in Medicine." *Computer Graphics Forum* 25, no. 1 (2006): 113-137. Excellent review of the state of the art. Discusses software development, diagnostic aids, educational tools, and computer-augmented reality.

WEB SITES
ACMSIGGRAPH (Association for Computing Machinery's Special Interest Group on Graphics and Interactive Techniques)
http://www.siggraph.org

Computer Society
http://www.computer.org

Institute of Electrical and Electronics Engineers
http://www.ieee.org

OpenGL Overview
http://www.opengl.org/about/overview

See also: Computer Engineering; Computer Science.

COMPUTER LANGUAGES, COMPILERS, AND TOOLS

FIELDS OF STUDY

Computer science; computer programming; information systems; information technology; software engineering.

SUMMARY

Computer languages are used to provide the instructions for computers and other digital devices based on formal protocols. Low-level languages, or machine code, were initially written using the binary digits needed by the computer hardware, but since the 1960's, languages have evolved from early procedural languages to object-oriented high-level languages, which are more similar to English. There are many of these high-level languages, with their own unique capabilities and limitations, and most require some type of compiler or other intermediate translator to communicate with the computer hardware. The popularity of the Internet has created the need to develop numerous applications and tools designed to share data across the Internet.

KEY TERMS AND CONCEPTS

- **application:** Computer program that completes a specific task.
- **basic input-output system (BIOS):** Computer program that allows the central processing unit of a computer to communicate with other computer hardware.
- **browser:** Software program used to view Web pages on the Internet.
- **compiler:** Program that converts source code in a text file into a format that can be executed by a computer.
- **graphical user interface (GUI):** Visual interface that allows a user to position a cursor over a displayed object or icon and click on that object or icon to make a selection.
- **interpreter:** Computer tool that is much faster and more efficient than a compiler.
- **mainframe:** Large stand-alone computer that completes batch processing (groups of computer instructions completed at once).
- **operating system:** Computer program written in a language so that it can function as an interface between a computer user and the hardware that runs the computer by managing resources.
- **portability:** Ability of a program to be downloaded from a remote location and executed on a variety of computers with different operating systems.
- **protocol:** Set of rules to be followed to allow for communication.
- **server:** Computer dedicated to managing network activities, such as printers and email, which are shared by many users (clients).
- **structured query language (SQL):** Computer language used to access databases.

DEFINITION AND BASIC PRINCIPLES

The traditional process of using a computer language to write a program has generally involved the initial design of the program using a flowchart based on the purpose and desired output of the program, followed by typing the actual instructions for the computer (the code) into a file using a text editor, and then saving this code in a file (the source code file). A text editor is used because it does not have the formatting features of a word processor. An intermediate tool called a compiler then has been used to convert this source code into a format that can be run (executed) by a computer.

However, as of 2011, there are new tools that are much faster and more efficient than compilers. Therefore, many compilers have been replaced by these new tools, called interpreters. Larger, more complex programs have evolved that have required an additional step to link external files. This process is called linking and it joins the main, executable program created by the compiler to other necessary programs. Finally, the executable program is run and its output is displayed on the computer monitor, printed, or saved to another digital file. If errors are found, the process of debugging is followed to go back through the code to make corrections.

BACKGROUND AND HISTORY

Early computers such as ENIAC (Electronic Numerical Integrator and Computer), the first general-purpose computer, were based on the use of switches that could be turned on or off. Thus, the binary digits of 0 and 1 were used to write machine code. In addition to being tedious for a programmer, the code had to be rewritten if used on a different type of machine, and it certainly could not be used to transmit data across the Internet, where different computers all over the world require access to the same code.

Assembly language evolved by using mnemonics (alphabetic abbreviations) for code instead of the binary digits. Because these alphabetic abbreviations of assembly language no longer used the binary digits, additional programs were developed to act as intermediaries between the human programmers writing the code and the computer itself. These additional programs were called compilers, and this process was initially known as compiling the code. This compilation process was still machine and vendor dependent, however, meaning, for example, that there were several types of compilers that were used to compile code written in one language. This was expensive and made communication of computer applications difficult.

The evolution of computer languages from the 1950's has accompanied technological advances that have allowed languages to become increasingly powerful, yet easier for programmers to use. FORTRAN and COBOL languages led the way for programmers to develop scientific and business application programs, respectively, and were dependent on a command-line user interface, which required a user to type in a command to complete a specific task. Several other languages were developed, including Basic, Pascal, PL/I, Ada, Lisp, Prolog, and Smalltalk, but each of these had limited versatility and various problems. The C and C++ languages of the 1970's and 1980's, respectively, have emerged as being the most useful and powerful languages, are still in use, and have been followed by development tools written in the Java and Visual Basic languages, including integrated development environments with editors, designers, debuggers, and compilers all built into a single software package.

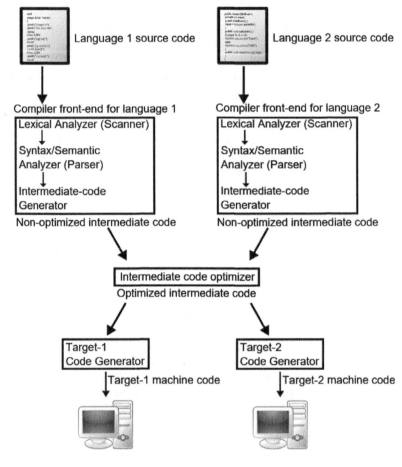

A diagram of the operation of a typical multi-language, multi-target compiler

How it Works

BIOS and operating system. The programs within the BIOS are the first and last programs to execute whenever a computer device is turned on or off. These programs interact directly with the operating system (OS). The early mainframe computers that were used in the 1960's and 1970's depended on several different operating systems, most of which are no longer in usage, except for UNIX and DOS. DOS (Disk Operating System) was used on the initial microcomputers of the 1980's and early 1990's, and it is still used for certain command-line specific instructions.

Graphical user interfaces (GUIs). Microsoft dominates the PC market with its many updated operating systems, which are very user-friendly with GUIs. These operating systems consist of computer programs and software that act as the management system for all of the computer's resources, including the various application programs most taken for granted, such as Word (for documents), Excel (for mathematical and spreadsheet operations), and Access (for database functions). Each of these applications is actually a program itself, and there are many more that are also available.

Since the 1980's, many programming innovations increasingly have been built to involve the client-server model, with less emphasis on large main frames and more emphasis on the GUIs for smaller microcomputers and handheld devices that allow consumers to have deep color displays with high resolution and voice and sound capabilities. However, these initial GUIs on client computers required additional upgrades and maintenance to be able to interact effectively with servers.

World Wide Web. The creation of the World Wide Web provided an improvement for clients to be able to access information, and this involvement of the Internet led to the creation of new programming languages and tools. The browser was developed to allow an end user (client) to be able to access Web information, and hypertext markup language (HTML) was developed to display Web pages. Because the client computer was manufactured by many different companies, the Java language was developed to include applets, which are mini-programs embedded into Web pages that could be displayed on any type of client computer. This was made possible by a special type of compiler-like tool called the Java Virtual Machine, which translated byte code. Java remains the primary computer language of the Internet.

APPLICATIONS AND PRODUCTS

FORTRAN and COBOL. FORTRAN was developed by a team of programmers at IBM and was first released in 1957 to be used primarily for highly numerical and scientific applications. It derived its name from formula translation. Initially, it used punched cards for input, because the text editors were not available in the 1950's. It has evolved but still continues to be used primarily in many engineering and scientific programs, including almost all programs written for geology research. Several updated versions have been released. FORTRAN77, released in 1980, had the most significant language improvements, and FORTRAN2000, released in 2002, is the most recent. COBOL (Common Business-Oriented Language) was released in 1959 with the goal of being used primarily for tracking retail sales, payroll, inventory control, and many other accounting-related activities. It is still used for most of these business-oriented tasks.

C and C++. The C computer language was the predecessor to the C++ language. Programs written in C were procedural and based on the usage of functions, which are small programming units. As programs grew in complexity, more functions were added to a C program. The problem was that eventually it became necessary to redesign the entire program, because trying to connect all of the functions, which added one right after the other, in a procedural way,

was too difficult. C++ was created in the 1980's based on the idea of objects grouped into classes as the building blocks of the programs, which meant that the order did not have to be procedural anymore. Object-oriented programming made developing complex programs much more efficient and is the current standard.

Microsoft.NET. In June, 2000, Microsoft introduced a suite of languages and tools named Microsoft.NET along with its new language called Visual C#. Microsoft.NET is a software infrastructure that consists of many programs that allow a user to write programs for a range of new applications such as server components and Web applications by using new tools. Although programs written in Java can be run on any machine, as long as the entire program is written in Java, Microsoft.NET allows various programs to be run on the Windows OS. Additional advantages of Microsoft.NET involve its new language of Visual C#. Visual C# provides services to help Web pages already in existence, and C# can be integrated with the Visual Basic and Visual C++ languages, which facilitate the work of Web programmers by allowing them to update existing Web applications, rather than having to rewrite them.

The Microsoft.NET framework uses a common type system (CTS) tool to compile programs written in Cobol.NET, PerlNET, Visual Basic.NET, Jscript, Visual C++, Fortran.NET, and Visual C# into an intermediate language. This common intermediate language (CIL) can then be compiled to a common language runtime (CLR). The result is that the .NET programming environment promotes interoperability to allow programs originally written in different languages to be executed on a variety of operating systems and computer devices. This interoperability is crucial for sharing data and communication across the Internet.

IMPACT ON INDUSTRY

Microsoft was founded in 1975 by Bill Gates, who dropped out of Harvard, and Paul Allen. Their first software sold was in the BASIC language, which was the first language to be used on personal computers (PCs). As PC prices decreased, the number of consumers able to purchase a PC increased, as did the revenue and market dominance of Microsoft. Microsoft is the largest software company in the world, with annual revenues of more than $32 billion.

Because it has gained such widespread dominance in the technology field, it offers its own certifications in order to maintain quality standards in its current and prospective employees. Among the most commonly held and useful certifications offered by Microsoft are the Microsoft Certified Applications Developer (MCAD), Microsoft Certified Solution Developer (MCSD), Microsoft Certified Professional Developer (MCPD), Microsoft Certified Systems Engineer (MCSE), Microsoft Certified Systems Administrator (MCSA), Microsoft Certified Database Administrator (MCDBA), and Microsoft Certified Trainer (MCT).

Sun Microsystems began with four employees in 1982 as a computer hardware manufacturer. With the rise of the Internet, the company seized the opportunity to create its computer language with its special tool, the Java Virtual Machine, which could be downloaded to any type of machine. This Java Virtual Machine contained a program called an interpreter, rather than a compiler, to act as an interface between the specific machine (platform) and the user. The result was a great enhancement in the interoperability of data on the Internet, and Java became the primary programming language on the Web.

Sun Microsystems also created Java servlets to allow more interaction with dynamic, graphical Web pages written in Java. However, although these servlets work very well with Java, they also must be compiled as classes before execution, which takes extra time.

Apple, one of the most successful American companies, has experienced phenomenal growth in the first decade of the twenty-first century. Its position as a market leader in consumer electronics is because of the success of its Macintosh computer accompanied by several Internet-enabled gadgets, such as the iPod, introduced in 2001, which uses various audio file formats, such as MP3 and WAV, to function as a portable media player. To further expand on the popularity of the iPod, Apple opened its own online media and applications stores, iTunes in 2003 and the App Store in 2008, to allow consumers to access music, video, and numerous applications via the Internet. In 2007, Apple expanded its market share by releasing the iPhone, which added to the phone the functions of text messaging, emailing of digital photos using its own camera capabilities, and a multi-touch screen for enhanced Internet access. In 2010, Apple further expanded with the release of the iPad, an electronic tablet that allows users access to the Internet to read, write, listen to, and view almost anything, including e-mail, movies, books, music, and magazines. The iPad also has a Global Positioning System (GPS) and a camera.

CAREERS AND COURSE WORK

Although it is becoming more common for entry-level job seekers to have a bachelor's degree in a computer-related field, either through a university computer-science or business-school department, it is possible to find rewarding employment as a programmer, software engineer, application developer, Web programmer or developer, database administrator, or software support specialist with just an associate's degree or relevant experience. This career field is unique in the large number of certifications that can be obtained to enhance job skills and increase the likelihood of finding employment, especially with just an associate's degree or related experience. Databases that can be accessed from within programming languages have also created the need for database administrators. There are many vendor-specific certifications for numerous job openings as database administrators who know computer languages such as SQL.

The future job prospects for software engineers are projected by the Bureau of Labor Statistics to be better than the job prospects for programmers for the period from 2008 to 2018. There is a great deal of similarity between these two careers, but generally programmers spend more time writing code in various languages, while the software engineers develop the overall design for the interaction of several programs to meet the needs of a customer. Software engineers are required more often to have a bachelor's or master's degree, usually in software engineering. The job growth for software engineers is expected to grow by 32 percent over the time period of 2008 to 2018, which is much faster than most other occupations.

SOCIAL CONTEXT AND FUTURE PROSPECTS

The Internet continues to bring the world together at a rapid pace, which has both positive and negative ramifications. Clearly, consumers have much easier access to many services, such as online education, telemedicine, and free search tools to locate doctors, learn more about any topic, comparison shop and purchase, and immediately access software, movies, pictures, and music. However, along with this increase in electronic commerce involving credit card purchases, bank accounts, and additional financial transactions has been the increase of cybercrime. Thousands of dollars are lost each year to various Internet scams and hackers being able to access private financial information. Some programmers even use computer languages to produce viruses and other destructive programs for purely malicious purposes, which have a negative impact on computer security.

Modern computer programs follow standard engineering principles to solve problems involving detail-driven applications ranging from radiation therapy and many medical devices to online banking, auctions, and stock trading. These online application programs require the special characteristics of Web-enabled software such as security and portability, which has given rise to the development of additional tools and will continue to produce the need for increased security features within computer languages and tools.

—Jeanne L. Kuhler, MS, PhD

FURTHER READING

Das, Sumitabha. *Your UNIX: The Ultimate Guide.* 2d ed. Boston: McGraw-Hill, 2006. This textbook provides background and instructions for using the UNIX operating system.

Dhillon, Gupreet. "Dimensions of Power and IS Implementation." *Information and Management* 41, no. 5 (May, 2004): 635-644. This article describes some of the work done by managers when choosing which computer languages and tools to implement.

Guelich, Scott, Shishir Gundavaram, and Gunther Birznieks. *CGI Programming with Perl.* 2d ed. Cambridge, Mass.: O'Reilly, 2000. This introductory overview text describes the use of the common gateway interface programming and its interactions with the Perl programming language.

Horstmann, Cay. *Big Java.* 4th ed. Hoboken, N.J.: John Wiley & Sons, 2010. This book describes the most recent version of the Java language, along with its history and special features of the Java interpreter.

Snow, Colin. "Embrace the Role and Value of Master Data Management," *Manufacturing Business Technology,* 26, no. 2 (February, 2008): 92-95. This article, written by the VP and research director of Ventana Research, covers how to implement data

---

FASCINATING FACTS ABOUT COMPUTER LANGUAGES, COMPILERS, AND TOOLS

- Job prospects for software engineers are expected to be excellent, with salaries typically in the range of $70,000 to $80,000 per year.
- As of October, 2010, Java is the most widely used computer language, followed by C++.
- Using computer languages to write programs for video games is expected to be one of the most in-demand skills in the gaming area.
- Computer languages are necessary for a wide variety of everyday applications, ranging from cell phones, microwaves, and security systems to banking, online purchases, Web sites, auction sites (such as eBay), social-networking sites (such as Facebook), and Internet search engines (such as Google).
- Businesses, including banks, almost exclusively use COBOL, which uses two-digit rather than four-digit years. The use of two-digit years was the basis for the Y2K fear, in which it was predicted that on January 1, 2000, all the banking and other business-related applications would default to the year 1900, documents would be lost, and chaos would ensue. January 1, 2000, passed without incident.
- In 2010, Mattel introduced a Barbie doll called Computer Software Engineer Barbie. She has her own laptop with the Linux operating system and an iPhone.

---

management and avoid the errors that cost businesses millions of dollars each year.

WEB SITES

Association of Information Technology Professionals
http://www.aitp.org

The Computer Society
http://www.computer.org

Computing Technology Industry Association
http://www.comptia.org/home.aspx

Microsoft Microsoft Certification Overview
http://www.microsoft.com/learning/en/us/ certification/cert-overview.aspx

See also: Computer Engineering; Computer Science.

COMPUTER SCIENCE

FIELDS OF STUDY

Computer science; computer engineering; electrical engineering; mathematics; artificial intelligence; computer programming; human-computer interaction; software engineering; databases and information management; bioinformatics; game theory; scheduling theory; computer networking; computer and data security; computer forensics; computer simulation and modeling; computation methods (including algorithms); ethics; computer graphics; multimedia.

SUMMARY

Computer science is the study of real and imagined computers, their hardware and software components, and their theoretical basis and application. Almost every aspect of modern life involves computers in some way. Computers allow people to communicate with people almost anywhere in the world. They control machines, from industrial assembly equipment to children's toys. Computers control worldwide stock market transactions and analyze and regulate those markets. They allow physicians to treat patients even when the physicians and patients cannot meet in the same location. Researchers endeavor to make the computers of science fiction everyday realities.

KEY TERMS AND CONCEPTS

- **algorithm:** Step-by-step procedure for solving a problem; a program can be thought of as a set of algorithms written for a computer.
- **artificial intelligence:** Study of how machines can learn and mimic natural human and animal abilities; it can involve enhancing or going beyond human abilities.
- **binary system:** Number system that has only two digits, 0 and 1; forms the basis of computer data representation.
- **computer:** Real or theoretical device that accepts data input, has a running program, stores and manipulates data, and outputs results.
- **computer network:** Group of computers and devices linked together to allow users to communi-

cate and share resources.
- **database:** System that allows for the fast and efficient storage and retrieval of vast amounts of data.
- **hardware:** Tangible part of a computer, consisting mostly of electronic and electrical components and their accessories.
- **programming:** Using an artificial programming language to instruct a computer to perform tasks.
- **software:** Programs running within a computer.
- **theoretical computer:** Imaginary computer made up by researchers to gain insights into computer science theory.

DEFINITION AND BASIC PRINCIPLES

Computer science is the study of all aspects of computers, applied and theoretical. However, considerable disagreement exists over the definition of such basic terms as computer science, computer, hardware, and software. This disagreement can be seen as a testament to the vitality and relative youth of this field. The Association for Computing Machinery's computing classification system, developed in 1998, is an attempt to define computer science.

The science part of computer science refers to the underlying theoretical basis of computers. Broadly speaking, computation theory, part of mathematics, looks at what mathematical problems are solvable. In computer science, it focuses on which problems can be solved using a computer. In 1936, English mathematician Alan Turing attempted to determine the limits of mechanical computation using a theoretical device, now called a Turing machine. Mathematics also forms the basis for research in programming languages, artificial languages developed for computers. Because computers do not have the ability to think like humans, these languages are very formal with strict rules on how programs using these languages can be written and used.

Another part of the underlying structure of computer science is engineering. The physical design of computers involves a number of disciplines, including electrical engineering and physics. The quest for smaller, faster, more powerful devices has led to research in fields such as quantum physics and nanotechnology.

Computer science is not just programming.

Computer scientists view programming languages as tools to research such issues as how to create better programs, how information is represented and used by computers, and how to do away with programming languages altogether and instead use natural languages (those used by people).

BACKGROUND AND HISTORY

Computer science can be seen as the coming together of two separate strands of development. The older strand is technology and machines, and the newer is the theoretical one.

Technology. The twentieth century saw the explosive rise of computers. Computers began to incorporate electronic components instead of the mechanical components that had gone into earlier machines, and they made use of the binary system rather than the decimal system. During World II, major developments in computer science arose from the attempt to build a computer to control antiaircraft artillery. For this project, the Hungarian mathematician John von Neumann wrote a highly influential draft paper on the design of computers. He described a computer architecture in which a program was stored along with data. The decades after World War II saw the development of programming languages and more sophisticated systems, including networks, which allow computers to communicate with one another.

Theory. The theoretical development of computer science has been primarily through mathematics. One early issue was how to solve various mathematical equations. In the eighteenth and nineteenth centuries, this blossomed into research on computation theory. At a conference in Germany in 1936 to investigate these issues, Turing presented the Turing machine concept.

The World War II antiaircraft project resulted not only in the development of hardware but also in extensive research on the theory behind what was being done as well as what else could be done. Out of this ferment eventually came such work as Norbert Wiener's cybernetics and Claude Shannon's information theory. Modern computer science is an outgrowth of all this work, which continues all around the world in industry, government, academia, and various organizations.

HOW IT WORKS

Computer organization and architecture. The most familiar computers are stand-alone devices based on the architecture that von Neumann sketched out in his draft report. A main processor in the computer contains the central processing unit, which controls the device. The processor also has arithmetic-processing capabilities.

Electronic memory is used to store the operating system that controls the computer, numerous other computer programs, and the data needed for running programs. Although electronic memory takes several forms, the most common is random access memory (RAM), which, in terms of size, typically makes up most of a computer's electronic memory. Because electronic memory is cleared when the computer is turned off, some permanent storage devices, such as hard drives and flash drives, were developed to retain these data.

Computers have an input/output (I/O) system, which communicates with humans and with other devices, including other computers. I/O devices include keyboards, monitors, printers, and speakers.

The instructions that computers follow are programs written in an artificial programming language. Different kinds of programming languages are used in a computer. Machine language, the only language that computers understand, is used in the processor and is composed solely of the binary digits 0 and 1.

Computers often have subsidiary processors that take some of the processing burden away from the main processor. For example, when directed by the main processor, a video processor can handle the actual placing of images on a monitor. This is an example of parallel processing, in which more than one action is performed simultaneously by a computer. Parallel processing allows the main processor to do one task while the subsidiary processors handle others. A number of computers use more than one main processor. Although one might think that doubling the number of processors doubles the processing speed, various problems, such as contention for the same memory, considerably reduce this advantage. Individual processors also use parallel processing to speed up processing; for example, some processors use multiple cores that together act much like individual processors.

Modern computers do not actually think. They are best at performing simple, repetitious operations incredibly fast and accurately. Humans can get bored and make mistakes, but not computers (usually).

Mathematics. Mathematics underlies a number of areas of computer science, including programming languages. Researchers have long believed that programming languages incorporating rules that are mathematically based will lead to programs that contain fewer errors. Further, if programs are mathematically based, then it should be possible to test programs without running them. Logic could be used to deduce whether the program works. This process is referred to as program proving.

Mathematics also underlies a number of algorithms that are used in computers. For example, computer games have a large math and physics component. Game action is often expressed in mathematical formulas that must be calculated as the game progresses. Different algorithms that perform the same task can be evaluated through an analysis of the relative efficiencies of different approaches to a problem solution. This analysis is mathematically based and independent of an actual computer.

Software. Computer applications are typically large software systems composed of a number of different programs. For example, a word-processing program might have a core set of programs along with programs for dictionaries, formatting, and specialized tasks. Applications depend on other software for a number of tasks. Printing, for example, is usually handled by the operating system. This way, all applications do not have to create their own basic printing programs. The application notifies the operating system (OS), which can respond, for example, with a list of all available printers. If an application user wishes to modify printer settings, the OS contacts the printer software, which can then display the settings. This way, the same interface is always used. To print the application, the user sends the work to the OS, which sends the work to a program that acts as a translator between the OS and the printer. This translator program is called a driver. Drivers work with the OS but are not part of it, allowing for new drivers to be developed and installed after an OS is released.

Machine language, since it consists only of 0's and 1's, is difficult for people to readily interpret. For human convenience, assembly language, which uses mnemonics rather than digits, was developed. Because computers cannot understand this language, a program called an assembler is used to translate assembly language to machine language.

For higher-level languages such as C++ and Java, a compiler program is used to translate the language statements first to assembly language and then to machine language.

APPLICATIONS AND PRODUCTS

Computers have penetrated into nearly every area of modern life, and it is hard to think of an area in which computer technology is not used. Some of the most common areas in which computers are used include communications, education, digitalization, and security.

Communication. In the early twentieth century, when someone immigrated to the United States, they knew that communication with those whom they were leaving behind would probably be limited. In the modern world, people across the world can almost instantly communicate with anyone else as long as the infrastructure is available. Products such as Skype and Vonage allow people to both see and talk to other people throughout the world. Instead of traveling to the other person's location, people can hold meetings through software such as Cisco's WebEx or its TelePresence, which allows for face-to-face meetings.

One of the most far-reaching computer applications is the Internet, a computer network that includes the World Wide Web. People are increasingly relying on the Internet to provide them with news and information, and traditional sources, such as printed newspapers and magazines are declining in popularity. Newer technologies such as radio and television have also been affected. The Internet seems to be able to provide everything: entertainment, information, telephone service, mail, business services, and shopping.

Telephones and the way people think of telephones have also been revolutionized. Telephones have become portable computing devices. Apple iPhones and Research in Motion (RIM) Blackberry phones are smart phones, which provide access to e-mail and the Internet; offer Global Positioning Systems (GPS) that guide the user to a selected destination; download motion pictures, songs, and other entertainment; and shoot videos and take photographs. Applications (apps) are a burgeoning industry for the iPhone. Apps range from purely entertaining applications to those that provide useful information, such as weather and medical data. Some people no longer have traditional land-line telephones and rely instead on their cell phones.

Smart phones demonstrate convergence, a trend toward combining devices used for individual activities into a single device that performs various activities and services. For example, increasingly high-definition televisions are connected to the Internet. Devices such as Slingbox allow users, wherever they are, to control and watch television on their desktop computer, laptop computer, or mobile phone. Digital video recorders (DVRs) allow users to record a television program and watch it at a later time or to pause live broadcasts. DVRs can be programmed to record shows based on the owner's specifications. Televisions are becoming two-way interactive devices, and viewers are no longer passive observers. Although televisions and DVRs are not themselves computers, they contain microprocessors, which are miniature computers.

Networking is possible through a vast network infrastructure that is still being extended and improved. Companies such as Belkin provide the cable, Cisco the equipment, and Microsoft the software that supports this infrastructure for providers and users. It is common for homes to have wireless networks with a number of different devices connected through a modem, which pulls out the Internet signal from the internet service provider (ISP), and a router, which manages the network devices.

Education. Distance education through the Internet (online education) is becoming more and more commonplace. The 2010 Horizon Report noted that people expect anywhere, anytime learning. Learning (or course) management systems (LMS or CMS) are the means by which the courses are delivered. These can be proprietary products such as Blackboard or Desire2Learn, or nonproprietary applications such as Moodle. Through these management systems, students can take entire courses without ever meeting their instructor in person. Applications such as Wimba and Elluminate allow classes to be almost as interactive as traditional classes and record those sessions for future viewing. Those participating can be anywhere as long as they can connect to the Internet. Through such software, students can ask questions of an instructor, student groups can meet online, and an instructor can hold review sessions.

Digitalization. This area is concerned with how data are translated for computer usage. Real-world experience is usually thought of as continuous phenomena. Long-play vinyl records (LPs) and magnetic tapes captured a direct analogy for the original sound; therefore, any change to the LP or tape meant a change to the sound quality, usually introducing distortions. Each copy (generation) that was made of that analogue signal introduced further distortions. LPs were several generations removed from the original recording, and additional distortions were added to accommodate the sound to the LP medium.

Rather than record an analogue of the original sound, digitalization translates the sound into discrete samples that can be thought of as snapshots of the original. The more samples (the higher the sampling rate) and the more information captured by each sample (the greater the bit depth), the better the result. Digitalization enables the storage of data of all types. The samples are encoded into one of the many binary file formats, some of which are open and some proprietary. MPEG-1 Audio Layer 3 (MP3) is an open encoding standard that is used in a number of different devices. It is a lossy standard, which means that the result is of a lower quality than the original. This tradeoff is made to gain a smaller file size, enabling more files to be stored. The multimedia equivalent is MPEG-4 Part 14 (MP4), also a lossy format. Lossless formats such as Free Lossless Audio Codec (FLAC) and Apple Lossless have better sound quality but require more storage space.

These formats and others have resulted in a proliferation of devices that have become commonplace. iPods and MP3 players provide portable sound in a relatively small container and, in some cases, play videos. Some iPods and computers can become media centers, allowing their users to watch television shows and films.

In a relatively short span of time, digital cameras have made traditional film cameras obsolete. Camera users have choices as to whether they want better quality (higher resolution) photographs or lower resolution photos but the capacity to store more of them in the same memory space. Similarly, DVDs have made tape videos obsolete. Blu-ray, developed by Sony, allows for higher quality images and better sound by using a purple laser rather than the standard red laser. Purple light has a higher frequency than red light, which means that the size (wavelength) of purple light is smaller than that of red light. This can be visualized as allowing the purple laser to get into smaller spaces than the red laser can; thus the 0's and 1's on a Blu-ray disc can be smaller than those on a standard DVD disc, so more data can be stored on the same size disc. Because the recordings are encoded in binary, they can be easily and exactly copied and manipulated without any distortion, unless that distortion is deliberately introduced, as with lossy file formats.

Security. The explosion of digital communication and products has caused some individuals to illegally, or at least unethically, exploit people's dependence on these technologies. These people write malware, programs designed to harm or compromise devices and send spam (digital junk mail). Malware includes viruses (programs that embed themselves in other programs and then make copies of themselves and spread), worms (programs that act like viruses but do not have to be embedded in another program), Trojan horses (malicious programs, often unknowingly downloaded with something else), and key loggers (programs that record all keystrokes and mouse movements, which might include user names and passwords). The key-stroke recorder is a type of spyware, programs that collect information about the device that they are on and those using it without their knowledge. These are often associated with Internet programs that keep track of people's browsing habits.

Spam can include phishing schemes, in which a person sends an e-mail that appears to come from a well-known organization such as a bank. The e-mail typically states that there is a problem with the recipient's account and to rectify it, the recipient must verify his or her identity by providing sensitive information such as his or her user name, password, and account number. The information, if given, is used to commit fraud using the recipient's identity.

Another type of spam is an advance-fee fraud, in which the sender of an e-mail asks the recipient to help him or her claim a substantial sum of money that is due the sender. The catch is that the recipient must first send money to "show good faith." The money that the recipient is to get in return never arrives. Another asks the recipient to cash a check and send the scammer part of the money. By the time the check is determined to be worthless, the recipient has sent the scammer a considerable sum of money. These scams are a form of social engineering, in which the scammer first obtains the trust of the e-mail recipient so that he or she will help the scammer.

These problems are countered by companies such as Symantec and McAfee, which produce security suites that contain programs to manage these threats. These companies keep databases of threats and require that their programs be regularly updated to be able to combat the latest threats. Part of these suites is a firewall, which is designed to keep the malware and spam from reaching the protected network.

IMPACT ON INDUSTRY

It is hard to think of any industry that has not felt the impact of computers in a major, perhaps transforming, way. The United States, the United Kingdom, and Japan led research and development in computer science, but Israel, the European Union, China, and Taiwan are increasingly engaging in research. Although India is not yet a major center of research, it has been very successful in setting up areas in Mumbai and elsewhere that support technology companies from around the world. Other countries, such as Pakistan, have tried to duplicate India's success. Indian and Chinese computer science graduates have been gaining a reputation worldwide as being very well prepared.

Industry and business sectors. Computer science has led to the creation of electronic retailing (etailing). Online sales rose 15.5 percent in 2009, while brick and mortar (traditional) sales were up 3.6 percent from the previous year. Online-only outlets such as Amazon.com have been carving out a significant part of the retail market. Traditional businesses, such as Macy's and Home Depot, are finding that they must have an online presence. Online sales for the 2009 Christmas season were estimated at $27 billion.

Some traditional industries are finding that survival has become difficult. Newspapers and magazines have been experiencing a circulation decline, as more and more people read the free online versions of these publications or get their news and information from other Internet sites. Some newspapers, such as *The Wall Street Journal*, sell subscriptions to their Web sites or offer subscriptions to enhanced Web sites. Travel agencies are losing business, as their clients instead use online travel sites such as Expedia.com or the Web sites of airlines and hotels to make their own travel arrangements. DVD rental outlets such as Blockbuster (which declared Chapter 11 bankruptcy in September, 2010) have been declining in the face of competition from Netflix, which initially offered a subscription service and sent DVDs through the U.S. Postal Service. As consumers have turned to streaming television programs and films, Netflix has entered this business, but faces competition from Amazon.com, Google, Apple, and cable and broadband television providers.

CD and DVD sales have been affected not only by online competition but also by piracy. Since digital sources can be copied without error, copying of CDs and DVDs has become rife. Online file sharing through such applications as BitTorrent has caused the industry to fight back through highly publicized lawsuits and legislation. The Digital Millennium Copyright Act of 1998 authorized, among many other measures, the use of Digital Rights Management (DRM) to restrict copying. Many DVDs and downloads use DRM protection. However, a thriving industry has sprung up to block the DRM features and allow copying. This is a highly contentious area, which engendered more controversy with the passage of the 2010 Digital Economy Act in the United Kingdom.

A thriving industry has been developed using virtual reality. Linden Research's Second Life is one of the most popular virtual reality offerings. Members immerse themselves in a world where they have the opportunity to be someone that they might never be in real life. A number of organizations use Second Life for various purposes, including education.

Professional organizations. Probably the two largest computer organizations are the Association for Computer Machinery (ACM), based in New York, and Computer Society of the Institute of Electrical and Electronic Engineers (IEEE), based in Washington,

D.C. The ACM Special Interest Groups explore all areas of computer science. When computer science higher education was a new frontier and the course requirements for a degree varied widely, it was the ACM that put forward a curriculum that has become the standard for computer science education in colleges and universities in the United States. Its annual Turing Award is one of the most prestigious awards for computing professionals.

The IEEE partners with the American National Standards Institute (ANSI) and the International Organization for Standardization on many of its wireless specifications, such as 1394 (FireWire) and Ethernet (802.3). Both computer organizations sponsor various contests for students from middle school to college.

Research. The National Science Foundation, a federal agency created in 1950, is responsible for about 20 percent of all federally funded basic research at colleges and universities in the United States. An example project is the California Regional Consortium for Engineering Advances in Technological Excellence (CREATE), a consortium of seven California community colleges, which works toward innovative technical education in community colleges. The consortium has funded a number of computer-networking initiatives.

The Defense Advanced Research Projects Agency (DARPA), part of the U.S. Department of Defense, funds research into a number of areas, including networks, cognitive systems, robotics, and high-priority computing. This agency initially funded the project that became the basis of the Internet.

CAREERS AND COURSE WORK

Computer science degrees require courses in mathematics (including calculus, differential equations, and discrete mathematics), physics, and chemistry as well as the usual set of liberal arts courses. Lower-division computer science courses are usually heavily weighted toward programming. These might include basic programming courses in C++ or Java, data structures, and assembly language. Other courses usually include computer architecture, networks, operating systems, and software engineering. A number of specialty areas, such as computer engineering, affect the exact mix of course work.

Careers in software engineering typically require a bachelor's degree in computer science. The U.S. Bureau of Labor Statistics sees probable growth in the need for software engineers. Computer skills are not all that employers want, however. They also want engineers to have an understanding of how technology fits in with their organization and its mission.

Those who wish to do research as computer scientists will generally require a doctorate in computer science or some branch of computer science. Computer scientists are often employed by universities, government agencies, and private industries such as IBM. AT&T Labs also has a tradition of Nobel Prize-winning research and was where the UNIX operating system and the C and C++ programming languages were developed.

SOCIAL CONTEXT AND FUTURE PROSPECTS

Computers are considered an essential part of modern life, and their influence continues to revolutionize society. The advantages and disadvantages of an always-wired society are being keenly debated. People not only are connected but also can be located by computer devices with great accuracy. Because people are doing more online, they are increasingly creating an online record of what they have done at various times in their lives. Many employers routinely search online for information on job candidates, and people are finding that comments they made online and the pictures they posted some years ago showing them partying are coming back to haunt them. People are starting to become aware of these pitfalls and are taking actions such as deleting these possibly embarrassing items and changing their settings on social network sites such as Facebook to limit access.

These issues bring up privacy concerns, including what an individual's privacy rights are on the Internet and who owns the data that are being produced. For example, in 1976, the U.S. Supreme Court ruled that financial records are the property of the financial institution not the customer. This would seem to suggest that a company that collects information about an individual's online browsing habits—not the individual whose habits are being recorded—owns that information. Most of these companies state that individuals are not associated with the data that they sell and all data are used in aggregate, but the capacity to link data with specific individuals and sell the information exists. Other questions include whether an employer has a right to know whether an employee visits a questionable Internet site. Certainly the technology to accomplish this is available.

These concerns lead to visions of an all-powerful and all-knowing government such as that portrayed in George Orwell's *Nineteen Eighty-Four* (1949). With the Internet a worldwide phenomenon, perhaps no single government can dictate regulations governing the World Wide Web and enforce them. The power of technology will only grow, and theses issues and many others will only become more pressing.

—Martin Chetlen, MCS

FURTHER READING

Brooks, Frederick P., Jr. *The Mythical Man-Month: Essays on Software Engineering.* Anniversary ed. Boston: Addison-Wesley, 2008. Brooks, who was the leader of the IBM 360 project, discusses computer and software development. The first edition of this book was published in 1975, and this 2008 edition adds four chapters.

Gaddis, Tony. *Starting Out With C++*. 6th ed. Boston: Addison-Wesley, 2009. A good introduction to a popular programming language.

Goldberg, Jan, and Mark Rowh. *Great Jobs for Computer Science Majors*. Chicago: VGM, 2003. Designed for a young and general audience, this career guide includes tools for self-assessment, researching computer careers, networking, choosing schools, and understanding a variety of career paths.

Henderson, Harry. *Encyclopedia of Computer Science and Technology*. Rev. ed. New York: Facts On File, 2009. An alphabetical collection of information about computer science. Contains bibliography, chronology, list of awards, and a list of computer organizations.

Kidder, Tracy. *The Soul of the New Machine*. 1981. Reprint. New York: Back Bay Books, 2000. Chronicles the development of a new computer and the very human drama and comedy that surrounded it.

Schneider, G. Michael, and Judith L. Gersting. *Invitation to Computer Science*. 5th ed. Boston: Course Technology, 2010. Gives a view of the breadth of computer science, including social and ethical issues.

WEB SITES

Association for Computer Machinery
http://www.acm.org

Computer History Museum
http://www.computerhistory.org

Computer Society
http://www.computer.org

Institute of Electrical and Electronic Engineers
http://www.ieee.org

See also: Communications; Computer Engineering; Computer Languages, Compilers, and Tools; Information Technology; Parallel Computing.

CRACKING

FIELDS OF STUDY

Chemistry; engineering; chemical engineering; chemical process modeling; fluid dynamics; heat transfer; distillation design; mechanical engineering; environmental engineering; control engineering; process engineering; industrial engineering; electrical engineering; safety engineering; plastics engineering; physics; thermodynamics; mathematics; materials science; metallurgy; business administration.

SUMMARY

In the petroleum industry, cracking refers to the chemical conversion process following the distillation of crude oil, by which fractions and residue with long-chain hydrocarbon molecules are broken down into short-chain hydrocarbons. Cracking is accomplished under pressure by thermal or catalytic means and by injecting extra hydrogen. Cracking is done because short-chain hydrocarbons, such as gasoline, diesel, and jet fuel, are more commercially valuable than long-chain hydrocarbons, such as fuel and bunker oil. Steam cracking of light gases or light naphtha is used in the petrochemical industry to obtain lighter alkenes, which are important petrochemical raw products.

KEY TERMS AND CONCEPTS

- **catalytic cracking:** Use of a catalyst to enhance cracking.
- **coking:** Most severe form of thermal cracking.
- **crude oil:** Liquid part of petroleum; has a wide mix of different hydrocarbons.
- **distillation:** Process of physically separating mixed components with different volatilities by heating them.
- **fluid catalytic cracker (FCC):** Cracking equipment that uses fluid catalysts.
- **fractionator:** Distillation unit in which product streams are separated and taken away; also known as fractionating tower, fractionating column, and bubble tower.
- **fractions:** Product streams obtained after each distillation; also known as cuts.
- **hydrocracking:** Special form of catalytic cracking that uses extra hydrogen to obtain the end prod-

ucts with the highest values.
- **regenerator:** Catalytic cracker in which accumulated carbon is burned off the catalyst.
- **residue:** Accumulated elements of crude oil that remain solid after distillation.
- **steam cracking:** Petrochemical process used to obtain lighter alkenes.
- **thermal cracking:** Oldest form of cracking; uses heat and pressure.

DEFINITION AND BASIC PRINCIPLES

Cracking is a key chemical conversion process in the petroleum and petrochemical industries. The process breaks down long-chain hydrocarbon molecules with high molecular weights and recombines them to form short-chain hydrocarbon molecules with lower molecular weights. This breaking apart, or cracking, is done by the application of heat and pressure and can be enhanced by catalysts and the addition of hydrogen. In general, products with short-chain hydrocarbons are more valuable. Cracking is a key process in obtaining the most valuable products from crude oil.

At a refinery, cracking follows the distillation of crude oil into fractions of hydrocarbons with different molecule chain lengths and the collection of the heavy residue. The heaviest fractions with the longest molecule chains and the residue are submitted to cracking. For petrochemical processes, steam cracking is used to convert light naphtha, light gases, or gas oil into short-chain hydrocarbons such as ethylene and propylene, crucial raw materials in the petrochemical industry.

Cracking may be done by various technological means, and the hydrocarbons cracked at a particular plant may differ. In general, the more sophisticated the cracking plant, the more valuable its end products will be but the more costly it will be to build. Being able to control and change a cracker's output to conform to changes in market demand has substantial economic benefits.

BACKGROUND AND HISTORY

By the end of the nineteenth century, demand rose for petroleum products with shorter hydrogen molecule chains, in particular, diesel and gasoline to fuel the new internal combustion engines. In Russia, engineer Vladimir Shukhov invented the thermal cracking process for hydrocarbon molecules and patented it on November 27, 1891. In the United States, the thermal cracking process was further developed and patented by William Merriam Burton on January 7, 1913. This doubled gasoline production at American refineries.

To enhance thermal cracking, engineers experimented with catalysts. American Almer McAfee was the first to demonstrate catalytic cracking in 1915. However, the catalyst he used was too expensive to justify industrial use. French mechanical engineer Eugene Jules Houdry is generally credited with inventing economically viable catalytic cracking in a process that started in a Paris laboratory in 1922 and ended in the Sun Oil refinery in Pennsylvania in 1937. Visbreaking, a noncatalytic thermal process that reduces fuel oil viscosity, was invented in 1939. On May 25, 1942, the first industrial-sized fluid catalytic cracker started operation at Standard Oil's Baton Rouge, Louisiana, refinery.

Research into hydrocracking began in the 1920's, and the process became commercially viable in the early 1960's because of cheaper catalysts such as zeolite and increased demand for the high-octane gasoline that hydrocracking could yield. By 2010, engineers and scientists worldwide sought to improve cracking processes by optimizing catalysts, using less energy and feedstock, and reducing pollution.

HOW IT WORKS

Thermal cracking. If hydrocarbon molecules are heated above 360 degrees Celsius, they begin to swing so vigorously that the bonds between the carbon atoms of the hydrocarbon molecule start to break apart. The higher the temperature and pressure, the more severe this breaking is. Breaking, or cracking, the molecules, creates short-chain hydrocarbon molecules as well as some new molecule combinations. Thermal cracking—cracking by heat and pressure only—is the oldest form of cracking at a refinery. In modern refineries, it is usually used on the heaviest residues from distillation and to obtain petrochemical raw materials.

Modern thermal crackers can operate at temperatures between 440 and 870 degrees Celsius. Pressure can be set from 10 to about 750 pound-force per square inch (psi). Heat and pressure inside different thermal crackers and the exact design of the crackers vary considerably.

Typically, thermal crackers are fed with residues

from the two distillation processes of the crude oil. Steam crackers are primarily fed with light naphtha and other light hydrocarbons. After their preheating, often feedstocks are sent through a rising tube into the cracker's furnace area. Furnace temperature and feedstock retention time is carefully set, depending on the desired outcome of the cracking process. Retention time varies from fractions of a second to some minutes.

After hydrocarbon molecules are cracked, the resulting vapor is either first cooled in a soaker or sent directly into the fractionator. There, the different fractions, or different products, are separated by distillation and extracted.

In cokers, severe thermal cracking creates an unwanted by-product. It is a solid mass of pure carbon, called coke. It is collected in coke drums, one of which supports the cracking process while the other is emptied of coke.

Catalytic cracking. Because it yields more of the desired short-chain hydrocarbon products with less use of energy than thermal cracking, catalytic cracking has become the most common method of cracking. Feedstocks are relatively heavy vacuum distillate hydrocarbon fractions from crude oil. They are preheated before being injected into the catalytic

cracking reactor, usually at the bottom of a riser pipe. There they react with the hot catalyst, commonly synthetic aluminum silicates called zeolites, which are typically kept in fluid form. In the reactor, temperatures are about 480 to 566 degrees Celsius and pressure is between 10 and 30 psi.

As feedstock vaporizes and cracks during its seconds-long journey through the riser pipe, feedstock vapors are separated from the catalyst at cyclones at the top of the reactor and fed into a fractionator. There, different fractions condense and are extracted as separate products. The catalyst becomes inactive as coke builds on its surface. Spent catalyst is collected and fed into the regenerator, where coke deposits are burned off. The regenerated catalyst is recycled into the reactor at temperatures of 650 to 815 degrees Celsius.

Hydrocracking. The most sophisticated and flexible cracking process, hydrocracking delivers the highest value products. It combines catalytic cracking with the insertion of hydrogen. Extra hydrogen is needed to form valuable hydrocarbon molecules with a low boiling point that have more hydrogen atoms per carbon atom than the less valuable, higher boiling point hydrocarbon molecules that are cracked. All hydrocracking involves high temperatures from 400 to 815 degrees Celsius and extremely high pressure from 1,000 to 2,000 psi, but each hydrocracker is basically custom designed.

In general, hydrocracking feedstock consists of middle distillates (gas oils), light and heavy cycle oils from the catalytic cracker, and coker distillates. Feedstock may also be contaminated by sulfur and nitrogen. Feedstocks are preheated and mixed with hydrogen in the first stage reactor. There, excess hydrogen and catalysts convert the contaminants sulfur and nitrogen into hydrogen sulfide and ammonia.

Some initial hydrogenating cracking occurs before the hydrocarbon vapors leave the reactor. Vapors are cooled,

Creating Ethene Gas through Cracking

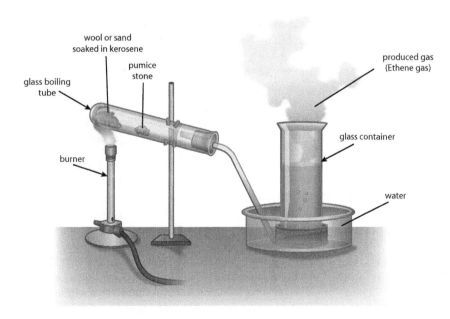

wool or sand soaked in kerosene

pumice stone

glass boiling tube

burner

produced gas (Ethene gas)

glass container

water

and liquefied products are separated from gaseous ones and hydrogen in the hydrocarbon separator. Liquefied products are sent to the fractionator, where desired products can be extracted.

The remaining feedstock at the bottom of the fractionator is sent into a second-stage hydrocracking reactor with even higher temperatures and pressures. Desired hydrocracked products are extracted through repetition of the hydrocracking process. Unwanted residues can go through the second stage again.

APPLICATIONS AND PRODUCTS

There are three modern applications of thermal cracking: visbreaking, steam cracking, and coking.

Visbreaking. Visbreaking is the mildest form of thermal cracking. It is applied to lower the viscosity (increasing the fluidity) of the heavy residue, usually from the first distillation of crude oil. In a visbreaker, the residue is heated no higher than 430 degrees Celsius. Visbreaking yields about 2 percent light gases such as butane, 5 percent gasoline products, 15 percent gas oil that can be catalytically cracked, and 78 percent tar.

Steam cracking. Steam cracking is used to turn light naphtha and other light gases into valuable petrochemical raw materials such as ethene, propene, or butane. These are raw materials for solvents, detergents, plastics, and synthetic fibers. Because light feed gases are cracked at very high temperatures between 730 and 900 degrees Celsius, steam is added before they enter the furnace to prevent their coking. The mix remains in the furnace for only 0.2 to 0.4 second before being cooled and fractionated to extract the cracked products.

Coking. Coking is the hottest form of cracking distillation residue. Because it leaves no residue, it has all but replaced conventional thermal cracking. Cracking at about 500 degrees Celsius also forms coke. Delayed coking moves completion of the cracking process, during which coke is created, out of the furnace area. To start cracking, feedstock stays in the furnace for only a few seconds before flowing into a coking drum, where cracking can take as long as one day.

In the coke drum, about 30 percent of feedstock is turned into coke deposits, while the valuable rest is sent to the fractionator. In addition to coke, delayed coking yields about 7 percent light gas, 20 percent light and heavy coker naphtha from which gasoline and gasoline products can be created, and 50 percent light and heavy gas oils. Gas oils are further processed through hydrocracking, hydrotreating, or subsequent fluid catalytic cracking, or used as heavy fuel oil. The coke drum has to be emptied of coke every half or full day. To ensure uninterrupted cracking, at least two are used. The coke comes in three kinds, in descending order of value: needle coke, used in electrodes; sponge coke, for part of anodes; and shot coke, primarily used as fuel in power plants.

Flexicoking. Flexicoking, continuous coking, and fluid coking are technological attempts to recycle coke as fuel in the cracking process. Although these cokers are more efficient, they are more expensive to build.

Fluid catalytic cracking. Because of its high conversion rate of vacuum wax distillate into far more valuable gasoline and lighter, olefinic gases, fluid catalytic crackers (FCCs) are the most important crackers at a refinery. By 2007, there were about four hundred FCCs worldwide working continuously at refineries. Together, they cracked about 10.6 million barrels of feedstock each day, about half of which was cracked in the United States. FCCs are essential to meet the global demand for gasoline.

Design of individual FCCs, while following the basic principle of fluid catalytic cracking, varies considerably, as engineers and scientists continuously seek to improve efficiency and lessen the environmental impact. By 2010, there were five major patents for FCCs that arranged the reactor, where cracking occurs, and the regenerator for the spent catalyst side by side. One major patent places the reactor atop the generator.

The typical products derived from vacuum distillate conversion in a FCC are about 21 percent olefinic gases, often called liquefied petroleum gas; 47 percent gasoline of high quality; 20 percent light cycle (or gas) oil, often called middle distillate; 7 percent heavy cycle (gas) oil, often called heavy fuel oil; and 5 percent coke. Gasoline is generally the most valuable. Light cycle oil is blended into heating oil and diesel, with highest demand for these blends in winter. The more a FCC can change the percentages of its outcome, the higher its economic advantage.

Hydrocracking. Hydrocrackers are the most flexible and efficient cracker, but they have high building and operation costs. High temperatures and pressure require significant energy, and the steel wall of a hydrocracker reactor can be as thick as 15 centimeters. Its use of hydrogen in the conversion process often requires a separate hydrogen generation plant. Hydrocrackers can accept a wide variety of feedstock ranging from crude oil distillates (gas oils or middle distillate) to light and heavy cycle oils from the FCC to coker distillates.

Hydrocracker output typically falls into flexible ranges for each product. Liquefied petroleum gas and other gases can make up 7 to 18 percent. Gasoline products, particularly jet fuel, one of the prime products of the hydrocracker, can be from 28 to 55 percent. Middle distillates, especially diesel and kerosene, can make up from 15 to 56 percent. Heavy distillates and residuum can range from 11 to 12 percent. Hydrocracking produces no coke but does have high carbon dioxide emissions. Its products have very low nitrogen and sulfur content.

IMPACT ON INDUSTRY

Cracking is a core activity at any modern refinery. Without modern cracking proesses, the world's demand for gasoline, diesel, kerosene, and jet fuel, as well as for basic petrochemicals, could not easily be met from processing crude oil. About 45 percent of the world's gasoline, for example, comes from FCCs or related units in 2006. Significant public and private research is focusing on improving cracking efficiency.

Government agencies. Throughout the industrialized and industrializing world, national governments and their agencies, as well as international entities, have promoted more environmentally friendly patterns of hydrocarbon production, processing, and consumption. This has increased the importance of and demand for cracking, as most of the heavy oil distillates and residues once used as heavy fuel oil in power plants are increasingly being replaced by cleaner burning natural gas. National legislation such as the United States Clean Air Act of 1963 with its crucial 1990 amendment imposes strict limits on emissions including those generated during the cracking process, as well as quality specifications for refined products such as cleaner burning gasoline and diesel. Crackers have come under governmental pressure to reduce their often high emissions of carbon dioxide and other gases through innovations or process changes, which means more costs.

Universities and research institutes. Particularly in industrialized societies, there is considerable public research into improving existing and developing new cracking processes. Of special interest are catalysts used by FCCs and hydrocrackers. Catalyst research looks into properties of and new materials for catalysts and experiments with optimizing their placement within the reactor, enhancing their recovery, and lengthening their life cycle.

However, a critical gap exists between what works in the laboratory or in small pilot plants and what can be implemented in an industrial setting. Occasionally, promising new catalysts have not risen to their perceived potential, and new reactor designs have proved commercially impractical. University and institute sciences tend to have a certain bias toward energy-saving and environmentally friendly discoveries, while industrial research tends to be more concerned with raising feedstock economy.

Industry and business sectors. Cracking provides a key process for the oil and gas and petrochemical industries. For example, in January, 2010, in the first step of atmospheric distillation, U.S. refineries processed a total of 17.8 million barrels of crude oil per stream day (almost identical to a calendar day). Of this, 8.5 million barrels were subjected to vacuum distillation. Among the distillates and residue from both steps, 2.5 million barrels (15 percent of the original crude oil entering the refinery) were processed in the thermal cracking process of delayed coking (fluid coking and visbreaking were negligible at U.S. refineries). Another 6.2 million barrels (35 percent of all products from the original crude oil) were submitted to an FCC for catalytic cracking, and 1.8 million barrels (10 percent of the base) were processed in a hydrocracker. Although the balance between FCC and hydrocracker processing is somewhat different in Europe and Asia, where demand for diesel and kerosene is higher than in the United States, the example is representative for the petroleum industry. Overall, about 60 percent of the crude oil that enters a refinery undergoes some form of cracking.

Major corporations. Because crackers are integral parts of a refinery, require a very large initial investment, and have relatively high operating costs, they are typically owned by the large international or national oil and gas companies that own the refinery. Their construction, however, may be executed by specialized engineering companies, and their design

---

FASCINATING FACTS ABOUT CRACKING

- If engineers had not invented cracking hydrocarbons, the world would need to produce almost double the amount of crude oil to meet the global demand of gasoline, diesel, and jet fuel.

- Crackers at refineries and petrochemical plants operate continuously for twenty-four hours and are typically staffed in three shifts.

- During World War II, newly invented catalytic cracking provided plenty of powerful high-octane gasoline for aircraft engines of the United States and Great Britain, giving their air forces an edge over German and Japanese air forces.

- Crackers are expensive chemical processing plants. The ethane steam cracker and high ole-fin fluid catalytic cracker constructed in 2009 at Rabigh Refining and Petrochemical Complex in Saudi Arabia cost $850 million to build.

- Reactors of catalytic and steam crackers are incredibly hot, reaching up to 900 degrees Celsius. At these temperatures, oil vapors are blown through the reactor in fractions of a second.

- New pharmaceuticals can contain some ingredients made from raw materials that came out of a steam cracker.

- Catalysts fundamentally support cracking processes. If operating staff mistreats or abuses catalysts, the whole cracker may have to be shut down for catalyst replacement.

---

can involve independent consultants. In addition, many cracking facilities are built under license from the company that holds the patent to their design.

Innovative or improved crackers can be added to existing refineries or built together with green field refineries that are increasingly erected in the Middle East, the source of much crude oil, or Asia, the source of much demand. Crackers follow economies of scale. In 2005, to build an FCC with a capacity of 10 million barrels of feedstock per stream day in a Gulf Coast refinery cost from $80 million to $120 million. An FCC with ten times more capacity, however, cost only between $230 million and $280 million. The economy-of-scale factor favors world-scale crackers at equally large refineries.

CAREERS AND COURSE WORK

Cracking is a key refining process, so good job opportunities should continue in this field. Students interested in designing and constructing or operating and optimizing a cracker at a refinery or petrochemical plant should take science courses, particularly chemistry and physics, in high school. The same is true for those who want to pursue research in the field. There also are many opportunities for technicians in building, operationing, and maintaining a cracker.

A bachelor of science degree in an engineering discipline, particularly chemical, electrical, computer, or mechanical engineering, is excellent job preparation in this field. A bachelor of science or arts degree in a major such as chemistry, physics, computer science, environmental science, or mathematics is also a good foundation. An additional science minor is useful.

For an advanced career, any master of engineering degree is a suitable preparation. A doctorate in chemistry or chemical or mechanical engineering is needed if the student wants a top research position, either with a corporation or at a research facility. Postdoctoral work in materials science (engaged in activities such as searching for new catalysts) is also advantageous.

Because cracking is closely related to selecting crude oil for purchase by the refinery, there are also positions for graduates in business or business administration. The same is true for members of the medical profession with an emphasis in occupational health and safety. Technical writers with an undergraduate degree in English or communication may also find employment at oil and engineering companies. As cracking is a global business, career advancement in the industry often requires willingness to work abroad.

SOCIAL CONTEXT AND FUTURE PROSPECTS

As fossilized hydrocarbons are a finite resource, they must be used as efficiently as possible. To this end, cracking at a refinery seeks to create the most valuable products out of its feedstocks derived from crude oil. This is not limited to fuels such as gasoline. The steam crackers of the petrochemical industry create raw materials for many extremely valuable and useful products such as pharmaceuticals, plastics, solvents, detergents, and adhesives, stretching the use of hydrocarbons for consumers.

At the same time, international concern with the negative environmental side effects of some hydrocarbon processes is increasing. Traditionally, crackers such as cokers or even hydrocrackers released a large amount of carbon dioxide or, in the case of cokers, other airborne pollutants as well. To make cracking more environmentally friendly, to save energy, and to convert feedstock efficiently are concerns shared by the public and the petroleum and petrochemical industries. This is especially so for companies when there are commercial rewards for efficient, clean operations.

The possible rise of alternative fuels, replacing some hydrocarbon-based fuels such as gasoline and diesel, would challenge refineries to adjust the output of their crackers. The more flexible hydrocrackers are best suited to meet this challenge.

—*R. C. Lutz, MA, PhD*

FURTHER READING

Burdick, Donald. Petrochemicals in Nontechnical Language. 3d ed. Tulsa, Okla.: Penn Well, 2010. Accessible introduction. Covers use of petrochemical materials gained from steam cracking. Figures and tables.

Conaway, Charles. The Petroleum Industry: A Nontechnical Guide. Tulsa, Okla.: Penn Well, 1999. Chapter 13 provides a short summary of cracking processes.

Gary, James, et al. Petroleum Refining: Technology and Economics. 5th ed. New York: CRC Press, 2007. Well-written textbook. Good presentation of all types of cracking in the petrochemical industry.

Includes economic aspects of cracking. Five appendixes, index, and photographs.

Leffler, William. Petroleum Refining in Nontechnical Language. Tulsa, Okla.: Penn Well, 2008. Covers all major aspects of cracking at a refinery, from chemical foundations to catalytic cracking and hydrocracking.

Meyers, Robert, ed. Handbook of Petroleum Refining Processes. 3d ed. New York: McGraw-Hill, 2004. Advanced-level technical compendium covers various industry methods for catalytic cracking, hydro-cracking, visbreaking, and coking. In-depth information for those considering a career in this field.

Sadeghbeigi, Reza. Fluid Catalytic Cracking Handbook. 2d ed. Houston, Tex.: Gulf Publishing, 2000. Detailed, technical look at this key cracking process. Figures, conversion table, glossary.

WEB SITES

American Association of Petroleum Geologists
http://www.aapg.org

American Petroleum Institute
http://www.api.org

National Petrochemical and Refiners Association
http://www.npra.org

National Petroleum Council
http://www.npc.org

See also: Detergents; Distillation.

CRYOGENICS

FIELDS OF STUDY

Astrophysics; cryogenic engineering; cryogenic electronics; nuclear physics; cryosurgery; cryobiology; high-energy physics; mechanical engineering; chemical engineering; electrical engineering; cryotronics; materials science; biotechnology; medical engineering; astronomy.

SUMMARY

Cryogenics is the branch of physics concerned with creation of extremely low temperatures and involves the observation and interpretation of natural phenomena resulting from subjecting various substances to those temperatures. At temperatures near absolute zero, the electric, magnetic, and thermal properties of most substances are greatly altered, allowing useful industrial, automotive, engineering, and medical applications.

KEY TERMS AND CONCEPTS

- **absolute zero:** Temperature measured 0 Kelvin (–273 degrees Celsius), where molecules and atoms oscillate at the lowest rate possible.
- **cryocooler:** Device that uses cycling gases to produce temperatures necessary for cryogenic work.
- **cryogenic processing:** Deep cooling of matter using cryogenic temperatures so the molecules and atoms of the matter slow or almost stop movement.
- **cryogenic tempering:** Onetime process using sensitive computerization to cool metal to cryogenic temperatures then tempering the metal to enhance performance, strength, and durability.
- **cryopreservation:** Cooling cells or tissues to subzero temperatures to preserve for future use.
- **evaporative cooling:** Process that allows heat in a liquid to change surface particles from a liquid to a gas.
- **heat conduction:** Technique where a substance is cooled by passing heat from matter of higher temperature to matter of lower temperature.
- **Joule-Thomson effect:** Technique where a substance is cooled by rapid expansion, which drops the temperature. So named for its discoverers, British physicists James Prescott Joule and William Thomson (Lord Kelvin).
- **Kelvin temperature scale (k):** Used to study extremely cold temperatures. On the Kelvin scale, water freezes at 273 K and boils at 373 K.
- **superconducting device:** Device known for its electrical properties and magnetic fields, such as magnetic resonance imaging (MRI) in medicine.
- **superconducting magnet:** Electromagnet with a superconducting coil where a magnetic field is maintained without any continuing power source.
- **superconductivity:** Absence of electrical resistance in metals, ceramics, and compounds when cooled to extremely low temperatures.
- **superfluidity:** Phase of matter, such as liquid helium, that is absent of viscosity and flows freely without friction at very low temperatures.

DEFINITION AND BASIC PRINCIPLES

Cryogenics comes from two Greek words: kryo, meaning "frost," and genic, "to produce." This science studies the implications of producing extremely cold temperatures and how these temperatures affect substances such as gases and metal. Cryogenic temperature levels are not found naturally on Earth. The usefulness of cryogenics is based on scientific principles. The three basic states of matter are gas, liquid, and solid. Matter moves from one state to another by the addition or subtraction of heat (energy). The molecules or atoms in matter move or vibrate at different rates depending on the level of heat. Extremely low temperatures, as achieved through cryogenics, slow the vibration of atoms and can change the state of matter. For example, cryogenic temperatures are used in the liquefaction of atmospheric gases such as oxygen, nitrogen, hydrogen, and methane for diverse industrial, engineering, automotive, and medical applications.

Sometimes cryogenics and cryonics are mistakenly linked, but use of subzero temperatures is the only thing these practices share. Cryonics is the practice of freezing a body right after death to preserve it for a future time when a cure for fatal illness or remedy for fatal injury may be available. The practice of cryonics is based on the belief that technology from cryobiology can be applied to cryonics.

If cells, tissues, and organs can be preserved by cryogenic temperatures, then perhaps whole body can be preserved for future thawing and life restoration. Facilities exist for interested persons or families, although the cryonic process is not considered reversible as of this writing.

BACKGROUND AND HISTORY

The history of cryogenics follows the evolution of low-temperature techniques and technology. Principles of cryogenics can be traced to 2500 BCE, when Egyptians and Indians evaporated water through porous earthen containers to produce cooling. The Chinese, Romans, and Greeks collected ice and snow from the mountains and stored it in cellars to preserve food. In the early 1800's, American inventor Jacob Perkins created a sulfuric-ether ice machine, a precursor to the refrigerator. By the mid-1800's, William Thomson, a British physicist known as Lord Kelvin, theorized that extremely cold temperatures could stop the motion of atoms and molecules. This became known as absolute zero, and the Kelvin scale of temperature measurement emerged.

Scientists of the time focused on liquefaction of permanent gases. By 1845, the work of British physicist Michael Faraday accomplished liquefaction of permanent gases by cooling immersion baths of ether and dry ice followed by pressurization. Six permanent gases— oxygen, hydrogen, nitrogen, methane, nitric oxide, and carbon monoxide—still resisted liquefaction. In 1877, French physicist Louis-Paul Cailletet and Swiss physicist Raoul Pictet produced drops of liquid oxygen, working separately and using completely different methods. In 1883, S. F. von Wroblewski at the University of Krakow in Poland, discovered oxygen would liquefy at 90 Kelvin (K) and nitrogen at 77 K. In 1898, Scottish chemist James Dewar discovered the boiling point of hydrogen at 20 K and its freezing point at 14 K.

Helium, with the lowest boiling point of all known substances, was liquefied in 1908 by Dutch physicist Heike Kamerlingh Onnes at the University of Leiden. Onnes was the first person to use the word "cryogenics." In 1892, Scottish physicist James Dewar invented the Dewar flask, a vacuum flask designed to maintain temperatures necessary for liquefying gases, which was the precursor to the Thermos. The liquefaction of gases had many important commercial applications, and many industries use Dewar's concept in applying cryogenics to their processes and products.

The usefulness of cryogenics continued to evolve, and by 1934 the concept was well established. During World War II, scientists discovered that metals became resistant to wear when frozen. In the 1950's, the Dewar flask was improved with the multilayer insulation (MLI) technique for insulating cryogenic propellants used in rockets. Over the next thirty years, Dewar's concept led to the development of small cryocoolers, useful to the military in national defense. The National Aeronautics and Space Administration (NASA) space program applies cryogenics to its programs. Cryogenics can be used to preserve food for long periods—this is especially helpful during natural disasters. Cryogenics continues to grow globally and serve a wide variety of industries.

HOW IT WORKS

Cryogenics is an ever-expanding science. The basic principle of cryogenics that the creation of extremely low temperatures will affect the properties of matter so the changed matter can be used for a number of applications. Four techniques can create the conditions necessary for cryogenics: heat conduction, evaporative cooling, rapid-expansion cooling (Joule-Thomson effect), and adiabatic demagnetization.

Creating low temperatures. With heat conduction, heat flows from matter of higher temperature to matter of lower temperature in what amounts, basically, to a transfer of thermal energy. As the process is repeated, the matter cools. This principle is used in cryogenics by allowing substances to be immersed in liquids with cryogenic temperatures or in an environment such as a cryogenic refrigerator for cooling.

Evaporative cooling is another technique employed in cryogenics. Evaporative cooling is demonstrated in the human body when heat is lost through liquid (perspiration) to cool the body via the skin. Perspiration absorbs heat from the body, which evaporates after it is expelled. In the early 1920's in Arizona during the summers, people hung wet sheets inside screened sleeping porches. Electric fans pulled air through the sheets to cool the sleeping space. In the same way, a container of liquid can evaporate, so the heat is removed as gas; the repetitive process drops the temperature of the liquid. An example is reducing the temperature of liquid nitrogen to its freezing point.

The Joule-Thomson effect occurs without the transfer of heat. Temperature is affected by the relationship between volume, mass, pressure, and temperature. Rapid expansion of a gas from high to low pressure results in a temperature drop. This principle was employed by Dutch physicist Heike Kamerlingh Onnes to liquefy helium in 1908 and is useful in home refrigerators and air conditioners.

Adiabatic demagnetization uses paramagnetic salts to absorb energy from liquid, resulting in a temperature drop. The principle in adiabatic demagnetization is the removal of the isothermal magnetized field from matter to lower the temperature. This principle is useful in application to refrigeration systems, which may include a superconducting magnet.

Cryogenic refrigeration. Cryogenic refrigeration, used by the military, laboratories, and commercial businesses, employs gases such as helium (valued for its low boiling point), nitrogen, and hydrogen to cool equipment and related components at temperatures lower than 150 K. The selected gas is cooled through pressurization to liquid or solid forms (dry ice used in the food industry is solidified carbon dioxide). The cold liquid may be stored in insulated containers until used in a cold station to cool equipment in an immersion bath or with sprayer.

Cryogenic processing and tempering. Cryogenic processing or treatment increases the length of wear of many metals and some plastics using a deep- freezing process. Metal objects are introduced to cooled liquid gases such as liquid nitrogen. The computer-controlled process takes about seventy-two hours to affect the molecular structure of the metal. The next step is cryogenic or heat tempering to improve the strength and durability of the metal object. There are about forty companies in the United States that provide cryogenic processing.

APPLICATIONS AND PRODUCTS

Early applications of cryogenics targeted the need to liquefy gases. The success of this process in the late 1800's paved the way for more study and research to apply cryogenics to developing life needs and products. Examples include applications in the auto and health care industries and in development of rocket fuels and methods of food preservation. Cryogenic engineering has applications related to commercial, industrial, aerospace, medical, domestic, and defense ventures.

Liquid nitrogen

Superconductivity applications. One property of cryogenics is superconductivity. This occurs when the temperature is dropped so low that the electrical current experiences no resistance. An example is electrical appliances, such as toasters, televisions, radios, or ovens, where energy is wasted trying to overcome electrical resistance. Another is with magnetic resonance imaging (MRI), which uses a powerful magnetic field generated by electromagnets to diagnosis certain medical conditions. High magnetic field strength occurs with superconducting magnets. Liquid helium, which becomes a free-flowing superfluid, cools the superconducting coils; liquid nitrogen cools the superconducting compounds, making cryogenics an integral part of this process. Another application is the use of liquefied gases that are sprayed on buried electrical cables to minimize wasted power and energy and to maintain cool cables with decreased electrical resistance.

Health care applications. The health care industry recognizes the value of cryogenics. Medical applications using cryogenics include preservation of cells or tissues, blood products, semen, corneas, embryos, vaccines, and skin for grafting. Cryotubes with liquid nitrogen are useful in storing strains of bacteria at low temperatures. Chemical reactions needed to release active ingredients in statin drugs, used for cholesterol control, must be completed at very low temperatures (−100 degrees Celsius). High-resolution imaging, like MRI, depends on cryogenic principles for the diagnosis of disease and medical conditions. Dermatologists uses cryotherapy to treat warts or skin lesions.

Food and beverage applications. The food industry uses cryogenic gases to preserve and transport mass amounts of food without spoilage. This is also useful in supplying food to war zones or natural-disaster areas. Deep-frozen food retains color, taste, and nutrient content while increasing shelf life.

FASCINATING FACTS ABOUT CRYOGENICS

- American businessman Clarence Birdseye revolutionized the food industry when he discovered that deep-frozen food tasted better than regular frozen food. In 1923, he developed the flash-freezing method of preserving food at below-freezing temperatures under pressure. The "Father of Frozen Food" first sold small-packaged foods to the public in 1930 under the name Birds Eye Frosted Foods.

- In cryosurgery, super-freezing temperatures as low as −200 Celsius are introduced through a probe of circulating liquid nitrogen to treat malignant tumors, destroy damaged brain tissue in Parkinson's patients, control pain, halt bleeding, and repair detached retinas.

- Cryogenics can be used to save endangered species from extinction. Smithsonian researcher Mary Hagedorn is using cryogenics to establish the first coral seed banks: She's collecting thousands of sample species and freezing them for the future. Hagedorn refers to this as an insurance policy for natural resources.

- The Joule-Thomson effect, discovered in 1852 by James Prescott Joule and William Thomson (Lord Kelvin), is responsible for the cooling used in home refrigerators and air conditioners.

- Helium's boiling point, 173 Kelvin, is the lowest of all known substances.

- Surgical tools and implants used by surgeons and dentists have increased strength and resistance to wear because of cryogenic processing.

- Cryogenic processing is 100 percent environmentally friendly with no use of harmful chemicals and no waste products.

- In 1988, microbiologist Curt Jones, who studied freezing techniques to preserve bacteria and enzymes for commercial use, created Dippin' Dots, a popular ice cream treat, using a quick-freeze process with liquid nitrogen.

Certain fruits and vegetables can be deep frozen for consumption out of season. Freeze-dried foods and beverages, such as coffee, soups, and military rations, can be safely stored for long periods without spoilage. Restaurants and bars use liquid gases to store beverages while maintaining the taste and look of the drink.

Automotive applications. The automotive industry employs cryogenics in diverse ways. One is through the use of thermal contraction. Because materials will contract when cooled, the valve seals of automobiles are treated with liquid nitrogen, which shrinks to allow insertion and then expands as it warms up, resulting in a tight fit. The automotive industry also uses cryogenics to increase strength and minimize wear of metal engine parts, pistons, cranks, rods, spark plugs, gears, axles, brake rotors and pads, valves, rings, rockers, and clutches. Cryogenic-treated spark plugs can increase an automobile's horsepower as well as its gasoline mileage. The use of cryogenics allows a race car to race as many as thirty times without a major rebuild on the motor compared with racing twice on an untreated car.

Aerospace industry applications. NASA's space program utilizes cryogenic liquids to propel rockets. Rockets carry liquid hydrogen for fuel and liquid oxygen for combustion. Cryogenic hydrogen fuel is what enables NASA's workhorse space shuttle to get into orbit. Another application is using liquid helium to cool the infrared telescopes on rockets.

Tools, equipment, and instrument applications. Metal tools can be treated with cryogenic applications that provide wear resistance. In surgery or dentistry, tools can be expensive, and cryogenic treatment can prolong usage. Sports equipment, such as golf clubs, benefits from cryogenics as it provides increased wear resistance and better performance. Another is the ability of a scuba diver to stay submerged for hours with an insulated Dewar flask of cryogenically cooled nitrogen and oxygen. Some claim musical instruments receive benefits from cryogenic treatment; in brass instruments, a crisper and cleaner sound is allegedly produced with cryogenic enhancement.

Other applications. Other applications are evolving as industries recognize the benefits of cryogenics to their products and programs. The military have used cryogenics in various ways, including infrared tracking systems, unmanned vehicles, and missile warning receivers. Companies can immerse discarded recyclables in liquid nitrogen to make them brittle, then these recyclables can be pulverized or grinded down to a more eco-friendly form. No doubt with continued research, many more applications will emerge.

IMPACT ON INDUSTRY

The science of cryogenics continues to impact the

quality and effectiveness of products of many industries. Various groups have initiated research studies to support the expanded use of cryogenics.

Government and university research. Government agencies and universities have conducted research on various aspects of cryogenics. The military have investigated the applications of cryogenics to national defense. The Air Force Research Laboratory (AFRL) Space Vehicles Directorate addressed applications of cryogenic refrigeration in the area of ballistic missile defense. The study looked at ground-based radars and space-based infrared sensors requiring cryogenic refrigeration. Future research targets the availability of flexible technology such as field cryocoolers. Such studies can be significant to a cost-effective national defense for the United States.

The health care industry has many possible applications for cryogenics. In 2005, Texas A&M University Health Science Center-Baylor College of Dentistry investigated the effect of cryogenic treatment of nickel-titanium on instruments used by dentists. Past research had been conducted on stainless-steel endodontic instruments with no significant increase in wear resistance. This research demonstrated an increase in microhardness but not in cutting efficiency.

Industry and business. In 2006, the Cryogenic Institute of New England (CINE) recognized that although cryogenic processing was useful in many industries, research validating its technologic advantages and business potential was scant. CINE located forty commercial companies that provided cryogenic processing services and conducted telephone surveys with thirty of them. The survey found that some $8 million were generated by these deep cryogenic services in the United States, with 75 percent coming from the services and 25 percent from the sales of equipment.

The survey asked participants to identify the list of top metals they worked with in cryogenic processing. These were given as cast irons, various steels (carbon, stainless, tool, alloy, mold), aluminum, copper, and others. The revenue was documented by market application. Some 42 percent of the cryogenic-processing-plant market was in the motor sports and automotive industry, where the goal was treatment of engine components to improve performance, extend life wear, or treat brake rotors. Thirty percent of the application market fell into the category of heavy metals, tooling, and cutting; examples of these include manufacturing machine tools, dies, piping, grinders, knives, food processing, paper and pulp processing, and printing. Ten percent were listed in heavy components such as construction, in-ground drilling, and mining, while 18 percent of the market was in areas such as recreational, firearms, electronics, gears, copper electrodes, and grinding wheels.

The National Institute of Standards and Technology (NIST) initiated the Cryogenic Technologies Project, a collaborative research effort between industry and government agencies to improve cryogenic processes and products. One goal is the investigation of cryogenic refrigeration.

The nonprofit Cryogenic Society of America (CSA) offers conferences on related work areas such as superconductivity, space cryogenics, cryocoolers, refrigeration, and magnet technology. It also has continuing-education courses and lists job postings on its Web site.

CAREERS AND COURSE WORK

Careers in cryogenics are as diverse as the applications of cryogenics. Interested persons can enter the profession in various ways, depending on their field of interest. Some secure jobs through additional education, while others learn on the job. In general, the jobs include engineers, technologists or technicians, and researchers.

A primary career track for those interested in working in cryogenics is cryogenic engineering. To become a cryogenic engineer requires a bachelor's or master's degree in engineering. Course work may include thermodynamics, production of low temperatures, refrigeration, liquefaction, solid and fluid properties, and cryogenic systems and safety.

In the United States, some four hundred academic institutions offer graduate programs in engineering with about forty committed to research and academic opportunities in cryogenics. Schools with graduate programs in cryogenics include the University of California, Los Angeles; University of Colorado; Cornell University; Georgia Institute of Technology; Illinois Institute of Technology; Florida International University; Iowa State University; Massachusetts Institute of Technology; Ohio State University; University of Wisconsin-Madison; Florida State University; Northwestern University; and University of Southampton in the United Kingdom.

SOCIAL CONTEXT AND FUTURE PROSPECTS

The economic and ecological impact of cryogenic research and applications holds global promise for the future. In 2009, Netherlands firm Stirling Cryogenics built a cooling system with liquid argon for the ICARUS project, which is being carried out by Italy's National Institute of Nuclear Physics. In China, the Cryogenic and Refrigeration Engineering Research Centre (CRERC) focuses on new innovations and technology in cryogenic engineering. Both private industry and government agencies in the United States are pursuing innovative ways to utilize existing applications and define future implications of cryogenics. Although cryogenics has proved useful to many industries, its full potential as a science has not yet been realized.

—*Marylane Wade Koch, MSN, RN*

FURTHER READING

Hayes, Allyson E., ed. *Cryogenics: Theory, Processes and Applications.* Hauppauge, N.Y.: Nova Science Publishers, 2010. Details global research on cryogenics and applications such as genetic engineering and cryopreservation.

Jha, A. R. *Cryogenic Technology and Applications.* Burlington, Mass.: Elsevier, 2006. Deals with most aspects of cryogenics and cryogenic engineering, including historical development and various laws, such as heat transfer, that make cryogenics possible.

Schwadron, Terry. "Hot Sounds From a Cold Trumpet? Cryogenic Theory Falls Flat." *New York Times*, November 18, 2003. Explains how two Tufts University researchers studied cryogenic freezing of trumpets and determined the cold did not improve the sound.

Ventura, Gugliemo, and Lara Risegari. *The Art of Cryogenics: Low-Temperature Experimental Techniques.* Burlington, Mass.: Elsevier, 2008. Comprehensive discussion of various aspects of cryogenics from heat transfer and thermal isolation to cryoliquids and instrumentation for cryogenics, such as the use of magnets.

WEB SITES

Cryogenic Society of America
http://www.cryogenicsociety.org

Help Mary Save Coral
http://www.helpmarysavecoral.org/obe

National Aeronautics and Space Administration Cryogenic Fluid Management
http://www.nasa.gov/centers/ames/research/ technology-onepagers/cryogenic-fluid-management. html

National Institute of Standards and Technology Cryogenic Technologies Project
http://www.nist.gov/mml/properties/cryogenics/ index.cfm

See also: Computer Science; Electrical Engineering; Electromagnet Technologies; Magnetic Resonance Imaging; Mechanical Engineering.

CRYPTOLOGY AND CRYPTOGRAPHY

FIELDS OF STUDY

History; computer science; mathematics; systems engineering; computer programming; communications; cryptographic engineering; security engineering; statistics.

SUMMARY

Cryptography is the use of a cipher to hide a message by replacing it with letters, numbers, or characters. Traditionally, cryptography has been a tool for hiding communications from the enemy during times of war. Although it is still used for this purpose, it is more often used to encrypt confidential data, messages, passwords, and digital signatures on computers. Although computer ciphers are based on manually applied ciphers, they are programmed into the computer using complex algorithms that include algebraic equations to encrypt the information.

KEY TERMS AND CONCEPTS

- **algorithm:** Mathematical steps to solve a problem

with a computer.

- **ciphertext:** Meaningless text produced by applying a cipher to a message.
- **cleartext:** Original message; also known as plaintext.
- **cryptosystem:** Computer program or group of programs that encrypt and decrypt a data file, message, password, or digital signature.
- **key:** Number that is used both in accessing encrypted data and in encrypting it.
- **polyalphabetic cipher:** Cipher that uses multiple alphabets or alphabetic arrangements.
- **prime number:** Number that can be divided only by itself and the numeral one.
- **proprietary information:** Information that is the property of a specific company.

DEFINITION AND BASIC PRINCIPLES

Cryptography is the use of a cipher to represent hidden letters, words, or messages. Cryptology is the study of ciphers. Ciphers are not codes. Codes are used to represent a word or concept, and they do not have a hidden meaning. An example is the international maritime signal flags, part of the International Code of Signals, known by all sailors and available to anyone else. Ciphers are schemes for changing the letters and numbers of a message with the intention of hiding the message from others. An example is a substitution cipher, in which the order of letters of the alphabet is rearranged and then used to represent other letters of the alphabet. Cryptography has been used since ancient times for communicating military plans or information about the enemy. In modern times, it is most commonly thought of in regard to computer security. Cryptography is critical to storing and sharing computer data files and using passwords to access information either on a computer or on the Internet.

BACKGROUND AND HISTORY

The most common type of cipher in ancient times was the substitution cipher. This was the type of cipher employed by Julius Caesar during the Gallic Wars, by the Italian Leon Battista Alberti in a device called the Alberti Cipher Disk described in a treatise in 1467, and by Sir Francis Bacon of Great Britain. In the 1400's, the Egyptians discovered a way to decrypt substitution ciphers by analyzing the frequency of the letters of the alphabet. Knowing the frequency of a letter made it easy to decipher a message.

German abbot Johannes Trithemius devised polyalphabetic ciphers in 1499, and French diplomat Blaise de Vigenère did the same in 1586. Both used the tableau, which consists of a series of alphabets written one below the other. Each alphabet is shifted one position to the left. De Vigenère added a key to his tableau. The key was used to determine the order in which the alphabets were used.

The Greeks developed the first encryption device, the scytale, which consisted of a wooden staff and a strip of parchment or leather. The strip was wrapped around the staff and the message was written with one letter on each wrap. When wrapped around another staff of the same size, the message appeared. In the 1780's, Thomas Jefferson invented a wheel cipher that used wooden disks with the alphabet printed around the outside. They were arranged side by side on a spindle and were turned to create a huge number of ciphers. In 1922, Jefferson's wheel cipher was adopted by the U.S. Army. It used metal disks and was named the M-94.

Four inventors—Edward H. Hebern, United States, in 1917; Arthur Scherbius, Germany, in 1918; Hugo Alexander Koch, Netherlands, in 1919; and Arvid Gerhard Damm, Sweden, in 1919—independently developed cipher machines that scrambled letters using a rotor or a wired code wheel. Scherbius, an electrical engineer, called his machine the Enigma, which looked like a small, narrow typewriter. The device changed the ciphertext with each letter that was input. The German navy and other armed forces tried the machine, and the German army redesigned the machine, developing the Enigma I in 1930 and using various models during World War II. The Japanese also developed an Enigma-like device for use during the war. The Allies were unable to decrypt Enigma ciphers, until the Polish built an Enigma and sold it to the British. The German military became careless in key choice and about the order of words in sentences, so the cipher was cracked. This was a factor in the defeat of Germany in World War II.

In the 1940's, British intelligence created the first computer, Colossus. After World War II, the British destroyed Colossus. Two Americans at the University of Pennsylvania are credited with creating the first American computer in 1945. It was named the Electronic Numerical Integrator and Calculator (ENIAC). It was able to easily decrypt manual and Enigma ciphers.

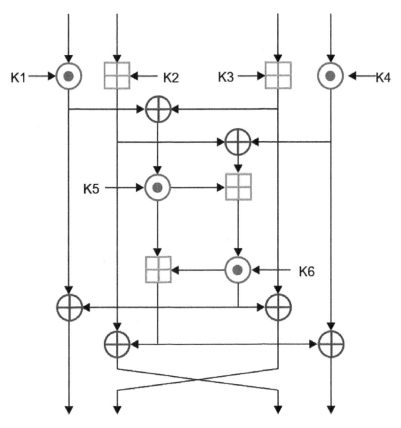

One round (out of 8.5) of the IDEA cipher, used in some versions of PGP for high-speed encryption of, for instance, e-mail

How it Works

The data, message, or password starts out as cleartext or plaintext. Once it is input into the computer, a computer algebra system (CAS) performs the actual encryption, using the key and selected algebraic equations, based on the cryptographic algorithm. It stores the data, message, or password as ciphertext.

Most encryption systems require a key. The length of the key increases the complexity of the cryptographic algorithm and decreases the likelihood that the cipher will be cracked. Modern key lengths range from 128 to 2,048 bits. There are two key types. The first is a symmetric or secret key that is used both to encrypt and to decrypt the data. A different key generates different ciphertext. It is critical to keep the key secret and to use a sufficiently complex cryptographic algorithm. The algebraic equations used with a symmetric key are two-way equations.

Symmetric key cryptography uses a polyalphabetic encryption algorithm, such as a block cipher, a stream cipher, or a hash function to convert the data into binary code. A block cipher applies the same key to each block of data. Stream ciphers encrypt the data bit by bit. They can operate in several ways, but two common methods are self-synchronizing and synchronous. The self-synchronizing cipher encrypts the data bit by bit, using an algebraic function applied to the previous bit. Synchronous ciphers apply a succession of functions independent of the data.

Public key or asymmetric cryptography uses two different keys: a public key and a private key. The public key can be distributed to all users, whereas the private key is unknown. The private key cannot be calculated from the public key, although there is a relationship between the two numbers. The public key is used for encryption and the private key is used for decryption. The algebraic equations used in public key cryptography can be calculated only one way, and different equations are used to encrypt and decrypt. The calculations are complex and often involve factorization of a large number or determining the specific logarithm of a large number. Public key algorithms use block ciphers and the blocks may be variable in size.

Digital signatures can be linked to the public key. A digital signature provides a way to identify the creator of the data and to authenticate the source. A digital signature is difficult to replicate. The typical components of a digital signature are the public key, the user's name and e-mail address, the expiration date of the public key, the name of the company that owns the signature, the serial number of the digital identification number (a random number), and the digital signature of the certification authority.

A hash function, or message digest, is an encryption algorithm that uses no key. A hash function takes a variable length record and uses a one-way function to calculate a shorter, fixed-length record. Hash functions are difficult to reverse, and so they are used to verify a digital signature or password. If the new hash is the same as the encrypted version, then the password, or digital signature, is accepted. There are a number of hash algorithms. Typically, they break the file into even-sized blocks and apply either a random number or a prime number to each block. Some hash functions act on all the values in the block, and others work on selected values. Sometimes they generate duplicate hashes, which are called collisions. If there is any change in the data, the hash changes.

There are specific security standards that cryptographic data must meet. They are authentication, privacy/confidentiality, integrity, and nonrepudiation. These standards not only protect the security of the data but also verify the identity of the user, the validity of the data, and that the sender provided the data.

APPLICATIONS AND PRODUCTS

Cryptographic software is created so that it can interact with a variety of computer systems. Some of this software is integrated into other computer programs, and some interfaces with other computer systems. Most cryptographic programs are written in the computer languages of Java, JavaScript, C, C+, or C++. The types of software used for cryptography include computer algebra systems, symmetric key algorithms, public key algorithms, hash functions schemes, and digital signature algorithms.

CAS software. A computer algebra system (CAS) is a software package that performs mathematical functions. Basic CAS software supports functions such as linear algebra, algebra of polynomials, calculus, nonlinear equations, functions, and graphics. More complex CAS software also supports command lines, animation, statistics, number theory, differential equations, networking, geometric functions, graphing, mathematical maximization, and a programming language. Some of the CAS software that is used for encryption of data includes FriCAS 1.0.3, Maple 12, Mathematica 6.0, Matlab 7.2.0.283, Maxima 5.19.1, and Sage 3.4.

Symmetric key algorithms. Symmetric key algorithms work best for storing data that will not be shared with others. This is because of the need to communicate the secret key to the receiver, which can compromise its secrecy. There are two types of secret key encryption algorithms: stream ciphers and block ciphers. The standard for secret key encryption is the Data Encryption Standard (DES) software that the National Bureau of Standards has adopted. Other secret key encryption algorithms include: IDEA (International Data Encryption Algorithm), Rivest Ciphers (RC1 through RC6), Blowfish, Twofish, Camellia, MISTYI, SAFER (Secure and Fast Encryption Routine), KASUMI, SEED, ARIA, and Skipjack. They are block ciphers, except for RC4, which is a stream cipher.

Public key algorithms. Asymmetric or public key algorithms support the review of digital signatures and variable length keys. They are used for data that are sent to another businesses or accessed by users, because there is no need to keep the public key secure. The U.S. National Institute of Standards and Technology (NIST) has adopted a new encryption cipher, called Advanced Encryption Standard (AES). AES uses a public key and a block algorithm and is considered to be more secure than DES. Some examples of public key algorithms are: RSA1, Diffie-Hellman, DSA, ElGamal, PGP (Pretty Good Privacy), ECC (Elliptic Curve Cryptography), Public-Key Cryptography Standards (PKCS through 15), Cramer-Shoup, and LUC.

Hash function algorithms. Hash function algorithms are not considered actual encryption, although they are often used with encryption algorithms. Hash functions are used to verify passwords or digital signatures, to look up a file or a table of data, to store data, to verify e-mail users, to verify data integrity, and to verify the parties in electronic funds transfer (EFT). Some examples of actual hash algorithms are: SHA-1, SHA-2, MD2, MD4, MD5, RIPEMD, Haval, Whirlpool, and Tiger.

Digital signatures. A digital signature scheme is used with a public key system and may be verified by a hash. It can be incorporated into the algorithms or just interface with them. If a digital signature algorithm is used, it requires that the user know both the public and the private keys. For hash functions, there may be no key. The digital key also verifies that the message or data were not altered during transmission. Digital signature schemes are used to verify a student's identity for access to an academic record, to verify the identity of a credit card user or a banking account owner, to verify the identity of an e-mail user, to verify the company identities in electronic funds transfers, to verify the source of data that is being transferred, and to verify the identity of the user who

FASCINATING FACTS ABOUT CRYPTOLOGY AND CRYPTOGRAPHY

- Codes and ciphers have been used since the beginning of recorded time, when humans drew figures on a cave wall to communicate ideas.
- During the Gallic Wars, Julius Caesar devised one of the first substitution ciphers by printing the ciphertext alphabet with the letters moved four places to the left under the plaintext alphabet.
- In 1795 Thomas Jefferson created the wheel cipher using thirty-six removable and numbered alphabetic disks on a spindle. The sender and receiver must arrange the disks in the same order to communicate a message.
- Hash means to cut up and mix, and hash programs do just that. They break files into pieces and mix them with random numbers.
- The scytale, which was invented by the Spartans during the Greek and Persian wars, was used in the 2006 film the *Da Vinci Code*.
- Random numbers generated by a computer program are not truly random. Computer programs use either a physical process such as clock drift to generate numbers and then try to compensate for any biases in the numbers, or they use a smaller number, or seed, to calculate a random number.

is storing data. Some digital signature algorithms include RSA, DSA, the elliptical curve variant ElGamel, variants of Schnorr and Pointcheval-Stern, Rabin, pairing-based schemes, undeniable signature, and aggregate signature.

IMPACT ON INDUSTRY

As society becomes more dependent on computers, data security by encryption becomes more important. Data need to be available to those who use them to perform their jobs, but they must be safe from hacking and unregistered users. Encryption of computerized data and passwords affects all types of businesses, including health care providers, the government, and schools.

Government. The U.S. government has standards for computer cryptography that are used to encrypt some of its own data. IBM designed Data Encryption Standard (DES) in the 1970's, and it has become the most commonly used secret key algorithm. DES has some shortcomings, but modifications have been made to strengthen it. In 2001, the National Institute of Standards and Technology adopted Advanced Encryption Standard, a more secure cryptosystem, for its applications. The U.S. government has to ensure the security of its own data, such as Internal Revenue records, personnel data, and confidential military data. In addition, it provides computers for its personnel. These require security of work data, messages, passwords, and e-mail. Some government agencies are responsible for gathering intelligence about foreign nations. This may involve attempting to decrypt messages sent by these countries.

Universities and schools. Universities require a secure computer system to store student demographic and health data; student grades and course work; computer access information; personnel information, including salaries; research data; and Internet use. Most students have their own personal computers and are able to access some areas of the university's computer system. Their digital signatures must be linked to a table that reflects their limited access. Faculty members, both on and off campus, are able to access the university's system, and their access also has limits. Researchers use the university system to store data for their projects and to perform analyses. Much of the research on encryption of data has been performed by university researchers. University staff have access to additional data based on their responsibilities. Even elementary and secondary schools have computerized data that include student records and grades, as well as personnel information. Some school systems permit parents to access the grades of their children online. Security is important to protect this data.

Industry and business. All big businesses and most small businesses have computers and use encryption algorithms. Computers have made it possible to store business data in a small space, to aggregate and analyze information, and to protect proprietary data through encryption. Banking has been changed to a primarily electronic business with the use of automated teller machines (ATMs), debit cards, electronic funds transfer, and credit cards. A key or password within the debit and credit cards is verified by a hash function. Access to an ATM is granted by a public key or credit card number, along with the PIN (personal identification number). Some banking transactions can even be done at home on a personal computer. Encryption keeps this information confidential. Health insurance companies record customer transactions on a computer, process claims, list participating health care providers, link employees to their employer's individual health insurance plans, perform aggregate reporting for income taxes (1099s), and report on employee claims. Actuaries use aggregate claims data to determine insurance premiums. All these functions require that only employees access computerized data. Similar types of data use are performed by most businesses.

CAREERS AND COURSE WORK

Most cryptographers have at least a bachelor's degree in computer science, mathematics, or engineering. Often they have either a master's degree or a doctorate. A number of universities and information technology institutes offer nondegree programs in cryptography. Cryptographers are persons who are knowledgeable in encryption programming, computer security software, data compression, and number theory, as well as firewalls and Windows and Internet security. People with this background may be employed in computer software firms, criminology, universities, or information technology.

There is a voluntary accreditation examination for cryptographers, which is administered by the International Information Systems Security Certification Consortium. The examination is called the Certified Information Systems Security Professional (CISSP) examination. To be certified, a cryptographer must have five years of relevant job experience and pass the CISSP examination. Recertification must be done every three years, either by retaking the examination or by earning 120 continuing professional education credits.

Position titles include cryptoanalysts, cryptosystem designers, cryptographic engineers, and digital rights professionals. A cryptoanalyst is involved in the examination and testing of cryptographic security systems. Cryptosystem designers develop the complex cryptographic algorithms. Cryptographic engineers work with the hardware of cryptographic computer systems. Digital rights professionals are responsible for the security of encrypted data, passwords, and digital signatures. They may be certificate administrators and be responsible for accepting new users and approving their digital signatures.

SOCIAL CONTEXT AND FUTURE PROSPECTS

Computer technology is an important part of contemporary life. Encryption of data, messages, passwords and digital signatures makes this possible. Without data and password security, both personal and professional users would find that anything that they loaded onto their computers would be available to hackers and spies, and they would be vulnerable to computer viruses that could damage their computers and corrupt their files. It would be difficult to use a computer under these circumstances because of the lack of dependability of the data and of the computer system. Many workers perform their jobs on a computer, and some are able to work from home. Wireless systems are increasingly being used. These capabilities require adequate security.

Despite the complexity of modern cryptography, there still are risks of attacks on computer data, messages, and passwords. No cryptography program is without its weak point. The likelihood of breaking a cryptographic algorithm is assessed by how long it would take to break the cipher with a high-speed computer. A common attack on key cryptography is brute force. This involves trying all the possible number combinations in order to crack the key. The longer the key is, the longer this will take. The only way to create an unbreakable cipher is to use a one-time pad, in which the secret key is encrypted using an input number that is used only once. Each time the data are accessed, another random number of the same length is used as the secret key.

Cryptosystems are not totally secure. As computer technology and cryptography knowledge advances, any particular encryption algorithm is increasingly likely to be broken. Cryptoanalysts must constantly evaluate and modify or abandon encryption algorithms as needed. A future risk to cryptographic systems is the development of a quantum computer, based on quantum theory. It is thought that a quantum computer could crack all the cryptographic ciphers in existence; however, no quantum computer has been created yet.

—*Christine M. Carroll, RN, MBA*

FURTHER READING

Blackwood, Gary. *Mysterious Messages: A History of Codes and Ciphers.* New York: Penguin Books, 2009. Provides a comprehensive history of the development of cryptography from ancient history through modern times.

Kahate, Atul. *Cryptography and Network Security.* 2d ed. Boston: McGraw-Hill Higher Education, 2009. Describes the concepts involved with the use of cryptography for network security.

Katz, Jonathan, and Yehuda Lindell. *Introduction to Modern Cryptography.* Boca Raton, Fla.: Chapman & Hall/CRC, 2008. Explores both public and private key cryptography. Looks at various models and systems.

Lunde, Paul, ed. *The Book of Codes: Understanding the World of Hidden Messages.* Los Angeles: University of California Press, 2009. Covers codes and ciphers used in society and the history of cryptography. Chapter 13 discusses the development of the computer and computer cryptography.

Seife, Charles. *Decoding the Universe.* New York: Penguin Group, 2006. Discusses information theory, which is the basis of computer cryptography.

WEB SITES
American Cryptogram Association
http://cryptogram.org

International Association for Cryptographic Research
http://www.iacr.org

International Financial Cryptography Association
http://ifca.ai

International Information Systems Security Certification Consortium
https://www.isc2.org

See also: Computer Science; Information Technology.

D

DETERGENTS

FIELDS OF STUDY

Biochemistry; biology; biotechnology; engineering; chemistry; environmental science; chemistry; materials science; pharmacy; physics and kinematics; polymer chemistry; process design; systems engineering; textile design.

SUMMARY

Synthetic detergents enable otherwise immiscible materials (water and oil) to form homogeneous dispersions. In addition to facilitating the breakdown and removal of stains or soil from textiles, hard surfaces, and human skin, detergents have found widespread application in food technology, oil-spill cleanup, and other industrial processes, such as the separation of minerals from their ores, the recovery of oil from natural deposits, and the fabrication of ceramic materials from powders. Detergents are used in the manufacture of thousands of products and have applications in the household, personal care, pharmaceutical, agrochemical, oil and mining, and automotive industries, as well as in the processing of paints, paper coatings, inks, and ceramics.

KEY TERMS AND CONCEPTS

- **critical micelle concentration (CMC):** Surfactant concentration at which appreciable micelle formation occurs, enabling the removal of soils.
- **hydrophilic-lipophilic balance (HLB):** Value expressing hydrophilic (polar/water-loving) and lipophilic (nonpolar/oil-loving) character of surfactants.
- **interface:** Surface that forms the boundary between two bodies of matter (liquid, solid, gas) or the area where two immiscible phases come in contact.
- **micelle:** Aggregate or cluster of individual surfactant molecules formed in solution, whereby polar ends are on the outside (toward the solvent) and nonpolar ends are in the middle.
- **surface tension:** Property of liquids to preferentially contract or expand at the surface, depending on the strength of molecular association and attraction forces.
- **surfactant:** Surface active agent containing both hydrophilic and hydrophobic groups, enabling it to lower surface tension and solubilize or disperse immiscible substances in water.

DEFINITION AND BASIC PRINCIPLES

Detergents comprise a group of synthetic water-soluble or liquid organic preparations that contain a mix of surfactants, builders, boosters, fillers, and other auxiliary constituents, the formulation of which is specially designed to promote cleansing action or detergency. Liquid laundry detergents are by far the largest single product category: Sales in the United States were reported at $3.6 billion for 2010. As of 2011, products are largely mixtures of surfactants, water softeners, optical brighteners and bleach substitutes, stain removers, and enzymes. They must be formulated with ingredients in the right proportion to provide optimum detergency without damaging the fabrics being washed.

Before the advent of synthetic detergents, soaps were used. Soaps are salts of fatty acids and made by alkaline hydrolysis of fats and oils. They consist of a long hydrocarbon chain with the carboxylic acid end group bonded to a metal ion. The hydrocarbon end is soluble in fats and oils, and the ionic end (carboxylic acid salt) is soluble in water. This structure gives soap surface activity, allowing it to emulsify or disperse oils and other water-insoluble substances. Because soaps are alkaline, they react with metal ions in hard water and form insoluble precipitates, decreasing their cleaning effectiveness. These precipitates became known as soap scum—the "gunk" that builds up and surrounds the bathtub and causes graying or yellowing in fabrics.

Three kinds of anionic detergents: a branched sodium dodecylbenzenesulfonate, linear sodium dodecylbenzenesulfonate, and a soap.

Once synthetic detergents were developed, this problem could be avoided. Detergents are structurally similar to soap and work much the same to emulsify oils and hold dirt in suspension, but they differ in their water-soluble portion in that their calcium, magnesium, and iron forms are more water soluble and do not precipitate out. This allowed detergents to work well in hard water, and thus, reduce the discoloration of clothes.

BACKGROUND AND HISTORY

Soap is the oldest cleaning agent and has been in use for 4,500 years. Ancient Egyptians were known to bathe regularly and Israelites had laws governing personal cleanliness. Records document that soaplike material was being manufactured as far back as 1500 BCE. Before soaps and detergents, clothes were cleansed by beating them on rocks by a stream. Plants such as soapwort and soapbark that contained saponins were known to produce soapy lather and probably served as the first crude detergent. By the 1800's, soap making was widespread throughout Europe and North America, and by the 1900's, its manufacture had grown into an industry. The chemistry of soap manufacturing remained primarily the same until 1907, when the first synthetic detergent was developed by Henkel. By the end of World War I, detergents had grown in popularity as people learned that they did not leave soap scum like their earlier counterparts.

The earliest synthetic detergents were short-chain alkyl naphthalene sulfonates. By the 1930's sulfonated straight-chain alcohols and long-chain alkyl and aryl sulfonates with benzene were being made. By the end of World War II, alkyl aryl sulfonates

dominated the detergent market. In 1946, the use of phosphate compounds in combination with surfactants was a breakthrough in product development and spawned the first of the "built" detergents that would prove to be much better-performing products. Sodium tripolyphosphate (STPP) was the main cleaning agent in many detergents and household cleaners for decades. By 1953, U.S. sales of detergents had surpassed those of soap. In the mid-1960's, it was discovered that lakes and streams were being polluted, and the blame was laid on phosphate compounds; however, the actual cause was found to be branching in their molecular structure, which prevented them from being degraded by bacteria. Detergent manufacturers then switched from commonly used compounds such as propylene tetramer benzene sulphonate to a linear alkyl version. Detergent manufacturers are still grappling with sustainability issues and are focusing much attention on developing products that are safe for the environment as well as consumers.

HOW IT WORKS

Just as forces exist between an ordered and disordered universe, so too do they between soil and cleanliness. People have been conditioned to believe that soil on the surface of an object is unwanted matter, but in reality, soil is being deposited continuously on all surfaces around us, and cleanliness itself is an unnatural, albeit desirable, state. In order to rid any surface of soil, one must work against nature and have an understanding of the concept of detergency—the act of cleaning soil from a surface (substrate).

The function of detergency. The cleaning action of detergents is based on their ability to emulsify or disperse different types of soil and hold it in suspension in water. The workhorse involved in this job is the surfactant, a compound used in all soaps and detergents. This ability comes from the surfactant's molecular structure and surface activity. When a soap or detergent product is added to water that contains insoluble materials like dirt, oil, or grease, surfactant molecules adsorb onto the substrate (clothes) and form clusters called micelles, which surround the immiscible droplets. The micelle itself is water soluble and allows the trapped oil droplets to be dispersed throughout the water and rinsed away. While this is a simplified explanation, detergency is a complex set of interrelated functions that relies on the diverse properties of surfactants, their interactions in solution, and their unique ability to disrupt the surface tension of water.

Surface tension. The internal attraction or association of molecules in a liquid is called surface tension. However simple this may seem, it is a complex phenomenon, and for many students can be hard to grasp. Examining the properties of water and the action of surfactants may dispel any confusion.

Water is polar in nature and very strongly associated, such that the surface tension is high. This is because of its nonsymmetrical structure, in which the double-atom oxygen end is more negative than the single-atom hydrogen end is positive. As a result, water molecules associate so strongly that they are relatively stable, with only a slight tendency to ionize or split into oppositely charged particles. This is why their boiling point and heat of vaporization are very high in comparison to their low molecular weight.

The surface tension of water can be explained by how the molecules associate. Water molecules in the liquid state, such as those in the center of a beaker full of water, are very strongly attracted to their neighboring molecules, and the pull is equal in all directions. The molecules on the surface, however, have no neighboring molecules in the air above; hence, they are directed inward and pulled into the bulk of the liquid where the attraction is greater. The result is a force applied across the surface, which contracts as the water seeks the minimum surface area per unit of volume. An illustration of this is the fact that one can spin a pail of water around without spilling the contents.

Surfactant compounds are amphiphilic, meaning their backbone contains at least one hydrophilic group attached to a hydrophobic group (called the hydrophobe), usually consisting of an 8 to 18 carbon-hydrocarbon chain. All surfactants possess the common property of lowering surface tension when added to water in small amounts, at which point the surfactant molecules are loosely integrated into the water structure. As they disperse, the hydrophilic portion of the surfactant causes an increased attraction of water molecules at the surface, leaving fewer sides of the molecule oriented toward the bulk of

the liquid and lessening the forces of attraction that would otherwise pull them into solution.

Micelle formation and critical micelle concentration (CMC). As surface active agents, surfactants not only have the ability to lower surface tension but also to form micelles in solution, unique behavior that is at the core of detergent action. Micelles are aggregate or droplet-like clusters of individual surfactant molecules, whose polar ends are on the outside, oriented toward the solvent (usually water), with the nonpolar ends in the middle. The driving force for micelle formation is the reduction of contact between the hydrocarbon chain and water, thereby reducing the free energy of the system. The micelles are in dynamic equilibrium, but the rate of exchange between surfactant molecule and micelle increase exponentially, depending on the structure of each individual surfactant. As surfactant concentration increases, surface tension decreases rapidly, and micelles proliferate and form larger units.

The concentration at which this phenomenon occurs is known as the critical micelle concentration (CMC). The most common technique for measuring this is to plot surface tension against surfactant concentration and determine the break point, after which surface tension remains virtually constant with additional increases in concentration. The corresponding surfactant concentration at this discontinuity point corresponds to the CMC. Every surfactant has its own characteristic CMC at a given temperature and electrolyte concentration.

APPLICATIONS AND PRODUCTS

The workhorse of detergents is the surfactant or more commonly, a mix of surfactants. The most important categories are the carboxylates (fatty acid soaps), the sulfates (alkyl sulfates, alkyl ether sulfates, and alkyl aryl sulfonates), and the phosphates.

Laundry products. The primary purpose of laundry products is the removal of soil from fabrics. As the cleaning agent, the detergent must fulfill three functions: wet the substrate, remove the soil, and prevent it from redepositing. This usually requires a mix of surfactants. For example, a good wetting agent is not necessarily a good detergent. For best wetting, the surface tension need only be lowered a little, but it must be done quickly. That requires surfactants with short alkyl chain lengths of 8 carbons and surfactants with an HLB of 7 to 9. For best detergency, the surface tension needs to be substantially lowered and that requires surfactants with higher chain lengths of 12 to 14 carbons and an HLB of 13 to 15. To prevent particles from redepositing, the particles must be stabilized in a solution, and that is done best by nonionic surfactants of the polyethylene oxide type. In general, nonionics are not as effective in removing dirt as anionic surfactants, which is one reason why a mixture of anionic and nonionic surfactants is used. However, nonionics are more effective in liquid dirt removal because they lower the oil-water interfacial tension without reducing the oil-substrate tension.

Skin-cleansing bars. As of 2011, the bar soap being manufactured is called superfatted soap and is made by incomplete saponification, an improved process over the traditional method in which superfatting agents are added during saponification, which prevents all of the oil or fat from being processed. Superfatting increases the moisturizing properties and makes the product less irritating. Transparent soaps are like traditional soap bars but have had glycerin added. Glycerin is a humectant (similar to a moisturizer) and makes the bar transparent and much milder.

Syndet bars are made using synthetic surfactants. Since they are not made by saponification, they are actually not soap. Syndet bars are very mild on the skin and provide moisturizing and other benefits. Dove was the first syndet bar on the market.

Mining and mineral processing. Because minerals are rarely found pure in nature, the desired material, called values, needs to be separated from the rocky, unwanted material, called gangue. Detergents are used to extract metals from their ores by a process called froth flotation. The ore is first crushed and then treated with a fatty material (usually an oil) which binds to the particles of the metal (the values), but not to the unwanted gangue. The treated ore is submerged in a water bath containing a detergent and then air is pumped over the sides. The detergent's foaming action produces bubbles, which pick up the water-repellant coated particles or values, letting them float to the top, flow over the sides, and be recovered. The gangue stays in the water bath.

Enhanced oil recovery (EOR). This process refers to the recovery of oil that is left behind after primary and secondary recovery methods have either been exhausted or have become uneconomical. Enhanced oil recovery is the tertiary recovery phase in which

surfactant-polymer (SP) flooding is used. SP flooding is similar to waterflooding, but the water is mixed with a surfactant-polymer compound. The surfactant literally cleans the oil off the rock and the polymer spreads the flow through more of the rock. An additional 15 to 25 percent of original oil in place (OOIP) can be recovered. Before this method is used, there is a great deal of evaluation and laboratory testing involved, but it has become a reliable and cost-effective method of oil recovery.

Ceramic dispersions. Ceramic is a nonmetallic inorganic material. Ceramic dispersions are the starting material for many applications. The use of detergents or surfactants enhances the wetting ability of the binder onto the ceramic particles and aids in the dispersion of ceramic powders in liquids. As dispersants, they reduce bulk viscosity of high-solid slurries and maintain stability in finely divided particle dispersions. Bi-block surfactants help agglomeration of the ceramic particles. In wastewater treatment, detergents are used in ceramic dispersions to reduce the amount of flocculents.

IMPACT ON INDUSTRY

Detergents are big business, not only in the United States, but around the world, and the industry and business sectors catering to it are expansive. These include the laundry and household products market; the industrial and institutional (I&I) market, which makes heavy-duty disinfectants, sanitizers, and cleaners; pharmaceuticals; the oil industry; food service and other service industries, such as hospitals and hospitality (hotels), as well as the personal-care industry.

The global picture. Worldwide, the detergent industry is estimated to be around $65 billion, $52 billion of which is the laundry-care category. Despite global cleaning products being a multibillion-dollar business, the global household products industry has seen slow growth over the last few years. The global I&I market is reported to be $30 billion.

Worldwide, the most growth for the detergent, household, and I&I sectors is expected to be in emerging markets, with China, India, and Brazil leading the pack. The three countries have been in the midst of a construction boom. China, for instance, has a new alkyl polyglucoside plant that is in full production. In contrast to the domestic market's slow-growth mode, experts say companies in these countries are seeing double-digit sales figures, while sales in other countries outside U.S. borders are rising by 5 to 10 percent. The consensus is that growth in the United States, Canada, Japan, and Western Europe will be slow compared with world averages. Some analysts predict that the growth in emerging markets will be explosive.

The local picture. The domestic market for detergents is not so rosy, say analysts. Market data for 2010 was mixed, as many companies are still feeling the pinch from the 2007-2009 recession. U.S. sales of liquid laundry detergents in food, drug, and mass merchant (FDMx) outlets (excluding Walmart) fell to $3.6 billion, a 3.14 percent drop from the previous year. However, growth in I&I as a whole rose 1.7 percent, amounting to about $11 billion. Projection data indicate that demand for disinfectants and sanitizers through 2014 will increase by more than 6 percent, outpacing the industry's overall growth for the same period. In the same vein, the personal-care sector saw FDMx sales of personal cleansers rise 22.22 percent,

an impressive jump to nearly $143 million. A 2010 observational study sponsored by the American Society for Microbiology and the American Cleaning Institute (formerly the Soap and Detergent Association), indicates that 85 percent of adults now wash their hands in public restrooms— the highest percentage observed since studies began in 1996. FDMx soap sales reached $2.08 billion in 2010, which represents a 5 percent increase over the previous year.

The supplier side. Surfactants are the primary ingredient in detergents and are made from two different types of raw materials. Those derived from agricultural feedstocks are called oleochemicals, and those derived from petroleum (crude oil), synthetic surfactants. The global market for surfactants by volume size is more than 18 million tons per year, with 80 percent of the demand represented by just ten different types of compounds. The global share of surfactant raw materials represents only about 0.1 percent of crude-oil consumption.

Major manufacturers. The top laundry detergent manufacturers in the United States are Procter & Gamble (P&G), Henkel, and Unilever, until 2008, when it sold its North American detergent business to Vestar Capital Partners, saying it could not compete with P&G. The former All, Wisk, Sunlight, Surf, and Snuggle brands are now part of a new company, the Sun Products Corporation. Vestar said Sun Products will have annual sales of more than $2 billion. Unilever NV of the Netherlands is still the largest detergent manufacturer in the United Kingdom.

P&G's detergent sales totaled $79 billion in 2010 and represent 70 percent of the company's annual sales. Tide continues to dominate the laundry detergent market, commanding more than 40 percent of sales and is the standout product of P&G's $1 Billion Club. Purex Complete was a big growth product for Henkel in the United States, but not in Europe. Besides the Purex brands, Henkel also markets Dial soaps, a brand it acquired in 2004. Henkel's worldwide sales grew 11 percent to $20.6 billion in 2010, but had stagnant sales in its U.S. division.

In the liquid laundry detergent category, the top five products by sales volume for the year ending October 31, 2010, were Tide (P&G), All (Sun Products), Arm & Hammer (Church & Dwight), Gain (P&G), and Purex (Henkel). Although ranked number eight, Cheer Brightclean (P&G) had the biggest jump in sales, reported at 41 percent. Reacting

---

FASCINATING FACTS ABOUT DETERGENTS

- According to biblical accounts, the ancient Israelites created a hair gel by mixing ashes and oil.

- Persil was the name of the first detergent made in 1907 by Germany's Henkel.

- Soap scum became immortalized with the ad campaign that coined the terms "ring around the tub" and "ring around the collar."

- Automatic dishwasher powders and liquid fabric softeners were invented in the 1950's, laundry powders with enzymes in the 1960's, and fabric softener sheets in the 1970's.

- Dawn dishwashing liquid was used to clean wildlife during the Exxon Valdez and BP oil spills in 1989 and 2010, respectively.

- The human lung contains a surface-active material called pulmonary surfactant that helps prevent the lung from collapsing after expiration.

- Enzymes used in detergents work much like they do in the body, since each has a personalized target soil it breaks down.

to the recession, P&G reduced the price of its line of laundry detergents by about 4.5 percent during 2010.

In the United States, Ecolab dominates the I&I sector. With a market share of more than 30 percent, it easily outdoes the number-two player, Diversey. Spartan Chemical is another big contender. Dial (Henkel), Purell (Gojo), and Gold Bond are top-selling brands in the personal cleanser and sanitizer market. Overall, in the personal-care market, private-label brands are by far the leader at mass merchandisers, with sales of $73.9 million in 2009.

CAREERS AND COURSE WORK

Science courses in the organic, inorganic, and physical chemistries, biology, and biochemistry, plus courses in calculus, physics, materials science, polymer technology, and analytical chemistry are typical requirements for students interested in pursuing careers in detergents. Other pertinent courses may include differential equations, instrumental analysis, statistics, thermodynamics, fluid mechanics, process design, quantitative analysis, and instrumental methods. Earning a bachelor of science degree in chemistry or chemical engineering is usually sufficient for entering the field or doing graduate work in a related area.

There are few degree programs in formulation chemistry, but students need to understand the chemistry involved, such as thermodynamics of mixing, phase equilibria, solutions, surface chemistry, colloids, emulsions, and suspensions. Even more important is how their dynamics affect such properties as adhesion, weather resistance, texture, shelf life, biodegradability, and allergenic response.

Students in undergraduate chemistry programs are encouraged to select a specialized degree track but are also being advised to take substantial course work in more than one of the primary fields of study related to detergent chemistry because product development requires skills drawn from multiple disciplines. Often a master's degree or doctorate is necessary for research and development.

There are a number of career paths for students interested in the detergent industry. Manufacturers of laundry detergents, household cleaning products, industrial and institutional (I&I) products, and personal-care products, as well as the ingredients suppliers of all these products, are the biggest employers of formulation chemists, technicians, and chemical operators. There are also career opportunities in research and development, marketing, or sales. Other areas where detergent chemists and technicians are needed include food technology, pharmaceuticals, oil drilling and recovery, mining and mineral processing, and ceramic powder production.

SOCIAL CONTEXT AND FUTURE PROSPECTS

Around the world, sustainability and environmental protection remain the buzzwords for detergents in the twenty-first century. The industry is very dependent on the price and availability of fats and oils, since these materials are needed to make the fatty acids and alcohols used in the manufacture of surfactants. The turmoil in the oil industry is not only impacting how much is paid at the gas pump but is also directly related to the price paid to keep the environment clean.

While the detergents industry, on the whole, has traditionally been relatively recession-proof, the economic slowdown and conflicts in the Middle East have taken a toll on this huge market. Robert McDonald, chairman of P&G, alluded to changes in consumer habits, volatility in commodity costs, and increasing complexity of the regulatory environment as rationales for the slowdown. A spokesperson for Kline and Company, a market research firm reporting on the industry, stated: "The industry's mantra is greener, cleaner, safer," but goes on to say that the challenge is a double-edge sword, as the industry is battling to hold down costs while trying to produce environmentally sustainable products.

Sustainability and environmental protection are global issues, but insights are becoming more astute and action is being taken. Henkel published its twentieth sustainability report in conjunction with its annual report for 2010, saying the company met its targets for 2012 early, namely to commit to principles and objectives relating to occupational health and safety, resource conservation, and emissions reduction. Important emerging markets such as China are no exception in this battle. The Chinese government puts great emphasis on environmental regulations. Analysts say this is a clear message that to expand their sales to other parts of the world, China needs to hone in on the green demand and manufacture products that offer innovative and sustainable solutions without compromising on performance.

—*Barbara Woldin*

FURTHER READING

Carter, C. Barry, and M. Grant Norton. *Ceramic Materials: Science and Engineering*. New York: Springer, 2007. Covers ceramic science, defects, and the mechanical properties of ceramic materials and how these materials are processed. Provides many examples and illustrations relating theory to practical applications; suitable for advanced undergraduate and graduate study.

Myers, Drew. *Surfactant Science and Technology*. 3d ed. Hoboken, N.J.: John Wiley & Sons, 2006. Written with the beginner in mind, this text clearly illustrates the basic concepts of surfactant action and application.

Rosen, Milton J. *Surfactants and Interfacial Phenomena*. 3d ed. Hoboken, N.J.: Wiley-Interscience, 2004. Easy-to-understand text on properties and applications of surfactants; covers many topics, including dynamic surface tension and other interfacial processes.

Tadros, Tharwat F. *Applied Surfactants: Principles and Applications*. Weinheim, Germany: Wiley-VCH Verlag, 2005. Author covers a wide range of topics on the preparation and stabilization of emulsion systems and highlights the importance of emulsion science in many modern-day industrial applications; discusses physical chemistry of emulsion systems, adsorption of surfactants at liquid/liquid interfaces, emulsifier selection, polymeric surfactants, and more.

Zoller, Uri, and Paul Sosis, eds. *Handbook of Detergents, Part F: Production*. Boca Raton, Fla.: CRC Press, 2009. One of seven in the Surfactant Science Detergents Series, the book discusses state of the art in the industrial production of the main players in detergent formulation—surfactants, builders, auxiliaries, bleaching ingredients, chelating agents, and enzymes.

WEB SITES

American Chemical Society
http://portal.acs.org/portal/acs/corg/content

American Chemistry Council
http://www.americanchemistry.com

American Oil Chemists' Society
http://www.aocs.org

Household and Personal Products Industry
http://www.happi.com

See also: Environmental Chemistry.

DIGITAL LOGIC

FIELDS OF STUDY

Mathematics; electronics; physics; analytical chemistry; chemical engineering; mechanical engineering; computer science; biomedical technology; cryptography; communications; information systems technology; integrated-circuit design; nanotechnology; electronic materials

SUMMARY

Digital logic is electronic technology constructed using the discrete mathematical principles of Boolean algebra, which is based on binary calculation, or the "base 2" counting system. The underlying principle is the relationship between two opposite states, represented by the numerals 0 and 1. The various combinations of inputs utilizing these states in integrated circuits permit the construction and operation of many devices, from simple on-off switches to the most advanced computers.

KEY TERMS AND CONCEPTS

- **Boolean algebra:** A branch of mathematics used to represent basic logic statements.
- **clock speed:** A specific frequency that controls the rate at which data bits are changed in a digital logic circuit.
- **gate:** A transistor assembly that combines input signals according to Boolean logic to produce a specific output signal.
- **interleaf:** To incorporate sections of different processes within the same body in such a way that they do not interfere with each other.
- **Karnaugh map:** A tabular representation of the

possible states resulting from combinations of a specific number of binary inputs.

- **sample:** To take discrete measurements of a specific quantity or property at a specified rate.

DEFINITION AND BASIC PRINCIPLES

Digital logic is built upon the result of combining two signals that can have either the same or opposite states, according to the principles of Boolean algebra. The mathematical logic is based on binary calculation, or the base 2 counting system. The underlying principle is the relationship between two opposite states, represented by the numerals 0 and 1.

The states are defined in modern electronic devices as the presence or absence of an electrical signal, such as a voltage or a current. In modern computers and other devices, digital logic is used to control the flow of electrical current in an assembly of transistor structures called gates. These gates accept the input signals and transform them into an output signal. An inverter transforms the input signal into an output signal of exactly opposite value.

An AND gate transforms two or more input signals to produce a corresponding output signal only when all input signals are present. An OR gate transforms two or more input signals to produce an output signal if any of the input signals are present. Combinations of these three basic gate structures in integrated circuits are used to construct NAND (or not-AND) and NOR (or not-OR) gates, accumulators, flip-flops, and numerous other digital devices that make up the functioning structures of integrated circuits and computer chips.

BACKGROUND AND HISTORY

Boolean algebra is named for George Boole (1815-1864), a self-taught English scientist. This form of algebra was developed from Boole's desire to express concrete logic in mathematical terms; it is based entirely on the concepts of true and false. The intrinsically opposite nature of these concepts allows the logic to be applied to any pair of conditions that are related as opposites. The modern idea of computing engines began with the work of Charles Babbage (1791-1871), who envisioned a mechanical "difference engine" that would calculate results from starting values. Babbage did not see his idea materialize, though others using his ideas were able to construct mechanical difference engines.

The development of the semiconductor junction transistor in 1947, attributed to William Shockley, John Bardeen, and Walter Brattain, provided the means to produce extremely small on-off switches that could be used to build complex Boolean circuits. This permitted electrical signals to carry out Boolean algebraic calculations and marked the beginning of what has come to be known as the digital revolution. These circuits helped produce the many modern-day devices that employ digital technology.

HOW IT WORKS

Boolean algebra. The principles of Boolean algebra apply to the combination of input signals rather than to the input signals themselves. If one associates one line of a conducting circuit with each digit in a binary number, it becomes easy to see how the presence or absence of a signal in that line can be combined to produce cumulative results. The series of signals in a set of lines provides ever larger numerical representations, according to the number of lines in the series. Because the representation is binary, each additional line in the series doubles the amount of information that can be carried in the series.

Bits and bytes. Digital logic circuits are controlled by a clock signal that turns on and off at a specific frequency. A computer operating with a CPU (central processing unit) speed of 1 gigahertz (10^9 cycles per second) is using a clock control that turns on and off 1 billion times each second. Each clock cycle transmits a new set of signals to the CPU in accord with the digital logic circuitry. Each individual signal is called a bit (plural byte) of data, and a series of 8 bits is termed one byte of data. CPUs operating on a 16bit system pass two bytes of data with each cycle, 32bit systems pass four bytes, 64-bit systems pass eight bytes, and 128-bit systems pass sixteen bytes with each clock cycle.

Because the system is binary, each bit represents two different states (system high or system low). Thus, two bits can represent four (or 2^2) different states, three bits represents eight (or 2^3) different states, four bits represents sixteen (or 2^4) different states, and so on. A 128-bit system can therefore represent 2^{128} or more than 3.40×10^{38} different system states.

Digital devices. All digital devices are constructed from semiconductor junction transistor circuits. This technology has progressed from individual transistors to the present technology in which millions of transistors can be etched onto a small silicon chip. Digital electronic circuits are produced in "packages" called integrated circuit, or IC, chips. The simplest digital logic device is the inverter, which converts an input signal to an output signal of the opposite value. The AND gate accepts two or more input signals such that the output signal will be high only if all of the input signals are high. The OR gate produces an output high signal if any one or the other of the input signals is high.

All other digital logic devices are constructed from these basic components. They include NAND gates, NOR gates, X-OR gates, flip-flops that produce two simultaneous outputs of opposite value, and counters and shift registers, which are constructed from series of flip-flops. Combinations of these devices are used to assemble accumulators, adders, and other components of digital logic circuits. One other important set of digital devices is the converters that convert a digital or analogue input signal to an analogue or digital output signal, respectively. These find extensive use in equipment that relies on analogue, electromagnetic, signal processing.

Karnaugh maps. Karnaugh maps are essential tools in designing and constructing digital logic circuits. A Karnaugh map is a tabular representation of all possible states of the system according to Boolean algebra, given the desired operating characteristics of the system. By using a Karnaugh map to identify the allowed system states, the circuit designer can select the proper combination of logic gates that will then produce those desired output states.

APPLICATIONS AND PRODUCTS

Digital logic has become the standard structural operating principle of most modern electronic devices, from the cheapest wristwatch to the most advanced supercomputer. Applications can be identified as programmable and nonprogrammable.

Nonprogrammable applications are those in which the device is designed to carry out a specific set of operations as automated processes. Common examples include timepieces, CD and DVD players, cellular telephones, and various household appliances. Programmable applications are those in which an operator can alter existing instruction sets or provide new ones for the particular device to carry out. Typical examples include programmable logic controllers, computers, and other hybrid devices into

which some degree of programmability has been incorporated, such as gaming consoles, GPS (global positioning system) devices, and even some modern automobiles.

Digital logic is utilized for several reasons. First, the technology provides precise control over the processes to which it is applied. Digital logic circuits function on a very precise clock frequency and with a rigorously defined data set in which each individual bit of information represents a different system state that can be precisely defined millions of times per second, depending on the clock speed of the system. Second, compared with their analogue counterparts, which are constructed of physical switches and relays, digital circuits require a much lower amount of energy to function. A third reason is the reduced costs of materials and components in digital technology. In a typical household appliance, all of the individual switches and additional wiring that would be required of an analogue device are replaced by a single small printed circuit containing a small number of IC chips, typically connected to a touchpad and LCD (liquid crystal display) screen.

Programmable logic controller. One of the most essential devices associated with modern production methods is the programmable logic controller, or PLC. These devices contain the instruction set for the automated operation of various machines, such as CNC (computer numerical control) lathes and milling machines, and all industrial robotics. In operation, the PLC replaces a human machinist or operator, eliminating the effects of human error and Digital logic fatigue that result in undesirable output variability. As industrial technology has developed, the precision with which automated machinery can meet demand has far exceeded the ability of human operators.

PLCs, first specified by General Motors Corporation (now Company) in 1968, are small computer systems programmed using a reduced-instruction-set programming language. The languages often use a ladder-like structure in which specific modules of instructions are stacked into memory. Each module consists of the instructions for the performance of a specific machine function. More recent developments of PLC systems utilize the same processors and digital logic peripherals as personal computers, and they can be programmed using advanced computer programming languages.

Digital communications. Digital signal processing is essential to the function of digital communications. As telecommunication devices work through the use of various wavelengths of the electromagnetic spectrum, the process is analogue in nature. Transmission of an analogue signal requires the continuous, uninterrupted occupation of the specific carrier frequency by that signal, for which analogue radio and television transmission frequencies are strictly regulated.

A digital transmission, however, is not continuous, being transmitted as discrete bits or packets of bits of data rather than as a continuous signal. When the signal is received, the bits are reassembled for audio or visual display. Encryption codes can be included in the data structure so that multiple signals can utilize the same frequency simultaneously without interfering with each other. Data reconstruction occurs at a rate that exceeds human perception so that the displayed signal is perceived as a continuous image or sound. The ability to interleaf signals in this way increases both the amount of data that can be transmitted in a limited frequency range and the efficiency with which the data is transmitted.

A longstanding application of digital logic in telecommunications is the conversion of analogue source signals into digital signals for transmission, and then the conversion of the digital signal back into an analogue signal. This is the function of digital-to-analogue converters (DACs), and analogue-to-digital converters (ACDs). A DAC uses sampling to measure the magnitude of the analogue signal, perhaps many millions of times per second depending upon the clock speed of the system. The greater the sampling rate, the closer its digital representation will be to the real nature of the analogue signal. The ACD accepts the digital representation and uses it to reconstruct the analogue signal as its output.

One important problem that exists with this method, however, is what is called aliasing, in which the DAC analogue output correctly matches the digital representation, but at the wrong frequencies. Present-day telecommunications technology is eliminating these steps by switching to an all-digital format that does not use analogue signals.

Servomechanisms. Automated processes controlled by digital logic require other devices by which the function of the machine can be measured. Typically, a measurement of some output property is automatically fed back into the controlling system and used to adjust functions so that desired output parameters are maintained. The adjustment is carried out through the action of a servomechanism, a mechanical device that performs a specific action in the operation of the machine. Positions and rotational speeds are the principal properties used to gauge machine function.

In both cases, it is common to link the output property to a digital representation such as Gray code, which is then interpreted by the logic controller of the machine. The specific code value read precisely describes either the position or the rotational speed of the component, and is compared by the controller to the parameters specified in its operating program. Any variance can then be corrected and proper operation maintained.

IMPACT ON INDUSTRY

Digital logic devices in industry, often referred to as embedded technology, represent an incalculable value, particularly in industrial production and fabrication, where automated processes are now the rule rather than the exception.

Machine shops once required the services of highly skilled machinists to control the operation of machinery for the production of quality parts. Typically, each machine required one master machinist per working shift, and while the work pieces produced were generally of excellent quality, dimensional tolerance was variable. In addition, human error and inconsistency caused by fatigue and other factors resulted in significant waste, especially given that master machinists represented the high end of the wage scale for skilled tradespersons.

Compare this earlier shop scenario to a modern machine shop, in which a single millwright ensures the proper operating condition of several machines, each of which is operated by a digital logic controller. The machines are monitored by unskilled laborers who must ensure that raw parts are supplied to the machine as needed and who must report any discrepancies to the millwright for needed maintenance. The actual function of the machine is automated through a PLC or similar device, providing output products that are dimensionally consistent, with minimal waste, and are produced at the same rate throughout the entire operating period.

It is easy to realize the economic value to industry

of even just this one example of change in production methods. Given that almost all industrial production methods that can be automated with digital logic have been, the effects on productivity and profitability are fairly obvious. Automation, however, reduces the requirement for skilled tradespersons. Machine operators now represent a much lower position on the wage scale than did their skilled predecessors. The reduced production costs in this area have led to the relocation of many company facilities to parts of the world where daily wages are typically much lower and willing workers are plentiful.

PLCs and other digital logic controls automate almost every industrial process. In many cases the function is entirely automatic, not even requiring the presence of a human overseer. Sophisticated machine design in these cases replaces each action that a human would otherwise be required to carry out. In other cases, human intervention is required, as some aspect of the process cannot be left to automatic control. For example, die-casting of parts from magnesium alloys involves a level of art over science that calls for continuous human attention to produce an acceptable cast and prevent a piece from jamming in the die. Between these two more extreme situations are those common ones in which digital logic replaces the need for an operator to carry out several minor actions with the press of a single button. Once that button is pressed, the operation proceeds automatically, and the operator can only stand and watch while preparing for the next cycle of the machinery.

CAREERS AND COURSEWORK

An understanding of digital logic and its applications is obtained through courses of study in electronics engineering technology. This will necessarily include the study of mathematics and physics as foundation courses. Because digital logic is used to control physical devices, students can also expect to study mechanics and mechanical engineering, control systems and feedback, and perhaps hydraulic and pneumatic systems technology.

The development of new electronic materials and production methods will draw students to the study of materials science and chemistry, while others will specialize in circuit design and layout. Graphic methods are extremely important in this area, and specialized design programs are essential components of careers involving logic circuit design.

FASCINATING FACTS ABOUT DIGITAL LOGIC

- Graphene, a one-atom-thick form of carbon that may allow transistors the size of single molecules, is one of the most exotic materials ever discovered, even though people have unknowingly been writing and drawing with it for centuries.
- The first functional transistor used two pieces of gold leaf that had been split in half with a razor blade.
- The term "transistor" was devised to indicate a resistor that could transfer electrical signals.
- Digital logic devices have automated nearly all modern production processes.
- An X-OR (Exclusive-OR) gate outputs a positive signal when just one or the other of its input signals is positive.
- Quantum dots can be thought of as artificial atoms containing a single electron with a precisely defined energy.
- Moore's law, stated by Gordon Moore of Intel Corporation in a 1965 paper, predicts that the number of transistors that can be put on a computer chip doubles about every eighteen months.
- The most sophisticated of digital logic devices, the CPU chips of advanced computers, are constructed from only three basic transistor gate designs, the NOT, AND, and OR gates. All other logic gates are constructed from these three.

Introductory and college-level courses will focus on providing a comprehensive understanding of logic gates and their functions in the design of relatively simple logic circuits. Special attention may be given to machine language programming, because this is the level of computer programming that works directly with the logical hardware. Typically, study progresses in these two areas, as students will design and build more complex logic circuits with the goal of interfacing a functional device to a controlling computer.

Postgraduate electronics engineering programs build on the foundation material covered in college-level programs and are highly specialized fields of study. This level will take the student into quantum mechanics and nanotechnology, as research continues to extend the capabilities of digital logic in the form of computer chips and other integrated circuits. This involves the development and study of new materials and methods, such as graphenes and fullerenes, superconducting organic polymers, and quantum dots, with exotic properties.

SOCIAL CONTEXT AND FUTURE PROSPECTS

At the heart of every consumer electronic device is an embedded digital logic device. The transistor quickly became the single most important feature of electronic technology, and it has facilitated the rapid development of everything from the transistor radio to space travel. Digital logic, as embedded devices, is becoming an ever more pervasive feature of modern technology; an entire generation has now grown up not knowing anything but digital computers and technology. Even the accoutrements of this generation are rapidly being displaced by newer versions, as tablet computers and smartphones displace more traditional desktop personal computers, laptops, and cell phones. The telecommunications industry in North America is in the process of making a government-regulated switch-over to digital format, making analogue transmissions a relic of the past.

Research to produce new materials for digital logic circuits and practical quantum computers is ongoing. The eventual successful result of these efforts, especially in conjunction with the development of nanotechnology, will represent an unparalleled advance in technology that may usher in an entirely new age for human society.

—*Richard M. Renneboog, MSc*

FURTHER READING

Brown, Julian. *Minds, Machines, and the Multiverse: The Quest for the Quantum Computer.* New York: Simon & Schuster, 2000. Discusses much of the history of digital logic systems, in the context of ongoing research toward the development of a true quantum computer.

Bryan, L. A., and E. A. Bryan. *Programmable Controllers: Theory and Application.* Atlanta: Industrial Text Company, 1988. Provides a detailed review of basic digital logic theory and circuits, and discusses the theory and applications of PLCs.

Holdsworth, Brian, and R. Clive Woods. *Digital Logic Design.* 4th ed. Woburn, Mass.: Newnes/Elsevier Science, 2002. A complete text providing a thorough treatment of the principles of digital logic and digital logic devices.

Jonscher, Charles. *Wired Life: Who Are We in the Digital Age?* London: Bantam Press, 1999. Discusses the real and imagined effects that the digital revolution will have on society in the twenty-first century.

Marks, Myles H. *Basic Integrated Circuits.* Blue Ridge Summit, Pa.: TAB Books, 1986. A very readable book that provides a concise overview of digital logic gates and circuits, then discusses their application in the construction and use of integrated circuits.

Miczo, Alexander. *Digital Logic Testing and Simulation.* 2d ed. Hoboken, N.J.: John Wiley & Sons, 2003. An advanced text that reviews basic principles of digital logic circuits, then guides the reader through various methods of testing and fault simulations.

WEB SITES

ASIC World
http://www.asic-world.com/

NobelPrize.org
http://www.nobelprize.org

See also: Algebra; Applied Mathematics; Applied Physics; Automated Processes And Servomechanisms; Communication; Computer Engineering; Computer Science; Electronics and Electronic Engineering; Engineering Mathematics; Information Technology; Integrated-Circuit Design; Liquid Crystal Technology; Mechanical Engineering; Nanotechnology; Transistor Technologies.

DIODE TECHNOLOGY

FIELDS OF STUDY

Electrical engineering; materials science; semiconductor technology; semiconductor manufacturing; electronics; physics; chemistry; nanotechnology; mathematics.

SUMMARY

Diodes act as one-way valves in electrical circuits, permitting electrical current to flow in only one direction and blocking current flow in the opposite direction. The original diodes used in circuits were constructed using vacuum tubes, but these diodes have been almost completely replaced by semiconductor-based diodes. Solid-state diodes, the most commonly used, are perhaps the simplest and most fundamental solid-state semiconductor devices, formed by joining two different types of semiconductors. Diodes have many applications, such as safety circuits to prevent damage by inadvertently putting batteries backward into devices and in rectifier circuits to produce direct current (DC) voltage output from an alternating current (AC) input.

KEY TERMS AND CONCEPTS

- **anode:** More positive side of the diode, through which current can flow into a diode while forward biased.
- **cathode:** More negative side of the diode, from which current flows out of the diode while forward biased.
- **forward bias:** Orientation of the diode in which current most easily flows through the diode.
- **knee Voltage:** Minimum forward bias voltage required for current flow, also sometimes called the threshold voltage, cut-in voltage, or forward voltage drop.
- **light-emitting diode (LED):** Diode that emits light when current passes through it in forward bias configuration.
- **p-n junction:** Junction between positive type (p-type) and negative type (n-type) doped semiconductors.
- **power dissipation:** Amount of energy per unit time dissipated in the diode.
- **rectifiers:** Diode used to make alternating current, or current flowing in two directions, into direct current, or current flowing in only one direction.
- **reverse bias:** Orientation of the diode in which current does not easily flow through the diode.
- **reverse breakdown voltage:** Maximum reverse biased voltage that can be applied to the diode without it conducting electrical current, sometimes just called the breakdown voltage.
- **thermionic diode:** Vacuum tube that permits current to flow in only one direction, also called a vacuum tube diode.
- **Zener diode:** Diode designed to operate in reverse bias mode, conducting current at a controlled breakdown voltage called the Zener voltage.

DEFINITION AND BASIC PRINCIPLES

A diode is perhaps the first semiconductor circuit element that a student learns about in electronics courses, though most early diodes were constructed using vacuum tubes. It is very simplistic in structure, and basic diodes are very simple to connect in circuits. They have only two terminals, a cathode and an anode. The very name diode was created by British physicist William Henry Eccles in 1919 to describe the circuit element as having only the two terminals, one in and one out.

Classic diode behavior, that for which most diodes are used, is to permit electric current to flow in only one direction. If voltage is applied in one direction across the diode, then current flows. This is called forward bias. The terminal on the diode into which the current flows is called the anode, and the terminal out of which current flows is called the cathode. However, if voltage is applied in the opposite direction, called reversed bias, then the diode prevents current flow. A theoretical ideal diode permits current to flow without loss in forward bias orientation for any voltage and prohibits current flow in reverse bias orientation for any voltage. Real diodes require a very small forward bias voltage in order for current to flow, called the knee voltage. The terms threshold voltage or cut-in voltage are also sometimes used in place of the term knee voltage. The electronic symbol for the diode signifies the classic diode behavior, with an arrow pointing in the direction of permitted current flow, and a bar on the other side of the diode signifying a block to current flow from the other direction.

Though most diodes are used to control the direction of current flow, there are many subtypes of diodes that have been developed with other useful properties, such as light-emitting diodes and even diodes designed to operate in reverse bias mode to provide a regulated voltage.

BACKGROUND AND HISTORY

Diode-like behavior was first observed in the nineteenth century. Working independently of each other in the 1870's, American inventors Thomas Alva Edison and Frederick Guthrie discovered that heating a negatively charged electrode in a vacuum permits current to flow through the vacuum but that heating a positively charged electrode does not produce the same behavior. Such behavior was only a scientific curiosity at the time, since there was no practical use for such a device.

At about the same time, German physicist Karl Ferdinand Braun discovered that certain naturally occurring electrically conducting crystals would conduct electricity in only one direction if they were connected to an electrical circuit by a tiny electrode connected to the crystal in just the right spot. By 1903, American electrical engineer Greenleaf Whittier Pickard had developed a method of detecting radio signals using the one-way crystals. By the middle of the twentieth century, homemade radio receivers using galena crystals had become quite popular among hobbyists.

As the electronics and the radio communication industries developed, it became apparent that there would be a need for human-made diodes to replace the natural crystals that were used in a trial-and-error manner. Two development paths were followed: solid-state diodes and vacuum tube diodes. By the middle of the twentieth century, inexpensive germanium-based diodes had been developed as solid-state devices. The problem with solid-state diodes was that they lacked the ability to handle large currents, so for high-current applications, vacuum tube diodes, or thermionic diodes, were developed. In the twenty-first century, most diodes are semiconductor devices, with thermionic diodes existing only for the rare very high-power applications.

HOW IT WORKS

Thermionic diodes. Though not used as frequently as they once were, thermionic diodes are the simplest type of diode to understand. Two electrodes are enclosed in an evacuated glass tube. Because the thermionic diode is a type of vacuum tube, it is often called a vacuum tube diode. The geometry of the electrodes in the tube depends on the manufacturer and the intended use of the tube. Heating one of the electrodes in some fashion permits electrons on that electrode to be thermally excited. If the electrode is heated past the work function of the material of which the electrode is fabricated, the electrons can come free of the electrode. If the heated electrode has a more negative voltage than the other electrode, then the electrons cross the space between the electrodes. More electrons flow into the negative electrode to replace the missing ones, and the electrons flow out of the positive electrode. Current flow is defined opposite to electron flow, so current would be defined as flowing into the positive electrode (labeled as the anode) and out of the negative electrode (labeled as the cathode). However, if the voltage is reversed, and the heated electrode is more positive than the other electrode, electrons liberated from the anode do not flow to the cathode, so no current flows, making the diode a one-way device for current flow.

Solid-state diodes. Thermionic diodes, or vacuum tube diodes, tend to be large and consume a lot of electricity. However, paralleling the development of vacuum tube diodes was the development of diodes based on the crystal structure of solids. The most important type of solid-state diodes are based on semiconductor technology.

Semiconductors are neither good conductors nor good insulators. The purity of the semiconductor determines, in part, its electrical properties. Extremely pure semiconductors tend to be poor conductors. However, all semiconductors have some impurities in them, and some of those impurities tend to improve conductivity of the semiconductor. Purposely adding impurities of the proper type and concentration into the semiconductor during the manufacturing process is called doping the semiconductor. If the impurity has one more outer shell electron than the number of electrons in atoms of the semiconductor, then extra electrons are available to move and conduct electricity. This is called a negative doped or n-type semiconductor. If the impurity has one fewer electrons than the atoms of the semiconductor, then electrons can move from one atom to another in the semiconductor. This acts as a positive charge moving in the semiconductor, though it is really a missing electron moving from atom to atom. Electrical engineers refer to this as a hole moving in the semiconductor. Semiconductors with this type of impurity are called positive doped or p-type semiconductor.

What makes a semiconductor diode is fabricating a device in which a p-type semiconductor is in contact with an n-type semiconductor. This is called a p-n junction. At the junction, the electrons from the n-type region combine with the holes of the p-type region, resulting in a depletion of charge carriers in the vicinity of the p-n junction. However, if a small positive voltage is applied across the junction, with the p-type region having the higher voltage, then additional electrons are pulled from the n-type region and additional holes are pulled from the p-type region into the depletion region, with electrons flowing into the n-type region from outside the device to make up the difference and out of the device from the p-type region to produce more holes. As with the thermionic device, current flows through the device, with the p-type side of the device being the anode and the n-type side of the device being the cathode. This is the forward bias orientation. When the voltage is reversed on the device, the depletion region simply grows larger and no current flows, so the device acts as a one-way valve for the flow of electricity. This is the reverse bias orientation. Though reverse bias diodes do not normally conduct electricity, a sufficiently high reverse voltage can create electric fields within the diode capable of moving charges through the depletion region and creating a large current through the diode. Because diodes act much like resistors in reverse bias mode, such a large current through the diode can damage or destroy the diode. However, two types of diodes, avalanche diodes and Zener diodes, are designed to be safely operated in reverse bias mode.

APPLICATIONS AND PRODUCTS
P-n junction devices, such as diodes, have a plethora of uses in modern technology.

Rectifiers. The classic application for a diode was to act as a one-way valve for electric current. Such a property makes diodes ideal for use in converting alternating current into direct-current circuits or circuits in which the current flows in only one direction. In fact, the devices were originally called rectifiers before the term diode was created to describe the function of these one-way current devices. Modern rectifier circuits consist of more than just a single diode, but they still rely heavily on diode properties.

Solid-state diodes, like most electronic components, are not 100 percent efficient, and so some energy is lost in their operation. This energy is typically dissipated in the diode as heat. However, semiconductor devices are designed to operate at only certain temperatures, and increasing the temperature

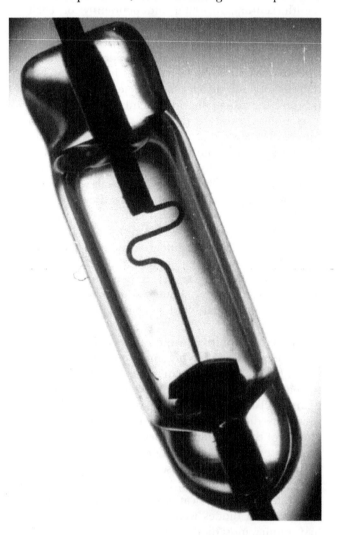

Extreme macro photo of a Chinese diode of the seventies.

beyond a specified range changes the electrical properties of the device. The more current that passed through the device, the hotter it gets. Thus, there is a limiting current that a solid-state diode can handle before it is damaged. Though solid-state diodes have been developed to handle higher currents, for the highest current and power situations, thermionic, or vacuum tube diodes, are still sometimes used, particularly in radio and television broadcasting.

Shottky diodes. All diodes require at least a small forward bias voltage in order to work. Shottky diodes are fabricated by using a metal-to-semiconductor junction rather than the traditional dual semiconductor p-n junction used with other diodes. Such a construction allows Shottky diodes to operate with extremely low forward bias.

Zener diodes. Though most diodes are designed to operate only in the forward bias orientation, Zener diodes are designed to operate in reverse bias mode. In such an orientation, they undergo a breakdown and conduct electric current in the reverse direction with a well-defined reverse voltage. Zener diodes are used to provide a stable and well-defined reference voltage.

Photodiodes. Operated in reverse bias mode, some p-n junctions conduct electricity when light shines on them. Such diodes can be used to detect and measure light intensity, since the more light that strikes the diodes, the more they conduct electricity.

Circuit protection. In most applications of diodes, they are used to take advantage of the properties of the p-n junction on a regular basis in circuits. For some applications, though, diodes are included in circuits in the hope that they will never be needed. One such application is for DC circuits, which are typically designed for current to flow in only one direction. This is automatically accomplished through a power supply with a particular voltage orientation such as a DC source, power converter, or battery. However, if the power supply were connected in reverse or if the batteries were inserted backward, then damage to the circuit could result. Diodes are often used to prevent current flow in such situations where voltage is applied in reverse, acting as a simple but effective reverse voltage protection system.

Light-emitting diodes (LEDs). For diodes with just the right kind of semiconductor and doping, the combination of holes and electrons at the p-n junction releases energy equal to that carried by photons of light. Thus, when current flows through these diodes in forward bias mode, the diodes emit light. Unlike most lighting sources, which produce a great deal of waste heat in addition to light (with incandescent lights often using energy to produce more heat than visible light), most of the energy dissipated in LEDs goes into light, making them far more energy-efficient light sources than most other forms of artificial lighting. Unfortunately, large high-power applications of light-emitting diodes are somewhat expensive, limiting them to uses where their small size and long life characteristics offset the cost associated with other forms of lighting.

Laser diodes. Very similar to light-emitting diodes are laser diodes, where the recombination of holes and electrons also produces light. However, with the laser diode, the p-n junction is placed inside a resonant cavity and the light produced stimulates more

---

FASCINATING FACTS ABOUT DIODE TECHNOLOGY

- The term diode comes from the Greek *di* (two) and *ode* (paths) signifying the two possible ways of connecting diodes in circuits.
- Naturally occurring crystals with diode-like properties were used in amateur crystal radio sets purchased by millions of hobbyists in the middle of the twentieth century.
- Despite their widely known property of allowing electricity to pass in only one direction, some diodes, such as Zener diodes, are made to be operated in reverse bias mode.
- Light-emitting diodes (LEDs) emit light only when current passes through them in the right direction.
- The laser light produced in laser pointers is made using laser diodes.
- Diodes are often used in battery-operated devices to protect the electronics in the device from users accidentally inserting batteries backward.
- Diodes are used with automobile alternators to produce the DC voltage required for automobile electrical systems.

light, producing coherent laser light. Laser diodes typically have much shorter operational lifetimes than other diodes, including LEDs, and are generally much more expensive. However, laser diodes cost much less than other methods of producing laser light, so they have become more common. Most lasers not requiring high-power application are based on laser diodes.

IMPACT ON INDUSTRY

The control of electric current is fundamental to electronics. Diodes, acting like one-way valves for electric current are, therefore, very important circuit elements in circuits. They can act as rectifiers, converting alternating current into direct current, but they have many other uses.

Government and university research. Though the basic concept of the p-n junction is understood, research continues into new applications for the junction. This research is conducted in both public and private laboratories. Much research goes into an effort to make diodes smaller, cheaper, and capable of higher power applications.

Industry and business. Most diodes are used in fabrication of devices. They are used in circuits and are part of almost every electronic device and will likely continue to play an important role in electronics. Specialized diodes, such as LEDs or laser diodes, have come to be far more important for their specialized properties than for the their current directionality. Laser diodes produce laser light inexpensively enough to permit the widespread use of lasers in many devices that would otherwise not be practical, such as DVD players. Laser diodes are easily modulated, so a great deal of fiber-optic communications use laser diodes. LEDs produce light much more efficiently than most other means of producing light. Many companies are working on developing LED lighting systems to replace more conventional lights.

Many uses of diodes exist outside of the traditional electronics industry. Diodes are, of course, used in electronic devices, but diodes are also used in automobiles, washing machines, and clocks. Almost all electronic devices use diodes, from televisions and radios to microwave ovens and DVD players.

Thermionic diodes are used in only a few industries, such as the power industry and radio and television broadcasting, where the extremely high-power applications would burn out semiconductor diodes.

Semiconductor diodes are used in all other applications, since they are more efficient to operate and less expensive to produce.

CAREERS AND COURSE WORK

The electronics field is vast and encompasses a wide variety of careers. Diodes exist in some form in most electronic devices. Thus, a wide range of careers come into contact with diodes, and therefore a wide range of background knowledge and preparation exists for the different careers.

Development of new types of diodes requires considerable knowledge of solid-state physics, materials science, and semiconductor manufacturing. Often advanced degrees in these fields would be required for research, necessitating students studying physics, electrical engineering, mathematics, and chemistry. However, diode technology is quite well evolved, so there are limited job prospects for developing new diodes or diode-like devices other than academic curiosity. Most of this area of study is simply determining how to manufacture or include smaller diodes in integrated circuits.

Electronics technicians repair electrical circuits containing diodes. So, knowledge of diodes and diode behavior is important in diagnosing failures in electronic circuits and circuit boards. Sufficient knowledge can be gained in basic electronics courses. A two-year degree in electronic technology is sufficient for many such jobs, though some jobs may require a bachelor's degree. Likewise, technicians designing and building circuits often do not need to know much about the physics of diodes— just the nature of diode behavior in circuits. Such knowledge can be gained through basic electronics courses or an associate's or bachelor's degree in electronics.

Manufacturing diodes does not actually require much knowledge about diodes for technicians who are actually making semiconductor devices. Such technicians need course work and training in operating the equipment used to manufacture semiconductors and semiconductor devices, and they must be able to follow directions meticulously in operating the machines. An associate's degree in semiconductor manufacturing is often sufficient for many such jobs. Manufacturing circuit boards with diodes, or any other circuit element, does not really require much knowledge of the circuit elements themselves, save for the ability to identify them by sight, though it would be helpful to understand basic diode behavior. Basic course work in circuits would be needed for such jobs.

Social Context and Future Prospects

Diodes exist in almost every electronic device, though most people do not realize that they are using diodes. Because electronics have been increasing in use in everyday life, diodes and diode technology will continue to play an important role in everyday devices. Diodes are very simple devices, however, and it is unlikely that the field will advance further in the development of basic diodes. Specialized devices using the properties of p-n junctions, such as laser diodes, continue to be important. It can be anticipated that additional uses of p-n junctions may be discovered and new types of diodes developed accordingly. Because the p-n junction is the basis of diode behavior and is the basis of semiconductor technology, diodes will continue to play an important role in electronics for the foreseeable future. LEDs produce light very efficiently, and work is proceeding to investigate the possibility of such devices replacing many other forms of lighting.

—*Raymond D. Benge, Jr., MS*

Further reading

Gibilisco, Stan. *Teach Yourself Electricity and Electronics.* 5th ed. New York: McGraw-Hill, 2011. Comprehensive introduction to electronics, with diagrams. A chapter on semiconductors includes a good description of the physics and use of diodes.

Held, Gilbert. *Introduction to Light Emitting Diode Technology and Applications.* Boca Raton, Fla.: Auerbach, 2009. A thorough overview of light-emitting diodes and their uses. The book also includes a good description of how diodes in general work.

Paynter, Robert T. *Introductory Electronic Devices and Circuits.* 7th ed. Upper Saddle River, N.J.: Prentice Hall, 2006. An excellent and frequently used introductory electronics textbook, with an excellent description of diodes, different diode types, and their use in circuits.

Razeghi, Manijeh. *Fundamentals of Solid State Engineering.* 3d ed. New York: Springer, 2009. An advanced undergraduate textbook on the physics of semiconductors, with a very detailed explanation of the physics of the p-n junction.

Schubert, E. Fred. *Light Emitting Diodes.* 2d ed. New York: Cambridge University Press, 2006. A very good and thorough overview of light-emitting diodes and their uses.

Turley, Jim. *The Essential Guide to Semiconductors.* Upper Saddle River, N.J.: Prentice Hall, 2003. A brief overview of the semiconductor industry and semiconductor manufacturing for the beginner.

Web Sites

The Photonics Society
http://photonicssociety.org

Schottkey Diode Flash Tutorial
http://cleanroom.byu.edu/schottky_animation.phtml

Semiconductor Industry Association
http://www.sia-online.org

University of CambridgeInteractive Explanation of Semiconductor Diode
http://www-g.eng.cam.ac.uk/mmg/teaching/ linearcircuits/diode.html

See also: Electrical Engineering; Electronic Materials Production; Light-Emitting Diodes; Nanotechnology.

DISTILLATION

FIELDS OF STUDY

Chemistry; chemical engineering; industrial studies; chemical hygiene; environmental chemistry.

SUMMARY

Distillation is a process for purifying liquid mixtures by collecting vapors from the boiling substance and condensing them back into the original liquid. Various forms of this technique, practiced since antiquity, continue to be used extensively in the petroleum, petrochemical, coal tar, chemical, and pharmaceutical industries to separate mixtures of mostly organic compounds as well as to isolate individual components in chemically pure form. Distillation has also been employed to acquire chemically pure water, including potable water through the desalination of seawater.

KEY TERMS AND CONCEPTS

- **condensation:** Phase transition in which gas is converted into liquid.
- **distilland:** Liquid mixture being distilled.
- **distillate:** Product collected during distillation.
- **forerun:** Small amount of low-boiling material discarded at the beginning of a distillation.
- **fraction:** Portion of distillate with a particular boiling range.
- **miscible:** Able to be mixed in any proportion without separation of phases.
- **petrochemical:** Chemical product derived from petroleum.
- **pot residue:** Oily material remaining in the boiling flask after distillation.
- **reflux:** To return condensed vapors (partially or totally) back to the original boiling flask.
- **reflux ratio:** Ratio of descending liquid to rising vapor during fractional distillation.
- **theoretical plate:** Efficiency of a fractionating column, being equal to the number of vapor-liquid equilibrium stages encountered by the distillate on passing through the column; often expressed as the height equivalent to a theoretical plate (HETP).
- **vaporization:** Phase transition in which liquid is converted into gas.
- **vapor pressure:** Pressure exerted by gas in equilibrium with its liquid phase.

DEFINITION AND BASIC PRINCIPLES

Matter commonly exists in one of three physical states: solid, liquid, or gas. Any phase of matter can be changed reversibly into another at a temperature and pressure characteristic of that particular sample. When a liquid is heated to a temperature called the boiling point, it begins to boil and is transformed into a gas. Unlike the melting point of a solid, the boiling point of a liquid is proportional to the applied pressure, increasing at high pressures and decreasing at low pressures.

When a mixture of several miscible liquids is heated, the component with the lowest boiling point is converted to the gaseous phase preferentially over those with higher boiling points, which enriches the vapor with the more volatile component. The distillation operation removes this vapor and condenses it back to the liquid phase in a different receiving flask. Thus, liquids with unequal boiling points can be separated by collecting the condensed vapors sequentially as fractions. Distillation also removes nonvolatile components, which remain behind as a residue.

BACKGROUND AND HISTORY

Applications of fundamental concepts such as evaporation, sublimation, and condensation were mentioned by Aristotle and others in antiquity; however, many historians consider distillation to be a discovery of Alexandrian alchemists (300 BCE to 200 CE), who added a lid (called the head) to the still and prepared oil of turpentine by distilling pine resin. The Arabians improved the apparatus by cooling the head (now called the alembic) with water, which allowed the isolation of a number of essential oils by distilling plant material and, by 800 CE, permitted the Islamic scholar Jbir ibn Hayyn to obtain acetic acid from vinegar. Alembic stills and retorts were widely employed by alchemists of medieval Europe. The first fractional distillation was developed by Taddeo Alderotti in the thirteenth century. The first comprehensive manual of distillation techniques was *Liber de arte distillandi, de simplicibus*, by Hieronymus Brunschwig, published in 1500 in Strasbourg, France.

The first account of the destructive distillation of coal was published in 1726. Large-scale continuous stills with fractionating towers similar to modern industrial stills were devised for the distillation of alcoholic beverages in the first half of the nineteenth century and later adapted to coal and oil refining. Laboratory distillation similarly advanced with the introduction of the Liebig condenser around 1850. The modern theory of distillation was developed by Ernest Sorel and reduced to engineering terms in his *Distillation et rectification industrielles* (1899).

HOW IT WORKS

Simple distillation. A difference in boiling point of at least 25 degrees Celsius is generally required for successful separations with simple distillation. The glass apparatus for laboratory-scale distillations consists of a round-bottomed boiling flask, a condenser, and a receiving flask. Vapors from the boiling liquid are returned to the liquid state by the cooling action of the condenser and are collected as distillate in the receiving flask. For high-boiling liquids, an air condenser may be sufficient, but often a jacketed condenser—in which a cooling liquid such as cold water is circulated—is required. The design of many styles of condensers (such as Liebig and Wes) enhances the cooling effect of the circulating liquid. An adapter called a still head connects the condenser to the boiling flask at a 45-degree angle and is topped with a fitting in which a thermometer is inserted to measure the temperature of the vapor (the boiling point). A second take-off adapter is often used to attach the receiving flask to the condenser at a 45-degree angle so that it is vertical and parallel to the boiling flask. One should never heat a closed system, so the take-off adapter contains a side-arm for connection to either a drying tube or a source of vacuum for distillations under reduced pressure. The apparatus was formerly assembled by connecting individual pieces with cork or rubber stoppers, but the ground-glass joints of modern glassware make these stoppers unnecessary.

Fractional distillation. When the boiling points of miscible liquids are within about 25 degrees Celsius, simple distillation does not yield separate fractions. Instead, the process produces a distillate whose composition contains varying amounts of the components, being initially enriched in the lower-boiling and more volatile one. The still assembly is modified to improve efficiency by placing a distilling column between the still head and boiling flask. This promotes multiple cycles of condensation and revaporization. Each of these steps is an equilibration of the liquid and gaseous phases and is, therefore, equivalent to a simple distillation. Thus, the distillate from a single fractional distillation has the composition of one obtained from numerous successive simple distillations. Still heads allowing higher reflux ratios and distilling columns having greater surface area permit more contact between vapor and liquid, which increases the number of equilibrations. Thus, a Vigreux column having a series of protruding fingers is more efficient than a smooth column. Even more efficient are columns packed with glass beads, single- or multiple-turn glass or wire helices, ceramic pieces, copper mesh, or stainless-steel wool. The

limit of efficiency is approached by a spinning-band column that contains a very rapidly rotating spiral of metal or Teflon over its entire length.

APPLICATIONS AND PRODUCTS

Batch and continuous distillation. Distilling very large quantities of liquids as a single batch is impractical, so industrial-scale distillations are often conducted by continuously introducing the material to be distilled. Continuous distillation is practiced in petrochemical and coal tar processing and can also be used for the low-temperature separation and purification of liquefied gases such as hydrogen, oxygen, nitrogen, and helium.

Vacuum distillation. Heating liquids to temperatures above about 150 degrees Celsius is generally avoided to conserve energy, to minimize difficulties of insulating the still head and distilling column, and to prevent the thermal decomposition of heat-sensitive organic compounds. A vacuum distillation takes advantage of the fact that a liquid boils when its vapor pressure equals the external pressure, which causes the boiling point to be lowered when the pressure decreases. For example, the boiling point of water is 100 degrees Celsius at a pressure of 760 millimeters of mercury (mmHg), but this drops to 79 degrees Celsius at 341 mmHg and rises to 120 degrees Celsius at 1,489 mmHg. Vacuum distillation can be applied to solids that melt when heated in the boiling flask; however, higher temperatures may be required in the condenser to prevent the distillate from crystallizing. The term "vacuum distillation" is actually a misnomer, for these distillations are conducted at a reduced pressure rather than under an absolute vacuum. A pressure of about 20 mmHg is obtainable with ordinary water aspirators and down to about 1 mmHg with a laboratory vacuum pump.

Molecular distillation. When the pressure of residual air in the still is lower than about 0.01 mmHg, the vapor can easily travel from the boiling liquid to the condenser, and distillate is collected at the lowest possible temperature. Distillation under high-vacuum conditions permits the purification of thermally unstable compounds of high molecular weight (such as glyceride fats and natural oils and waxes) that would otherwise decompose at temperatures encountered in an ordinary vacuum distillation. Molecular stills often have a simple design that minimizes refluxing and accelerates condensation. For example, the high-vacuum short-path still consists of two plates, one heated and one cooled, that are separated by a very short distance. Industrially, the distillate can be condensed on a rapidly rotating cone and removed quickly by centrifugal force.

Steam distillation. Another method to lower boiling temperature is steam distillation. When a homogeneous mixture of two miscible liquids is distilled, the vapor pressure of each liquid is lowered according to Raoult's and Dalton's laws; however, when a heterogeneous mixture of two immiscible liquids is distilled, the boiling point of the mixture is lower than that of its most volatile component because the vapor pressure of each liquid is now independent of the other liquid. A steam distillation occurs when one of these components is water and the other an immiscible organic compound. The steam may be introduced into the boiling flask from an external source or may be generated internally by mixing water with the material to be distilled. Steam distillation is especially useful in isolating the volatile oils of plants.

Azeotropic extractive distillation. Certain nonideal solutions of two or more liquids form an azeotrope, which is a constant-boiling mixture whose composition does not change during distillation. Water (boiling point of 100.0 degrees Celsius) and ethanol (boiling point of 78.3 degrees Celsius) form a binary azeotrope (boiling point of 78.2 degrees Celsius) consisting of 4 percent water and 96 percent ethanol. No amount of further distillation will remove the remaining water; however, addition of benzene (boiling point 80.2 degrees Celsius) to this distillate forms a tertiary benzene-water-ethanol azeotrope (boiling point of 64.9 degrees Celsius) that leaves pure ethanol behind when the water is removed. This is an example of azeotropic drying, which is a special case of azeotropic extractive distillation.

Microscale distillation. Microscale organic chemistry, with a history that spans more than a century, is not a new concept to research scientists; however, the traditional 5- to 100-gram macroscale of student laboratories was reduced one hundred to one thousand times by the introduction of microscale glassware in the 1980's to reduce the risk of fire and explosion, limit exposure to toxic substances, and minimize hazardous waste. Microscale glassware comes in a variety of configurations, such as Mayo-Pike or Williamson styles. Distillation procedures are especially troublesome in microscale because the ratio of wetted-glass

surface area to the volume of distillate increases as the sample size is reduced, thereby causing significant loss of product. Specialized microscale glassware such as the Hickman still head has been designed to overcome this difficulty.

Analytical distillation. The composition of liquid mixtures can be quantitatively determined by weighing the individual fractions collected during a carefully conducted fractional distillation; however, this technique has been largely replaced by instrumental methods such as gas and liquid chromatography.

IMPACT ON INDUSTRY

Chemical and petrochemical industries. Distillation is one of the fundamental unit operations of chemical engineering and is an integral part of many chemical manufacturing processes. Modern industrial chemistry in the twentieth century was based on the numerous products obtainable from petrochemicals, especially when thermal and catalytic cracking is applied. Industrial distillations are performed in large, vertical distillation towers that are a common sight at chemical and petrochemical plants and petroleum refineries. These range from about 2 to 36 feet in diameter and 20 to 200 feet or more in height. Chemical reaction and separation can be combined in a process called reactive distillation, where the removal of a volatile product is used to shift the equilibrium toward completion.

Petroleum industry. Distillation is extensively used in the petroleum industry to separate the hydrocarbon components of petroleum. Crude oil is a complex mixture of a great many compounds, so initial refining yields groups of substances within a range of boiling points: natural gas below 20 degrees Celsius (C_1 to C_4 hydrocarbons), petroleum ether from 20 to 60 degrees Celsius (C_5 to C_6 hydrocarbons), naphtha (or ligroin) from 60 to 100 degrees Celsius (C_5 to C_9 hydrocarbons), gasoline from 40 to 205 degrees Celsius (C_5 to C_{12} hydrocarbons and cycloalkanes), kerosene from 175 to 325 degrees Celsius (C_{10} to C_{18} hydrocarbons and aromatics), gas oil (or diesel) from 250 to 400 degrees Celsius (C_{12} and higher hydrocarbons). Lubricating oil is distilled at reduced pressure, leaving asphalt behind as a residue

Destructive distillation. When bituminous or soft coal is heated to high temperatures in an oven that excludes air, destructive distillation converts the coal

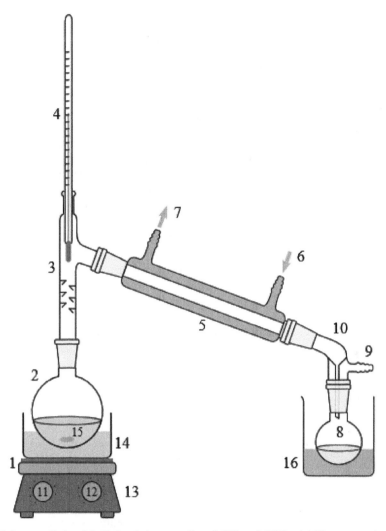

Laboratory display of distillation: 1: A source of heat 2: Still pot 3: Still head 4: Thermometer/ Boiling point temperature 5: Condenser 6: Cooling water in 7: Cooling water out 8: Distillate/ receiving flask 9: Vacuum/gas inlet 10: Still receiver 11: Heat control 12: Stirrer speed control 13: Stirrer/heat plate 14: Heating (Oil/sand) bath 15: Stirring means e.g. (shown), boiling chips or mechanical stirrer 16: Cooling bath

to light street lamps and were the main source of phenol, toluene, ammonia, and naphthalene during World War I. When wood is similarly heated in a closed vessel, destructive distillation converts the wood into charcoal and yields methyl alcohol (called wood alcohol), together with acetic acid and acetone as liquid distillates.

Essential oils. Essential oils are generally obtained by steam distillation of plant materials (flowers, leaves, wood, bark, roots, seeds, and peel) and are used in perfumes, flavorings, cosmetics, soap, cleaning products, pharmaceuticals, and solvents. Examples include turpentine and oils of cloves, eucalyptus, lavender, and wintergreen.

Distilled water and desalination. Tap water commonly contains dissolved salts that contribute to its overall hardness. These are removed through distillation to provide distilled water for use in automobile batteries and radiators, steam irons, and other applications where pure water is beneficial. Specialized stills fed with continuously flowing tap water are used to prepare distilled water in chemistry laboratories; however, deionized water is becoming an increasingly popular and more convenient alternative. Seawater can be similarly distilled to provide potable drinking water. Desalination of seawater is especially important in arid and semi-arid regions of the Middle East, where the abundant sunshine is used in solar distillation facilities.

into coke as gases and liquids form and are distilled over. The coke is chiefly employed in the iron and steel industry, and coal gas is used for heating. The liquid fraction, called coal tar, contains some compounds originally in the coal and others produced by chemical reactions during heating. Coal tar yields numerous organic compounds on repeated distilling and redistilling of the crude fractions. It provided the raw material for the early synthetic organic chemical industry in the nineteenth century. In the early twentieth century, coal-derived chemicals provided gas

CAREERS AND COURSE WORK

The art of distillation is most commonly practiced by chemists whose college majors were chemistry or chemical engineering, the former distilling samples on a laboratory scale and the latter conducting distillations on larger pilot plant and considerably larger industrial scales. The difference between chemistry and chemical engineering majors occurs in advanced and elective course work. Chemistry majors concentrate on molecular structure to better understand the chemical and physical properties of matter, whereas

FASCINATING FACTS ABOUT DISTILLATION

- The word "distill" in the late fourteenth century was originally applied to the separation of alcoholic liquors from fermented materials and comes from the Middle English "distillen," which comes from the Old French *distiller*, which comes from the Late Latin *distillare*, which is an alteration of *destillare* (*de* and *stillare*) meaning "to drip."
- In *Aristotelous peri geneses kai phthoras* (335-323 BCE; *Meteoroligica*, 1812), Aristotle described how potable water was obtained as condensation in a sponge suspended in the neck of a bronze vessel containing boiling seawater.
- Distilled alcoholic beverages first appeared in Europe in the late twelfth century, developed by alchemists.
- The Celsius scale of temperature was originally defined in the eighteenth century by the freezing and boiling points of water at 0 and 100 degrees Celsius, respectively, at a pressure of one standard atmosphere (760 millimeters of mercury).
- The first vertical continuous distillation column was patented in France by Jean-Baptiste Cellier Blumenthal in 1813 for use in alcohol distilleries.
- The first modern book on the fundamentals of distillation, *La Rectification de l'alcool* (*The Rectification of Alcohol*), was published by Ernest Sorel in 1894.
- The gas lights of Sherlock Holmes's London burned coal gas from the destructive distillation of coal.
- In 1931, M. R. Fenske separated two isomeric hydrocarbons that boiled at only 3 degrees Celsius apart using a 52-foot fractionating column mounted in an airshaft at Pond Laboratory of Pennsylvania State College.

economical, and safe manufacture of useful products. Advanced graduate study can be pursued in both disciplines at the master's, and doctoral levels, the latter degree being very common among professors of chemistry and chemical engineering at colleges and universities.

Both chemistry and chemical engineering majors must have strong backgrounds in physics and mathematics and take a year of general chemistry followed by a year of organic chemistry and another year of physical chemistry. The ability to infer the relative boiling points of chemical substances by predicting the strength of intermolecular forces from molecular structure begins in general chemistry and is pursued in detail in organic chemistry. Student laboratories employ distillation to isolate and purify products of synthesis, often with very small amounts using microscale glassware. The theoretical and quantitative aspects of phase transitions are studied in depth in the physical chemistry lecture and laboratory. Participation in research by working with an established research group is an important way for students to gain practical experience in laboratory procedures, methodologies, and protocols. Knowledge of correct procedures for handling toxic materials and the disposal of hazardous wastes are vital for all practicing chemists.

SOCIAL CONTEXT AND FUTURE PROSPECTS
The process and apparatus of distillation, more than any other technology, gave birth to modern chemical industry because of the numerous chemical products derived first from coal tar and later from petroleum. The continued role played by distillation in modern technology will depend on several factors including the sustainability of raw materials and energy conservation. Increased demand and diminishing supplies of raw materials, together with the accumulation of increasing amounts of hazardous waste, have made recycling economically feasible on an industrial scale, and distillation has a role to play in many of these processes. Likewise, the increasing cost of crude oil because of diminishing supplies of this finite resource encourages the use of alternate sources of oil such as coal (nearly 75 percent of total fossil fuel reserves), and distillation would be expected to play the same central role as it does in refining petroleum. However, distillation is also an energy-intensive technology in the requirements of both heating liquids to

chemical engineering students focus more on the properties of bulk matter involved in the large-scale, boiling and cooling the resulting vapors so that they condense back to liquid products. Thus, one would expect that the chemical industry of the future would seek alternate energy sources such as solar power as well as ways to conserve energy through improvements in distillation efficiency.

—Martin V. Stewart, PhD

FURTHER READING

Donahue, Craig J. "Fractional Distillation and GC Analysis of Hydrocarbon Mixtures." *Journal of Chemical Education* 79, no. 6 (June, 2002): 721-723. Demonstrates analytical and physical methods employed in petroleum refining.

El-Nashar, Ali M. *Multiple Effect Distillation of Seawater Using Solar Energy.* New York: Nova Science, 2008. Tells the story of the Abu Dhabi, United Arab Emirates, desalinization plant that uses solar energy to distill seawater.

Forbes, R. J. *Short History of the Art of Distillation from the Beginnings Up to the Death of Cellier Blumenthal.* 1948. Reprint. Leiden, Netherlands: E. J. Brill, 1970. Discusses the history of distillation, focusing on earlier periods.

Kister, Henry Z. *Distillation Troubleshooting.* Hoboken, N.J.: Wiley-Interscience, 2006. Examines distillation operations and how they are maintained and repaired.

Owens, Bill, and Alan Dikty, eds. *The Art of Distilling Whiskey and Other Spirits: An Enthusiast's Guide to the Artisan Distilling of Potent Potables.* Beverly, Mass.: Quarry Books, 2009. Photographs from two cross-country road trips illustrate this paperback guide to the small-scale distillation of whiskey, vodka, gin, rum, brandy, tequila, and liquors.

Stichlmair, Johann G., and James R. Fair. *Distillation: Principles and Practices.* New York: Wiley-VCH, 1998. This 544-page comprehensive treatise for distillation technicians contains chapters on modern industrial distillation processes and energy savings during distillation.

Towler, Gavin P., and R. K. Sinnott. *Chemical Engineering Design: Principles, Practice, and Economics of Plant and Process Design.* Boston: Elsevier/Butterworth-Heinemann, 2008. Chapter 11 examines continuous and multi-component distillation, looking at the principles involved and design variations. Also discusses other distillation systems and components of a system.

WEB SITES

American Institute of Chemical Engineers
http://www.aiche.org

American Petroleum Institute
http://www.api.org

Fractionation Research
http://www.fri.org

Institution of Chemical Engineers
http://unified.icheme.org

See also: Cracking; Petroleum Extraction and Processing.

EARTHQUAKE ENGINEERING

FIELDS OF STUDY

Civil engineering; earthquake engineering; engineering seismology; geology; geotechnical engineering.

SUMMARY

Earthquake engineering is a branch of civil engineering that deals with designing and constructing buildings, bridges, highways, railways, and dams to be more resistant to damage by earthquakes. It also includes retrofitting existing structures to make them more earthquake resistant.

KEY TERMS AND CONCEPTS

- **asperity:** Surface roughness that projects outward from the surface.
- **epicenter:** Point on the Earth's surface directly above the hypocenter.
- **hypocenter:** Point beneath the Earth's surface where an earthquake originates.
- **love wave:** Wave formed by the combination of secondary waves and primary waves on the surface; causes the ground to oscillate from side to side perpendicular to the propagation direction of the wave and is the most destructive.
- **P wave:** First wave from an earthquake to reach a seismograph; travels through the body of the Earth, including through a liquid; also called a primary wave.
- **rammed earth:** Mixture of damp clay, sand, and a binder such as crushed limestone that is poured into a form and then rammed down by thrusting with wooden posts; after it dries the forms are removed.
- **Rayleigh wave:** Wave formed by the combination of secondary waves and primary waves on the surface; causes the ground to oscillate in a rolling motion parallel with the direction of the wave.
- **wave:** Wave that reaches a seismic station after a primary wave; travels through the body of the Earth, but not through a liquid; also called secondary wave.

DEFINITION AND BASIC PRINCIPLES

Worldwide, each year there are about eighteen major earthquakes (magnitude 7.0 to 7.9) strong enough to cause considerable damage and one great earthquake (magnitude 8.0 or greater) strong enough to destroy a city. The outermost layer of the Earth is the rocky crust where humans live. The continental crust of the Earth is 30 to 50 kilometers thick, while the oceanic crust is 5 to 15 kilometers thick. The crust is broken up into about two dozen plates that fit together like pieces of a jigsaw puzzle, with the larger plates being hundreds to thousands of kilometers across. As these plates move about on the underlying mantle at rates of a few to several centimeters per year, they rub against neighboring plates. Asperities (irregularities) from one plate lock with those of an adjacent plate and halt the motion. While the plate boundary is held motionless, the rest of the plate continues in motion in response to the forces on it, and this action builds up stress in the boundary rocks.

Finally, when the stress on the boundary rocks is too great the asperities are sheared off as the boundary rock surges ahead several centimeters to several meters. It is this sudden lurching of the rock that produces earthquake waves. The point of initial rupture produces the most waves and is called the hypocenter, while the point on the surface directly above the hypocenter is called the epicenter. The epicenter is usually the site of the worst damage on the surface. Earthquake engineers can design structures to reduce the damage and the number of deaths, but the limited resources available means that not everything that might be done is done. The philosophy generally adopted is that while a strong earthquake may damage most structures they should remain standing at least long enough for the people in them to evacuate safely. Essential structures such as hospitals should not only remain standing but should be still usable after the quake.

BACKGROUND AND HISTORY

Earthquakes have plagued mankind throughout history. The Antioch (now in Turkey) earthquake in 526 killed an estimated 250,000 people. A thousand years later in 1556, an estimated 830,000 died in Shaanxi, China. In ancient times, earthquakes were ascribed to various fanciful causes such as air rushing out of deep caverns. Chinese mathematician Zhang Heng (78-139) is credited with the invention of a Chinese earthquake detector consisting of a large, nearly spherical vessel with eight dragon heads projecting outward from its circumference. A brass ball is loosely held in each dragon's mouth. A pendulum is suspended inside the vessel so that if an earthquake sets it swinging, it will strike a dragon causing the ball to fall from its mouth and into the waiting mouth of a toad figure. The sound of the ball striking the metal toad alerts the operator that an earthquake has occurred, and whichever toad has the ball indicates the direction to the epicenter.

Earthquake 'Shaking Table' Test

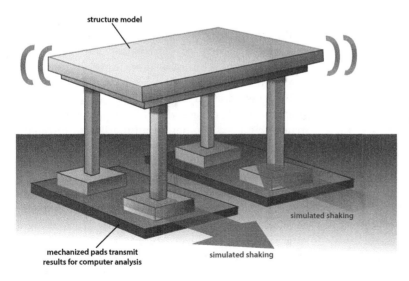

An earthquake shake table is used to test the resistance of certain components or structures to seismic activity.

The scientific study of earthquakes did not blossom until the twentieth century. In 1935, American seismologist Charles Richter with German seismologist Beno Gutenberg developed the Richter scale for measuring the intensity of an earthquake based on the amplitude of the swinging motion of the needle on a seismometer. The Richter scale was superseded in 1979 by the moment magnitude scale based on the energy released by the quake. This scale was developed by Canadian seismologist Thomas C. Hanks and Japanese American seismologist and Kyoto Prize winner Hiroo Kanamori and is the same as the Richter scale for quakes of magnitude 3 through 7.

Following the 1880 Yokohama earthquake, the Seismological Society of Japan was founded to see what might be done to reduce the consequences of earthquakes. It was the world's first such society and marks the beginning of earthquake engineering. The Japanese were forced into such a leadership role by being an industrial society sitting on plate boundaries and therefore subject to frequent earthquakes. In 1893, Japanese seismologist Fusakichi Omori and British geologist John Milne studied the behavior of brick columns on a shake table (to simulate earthquakes). Toshikata Sano, a professor of structural engineering at the Imperial University of Tokyo, published "Earthquake Resistance of Buildings" in 1914, and by the 1930's, several nations had adopted seismic building codes. Knowledge from earthquake engineering was beginning to be put into practice.

The early twenty-first century has seen several devastating quakes. On January, 12, 2010, a magnitude-7 earthquake struck Haiti, killing an estimated 316,000 people, injuring 300,000, and leaving more than one million people homeless. On March 11, 2011, a magnitude-8.9 quake occurred off the east coast of Honshu, Japan, causing a massive tsunami that destroyed entire villages and also affected places as far away as Australia and the West Coast of the United States. This was the strongest quake in Japanese history. In addition to an estimated 10,000 deaths and almost 8,000 people reported missing,

Japanese officials had to deal with subsequent leaks at three nuclear reactors in the affected region.

HOW IT WORKS

Earthquakes may be classified by their depth. Shallow-focus quakes have a hypocenter less than 70 kilometers deep and are the most destructive. Mid-focus quakes originate between 70 kilometers and 300 kilometers deep, and deep-focus quakes originate deeper than 300 kilometers and are the least destructive.

P waves and S waves. Underground quakes emit two kinds of waves, P waves (primary waves) and S waves (secondary waves). P waves are longitudinal or compression waves just like sound waves. The rock atoms vibrate along the direction in which the wave moves. P waves travel about 5,000 meters per second in granite. S waves are transverse waves, so atoms vibrate up and down perpendicular to the direction of wave travel. The speed of an S wave is about 60 percent that of a P wave. The difference in arrival times of P waves and S waves at a seismic station provides an estimate of the distance to the hypocenter. Therefore during a site evaluation, earthquake engineers must locate any nearby faults and the location of past hypocenters. Then they must try to determine the most likely hypocenters of future earthquakes, which they hope will be deep and far away.

Love waves and Rayleigh waves. S waves and P waves interact at the Earth's surface to produce two new types of surface waves: Love waves and Rayleigh waves. These are the waves that destroy buildings and knock people off their feet. Love waves are named for British mathematician Augustus Edward Hough Love, who developed the theory about the waves in his book *Some Problems About Geodynamics* (1911). They cause the ground to oscillate from side to side perpendicular to the propagation direction of the wave. Love waves are the greatest source of destruction outside the epicenter. Rayleigh waves are named for British physicist and Nobel Prize winner John William Strutt (Lord Rayleigh). They cause the ground to oscillate in a rolling motion parallel Earthquake Engineering with the direction of the wave. The greater the wave amplitude, the more violent the shaking. A careful examination of the rocks and soil underlying a site should give information on propagation, damping, and direction of likely earthquake waves.

Core sampling. To complete the site investigation, core samples may need to be taken to look for damp and insufficiently compacted soil. The shaking of an earthquake can turn damp sand into jelly, a process called liquefaction. Sand grains are surrounded by liquid and cannot cling together. Buildings sink into liquified soil. If it looks like liquefaction may be a problem, support piers must go from the building's foundation down into bedrock. If that is not possible, densification and grout injection may stabilize the soil. Liquefaction caused a segment of roadway to drop 2.4 meters in the 1964 Alaska earthquake. It caused great destruction in the Marina District of the 1989 Loma Prieta earthquake, and contributed to the destruction of Christchurch, New Zealand, in the February 22, 2011, magnitude-6.3 earthquake.

If the site involves a slope, or if the site is a railroad embankment, an earthquake might cause the slope to collapse. In this case the angle of repose of the slope could be reduced. If that is not possible, a retaining wall may be required, or a geotextile (made from polyester or polypropylene) covered with sand could be used to stabilize the slope.

APPLICATIONS AND PRODUCTS

Geologists and geophysicists have learned enough about earthquakes to be able to identify ways in which the damage they cause can be minimized by appropriate human behavior. In general, there are two main approaches to mitigating the effects of earthquakes: Build the structure to withstand a quake, and make the structure invisible to the earthquake waves, but if that cannot be done, dampen the waves as quickly as possible.

Designing for earthquake safety. Earthquake engineers can test design ideas by carefully modeling a structure on a computer, inputting the location and strength of the various materials that will be used to construct the building. They can then use the computer to predict the results of various stresses. They can also build a scaled-down model, or even a full-scale model, on a shake table, a platform that can be shaken to simulate an earthquake. When using a scale model, care must be taken so that it is not only geometrically similar to the full-size structure but that other factors such as the velocity of waves moving through the structure are also scaled down. The ultimate test is to build the structure and see how it fares.

Retrofitting old structures. The single act that has

the potential to save the most lives is to fortify adobe houses against earthquakes. About 50 percent of the population in developing countries live in houses made of adobe or rammed earth, since dirt is cheap and is the ubiquitous building material. Adobe bricks are made by mixing water into 50 percent sand, 35 percent clay, and 15 percent binder-either straw or manure (said to repel insects). The mixture is poured and patted into a mold and then the mold is turned upside down so that the new brick will fall onto the ground to dry in the sun. The bricks may be assembled into a wall using a wet mixture of sand and clay as mortar. Mortar joints should be no more than 20 millimeters thick to avoid cracking. Rammed earth uses a similar mixture of sand and clay but uses lime, cement, or asphalt as a stabilizer. The wet mixture is poured into a form and then tamped down, or rammed, by workers with thick poles. After it sets, the forms are removed.

In an earthquake, both rammed earth and adobe crack and shatter. Walls tumble down, and roofs that had rested on the walls collapse onto people. The magnitude-7 quake that struck Haiti in 2011 flattened a large part of its capital, Port-Au-Prince. It was so devastating because its hypocenter was shallow (only 13 kilometers deep) and only 25 kilometers away, and the poorly constructed adobe houses fell on people and buried them.

Earthquake engineers have figured out relatively inexpensive ways to strengthen adobe houses. Laying bamboo lengthwise in the mortar strengthens the wall as does drilling holes in the bricks and inserting vertical bamboo sticks so that they tie rows of blocks together. A continuous wooden or cement ring should go all around the top of the walls to tie the walls together and to provide a way to fasten ceiling and roof joists to the walls. If an existing house is being retrofitted, vertical and horizontal strips of wire mesh should be nailed onto the walls both inside and outside at corners and around windows and doors. In practice, the mesh strips used range from seven centimeters to sixty centimeters in width. The strips are attached by driving nails through metal bottle caps and into the adobe.

Bridges, dams, and isolation bearings. Structures can be protected by strengthening them to withstand an earthquake or by isolating them from the ground so that earthquake energy does not enter the structure. Dams are built to the first plan, and bridges

follow the second plan. To ensure that a concrete dam survives an earthquake, extra reinforcing steel bars (rebars) would be used. The dam must either rest directly on bedrock or on massive pillars that extend downward to bedrock. If there are known fault lines in the area so that it is known in which direction land will move, it may be possible to construct a slip joint. A slip joint works like sliding closet doors, where one door can slide in front of the other. The Clyde Dam in Central Otago, New Zealand, has a slip joint that will allow the land on either side to slip up to two meters horizontally and one meter vertically. Finally, the dam should be built several meters higher than originally planned so that if an earthquake causes the impounded water to slosh back and forth it will not overtop the dam.

Bridges have long used bearings in the form of several hardened steel cylinders between a flat bridge plate and a flat supporting plate. These bearings allow the bridge to move as it expands or contracts with temperature. The same method can be used to allow motion during an earthquake without damaging the bridge, but now the bridge is isolated from the pier since the pier can move back and forth while the bridge remains stationary.

Waves that shake the foundation of a building send some of that vibrational energy up into the building. Placing a bridge or a building on lead and rubber bearings lessens the energy transmitted to the bridge or building. A typical bearing consists of a large block of alternating steel and rubber layers surrounding a vertical lead cylinder. It is quite rigid in the vertical direction but allows considerable motion in the horizontal direction. Since the rubber heats up as it is deformed, it converts the horizontal motion into heat and thereby damps this motion. The Museum of New Zealand and the New Zealand Parliament buildings stand on lead-rubber bearings and are thereby partially isolated from ground motion. Ironically, Christchurch, New Zealand, was considering reinforcing a large number of buildings prior to the February, 2011, quake, but had not proceeded very far because of the cost. That cost was a pittance compared with the rebuilding cost.

Mass dampers. A building of a certain height, mass, and stiffness will tend to oscillate or resonate at a given frequency. Small oscillations can quickly build up into large oscillations, just as repeatedly giving a small push at the right frequency to a child in a playground swing will cause the swing to move in a large arc. If the upper floors of a tall building sway too much, people get motion sickness, and the structure gets weakened and may eventually fall. Mass dampers are usually huge concrete blocks mounted on tracks on an upper floor. When sensors detect lateral motion of the building's upper floors, motors drive the block in the opposite direction to the building's motion. This pushes the building in the opposite direction from its motion and causes that motion to die out. A tuned mass damper oscillates at the natural frequency of the building. This technique also works with motion caused by earthquake waves.

Taipei 101 in Taiwan is the world's second-tallest building. It is 101 stories (508 meters) tall, and its mass wind damper is a 660-metric ton metal sphere suspended like a pendulum between the eighty-seventh and ninety-second floors. If the building sways, the pendulum is driven in the opposite direction. This passively tuned mass wind damper reduces the building's lateral motion by more than half. Two sixmetric-ton pendula are positioned in the tower to control its motion.

X-braces and pneumatic dampers. The vertical and horizontal beams of a building's steel framework form rectangles. Consider a vertical rectangle in the wall at the bottom of the building. The bottom of the rectangle is fastened to the foundation and will move with a seismic wave, but the top of the rectangle is fastened to the building above it so that inertia will tend to hold it fixed. As the foundation moves laterally, the rectangle will be deformed into a parallelogram. Two diagonal beams making an "X" in the rectangle will keep it from deforming very much. The beautiful seventy-one story Pearl River Tower in Guangzhou, China, uses massive X-braces to keep the tower from swaying in the wind or an earthquake. These beams can clearly be seen in construction photographs of the tower. The tower uses integral wind turbines and solar cells to be largely energy self-sufficient. Rather than using the X-braces to stiffen the tower, the centers of the diagonal beams could have been clamped together with break-lining material between them, then they would damp the horizontal motion. Diagonally mounted massive hydraulic pistons can also be used to strengthen a structure and simultaneously to damp out the earthquake energy in a building.

Pyramids. Large amplitude horizontal motion can also be avoided if earthquake motion is not concentrated at the building's natural frequency, but is spread over many frequencies. A building can be designed not to amplify waves of certain frequencies and to deflect some waves and absorb others. The speed of a wave traveling up a building depends on the amount of stress present, the amount of mass per meter of height, and the frequency of the wave. A bullwhip made of woven leather thongs makes a good analogue. Near the handle, the whip is as thick as the handle, but it tapers to a single thin thong at the far end. When the handle is given a quick backward and forward jerk, a wave speeds down the length of the whip. If the momentum (mass times velocity) of a whip segment is to remain constant, as the mass of a segment decreases (because of the taper), the speed must increase. The distinctive whip crack occurs as the whip end exceeds the speed of sound. In a similar fashion, the speed of a wave traveling up a pyramid-shaped building changes, and as the speed changes, the frequency changes. If a pyramidal building is properly designed, earthquake waves will attenuate as they try to pass upward. This is the idea behind the design of the forty-eight story Transamerica Pyramid building in San Francisco. It is not essential that the shape of the building be a pyramid since changing the mass density or the tension in the steel structure can have a similar effect.

IMPACT ON INDUSTRY

Government and university research. Most earthquake engineering work is done by consulting companies, while research is carried out by universities and research institutes. The United States Geological Survey (USGS) is the federal agency tasked with recording and reporting earthquake activity nationwide. Data is provided by the Advanced National Seismic System (ANSS), a nationwide array of seismic stations. USGS maintains several active research projects. The Borehole Geophysics and Rock Mechanics project drills deep into fault zones to measure heat flow, stress, fluid pressure, and the mechanical properties of the rocks. The Earthquake Geology and Paleoseismology project seeks out and analyzes the rocks pertaining to historic and prehistoric earthquakes.

Earthquake Engineering Research Institute (EERI) of Oakland, California, carries out various research projects. One project involves surveying concrete buildings that failed during an earthquake in an effort to discover the top ten reasons for the failure of these buildings. The Pacific Earthquake Engineering Research Center (PEER) at the University of California, Berkeley, has a 6-meter-by-6 meter shaking table. It can move horizontally and vertically and can rotate about three different axes. It can carry structures weighing up to 45 tons and subject them to horizontal accelerations of 1.5 times gravity. Recent projects include the seismic-qualification testing of three types of 245-kilovolt disconnect switches, testing a friction pendulum system (for damping vibrations), testing a two-story wood-frame house, and testing a reinforced concrete frame.

The John A. Blume Earthquake Engineering Center at Stanford University pursues the advancement of research, education, and practice of earthquake engineering. Scientists there did the research into the earthquake risk for the 2-mile-long Stanford Linear Accelerator, the Diablo Canyon Nuclear Power Plant, and many other sites. The founder, John Blume, is quoted as reminding a reporter that the center designed "earthquake-resistant" buildings not "earthquake-proof" buildings, and added, "Don't say 'proof' unless you're talking about whiskey."

Feeling that more coordinated efforts were needed, the Japanese formed the Japan Association for Earthquake Engineering (JAEE) in January, 2001. The association was to be involved with the evaluation of seismic ground motion and active faults, resistance measures before an earthquake, education on earthquake disaster reduction, and sponsoring meetings and seminars where new techniques could be shared and analyzed. After an earthquake, they hope to aid in damage assessment, emergency rescue and medical care, and in evaluating what building techniques worked and what did not work.

Major corporations. ABS Consulting, with headquarters in Houston, Texas, is a worldwide risk-management company. It has 1,400 employees and uses earthquake engineers when it evaluates the earthquake risk for a site.

Air Worldwide provides risk analysis and catastrophe modeling software and consulting services. It has offices in Boston, San Francisco, and several major cities in other countries. It hires civil engineers to perform seismic design studies and to prepare plans for structural engineering.

ARUP, an engineering consulting company headquartered in London, employs 10,000 people worldwide. To reduce the lateral movement of the upper stories of buildings, it employs its damped outrigger system. It uses large hydraulic cylinders to tie the central pillar of the building to the outer walls. The alternative is to make the building stiffer with more concrete and steel (which costs several million dollars) and then add a tuned mass damper (which ties up a great deal of space). The company used a few dozen of these dampers in the beautiful twin towers of the St. Francis Shangri-La Place in Manila, Philippines.

International Seismic Application Technology (ISAT), with headquarters in La Mirada, California, uses earthquake engineers to do site studies and to design seismic-restraint systems for plumbing, air-conditioning ducts, and electrical systems in buildings.

Miyamoto International is a global earthquake and structural engineering firm that specializes

---

FASCINATING FACTS ABOUT EARTHQUAKE ENGINEERING

- San Francisco is moving toward Los Angeles at about 5 centimeters per year. They should be across from each other in about 12 million years.
- There are about 500,000 earthquakes each year, but only about 100,000 are strong enough to be felt by people.
- Only about 100 earthquakes each year are strong enough to do any damage.
- The largest recorded earthquake was a magnitude 9.5 in Chile on May 22, 1960.
- On March 28, 1964, a woman in Anchorage, Alaska, was trying to remove a stuck lid from a jar of fruit. At the instant she tapped the lid against the corner of the kitchen counter, a 9.2 magnitude earthquake struck. For a few moments she feared she had caused the quake.
- Without a tuned mass damper, the top floors of very tall buildings can sway back-and-forth 30 centimeters or more in a strong wind, let alone an earthquake.
- The 2004 Indian Ocean 9.0-magnitude earthquake released energy equivalent to 9,560,000 megatons of TNT, about 1,000 times the world supply of strategic nuclear warheads.

in designing earthquake engineering solutions. It has offices throughout California, and in Portland-Vancouver and Tokyo, and specializes in viscous and friction cross-bracing dampers.

The Halcrow Group based in London does seismic-hazard analysis, design, and remediation. It also does earthquake site response analysis and liquefaction assessment and remediation. It has done site evaluations for nuclear power plants, dams, and intermediate level nuclear waste storage. It is a large company with 8,000 employees and many interests.

CAREERS AND COURSE WORK

Earthquake engineering is a subset of geotechnical engineering, which itself is a branch of civil engineering. Perhaps the most direct route would be to attend a university such as the University of California, Los Angeles, which offers both graduate and undergraduate degrees in engineering, and get an undergraduate degree in civil engineering. An undergraduate should take principles of soil mechanics, design of foundations and earth structures, advanced geotechnical design, fundamentals of earthquake engineering, soil mechanics laboratory, and engineering geomatics. Graduate courses should include advanced soil mechanics, advanced foundation engineering, soil dynamics, earth retaining structures, advanced cyclic and monotonic soil behavior, geotechnical earthquake engineering, geoenvironmental engineering, numerical methods in geotechnical engineering, and advanced soil mechanics laboratory.

Other schools will have their own version of the program. For example, Stanford University offers a master's degree in structural engineering and geomechanics. The program requires a bachelor's degree in civil engineering including courses in mechanics of materials, geotechnical engineering, structural analysis, design of steel structures, design of reinforced concrete structures, and programming methodology. The University of Southern California; the University of California, San Diego; the University of California, Berkeley; and the University of Alaska at Anchorage all have earthquake engineering programs. On the east coast, the Multidisciplinary Center for Earthquake Engineering Research at the State University of New York at Buffalo and the Center for Earthquake Engineering Simulation at Rensselaer Polytechnic Institute in Troy, New York,

are centers for earthquake engineering.

SOCIAL CONTEXT AND FUTURE PROSPECTS

Although earthquake engineering has made a lot of progress, some areas of society have been surprisingly slow to implement proven measures. Many of the buildings that collapsed in the New Zealand earthquake in February, 2011, would have remained standing had they been reinforced to the recommended standard. They were not reinforced because of cost, but that cost was a small fraction of what it will now cost to rebuild. The January, 2010, Haitian earthquake was so deadly because there are no national building standards. The December, 2003, earthquake in Bam, Iran, was so devastating because building codes were not followed. In particular, enough money was budgeted to build the new hospitals to standards that would have kept them standing. The hospitals collapsed into piles of rubble while corrupt officials (according to expatriates) enriched themselves. Building codes will do no good until they are enforced, and they will not be enforced without honest officials.

On a more positive note, an exciting, recent proposal is to make a building invisible to earthquake waves. Earthquake surface waves cause the damage. The speed of such waves depends upon the density and rigidity of the rock and soil they traverse. Consider a wave coming toward a building almost along a radius, and suppose that the building's foundation is surrounded by a doughnut-shaped zone in which the speed of an earthquake wave is increased above that of the surrounding terrain. The incoming wave will necessarily bend away from the radius. One or more properly constructed doughnuts, or rings, should steer the earthquake waves around the building. No doubt there will be problems in implementing this method, but it seems promising. It may even be possible to surround a town with such rings and thereby protect the whole town.

—Charles W. Rogers, MS, PhD

FURTHER READING

Bozorgnia, Yousef, and Vitelmo V. Bertero, eds. *Earthquake Engineering: From Engineering Seismology to Performance-Based Engineering*. Boca Raton, Fla.: CRC Press, 2004. Provides a good overview of the problems encountered in earthquake engineering and ways to solve them. Requires a good science and math background.

Building Seismic Safety Council for the Federal Emergency Management Agency of the Department of Homeland Security. *Homebuilder's Guide to Earthquake-Resistant Design and Construction.* Washington, D.C.: National Institute of Building Sciences, 2006. A gold mine for the non-engineer or the prospective engineer. Introduces the terms and techniques of earthquake-resistant structures in an understandable fashion.

Kumar, Kamalesh. *Basic Geotechnical Earthquake Engineering.* New Delhi, India: New Age International, 2008. Emphasizes site properties and preparation, when to expect liquefaction and what to do about it. Easily read by the science-savvy layperson.

Stein, Ross S. "Earthquake Conversation." *Scientific American* 288 (January, 2003): 72-79. Active faults are responsive to even a small increase in stress that they acquire when there is a quake in a nearby fault. This may make earthquake prediction more accurate.

Yanev, Peter I., and Andrew C. T. Thompson. *Peace of Mind in Earthquake Country: How to Save Your Home, Business, and Life.* 3d ed. San Francisco: Chronicle Books, 2008. An excellent introductory treatment of earthquakes, how they damage structures, and what may be done beforehand to reduce damage. Discusses building sites and possible problems such as liquefaction.

WEB SITES
ArchitectJaved.com Earthquake Resistant Structures
http://articles.architectjaved.com

Earthquake Engineering Research Institute
http://www.eeri.org/site

Seismological Society of America
http://www.seismosoc.org

See also: Bridge Design and Barodynamics; Civil Engineering.

ELECTRICAL ENGINEERING

FIELDS OF STUDY

Physics; quantum physics; thermodynamics; chemistry; calculus; multivariable calculus; linear algebra; differential equations; statistics; electricity; electronics; computer science; computer programming; computer engineering; digital signal processing; materials science; magnetism; integrated circuit design engineering; biology; mechanical engineering; robotics; optics.

SUMMARY

Electrical engineering is a broad field ranging from the most elemental electrical devices to high-level electronic systems design. An electrical engineer is expected to have fundamental understanding of electricity and electrical devices as well as be a versatile computer programmer. All of the electronic devices that permeate modern living originate with an electrical engineer. Items such as garage-door openers and smart phones are based on the application of electrical theory. Even the computer tools,

fabrication facilities, and math to describe it all is the purview of the electrical engineer. Within the field there are many specializations. Some focus on high-power analogue devices, while others focus on integrated circuit design or computer systems.

KEY TERMS AND CONCEPTS

- **alternating current (AC):** Current that alternates its potential difference, changes its rate of flow, and switches direction periodically.
- **analogue:** Representation of signals as a continuous set of numbers such as reals.
- **binary:** Counting system where there are only two digits, 0 and 1, which are best suited for numerical representations in digital applications.
- **capacitance:** Measure of potential electrical charge in a device.
- **charge:** Electrical property carried by all atomic particles (protons, neutrons, and electrons).
- **current:** Flow of electrical charge from one region to another.
- **digital:** Representation of signals as a discrete

271

number, such as an integer.

- **digital signal processing (DSP):** Mathematics that describes the processing of digital signals.
- **direct current (DC):** Current that flows in one direction only and does not change its potential difference.
- **inductance:** Measure of a device's ability to store magnetic flux.
- **integrated circuit (IC):** Microscopic device where many transistors have been etched into the surface and then connected with wire.
- **resistance:** Measure of how easily current can flow through a material.
- **transistor:** Three-terminal device where one terminal controls the rate of flow between the other two.
- **voltage:** Measure of electrical potential energy between two regions.

DEFINITION AND BASIC PRINCIPLES

Electrical engineering is the application of multiple disciplines converging to create simple or complex electrical systems. An electrical system can be as simple as a light bulb, power supply, or switch and as complicated as the Internet, including all its hardware and software subcomponents. The spectrum and scale of electrical engineering is extremely diverse. At the atomic scale, electrical engineers can be found studying the electrical properties of electrons through materials. For example, silicon is an extremely important semiconductive material found in all integrated circuit (IC) devices, and knowing how to manipulate it is extremely important to those who work in microelectronics.

While electrical engineers need a fundamental background in basic electricity, many (if not most) electrical engineers do not deal directly with wires and devices, at least on a daily basis. An important subdiscipline in electrical engineering includes IC design engineering: A team of engineers are tasked with using computer software to design IC circuit schematics. These schematics are then passed through a series of verification steps (also done by electrical engineers) before being assembled. Because computers are ubiquitous, and the reliance on good computer programs to perform complicated operations is so important, electrical engineers are adept computer programmers as well. The steps would be the same in any of the subdisciplines of the field.

BACKGROUND AND HISTORY

Electrical engineering has its roots in the pioneering work of early experimenters in electricity in the eighteenth and nineteenth centuries, who lent their names to much of the nomenclature, such as French physicist André-Marie Ampère and Italian physicist Alessandro Volta. The title electrical engineer began appearing in the late nineteenth century, although to become an electrical engineer did not entail any special education or training, just ambition. After American inventor Thomas Edison's direct current (DC) lost the standards war to Croatian-born inventor Nicola Tesla's alternating current (AC), it was only a matter of time before AC power became standard in every household.

Vacuum tubes were used in electrical devices such as radios in the early twentieth century. The first computers were built using warehouses full of vacuum tubes. They required multiple technicians and programmers to operate because when one tube burst, computation could not begin until it had been identified and replaced.

The transistor was invented in 1947 by John Bardeen, Walter Brattain, and William Shockley, employees of Bell Laboratories. By soldering together boards of transistors, electrical engineers created the first modern computers in the 1960's. By the 1970's, integrated circuits were shrinking the size of computers and the purely electrical focus of the field. As of 2011, electrical engineers dominate IC design and systems engineering, which include mainframes, personal computers, and cloud computing. There is, of course, still a demand for high-energy electrical devices, such as airplanes, tanks, and power plants, but because electricity has so many diverse uses, the field will continue to diversify as well.

HOW IT WORKS

In a typical scenario, an electrical engineer, or a team of electrical engineers, will be tasked with designing an electrical device or system. It could be a computer, the component inside a computer, such as a central processing unit (CPU), a national power grid, an office intranet, a power supply for a jet, or an automobile ignition system. In each case, however, the electrical engineer's grasp on the fundamentals of the field are crucial.

Electricity. For any electrical application to work, it needs electricity. Once a device or system has been identified for assembly, the electrical engineer must know how it uses electricity. A computer will use low voltages for sensitive IC devices and higher ones for fans and disks. Inside the IC, electricity will be used as the edges of clock cycles that determine what its logical values are. A power grid will generate the electricity itself at a power plant, then transmit it at high voltage over a grid of transmission lines.

Electric power. When it is determined how the device or application will use electricity, the source of that power must also be understood. Will it be a standard AC power outlet? Or a DC battery? To power a computer, the voltage must be stepped down to a lower voltage and converted to DC. To power a jet, the spinning turbines (which run on jet fuel) generate electricity, which can then be converted to DC and will then power the onboard electrical systems. In some cases, it's possible to design for what happens in the absence of power, such as the battery backup on an alarm clock or an office's backup generator. An interesting case is the hybrid motor of certain cars such as the Toyota Prius. It has both an electromechanical motor and an electric one. Switching the drivetrain seamlessly between the two is quite a feat of electrical and mechanical engineering.

Circuits. If the application under consideration has circuit components, then its circuitry must be designed and tested. To test the design, mock-ups are often built onto breadboards (plastic rows of contacts that allow wiring up a circuit to be done easily and quickly). An oscilloscope and voltmeter can be used to measure the signal and its strength at various nodes. Once the design is verified, if necessary the schematic can be sent to a fabricator and mass manufactured onto a circuit board.

Digital logic. Often, an electrical engineer will not need to build the circuits themselves. Using computer design tools and tailored programming languages, an electrical engineer can create a system using logic blocks, then synthesize the design into a circuit. This is the method used for designing and fabricating application-specific integrated circuits (ASICs) and field-programmable gate arrays (FPGAs).

Digital signal processing (DSP). Since digital devices require digital signals, it is up to the electrical engineer to ensure that the correct signal is coming in and going out of the digital circuit block. If the incoming signal is analogue, it must be converted to digital via an analogue-to-digital converter, or if the circuit block can only process so much data at a time, the circuit block must be able to time slice the data into manageable chunks. A good example is an MP3 player: The data must be read from the disk while it is moving, converted to sound at a frequency humans can hear, played back at a normal rate, then converted to an analogue sound signal in the headphones. Each one of those steps involves DSP.

Computer programming. Many of the steps above can be abstracted out to a computer programming language. For example, in a logical programming language such as Verilog, an electrical engineer can write lines of code that represent the logic. Another program can then convert it into the schematics of an IC block. A popular programming language called SPICE can simulate how a circuit will behave, saving the designer time by verifying the circuit works as expected before it is ever assembled.

APPLICATIONS AND PRODUCTS

The products of electrical engineering are an integral part of our everyday life. Everything from cell phones and computers to stereos and electric lighting encompass the purview of the field.

For example, a cell phone has at every layer the mark of electrical engineering. An electrical engineer designed the hardware that runs the device. That hardware must be able to interface with the established communication channels designated for use. Thus, a firm knowledge of DSP and radio waves went into its design. The base stations with which the cell phone communicates were designed by electrical engineers. The network that allows them to work in concert is the latest incarnation of a century of study in electromagnetism. The digital logic that allows multiple phone conversations to occur at the same time on the same frequency was crafted by electrical engineers. The whole mobile experience integrates seamlessly into the existing landline grid. Even the preexisting technology (low voltage wire to every home) is an electrical engineering accomplishment—not to mention the power cable that charges it from a standard AC outlet.

One finds the handiwork of electrical engineers in such mundane devices as thermostats to the ubiquitous Internet, where everything from the network cards to the keyboards, screens, and software are crafted by electrical engineers. Electrical engineers are historically involved with electromagnetic devices as well, such as the electrical starter of a car or the turbines of a hydroelectric plant. Many devices that aid artists, such as sound recording and electronic musical instruments, are also the inspiration of electrical engineers.

Below is a sampling of the myriad electrical devices that are designed by electrical engineers.

Computers. Computer hardware and often computer software are designed by electrical engineers. The CPU and other ICs of the computer are the product of hundreds of electrical engineers working together to create ever-faster and more miniature devices. Many products can rightfully be considered computers, though they are not often thought of as such. Smart phones, video-game consoles, and even the controllers in modern automobiles are computers, as they all employ a microprocessor. Additionally, the peripherals that are required to interface with a computer have to be designed to work with the computer as well, such as printers, copiers, scanners, and specialty industrial and medical equipment.

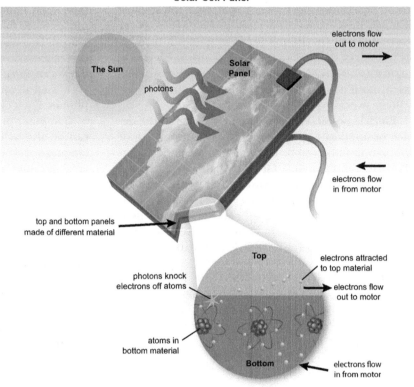

Solar panels are a collection of solar cells that convert light into electricity.

Test equipment. Although these devices are seldom seen by the general public, they are essential to keeping all the other electrical devices in the world working. For example, an oscilloscope can help an electrical engineer test and debug a failing circuit because it can show how various nodes are behaving relative to each other over time. A carpenter might use a wall scanner to find electrical wire, pipes, and studs enclosed behind a wall. A multimeter, which measures voltage, resistance, and current, is handy not just for electrical engineers but also for electricians and hobbyists.

Sound amplifiers. Car stereos, home theaters, and electric guitars all have one thing in common: They all contain an amplifier. In the past, these have been purely analogue devices, but since the late twentieth century, digital amplifiers have supplanted their

analogue brethren due to their ease of operation and size. Audiophiles, however, claim that analogue amplifies sound better.

Power supplies. These can come in many sizes, both physically and in terms of power. Most people encounter a power supply as a black box plugged into an AC outlet with a cord that powers electrical devices such as a laptop, radio, or television. Inside each is a specially designed power inverter that converts AC power to the required volts and amperes of DC power.

Batteries. Thomas Edison is credited with creating the first portable battery, a rechargeable box that required only water once a week. Batteries are an electrochemical reaction, that is the realm of chemistry, and demonstrate how far afield electrical engineering can seem to go while remaining firmly grounded in its fundamentals. Battery technology is entering a new renaissance as the charge life is extending and the size is shrinking. Edison first marketed his "A" battery for use in electric cars before they went out of fashion. Electric cars that run on batteries may be making a comeback, and their cousin, the hybrid, runs on both batteries and combustion.

The power grid. This is one of the oldest accomplishments of electrical engineering. A massive nationwide interdependent network of transmission lines delivers power to every corner of the country. The power is generated at hydroelectric plants, coal plants, nuclear plants, and wind and solar plants. The whole thing works such that if any one section fails, the others can pick up the slack. Wind and solar pose particular challenges to the field, as wind and sunshine do not flow at a constant rate, but the power grid must deliver the same current and voltage at all times of day.

ELECTRICAL ENGINEERING

Electric trains and buses. Many major cities have some kind of public transportation that involves either an electrified rail, or bus wires, or both. These subways, light rails, and trolleys are an important part of municipal infrastructure, built on many of the same principles as the power grid, except that it is localized.

Automobiles. There are many electronic parts in a car. The first to emerge historically is the electric starter, obviating the hand crank. Once there was a battery in the car to power the starter, engineers came up with all sorts of other uses for it: headlamps, windshield wipers, interior lighting, a radio (and later tape and CD players), and the dubious car alarm, to name a few. The most important electrical component of modern automobiles is the computer-controlled fuel injector. This allows for the right amount of oxygen and fuel to be present in the engine for maximum fuel efficiency (or for maximum horsepower). The recent success of hybrids, and the potentially emerging market of all electric vehicles, means that there is still more electrical innovation to be had inside a century-old technology.

Medical devices. Though specifically the domain of biomedical engineering, many, if not most, medical devices are designed by electrical engineers who have entered this subdiscipline. Computed axial tomography (CAT) scanners, X rays, ultrasound, and magnetic resonance imaging (MRI) machines all rely

on electromagnetic and nuclear physics applied in an electrical setting (and controlled by electronics). These devices can be used to look into things other than human bodies as well. Researchers demonstrated that an MRI could determine if a block of cheese had properly aged.

Telecommunications. This used to be an international grid of telephone wires and cables connecting as many corners of the globe where wire could be strung. As of 2011, even the most remote outposts can communicate voice, data, and video thanks to advances in radio technology. The major innovation in this field has been the ability for multiple connections to ride the same signal. The original cell phone technology picked a tiny frequency for each of its users, thus limiting the number of total users to a fixed division in that band. Mobile communication has multiple users on the same frequency, which opens up the band to more users.

Broadcast television and radio. These technologies are older but still relevant to the electrical engineer. Radio is as vibrant as ever, and ham radio is even experiencing a mini renaissance. While there may not be much room for innovation, electrical engineers must understand them to maintain them, as well as understand their derivative technologies.

Lighting. Light-emitting diodes (LEDs), are low-power alternatives to incandescent bulbs (the lightbulb that Thomas Edison invented). They are just transistors, but as they have grown smaller and more colors have been added to their spectrum, they have found their way into interior lighting, computer monitors, flashlights, indicator displays, and control panels.

IMPACT ON INDUSTRY

New electrical devices are being introduced every day in a quantity too numerous to document. In 2006, consumer electronics alone generated $169 trillion in revenue. Nonetheless, electrical engineering has a strong public research and development component. Increasingly, businesses are partnering with universities to better capitalize on independent research. As of 2011, most IC design is done in the United States, Japan, and Europe, and most of the manufacturing is outsourced to Taiwan and Singapore. As wind and solar power become more popular, so does the global need for electrical engineers. Spain, Portugal, and Germany lead the European Union in solar panel use and production. In the United States, sunny states such as California give financial incentives to homes and businesses that incorporate solar.

University research. University research is funded by the United States government in the forms of grants from organizations such as the Defense Advanced Research Projects Agency (DARPA) and the National Aeronautics and Space Administration (NASA). The National Science Foundation (NSF) indirectly supports research through fellowships. The rest of the funding comes from industry. Research can be directed at any of the subdisciplines of the field: different material for transistors or new mathematics for better DSP or unique circuit configuration that optimizes a function. Often, the research is directed at combining disciplines, such as circuits

that perform faster DSP. DARPA is interested in security and weapons, such as better digital encryption and spy satellites. Solar power is also a popular area of research. There is a race to increase the performance of photovoltaic devices so that solar power can compete with gas, coal, and petroleum in terms of price per kilowatt hour. Universities that are heavily dependent on industry funding tend to research in areas that are of concern to their donors.

For example, Intel, the largest manufacturer of microprocessors, is a sponsor of various University of California electrical engineering departments and in 2011 announced their new three-dimensional transistor. The technology is based on original research first described by the University of California, Berkeley, in 2000 and funded by DARPA.

The Internet was originated from a Cold War era DARPA project. The United States wanted a communications network that would survive a first-strike nuclear assault. When the Cold War ended, the technology found widespread use in the civilian sphere. The Internet enabled universities to share research data and libraries a decade before it became a household commodity.

Business sector. More than half of the electrical engineers employed in the United States are working in electronics. Consumer electronics include companies such as Apple, Sony, and Toshiba that make DVD players, video-game consoles, MP3 players, laptops, and computers. But the majority of the engineering takes place at the constituent component level. Chip manufacturers such as Intel and Advanced Micro Devices (AMD) are household names, but there are countless other companies producing all the other kinds of microchips that find their way into refrigerators, cars, network storage devices, cameras, and home lighting. The FPGA market alone was $2.75 billion, though few end consumers will ever know that they are in everything from photocopiers to cell phone base stations.

In the non-chip sector, there are behemoths such as General Electric, which make everything from light bulbs to household appliances to jet engines. There are about 500 electric power companies in the contiguous forty-eight states of the United States. Because they are all connected to each other and the grid is aging, smart engineering is required to bring new sources online such as solar and wind. In 2009, the Federal Communications Commission (FCC) issued the National Broadband Plan, the goal of which is to bring broadband Internet access to every United States citizen. Telecommunications companies such as AT&T and cable providers are competing fiercely to deliver ever-faster speeds at lower prices to fulfill this mandate.

CAREERS AND COURSE WORK

Electrical engineering requires a diverse breadth of background course work—math, physics, computer science, and electrical theory—and a desire to specialize while at the same time being flexible to work with other electrical engineers in their own areas of expertise. A bachelor of science degree in electrical engineering usually entails specialization after the general course work is completed. Specializations include circuit design, communications and networks, power systems, and computer science. A master's degree is generally not required for an electrical engineer to work in the industry, though it would be required to enter academia or to gain a deeper understanding in the specialization. An electrical engineer wishing to work as an electrical systems contractor will probably require professional engineer (PE) certification, which is issued by the state after one has several years of work experience and has passed the certification exam.

Careers in the field of electrical engineering are as diverse as its applications. Manufacturing uses electrical engineers to design and program industrial equipment. Telecommunications employs electrical engineers because of their understanding of DSP. More than half of all electrical engineers work in the microchip sector, which uses legions of electrical engineers to design, test, and fabricate ICs on a continually shrinking scale. Though these companies seem dissimilar—medical devices, smart phones, computers (any device that uses an IC)— they have their own staffs of electrical engineers that design, test, fabricate, and retest the devices.

Electrical engineers are being seen more and more in the role of computer scientist. The course work has been converging since the twentieth century. University electrical engineering and computer science departments may share lecturers between the two disciplines. Companies may use electrical engineers to solve a computer-programming problem in the hopes that the electrical engineer can debug both the hardware and software.

SOCIAL CONTEXT AND FUTURE PROSPECTS

Electrical engineering may be the most under-recognized driving force behind modern living. Everything from the electrical revolution to the rise of the personal computer to the Internet and social networking has been initiated by electrical engineers. This field first brought electricity into our homes and then ushered in the age of transistors. Much of the new technology being developed is consumed as software and requires computer programmers. But the power grid, hardware, and Internet that powers it were designed by electrical engineers and maintained by electrical engineers.

As the field continues to diversify and the uses for electricity expands, the need for electrical engineers will expand, as will the demands placed on the knowledge base required to enter the field. Electrical engineers have been working in the biological sciences, a field rarely explored by the electrical engineer. The neurons that comprise the human brain are an electrical system, and it makes sense for both fields to embrace the knowledge acquired in the other.

Other disciplines rely on electrical engineering as the foundation. Robotics, for example, merge mechanical and electrical engineering. As robots move out of manufacturing plants and into our offices and homes, engineers with a strong understanding of the underlying physics are essential. Another related field, biomedical engineering, combines medicine and electrical engineering to produce life-saving devices such pacemakers, defibrillators, and CAT scanners. As the population ages, the need for more advanced medical treatments and early detection devices becomes paramount. Green power initiatives will require electrical engineers with strong mechanical engineering and chemistry knowledge. If recent and past history are our guides, the next scientific revolution will likely come from electrical engineering.

—*Vincent Jorgensen, BS*

FURTHER READING

Adhami, Reza, Peter M. Meenen III, and Dennis Hite. *Fundamental Concepts in Electrical and Computer Engineering with Practical Design Problems.* 2d ed. Boca Raton, Fla.: Universal Publishers, 2005. A well-illustrated guide to the kind of math required to analyze electrical circuits, followed by sections on circuits, digital logic, and DSP.

Davis, L. J. *Fleet Fire: Thomas Edison and the Pioneers of the Electric Revolution.* New York: Arcade, 2003. The stunning story of the pioneer electrical engineers, many self-taught, who ushered in the electric revolution.

Gibilisco, Stan. *Electricity Demystified.* New York: Mc-Graw-Hill, 2005. A primer on electrical circuits and magnetism.

Mayergoyz, I. D., and W. Lawson. *Basic Electric Circuit Theory: A One-Semester Text.* San Diego: Academic Press, 1997. Introductory textbook to the fundamental concepts in electrical engineering. Includes examples and problems.

McNichol, Tom. *AC/DC: The Savage Tale of the First Standards War.* San Francisco: Jossy-Bass, 2006. The riveting story of the personalities in the AC/DC battle of the late nineteenth century, focusing on Thomas Edison and Nicola Tesla.

Shurkin, Joel N. *Broken Genius: The Rise and Fall of William Shockley, Creator of the Electronic Age.* New York: Macmillian, 2006. Biography of the Nobel Prize-winning electrical engineer and father of the Silicon Valley, who had the foresight to capitalize on invention of the transistor but ultimately went down in infamy and ruin.

WEB SITES
Association for Computing Machinery
http://www.acm.org

Computer History Museum
http://www.computerhistory.org

Institute of Electrical and Electronics Engineers
http://www.ieee.org

National Society of Professional Engineers
http://www.nspe.org/index.html

See also: Bionics and Biomedical Engineering; Computer Engineering; Computer Science; Mechanical Engineering.

ELECTRICAL MEASUREMENT

FIELDS OF STUDY

Electronics; electronics technology; instrumentation; industrial machine maintenance; electrical engineering; metrology; avionics; physics; electrochemistry; robotics; electric power transmission and distribution services.

SUMMARY

Electrical measurement has three primary aspects: the definition of units to describe the electrical properties being measured; the modeling, design, and construction of instrumentation by which those units may be applied in the measurement process; and the use of measurement data to analyze the functioning of electric circuits. The measurement of any electrical property depends on the flow of electric current through a circuit. A circuit can exist under any conditions that permit the movement of electric charge, normally as electrons, from one point to another. In a controlled or constructed circuit, electrons move only in specific paths, and their movement serves useful functions.

KEY TERMS AND CONCEPTS

- **ampere:** Measure of the rate at which electrons are passing through an electric circuit or some component of that circuit.
- **analogue:** Electric current that is continuous and continuously variable in nature
- **capacitance:** Measure of the ability of a nonconducting discontinuous structure to store electric charge in an active electric circuit.
- **digital:** Electric current that flows in packets or bits, with each bit being of a specific magnitude and duration.
- **electromagnetism:** Property of generating a magnetic field by the flow of electricity through a conductor.
- **hertz:** Unit of measurement defined as exactly one complete cycle of any process over a time span of exactly one second.
- **inductance:** Measure of the work performed by an electric current in generating a magnetic field as it

passes through an induction coil.

- **left-hand rule:** Rule of thumb in determining the direction of the north pole of the magnetic field in a helically wound electromagnetic coil.
- **phase difference:** Measure of the extent to which a cyclic waveform such as an alternating voltage follows or precedes another.
- **resistance:** Measure of the ability of electric current to flow through a circuit or some component part of a circuit or device; the inverse of conductance.

DEFINITION AND BASIC PRINCIPLES

Electrical measurement refers to the quantification of electrical properties. As with all forms of measurement, these procedures provide values relative to defined standards. The basic electrical measurements are voltage, resistance, current, capacitance, and waveform analysis. Other electrical quantities such as inductance and power are generally not measured directly but are determined from the mathematical relationships that exist among actual measured properties of an electric circuit.

An electric circuit exists whenever physical conditions permit the movement of electrons from one location to another. It is important to note that the formation of a viable electric circuit can be entirely accidental and unexpected. For example, a bolt of lightning follows a viable electric circuit in the same way that electricity powering the lights in one's home follows a viable electric circuit.

Electrons flow in an electric circuit because of differences in electrical potential between one end of the circuit and the other. The flow of electrons in the circuit is called the current. When the current flows continuously in just one direction, it is called direct current, or DC. In direct current flow, the potential difference between ends of the circuit remains the same, in that one end is always relatively positive and the other is relatively negative. A second type of electric current is called alternating current, or AC. In alternating current, the potential difference between ends of the circuit alternates signs, switching back and forth from negative to positive and from positive to negative. The electrons in alternating current do not flow from one end of the circuit to the other but instead oscillate back and forth between the ends at a specific frequency.

The movement of electrons through an electric circuit is subject to friction at the atomic level and to other effects that make it more or less difficult for the electrons to move about. These effects combine to restrict the flow of electrons in an overall effect called the resistance of the circuit. The current, or the rate at which electrons can flow through a circuit, is directly proportional to the potential difference (or applied voltage) and inversely proportional to the resistance. This basic relationship is the foundation of all electrical measurements and is known as Ohm's law.

Another basic and equally important principle is Kirchoff's current law, which states that the electric current entering any point in a circuit must always be equal to the current leaving that point in the circuit. In the light of the definition of electric current as the movement of electrons from point to point through a conductive pathway, this law seems concrete and obvious. It is interesting to note, however, that it was devised in 1845, well before the identification of electrons as discrete particles and the discovery of their role in electric current.

Electrical measurement, like all measurement, is a comparative process. The unit of potential difference—called the volt—defines the electrical force required to move a current of one ampere through a resistance of one ohm. Devices that measure voltage are calibrated against this standard definition. This definition similarly defines the ohm but not the ampere. The ampere is defined in terms of electron flow, such that a current of one ampere represents the movement of one coulomb of charge (equivalent to 6.24×10^{18} electrons) past a point in a period of one second.

The capacitance of a device in an electric circuit is defined as the amount of charge stored in the device relative to the voltage applied across the device. The inductance of a device in an electric circuit is more difficult to define but may be thought of as the amount of current stored in the device relative to the voltage applied across the device. In both cases, the current flow is restricted as an accumulation of charge within the device but through different methods. Whereas a capacitor restricts the current flow by presenting a physical barrier to the movement of electrons, an inductor restricts current flow by effectively trapping a certain amount of flowing current within the device.

BACKGROUND AND HISTORY

Wild electricity—lightning and other natural

phenomena that result from differences in the oxidation potentials of different materials—has been observed and known for ages. Artificially produced electricity may have been known thousands of years ago, although this has not been proven conclusively. For example, artifacts recovered from some ancient Parthian tombs near Baghdad, Iraq, bear intriguing similarities in construction to those of more modern electrochemical cells or batteries. Reconstructions of the ancient device have produced electric current at about 0.87 volts, and other observations indicate that the devices may have been used to electroplate metal objects with gold or silver.

The modern battery, or voltaic pile, began with the work of Alessandro Volta in 1800. During the nineteenth century, a number of other scientists investigated electricity and electrical properties. Many of the internationally accepted units of electrical measurement were named in their honor.

HOW IT WORKS

Ohm's law. The basis of electrical measurement is found in Ohm's law, derived by Georg Simon Ohm in the nineteenth century. According to Ohm's law, the current flowing in an electric circuit is directly proportional to the applied voltage and inversely proportional to the resistance of the circuit. In other words, the greater the voltage applied to the circuit, the more current will flow. Conversely, the greater the resistance of the circuit, the less current will flow. This relationship can be stated mathematically as $E = I \times R$ (where E = voltage, I = current, and R = resistance), in which voltage is represented as the product of current and resistance.

Given this relationship, it is a relatively simple matter to design a device that uses two specific properties to determine the third. By constructing a device that employs set values of voltage and resistance, one can measure current. Similarly, by constructing a device that describes a system in which current and resistance are constant, one can measure voltage, and by devising a system in which current and voltage are regulated, one can measure resistance.

If the three primary properties of a circuit are known, then all other properties can be determined by arithmetic calculations. The capacitance of a circuit or circuit component, for example, is the amount of charge stored in the device at a given applied voltage, and the amount of charge is proportional to the current in the device. Similarly, the inductance in a circuit or circuit component depends on the current passing through the device at a given voltage.

Units of measurement. All electrical properties must have an associated defined standard unit to be measurable. To that end, current is measured in amperes, named after Louis Ampere. Potential difference, sometimes called electromotive force, is measured in volts, named after Volta. Resistance is measured in ohms, named after Ohm. Power is measured in watts, named after James Watt. Capacitance is measured in farads, named after Michael Faraday. Inductance is measured in henrys, named after Joseph Henry. Conductance, the reciprocal of resistance, is measured in siemens, named after Ernst W. von Siemens. Frequencies are measured in hertz, or cycles per second, named after Gerhard Hertz.

Basic electricity concepts. Electricity can be produced in a continuous stream known as direct current (DC), in which electrons flow continuously from a negative source to a positive sink. The potential difference between the source and the sink is the applied voltage of the circuit, and it does not change. Electricity can also be produced in a varying manner called alternating current (AC), in which electron flow oscillates back and forth within the circuit. The applied voltage in such a system varies periodically between positive and negative values that are equal in magnitude. It is important to understand that circuits designed to operate with one type of applied voltage do not function when the other type of voltage is applied. In other words, a circuit designed to perform certain functions when supplied with a constant voltage and direct current will not perform those functions when supplied with a varying voltage and alternating current. The fundamental concept of Ohm's law applies equally to both cases, but other characteristics such as phase and frequency differences and voltage waveform make the relationships more complex in alternating current applications. Electrical measurement devices are designed to accommodate these characteristics and are capable of extremely fine differentiation and precision.

APPLICATIONS AND PRODUCTS

The easiest electrical properties to measure accurately are voltage and resistance. Thus, the primary application tool of electrical measurement is the common volt-ohm meter (VOM), either as an analogue device or as its digital counterpart, the digital volt-ohm meter (DVOM).

Basic analogue measuring devices. Two systems are required for any measuring device. One system is the structure by which the unknown value of a property is measured, and the other is the method of indicating that value to the user of the device. This latter feature was satisfied through the application of electromagnetic induction in the moving coil to produce a D'Arsonval movement. The strength of a magnetic field produced by current flowing through a coil is directly proportional to the magnitude of that current. In a basic analogue measuring device, the small coil is allowed to pivot freely within a permanent magnetic field. The amount by which the coil pivots is determined by the amount of current flowing through it; a needle attached to the coil indicates the appropriate value being measured on a calibrated scale. Analogue meters used to measure most electrical properties employ the moving-coil system to indicate the value of the property being measured.

Basic digital measuring devices. The advent of digital electronics brought about a revolution in electrical measurement devices. Digital meters use a number of different systems to produce the desired information. The basic operation of digital devices is controlled by a central clock cycle in such a way that the value of the inputs are effectively measured thousands of separate times per second rather than continuously. This is known as sampling. Because the flow of electricity is a continuous or analogue process, sampling converts the analogue input value to a digital data stream. The data values can then be manipulated to be displayed directly as a numerical readout, eliminating the guesswork factor involved in reading a needle scale indicator. Another advantage of digital measurement devices is their inherent sensitivity. Through the use of an operational amplifier, or op-amp integrated circuits (IC), input signals can be amplified by factors of hundreds of thousands. This allows extremely small electrical values to be measured with great accuracy.

Other measuring devices. One of the most valuable devices in the arsenal of electrical measurement is the oscilloscope, which is available in both analogue and digital models. Like a typical meter, the oscilloscope measures electrical inputs, but it also has the capability to display the reading as a dynamic trace on a display screen. The screen is typically a cathode-ray tube (CRT) display, but later versions use liquid crystal display (LCD) screens and even can be used with desktop and laptop computers.

Another highly useful device for electrical measurement is the simple logic probe used with digital circuitry. In digital electronics, the application of voltages, and therefore the flow of current, is not continuous but appears in discrete bits as either an applied voltage or no applied voltage. These states are known as logic high and logic low, corresponding to on and off. The manipulation of these bits of data is governed by strict Boolean logic (system for logical operations on which digital electronics is based). Accordingly, when a digital device is operating properly, certain pins (or leads) of an integrated-circuit connection must be in the logic high state while others are in the logic low state. The logic probe is used to indicate which state any particular pin is in, generally by whether an indicator light in the device is on or off. Unlike a meter, the logic probe does not provide quantitative data or measurements, but it is no less invaluable as a diagnostic tool for troubleshooting digital circuitry.

Ancillary devices. A more specific analytical device is the logic analyzer. Designed to read the input or pin signals of specific central processing unit (CPU) chips in operation, the logic analyzer can provide a running record of the actual programming used in the function of a digital electronic circuit. Another device that is used in electrical measurement, more typically as an electrical source than as a measuring device, is the waveform generator. This device is used to provide a specific shape of input voltage for an electric circuit to verify or test the function of the circuit.

Indirect applications. Because of the high sensitivity that can be achieved in electrical measurement, particularly in the application of digital electronic devices, the measurement of certain electrical properties is widely used in analytical devices. The most significant electrical property employed in this way is resistance measurement. Often, this is measured as its converse property—conductivity. Gas-phase detectors on analytical devices such as gas chromatographs, high-performance liquid chromatographs, and combination devices with mass spectrometers are designed to measure the resistance, or conductivity, of the output stream from the device. For example, in gas chromatography, a carrier gas is passed through a heated column packed with a chromatography medium. When a mixture of compounds is injected into the gas stream, the various components become separated as they pass through the column. As the materials exit the column, they pass through a detector that measures changes in conductivity that occur with each different material. The changes are recorded either as a strip chart or as a collection of data.

The sensitive measurement of electrical resistance is made possible by the use of specific electric circuits, most notably the Wheatstone bridge circuit. In a Wheatstone bridge circuit, four resistance values are used, connected in a specific order. Two sets of resistance values are connected parallel to each other, with each set containing two resistance values connected in series with each other. One of the resistance values is the unknown quantity to be measured, while the other three are precisely known values. The voltage between the two midpoints of the series circuits of the bridge changes very precisely with any change in one of the resistance values. In addition, the output voltage signal can be amplified by several orders of magnitude, making even very small changes in output voltage meaningful. The role of digital electronics and op-amps cannot be overstated in this area.

IMPACT ON INDUSTRY

Electrical measurement has had a tremendous impact on the manner in which industry is carried out, particularly in regard to automation and robotic control, and in process control. The electrical measurement of process variables in manufacturing and processing can be applied to control circuitry that automatically adjusts and maintains the correct and proper functioning of the particular system. This both renders the process more precise and eliminates problems that are caused by human error, as well as reduces the number of personnel required for hands-on operational checking and process maintenance.

Electrical measurement is central to essentially all modern methods of testing and analysis and can be attributed to several factors. The technology is superbly adaptable to such functions and is fundamental to the operations of detection and electronic control. It is also extremely sensitive and precise, with the capability of measuring extremely small values of voltage, resistance, and current with precision. A

thriving industry has subsequently developed to provide the various control mechanisms needed for the operation of modern methods of production and analysis. The procedures of electrical measurement also have a central role in medical analysis and other high-tech sciences such as physics and chemistry. In these fields, the devices used to carry out everything from the routine analysis of large groups of

---

FASCINATING FACTS ABOUT ELECTRICAL MEASUREMENT

- Devices nearly 2,000 years old that were recovered from tombs near Baghdad, Iraq, were almost certainly batteries used to gold plate other metal objects. The technology of the battery was lost and not rediscovered until 1800.
- Henry Cavendish discovered what would be known as Ohm's law about fifty years earlier than Georg Simon Ohm did. However, as Cavendish did not publish his observations, the law has been attributed to Ohm.
- The most commonly and easily measured electrical quantity is voltage, or potential difference.
- Essentially all electrical measurements are determined by the relationship between voltage, resistance, and current (Ohm's law).
- Kirchoff's current law was stated before electrons were discovered and identified, and electricity was still regarded as some kind of mysterious fluid that certain materials contained.
- A bolt of lightning results when electrons flow through a regular electric circuit formed by the presence of charged particles (ions) in the atmosphere between clouds and the surface of the ground.
- An average lightning bolt transfers only about 5 coulombs of charge, but the transfer takes place so quickly that the electric current is about 30,000 amperes.
- Power, as measured in watts, is used to train elite cyclists. Over the course of an hour, professional cyclist Lance Armstrong can consistently generate 350 watts of electricity; the average in-shape cyclist, only about 100 watts.
- A static shock that a person can hear, feel, and see is about 250 volts of electricity.
- Conductance is the reciprocal of resistance; the unit of conductance was originally called the mho, as the reciprocal of the ohm.

urine samples in a biomedical analytical laboratory to single-run experiments in the newest, largest subatomic particle colliders depend on electrical measurement for their functioning.

Nanoscience and nanotechnology. Scientists continuously discover new materials and develop them for applications that capitalize on their specific electrical properties. They also modify existing measurement devices and design completely new ones. This often requires the redesign of electric circuitry, particularly the miniaturization of the devices. In the field of nanotechnology, for example, control circuitry for such small devices may consist of no more than a network of metallic tracings on a surface that measures correspondingly small electrical values.

Basic electric service. The mainstay of electrical measurement in industry is the service industry that maintains the operation of the continental electric grid and the power supplied to individual locations. Distribution of electric power calls for close monitoring of the network by which electric power is carried to individual residences, factories, and other locations. It also requires the installation and operation of new sources of electric power. This requires a skilled workforce that is capable of using basic electrical measurement practices. Accordingly, a significant component of training programs provided by colleges, universities, and vocational schools focuses on the practical aspects of the physical delivery of electric power, where measurement is a foundation skill. Simple economics demands that the amount of electricity being produced and the amount being consumed must be known. The effective distribution of that electricity, both generally and within specific installations, is vitally important to its continued functioning.

CAREERS AND COURSE WORK

Studies related to the use of electrical devices provide a good, basic knowledge of electrical measurement. A sound basis in physics and mathematics will also be required. College and university level course work will depend largely on the chosen area of specialization. Options at this level range from basic electrical service technician to fundamental physics. At a minimum, students will pursue studies in mathematics, physical sciences, industrial technologies (chemical, electrical, and mechanical), and business at the undergraduate level or as trade students.

More advanced studies usually consist of specialized courses in a chosen field. The study of applied mathematics will be particularly appropriate in advanced studies, as this branch of mathematics provides the mathematical basis for phase relationships, quantum boundaries, and electron behavior that are the central electrical measurement features of advanced practices.

In addition, as technologies of electrical measurement and application and regulations governing the distribution of electric energy change, those working in the field can expect to be required to upgrade their working knowledge on an almost continual basis to keep abreast of changes.

SOCIAL CONTEXT AND FUTURE PROSPECTS

Economics drives the production of electricity for consumption, and in many ways, the developed world is very much an electrical world. A considerable amount of research and planning has been devoted to the concept of the smart grid, in which the grid itself would be capable of controlling the distribution of electricity according to demand. Effective electrical measurement is absolutely necessary for this concept to become a working reality.

At a very basic level, the service industry for the maintenance of electrically powered devices will continue to provide employment for many, particularly as the green movement tends to shift society toward recycling and recovering materials rather than merely disposing of them. Accordingly, repair and refurbishment of existing units and the maintenance of residential wiring systems would focus more heavily on effective troubleshooting methods to determine the nature of flaws and correct them before any untoward incident should occur.

The increased focus on alternative energy sources, particularly on the development of solar energy and fuel cells, will also place higher demands on the effectiveness of electrical measurement technology. This will be required to ensure that the maximum amount of usable electric energy is being produced and that the electric energy being produced is also being used effectively.

—*Richard M. J. Renneboog, MSc*

FURTHER READING

Clark, Latimer. *An Elementary Treatise on Electrical*

Measurement for the Use of Telegraph Inspectors and Operators. London: E. and F. N. Spon, 1868. An interesting historical artifact that predates the identification of electrons and their role in electric current and measurement.

Herman, Stephen L. *Delmar's Standard Textbook of Electricity.* 5th ed. Clifton Park, N.Y.: Delmar Cengage Learning, 2009. A good basic presentation of the fundamental principles of electricity and electrical measurement.

Herman, Stephen L., and Orla E. Loper. *Direct Current Fundamentals.* 7th ed. Clifton Park, N.Y.: Thomson Delmar Learning, 2007. An introductory level textbook offering basic information regarding direct current electricity and electrical measurement.

Lenk, John D. *Handbook of Controls and Instrumentation.* Englewood Cliffs, N.J.: Prentice Hall, 1980. Describes general electrical measurement applications in principle and practice, as used for system and process control purposes.

Malvino, Albert Paul. *Malvino Electronic Principles.* 6th ed. New York: Glencoe/McGraw-Hill Books, 1999. Presents a detailed analysis of the electronic principles and semiconductor devices behind digital electronic applications in electrical measurement.

Strobel, Howard A., and William R. Heineman. *Chemical Instrumentation: A Systematic Approach.* 3d ed. New York: John Wiley & Sons, 1989. Describes the fundamentals of design and the operation of instrumentation used in chemistry and chemical applications, all of which ultimately use and depend on electrical measurement.

Tumanski, Slawomir. *Principles of Electrical Measurement.* New York: Taylor and Francis, 2006. Provides a thorough, descriptive introduction to the principles of electrical measurement and many of the devices that are used to manipulate and measure electric signals.

WEB SITES
U.S. Department of Energy Electric Power
http://www.energy.gov/energysources/electricpower.htm

U.S. Energy Information Administration Electricity
http://www.eia.doe.gov/fuelelectric.html

See also: Electrical Engineering; Electrochemistry; Electrometallurgy; Electronics and Electronic Engineering; Fossil Fuel Power Plants; Integrated-Circuit Design; Nanotechnology.

ELECTRIC AUTOMOBILE TECHNOLOGY

FIELDS OF STUDY

Automotive engineering; chemical engineering; clean-energy technologies; electrical engineering; informatics; information technologies; materials engineering; mathematics; physics; resource management.

SUMMARY

Electric vehicles have been around even longer than internal combustion engine cars. With health issues resulting from the modern use of internal combustion engines, the automotive industry is intensifying its efforts to produce novel machines that run on electricity. Many cars come with drivetrains that can accept electric propulsion, offering quieter, healthier transportation options. Although consumers still seem to shy away from completely electric vehicles, hybrid vehicles that use both internal combustion engines and electric power are in use in many cities around the world.

KEY TERMS AND CONCEPTS

- **aromatic compound:** Carbon-based chemical such as benzene or toluene; sometimes harmful.
- **battery:** Combination of one or more electrochemical cells used to convert stored chemical energy into an electric current.
- **combustion:** Process by which a substance (fuel) reacts with oxygen to form carbon dioxide, water, heat, and light.
- **current:** Flow of electricity through a material.
- **drivetrain:** Mechanical system that transmits power or torque from one place (such as the motor) to another (such as the wheels).
- **electrochemical cell:** Device that can derive elec-

trical energy from chemical reactions or facilitate chemical reactions by applying electricity.

- **fuel:** Chemical compound that can provide energy by its conversion to water or carbon dioxide, either by combustion or by electrochemical conversion.
- **fuel cell:** Device that combines oxygen from the air with a fuel such as hydrogen to form water, heat, and electricity by splitting off protons from the fuel and allowing only those protons to pass through a dense membrane.
- **hybrid vehicle:** Car using a combination of several essentially different drive mechanisms, usually an internal combustion engine in combination with an electric motor, which in turn can derive its power from a battery or a fuel cell.
- **lithium-ion battery:** Modern battery that involves lithium in storing and converting energy, has a high energy density, but degrades fast at elevated temperatures.
- **membrane:** Dense dividing wall between two compartments.
- **polarization:** Magnetization of a material with a defined polarity.
- **proton exchange membrane fuel cell:** Polymer-based electrochemical device that converts various fuels and air into water, heat, and electric energy at low operating temperatures (70 to 140 degrees Celsius).
- **solid oxide fuel cell:** Ceramic electrochemical device that converts various fuels and air into water, heat, and electric energy at high operating temperatures (600 to 1,000 degrees Celsius).
- **torque:** Tendency of a force to rotate an object around a defined axis or pivot; also known as moment force.
- **voltage:** Electromotive force of electricity; also defined as the work required to move a charged object through an electric field.

DEFINITION AND BASIC PRINCIPLES

Electric vehicles are driven by an electric motor. The electricity for this motor can come from different sources. In vehicle technology, electrical power is usually provided by batteries or fuel cells. The main advantages of these devices are that they are silent, operate with a high efficiency, and do not have tailpipe emissions harmful to humans and the environment. In hybrid vehicles, two or more motors coexist in the vehicle. When large quantities of power are required rapidly, the power is provided by combusting fuels in the internal combustion engine; when driving is steady, or the car is idling at a traffic light, the car is entirely driven by the electric motor, thereby cutting emissions while providing the consumer with the normal range typically associated with traditional cars that would rely entirely on internal combustion engines. Electric vehicles make it possible for drivers to avoid having to recharge at a station. Recharging can occur at home, at work, and in parking structures, quietly, cleanly, and without involving potentially carcinogenic petroleum products.

BACKGROUND AND HISTORY

Electric vehicles have been around since the early 1890's. Early electric vehicles had many advantages over their combustion-engine competitors. They had no smell, emitted no vibration, and were quiet. They also did not need gear changes, a mechanical problem that made early combustion-engine cars cumbersome. In addition, the torque exhibited by an electric engine is superior to the torque generated by an equivalent internal combustion engine. However, battery technology was not yet sufficiently developed, and consequently, as a result of low charge, storage capacity in the batteries, and rapid developments in internal combustion engine vehicle technology, electric vehicles declined on the international markets in the early 1900's.

At the heart of electric vehicles is the electric motor, a relatively simple device that converts electric energy into motion by using magnets. This technology is typically credited to the English chemist and physicist Michael Faraday, who discovered electromagnetic induction in 1831. These motors require some electrical power, which is typically provided by batteries, as it was done in the early cars, or by fuel cells. Batteries were described as early as 1748 by one of the founding fathers of the United States, Benjamin Franklin. These devices can convert chemically stored energy into a flow of electrons by converting the chemicals present in the battery into different chemicals. Depending on the materials used, some of these reactions are reversible, which means that by applying electricity, the initial chemical can be recreated and the battery reused. The development of fuel cells is typically credited to the English lawyer and physicist Sir William Grove in 1839, who discovered that flowing hydrogen and air over the surfaces of platinum rods in sulfuric acid creates a current of electricity and leads to the formation of water. The devices necessary to develop an electric car had been around for many decades before they were first assembled into a vehicle.

A major circumstance that led to the commercial success of combustion-engine vehicles over electric vehicles was the discovery and mining of cheap and easily available oil. Marginal improvements in battery technology compared with internal combustion engine technology occurred during the twentieth century. As a result of stricter emissions standards near the end of the twentieth century, global battery research began to reemerge and has significantly accelerated in the 2010's. Some early results of this research are on the market.

During the 1990's, oil was still very cheap, and consumers, especially in North America, demanded heavier and larger cars with stronger motors. During this decade, General Motors (GM) developed an electric vehicle called the EV1, which gained significant, though brief, international positive attention before it was taken off the market shortly after its introduction. All produced new cars were destroyed, and the electric-vehicle program was shut down. The development of electric vehicles was then left to other companies.

Barely twenty years later, following a significant negative impact on the car manufacturing companies in North America from the 2009 financial crisis and their earlier abandonment of research and development of electric car technology, North American companies tried to catch up with the electric vehicle technology of the global vehicle manufacturing industry. During the hiatus, global competitors surpassed North American companies by creating modern, fast, useful electrical vehicles such as the Nissan Cube from Japan and the BMW ActiveE models from Germany. GM, at least, has made a full turnaround after receiving government financial incentives to develop battery-run vehicles and has actively pursued a new electric concept called Chevrolet Volt. Similar to hybrid cars, the Volt has a standard battery, but because early twenty-first century batteries are not yet meeting desired performance levels, the Volt also has a small engine to extend its range. Installing two different power sources in a vehicle, one electric and one combustion based, makes sense in order to develop a product that has lower emissions but the same range as combustion-enginebased vehicles.

HOW IT WORKS

Power source. Gasoline, which is mainly a chemical called octane, is a geologic product of animals and plants that lived many millions of years ago. They stored the energy of the Sun either directly from photosynthesis or through digestion of plant matter. The solar energy that is chemically stored in gasoline is released during combustion.

The storage of energy in batteries occurs through different chemicals, depending on the type of battery. For example, typical car-starter lead-acid batteries use the metal lead and the ceramic lead oxide to store energy. During discharge, both these materials convert into yet another chemical called lead sulfate. When a charge is applied to the battery, the original lead and lead oxide are re-created from the lead sulfate. Over time, some of the lead and lead oxide are lost in the battery, as they separate from the main material. This can be seen as a black dust swimming in the sulfuric acid of a long-used battery and indicates that the battery can no longer be recharged to its full initial storage capacity. This happens to all types of batteries. Modern lithium-ion batteries used in anything from vehicles to mobile phones use lithium-cobalt/nickel/manganese oxide, and lithium graphite. These batteries use lithium ions to transport the charges around, while allowing the liberated electrons to be used in an electric motor. Other batteries use zinc to store energy—for example, in small button cells. Toxic materials such as mercury and cadmium have for some years been used in specific types of batteries, but have mostly been phased out because of the potential leaching of these materials into groundwater after the batteries' disposal.

Fuel cells do not use a solid material to store their charge. Instead, low-temperature proton exchange membrane fuel cells use gases such as hydrogen and liquid ethanol (the same form of alcohol found in vodka) or methanol as fuels. These materials are pumped over the surface of the fuel cells, and in the presence of noble-metal catalysts, the protons in these fuels are broken away from the fuel molecule and transported through the electrolyte membrane to form water and heat in the presence of air. The liberated electrons can, just as in the case of batteries, be used to drive an electric motor. Other types of fuel cells, such as molten carbonate fuel cells and solid oxide fuel cells, can use fuels such as carbon in the form of coal, soot, or old rubber tires and operate at 800 degrees Celsius with a very high efficiency.

Converting electricity into motion. Most electric motors use a rotatable magnet the polarity of which can be reversed inside a permanent magnet. Once electricity is available to an electric motor, electrons, traveling through an electric wire and coiled around a shaft that can be magnetized, generate an electrical field that polarizes the shaft. As a result, the shaft is aligned within the external permanent magnet, since reverse polarities in magnets are attracted to each other. If the polarity of the rotatable shaft is now reversed by changing the electron flow, the magnet reverses polarity and rotates 180 degrees. If the switching of the magnetic polarity is precisely timed, constant motion will be created. Changes in rotational speed can be achieved by changing the frequency of the change in polarization. The rotation generated by an electric motor can then be used like the rotation generated by an internal combustion engine by transferring it to the wheels of the vehicle.

Research and development. Many components of electric vehicles can be improved by research and development. In the electric motor, special magnetic glasses can be used that magnetize rapidly with few losses to heat, and the magnet rotation can occur in a vacuum and by using low-friction bearings. Materials research of batteries has resulted in higher storage capacities, lower overall mass, faster recharge cycles, and low degradation over time. However, significant further improvements can still be expected from this type of research as the fundamental understanding of the processes occurring in batteries become better understood.

Novel fuel cells are being developed with the goal of making them cheaper by using non-precious-metal catalysts that degrade slowly with time and are reliable throughout the lifetime of the electric motor. The U.S. Department of Energy has set specific lifetime and performance targets to which all

Early electric car, built by Thomas Parker, photo from 1895

these devices have to adhere to be useful on the commercial market. In fact, electric car prototypes are already available and comparable to vehicles using internal combustion engines. As of 2011, the main factor preventing deep mass-market penetration is cost, but that is continuously addressed by research and development of novel batteries and fuel cells that are lighter, use less expensive precious-metal catalysts, last longer, and are more reliable than previous devices. Since the 2010's, the development of these devices has significantly accelerated, especially because of international funding that is being poured into clean-energy technologies.

There are, however, disadvantages to all energy-conversion technologies. Internal combustion engines require large amounts of metals, including iron, chromium, nickel, manganese, and other alloying elements. They also require very high temperatures in forges during production. Additionally, the petroleum-based fuels contain carcinogenic chemicals, and the exhausts are potentially dangerous to humans and the environment, even when catalytic converters are used; to function well, these devices require large amounts of expensive and rare noble metals such as palladium and platinum. The highest concentrations of oil deposits have been found in politically volatile regions, and oil developments in those regions have been shown to increase local poverty and to cause severe local environmental problems.

Batteries require large quantities of rare-earth elements such as lanthanum. Most of these elements are almost exclusively mined in China, which holds a monopoly on the pricing and availability of these elements. Some batteries use toxic materials such as lead, mercury, or cadmium, although the use of these elements is being phased out in Europe. Lithium-ion batteries can rapidly and explosively discharge when short-circuited and are also considered a health risk. Electricity is required to recharge batteries, and it is often produced off-site in reactors whose emissions and other waste can be detrimental to human health and the environment.

Fuel cells require catalysts that are mostly made from expensive noble metals. Severe price fluctuations make it difficult to identify a stable or predictable cost for these devices. The fuels used in fuel cells, mostly hydrogen and methanol or ethanol, have to be produced, stored, and distributed. As of 2011, the majority of the hydrogen used is derived via a water-gas shift reaction, where oxygen is stripped off the water molecules and binds with carbon molecules from methane gas, producing hydrogen with carbon dioxide as a by-product; the process requires large quantities of natural gas. Methanol or ethanol can be derived from plant matter, but if it is derived from plants originally intended as food, food prices may increase, and arable land once used for food production then produces fuels instead.

Nevertheless, while the advantages and disadvantages of cleaner energy technologies such as fuel cells and batteries must be weighed against their ecological and economic impacts, it is important to remember that they are significantly cleaner than current internal combustion engine technologies.

APPLICATIONS AND PRODUCTS

Batteries. Battery technology still needs to be developed to be lighter without reducing the available charge. This means that the energy density of the battery (both by mass and by volume) needs to increase in order to improve a vehicle's range. Furthermore, faster recharge cycles have to be developed that will not negatively impact the degradation of the batteries. Overnight recharge cycles are possible, and good for home use, but a quick recharge during a shopping trip should allow the car to

regain a significant proportion of its original charge. Repeated recharge cycles at different charge levels as well as longtime operation with large temperature fluctuations should not detrimentally affect the microstructure of the batteries, so the power density of the batteries will remain intact. Furthermore, operation in very cold environments, in which the charge carriers inside the battery are less mobile, should be realized for a good market penetration. The introduction of the Chevrolet Volt into the North American market in March, 2011, resulted in a disappointment, as customers appeared unwilling to pay a premium for battery-operated cars. Stricter policies enforcing the conversion of a more significant proportion of cars into electric vehicles are necessary to change the market, especially in North America.

Personal vehicles. GM's EV1 was an attempt to market electric vehicles in North America in the 1990's. It was fast and lightweight and had all the amenities required by consumers but was discontinued by the manufacturer because it was not commercially viable. The 2011 edition of GM's Chevrolet Volt was almost indistinguishable from other GM station wagons, but the cost for the battery-powered car proved too high for a market that was used to very cheap vehicles with internal combustion engines. Electric vehicle technology is arguably much more advanced in Asia. Asian vehicle manufacturers were up to ten years ahead of the rest of the world in producing hybrid-electric vehicles, and they are set up to be ahead in the manufacturing of completely electric vehicles as well. For example, battery-only vehicles such as the Toyota iQ and the Nissan Leaf can drive up to 100 miles on a single battery charge with performance similar to that of an internal combustion engine car. European manufacturers, such as Renault, have teamed up with Nissan so as to not be left behind in the electric vehicle business and offered their first electric vehicle lineup to the European markets in 2011. Volkswagen has produced several studies including a concept called SpaceUp, but Volkswagen CEO Martin Winterkorn said in 2011 that battery technology was not mature enough for vehicles, and that Volkswagen would not produce any mass-market devices before 2013. Car manufacturer Fisker developed two plug-in hybrid electric vehicles, Nina and Karma, and planned to sell 1,000 units in the United States in 2011. As of 2011, Tesla Motors had two electric vehicles ready for the North American market:

Model S and Roadster. Think City also manufactures small electric vehicles. However, while demand outside North America is large, demand in the United States is low. With major international governmental tax incentives in place, all vehicle manufacturers are developing at least some studies of electric vehicles for auto shows. Whether these models will actually go on to be developed for commercial markets remains to be seen.

Utility vehicles and trucks. To develop a green image, some municipalities considered switching their fleets to electric vehicles based on fuel cells or batteries. Ford has created a model called Transit Connect, which is an electrified version of its Ford Transporter. Navistar has developed an electric truck, the eStar, and expects to sell a maximum of 1,000 units each year until 2015 in the United States. Smith Electric Vehicles has developed several models, such as the Newton, for the expected demand in electric-utility vehicles. Additionally, there are many small companies producing small utility trucks, such as the Italian manufacturer Alkè. The products of these companies are small, practical multi-purpose vehicles for cities and municipalities.

Bicycles and scooters. Small electric motor-assisted bicycles and electric scooters have been in use since the early twentieth century. Other small electric vehicles include wheelchairs, skateboards, golf carts, lawn-mowers, and other equipment that typically does not require much power. For customers looking for faster vehicles, Oregon-based Brammo produces electric motorcycles that have a range of 100 miles at a speed of 100 miles per hour. These electric vehicles have comparable performance to any internal combustion motorcycle but lack any tailpipe or noise emissions.

Mass transit. In North America, many cities had electric public transit similar to the San Francisco cable cars until they were sold to car manufacturers who decommissioned them. As a result, most public transit systems rely heavily on diesel engine buses. Some of the public transit companies have considered testing fuel-cell- or battery-powered electric buses, but all these efforts have remained very small, with a handful of buses running at any time throughout Europe and North America. For example, during the 2010 Olympics, the Canadian Hydrogen and Fuel Cell Association ran fuel-cell electric vehicle busses between the cities of Vancouver and Whistler. United Kingdom-based manufacturer Optare has produced battery-powered buses since 2009. Cities, such as Seattle, that use electric overhead lines to power trolley buses, trams, and trains have had much higher impact in terms of actual transported passengers. All these systems constitute electric vehicles, but all of them are dependent on having electric wires in place before they can operate. On the other hand, once the wires are in place, the public transit systems can operate silently and cleanly, using electricity provided through an electric grid instead of a battery or a fuel cell.

Forklifts. In spaces with little ventilation, the exhaust of internal combustion engines can be harmful and potentially toxic to humans, which is why warehouse forklifts are typically powered by electric engines. Traditionally, these engines are powered by batteries, but the recharging time of several hours often requires the purchase of at least twice as many batteries as forklifts—or twice as many forklifts as drivers—to be able to work around the clock. Using fuel cells as a power source, forklifts such as the ones produced by Vancouver-based company Cellex require only a short time at a hydrogen refueling station before being ready for use. Such a short downtime of fuel-cell powered electric forklifts compared with battery-powered forklifts allows warehouses to operate with less machinery, cutting back on the initial capital cost of operation.

IMPACT ON INDUSTRY

Government research. Globally, most governments have introduced specific targets for the electrification of vehicles to reduce pollution. In Germany, for example, the federal government plans to have at least one million electric vehicles on the streets by 2020. This is an ambitious target that is backed by significant funding for government, university, and private-sector research. However, in 2009-2010, the German government offered car wreckage flat-rate premium to all car owners, many of whom traded their used but still very usable cars for newer cars with larger motors and increased emissions. Although the funding of this incentive program significantly benefited automakers worldwide, incentives to improve the performance of electric motors and the market penetration of electric vehicles were not included. As a consequence, no net reduction in vehicle-emission pollution was achieved, and the number of electric cars actually on the road remained negligible. About 1,500 electric vehicles were registered in Germany in 2010 out of a total of about 41 million cars. With no incentives in place as of 2011, the number of registered electric vehicles has not significantly changed either. Asian countries appear to invest more into an electric-vehicle infrastructure, with China set to place a half million electric vehicles on its roads in 2012, although it remains doubtful whether this target can be met completely. The U.S. Department of Energy's February, 2011, Status Report stated the ambitious target of having one million electric vehicles on the road by 2015. This is based on 2010 statistics in which 97 percent of cars sold had conventional internal combustion engines and 3 percent had hybrid electric engines. While one million cars sounds like a lot, in the actual car market, this amount will not represent any significant step in reducing pollution from internal combustion engines. Significantly higher numbers of electric vehicle usage are required to make cleaner inner-city air a reality.

University research. Universities have always been involved in battery research, but only since the 2010's has there been a significant increase in attention to clean-energy storage technologies from super capacitors to flywheels, hydrogen, and battery technologies. Although university projects suffered some detrimental impact from the 2009 world economic crisis, the research, especially in batteries, was minimally affected and has since seen significant increases in funding, scope, and technology-focused applications. One example is the Institute for Electrochemical Energy Storage and Conversion at the University of Ulm in Germany, which has significantly expanded its activities in battery research. Similar expansions in battery-technology research can be seen around the globe.

Industry and business. One of the main business sectors for electronic vehicles includes forklifts and other indoor equipment for confined or explosive environments, such as mines. Other sectors that have made profits over the past years include electric scooters and small electric bicycles. And while there is significant media attention on electric cars, both fuel cell and battery based, as well as hybrid-electric vehicles, their contribution to the global automobile market is negligibly small and requires significant governmental incentives to become a larger part of the global automotive economy.

Major corporations. Although North America was a global leader in electric vehicles and battery research, the car industry mostly dismissed electric vehicle developments in the 1990's, as North American consumers demanded larger, heavier, and more powerful vehicles with internal combustion engines. As of 2011, North American car manufacturers lag behind the rest of the world in developing mature vehicles that will be accepted in the market. Both Ford, with its Focus electric vehicle, and GM, with its Chevrolet Volt, have produced early studies of electric vehicles that they intend to develop for the North American market. They compete with all major international car manufacturers that are introducing the electric vehicles worldwide, as well as domestic corporations that focus on electric vehicles, such as the California-based luxury-car manufacturer Tesla Motors and Missouri-based electric truck and commercial vehicle manufacturer Smith Electric Vehicles. Internationally, there is a large number of electric car manufacturing businesses starting up, reviving a business that had become the monopoly of very few corporations. Because of more stringent emissions targets and significant tax incentives, North American car manufacturers have returned to some electric car developments in the 2010's. Whether these are just greening initiatives and convenient tax-reduction programs or whether these cars can become a major business in the North American market remains to be seen.

Most car manufacturers have created spin-off companies that develop customer-specific batteries for modern vehicles. For example, the European car manufacturer Daimler and the international chemical corporation Evonik and its subsidiary Li-Tec formed a new company called Deutsche Accumotive, with the sole aim of producing better batteries for the international vehicle market. This comes as a result of the dominance of lithium-ion batteries in the low-end markets by Chinese manufacturers and high-end markets by Japan and South Korea. Most batteries are used in power tools and small handheld devices such as mobile phones and laptop computers. The clean technology consulting firm Pike Research estimates that the global annual market for lithium-ion batteries for vehicles is $8 billion.

CAREERS AND COURSE WORK

As gasoline prices increase, it becomes more

FASCINATING FACTS ABOUT ELECTRIC AUTOMOBILE TECHNOLOGY

- The average power capacity of rechargeable batteries has tripled since the 1970's.
- Fuel cells produce electricity with clean water as the only exhaust. During space missions, this water has been used for drinking.
- The clean and quiet Brammo Enertia electric motorcycle can go 100 miles at a velocity of 100 miles per hour before requiring a recharge.
- Hitting the accelerator in a car with an electric motor results in a much faster acceleration than in a comparable gasoline-run car, since the torque of the electric motor far exceeds that of the internal combustion engine.
- Modern electric vehicles from major carmakers are indistinguishable from their standard line of models, as this reduces potential customer barriers to purchasing electric or electric-hybrid vehicles and lowers manufacturing cost.
- Non-rechargeable batteries that are not properly disposed of may leach toxic chemicals into the ground over time.
- It takes significantly more energy to create a new car than it does to keep an old car in good shape and run it—especially if the new car uses an internal combustion engine.

important to have lighter vehicles that require less material during manufacturing, as these have to be mined and transported around the world and machined using energy coming primarily from fossil fuels. Additionally, vehicles should become more efficient, to reduce the operating cost for vehicle owners. All these issues are addressed by selecting and designing better, novel materials. Those interested in a career in electric vehicle manufacturing or design would do well studying materials, mechanical, chemical, mining, or environmental engineering for designing novel cars, highly efficient motors, better batteries, and cheaper, more durable fuel cells. The mathematical modeling of the electrochemistry involved in electric motors is also very important to understand how to improve electric devices, and studies in chemistry and physics may lead to improvements in the efficiency of vehicles.

After earning a bachelor's degree in one of the above-mentioned areas, an internship would be ideal. After an internship, one's career path can be

extremely varied. In the research sector, for example, working on catalysts for batteries and fuel cells in a chemical company could include developing new materials that involve inexpensive, nontoxic, durable noble metals that are at least as efficient as traditional catalysts. This is only one example of many potential careers in the global electric vehicle market.

SOCIAL CONTEXT AND FUTURE PROSPECTS

Energy consumption per capita is increasing continuously. The majority of power production uses the combustion of fossil fuels with additional contributions from hydroelectric and nuclear energy conversion. These energy-conversion methods create varying kinds of pollution and dangers to the environment such as habitat destruction, toxic-waste production, or radiation, as seen in nuclear reactors hit by earthquakes, equipment malfunction, or operator errors. The increasing demand for a finite quantity of fossil fuels has the potential to increase the cost of these resources significantly. Another undesirable consequence of the thermochemical conversion of fossil fuels by combustion is environmental contamination. The reaction products from combustion can be harmful to humans on a local scale and have been cited as contributing to global climate change. The remaining ash of coal combustion contains heavy metals and radioactive isotopes that can be severely damaging to health, as seen in the 2008 Kingston Fossil Plant coal ash slurry spill in Tennessee.

Furthermore, fossil fuel resources are unevenly distributed over the globe, leading to geopolitical unrest as a result of the competition for resource access. As a consequence of upheavals in the Middle East and North Africa in 2011, oil and food prices have soared, and may continue doing so.

Clearly, the energy demands of society need to be satisfied in a more appropriate, sustainable, and efficient way. Cleaner devices for energy conversion are batteries and fuel cells. They operate more efficiently, produce less pollution, are modular, and are less likely to fail mechanically since they have fewer moving parts than energy conversion based on combustion.

The advantages of electric vehicles are clear: a world in which all or most vehicles are quiet, with no truck engine brakes to rattle windows from a mile away and no lawnmowers disturbing the quiet or fresh air of a neighborhood; a society with no harmful local emissions from any of the machines being used, allowing people to walk by a leaf blower without having to hold their breath and to live next to major roads without risking chronic diseases from continually breathing in harmful emissions. All this could already be humanity's present-day reality if people were willing to change their habits and simply use electric motors instead of combustion engines.

—*Lars Rose, MSc, PhD*

FURTHER READING

Cancilla, Riccardo, and Monte Gargano, eds. *Global Environmental Policies: Impact, Management and Effects.* Hauppauge, N.Y.: Nova Science Publishers, 2010. Features multifaceted chapters on various international aspects of environmental policies, laying the groundwork for a change in consumer attitude toward new clean energy and infrastructure.

Hoel, Michael, and Snorre Kverndokk. "Depletion of Fossil Fuels and the Impact of Global Warming." *Resource and Energy Economics* 18, no. 2 (June, 1996): 115-136. Gives an explanation of the calculation of global oil depletion.

Husain, Iqbal. *Electric and Hybrid Vehicles: Design Fundamentals.* 2d ed. Boca Raton, Fla.: CRC Press, 2011. A very technical but comprehensive book that provides an overview of modern electric and electric-hybrid vehicle technologies.

Root, Michael. *The TAB Battery Book: An In-Depth Guide to Construction, Design, and Use.* New York: McGraw-Hill, 2011. Provides a good, readable background for all different types of batteries and the challenges in research and development.

Taylor, Peter J., and Frederick H. Buttel. "How Do We Know We Have Global Environmental Problems? Science and the Globalization of Environmental Discourse." *Geoforum* 23, no. 3 (1992): 405-416. Outlines the science behind global environmental issues.

Army Center for Health Promotion and Preventive Medicine. "Engine Emissions—Health and Medical Effects." http://phc.amedd.army.mil/PHC%20Resource%20Library/FS65-039-1205.pdf. Outlines the detrimental acute and chronic effects of engine exhaust intake via the lungs and the skin.

Department of Energy. "One Million Electric Vehicles By 2015." http://www.energy.gov/media/1_Million_Electric_Vehicle_Report_Final.pdf. Details the U.S. plan to achieve a sale rate of at least 1.7 percent electric vehicles every year until 2015, also indicating preferred companies involved for the North American electric vehicle market.

WEB SITES

American Society for Engineering Education
http://www.asee.org

Electric Auto Association
http://www.electricauto.org

Electric Drive Transportation Organization
http://www.electricdrive.org

European Association for Battery, Hybrid and Fuel Cell Electric Vehicles
http://www.avere.org/www/index.php

See also: Electrical Engineering; Hybrid Vehicle Technologies.

ELECTROCHEMISTRY

FIELDS OF STUDY

Physical chemistry; thermodynamics; organic chemistry; inorganic chemistry; quantitative analysis; chemical kinetics; analytical chemistry; metallurgy; chemical engineering; electrical engineering; industrial chemistry; electrochemical cells; fuel cells; electrochemistry of nanomaterials; advanced mathematics; physics; electroplating; nanotechnology; quantum chemistry; electrophoresis; biochemistry; molecular biology.

SUMMARY

Electrochemists study the chemical changes produced by electricity, but they are also concerned with the generation of electric currents due to the transformations of chemical substances. Whereas traditional electrochemists investigated such phenomena as electrolysis, modern electrochemists have broadened and deepened their interdisciplinary field to include theories of ionic solutions and solvation. This theoretical knowledge has led to such practical applications as efficient batteries and fuel cells, the production and protection of metals, and the electrochemical engineering of nanomaterials and devices that have great importance in electronics, optics, and ceramics.

KEY TERMS AND CONCEPTS

- **anode:** Positive terminal (or electrode) of an electrochemical cell to which negatively charged ions travel with the passage of an electric current.
- **battery:** Electrochemical device that converts chemical energy into electrical energy.
- **cathode:** Negative electrode of an electrochemical cell in which positively charged ions migrate under the influence of an electric current.
- **electrolysis:** Process by which an electric current causes chemical changes in water, solutions, or molten electrolytes.
- **electrolyte:** Substance that generates ions when molten or dissolved in a solvent.
- **electrophoresis:** Movement of charged particles through a conducting medium due to an applied electric field.
- **electroplating:** Depositing a thin layer of metal on an object immersed in a solution by passing an electric current through it.
- **Faraday's law:** Magnitude of the chemical effect of an electrical current is directly proportional to the amount of current passing through the system.
- **fuel cell:** Electrochemical device for converting a fuel with an oxidant into direct-current electricity.
- **ion:** Atom or group of atoms carrying a charge, either positive (cation) or negative (anion).
- **nanomaterials:** Chemical substances or particles the masses of which are measured in terms of billionths of a gram.
- **pH:** Numerical value, extending from 0 to 14, representing the acidity or alkalinity of an aqueous solution (the number decreases with increasing acidity and increases with increasing alkalinity).

DEFINITION AND BASIC PRINCIPLES

As its name implies, electrochemistry concerns all systems involving electrical energy and chemical processes. More specifically, this field includes the study of chemical reactions caused by electrical forces as well as the study of how chemical processes give rise to electrical energy. Some electrochemists investigate the electrical properties of certain chemical substances, for instance, these substances' ability to serve as insulators or conductors. Because the atomic structure of matter is fundamentally electrical, electrochemistry is intimately involved in all fields of chemistry from physical and inorganic through organic and biochemistry to such new disciplines as nanochemistry. No matter what systems they study, chemists in some way deal with the appearance or disappearance of electrical energy into the surroundings. On the other hand, electrochemists concentrate on those systems consisting of electrical conductors, which can be metallic, electrolytic, or gaseous.

Because of its close connection with various branches of chemistry, electrochemistry has applications that are multifarious. Early applications centered on electrochemical cells that generated a steady current. New metals such as potassium, sodium, calcium, and strontium were discovered by electrolysis of their molten salts. Commercial production of such metals as magnesium, aluminum, and zinc were mainly accomplished by the electrolyhsis of solutions or melts. An understanding of the electrical nature of chemical bonding led chemists to create many new dyes, drugs, plstics, and artificial plastics. Electroplating has served both aesthetic and practical purposed, and it has certainly decreased the corrosion of several widely used metals.

Electrochemistry played a significant part in the research and development of such modern substances as silicones, fluorinated hydrocarbons, synthetic rubbers, and plastics. Even tough semiconductors such as germanium and silicon do not conduct electricity as well as copper, an understanding of electrochemical principles has been important in the invention of various solid-state devices that have revolutionized the electronics industries, from radio and television to computers. Electrochemistry, when it has been applied in the life sciences, has resulted in an expanded knowledge of biological molecules. For example, American physical chemist Linus Pauling used eletrophoretic techniques to discover the role of a defective hemoglobin molecule in sickle-cell anemia. A grasp of electrochemical phenomena occurring in the human heart and brain has led to diagnostic and palliative technologies that have improved the quality and length of human lives. Much research and development are being devoted to increasingly sophisticated electrochemical devices for implantation in the human body, and some even predict, such as American inventory Ray Kruzweil, that these "nanobots" will help extend human life indefinitely.

BACKGROUND AND HISTORY

Most historians of science trace the origins of electrochemistry to the late eighteenth and early nineteenth centuries, when Italian physician Luigi Galvani studied animal electricity and Italian physicist Alessandro Volta invented the first battery. Vota's device consisted of a pile of dissimilar metals such as zinc and silver separated by a moist conductor. This "Voltaic pile" produced a continuous current, and applications followed quickly. Researchers showed that a Voltaic pile could decompose water into hydrogen and oxygen by a process later called electrolysis. English chemist Sir Humphrey Davy used the electrolysis of melted inorganic compounds to discover several new elements. The Swedish chemist Jöns Jacob Berzelius used those electrochemical studies to formulate a new theory of chemical combination . In

his dualistic theory atoms are held together in compounds by opposite charges, but his theory declined in favor when it was unable to explain organic compounds, or even such a simple molecule as diatomic hydrogen.

Though primarily a physicist, Michael Faraday made basic discoveries in electrochemistry, and, with the advice of others, he developed the terminology of this new science. For example, he introduced the terms "anode" and "anion," "cathode" and "cation," "electrode," "electrolyte," as well as "electrolysis." In the 1830s his invention of a device to measure the quantity of electric current resulted in his discovery of a fundamental law of electrochemistry—that the quantity of electric current that leads to the formation of a certain amount of a particular chemical substance also leads to chemically equivalent amounts of other substances. Even though Faraday's discovery of the relationship between quantity of electricity and electrochemical equivalents was extraordinarily significant, it was not properly appreciated until much later. Particularly helpful was the work of the Swedish chemist Svante August Arrhenius, whose ionic theory, proposed toward the end of the nineteenth century, contained the surprising new idea that anions and cations are present in dilute solutions of electrolytes.

In the twentieth century, Dutch-American physical chemist Peter Debye, together with German chemist Erich Hückel, corrected and extended the Arrhenius theory by taking into account that, in concentrated solutions, cations have a surrounding shell of anions, and vice versa, causing these ions' movements to be retarded in an electric field. Norwegian-American chemist Lars Onsager further refined this theory by taking into account Brownian motion, the movement of these ionic atmospheres due to heat. Other scientists used electrochemical ideas to understand the nature of acids and bases, the interface between dissimilar chemicals in electrochemical cells, and the complexities of oxidation-reduction reactions, whether they occur in electrochemical cells or in living things.

Electrochemistry

Connecting copper and zinc creates an electrochemical cell.

How it Works

Primary and secondary cells. The basic device of electrochemistry is the cell, generally consisting of a container with electrodes and an electrolyte, designed to convert chemical energy into electrical energy. A primary cell, also known as a galvanic or Voltaic cell, is one that generates electrical current via an irreversible chemical reaction. This means that a discharged primary cell cannot be recharged from an external source. By taking measurements at different temperatures, chemists use primary cells to calculate the heat of reactions, which have both theoretical and practical applications. Such cells can also be used to determine the acidity and alkalinity of solutions. Every primary cell has two metallic electrodes, at which electrochemical reactions occur. In one of these reactions the electrode gives up electrons, and at the other electrode electrons are absorbed.

In a secondary cell, also known as a rechargeable or storage cell, electrical current is created by chemical reactions that are reversible. This means that a discharged secondary cell may be recharged by circulating through the cell in a quantity of electricity equal to what had been withdrawn. This process can be repeated as often as desired. The manufacture of secondary cells has grown into an immense industry, with such commercially successful products as lead-acid cells and alkaline cells with either nickel-iron or nickel-cadmium electrodes.

Electrolyte pocesses. Electrolysis, one of the first electrochemical processes to be discovered, has increased in importance in the twentieth and twenty-first centuries. Chemists investigating electrolysis soon discovered that chemical reactions take place at the two electrodes, but the liquid solution between them remains unchanged. An early explanation was that with the passage of electric current, ions in the solution alternated decompositions and recombinations of the electrolyte. This theory had to be later revised in the light of evidence that chemical components had different motilities in solution.

For more than two hundred years, the electrolysis of water has been used to generate hydrogen gas. In an electrolytic cell with a pair of platinum electrodes, to the water of which a small amount of sulfuric acid has been added (to reduce the high voltage needed), electrolysis begins with the application of an external electromotive force, with bubbles of oxygen gas appearing at the anode (due to an oxidation reaction) and hydrogen gas at the cathode (due to a reduction reaction). If sodium chloride is added to the water, the electrochemical reaction is different, with sodium metal and chlorine gas appearing at the appropriate electrodes. In both these electrolyses the amounts of hydrogen and sodium produced are in accordance with Faraday's law, the mass of the products being proportional to the current applied to the cell.

Redox ractions. For many electrochemists the paramount concern of their discipline is the reduction and oxidation (redox) reaction that occurs in electrochemical cells, batteries, and many other devices and applications. Reduction takes place when an element or radical (an ionic group) gains electrons, such as when a double positive copper ion in solution gains two electrons to form metallic copper. Oxidation takes place when an element or radical loses electrons, such as when a zinc electrode loses two electrons to form a doubly positive zinc ion in solution. In electrochemical research and applications the sites of oxidation and reduction are spatially separated. The electrons produced by chemical processes can be forced to flow through a wire, and this electric current can be used in various applications.

Electrodes. Electrochemists employ a variety of electrodes, which can consist of inorganic or organic materials. Polarography, a subdiscipline of electrochemistry dealing with the measurement of current and voltage, uses a dropping mercury electrode, a technique enabling analysts to determine such species as trace amounts of metals, dissolved oxygen, and certain drugs. Glass electrodes, whose central feature is a thin glass membrane, have been widely used by chemists, biochemists, and medical researchers. A reversible hydrogen electrode plays a central role in determining the pH of solutions. The quinhydrone electrode, consisting of a platinum electrode immersed in a quinhydrone solution, can also be used to measure pH (it is also known as an indicator electrode because it can indicate the concentration of certain ions in the electrolyte). Also widely used, particularly in industrial pH measurements, is the calomel electrode, consisting of liquid mercury covered by a layer of calomel (mercurous chloride), and immersed in a potassium chloride solution. Electrochemists have also created electrodes with increasing (or decreasing) power as oxidizing or reducing agents. With this quantitative information they are then able to choose a particular electrode material to suit a specific purpose.

APPLICATIONS AND PRODUCTS

Batteries and fuel cells. Soon after Volta's invention of the first electric battery, investigators found applications, first as a means of discovering new elements, then as a way to deepen understanding of chemical bonding. By the 1830's, when new batteries were able to serve as reliable sources of electric current, they began to exhibit utility beyond their initial value for experimental and theoretical science. For example, the Daniell cell was widely adopted by the telegraph industry. Also useful in this industry was the newly invented fuel cell that used the reaction of a fuel such as hydrogen and an oxidant such as oxygen to produce direct-current electricity. However, its requirement of expensive platinum electrodes led to its replacement, in the late nineteenth and throughout the twentieth century, with the rechargeable lead-acid battery, which came to be extensively used in the automobile industry. In the late twentieth and early twenty-first centuries many electrochemists predicted a bright future for fuel cells based on hydrogen and oxygen, especially with the pollution problems associated with widespread fossil-fuel use.

Electrodeposition, electroplating, and electrorefining. When an electric current passes through a solution (for instance, silver nitrate), the precipitation of a material (silver) at an electrode (the cathode) is called electrodeposition. A well-known category of electrodeposition is electroplating, when a thin layer of one metal is deposited on another. In galvanization, for example, iron or steel objects are coated with rust-resistant zinc. Electrodeposition techniques have the advantage of being able to coat objects thoroughly, even those with intricate shapes. An allied technique, electrorefining, transforms metals contaminated with impurities to very pure states by anodic dissolution and concomitant redeposition of solutions of their salts. Some industries have used so-called electrowinning techniques to produce salable metals from low-grade ores and mine tailings.

Advances in electrochemical knowledge and techniques have led to evermore sophisticated applications of electrodeposition. For example, knowledge of electrode potentials has made the electrodeposition of alloys possible and commercial. Methods have also been discovered to provide plastics with metal coatings. Similar techniques have been discovered to coat such rubber articles as gloves with a metallic layer. Worn or damaged metal objects can be returned to pristine condition by a process called electroforming. Some commercial metal objects, such as tubes, sheets, and machine parts, have been totally manufactured by electrodeposition (sometimes called electromachining).

Electrometallurgy. A major application of electrochemical principles and techniques occurs in the manufacture of such metals as aluminum and titanium. Plentiful aluminum-containing bauxite ores exist in large deposits in several countries, but it was not until electrochemical techniques were developed in the United States and France at the end of the nineteenth century that the cost of manufacturing this light metal was sufficiently reduced to make it a commercially valuable commodity. This commercial process involved the electrolysis of alumina (aluminum oxide) dissolved in fused cryolite (sodium aluminum fluoride). During the century that followed this process's discovery, many different uses for this lightweight metal ensued, from airplanes to zeppelins.

Corrosion control and dielectric materials. The destruction of a metal or alloy by oxidation is itself an electrochemical process, since the metal loses electrons to the surrounding air or water. A familiar example is the appearance of rust (hydrated ferric oxide) on an iron or steel object. Electrochemical knowledge of the mechanism of corrosion led researchers to ways of preventing or delaying it. Keeping oxidants away from the metallic surface is an obvious means of protection. Substances that interfere with the oxidizing of metals are called inhibitors. Corrosion inhibitors include both inorganic and organic materials, but they are generally categorized by whether the inhibitor obstructs corrosive reactions at the cathode or anode. Cathodic protection is used extensively for such metal objects as underground pipelines or such structures as ship hulls, which have to withstand the corrosive action of seawater. Similarly, dielectric materials with low electrical conductivity, such as insulators, require long-term protection from high and low temperatures as well as from corrosive forces. An understanding of electrochemistry facilitates the construction of such electrical devices as condensers and capacitors that involve dielectric substances.

Electrochemistry, molecular biology, and medicine. Because of the increasing understanding of electrochemistry as it pertains to plant, animal, and human life, and because of concerns raised by the modern environmental movement, several significant applications have been developed, with the promise of many more to come. For example, electrochemical devices have been made for the analysis of proteins and deoxyribonucleic acid (DNA). Researchers have fabricated DNA sensors as well as DNA chips. These DNA sensors can be used to detect DNA damage. Electrochemistry was involved in the creation of implantable pacemakers designed to regulate heart beats, thus saving lives. Research is under way to create an artificial heart powered by electrochemical processes within the human body. Neurologists electrically stimulate regions of the brain to help mitigate or even cure certain psychological problems. Developments in electrochemistry have led to the creation of devices that detect various environmental pollutants in air and water. Photoelectrochemistry played a role in helping to understand the dramatic depletion of the ozone layer in the stratosphere and the role that chlorofluorocarbons (CFCs) played in exacerbating this problem. Because a large hole in the ozone layer allows dangerous solar radiation to damage plants, animals, and humans, many countries have banned the use of CFCs.

Nanomaterials in electrochemistry. Miniaturization of electronic technologies became evident and important in the computer industry, where advances have been enshrined in Moore's law, which states that transistor density in integrated circuits doubles every eighteen months. Electrodeposition has proved to be a technique well-suited to the preparation of metal nanostructures, with several applications in electronics, semiconductors, optics, and ceramics. In particular, electrochemical methods have contributed to the understanding and applications of quantum dots, nanoparticles that are so small that they follow quantum rather than classical laws. These quantum dots can be as small as a few atoms, and in the form of ultrathin cadmium-sulfide films they have been shown to generate high photocurrents in solar cells. The electrochemical synthesis of such nanostructured products as nanowires, biosensors, and microelectroanalytical devices has led researchers to predict the ultimate commercial success of these highly efficient contrivances.

IMPACT ON INDUSTRY

Because electrochemistry itself an interdisciplinary field, and is a part of so many different scientific disciplines and commercial applications, it is difficult to arrive at an accurate figure for the economic worth and annual profits of the global electrochemical industry. More reliable estimates exist for particular segments of this industry in specific countries. For example, in 2008, the domestic revenues of the United States battery and fuel cell industry were about $4.9 billion, the lion's share of which was due to the battery business (in 2005, the United States fuel cell industry had revenues of about $266 million). During the final decades of the twentieth century, the electrolytic production of aluminum in the United States was about one-fifth of the world's total, but in the twenty-first century competition from such countries as Norway and Brazil, with their extensive and less expensive hydropower, has reduced the American share.

Government and university research. In the United States during the decades after World War II, the National Science Foundation provided support for many electrochemical investigations, especially those projects that, because of their exploratory nature, had no guarantee of immediate commercial success. An example is the 1990's electrochemical research on semiconducting nanocrystals. The United States' Office of Naval Research has supported projects on nanostructured thin films. It is not only the federal government that has seen fit through various agencies to support electrochemical research but state governments as well. For example, the New York State Foundation for Science, Technology and Innovation has invested in investigations of how ultrathin films can be self-assembled from polyelectrolytes, nanoparticles, and nanoplatelets with the hope that these laboratory-scale preparations may have industrial-scale applications.

Government agencies and academic researchers have often worked together to fund basic research in electrochemistry. The Department of Energy through its Office of Basic Energy Sciences has supported research on electrochromic and photo-chromic effects, which has led to such commercial products as switchable mirrors in automobiles. Researchers at the Georgia Institute of Technology have contributed to the improvement of proton exchange membrane fuel cells (PEMFCs), and scientists at the University of Dayton in Ohio have shown the value of carbon nanotubes in fuel cells. Hydrogen fuel cells became the focus of an initiative promoted by President George W. Bush in 2003, which was given direction and financial support in the 2005 Energy Policy Act and the 2006 Advanced Energy Initiative. However, a few years later, the Obama administration chose to de-emphasize hydrogen fuel cells and emphasize other technologies that will create high energy efficiency and less polluting automobiles in a shorter period of time.

Industry and business. Because of the widespread need for electrochemical cells and batteries, companies manufacturing them have devoted extensive human and financial resources to the research, development, and production of a variety of batteries. Some companies, such as Exide Technologies, have emphasized lead-acid batteries, whereas General Electric, whose corporate interest in batteries goes back to its founder, Thomas Alva Edison, has made fuel cells a significant part of its diversified line of products. Some businesses, such as Alcoa, the world's leading producer of aluminum, were based on the discovery of a highly efficient electrolytic process, which led to a dramatic decrease in the cost of aluminum and, in turn, to its widespread use (it is second only to steel as a construction metal).

CAREERS AND COURSE WORK

Electrochemistry is an immense field with a large variety of specialties, though specialized education generally takes place at the graduate level. Undergraduates usually major in chemistry, chemical engineering, or materials science engineering, the course work of which involves introductory and advanced physics courses, calculus and advanced mathematics courses, and elementary and advanced chemistry courses. Certain laboratory courses, for example, qualitative, quantitative and instrumental analysis, are often required. Because of the growing sophistication of many electrochemical disciplines, those interested in becoming part of these fields will need to pursue graduate degrees. Depending on their specialty, graduate students need to satisfy core courses, such as electrochemical engineering, and a certain number of electives, such as semiconductor devices. Some universities, technical institutes, and engineering schools have programs for students interested in theoretical electrochemistry, electrochemical cells, electrodeposition, nanomaterials, and many others. For doctoral and often for a master's degree, students are required to write a thesis under the supervision of a faculty director.

Career opportunities for electrochemists range from laboratory technicians at small businesses to research professors at prestigious universities. The battery business employs many workers with bachelor of science degrees in electrochemistry to help manufacture, service, and improve a variety of products. Senior electrochemical engineers with advanced degrees may be hired to head research programs to develop new products or to supervise the production of the company's major commercial offerings. Electrochemical engineers are often hired to manage the manufacture of electrochemical components or oversee the electrolytic production of such metals as aluminum and magnesium. Some electrochemists are employed by government agencies, for example, to design and develop fuel cells for the National Aeronautics and Space Administration (NASA), while others may be hired by pharmaceutical companies to develop new drugs and medical devices.

SOCIAL CONTEXT AND FUTURE PROSPECTS

Even though batteries, when compared with other energy sources, are too heavy, too big, too inefficient, and too costly, they will continue to be needed in the twenty-first century, at least until suitable substitutes are found. Although some analysts predict a bright future for fuel cells, others have been discouraged by their slow rate of development. As advanced industrialized societies continue to expand, increasing demand for such metals as beryllium, magnesium, aluminum, titanium, and zirconium will necessarily follow, forcing electrochemists to improve the electrolytic processes for deriving these metals from dwindling sources. If Moore's law holds well into the future, then computer engineers, familiar with electrochemical principles, will find new ways to populate integrated circuits with more and better micro-devices.

Some prognosticators foresee significant progress in the borderline field between electrochemistry and organic chemistry (sometimes called electro-organic chemistry). When ordinary chemical methods have proved inadequate to synthesize desired compounds of high purity, electrolytic techniques have been much better than traditional ones in accomplishing this, though these successes have occurred at the laboratory level and the development of industrial processes will most likely take place in the future. Other new fields, such as photoelectrochemistry, will mature in the twenty-first century, leading to important applications. The electrochemistry of nanomaterials is already well underway, both theoretically and practically, and a robust future has been envisioned

as electrochemical engineers create new nanophase materials and devices for potential use in a variety of applications, from electronics to optics.

—Robert J. Paradowski, MS, PhD

FURTHER READING

Bagotsky, Vladimir, ed. *Fundamentals of Electrochemistry.* 2d ed. New York: Wiley-Interscience, 2005. Provides a good introduction to this field for those unfamiliar with it, though later chapters contain material of interest to advanced students. Index.

Bard, Allen J., and Larry R. Faulkner. *Electrochemical Methods: Fundamentals and Applications.* 2d ed. Hoboken, N.J.: John Wiley & Sons, 2001. For many years this "gold standard" of electrochemistry textbooks was the most widely used such book in the world. Its advanced mathematics may daunt the beginning student, but its comprehensive treatment of theory and applications rewards the extra effort needed to understand this field's fundamentals; index.

Brock, William H. *The Chemical Tree: A History of Chemistry.* New York: Norton, 2000. This work, previously published as *The Norton History of Chemistry,* was listed as a *New York Times* "Notable Book" when it appeared. The development of electrochemistry forms an important part of the story of chemistry; includes an extensive bibliographical essay, notes, and index.

Ihde, Aaron J. *The Development of Modern Chemistry.* New York: Dover, 1984. Makes available to general readers a well-organized treatment of chemistry from the eighteenth to the twentieth century, of which electrochemical developments form an essential part. Illustrated, with extensive bibliographical essays on all the chapters; author and subject indexes.

MacInnes, Duncan A. *The Principles of Electrochemistry.* New York: Dover, 1961. This paperback reprint brings back into wide circulation a classic text that treats the field as an integrated whole; author and subject indexes.

Schlesinger, Henry. *The Battery: How Portable Power Sparked a Technological Revolution.* New York: HarperCollins, 2010. The author, a journalist specializing in technology, provides an entertaining, popular history of the battery, with lessons for readers familiar only with electronic handheld devices.

Zoski, Cynthia G., ed. *Handbook of Electrochemistry.* Oxford, England: Elsevier, 2007. After an introductory chapter, this book surveys most modern research areas of electrochemistry, such as reference electrodes, fuel cells, corrosion control, and other laboratory techniques and practical applications.

WEB SITES
Electrochemical Science and Technology Information Resource (ESTIR)

http://electrochem.cwru.edu/estir

The Electrochemical Society
http://www.electrochem.org

International Society of Electrochemistry
http://www.ise-online.org

See also: Electrical Engineering; Electrometallurgy; Metallurgy; Nanotechnology.

ELECTROMAGNET TECHNOLOGIES

FIELDS OF STUDY

Mathematics; physics; electronics; materials science.

SUMMARY

Electromagnetic technology is fundamental to the maintenance and progress of modern society. Electromagnetism is one of the essential characteristics of the physical nature of matter, and it is fair to say that everything, including life itself, is dependent upon it. The ability to harness electromagnetism has led to the production of most modern technology.

KEY TERMS AND CONCEPTS

- **Curie temperature:** The temperature at which a ferromagnetic material is made to lose its magnetic properties.
- **induction:** The generation of an electric current in a conductor by the action of a moving magnetic field.
- **LC oscillation:** Alternation between stored electric and magnetic field energies in circuits containing both an inductance (L) and a capacitance (C).
- **magnetic damping:** The use of opposing magnetic fields to damp the motion of a magnetic oscillator.
- **sinusoidal:** Varying continuously between maximum and minimum values of equal magnitude at a specific frequency, in the manner of a sine wave.
- **toroid:** A doughnut-shaped circular core about which is wrapped a continuous wire coil carrying an electrical current.

DEFINITION AND BASIC PRINCIPLES

Magnetism is a fundamental field effect produced by the movement of an electrical charge, whether that charge is within individual atoms such as iron or is the movement of large quantities of electrons through an electrical conductor. The electrical field effect is intimately related to that of magnetism. The two can exist independently of each other, but when the electrical field is generated by the movement of a charge, the electrical field is always accompanied by a magnetic field. Together they are referred to as an electromagnetic field. The precise mathematical relationship between electric and magnetic fields allows electricity to be used to generate magnetic fields of specific strengths, commonly through the use of conductor coils, in which the flow of electricity follows a circular path. The method is well understood and is the basic operating principle of both electric motors and electric generators.

When used with magnetically susceptible materials, this method transmits magnetic effects that work for a variety of purposes. Bulk material handling can be carried out in this way. On a much smaller scale, the same method permits the manipulation of data bits on magnetic recording tape and hard-disk drives. This fine degree of control is made possible through the combination of digital technology and the relationship between electric and magnetic fields.

BACKGROUND AND HISTORY

The relationship between electric current and magnetic fields was first observed by Danish physicist and chemist Hans Christian Oersted (1777-1851), who noted, in 1820, how a compass placed near an electrified coil of wires responded to those wires. When electricity was made to flow through the coil, these wires changed the direction in which the compass pointed. From this Oersted reasoned that a magnetic field must exist around an electrified coil.

In 1821, English chemist and physicist Michael Faraday (1791-1867) found that he could make an electromagnetic field interact with a permanent magnetic field, inducing motion in one or the other of the magnetized objects. By controlling the electrical current in the electromagnet, the permanent magnet can be made to spin about. This became the operating principle of the electric motor. In 1831, Faraday found that moving a permanent magnetic field through a coil of wire caused an electrical current to flow in the wire, the principle by which electric generators function. In 1824, English physicist and inventor William Sturgeon (1783-1850) discovered that an electromagnet constructed around a core of solid magnetic material produced a much stronger magnetic field than either one alone could produce.

Since these initial discoveries, the study of electromagnetism has refined the details of the mathematical relationship between electricity and magnetism as it is known today. Research continues to refine understandings of the phenomena, enabling its use in new and valuable ways.

HOW IT WORKS

The basic principles of electromagnetism are the same today as they have always been, because electromagnetism is a fundamental property of matter. The movement of an electrical charge through a conducting medium induces a magnetic field around the conductor. On an atomic scale, this occurs as a function of the electronic and nuclear structure of the atoms, in which the movement of an electron in a specific atomic orbital is similar in principle to the movement of an electron through a conducting material. On larger scales, magnetism is induced by the movement of electrons through the material as an electrical current.

Although much is known about the relationship of electricity and magnetism, a definitive understanding has so far escaped rigorous analysis by physicists and mathematicians. The relationship is apparently related to the wave-particle duality described by quantum mechanics, in which electrons are deemed to have the properties of both electromagnetic waves and physical particles. The allowed energy levels of electrons within any quantum shell are determined by two quantum values, one of which is designated the magnetic quantum number. This fundamental relationship is also reflected in the electromagnetic spectrum, a continuum of electromagnetic wave phenomena that includes all forms of light, radio waves, microwaves, X rays, and so on. Whereas the electromagnetic spectrum is well described by the mathematics of wave mechanics, there remains no clear comprehension of what it actually is, and the best theoretical analysis of it is that it is a field effect consisting of both an electric component and a magnetic component. This, however, is more than sufficient to facilitate the physical manipulation and use of electromagnetism in many forms.

Ampere's circuit law states that the magnetic field intensity around a closed circuit is determined by the sum of the currents at each point around that circuit. This defines a strict relationship between electrical current and the magnetic field that is produced around the conductor by that current. Because electrical current is a physical quantity that can be precisely controlled on scales that currently range from single electrons to large currents, the corresponding magnetic fields can be equally precisely controlled.

Electrical current exists in two forms: direct current and alternating current. In direct current, the movement of electrons in a conductor occurs in one direction only at a constant rate that is determined by the potential difference across the circuit and the resistance of the components that make up that circuit. The flow of direct current through a linear conductor generates a similarly constant magnetic field around that conductor.

In alternating current, the movement of electrons alternates direction at a set sinusoidal wave frequency. In North America this frequency is 60 hertz, which means that electrons reverse the direction of their movement in the conductor 120 times each second, effectively oscillating back and forth through the wires. Because of this, the vector direction of the magnetic field around those conductors also reverses at the same rate. This oscillation requires that the phase of the cycle be factored into design and operating principles of electric motors and other machines that use alternating current.

Both forms of electrical current can be used to induce a strong magnetic field in a magnetic material that has been surrounded by electrical conductors.

The classic example of this effect is to wrap a large steel nail with insulated copper wire connected to the terminals of a battery so that as electrical current flows through the coil, the nail becomes magnetic. The same principle applies on all scales in which electromagnets are utilized to perform a function.

APPLICATIONS AND PRODUCTS

The applications of electromagnetic technology are as varied and widespread as the nature of electromagnetic phenomena. Every possible variation of electromagnetism provides an opportunity for the development of some useful application.

Electrical power generation. The movement of a magnetic field across a conductor induces an electrical current flow in that conductor. This is the operating principle of every electrical generator. A variety of methods are used to convert the mechanical energy of motion into electrical energy, typically in generating stations.

Hydroelectric power uses the force provided by falling water to drive the magnetic rotors of generators. Other plants use the combustion of fuels to generate steam that is then used to drive electrical generators. Nuclear power plants use nuclear fission for the same purpose. Still other power generation projects use renewable resources such as wind and ocean tides to operate electrical generators.

Solar energy can be used in two ways to generate electricity. In regions with a high amount of sunlight, reflectors can be used to focus that energy on a point to produce steam or some other gaseous material under pressure; this material can then drive an electrical generator. Alternatively, and more commonly, semiconductor solar panels are used to capture the electromagnetic property of sunlight and drive electrons through the system to generate electrical current.

Material handling. Electromagnetism has long been used on a fairly crude scale for the handling and manipulation of materials. A common site in metal recycling yards is a crane using a large electromagnet to pick up and move quantities of magnetically susceptible materials, such as scrap iron and steel, from one location to another. Such electromagnets are powerful enough to lift a ton or more of material at a time. In more refined applications, smaller electromagnets operating on exactly the same principle are often incorporated into automated processes under robotic control to manipulate and move individual metal parts within a production process.

Machine operational control. Electromagnetic technology is often used for the control of operating machinery. Operation is achieved normally through the use of solenoids and solenoid-operated relays. A solenoid is an electromagnet coil that acts on a movable core, such that when current flows in the coil, the core responds to the magnetic field by shifting its position as though to leave the coil. This motion can be used to close a switch or to apply pressure according to the magnetic field strength in the coil.

When the solenoid and switch are enclosed together in a discrete unit, the structure is known as a relay and is used to control the function of electrical circuitry and to operate valves in hydraulic or pneumatic systems. In these applications also, the extremely fine control of electrical current facilitates the design and application of a large variety of solenoids. These solenoids range in size from the very small (used in micromachinery) and those used in typical video equipment and CD players, to very large ones (used to stabilize and operate components of large, heavy machinery).

A second use of electromagnetic technology in machine control utilizes the opposition of magnetic fields to effect braking force for such precision machines as high-speed trains. Normal frictional braking in such situations would have serious negative effects on the interacting components, resulting in warping, scarring, material transfer welding, and other damage. Electromagnetic braking forces avoid any actual contact between components and can be adjusted continuously by control of the electrical current being applied.

Magnetic media. The single most important aspect of electromagnetic technology is also the smallest and most rigidly controlled application of all. Magnetic media have been known for many years, beginning with the magnetic wire and progressing to magnetic recording tape. In these applications, a substrate material in tape form, typically a ribbon of an unreactive plastic film, is given a coating that contains fine granules of a magnetically susceptible material such as iron oxide or chromium dioxide. Exposure of the material to a magnetic field imparts the corresponding directionality and magnitude of that magnetic field to the individual granules in the coating. As the medium moves past an electromagnetic

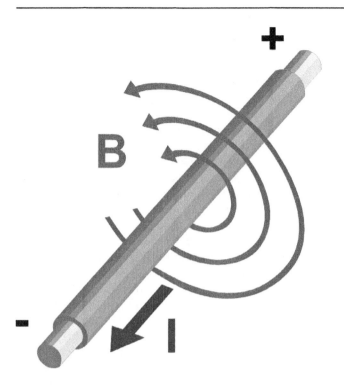

Current (I) through a wire produces a magnetic field (B). The field is oriented according to the right-hand rule.

"read-write head" at a constant speed, the variation of the magnetic properties over time is embedded into the medium. When played back through the system, the read-write head senses the variation in magnetic patterns in the medium and translates that into an electronic signal that is converted into the corresponding audio and video information. This is the analogue methodology used in the operation of audio and videotape recording.

With the development of digital technology, electromagnetic control of read-write heads has come to mean the ability to record and retrieve data as single bits. This was realized with the development of hard drive and floppy disk-drive technologies. In these applications, an extremely fine read-write head records a magnetic signal into the magnetic medium of a spinning disk at a strictly defined location. The signal consists of a series of minute magnetic fields whose vector orientations correspond to the 1s and 0s of binary code; one orientation for 1, the opposite orientation for 0. Also recorded are identifying codes that allow the read-write head to locate the position

of specific data on the disk when requested. The electrical control of the read-write head is such that it can record and relocate a single digit on the disk. Given that hard disks typically spin at 3,000 rpm (revolutions per minute) and that the recovery of specific data is usually achieved in a matter of milliseconds, one can readily grasp the finesse with which the system is constructed.

Floppy disk technology has never been the equal of hard disk technology, instead providing only a means by which to make data readily portable. Other technologies, particularly USB (universal serial bus) flash drives, have long since replaced the floppy drive as the portable data medium, but the hard disk drive remains the staple data storage device of modern computers. New technology in magnetic media and electromagnetic methods continues to increase the amount of data that can be stored using hard disk technology. In the space of twenty years this has progressed from merely a few hundred megabytes (10^6 bytes) of storage to systems that easily store 1 terabyte (10^{12} bytes) or more of data on a single floppy disk. Progress continues to be made in this field, with current research suggesting that a new read-write head design employing an electromagnetic "spin-valve" methodology will allow hard disk systems to store as much as 10 terabytes of data per square inch of disk space.

IMPACT ON INDUSTRY

It is extremely difficult to imagine modern industry without electromagnetic technologies in place. Electric power generation and electric power utilization, both of which depend upon electromagnetism, are the two primary paths by which electromagnetism affects not only industry but also modern society as a whole. The generation of alternating current electricity, whether by hydroelectric, combustion, nuclear, or other means, energizes an electrical grid that spans much of North America. This, in turn, provides homes, businesses, and industry with power to carry out their daily activities, to the extent that the system is essentially taken for granted and grinds to an immediate halt when some part of the system fails. There is a grave danger to society in this combination of complacency and dependency, and efforts are expended both in recovering from any failure and in preventing failures from occurring.

That modern society is entirely dependent upon a continuous supply of electrical energy is an unavoidable conclusion of even the most cursory of examinations. The ability to use electricity at will is a fundamental requirement of many of the mainstays of modern industry. Aluminum, for example, is the most abundant metal known, and its use is essentially ubiquitous today, yet its ready availability depends on electricity. This metal, as well as magnesium, cannot be refined by heat methods used to refine other metals because of their propensity to undergo oxidation in catastrophic ways. Without electricity, these materials would become unavailable.

Paradoxically, the role of electromagnetism in modern electronic technology, although it has been very large, has not been as essential to the development of that technology as might be expected. It is a historical fact that the development of portable storage and hard disk drives greatly facilitated development and spawned whole new industries in the process. Nevertheless, it is also true that portable storage, however inefficient in comparison, was already available in the form of tape drives and nonmagnetic media, such as punch cards. It is possible that other methods, such as optical media (CDs and DVDs) or electronic media (USB flash drives), could have developed without magnetic media.

One area of electromagnetic technology that would be easy to overlook in the context of industry is that of analytical devices used for testing and research applications. Because light is an electromagnetic phenomenon, any device that uses light in specific ways for measurement represents a branch of electromagnetic technology. Foremost of these are the spectrophotometers that are routinely used in medical and technical analysis. These play an important role in a broad variety of fields. Forensic analysis utilizes spectrophotometry to identify and compare materials. Medical analytics uses this technology in much the same way to quantify components in body fluids and other substances. Industrial processes often use some form of spectrophotometric process to monitor system functions and obtain feedback data to be used in automated process control.

In basic research environments, electromagnetic technologies are essential components, providing instrumentation such as the mass spectrometer and the photoelectron spectrometer, both of which function strictly on the principles of electric and magnetic fields. Nuclear magnetic resonance and electron spin resonance also are governed by those same principles. In the former case, these principles have been developed into the diagnostic procedure known as magnetic resonance imaging (MRI). Similar methods have application in materials testing, quality control, and nondestructive structural testing.

Careers and Coursework

Because electromagnetic technology is so pervasive in modern society and industry, students are faced with an extremely broad set of career options. It is not an overstatement to say that almost all possible careers rely on or are affected in some way by electromagnetic technology. Thus, those persons who have even a basic awareness of these principles will be somewhat better prepared than those who do not.

Students should expect to take courses in mathematics, physics, and electronics as a foundation for understanding electromagnetic technology. Basic programs will focus on the practical applications of electromagnetism, and some students will find a rewarding career simply working on electric motors and generators. Others may focus on the applications of electromagnetic technology as they apply to basic computer technology. More advanced careers, however, will require students to undertake more advanced studies that will include materials science, digital electronics, controls and feedback theory, servomechanism controls, and hydraulics and pneumatics, as well as advanced levels of physics and applied mathematics. Those who seek to understand the electromagnetic nature of matter will specialize in the study of quantum mechanics and quantum theory and of high-energy physics.

Social Context and Future Prospects

Electromagnetic technologies and modern society are inextricably linked, particularly in the area of communications. Despite modern telecommunications swiftly becoming entirely digital in nature, the transmission of digital telecommunications signals is still carried out through the use of electromagnetic carriers. These can be essentially any frequency, although regulations control what range of the electromagnetic spectrum can be used for what purpose. Cellular telephones, for example, use the microwave region of the spectrum only, while other devices are restricted to operate in only the infrared region or

FASCINATING FACTS ABOUT ELECTROMAGNET TECHNOLOGIES

- It may soon be possible for hard disk drives to store as much as 10 terabytes of data per square inch of disk space.
- The giant electromagnet at work in a junkyard works on the same principle as the tiny read-write heads in computer disk drives.
- What came first: electricity or magnetism?
- An electron moving in an atomic orbital has the same magnetic effect as electric current moving through a wire.
- Before generators were invented to make electricity readily available for the refining of aluminum metal from bauxite ore, the metal was far more valuable than gold, even though it is the most common known metal.
- Electromagnetic "rail gun" tests have accelerated a projectile from 0 to 13,000 miles per hour in 0.2 seconds.
- Non-ionizing electromagnetic fields are being investigated as treatments for malaria and cancer.
- Coupling an electric field and a magnet produces a magnetic field that is stronger than either one alone could produce.
- In 1820, Oerstad first observed the existence of a magnetic field around an electrified coil.
- Because of the electrical activity occurring inside the human body, people generate their own magnetic fields.
- Mass spectrometers use the interaction of an electric charge with a magnetic field to analyze the mass of fragments of molecules.

in the visible light region of the electromagnetic spectrum.

One cannot overlook the development of electromagnetic technologies that will come about through the development of new materials and methods. These will apply to many sectors of society, including transportation and the military. In development are high-speed trains that use the repulsion between electrically generated magnetic fields to levitate the vehicle so that there is no physical contact between the machine and the track. Electromagnetic technologies will work to control the acceleration, speed, deceleration, and other motions of the machinery. High-speed transit by such machines has the potential to drastically change the transportation industry.

In future military applications, electromagnetic technologies will play a role that is as important as its present role. Communications and intelligence, as well as analytical reconnaissance, will benefit from the development of new and existing electromagnetic technologies. Weaponry, also, may become an important field of military electromagnetic technology, as experimentation continues toward the development of such things as cloaking or invisibility devices and electromagnetically powered rail guns.

—*Richard M. Renneboog, MSc*

FURTHER READING

Askeland, Donald R. "Magnetic Behaviour in Materials." In *The Science and Engineering of Materials*, edited by Donald Askeland. New York: Chapman & Hall, 1998. Provides an overview of the behavior of magnetic materials when acted upon by electromagnetic fields.

Dugdale, David. *Essentials of Electromagnetism.* New York: American Institute of Physics, 1993. A thorough discussion of electromagnetism through electric and magnetic properties, from first principles to practical applications and materials in electrical circuits and magnetic circuits.

Funaro, Daniele. *Electromagnetism and the Structure of Matter.* Hackensack, N.J.: World Scientific, 2008. An advanced treatise discussing deficiencies of Maxwell's theories of electromagnetism and proposing amended versions.

Han, G. C., et al. "A Differential Dual Spin Valve with High Pinning Stability." *Applied Physics Letters* 96 (2010). Discusses how hard disk drives could include ten terabytes of data per square inch using a new read-write head design.

Sen, P. C. *Principles of Electric Machines and Power Electronics.* 2d ed. New York: John Wiley & Sons, 1997. An advanced textbook providing a thorough treatment of electromagnetic machine principles, including magnetic circuits, transformers, DC machines, and synchronous and asynchronous motors.

See also: Applied Physics; Computer Engineering; Electrical Engineering; Electronics and Electronic Engineering; Magnetic Resonance Imaging; Magnetic Storage.

ELECTROMETALLURGY

FIELDS OF STUDY

Chemistry; physics; engineering; materials science.

SUMMARY

Electrometallurgy includes electrowinning, electroforming, electrorefining, and electroplating. The electrowinning of aluminum from its oxide (alumina) accounts for essentially all the world's supply of this metal. Electrorefining of copper is used to achieve levels of purity needed for its use as an electrical conductor. Electroplating is used to protect base metals from corrosion, to increase hardness and wear resistance, or to create a decorative surface. Electroforming is used to produce small, intricately shaped metal objects.

KEY TERMS AND CONCEPTS

- **anode:** Electrode where electrons return to the external circuit.
- **cathode:** Electrode where electrons react with the electrolyte.
- **electrode:** Interface between an electrolyte and an external circuit; usually a solid rod.
- **electroforming:** Creation of metal objects of a desired shape by electrolysis.
- **electrolysis:** Chemical change produced by passage of an electric current through a substance.
- **electrolyte:** Water solution or molten salt that conducts electricity.
- **electroplating:** Creation of an adherent metal layer on some substrate by electrolysis.
- **electrorefining:** Increasing the purity of a metal by electrolysis.
- **electrowinning:** Obtaining a metal from its compounds by the use of electric current.
- **equivalent weight:** Atomic weight of a metal divided by the integer number of electron charges on each of its ions.
- **Faraday:** Least amount of electric charge needed to plate out one equivalent weight of a metal.
- **pyrometallurgy:** Reduction of metal ores by heat and chemical reducing agents such as carbon.

DEFINITION AND BASIC PRINCIPLES

Electrometallurgy is that part of metallurgy that involves the use of electric current to reduce compounds of metals to free metals. It includes uses of electrolysis such as metal plating, metal refining, and electroforming. Electrolysis in this context does not include the cosmetic use of electrolysis in hair removal.

BACKGROUND AND HISTORY

The field of electrometallurgy started in the late eighteenth century and was the direct result of the development of a source of electric current, the cell invented by Alessandro Volta of Pavia, Italy. Volta placed alternating disks of zinc and copper separated by brine-soaked felt disks into a long vertical stack. In March, 1800, Volta described his "pile" in a letter to the Royal Society in London. His primitive device could produce only a small electric current, but it stimulated scientists all over Europe to construct bigger and better batteries and to explore their uses. Luigi Valentino Brugnatelli, an acquaintance of Volta, used a voltaic pile in 1802 to do electroplating. British surgeon Alfred Smee published *Elements of Electrometallurgy* in 1840, using the term "electrometallurgy" for the first time.

Michael Faraday discovered the basic laws of electrolysis early in the nineteenth century and was the first to use the word "electrolysis." Both Sir Humphry Davy and Robert Bunsen used electrolysis to prepare chemical elements. The development of dynamos based on Faraday's ideas made possible larger currents and opened up industrial uses of electrolysis, most notably the Hall-Héroult process for the production of aluminum around 1886.

The Aluminum Company of America (known as Alcoa since 1999) was founded in 1909 and became a major producer of aluminum by the Hall-Héroult process. Canada-based company Rio Tinto Alcan and Norway-based Norsk Hydro are also major producers of aluminum. Magnesium was first made commercially in Germany by I. G. Farben in the 1890's, and later in the United States by Dow Chemical. In both instances, electrolysis of molten magnesium chloride was used to produce the magnesium.

How it Works

Electrolysis involves passage of an electric current through a circuit containing a liquid electrolyte, which may be a water solution or a molten solid. The current is carried through the electrolyte by migration of ions (positively or negatively charged particles). At the electrodes, the ions in the electrolyte undergo chemical reactions. For example, in the electrolysis of molten sodium chloride, positively charged sodium ions migrate to one electrode (the cathode) and negatively charged chloride ions to the other electrode (the anode). As the sodium ions interact with the cathode, they absorb electrons from the external circuit, forming sodium metal. This absorption of electrons is known as reduction of the sodium ions. At the anode, chloride ions lose electrons to the external circuit and are converted to chlorine gas—an oxidation reaction.

The chemical reaction occurring in electrolysis requires a minimum voltage. Sufficient voltage also must be supplied to overcome internal resistance in the electrolyte. Because an electrolyte may contain a variety of ions, alternative reactions are possible. In aqueous electrolytes, positively charged hydrogen ions are always present and can be reduced at the cathode. In a water solution of sodium chloride, hydrogen ions are reduced to the exclusion of sodium ions, and no sodium metal forms. The only metals that can be liberated by electrolysis from aqueous solution are those such as copper, silver, cadmium, and zinc, whose ions are more easily reduced than hydrogen ions.

An electrochemical method of liberating active metals depends on finding a liquid medium more resistant to reduction than water. The use of molten salts as electrolytes often solves this problem. Salts usually have high melting points, and it can be advantageous to choose mixtures of salts with lower melting temperatures, as long as the new cations introduced do not interfere with the desired reduction reaction. Choice of electrode materials is important because it can affect the purity of the metal being liberated. Usually inert electrodes are preferable.

In electroplating, it is necessary to produce a uniform adherent coating on the object that forms the cathode. Successful plating requires careful control of several factors such as temperature, concentration of the metal ion, current density at the cathode, and cleanliness of the surface to be plated. The metal

Electrometallurgy

ions in the electrolyte may need to be replenished as they are being depleted. Current density is usually kept low and plating done slowly. Sometimes additives that react with the metal ions or modify the viscosity or surface tension of the medium are used in the electrolyte. Successful plating conditions are often discovered by experiment and may not be understood in a fundamental sense.

Electroforming consists of depositing a metal plate on an object with the purpose of preparing a duplicate of the object in all its detail. Nonconducting objects can be rendered conductive by coating them with a conductive layer. The plated metal forms on a part called a mandrel, which forms a template and from which the new metal part is separated after the operation. The mandrel may be saved and used again or dissolved away with chemicals to separate it from the new object.

In electrorefining, anodes of impure metals are immersed in an electrolyte that contains a salt of the metal to be refined—often a sulfate—and possibly sulfuric acid. When current is passed through, metal dissolves from the anode and is redeposited on the cathode in purer form. Control of the applied voltage is necessary to prevent dissolution of less easily oxidized metal impurities. Material that is not oxidized at the anode falls to the bottom of the electrolysis cell as anode slime. This slime can be further processed for any valuable materials it may contain.

Applications and Products

Electrowinning of aluminum. Aluminum, the most abundant metal in the Earth's crust, did not become readily available commercially until the development of the Hall-Héroult process. This process involves electrolysis of dry aluminum oxide (alumina) dissolved in cryolite (sodium aluminum hexafluoride). Additional calcium fluoride is used to lower the melting point of the cryolite. The process runs at about 960 degrees Celsius and uses carbon electrodes. The alumina for the Hall-Héroult process is obtained from an ore called bauxite, an impure aluminum oxide with varying amounts of compounds such as iron oxide and silica. The preparation of pure alumina follows the Bayer process: The alumina is extracted from the bauxite as a solution in sodium hydroxide (caustic soda), reprecipitated by acidification, filtered, and dried. The electrolysis cell has a carbon coating at the bottom that forms a cathode, while the anodes are graphite rods extending down into the molten salt electrolyte. As aluminum forms, it forms a pool at the bottom of the cell and can be removed periodically

The anodes are consumed as the carbon reacts with the oxygen liberated by the electrolysis. The alumina needs to be replenished from time to time in the melt. The applied voltage is about 4.5 volts, but it rises sharply if the alumina concentration is too low. The electrolysis cells are connected in series, and there may be several hundred cells in an aluminum plant. The consumption of electric power amounts to about 15,000 kilowatt-hours per ton of aluminum. (A family in a home might consume 150 kilowatt-hours of power in a month.) Much of the power goes for heating and melting the electrolyte. The cost of electric power is a significant factor in aluminum manufacture and makes it advantageous to locate plants where power is relatively inexpensive, for example, where hydropower is available. Although some other metals are produced by electrolysis, aluminum is the metal produced in the greatest amount, tens of millions of metric tons per year.

Aluminum is the most commonly used structural metal after iron. As a low-density strong metal, aluminum tends to find uses where weight saving is important, such as in aircraft. When automobile manufacturers try to increase gas mileage, they replace the steel in vehicles with aluminum to save weight. Aluminum containers are commonly used for foods and beverages and aluminum foil for packaging.

Alcoa has developed a second electrolytic aluminum process that involves electrolysis of aluminum chloride. The aluminum chloride is obtained by chlorinating aluminum oxide. The chlorine liberated at the anode can be recycled. Also the electric power requirements of this process are less than for the Hall-Héroult process. This aluminum chloride reduction has not been used as much as the oxide reduction.

Electrowinning of other metals. Magnesium, like aluminum, is a low-density metal and is also manufactured by electrolysis. The electrolysis of molten magnesium chloride (in the presence of other metal chlorides to lower the melting point) yields magnesium at the cathode. The scale of magnesium production is not as large as that of aluminum, amounting to several hundred thousand metric tons per year. Magnesium is used in aircraft alloys, flares, and chemical syntheses.

Sodium metal comes from the electrolysis of molten sodium chloride in an apparatus called the Downs cell after its inventor J. C. Downs. Molten sodium forms at the cathode. Sodium metal was formerly very important in the process for making the gasoline additive tetraethyl lead. As leaded fuel is no longer sold in the United States, this use has

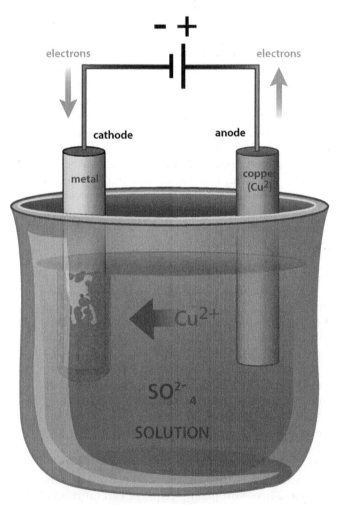

ELECTROPLATING

Electroplating is the creation of an adherent metal layer on some substrate by electrolysis.

declined, but sodium continues to be used in organic syntheses, the manufacture of titanium, and as a component (with potassium) in high-temperature heat exchange media for nuclear reactors.

Lithium and calcium are obtained in relatively small quantities by the electrolysis of their chlorides. Lithium is assuming great importance for its use in high-performance batteries for all types of applications but particularly for powering electric automobiles. A lightweight lithium-aluminum alloy is used in the National Aeronautics and Space Administration's Ares rocket and the external tank of the space shuttle.

Electrorefining applications. Metals are often obtained from their ores by pyrometallurgy (sequences of heating) and the use of reducing agents such as carbon. Metals obtained this way include iron and copper. Iron can be refined electrolytically, but much iron is used for steel production without refining.

Copper, however, is used in applications where purity is important. Copper, when pure, is ductile and an excellent electrical conductor, so it needs to be refined to be used in electrical wiring. Copper anodes (blister copper) are suspended in a water solution containing sulfuric acid and copper sulfate with steel cathodes. Electrolysis results in dissolution of copper from the anode and migration of copper ions to the cathode, where purified metal is deposited. The result is copper of 99.9 percent purity. A similar procedure may be used in recycling copper. Other metals that are electrorefined include aluminum, Electrometallurgy zinc, nickel, cobalt, tin, lead, silver, and gold. Materials are added to the electrolyte to make the metal deposit more uniform: glue, metal chlorides, levelers, and brighteners may be included. The details of the conditions for electrorefining vary depending on the metal. The electrorefining done industrially worldwide ranges from 1,000 to 100,000 metric tons of metals per year.

Electroforming applications. The possibility of reproducing complicated shapes on both large and small scale is an advantage of electroforming. Objects that would be impossible to produce because of their intricate shapes or small sizes are made by electroforming. The metal used may be a single metal or an alloy. The metals used most often are nickel and copper. The manufacture of compact discs for recording sound makes use of electroforming in the reproduction of the bumps and grooves in a studio disk of glass, which is a negative copy, and is then used to make a mold from which plastic discs can be cast. Metal foil can be electroformed by using a rotating mandrel surrounded by a cylindrical anode. The foil is peeled off the mandrel in a continuous sheet. The electroforming of copper foil for electronic applications is the largest application of electroforming. About $2 billion per year is spent on electroforming.

Electroplating applications. Plating is done to protect metal surfaces from corrosion, to enhance the appearance of a surface, or to modify its properties in other ways such as to increase hardness or reflectivity. Familiar applications include chromium plating of automobile parts, silver plating of tableware and jewelry, and gold plating of medals and computer parts. Plating can also be done on nonmetallic surfaces such as plastic or ceramics. The manufacture of circuit boards requires a number of steps, some of which involve the electroplating of copper, lead, and tin. Many switches and other electrical contacts are plated with gold to prevent corrosion. The metals involved in commercial electroplating are mostly deposited from an aqueous electrolyte. This excludes metals such as aluminum or magnesium, which cannot be liberated in water solution. If a molten salt electrolyte is used, aluminum plating is possible, but it is seldom done.

IMPACT ON INDUSTRY

In 2006, 33.4 million tons of primary aluminum was produced worldwide, with the majority of it being made in China, the United States, Russia, Canada, Brazil, Australia, Norway, and India. Magnesium production in 2008 was 719,000 tons. These two metals account for most of the electrochemical metal production. Because of the high economic value of these industries, the processes involved are the subjects of research at government institutions and universities and at major corporations. Continuing research is done to improve the energy efficiency of processes and to mitigate pollution problems. A continuing problem with the Hall-Héroult process is emission of fluorine-containing compounds that are either toxic or are greenhouse gases. The high temperature of the process leads to corrosion problems in the cells, but attempts to lower the temperature can lead to reduced solubility of alumina in the electrolyte.

Alcoa and Rio Tinto Alcan are two of the largest aluminum producers in the world. Each has tens of thousands of employees spread over forty countries.

FASCINATING FACTS ABOUT ELECTROMETALLURGY

- The engine block of the airplane built by Orville and Wilbur Wright in 1903 was made of aluminum obtained by electrolysis.
- In the early nineteenth century, aluminum was so expensive that it was used in jewelry.
- Roughly 5 percent of the electricity in the United States is used to make aluminum.
- The price of aluminum fell to its lowest recorded level in 1993; $0.53 per pound.
- The U.S. industry that uses the most aluminum is the beverage industry. Americans use 80 billion aluminum cans per year.
- When Prince Charles of England was invested as Prince of Wales in 1969, the coronet he received was made partly by electroforming.
- The Voyager space vehicle launched in 1977 contained a gold-plated copper disk electroplated with a patch of pure uranium-238 isotope. If this disk were recovered by some advanced alien civilization, the disk's age could be determined from the radioactive decay rate of the isotope.

Alcoa maintains a technical research center near Pittsburgh. In 2007, the company signed an agreement with a consortium of universities in Russia to sponsor research and development. Similar agreements were made in both India and China. Rio Tinto Alcan makes grants to several Canadian universities for research on aluminum production.

The National Research Council of Canada has an aluminum technical center in Saguenay, Quebec, and the E. B. Yeager Center for Electrochemical Science at Case Western University is in Cleveland, Ohio. The center sponsors workshops, lectures, and research in all areas of electrochemistry, including the use of computer-controlled processes to achieve optimum conditions in manufacturers. U.S. government laboratories such as Argonne National Laboratory in Illinois also do research on electrochemistry. The U.S. Department of Energy supports university research at various institutions, including Michigan Technological University, which received a $2 million grant for research on uses for aluminum smelting wastes.

CAREERS AND COURSE WORK

The path to a career in electrometallurgy is through a bachelor's or master's degree in chemistry or chemical engineering, although careers in research require a doctorate. The course work involves a thorough grounding in physical science (chemistry, physics), two years of calculus, and courses in computer science and engineering principles. Most large state universities offer programs in chemistry and chemical engineering. A few universities, including the University of Utah and the University of Nevada at Reno, offer specialized work in electrometallurgy. Rio Tinto Alcan and Alcoa offer summer internships for students and Rio Tinto Alcan also offers scholarships.

Multiday, personal development courses in electroplating are widely available through groups such as the National Society for Surface Finishing.

SOCIAL CONTEXT AND FUTURE PROSPECTS

The electometallurgy industry, like many industries, poses challenges for society. Metals have great value and many uses, which are an essential part of modern life. Unfortunately, electrometallurgy consumes huge amounts of energy and uses tons of unpleasant chemicals. In addition, aluminum plants emit carbon dioxide and fluorine compounds. However, the use of electricity to produce metals probably remains the cleanest and most efficient method.

In the future, the process of electrometallurgy probably will become more efficient and less polluting. Also, new techniques are likely to be developed to permit additional metals to be obtained by electrometallurgy. Titanium metal continues to be made by reduction of its chloride by sodium metal, but the Fray Farthing Chen (FFC) Cambridge process announced in 2000 shows promise as an electrolytic method for obtaining titanium. The process involves reduction of titanium oxide by electrically generated calcium in a molten calcium chloride medium. Titanium is valued for its strength, light weight, high temperature performance, and corrosion resistance. These qualities make it essential in jet engine turbines. Titanium is stable in the human body and can be used for making artificial knees and hips for implantation.

—*John R. Phillips, PhD*

FURTHER READING

Chen, G. Z., D. J. Fray, and T. W. Farthing. "Direct Electrochemical Reduction of Titanium Dioxide to Titanium in Molten Calcium Chloride." *Nature* 407 (2000): 361-364. First description of the FFC-Cambridge method—a possibly game-changing procedure for reducing metal oxides.

Curtis, Leslie. *Electroforming*. London: A & C Black, 2004. A simple description of electroforming with practical directions for applications.

Geller, Tom. "Common Metal, Uncommon Past." *Chemical Heritage* 25 no. 4 (Winter, 2007): 32-36. Historical details on the discovery and manufacture of aluminum.

Graham, Margaret B. W., and Bettye H. Pruitt. *R&D for Industry: A Century of Technical Innovation at Alcoa*. New York: Cambridge University Press, 1990. An extensive history that is mostly nontechnical. Much discussion of organizational and management matters.

Kanani, Nasser. *Electroplating: Basic Principles, Processes and Practice*. New York: Elsevier, 2005. A thorough discussion of electroplating processes including measurements of thickness and adherence using modern instruments.

Mertyns, Joost. "From the Lecture Room to the Workshop: John Frederic Daniell, the Constant Battery and Electrometallurgy Around 1840." *Annals of Science* 55, no. 3 (July, 1998): 241-261. Early developments in batteries, electroplating, and electroforming are described.

Pletcher, Derek, and Frank C. Walsh. *Industrial Electrochemistry*. 2d ed. New York: Blackie Academic & Professional, 1992. Particularly good discussion of metal plating and electroforming with both theoretical and practical aspects treated. List of definitions and units.

Popov, Konstantin I., Stojan S. Djokić, and Branimir N. Grgur, eds. *Fundamental Aspects of Metallurgy*. New York: Kluwer Academic/Plenum, 2002. Provides background on metallurgy, then the specifics of metal disposition.

WEB SITES

The Electrochemical Society
http://www.electrochem.org

International Society of Electrochemistry
http://www.ise-online.org

See also: Aeronautics and Aviation; Electrochemistry; Metallurgy.

ELECTRONIC MATERIALS PRODUCTION

FIELDS OF STUDY

Mathematics; physics; chemistry; crystallography; quantum theory; thermodynamics

SUMMARY

While the term "electronic materials" commonly refers to the silicon-based materials from which computer chips and integrated circuits are constructed, it technically includes any and all materials upon which the function of electronic devices depends. This includes the plain glass and plastics used to house the devices to the exotic alloys and compounds that make it possible for the devices to function. Production of many of these materials requires not only rigorous methods and specific techniques but also requires the use of high-precision analytical methods to ensure the structure and quality of the devices.

KEY TERMS AND CONCEPTS

- **biasing:** The application of a voltage to a semiconductor structure (transistor) to induce a directional current flow in the structure.
- **Czochralski method:** A method of pulling material from a molten mass to produce a single large crystal.
- **denuded zone:** Depth and area of a silicon wafer that contains no oxygen precipitates or interstitial oxygen.
- **epi reactor:** A thermally programmable chamber in which epitaxial growth of silicon chips is carried out.
- **gettering:** A method of lowering the potential for precipitation from solution of metal contaminants in silicon, achieved by controlling the locations at which precipitation can occur.
- **polysilicon (metallurgical grade silicon):** A form of

silicon that is 99 percent pure, produced by the reaction of silicon dioxide (SiO_2) and carbon (C) to produce silicon (Si) and carbon monoxide (CO) at a temperature of 2000 degrees Celsius.

DEFINITION AND BASIC PRINCIPLES

Electronic materials are those materials used in the construction of electronic devices. The major electronic material today is the silicon wafer, from which computer chips and integrated circuits (ICs) are made. Silicon is one of a class of elements known as semiconductors. These are materials that do not conduct electrical currents appreciably unless acted upon, or "biased," by an external voltage. Another such element is germanium.

The construction of silicon chips requires materials of high purity and consistent internal structure. This, in turn, requires precisely controlled methods in the production of both the materials and the structures for which they are used. Large crystals of ultrapure silicon are grown from molten silicon under strictly controlled environmental conditions. Thin wafers are sliced from the crystals and then polished to achieve the desired thickness and mirror-smooth surface necessary for their purpose. Each wafer is then subjected to a series of up to five hundred, and sometimes more, separate operations by which extremely thin layers of different materials are added in precise patterns to form millions of transistor structures. Modern CPU (central processing unit) chips have between 10^7 and 10^9 separate transistors per square centimeter etched on their surfaces in this way.

One of the materials added by the thin-layer deposition process is silicon, to fill in spaces between other materials in the structures. These layers must be added epitaxially, in a way that maintains the base crystal structure of the silicon wafer.

Other materials used in electronic devices are also formed under strictly controlled environmental conditions. Computers could not function without some of these materials, especially indium tin oxide (ITO) for what are called transparent contacts and indium nitride for light-emitting diodes in a full spectrum range of colors.

BACKGROUND AND HISTORY

The production of modern electronic materials began with the invention of the semiconductor bridge transistor in 1947. This invention, in turn, was made possible by the development of quantum theory and the vacuum tube technology with which electronic devices functioned until that time.

The invention of the transistor began the development of electronic devices based on the semiconducting character of the element silicon. Under the influence of an applied voltage, silicon can be induced to conduct an electrical current. This feature allows silicon-based transistors to function somewhat like an on-off switch according to the nature of the applied voltage.

In 1960, the construction of the functional laser by American physicist and Nobel laureate Arthur Schawlow began the next phase in the development of semiconductor electronics, as the assembly of transistors on silicon substrates was still a tedious endeavor that greatly limited the size of transistor structures that could be constructed. As lasers became more powerful and more easily controlled, they were applied to the task of surface etching, an advance that has produced ever smaller transistor structures. This development has required ever more refined methods of producing silicon crystals from which thin wafers can be cut for the production of silicon semiconductor chips, the primary effort of electronic materials production (though by no means the most important).

HOW IT WORKS

Melting and crystallization. Chemists have long known how to grow large crystals of specific materials from melts. In this process, a material is heated past its melting point to become liquid. Then, as the molten material is allowed to cool slowly under controlled conditions, the material will solidify in a crystalline form with a highly regular atomic distribution.

Now, molten silicon is produced from a material called polysilicon, which has been stacked in a closed oven. Specific quantities of doping materials such as arsenic, phosphorus, boron, and antimony are added to the mixture, according to the conducting properties desired for the silicon chips that will be produced. The polysilicon melt is rotated in one direction (clockwise); then, a seed crystal of silicon, rotating in the opposite direction (counterclockwise), is introduced. The melt is carefully cooled to a specific temperature as the seed crystal structure is drawn out of the molten mass at a rate that determines the diameter of the resulting crystal.

To maintain the integrity of the single crystal that results, the shape is allowed to taper off into the form of a cone, and the crystal is then allowed to cool completely before further processing. The care with which this procedure is carried out produces a single crystal of the silicon alloy as a uniform cylinder, whose ends vary in diameter first as the desired extraction rate was achieved and then due to the formation of the terminal cone shape.

Wafers. In the next stage of production, the non-uniform ends of the crystal are removed using an inner diameter saw. The remaining cylinder of crystal is called an ingot, and is then examined by X ray to determine the consistency and integrity of the crystal structure. The ingot then will normally be cut into smaller sections for processing and quality control.

To produce the rough wafers that will become the substrates for chips, the ingot pieces are mounted on a solid base and fed into a large wire saw. The wire saw uses a single long moving wire to form a thick network of cutting edges. A continuous stream of slurry containing an extremely fine abrasive provides the cutting capability of the wire saw, allowing the production of many rough wafers at one time. The rough wafers are then thoroughly cleaned to remove any residue from the cutting stage.

Another procedure rounds and smooths the edges of each wafer, enhancing its structural strength and resistance to chipping. Each wafer is also laser-etched with identifying data. They then go on to a flat lapping procedure that removes most of the machining marks left by the wire saw, and then to a chemical etching process that eliminates the marking that the lapping process has left. Both the lapping process and the chemical etching stage are used to reduce the thickness of the wafers.

Polishing. Following lapping and rigorous cleaning, the wafers move into an automated chemical-mechanical polishing process that gives each wafer an extremely smooth mirror-like and flat surface. They are then again subjected to a series of rigorous chemical cleaning baths, and are then either packaged for sale to end users or moved directly into the epitaxial enhancement process.

Epitaxial enhancement. Epitaxial enhancement is used to deposit a layer of ultrapure silicon on the surface of the wafer. This provides a layer with different properties from those of the underlying wafer

material, an essential feature for the proper functioning of the MOS (metal-oxide-semiconductor) transistors that are used in modern chips. In this process, polished wafers are placed into a programmable oven and spun in an atmosphere of trichlorosilane gas. Decomposition of the trichlorosilane

deposits silicon atoms on the surface of the wafers. While this produces an identifiable layer of silicon with different properties, it also maintains the crystal structure of the silicon in the wafer. The epitaxial layer contains no imperfections that may exist in the wafer and that could lead to failure of the chips in use.

From this point on, the wafers are submitted to hundreds more individual processes. These processes build up the transistor structures that form the functional chips of a variety of integrated circuit devices and components that operate on the principles of digital logic.

APPLICATIONS AND PRODUCTS

Microelectronics. The largest single use of silicon chips is in the microelectronics industry. Every digital device functions through the intermediacy of a silicon chip of some kind. This is as true of the control pad on a household washing machine as it is of the most sophisticated and complex CPU in an ultramodern computer.

Digital devices are controlled through the operation of digital logic circuits constructed of transistors built onto the surface of a silicon chip. The chips can be exceedingly small. In the case of integrated circuit chips, commonly called ICs, only a few transistors may be required to achieve the desired function.

The simplest of these ICs is called an inverter, and a standard inverter IC provides six separate inverter circuits in a dual inline package (DIP) that looks like a small rectangular block of black plastic about 1 centimeter wide, 2 centimeters long, and 0.5 centimeters thick, with fourteen legs, seven on each side. The actual silicon chip contained within the body of the plastic block is approximately 5 millimeters square and no more than 0.5 millimeters thick. Thousands of such chips are cut from a single silicon wafer that has been processed specifically for that application.

Inverters require only a single input lead and a single output lead, and so facilitate six functionalities on the DIP described. However, other devices typically use two input leads to supply one output lead. In those devices, the same DIP structure provides only four functionalities. The transistor structures are correspondingly more complex, but the actual chip size is about the same. Package sizes increase according to the complexity of the actual chip and the number of leads that it requires for its function and application to physical considerations such as dissipation of the heat that the device will generate in operation.

In the case of a modern laptop or desktop computer, the CPU chip package may have two hundred leads on a square package that is approximately 4 centimeters on a side and less than 0.5 centimeters in thickness. The actual chip inside the package is a very thin sheet of silicon about 1 square centimeter in size, but covered with several million transistor structures that have been built up through photo-etching and chemical vapor deposition methods, as described above. Examination of any service listing of silicon chip ICs produced by any particular manufacturer will quickly reveal that a vast number of different ICs and functionalities are available.

Solar technology. There are several other current uses for silicon wafer technology, and new uses are yet to be realized. Large quantities of electronic-grade silicon wafers are used in the production of functional solar cells, an area of application that is experiencing high growth, as nonrenewable energy resources become more and more expensive. Utilizing the photoelectron effect first described by Albert Einstein in 1905, solar cells convert light energy into an electrical current. Three types are made, utilizing both thick (> 300 micrometers [μm]) and thin (a few μm) layers of silicon. Thick-layer solar cells are constructed from single crystal silicon and from large-grain polycrystalline silicon, while thin-layer solar cells are constructed by using vapor deposition to deposit a layer of silicon onto a glass substrate.

Microelectronic and mechanical systems. Silicon chips are also used in the construction of microelectronic and mechanical systems (MEMS). Exceedingly tiny mechanical devices such as gears and single-pixel mirrors can be constructed using the technology developed for the production of silicon chips. Devices produced in this way are by nature highly sensitive and dependable in their operation, and so the majority of MEMS development is for the production of specialized sensors, such as the accelerometers used to initiate the deployment of airbag restraint systems in automobiles. A variety of other products are also available using MEMS technology, including biosensors, the micronozzles of inkjet printer cartridges, microfluidic test devices, microlenses and arrays of microlenses, and microscopic versions of tunable capacitors and resonators.

Other applications. Other uses of silicon chip technology, some of which is in development, include mirrors for X-ray beams; mirrors and prisms for application in infrared spectroscopy, as silicon is entirely transparent to infrared radiation; and the material called porous silicon, which is made electro-chemically from single-crystal silicon and has itself presented an exceptionally varied field of opportunity in materials science.

As mentioned, there are also many other materials that fall into the category of electronic materials. Some, such as copper, gold, and other pure elements, are produced in normal ways and then subjected to methods such as zone refining and vapor deposition techniques to achieve high purity and thin layers in the construction of electronic devices. Many exotic elements and metallic alloys, as well as specialized plastics, have been developed for use in electronic devices. Organic compounds known as liquid crystals, requiring no extraordinary synthetic measures, are normally semisolid materials that have properties of both a liquid and a solid. They are extensively used as the visual medium of thin liquid crystal display (LCD) screens, such as would be found in wristwatches, clocks, calculators, laptop and tablet computers, almost all desktop monitors, and flat-screen televisions.

Another example is the group of compounds made up of indium nitride, gallium nitride, and aluminum nitride. These are used to produce light-emitting diodes (LEDs) that provide light across the full visible spectrum. The ability to grow these LEDs on the same chip now offers a technology that could completely replace existing CRT (cathode ray tube) and LCD technologies for visual displays.

IMPACT ON INDUSTRY

Electronic materials production is an entire industry unto itself. While the products of this industry are widely used throughout society, they are not used in the form in which they are produced. Rather, the products of the electronic materials industry become input supplies for further manufacturing processes. Silicon chips, for example, produced by any individual manufacturer, are used for in-house manufacturing or are marketed to other manufacturers, who, in turn, use the chips to produce their own particular products, such as ICs, solar cells, and microdevices.

This intramural or business-to-business market aspect of the electronic materials production industry, with its novel research and development efforts and especially given the extent to which society now relies on information transfer and storage, makes ascribing an overall economic value to the industry impossible. One has only to consider the number of computing devices produced and sold each year around the world to get a sense of the potential value of the electronic materials production industry.

Ancillary industries provide other materials used by the electronic materials production industry, many of which must themselves be classified as

---

FASCINATING FACTS ABOUT ELECTRONIC MATERIALS PRODUCTION

- Large single crystals of silicon are grown from a molten state in a process that literally pulls the molten mass out into a cylindrical shape.
- About 70 percent of silicon chips fail during the manufacturing process, leaving only a small percentage of chips that are usable.
- Silicon is invisible to infrared light, making it exceptionally useful for infrared spectroscopy and as mirrors for X rays.
- Quantum dot and graphene-based transistors will produce computers that are orders of magnitude more powerful than those used today.
- The photoelectric effect operating in silicon allows solar cells to convert light energy into an electrical current.
- Semiconductor transistors were invented in 1947 and integrated circuits in 1970, and the complexity of electronic components has increased by about 40 percent each year.
- Copper and other metals dissolve very quickly in liquid silicon, but precipitate out as the molten material cools, often with catastrophic results for the silicon crystal.
- Porous silicon is produced electrochemically from single crystals of silicon. Among its other properties, porous silicon is highly explosive.

---

electronic materials. An electric materials company, for example, may provide polishing and surfacing materials, photovoltaic materials, specialty glasses, electronic packaging materials, and many others.

Given both the extremely small size and sensitivity of the structures created on the surface of silicon chips and the number of steps required to produce those structures, quality control procedures are stringent. These steps may be treated as part of a multistep synthetic procedure, with each step producing a yield (as the percentage of structures that meet functional requirements). In silicon-chip production, it is important to understand that only the chips that are produced as functional units at the end of the process are marketable. If a process requires two hundred individual construction steps, even a 99 percent success rate for each step translates into a final yield of functional chips of only 0.99^{200}, or 13.4 percent. The majority of chip structures fail during construction, either through damage or through a step failure. It is therefore imperative that each step in the construction of silicon chips be precisely carried out.

To that end, procedures and quality control methods have been developed that are applicable in other situations too. Clean room technology that is essential for maximizing usable chip production is equally valuable in biological research and medical treatment facilities, applied physics laboratories, space exploration, aeronautics repair and maintenance facilities, and any other situations in which steps to protect either the environment or personnel from contamination must be taken.

CAREERS AND COURSEWORK

Electronic materials production is a specialist field that requires interested students to take specialist training in many subject areas. For many such careers, a university degree in solid state physics or electronic engineering is required. For those who will specialize in the more general field of materials science, these subject areas will be included in the overall curriculum. Silicon technology and semiconductors are also primary subject areas. The fields of study listed here are considered prerequisites for specialist study in the field of electronic materials production, and students can expect to continue studies in these subjects as new aspects of the field develop.

Researchers are now looking into the development of transistor structures based on graphene. This represents an entirely new field of study and application, and the technologies that develop from it will also set new requirements for study. High-end spectrometric methodologies are essential tools in the study and development of this field, and students can expect to take advanced study and training in the use of techniques such as scanning probe microscopy.

SOCIAL CONTEXT AND FUTURE PROSPECTS

Moore's law has successfully predicted the progression of transistor density that can be inscribed onto a silicon chip. There is a finite limit to that density, however, and the existing technology is very near or at that limit. Electronic materials research continues to improve methods and products in an effort to push the Moore limit.

New technologies must be developed to make the use of transistor logic as effective and as economic as possible. To that end, there exists a great deal of research into the application of new materials. Foremost is the development of graphene-based transistors and quantum dot technology, which will drive the level of technology into the molecular and atomic scales.

—*Richard M. Renneboog, MSc*

FURTHER READING

Akimov, Yuriy A., and Wee Shing Koh. "Design of Plasmonic Nanoparticles for Efficient Subwavelength Trapping in Thin-Film Solar Cells." Plasmonics 6 (2010): 155-161. This paper describes how solar cells may be made thinner and lighter by the addition of aluminum nanoparticles on a surface layer of indium tin oxide to enhance light absorption.

Askeland, Donald R. The Science and Engineering of Materials. London: Chapman & Hall, 1998. A recommended resource, this book provides a great deal of fundamental background regarding the physical behavior of a wide variety of materials

and processes that are relevant to electronic materials production.

Falster, Robert. "Gettering in Silicon: Fundamentals and Recent Advances." Semiconductor Fabtech 13 (2001). This article provides a thorough description of the effects of metal contamination in silicon and the process of gettering to avoid the damage that results from such contamination.

Zhang, Q., et al. "A Two-Wafer Approach for Integration of Optical MEMS and Photonics on Silicon Electronic Materials Production Substrate." IEEE Photonics Technology Letters 22 (2010): 269-271. This paper examines how photonic and micro-electromechanical systems on two different silicon chips can be precisely aligned.

Zheng, Y., et al. "Graphene Field Effect Transistors with Ferroelectric Gating." Physical Review Letters 105 (2010). This paper discusses the experimental development and successful testing of a graphene-based field-effect transistor system using gold and graphene electrodes with SiO_2 gate structures on a silicon substrate.

WEB SITES

SCP Symposium (June 2005) "Silicon Starting Materials for Sub-65nm Technology Nodes."
http://www.memc.com/assets/file/technology/ papers/SCP-Symposium-Seacrist.pdf

University of Kiel "Electronic Materials Course."
http://www.tf.uni-kiel.de/matwis/amat/elmat_en/index.html

See also: Applied Physics; Computer Engineering; Computer Science; Electrochemistry; Electronics and Electronic Engineering; Integrated-Circuit Design; Liquid Crystal Technology; Nanotechnology; Surface and Interface Science; Transistor Technologies.

ELECTRONICS AND ELECTRONIC ENGINEERING

FIELDS OF STUDY

Mathematics; physics; electronics; electrical engineering; automotive mechanics; analytical technology; chemical engineering; aeronautics; avionics; robotics; computer programming; audio/video technology; metrology; audio engineering; telecommunications; broadcast technology; computer technology; computer engineering; instrumentation

SUMMARY

A workable understanding of the phenomenon of electricity originated with proof that atoms were composed of smaller particles bearing positive and negative electrical charges. The modern field of electronics is essentially the science and technology of devices designed to control the movement of electricity to achieve some useful purpose. Initially, electronic technology consisted of devices that worked with continuously flowing electricity, whether direct or alternating current. Since the development of the transistor in 1947 and the integrated circuit in 1970, electronic technology has become digital, concurrent with the ability to assemble millions of transistor structures on the surface of a single silicon chip.

KEY TERMS AND CONCEPTS

- **cathode rays:** Descriptive term for energetic beams emitted from electrically stimulated materials inside of a vacuum tube, identified by J. J. Thomson in 1897 as streams of electrons.
- **channel rays:** Descriptive term for energetic beams having the opposite electrical charge of cathode rays, emitted from electrically stimulated materials inside a vacuum tube, also identified by J. J. Thomson in 1897.
- **gate:** A transistor structure that performs a specific function on input electrical signals to produce specific output signals.
- **operational amplifier (op-amp):** An integrated circuit device that produces almost perfect signal reproduction with high gains of amplification and precise, stable voltages and currents.
- **sampling:** Measurement of a specific parameter such as voltage, pressure, current, and loudness at a set frequency determined by a clock cycle such as 1 MHz.
- **semiconductor:** An element that conducts electricity effectively only when subjected to an applied voltage.
- **Zener voltage:** The voltage at which a Zener diode is designed to operate at maximum efficiency to produce a constant voltage, also called the breakdown voltage.

DEFINITION AND BASIC PRINCIPLES

The term "electronics" has acquired different meanings in different contexts. Fundamentally, "electronics" refers to the behavior of matter as affected by the properties and movement of electrons. More generally, electronics has come to mean the technology that has been developed to function according to electronic principles, especially pertaining to basic digital devices and the systems that they operate. The term "electronic engineering" refers to the practice of designing and building circuitry and devices that function on electronic principles.

The underlying principle of electronics derives from the basic structure of matter: that matter is composed of atoms composed of smaller particles. The mass of atoms exists in the atomic nucleus, which is a structure composed of electrically neutral particles called neutrons and positively charged particles called protons. Isolated from the nuclear structure by a relatively immense distance is an equal number of negatively charged particles called electrons. Electrons are easily removed from atoms, and when a difference in electrical potential (voltage) exists between two points, electrons can move from the area of higher potential toward that of lower potential. This defines an electrical current.

Devices that control the presence and magnitude of both voltages and currents are used to bring about changes to the intrinsic form of the electrical signals so generated. These devices also produce physical changes in materials that make comprehensible the information carried by the electronic signal.

BACKGROUND AND HISTORY

Archaeologists have found well-preserved Parthian relics that are now believed to have been rudimentary, but functional, batteries. It is believed that these ancient devices were used by the Parthians to plate objects with gold. The knowledge was lost until 1800, when Italian physicist Alessandro Volta reinvented the voltaic pile. Danish physicist and chemist Hans Christian Oersted demonstrated the relationship between electricity and magnetism in 1820, and in 1821, British physicist and chemist Michael Faraday used that relationship to demonstrate the electromagnetic principle on which all electric motors work. In 1831, he demonstrated the reverse relationship, inventing the electrical generator in the process.

Electricity was thought, by American statesman and scientist Benjamin Franklin and many other scientists of the eighteenth and nineteenth centuries, to be some mysterious kind of fluid that might be captured and stored. A workable concept of electricity was not developed until 1897, when J. J. Thomson identified cathode rays as streams of light electrical particles that must have come from within the atoms of their source materials. He arbitrarily ascribed their electrical charge as negative. Thomson also identified channel rays as streams of massive particles from within the atoms of their source materials that are endowed with the opposite electrical charge of the electrons that made up cathode rays. These observations essentially proved that atoms have substructures. They also provided a means of explaining electricity as the movement of charged particles from one location to another.

With the establishment of an electrical grid, based on the advocacy of alternating current by Serbian American engineer and inventor Nikola Tesla (18561943) , a vast assortment of analogue electrical devices were soon developed for consumer use, though initially these devices were no more than electric lights and electromechanical applications based on electric motors and generators.

As the quantum theory of atomic structure came to be better understood and electricity better controlled, electronic theory became much more important. Spurred by the success of the electromagnetic telegraph of American inventor Samuel Morse (17911872), scientists sought other applications. The first major electronic application of worldwide importance was wireless radio, first demonstrated by Italian inventor Guglielmo Marconi (1874-1937). Radio depended on electronic devices known as vacuum tubes, in which structures capable of controlling currents and voltages could operate at high temperatures in an evacuated tube with external contacts. In 1947, American physicist William Shockley and colleagues invented the semiconductor-based transistor, which could be made to function in the same manner as vacuum tube devices, but without the high temperatures, electrical power consumption, and vacuum construction of those analogue devices.

In 1970, the first integrated circuit "chips" were made by constructing very small transistor structures on the surface of a silicon chip. This gave rise to the entire digital technology that powers the modern world.

APPLICATIONS AND PRODUCTS

Electronics are applied in practically every conceivable manner today, based on their utility in converting easily-produced electrical current into mechanical movement, sound, light, and information signals.

Basic electronic devices. Transistor-based digital technology has replaced older vacuum tube technology, except in rare instances in which a transistorized device cannot perform the same function. Electronic circuits based on vacuum tubes could carry out essentially the same individual operations as transistors, but they were severely limited by physical size, heat production, energy consumption, and mechanical failure. Nevertheless, vacuum tube technology was the basic technology that produced radio, television, radar, X-ray machines, and a broad variety of other electronic applications.

Electronic devices that did not use vacuum tube technology, but which operated on electronic and electromagnetic principles, were, and still are, numerous. These devices include electromagnets and all electric motors and generators. The control systems for many such devices generally consisted of nothing more than switching circuits and indicator lights. More advanced and highly sensitive devices required control systems that utilized the more refined and correspondingly sensitive capabilities available with vacuum tube technology.

Circuit boards. The basic principles of electricity, such as Ohm's resistance law and Kirchoff's current law and capacitance and inductance, are key features in the functional design and engineering of analogue electronic systems, especially for vacuum-tube control systems. An important application that facilitated the general use and development of electronic systems of all kinds is printed circuit board technology. A printed circuit board accepts standardized components onto a nonconducting platform made initially of compressed fiber board, which was eventually replaced by a resin-based composite board. A circuit design is photo-etched onto a copper sheet that makes up one face of the circuit board, and all nonetched copper is chemically removed from the surface of the board, leaving the circuit pattern. The leads of circuit components such as resistors, capacitors, and inductors are inserted into the circuit pattern and secured with solder connections.

Mass production requirements developed the

flotation soldering process, whereby preassembled circuit boards are floated on a bed of molten solder, which automatically completes all solder connections at once with a high degree of consistency. This has become the most important means of circuit board production since the development of transistor technology, being highly compatible with mechanization and automation and with the physical shapes and dimensions of integrated circuit (IC) chips and other components.

Digital devices. Semiconductor-based transistors comprise the heart of modern electronics and electronic engineering. Unlike vacuum tubes, transistors do not work on a continuous electrical signal. Instead, they function exceedingly well as simple on-off switches that are easily controlled. This makes them well adapted to functions based on Boolean algebra. All transistor structures consist of a series of "gates" that perform a specific function on the electronic signals that are delivered to them.

Printed circuit board

Digital devices now represent the most common (and rapidly growing) application of electronics and electronic engineering, including relatively simple consumer electronic devices such as compact fluorescent light bulbs and motion-detecting air fresheners to the most advanced computers and analytical instrumentation. All applications, however, utilize an extensive, but limited, assortment of digital components in the form of IC chips that have been designed to carry out specific actions with electrical or electromagnetic input signals.

Input signals are defined by the presence or absence of a voltage or a current, depending upon the nature of the device. Inverter gates reverse the sense of the input signal, converting an input voltage (high input) into an output signal of no voltage (low output), and vice versa. Other transistor structures (gates) called AND, NAND, OR, NOR and X-OR function to combine input signals in different ways to produce corresponding output signals. More advanced devices (for example, counters and shift registers) use combinations of the different gates to construct various functional circuits that accumulate

signal information or that manipulate signal information in various ways.

One of the most useful of digital IC components is the operational amplifier, or Op-Amp. Op-Amps contain transistor-based circuitry that boosts the magnitude of an input signal, either voltage or current, by five orders of magnitude (100,000 times) or more, and are the basis of the exceptional sensitivity of the modern analytical instruments used in all fields of science and technology.

Electrical engineers are involved in all aspects of the design and development of electronic equipment. Engineers act first as the inventors and designers of electronic systems, conceptualizing the specific functions a potential system will be required to carry out. This process moves through the specification of the components required for the system's functionality to the design of new system devices. The design parameters extend to the infrastructure that must support the system in operation. Engineers determine the standards of safety, integrity, and operation that must be met for electronic systems.

Consumer electronics. For the most part, the term "electronics" is commonly used to refer to the electronic devices developed for retail sale to consumers. These devices include radios, television sets, DVD and CD players, cell phones and messaging devices, cameras and camcorders, laptops, tablets, printers, computers, fax and copy machines, cash registers, and scanners. Millions such devices are sold around the world each day, and numerous other businesses have formed to support their operation.

IMPACT ON INDUSTRY

With electrical and electronic technology now intimately associated with all aspects of society, the impact of electronics and electronic engineering on industry is immeasurable. It would be entirely fair to say that modern industry could not exist without electronics. Automated processes, which are ubiquitous, are not possible without the electronic systems that control them.

The transportation industry, particularly the automotive industry, is perhaps the most extensive user of electronics and electronic engineering. Modern automobiles incorporate an extensive electronic network in their construction to provide the ignition and monitoring systems for the operation of their internal combustion engines and for the many monitoring and control systems for the general safe operation of the vehicle; an electronic network also informs and entertains the driver and passengers. In some cases, electronic systems can completely take control of the vehicle to carry out such specific programmable actions as speed control and parallel parking. Every automated process in the manufacture of automobiles and other vehicles serves to reduce the labor required to carry out the corresponding tasks, while increasing the efficiency and precision of the process steps. Added electronic features also increase the marketability of the vehicles and, hence, manufacturer profits.

Processes that have been automated electronically also have become core components of general manufacturing, especially in the control of production machinery. For example, shapes formed from bent tubing are structural components in a wide variety of applications. While the process of bending the tubing can be carried out by the manual operation of a suitably equipped press, an automated process will produce tube structures that are bent to exact angles and radii in a consistent manner. Typically, a human operator places a straight tube into the press, which then positions and repositions the tube for bending over its length, according to the program that has

been entered into the manufacturing system's electronic controller. Essentially, all continuous manufacturing operations are electronically controlled, providing consistent output.

Electronics and electronic engineering make up the essence of the computer industry; indeed, electronics is an industry worth billions of dollars annually. Electronics affects not only the material side of industry but also the theoretical and actuarial side. Business management, accounting, customer relations, inventory and sales data, and human resources management all depend on the rapid information-handling that is possible through electronics.

XML (extensible markup language) methods and applications are being used (and are in development) to interface electronic data collection directly to physical processes. This demands the use of specialized electronic sensing and sampling devices to convert measured parameters into data points within the corresponding applications and databases. XML is an application that promises to facilitate information exchange and to promote research using large-scale databases. The outcome of this effort is expected to enhance productivity and to expand knowledge in ways that will greatly increase the efficiency and effectiveness of many different fields.

An area of electronics that has become of great economic importance in recent years is that of electronic commerce: the exclusive use of electronic communication technology for the conduct of business between suppliers and consumers. Electronic communications encompasses interoffice faxing, e-mail exchanges, and the Web commerce of companies such as eBay, Amazon, Google, and of the New York and other stock exchanges. The commercial value of these undertakings is measured in billions of dollars annually, and it is expected to continue to increase as new applications and markets are developed.

The fundamental feature here is that these enterprises exist because of the electronic technology that enables them to communicate with consumers and with other businesses. The electronics technology and electronics engineering fields have thus generated entirely new and different daughter industries, with the potential to generate many others, all of which will depend on persons who are knowledgeable in the application and maintenance of electronics and electronic systems.

CAREERS AND COURSEWORK

Many careers depend on knowledge of electronics and electronic engineering because almost all machines and devices used in modern society either function electronically or utilize some kind of electronic control system. The automobile industry is a prime example, as it depends on electronic systems at all stages of production and in the normal operation of a vehicle. Students pursuing a career in automotive mechanics can therefore be expected to study electronic principles and applications as a significant part of their training. The same reasoning applies in all other fields that have a physical reliance on electronic technology.

Knowledge of electronics has become so essential that atomic structure and basic electronic principles, for example, have been incorporated into the elementary school curriculum. Courses of study in basic electronics in the secondary school curriculum are geared to provide a more detailed and practicable knowledge to students.

Specialization in electronics and other fields in which electronics play a significant role is the province of a college education. Interested students can expect to take courses in advanced mathematics, physics, chemistry, and electronics technology as part of the curriculum of their specialty programs. Normally, a technical career, or a skilled trade, requires a college-level certification and continuing education. In some cases, recertification on a regular schedule is also required to maintain specialist standing in that trade.

Students who plan to pursue a career in electronic engineering at a more theoretical level will require, at minimum, a bachelor's degree. A master's degree can prepare a student for a career in forensics, law, and other professions in which an intimate or specialized knowledge of the theoretical side of electronics can be advantageous. (The Vocational Information Center provides an extensive list of careers involving electronics at http://www.khake.com/page19.html.)

SOCIAL CONTEXT AND FUTURE PROSPECTS

It is difficult, if not impossible, to imagine modern society without electronic technology. Electronics has enabled the world of instant communication, wherein a person on one side of the world can communicate directly and almost instantaneously with someone on the other side of the world. As a social tool, such facile communication has the potential to bring about understanding between peoples in a way that has until now been imagined only in science fiction.

Consequently, this facility has also resulted in harm. While social networking sites, for example, bring people from widely varied backgrounds together peacefully to a common forum, network hackers and so-called cyber criminals use electronic technology to steal personal data and disrupt financial markets.

Electronics itself is not the problem, for it is only a tool. Electronic technology, though built on a foundation that is unlikely to change in any significant way, will nevertheless be transformed into newer and better applications. New electronic principles will

FASCINATING FACTS ABOUT ELECTRONICS AND ELECTRONIC ENGINEERING

- In 1847, George Boole developed his algebra for reasoning that was the foundation for first-order predicate calculus, a logic rich enough to be a language for mathematics.
- In 1950, Alan Turing gave an operational definition of artificial intelligence. He said a machine exhibited artificial intelligence if its operational output was indistinguishable from that of a human.
- In 1956, John McCarthy and Marvin Minsky organized a two-month summer conference on intelligent machines at Dartmouth College. To advertise the conference, McCarthy coined the term "artificial intelligence."
- Digital Equipment Corporation's XCON, short for eXpert CONfigurer, was used in-house in 1980 to configure VAX computers and later became the first commercial expert system.
- In 1989, international chess master David Levy was defeated by a computer program, Deep Thought, developed by IBM. Only ten years earlier, Levy had predicted that no computer program would ever beat a chess master.
- In 2010, the Haystack group at the Computer Science and Artificial Intelligence Laboratory at the Massachusetts Institute of Technology developed Soylent, a word-processing interface that lets users edit, proof, and shorten their documents using Mechanical Turk workers.

come to the fore. Materials such as graphene and quantum dots, for example, are expected to provide entirely new means of constructing transistor structures at the atomic and molecular levels. Compared with the 50 to 100 nanometer size of current transistor technology, these new levels would represent a difference of several orders of magnitude. Researchers suggest that this sort of refinement in scale could produce magnetic memory devices that can store as much as ten terabits of information in one square centimeter of disk surface. Although the technological advances seem inevitable, realizing such a scale will require a great deal of research and development.

—*Richard M. Renneboog, MSc*

FURTHER READING

Gates, Earl D. *Introduction to Electronics.* 5th ed. Clifton Park, N.Y.: Cengage Learning, 2006. This book presents a serious approach to practical electronic theory beginning with atomic structure and progressing through various basic circuit types to modern digital electronic devices. Also discusses various career opportunities for students of electronics.

Mughal, Ghulam Rasool. "Impact of Semiconductors in Electronics Industry." *PAF-KIET Journal of Engineering and Sciences* 1, no. 2 (July-December, 2007): 91-98. This article provides a learned review of the basic building blocks of semiconductor devices and assesses the effect those devices have had on the electronics industry.

Petruzella, Frank D. *Introduction to Electricity and Electronics 1.* Toronto: McGraw-Hill Ryerson, 1986. A high-school level electronics textbook that provides a beginning-level introduction to electronic principles and practices.

Platt, Charles. *Make: Electronics.* Sebastopol, Calif.: O'Reilly Media, 2009. This book promotes learning about electronics through a hands-on experimental approach, encouraging students to take things apart and see what makes those things work.

Robbins, Allen H., and Wilhelm C. Miller. *Circuit Analysis Theory and Practice.* Albany, N.Y.: Delmar, 1995. This textbook provides a thorough exposition and training in the basic principles of electronics, from fundamental mathematical principles through the various characteristic behaviors of complex circuits and multiphase electrical currents.

Segura, Jaume, and Charles F. Hawkins. *CMOS Electronics: How It Works, How It Fails.* Hoboken, N.J.: John Wiley & Sons, 2004. The introduction to basic electronic principles in this book leads into detailed discussion of MOSFET and CMOS electronics, followed by discussions of common failure modes of CMOS electronic devices.

Singmin, Andrew. *Beginning Digital Electronics Through Projects.* Woburn, Mass.: Butterworth-Heinemann, 2001. This book presents a basic introduction to electrical properties and circuit theory and guides readers through the construction of some simple devices.

Strobel, Howard A., and William R. Heineman. *Chemical Instrumentation: A Systematic Approach.* 3d ed. New York: John Wiley & Sons, 1989. This book provides an exhaustive overview of the application of electronics in the technology of chemical instrumentation, applicable in many other fields as well.

WEB SITES

Institute of Electrical and Electronics Engineers
http://www.ieee.org

See also: Automated Processes and Servomechanisms; Communication; Computer Engineering; Electrical Engineering; Electrical Measurement; Electromagnet Technologies; Electronic Commerce; Electronic Materials Production; Electronics and Electronic Engineering; Information Technology; Integrated-Circuit Design; Nanotechnology; Radio; Transistor Technologies.

ELECTRON SPECTROSCOPY ANALYSIS

FIELDS OF STUDY

Mathematics; physics; chemistry; biology; materials science; biochemistry; biomedical technology; metallurgy

SUMMARY

Electron spectroscopy analysis is a scientific method that uses ionizing radiation, such as ultraviolet, X-ray, and gamma radiation, to eject electrons from atomic and molecular orbitals in a given material. The properties of these electrons are then interpreted to provide information about the system from which they were ejected.

KEY TERMS AND CONCEPTS

- **auger electron:** An electron emitted from a valence orbital as a result of an energy cascade initiated by the photoemission of an electron from a core orbital.
- **Balmer series:** A set of absorption lines in the electromagnetic spectrum of the hydrogen molecule, corresponding to the specific frequencies of the energy differences between molecular orbitals in the H molecule.
- **bond strength (or bond dissociation energy):** The amount of energy required to overcome the bond between two atoms and separate them from each other.
- **electron-volt:** The energy acquired by any charged particle with a unit charge on passing through a potential difference of one volt, equal to 23,053 calories per mole.
- **ionization potential:** The amount of work required to completely remove a specific electron from an atomic or molecular orbital.
- **paramagnetism:** A measurable increase in the strength of an applied magnetic field caused by alignment of electron orbits in the material.
- **photoelectron emission:** The emission of an electron from an orbital caused by the impingement of a photon.

DEFINITION AND BASIC PRINCIPLES

The quantum mechanical theory of matter describes the positions and energies of electrons within atoms and molecules. When ionizing radiation is applied to a sample of a material, electrons are ejected from atomic and molecular orbitals in that material. The measured energies of those ejected electrons provide information that corresponds to the chemical identity and molecular structure of the material.

The analytical methods that employ this technique, such as mass spectrometry, typically study the properties of the molecular ions themselves rather than the electrons that were removed. The two processes are related, however, because the energies observed for one technique are often identical to those observed for the other. This can be understood at a rudimentary level by considering the law of conservation of energy as it must apply to the overall process of rearrangement. Electrons move from one orbital to another after one has been removed from an inner orbital and rearrangement of the electron distribution takes place to "fill in the hole."

Electron spectroscopic methods require that the electron emission process be carried out under high vacuum and with the use of sensitive electronic equipment to capture and measure the emitted electrons and their properties. Each technique utilizes unique methods, but similar devices, to carry out its tasks.

BACKGROUND AND HISTORY

The beginning of the science of electron spectroscopy can only be equated to the experiments of British physicist and Nobel laureate J. J. Thomson in 1897. These experiments first identified electrons and protons as the electrically charged particles of which atoms were composed, according to the atomic model propounded by British chemist and physicist Ernest Rutherford. Thomson's experiments were also the first to measure the ratio of the charge of the electron to the mass of the electron. This feature must be known to utilize the interaction of electrons and electromagnetic fields quantitatively.

In 1905, Albert Einstein identified and explained the photoelectron effect, in which light is observed to provide the energy by which electrons are ejected from within atoms. This work, one of only a handful of scientific papers actually published by Einstein, earned him the Nobel Prize in Physics in 1921.

German physicist Wilhelm Röntgen's discovery of X rays in 1895, and the subsequent development of the means to precisely control their emission, provided an important way to probe the nature of matter. X rays are designated in the electromagnetic spectrum as intermediate between ultraviolet light and gamma rays. High-vacuum technology and, most recently, digital electronic technology, all combine in the construction of devices that permit the precise measurement of minute changes in the properties of electrons in atoms and molecules.

HOW IT WORKS

Photoelectron spectroscopy. Two general categories of photoelectron spectroscopy are commonly used. These are ultraviolet photoelectron spectroscopy (UPS) and X-ray photoelectron spectroscopy (XPS). Both methods function in precisely the same manner, and both utilize the same devices. The difference between them is that UPS uses ultraviolet radiation as the ionizing method, while XPS uses X rays to effect ionization.

A typical photoelectron spectrometer consists of a high-vacuum chamber containing a sample target; both are connected to an ionizing radiation emitter and a detection system constructed around a magnetic field. In operation, the vacuum chamber is placed under high vacuum. When the system has been evacuated, the sample is introduced and the emitter irradiates the sample, bringing about the emission of electrons from atomic or molecular orbitals in the material. The emitted electrons are then free to move through the magnetic field, where they impinge upon the detector.

The ability to precisely control the strength of the magnetic field allows an equally precise measurement of the energy of the emitted electrons. This measured energy must correspond to the energy of the electrons within the atomic or molecular orbitals of the target material, according to the mathematics of quantum mechanical theory, and so provides information about the intimate internal structure of the atoms and molecules in the material. The methodology has been developed such that measurements are obtainable using matter in any phase as a solid, liquid, or gas. Each phase requires its own modification of the general technique.

The direct measurement of emitted electron energies through the use of photomultiplying devices is displacing more complex methods based on magnetic field because of the inherent difficulties of providing adequate magnetic shielding to the ever more sensitive components of the devices.

XPS is also known as electron spectroscopy for chemical analysis, or ESCA. The use of this identifier, however, is becoming less common in practice and in the chemical literature.

Auger electron spectroscopy (AES). The Auger electron process is a secondary electron emission process that begins with the normal ejection of a core electron by ultraviolet or X-ray radiation. In the Auger process, electrons from higher energy levels shift to lower levels to fill in the gap left by the emission of the core electron. Excess energy that accrues from the difference in orbital energies as the electrons shift then brings about the secondary emission of an electron from a valence shell. The overall process is in accord with both quantum mechanics and the law of conservation of energy, which requires the total energy of a system before a change occurs to be exactly the same as the total energy of the system after a change occurs.

Unlike UPS and XPS, AES generally utilizes an electron beam to effect core electron emission. Detection of emitted electrons is entirely by direct measurement through photomultiplying devices rather than by any magnetic field methods. The methodology of the technique is otherwise similar to that of UPS and XPS. It is amenable to the study of matter in all phases, except for hydrogen and helium, but is generally valuable for use only with solids as a surface analysis technique. This is true because sample materials must be stable under vacuum at pressures of 10^{-}Torr. Also, AES is known to be highly sensitive and capable of fast response.

Electron spin resonance (ESR). The principles of ESR are based on an entirely different physical property of electrons in their atomic or molecular orbitals. In quantum mechanics, each electron is allowed to exist only in very specific states with very specific energies within an atom or molecule.

One of the allowed states is designated as "spin." In this state, the electron can be thought of as an electrical charge that is physically spinning about an axis, thus generating a magnetic field. Only two orientations are allowed for the magnetic fields generated in this way, and according to the Pauli exclusion principle, pairs of electrons must occupy both states in opposition. This requirement means that ESR can be used only with materials that contain single or unpaired electrons, including ions. When placed in an external magnetic field, the magnetic fields of the single electrons align with the external magnetic field.

Subsequent irradiation with an electromagnetic field fluctuating at microwave frequencies acts to invert the magnetic fields of the electrons. Measurement of the frequencies at which inversion takes place provides specific information about the structure of the particular material being examined. The precise locations of inversion signals depend upon the atomic or molecular structure of the material, as these environments affect the nature of the magnetic field surrounding the electron.

APPLICATIONS AND PRODUCTS

In application, electron spectroscopy is strictly an analytical methodology, and it serves only as a probe of material composition and properties. It does not serve any other purpose, and all applications and products related to electron spectroscopy are the corresponding spectroscopic analyzers and the ancillary products that support their operation.

Spectroscopic analyzers come in a variety of forms and designs, according to the environment in which they will be expected to function, but more with respect to the nature of the use to which they will be put. These range from machines for routine analysis of a limited range of materials and properties at the low end of the scale, to the complex machines used in high-end research that must be capable of extreme sensitivity and finely detailed analysis.

The applications of electron spectroscopic analysis are, in contrast, wide ranging and are applicable in many fields. In its roles in those fields, the methodology has enabled some of the most fundamental technology to be found in modern society.

One application in which electron spectroscopy has proven unequaled in its role is submicroscopic surface analysis. Both XPS and AES are the methods of choice in this application, because each can probe to a depth of about 30 microns below the actual surface of a solid material, enabling analysts to see and understand the physical and chemical changes that occur in that region.

The surface of a solid typically represents the point of contact with another solid, and physical interaction between the two normally effects some kind of change to those surfaces because of friction, impact, or electrochemical interaction. A good example of this is the tribological study of moving parts in internal combustion engines. In normal operation, a piston fitted with sealing rings moves with a reciprocating motion within a closely fitted cylinder. The rings physically interact with the wall of the cylinder with intense friction, even though well lubricated, under the influence of the high heat produced through the combustion of fuel. At the same time, the top of the piston is subjected to intense pressures and heat from the explosive combustion of the fuel. At an engine revolution rate of 2,400 revolutions per minute (rpm), each cylinder in a four-cylinder internal combustion engine goes through its reciprocating motion six hundred times each minute, or ten times each second.

In turbine and jet engines, for example, parts are subjected to such stress and friction at a rate of hundreds and even thousands of times per second. Engine and automobile manufacturers and developers must understand what happens to the materials used in the corresponding parts under the conditions of operation. Both XPS and AES are used to probe the material effects at these surfaces for the development of better formulations and materials, and for understanding the weaknesses and failure modes of existing materials.

ESR, on the other hand, is used entirely for the study of the chemical nature of materials in the liquid or gaseous phase. In this role, researchers and analysts use the methodology to study the reactions and mechanisms involving single-electron chemical species. This includes the class of compounds known as radicals, which are essentially molecules containing their full complement of electrons but not of atoms. The methyl radical, for example, is basically a molecule of methane (CH) that has lost one hydrogen atom. The remaining $\dot{C}H$ portion is electrically neutral, because it has all of the electrons that would normally be present in a neutral molecule of CH, but with one of its electrons free to latch on to the first available molecule that comes along.

Radical reactions are understood to be responsible for many effects: aging in living systems, especially humans; atmospheric reactions, especially in the upper atmosphere and ozone layer; the detrimental effects of singlet oxygen; combustion processes; and many others. In biological systems, specially designed molecules are used to tag other nonparamagnetic molecules so that they can be studied by ESR. Such molecules often include a "nitroso" functional group in their structures to provide a paramagnetic radical site that can be monitored by ESR. The production and testing of these specialty chemicals is another area of application.

IMPACT ON INDUSTRY

The impact on industry of electron spectroscopy is not obvious. The methodology plays very much a behind-the-scenes or supportive role that is not apparent from outside any industry that uses electron spectroscopy. Although this is true, the role played by electron spectroscopy in the development and

---

FASCINATING FACTS ABOUT ELECTRON SPECTROSCOPY ANALYSIS

- The photoelectron effect, one of the basic principles of electron spectroscopy, was first explained by Albert Einstein in 1905, for which he received the 1921 Nobel Prize in Physics.
- Spinning electrons generate a magnetic field around themselves in the same way that moving electrons through a wire produces a magnetic field around the wire.
- X rays and ultraviolet light can both eject electrons from the inner orbitals of atoms.
- Auger electrons are electrons emitted by the extra energy released when electrons cascade into lower orbitals to replace electrons that have been ejected by X rays or ultraviolet light.
- Electron spectroscopy can measure the chemical and physical properties of materials up to 30 microns below the surface of a solid.
- Electron spin resonance measures the frequencies required to switch the orientation of the magnetic fields of single electrons.
- Scanning Auger microscopy can produce detailed maps of the distribution of specific metal atoms at the surface of an alloy, allowing metallurgists to see how the material is structured.

---

improvement of products and materials has been valuable to those same industries.

Many advances in metallurgy and tribology, the study of friction and its effects, have been made possible through knowledge obtained by electron spectroscopy, especially AES and XPS. Given the untold millions of moving parts–bearings, pistons, slides, shafts, link chains–that are in operation around the world every single minute of every single day, lubrication and lubricating materials represents a huge world-wide industry. With the vast majority of lubricating materials (oils and greases) being derived from nonrenewable resources, the need to enhance the performance of those materials through better understanding of material interactions has been one driving force behind the application of electron spectroscopy in industry.

Other industrial processes require that materials undergo a chemical process called passivation, which is essentially the rendering of the surface of a material inert to chemical reaction through the formation of a thin coating layer of oxide, nitride, or some other suitable chemical form. With its ability to accurately measure the thickness and properties of thin films such as oxide layers on a surface, electron spectroscopy is uniquely appropriate to use in industries that rely on passivation or on the formation of thin layers with specific properties. One such industry is the semiconductor industry, upon which the computer and digital electronics fields have been built. AES and XPS are commonly used to monitor the quality and properties of thin layers of semiconductor materials used to construct computer chips and other integrated circuits.

CAREERS AND COURSEWORK

Electron spectroscopy is used to examine the intimate details of the atomic and molecular structure of matter, making electron spectroscopy an advanced career. Students who look to a career in this field will undertake highly technical foundation courses in mathematics, physics, inorganic chemistry, organic chemistry, surface chemistry, physical chemistry, chemical physics, and electronics. The minimum requirement for a career in this field is an associate's degree in electronics technology or an honors (four-year) bachelor's degree in chemistry, which will allow a student to specialize in the practice as a technician in a research or analytical facility. More advanced positions will require a postgraduate degree (master's or doctorate).

As the field finds more application in materials research and forensic investigation, general opportunities should further develop. These applications will require, however, that those wishing a career involving electron spectroscopy have specialist training. The vast majority of opportunities in the field are to be found in such academic research facilities as surface science laboratories and in materials science research. Forensic analysis also holds a number of opportunities for electron spectroscopists.

SOCIAL CONTEXT AND FUTURE PROSPECTS

Electron spectroscopy analysis is a methodology with an important role behind the scenes. The field is neither well known nor readily recognized. Nevertheless, it is a critical methodology for advancing the understanding of materials and the nature of matter. As such, electron spectroscopy adds to the general wealth of knowledge in ways that permit the development of new materials and processes and to advancing the understanding of how existing materials function.

One development of electron spectroscopy, known as scanning Auger microscopy (SAM) has the potential to become an extremely valuable technique because of its ability to generate detailed maps of the surface structure of materials at the atomic and molecular level. By tuning SAM to focus on specific elements, the precise distribution of those elements in the surface being examined can be identified and mapped, providing detailed knowledge of the granularity, crystallinity, and other structural details of the material. This is especially valuable in such widely varied fields as metallurgy, geology, and advanced composite materials.

XPS and AES have been applied in a variety of different fields and are themselves becoming very important surface analytical methods in those fields. These areas include the aerospace and automotive industries, biomedical technology and pharmaceuticals, semiconductors and electronics, data storage, lighting and photonics, telecommunications, polymer science, and the rapidly growing fields of solar cell and battery technology.

—*Richard M. Renneboog, MSc*

FURTHER READING

Chourasia, A. R., and D. R. Chopra. "Auger Electron Spectroscopy." In *Handbook of Instrumental Techniques for Analytical Chemistry*, edited by Frank Settle. New York: Prentice Hall Professional Reference, 1997. This chapter provides a thorough and systematic description of the principles and practical methods of Auger spectroscopy, including its common applications and limitations.

Kolasinski, Kurt W. *Surface Science: Foundations of Catalysis and Nanoscience*. 2d ed. Chichester, England: John Wiley & Sons, 2008. Provides a complete study of the utility of electron spectroscopy as applied to the study of processes that occur on material surfaces.

Merz, Rolf. "Nano-analysis with Electron Spectroscopic Methods: Principle, Instrumentation, and Performance of XPS and AES." In *NanoS Guide 2007*. Weinheim, Germany: Wiley-VCH, 2007. A lucid and readable presentation of the principles, capabilities, and limitations of XPS and AES based on actual applications and comparisons with methods such as scanning electron microscopy.

Strobel, Howard A., and William R. Heineman. *Chemical Instrumentation: A Systematic Approach*. 3d ed. New York: John Wiley & Sons, 1989. This book provides a valuable resource for the principles and practices of electron spectroscopy and for many other analytical methods. Geared toward the operation and maintenance of the devices used in those practices.

WEB SITES

Farach, H. A., and C. P. Poole "Overview of Electron Spin Resonance and Its Applications"
http://www.uottawa.ca/publications/interscientia/inter.2/spin.html

Molecular Materials Research Center, California Institute of Technology "Overview of Electron Spin Resonance and Its Applications"
http://mmrc.caltech.edu/SS_XPS/XPS_PPT/XPS_Slides.pdf

See also: Applied Physics; Computer Engineering; Electrical Engineering; Electrochemistry; Electromagnet Technologies; Electrometallurgy; Electronics and Electronic Engineering.

ENGINEERING

FIELDS OF STUDY

Physics; chemistry; computer science; mathematics; calculus; design; systems; processes; materials; circuitry; electronics; environmental science; miniaturization; biology; aeronautics; fluids; gases; technical communication.

SUMMARY

Engineering is the application of scientific and mathematical principles for practical purposes. Engineering is subdivided into many disciplines; all create new products and make existing products or systems work more efficiently, faster, safer, or at less cost. The products of engineering are ubiquitous and range from the familiar, such as microwave ovens and sound systems in movie theaters, to the complex, such as rocket propulsion systems and genetic engineering.

KEY TERMS AND CONCEPTS

- **analogue:** Technology for recording a wave in its original form.
- **design:** Series of scientifically rigorous steps engineers use to create a product or system.
- **dDigital:** Technology for sampling analogue waves and turning them into numbers; these numbers are turned into voltage.
- **energy:** Capacity to do work; can be chemical, electrical, heat, kinetic, nuclear, potential, radiant, radiation, or thermal in nature.
- **feasibility study:** Process of evaluating a proposed design to determine its production difficulty in terms of personnel, materials, and cost.
- **force:** Anything that produces or prevents motion; a force can be precisely measured.
- **matter:** Anything that occupies space and has weight.
- **power:** Time rate of doing work.
- **prototype:** Original full-scale working model of a product or system.
- **quantum mechanics:** Science of understanding how a particle can act like both a particle and a wave.

- **specifications:** Exact requirements engineers must comply with to create products or services.
- **work:** Product of a displacement and the component of the force in the direction of the displacement.

DEFINITION AND BASIC PRINCIPLES

Engineering is a broad field in which practitioners attempt to solve problems. Engineers work within strict parameters set by the physical universe. Engineers first observe and experiment with various phenomena, then express their findings in mathematical and chemical formulas. The generalizations that describe the physical universe are called laws or principles and include gravity, the speed of light, the speed of sound, the basic building or subatomic particles of matter, the chemical construction of compounds, and the thermodynamic relationship that to produce energy requires energy. The fundamental composition of the universe is divided into matter and energy. The potential exists to convert matter into energy and vice versa. The physical universe sets the rules for engineers, whether the project is designing a booster rocket to lift thousands of tons into outer space or creating a probe for surgery on an infant's heart.

Engineering is a rigorous, demanding discipline because all work must be done with regard to the laws of the physical universe. Products and systems must withstand rigorous independent trials. A team in Utah, for example, must be able to replicate the work of a team in the Ukraine. Engineers develop projects using the scientific method, which has four parts: observing, generalizing, theorizing, and testing.

BACKGROUND AND HISTORY

The first prehistoric man to use a branch as a lever might be called an engineer although he never knew about fulcrums. The people who designed and built the pyramids of Giza (2500 BCE) were engineers. The term "engineer" derives from the medieval Latin word *ingeniator*, a person with "ingenium," connoting curiosity and brilliance. Leonardo da Vinci, who used mathematics and scientific principles in everything from his paintings to his designs for military fortifications, was called the Ingegnere Generale (general engineer). Galileo is credited with seeking a systematic explanation for phenomena and adopting a scientific approach to problem solving. In 1600, William Gilbert, considered the first electrical engineer, published *De magnete, magneticisque corporibus et de magno magnete tellure* (*A New Natural Philosophy of the Magnet, Magnetic Bodies, and the Great Terrestrial Magnet*, 1893; better known as *De magnete*) and coined the term "electricity." Until the Industrial Revolution of the eighteenth and nineteenth centuries, engineering was done using trial and error. The British are credited with developing mechanical engineering, including the first steam engine prototype developed by Thomas Savery in 1698 and first practical steam engine developed by James Watt in the 1760's.

Military situations often propel civilian advancements, as illustrated by World War II. The need for advances in flight, transportation, communication, mass production, and distribution fostered growth in the fields of aerospace, telecommunication, computers, automation, artificial intelligence, and robotics. In the twenty-first century, biomedical engineering spurred advances in medicine with developments such as synthetic body parts and genetic testing.

HOW IT WORKS

Engineering is made up of specialties and sub-specialties. Scientific discoveries and new problems constantly create opportunities for additional sub-specialties. Nevertheless, all engineers work the same way. When presented with a problem to solve, they research the issue, design and develop a solution, and test and evaluate it. For example, to create tiles for the underbelly of the space shuttle, engineers begin by researching the conditions under which the tiles must function. They examine the total area covered by the tiles, their individual size and weight, and temperature and frictional variations that affect the stability and longevity of the tiles. They decide how the tiles will be secured and interact with the materials adjacent to them. They also must consider budgets and deadlines.

Collaboration. Engineering is collaborative. For example, if a laboratory requires a better centrifuge, the laboratory needs designers with knowledge in materials, wiring, and metal casting. If the metal used is unusual or scarce, mining engineers need to determine the feasibility of providing the metal. At the assembly factory, an industrial engineer alters the

assembly line to create the centrifuge. Through this collaborative process, the improved centrifuge enables a biomedical engineer to produce a life-saving drug.

Communication. The collaborative nature of engineering means everyone relies on proven scientific knowledge and symbols clearly communicated among engineers and customers. The increasingly complex group activity of engineering and the need to communicate it to a variety of audiences has resulted in the emergence of the field of technical communications, which specializes in the creation of written, spoken, and graphic materials that are clear, unambiguous, and technically accurate.

Design and development. Design and development are often initially at odds with each other. For example, in an architectural team assigned with creating the tallest building in the world, the design engineer is likely to be very concerned with the aesthetics of the building in a desire to please the client and the city's urban planners. However, the development engineer may not approve the design, no matter how beautiful, because the forces of nature (such as wind shear on a mile-high building) might not allow for facets of the design. The aesthetics of design and the practical concerns of development typically generate a certain level of tension. The ultimate engineering challenge is to develop materials or methods that withstand these forces of nature or otherwise circumvent them, allowing designs, products, and processes that previously were impossible.

Testing. With computers, designs that at one time took days to draw can be created in hours. Similarly, computers allow a prototype (or trial product) to be quickly produced. Advances in computer simulation make it easier to conduct tests. Testing can be done multiple times and under a broad range of harsh conditions. For example, computer simulation is used to test the composite materials that are increasingly used in place of wood in building infrastructures. These composites are useful for a variety of reasons, including fire retardation. If used as beams in a multistory building, they must be able to withstand tremendous bending and heat forces. Testing also examines the materials' compatibility with the ground conditions at the building site, including the potential for earthquakes or other disasters.

Financial considerations. Financial parameters often vie with human cost, as in biomedical

The International Space Station represents a modern engineering challenge from many disciplines.

advancements. If a new drug or stent material is rushed into production without proper testing to maximize the profit of the developing company, patients may suffer. Experimenting with new concrete materials without determining the proper drying time might lower the cost of their development, but buildings or bridges could collapse. Dollars and humanity are always in the forefront of any engineering project.

APPLICATIONS AND PRODUCTS

The collaborative nature of engineering requires the cooperation of engineers with various types of knowledge to solve any single problem. Each branch of engineering has specialized knowledge and expertise.

Aerospace. The field of aerospace engineering is divided into aeronautical engineering, which deals with aircraft that remain in the Earth's atmosphere, and astronautical engineering, which deals with spacecraft. Aircraft and spacecraft must endure extreme changes in temperature and atmospheric pressure and withstand massive structural loads. Weight and cost considerations are paramount, as is reliability. Engineers have developed new composite materials to reduce the weight of aircraft and enhance fuel efficiency and have altered spacecraft design to help control the friction generated when spacecraft leave and reenter the Earth's atmosphere. These developments have influenced earthbound transportation from cars to bullet trains.

Architectural. The field of architectural engineering applies the principles of engineering to the design and construction of buildings. Architectural engineers address the electrical, mechanical, and structural aspects of a building's design as well as its appearance and how it fits in its environment. Areas of concern to architectural engineers include plumbing, lighting, acoustics, energy conservation, and heating, ventilation, and air conditioning (HVAC). Architectural engineers must also make sure that buildings they design meet all regulations regarding accessibility and safety in addition to being fully functional.

Bioengineering. The field of bioengineering involves using the principles of engineering in biology, medicine, environmental studies, and agriculture. Bioengineering is often used to refer to biomedical engineering, which involves the development of artificial limbs and organs, including ceramic knees and hips, pacemakers, stents, artificial eye lenses, skin grafts, cochlear implants, and artificial hands. However, bioengineering also has many other applications, including the creation of genetically modified plants that are resistant to pests, drugs that prevent organ rejection after a transplant operation, and chemical coatings for a stent placed in a heart blood vessel that will make the implantation less stressful for the body. Bioengineers must concern themselves with not only the biological and mechanical functionality of their creations but also financial and social issues such as ethical concerns.

Chemical. Everything in the universe is made up of chemicals. Engineers in the field of chemical engineering develop a wide range of materials, including fertilizers to increase crop production, the building materials for a submarine, and fabric for everything from clothing to tents. They may also be involved in finding, mining, processing, and distributing fuels and other materials. Chemical engineers also work on processes, such as improving water quality or developing less-polluting, readily available, inexpensive fuels.

Civil. Some of the largest engineering projects are in the field of civil engineering, which involves the design, construction, and maintenance of infrastructure such as roads, tunnels, bridges, canals, dams, airports, and sewage and water systems. Examples include the interstate highway system, the Hoover Dam, and the Brooklyn Bridge. Completion of civil engineering projects often results in major shifts in population distribution and changes in how people live. For example, the highway system allowed fresh produce to be shipped to northern states in the wintertime, improving the diets of those who lived there. Originally, the term "civil engineer" was used to distinguish between engineers who worked on public projects and "military engineers" who worked on military projects such as topographical maps and the building of forts. The subspecialties of civil engineering include construction engineering, irrigation engineering, transportation engineering, soils and foundation engineering, geodetic engineering, hydraulic engineering, and coastal and ocean engineering

Computer. The field of computer engineering has two main focuses, the design and development of hardware and of the accompanying software. Computer hardware refers to the circuits and architecture of the computer, and software refers to the computer programs that run the computer. The hardware does only what the software instructs it to do, and the software is limited by the hardware. Computer engineers may research, design, develop, test, and install hardware such as computer chips, circuit boards, systems, modems, keyboards, printers, or computers embedded in various electronic products, such as the tracking devices used to monitor parolees. They may also create, maintain, test, and install software for mainframes, personal computers, electronic devices, and smartphones. Computer programs range from simple to complex and from familiar to unfamiliar. Smartphone applications are extremely numerous, as are applications for personal computers. Software is used to track airplanes and other transportation, to browse the Web, to provide security for financial transactions and corporations, and to direct unmanned missiles to a precisely defined target. Computers can operate from a remote location. For example, anaerobic manure digesters are used to convert cattle manure to biogas that can be converted to energy, a biosolid that can be used as bedding or soil amendment, and a nonodorous liquid stream that can be used as fertilizer. These digesters can be placed on numerous cattle farms in

different states and operated and controlled by computers miles away.

Electrical. Electrical engineering studies the uses of electricity and the equipment to generate and distribute electricity to homes and businesses. Without electrical engineering, digital video disc (DVD) players, cell phones, televisions, home appliances, and many life-saving medical devices would not exist. Computers could not turn on. The Global Positioning System (GPS) in cars would be useless, and starting a car would require using a hand crank. This field of engineering is increasingly is involved in investigating different ways to produce electricity, including alternative fuels such as biomass and solar and wind power.

Environmental. The growth in the population of the world has been accompanied by increases in consumption and the production of waste. Environmental engineering is concerned with the reduction of existing pollution in the air, in the water, and on land, and the prevention of future harm to the environment. Issues addressed include pollution from manufacturing and other sources, the transportation of clean water, and the disposal of nonbiodegradable materials and hazardous and nuclear waste. Because pollution of the air, land, and water crosses national borders, environmental engineers need a broad, global perspective.

Industrial. Managing production and delivery of any product is the expertise of industrial engineers. They observe the people, machines, information, and technology involved in the process from start to finish, looking for any areas that can be improved. Increasingly, they use computer simulations and robotics. Their goals are to increase efficiency, reduce costs, and ensure worker safety. For example, worker safety can be improved through ergonomics and the use of less-stressful, easier-to-manipulate tools. The expertise of industrial engineers can have a major impact on the profitability of companies.

Manufacturing. Manufacturing engineering examines the equipment, tools, machines, and processes involved in manufacturing. It also examines how manufacturing systems are integrated. Its goals are to increase product quality, safety, output, and profitability by making sure that materials and labor are used optimally and waste—whether of time, labor, or materials—is minimized. For example, engineers may improve machinery that folds disposable diapers or that machines the gears for a truck, or they may reconfigure the product's packaging to better protect it or facilitate shipping. Increasingly, robots are used to do hazardous, messy, or highly repetitive work, such as painting or capping bottles.

Mechanical. The field of mechanical engineering is the oldest and largest specialty. Mechanical engineers create the machines that drive technology and industry and design tools used by other engineers. These machines and tools must be built to specifications regarding usage, maintenance, cost, and delivery. Mechanical engineers create both power-generating machinery such as turbines and power-using machinery such as elevators by taking advantage of the compressibility properties of fluids and gases.

Nuclear. Nuclear engineering requires expertise in the production, handling, utilization, and disposal of nuclear materials, which have inherent dangers as well as extensive potential. Nuclear materials are used in medicine for radiation treatments and diagnostic testing. They also function as a source of energy in nuclear power plants. Because of the danger of nuclear materials being used for weapons, nuclear engineering is subject to many governmental regulations designed to improve security.

IMPACT ON INDUSTRY
The U.S. economy and national security are closely linked to engineering. For example, engineering is vital for developing ways to reduce the cost of energy, decrease American reliance on foreign sources of energy, and conserve existing natural resources. Because of its importance, engineering is supported and highly regulated by government agencies, universities, and corporations. However, some experts question whether the United States is educating enough engineers, especially in comparison with China and India. The number of engineering degrees awarded each year in the United States is not believed to be keeping pace with the demand for new engineers.

Government research. The U.S. government has a vested interest in the commerce, safety, and military preparedness of the nation. It both funds and regulates development through subcontractors, laws, guidelines, and educational initiatives. For example, the Americans with Disabilities Act of 1990 made it mandatory that public buildings be accessible for all American citizens. Its passage spurred innovations in engineering such as kneeling buses, which make it possible for those in wheelchairs to board a bus. The

U.S. Department of Defense (DOD) is the largest contractor and the largest provider of funds for engineering research. For example, it issued 80,000 specifications for the creation of a synthetic jet fuel. The Federal Drug Administration (FDA) concentrates its efforts on supplier control and testing of products and materials. Funding for research in engineering is also provided by the National Science Foundation. The government has also sponsored educational initiatives in science, technology, engineering, and mathematics, an example of which is the America Competes Act of 2007.

Academic research. Universities, often in collaboration with governmental agencies, conduct research in engineering. Also, universities have entered into partnerships with private industry as investors have sought to capitalize on commercial possibilities presented by research, as in the case of stem cell research in medicine. Universities are also charged with providing rigorous, up-to-date education for engineers. Numerous accrediting agencies, including the Accreditation Board for Engineering and Technology, ensure that graduates from engineering programs have received an adequate and appropriate education. Attending an institution without accreditation is not advisable.

Industry and business. Engineering has a role in virtually every company in every industry, including nonprofits in the arts, if only because these companies use computers in their offices. Consequently, some fields of engineering are sensitive to swings in the economy. In a financial downturn, no one develops office buildings, so engineers working in architecture and construction are downsized. When towns and cities experience a drop in tax income, projects involving roads, sewers, and environmental cleanup are delayed or canceled, and civil, mechanical, and environmental engineers lose their jobs. However, some economic problems can actually spur developments in engineering. For example, higher energy costs have led engineers to create sod roofs for factories, which keep the building warmer in winter and cooler in summer, and to develop lighter, stronger materials to use in a airplanes.

CAREERS AND COURSE WORK
To pursue a career in engineering, one must obtain a degree from an accredited college in any of the major

FASCINATING FACTS ABOUT ENGINEERING

- The Bering Strait is 53 miles of open water between Alaska and Russia. In the 2009 Bering Strait Project Competition, people submitted plans for bridging the strait, including a combination bridge-tunnel with passageways for migrating whales and the capability of circulating the frigid arctic waters of the north to help fend off global warming.

- The Gotthard Base Tunnel network is being built under the Alps from Switzerland to Milan, Italy. At about 35 miles, the two-way tunnel will be the longest in the world and will reduce the time to make the trip by automobile from 3.5 hours to 1 hour. High-speed trains traveling at speeds of 155 miles per hour will make more than two hundred trips through the tunnel per day.

- · A typical desktop computer can handle 100 million instructions per second. As of June, 2010, the fastest supercomputer was the Cray Jaguar at Oak Ridge National Laboratory. Its top speed is

- 1.75 petaflops (1 quadrillion floating point operations) per second.

- The world's smallest microscope weighs about as much as an egg. Instead of using a lens to magnify, it generates holographic images of microscopic particles or cells using a light emitting diode (LED) to illuminate and a digital sensor to capture the image.

- The National Institutes of Health has developed an implant made of silk and metal that when placed in the brain can detect impending seizures and send out electric pulses to halt them. It can also send out electric signals to prostheses used by people with spinal cord injuries.

- Stanford University is developing the computers and technology for a driverless car. The 2010 autonomous race car, dubbed Shelley, is an Audi TTS equipped with a differential Global Positioning System accurate to an inch. It calculates the right times to brake and accelerate while turning.

fields of engineering. A bachelor's degree is sufficient for some positions, but by law, each engineering project must be approved by a licensed professional engineer (P.E.). To gain P.E. registration, an engineer must pass the comprehensive test administered by the National Society of Professional Engineers and work for a specified time period. In addition, each state has its own requirements for being licensed, including an exam specific to the state. An engineer with a bachelor's degree may work as an engineer with or without P.E. registration, obtain a master's degree or doctorate to work in a specialized area of engineering or pursue an academic career, or obtain an M.B.A. in order to work as a manager of engineers and products.

SOCIAL CONTEXT AND FUTURE PROSPECTS

Engineering can both prolong life through biomedical advances such as neonatal machinery and destroy life through unmanned military equipment and nuclear weaponry. An ever-increasing number of people and their concentration in urban areas means that ways must be sought to provide more food safely and to ensure an adequate supply of clean, safe drinking water. These needs will create projects involving genetically engineered crops, urban agriculture, desalination facilities, and the restoration of contaminated rivers and streams. The never-ending quest for energy will remain a fertile area for research and development. Heated political debates about taxing certain fuels and subsidizing others are part of the impetus behind solar, wind, and biomass development and renewed discussions about nuclear power.

The lack of minorities, including women, Hispanics, African Americans, and Native Americans, in engineering is being addressed through education initiatives. Women's enrollment in engineering schools has hovered around 20 percent since about 2000. African Americans make up about 13 percent of the U.S. population yet only about 3,000 blacks earn bachelor's degrees in engineering each year. About 4,500 Hispanics, who represent about 15 percent of the U.S. population, earn bachelor's degrees in engineering each year.

—*Judith L. Steininger, MA*

FURTHER READING

Addis, Bill. *Building: Three Thousand Years of Design Engineering and Construction*. New York: Phaidon Press, 2007. Traces the history of building engineering in the Western world, covering the people, buildings, classic texts, and theories. Heavily illustrated.

Baura, Gail D. *Engineering Ethics: An Industrial Perspective*. Boston: Elsevier Academic Press, 2006.

Thirteen case studies examine problems with products, structures, and systems and the role of engineers in each. Chapters cover the Ford Explorer rollovers, the San Francisco-Oakland Bay Bridge earthquake collapse, the Columbia space shuttle explosion, and the 2003 Northeast blackout.

Dieiter, George E., and Linda C. Schmidt. *Engineering Design*. 4th ed. Boston: McGraw-Hill Higher Education, 2009. Looks at design as the central activity of engineering. Provides a broad overview of basic topics and guidance on the design process, including materials selection and design implementation.

Nemerow, Nelson L., et al., eds. *Environmental Engineering: Environmental Health and Safety for Municipal Infrastructure, Land Use and Planning, and Industry*. Hoboken, N.J.: John Wiley & Sons, 2009. Covers environmental issues such as waste disposal for industry, the residential and institutional environment, air pollution, and surveying and mapping for environmental engineering.

Petroski, Henry. *Success Through Failure: The Paradox of Design*. 2006. Reprint. Princeton, N.J.: Princeton University Press, 2008. Examines failure as a motivator for engineering and defines success as "anticipating and obviating failure." Chapters deal with bridges, buildings, and colossal failures.

Yount, Lisa. *Biotechnology and Genetic Engineering*. 3d ed. New York: Facts On File, 2008. Covers the history of genetic engineering and biotechnology, including the important figures. Contains a bibliography and index.

WEB SITES

American Association of Engineering Societies
http://www.aaes.org

American Engineering Association
http://www.aea.org

American Society for Engineering Education
http://www.asee.org

Institute of Electrical and Electronics Engineers (IEEE)
http://www.ieee.org

National Society of Professional Engineers
http://www.nspe.org

See also: Bioengineering; Biomechanical Engineering; Civil Engineering; Computer Engineering; Electrical Engineering; Electronics and Electronic Engineering; Engineering Mathematics; Environmental Engineering.

ENGINEERING MATHEMATICS

FIELDS OF STUDY

Algebra; geometry; trigonometry; calculus (including vector calculus); differential equations; statistics; numerical analysis; algorithmic science; computational methods; circuits; statics; dynamics; fluids; materials; thermodynamics; continuum mechanics; stability theory; wave propagation; diffusion; heat and mass transfer; fluid mechanics; atmospheric engineering; solid mechanics.

SUMMARY

Engineering mathematics focuses on the use of mathematics as a tool within the engineering design process. Such use includes the development and application of mathematical models, simulations, computer systems, and software to solve complex engineering problems. Thus, the solution of the engineering problem might be a component, system, or process.

KEY TERMS AND CONCEPTS

- **algorithmic science:** Study, implementation, and application of real-number algorithms for solving problems of continuous mathematics that arise in the realms of optimization and numerical analysis.
- **applied mathematics:** Branch of mathematics devoted to developing and applying mathematical methods, including math-modeling techniques, to solve scientific, engineering, industrial, and social problems. Areas of focus include ordinary and partial differential equations, statistics, probability, operational analysis, optimization theory,

solid mechanics, fluid mechanics, numerical analysis, and scientific computing.

- **bioinformatics:** Field that uses sophisticated mathematical and computational tools for problem solving in diverse biological disciplines.
- **computational science:** Broad field blending applied mathematics, computer science, engineering, and other sciences that uses computational methods in problem solving.

Definition and Basic Principles

Engineering mathematics entails the development and application of mathematics (such as algorithms, models, computer systems, and software) within the engineering design process. In engineering mathematics course work and professional research, a variety of tools may be used in the collection, analysis, and display of data. Standard tools of measurement include rules, spirit levels, micrometers, calipers, and gauges. Software tools include Maplesoft, Mathematica, MATLAB, and Excel. Engineering mathematics has roots and applications in many areas, including algorithmic science, applied mathematics, computational science, and bioinformatics.

Background and History

Since engineering is such a broad area, engineering mathematics includes a variety of applications. In general, a link between engineering and mathematics is established when mathematical descriptions of physical systems are formulated.

Links may involve precise mathematical relationships or formulas. For example, Galileo Galilei's pioneering work on the study of the motion of physical objects led to the equations of accelerated motion, $v = at$ and $d = \frac{1}{2} at$, in which velocity is v, acceleration a, time t, and distance d. His work paved the way for Newtonian physics.

Other links are established through empirical relationships that have the status of laws. French physicist Charles-Augustin de Coulomb discovered that the force between two electrical charges is proportional to the product of the charges and inversely proportional to the square of the distance between them: $F = kq1q2/d2$ (force is F, constant of variation k, charges $q1$ and $q2$, and distance d). The coulomb, a measure of electrical charge was named for him. Coulomb's work was the first in a sequence of related discoveries by other notable scientists, many of whose findings

led to additional laws. The list includes Danish physicist Hans Christian Ørsted, French physicists André-Marie Ampère, Jean-Baptiste Biot, and Félix Savart, British physicist and chemist Michael Faraday, and Russian physicist Heinrich Lenz.

Sometimes mathematical expressions of principles apply almost universally. In physics, for example, the conservation laws indicate that in a closed system certain measurable quantities remain constant: mass, momentum, energy, and mass-energy. Lastly, systems of equations are required to describe physical phenomena of various levels of complexity. Examples include English astronomer and mathematician Sir Isaac Newton's equations of motion, Scottish physicist and mathematician James Clerk Maxwell's equations for electromagnetic fields, and Swiss mathematician and physicist Leonhard Euler's and French engineer Claude-Louis Navier and British mathematician and physicist George Gabriel Stokes's (Navier-Stokes) equations in fluid mechanics.

Further links between engineering and mathematics are discovered through the ongoing development, extension, modification, and generalization of equations and models in broader physical systems. For example, the Euler equations used in fluid mechanics can be connected to the conservation laws of mass and momentum.

How it Works

Many problems in engineering mathematics lead to the construction of models that can be used to describe physical systems. Because of the power of technology, a model may be derived from a system of a few equations that may be linear, quadratic, exponential, or trigonometric—or a system of many equations of even greater complexity. In engineering, such equations include ordinary differential equations, differential algebraic equations, and partial differential equations.

As the system of equations is solved, the mathematical model is formulated. Models are expressed in terms of mathematical symbols and notation that represent objects or systems and the relationships between them. Computer software, such as Maplesoft, Mathematica, MATLAB, and Excel, facilitates the process.

Engineers have available many models of physical systems. The development, extension, and modification of existing models, and the development of new

- Albert Einstein proposed his general theory of relativity, incorporating the Riemannian or elliptical geometry concept that space can be unbounded without being infinite. The theory of relativity, along with later nuclear-energy research, led to development of the atomic and hydrogen bombs and theoretical understanding of thermonuclear fusion as the energy source powering the Sun and stars.
- Among the prominent figures involved in the design and development of electronic computers is American mathematician Norbert Wiener. Wiener is the founder of the science of cybernetics, the mathematical study of the structure of control systems and communications systems in living organisms and machines.

models, are the subject of ongoing research. In this way, engineering mathematics continues to advance.

The ultimate test of a mathematical model is whether it truly reflects the behavior of the physical system under study. Computational experiments can be run to test the model for unexpected characteristics of the system and possibly optimize its design. However, models are approximations, and the accuracy of computed results must be evaluated through some form of error analysis.

The level of complexity of the construction and use of models depends on the engineering application. Further appreciation of the utility of a model may be gained by examining the use of a new model in engineering mathematics that impacts several scientific and technological areas. A mathematical model can now be used to investigate how materials break: One led to a new law of physics that depicts fracturing before it happens, or even as it occurs. In addition to the breakage of materials such as glass and concrete used in construction, the model enables better examination of bone breakage in patients with pathologies such as osteoporosis.

APPLICATIONS AND PRODUCTS

Cell biology. Advances in research in cell growth and division have proved helpful in disease detection, pharmaceutical research, and tissue engineering. Biologists have extensively explored cell growth and mass and the relationship between them. Using microsensors, bioengineers can now delineate colon cancer cell masses and divisions over given time

periods. They have found that such cells grow faster as they grow heavier. With additional cell measurements and mathematical modeling, the scientists examined other properties such as stiffness. They also performed simulations to study the relationship between cell stiffness, contact area, and mass measurement.

Genetics. New genes involved in stem-cell development can be found, quickly and inexpensively, along the same pathway as genes already known. When searching for genes involved in a particular biological process, scientists try to find genes with a symmetrical correlation. However, many biological relationships are asymmetric and can now be found using Boolean logic in data-mining techniques. Engineering and medical researchers can then examine whether such genes become active, such as those in developing cancers. This research is expected to lead to advances in disease diagnosis and cancer therapy.

Energy. A new equation could help to further the use of organic semiconductors. The equation represents the relationship between current and voltage at the junctions of the organic semiconductors. Research in the use of organic semiconductors may lead to advances in solar cells, displays, and lighting. Engineers have been studying organic semiconductors for about 75 years but have only recently begun to discover innovative applications.

IMPACT ON INDUSTRY

Government and university research. Engineering mathematics has roots and applications in many research areas, including algorithmic science, applied mathematics, computational science, and bioinformatics. Theoretical, empirical, and computational research includes the development of effective and efficient mathematical models and algorithms. Professional research includes diverse medical, scientific, and research in university-level engineering mathematics includes work that challenges students to integrate and apply course-work knowledge from many disciplines as they examine cracks in solids, mixing in small geometries, the crumpling of paper, lattice packings in curved geometries, materials processes, and predict optimal mechanics for device applications.

Engineering and technology. Engineering mathematics is among the tools used to create, develop, and maintain products, systems, and processes.

Applications are found in many areas, including nanotechnology. For example, radio-frequency designs are approaching higher frequency ranges and higher levels of complexity. The underlying mathematics of such designs is also complex. It requires new modeling techniques, mathematical methods, and simulations with mixed analogue and digital signals. Ordinary differential equations, differential algebraic equations, and partial differential algebraic equations are used in the analyses. The goal is to predict the circuit behavior before costly production begins. In such research, algorithms have been modified and new algorithms created to meet simulation demands.

Other applied sciences. There has been a dramatic rise in the power of computation and information technology. With it have come vast amounts of data in various fields of applied science and engineering. The challenge of understanding the data has led to new tools and approaches, such as data mining. Applied mathematics includes the use of mathematical models and control theory that facilitate the study of epidemics, pharmacokinetics, and physiologic systems in the medical industry. In telematics, models are developed and used in the enhancement of wireless mobile communications.

CAREERS AND COURSE WORK

Data analyst or data miner. Data mining involves the discovery of hidden but useful information in large databases. In applications of data mining, career opportunities emerge in medicine, science, and engineering. Data mining involves the use of algorithms to identify and verify previously undiscovered relationships and structure from rigorous data analysis. Course work should include a focus on higher-level mathematics in such areas as topology, combinatorics, and algebraic structures.

Materials science. Materials science is the research, development, and manufacture of such items as metallic alloys, liquid crystals, and biological materials. There are many career opportunities in aerospace, electronics, biology, and nanotechnology. Research and development uses mathematical models and computational tools. Course work should include a focus on applied mathematics, including differential equations, linear algebra, numerical analysis, operations research, discrete mathematics, optimization, and probability.

FASCINATING FACTS ABOUT ENGINEERING MATHEMATICS

- In aircraft design, computational simulations have been used extensively in the analysis of lift and drag. Advanced computation and simulation are now essential tools in the design and manufacture of aircraft.

- Auto-engineering researchers have developed a simulation model that can significantly decrease the time and cost of calibrating a new engine. Unlike statistics-based models, the new physics-based model can generate data for transient behavior (acceleration or deceleration between different speeds).

- Although the number of nuclear weapons held by each country in the world is securely guarded information, the rising demands of regulatory oversight require computing technology that exceeds current levels. A knowledge of the mathematical development and use of uncertainty models is critical for this application.

- Researchers have used nontraditional mathematical analyses to identify evolving drug resistance in strains of malaria. Their goal is to enable the medical community to react quickly to inevitable drug resistance and save lives. They also want to increase the life span of drugs used against malaria. The researchers used mathematical methods that involved graphing their data in polar coordinates and in rectangular coordinates. The results of the study are of particular interest to biomedical engineers focusing on genetics research.

- Bioengineers studying Alzheimer's disease found that amyloid beta peptides generate calcium waves. Following these waves in brain-cell networks, the scientists discovered voltage changes that signify intracellular communication. This research involves mathematical modeling of brain networks.

- Medical and engineering researchers have developed a mathematical model reflecting how red blood cells change in size and hemoglobin content during their four-month life span. This model may provide valuable clinical information that can be used to predict who is likely to become anemic.

- Mathematically gifted philosophers Sir Isaac Newton and Gottfried Leibniz invented the calculus that is now part of engineering programs.

- French mathematician Joseph Fourier introduced a mathematical series using sines and cosines to model heat flow.

Ecological and environmental engineering. The work of professionals in these fields covers many areas that transcend pollution control, public health, and waste management. It might, for example, involve the design, construction, and management of an aquatic ecosystem or the research and development of appropriate sustainable technologies.

Course work should include a focus on higher-level mathematics in such areas as calculus, linear algebra, differential equations, and statistics.

Meteorology and climatology. These career areas incorporate not only atmospheric, hydrologic, and oceanographic sciences but modeling, forecasting, geoengineering, and geophysics. In general, historical weather data and current data from satellites, radar, and monitoring equipment are combined with other measurements to develop, process, and analyze complex models using high-performance computers. Current research areas include global warming and the impact of atmospheric radiation and industrial pollutants. Mathematics courses in meteorology and atmospheric science programs include calculus, differential equations, linear algebra, statistics, computer science, numerical analysis, and matrix algebra or computer systems.

SOCIAL CONTEXT AND FUTURE PROSPECTS

Within engineering mathematics, an interdisciplinary specialty has emerged, computational engineering. Computational engineering employs mathematical models, numerical methods, science, engineering, and computational processes that connect various fields of engineering science. Computational engineering emerged from the impact of supercomputing on engineering analysis and design. Computational modeling and simulation are vitally important for the development of high-technology products in a globally competitive marketplace. Computational engineers develop and use advanced software for real-world engineering analysis and design problems. The research work of engineering professionals and academics has potential for applications in several engineering disciplines.

—*June Gastón, MSEd, MEd, EdD*

FURTHER READING

Gribbin, John. *The Scientists: A History of Science Told Through the Lives of Its Greatest Inventors.* New York: Random House, 2003. This compelling text on the history of modern science marks the subject's discoveries and milestones through the great thinkers who were integral to its creation, both well known and not as well known.

Merzbach, Uta C., and Carl B. Boyer. *A History of Mathematics.* 3d ed. Hoboken, N.J.: John Wiley & Sons, 2011. An excellent and highly readable history of the subject that chronicles the earliest principles as well as the latest computer-aided proofs.

Schäfer, M. *Computational Engineering, Introduction to Numerical Methods.* New York: Springer, 2006. Schäfer includes applications in fluid mechanics, structural mechanics, and heat transfer for newer fields such as computational engineering and scientific computing, as well as traditional engineering areas.

Shiflet, Angela B., and George W. Shiflet. *Introduction to Computational Science: Modeling and Simulation for the Sciences.* Princeton, N.J.: Princeton University Press, 2006. Two approaches to computational science problems receive focus: system dynamics models and cellular automaton simulations. Other topics include rate of change, errors, simulation techniques, empirical modeling, and an introduction to high-performance computing. Numerous examples, exercises, and projects explore applications.

Stroud, K.A. *Engineering Mathematics.* 6th ed. New York: Industrial Press, 2007. Providing a broad mathematical survey, this innovative volume covers a full range of topics from basic arithmetic and algebra to challenging differential equations, Laplace transforms, and statistics and probability.

Velten, Kai. *Mathematical Modeling and Simulation: Introduction for Scientists and Engineers.* Weinheim, Germany: Wiley-VCH, 2009. Velten explains the principles of mathematical modeling and simulation. After treatment of phenomenological or data-based models, the remainder of the book focuses on mechanistic or process-oriented models and models that require the use of differential equations.

WEB SITES

American Mathematical Society
http://www.ams.org/home/page

National Society of Professional Engineers
http://www.nspe.org

Society for Industrial and Applied Mathematics
http://www.siam.org

See also: Algebra; Applied Mathematics; Calculus; Geometry; Numerical Analysis.

ENGINEERING SEISMOLOGY

FIELDS OF STUDY

Engineering; geology; physics; geophysics; earth science; electrical engineering; volcanology; volcanic seismology; plate tectonics; geodynamics; mineral physics; tectonic geodesy; mantle dynamics; seismic modeling; seismic stratigraphy; statistical seismology; computer science; mathematics; earthquake engineering; paleoseismology; archeoseismology; historical seismology; structural engineering; geography

SUMMARY

Engineering seismology is a scientific field focused on studying the likelihood of future earthquakes and the potential damage such seismic activity can cause to buildings and other structures. Engineering seismology utilizes computer modeling, geological surveys, existing data from historical earthquakes, and other scientific tools and concepts. Engineering seismology is particularly useful for the establishment of building codes and for land-use planning.

KEY TERMS AND CONCEPTS:

- **duration:** The length of time ground motion occurs during an earthquake.
- **epicenter:** The surface-level geographic point located directly above an earthquake's hypocenter.
- **focal depth:** The depth of an earthquake's hypocenter.
- **ground motion:** The ground-level shaking that occurs during an earthquake.
- **hypocenter:** The point of origin of an earthquake.
- **love waves:** Seismic waves that occur in a side-to-side motion.
- **magnitude:** An earthquake's size and relative strength.
- **Rayleigh waves:** Seismic waves that occur in a circular, rolling fashion.
- **Richter scale:** Logarithmic scale used to assign a numerical value to the magnitude of an earthquake.
- **source parameters:** A series of earthquake characteristics, including distance, duration, energy, and the types of waves that occur.
- **stress drop:** The amount of energy released when locked tectonic plates separate, causing an earthquake.
- **wave propagation path:** The directions in which seismic waves travel in an earthquake.

DEFINITION AND BASIC PRINCIPLES

Engineering seismology (also known as earthquake engineering) is a multidisciplinary field that assesses the effects of earthquakes on buildings, bridges, roads, and other structures. Seismology engineers work in the design and construction of structures that can withstand seismic activity. They also assess the damages and effects of seismic activity on existing structures. Engineering seismologists analyze such factors as quake duration, ground motion, and focal depth in assessing the severity of seismic events and how those events affect fabricated structures. They also consider source parameters, which help seismologists zero in on a seismic event's location and the speed and trajectory at which the quake's resulting waves are traveling.

Earthquake engineers also study theoretical concepts and models related to potential earthquakes and historical seismic events. Such knowledge can help engineers and architects design structures that can withstand as powerful an earthquake as the geographic region has produced (or possibly will produce). Mapping systems and programs and mathematical and computer-based models are essential to engineering seismologists' work. Such techniques are also useful for archeologists and paleontologists, both of whom may use engineering seismology concepts to understand how the earth has evolved over millions of years and how ancient civilizations were affected by seismic events.

BACKGROUND AND HISTORY

Throughout human history, people have struggled to understand the nature of earthquakes and, as a result, have faced the challenges of preparing for these seismic events. Some ancient civilizations attributed earthquakes to giant snakes, turtles, and other creatures living and moving beneath the earth's surface. In the fourth century BCE, Aristotle was the first to speculate that earthquakes were not caused by supernatural forces but rather were natural events. However, little scientific study on earthquakes took place for hundreds of years, despite the occurrence of many major seismic events (including the eruption of Mount Vesuvius in Italy in 79 CE, which was preceded by a series of earthquakes).

In the mid-eighteenth century, however, the British Isles experienced a series of severe earthquakes, which created a tsunami that destroyed Lisbon, Portugal, killing tens of thousands of people. Scientists quickly developed an interest in cataloging and understanding seismic events. In the early nineteenth century, Scottish physicist and glaciologist James D. Forbes invented the inverted pendulum seismometer, which gauged not only the severity of an earthquake but also its duration.

Throughout history, seismology has seen advances that immediately followed significant seismic events. Engineering seismology, which is proactive, represents a departure from reactionary approaches to the study of earthquakes. Today, engineering seismology uses seismometers, computer modeling, and other advanced technology and couples it with historical data for a given site. The resulting information helps civil engineers and architects construct durable buildings, bridges, and other structures and assess the risks to existing structures posed by an area's seismic potential.

HOW IT WORKS

To understand engineering seismology, one must understand the phenomenon of earthquakes. Earthquakes may be defined as the sudden shaking of the earth's surface as caused by the movement of subterranean rock. These massive rock formations (plates), resting on the earth's superheated core, experience constant movement caused predominantly by gravity. While some plates move above and below one another, others come into contact with one another as they pass. The boundaries formed by these passing plates are known as faults. When passing plates lock together, stored energy builds up gradually. The plates eventually give, causing that energy to be released and sent from the quake's point of origin (the hypocenter) outward to the surface in the form of seismic (or surface) waves. Such waves occur either in a circular, rolling fashion (Rayleigh waves) or in a twisting, side-to-side motion (Love waves).

The field of seismology has developed only over the last few centuries, largely because of major, devastating seismic events. The practice of engineering seismology has grown in demand in recent years, mainly because of the modern world's dependency on major cities, infrastructure (such as bridges, roadways, and rail systems), and energy resources (including nuclear power plants and offshore oil rigs). Earthquake engineers, therefore, have two main areas of focus: studying seismology and developing structures that can withstand the force of an earthquake.

To study seismic activity and earthquakes, engineering seismologists may use surface-based detection systems, such as seismometers, to monitor and catalog tremors. They also employ equipment—including calibrators and accelerometers—that is lowered into deep holes. Such careful monitoring practices help seismologists and engineering seismologists better understand a region's potential for seismic activity.

When earthquakes occur, engineering seismologists quickly attempt to locate the hypocenter and the epicenter (the surface point that lies directly above the hypocenter). They are able to do so by monitoring two types of waves—P and S waves—that move much quicker than surface waves and, therefore, act as precursors to surface waves. These engineers also work to determine the magnitude (a measurement of an earthquake's size) of the event.

Magnitude may be based on a number of key factors (or source parameters), including duration, distance to the epicenter and hypocenter, the size and speed of the surface waves, the amount of energy (known as the stress drop) that is released from the hypocenter, P and S waves, and the directions in which surface waves move (the wave propagation path). Analyzing an earthquake's magnitude provides an accurate profile of the quake and the conditions that caused it.

In addition to developing a profile of a region's

past seismic activity, earthquake engineers use such information to ascertain the type of activity a geographic region may experience in the future. For example, scientific evidence suggests that the level of stress drop is a major contributor to the severity of seismic activity that can cause massive destruction in major urban centers. Similarly, studies show that the duration of ground motion (the "shaking" effects of an earthquake) may be more of a factor in the amount of damage to buildings and other structures than stress drop.

The field of engineering seismology is less than one century old, but in the twenty-first century, it plays an important role in urban development and disaster prevention. Earthquake seismologists work with civil engineers and architects to design buildings, roads, bridges, and tunnels that may withstand the type of seismic activity that has occurred in the past.

APPLICATIONS AND PRODUCTS
Engineering seismology applies knowledge of seismic conditions, events, and potential to the design and development of new and existing fabricated

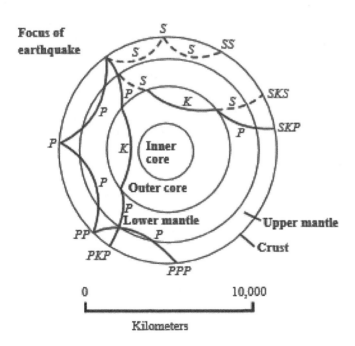

Seismic velocities and boundaries in the interior of the Earth sampled by seismic waves

structures. Among the methods and applications employed by earthquake seismologists are the following:

Experimentation. Engineering seismologists may construct physical scale models of existing structures or proposed structures. Using data from a region's known seismic history, the engineers attempt to recreate an earthquake by placing these models on so-called shake tables, large mechanical platforms that simulate a wide range of earthquake types. After the "event," engineering seismologists examine the simulation's effects on the model structure, including its foundations, support beams, and walls. This approach enables the engineers and architects to directly examine the pre- and post-simulation structure and determine what sort of modifications may be warranted.

Computer models. One of the most effective tools utilized by engineering seismologists is computer modeling. Through the application of software, engineering seismologists can input a wide range of source parameters, ground motion velocities, wave types, and other key variables. They also can view how different structural components withstand (or fail to withstand) varying degrees of seismic activities without the expense and construction time of a shake table. Computer modeling has become increasingly useful when attempting to safeguard against earthquake damage to dams, nuclear power plants, and densely developed urban centers. Computer modeling is also used by engineering seismologists to predict the path of destruction that often occurs after an earthquake, destruction such as that caused by fires or flooding.

Seismic design software. Earthquake engineers study seismic activity in terms of how it affects structures. To this end, engineers must attempt to predict how earthquakes will strike an area. Seismic design software is used to create a map of a region's seismic activity and how those conditions will potentially cause structural damage. The software enables government officials to establish formalized building codes for buildings, bridges, power plants, and other structures. This software is easily obtained on the Web and through the U.S. Geological Survey (USGA) and other organizations.

Mathematics. Engineering seismology is an interdisciplinary field that relies heavily on an understanding of physics and mathematics. To quantify the severity of earthquakes, to calculate the scope of seismic activity, and, in general, to create a profile of a region's seismic environment, engineering seismologists utilize a number of mathematical formulae. One of the most-recognized of these formulae is the Richter scale, which was developed in 1935 by American seismologist and physicist Charles Richter. The Richter scale uses a logarithm to assign a numerical value (with no theoretical limit) to establish the magnitude of an earthquake. The Richter scale takes into account the amplitude (the degree of change) between seismic waves and the distance between the equipment that detects the quake and the quake's epicenter.

Earthquake engineers use such mathematical data as part of their analyses when working with civil engineers on construction projects. Earthquake engineers also are increasingly called upon by government officials to use this data to assess individual structure and citywide structural deficiencies that resulted in earthquake destruction. Forensic engineering was called into service in 2009, when Australian emergency officials intervened in Padang, Indonesia, after a magnitude 7.6 quake devastated that city. Engineers used mathematical formulae and statistical data to assess system-wide structural deficiencies in Padang rather than analyzing damage on a structure-by-structure basis. In light of the countless variables involved with studying earthquakes and their effects on fabricated structures, the use of established logarithms, data sets, and mathematical formulae is a time-honored practice of engineering seismologists.

Sensors. Not all earthquakes cause immediate and significant damage to affected structures. According to the USGA, the Greater San Francisco area experienced more than eighty earthquakes in 2011 alone, with none of those quakes registering higher than a 2.3 on the Richter scale. However, seismic activity on a small but frequent basis can cause long-term damage to structures. For example, seismic events can shift soil pressure on underground structures (such as pipes and foundations). Earthquake engineers are therefore highly reliant on sensor equipment, which enables them to gauge the effects of frequent seismic activity not only on above-ground structures but also on the ground itself.

To examine shifts in soil pressure caused by seismic activity, seismology engineers used an array of tactile pressure sensors, which were originally designed for artificial intelligence systems but were later utilized for the purposes of designing car seats and brake pad systems.. The use of such equipment helps engineers study the long-term effects of seismic activity on water pipes, underground cables, and underground storage tanks.

Arguably one of the best-known types of seismic detector systems is the seismograph. The seismograph uses a pendulum-based system to detect ground motion from seismic activity. Originally, the modern seismograph was designed to detect only significant earthquakes and tremors, and it could be found only in stable environments (such as a laboratory). Today, however, there are many different types of seismographs; some may be placed underground, others can be used in the field, while others are so sensitive that they can detect distant explosions or minute tremors.

IMPACT ON INDUSTRY
Engineering seismology is a relatively new combination of civil engineering and seismology, along with other fields (such as geology, emergency management, and risk management). Its uses have proven invaluable, however, as many urban centers in earthquake-prone regions, such as Tokyo and San Francisco, have benefited from the careful application of safe building practices and disaster mitigation programs that are borne of engineering seismology. Japan, the United States, and Switzerland are among the leaders in earthquake engineering, along with Australia and New Zealand. Engineering seismology also involves a range of public and private organizations, such as the following:

Governments. National and regional governments play an important role in the application of the findings of engineering seismologists. The USGA, for example, offers a wide range of resources and services for studying seismic activity and the dangers it poses. Additionally, the Federal Emergency Management Agency (FEMA), the National Science Foundation, the National Institute of Standards, and the USGA combine their resources to operate the National Earthquake Hazards Reduction Program, which seeks to reduce property losses and human casualties caused by earthquakes through careful engineering seismology practices (including mapping seismically active areas and generating building codes).

Engineering seismology societies. Engineering seismologists share information and theories with their peers through professional associations and societies, many of which are global in nature. The International Association for Earthquake Engineering is one such network, holding worldwide conferences on engineering seismology every four years. This organization has branch societies in Japan, Europe, and the United States, each working locally but also contributing to the larger association.

Universities. Because of the contributions engineering seismology provides to the field of civil engineering, earthquake engineering continues to evolve within this educational discipline. Universities such as Stanford and the University of California, Berkeley, offer such programs. Many other universities feature coursework in environmental engineering, which includes earthquake engineering and seismology studies. Furthermore, a large number of universities house research laboratories, shake tables, and seismograph stations covering seismic activity throughout a given geographic area.

Consulting firms. There is a considerable financial benefit to constructing a building, bridge, or other structure that will survive in a seismically active environment. Oil companies and other energy corporations, mining operations, and other businesses frequently seek the advice of private engineering consultants who offer seismic monitoring services. Construction companies also look to these consultants, seeking structural analyses and other services. The person who introduced the Richter scale, Charles Richter, founded a consulting firm upon his retirement from the California Institute of Technology in 1970.

CAREERS AND COURSEWORK

Students interested in engineering seismology should pursue and complete a bachelor's degree program in a related field, such as geology or civil engineering. They should also obtain a master's degree, preferably in a field of relevance to earthquake engineering, such as environmental engineering, structural engineering, geology, or seismology. Engineering seismologists' competitiveness as job candidates is improved greatly when they also earn a doctorate.

Engineering seismologists must receive training in the geosciences, including seismology, geology, and physics. These fields include courses in geodynamics, plate tectonics, statistical seismology, and mineral dynamics. They must also study civil engineering, structural design, and computer science (which must include training in computer modeling, digital mapping systems, and design software, which are essential in this arena). Furthermore, engineering seismologists must demonstrate excellent mathematical skills, particularly in geometry, algebra, and calculus. Finally, earthquake engineers must be trained in the use of many of the technical systems and devices that seismologists must use to monitor earthquakes.

FASCINATING FACTS ABOUT ENGINEERING SEISMOLOGY

- There are approximately 500,000 detectable earthquakes in the world annually, 100,000 of which may be felt and only 100 that cause damage.
- Between 1975 and 1995, only the U.S. states of Florida, North Dakota, Iowa, and Wisconsin did not experience any earthquakes.
- The earliest known "seismograph" was introduced in 132 CE in China. It took the shape of a hollow urn (with a hidden pendulum inside) with dragon heads adorning the sides and frogs at the base directly under each dragon (which held a single ball). During an earthquake, a dragon would drop its ball into the mouth of the frog below, revealing the direction of the waves. The device once detected an earthquake four hundred miles away.
- Scientists cannot presently predict earthquakes; they can only calculate the potential for a quake to strike.
- The largest recorded earthquake in the world was a magnitude 9.5 quake in Chile in 1960.
- In 1931, there were 350 operating seismic stations in the world. In 2011, there existed more than 4,000 .
- The largest recorded earthquake in the United States was a magnitude 9.2 quake in Alaska in 1964.
- A magnitude 8.8 earthquake in Chile in 2010 shifted that country's coastline approximately 1,640 feet toward the Pacific Ocean.

Social Context and Future Prospects

The study of earthquakes is a practice that dates back hundreds of years. Earthquake engineering specifically, however, represents an evolution toward a practical application of the study of seismic activity to the design and construction of large buildings, power plants, and other structures.

Engineering seismologists work closely with seismologists and civil engineers. On the former front, these engineers help design and operate detection equipment and systems to help explain seismic activity. This collaboration is critical, as improved knowledge of seismic activity can save lives and property.

For example, Japan has long utilized engineering seismology practices in its urban centers. The magnitude 8.9 earthquake in that country in March, 2011, did not devastate Tokyo because of strong building codes that, among other things, cause skyscrapers to sway with the region's seismic waves rather than stand in rigid fashion. Comparatively, the magnitude 7.0 Haiti earthquake in 2010 virtually flattened the country's capital, Port-au-Prince, and outlying areas, largely because Haiti did not have earthquake-safe building codes. Its buildings were built using insufficient steel and on slopes with no reinforcing foundation or support systems. One observer in Haiti reported that Port-au-Prince would likely not have survived even a magnitude 2.0, much less the 7.0 quake it did have.

The significance of the 2011 Japan disaster and of the rare 5.8 Virginia earthquake that struck the East Coast of the United States in August, 2011, continues to cast light on the need for earthquake engineering in structural design and construction. As regions with a history of major seismic activity (and those regions with the potential for such activity) continue to grow in size and population, engineering seismologists are likely to remain in high demand.

—*Michael P. Auerbach, MA*

Further reading

Chopra, Anil K. *Dynamics of Structures: Theory and Applications to Earthquake Engineering.* 3d ed. Upper Saddle River, N.J.: Prentice Hall, 2007. This book analyzes the theories of structural dynamics and applies the effects of earthquakes and seismology to them, including structural design and energy dissipation models.

Griffith, M. C., et al. "Earthquake Reconnaissance: Forensic Engineering on an Urban Scale." *Australian Journal of Structural Engineering* 11, no. 1 (2010): 63

74. This article describes the observations of a team of Australian aid workers who were dispatched to Indonesia to assess structural damages caused by a magnitude 7.6 earthquake there in 2009.

Saragoni, G. Rudolfo. "The Challenge of Centennial Earthquakes to Improve Modern Earthquake Engineering." *AIP Conference Proceedings,* 1020, no. 1 (July 8, 2008): 1113-1120. This article uses the 1906 San Francisco earthquake as a point of reference for reviewing the evolution of modern engineering seismology.

Slak, Tomaz, and Vojko Kilar. "Development of Earthquake Resistance in Architecture from an Intuitive to an Engineering Approach." *Prostor* 19, no. 1 (2011): 252-263. This article reviews the application of earthquake engineering concepts to the design of modern structures.

Stark, Andreas. *Seismic Methods and Applications.* Boca Raton, Fla.: BrownWalker Press, 2008. This book first reviews the principles of seismology and relevant geosciences and then proceeds to the practical application of these principles to structural design and engineering.

Villaverde, Roberto. *Fundamental Concepts of Earthquake Engineering.* New York: CRC Press, 2009. This book features a review of the history of the field of engineering seismology and includes some examples of how certain seismic wave types and other conditions are taken into account in building design.

See also: Applied Physics; Civil Engineering; Earthquake Engineering; Engineering; Environmental Engineering.

ENVIRONMENTAL CHEMISTRY

FIELDS OF STUDY

Chemistry; chemical engineering; bioengineering; physics; physical chemistry; organic chemistry; biochemistry; molecular biology; electrochemistry; analytical chemistry; photochemistry; atmospheric chemistry; agricultural chemistry; industrial ecology; toxicology.

SUMMARY

Environmental chemistry is an interdisciplinary subject dealing with chemical phenomena in nature. Environmental chemists are concerned with the consequences of anthropogenic chemicals in the air people breathe and the water they drink. They have become increasingly involved in managing the effects of these chemicals through both the creation of ecologically friendly products and efforts to minimize the pollution of the land, water, and air.

KEY TERMS AND CONCEPTS

- **anthrosphere:** Artificial environment created, modified, and used by humans for their purposes and activities, from houses and factories to chemicals and communications systems.
- **biomagnification:** Increase in the concentration of such chemicals as dichloro-diphenyl-trichloroethane (DDT) in different life-forms at successively higher trophic levels of a food chain or web.
- **barcinogen:** Chemical that causes cancer in organisms exposed to it.
- **chlorofluorocarbon (CFC):** Organic chemical compound containing carbon, hydrogen, chlorine, and florine, such as Freon (a trademark for several CFCs), once extensively used as a refrigerant.
- **dichloro-diphenyl-trichloroethane (DDT):** Organic chemical that was widely used for insect control during and after World War II but has been banned in many countries.
- **ecological footprint:** Measure of how much land and water a human population needs to regenerate consumed resources and to absorb wastes.
- **greenhouse effect:** Phenomenon in which the atmosphere, like a greenhouse, traps solar heat, with the trapping agents being such gases as carbon dioxide, methane, and water vapor.
- **hazardous waste:** Waste, or discarded material, that contains chemicals that are flammable, corrosive, toxic, or otherwise pose a threat to the health of humans or other organisms or a hazard to the environment.
- **heavy metal:** Any metal with a specific gravity greater than 5; usually used to refer to any metal that is poisonous to humans and other organisms, such as cadmium, lead, and mercury.
- **mutagen:** Chemical that causes or increases the frequency of changes (mutations) in the genetic material of an organism.
- **ozone:** Triatomic form of oxygen produced when diatomic oxygen is exposed to ultraviolent radiation (its presence in a stratospheric layer protects life-forms from harmful solar radiation).
- **pesticide:** Chemical used to kill or inhibit the multiplication of organisms that humans consider undesirable, such as certain insects.

DEFINITION AND BASIC PRINCIPLES

Environmental chemistry is the science of chemical processes in the environment. It is a profoundly interdisciplinary and socially relevant field. What environmental chemists do has important consequences for society because they are concerned with the effect of pollutants on the land, water, and air that humans depend on for their life, health, and proper functioning. Environmental chemists are forced to break down barriers that have traditionally kept chemists isolated from other fields.

Environmental chemistry needs to be distinguished from its later offshoot, green chemistry. As environmental chemistry developed, it tended to emphasize the detection and mitigation of pollutants, the study of the beneficial and adverse effects of various chemicals on the environment, and how the beneficial effects could be enhanced and the adverse eliminated or attenuated. Green chemistry, however, focuses on how to create sustainable, safe, and non-polluting chemicals in ways that minimize the ecological footprint of the processes. Some scholars define this field simply as sustainable chemistry.

The work of environmental chemists is governed by several basic principles. For example, the prevention principle states that when creating chemical products, it is better to minimize waste from the start than to later clean up wastes that could have been eliminated. Another principle declares that in making products, chemists should avoid using substances that could harm humans or the environment. Furthermore, chemists should design safe chemicals with the lowest practicable toxicity. In manufacturing products, chemists must minimize energy use and maximize energy efficiency; they should also use, as much as possible, renewable materials and energy resources. The products chemists make should be, if possible, biodegradable. Environmental chemists should use advanced technologies, such as computers, to monitor and control hazardous wastes. Finally, they must employ procedures that minimize accidents.

BACKGROUND AND HISTORY

Some scholars trace environmental chemistry's roots to the industrial revolutions in Europe and the United States in the eighteenth and nineteenth centuries. The newly created industries accelerated the rate at which chemicals were produced. By the late nineteenth century, some scientists and members of the general public were becoming aware of and concerned about certain negative consequences of modern chemical technologies. For example, the Swedish chemist Svante August Arrhenius recognized what later came to be known as the greenhouse effect, and a Viennese physician documented the health dangers of asbestos, which had become an important component in more than 3,000 products.

Modern industrialized societies were also generating increasing amounts of wastes, and cities and states were experiencing difficulties in discovering how to manage them without harm to the environment. World War I, often called the chemists' war, revealed the power of scientists to produce poisonous and explosive materials. During World War II, chemists were involved in the mass production of penicillin and DDT, substances that saved thousands of lives. However, the widespread and unwise use of DDT and other pesticides after the war prompted Rachel Carson to write *Silent Spring* (1962), which detailed the negative effects that pesticides were having on birds and other organisms. Many associate the

start of the modern environmental movement with the publication of Carson's book.

The warnings Carson issued played a role in the establishment of the Environmental Protection Agency (EPA) in 1971 and the EPA's ban of DDT in 1972. Reports in 1974 that CFCs were destroying the Earth's ozone layer eventually led many countries to halt, then ban their production. From the 1970's on, environmental chemists devoted themselves to the management of pollutants and participation in government policies and regulations that attempted to prevent or mitigate chemical pollution. In the 1990's, criticism of this command-and-control approach led to the formation of the green chemistry movement by Paul Anastas and others. These chemists fostered a comprehensive approach to the production, utilization, and termination of chemical materials that saved energy and minimized wastes. By the first decade of the twenty-first century, environmental chemistry had become a thriving profession with a wide spectrum of approaches and views.

HOW IT WORKS

Chemical analysis and the atmosphere. Indispensable to the progress of environmental chemistry is the ability to measure, quantitatively and qualitatively, certain substances that even at very low concentrations, pose harm to humans and the environment. By using such techniques as gravimetric and volumetric analysis, various types of spectroscopy, electroanalysis, and chromatography, environmental chemists have been able to accurately measure such atmospheric pollutants as sulfur dioxide, carbon monoxide, hydrogen sulfide, nitrogen oxides, and several hydrocarbons. Many governmental and nongovernmental organizations require that specific air contaminants be routinely monitored. Because of heavy demands on analytic chemists, much monitoring has become computerized and automatic.

Atmospheric particles range in size from a grain of sand to a molecule. Nanoparticles, for example, are about one-thousandth the size of a bacterial cell, but environmental chemists have discovered that they can have a deleterious effect on human health. Epidemiologists have found that these nanoparticles adversely affect respiratory and cardiovascular functioning.

Atmospheric aerosols are solid or liquid particles smaller than a hundred millimicrons in diameter.

These particles undergo several possible transformations, from coagulation to phase transitions. For example, particles can serve as nuclei for the formation of water droplets, and some chemists have experimented with particulates in forming rain clouds. Human and natural biological sources contribute to atmospheric aerosols. The CFCs in aerosol cans have been factors in the depletion of the Earth's ozone layer. Marine organisms produce such chemicals as halogen radicals, which in turn influence reactions of atmospheric sulfur, nitrogen, and oxidants. As marine aerosol particles rise from the ocean and are oxidized in the atmosphere, they may react with its components, creating a substance that may harm human health. Marine aerosols contain carbonaceous as well as inorganic materials, and when an organic aerosol interacts with atmospheric oxygen, its inert hydrophobic (water-repelling) film is transformed into a reactive hydrophilic (water-absorbing) layer. A consequence of this process is that organic aerosols serve as a conduit for organic compounds to enter the atmosphere.

Carbon dioxide is an example of a molecular atmospheric component, and some environmental chemists have been devoting their efforts to determining its role in global warming, but others have studied the Earth's prebiotic environment to understand how inorganic carbon initially formed the organic molecules essential to life. Using photoelectrochemical techniques (how light affects electron transfers in chemical reactions), researchers discovered a possible metabolic pathway involving carbon dioxide fixation on mineral surfaces.

Water pollution. Because of water's vital importance and its uneven distribution on the Earth's surface, environmental chemists have had to spend a great deal of time and energy studying this precious resource. Even before the development of environmental chemistry as a profession, many scientists, politicians, and citizens were concerned with water management. Various governmental and nongovernmental organizations were formed to monitor and manage the quality of water. Environmental chemists have been able to use their expertise to trace the origin and spread of water pollutants throughout the environment, paying special attention to the effects of water pollutants on plant, animal, and human life.

A particular interest of environmental chemists has been the interaction of inorganic and organic matter with bottom sediments in lakes, rivers, and oceans. These surface sediments are not simply unreactive sinks for pollutants but can be studied quantitatively in terms of how many specific chemicals are bonded to a certain amount of sediment, which in turn is influenced by whether the conditions are oxidizing or reducing. Chemists can then study the bioavailability of contaminants in sediments. Furthermore, environmental chemists have studied dissolution and precipitation, discovering that the rates of these processes depend on what happens in surface sediments. Using such techniques as scanning polarization force microscopy, they have been able to quantify pollutant immobilization and bacterial attachment on surface sediments. Specifically, they have used these methods to understand the concentrations and activities of heavy metals in aquatic sediments.

Hazardous waste. One of the characteristics of advanced industrialized societies has been the creation of growing amounts of solid and liquid wastes, an important proportion of which pose severe dangers to the environment and human health. These chemicals can be toxic, corrosive, or flammable. The two largest categories of hazardous wastes are organic compounds such as polychlorinated biphenyls (PCBs) and dioxin, and heavy metals, such as lead and mercury. Environmental chemists have become involved in research on the health and environmental effects of these hazardous substances and the development of techniques to detect, monitor, and control them. For example, they have studied the rapidly growing technology of incineration as a means of reducing and disposing of wastes. They have also studied the chemical emissions from incinerators and researched methods for the safe disposal of the ash and slag produced. Because of the passage of the Resource Recovery Act of 1970, environmental chemists have devoted much attention to finding ways of reclaiming and recycling materials from solid wastes.

Pesticides. The mismanagement of pesticides inspired Carson to write *Silent Spring*, and pesticides continue to be a major concern of industrial and environmental chemists. One reason for the development of pesticides is the great success farmers had in using them to control insects, thereby dramatically increasing the quantity and quality of various agricultural products. By 1970, more than 30,000 pesticide products were being regularly used. This expansion in pesticides is what alarmed Carson, who was not in favor of a total ban of pesticides but rather their reduction and integration with biological and cultural controls.

Because pesticides are toxic to targeted species, they often cause harm to beneficial insects and, through biomagnification, to birds and other animals. Pesticide residues on agricultural products have also been shown to harm humans. Therefore, environmental chemists have become involved in monitoring pesticides from their development and use to their effects on the environment. They have also helped create pesticide regulations and laws. This regulatory system has become increasingly complex, costing companies, the government, and customers large amounts of money. The hope is that integrated control methods will prove safer and cheaper than traditional pesticides.

APPLICATIONS AND PRODUCTS

Anthrosphere. Environmental chemists are concerned with how humans and their activities, especially making and using chemicals, affect the environment. Building homes and factories, producing food and energy, and disposing of waste all have environmental consequences. Whereas some environmentalists study how to create ecologically friendly dwellings, environmental chemists study how chemical engineers should design factories that cause minimum harm to the environment. Specific examples of applications of environmental chemistry to industry include the creation of efficient catalysts that speed up reactions without themselves posing health or environmental hazards.

Because many problems arise from the use of hazardous solvents in chemical processes, environmental chemists try to develop processes that use only safe solvents or avoid their use altogether. Because the chemical industry depends heavily on petroleum resources, which are nonrenewable and becoming drastically diminished, environmental chemists study how renewable resources such as biomass may serve as substitutes for fossil fuels. They are also creating products that degrade rapidly after being discarded, so that their environmental impact is transient. Following the suggestions of environmental chemists, some companies are developing long-lasting, energy-saving batteries, and others are selling their products with less packaging than previously.

Hydrosphere. Throughout history, water has been essential in the development of human civilizations, some of which have declined and disappeared because of deforestation, desertification, and drought. Water has also been a vehicle for the spread of diseases and pollutants, both of which have caused serious harm to humans and their environment. Environmental chemists have consequently been involved in such applications as the purification of water for domestic use, the monitoring of water used in the making of chemicals, and the treatment of wastewater so that its release and reuse will not harm humans or the environment. For example, such heavy metals as cadmium, mercury, and lead are often found in wastewater from various industries, and environmental chemists have developed such techniques as electrodeposition, reverses osmosis, and ion exchange to remove them. Many organic compounds are carcinogens and mutagens, so chemists want to remove them from water. Besides such traditional methods as powdered activated carbon, chemists have used adsorbent synthetic polymers to attract insoluble organic compounds. Detergents in wastewater can contribute to lake eutrophication and cause harm to wildlife, and some companies have created detergents specially formulated to cause less environmental damage.

Atmosphere. In the twentieth century, scientists discovered that anthropogenic greenhouse gases have been contributing to global warming that could have catastrophic consequences for island nations and coastal cities. Industries, coal-burning power plants, and automobiles are major air polluters, and environmental chemical research has centered on finding ways to reduce or eliminate these pollutants. The Clean Air Act of 1970 and subsequent amendments set standards for air quality and put pressure on air polluters to reduce harmful emissions. For example, chemists have helped power plants develop desulfurization processes, and other scientists developed emission controls for automobiles.

The problem of global warming has proved difficult to solve. Some environmental chemists believe that capturing and storing carbon dioxide is the answer, whereas others believe that government regulation of carbon dioxide and methane by means of energy taxes will lead to a lessening of global warming. Some think that the Kyoto Protocol, an international agreement that went into effect in 2005, is a small but important first step, whereas others note that the lack of participation by the United States and the omission of a requirement that such countries as China and India reduce greenhouse gas emissions seriously weakened the agreement. On the other hand, the Montreal and Copenhagen Protocols did foster global cooperation in the reduction of and phasing out of CFCs, which should lead to a reversal in ozone layer depletion.

Agricultural and industrial ecology. Agriculture, which involves the production of plants and animals as food, is essential in ministering to basic human needs. Fertilizers and pesticides developed by chemists brought forth the green revolution, which increased crop yields in developed and developing countries. Some believe that genetic engineering techniques will further revolutionize agriculture. Initially, chemists created such highly effective insecticides as DDT, but DDT proved damaging to the environment. Activists then encouraged chemists to develop biopesticides from natural sources because they are generally more ecologically friendly than synthetics.

Industrial ecology is a new field based on chemical engineering and ecology; its goal is to create products in a way that minimizes environmental harm. Therefore, environmental chemical engineers strive to build factories that use renewable energy as much as possible, recycle most materials, minimize wastes, and extract useful materials from wastes. In general, these environmental chemists act as wise stewards of their facilities and the environment. A successful example of ecological engineering is phytoremediation, or the use of plants to remove pollutants from contaminated lands. Artificially constructed wetlands have also been used to purify wastewater.

IMPACT ON INDUSTRY

Industries are involved in a wide variety of processes that have environmental implications, from food production and mineral extraction to manufacturing and construction. In some cases, such as the renewable energy industries, the environmental influence is strong, but even in traditional industries such as utilities and transportation companies, problems such as air and water pollution have become corporate concerns.

Government and university research. Since the beginning of the modern environmental movement,

---

FASCINATING FACTS ABOUT ENVIRONMENTAL CHEMISTRY

- At life's beginning, single-celled cyanobacteria made possible the evolution of millions of new species; however, what modern humans are doing to the atmosphere will result in the extinction of hundreds of thousands of species.
- More than 99 percent of the total mass of the Earth's atmosphere is found within about 30 kilometers (about 20 miles) of its surface.
- Although 71 percent of the Earth's surface is covered by water, only 0.024 percent of this water is available as freshwater.
- According to a National Academy of Sciences study, legally permitted pesticide residues in food cause 4,000 to 20,000 cases of cancer per year in the United States.
- From 1980 to 2010, the quality of outdoor air in most developed countries greatly improved.
- From 1980 to 2008, the EPA placed 1,569 hazardous-waste sites on its priority list for cleanup.
- According to the U.S. Geological Survey, even though the population of the United States grew by 16 percent from 1980 to 2004, total water consumption decreased by about 9 percent.
- Because of global warming, in February and March, 2002, a mass of ice larger than the state of Rhode Island separated from the Antarctic Peninsula.
- The United States leads the world in producing solid waste. With only 4.6 percent of the world's population, it produced about one-third of the world's solid waste.
- Each year, 12,000 to 16,000 American children under nine years of age are treated for acute lead poisoning, and about 200 die.

---

state and federal governments as well as universities have increased grants and fellowships for projects related to environmental chemistry. For example, the EPA's Green Chemistry Program has supported basic research to develop chemical products and manufacturing techniques that are ecologically benign. Sometimes government agencies cooperate with each other in funding environmental chemical projects; for instance, in 1992, the EPA's Office of Pollution Prevention and Toxics collaborated with the National Science Foundation to fund several green chemical proposals. These grants were significant, totaling tens of millions of dollars. Besides government and academia, professional organizations have also sponsored green research. For example, the American Chemical Society has established the Green Chemistry Institute, whose purpose is to encourage collaboration with scientists in other disciplines to discover chemical products and processes that reduce or eliminate hazardous wastes.

Industry and business. Environmentalists and government regulations have forced leaders in business and industry to make sustainability a theme in their plans for future development. In particular, the U.S. chemical industry, the world's largest, has directly linked its growth and competitiveness to a concern for the environment. Industrial leaders realize that they will have to cooperate with officials in government and academia to realize this vision. They also understand that they will need to join with such organizations as the Environmental Management Institute and the Society of Environmental Toxicology and Chemistry to minimize the environmental contamination that has at times characterized the chemical industry of the past.

Major corporations. Top American chemical companies, such as Dow, DuPont, Eastman Chemical, and Union Carbide, have gone on record as vowing to use resources more efficiently, deliver products to consumers that meet their needs and enhance their quality of life, and preserve the environment for future generations. Nevertheless, these promised changes must be seen against the background of past environmental depredations and disasters. For example, Dow is responsible for ninety-six of the worst Superfund toxic-waste dumps, and Union Carbide shared responsibility for the deaths of more than 2,000 people in a release of toxic chemicals in Bhopal, India. Eastman Chemical, along with other industries, is a member of Responsible Care, an organization devoted to the principles of green chemistry, and the hope is that the member industries will encourage the production of ecologically friendly chemicals without the concomitant of dangerous wastes.

CAREERS AND COURSE WORK

Students in environmental chemistry need to take many chemistry courses, including general chemistry, organic chemistry, quantitative and qualitative analysis, instrumental analysis, inorganic chemistry,

physical chemistry, and biochemistry. They also should study advanced mathematics, physics, and computer science. Although job opportunities exist for students with a bachelor's degree, there are greater opportunities for those who obtain a master's degree or a doctorate. Graduate training allows students to specialize in such areas as soil science, hazardous-waste management, air-quality management, water-quality management, environmental education, and environmental law. Because of increasing environmental concerns in industries, governments, and academia, numerous careers are possible for environmental chemistry graduates. They have found positions in business, law, marketing, public policy, government agencies, laboratories, and chemical industries. Some graduates pursue careers in such government agencies as the EPA, the Food and Drug Administration, the Natural Resource Conservation Services, the Forest Service, and the Department of Health and Human Services. After obtaining a doctorate, some environmental chemists become teachers and researchers in one of the many academic programs devoted to their field.

SOCIAL CONTEXT AND FUTURE PROSPECTS

In a world increasingly concerned with environmental quality, the sustainability of lifestyles, and environmental justice, the future for environmental chemistry appears bright. For example, analysts have predicted that environmental chemical engineers will have a much faster employment growth than the average for all other occupations. Environmental chemists will be needed to help industries comply with regulations and to develop ways of cleaning up hazardous wastes. However, other analysts warn that, in periods of economic recession, environmental concerns tend to be set aside, and this could complicate the employment forecast for environmental chemists.

Some organizations, such as the Environmental Chemistry Group in England, have as a principal goal the promotion of the expertise and interests of their members, and the American Chemical Society's Division of Environmental Chemistry similarly serves its members with information on educational programs, job opportunities, and awards for significant achievement, such as the Award for Creative Advances in Environmental Chemistry. These organizations also issue reports on their social goals, and documents detailing their social philosophy emphasize that environmental chemists should be devoted to the safe operation of their employers' facilities. Furthermore, they should strive to protect the environment and make sustainability an integral part of all business activities.

—Robert J. Paradowski, MS, PhD

FURTHER READING

Baird, Colin, and Michael Cann. *Environmental Chemistry.* 4th ed. New York: W. H. Freeman, 2008. A clear and comprehensive survey of the field. Each chapter has further reading suggestions and Web sites of interest. Index.

Carson, Rachel. *Silent Spring.* 1962. Reprint. Boston: Houghton Mifflin, 2002. Originally serialized in *The New Yorker* magazine, this book has been honored as one of the best nonfiction works of the twentieth century. Its criticism of the chemical industry and the overuse of pesticides generated Environmental Chemistry controversy, and most of its major points have stood the test of time.

Girard, James E. *Principles of Environmental Chemistry.* Sudbury, Mass.: Jones and Bartlett, 2010. Emphasizes the chemical principles undergirding environmental issues as well as the social and economic contexts in which they occur. Five appendixes and index.

Howard, Alan G. *Aquatic Environmental Chemistry.* 1998. Reprint. New York: Oxford University Press, 2004. Analyzes the chemistry behind freshwater and marine systems. Also includes useful secondary material that contains explanations of unusual terms and advanced chemical and mathematical concepts.

Manahan, Stanley E. *Environmental Chemistry.* 9th ed. Baca Raton, Fla.: CRC Press, 2010. Explores the anthrosphere, industrial ecosystems, geochemistry, and aquatic and atmospheric chemistry. Each chapter has a list of further references and cited literature. Index.

Schwedt, Georg. *The Essential Guide to Environmental Chemistry.* 2001. Reprint. New York: John Wiley & Sons, 2007. Provides a concise overview of the field. Contains many color illustrations and an index.

WEB SITES

American Chemical Society Division of Environmental Chemistry
http://www.envirofacs.org

Environmental Protection Agency National Exposure Research Library, Environmental Sciences Division
http://www.epa.gov/nerlesd1

Royal Society of Chemistry Environmental Chemistry Group
http://www.rsc.org/membership/networking/interestgroups/environmental/index.asp

Society of Environmental Toxicology and Chemistry
http://www.setac.org

See also: Air-Quality Monitoring; Environmental Engineering.

ENVIRONMENTAL ENGINEERING

FIELDS OF STUDY

Mathematics; chemistry; physics; biology; geology; engineering mechanics; fluid mechanics; soil mechanics; hydrology.

SUMMARY

Environmental engineering is a field of engineering involving the planning, design, construction, and operation of equipment, systems, and structures to protect and enhance the environment. Major areas of application within the field of environmental engineering are wastewater treatment, water-pollution control, water treatment, air-pollution control, solid-waste management, and hazardous-waste management. Water-pollution control deals with physical, chemical, biological, radioactive, and thermal contaminants. Water treatment may be for the drinking water supply or for industrial water use. Air-pollution control is needed for stationary and moving sources. The management of solid and hazardous wastes includes landfill and incinerators for disposal of solid waste and identification and management of hazardous wastes.

KEY TERMS AND CONCEPTS

- **activated sludge process:** Biological wastewater-treatment system for removing waste organic matter from wastewater.
- **baghouse:** Air-pollution control device that filters particulates from an exhaust stream; also called a bag filter.
- **biochemical oxygen demand (BOD):** Amount of oxygen needed to oxidize the organic matter in a water sample.
- **catch basin:** Chamber to retain matter flowing from a street gutter that might otherwise obstruct a sewer.
- **digested sludge:** Wastewater biosolids (sludge) that have been stabilized by an anaerobic or aerobic biological process.
- **effluent:** Liquid flowing out from a wastewater-treatment process or treatment plant.
- **electrostatic precipitator:** Air-pollution control device to remove particulates from an exhaust stream by giving the particles a charge.
- **nonvisual uses for LEDs:** Pollution source that cannot be traced back to a single emission source, such as storm-water runoff.
- **oxidation pond:** Large shallow basin used to treat wastewater, using sunlight, bacteria, and algae.
- **photochemical smog:** Form of air pollution caused by nitrogen oxides and hydrocarbons in the air that react to form other pollutants because of catalysis by sunlight.
- **pollution prevention:** Use of conscious practices or processes to reduce or eliminate the creation of wastes at the source.
- **primary treatment:** Wastewater treatment to remove suspended and floating matter that will settle from incoming wastewater.
- **sanitary landfill:** Site used for disposal of solid waste that uses liners to prevent groundwater contamination and is covered daily with a layer of earth.
- **secondary treatment:** Wastewater treatment to remove dissolved and fine suspended organic matter that would exert an oxygen demand on a receiving stream.

- **trickling filter:** Biological wastewater-treatment process in which wastewater trickles through a bed of rocks with a coating containing microorganisms.

DEFINITION AND BASIC PRINCIPLES

Environmental engineering is a field of engineering that split off from civil engineering as the importance of the treatment of drinking water and wastewater was recognized. This field of engineering was first known as sanitary engineering and dealt almost exclusively with the treatment of water and wastewater. As awareness of other environmental concerns and the need to do something about them grew, this field of engineering became known as environmental engineering, with the expanded scope of dealing with air pollution, solid wastes, and hazardous wastes, in addition to water and wastewater treatment.

Environmental engineering is an interdisciplinary field that makes use of principles of chemistry, biology, mathematics, and physics, along with engineering sciences (such as soil mechanics, fluid mechanics, and hydrology) and empirical engineering correlations and knowledge to plan for, design, construct, maintain, and operate facilities for treatment of liquid and gaseous waste streams, for prevention of air pollution, and for management of solid and hazardous wastes. The field also includes investigation of sites with contaminated soil and/or groundwater and the planning and design of remediation strategies. Environmental engineers also provide environmental impact analyses, in which they assess how a proposed project will affect the environment.

BACKGROUND AND HISTORY

When environmental engineering, once a branch of civil engineering, first became a separate field in the mid-1800's, it was known as sanitary engineering. Initially, the field involved the water supply, water treatment, and wastewater collection and treatment.

In the middle of the twentieth century, people began to become concerned about environmental quality issues such as water and air pollution. As a consequence, the field of sanitary engineering began to change to environmental engineering, expanding its scope to include air pollution, solid- and hazardous-waste management, and industrial hygiene.

Several pieces of legislation have affected and helped define the work of environmental engineers.

Some of the major laws include the Clean Air Act of 1970, the Safe Drinking Water Act of 1974, the Toxic Substances Control Act of 1976, the Resource Recovery and Conservation Act (RCRA) of 1976, and the Clean Water Act of 1977.

HOW IT WORKS

Environmental engineering uses chemical, physical, and biological processes for the treatment of water, wastewater, and air, as well as in-site remediation processes. Therefore, knowledge of the basic sciences—chemistry, biology, and physics—is important along with knowledge of engineering sciences and applied engineering.

Chemistry. Chemical processes are used to treat water and wastewater, to control air pollution, and for site remediation. These chemical treatments include chlorination for disinfection of both water and wastewater, chemical oxidation for iron and manganese removal in water-treatment plants, chemical oxidation for odor control, chemical precipitation for removal of metals or phosphorus from wastewater, water softening by the lime-soda process, and chemical neutralization for pH (acidity) control and for scaling control.

The chemistry principles and knowledge that are needed for these treatment processes include the ability to understand and work with chemical equations, to make stoiciometric calculations for dosages, and to determine size and configuration requirements for chemical reactors to carry out the various processes.

Biology. The major biological treatment processes used in wastewater treatment are biological oxidation of dissolved and fine suspended organic matter in wastewater (secondary treatment) and stabilization of biological wastewater biosolids (sludge) by anaerobic digestion or aerobic digestion. Biological principles and knowledge that are useful in designing and operating biological wastewater treatment and biosolids digestion processes include the kinetics of the biological reactions and knowledge of the environmental conditions required for the microorganisms. The required environmental conditions include the presence or absence of oxygen and the appropriate temperature and pH.

Physics. Physical treatment processes used in environmental engineering include screening, grinding, comminuting, mixing, flow equalization, flocculation, sedimentation, flotation, and granular filtration. These processes are used to remove materials that can be screened, settled, or filtered out of water or wastewater and to assist in managing some of the processes. Many of these physical treatment processes are designed on the basis of empirical loading factors, although some use theoretical relationships such as the use of estimated particle settling velocities for design of sedimentation equipment.

Soil mechanics. Topics covered in soil mechanics include the physical properties of soil, the distribution of stress within the soil, soil compaction, and water flow through soil. Knowledge of soil mechanics is used by environmental engineers in connection with design and operation of sanitary landfills for solid waste, in storm water management, and in the investigation and remediation of contaminated soil and groundwater.

Fluid mechanics. Principles of fluid mechanics are used by environmental engineers in connection with the transport of water and wastewater through pipes and open channels. Such transport takes place in water distribution systems, in sanitary sewer collection systems, in storm water sewers, and in wastewater-treatment and water-treatment plants. Design and sizing of the pipes and open channels make use of empirical relationships such as the Manning equation for open channel flow and the Darcy-Weisbach equation for frictional head loss in pipe flow. Environmental engineers also design and select pumps and flow measuring devices.

Hydrology. The principles of hydrology (the science of water) are used to determine flow rates for storm water management when designing storm sewers or storm water detention or retention facilities. Knowledge of hydrology is also helpful in planning and developing surface water or groundwater as sources of water.

Practical knowledge. Environmental engineers make use of accumulated knowledge from their work in the field. Theoretical equations, empirical equations, graphs, nomographs, guidelines, and rules of thumb have been developed based on experience. Empirical loading factors are used to size and design many treatment processes for water and wastewater. For example, the design of rapid sand filters to treat drinking water was based on a specified loading rate in gallons per minute of water per square foot of sand filter. Also the size required for a rotating biological contactor to provide secondary treatment of wastewater was determined based on a loading rate in pounds of biochemical oxygen demand (BOD) per day per 1,000 square feet of contactor area.

Engineering tools. Tools such as engineering graphics, computer-aided drafting (CAD), geographic information systems (GIS), and surveying are available for use by environmental engineers. These tools are used for working with plans and drawings and for laying out treatment facilities or landfills.

Codes and design criteria. Much environmental engineering work makes use of codes or design criteria specified by local, state, or federal government agencies. Examples of such design criteria are the storm return period to be used in designing storm sewers or storm water detention facilities and the loading factor for rapid sand filters. Design and operation of treatment facilities for water and wastewater are also based on mandated requirements for the finished water or the treated effluent.

APPLICATIONS AND PRODUCTS

Environmental engineers design, build, operate, and maintain treatment facilities and equipment for the treatment of drinking water and wastewater, air-pollution control, and the management of solid and hazardous wastes.

Air-pollution control. Increasing air pollution from industries and power plants as well as automobiles led to passage of the Clean Air Act of 1970. This law led to greater efforts to control air pollution. The two major ways to control air pollution are the treatment of emissions from fixed sources and from moving sources (primarily automobiles). The fixed sources of air pollution are mainly the smokestacks of industrial facilities and power plants.

Devices used to reduce the number of particulates emitted include settling chambers, baghouses, cyclones, wet scrubbers, and electrostatic precipitators. Electrostatic precipitators impart the particles with an electric charge to aid in their removal. They are often used in power plants, at least in part because of the readily available electric power to run them. Water-soluble gaseous pollutants can be removed by wet scrubbers. Other options for gaseous pollutants are adsorption on activated carbon or incineration of combustible pollutants. Because sulfur is contained in the coal used as fuel, coal-fired power plants produce sulfur oxides, particularly troublesome pollutants.

The main options for reducing these sulfur oxides are desulfurizing the coal or desulfurizing the flue gas, most typically with a wet scrubber using lime to precipitate the sulfur oxides. Legislation has greatly reduced the amount of automobile emissions, the main moving source of air pollution. The reduction in emissions has been accomplished through catalytic converters to treat exhaust gases and improvements in the efficiency of automobile engines.

Water treatment. The two main sources for the water supply are surface water (river, lake, or reservoir) and groundwater. The treatment requirements for these two sources are somewhat different.

For surface water, treatment is aimed primarily at removal of turbidity (fine suspended matter) and perhaps softening the water. The typical treatment processes for removal of turbidity involve the addition of chemicals such as alum or ferric chloride. The chemicals are rapidly mixed into the water so that they react with alkalinity in the water, then slowly mixed (flocculation) to form a settleable precipitate. After sedimentation, the water passes through a sand filter and finally is disinfected with chlorine. If the water is to be softened as part of the treatment, lime, $Ca(OH)$, and soda ash, $NaCO$, are used in place of alum or ferric chloride, and the water hardness (calcium and magnesium ions) is removed along with its turbidity. Groundwater is typically not turbid (cloudy), so it does not require the type of treatment used for surface water. At minimum, it requires disinfection. Removal of iron and manganese by aeration may be needed, and if the water is very hard, it may be softened by ion exchange.

Wastewater treatment. The Clean Water Act of 1977 brought wastewater treatment to a new level by requiring that all wastewater discharged from municipal treatment plants must first undergo at least secondary treatment. Before the passage of the legislation, many large cities located on a river or along the ocean provided only primary treatment in their wastewater-treatment plants and discharged effluent with only settleable solids removed. All dissolved and fine suspended organic matter remained in the effluent. Upgrading treatment plants involved added a biological treatment to remove dissolved and fine suspended organic matter that would otherwise exert an oxygen demand on the receiving stream, perhaps depleting the oxygen enough to cause problems for fish and other aquatic life.

Solid-waste management. The main options for solid-waste management are incineration, which reduces the volume for disposal to that of the ash that is produced, and disposal in a sanitary landfill. Some efforts have been made to reuse and recycling materials to reduce the amount of waste sent to incinerators or landfills. A sanitary landfill is a big improvement over the traditional garbage dump, which was simply an open dumping ground. A sanitary landfill uses liners to prevent groundwater contamination, and each day, the solid waste is covered with soil.

Hazardous-waste management. The Resource Conservation and Recovery Act (RCRA) of 1976 provides the framework for regulating hazardous-waste handling and disposal in the United States. One very useful component of RCRA is that it specifies a very clear and organized procedure for determining if a particular material is a hazardous waste and therefore subject to RCRA regulations. If the material of interest is indeed a waste, then it is defined to be a hazardous waste if it appears on one of RCRA's lists of hazardous wastes, if it contains one or more hazardous chemicals that appear on an RCRA list, or if it has one or more of the four RCRA hazardous waste characteristics as defined by laboratory tests. The four RCRA hazardous waste characteristics are flammability, reactivity, corrosivity, and toxicity. The RCRA regulations set standards for secure landfills and treatment processes for disposal of hazardous waste.

Much work has been done in investigating and cleaning up sites that have been contaminated by hazardous wastes in the past. In some cases, funding is available for cleanup of such sites through the Comprehensive Environmental Response, Compensation, and Liability Act of 1980 (known as CERCLA or Superfund) or its amendment, the Superfund Amendments and Reauthorization Act (SARA) of 1986.

IMPACT ON INDUSTRY

Increased interest in environmental issues in the last quarter of the twentieth century has made environmental engineering more prominent. The U.S. Bureau of Labor Statistics shows environmental engineering as the eighth largest field of engineering, with an estimated 54,300 environmental engineers employed in the United States in 2008. The bureau projects a 31 percent rate of growth in environmental engineering employment through much of the 2010's, which is much higher than the average for all occupations. Environmental engineers are employed by consulting engineering firms, industry, universities, and federal, state, and local government agencies.

Consulting engineering firms. Slightly more than half of all environmental engineers in the United States are employed by firms that engage in consultng in architecture, engineering, management, and scientific and technical issues. Some engineering consulting companies specialize in environmental projects, while others have an environmental division or simply have some environmental engineers on staff. The U.S. environmental consulting industry is made up of about 8,000 companies, ranging in size from one-person shops to huge multinational corporations. Some of the largest environmental consulting firms are CH2M Hill, Veolia Environmental Services North America, and Tetra Tech. Two engineering and construction firms with large environmental divisions are Bechtel and URS.

Government agencies. Environmental engineers are employed by government agencies at the local, state, and federal levels. The U.S. Environmental Protection Agency and state environmental agencies employ the most environmental engineers, but many other government agencies, such as the Army Corps of Engineers, the Bureau of Reclamation, the Department of Agriculture, the Department of Defense, the Federal Emergency Management Agency, and the Natural Resources Conservation Service also have environmental engineers on their staffs. At the local government level, environmental engineers are used by city and county governments, in city or county engineering offices, and in public works departments.

University research and teaching. Colleges and universities employ environmental engineers to teach environmental engineering and the environmental component of civil engineering programs. Civil engineering is one of the largest engineering specialties, and its former subspecialty, environmental engineering, is taught at numerous colleges and universities around the world.

Industry. A small percentage of environmental engineers work in industry, at companies such as 3M, Abbott Laboratories, BASF, Bristol-Myers Squibb, Chevron, the Dow Chemical Company, DuPont, and IBM.

CAREERS AND COURSE WORK

An entry-level environmental engineering position can be obtained with a bachelor's degree in environmental engineering or in civil or chemical engineering with an environmental specialization. However, because many positions require registration as an engineer in training or as a professional engineer, it is important that the bachelor's degree program is accredited by the Accreditation Board for Engineering and Technology (ABET). Students must first graduate from an accredited program before taking the exam to become a registered engineer in training. After four years of experience, the engineer in training can take another exam for registration as a professional engineer.

A typical program of study for an environmental engineering degree at the undergraduate level includes the chemistry, calculus-based physics, and mathematics that is typical of almost all engineering programs in the first two years of study. It also may include biology, additional chemistry, and engineering geology. The last two years of study will typically include hydrology, soil mechanics, an introductory course in environmental engineering, and courses in specialized areas of environmental engineering such as water treatment, wastewater treatment, air-pollution control, and solid- and hazardous-waste management.

Master's degree programs in environmental engineering fall into two categories: those designed primarily for people with an undergraduate degree in environmental engineering and those for people with an undergraduate degree in another type of engineering. Some environmental engineering positions require a master's degree. A doctoral degree in environmental engineering is necessary for a position in research or teaching at a college or university.

SOCIAL CONTEXT AND FUTURE PROSPECTS

Many major areas of concern in the United States and around the world are related to the environment. Issues such as water-pollution control, air-pollution control, global warming, and climate change all need the work of environmental engineers. These issues, as well as the need for environmental engineers, are likely to remain concerns for much of the twenty-first century. Water supply, wastewater treatment, and solid-waste management all involve infrastructure, needing Environmental Engineering repair,

---

FASCINATING FACTS ABOUT ENVIRONMENTAL ENGINEERING

- In March, 1987, a barge loaded with municipal solid waste departed from Islip, New York, headed to a facility in North Carolina, where state officials turned it away. It traveled to six states and three countries over seven months trying to find a place to unload its cargo before it was allowed to return to New York.

- The activated sludge process for treating wastewater was invented in England in 1914. Interest in the activated sludge process spread rapidly, and it soon became the most widely used biological wastewater-treatment process in the world.

- The use of chlorine to disinfect drinking water supplies began in the late 1800's and early 1900's. It dramatically reduced the incidence of waterborne diseases such as cholera and typhoid fever.

- In the book *Silent Spring* (1962), Rachel Carson described the negative effect of the pesticide dichloro-diphenyl-trichloroethane (DDT) on birds. This book increased environmental awareness and is often cited as the beginning of the environmental movement.

- The first comprehensive sewer system in the United States was built in Chicago in 1850. The city level was raised 10 to 15 feet so that gravity would drain the sewers into the Chicago River, which emptied into Lake Michigan.

- A "solid waste" as defined by the U.S. Resource Conservation and Recovery Act may be solid, liquid, or semi-solid in form.

---

maintenance, and upgrading, which are all likely to need the help of environmental engineers.

—Harlan H. Bengtson, MS, PhD

FURTHER READING

Anderson, William C. "A History of Environmental Engineering in the United States." In *Environmental and Water Resources History*, edited by Jerry R. Rogers and Augustine J. Fredrich. Reston, Va.: American Society of Civil Engineers, 2003. Describes the development of environmental engineering in the United States, starting in the 1830's, through the growth of environmental awareness in the 1970's, and into the twenty-first century.

Identifies and discusses significant pioneers in the field.

Davis, Mackenzie L., and Susan J. Masten. *Principles of Environmental Engineering and Science.* 2d ed. Boston: McGraw-Hill Higher Education, 2009. An introduction to the field of environmental engineering. Includes illustrations and maps.

Juuti, Petri S., Tapio S. Katko, and Heikki Vuorinen. *Environmental History of Water: Global Views on Community Water Supply and Sanitation.* London: IWA, 2007. Provides information on the history of the water supply and sanitation around the world.

Leonard, Kathleen M. "Brief History of Environmental Engineering: 'The World's Second Oldest Profession.'" *ASCE Conference Proceedings* 265, no. 47 (2001): 389-393. Describes the evolution of environmental engineering from its earliest beginnings.

Spellman, Frank R., and Nancy E. Whiting. *Environmental Science and Technology: Concepts and Applications.* 2d ed. Lanham, Md.: Government Institutes, 2006. Provides background basic science and engineering science information as well as an introduction to the different areas of environmental engineering.

Vesilind, P. Aarne, Susan M. Morgan, and Lauren G. Heine. *Introduction to Environmental Engineering.* 3d ed. Stamford, Conn.: Cengage Learning, 2010. A holistic approach to solving environmental problems with two unifying themes—material balances and environmental ethics.

WEB SITES
American Academy of Environmental Engineers
http://www.aaee.net

American Society of Civil Engineers
http://www.asce.org

U.S. Environmental Protection Agency
http://www.epa.gov

See also: Civil Engineering.

F

FIBER-OPTIC COMMUNICATIONS

FIELDS OF STUDY

Physics; chemistry; solid-state physics; electrical engineering; electromechanical engineering; mechanical engineering; materials science and engineering; telecommunications; computer programming; broadcast technology; information technology; electronics; computer networking; mathematics; network security.

SUMMARY

The field of fiber optics focuses on the transmission of signals made of light through fibers made of glass, plastic, or other transparent materials. The field includes the technology used to create optic fibers as well as modern applications such as telephone networks, computer networks, and cable television. Fiber optics are used in almost every part of daily life in technologies such as fax machines, cell phones, television, computers, and the Internet.

KEY TERMS AND CONCEPTS

- **attenuation:** Loss of light power as the signal travels through fiber-optic cable.
- **bandwidth:** Range of frequencies within which a fiber-optic transmitting device can transmit data or information.
- **broadband:** Telecommunications signal with a larger-than-usual bandwidth.
- **dispersion:** Spreading of light-signal pulses as they travel through fiber-optic cable.
- **endoscope:** Fiber optic medical device that is used to see inside the human body without surgery.
- **fiber-optic cable:** Cable consisting of numerous fiber-optic fibers lined with a reflective core medium to direct light.
- **fiberscope:** First device that was able to transmit images over a glass fiber.
- **light-emitting diodes (LED):** Light source some-times used in fiber-optic systems to transmit data.
- **receiver:** In fiber-optics systems, a device that captures the signals transmitted through the fiber-optic cable and then translates the light back into electronic data.
- **semiconductor laser:** Laser sometimes used in fiber-optic systems to transmit data.
- **transmitter:** In fiber optic systems, a device that codes electronic data into light signals.

DEFINITION AND BASIC PRINCIPLES
The field of fiber optics focuses on the transmission of signals made of light through fibers made of glass, plastic, or other transparent media. Using the principles of reflection, optical fibers transmit images, data, or voices and provide communications links for a variety of applications such as telephone networks, computer networks, and cable television.

BACKGROUND AND HISTORY
The modern field of fiber optics developed from a series of important scientific discoveries, principles, technologies, and applications. Early work in use of light as a signal by French engineer Claude Chappe, British physicist John Tyndall, Scottish physicist Alexander Graham Bell, and American engineer William Wheeler in the eighteenth and nineteenth centuries laid the foundation for harnessing light through conductible materials such as glass. These experiments also served as proof of the concept that sound could be transmitted as light. The failures in the inventions indicated further areas of work before use in practical applications: The main available light source was the Sun, and the light signal was reduced by travel through the conductible substance. For example, in 1880 Bell created a light-based system of sound transmission or photophone that was abandoned for being too affected by the interruption of the light transmission beam. In the 1920's, the transmission of facsimiles (faxes) or television images through light signals via glass or plastic rods or pipes

was patented by Scottish inventor John Logie Baird and American engineer Clarence Hansell. The fiber-scope, developed in the 1950's, was able to transmit low- resolution images of metal welds over a glass fiber. In the mid1950's, Dutch scientist Abraham Van Heel reported a method of gathering fibers into bundles and coating them in a clear coating or cladding that decreased interference between the fibers and reduced distortion effects from the outside. In 1966, English engineer George Hockham and Chinese physicist Charles Kao published a theoretical method designed to dramatically decrease the amount of light lost as it traveled through glass fibers. By 1970, scientists at Corning Glass Works created fibers that actualized Hockham and Kao's theoretical method. In the mid-1970's, the first telephone systems using fiber optics were piloted in Atlanta and Chicago. By 1984, other major cities on the Eastern seaboard were connected by AT&T's fiber-optic systems. In 1988, the first transatlantic fiber-optic cable connected the United States to England and France.

By the late 1980's, fiber-optic technology was in use for such medical applications as the gastroscope, which allowed doctors to look inside a patient and see the image transmitted along the fibers. However, more work was still needed to allow effective and accurate transmission of electronic data for computer work.

In the mid-twentieth century, the use of fiber optics accelerated in number of applications and technological advances. Scientists found a way to create a glass fiber coated in such a way that the light transmitted moved forward at full strength and signal. Coupled with the development of the semiconductor laser, which could emit a high-powered, yet cool and energy-efficient, targeted stream of light, fiber optics quickly became integrated into existing and new technology associated with computer networking, cable television, telephone networks, and other industry applications that benefited from high-speed and long-distance data transfer.

HOW IT WORKS

The major elements required for fiber-optics transmission include: long flexible fibers made of transparent materials such as glass, plastic, or plastic-clad silica; a light-transmittal source such as a laser of light-emitting diode (LED); cables or rods lined with a reflective core medium to direct light; and a receiver to capture the signal. Many systems also include a signal amplifier or optoelectronic repeater to increase the transmission distance of a signal. Electronic data is coded into light signals using the transmitter. The light signals then move down the fibers bouncing off the reflective core of the fibers to the receiver. The receiver captures the signals and then translates the light back into electronic data. This process is used to transmit data in the form of images, sound, or other signals down the rods at the speed of light.

APPLICATIONS AND PRODUCTS

Information transmittal. Fiber-optics technology revolutionized the ability to transfer data between computers. Networked computers share and distribute information via a main computer (a server) and its connected computers (nodes). The use of fiber optics exponentially increases the data-transmission speed and ability of computers to communicate. In addition, fiber-optics data transfer is more secure than lines affected by magnetic interference. Industries that use information and data transmission through networks include banking, communications, cable television, and telecommunications. Fiber-optic information transmission has advantages over copper-cable transmission in that it is relatively easy to install, is lighter weight, very durable, can transmit for long distances at a higher bandwidth, and is not influenced by electromagnetic disruptions such as lightning or fluorescent lighting fixture transformers.

Modern communications. The use of fiber-optics technology in telephone communication has increased the capacity, ease, and speed of standard copper-wired phones. The quality of voices over the phone is improved, as the sound signal is no longer distorted by distance or is subject to time delay. Fiber-optic lines are not affected by electromagnetic interference and are less subject to security breaches related to unauthorized access to phone calls and data transfer over phone lines. Additionally, fiber-optic cables are less expensive and easier to install than copper wire or coaxial cables, and since the 1980's they have been installed in many areas. Fiber-optic cabling can be used to provide high-speed Internet access, cable television, and regular telephone service over one line. In addition to traditional phone lines and home-based services, fiber-optic links between mobile towers and networks also allow the use of smart phones, which can be used to send and receive e-mails, surf the Internet, and have device-specific applications such as Global Positioning Systems.

Manufacturing. The increased globalization of the manufacturing of goods requires information, images, and data to be transmitted quickly from one location to another (known as point-to-point connections). For example, a car may be assembled in Detroit, but one part may be made in Mexico, another in Taiwan, and a third in Alabama. The logistics to make sure all the parts are of appropriate quality and quantity to be shipped to Detroit for assembly are coordinated through networked computer systems and fiber-optic telephone lines. In addition, the ability to use fiber optics to capture and transmit images down a very small cable allows quality-control personnel to "see" inside areas that the human eye cannot. As an example, a fiberscope can be used to inspect a jet engine's welding work within combustion chambers and reactor vessels.

The I=nternet. According to the United Nations' International Telecommunication Union (ITU), the number of Internet users across the world met the two billion mark in January 2011. Much like standard telephone service, the capacity, ease, and speed to the Internet has been greatly increased by the replacement of phone-based modem systems, cable modems, and digital subscriber line (DSL) by fiber-optics wired systems. Although fiber-optic connections directly to homes in the United States are not available in all areas, some companies use fiber-optic systems down major networking lines and then split to traditional copper wiring for houses.

Medicine. Fiber optics have significantly altered medical practice by allowing physicians to see and work within the human body using natural or small surgical openings. The fiberscopes or endoscopes are fiber-optics-based instruments that can image and illuminate internal organs and tissues deep within the human body. A surgeon is able to visualize an area of concern without performing large-scale exploratory surgery. In addition to viewing internal body surfaces, laproscopic surgery using fiber-optic visualization allows the creation of very small cuts to target and perform surgery reducing overall surgical risks and recovery time in many cases. Beyond the use of endoscopes, fiber-optic technology has been used to update standard medical equipment so that it may be used in devices that emit electromagnetic fields. As an example, companies have developed a fiber-optic

pulse oximeter to be used to measure heart rate and oxygen saturation during magnetic resonance imaging (MRI).

Broadcast industry. The broadcast industry has moved much of its infrastructure to fiber-optics technology. This change has also allowed the creation and transmission of television signals with increased clarity and picture definition known as high-definition television (HDTV). The use of fiber optics and its increased data-transmission ability was key in 2009, when all television stations changed from analogue to digital signals for their content broadcasts.

Military. The military began using fiber optics as a reliable method of communications early in the development of the technology. This quick implementation was due to recognition that fiber optics cables were able to withstand demanding conditions and temperature extremes while still transferring information accurately and quickly. Programs such as the Air Force's Airborne Light Optical Fiber Technology (ALOFT) program helped move fiber-optic technology along even as it served as proof of concept: fiber-optic signal transmission could transmit data reliably even in outer space. Beyond communications, the military uses fiber-optic gyroscopes (FOGs) in navigation systems to direct guided missiles accurately. Additionally, fiber optics have been used to increase the accuracy of rifle-bullet targeting by using sensitive laser-based fiber-optic sensors that adjust crosshairs on the scope based on the precise measurement of the barrel's deflection.

Traffic control. According to the United States Department of Transportation, traffic signals that are not synchronized result in nearly 10 percent of all traffic delays and waste nearly 300 million vehicle-hours nationwide each year. Fiber optics has been used as part of intelligent transport systems to help coordinate traffic signals and improve the flow of cars via real-time monitoring of congestion, accidents, and traffic flow. Beyond traffic congestion, some cities capture data on cars running red lights, paying tolls, and the license plates moving through toll roads, tunnels, and bridges.

IMPACT ON INDUSTRY

The total value of the fiber-optic communications industry is difficult to estimate as different aspects of the industry are divided and captured under different financial projections. For example, according to market-research firm BCC Research, the total estimated global market value for fiber-optic connectors in 2010 was an estimated $1.9 billion with an annual average growth rate of 9.6 percent. These projections are different from the total estimated global market value for fiber-optic circulators, which, as of 2011, is forecast to increase annually at 14.29 percent according to technology-forecasting firm ElectroniCast. Further projections are divided into categories such as medical fiber optics, glass and fiber manufacturing, and fiber optic components.

The fiber-optics industry has a global presence; however, the United States is a major player, dominating the optical cable and fiber markets internationally. According to market-research firm First Research, the U.S. glass and fiber-optic manufacturing industry includes about 2,000 companies with combined annual revenue of $20 billion. The majority of the market is concentrated on companies such as Corning and PPG Industries, and 80 percent of the market is captured by the fifty largest companies. Japan and the countries of the European Union are also instrumental in the fiber-optic communications industries as both manufacturers and consumers.

Government and university research. The United States federal government funds fiber-optics research through branches such as the Department of Defense (DOD). One main category of basic research funded through the DOD includes lasers and fiber optics in communications and medicine. This research funding encourages joint ventures between universities and corporations such as the Lockheed Martin-Michigan Technological University project to develop a fiber-optic-based circuit board manufacturing process.

Industry and business. Research and development programs in the industry and business sector continue to search for the next technological innovation in fiber-optics access and technology. Upcoming updates in speed and improved processing from industry will soon be seen in the next generation of fiber-optic cables. In addition, businesses are working to connect phone services, the Internet, and cable directly to an increasing number of private homes via fiber-optic cables.

Military. The military continues to fund research and implement new fiber-optics technologies related to several applications. Monitoring systems based on

fiber optics are being developed to detect chemical weapons, explosives, or biohazardous substances based on a specific wavelength emitted by the substance or device. The military has also integrated use of smart phones into its operation activities with applications such as BulletFlight, which helps snipers determine the most effective angle from which to fire and input data to account for changes based on altitude and weather conditions.

CAREERS AND COURSE WORK

There are many careers in the fiber-optics industry and entry-level requirements vary significantly by position. Given the wide spectrum of difference between the careers, a sampling of careers and course work follows.

Professional, management, and sales occupations generally require a bachelor's degree. Technical occupations often require specific course work but not necessarily a bachelor's degree. However, it is easier to obtain employment and gain promotions with a degree, especially in larger, more competitive markets. Advanced schooling usually is required for supervisory positions—including technical occupations—which have greater responsibility and higher salaries. These positions comprise about 19 percent of the fiber-optics communications industry careers.

Engineering roles in the fiber-optics industry range from cable logistics and installation planning to research and development positions in fiber optics and lasers. Positions may be found in universities, corporations, and the military. Engineers may specialize in a particular area of fiber optics such as communication systems, telecommunications design, or computer network integration with fiber-optic technology. Education requirements for entry-level positions begin with a bachelor's degree in engineering, computer science, or a related field.

Telecommunications equipment installers and repairers usually acquire their skills through formal training at technical schools or college, where they major in electronics, communications technology, or computer science. Military experience in the field, on-the-job training with a software manufacturer, or prior work as a telecommunications line installer may also provide entry into more complicated or complex positions.

Optics physicists work in the fiber-optics industry in research and development. The role of the optics physicist is to develop solutions to fiber-optics communications quandaries using the laws of physics. Most optics physicists have a doctorate in physics, usually with a specialization in optics. They also tend to spend several years after obtaining their doctorate performing academic research before moving to industry positions.

Specialized roles in computer software engineering and networking in the fiber-optic telecommunications industry also exist. Much like the engineering roles, individuals may specialize in a particular area of fiber optics such as networking, communication systems, telecommunications design, data communications, or computer software. Education requirements for entry-level positions begin with a bachelor's degree with a major in engineering, computer science, or a related field.

SOCIAL CONTEXT AND FUTURE PROSPECTS

Fiber-optic communications technologies are constantly changing and integrating new innovations and applications. Some countries, such as Japan, have fully embraced use of fiber optics in the home as well as in business; however, the investment in infrastructure is not as fully actualized in other areas. The consumer demand for faster, better access to the Internet and related data-transmittal applications is driving the move from standard copper wiring to fiber optics. New types of fibers will increase fiber-optic application beyond telecommunications into more medical, military, and industrial uses. Though wireless technology use could negatively check industry growth, the strong consumer demand and increasing number of fiber-optics applications suggest that the fiber-optic industry will continue to grow. However, the industry may have more moderate growth as the telecommunication industry experiences decreased growth. This was seen during the economic recession of 2008 to 2010, as consumers held off upgrading from copper cabling to fiber optics. According to a report by Global Industry Analysts, the recession's impact on fiber-optic cabling ended in 2011. Overall, the report anticipates significant growth in the industry as more fiber-optic cable networks are installed and businesses, consumers, and telecom providers invest in advanced tools to facilitate the new networks. Employment in the wired telecommunications industry is expected to decline by 11 percent during the period from 2008 to 2018; however, telecommunications jobs focused on fiber optics are expected to rise.

—*Dawn A. Laney, MS, CGC, CCRC*

FURTHER READING

Allen, Thomas B. "The Future Is Calling," *National Geographic* 200, Issue 6 (December 2001): 76. An interesting and well-written description of the growth of the fiber-optics industry.

Belson, Ken. "Unlike U.S., Japanese Push Fiber Over Profit." *The New York Times.* October 3, 2007. Compares the United States' and Japan's approach to updating infrastructure with fiber-optic cabling.

Crisp, John, and Barry Elliott. *Introduction to Fiber Optics.* 3d ed. Burlington, Mass.: Elsevier, 2005. An excellent text for anyone, of any skill level, who wants to learn more about fiber optics from the ground up. Each chapter ends with review questions.

Goff, David R. *Fiber Optic Reference Guide: A Practical Guide to Communications Technology.* 3d ed. Burlington, Mass.: Focal Press, 2002. An excellent review of the history of fiber optics, the basic principles of the technology, and information on practical applications particularly in communications.

Hayes, Jim. *FOA Reference Guide to Fiber Optics: Study Guide to FOA Certification.* Fallbrook, Calif.: The

---

FASCINATING FACTS ABOUT FIBER-OPTIC COMMUNICATIONS

- As of 2011, all new undersea cables are made of optical fibers.
- Airplanes use fiber-optic cabling in order to keep overall weight down and increase available capacity.
- Fiber optics were an integral part of the television cameras sent to film the first Moon walk in 1969.
- Industry analysts predict that sometime in the early twenty-first century 98 percent of copper wire will have been replaced by fiber-optic cable.
- A fiber-optic fiber is thinner than a human hair.
- A fiber-optics system is capable of transmitting more than the equivalent of a twenty-four-volume encyclopedia worth of information in one second.

Fiber Optic Association, 2009. A useful guide that details the design and installation of fiber optic cabling networks including expansive coverage of the components and processes of fiber optics.

Hecht, Jeff. *City of Light: The Story of Fiber Optics.* Rev. ed. New York: Oxford University Press, 1999. A readable history of the development of fiber optics.

WEB SITES

Fiber Optic Association
http://www.thefoa.org

International Telecommunication Union (ITU)
http://www.itu.int

U.S. Bureau of Labor Statistics
Career Guide to Industries: Telecommunications
http://www.bls.gov/oco/cg/cgs020.htm

See also: Computer Engineering; Electrical Engineering; Mechanical Engineering; Telephone Technology and Networks.

FIBER TECHNOLOGIES

FIELDS OF STUDY

Chemistry; physics; mathematics; chemical engineering; polymer chemistry; agriculture; agronomy; mechanical engineering; industrial management; waste management; business management

SUMMARY

Fibers have been used for thousands of years, but not until the nineteenth and twentieth centuries did chemically modified natural fibers (cellulose) and synthetic plastic or polymer fibers become extremely important, opening new fields of application. Advanced composite materials rely exclusively on synthetic fibers. Research has also produced new applications of natural materials such as glass and basalt in the form of fibers. The current "king" among fibers is carbon, and new forms of carbon, such as carbon nanotubes, promise to advance fiber technology even further.

KEY TERMS AND CONCEPTS

- **denier:** A unit indicating the fineness of a filament; a filament 9,000 meters in length weighing 1 gram has a fineness of 1 denier.
- **roving:** Nonwoven fiber fabric whose strands all have the same absolute orientation.
- **warp clock:** A visual guide to the orientation of the warp of woven fiber fabrics, used in the laying-up of composite materials to provide a quasi-isotropic

character to the final product.

DEFINITION AND BASIC PRINCIPLES

A fiber is a long, thin filament of a material. Fiber technologies are used to produce fibers from different materials that are either obtained from natural sources or produced synthetically. Natural fibers are either cellulose-based or protein-based, depending on their source. All cellulosic fibers come from plant sources, while protein-based fibers such as silk and wool are exclusively from animal sources; both fiber types are referred to as biopolymers. Synthetic fibers are manufactured from synthetic polymers, such as nylon, rayon, polyaramides, and polyesters. An infinite variety of synthetic materials can be used for the production of synthetic fibers.

Production typically consists of drawing a melted material through an orifice in such a way that it solidifies as it leaves the orifice, producing a single long strand or fiber. Any material that can be made to melt can be used in this way to produce fibers. There are also other ways in which specialty fibers also can be produced through chemical vapor deposition. Fibers are subsequently used in different ways, according to the characteristics of the material.

BACKGROUND AND HISTORY

Some of the earliest known applications of fibers date back to the ancient Egyptian and Babylonian civilizations. Papyrus was formed from the fibers of the papyrus reed. Linen fabrics were woven from flax fibers. Cotton fibers were used to make sail fabric. Ancient

China produced the first paper from cellulose fiber and perfected the use of silk fiber.

Until the nineteenth century, all fibers came from natural sources. In the late nineteenth century, nitrocellulose was first used to develop smokeless gunpowder; it also became the first commercially successful plastic: celluloid.

As polymer science developed in the twentieth century, new and entirely synthetic materials were discovered that could be formed into fine fibers. Nylon-66 was invented in 1935 and Teflon in 1938. Following World War II, the plastics industry grew rapidly as new materials and uses were invented. The immense variety of polymer formulations provides an almost limitless array of materials, each with its own unique characteristics. The principal fibers used today are varieties of nylons, polyesters, polyamides, and epoxies that are capable of being produced in fiber form. In addition, large quantities of carbon and glass fibers are used in an ever-growing variety of functions.

HOW IT WORKS

The formation of fibers from natural or synthetic materials depends on some specific factors. A material must have the correct plastic characteristics that allow it to be formed into fibers. Without exception, all natural plant fibers are cellulose-based, and all fibers from animal sources are protein-based. In some cases, the fibers can be used just as they are taken from their source, but the vast majority of natural fibers must be subjected to chemical and physical treatment processes to improve their properties.

Cellulose fibers. Cellulose fibers provide the greatest natural variety of fiber forms and types. Cellulose is a biopolymer; its individual molecules are constructed of thousands of molecules of glucose chemically bonded in a head-to-tail manner. Polymers in general are mixtures of many similar compounds that differ only in the number of monomer units from which they are constructed. The processes used to make natural and synthetic polymers produce similar molecules having a range of molecular weights. Physical and chemical manipulation of the bulk cellulose material, as in the production of rayon, is designed to provide a consistent form of the material that can then be formed into long filaments, or fibers.

Synthetic polymers. Synthetic polymers have greatly expanded the range of fiber materials that are available, and the range of uses to which they can be applied. Synthetic polymers come in two varieties: thermoplastic and thermosetting. Thermoplastic polymers are those whose material becomes softer and eventually melts when heated. Thermosetting polymers are those whose the material sets and becomes hard or brittle through heating. It is possible to use both types of polymers to produce fibers, although thermoplastics are most commonly used for fiber production.

The process for both synthetic fibers is essentially the same, but with reversed logic. Fibers from thermoplastic polymers are produced by drawing the liquefied material through dies with orifices of the desired size. The material enters the die as a viscous liquid that is cooled and solidifies as it exits the die. The now-solid filament is then pulled from the die, drawing more molten material along as a continuous fiber. This is a simpler and more easily controlled method than forcing the liquid material through the die using pressure, and it produces highly consistent fibers with predictable properties.

Fibers from thermosetting polymers are formed in a similar manner, as the unpolymerized material is forced through the die. Rather than cooling, however, the material is heated as it exits the die to drive the polymerization to completion and to set the polymer. Other materials are used to produce fibers in the manner used to produce fibers from thermoplastic polymers. Metal fibers were the first of these materials. The processes used for their production provided the basic technology for the production of fibers from polymers and other nonmetals. The best-known of these fibers is glass fiber, which is used with polymer resins to form composite materials. A somewhat more high-tech variety of glass fiber is used in fiber optics for high-speed communications networks. Basalt fiber has also been developed for use in composite materials. Both are available commercially in a variety of dimensions and forms.

Production of carbon fiber begins with fibers already formed from a carbon-based material, referred to as either pitch or PAN. Pitch is a blend of polymeric substances from tars, while PAN indicates that the carbon-based starting material is polyacrylonitrile. These starting fibers are then heat-treated in such a way that essentially all other atoms in the material are driven off, leaving the carbon skeletons of the original polymeric material as the end-product fiber.

Boron fiber is produced by passing a very thin filament of tungsten through a sealed chamber, during which the element boron is deposited onto the tungsten fiber by the process of chemical vapor deposition.

APPLICATIONS AND PRODUCTS

All fiber applications derive from the intrinsic nature of the material from which the fibers are formed. Each material, and each molecular variation of a material, produces fibers with unique characteristics and properties, even though the basic molecular formulas of different materials are very similar. As well, the physical structure of the fibers and the manner in which they were processed work to determine the properties of those fibers. The diameter of the fibers is a very important consideration. Other considerations are the temperature of the melt from which fibers of a material were drawn; whether the fibers were stretched or not, and the degree by which they were stretched; whether the fibers are hollow, filled, or solid; and the resistance of the fiber material to such environmental influences as exposure to light and other materials.

Structural fibers. Loosely defined, all fibers are structural fibers in that they are used to form various structures, from plain weave cloth for clothing to advanced composite materials for high-tech applications. That they must resist physical loading is the common feature identifying them as structural fibers. In a stricter sense, structural fibers are fibers (materials such as glass, carbon, aramid, basalt, and boron) that are ordinarily used for construction purposes. They are used in normal and advanced composite materials to provide the fundamental load-bearing strength of the structure.

A typical application involves "laying-up" a structure of several layers of the fiber material, each with its own orientation, and encasing it within a rigid matrix of polymeric resin or other solidifying material. The solid matrix maintains the proper orientation of the encased fibers to maintain the intrinsic strength of the structure. Materials so formed have many structural applications. Glass fiber, for example, is commonly used to construct different fiberglass shapes, from flower pots to boat hulls, and is the most familiar of composite fiber materials. Glass fiber is also used in the construction of modern aircraft,

such as the Airbus A-380, whose fuselage panels are composite structures of glass fibers embedded in a matrix of aluminum metal.

Carbon and aramid fibers such as Kevlar are used for high-strength structures. Their strength is such that the application of a layer of carbon fiber composite is frequently used to prolong the usable lifetime of weakened concrete structures, such as bridge pillars and structural joists, by several years. While very light, Kevlar is so strong that high-performance automotive drive trains can be constructed from it. It is the material of choice for the construction of modern high-performance military and civilian aircraft, and for the remote manipulators that were used aboard the space shuttles of the National Aeronautics and Space Administration. Kevlar is recognizable as the high stretch-resistance cord used to reinforce vehicle tires of all kinds and as the material that provides the impact-resistance of bulletproof vests.

In fiber structural applications, as with all material applications, it is important to understand the manner in which one material can interact with another. Allowing carbon fiber to form a galvanic connection to another structural component such as aluminum, for example, can result in damage to the overall structure caused by the electrical current that naturally results.

Fabrics and textiles. The single most recognized application of fiber technologies is in the manufacture of textiles and fabrics. Textiles and fabrics are produced by interweaving strands of fibers consisting of single long fibers or of a number of fibers that have been spun together to form a single strand. There is no limit to the number of types of fibers that can be combined to form strands, or on the number of types of strands that can be combined in a weave.

The fiber manufacturing processes used with any individual material can be adjusted or altered to produce a range of fiber textures, including those that are soft and spongy or hard and resilient. The range of chemical compositions for any individual polymeric material, natural or synthetic, and the range of available processing options, provides a variety of properties that affect the application of fabrics and textiles produced.

Clothing and clothing design consume great quantities of fabrics and textiles. Also, clothing designers seek to find and utilize basic differences in fabric and textile properties that derive from variations in

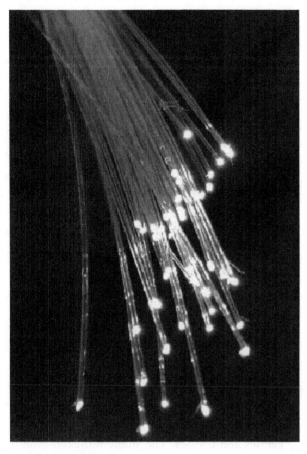

A bundle of optical fibers

chemical composition and fiber processing methods.

Fibers for fabrics and textiles are quantified in units of deniers. Because the diameter of the fiber can be produced on a continuous diameter scale, it is therefore possible to have an essentially infinite range of denier weights. The effective weight of a fiber may also be adjusted by the use of sizing materials added to fibers during processing to augment or improve their stiffness, strength, smoothness, or weight. The gradual loss of sizing from the fibers accounts for cotton denim jeans and other clothing items becoming suppler, less weighty, and more comfortable over time.

The high resistance of woven fabrics and textiles to physical loading makes them extremely valuable in many applications that do not relate to clothing. Sailcloth, whether from heavy cotton canvas or light nylon fabric, is more than sufficiently strong to move the entire mass of a large ship through water by resisting the force of wind pressing against the sails. Utility covers made from woven polypropylene strands are also a common consumer item, though used more for their water-repellent properties than for their strength. Sacks made from woven materials are used worldwide to carry quantities of goods ranging from coffee beans to gold coins and bullion. One reason for this latter use is that the fiber fabric can at some point be completely burned away to permit recovery of miniscule flakes of gold that chip off during handling.

Cordage. Ropes, cords, and strings in many weights and winds traditionally have been made from natural fibers such as cotton, hemp, sisal, and manila. These require little processing for rough cordage, but the suppleness of the cordage product increases with additional processing. Typically, many small fibers are combined to produce strands of the desired size, and these larger strands can then be entwined or plaited to produce cordage of larger sizes. The accumulated strength of the small fibers produces cordage that is stronger than cordage of the same size consisting of a single strand. The same concept is applied to cordage made from synthetic fibers.

Ropes and cords made from polypropylene can be produced as a single strand. However, the properties of such cordage would reflect the properties of the bulk material rather than the properties of combined small fibers. It would become brittle when cold, overly stretchy when warm, and subject to failure by impact shock. Combined fibers, although still subject to the effects of heat, cold, and impact shock, overcome many of these properties as the individual fibers act to support each other and provide superior resistance.

IMPACT ON INDUSTRY

Industries based on fiber technologies can be divided into two sectors: those that produce fibers and those that consume fibers; both are multibillion dollar sectors. Fiber production industries are agricultural and technical. Forestry and other agricultural industries produce large amounts of cellulosic fiber each year. Production of cellulosic fiber peaked at some 3 million metric tons in 1982, representing 21 percent of the fiber market share, but this heavy production has steadily declined since 1982. By 2002, cellulosic fiber production had decreased to 6 percent of the world fiber-market share. Production in Eastern Europe dropped from 1.1 million metric tons to just 92,000 metric tons, and production in Asia increased by 660,000 metric tons over the same period, accounting for fully 69 percent of global production in 2002. Most of the cellulosic fiber that is produced by the forestry industry is used in the manufacture of paper and paper products, while the other cellulosic fiber types—cotton, hemp, sisal, and manila—are used primarily for fabrics, textiles, and cordage.

Synthetic fiber production is the principal driving force behind the decline in cellulosic fiber production. Industrial polymerization processes provide a much greater variety of fiber-forming materials with a lower requirement for process control. In the period from 1982 to 2002, manufactured fiber production increased by 155 percent, with the rise attributed to synthetic fiber as the production of cellulosic fiber decreased. Fibers from synthetic materials account for 94 percent of total global fiber production.

A great deal of research has been expended in the field of polymerization chemistry to identify effective catalysts and processes for the reactions involved, so it has become ever easier to control the nature of the materials being produced. Polymerization reactions have become controllable such that their products now span a much narrower range of monomer weights. This specificity of control has enabled the production of more specific fiber materials, which has in turn driven research and development of more specific uses of those fibers.

The greatest increase in synthetic fiber materials has been an increase in the polyesters, which now account for almost two-thirds of total synthetic fiber production. It must be remembered that terms such as "polyester" and "polyamide" refer to broad classes of compounds and not to specific materials. Each class contains untold thousands of possible variations in molecular structure, both from the chemical identity of the monomers used and from the order in which they react during polymerization.

Technical industries that produce fibers from both natural and synthetic materials make use of several classes of machinery. One class of machinery (for example, pelletizers and masticators) manipulates raw materials into a form that is readily transported and usable in the fiber-forming process. Another class of machinery (for example, injection molders and spinnerets) carries out the fiber-forming process, while

still another class of machinery (for example, sizers and mercerizers) uses the raw fiber to provide a finished fiber product ready for consumers.

Industries that consume finished fibers include weaving mills that produce fabrics and textiles from all manner of fibers, including glass, basalt, and carbon. These fibers are then used accordingly, for clothing, composites, or other uses. Cordage manufacturers use finished fibers as an input feedstock to produce every type of cordage, from fine thread to coarse rope.

A more recent development in the fiber industry is the utilization of recycling wastes as feedstock for fiber-forming processes. Synthetic materials such as nylon, polyethylene, and polyethylene terephalate (PET), which have historically been condemned to landfills, are now accepted for reprocessing through recycling programs and formed into useful fibers for many different purposes. The versatility of the materials themselves lends to the development of new industries based on their use.

FASCINATING FACTS ABOUT FIBER TECHNOLOGIES

- In 1845, German-Swiss chemist C. F. Schonbein accidentally discovered the explosive properties of nitrocellulose fibers when he used his wife's cotton apron to mop up some nitric acid, then hung the apron by a stove to dry.
- Nitrocellulose became the first commercially successful plastic as celluloid, the basis of the photographic film industry.
- French chemist and microbiologist Louis Pasteur and French engineer Count Hilaire de Chardonnet tried to develop a synthetic alternative for silk. The cellulose fiber rayon was the result, patented by Chardonnet in 1885.
- Any material that can be melted, even the volcanic lava known as basalt, can be formed into a usable fiber.
- Many materials, such as PET bottles, which have long been resigned to landfills after a single use, are now being recycled as feedstock for synthetic fiber production.
- A fiber with a textile weight of 1 denier is 9,000 meters long, but weighs only 1 gram.
- The synthetic fiber industry has an economic value of about US$100 billion annually, worldwide.

CAREERS AND COURSEWORK

Careers in textile production depend on a sound basic education in chemistry, physics, mathematics, and materials science. Students anticipating such a career should be prepared to take advanced courses in these areas at the college and university level, with specialization in organic chemistry, polymer chemistry, and industrial chemistry. Postsecondary programs specializing in textiles and the textile industry also are available, and provide the specialist training necessary for a career in the manufacture and use of fibers and textiles. The chemistry of color and dyes is another important aspect of the field, representing a distinct area of specialization.

Composite-materials training and advanced composite-materials specialist training can be obtained through only a limited number of public and private training facilities, although considerable research in this field is carried out in a number of universities and colleges. Private industries that require this kind of specialization, particularly aircraft manufacturers, often have their own patented fabrication processes and so prefer to train their personnel on-site rather than through outside agencies.

SOCIAL CONTEXT AND FUTURE PROSPECTS

One could argue that the fiber industry is the principal industry of modern society, solely on the basis that everyone wears clothes of some kind that have been made from natural or synthetic fibers. As this is unlikely ever to change, given the climatic conditions that prevail on this planet and given the need for protective outerwear in any environment, there is every likelihood that there will always be a need for specialists who are proficient in both fiber manufacturing and fiber utilization.

—Richard M. Renneboog, MSc

FURTHER READING

Fenichell, Stephen. *Plastic: The Making of a Synthetic Century.* New York: HarperCollins, 1996. A well-researched account of the plastics industry, focusing on the social and historical contexts of plastics, their technical development, and the many uses for synthetic fibers.

Morrison, Robert Thornton, and Robert Nielson Boyd. *Organic Chemistry..* 5th ed. Newton, Mass.: Allyn & Bacon, 1987. Provides one of the best and most readable introductions to organic chemistry and polymerization.

Selinger, Ben. *Chemistry in the Marketplace.* 5th ed. Sydney: Allen & Unwin, 2002. The seventh chapter of this book provides a concise overview of many fiber materials and their common uses and properties.

Weinberger, Charles B. "Instructional Module on Synthetic Fiber Manufacturing." Gateway Engineering Education Coalition: 30 Aug. 1996. This article presents an introduction to the chemical engineering of synthetic fiber production, giving an idea of the sort of training and specialization required for careers in this field.

WEB SITES

University of Tennessee Space Institute
http://www.utsi.edu/research/carbonfiber

U.S. Environmental Protection Agency
http://www.epa.gov

See also: Fiber-Optic Communications.

FLUID DYNAMICS

FIELDS OF STUDY

Physics; mathematics; engineering; chemistry; suspension mechanics; hydrodynamics; computational fluid dynamics; microfluidic systems; coating flows; multiphase flows; viscous flows.

SUMMARY

Fluid dynamics is an interdisciplinary field concerned with the behavior of gases, air, and water in motion. An understanding of fluid dynamic principles is essential to the work done in aerodynamics. It informs the design of air and spacecraft. An understanding of fluid dynamic principles is also essential to the field of hydromechanics and the design of oceangoing vessels. Any system with air, gases, or water in motion incorporates the principles of fluid dynamics.

KEY TERMS AND CONCEPTS

- **aerodynamics:** Study of air in motion.
- **boundary layer:** Region between the wall of a flowing fluid and the point where the flow speed is nearly equal to that of the fluid.
- **continuum:** Continuous flow of fluid.
- **fluid:** State of matter in which a substance cannot maintain a shape on its own.
- **hydrodynamics:** Study of water in motion.
- **ideal fluids:** Fluids without any internal friction (viscosity).
- **incompressible flows:** Those in which density does not change when pressure is applied.
- **inviscid fluid:** Fluid without viscosity.
- **Newtonian fluids:** Fluids that quickly correct for shear strain.
- **shear strain:** Stress in a fluid that is parallel to the fluid motion velocity or streamline.
- **streamline:** Manner in which a fluid flows in a continuum with unbroken continuity.
- **viscosity:** Internal friction in a fluid.

DEFINITION AND BASIC PRINCIPLES

Fluid dynamics is the study of fluids in motion. Air, gases, and water are all considered to be fluids. When the fluid is air, this branch of science is called aerodynamics. When the fluid is water, it is called hydrodynamics.

The basic principles of fluid dynamics state that fluids are a state of matter in which a substance cannot maintain an independent shape. The fluid will take the shape of its container, forming an observable surface at the highest level of the fluid when it does not completely fill the container. Fluids flow in a continuum, with no breaks or gaps in the flow. They are said to flow in a streamline, with a series of particles following one another in an orderly fashion in parallel with other streamlines. Real fluids have some amount of internal friction, known as viscosity. Viscosity is the phenomenon that causes some fluids to flow more readily than others. It is the reason that molasses flows more slowly than water at room temperature.

Fluids are said to be compressible or incompressible. Water is an incompressible fluid because its density does not change when pressure is applied. Incompressible fluids are subject to the law of continuity, which states that fluid flows in a pipe are constant. This theory explains why the rate of flow increases when the area of the pipe is reduced and vice versa. The viscosity of a fluid is an important consideration when calculating the total resistance on an object.

The point where the fluid flows at the surface of an object is called the boundary layer. The fluid "sticks" to the object, not moving at all at the point of contact. The streamlines further from the surface are moving, but each is impeded by the streamline between it and the wall until the effect of the streamline closest to the wall is no longer a factor. The boundary layer is not obvious to the casual observer, but it is an important consideration in any calculations of fluid dynamics.

Most fluids are Newtonian fluids. Newtonian fluids have a stress-strain relationship that is linear. This means that a fluid will flow around an object in its path and "come together" on the other side without a delay in time. Non-Newtonian fluids do not have a linear stress-strain relationship. When they encounter shear stress their recovery varies with the type of non-Newtonian fluid.

A main consideration in fluid dynamics is the amount of resistance encoutenered by an object moving moving through a fluid. Resistance, also known as drag, is made up of several components but all have in common that they occur at the point where the object meets the fluid. The area can be quite large as in the wetted surface of a ship, the portion of a ship that is below the waterline. For an airplane, the equivalent is the body of the plane as it moves through the air. The goal for those who work in the field of fluid dynamics is to understand the effects of fluid flows and minimize their effect on the object in question.

BACKGROUND AND HISTORY

Swiss mathematician Daniel Bernoulli introduced the term "hydrodynamics" with the publication of his book *Hydrodynamica* in 1738. The name referred to water in motion and gave the field of fluid dynamics its first name, but it was not the first time water in action had been noted and studied. Leonardo da Vinci made observations of water flows in a river and was the one who realized that water was an incompressible flow and that for an incompressible flow, V = constant. This law of continuity states that fluid flow in a pipe is constant. In the late 1600's, French physicist Edme Mariotte and Dutch mathematician Christiaan Huygens contributed the velocity-squared law to the science of fluid dynamics. They did not work together but they both reached the conclusion that resistance is not proportional to velocity; it is instead the square of the velocity.

Sir Isaac Newton put forth his three laws in the 1700's. These laws play a fundamental part in many branches of science, including fluid dynamics. In addition to the term hydrodynamics, Bernoulli's contribution to fluid dynamics was the realization that pressure decreases as velocity increases. This understanding is essential to the understanding of lift. Leonhard Euler, the father of fluid dynamics, is considered by many to be the preeminent mathematician of the eighteenth century. He is the one who derived what is today known as the Bernoulli equation from the work of Daniel Bernoulli. Euler also developed equations for inviscid flows. These equations were based on his own work and are still used for compressible and incompressible fluids.

The Navier-Stokes equations result from the work of French engineer Claude-Louis Navier and British physicist George Gabriel Stokes in the mid-nineteenth century. They did not work together, but their equations apply to incompressible flows. The Navier-Stokes equations are still used. At the end of the nineteenth century, Scottish engineer William John Macquorn Rankine changed the understanding of the way fluids flow with his streamline theory, which states that water flows in a steady current of parallel flows unless disrupted. This theory caused a fundamental shift in the field of ship design because it changed the popular understanding of resistance in oceangoing vessels. Laminar flow is measured today by use of the Reynolds number, developed by British engineer and physicist Osborne Reynolds in 1883. When the number is low, viscous forces dominate. When the number is high, turbulent flows are dominant.

American naval architect David Watson Taylor designed and operated the first experimental model basin in the United States at the start of the twentieth century. His seminal work, *The Speed and Power of Ships*, first published in 1910, is still read. Taylor played a role in the use of bulbous bows on vessels

of the navy. He also championed the use of airplanes that would be launched from naval craft underway in the ocean.

The principles of fluid dynamics took to the air in the eighteenth century with the work done by aviators such has the Montgolfier brothers and their hot-air balloons and French physicist Louis-Sébastien Lenormand's parachute. It was not until 1799 when English inventor Sir George Cayley designed the first airplane with an understanding of the roles of lift, drag, and propulsion, that aerodynamics came under scrutiny. Cayley's work was soon followed by the work of American engineer Octave Chanute. In 1875, he designed several biplane gliders, and with the publication of his book *Progress in Flying Machines* in 1894, he became internationally recognized as an aeronautics expert.

The Wright brothers are rightfully called the first aeronautical engineers because of the testing they did in their wind tunnel. By using balances to test a variety of different airfoil shapes, they were able to correctly predict the lift and drag of different wing shapes. This work enabled them to fly successfully at Kitty Hawk, North Carolina, on December 17, 1903.

German physicist Ludwig Prandtl identified the boundary layer in 1904. His work led him to be known as the father of modern aerodynamics. Russian scientist Konstantin Tsiolkovsky and American physicist Robert Goddard followed, and Goddard's first successful liquid propellant rocket launch in 1926 earned him the title of the father of modern rocketry.

All of the principles that applied to hydrodynamics—the study of water in motion—applied to aerodynamics: the study of air in motion. Together these principles comprise the field of fluid dynamics.

HOW IT WORKS

When an object moves through a fluid such as gas or water, it encounters resistance. How much resistance depends upon the amount of internal friction in the fluid (the viscosity) as well as the shape of the object. A torpedo, with its streamlined shape, will encounter less resistance than a two-by-four that is neither sanded nor varnished. A ship with a square bow will encounter more resistance than one with a bulbous bow and V shape. All of this is important because with greater resistance comes the need for greater power to cover a given distance. Since power requires a fuel source and a way to carry that fuel, a vessel that can travel with a lighter fuel load will be more efficient. Whether the design under consideration is for a tractor trailer, an automobile, an ocean liner, an airplane, a rocket, or a space shuttle, these basic considerations are of paramount importance in their design.

APPLICATIONS AND PRODUCTS

Fluid dynamics plays a part in the design of everything from automobiles to the space shuttle. Fluid dynamic principles are also used in medical research by bioengineers who want to know how a pacemaker will perform or what effect an implant or shunt will have on blood flow. Fire flows are also being studied to aid in the science of wildfire management. Until now the models have focused on heat transfer but new studies are looking at fire systems and their fluid dynamic properties. Sophisticated models are used to predict fluid flows before model testing is done. This lowers the cost of new designs and allows the people involved to gain a thorough understanding of the trade-off between size and power given a certain design and level of resistance.

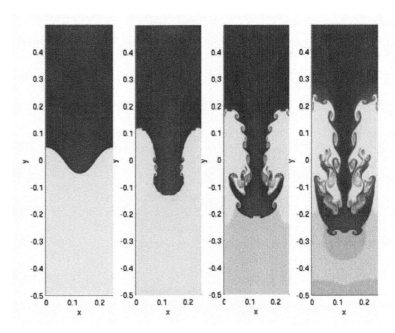

Hydrodynamics simulation of the Rayleigh–Taylor instability

IMPACT ON INDUSTRY

Before English engineer William Froude, ships were built based on what had worked and what should work. Once Froude proved that scale-model testing could reliably predict the performance of full-scale vessels, the entire process of vessel design was forever altered. The experience with scale models in a model basin transferred to the testing of scale models of airplanes and automobiles in wind tunnels. Computational fluid dynamic models are used for testing key elements of everything from ships to skyscrapers. The use of fluid dynamic theories to predict performance is ongoing and continues to be vital to engineers.

CAREERS AND COURSE WORK

Fluid dynamics plays a part in a host of careers. Naval architects use fluid dynamic principles to design vessels. Aeronautical engineers use the principles to design aircraft. Astronautical engineers use fluid dynamic principles to design spacecraft. Weapons are constructed with and understanding of fluids in motion. Automotive engineers must understand fluid dynamics to design fuel-efficient cars. Architects must take the motion of air into their design of skyscrapers and other large buildings. Bioengineers use fluid dynamic principles to their advantage in the design of components that will interact with blood flow in the human body. Land-management professionals can use their understanding of fluid flows to develop plans for protecting the areas under their care from catastrophic loss due to fires. Civil engineers take the principles of fluid dynamics into consideration when designing bridges. Fluid dynamics also plays a role in sports: from pitchers who want to improve their curveballs to quarterbacks who are determined to increase the accuracy of their passes.

Students should take substantial course work in more than one of the primary Fields of Study related to fluid dynamics (physics, mathematics, computer science, and engineering), because the fields that depend upon knowledge of fluid dynamic principles draw from multiple disciplines. In addition, anyone desiring to work in fluid dynamics should possess skills that go beyond the academic, including an aptitude for mechanical details and the ability to envision a problem in more than one dimension. A collaborative mind-set is also an asset, as fluid dynamic applications tend to be created by teams.

SOCIAL CONTEXT AND FUTURE PROSPECTS

The science of fluid dynamics touches upon a number of career fields that range from sports to bioengineering. Anything that moves through liquids such as air, water, or gases is subject to the principles of fluid dynamics. The more thorough the understanding, the more efficient vessel and other designs will be. This will result in the use of fewer resources in the form of power for inefficient designs and help create more efficient aircraft and launch vehicles as well as medical breakthroughs.

—*Gina Hagler, MBA*

FURTHER READING

Anderson, John D., Jr. *A History of Aerodynamics and Its Impact on Flying Machines.* New York: Cambridge University Press, 1997. Includes several chapters that deal with the theories of fluid dynamics and their application to flight.

Carlisle, Rodney P. *Where the Fleet Begins: A History of the David Taylor Research Center.* Washington, D.C.: Naval Historical Center, 1998. A detailed history of the work done at the David Taylor Research Center.

Çengel, Yunus A., and John M. Cimbala. *Fluid Mechanics: Fundamentals and Applications.* Boston: McGraw-Hill, 2010. An essential text for those seeking familiarity with the principles of fluid dynamics.

Darrigol, Olivier. *Worlds of Flow: A History of Hydrodynamics from the Bernoullis to Prandtl.* New York: Oxford University Press, 2005. A thorough account of the progress in hydrodynamic and fluid dynamic theory.

Eckert, Michael. *The Dawn of Fluid Dynamics: A Discipline Between Science and Technology.* Weinheim, Germany: Wiley-VCH, 2006. An introduction to fluid dynamics and its applications.

Ferreiro, Larrie D. *Ships and Science: The Birth of Naval Architecture in the Scientific Revolution, 1600-1800.* Cambridge, Mass.: MIT Press, 2007. A fully documented account of the transition from art to science in the field of naval architecture.

Johnson, Richard W., ed. *The Handbook of Fluid Dynamics.* Boca Raton, Fla.: CRC Press, 1998. A definitive text on the principles of fluid dynamics.

Mahan, A. T. *The Influence of Sea Power Upon History, 1660-1783.* 1890. Reprint. New York, Barnes & Noble Books, 2004. Mahan wrote the book that changed the way nations viewed the function of their navies.

FASCINATING FACTS ABOUT FLUID DYNAMICS

- The pitot tube is a simple device invented by French hydraulic engineer Henri Pitot in the 1700's. It is used to measure air speed in wind tunnels and on aircraft.
- The Bernoulli principle, that pressure decreases as velocity increases, explains why an airplane wing produces lift.
- English engineer William Froude was the first to prove the validity of scale-model tests in the design of full-size vessels. He did this in the 1870's, building and operating a model basin for this purpose.
- The Froude number is a dimensionless number that measures resistance. The greater the Froude number, the greater the resistance.
- Froude performed the seminal work on the rolling of ships.
- Alfred Thayer Mahan's book *The Influence of Sea Power Upon History:1660-1783*, published in 1890, made such a powerful case for the importance of a strong navy that it caused the major powers of that time to invest heavily in new technology for their fleets.
- American naval architect David Watson Taylor was a rear admiral in the U.S. Navy. He was meticulous in his work and developed procedures still in use in model basins. The Taylor Standard Series was a series of trials run with specific models. The results could be used to estimate the resistance of a ship effectively before it was built.
- The bulbous bow is a torpedo-shaped area of the bow of a ship. It is below the waterline. It reduces the resistance of a ship by reducing the impact of the waves on the front bow of a ship underway.
- David Watson Taylor studied the phenomenon of suction between two vessels moving close together in a narrow channel. He was called as an expert witness for the Olympic-Hawke trial in 1911.
- Bioengineers examine fluid flows when designing pacemakers and other medical equipment that will be implanted in the human body.

FUEL CELL TECHNOLOGIES

FIELDS OF STUDY

Physics; chemistry; electrochemistry; thermody-namics; heat and mass transfer; fluid mechanics; com-bustion; materials science; chemical engineering; mechanical engineering; electrical engineering; sys-tems engineering; advanced energy conversion.

SUMMARY

The devices known as fuel cells convert the chemical energy stored in fuel materials directly into electrical energy, bypassing the thermal-energy stage. Among the many technologies used to convert chemical en-ergy to electrical energy, fuel cells are favored for their high efficiency and low emissions. Because of their high efficiency, fuel cells have found ap-plications in spacecraft and show great potential as sources of energy in generating stations.

KEY TERMS AND CONCEPTS

- **anode:** Electrode through which electric current flows into a polarized electrical device.
- **Carnot efficiency:** Highest efficiency at which a heat engine can operate between two tempera-tures: that at which energy enters the cycle and that at which energy exits the cycle.
- **cathode:** Electrode through which electric current flows out of a polarized electrical device.
- **cogeneration:** Using a heat engine to generate both electricity and useful heat simultaneously.
- **electrocatalysis:** Using a material to enhance elec-trode kinetics and minimize overpotential.
- **electrode:** Electrical conductor used to make con-tact with a nonmetallic part of a circuit.

- **electrolyte:** Substance containing free ions that make the substance electrically conductive.
- **electron:** Subatomic particle carrying a negative electric charge.
- **in situ:** Latin for "in position"; here, it refers to being in the reaction mixture.
- **proton:** Subatomic particle carrying a positive electric charge.

DEFINITION AND BASIC PRINCIPLES

Fuel cells provide a clean and versatile means to convert chemical energy to electricity. The reaction between a fuel and an oxidizer is what generates electricity. The reactants flow into the cell, and the products of that reaction flow out of it, leaving the electrolyte behind. As long as the necessary reactant and oxidant flows are maintained, they can operate continuously. Fuel cells differ from electrochemical cell batteries in that they use reactant from an ex-ternal source that must be replenished. This is known as a thermodynamically open system. Batteries store electrical energy chemically and are considered a thermodynamically closed system. In general, fuel cells consist of three components: the anode, where oxidation of the fuel occurs; the electrolyte, which al-lows ions but not electrons to pass through; and the cathode, which consumes electrons from the anode.

A fuel cell does not produce heat as a primary en-ergy conversion mode and is not considered a heat engine. Consequently, fuel cell efficiencies are not limited by the Carnot efficiency. They convert chem-ical energy to electrical energy essentially in an iso-thermal manner.

Fuel cells can be distinguished by: reactant type (hydrogen, methane, carbon monoxide, methanol for a fuel and oxygen, air, or chlorine for an oxidizer);

electrolyte type (liquid or solid); and working temperature (low temperature, below 120 degrees Celsius, intermediate temperature, 120 degrees to 300 degrees Celsius, or high temperature, more than 600 degrees Celsius).

BACKGROUND AND HISTORY

The first fuel cell was developed by the Welsh physicist and judge Sir William Robert Grove in 1839, but fuel cells did not receive serious attention until the early 1960's, when they were used to produce water and electricity for the Gemini and Apollo space programs. These were the first practical fuel cell applications developed by Pratt & Whitney. In 1989, Canadian geophysicist Geoffrey Ballard's Ballard Power Systems and Perry Oceanographics developed a submarine powered by a polymer electrolyte membrane or proton exchange membrane fuel cell (PEMFC). In 1993, Ballard developed a fuel-cell-powered bus and

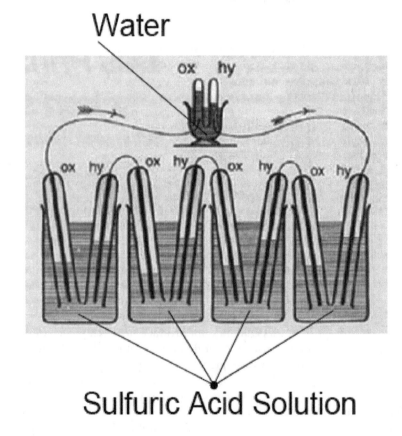

Water

Sulfuric Acid Solution

Sketch of William Grove's 1839 fuel cell

374

later a PEMFC-powered passenger car. Also in the late twentieth century, United Technologies (UTC) manufactured a large stationary fuel cell system for the cogeneration power plant, while continuously developing the fuel cells for the U.S. space program. UTC is also developing fuel cells for automobiles. Siemens Westinghouse has successfully operated a 100-kilowatt (kW) cogeneration solid oxide fuel cell (SOFC) system, and 1-megawatt (MW) systems are being developed.

HOW IT WORKS

Polymer electrolyte membranes or proton exchange membrane fuel cells (PEMFCs). PEMFCs use a proton conductive polymer membrane as an electrolyte. At the anode, the hydrogen separates into protons and electrons, and only the protons pass through the proton exchange membrane. The excess of electrons on the anode creates a voltage difference that can work across an exterior load. At the cathode, electrons and protons are consumed and water is formed.

For PEMFC, the water management is critical to the fuel cell performance: Excess water at the positive electrode leads to flooding of the membrane; dehydration of the membrane leads to the increase of ohmic resistance. In addition, the catalyst of the membrane is sensitive to carbon monoxide poisoning. In practice, pure hydrogen gas is not economical to mass produce. Thus, hydrogen gas is typically produced by steam reforming of hydrocarbons, which contains carbon monoxide.

Direct methanol fuel cells (DMFCs). Like PEMFCs, DMFCs also use a proton exchange membrane. The main advantage of DMFCs is the use of liquid methanol, which is more convenient and less dangerous than gaseous hydrogen. As of 2011, the efficiency is low for DMFCs, so they are used where the energy and power density are more important than efficiency, such as in portable electronic devices.

At the anode, methanol oxidation on a catalyst layer forms carbon dioxide. Protons pass through the proton exchange membrane to the cathode. Water is produced by the reaction between protons and oxygen at the cathode and is consumed at the anode. Electrons are transported through an external circuit from anode to cathode, providing power to connected devices.

Solid oxide fuel cells (SOFCs). Unlike PEMFCs, SOFCs can use hydrocarbon fuels directly and do not require fuel preprocessing to generate hydrogen prior to utilization. Rather, hydrogen and carbon monoxide are generated in situ, either by partial oxidation or, more typically, by steam reforming of the hydrocarbon fuel in the anode chamber of the fuel cell. SOFCs are all-solid electrochemical devices. There is no liquid electrolyte with its attendant material corrosion and electrolyte management problems. The high operating temperature (typically 500-1,000 degrees Celsius) allows internal reforming, promotes rapid kinetics with nonprecious materials, and yields high-quality byproduct heat for cogeneration. The total efficiency of a cogeneration system can be 80 percent—far beyond the conventional power-production system.

The function of the fuel cell with oxides is based on the activity of oxide ions passing from the cathode region to the anode region, where they combine with hydrogen or hydrocarbons; the freed electrons flow through the external circuit. The ideal performance of an SOFC depends on the electrochemical reaction that occurs with different fuels and oxygen.

Molten carbonate fuel cells (MCFCs). MCFCs use an electrolyte composed of a molten carbonate salt mixture suspended in a porous, chemically inert ceramic matrix. Like SOFCs, MCFCs do not require an external reformer to convert fuels to hydrogen. Because of the high operating temperatures, these fuels are converted to hydrogen within the fuel cell itself by an internal re-forming process.

MCFCs are also able to use carbon oxides as fuel. They are not poisoned by carbon monoxide or carbon dioxide, thus MCFCs are advanced to use gases from coal so that they can be integrated with coal gasification.

APPLICATIONS AND PRODUCTS

Hydrogen fuel cell vehicles. In recent years, both the automobile and energy industries have had great interest in the fuel cell powered vehicle as an alternative to internal combustion engine vehicles, which

are driven by petroleum-based liquid fuels. Many automobile manufacturers, such as General Motors, Renault, Hyundai, Toyota, and Honda, have been developing prototype hydrogen fuel cell vehicles. Energy industries have also been installing prototype hydrogen filling stations in large cities, including Los Angeles; Washington, D.C.; and Tokyo.

The first hydrogen fuel cell passenger vehicle for a private individual was leased by Honda in 2005. However, public buses provide better demonstrations of hydrogen fuel cell vehicles compared with passenger vehicles, since public buses are operated and maintained by professionals and they have more volume for the hydrogen fuel storage than passenger vehicles. A number of bus manufacturers such as Toyota, Man, and Daimler have developed hydrogen fuel cell buses and they have been in service in Palm Springs, California; Nagoya, Japan; Vancouver; and Stockholm.

Despite many advantages of hydrogen fuel cell vehicles, this technology still faces substantial challenges such as high costs of novel metal catalyst, safety of hydrogen fuel, effective storage of hydrogen onboard, and infrastructure needed for public refueling stations.

Stationary power plants and hybrid power systems. Siemens Westinghouse and UTC have produced a number of power plant units in the range of about 100 kW by using SOFCs, MCFCs, and phosphoric acid fuel cells (PAFCs). Approximately half of the power plants were MCFC-based plants. They showed that these fuel cell systems have exceeded the research-and-discovery level and already produced an economic benefit. These systems generate power with less fossil fuel and lower emissions of greenhouse gases and other harmful products. Just a small number PEMFC-based power plants were built as the cost of fuel cell materials was prohibitive. In many cases, the fuel-cell-based stationary power plants are used for heat supply in addition to power production, enabling so-called combined heat and power systems. Such systems increase the total efficiency of the power plants and offer an economic benefit.

More recently, many efforts to develop hybrid power plants combining fuel cells and gas turbines were made. While the high-temperature fuel cells, such as SOFCs and MCFCs, produce electrical power, the gas turbines produce additional electrical power from the heat produced by the fuel cells' operation.

At the same time, the gas turbines compress the air fed into the fuel cells. The expected overall efficiency for the direct conversion of chemical energy to electrical energy is up to 80 percent.

Small power generation for the portable electronic devices. At the end of the twentieth century, the demand for electricity continued to increase in many applications, including portable electronics. Batteries have seen significant advances, but their power density is still far inferior to combustion

---

FASCINATING FACTS ABOUT FUEL CELL TECHNOLOGIES

- In 2003, U.S. president George Bush launched the Hydrogen Fuel Initiative (HFI), which was later implemented by legislation through the 2005 Energy Policy Act and the 2006 Advanced Energy Initiative. President Bush stated that "the first car driven by a child born today could be powered by hydrogen and pollution free."

- The Department of Energy is the largest funder of fuel cell science and technology in the United States.

- As of 2011, 191 states have signed Kyoto Protocol, which is a legally binding international agreement to reduce greenhouse-gas emissions by 5.2 percent of 1990 levels by the year 2012.

- In 2008, Boeing announced that it has, for the first time in aviation history, flown a manned airplane powered by hydrogen fuel cells. The Fuel Cell Demonstrator Airplane used a proton exchange membrane fuel cell and lithium-ion battery hybrid system to power an electric motor, which was coupled to a conventional propeller.

- In 2002, typical fuel cell systems cost $1,000 per kilowatt. But, by 2009, the fuel cell system costs had been reduced with volume production (estimated at 500,000 units per year) to $61 per kilowatt.

- Top international universities built their own hydrogen fuel cell racing vehicle to compete against one another on a mobile track in a race called Formula Zero Championship. More advanced races are planned for 2011 Street Edition: The race class will scale up to hydrogen racers, which will compete globally on street circuits in city centers. In the 2015 Circuit Edition, full-size hydrogen fuel cell racing cars built by car manufacturers will compete on racing circuits around the world.

---

devices. Typically, hydrocarbon fuels have 50 to 100 times more energy storage density than commercially available batteries. Even with low conversion efficiencies, fuel-driven generators will still have superior energy density. There is considerable interest in miniaturizing thermochemical systems for electrical power generation for remote sensors, micro-robots, unmanned vehicles (UMVs), unmanned aerial vehicles (UAVs), even portable electronic devices such as laptop computers and cell phones.

Much work on such systems has been developed by the military. The Defense Advanced Research Projects Agency (DARPA) has initiated and developed many types of portable power concepts using the fuel cells. Industries such as Samsung, Sony, NEC, Toshiba, and Fujitsu have developed fuel cells based portable power generation. Most (about 90 percent) devices were based on PEMFCs or DMFCs, which require lower operating temperatures than SOFC. However, development of SOFC-based portable power generation under the DARPA Microsystems Technology Office showed the feasibility of employing high-temperature fuel cells with appropriate thermal management.

The military. In addition to the portable power generation for the foot soldiers, the military market has been interested in developing medium-size power plants (a few hundred watts) for recharging various types of storage batteries and high stationary power plants (more than a few kW) for the auxiliary power units.

Military programs in particular have been interested in the direct use of logistic fuel (for example, Jet Propellant 8) for the fuel cells, because of the complexities and difficulties of the re-forming processes. While the new and improved re-forming processes of logistic fuel were being developed to feed hydrogen into the fuel cells, direct jet-fuel SOFCs were also demonstrated by developing new anode materials that had a high resistance to coking and sulfur poisoning.

IMPACT ON INDUSTRY

Government and university research. One of the biggest sources of funding for fuel cell research in the United States is the Department of Energy (DOE). DOE has developed many programs for the fuel cells and hydrogen. For example, DOE formed Solid State Energy Conversion Alliance (SECA) in 1999 and formulated a program with funding of $1 billion for 10 years. Other government agencies, such as the Department of Defense (DOD), DARPA, Air Force Office of Scientific Research (AFOSR), Office of Naval Research (ONR), and Army Research Laboratory (ARL) have also funded a number of the fuel cell projects taking place in academic and corporate settings to bring about the transfer of the energy technologies to those fighting wars.

Professional societies have also noticed the importance of the energy security and advanced energy technologies. In 2003, the American Institute of Aeronautics and Astronautics (AIAA) and American Society of Mechanical Engineers (ASME) brought in new international conferences: AIAA International Energy Conversion Engineering Conference (IECEC) and ASME International Fuel Cell Science, Engineering, and Technology Conference. The conferences' goals are to expand international cooperation, understanding, promotion of efforts, and disciplines in the area of energy conversion technology, advanced energy and power systems and devices, and the policies, programs, and environmental impacts associated with the development and utilization of energy technologies.

As of 2011, the Korean fuel cell market is in a nascent stage and is expected to witness rapid growth as a result of government-supported policies. Korea has nine fuel cell units installed in various regions. The major driver behind the future development of the hydrogen and fuel cell industry in Korea is the country's need to achieve energy security.

Industry and business. Almost every car manufacturer has developed a fuel cell vehicle powered by PEMFCs or SOFCs. They hope that the fuel cell vehicles double the efficiency of internal combustion engine vehicles. Some (Honda, General Motors, Toyota) are using their own developed fuel cells, but most companies buy the fuel cell systems from the fuel cell manufacturers such as UTC and De Nora. Many electronic companies such as Motorola, NEC, Toshiba, Samsung, and Matsushita are rushing to develop their own small fuel cells that will provide power up to ten times longer on a single charge than conventional batteries for small portable electronic devices.

The U.S. fuel cell market is growing rapidly. By the end of 2009, 620 fuel cell power units were installed. Government-supported promotion of clean energy is responsible for this, and the tax credits permitted under Energy Policy Act of 2005 continue to drive the U.S. fuel cell market. Fuel cell manufacturers are also working on development of small fuel cell power systems intended to be used in homes and office buildings. For example, a 200 kW PAFC system was installed to power a remote police station in New York City's Central Park.

CAREERS AND COURSE WORK

Courses in chemistry, physics, electrochemistry, materials science, chemical engineering, and mechanical engineering make up the foundational requirements for students interested in pursuing careers in fuel cell research. Earning a bachelor of science degree in any of these fields would be appropriate preparation for graduate work in a similar area. In most circumstances, either a master's or doctorate degree is necessary for the most advanced career opportunities in both academia and industry. Careers in the fuel cells field can take several different shapes. Fuel cell industries are the biggest employers of fuel cell engineers, who focus on developing and manufacturing new fuel cell units as well as maintaining or repairing fuel cell units. Other industries in which fuel cell engineers often find work include aviation, automotive, electronics, telecommunications, and education.

Many fuel cell engineers prefer employment within the national laboratories and government agencies such as the Pacific Northwest National Laboratory, the National Renewable Energy Laboratory, the Argonne National Laboratory, the National Aeronautics and Space Administration (NASA), DOE, and DARPA. Others find work in academia. Such professionals divide their time between teaching university classes on fuel cells and conducting their own research.

SOCIAL CONTEXT AND FUTURE PROSPECTS

In the future, it is not likely that sustainable transportation will involve use of conventional petroleum. Transportation energy technologies should be developed with both the goal of providing an alternative to the petroleum-based internal combustion engine vehicles. People evaluate vehicles not only on the basis of fuel economy but also performance. Vehicles using an alternative energy source should be designed with these parameters.

One of the most promising energy sources for the

future will be hydrogen. The hydrogen fuel cell vehicles face cost and technical challenges, especially the fuel cell stack and onboard hydrogen storage.

For the fuel cell power plants, the economic and lifetime related issues hinder the acceptance of fuel cell technologies. Such problems were not associated with fuel cells but with auxiliary fuel cell units such as thermal management, reactant storage, and water management. Therefore, the auxiliary units of fuel cell systems should be further developed to address these issues.

The fundamental problems of fuel cells related to electrocatalysis also need to be addressed for improvement in performance, as highly selective catalysts will provide better electrochemical reactions.

Lastly, once the new fuel cell technologies are successfully developed and meet the safety requirements, the infrastructure to distribute and to recycle fuel cells will also be necessary.

—*Jeongmin Ahn, MS, PhD*

FURTHER READING

Bagotsky, Vladimir S. Fuel Cells: Problems and Solutions. Hoboken, N.J.: John Wiley & Sons, 2009. Provides extensive explanations of the various types of fuel cells operation.

Hoogers, Gregor, ed. Fuel Cell Technology Handbook. Boca Raton, Fla.: CRC Press, 2003. Recognizes the part played by the change in Gibb's potential.

Kotas, T.J. The Exergy Method of Thermal Plant Analysis. Malabar, Fla.: Krieger Publications, 1995. Proves that the fuel chemical exergy and the lower calorific value of the fuel, with different units, are numerically equal.

Larminie, James, and Andrew Dicks. Fuel Cell Systems Explained. 2d ed. Hoboken, N.J.: John Wiley & Sons, 2000. This text provides construction details of the various types of fuel cells.

O'Hayre, Ryan P., et al. Fuel Cell Fundamentals. 2d ed. Hoboken, N.J.: John Wiley & Sons, 2008. Includes extensive discussions on thermodynamics, transport science, and chemical kinetics in the early chapters with a supporting appendix on quantum-mechanical issues. Also addresses modeling and characterization of fuel cells and fuel cell systems and their environmental impact.

Reddy, Thomas B., ed. Linden's Handbook of Batteries. 4th ed. New York: McGraw-Hill, 2011. Includes detailed technical descriptions of chemistry, electrical characteristics, construction details, applications, and pros and cons charts.

WEB SITES
Battery Council International
http://www.batterycouncil.org

Fuel Cell and Hydrogen Energy Association
http://www.fchea.org

Fuel Cell Europe
http://www.fuelcelleurope.org

See also: Electrical Engineering; Electric Automobile Technology; Electronics and Electronic Engineering; Hybrid Vehicle Technologies; Mechanical Engineering.

G

GEOMETRY

FIELDS OF STUDY

Civil engineering; architecture; surveying; agriculture; environmental and conservation sciences; mechanical engineering; computer-aided design; molecular design; nanotechnology; physical chemistry; biology; physics; graphics; computer game programming; textile and fabric arts; fine arts; cartography; geographic information systems; Global Positioning Systems; medical imaging; astronomy; robotics.

SUMMARY

Geometry, which literally means "earth measurement," is absolutely critical to most fields of physical science and technology and especially to any application that involves surfaces and surface measurement. The term applies on all scales, from nanotechnology to deep space science, where it describes the relative physical arrangement of things in space. Although the first organized description of the principles of geometry is ascribed to the ancient Greek philosopher Euclid, those principles were known by others before him, and certainly had been used by the Egyptians and the Babylonians. Euclidean, or plane, geometry deals with lines and angles on flat surfaces (planes), while non-Euclidean geometry applies to nonplanar surfaces and relationships.

KEY TERMS AND CONCEPTS

- **conic section:** Geometric form that can be defined by the intersection of a plane and a cone; examples are a circle, an ellipse, a parabola, a hyperbola, a line, and a point.
- **Euclidean:** Describing geometric principles included in the range of principles described by Euclid.
- **Fibonacci series:** Geometric sequence described by Italian mathematician Leonardo of Pisa (known as Fibonacci) in about 1200, beginning with zero and in which each subsequent number is the sum of the previous two: 0, 1, 1, 2, 3, 5, 8, 13, 21, 34, 55, . . .
- **geometric isomers:** Chemical term used to denote molecular structures that differ only by the
- **relative arrangement of atoms in different molecules.**
- **golden ratio:** Represented as φ or Φ, a seemingly ubiquitous naturally occurring ratio or proportion having the approximate value 1.618018513; also known as the golden mean, the golden section, and the divine proportion.
- **pi:** Represented as π, an irrational number that is equivalent to the ratio of the circumference of a circle to its diameter, equal to about 3.14159265358979323846264 (the decimal part is believed to be nonrepeating and indeterminate, or unending).
- **plane:** Flat surface that would be formed by the translation of a line in one direction.
- **polygon:** Two-dimensional, or plane, geometric shape having a finite number of sides; designated as a regular polygon if all sides are of equal length.
- **polyhedron:** Three-dimensional geometric shape having a finite number of planar faces; designated as a regular polyhedron if all faces are equivalent.
- **postulate:** Rule that is accepted as true without proof.
- **Pythagorean:** Describing a relation to the theorem and principles ascribed to Pythagoras, especially the properties of right triangles.
- **theorem:** Rule that is accepted as true but requires a rigorous proof.
- **torus:** Structure formed by the translation of a circle through space along a path defined by a second circle whose plane is orthogonal to that of the first circle.

DEFINITION AND BASIC PRINCIPLES

Geometry is the branch of mathematics concerned with the properties and relationships of points, lines, and surfaces, and the space contained by those entities. A point translated in a single direction describes a line, while a line translated in a single direction describes a plane. The intersections and rotations of various identities describe corresponding structures that have specific mathematical relationships and properties. These include angles and numerous two- and three-dimensional forms. Geometric principles can also be extended into realms encompassing more than three dimensions, such as with the incorporation of time as a fourth dimension in Albert Einstein's space-time continuum. Higher dimensional analysis is also possible and is the subject of theoretical studies.

The basic principles of geometry are subject to various applications within different frames of reference. Plane geometry, called Euclidean geometry after the ancient Greek mathematician Euclid, deals with the properties of two-dimensional constructs such as lines, planes, and polygons. The five basic principles of plane geometry, called postulates, were described by Euclid. The first four are accepted as they are stated and require no proof, although they can be proven. The fifth postulate differs considerably from the first four in nature, and attempts to prove or disprove it have consistently failed. However, it gave rise to other branches of geometry known as non-Euclidean. The first four postulates apply equally to all branches of Euclidean and non-Euclidean geometry, while each of the non-Euclidean branches uses its own interpretation of the fifth postulate. These are known as hyperbolic, elliptic, and spherical geometry.

The point, defined as a specific location within the frame of reference being used, is the common foundation of all branches of geometry. Any point can be uniquely and unequivocally defined by a set of coordinates relative to the central point or origin of the frame of reference. Any two points within the frame of reference can be joined with a single line segment, which can be extended indefinitely in that direction to produce a line. Alternatively, the movement of a point in a single direction within the frame of reference describes a line. Any line can be translated in any orthogonal direction to produce a plane. Rotation of a line segment in a plane about one of its end points describes a circle whose radius is equal to the length of that line segment. In any plane, two lines that intersect orthogonally produce a right angle (an angle of 90 degrees). These are the essential elements of the first four Euclidean postulates. The fifth, which states that nonparallel lines must intersect, could not be proven within Euclidean geometry, and its interpretation under specific conditions gives rise to other frames of reference.

BACKGROUND AND HISTORY
The Greek historian Herodotus soundly argued that geometry originated in ancient Egypt. During the time of the legendary pharaoh Sesostris, the farmland of the empire was apportioned equally among the people, and taxes were levied accordingly. However, the annual inundation of the Nile River tended to wash away portions of farmland, and farmers who lost land in this way complained that it was unfair for them to pay taxes equal to those whose farms were complete. Sesostris is said to have sent agents to measure the loss of land so that taxation could be made fair again. The agents' observations of the relationships that existed gave rise to an understanding of the principles of geometry.

It is well documented that the principles of geometry were known to people long before the ancient Greeks described them. The Rhind papyrus, an Egyptian document dating from 2000 BCE, contains a valid geometric approximation of the value of pi, the ratio of the circumference of a circle to its diameter. The ancient Babylonians were also aware of the principles of geometry, as is evidenced by the inscription on a clay tablet that is housed in Berlin. The inscription has been translated as an explanation of the relationship of the sides and hypotenuse of a right triangle in what is known as the Pythagorean theorem, although the tablet predates Pythagoras by several hundred years.

From these early beginnings to modern times, studies in geometry have evolved from being simply a means of describing geometric relationships toward encompassing a complete description of the workings of the universe. Such studies allow the behavior of materials, structures, and numerous processes to be predicted in a quantifiable way.

HOW IT WORKS
Geometry is concerned with the relationship between points, lines, surfaces, and the spaces enclosed or bounded by those entities.

Points. A point is any unique and particular location within a frame of reference. It exists in one

Woman teaching geometry. Illustration at the beginning of a medieval translation of Euclid's Elements, (c. 1310)

the number of distinct dimensions assigned to the system. For the purposes of all but the most theoretical of applications, however, three dimensions are sufficient for normal physical representations.

Points can also be identified as corresponding to specific distances and angles, also relative to a coordinate system origin. Thus, a point located in a spherical coordinate system is defined by the radius (the straight-line distance from the origin to the point), the angle that the radius is swept through a plane, and the angle that the radius is swept through an orthogonal plane to achieve the location of the point in space.

Lines. A line can be formed by the translation of a point in a single direction within the reference system. In its simplest designation, a line is described in a two-dimensional system when one of the coordinate values remains constant. In a three-dimensional system, two of the three coordinate values must remain constant. The lines so described are parallel to the reference coordinate axis. For example, the set of points (x, y) in a two-dimensional Cartesian system that corresponds to the form $(x, 3)$—so that the value of the y-coordinate is 3 no matter what the value of x—defines a line that is parallel to the x-axis and always separated from it by 3 units in the positive y direction.

Lines can also be defined by an algebraic relationship between the coordinate axes. In a two-dimensional Cartesian system, a line has the general algebraic form $y = mx + b$. In three-dimensional systems, the relationship is more complex but can be broken down into the sum of two such algebraic equations involving only two of the three coordinate axes.

Planes and surfaces. Planes are described by the translation of a line through the reference system, or by the designation of two of the three coordinates having constant values while the third varies. A plane can be thought of as a flat surface. A curved surface can be formed in an analogous manner by translating a curved line through the reference system, or by definition, as the result of a specific algebraic relationship between the coordinate axes.

Angles. Intersecting lines have the property of defining an angle that exists between them. The angle can be thought of as the amount by which one line must be rotated about the intersection point in order to coincide with the other line. The magnitude, or value, of angles rigidly determines the shape of

dimension only, having neither length nor width nor breadth but only location. The location of any point can be uniquely and unequivocally defined by a set of coordinate values relative to the central point of the particular reference system being used. For example, in a Cartesian coordinate system—named after French mathematician René Descartes although the method was described at the same time by Pierre de Fermat—the location of any point in a two-dimensional plane is described completely by an x-coordinate and a y-coordinate, as (x, y), relative to the origin point at $(0, 0)$. Thus, a point located at $(3, 6)$ is 3 units away from the origin in the direction corresponding to positive values of x, and 6 units away from the origin in the direction corresponding to positive values of y. Similarly, a point in a three-dimensional Cartesian system is identified by three coordinates, as (x, y, z). In Cartesian coordinate systems, each axis is orthogonal to the others. Because orthogonality is a mathematical property, it can also be ascribed to other dimensions as well, allowing the identification of points in theoretical terms in n-space, where n is structures, especially when the structures are formed by the intersection of planes.

Conic sections. A cone is formed by the rotation of a line at an angle about a point. Conic sections are described by the intersection of a plane with the cone structure. As an example, consider a cone formed in the three-dimensional Cartesian system by rotating the line $x = y$ about the y-axis, forming both a positive and a negative cone shape that meet at the origin point. If this is intersected by a plane parallel to the x-z plane, the intersection describes a circle. If the plane is canted so that it is not parallel to the x-z plane and intersects only one of the cone ends, the result is an ellipse. If the plane is canted further and positioned so that it intersects the positive cone on one side of the y-axis and the negative cone on the other side of the y-axis, the intersection defines a hyperbola. If the plane is canted still further so that it intersects both cones on the same side of the y-axis, then a parabola is described.

APPLICATIONS AND PRODUCTS

It is impossible to describe even briefly more than a small portion of the applications of geometry because geometry is so intimately bound to the structures and properties of the physical universe. Every physical structure, no matter its scale, must and does adhere to the principles of geometry, since these are the properties of the physical universe. The application of geometry is fundamental to essentially every field, from agriculture to zymurgy.

GIS and GPS. Geographical information systems (GIS) and Global Positioning Systems (GPS) have been developed as a universal means of location identification. GPS is based on a number of satellites orbiting the planet and using the principles of geometry to define the position of each point on the Earth's surface. Electronic signals from the various satellites triangulate to define the coordinates of each point. Triangulation uses the geometry of triangles and the strict mathematical relationships that exist between the angles and sides of a triangle, particularly the sine law, the cosine law, and the Pythagorean theorem.

GIS combines GPS data with the geographic surface features of the planet to provide an accurate "living" map of the world. These two systems have revolutionized how people plan and coordinate their movements and the movement of materials all over the world. Applications range from the relatively

simple GPS devices found in many modern vehicles to precise tracking of weather systems and seismic activity. An application that is familiar to many through the Internet is GoogleEarth, which presents a satellite view of essentially any place on the planet at a level of detail that once was available from only the most top secret of military reconnaissance satellites. The system also allows a user to view traditional map images in a way that allows them to be scaled as needed and to add overlays of specific buildings, structures, and street views. Anyone with access to the Internet can quickly and easily call up an accurate map of almost any desired location on the planet.

GIS and GPS have provided a whole new level of security for travelers. They have also enabled the development of transportation security features such as General Motors' OnStar system, the European Space Agency's Satellite Based Alarm and Surveillance System (SASS), and several other satellite-based security applications. They invariably use the GPS and GIS networks to provide the almost instantaneous location of individuals and events as needed.

CAD. Computer-aided design (CAD) is a system in which computers are used to generate the design of a physical object and then to control the mechanical reproduction of the design as an actual physical object. A computer drafting application such as AutoCAD is used to produce a drawing of an object in electronic format. The data stored in the drawing file include all the dimensions and tolerances that define the object's size and shape. At this point, the object itself does not exist; only the concept of it exists as a collection of electronic data. The CAD application can calculate the movements of ancillary robotic machines that will then use the program of instructions to produce a finished object from a piece of raw material. The operations, depending on the complexity and capabilities of the machinery being directed, can include shaping, milling, lathework, boring or drilling, threading, and several other procedures. The nature of the machinery ranges from basic mechanical shaping devices to advanced tooling devices employing lasers, high-pressure jets, and plasma- and electron-beam cutting tools.

The advantages provided by CAD systems are numerous. Using the system, it is possible to design and produce single units, or one offs, quickly and precisely to test a physical design. Adjustments to production steps are made very quickly and simply by adjusting the object data in the program file rather than through repeated physical processing steps with their concomitant waste of time and materials. Once perfected, the production of multiple pieces becomes automatic, with little or no variation from piece to piece.

Metrology. Closely related to CAD is the application of geometry in metrology, particularly through the use of precision measuring devices such as the measuring machine. This is an automated device that uses the electronic drawing file of an object, as was produced in a CAD procedure, as the reference standard for objects as they are made in a production facility. Typically, this is an integral component of a statistical process control and quality assurance program. In practice, parts are selected at random from a production line and submitted to testing for the accuracy of their construction during the production process. A production piece is placed in a custom jig or fixture, and the calibrated measuring machine then goes through a series of test measurements to determine the correlation between the features of the actual piece and the features of the piece as they are designated in the drawing file. The measuring machine is precisely controlled by electronic mechanisms and is capable of highly accurate measurement.

Game programming and animation. Basic and integral parts of both the video game and the motion-picture industry are described by the terms "polygon," "wire frame," "motion capture," and "computer-generated imagery" (CGI). Motion capture uses a series of reference points attached to an actor's body. The reference points become data points in a computer file, and the motions of the actor are recorded as the geometric translation of one set of data points into another in a series. The data points can then be used to generate a wire-frame drawing of a figure that corresponds to the character whose actions have been imitated by the actor during the motion-capture process. The finished appearance of the character is achieved by using polygon constructions to provide an outward texture to the image. The texture can be anything from a plain, smooth, and shiny surface to a complex arrangement of individually colored hairs. Perhaps the most immediately recognizable application of the polygon process is the generation of dinosaur skin and aliens in video games and films.

The movements of the characters in both games and films are choreographed and controlled through strict geometric relationships, even to the play of light over the character's surface from a single light source. This is commonly known as ray tracing and is used to produce photorealistic images.

IMPACT ON INDUSTRY

One cannot conceive of modern agriculture without the economic assessment scales of yield, fertilizer rates, pesticide application rates, and seed rates, all on a per hectare or per acre basis. Similarly, one cannot envisage scientific applications and research programs that do not rely intimately on the principles of geometry in both theory and practice. Processes such as robotics and CAD have become fundamental aspects of modern industry, representing entirely different paradigms of efficiency, precision, and economics than existed before. The economic value of computer-generated imagery and other computer graphics applications is similarly inestimable. For example, James Cameron's film *Avatar* (2009), which featured computer-generated imagery, returned more than $728 million within two weeks of its release date. Indeed, many features simply would not exist without computer-generated imagery, which in turn would not exist without the application of geometry. Similarly, computer gaming, from which computer-aided imagery developed, is a billion-dollar industry that is relies entirely on the mathematics of geometry.

Academic research. All universities and colleges maintain a mathematics department in which practical training and theoretical studies in the applications of geometric principles are carried out. Similarly, programs of research and study in computer science departments examine new and more effective ways to apply geometry and geometric principles in computing algorithms. Graphic arts programs in particular specialize in the development and application of computer-generated imagery and other graphics techniques relying on geometry. Training in plane surveying and mechanical design principles are integral components of essentially all programs of training in civil and mechanical engineering.

Industry and business. A wide variety of businesses are based almost entirely on the provision of goods and services that are based on geometry. These include contract land-surveying operations that serve agricultural needs; environmental conservation and management bodies; forestry and mining companies; municipal planning bodies; transportation infrastructure and the construction industries; data analysis; graphics programming; advertising; and cinematographic adjuncts, to name but a few. The opportunity exists for essentially anyone with the requisite knowledge and some resources to establish a business that caters to a specific service need in these and other areas.

Another aspect of these endeavors is the provision of materials and devices to accommodate the services. Surveyors, for example, require surveying equipment such as transit levels and laser source-detectors in order to function. Similarly, construction contractors typically require custom-built roof trusses and other structures. Graphics and game design companies often subcontract needed programming expertise. Numerous businesses and industries exist, or can be initiated, to provide needed goods.

CAREERS AND COURSE WORK

Geometry is an essential and absolutely critical component of practically all fields of applied science. A solid grounding in mathematics and basic geometry is required during secondary school studies. In more advanced or applied studies at the post-secondary

level, in any applied field, mathematical training in geometrical principles will focus more closely on the applications that are specific to that field of study. Any program of study that integrates design concepts will include subject-specific applications of geometric principles. Applications of geometric principles are used in mechanical engineering, manufacturing, civil engineering, industrial plant operations, agricultural development, forestry management, environmental management, mining, project management and logistics, transportation, aeronautical engineering, hydraulics and fluid dynamics, physical chemistry, crystallography, graphic design, and game programming.

The principles involved in any particular field of study can often be applied to other fields as well. In economics, for example, data mining uses many of the same ideological principles that are used in the mining of mineral resources. Similarly, the generation of figures in electronic game design uses the same geometric principles as land surveying and topographical mapping. Thus, a good grasp of geometry and its applications can be considered a transferable skill usable in many different professions.

SOCIAL CONTEXT AND FUTURE PROSPECTS

Geometry is set to play a central role in many fields. Geometry is often the foundation on which decisions affecting individuals and society are made. This is perhaps most evident in the establishment and construction of the most basic infrastructure in every country in the world, and in the most high-tech advances represented by the satellite networks for GPS and GIS. It is easy to imagine the establishment of similar networks around the Moon, Mars, and other planets, providing an unprecedented geological and geographical understanding of those bodies. In between these extremes that indicate the most basic and the most advanced applications of geometry and geometrical principles are the typical everyday applications that serve to maintain practically all aspects of human endeavor. The scales at which geometry is applied cover an extremely broad range, from the ultrasmall constructs of molecular structures and nanotechnological devices to the construction of islands and buildings of novel design and the ultra-large expanses of interplanetary and even interstellar space.

—Richard M. J. Renneboog, MSc

FURTHER READING

Bar-Lev, Adi. "Big Waves: Creating Swells, Wakes and Everything In-Between." *Game Developer* 15, no. 2, (February, 2008): 14-24. Describes the application of geometry in the modeling of liquid water actions in the computer graphics of gaming and simulations.

Bonola, Roberto. *Non-Euclidean Geometry: A Critical and Historical Study of Its Development.* 1916. Reprint. Whitefish, Mont.: Kessinger, 2007. A facsimile reproduction of a classic work on the history of alternate geometries.

Boyer, Carl B. *History of Analytic Geometry.* Mineola, N.Y.: Dover Publications, 2004. Traces the history of analytic geometry from ancient Mesopotamia, Egypt, China, and India to 1850.

Darling, David. *The Universal Book of Mathematics: From Abracadabra to Zeno's Paradoxes.* 2004. Reprint. Edison, N.J.: Castle Books, 2007. An encyclopedic account of things mathematical.

Heilbron, J. L. *Geometry Civilized: History, Culture, and Technique.* Reprint. New York: Oxford University Press, 2003. A very readable presentation of the history of geometry, including geometry in cultures other than the Greek, such as the Babylonian, Indian, Chinese, and Islamic cultures.

Herz-Fischler, Roger. *A Mathematical History of the Golden Number.* Mineola, N.Y.: Dover Publications, 1998. A well-structured exploration of the golden number and its discovery and rediscovery throughout history.

Holme, Audun. *Geometry: Our Cultural Heritage.* New York: Springer, 2002. An extensive discussion of the historical development of geometry from prehistoric to modern times, focusing on many major figures in its development.

Livio, Mario. *The Golden Ratio: The Story of Phi, the World's Most Astonishing Number.* New York: Broadway Books, 2002. Describes how the occurrence of the golden ratio, also known as the divine ratio, defined by the ancient mathematician Euclid, seems to be a fundamental constant of the physical world.

Szecsei, Denise. *The Complete Idiot's Guide to Geometry.* 2d ed. Indianapolis: Alpha Books, 2007. A fun, step-by-step presentation that walks the reader through the mathematics of geometry and assumes absolutely no prior knowledge beyond basic arithmetic.

West, Nick. "Practical Fluid Dynamics: Part 1." *Game Developer* 14, no. 3 (March, 2007): 43-47. Introduces the application of geometric principles in the modeling of smoke, steam, and swirling liquids in the graphics of computer games and simulations.

WEB SITES

American Mathematical Society
http://www.ams.org

Mathematical Association of America
http://www.maa.org

See also: Applied Mathematics; Computer Graphics; Engineering; Engineering Mathematics.

GLASS AND GLASSMAKING

FIELDS OF STUDY

Optics; physics; mechanical engineering; electrical engineering; materials science; ceramics; chemistry; architecture; mathematics; thermodynamics; microscopy; spectroscopy.

SUMMARY

Glassmaking is a diverse field with applications ranging from optics to art. Most commercially produced glass is used to make windows, lenses, and food containers such as bottles and jars. Glass is also a key component in products such as fiber-optic cables and medical devices. Because many types of glass are recyclable, the demand for glass is expected to increase. Most careers in glassmaking require significant hands-on experience as well as formal education.

KEY TERMS AND CONCEPTS

- **ceramic:** Material made from nonorganic, nonmetallic compounds subjected to high levels of heat.
- **crystal:** Solid formed from molecules in a re-

peating, three-dimensional pattern; also refers to a high-quality grade of lead glass.
- **glazier:** Professional who makes, cuts, installs, and replaces glass products.
- **pellucidity:** Rate at which light can pass through a substance; also known as transparency.
- **silica:** Silicon dioxide, the primary chemical ingredient in common glass.
- **tempered glass:** Glass strengthened by intensive heat treatment during manufacturing.
- **viscosity:** Rate at which a fluid resists flow; sometimes described as thickness.
- **vitreous:** Being glasslike in texture or containing glass.

DEFINITION AND BASIC PRINCIPLES

Glass is one of the most widely used materials in residential and commercial buildings, vehicles, and many different devices. Glass windows provide natural light while protecting indoor environments from changes in the outside temperature and humidity. The transparency of glass also makes it a good choice for lightbulbs, light fixtures, and lenses for items such as eyeglasses. A nonporous and nonreactive substance, glass is an ideal material for food packaging and preparation. Most types of glass can be recycled with no loss in purity or quality, a factor that has increased its appeal.

Although glass resembles crystalline substances found in nature, it does not have the chemical properties of a crystal. Glass is formed through a fusion process involving inorganic chemical compounds such as silica, sodium carbonate, and calcium oxide (also known as lime). The compounds are heated to form a liquid. When the liquid is cooled rapidly, it becomes a solid but retains certain physical characteristics of a liquid, a process known as glass transformation.

BACKGROUND AND HISTORY

The earliest known glass artifacts come from Egypt and eastern Mesopotamia, where craftspeople began making objects such as beads more than five thousand years ago. Around 1500 BCE, people began dipping metal rods into molten sand to create bottles. In the third century BCE, craftspeople in Babylon discovered that blowing air into molten glass was a rapid, inexpensive way to make hollow shapes.

Glassmaking techniques quickly spread throughout Europe with the expansion of the Roman Empire in the first through fourth centuries. The region that would later be known as Italy, led by the city of Venice, dominated the glass trade in Europe and the Americas for several hundred years.

With the development of the split mold in 1821, individual glass objects no longer needed to be blown and shaped by hand. The automation of glass manufacturing allowed American tinsmith John L. Mason to introduce the Mason jar in 1858. Over the next few decades, further innovations in glass-making led to a wide range of glass applications at increasingly lower costs. One of the most noteworthy advancements in glass use came in the 1960's and 1970's with the design and rollout of fiber-optic cable for long-range communications.

HOW IT WORKS

Most types of glass have translucent properties, which means that certain frequencies, or colors, of light can pass through them. Transparent glass can transmit all frequencies of visible light. Other types of glass act as filters for certain colors of light so that when objects are viewed through them, the objects appear to be tinted a specific shade. Some types of glass transmit light but scatter its rays so that objects on the other side are not visible to the human eye. Glass is often smooth to the touch because surface tension, a feature similar to that of water, binds its molecules together during the cooling process. Unless combined with certain other compounds, glass is brittle in texture.

There is a widespread but incorrect belief that glass is a liquid. Many types of glass are made by heating a mixture to a liquid state, then allowing it to reach a supercooled state in which it cannot flow. The process, known in glassmaking as the glass transition, causes the molecules to organize themselves into a form that does not follow an extended pattern. In this state, known as the vitreous or amorphous state, glass behaves like a solid because of its hardness and its tendency to break under force.

Chemical characteristics. The most common type of glass is known as soda-lime glass and is made primarily of silica (60 to 75 percent). Sodium carbonate, or soda ash, is added to the silica to lower its melting point. Because the presence of soda ash makes it possible for water to dissolve glass, a third compound such as calcium carbonate, or limestone, must be added to increase insolubility and hardness. For some types of glass, compounds such as lead oxide or boric oxide are added to enhance properties such as brilliance and resistance to heat.

Manufacturing processes. The melting and cooling of a mixture to a liquid, then supercooled state is the oldest form of glassmaking. Most glass products, including soda-lime, are made using this type of process. As the liquid mixture reaches the supercooled state, it is often treated to remove stresses that could weaken the glass item in its final form. This process is known as annealing. The item's surface is smoothed and polished through a stream of pressurized nitrogen. Once the glass transformation is complete, the item can be coated or laminated to increase traits such as strength, electrical conductivity, or chemical durability.

Specialized glass can be made by processes such as vapor deposition or sol-gel (solution-gel). Under vapor deposition, chemicals are combined without being melted. This approach allows for the creation of thin films that can be used in industrial settings. Glass created through a sol-gel process is made by combining a chemical solution, often a metal oxide, with a compound that causes the oxide to convert to a gel. High-precision lenses and mirrors are examples of sol-gel glass.

APPLICATIONS AND PRODUCTS

Glass is one of the most widely used materials in the world. Some of the most common uses for glass include its incorporation into buildings and other structures, vehicles such as automobiles, fiberglass packaging materials, consumer and industrial optics, and fiber optics.

Construction. The homebuilding industry relies on glass for the design and installation of windows, external doors, skylights, sunrooms, porches, mirrors, bathroom and shower doors, shelving, and display cases. Many office buildings have floor-to-ceiling windows or are paneled with glass. Glass windows allow internal temperatures and humidity levels to be controlled while still allowing natural light to enter a room. Light fixtures

frequently include glass components, particularly bulbs and shades, because of the translucent quality of glass and its low manufacturing costs.

Automotive glass. Windshields and windows in automobiles, trucks, and other vehicles are nearly always made of glass. Safety glass is used for windshields. Most safety glass consists of transparent glass layered with thin sheets of a nonglass substance such as polyvinyl butyral. The nonglass layer, or laminate, keeps the windshield from shattering if something strikes it and protects drivers and passengers from injury in an accident. A car's side and back windows generally are not made of safety glass but rather of tempered glass treated to be heat resistant and to block frequencies of light such as ultraviolet rays.

Fiberglass. Russell Games Slayter, an employee of Owens-Corning, invented a glass-fiber product in 1938 that the company named Fiberglas. The product was originally sold as thermal insulation for buildings and helped phase out the use of asbestos, a carcinogen. The tiny pockets of air created by the material's fabriclike glass filaments are the source of its insulating properties. Fiberglass is also used in the manufacturing of vehicle bodies, boat hulls, and sporting goods such as fishing poles and archery bows because of its light weight and durability. Because recycled glass makes up a significant portion of many

Glass container forming

fiberglass products, the material is considered environmentally friendly.

Packaging. Processed foods and beverages are often sold in glass bottles and jars. The clarity of glass allows consumers to see the product and judge its quality. Glass is a nonporous, nonreactive material that does not affect the flavor, aroma, or consistency of the product it contains. Consumers also prefer glass because it can be recycled. A disadvantage of glass packaging is the ease with which containers can be shattered. Glass also is heavier than competing packaging materials such as plastic, aluminum, and paper.

Consumer optics. Glass lenses can be found in a wide range of consumer-oriented optical products ranging from eyeglasses to telescopes. The use of glass in eyeglasses is diminishing as more lenses are made from specialized plastics. Cameras, however, rarely use plastic lenses, as they tend to create lower quality images and are at greater risk of being scratched. Binoculars, microscopes, and telescopes designed for consumer use are likely to contain a series of glass lenses. These lenses are often treated with coatings that minimize glare and improve the quality of the image being viewed.

Industrial optics. For the same reasons that glass is preferred in consumer optics, it is the material of choice for industrial applications. Many specialized lenses and mirrors used for telescopes, microscopes, and lasers are asymmetrical or aspherical in shape. The manufacture of aspherical lenses for high-precision equipment was difficult and expensive until the 1990's, when glass engineers developed new techniques based on the processing of preformed glass shapes at relatively low temperatures. These techniques left the surface of the glass smooth and made it more cost-effective to manufacture highly customized lenses.

Fiber optics. The optical fibers in fiber-optic communication infrastructures are made from glass. The fibers carry data in the form of light pulses from one end of a glass strand to the other. The light pulses then jump through a spliced connection to the next glass fiber. Fiber-optic cable can transmit information more quickly and over a longer distance than electrical wire or cable. Glass fibers can carry more than one channel of data when multiple frequencies of light are used. They also are not subject to electromagnetic interference such as lightning strikes,

---

FASCINATING FACTS ABOUT GLASS AND GLASSMAKING

- The oldest known glassmaking handbook was etched onto tablets and kept in the library of Ashurbanipal, king of Assyria in the seventh century BCE.

- In 1608, settlers in Jamestown, Virginia, opened the first glassmaking workshop in the American colonies. It produced bottles used by pharmacists for medicines and tonics.

- Nearly 36 percent of glass soft drink and beer bottles and 28 percent of all glass containers were recycled in 2008. About 80 percent of recovered glass is used for new bottles.

- Toledo, Ohio, is nicknamed the Glass City because of its industrial heritage of glassmaking. In the early twentieth century, Toledo began providing glass to automobile factories in nearby Detroit.

- China manufactures about 45 percent of all glass in the world. Most Chinese glass is installed in automobiles made locally.

- The first sunglasses were built from lenses made out of quartz stone and tinted by smoke. The glasses were worn by judges in fourteenth-century China to hide their facial expressions during court proceedings.

- The National Aeronautics and Space Administration has developed NuStar (Nuclear Spectroscopic Telescope Array), a set of two space telescopes carrying a new type of glass lens made from 133 glass layers. NuStar is about one thousand times as sensitive as space telescopes such as Hubble.

---

an advantage when fiber-optic cable is used to wire offices in skyscrapers.

IMPACT ON INDUSTRY

The worldwide market for glass is divided by regions and product types. By its nature, glass is heavy and fragile. It is subject to breaking when shipped over long distances. For this reason, most glass is manufactured in plants located as close to customers as possible.

Flat glass is one of the largest glass markets globally. An estimated 53 million tons of flat glass, valued at $66 billion, were made in 2008. The majority of flat glass, about 70 percent of the total market, was used for windows in homes and industrial buildings. Furniture, light fixtures, and other interior products made up 20 percent of the market, while the rest (10 percent) was used primarily in the automotive industry. The manufacturing of flat glass requires significant capital investment in large plants, so the market is dominated by a few large companies. In 2009, the world's leading makers of flat glass were the NSG Group (based in Japan), Saint-Gobain (France), Asahi Glass (Japan), and Guardian Industries (United States). The market is growing at about 4 to 5 percent each year. Much of this growth is fueled by demand from China, where the construction and automotive industries are expanding at a faster rate than in North America and Europe. New flat glass products incorporate nonglass materials such as polymers, which enhance safety and the insulating properties of windows.

The glass packaging industry worldwide is led by bottles used for beverage products. In 2009, glass container manufacturers produced more than 150 billion beer bottles and more than 60 billion bottles for spirits, wines, and carbonated drinks. Research firm Euromonitor International predicts that demand for glass packaging will grow, but growth rates across package types will vary significantly. Glass bottles face competition from polyethylene terephthalate (PET) bottles, which are much lighter and more durable. A rise in median consumer income across the Asia-Pacific region in the 2000's has increased demand for processed food products sold in jars and bottles.

Niches such as glass labware are also facing pressure from items made with nonglass materials. The global market for glass labware could be as high as $3.87 billion by 2015, according to Global Industry Analysts. New plastics developed for specialized applications such as the containment of hazardous chemicals may slow down the rate at which laboratories purchase glass containers. However, glass will continue to be a laboratory staple as long as it remains inexpensive and widely available.

Fiber-optic cable continues to fuel growth for glass manufacturing. In late 2010, many telecommunications firms were preparing to invest in expanding their fiber-optic networks. Telecom research firm CRU estimated that companies worldwide ordered more than 70 million miles of new fiber-optic cable between July, 2009, and June, 2010. The growth in

demand in Europe was nearly twice that of the United States in 2009 and 2010.

CAREERS AND COURSE WORK

The skills required to work in the glassmaking industry depend on the nature of the glass products being made. Manufacturing and installing glass windows in buildings is a career within the construction industry, while the design of glass homewares involves the consumer products industry. The making of lenses for eyeglasses is a career track that differs significantly from the production of specialty lenses for precision equipment such as telescopes.

Glaziers in construction gain much of their knowledge from on-the-job training. Many glaziers enter apprenticeships of about three years with a contractor. The apprentice glazier learns skills such as glass cutting by first practicing on discarded glass. Tasks increase in difficulty and responsibility as the apprentice's skills grow. This work is supplemented by classroom instruction on topics such as mathematics and the design of construction blueprints. Similarly, makers of stemware and high-end glass products used in homes learn many of their skills through hands-on work with experienced designers.

Careers in optical glass and lens design, on the whole, require more formal education than other areas of glass production do. Several universities offer course concentrations in glass science, optics, or ceramics within undergraduate engineering programs. Graduate programs can focus on fields as narrow as electro-optics (fewer than ten American schools offer this option).

Students planning careers in optical glass take general classes such as calculus, physics, and chemistry. Course work on topics such as thermodynamics, microscopy, spectroscopy, and computer-aided design are also standard parts of an optical engineer's education.

SOCIAL CONTEXT AND FUTURE PROSPECTS

Demand for glaziers in the construction industry rises and falls with the rate of new buildings being built. The early 2000's saw a sharp increase in demand for glaziers, particularly those with experience in installing windows in new houses. This trend reversed with the end of the construction boom in the late 2000's. At the same time, the development of window glass with energy-efficient features led to the upgrading of windows in office buildings and homes, which has increased the need for skilled glaziers. The U.S. Bureau of Labor Statistics predicts that jobs in this field will grow at about 8 percent from 2008 to 2018, an average rate compared to all other occupations.

Within the field of optical glass, manufacturers have developed a wide range of glass types for specialized applications such as lasers. The need for increasingly precise lenses has grown steadily with developments in fields ranging from microsurgery to astronomy.

—*Julia A. Rosenthal, MS*

FURTHER READING

Beretta, Marco. The Alchemy of Glass: Counterfeit, Imitation, and Transmutation in Ancient Glassmaking. Sagamore Beach, Mass.: Science History Publica-tions/USA, 2009. Examines the history of glass-making, from Egypt and Babylonia to modern times, looking at the effects of technology.

Hartmann, Peter, et al. "Optical Glass and Glass Ceramic Historical Aspects and Recent Developments: A Schott View." Applied Optics (June 1, 2010): D157D176. A brief but informative overview of optical glass from the Italian Renaissance to the modern times.

Le Bourhis, Eric. Glass: Mechanics and Technology. Weinheim, Germany: Wiley-VCH, 2008. Explores the properties of glass and their physical and chemical sources.

Opie, Jennifer Hawkins. Contemporary International Glass. London: Victoria and Albert Museum, 2004. Photographs and artistic statements from sixty international glass artists, based on pieces in the collection of the Victoria and Albert Museum in London.

Shelby, J. E. Introduction to Glass Science and Technology. Cambridge, England: Royal Society of Chemistry, 2005. A college-level textbook with detailed information on the physical and chemical properties of glass.

Skrabec, Quentin, Jr. Michael Owens and the Glass Industry. Gretna, La.: Pelican, 2007. A biography of American industrialist Michael Owens, whose mass production techniques for glass bottles and cans led to major changes in the lives of consumers.

WEB SITES

American Scientific Glassblowers Society
http://www.asgs-glass.org

Association for the History of Glass
http://www.historyofglass.org.uk

Corning Museum of Glass
http://www.cmog.org

Society of Glass Technology
http://www.societyofglasstechnology.org.uk

See also: Mirrors and Lenses; Optics.

GRAPHENE

FIELDS OF STUDY

Electronics; Nanotechnology

SUMMARY

Graphene is an allotrope of elemental carbon, with a planar molecular structure that can theoretically extend to infinite size in two dimensions. While only recently recognized, graphene is a naturally-occurring material found as graphite. Closure of graphene sheets produces carbon nanotubes and other fullerenes. The versatile structural and unique electronic characteristics of graphene present entirely new fields of application and may see the material replace silicon as the basic material of transistor devices.

PRINCIPAL TERMS

- **continuity:** a clear path for electricity from point A to point B
- **covalent bond:** a type of chemical bond in which electrons are shared between two adjacent atoms.
- **element:** a form of matter consisting only of atoms of the same atomic number

- **molecular formula:** a chemical formula that indicates how many atoms of each element are present in one molecule of a substance.
- **standard model:** a generally accepted unified framework of particle physics that explains electromagnetism, the weak interaction, and the strong interaction as products of interactions between different types of elementary particles.
- **structural formula:** a graphical representation of the arrangement of atoms and bonds within a molecule.

THE UNIQUE CHARACTER OF CARBON

In the standard model, atoms consist of a very small, dense nucleus composed of protons and neutrons that is surrounded by a very large, and very diffuse, cloud of electrons. The identity of any particular atom as an element is defined by the number of protons that are contained within its nucleus. The various forces at work within the nucleus of an atom define the specific energies that the surrounding electrons are 'allowed' to have and the specific regions about the nucleus that are generally called 'orbitals'. This is an important factor for understanding the nature of carbon, and therefore of graphene. The carbon atom contains six protons, and therefore has six electrons occupying the orbitals about the carbon nucleus. Two of these fully occupy the lowest energy atomic orbital, but the remaining four electrons have four orbitals available at the next higher quantum level, and this permits the carbon atom to form different kinds and numbers of covalent bonds in the formation of molecules. In one arrangement, three of the four slightl;y different orbitals combine to form three identical 'hybrid' orbitals that lie in a plane and are equally spaced at 120° about the nucleus. The fourth orbital keeps its original shape and is perpendicular to the plane of the three hybrid orbitals. This geometry is perfectly suited to the formation of very stable rings of six carbon atoms, which is the essential feature of the structural formula of graphene. It can be easily visualized as a sheet of 'chicken wire'. The variable size of the polymeric structure of graphene makes identification of an individual molecular formula problematic, except in the case of specific fullerenes such as carbon nanotubes and 'buckyball' molecules.

PROPERTIES OF GRAPHENE

Graphene is perhaps the most advanced material known, yet as naturally occurring graphite as is used in common pencils, people have been using it to write and draw with from prehistoric times. Graphite is composed of layer upon layer of graphene molecules that have been formed from carbonaceous matter under the effects of heat and pressure for hundreds of thousands of years, and has been known since 1947, but only in the context of a macroscopic complex material. Actual molecules of graphene were first isolated in 2004 at Manchester University, by Andre Geim and Konstantin Novoselov, who were later awarded the Nobel Prize in Physics for this work. The actual isolation of the material was achieved by using sticky tape to repeatedly separate layers of graphene until just the single one-atom-thick molecule remained. Graphene is the thinnest material known, about one million times thinner than a single human hair, yet it is two hundred times stronger than steel. It is transparent, and impermeable. It has a high degree of flexibility, as is demonstrated by its relationship to carbon nanotubes, which are essentially just tightly rolled sheets of graphene whose edge carbon atoms have bonded to each other. The uniform field of perpendicular non-bonding atomic orbitals, one on each carbon atom, overlap in such a way that they provide perfect continuity for the movement of electrons in an electrical current, making graphene the most highly conductive material known. Being composed of just carbon atoms, it is very light in weight relative to materials of similar strength properties.

POTENTIAL OF GRAPHENE

The unique nature of graphene opens up entirely new fields of possible applications based on both its structural properties and its electronic properties. Structurally, graphene is a perfect material to use as a substrate for the construction of atom-sized structures such as transistors. As a scaffold for "nanodots" - structures composed of just a few single atoms – graphene has numerous potential applications in such technologies as solar energy, optics and surface science. As a scaffold for new transistor designs, graphene may replace silicon as the material of choice for the fabrication of digital electronic devices. Transistors constructed at the atomic scale rather than the current nanometer scale would be thousands of times faster and more reliable, and consume much less energy than computers based on silicon technology. The material is in actuality so versatile that it may impossible to know all of the applications in which it may be used and the new technologies that may be developed from it.

Further Reading

Jiang, De-en and Chen, Zhomgfanf, eds. *Graphene Chemistry. Theoretical Perspectives* Hoboken, NJ: John Wiley & Sons, 2013.

Koratkar, Nikhil A., *Graphene in Composite Materials* Lancaster, PA: DESTech Publications, 2013.

Torres, Luis E.F. Foa, Roche, Stephan and Charlier, Jean-Christophe *Introduction to Graphene-Based Nanomaterials. From Electronic Structure to Quantum Transport* New York, NY: Cambridge University Press, 2014.

Warner, Jamie H., Schäffel, Franziska, Bachmatiuk, Alicja and Rümmeli, Mark H., *Graphene. Fundamentals and Emergent Applications* Wlatham, MA: Elsevier, 2013

Wolf, E.L. *Graphene. A New Paradigm in Condensed Matter and Device Physics* New York, NY: Oxford University Press, 2014.

—*Richard M. Renneboog MSc*

GRAVITATIONAL RADIATION

FIELDS OF STUDY

Relativity, Cosmology

SUMMARY

Of all the fundamental forces known only two have infinite range, the electromagnetic and the gravitational. The two forces are both inverse square forces, but the electromagnetic force can be attractive or repulsive where the gravitational force is, as far as we know, always attractive. The electromagnetic force is responsible for the production of light or electromagnetic radiation whenever an electrical charge is accelerated. The electromagnetic field is described by four coupled differential equations, commonly called Maxwell's equations, after the British theoretical physicist James Clerk Maxwell (1831-1879) who systematized them around 1865. Maxwell's equations gave a unified treatment of electric and magnetic phenomena and made predictions about light and electromagnetism that were quickly verified. Maxwell's discoveries prompted a search for a field theory of gravitation and possibly a unified field theory of the electromagnetic and gravitational field which attracted the efforts of Albert Einstein (1879-1955) among others. Einstein realized that the acceleration of two massive bodies with respect to each other would give rise to gravitational waves that could be detected by their action on large bodies. Nonetheless, the gravitational force is so weak compared to electromagnetism that the gravitational waves remained undetected for one hundred years. Their detection was an engineering tour de force involving signals detected 2000 miles apart.

PRINCIPAL TERMS

- **black hole:** A celestial body whose gravitational field is so strong that even light cannot escape.
- **Coulomb's law:** The basic law of force between nonmoving charges. ($F=kq_1q_2/r^2$)
- **interferometry:** Using the interference of waves of any type to measure distances.
- **signal:** A measurable meaningful physical quantity.
- **noise:** Unavoidable fluctuations in a measured physical quantity.
- **tensor calculus:** a mathematical technique for studying the transformation from one set of tensor components to another, particularly valuable when the curvature of space time cannot be ignored.
- **universal gravitational constant:** The constant $G = 6.67 \times 10^{-11}$ Nm2/kg2 appearing in Newton's Law of Universal Gravitation ($F = GM_1M_2/r^2$)

THEORY OF RELATIVITY

Albert Einstein launched a revolution in physics in 1905 when he pointed out that two observers in relative motion might disagree on the spatial separation of two events, r and their separation in time t and yet agree on the space time interval $(ct)^2-(r)^2$ between them. The special theory of relativity allows discussion of measurements made by observers in relative uniform motion. It is itself relatively simple and can be taught to junior high school students.

Einstein did not stop there of course. A decade later he announced the general theory of relativity which allowed the discussion of measurements made in reference frames which are accelerated with respect to each other. While the special theory did not place heavy mathematical demands on the student the general theory does, requiring tensor calculus, as one must deal with the mixing of space and time coordinates. The general theory assumes the equivalence of gravitation and uniform acceleration. Unlike special relativity which is tested numerous times each day in all manner of experiments, there are a relatively small number of predictions of general relativity that can be tested because of the different scale of gravitational and electromagnetic phenomena. Nonetheless general relativity has passes each experimental test to which it has been subjected.

General relativity deviates from classical Newtonian mechanics particularly in the realm of very dense material. One of the first of its predictions to be verified concerned the bending of starlight as it grazed the surface of our sun during a solar eclipse. It was in fact the success of this prediction that led to Einstein's position and fame. But a stellar produced electromagnetic radiation one expects the acceleration of masses to produce gravitational radiation. However, because the gravitational force is so much weaker than the electromagnetic one requires the collision of very large masses (say two black holes) and very sensitive detectors to find evidence for it.

Detecting gravitational radiation is no tabletop experiment. One requires a very large baseline, say 2000 miles and massive receivers, masses of several tons, and then one must somehow extract the signal from random vibrations, such as trucks driving past the laboratory. Detection was reported in 2016 by the LIGO (Laser Interferometry Gravitational-Wave Observatory) collaboration and the Virgo collaboration and reported in the Journal Physical Review

---

An Example: Calculate the ratio of gravitational force to the electrical force of attraction between the proton and the electron in a hydrogen atom.

Data: m_{proton} = 1.67 x 10-27 kg

$m_{electron}$ = 9.1 x 10-31kg

q_{proton} = -$q_{electron}$=1.6x10-19C

ratio = (6.67x10-11Nm2kg-2x9.1x10-31kgx1.67 x 10-27 kg)/(9x109 Nm2C-2 x (1.6x10-19C)2

= 4x10-40, a much weaker force indeed.

Letters in a paper sixteen pages long, with nearly a thousand authors affiliated with over 60 institutions. Needless to say, coordinating the work of so many individuals is quite expensive and not surprisingly subject to political influence, thus the experiment connected scientists in Hanford, Washington with those as a site in Louisiana, rather than on the East coast which would have made for a longer baseline and slightly more accurate result.

—*Donald R. Franceschetti, PhD*

FOR FURTHER READING

Moerchen, Margaret and Coontz, Robert, Einstein's Vision: General Relativity turns 100, Science, 21 March 2015.

Taylor, Edwin F. and Wheeler, J. Archibald, Spacetime Physics (Freeman, San Francisco, 1963)

Misner, Charles W., Thorne, Kip S., Wheeler, J. Archibald Gravitation (Freeman, San Francisco, 1970)

HEAT-EXCHANGER TECHNOLOGIES

FIELDS OF STUDY

Thermodynamics; fluid mechanics; heat transfer; mechanics of materials; calculus; differential equations.

SUMMARY

A heat exchanger transfers thermal energy from one flowing fluid to another. A car radiator transfers thermal energy from the engine-cooling water to the atmosphere. A nuclear power plant contains very large heat exchangers called steam generators, which transfer thermal energy out of the water that circulates through the reactor core and makes steam that drives the turbines. In some heat exchangers, the two fluids mix together, while in others, the fluids are separated by a solid surface such as a tube wall.

KEY TERMS AND CONCEPTS

- **closed heat exchanger:** Exchanger in which the two fluids are separated by a solid surface.
- **counterflow:** Fluids flowing in opposite directions.
- **cross-flow:** Fluids flowing perpendicular to each other.
- **deaerator (direct-contact heater):** Open heat exchanger in which steam is mixed with water to bring the water to its boiling point.
- **open heat exchanger:** Exchanger in which the two fluids mix together and exit as a single stream.
- **parallel flow (co-current):** Both fluids flowing in the same direction.
- **plate heat exchanger:** Exchanger in which corrugated flat metal sheets separate the two flowing fluids.
- **shell:** Relatively large, usually cylindrical, enclosure with many small tubes running through it.
- **shell and tube heat exchanger:** Exchanger composed of many relatively small tubes running through a much larger enclosure called a shell.
- **tube sheet:** Flat plate with many holes drilled in it. Tubes are inserted through the holes.

DEFINITION AND BASIC PRINCIPLES

There are three modes of heat transfer. Conduction is the method of heat transfer within a solid. In closed heat exchangers, conduction is how thermal energy moves through the solid boundary that separates the two fluids. Convection is the method for transferring heat between a fluid and a solid surface. In a heat exchanger, heat moves out of the hotter fluid into the solid boundary by convection. That is also the way it moves from the solid boundary into the cooler fluid. The final mode of heat transfer is radiation. This is how the energy from the Sun is transmitted through space to Earth.

The simplest type of closed heat exchanger is composed of a small tube running inside a larger one. One fluid flows through the inner tube, while the other fluid flows in the annular space between the inner and outer tubes. In most applications, a double-pipe heat exchanger would be very long and narrow. It is usually more appropriate to make the outer tube large in diameter and have many small tubes inside it. Such a device is called a shell and tube heat exchanger. When one of the fluids is a gas, fins are often added to heat-exchanger tubes on the gas side, which is usually on the outside.

Plate heat exchangers consist of many thin sheets of metal. One fluid flows across one side of each sheet, and the other fluid flows across the other side.

BACKGROUND AND HISTORY

Boilers were probably the first important heat exchangers. One of the first documented boilers was invented by Hero of Alexandria in the first century. This device included a crude steam turbine, but Hero's engine was little more than a toy. The first truly useful boiler may have been invented by the Marquess of Worcester in about 1663. His boiler provided steam to drive a water pump. Further developments were made by British engineer Thomas Savery and British blacksmith Thomas Newcomen, though many people mistakenly believe that James Watt invented the steam engine. Watt invented the condenser, another kind of heat exchanger. Combining the condenser with existing engines made them much more efficient. Until the late nineteenth century boilers and condensers dominated the heat-exchanger scene.

The invention of the diesel engine in 1897 by Bavarian engineer Rudolf Diesel gave rise to the need for other heat exchangers: lubricating oil coolers, radiators, and fuel oil heaters. During the twentieth century, heat exchangers grew rapidly in number, size, and variety. Plate heat exchangers were invented. The huge steam generators used in nuclear plants were produced. Highly specialized heat exchangers were developed for use in spacecraft.

HOW IT WORKS

Thermal calculations. There are two kinds of thermal calculations–design calculations and rating calculations. In design calculations, engineers know what rate of heat transfer is needed in a particular application. The dimensions of the heat exchanger that will satisfy the need must be determined. In rating calculations, an existing heat exchanger is to be used in a new situation. The rate of heat transfer that it will provide in this situation must be determined.

In both design and rating analyses, engineers must deal with the following resistances to heat transfer: the convection resistance between the hot fluid and the solid boundary, the conduction resistance of the solid boundary itself, and the convection resistance between the solid boundary and the cold fluid. The conduction resistance is easy to calculate. It depends only on the thickness of the boundary and the thermal conductivity of the boundary material. Calculation of the convection resistances is much more complicated. They depend on the velocities of the fluid flows and on the properties of the fluids such as viscosity, thermal conductivity, heat capacity, and density. The geometries of the flow passages are also a factor.

Because the convection resistances are so complicated, they are usually determined by empirical methods. This means that research engineers have conducted many experiments, graphed the results, and found equations that represent the lines on their graphs. Other engineers use these graphs or equations to predict the convection resistances in their heat exchangers.

In liquids, the convection resistance is usually low,

but in gases, it is usually high. In order to compensate for high convection resistance, fins are often installed on the gas side of heat-exchanger tubes. This can increase the amount of heat transfer area on the gas side by a factor of ten without significantly increasing the overall size of the heat exchanger.

Hydraulic calculations. Fluid friction and turbulence within a heat exchanger cause the exit pressure of each fluid to be lower than its entrance pressure. It is desirable to minimize this pressure drop. If a fluid is made to flow by a pump, increased pressure drop will require more pumping power. As with convection resistance, pressure drop depends on many factors and is difficult to predict accurately. Empirical methods are again used. Generally, design changes that reduce the convection resistance will increase the pressure drop, so engineers must reach a compromise between these issues.

Strength calculations. The pressure of the fluid flowing inside the tubes is often significantly different from the pressure of the fluid flowing around the outside. Engineers must ensure that the tubes are strong enough to withstand this pressure difference so the tubes do not burst. Similarly, the pressure of the fluid in the shell (outside the tubes) is often significantly different from the atmospheric pressure outside the shell. The shell must be strong enough to withstand this.

Fouling. In many applications, one or both fluids may cause corrosion of heat-exchanger tubes, and they may deposit unwanted material on the tube surfaces. River water may deposit mud. Seawater may deposit barnacles and other biological contamination. The general term for all these things is fouling. The tubes may have a layer of fouling on the inside surface and another one on the outside. In addition to the two convection resistances and the conduction resistance, there may be two fouling resistances. When heat exchangers are designed, a reasonable allowance must be made for these fouling resistances.

APPLICATIONS AND PRODUCTS

Heat exchangers come in an amazing variety of shapes and sizes. They are used with fluids ranging from liquid metals to water and air. A home with hot-water heat has a heat exchanger in each room. They are called radiators, but they rely on convection, not radiation. A room air conditioner has two heat exchangers in it. One transfers heat from room air to the refrigerant, and the other transfers heat from the refrigerant to outside air. Cars with water-cooled engines have a heat exchanger to get rid of engine heat. It is called a radiator, but again it relies on convection.

Boilers. Boilers come in two basic types: fire tube and water tube. In both cases, the heat transfer is between the hot gases produced by combustion of fuel and water that is turning to steam. As the name suggests, a fire-tube boiler has very hot gas, not actually fire, inside the tubes. These tubes are submerged in water, which absorbs heat and turns into steam. In a water-tube boiler, water goes inside the tubes and hot gases pass around them. Water-tube boilers often include superheaters. These heat exchangers allow the steam to flow through tubes that are exposed to the hot combustion gases. As a result the final temperature of the steam may reach 1,000 degrees Fahrenheit or higher. An important and dangerous kind of fouling in boilers is called scale. Scale forms when minerals in the water come out of solution and form a layer of fouling on the hot tube surface. Scale is dangerous because it causes the tube metal behind it to get hotter. In high-performance boilers, this can cause a tube to overheat and burst.

Condensers. Many electric plants have generators driven by steam turbines. As steam leaves a turbine it is transformed back into liquid water in a condenser. This increases the efficiency of the system, and it recovers the mineral-free water for reuse. A typical condenser has thousands of tubes with cooling water flowing inside them. Steam flows around the outsides, transfers heat, and turns back into liquid water. The cooling water may come from a river or ocean. When a source of a large amount of water is not available, cooling water may be recirculated through a cooling tower. Hot cooling water leaving the condenser is sprayed into a stream of atmospheric air. Some of it evaporates, which lowers the temperature of the remaining water. This remaining water can be reused as cooling water in the condenser. The water that evaporates must be replaced from some source, but the quantity of new water needed is much less than when cooling water is used only once. When

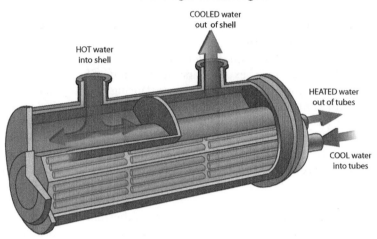

Heat-Exchanger Technologies

COOLED water out of shell

HOT water into shell

HEATED water out of tubes

COOL water into tubes

river water or seawater is used for cooling, there may be significant fouling on the insides of the tubes. Because the steam leaving a turbine contains very small droplets of water moving at high speed, erosion on the outsides of the tubes is a problem. Eventually a hole may develop, and cooling water, which contains dissolved minerals, can leak into the condensing steam.

Steam generators. In a pressurized-water nuclear power plant, there is a huge heat exchanger called a steam generator. The primary loop contains water under high pressure that circulates through the reactor core. This water, which does not boil, then moves to the steam generator, where it flows inside a large number of tubes. Secondary water at lower pressure surrounds these tubes. As the secondary water absorbs heat it turns into steam. This steam is used to drive the turbines. Steam generators are among the largest heat exchangers in existence.

Deaerators. Because the condensers in steam systems operate with internal pressures below one atmosphere, air may leak in. Some of this air dissolves in the water that forms as steam condenses. If this air remained in the water as it reached the boiler, rapid rusting of boiler surfaces would result. To prevent this, an open heat exchanger, called a deaerator, is installed. Water is sprayed into the deaerator as a fine mist, and steam is also admitted. As the steam and water mix, the water droplets are heated to their boiling point but not actually boiled. The solubility of air in water goes to zero as the water temperature approaches the boiling point, so nearly all air is forced out of solution in the deaerator. Once the air is in gaseous form, it is removed from the system.

Feedwater heaters. Leaving the deaerator, the water in a steam plant is on its way to the boiler. The system can be made more efficient by preheating the water along the way. This is done in a feedwater heater. Steam is extracted from the steam turbines to serve as the heat source in feedwater heaters. Feedwater flows inside the tubes and steam flows around the tubes. Feedwater heaters are often multi-pass heat exchangers. This means that the feedwater passes back and forth through the heat exchanger several times. This makes the heat exchanger shorter and fatter, which is a more convenient shape.

Intercoolers. Many diesel engines have turbochargers that pressurize the air being fed to the cylinders. As air is compressed, its temperature rises. It is desirable to lower the temperature of the air before it enters the cylinders, because that means a greater mass of air can occupy the same space. More air in the cylinder means more fuel can be burned, and more power can be produced. An intercooler is a closed heat exchanger between the turbocharger and the engine cylinders. In this device, air passes around the tubes, and cooling water passes inside them. There are usually fins on the outsides of the tubes to provide increased heat transfer area.

Industrial air compressors also have intercoolers. These compressors are two-stage machines. That means air is compressed part way in one cylinder and the rest of the way in another. As with turbochargers, the first compression raises the air temperature. An intercooler is often installed between the cylinders. Compressed air flows through the intercooler tubes. Either atmospheric air or cooling water flows around the outside. Cooling the air before the second compression reduces the power required there.

IMPACT ON INDUSTRY

By 1880 the use of boilers was widespread, but there was little or no regulation of their design and construction. As a result, boiler explosions were commonplace, and many lives were lost in that way every year. The American Society of Mechanical Engineers (ASME) was founded that year to develop safety standards for boilers.

Early in the twenty-first century, an industry group called Tubular Exchanger Manufacturers Association (TEMA) published standards for shell-and-tube heat exchangers. This organization certifies heat exchangers that are built in accordance with its standards. Purchasers of heat exchangers with TEMA certification are assured that these devices are properly designed and safe to operate. More than twenty manufacturers are members of TEMA. Standards for shell-and-tube heat exchangers are also published by ASME, Heat Transfer Research, and the American Petroleum Institute.

Manufacturers of small heat exchangers include Exergy LLC and Springboard Manufacturing. These heat exchangers are small enough to hold in one's hand. Basco/Whitlock manufactures medium-size shell-and-tube heat exchangers from two inches to eight inches in diameter and eight inches to ninety-six inches in length. The shells are made of brass or stainless steel, and the tubes are made of copper, Admiralty metal (an alloy of copper, zinc, and tin), or stainless steel.

Westinghouse, Combustion Engineering, and Babcock & Wilcox have built many of the huge steam generators used in U.S. nuclear power plants. Heat Transfer Equipment Company manufactures large steel heat exchangers. It specializes in heat exchangers for the oil and gas industry. In Japan, Mitsubishi Heavy Industries has manufactured about one hundred steam generators for use in nuclear power plants there.

The Electric Power Research Institute, an organization composed of companies that generate electricity, funds and conducts research on all aspects of electric power generation, including the heat exchangers that are involved. Much of the basic research on convection resistance in heat exchangers was conducted in university laboratories during the middle of the twentieth century. Although this remains an active research area, the level of activity is lower in the early twenty-first century.

CAREERS AND COURSE WORK

Heat exchangers are usually designed by mechanical engineers who hold bachelor or master of science degrees in this field. Students of mechanical engineering take courses in advanced mathematics, mechanics of materials, thermodynamics, fluid

FASCINATING FACTS ABOUT HEAT-EXCHANGER TECHNOLOGIES

- Heat exchangers that have seawater flowing through them usually have small zinc blocks in the flow. These blocks help protect the exchanger surfaces from corrosion.

- Steam condensers operate below atmospheric pressure. Any leakage that occurs is air going in rather than steam coming out. These heat exchangers have air-removal devices called air ejectors connected to them.

- Heat exchanger tubes are usually attached to tube sheets by "rolling," a process that expands the tube to make it fit tightly in its hole in the tube sheet.

- Shell and tube heat exchangers have baffle plates that direct the fluid flowing over the outsides of the tubes. The baffle plates also help support the tubes.

- Electronic sensing devices called eddy-current sensors are passed through the tubes of a shell heat exchanger to detect potential problems before tubes fail.

- Steam generators in nuclear power plants can be as tall as 70 feet and weigh as much as 800 tons when empty. These are among the largest heat exchangers in existence.

- The slang "crud" is said to be an acronym of Chalk River unidentified deposits, which refers to radioactive fouling found in the steam generators of the Chalk River Laboratories in Canada.

- Radiators in cars and home-heating systems make use of convection heat transfer, which refers to the transfer of heat into a fluid. In both cases, heat is transferred to air. They do not make significant use of radiation heat transfer.

- Heat exchangers in spacecraft make use of radiation heat transfer, because there is no air or water to absorb the heat.

mechanics, and heat transfer. An M.S. degree provides advanced understanding of the physical phenomena involved in heat exchangers. Research into the theory of heat transfer is normally carried out by mechanical engineers with doctoral degrees. They conduct research in laboratories, at universities, private research companies, or large corporations that build heat exchangers. As mentioned earlier, convection heat transfer calculations rely on equations that are derived from extensive experiments. Much research work continues to be devoted to improving the accuracy of these equations.

Construction of heat exchangers is executed by companies large and small. The work is carried out by skilled craftsmen using precise machine tools and other equipment. Machinists, welders, sheet-metal, and other highly trained workers are involved. Students who pursue such careers may begin with vocational-technical training at the high school level. They become apprentices to one of these trades. During apprenticeship, the workers receive formal training in classrooms and on-the-job training. As their skills develop they become journeymen and then master mechanics.

Workers who operate, maintain, and repair heat exchangers have a variety of backgrounds. Some have engineering or engineering technology degrees. Others have vocational-technical and on-the-job training. At nuclear power plants, the Nuclear Regulatory Commission requires a program of extensive testing of the vital heat exchangers. This is carried out by engineers with B.S. or M.S. degrees, assisted by skilled craftsmen.

SOCIAL CONTEXT AND FUTURE PROSPECTS

Although heat exchangers are not glamorous, they are an essential part of people's lives. Every home has several, as does every car and truck. Without heat exchangers, people would still be heating their homes with fireplaces, and engines of all types and sizes would not be possible. Heat exchangers are essential in all manner of industries. In particular, they play a key role in the generation of electricity.

The design of heat exchangers is based on empirical methods rather than basic principles. While empirical methods are reasonably effective, design from basic principles would be preferred. In the early twenty-first century, extensive research projects are under way with the goal of solving the very complicated equations that represent the basic principles of heat transfer. These projects make use of very powerful computers. As the cost of computers continues to drop and their power continues to increase, heat exchangers may come to be designed from basic principles.

—*Edwin G. Wiggins, MS, PhD*

FURTHER READING

Babcock and Wilcox Company. *Steam: Its Generation and Use.* Reprint. Whitefish, Mont.: Kessinger Publishing, 2010. This easy-to-read book provides extensive information about all manner of fossil-fueled boilers as well as nuclear steam generators.

Blank, David A., Arthur E. Bock, and David J. Richardson. *Introduction to Naval Engineering*. 2d ed. Annapolis, Md.: Naval Institute Press, 2005. Intended for use by freshmen at the U.S. Naval Academy, this book provides a simple, nonmathematical discussion of heat exchangers used on naval ships.

McGeorge, H. D. *Marine Auxiliary Machinery*. 7th ed. Burlington, Mass.: Butterworth-Heinemann, 1995. Excellent descriptions of shell-and-tube heat exchangers, plate heat exchangers, and deaerators are provided in very readable form, including a simple presentation of heat exchanger theory.

Thurston, Robert Henry. *A History of the Growth of the Steam-Engine*. Ithaca, N.Y.: Cornell University Press, 1939. Although very old, this book is useful because it provides a comprehensive history of the development of steam engines, boilers, and condensers.

Web Sites

American Petroleum Institute
http://www.api.org

American Society of Mechanical Engineers
http://www.asme.org

Electric Power Research Institute
http://my.epri.com/portal/server.pt?

Heat Transfer Research, Inc.
http://www.htri.net

Tubular Exchanger Manufacturers Association
http://tema.org

See also: Calculus; Fluid Dynamics; Mechanical Engineering.

HYBRID VEHICLE TECHNOLOGIES

FIELDS OF STUDY

Electrical engineering; mechanical engineering; chemistry; ecology.

SUMMARY

Hybrid vehicle technologies use shared systems of electrical and gas power to create ecologically sustainable industrial and passenger vehicles. With both types of vehicles, the main goals are to reduce hazardous emissions and conserve fuel consumption.

KEY TERMS AND CONCEPTS

- **driving range:** Distances, both in city driving and open-road driving, that can be covered by HEVs before refueling or recharging is necessary.
- **fuel cells:** Very advanced form of electrochemical cell (instead of a conventional battery) that produces electricity following a reaction between an externally supplied fuel, frequently hydrogen, which becomes positively charged ions following contact with an oxidant, frequently oxygen or chlorine, which "strips" its electrons.
- **internal combustion engine (ICE):** Produces power by the pressure of expanding fossil fuel gases after they are ignited in cylinder chambers.
- **lithium-ion battery:** Lightweight rechargeable battery with energy-efficient qualities tied to the fact that lithium ions travel from the negative to positive electrode when the battery discharges, returning to the negative during charging.
- **parallel system hybrid:** Main power comes from the gas engine; electric motor steps in only for power to accelerate.
- **series system hybrid:** Power is delivered to the wheels from the electric motor; the gas engine's role is to generate electricity.
- **third-generation hybrid:** Improved vehicle models using newer technology after the previous (first and second) generations.

DEFINITION AND BASIC PRINCIPLES

As the word "hybrid" suggests, hybrid vehicle technology seeks to develop an automobile (or, more broadly defined, any power-driven mechanical system) using power from at least two different sources. Before and during the first decade of the twenty-first century, hybrid technology emphasized the combination of an internal combustion engine working with an electric motor component.

BACKGROUND AND HISTORY

Before technological development of what is now called a hybrid vehicle, the automobile industry, by necessity, had to have two existing forms of motor energy to hybridize–namely internal combustion in combination with some form of electric power. Early versions of cars driven with electric motors emerged in the 1890's and seemed destined to compete very seriously with both gasoline (internal combustion engines) and steam engines at the turn of the twentieth century.

Although development of commercially attractive hybrid vehicles would not occur until the middle of the twentieth century, the Austrian engineer Ferdinand Porsche made a first-series hybrid automobile in 1900. Within a short time, however, the commercial attractiveness of mass-produced internal combustion engines became the force that dominated the automobile industry for more than a half century. Experimentation with hybrid technology as it could be applied to other forms of transport, especially motorcycles, however, continued throughout this early period.

By the 1970's the main emerging goal of hybrid car engineering was to reduce exhaust emissions; conservation of fuel was a secondary consideration. This situation changed when, in the wake of the 1973 Arab-Israeli War, many petroleum-producing countries supporting the Arab cause cut exports drastically, causing a nationwide gas shortage and worldwide fears that oil would be used as a political weapon.

Until 1975 government support for research and development of hybrid cars was tied to the Environmental Protection Agency (EPA). In that year (and after at least two unsatisfactory results of EPA-supported hybrid car projects), this role was shifted to the Energy Research and Development Administration, which later became the Department of Energy (DOE).

During the decade that followed the introduction of Honda's Insight hybrid car in 1999, the most widely recognized commercially marketed hybrid automobile was Toyota's Prius. Despite some setbacks in sales in 2010 following major recalls connected with (among other less dangerous problems) the malfunctioning anti-lock braking system and accelerator

devices, the third-generation Prius still held a strong position in total hybrid car sales globally going into 2011.

HOW IT WORKS

"Integrated motor assist," a common layperson's engineering phrase borrowed from Honda's late1990's technology, suggests a simple explanation of how a hybrid vehicle works. The well-known relationship between the electrical starter motor and the gas-driven engine in an internal combustion engine (ICE) car provides a (technically incomplete) analogy: The electric starter motor takes the load needed to turn the crankcase (and the wheels if gears are engaged) until the ICE itself kicks in. This overly general analogy could be carried further by including the alternator in the system, since it relieves the battery of the job of supplying constant electricity to the running engine (recharging the battery at the same time).

In a hybrid system, however, power from the electric motor (or the gas engine) enters and leaves the drivetrain as the demand for power to move the vehicle increases or decreases. To obtain optimum results in terms of carbon dioxide emissions and overall fuel efficiency, the power train of most hybrid vehicles is designed to depend on a relatively small internal combustion engine with various forms of rechargeable electrical energy. Although petroleum-driven ICE's are commonly used, hybrid car engineering is not limited to petroleum. Ethanol, biodiesel, and natural gas have also been used.

In a parallel hybrid, the electric motor and ICE are installed so that they can power the vehicle either individually or together. These power sources are integrated by automatically controlled clutches. For electric driving, the clutch between the ICE and the gearbox is disengaged, while the clutch connecting the electric motor to the gearbox is engaged. A typical situation requiring simultaneous operation of the ICE and the electric motor would be for rapid acceleration (as in passing) or in climbing hills. Reliance on the electric motor would happen only when the car is braking, coasting, or advancing on level surfaces.

It is extremely important to note that one of the most vital challenges for researchers involved in hybrid-vehicle technology has to do with variable options for supplying electricity to the system. It is far too simple to say that the electrical motor is run by a rechargeable battery, since a wide range of batteries (and alternatives to batteries) exists. A primary and obvious concern, of course, will always be reducing battery weight. To this aim, several carmakers, including Ford, have developed highly effective first-, second-, and even third-generation lithium-ion batteries. Many engineers predict that, in the future, hydrogen-driven fuel cells will play a bigger role in the electrical components of hybrids.

Selection of the basic source of electrical power ties in with corollary issues such as calculation of the driving range (time elapsed and distances covered before the electrical system must be recharged) and optimal technologies for recharging. The simplest scenario for recharging, which is an early direct borrowing from pure-electric car technology, involves plugging into a household outlet (either 110 volt or 220 volt) overnight. But, hybrid-car engineers have developed several more sophisticated methods. One is a "sub-hybrid" procedure, which uses very lightweight fuel cells, mentioned above, in combination with conventional batteries (the latter being recharged by the fuel cells while the vehicle is underway). Research engineers continue to look at any number of ways to tweak energy and power sources from different phases of hybrid vehicle operation. One example, which has been used in Honda's Insight, is a process that temporarily converts the motor into a generator when the car does not require application of the accelerator. Other channels are being investigated for tapping kinetic-energy recovery during different phases of simple mechanical operation of hybrid vehicles.

APPLICATIONS AND PRODUCTS

Some countries, especially Japan, have begun to use the principle of the hybrid engine for heavy-duty transport or construction-equipment needs, as well as hybrid systems for diesel road graders and new forms of diesel-powered industrial cranes. Hybrid medium-power commercial vehicles, especially urban trolleys and buses, have been manufactured, mainly in Europe and Japan. Important for broad ecological planning, several countries, including China and Japan, have incorporated hybrid (diesel combined with electric) technology into their programs for rail transport. The biggest potential consumer market for hybrid technology, however, is probably in the private automobile sector.

By the second decade of the twenty-first century, a wide variety of commercially produced hybrid automobiles were on European, Asian, and American markets. Among U.S. manufacturers, Ford has developed the increasingly popular Escape, and General Motors produces about five models ranging from Chevrolet's economical Volt to Cadillac's more expensive Escalade. Japanese manufacturers Nissan, Honda, and Toyota have introduced at least one, three, and two standard hybrid models, respectively, to which one should add Lexus's RX semi-luxury and technologically more advanced series of cars. Korea's Hyundai Elantra and Germany's Volkswagen Golf also competed for some share of the market.

One of the chief attractions to Toyota's hybrid technology has usually been its primary goal of using electric motors in as many operational phases as possible. The closest sales competitor in the United States to the Prius (mainly for mileage efficiency) was Chevrolet's Volt.

At the outset of 2011, Lexus launched an ambitious campaign to attract attention to what it called its full hybrid technology (as compared with mild hybrid) in its high-end RX models. A main feature of the full hybrid system, according to Lexus, is a combination of both series and parallel hybrid power in one vehicle. Such a combination aims at transferring a variable but continuously optimum ratio of gas-engine and electric-motor power to the car. Another advance claimed by Lexus's full hybrid over parallel hybrids is its reliance on the electric motor only at lower speeds.

Early in 2011, Mercedes-Benz also announced its intention to capture more sales of high-end hybrids by dedicating, over a three-year period, more research to improve the technology used in its S400 model. Audi, a somewhat latecomer, unveiled plans for its first hybrid, the Q5, to appear in European markets late in 2011.

As fuel alternatives continue to be added to the ICE components of HEVs, advanced fuel-cell technology could transform the technological field that supplies electrical energy to the combined system.

IMPACT ON INDUSTRY

Given the factors of added cost associated with designing and producing hybrid vehicles, private companies, both manufacturers and research institutions, are more likely to enter the field if they can receive some form of governmental financial assistance. This can be in the form of direct subsidies, tax reductions, or grants. Similarly, institutions of higher learning, both public and private, frequently seek outside sources of funding for specially targeted research activities. Major national laboratories that are not tied to academic institutions, such as the Argonne National Laboratory near Chicago, also submit research proposals for grants. Private foundations that favor ecological research, particularly reduction of carbon dioxide emissions and alternative methods for producing electrical energy, can also be approached for seed money.

In the United States in 2010, hybrid vehicles made up only 3 percent of the total car sales. If this is a valid indicator, only a much bigger sales potential is likely to induce manufacturers to fund research that could bring to the market more fuel-efficient hybrid-technology cars at increasingly attractive prices. Major vested interests on the potentially negative side are: the gigantic ICE automobile industry itself, which is resistant to changes that require major new investment costs, and the fossil-fuel (petroleum) industry, which holds a near monopoly on fuels supplying power to automobiles, including diesel, and, in some cases, ethanol, around the world.

Future expansion in the number of hybrid cars might also cause important changes in the nature of equipment needed for various aspects of hybrid refueling and recharging–equipment that will eventually have to be made available for commercial distribution. At an even more local level, hiring specialized and, perhaps, higher-paid mechanics capable of dealing with the more advanced technical components of hybrid vehicles may also become part of automotive shops' planning for as-yet unpredictable new directions in their business.

CAREERS AND COURSE WORK

Academic preparation for careers tied to HEV technology is, of course, closely tied to the fields of electrical and mechanical engineering and, perhaps to a lesser degree, chemistry. All of these fields demand course work at the undergraduate level to develop familiarity not only with engineering principles but with basic sciences and mathematics used, especially those used by physicists. Beyond a bachelor's degree, graduate-level preparation would include continuation of

all of the above subjects at more advanced levels, plus an eventual choice for specialization, based on the assumption that some subfields of engineering are more relevant to HEV technology than others.

The most obvious employment possibilities for engineers interested in HEV technology is with actual manufacturers of automobiles or heavy equipment. Depending on the applicant's academic background, employment with manufacturing firms can range from hands-on engineering applications to more conceptually based research and design functions.

Employment openings in research may be found with a wide variety of private- sector firms, some involving studies of environmental impact, others embedded in actual hybrid-engineering technology. These are too numerous to list here, but one outstanding example of a major private firm that is engaged on an international level in environmentally sustainable technology linked to hybrid vehicle research is ABB. ABB grew from late-nineteenth-century origins in electrical lighting and generator manufacturing in Sweden (ASEA), merging in 1987 with the Swiss firm Brown Boveri. ABB carries on operations in many locations throughout the world.

Internationally known U.S. firm Argonne National Laboratory not only produces research data but also serves as a training ground for engineers who either move on to work with smaller ecology-sensitive engineering enterprises or enter government agencies and university research programs.

Finally, employment with government agencies, especially the EPA, the DOE, and Department of Transportation, represents a viable alternative for applicants with requisite advanced engineering and managerial training.

SOCIAL CONTEXT AND FUTURE PROSPECTS

Although obvious ecological advantages can result as more and more buyers of new vehicles opt for hybrid cars, a variety of potentially negative socioeconomic factors could come into play, certainly over the short to medium term. The higher sales price of hybrids that were available toward the end of 2010 already raised the question of consumer ability (or willingness) to pay more at the outset for fuel-economy savings that would have to be spread out over a fairly long time frame–possibly even longer than the owner kept the vehicle. It is nearly impossible to predict the number of potential buyers whose statistically lower

FASCINATING FACTS ABOUT HYBRID VEHICLE TECHNOLOGIES

- The basic technology that is being used, albeit in perfected form, in post-2000 hybrid vehicles was first used to manufacture a working hybrid car more than one hundred years ago.
- Consideration of total weight of a hybrid vehicle is so important that engineers devote major attention to possible innovations for any and all hybrid electric vehicle (HEV) components. The most obvious component that undergoes changes from one generation of HEV to the next involves ever-more efficient modes of supplying electrical energy.
- Technological use of hybrid power systems need not, probably will not, be limited to land transportation. It is possible that aviation--a transport sector that went from conventional to jet engines in the middle of the 20th century--could become considerably more economical and ecological by a combination of power sources.
- Many major cities, especially in Europe (most notably Paris) have fleets of municipal bicycles that can be checked in and out for inner-city use by individuals physically capable and desirous of peddling. It is to be hoped that a next stage-hourly rentals of small HEVs, especially those with major electrical power sources--will follow when production of "basic" (markedly less expensive) hybrid vehicles becomes feasible.
- As more and more sophisticated hybrid-power procedures are developed, the possibility of using different forms of fuel, and eventually bypassing dependence on petroleum, and even biofuels, is a long-range goal of hybrid technology.
- Using regenerative braking, engineers are able to recover electric energy from the magnetic field created when braking results and store it in the HEV's battery for future use.

purchasing ability prevents them from paying higher prices for hybrids. Continued unwillingness or inability to purchase hybrids would mean that a proportionally large number of used older-model ICE's (or brand-new models of older-technology vehicles) would remain on the roads. This socioeconomic potentiality remains linked, of course, to any investment strategies under consideration by industrial producers of cars.

How is one to know which companies worldwide are developing new, economically attractive applications for forthcoming hybrid cars?

The European digital news service EIN News, established in the mid-1990's, provides (among dozens of other categories of information) a specific subsection on hybrid vehicle technology and marketing events, including exhibitions, to its subscribers.

Subscribers from all over the world can obtain up-to-date information on hybrid technology from the Detroit publication *Automotive News* and *Automotive News Europe* and *Automotive News China*. There are, of course, many different marketing congresses (popularly labeled automotive shows) all over the globe, where the latest hybrid technology is introduced and different manufacturers' models can be compared.

In the United States, the Society of Automotive Engineers (SAE) is an important source of up-to-date information for ongoing hybrid vehicle research for both engineering specialists and well-informed general readers.

—Byron D. Cannon, MA, PhD

FURTHER READING

Bethscheider-Kieser, Ulrich. *Green Designed: Future Cars.* Ludwigsburg, Germany: Avedition, 2008. Presents European estimates of technologies that should be compared with the hybrid gas-electric approach to fuel economy.

Clemens, Kevin. *The Crooked Mile: Through Peak Oil, Hybrid Cars and Global Climate Change to Reach a Brighter Future.* Lake Elmo, Minn.: Demontreville Press, 2009. As the title suggests, issues of hybrid car technology need to be placed in a very broad ecological context, where even bigger issues (downward decline in world oil reserves, climate change) may necessitate emphasis on new possible technological solutions.

Lim, Kieran. *Hybrid Cars, Fuel-cell Buses and Sustainable Energy Use.* North Melbourne, Australia: Royal Australian Chemical Institute, 2004. Provides an idea of technologies and programs imagined in other countries.

Society of Automotive Engineers. *1994 Hybrid Electric Vehicle Challenge.* Warrendale, Penn.: Society of Automotive Engineers, 1995. Reports published by thirty American and Canadian college and university engineering laboratories on their respective HEV research programs.

WEB SITES
Electric Auto Association
http://www.electricauto.org

Electric Drive Transportation Association
http://www.electricdrive.org

Society of Automotive Engineers
http://www.sae.org

U.S. Department of Energy Clean Cities
http://www1.eere.energy.gov/cleancities

See also: Biofuels and Synthetic Fuels; Electrical Engineering; Electric Automobile Technology; Mechanical Engineering.

HYDRAULIC ENGINEERING

FIELDS OF STUDY

Physics; mathematics; computer programming; numerical analysis and modeling; material science; civil engineering; water resources engineering; fluid mechanics; hydrostatics; fluid kinematics; hydrodynamics; hydraulic structures; reservoir operations; dam design; open-channel flow; channel design; bridge design; river navigation; coastal engineering; water supply; hydraulic transients; pipeline design; storm drainage; irrigation; water reclamation and recycling; sanitary engineering; environmental engineering; hydraulic machinery; hydroelectric power; pump design.

SUMMARY

Hydraulic engineering is a branch of civil engineering concerned with the properties, flow, control, and uses of water. Its applications are in the fields of water supply, sewerage evacuation, water recycling, flood management, irrigation, and the generation of electricity. Hydraulic engineering is an essential element in the design of many civil and environmental engineering projects and structures, such as water distribution systems, wastewater management systems, drainage systems, dams, hydraulic turbines, channels, canals, bridges, dikes, levees, weirs, tanks, pumps, and valves.

KEY TERMS AND CONCEPTS

- **Froude number:** Dimensionless number for open-channel flow, defined as the ratio of the velocity to the square root of the product of the hydraulic depth and the gravitational acceleration. The Froude number equals 1 for critical flow.
- **gravity flow:** Flow of water with a free surface, as in open channels or partially full pipes.
- **ideal fluid flow:** Hypothetical flow that assumes a frictionless flow and no fluid viscosity; also known as inviscid flow.
- **incompressible fluid:** Fluid that assumes constant fluid density; applies to fluids such as water and oil, but not gases, which are compressible.
- **laminar flow:** Streamlined flow in a pipe in which a fluid particle follows an observable path. It occurs at low velocities, in high-viscosity fluids, and at low Reynolds numbers, typically less than 2,000.
- **pressure flow:** Flow under pressure, such as water in a pipe flowing at capacity.
- **real fluid flow:** Flow that assumes frictional (viscous) effects; also known as viscid flow.
- **Reynolds number:** Dimensionless number defined by the product of the pipe diameter and the flow velocity, divided by the fluid's kinematic viscosity.
- **steady flow:** Flow that remains constant over time at every point.
- **subcritical flow:** Relatively low-velocity and high-depth flow in an open channel. Its Froude number is smaller than 1.
- **supercritical flow:** Relatively high-velocity and small-depth flow in an open channel. Its Froude number is larger than unity.
- **turbulent flow:** Flow characterized by the irregular motion of particles following erratic paths. It may be found in streams that appear to be flowing very smoothly and in swirls and large eddies caused by a disturbance source.
- **uniform flow:** Flow that occurs when the cross section and the velocity remain constant along the channel or pipe.

- **unsteady flow:** Flow that changes with time; also known as hydraulic transient flow.
- **varied flow:** Flow that occurs when the free water surface and the velocity vary along the flow path; also known as nonuniform flow.

DEFINITION AND BASIC PRINCIPLES

Hydraulic engineering is a branch of civil engineering that focuses on the flow of water and its role in civil engineering projects. The principles of hydraulic engineering are rooted in fluid mechanics. The conservation of mass principle (or the continuity principle) is the cornerstone of hydraulic analysis and design. It states that the mass going into a control volume within fixed boundaries is equal to the rate of increase of mass within the same control volume. For an incompressible fluid with fixed boundaries, such as water flowing through a pipe, the continuity equation is simplified to state that the inflow rate is equal to the outflow rate. For unsteady flow in a channel or a reservoir, the continuity principle states that the flow rate into a control volume minus the outflow rate is equal to the time rate of change of storage within the control volume.

Energy is always conserved, according to the first law of thermodynamics, which states that energy can neither be created nor be destroyed. Also, all forms of energy are equivalent. In fluid mechanics, there are mainly three forms of head (energy expressed in unit of length). First, the potential head is equal to the elevation of the water particle above an arbitrary datum. Second, the pressure head is proportional to the water pressure. Third, the kinetic head is proportional to the square of the velocity. Therefore, the conservation of energy principle states that the potential, pressure, and kinetic heads of water entering a control volume, plus the head gained from any pumps in the control volume, are equal to the potential, pressure, and kinetic heads of water exiting the control volume, plus the friction loss head and any head lost in the system, such as the head lost in a turbine to generate electricity.

Hydraulic engineering deals with water quantity (flow, velocity, and volume) and not water quality, which falls under sanitary and environmental engineering. However, hydraulic engineering is an essential element in designing sanitary engineering facilities such as wastewater-treatment plants.

Hydraulic engineering is often mistakenly thought to be petroleum engineering, which deals with the

Hydraulic Flood Retention Basin (HFRB)

flow of natural gas and oil in pipelines, or the branch of mechanical engineering that deals with a vehicle's engine, gas pump, and hydraulic breaking system. The only machines that are of concern to hydraulic engineers are hydraulic turbines and water pumps.

BACKGROUND AND HISTORY

Irrigation and water supply projects were built by ancient civilizations long before mathematicians defined the governing principles of fluid mechanics. In the Andes Mountains in Peru, remains of irrigation canals were found, radiocarbon dating from the fourth millennium BCE. The first dam for which there are reliable records was built before 4000 BCE on the Nile River in Memphis in ancient Egypt. Egyptians built dams and dikes to divert the Nile's floodwaters into irrigation canals. Mesopotamia (now Iraq and western Iran) has low rainfall and is supplied with surface water by two major rivers, the Tigris and the Euphrates, which are much smaller than the Nile but have more dramatic floods in the spring. Mesopotamian engineers, concerned about water storage and flood control as well as irrigation, built diversion dams and large weirs to create reservoirs and to supply canals that carried water for long distances. In the Indus Valley civilization (now Pakistan and northwestern India), sophisticated irrigation and storage systems were developed.

One of the most impressive dams of ancient times is near Marib, an ancient city in Yemen. The 1,600-foot-long dam was built of masonry strengthened by copper around 600 BCE. It holds back some of the annual floodwaters coming down the valley and diverts the rest of that water out of sluice gates and into a canal system. The same sort of diversion dam system was independently built in Arizona by the Hohokam civilization around the second or third century CE.

In the Szechwan region of ancient China, the Dujiangyan irrigation system was built around 250 BCE and still supplies water in modern times. By the second century CE, the Chinese used chain pumps, which lifted water from lower to higher elevations, powered by hydraulic waterwheels, manual foot pedals, or rotating mechanical wheels pulled by oxen.

The Minoan civilization developed an aqueduct system in 1500 BCE to convey water in tubular conduits in the city of Knossos in Crete. Roman aqueducts were built to carry water from large distances to Rome and other cities in the empire. Of the 800 miles of aqueducts in Rome, only 29 miles were above ground. The Romans kept most of their aqueducts underground to protect their water from enemies and diseases spread by animals.

The Muslim agricultural revolution flourished during the Islamic golden age in various parts of Asia and Africa, as well as in Europe. Islamic hydraulic engineers built water management technological complexes, consisting of dams, canals, screw pumps, and *norias*, which are wheels that lift water from a river into an aqueduct.

The Swiss mathematician Daniel Bernoulli published *Hydrodynamica* (1738; *Hydrodynamics by Daniel Bernoulli*, 1968), applying the discoveries of Sir Isaac Newton and Gottfried Wilhelm Leibniz in mathematics and physics to fluid systems. In 1752, Leonhard Euler, Bernoulli's colleague, developed the more generalized form of the energy equation.

In 1843, Adhémar-Jean-Claude Barré de Saint Venant developed the most general form of the differential equations describing the motion of fluids, known as the Saint Venant equations. They are sometimes called Navier-Stokes equations after Claude-Louis Navier and Sir George Gabriel Stokes, who were working on them around the same time.

The German scientist Ludwig Prandtl and his students studied the interactions between fluids and solids between 1900 and 1930, thus developing the boundary layer theory, which theoretically explains the drag or friction between pipe walls and a fluid.

How it Works

Properties of water. Density and viscosity are important properties in fluid mechanics. The density of a fluid is its mass per unit volume. When the temperature or pressure of water changes significantly, its density variation remains negligible. Therefore, water is assumed to be incompressible. Viscosity, on the other hand, is the measure of a fluid's resistance to shear or deformation. Heavy oil is more viscous than water, whereas air is less viscous than water. The viscosity of water increases with reduced temperatures. For instance, the viscosity of water at its freezing point is six times its viscosity at its boiling temperature. Therefore, a flow of colder water assumes higher friction.

Hydrostatics. Hydrostatics is a subdiscipline of fluid mechanics that examines the pressures in water at rest and the forces on floating bodies or bodies submerged in water. When water is at rest, as in a tank or a large reservoir, it does not experience shear stresses; therefore, only normal pressure is present. When the pressure is uniform over the surface of a body in water, the total force applied on the body is a product of its surface area times the pressure. The direction of the force is perpendicular (normal) to the surface. Hydrostatic pressure forces can be mathematically determined on any shape. Buoyancy, for instance, is the upward vertical force applied on floating bodies (such as boats) or submerged ones (such as submarines). Hydraulic engineers use hydrostatics to compute the forces on submerged gates in reservoirs and detention basins.

Fluid kinematics. Water flowing at a steady rate in a constant-diameter pipe has a constant average velocity. The viscosity of water introduces shear stresses between particles that move at different velocities. The velocity of the particle adjacent to the wall of the pipe is zero. The velocity increases for particles away from the wall, and it reaches its maximum at the center of the pipe for a particular flow rate or pipe discharge. The velocity profile in a pipe has a parabolic shape. Hydraulic engineers use the average velocity of the velocity profile distribution, which is the flow rate over the cross-sectional area of the pipe.

Bernoulli's theorem. When friction is negligible and there are no hydraulic machines, the conservation of energy principle is reduced to Bernoulli's equation, which has many applications in pressurized flow and open-channel flow when it is safe to neglect the losses.

APPLICATIONS AND PRODUCTS

Water distribution systems. A water distribution network consists of pipes and several of the following components: reservoirs, pumps, elevated storage tanks, valves, and other appurtenances such as surge tanks or standpipes. Regardless of its size and complexity, a water distribution system serves the purpose of transferring water from one or more sources to customers. There are raw and treated water systems. A raw water network transmits water from a storage reservoir to treatment plants via large pipes, also called transmission mains. The purpose of a treated water network is to move water from a water-treatment plant and distribute it to water retailers through transmission mains or directly to municipal and industrial customers through smaller distribution mains.

Some water distribution systems are branched, whereas others are looped. The latter type offers more reliability in case of a pipe failure. The hydraulic engineering problem is to compute the steady velocity or flow rate in each pipe and the pressure at each junction node by solving a large set of continuity equations and nonlinear energy equations that characterize the network. The steady solution of a branched network is easily obtained mathematically; however, the looped network initially offered challenges to engineers. In 1936, American structural engineer Hardy Cross developed a simplified method that tackled networks formed of only pipes. In the 1970's and 1980's, three other categories of numerical methods were developed to provide solutions for complex networks with pumps and valves. In 1996, engineer Habib A. Basha and his colleagues offered a perturbation solution to the nonlinear set of equations in a direct, mathematical fashion, thus eliminating the risk of divergent numerical solutions.

Hydraulic transients in pipes. Unsteady flow in pipe networks can be gradual; therefore, it can be modeled as a series of steady solutions in an extended period simulation, mostly useful for water-quality analysis. However, abrupt changes in a valve position, a sudden shutoff of a pump because of power failure, or a rapid change in demand could cause a hydraulic transient or a water hammer that travels back and forth in the system at high speed, causing large pressure fluctuations that could cause pipe rupture or collapse.

The solution of the quasi-linear partial differential equations that govern the hydraulic transient problem is more challenging than the steady network solution.

The Russian scientist Nikolai Zhukovsky offered a simplified arithmetic solution in 1904. Many other methods–graphical, algebraic, wave-plane analysis, implicit, and linear methods, as well as the method of characteristics–were introduced between the 1950's and 1990's. In 1996, Basha and his colleagues published another paper solving the hydraulic transient problem in a direct, noniterative fashion, using the mathematical concept of perturbation.

Open-channel flow. Unlike pressure flow in full pipes, which is typical for water distribution systems, flow in channels, rivers, and partially full pipes is called gravity flow. Pipes in wastewater evacuation and drainage systems usually flow partially full with a free water surface that is subject to atmospheric pressure. This is the case for human-built canals and channels (earth or concrete lined) and natural creeks and rivers.

The velocity in an open channel depends on the area of the cross section, the length of the wetted perimeter, the bed slope, and the roughness of the channel bed and sides. A roughness factor is estimated empirically and usually accounts for the material, the vegetation, and the meandering in the channel.

Open-channel flow can be characterized as steady or unsteady. It also can be uniform or varied flow, which could be gradually or rapidly varied flow. A famous example of rapidly varied flow is the hydraulic jump.

When high-energy water, gushing at a high velocity and a shallow depth, encounters a hump, an obstruction, or a channel with a milder slope, it cannot sustain its supercritical flow (characterized by a Froude number larger than 1). It dissipates most of its energy through a hydraulic jump, which is a highly turbulent transition to a calmer flow (subcritical flow with a Froude number less than 1) at a higher depth and a much lower velocity. One way to solve for the depths and velocities upstream and downstream of the hydraulic jump is by applying the conservation of momentum principle, the third principle of fluid mechanics and hydraulic engineering. The hydraulic jump is a very effective energy dissipater that is used in the designs of spillways.

Hydraulic structures. Many types of hydraulic structures are built in small or large civil engineering projects. The most notable by its size and cost is the dam. A dam is built over a creek or a river, forming a reservoir in a canyon. Water is released through an outlet structure into a pipeline for water supply or into the river or creek for groundwater recharge and environmental reasons (sustainability of the biological life in the river downstream). During a large flood, the reservoir fills up and water can flow into a side overflow spillway–which protects the integrity of the face of the dam from overtopping–and into the river.

The four major types of dams are gravity, arch, buttress, and earth. Dams are designed to hold the immense water pressure applied on their upstream face. The pressure increases as the water elevation in the reservoir rises.

Hydraulic machinery. Hydraulic turbines transform the drop in pressure (head) into electric power. Also, pumps take electric power and transform it into water head, thereby moving the flow in a pipe to a higher elevation.

There are two types of turbines, impulse and reaction. The reaction turbine is based on the steam-powered device that was developed in Egypt in the first century CE by Hero of Alexandria. A simple example of a reaction turbine is the rotating lawn sprinkler. Pumps are classified into two main categories, centrifugal and axial flow. Pumps have many industrial, municipal, and household uses, such as boosting the flow in a water distribution system or pumping water from a groundwater well.

IMPACT ON INDUSTRY

The vast field of hydraulics has applications ranging from household plumbing to the largest civil engineering projects. The field has been integral to the history of humankind. The development of irrigation, water supply systems, and flood protection shaped the evolution of societies.

Hydraulic engineering is not a new field. Its governing principles were established starting in the eighteenth century and were refined through the twentieth century. Modern advances in the industry have been mainly the development of commercial software that supports designers in their modeling.

Since the 1980's, water distribution systems software has been evolving. Software can handle steady flow, extended period simulations, and hydraulic transients. Wastewater system and storm drainage software are also being developed.

One-dimensional open-channel software can be used for modeling flow in channels and rivers and even to simulate flooding of the banks. However,

FASCINATING FACTS ABOUT HYDRAULIC ENGINEERING

- An interesting application of Bernoulli's theorem outside the field of hydraulic engineering is the lift force on an airplane wing. The air velocity over the longer, top side of the wing is faster than the velocity along the shorter underside. Bernoulli's theorem proves that the pressure on the top of the wing is lower than the pressure on the bottom, which results in a net upward force that lifts the plane in the air.

- Hoover Dam, built between 1931 and 1936, is a 726-foot-tall concrete arch-gravity dam on the border between Nevada and Arizona. It is the second tallest dam in the United States after the 770-foot-tall Oroville Dam in California, built between 1961 and 1968. The world's tallest dam is the 984-foot-tall Nurek Dam in Tajikistan, an earth-fill embankment dam constructed between 1961 and 1980.

- One of the first recorded dam failures was in Grenoble, France, in 1219. Since 1865, twenty-nine dam failures have been recorded worldwide, ten of which occurred in the United States.

- The 305-foot-tall Teton Dam in Idaho became the highest dam to fail in June, 1976, a few months after construction was complete. Spring runoff had filled the reservoir with 80 million gallons of water, which gushed through the mostly evacuated downstream towns, killing eleven people and causing $1 billion in damages.

- Hurricane Katrina hit New Orleans in August, 2005, causing the failure of levees and flood walls at about fifty locations. Millions of gallons of water flooded 85 percent of the coastal city, and more than 1,800 people died.

- A tidal bore is a moving hydraulic jump that occurs when the incoming high tide forms a wave that travels up a river against the direction of the flow. Depending on the water level in the river, the tidal bore can vary from an undular wave front to a shock wave that resembles a wall of water.

- A water hammer is hydraulic transient in a pipe, characterized by dangerously large pressure fluctuations caused by the sudden shutdown of a pump or the rapid opening or closing of a valve. The high-velocity wave travels back and forth in the pipe, causing the pressure to fluctuate from large positive values that could burst the pipe to very low negative values that could cause the walls of the pipe to collapse.

two-dimensional software is better for modeling flow in floodplains and estuaries, although most software still has convergence problems. Three-dimensional software is used for modeling the flow in bays and lakes.

CAREERS AND COURSE WORK

Undergraduate students majoring in civil or environmental engineering usually take several core courses in hydraulic engineering, including fluid mechanics, water resources, and fluid mechanics laboratory. Advanced studies in hydraulic engineering lead to a master of science or a doctoral degree. Students with a bachelor's degree in a science or another engineering specialty could pursue an advanced degree in hydraulic engineering, but they may need to take several undergraduate level courses before starting the graduate program.

Graduates with a bachelor's degree in civil engineering or advanced degrees in hydraulics can work for private design firms that compete to be chosen to work on the planning and design phases of large governmental hydraulic engineering projects. They can work for construction companies that bid on governmental projects to build structures and facilities that include hydraulic elements, or for water utility companies, whether private or public.

To teach or conduct research at a university or a research laboratory requires a doctoral degree in one of the branches of hydraulic engineering.

SOCIAL CONTEXT AND FUTURE PROSPECTS

In the twenty-first century, hydraulic engineering has become closely tied to environmental engineering. Reservoir operators plan and vary water releases to keep downstream creeks wet, thus protecting the biological life in the ecosystem.

Clean energy is the way to ensure sustainability of the planet's resources. Hydroelectric power generation is a form of clean energy. Energy generated by ocean waves is a developing and promising field, although wave power technologies still face technical challenges.

—*Bassam Kassab, MSc*

FURTHER READING

Basha, Habib A. "Nonlinear Reservoir Routing: Particular Analytical Solution." *Journal of Hydraulic Engineering* 120, no. 5 (May, 1994): 624-632. Presents a mathematical solution for the flood routing equations in reservoirs.

Basha, Habib A., and W. El-Asmar. "The Fracture Flow Equation and Its Perturbation Solution." *Water Resources Research* 39, no. 12 (December, 2003): 1365. Shows that the perturbation method could be used not only on steady and transient water distribution problems but also on any nonlinear problem.

Boulos, Paul F. "H2ONET Hydraulic Modeling." *Journal of Water Supply Quarterly, Water Works Association of the Republic of China (Taiwan)* 16, no. 1 (February, 1997): 17-29. Introduces the use of modeling software (the H2ONET) in water distribution networks.

Boulos, Paul F., Kevin E. Lansey, and Bryan W. Karney. *Comprehensive Water Distribution Systems Analysis Handbook for Engineers and Planners.* 2d ed. Pasadena, Calif.: MWH Soft Press, 2006. Includes chapters on master planning and water-quality simulation.

Chow, Ven Te. *Open-Channel Hydraulics.* 1959. Reprint. Caldwell, N.J.: Blackburn, 2008. Contains chapters on uniform flow, varied flow, and unsteady flow.

Finnemore, E. John, and Joseph B. Franzini. *Fluid Mechanics With Engineering Applications.* 10th international ed. Boston: McGraw-Hill, 2009. Features chapters on kinematics, energy principles, hydrodynamics, and forces on immersed bodies.

Walski, Thomas M. *Advanced Water Distribution Modeling and Management.* Waterbury, Conn.: Haestad Methods, 2003. Examines water distribution modeling, has a chapter on hydraulic transients.

WEB SITES

American Society of Civil Engineers
http://www.asce.org

International Association of Hydro-Environment Engineering and Research
http://www.iahr.net

International Association of Hydrological Sciences
http://iahs.info

United States Society on Dams
http://www.ussdams.org

U.S. Bureau of Reclamation Waterways and Concrete Dams Group
http://www.usbr.gov/pmts/waterways

See also: Civil Engineering.

I

INFORMATION TECHNOLOGY

FIELDS OF STUDY

Information systems; computer science; software engineering; systems engineering; networking; computer security; Internet and Web engineering; computer engineering; computer programming; usability engineering; business; mobile computing; project management; information assurance; computer-aided design and manufacturing; artificial intelligence; knowledge engineering; mathematics; robotics.

SUMMARY

Information technology is a new discipline with ties to computer science, information systems, and software engineering. In general, information technology includes any expertise that helps create, modify, store, manage, or communicate information. It encompasses networking, systems management, program development, computer hardware, interface design, information assurance, systems integration, database management, and Web technologies. Information technology places a special emphasis on solving user issues, including helping everyone learn to use computers in a secure and socially responsible way.

KEY TERMS AND CONCEPTS

- **application software:** Software, such as word processors, used by individuals at home and at work.
- **cloud computing:** Software developed to execute on multiple servers and delivered by a Web service.
- **computer hardware:** Computing devices, networking equipment, and peripherals, including smartphones, digital cameras, and laser printers.
- **human-computer interaction:** Ways in which humans use or interact with computer systems.
- **information assurance:** Discipline that not only secures information but also guarantees its safety and accuracy.

- **information management:** Storage, processing, and display of information that supports its use by individuals and businesses.
- **networking:** Sharing of computer data and processing through hardware and software.
- **programming:** Development of customized applications, using a special language such as Java.
- **software integration:** Process of building complex application programs from components developed by others.
- **systems integration:** Technique used to manage the computing systems of companies that use multiple types of software, such as an Oracle database and Microsoft Office.
- **Web systems:** Database-driven Web applications, such as an online catalog, as well as simple Web sites.

DEFINITION AND BASIC PRINCIPLES

Information technology is a discipline that stresses systems management, use of computer applications, and end-user services. Although information technology professionals need a good understanding of networking, program development, computer hardware, systems development, software engineering, Web site design, and database management, they do not need as complete an understanding of the theory of computer science as computer scientists do. Information technology professionals need a solid understanding of systems development and project management but do not need as extensive a background in this as information systems professionals. In contrast, information technology professionals need better interpersonal skills than computer science and information systems workers, as they often do extensive work with end-users.

During the second half of the twentieth century, the world moved from the industrial age to the computer age, culminating in the development of the World Wide Web in 1990. The huge success of the Web as a means of communications marked a transition from the computer age, with an emphasis on technology, to the information age, with an emphasis on how technology enhances the use of information. The Web is used in many ways to enhance the use and transfer of information, including telephone service, social networking, e-mail, teleconferencing, and even radio and television programs. Information technology contains a set of tools that make it easier to create, organize, manage, exchange, and use information.

BACKGROUND AND HISTORY

The first computers were developed during World War II as an extension of programmable calculating machines. John von Neumann added the stored program concept in 1944, and this set off the explosive growth in computer hardware and software during the remainder of the twentieth century. As computing power increased, those using the computers began to think less about the underlying technology and more about what the computers allowed them to do with data. In this way, information, the organization of data in a way that facilitates decision making, became what was important, not the technology that was used to obtain the data. By 1984, organizations such as the Data Processing Management Association introduced a model curriculum defining the training required to produce professionals in information systems management, and information systems professionals began to manage the information of government and business.

In 1990, Computer scientist Tim Berners-Lee developed a Web browser, and the Web soon became the pervasive method of sharing information. In addition to businesses and governmental agencies, individuals became extensive users and organizers of information, using applications such as Google (search engine) and Facebook (social networking). In the early 2000's, it became clear that a new computer professional, specializing in information management, was needed to complement the existing information systems professionals. By 2008, the Association of Computing Machinery and the IEEE Computer Society released their recommendations for educating information technology professionals, authenticating the existence of this new computing field. The field of information technology has become one of the most active areas in computer science.

HOW IT WORKS

The principal activity of information technology is the creation, storage, manipulation, and communication of information. To accomplish these activities, information technology professionals must have a background from a number of fields and be able to use a wide variety of techniques.

Networking and Web systems. Information is stored and processed on computers. In addition, information is shared by computers over networks. Information technology professionals need to have a good working knowledge of computer and network systems to assist them in acquiring, maintaining, and managing these systems. Of the many tasks performed by information technology professionals, none is more important than installing, updating, and training others to use applications software. The Web manages information by storing it on a server and distributing it to individuals using a browser, such as Internet Explorer or Foxfire. Building Web sites and applications is a big part of information technology and promises to increase in importance as more mobile devices access information provided by Web services.

Component integration and programming. Writing programs in a traditional language such as Java will continue to be an important task for information technology professionals, as it had been for other computer professionals in the past. However, building new applications from components using several Web services appears to be poised to replace traditional programming. For all types of custom applications, the creation of a user-friendly interface is important. This includes the careful design of the screen elements so that applications are easy to use and have well thought-out input techniques, such as allowing digital camera and scanner input at the appropriate place in a program, as well as making sure the application is accessible to visually and hearing impaired people.

Databases. Data storage is an important component of information technology. In the early days of computers, most information was stored in files. The difficulty of updating information derived from a file led to the first database systems, such as the information management system created by IBM in the 1960's. In the 1970's, relational databases became the dominant method of storing data, although a number of competing technologies are also in use.

For example, many corporate data are stored in large spreadsheets, many personal data are kept in word-processing documents, and many Web data are stored directly in Web pages.

Information security, privacy, and assurance. Regardless of how information is collected, stored, processed, or transferred, it needs to be secure. Information security techniques include developing security plans and policies, encrypting sensitive data during storage and transit, having good incident response and recovery teams, installing adequate security controls, and providing security training for all levels of an organization. In addition to making sure that an organization's data are secure, it is also important to operate the data management functions of an organization in such a way that each individual's personal data are released only to authorized parties. Increasingly, organizations want information handled in such a way that the organization can be assured that the data are accurate and have not been compromised.

Professionalism. Information technology professionals need to inform themselves of the ethical principles of their field (such as the Association of Computing Machinery's code of ethical and professional conduct) and conscientiously follow this code. Increasingly, laws are being passed about how to handle information, and information technology professionals need to be aware of these laws and to follow them.

APPLICATIONS AND PRODUCTS

There are more useful applications of information technology than can be enumerated, but they all involve making information accessible and usable.

End-user support. One of the most important aspects of information technology is its emphasis on providing support for a wide variety of computer users. In industrial, business, and educational environments, user support often starts immediately after someone is given access to a computer. An information technology professional assists the user with the login procedure, shows him or her how to use e-mail and various applications. After this initial introduction, technology professionals at a help desk answer the user's questions. Some companies also provide training courses. When hardware and software problems develop, users contact the information technology professionals in the computer

support department for assistance in correcting their problems.

Information technology educational programs usually cover the theory and practice of end-user support in their general courses, and some offer courses dedicated to end-user support. This attention to end-user support is quite different from most programs in computer science, information systems, or software engineering and is one of the most important differences between information technology programs and other areas of computer science.

The electronic medical record. A major use of information technology is in improving the operation of hospitals in providing medical services, maintaining flexible schedules, ensuring more accurate billing, and reducing the overall cost of operation. In much the same way, information technology helps doctors in clinics and individual practices improve the quality of their care, scheduling, billing, and operations. One of the early successes in using computers in medicine was the implementation of e-prescribing. Doctors can easily use the Internet to determine the availability of a drug needed by a patient, electronically send the prescription to the correct pharmacy, and bill the patient, or their insurance company, for the prescription. Another success story for medical information technology is in the area of digital imaging. Most medical images are created digitally, and virtually all images are stored in a digital format. The Digital Imaging and Communications in Medicine (DICOM) standard for storage and exchange of digital images makes it possible for medical images to be routinely exchanged between different medical facilities. Although the process of filling prescriptions and storing images is largely automated, the rest of medicine is still in the process of becoming fully automated. However, computers are likely to become as much a part of medical facilities as they are of financial institutions.

One of the major goals of information technology is the development of a true electronic medical record (EMR). The plan is to digitize and store all the medical information created for an individual by such activities as filling out forms, taking tests, and receiving care. This electronic medical record would be given to individuals in its entirety and to hospital and insurance companies as needed. Electronic medical records would be available to the government for data mining to determine health policies. Many professionals in information technology are likely to be involved in the collection, storage, and securitization of electronic medical records.

Geographic information management. Maps have been used by governments since the beginning of recorded history to help with the process of governing. Geographic information systems (GIS) are computerized systems for the storage, retrieval, manipulation, analysis, and display of geographic data and maps. In the 1950's, the first GIS were used by government agencies to assist in activities such as planning, zoning, real estate sales, and highway construction. They were developed on mainframes and accessed by government employees on behalf of those desiring the information contained in the systems. The early GIS required substantial computing resources because they used large numbers of binary maps but were relatively simple as information retrieval systems.

Geographic information systems still serve the government, but they are also used by industry and educational institutions and have gained many applications, including the study of disease, flood control, census estimates, and oil discovery. GIS information is easily accessed over the Internet. For example, zoning information about Dallas, Texas, is readily available from the city's Web site. Many general portals also provide GIS information to the public. For example, Google Maps allows travelers to print out their route and often provides a curbside view of the destination. Modern geographic information systems are complex, layered software systems that require expertise to create and maintain.

Network management. Computer networking developed almost as quickly as the computing field itself. With the rapid acceptance of the Internet in the 1990's, computer connectivity through networking became as common as the telephone system. As computer networks developed, a need also developed for professionals to manage these networks. The first network managers were generally hardware specialists who were good at running cable through a building, adding hardware to a computer, and configuring network software so that it would work. Modern network specialists need all these early capabilities but also must be network designers, managers, and security experts.

Database management. One of the major functions of information technology is the storage of information. Information consists of data organized for meaningful usage. Both data and information regarding how the data are related can be stored in many ways. For example, many corporate data are stored in word-processing documents, spreadsheets, and e-mails. Even more corporate data are stored in relational databases such as Oracle. Most businesses, educational institutions, and government agencies have database management specialists who spend most of their time determining the best logical and physical models for the storage of data.

Mobile computing and the cloud. Mobile phones have become so powerful that they rival small computers, and Web services and applications for computers or mobile devices are being developed at a breathtaking rate. These two technological developments are working together to provide many Web applications for smartphones. For example, any cell phone can play MP3 music. Literally dozens of Web services have been developed that will automatically download songs to cell phones and bill the owner a small fee. In addition, many smartphone applications will download and play songs from Web services.

Cloud computing and storage is beginning to change how people use computers in their homes. Rather than purchasing computers and software, some home users are paying for computing as a service and storing their data in an online repository. Microsoft's Live Office and Google's Apps are two of the early entrants in the software and storage as a service market.

Computer integrated manufacturing. Computer integrated manufacturing (CIM) provides complete support for the business analysis, engineering, manufacturing, and marketing of a manufactured item, such as a car. CIM has a number of key areas including computer-aided design (CAD), supply chain management, inventory control, and robotics.

Many of the areas of CIM require considerable use of information technology. For example, CAD programs require a good data management program to keep track of the design changes, complex algorithms to display the complicated graphical images, fast computer hardware to display the images quickly, and good project management tools to assist in completing the project on time. Robots, intelligent machines used to build products, require much innovative computer and machine hardware, very complex artificial intelligence algorithms simulating human operation of machine tools, and sophisticated computer networks for connecting the robotic components on the factory floor.

Computer security management. One of the most active areas of information technology is that of computer security management. As theoretical computer scientists develop new techniques to protect computers, such as new encryption algorithms, information technology specialists work on better ways to implement these techniques to protect computers and computer networks. Learning how to acquire the proper hardware and software, to do a complete risk analysis, to write a good security policy, to test a computer network for vulnerabilities, and to accredit a business's securing of its computers requires all the talents of an information technologist.

Web site development. The World Wide Web was first introduced in 1990, and since that time, it has become one of the most important information distribution technologies available. Information technology professionals are the backbone of most Web site development teams, providing support for the setup and maintenance of Web servers, developing the HTML pages and graphics that provide the Web content, and assisting others in a company in getting their content on the Web.

IMPACT ON INDUSTRY

The use of computers became pervasive in companies, government agencies, and universities in the second half of the twentieth century. Information technology gradually developed as a separate discipline in the late 1970's. By 1980, companies, government agencies, and universities recognized that it was the information provided by the computer that was important, not the computers themselves. Although access to and the ability to use the information was what people wanted, getting the information was difficult and required a number of skills. These included the building and maintenance of computers and networks, the development of computer software, providing end-user support, and using management information and decision support systems to add value to the information. Government and industry made their information needs known; universities and some computer manufacturers did basic and applied research developing hardware and software to meet these needs, and information technology professionals provided support to allow those in government and industry to actually use the information.

Although estimates of the total value of the goods and services provided by information technology in a year are impossible to determine, it is estimated to be close to $1 trillion per year.

Government, industry, and university uesearch. The United States government exerts a great influence on the development of information technology because it purchases a large amount of computer hardware and software and makes substantial use of the information derived from its computers and networks. Companies such as IBM, SAP, and Oracle have geared much of their research and development toward meeting the needs of their government customers. For example, IBM's emphasis on COBOL for its mainframes was a result of a government requirement that all computers purchased by the government support COBOL compilers. SAP's largest customer base in the United States is federal, state, and local government agencies. One of the main reasons for Oracle's acquisition of PeopleSoft was to enhance its position in the educational enterprise resource planning market.

The National Science Foundation, the National Institutes of Health, and the U.S. Department of Defense have always provided support for university research for information technology. For example, government agencies supporting cancer research gave a great deal of support to Oracle and to university researchers in the development of a clinical research data management system to better track research on cancer. The government also has been a leader in developing educational standards for government information technology professionals, and these standards have greatly influenced university programs in information technology. For example, after the September 11, 2001, terrorist attacks on the United States, the National Security Agency was charged with coming up with security standards that all government employees needed to meet. The agency created an educational branch, the Committee on National Security Systems, which defined the standards and developed a set of security certifications to implement the standards. Its certifications have had a significant influence on the teaching of computer security.

The computer industry has been very active in developing better hardware and software to support information technology. For example, one of the original relational database implementations, System R and SQL/Data System, was developed at the IBM Research Laboratory in San Jose, California, in the 1970's, and this marked the beginning of the dominance of relational databases for data storage. There have also been a number of successful collaborations between industry and universities to improve information technology. For example, the Xerox PARC (later PARC) laboratories, established as a cooperative effort between Xerox and several universities, developed the Ethernet, the graphical user interface, and the laser printer. Microsoft has one of the largest industry research programs, which supports information technology development centers in Redmond, Washington; Cambridge, England; and six other locations. In 2010, Microsoft announced Azure, its operating system to support cloud computing. Azure will allow individuals to access information from the cloud from almost anywhere on a wide variety of devices, providing information on demand.

Information technology companies. Many successful large information technology companies are operating in the United States. Dell and HP (Hewlett Packard) concentrate on developing the computing hardware needed to store and process information. IBM produces both the hardware and software needed to store and process information. In the early 2010's, Microsoft produced more information technology software than any other company, but SAP was reported to have the greatest worldwide sales volume. Oracle, the largest and most successful database company, was the leader in providing on-site storage capacity for information, but Microsoft and Google had begun offering storage as a service over the Internet, a trend predicted to continue. Smaller but important information technology companies include Adobe Systems, which developed portable document format (PDF) files for transporting information and several programs to create PDF files.

CAREERS AND COURSE WORK
The most common way to prepare for a career in information technology is to obtain a degree in information systems. Students begin with courses in mathematics and business and then take about thirty hours of courses on information systems development. Those getting degrees in information systems often take information technology jobs as systems analysts,

data modelers, or system managers, performing tasks such as helping implement a new database management system for a local bank.

A degree in computer science or software engineering is another way to prepare for an information technology career. The courses in ethics, mathematics, programming, and software management that are part of these majors provide background for becoming an information technology professional. Those getting degrees in these areas often take information technology jobs as programmers or system managers, charged with tasks such as helping write a new program to calculate the ratings of a stock fund.

Degree programs in information technology are available at some schools. Students begin with courses in problem solving, ethics, communications, and management, then take about thirty-six hours of courses in programming, networking, human-computer communications, databases, and Web systems. Those getting an information technology degree often take positions in network management, end-user support, database management, or data modeling. A possible position would be a network manager for a regional real estate brokerage.

In addition to obtaining a degree, many information technology professionals attend one of the many professional training and certification programs. One of the original certification programs was the Novell Certified Engineer, followed almost immediately by a number of Microsoft certification programs. These programs produced many information technology professionals for network management. Cisco has created a very successful certification program for preparing network specialists in internetworking (connecting individual local-area networks to create wide-area networks).

SOCIAL CONTEXT AND FUTURE PROSPECTS
The future of information technology is bright, with good jobs available in end-user support, network management, programming, and database management. These areas, and the other traditional areas of information technology, are likely to remain important. Network management and end-user support both appear to be poised for tremendous growth over the next few years. The growth in the use of the Internet and mobile devices requires the support of robust networks, and this, in turn, will require a large number of information technology professionals to

information technology are being developed. One of the areas of development is in medical informatics. This includes the fine-tuning of existing hospital software systems, development of better clinical systems, and the integration of all of these systems. The United States and many other nations are committed to the development of a portable electronic health record for each person. The creation of these electronic health records is a massive information technology project and will require a large workforce of highly specialized information technology professionals. For example, to classify all of the world's medical information, a new language, Health Level 7, has been created, and literally thousands of specialists have been encoding medical information into this language.

The use of mobile devices to access computing from Web services is another important area of information technology development. Many experts believe the use of Web-based software and storage, or cloud computing, may be the dominant form of computing in the future, and it is likely to employ many information technology professionals.

Another important area of information technology is managing information in an ethical, legal, and secure way, while ensuring the privacy of the owners of the information. Security management specialists must work with network and database managers to ensure that the information being processed, transferred, and stored by organizations is properly handled.

—*George M. Whitson III, MS, PhD*

FURTHER READING
Miller, Michael. *Cloud Computing: Web-Based Applications That Change the Way You Work and Collaborate.* Indianapolis, Ind.: Que, 2009. Provides an excellent overview of the new method of using computing and storage on demand over the cloud.

Reynolds, George. *Information Technology for Managers.* Boston: Course Technology, 2010. A short but very complete introduction to information technology by one of the leading authors of information technology books.

Senn, James. *Information Technology: Principles, Practices, and Opportunities.* 3d ed. Upper Saddle River, N.J.: Prentice Hall, 2004. An early information technology book that emphasizes the way information technology is used and applied to problem

install, repair, update, and manage networks. The greater use of the Internet and mobile devices also means that there will be a large number of new, less technically aware people trying to use the Web and needing help provided by information technology end-user specialists.

A large number of new applications for

solving.

Shneiderman, Ben, and Catherine Plaisant. *Designing the User Interface.* Boston: Addison-Wesley, 2010. An excellent book about developing good user interfaces. In addition to good coverage on developing interfaces for desktop applications, it covers traditional Web applications and applications for mobile devices.

Stair, Ralph, and George Reynolds. *Principles of Information Systems.* 9th ed. Boston: Course Technology, 2010. Contains not only a complete introduction to systems analysis and design but also a very good description of security, professionalism, Web technologies, and management information systems.

WEB SITES
Association for Computing Machinery
http://www.acm.org

CompTIA
http://www.comptia.org/home.aspx

IEEE Computer Society
http://sites.computer.org/ccse

Information Technology Industry Council
http://www.itic.org

TechAmerica
http://www.itaa.org

See also: Communication; Computer Science; Telephone Technology and Networks.

INTEGRATED-CIRCUIT DESIGN

FIELDS OF STUDY

Mathematics; physics; electronics; quantum mechanics; mechanical engineering; electronic engineering; graphics; electronic materials science.

SUMMARY

The integrated circuit (IC) is the essential building block of modern electronics. Each IC consists of a chip of silicon upon which has been constructed a series of transistor structures, typically MOSFETs, or metal oxide semiconductor field effect transistors. The chip is encased in a protective outer package whose size facilitates use by humans and by automated machinery. Each chip is designed to perform specific electronic functions, and the package design allows electronics designers to work at the system level rather than with each individual circuit. The manufacture of silicon chips is a specialized industry, requiring the utmost care and quality control.

KEY TERMS AND CONCEPTS

- **clock:** A multitransistor structure that generates square-wave pulses at a fixed frequency; used to regulate the switching cycles of electronic circuitry.
- **fan-out:** The number of transistor inputs that can be served by the output from a single transistor.
- **flip-flop:** A multitransistor structure that stores and maintains a single logic state while power is applied or until it is instructed to change.
- **vibrator:** A multitransistor structure that outputs a constant sinusoidal pulse by switching between two unstable states, rather than by modifying an input signal.

DEFINITION AND BASIC PRINCIPLES

An integrated circuit, or IC, is an interconnected series of transistor structures assembled on the surface of a silicon chip. The purpose of the transistor assemblages is to perform specific operations on electrical signals that are provided as inputs. All IC devices can be produced from a structure of four transistors that function to invert the value of the input signal, called NOT gates or inverters. All digital electronic circuits are constructed from a small number of different transistor assemblages, called gates, that are built into the circuitry of particular ICs. The individual ICs are used as the building blocks of the digital electronic circuitry that is the functional heart of modern digital electronic technology.

The earliest ICs were constructed from bipolar transistors that function as two-state systems, a system high and a system low. This is not the same as "on" and "off," but is subject to the same logic. All digital

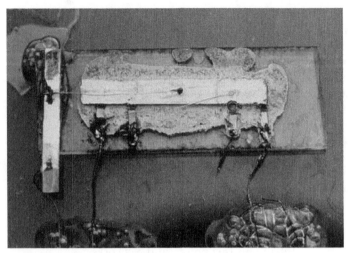
Jack Kilby's original integrated circuit

systems function according to Boolean logic and binary mathematics.

Modern ICs are constructed using metal oxide semiconductor field effect transistor (MOSFET) technology, which has allowed for a reduction in size of the transistor structures to the point in which, at about 65 nanometers (nm) in size, literally millions of them can be constructed per square centimeter (cm^2) of silicon chip surface. The transistors function by conducting electrical current when they are biased by an applied voltage. Current ICs, such as the central processing units (CPUs) of personal computers, can operate at gigahertz frequencies, changing that state of the transistors on the IC billions of times per second.

BACKGROUND AND HISTORY

Digital electronics got its start in 1906, when American inventor Lee de Forest constructed the triode vacuum tube. Large, slow, and power-hungry as they were, vacuum tubes were nevertheless used to construct the first analogue and digital electronic computers. In 1947, American physicist William Shockley and colleagues constructed the first semiconductor transistor junction, which quickly developed into more advanced silicon-germanium junction transistors.

Through various chemical and physical processes, methods were developed to construct small transistor structures on a substrate of pure crystalline silicon. In 1958, American physicist and Nobel laureate Jack Kilby first demonstrated the method by constructing germanium-based transistors as an IC chip, and

American physicist Robert Noyce constructed the first silicon-based transistors as an IC chip. The transistor structures were planar bipolar in nature, until the CMOS (complementary metal oxide semiconductor) transistor was invented in 1963. Methods for producing CMOS chips efficiently were not developed for another twenty years.

Transistor structure took another developmental leap with the invention of the field effect transistor (FET), which was both a more efficient design than that of semiconductor junction transistors and easier to effectively manufacture. The MOSFET structure also is amenable to miniaturization and has allowed designers to engineer ICs that have one million or more transistor structures per centimeters squared.

HOW IT WORKS

The electronic material silicon is the basis of all transistor structures. It is classed as a pure semiconductor. It is not a good conductor of electrical current or insulate well against electrical current. By adding a small amount of some impurity to the silicon, its electrical properties can be manipulated such that the application of a biasing voltage to the material allows it to conduct electrical current. When the biasing voltage is removed, the electrical properties of the material revert to their normal semiconductive state.

Silicon manufacture. Integrated-circuit design begins with the growth of single crystals of pure silicon. A high-purity form of the material, known as polysilicon, is loaded into a furnace and heated to melt the material. At the proper stage of melt, a seed crystal is attached to a slowly turning rod and introduced to the melt. The single crystal begins to form around the seed crystal and the rotating crystal is "pulled" from the melt as it grows. This produces a relatively long, cylindrical, single crystal that is then allowed to cool and set.

Wafers. From this cylinder, thin wafers or slices are cut using a continuous wire saw that produces several uniform slices at the same time. The slices are subjected to numerous stages of polishing, cleaning, and quality checking, the end result of which is a consistent set of silicon wafers suitable for use as substrates for integrated circuits.

Circuitry. The integrated circuit itself begins as a complex design of electronic circuitry to be constructed from transistor structures and "wiring" on the surface of the silicon wafer. The circuits can be no more than a series of simple transistor gates (such as invertors, AND-gates, and OR-gates), up to and including the extremely complex transistor circuitry of advanced CPUs for computers.

Graphics technology is extremely important in this stage of the design and production of integrated circuits, because the entire layout of the required transistor circuitry must be imaged. The design software also is used to conduct virtual tests of the circuitry before any ICs are made. When the theoretical design is complete and imaged, the process of constructing ICs can begin.

Because the circuitry is so small, a great many copies can be produced on a single silicon wafer. The actual chips that are housed within the final polymer or ceramic package range in size from two to five cm^2. The actual dimensions of the circa 1986 Samsung KS74AHCT240 chip, for example, are just 1 cm x 2 cm. The transistor gate sizes used in this chip are 2 micrometers (um) (2×10^{-6} meters [m]), and each chip contains the circuitry for two octal buffers, constructed from hundreds of transistor gate structures. Transistor construction methods have become much more efficient, and transistor gate sizes are now measured in nanometers (10^{-9} m) rather than um, so that actual chip sizes have also become much smaller, in accord with Moore's law. The gate structures are connected through the formation of aluminum "wires" using the same chemical vapor deposition methodology used to form the silicon oxide and other layers needed.

Photochemical etching. The transistor structures of the chip are built up on the silicon wafer substrate through a series of steps in which the substrate is photochemically etched and layers of the necessary materials are deposited. Absolute precision and an ultraclean environment are required at each step. The processes are so sensitive that any errant speck of dust or other contaminant that finds its way to the wafer's surface renders that part of the structure useless.

Accounting for losses of functional chips at each stage of the multistep process, it is commonly the case that as little as 10 percent of the chips produced from any particular wafer will prove viable at the end of their construction. If, for example, the procedure requires one hundred individual steps, not including quality-testing steps, to produce the final

product, and a mere 2 percent of the chips are lost at each step, then the number of viable chips at the end of the procedure will be 0.98^{100}, or 13.26 percent of the number of chips that could ideally have been produced.

Each step in the formation of the IC chip must be tested to identify the functionality of the circuitry as it is formed. Each such step and test procedure adds significantly to the final cost of an IC. When the ICs are completed, the viable ones are identified, cut from the wafer, and enclosed within a protective casing of resin or ceramic material. A series of leads are also built into this "package" so that the circuitry of the IC chip can be connected into an electronic system.

APPLICATIONS AND PRODUCTS

Bipolar transistors and MOSFETs. Transistors are commonly pictured as functioning as electronic on-off switches. This view is not entirely correct. Transistors function by switching between states according to the biasing voltages that are applied. Bipolar switching transistors have a cut-off state in which the applied biasing voltage is too low to make the transistor function. The normal operating condition of the transistor is called the linear state. The saturation state is achieved when the biasing voltage is applied to both poles of the transistor, preventing them from functioning. MOSFET transistors use a somewhat different means, relying on the extent of an electric field within the transistor substrate, but the resulting functions are essentially the same.

The transistor structures that form the electronic circuitry of an IC chip are designed to perform specific functions when an electrical signal is introduced. For simple IC circuits, each chip is packaged to perform just one function. An inverter chip, for example, contains only transistor circuitry that inverts the input signal from high to low or from low to high. Typically, six inverter circuits are provided in each package through twelve contact points. Two more contact points are provided for connection to the biasing voltage and ground of the external circuit. It is possible to construct all other transistor logic gates using just inverter gates. All ICs use this same general package format, varying only in their size and the number of contact points that must be provided.

MOSFETS have typical state switching times of

something less than 100 nanoseconds, and are the transistor structures of choice in designing ICs, even though bipolar transistors can switch states faster. Unlike bipolars, however, MOSFETS can be constructed and wired to function as resistors and can be made to a much smaller scale than true resistors in the normal production process. MOSEFTS are easier to manufacture than are bipolars as they can be made much smaller in VLSI (very large scale integration) designs. MOSFETS also cost much less to produce.

NOTs, ANDs, ORs, and other gates. All digital electronic devices comprise just a few basic types of circuitry—called logic elements—of which there are two basic types: decision-making elements and storage elements. All logic elements function according to the Boolean logic of a two-state (binary) system. The only two states that are allowed are "high" and "low," representing an applied biasing voltage that either does or does not drive the transistor circuitry to function. All input to the circuitry is binary, as is all output from the circuitry.

Decision-making functions are carried out by logic gates (the AND, OR, NOT gates and their constructs) and memory functions are carried out by combination circuitry (flip-flops) that maintains certain input and output states until it is required to change states. All gates are made up of a circuit of interconnected transistors that produces a specific output according to the input that it receives.

A NOT gate, or inverter, outputs a signal that is the opposite of the input signal. A high input produces a low output, and vice versa. The AND gate outputs a high signal only when all input signals are high, and a low output signal only if any of the input signals are low. The OR gate functions in the opposite sense, producing a high output signal if any of the input signals are high and a low output signal only when all of the input signals are low. The NAND gate, which can be constructed from either four transistors and one diode or five diodes, is a combination of the AND and NOT gates. The NOR gate is a combination of the OR and NOT gates. These, and other gates, can have any number of inputs, limited only by the fan-out limits of the transistor structures.

Sequential logic circuits are used for timing, sequencing, and storage functions. The flip-flops are the main elements of these circuits, and memory functions are their primary uses. Counters consist of a series of flip-flops and are used to count the number of applied input pulses. They can be constructed to count up or down, as adders, subtracters, or both. Another set of devices called shift registers maintains a memory of the order of the applied input pulses, shifting over one place with each input pulse. These can be constructed to function as series devices, accepting one pulse (one data bit) at a time in a linear fashion, or as parallel devices, accepting several data bits, as bytes and words, with each input pulse. The devices also provide the corresponding output pulses.

Operational amplifiers, or OP-AMPs, represent a special class of ICs. Each OP-AMP IC contains a self-contained transistor-based amplifying circuit that provides a high voltage gain (typically 100,000 times or more), a very high input impedance and low output impedance, and good rejection of common-mode signals (that is, the presence of the same signal on both input leads of the OP-AMP).

Combinations of all of the gates and other devices constructed from them provide all of the computing power of all ICs, up to and including the most cutting-edge CPU chips. Their manufacturing process begins with the theoretical design of the desired functions of the circuitry. When this has been determined, the designer minimizes the detailed transistor circuitry that will be required and develops the corresponding mask and circuitry images that will be required for the IC production process. The resulting ICs can then be used to build the electronic circuitry of all electronic devices.

IMPACT ON INDUSTRY

Digital-electronic control mechanisms and appliances are the mainstays of modern industry, mainly because of the precision with which processes and quality can be monitored through digital means. Virtually all electronic systems now operate using digital electronics constructed from ICs.

Digital electronics has far-reaching consequences. For example, in September, 2011, the telecommunications industry in North America ceased using analogue systems for broadcasting and receiving commercial signals. This development can be considered a continuation of a process that began with the invention of the Internet, with its entirely digital protocols and access methods. As digital transmission methods and infrastructure have developed to facilitate the transmission of digital communications between computer systems, the feasibility of supplanting traditional analogue communications systems with their digital counterparts became apparent. Electronic technology and the digital format enable the transmission of much greater amounts of information within the same bandwidth, with much lower energy requirements.

Modern industry without the electronic capabilities provided by IC design is almost inconceivable, from both the producer and the consumer points of view. The production of ICs themselves is an expensive proposition because of the demands for ultraclean facilities, special production methods, and the precision of the tools required for producing the transistor structures.

FASCINATING FACTS ABOUT INTEGRATED-CIRCUIT DESIGN

- The ENIAC computer, built in 1947, weighed more than 30 tons; used 19,000 vacuum tubes, 1,500 relays, about one-half-million resistors, capacitors, and inductors; and consumed almost 200 kilowatts of electricity per hour.
- In 1997, the ENIAC computer was reproduced on a single triple-layer CMOS chip that was 7.44 millimeters (mm) x 5.29 mm in size and contained 174,569 transistors.
- In 1954, RAND scientists predicted that by the year 2000, some homes in the United States could have their own computer room, including a teletype terminal for FORTRAN programming, a large television set for graphic display, rows of blinking lights, and a double "steering wheel."
- The point transistor was invented in 1947, and the first integrated circuits were demonstrated in 1950 by two different researchers, leading directly to the first patent infringement litigation involving transistor technology.
- The flow of electricity in a bipolar transistor, and hence its function, is controlled by a bias voltage applied to one or the other pole.
- The flow of electricity, and hence its function, of a MOSFET is controlled by the extent of the electrical field formed within the substrate materials.
- Silicon wafers sliced from cylindrical single crystals of silicon undergo dozens of production steps before they are suitable for use as substrates for ICs.
- The production of ICs from polished silicon wafers can take hundreds of steps, after which as few as 10 percent of the ICs will be usable.

The transistors are produced on the surface of silicon wafers through photoetching and chemical vapor-deposition methodologies. Photoetching requires the use of specially designed masks to provide the shape and dimensions of the transistors' component structures. As transistor dimensions have decreased from 35 um to 65 nm, the cost of the masks has risen exponentially, from about \$100,000 to more than \$3 million. At the same time, however, semiconductor industry revenues have increased from about \$20 million in 1955 to more than \$200 billion in the early twenty-first century, and they continue to increase at a rate of about 6 percent per year. This is mirrored by the amount of investment in research and development. In 2006, for example, such investment amounted to some \$19.9 billion.

The consumer aspect of IC design is demonstrated by the market proliferation of consumer electronic devices and appliances. It seems as though everything, from compact fluorescent light bulbs and motion-sensing air fresheners to high-definition television monitors, personal computers, cellular telephones, and even automobiles depends on IC technology. This is a quantity whose value rivals the gross domestic product of the United States.

CAREERS AND COURSE WORK

IC design and manufacturing is a high-precision field. Students who are interested in pursuing a career working with devices constructed with ICs will require a sound basic education in mathematics and physics, as the basis for the study of electronics technology and engineering. As specialists in these fields, the focus will be on using ICs as the building blocks of electronic circuits for numerous devices and appliances. A recognized college degree or the equivalent is the minimum requirement for qualification as an electronics technician or technologist. The minimum qualification for a career in electronics engineering is a bachelor's degree in that field from a recognized university. In addition, membership in professional associations following graduation, if maintained, will require regular upgrading of qualifications to meet standards and keep abreast of changing technology.

Electronics technology is an integral component of many technical fields, especially for automotive mechanics and transportation technicians, as electronic systems have been integrated into transportation and vehicle designs of all kinds.

For those pursuing a career in designing integrated circuitry and the manufacture of IC chips, a significantly more advanced level of training is required. Studies at this level include advanced mathematics and physics, quantum mechanics, electronic theory, design principles, and graphics. Integrated circuitry is cutting-edge technology, and the nature of ICs is likely to undergo rapid changes that will, in turn, require the IC designer to acquire knowledge of entirely new concepts of transistors and electronic materials. A graduate degree (master's or doctorate) from a recognized university will be the minimum qualification in this field.

SOCIAL CONTEXT AND FUTURE PROSPECTS

IC technology is on the verge of extreme change, as new electronic materials such as graphene and carbon nanotubes are developed. Research with these materials indicates that they will be the basic materials of molecular-scale transistor structures, which will be thousands of times smaller and more energy-efficient than VLSI technology based on MOSFETs. The computational capabilities of computers and other electronic devices are expected to become correspondingly greater as well.

Such devices will utilize what will be an entirely new type of IC technology, in which the structural features are actual molecules and atoms, rather than what are by comparison mass quantities of semiconductor materials and metals. As such, future IC designers will require a comprehensive understanding of both the chemical nature of the materials and of quantum physics to make the most effective use of the new concepts.

The scale of the material structures as well will have extraordinary application in society. It is possible, given the molecular scale of the components, that the technology could even be used to print ultrahigh resolution displays and computer circuitry that would make even the lightest and thinnest of present-day appliances look like the ENIAC (Electronic Numerical Integrator and Computer) of 1947, which was the first electronic computer.

The social implications of such miniaturized technology are far-reaching. The RFID (radio-frequency identification) tag is now becoming an important means of embedding identification markers directly into materials. RFID tags are tiny enough to be included as a component of paints, fuels, explosives, and other materials, allowing identification of the exact source of the material, a useful implication for forensic investigations and other purposes. Even the RFID tag, however, would be immense compared with the molecular scale of graphene and nanotubebased devices that could carry much more information on each tiny particle.

The ultimate goal of electronic development, in current thought, is the quantum computer, a device that would use single electrons, or their absence, as data bits. The speed of such a computer would be unfathomable, taking seconds to carry out calculations that would take present-day supercomputers billions of years to complete. The ICs used for such a device would bear little resemblance, if any, to the MOSFET-based ICs used now.

—*Richard M. Renneboog, MSc*

FURTHER READING

Brown, Julian R. *Minds, Machines, and the Multiverse: The Quest for the Quantum Computer.* New York: Simon & Schuster, 2000. Provides historical insight and speculates about the future of computer science and technology.

Kurzweil, Ray. *The Age of Spiritual Machine: When Computers Exceed Human Intelligence.* New York: Penguin, 2000. Offers some speculations about computers and their uses through the twenty-first century.

Marks, Myles H. *Basic Integrated Circuits.* Blue Ridge Summit, Pa.: Tab Books, 1986. An introduction to functions and uses of integrated circuits.

Zheng, Y., et al. "Graphene Field Effect Transistors with Ferroelectric Gating." *Physical Review Letters* 105 (2010). Reports on the experimental development and successful testing of a graphene-based field-effect transistor system using gold and graphene electrodes with SiO_2 gate structures on a silicon substrate.

See also: Applied Mathematics; Applied Physics; Computer Engineering; Computer Science; Electrochemistry; Electronic Materials Production; Electronics and Electronic Engineering; Liquid Crystal Technology; Nanotechnology; Surface and Interface Science; Transistor Technologies.

L

LASER INTERFEROMETRY

FIELDS OF STUDY

Control engineering; electrical engineering; engineering metrology; interferometry; laser science; manufacturing engineering; materials science; mechanical engineering; nanometrology; optical engineering; optics; physics.

SUMMARY

Laser interferometry includes many different measurement methods that are all based on the unique interference properties of laser lights. These techniques are used to measure distance, velocity, vibration, and surface roughness in industry, military, and scientific research.

KEY TERMS AND CONCEPTS

- **beam splitter:** Partially reflecting and partially transmitting mirror.
- **constructive interference:** Addition of two or more waves that are in phase, leading to a larger overall wave.
- **destructive interference:** Addition of two or more waves that are out of phase, leading to a smaller overall wave.
- **heterodyne detection:** Mixing of two different frequencies of light to create a detectable difference in their interference pattern.
- **homodyne detection:** Mixing of two beams of light at the same frequency, but different relative phase, to create a detectable difference in their interference pattern.
- **monochromatic:** Containing a single wavelength.
- **spatial coherence:** Measure of the phase of light over a defined space.
- **temporal coherence:** Measure of the phase of light as a function of time.

DEFINITION AND BASIC PRINCIPLES

Laser interferometry is a technique that is used to make extremely precise difference measurements between two beams of light by measuring their interference pattern. One beam is reflected off a reference surface and the other either reflects from or passes through a surface to be measured. When the beams are recombined, they either add (constructive interference) or subtract (destructive interference) from each other to yield dark and light patterns that can be read by a photosensitive detector. This interference pattern changes as the relative path length changes or if the relative wavelength or frequency of the two beams changes. For instance, the path lengths might vary because one object is moving, yielding a measurement of vibration or velocity. If the path lengths vary because of the roughness of one surface, a "map" of surface smoothness can be recorded. If the two beams travel through different media, then the resulting phase shift of the beams can be used to characterize the media.

Lasers are not required for interferometric measurements, but they are often used because laser light is monochromatic and coherent. It is principally these characteristics that make lasers ideal for interferometric measurements. The resulting interference pattern is stable over time and can be easily measured, and the precision is on the order of the wavelength of the laser light.

BACKGROUND AND HISTORY

The interference of light was first demonstrated in the early 1800's by English physicist Thomas Young in his double-slit experiment, in which he showed that two beams of light can interact like waves to produce alternating dark and light bands. Many scientists believed that if light were composed of waves, it would require a medium to travel through, and this medium (termed "ether") had never been detected. In the late 1800's, American physicist Albert Michelson designed an interferometer to measure the effect of the ether on the speed of light. His experiment was considered a failure in that he was not able to provide proof of the existence of the ether. However, the utility of the interferometer for measuring a precise distance was soon exploited. One of Michelson's first uses of his interferometer was to measure the international unit of a meter using a platinum-iridium metal bar, paving the way for modern interferometric methods of measurement. Up until the mid-twentieth century, atomic sources were used in interferometers, but their use for measurement was limited to their coherence length, which was less than a meter. When lasers were first developed in the 1960's, they quickly replaced the spectral line sources used for interferometric measurements because of their long coherence length, and the modern field of laser interferometry was born.

HOW IT WORKS

The most common interferometer is the Michelson interferometer, in which a laser beam is divided in two by use of a beam splitter. The split beams travel at right angles from each other to different surfaces, where they are reflected back to the beam splitter and redirected into a common path. The interference between the recombined beams is recorded on a photosensitive detector and is directly correlated with the differences in the two paths that the light has traveled.

In the visible region, one of the most commonly available lasers is the helium-neon laser, which produces interference patterns that can be visually observed, but it is also possible to use invisible light lasers, such as those in the X-ray, infrared, or ultraviolet regions. Digital cameras and photodiodes are routinely used to capture interference patterns, and these can be recorded as a function of time to create a movie of an interference pattern that changes with time. Mathematical methods, such as Fourier analysis, are often used to help resolve the wavelength composition of the interference patterns. In heterodyne detection, one of the beams is purposefully phase shifted a small amount relative to the other, and this gives rise to a beat frequency, which can be measured to even higher precision than in standard homodyne detection. Fiber optics can be used to direct the light beams, and these are especially useful to control the environment through which the light travels. In this case, the reflection from the ends of the fiber optics

have to be taken into account or used in place of reflecting mirrors. Polarizers and wave-retarding lenses can be inserted in the beam path to control the polarization or the phase of one beam relative to the other.

While Michelson interferometers are typically used to measure distance differences between the two reflecting surfaces, there are many other configurations. Some examples are the Mach-Zehnder and Jamin interferometers, in which two beams are reflected off of identical mirrors but travel through different media. For instance, if one beam travels through a gas, and the other beam travels through vacuum, the beams will be phase shifted relative to the other, causing an interference pattern that can be interpreted to give the index of refraction of the gas. In a Fabry-Perot interferometer, light is directed into a cavity consisting of two highly reflecting surfaces. The light bounces between the surfaces multiple times before exiting to a detector, creating an interference pattern that is much more highly resolved than in a standard Michelson interferometer. Several other types of interferometers are described below in relation to specific applications.

APPLICATIONS AND PRODUCTS

Measures of standards and calibration. Because of the accuracy possible with laser interferometry, it is widely used for calibration of length measurements. The National Institute of Standards and Technology (NIST), for example, offers measurements of gauge blocks and line scales for a fee using a modified Michelson-type interferometer. Many commercial companies also offer measurement services based on laser interferometer technology. Typical services are for precise measurement of mechanical devices, such as bearings, as well as for linear, angular, and flatness calibration of other tools such as calipers, micrometers, and machine tools. Interferometers are also used to measure wavelength, coherence, and spectral purity of other laser systems.

Dimensional measurements. Many commercial laser interferometers are available for purchase and can be used for measurements of length, distance, and angle. Industries that require noncontact measurements of complex parts use laser interferometers to test whether a part is good or to maintain precise positioning of parts during fabrication. Laser interferometers are widely used for these purposes in the automotive, semiconductor, machine tool, and medical- and scientific-parts industries.

Vibrational measurements. Laser vibrometers make use of the Doppler shift, which occurs when one laser beam experiences a frequency shift relative to the other because of the motion of the sample. These interferometers are used in many industries to measure vibration of moving parts, such as in airline or automotive parts, or parts under stress, such as those in bridges.

Optical metrology. Mirrors and lenses used in astronomy require high-quality surfaces. The Twyman-Green and Fizeau interferometers are variations on the Michelson interferometer, in which an optical lens or mirror to be tested is inserted into the path of one of beams, and the measured interference pattern is a result of the optical deviations between the two surfaces. Other industrial optical testing applications include the quality control of lenses in glasses or microscopes, the testing of DVD reader optical components, and the testing of masks used in lithography in the semiconductor industry.

Ring lasers and gyroscopes. In the last few decades, laser interferometers have started to replace mechanical gyroscopes in many aircraft navigation systems. In these interferometers, the laser light is reflected off of mirrors such that the two beams travel in opposite directions to each other in a ring, recombining to produce an interference pattern at the starting point. If the entire interferometer is rotated, the path that the light travels in one direction is longer than the path length in the other direction, and this results in the Sagnac effect: an interference pattern that changes with the angular velocity of the apparatus. These ring interferometers are widely available from both civilian and military suppliers such as Honeywell or Northrop Grumman.

Ophthalmology. A laser interferometry technique using infrared light for measuring intraocular distances (eye length) is widely used in ophthalmology. This technique has been developed and marketed primarily by Zeiss, which sells an instrument called the IOL Master. The technique is also referred to as partial coherence interferometry (PCI) and laser Doppler interferometry (LDI) and is an area of active research for other biological applications.

Sensors. Technology based on fiber-optic acoustic sensors to detect sound waves in water have been developed by the Navy and are commercially available from manufacturers such as Northrop Grumman.

Gravitational wave detection. General relativity predicts that large astronomical events, such as black hole formation or supernova, will cause "ripples" of gravitational waves that spread out from their source. Several interferometers have been built to try to measure these tiny disturbances in the local gravitational fields around the earth. These interferometers typically have arm lengths on the order of miles and require a huge engineering effort to achieve the necessary mechanical and vibrational stability of the lasers and the mirrors. There are currently efforts to build space-based gravity-wave-detecting interferometers, which would not be subject to the same seismic instability as Earth-based systems.

Research applications. Laser interferometers are used in diverse form in many scientific experiments. In many optical physics applications, laser interferometers are used to align mirrors and other experimental components precisely. Interferometers are also used for materials characterization in many basic research applications, while ultrasonic laser interferometers are used to characterize velocity distributions and structures in solids and liquids. A more recent technology development is interferometric sensors, which are used to monitor chemical reactions in real time by comparing laser light directed through waveguides. The

Laser Interferometry

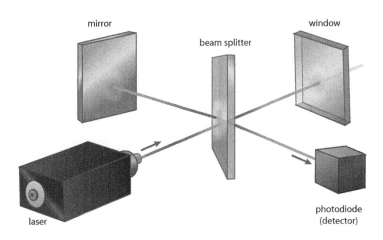

interference pattern from the two beams changes as the chemical reaction progresses. This technique is often referred to as dual-polarization interferometry.

IMPACT ON INDUSTRY

The invention of the laser in 1960 opened up the field of interferometry to a huge range of applications, allowing measurements both at long distances and with extremely high precision. The following decades saw the initial application of laser interferometry to vibration and dimensional measurements in industrial processes. The invention of cheap semiconductor diode lasers in the last decade has further decreased the price of laser interferometers and has widened the range of accessible wavelengths. Many laser interferometer systems are now available commercially, and they are used in a large segment of the semiconductor, automotive, and measurement industries. Systems now sold by many companies give compact, reliable ways to measure surface roughness, calibrate mechanical components, align or position parts during manufacturing, or for precision machining. Lasers are a multibillion-dollar industry, and laser interferometers are a constantly growing segment of this market.

Laser interferometers have hugely affected the quality control of lenses in the optical-manufacturing industries. Interferometers are the method of choice to measure curvature and smoothness of lenses used in microscopes, telescopes, and eyeglasses. Companies specializing in optical measurements are available for on- or off-site testing of optical components. Astronomy, in particular, has benefited from the precision testing of polished surfaces made possible by laser interferometers; mirrors such as the ones used in the Hubble Space Telescope would not be possible without a laser interferometry testing system.

The National Science Foundation (NSF) has invested substantially in ground-based interferometers for the measurement of gravity waves to test the predictions of general relativity. In the 1990's, the NSF funded the building of two very large-scale Michelson interferometers in the United States—one in Hanford, Washington, and the other in Livingston, Louisiana. Together they make up the Laser Interferometer Gravitational-Wave Observatory (LIGO) and consist of interferometer arms that are miles long. Though gravity waves have not yet been detected,

---

FASCINATING FACTS ABOUT LASER INTERFEROMETRY

- When the first laser was demonstrated in 1960, it was called a "solution looking for a problem."
- The Laser Interferometer Gravitational-Wave Observatory (LIGO) is sensitive to disturbances in the local gravitational fields that are caused by astronomical events as far back in time as 70 million years.
- Sensors based on laser interferometers are being developed to detect acoustic signals for surveillance. A buried sensor can detect the sound of footsteps from as far away as 30 feet.
- The Laser Interferometer Space Antenna (LISA) will consist of three spacecraft orbiting at 5 million kilometers apart and will provide information on the growth and formation of black holes and other events never before seen.
- Laser interferometers will be used to measure the optical quality of the mirror used in the James Webb Space Telescope, scheduled to launch in 2014. The surface smoothness must be unprecedented: If the mirror were scaled up to a size of 3,000 miles across, the surface height would be allowed to vary only by a foot at most.
- Quantum interferometers have been built to recreate the classic double-slit interference experiment, using single atoms instead of a light beam.

---

government-funded efforts around the world continue with large-scale interferometers currently operating in a half-dozen countries.

Sensor technology is a growing field in laser interferometry applications. Sensors based on interference of laser light using fiber optics have been developed to detect environmental changes, such as temperature, moisture, pressure, strain on components, or chemical composition of an environment. These types of sensors are not yet widely commercially available but are in development by industry and by the military for applications such as chemical-agent detection in the field or for use in harsh or normally inaccessible environments. Combined with wireless technology, they could be used, for instance, to monitor environmental conditions far underground or within the walls of buildings.

CAREERS AND COURSE WORK

Basic research on laser interferometry and its applications is conducted in academia, in many metrology industries, and in government laboratories and agencies. For research careers in new and emerging interferometry methods in academia or as a primary investigator in industry, a doctorate degree is generally required. Graduate work should be in the area of physics or engineering. The undergraduate program leading into the graduate program should include classes in mathematics, engineering, computer, and materials science.

For careers in industries that provide or use commercial laser interferometers but do not conduct basic research, a master's or bachelor's degree would be sufficient, depending on the career path. Senior careers in these industries involve leading a team of engineers in new designs and applications or guiding a new application into the manufacturing field. In this case, the focus of course work should be in engineering.

Mechanical, electrical, optical, or laser engineering will provide a solid background and an understanding of the basic theory of interferometer science. Additional courses should include physics, mathematics, and materials science. A bachelor's degree would also be required for a marketing position in laser interfer-_ ometry industries. In this case, focus should be on business, but a strong background in engineering or physics will make a candidate much more competitive. Technical jobs that do not require a bachelor's degree could involve maintenance, servicing, or calibration of laser interferometers for measurements in industry. They could involve assembly of precision optomechanical systems or machining of precision parts.

SOCIAL CONTEXT AND FUTURE PROSPECTS

The development of increasingly precise interferometers in the last few decades, such as for gravitational-wave measurement, has spurred corresponding leaps in mechanical and materials engineering, since these systems require unprecedented mechanical and vibrational stability. Laser interferometers are beginning to be used in characterization of nanomaterials, and this will push the limits of resolution of laser interferometers even further. As the cost of lasers and optical components continues to decrease, the use of laser interferometers in many industrial

manufacturing applications will likely increase. They are an ideal measurement system in that they do not contain moving parts, so there is no wear on parts, and they do not mechanically contact the sample being measured.

Active research is conducted in the field of laser interferometric sensors, with potential applications in military and manufacturing industries. Oil and gas companies may also drive development of sensors for leak and gas detection during drilling. In addition, commercial applications for laser sensors will open up in areas of surveillance as acoustic laser interferometry technology is developed.

Research continues in academic labs in areas that may someday become real-world applications, such as using laser interferometers for the detection of seismic waves, or in the area of quantum interferometry in which single photons are manipulated to interfere with each other in a highly controlled manner.

—Corie Ralston, PhD

FURTHER READING

Beers, John S., and William B. Penzes. "The NIST Length Scale Interferometer." *Journal of Research of the National Institute of Standards and Technology* 104, no. 3 (May/June, 1999): 225-252. Detailed description of the original NIST Michelson interferometer, with historical context.

Halsey, David, and William Raynor, eds. *Handbook of Interferometers: Research, Technology and Applications.* New York: Nova Science Publishers, 2009. A comprehensive text on current interferometry technologies. Designed for readers with a strong physics background.

Hariharan, P. *Basics of Interferometry.* 2d ed. Burlington, Mass.: Academic Press, 2007. Covers most of the current interferometer configurations and includes sections on applications.

_____. *Optical Interferometry.* 2d ed. San Diego: Academic Press, 2003. Overview of the basic theory of interferometry methods, including sections on laser interferometry.

Malacara, Daniel, ed. *Optical Shop Testing.* 3d ed. Hoboken, N.J.: John Wiley & Sons, 2007. An introduction to interferometers used in testing optical components.

Sirohi, Rajpal S. *Optical Methods of Measurement:*

Whole-field Techniques. 2d ed. Boca Raton, Fla.: CRC Press, 2009. Covers the basics of wave equations and interference phenomena. Includes multi-wavelength techniques of interferometry.

Tolansky, Samuel. *An Introduction to Interferometry.* 2d ed. London: Longman, 1973. Includes many interferometry methods that do not use lasers.

WEB SITES
Laser Interferometer Gravitational-Wave Observatory
http://www.ligo.caltech.edu

National Aeronautics and Space Administration James Webb Space Telescope
http://www.jwst.nasa.gov

National Institute of Standards and Technology
http://www.nist.gov

See also: Computer Science; Electrical Engineering; Engineering; Laser Technologies; Mechanical Engineering; Optics.

LASER TECHNOLOGIES

FIELDS OF STUDY

Cosmetic surgery; dermatology; fiber optics; laparoscopic surgery; medicine; ophthalmology; optical disk manufacturing; physics; printer manufacturing; spectroscopy; surgery.

SUMMARY

The term "laser" is an acronym for light amplification by stimulated emission of radiation. A laser device emits light via a process of optical amplification; the process is based on the stimulated emission of photons (particles of electromagnetic energy). Laser beams exhibit a high degree of spatial and temporal coherence, which means that the beam does not widen or deteriorate over a distance. Laser applications are numerous and include surgery, skin treatment, eye treatment, kidney-stone treatment, light displays, optical disks, bar-code scanners, welding, and printers.

KEY TERMS AND CONCEPTS

- **fiber optics:** Transmission of light through extremely fine, flexible glass or plastic fibers.
- **laparoscopic surgery:** Minimally invasive surgery that is accomplished with a laparoscope and the use of specialized instruments, including laser devices, through small incisions.
- **laser guidance:** Guidance via a laser beam, which continuously illuminates a target; missiles, bombs, or projectiles (for example, bullets) home on the beam to strike the target.
- **optical disk:** Plastic-coated disk that stores digital data, such as music images, or text; tiny pits in the disk are read by a laser beam.
- **photon:** Electromagnetic energy, which is regarded to be a discrete particle that has zero mass, no electrical charge, and an indefinitely long lifetime.
- **spectroscopy:** Analysis of matter via a light beam.

DEFINITION AND BASIC PRINCIPLES
A laser consists of a highly reflective optical cavity, which contains mirrors and a gain medium. The gain medium is a substance with light-amplification properties. Energy is applied to the gain medium via an electrical discharge or an optical source such as another laser or a flash lamp. The process of applying energy to the gain medium is known as pumping. Light of a specific wavelength is amplified in the optical cavity. The mirrors ensure that the light bounces repeatedly back and forth in the chamber. With each bounce, the light is further amplified. One mirror in the chamber is partially transparent; amplified light escapes through this mirror as a light beam.

Many types of gain media are employed in lasers, they include: gases (carbon dioxide, carbon monoxide, nitrogen, argon, and helium/neon mixtures); silicate or phosphate glasses; certain crystals (yttrium aluminum garnet); and semiconductors (gallium arsenide and indium gallium arsenide). A basic concept of laser technology is "population inversion."

Normally, most of the particles comprising a gain medium lack energy and are in the ground state. Pumping the medium places most or all the particles in an excited state. This results in a powerful, focused laser beam. Lasers are classified as operating in either continuous or pulsed mode. In continuous mode, the power output is continuous and constant; in pulse mode, the laser output takes on the form of intermittent pulses of light.

BACKGROUND AND HISTORY

In 1917, Albert Einstein theorized about the process of stimulated emission, which makes lasers possible. The precursor of the laser was the maser (microwave amplification by stimulated emission of radiation). A patent for the maser was granted to American physicists Charles Hard Townes and Arthur L. Schawlow on March 24, 1959. The maser did not emit light; rather it amplified radio signals for space research. In 1958, Townes and Schawlow published scientific papers, which theorized about visible lasers. Also in 1958, Gordon Gould, a doctoral student at Columbia University under Townes, began building an optical laser. He did not obtain a patent until 1977. In 1960, the first gas laser (helium-neon) was invented by Iranian-American physicist Ali Javan. The device converted electrical energy to light, and this type of laser has many practical applications such as laser surgery. In 1962, American engineer Robert Hall invented the semiconductor laser, which has many applications for communications systems and electronic appliances. In 1969, Gary Starkweather, a Xerox researcher, demonstrated the use of a laser beam to print. Laser printers were a marked improvement over the dot-matrix printer. The print quality was much better and it could print on a single sheet of paper rather than a continuous pile of fan-folded paper. In September, 1976, Sony first demonstrated an optical audio disk. This was the precursor to compact discs (CDs) and digital versatile discs (DVDs). In 1977, telephone companies began trials using fiber-optic cables to carry telephone traffic. In 1987, New York City ophthalmologist Stephen Trokel performed the first laser surgery on a patient's eyes.

HOW IT WORKS

Laser ablation. Laser ablation involves removing material from a surface (usually a solid but occasionally a liquid). Pulsed laser is most commonly used;

however, at a high intensity, continuous laser can ablate material. At lower levels of laser energy, the material is heated and evaporates. At higher levels, the material is converted to plasma. Plasma is similar to gas, but it differs from gas in that some of the particles are ionized (a loss of electrons). Because of this ionization, plasma is electrically conductive.

Laser cutting. Laser cutting uses laser energy to cut materials either by melting, burning, or vaporizing. Laser cutting is extremely focusable to about 25 microns (one-quarter the width of a human hair); thus, a minimal amount of material is removed. The three common types of lasers used for cutting are: carbon dioxide (CO_2), neodymium (Nd), and neodymium-yttrium aluminum garnet (Nd-YAG).

Laser guidance. Laser guidance involves the use of a laser beam to guide a projectile (a bomb or a bullet) to a target. In its simplest form, such as a beam emitted from a rifle, the shooter points the laser beam so that the bullet will hit the target. A much more complex process is the guidance of a missile or a bomb. In some cases, the missile contains a laser homing device in which the projectile "rides" the laser beam to the target. More commonly, a technique referred to as semi-active laser homing (SALH) is employed. With SALH, the laser beam is kept pointed at the target after the projectile is launched. Laser energy is scattered from the target, and as the missile approaches the target, heat sensors home in on this energy. If a target does not reflect laser energy well, the beam is aimed at a reflective source near the target.

Laser lighting displays. The focused beam emitted by a laser makes it useful for light shows. The bright, narrow beam is highly visible in the night sky. The beam can also be used to draw images on a variety of surfaces, such as walls or ceilings, or even theatrical smoke. The image can be reflected from mirrors to produce laser sculptures. The beam can be moved at any speed in different directions by the use of a galvanometer, which deflects the beam via an electrical current. The variety of vivid colors available with lasers enhances the visual effects.

Laser printing. Laser printing involves projecting an image onto an electrically charged, rotating drum that is coated with a photoconductor. When exposed to light, the electrical conductivity of the photoconductor increases. The drum then picks up particles of dry ink (toner). These particles are picked up by varying degrees depending on the amount of charge. The toner is then applied to a sheet of paper. The process involves rapidly "painting" the image line by line. Fast (and more expensive) laser printers can print up to 200 pages per minute. Printers are both monochrome (one color–usually black) or color. Color printers contain toners in four colors: cyan, magenta, yellow, and black. A separate drum is used for each toner. The mixture of the colors on the toners produces a crisp, multicolored image. Duplex laser printers are available, which print on both sides of a sheet of paper. Some duplex printers are manual devices, which require the operator to flip one or more pages manually when indicated. Automatic duplexers mechanically turn each sheet of paper and feed it past the drum twice.

Optical disks. Optical disks are flat, circular disks that contain binary data in the form of microscopic pits, which are non-reflective and form a binary value of 0. Smooth areas are reflective and form a binary value of 1. Optical disks are both created and read with a laser beam. The disks are encoded in a continuous spiral running from the center of the disk to the perimeter. Some disks are dual layer; with these disks, after reaching the perimeter, a second spiral track is etched back to the center. The amount of data storage is dependent on the wavelength of the laser beam. The shorter the wavelength, the greater the storage capacity (shorter-wavelength lasers can read a smaller pit on the disk surface). For example, the high-capacity Blu-ray Disc uses short-wavelength blue light. Laser can be used to create a master disk from which duplicates can be made by a stamping process.

Laser 3-D scanners. Laser 3-D scanners analyze an object via a laser beam. The collected data is used to construct a digital, three-dimensional object of the model. In addition to shape, some scanners can replicate color.

APPLICATIONS AND PRODUCTS

Laser technology includes a vast number of applications for business, entertainment, industrial, medical, and military use. The reflective ability of a laser beam has one major drawback: It can inadvertently strike an unintended target. For example, a reflected laser beam could damage an eye, so protective goggles are worn by laser operators. Products range from inexpensive laser pointers and CDs to surgical, industrial, and military devices costing hundreds of thousands

of dollars. The following applications and products are a representative sample and are by no means comprehensive.

Optical disks. Most optical disks are read-only; however, some are rewritable. They are used for the storage of data, computer programs, music, graphic images, and video games. Since the first CD was introduced in 1982, this technology has evolved markedly. Optical data storage has in a large part supplanted storage on magnetic tape. Although optical storage media can degrade over time from environmental factors, they are much more durable than magnetic tape, which loses its magnetic charge over time. Magnetic tape is also subject to wear as it passes through the rollers and recording head. This is not the case for optical media in which the only contact with the recording surface is the laser beam. CDs are primarily used for the storage of music: A five-inch CD can hold an entire recorded album, replacing the vinyl record, which was subject to wear and degradation. A limitation of the CD is its storage capacity: 700 megabytes of data (eighty minutes of music). Actually, three years before the introduction of the CD, the larger LaserDisc appeared for home video use. However, it never attained much popularity in the United States. The DVD, which appeared in 1996, rapidly gained popularity and soon outpaced VHS tape for the storage of feature-length movies. The DVD can store 4.7 gigabytes of data in single-layer format and 8.5 gigabytes in dual-layer format. The development of high-definition (HD) television fueled the development of higher-capacity storage media. After a format war of several year's duration, the Blu-ray DVD won out over the HD disc. The Bluray Disc can store about six times the amount of data as a standard DVD: 25 gigabytes of data in single-layer format and 50 gigabytes of data in dual-layer format.

Medical uses. Medical applications of laser technology exist for abdominal surgery, ophthalmic surgery, vascular surgery, and dermatology. The narrow laser beam can cut through tissue and cauterize small blood vessels at the same time. Laser can be used with an open surgical incision as well as laparoscopic procedures in which the surgery is performed through small incisions for passage of the laparoscope and surgical instruments. The surgeon and his or her assistants can view images on a video monitor. Mirrors can be used to deflect the laser beam in the desired direction. However, the surgeon must be extremely careful when directing a laser beam at an internal structure–inadvertent reflection of the beam can damage tissue (for example, puncture the bowel).

A common ophthalmic procedure is the radial keratotomy. With this procedure, fine laser cuts are made in a radial fashion around the cornea. These precision cuts can correct myopia (nearsightedness) and hyperopia (farsightedness). Laser is also used to treat a number of eye problems such as a detached retina (separation of the imaging surface of the retina from the back of the eye), glaucoma (increased pressure within the eyeball), and cataracts (clouding of the lens).

Atherosclerosis (a narrowing of the blood vessels due to plaque formation) is a common disease. Laser can be used to vaporize the plaque. It also can be used to bore small holes within the heart to improve circulation within the heart muscle.

Dermatology procedures can be performed with the laser–many of which are cosmetic procedures done to improve appearance. Some laser dermatologic applications are: acne treatment and acne scar removal, removal of age spots, skin resurfacing, scar removal, tattoo removal, spider vein removal, and hair removal.

Innovative uses of laser for medical purposes are being reported frequently. For example, in December, 2010, French researchers reported a noninvasive technique to diagnose cystic fibrosis prenatally. Cystic fibrosis is a genetic disease in which thick mucus secretions form in the lungs and glands such as the pancreas, which result in progressive deterioration of the affected organs. The technology involves the identification of affected fetuses by laser microdissection of a fetal cell, which was circulating in the mother's bloodstream. The procedure avoids the risk of a miscarriage when either chorionic villus sampling or amniocentesis is used. Another innovative use of laser was also reported in December, 2010. Medical devices, which are made of plastic or metal, are surgically placed within the body. Examples of medical devices are stents, which are inserted to improve blood flow to the heart, and hip-replacement prosthetics. These devices can become coated with biofilm, which is of bacterial origin. This biofilm is resistant to antibiotics; however, a laser technique was reported to be successful in removing biofilm.

Industrial uses. Lasers can make precision cuts on metal or other materials. A minimal amount of material is removed, leaving a smooth, polished surface on both sides of the cut. It also can weld and heat-treat materials. Other industrial uses include marking and measuring parts.

Business uses. The supermarket bar-code scanner was one of the earliest laser applications–it appeared in 1974. Laser printers are ubiquitous in all but the smallest business offices. They range in price from less than $100 to more than $10,000. The higher-priced models have features such as duplexing capability, color, high-speed printing, and collating.

Military uses. In addition to marking targets and weapons guidance, laser is used by the military for defensive countermeasures. It also can produce temporary blindness, which can temporarily impair an enemy's ability to fire a weapon or engage in another harmful activity. The military also uses laser guidance systems. Defensive countermeasure applications include small infrared lasers, which confuse heat-seeking missiles to intercept lasers, and power boost-phase intercept laser systems, which contain a complex system of lasers that can locate, track, and destroy intercontinental ballistic missiles (ICBMs). Intercept systems are powered by chemical lasers. When deployed, a chemical reaction results in the quick release of large amounts of energy.

Laser lighting displays. Laser light displays are popular worldwide. The displays range from simple to complex and are often choreographed to music. A popular laser multimedia display is Hong Kong's Symphony of Lights, which is presented nightly. The display is accentuated during Tet, the Vietnamese New Year. The exteriors of 44 buildings on either side of Victoria Harbour are illuminated with a synchronized, multicolored laser display, which is accompanied by music. More than four million visitors and locals have viewed the display.

IMPACT ON INDUSTRY

Laser technology is a major component of many industries, including electronics, medical devices, data storage, data transmission, research, and entertainment. Laser products enjoy significant repeat business. Over time the devices fail and need to be repaired or replaced. Often, before device failure occurs, the consumer will be attracted to an improved rendition of the product. For example the original LaserDisc, which never achieved popularity in the United States, has evolved to the Blu-ray Disc, which boasts vastly improved storage capacity as well as image and sound quality. In addition, many laser devices come with a hefty price tag–ranging from several thousand dollars to well in excess of 1 million dollars.

Industry and business. The average home contains laser devices such as CDs (discs and players) and DVDs (discs and players); and home computers, televisions, and telephones that are often connected to a fiber-optic cable network. Many homes and businesses use one or more laser printers. Virtually every retail establishment has a laser bar-code scanner. The military uses laser devices ranging from inexpensive

---

FASCINATING FACTS ABOUT LASER TECHNOLOGIES

- In early 1942, long before laser guidance systems appeared, psychologist B. F. Skinner at the University of Minnesota investigated weapons guidance by trained pigeons. The birds were trained with slides of aerial photographs of the target. They were placed in a harness inside the guidance system and kept the target in the crosshairs by pecking. Skinner never overcame official skepticism, and the project was abandoned.

- The radial keratotomy procedure originated in Soviet Russia with physician Svyatoslav Fyodorov in the 1970's. Fyodorov treated a boy whose glasses had been broken in a fight and cut his cornea. After the boy recovered, his eyesight had improved—his myopia was significantly lessened.

- Physician Stephen Trokel, who patented the excimer laser for vision correction, adapted the device from its original use: etching silicone computer chips.

- In 1960, American inventor Hildreth Walker, Jr., developed a ruby laser, which precisely measured the distance from the Earth to the Moon during the Apollo 11 mission.

- The original Hewlett Packard (HP) LaserJet printer retailed for more than $3,000 and could print eight pages per minute at a resolution of 300 dots per inch. The only font available was Courier, and it could print only a quarter-page of graphics at that resolution. As of 2011, much smaller laser printers with a multitude of features, including higher speed and print quality, retail for around $100.

- Laboratory lasers can measure distances accurate to less than a thousandth of the wavelength of light, which is less than the size of an atom.

---

pointers to guidance systems, which can cost many thousands of dollars. These expensive devices offer a continuous revenue stream for a vendor as they are only used once. The United States automobile industry has been using laser to cut the steel for its vehicles since the 1980's. As of 2011, many manufacturing industries use laser for the cutting and welding of metal parts. The medical-device industry is responsible for many high-end products. Laparoscopic surgery is currently popular, and laser instruments are well-adapted to operating by remote control through small incisions. Ophthalmologists are heavy users of laser instruments. Dermatologic procedures, particularly in the realm of cosmetic surgery, often employ expensive laser devices. Scientific procedures such as spectroscopy are commonplace in all industrialized nations.

Government and university research. One of the biggest sources of funding for laser research in the United States is the Defense Advanced Research Projects Agency (DARPA). It is also the biggest client for certain kinds of laser applications. The agency is primarily concerned with lasers with a military focus, such as guidance systems, missile countermeasures, and laser-based weapons. Another source of funding for laser research is the United States Department of Energy (DOE).The DOE is currently focused on energy efficiency and renewable energy.

DARPA and the DOE supply funds to many universities in the United States for laser research and development. For example, the two major universities in the Los Angeles area, University of California, Los Angeles (UCLA), and the University of Southern California (USC), are actively engaged in highly technical laser research. UCLA is the home to the Plasma-Accelerator Group, which publishes papers such as "Highest Power CO_2 Laser in the World" (2010). The USC Center for Laser Studies contains a number of research groups: Optical Devices Research Group, Research at High Speed Technology, Semiconductor Device Technology, Solid State Lasers, and Theoretical Research.

CAREERS AND COURSE WORK

The laser technology industry offers careers ranging from entry-level positions, such as lighting display operators, to high-level and highly technical positions, which require a scientific or engineering degree. The high-level positions require at least a bachelor's degree with course work in several fields related to laser technology: engineering, physics, computer science, mathematics, and robotics. Many of the positions require a master's or doctorate degree.

Positions are available for individuals with a degree in laser engineering in both the government and private sectors. The ability to be a team player is often of value for these positions because ongoing research is often a collaborative effort. University research positions are also available. In this arena, the employee is expected to divide his or her time between research and teaching.

Laser technicians are needed in a variety of fields including medical, business, and entertainment. Many of these positions require some training beyond high school at a community college or trade school. Dermatologists and other medical specialists often employ technicians to perform laser procedures. If a company employs a number of laser technicians, supervisory positions may be available.

SOCIAL CONTEXT AND FUTURE PROSPECTS

As of 2011, laser applications have become innumerable and ubiquitous. Many aspects of daily living that are taken for granted, such as ringing up groceries, making a phone call, or playing a video, are dependent on this technology. Virtually every branch of laser technology is continually evolving. Furthermore, existing technologies are being implemented in increasing numbers. For example, use of the CD for data storage is being replaced with the much higher capacity DVD. Research in the realm of military applications is particularly vigorous. The science-fiction weapon, the laser blaster, will soon become a reality. A primary obstacle to developing lasers as weapons has been the generation of enough power to produce a laser blast with a sufficient level of destruction. That power level is rapidly approaching. In March, 2009, Northrop Grumman first fired up their 100 kilowatt (100,000 watts) weapons-grade laser. (One hundred kilowatts is enough energy to power about six U.S. homes for a month). Tests are ongoing for the device. Grumman announced in late 2010 that the device would soon be transferred to the White Sands Missile Range in New Mexico for further testing. Medical applications for laser are expanding rapidly. The consumer, however, must be aware that adding the "laser" adjective to a procedure does not necessarily make it superior to previous techniques. The laser scalpel,

for example, is basically just another means of cutting tissue..00

—*Robin L. Wulffson, MD, FACOG*

FURTHER READING

Bone, Jan. *Opportunities in Laser Technology Careers.* New York: McGraw-Hill, 2008. Offers a comprehensive overview of career opportunities in laser technology; includes salary figures as well as experience required to enter the field.

Hecht, Jeff. *Understanding Lasers: An Entry-Level Guide.* 3d ed. Piscataway, N.J.: IEEE Press, 2008. This introductory text is suitable for students at the advanced high school level.

Lele, Ajey. *Strategic Technologies for the Military: Breaking New Frontiers.* Thousand Oaks, Calif.: Sage, 2009. Describes the nuances of technological development in a purely scientific manner and provides a social perspective to their relevance for future warfare and for issues such as disarmament and arms control, as well as their impact on the environment.

Sarnoff, Deborah S., and Joan Swirsky. *Beauty and the Beam: Your Complete Guide to Cosmetic Laser Surgery.* New York: St. Martin's, 1998. Provides information for the consumer on laser dermatology procedures and aids the reader in selecting the safest and most experienced practitioner. It also describes what each surgical procedure entails and details costs in different regions.

Silfvast, William Thomas. *Laser Fundamentals.* 2d ed. New York: Cambridge University Press, 2004. Covers topics from laser basics to advanced laser physics and engineering.

Townes, Charles H. *How the Laser Happened: Adventures of a Scientist.* New York: Oxford University Press, 1999. A personal account by a Nobel laureate that describes some of the leading events in twentieth-century physics.

WEB SITES

American Society for Laser Medicine and Surgery
http://www.aslms.org

Defense Advanced Research Projects Agency
http://www.darpa.mil

USC Center for Laser Studies
http://www.usc.edu/dept/CLS/page.html

LIGHT-EMITTING DIODES

FIELDS OF STUDY

Electrical engineering; materials science; semiconductor technology; semiconductor manufacturing; electronics; physics; chemistry; mathematics; optics; lighting; environmental studies; physics.

SUMMARY

Light-emitting diodes (LEDs) are diodes, semiconductor devices that pass current easily in only one direction, that emit light when current is passing through them in the proper direction. LEDs are small and are easier to install in limited spaces or where small light sources are preferred, such as indicator lights in devices. LEDs are also generally much more efficient at producing visible light than other light sources. As solid-state devices, when used properly, LEDs also have very few failure modes and have longer operational lives than many other light sources. For these reasons, LEDs are gaining popularity as light sources in many applications, despite their higher cost compared with other more traditional light sources.

KEY TERMS AND CONCEPTS

- **anode:** More positive side of the diode, through which current can easily flow into the device when forward biased.
- **cathode:** More negative side of the diode, through which current can flow out of the device when forward biased.
- **color temperature:** Temperature of a blackbody radiating thermal energy having the same color as the light emitted by the LED.
- **forward bias:** Orientation of the diode in which current most easily flows through the device.
- **photon:** Quantum mechanical particle of light.
- **p-n junction:** Junction between positive type (p-type) and negative type (n-type) doped semiconductors on which all diodes are based.
- **radiant efficiency:** Ratio of optical power output to the electrical power input of the device.
- **reverse bias:** Orientation of the diode in which current does not easily flow through the device.

- **reverse breakdown voltage:** Maximum reverse biased voltage that can be applied to the device before it begins to conduct electricity, often in an uncontrolled manner; sometimes simply called the breakdown voltage.
- **thermal power dissipation:** Rate of energy per unit time dissipated in the device in the form of heat.

DEFINITION AND BASIC PRINCIPLES

Diodes act as one-way valves for electrical current. Current flows through a diode easily in one direction, and the ideal diode blocks current flow in the other direction. The very name diode comes from the Greek meaning two pathways. The diode-like behavior comes from joining two types of semiconductors, one that conducts electricity using electrons (ntype semiconductor) and one that conducts electrons using holes, or the lack of electrons (p-type semiconductor). The electrons will try to fill the holes, but applying voltage in the proper direction ensures a constant supply of holes and electrons to conduct electricity through the diode. The electrons and holes have different energies, so when the electrons combine with holes, they release energy. For most diodes this energy heats the diode. However, by adjusting the types and properties of the semiconductors making up the diodes, the energy difference between holes and electrons can be made larger or smaller. If the energy difference corresponds to the energy of a photon of light, then the energy is given off in the form of light. This is the basis of how LEDs work.

LEDs are not 100 percent efficient, and some energy is lost in current passing through the device, but the majority of energy consumed by LEDs goes into the production of light. The color of light is determined by the semiconductors making up the device, so LEDs can be fabricated to make light only in the range of wavelengths desired. This makes LEDs among the most energy-efficient sources of light.

BACKGROUND AND HISTORY

In 1907, H. J. Round reported that light could be emitted by passing current through a crystal rectifier junction under the right circumstances. This was the ancestor of the modern LED, though the term diode had not yet been invented. Though research continued on these crystal lamps, as they were called, they were seen as impractical alternatives to incandescent and other far less expensive means of producing light. By 1955, Rubin Braunstein, working at RCA, had shown that certain semiconductor junctions produced infrared light when current passed through them. Scientists Robert Biard and Gary Pittman, however, managed to produce a usable infrared LED, receiving a patent for their device. Nick Holonyak, Jr., a scientist at General Electric, then created a red LED–the viable and useful visual spectrum LED–in 1961. Though these early LEDs were usable, they were far too expensive for widespread adoption. By the 1970's, Fairchild Semiconductor had developed inexpensive red LEDs. These LEDs were soon incorporated into seven segment numeric indicators for calculators produced by Hewlett Packard and Texas Instruments. Red LEDs were also used in digital watch displays and as red indicator lights on various pieces of equipment.

Early LEDs were limited in brightness, and only the red ones could be fabricated inexpensively. Eventually, other color LEDs and LEDs capable of higher light output were developed. As the capabilities of LEDs expanded, they began to see more uses. By the early twenty-first century, LEDs began to compete with other forms of artificial lighting based on their energy efficiency.

HOW IT WORKS

An LED is a specific type of solid-state diode, but it still retains the other properties typical of diodes. Solid-state diodes are formed at the junction of two semiconductors of different properties. Semiconductors are materials that are inherently neither good conductors nor good insulators. The electrical properties of the materials making up semiconductors can be altered by the addition of impurities into the crystal structure of the material as it is fabricated. Adding impurities to semiconductors to achieve the proper electrical nature is called doping. If the added impurity has one more electron in its outermost electron shell compared with the semiconductor material, then extra electrons are available to conduct electricity. This is a negative doped, or n-type, semiconductor. However, if the impurity has one fewer electron in its outermost electron shell compared with the semiconductor material, then there are too few

electrons in the crystal structure, and an electron can move from atom to atom to fill the void. This results in a missing electron moving from place to place and acts like a positive charge moving through the semiconductor. Engineers call this missing electron a hole, and semiconductors in which holes dominate are called positive doped, or p-type, semiconductors.

The p-n junction. To make a diode, a device is fabricated in which a p-type semiconductor is placed in contact with a n-type semiconductor. The shared boundary between the two types of semiconductors is called a p-n junction. In the vicinity of the junction, the extra electrons in the n-type region combine with the holes of the p-type region. The results in the removal of charge carriers in the vicinity of the p-n junction and the area of few charge carriers is called the depletion region.

When a voltage is applied across the p-n junction, with the p-type region having the higher voltage, then electrons are pulled from the n-type region and holes are pulled from the p-type region into the depletion region. Additionally, electrons are pulled into cathode (the exterior terminal connecting to the n-type region) replenishing the supply of electrons in the n-type region, and electrons are pulled from the anode (the exterior terminal connecting to the p-type region) replenishing the holes in the p-type region. This is the forward-bias orientation of the diode, and current flows through the diode when voltage is applied in this manner. However, when voltage is applied in the reverse direction, electrons are pulled from the n-type region and into the p-type region, resulting in a larger depletion region and fewer available charge carriers. Electric current does not flow through the diode in this reverse-bias orientation.

Electroluminescence. Electrons and holes have different energy levels. When the electrons and holes combine in the depletion region, therefore, they release energy. For most diodes, the energy difference between the p-type holes and the n-type electrons is fairly small, so the energy released is correspondingly small. However, if the energy difference is sufficiently large, then when an electron and hole combine, the amount of energy released is the same as that of a photon of light, and the energy is released in the form of light. This is called electroluminescence. Different wavelengths or colors of light have different energies, with infrared light having less energy than visual

A bulb-shaped modern retrofit LED lamp with aluminum heat sink, a light diffusing dome and E27 screw base, using a built-in power supply working on mains voltage

light, and red light having less energy than other forms of visual light. Blue light has more energy than other forms of visual light. The color of light emitted by the recombination of electrons and holes is determined by the energy difference between the electrons and holes. Different semiconductors have different energies of electrons and holes, so p-n junctions made of different kinds of semiconductors with different doping result in different colors of light emitted by the LED.

Efficiency. Most light sources emit light over a wide range of wavelengths, often including both visual and nonvisual light as well as heat. Therefore, only a portion of the energy used goes into the form of light desired. For an idealized LED, all of the light goes into one color of light, and that color is determined by the composition of the semiconductors making the p-n junction. For real LEDs, not all of the light makes it out of the material. Some of it is internally reflected and absorbed. Furthermore, there is some electrical resistance to the device, so there is some energy lost in heat in the LED—but nowhere near as much as with many other light sources. This makes LEDs very efficient as light sources. However, LED efficiency is temperature dependent, and they are most efficient at lower temperatures. High temperatures tend to reduce LED efficiency and shorten the lifetime of the devices.

APPLICATIONS AND PRODUCTS

LEDs produce light, like any other light source, and they can be used in applications where other light sources would have been used. LEDs have certain properties, however, that sometimes make their use preferable to other artificial light sources.

Indicator lights. Among the first widespread commercial use of LEDs for public consumption was as indicator lights. The early red LEDs were used as small lights on instruments in place of small incandescent lights. The LEDs were smaller and less likely to burn out. LEDs are still used in a similar way, though not with only the red LEDs. They are used as the indicator lights in automobile dashboards and in aircraft instrument panels. They are also used in many other applications where a light is needed and there is little room for an incandescent bulb.

Another early widespread commercial use of LEDs was the seven-segment numeric displays used to show digits in calculators and timepieces. However, LEDs require electrical current to operate, and calculators and watches would rather quickly discharge the batteries of these devices. Often the display on the watches was visible only when a button was pressed to light up the display. However, the advent of liquid crystal displays (LCDs) has rendered these uses mostly obsolete since they require far less energy to operate, and LEDs are needed to light the display at night only.

Replacements for colored incandescent lights. Red LEDs have become bright enough to be used as brake lights in automobiles. Red, green, and yellow LEDs are sometimes used for traffic lights and for runway lights at airports. LEDs are even used in Christmas-decoration lighting. They are also used in message boards and signs. LEDs are sometimes used for backlighting LCD screens on televisions and laptops. Colored LEDs are also frequently used in decorative or accent lighting, such as lighting in aquariums to accentuate the colors of coral or fish. Some aircraft use LED lighting in their cabins because of energy efficiency. Red LEDs are also used in pulse oximeters used in a medical setting to measure the

FASCINATING FACTS ABOUT LIGHT-EMITTING DIODES

- Most of the device often called a light-emitting diode is really just the packaging for the LED. The actual diode is typically very tiny and embedded deep inside the packaging.
- The fist inexpensive digital watches marketed to the public used red LEDs for displays.
- Most remote controls for televisions, DVD players, and similar devices use an infrared LED to communicate between the remote control and the devices.
- LEDs operate more efficiently when they are cold than when they are hot.
- Many automobile manufacturers use LEDs in brake lights, particularly the center third brake light.
- An early term for the LED was "crystal lamp." This term eventually gave way to the term "light-emitting diode," later commonly abbreviated LED.
- LEDs, if properly cared for and operated under specified conditions, can last for upward of 40,000 hours of operation.
- The first LEDs produced were infrared-emitting diodes. The first mass-produced visible-light LEDs emitted red light.
- LEDs switch on and off far more quickly than most other light sources, often being able to go from off to fully on in a fraction of a microsecond.

oxygen saturation in a patient's blood.

The biggest obstacle to replacing incandescent lights with LEDs for room lighting or building lighting is that they produce light of only one color. Several strategies are in development for producing white light using LEDs. One strategy is to use multiple-colored LEDs to simulate the broad spectrum of light produced by incandescent lights or fluorescent lights. However, arrays of LEDs produce a set of discrete colors of light rather than all colors of the rainbow, thus distorting colors of objects illuminated by the LED arrays. This is aesthetically unpleasing to most people. Another strategy for producing white light from LEDs is to include a phosphorescent coating in the casing around the LED. This coating would provide the different colors of light that would mimic the light of fluorescent bulbs; however, such

a strategy removes much of the efficiency of LEDs. Research continues to produce a pleasing white light from LEDs.

Despite the color problems and the high initial cost of LEDs, LEDs have many properties that make them attractive replacements for incandescent or fluorescent lights. LEDs typically have no breakable parts and being solid-state devices are very durable and have low susceptibility to vibrational damage. LEDs are very energy efficient, but they tend to be less efficient at high power and high light output. LEDs are slightly more efficient, and far more expensive, than high-efficiency fluorescent lights, but research continues.

Nonvisual uses for LEDs. Infrared LEDs are often used as door sensors or for communication by remote controls for electronic devices. They can also be used in fiber optics. The rapid switching capabilities of LEDs makes them well suited for high-speed communication purposes. Ultraviolet LEDs are being investigated as replacements for black lights for purposes of sterilization, since many bacteria are killed by ultraviolet light.

IMPACT ON INDUSTRY

Government and university research. Research continues to produce less expensive and more capable LEDs. The nature of LEDs makes them potentially among the most energy-efficient light sources. The United States Department of Energy is a significant driving force behind development of energy-efficient solid-state lighting systems, specifically the development of LED technology for widespread use. University researchers, particularly faculty in engineering and materials science, receive government grants to study improving LED performance.

Industry and business. There are two different ways that the private sector is impacted by advances in LEDs and LED uses. Major corporations, particularly those in semiconductor manufacturing, are doing research alongside government and university researchers to develop more efficient LEDs with greater capabilities. The increased demand for LED lighting from energy-conscious consumers spurs companies to compete to develop new LEDs for consumer use.

However, no matter what types of new LEDs are developed, they must be used in products that consumers want to purchase in order to be commercially viable. In the early twenty-first century, environmentally conscious consumers began looking for more energy-efficient lights. A great many companies began to make products that used LED lights in place of incandescent lights for these energy-conscious consumers who were willing to pay more for a product that was perceived as being more environmentally friendly and energy efficient. A prime example of this sort of development was the advent of LED Christmas lights; however, many other applications have been developed. The long operational life of LEDs is also attractive to consumers, since they are less likely to need a replacement compared with other light sources. The continued development of increasingly inexpensive LEDs with more different colors of light possible is spurring more companies to develop more products using LEDs.

CAREERS AND COURSE WORK
LEDs are used in many industries, not just in electronics, which means that there are many different degree and course-work pathways to working with LEDs.

The development of new types of LEDs requires detailed understanding of semiconductor physics, chemistry, and materials science. Typically, such research requires an advanced degree in physics, materials science, or electrical engineering. Such degrees require courses in physics, mathematics, chemistry, and electronics. The different degrees will have different proportions of those courses.

Utilization of LEDs in circuits, however, requires a quite different background. Technicians and assembly workers need only basic electronics and circuits courses to incorporate the LEDs into circuits or devices.

Lighting technicians and lighting engineers also work with LEDs in new applications. Such careers could require bachelor's degrees in their field. New LED lamps are being developed and LEDs are seen as a possible energy-efficient alternative to other types of lighting. They also have long operational lives, so there is continual development to include LEDs in any type of application where light sources of any sort are used.

SOCIAL CONTEXT AND FUTURE PROSPECTS
At first, LEDs were a niche field, with limited uses. However, as LEDs with greater capabilities and different colors of emitted light were produced, uses began to grow. LEDs have evolved past the point of simply being indicator lights or alphanumeric displays. Developments in semiconductor manufacturing have driven down the cost of many semiconductor devices, including LEDs. The reducing cost combined with the energy efficiency of LEDs has led these devices to become more prominent, particularly where colored lights are desired. Research continues to produce newer LEDs with different colors, different power requirements, and different intensities. Newer techniques are being developed to produce white light using LEDs. These technological developments will make LEDs even more practical replacements for current light sources, despite their higher initial up-front costs.

Research continues on LEDs to make them more commercially and aesthetically viable as alternatives to more traditional light sources. However, research is also continuing on other alternative light sources. The highest-efficiency fluorescent lights have similar efficiencies to standard LEDs, but they cost less and are able to produce pleasing white light that LEDs do not yet produce. LEDs will continue to play an increasing role in their current uses, but it is unclear if they will eventually become wide-scale replacements for incandescent or fluorescent lights.

—*Raymond D. Benge, Jr., MS*

FURTHER READING
Held, Gilbert. *Introduction to Light Emitting Diode Technology and Applications.* Boca Raton, Fla.: Auerbach, 2009. A comprehensive overview of light-emitting diode technology and applications of LEDs.
Mottier, Patrick, ed. *LEDs for Lighting Applications.* Hoboken, N.J.: John Wiley & Sons, 2009. A detailed book about LEDs, the manufacture of LEDs, and use of the devices in artificial lighting.
Paynter, Robert T. *Introductory Electronic Devices and Circuits: Conventional Flow Version.* 7th ed. Upper Saddle River, N.J.: Prentice Hall, 2006. An excellent and frequently used introductory electronics textbook, containing several sections on diodes, with a very good description of light-emitting diodes.
Razeghi, Manijeh. *Fundamentals of Solid State Engineering.* 3d ed. New York: Springer, 2009. An advanced undergraduate textbook on the physics of

semiconductors, with a very detailed explanation of the physics of the p-n junction, which is the heart of diode technology.

Schubert, E. Fred. *Light-Emitting Diodes.* 2d ed. New York: Cambridge University Press, 2006. A very good and thorough overview of light-emitting diodes and their uses.

Žukauskas, Artūras, Michael S. Shur, and Remis Gaska. *Introduction to Solid-State Lighting.* New York: John Wiley & Sons, 2002. A fairly advanced and very thorough treatise on artificial lighting technologies, particularly solid-state lighting, such as LEDs.

WEB SITES
LED Magazine
http://www.ledsmagazine.com

The Photonics Society
http://photonicssociety.org

Schottkey Diode Flash Tutorial
http://cleanroom.byu.edu/schottky_animation.phtml

University of Cambridge Interactive Explanation of Semiconductor Diode
http://www-G.eng.cam.ac.uk/mmg/teaching/ linearcircuits/diode.html

U.S. Department of Energy Solid-State Lighting
http://www1.eere.energy.gov/buildings/ssl/about.html

See also: Diode Technology; Electrical Engineering; Electronics and Electronic Engineering; Optics.

LIQUID CRYSTAL DISPLAYS

FIELDS OF STUDY

Electronics

SUMMARY

Liquid crystal displays (LCDs) are used on numerous devices from cell phones to high definition television sets. Their function depends on the response of liquid crystal compounds to an electrical field. LCDs have three principal design modes called passive matrix, active matrix and OLED. When electrically stimulated the molecules of a liquid crystal alter their shape and, orientation and can thus act as shutters for reflected or transmitted light in passive and active matrix displays. In OLED displays, the individual pixels are activated directly and emit light.

PRINCIPAL TERMS

- **capacitor:** an electrical component consisting of two conductors separated from each other by a nonconductor, allowing it to store electrical charge temporarily.
- **compound:** a chemically unique material whose molecules consist of several atoms of two or more different elements

- **crystal:** a solid consisting of atoms, molecules, or ions arranged in a regular, periodic pattern in all directions, often resulting in a similarly regular macroscopic appearance
- **phase:** a region of matter having specific defined properties
- **polarity:** a characteristic of a molecule or functional group in which there is a difference in the distribution of electronic charge, causing one part of the molecule or group to be relatively electrically positive and another part to be relatively electrically negative.
- **synthetic:** produced by artificial means or manipulation rather than by naturally occurring processes.
- **visible light:** electromagnetic radiation that human eyes can see, with a wavelength between 4 X 10^{-7} and 7 X 10^{-7} meters.

LIQUID CRYSTALS
Typically, a material that is classed as a crystal is a solid with rigidly defined shape and volume. The key characteristic of a crystal is that the molecules of the particular compound are arranged in a strict three-dimensional array that is held together by intermolecular forces that often are due to the polarity of the molecule. In most cases this can be visualized as a set

of tiny bar magnets that align themselves by the attraction of their north and south poles. In a similar way, the positive and negative regions of a polar molecule are attracted to each other and the molecules align themselves accordingly. A liquid crystal is a substance that can exist in a state that is somewhere between the liquid and solid states. The molecules of the material can line up in much the same way as in a solid crystalline material, but without forming a rigid three-dimensional structure. This allows the material to flow as a liquid, even though it has molecular properties that normally characterize a solid. The most important aspect of liquid crystals is that they are affected in a predictable and controllable way by an electric eield. In the "nematic phase" the molecules of a liquid crystal compound have a random orientation. There is no spatial ordering of the molecules as there would be in a solid crystal, and the material is a true liquid. When an electrical field is applied to the material, however, the molecules assume an ordered arrangement depending on the polarity of the current flow. The shape of the molecule changes predictably under specific conditions, and this feature allows them to be used like tiny 'open-and-close' shutters in the function of a liquid crystal display (LCD). Many liquid ccrystal materials are synthetic, but there are also a great many naturally occurring materials such as cell membranes and vanadium (V) oxide that exhibit liquid crystal behavior.

Passive Matrix and Active Matrix LCDs

LCDs are the principal visual display component of almost all electronic devices, and are used in cell phones, laptops, flat-screen televisions and computer monitors, vehicle information displays, watches, clocks, industrial and instrumentation read-outs, and any number of other such devices. The manner in which the liquid crystal material is activated determines if the LCD is of the passive matrix design or the active matrix design. In the passive matrix design, the liquid crystal material is sandwiched between two conductive plates, one of which determines the columns of the matrix array while the other determines the rows of the array. As the appropriate intersections of the grid are activated, the liquid crystal material at that location responds and allows the appropriate visible light to pass through either from a reflective surface or from a lighted surface behind the matrix. The actual process is somewhat more complicated

than this would suggest, as the structure of the LCD includes polarizing filters so that the colored visible light that is seen is plane-polarized. This can be demonstrated by looking at an LCD screen through a polarizing filter. Rotation of the filter to an orientation that is 90° to the plane of polarization will block the light completely so that only a black disc can be seen. The shuttering effect of the liquid crystal works specifically on the light being transmitted perpendicularly to the screen, so image integrity is quickly lost when viewed at an angle and may even be seen as a negative image. Active matrix LCDs appear similar in structure, but operate on a very different principle. In an active matrix LCD, the nematic liquid crystal is sandwiched between an printed circuit containing millions of very small transistors, each of which is combined with a capacitor to prevent charge leakage and maintain the brightness of the pixel at that location. Each pixel in the display is activated directly by its associated transistor-capacitor pair, rather than by the passive circuit structure of a passive matrix LCD. This allows active matrix LCDs to be viewed from a wider angle without loss of the image integrity.

OLED LCDs

The acronym OLED stands for Organic Light Emitting Diode. OLED LCDs use specific organic compounds that emit particular wavelengths of light when electrically stimulated. Because the phosphors of OLED screens actually emit light, OLED LCDs do not use reflected or transmitted light for the display. This produces a much brighter display that has the added feature of being viewable at any angle with the only detriment being loss of perspective when viewed at a significant angle. Perspective loss is what happens when the screen in a movie theater is viewed from an angle rather than from seven rows back in the middle of the theater. Perspective loss on OLED LCDs works the same way. OLED LCD screens have quickly become the most popular for high definition screens because of their functional characteristics. The trade-off, however, is that construction of an OLED LCD screen is a high precision operation in regard to the placement of the phosphors and the demanding nature of the transistor formation. Any transistor that does not function flawlessly will result in a permanent black or colored spot at the corresponding location on the screen.

Further Reading

Yang, Deng-Ke and Wu, Shin-Tson *Fundamentals of Liquid Crystal Devices* Hoboken, NJ: John Wiley & Sons, 2006.

Kelly, S.M. *Flat Panel Displays. Advanced Organic Materials* Cambridge, UK: The Royal Society of Chemistry, 2000.

Ge, Zhebing and Wu, Shin-Tson *Transflective Liquid Crystal Displays.* Hoboken, NJ: John Wiley & Sons, 2010.

Li, Quan, ed. *Liquid Crystals Beyond Displays. Chemistry, Physics, and Applications* Hoboken, NJ: John Wiley & Sons, 2012.

Collings, Peter J. *Liquid Crystals. Nature's Delicate Phase of Matter* Princeton, NJ: Princeton University Press, 2002.

—*Richard M. Renneboog MSc*

LIQUID CRYSTAL TECHNOLOGY

FIELDS OF STUDY

Electronics; chemistry; organic chemistry; physics; optics; materials science; mathematics

SUMMARY

Liquid crystal devices are the energy efficient, low-cost displays used in a variety of applications in which information or images are presented. The operation of the devices is based on the unique electrical and optical properties of liquid crystal materials.

KEY TERMS AND CONCEPTS

- **active matrix liquid crystal display:** A display with transistors to control voltage at each pixel, which allows for a sharper picture.
- **anisotropic:** Molecules that are not similar in all directions but have a longer axis in certain planes, so certain measurements may depend on direction.
- **calamitic liquid crystal:** An anisotropic liquid crystal with a longer axis in one direction, giving it a rodlike shape (also referred to as prolate).
- **discotic liquid crystal:** An anisotropic liquid crystal with one axis shorter than the other two, giving it a disclike shape (also referred to as oblate).
- **lyotropic liquid crystal:** Liquid crystals achieved when the molecule is combined to a high enough concentration with a solvent (rather than through temperature change). Soap is one example.
- **nematic state:** Liquid crystals that are not distinctly layered but can be oriented and aligned by a charge.
- **smectic state:** Liquid crystals that are somewhat disordered but maintain distinct layers.
- **thermotropic liquid crystals:** Liquid crystals in which the specific properties are maintained within a specific temperature range.
- **thin film transistor (TFT):** Transistors in a liquid crystal display (LCD) that help control the voltage, and therefore, the coloring, at individual pixels.
- **twisted-nematic effect (TNE):** The effect in which, in a turned-on LCD, light travels through the display changing from being polarized in one direction to being twisted 90 degrees and polarized in that direction.
- **twisted-nematic liquid crystal display (TN LCD):** The dominant type of LCD display, which utilizes the TNE.

DEFINITION AND BASIC PRINCIPLES

Liquid crystal technology is the use of a unique property of matter to create visual displays that have become the standard for modern technology.

Originally discovered as a state existing between a solid and a liquid, liquid crystals were later found to have applications for visual display. While liquid crystals are less rigid than something in a solid state of matter, they also are ordered in a manner not found in liquids. As anisotropic molecules, liquid crystals can be polarized to a specific orientation to achieve the desired lighting effects in display technologies.

Liquid crystals themselves can exist in several states. These range from a well-ordered crystal state to a disordered liquid state. In-between states are known as the smectic phase, which have layering, and the nematic phase, in which the separate layers no longer exist but the molecules can still be ordered.

Liquid crystal displays (LCDs) at this point typically use crystals in the nematic state. They also use

441

calamitic liquid crystals, whose rodlike shape and orientation along one axis allow the display to lighten and darken.

Over time, researchers have made advances in the materials used for LCDs and the route of power for manipulating the crystals, allowing for the low-cost, high-resolution, energy-efficient displays that have become the dominant technology for displays such as computers and television sets. Future research should allow for improvements in the response time of the displays and for better viewing from different angles.

BACKGROUND AND HISTORY

Liquid crystals were discovered by Austrian botanist and chemist Friedrich Reinitzer in 1888. While working with cholesterol, he discovered what appeared to be a phase of matter between the solid (crystal) state and the liquid state. While attempting to find the melting point, Reinitzer observed that within a certain temperature range he had a cloudy mixture, and only at a higher temperature did that mixture become a liquid. Reinitzer wrote of his discovery to his friend, German physicist Otto Lehmann, who not only confirmed Reinitzer's discovery—that the liquid crystal state was unique and not simply a mixture of solid and liquid states—but also noted some distinct visual properties, namely that light can travel in one of two different ways through the crystals, a property known as birefringence.

After the discovery of liquid crystals, the field saw a lengthy period of dormancy. Modern display applications have their roots in the early 1960s, in part because of the work of French physicist and Nobel laureate Pierre-Gilles de Gennes, who connected research in liquid crystals with that in superconductors. He found that applying voltages to liquid crystals allowed for control of their orientation, thus allowing for control of the passage of light through them.

In the early 1970s, researchers, including Swiss physicist and inventor Martin Schadt at the Swiss company Hoffman-LaRoche, discovered the twisted-nematic effect—a central idea in LCD technology. (The year of invention is typically said to be 1971, although patents were awarded later.) The idea was patented in the United States at the same time by the International Liquid Xtal Company (now LXD), which was founded by American physicist and inventor James Fergason in Kent, Ohio. (Fergason

was part of the Liquid Crystal Institute at Kent State University. The institute was founded by American chemist Glenn H. Brown in 1965.) Licensing the patents to outside manufacturers allowed for the production of simple LCDs in products such as calculators and wristwatches.

In the 1980s, LCD technology expanded into computers. LCDs became critical components of laptop computers and smaller television sets. With research continuing on liquid crystals into the twenty-first century, LCD televisions overtook cathode ray tubes (CRTs) as the dominant technology for television sets.

HOW IT WORKS

LCDs have a similar structure, whether in a digital watch or in a 40-inch television. The liquid crystals are held between two layers of glass. A layer of transparent conductors on the liquid crystal side of the glass allows the liquid crystal layer to be manipulated. Polarized film layers are placed on the outside ends of the glass, one of which will face the viewer and the other will remain at the back of the display.

Polarizers alter the course of light. Typically, light travels outward in random directions. Polarizers present a barrier, blocking light from traveling in certain directions and preventing glare. The polarizers in an LCD are oriented at 90-degree angles from each other. With the polarizers in place alone, all light would be blocked from traveling through an LCD, but the workings of liquid crystals allow that light to come through.

The electrical current running through the liquid crystals controls their orientation. The rodlike crystals, without voltage, are oriented perpendicular to the glass of the screens. In this state, the crystals do not alter the direction of the light passing through. As voltage is applied, the crystals turn parallel to the direction of the screen. Like the polarized films, the conductors are oriented at 90-degree angles to each other, as are the crystals next to each screen (that is, crystals on one end are at a 90-degree angle from those on the other end). Between, however, the crystals orient in a twisting pattern, so light polarized in one direction will be redirected and turned 90 degrees when it emerges at the other end of the display. This is known as the twisted-nematic effect.

Thus, the voltage applied to the crystals controls the light coming through the LCD screen. At lower

Otto Lehmann coined the term cholesteric liquid crystals in 1904

However, materials that do one of these things better may make other features of a display worse. In some cases, a mixture of compounds for the liquid crystals in a device may be used.

Simple LCDs. The simplest LCD displays, such as calculators and watches, typically do not have their own light sources. Instead, they have what is known as passive display. In back of the LCD display is a reflective surface. Light enters the display and then bounces off the reflective surface to allow for the screen display. Simple LCDs are monochromatic and have specific areas (typically bars or dots) that become light or dark. While these devices are lower-powered, some do still use a light source of their own. Alarm clocks, for example, have light-emitting diodes (LEDs) as part of their display so that they can be seen in the dark.

Personal computers and televisions. For larger monitors that display complex images in color, the setup for an LCD becomes more complicated. Multi-color LCDs need a significant light source at the back of the display.

The glass used for more sophisticated LCD displays will have microscopic etchings on the glass plates at the front and back of the display. As with the polarizing filters and the conductors, the etchings are at 90-degree angles from each other, vertical on one plate and horizontal on the other. This alignment forms a matrix of points in each location where the horizontal and vertical etchings cross, resulting in what are known as pixels. Each pixel has a unique "address" for the electronic workings of the display. Many television sets are marketed as having 1080p, referring to 1,080 horizontal lines of pixels.

An active matrix (AM) display will have individual thin film transistors (TFTs) added at each pixel to allow for control of those sites. Three transistors are actually present at each pixel, each accompanied by an additional filter of red, green, or blue. Each of those transistors has 256 power levels. The blending of the different levels of those three colors (256^3, or 16,777,216 possible combinations) and the number of pixels allows for the full-color LCD displays.

While light is displayed as a combination of red, green, and blue on screens, printing is typically done on a scale that uses cyan, magenta, yellow, and black as base colors. This accounts for some discrepancy between colors that appear on screen and those that

voltage levels, some, but not all, light is allowed through the display. By manipulating the intensity of the incoming light, the LCD can display in a gray scale.

Because liquid crystals are a state of matter, they exist only at a certain temperature: between the melting and freezing points of the material. Thus, LCD displays may have trouble working in extreme heat or extreme cold. One of the primary challenges of LCD display is finding materials that remain in liquid crystal forms at the temperatures in which the devices are likely to be used. Some of the challenge also lies in finding materials that may display better color or allow for lower energy consumption.

show up on paper.

APPLICATIONS AND PRODUCTS

Liquid crystals are used in displays for a number of products. Early uses included digital thermometers, digital wristwatches, electronic games, and calculators. As the power needed for an LCD display and resolution improved, LCDs came to be used in computer monitors, television sets, car dashboards, and cellphones.

Calculators and digital watches. Watches and calculators use what is known as a seven-segment display, wherein each of the seven segments that make up a number are "lit" or "unlit" to represent the ten digits. Looking closely at an LCD will reveal that most numbers come from seven segments, which can be lit to display the ten different digits. Without the polarizing layer, the display would not work. Placing the polarized layers in parallel on the surface would, for example, cause the outlines of all the numbers and other areas on the display to illuminate (appearing as 8s) and leave as blanks the rest of the display.

Early electronic games also used a segment display. Fixed places on the display would be either lit or unlit, allowing game characters to appear to move across the screen.

Temperature monitors. Because of their sensitivity to heat, liquid crystals have been studied for their use as temperature monitors. Molecules in the smectic liquid crystal state rotate around their axes, and the angle at which they rotate (the pitch) can be temperature sensitive. At different temperatures, the wavelength of light given off will change. Some liquid crystal mixtures are fairly temperature sensitive, and so the mixture of the colors will change with relatively small changes in the temperature. Because of this, they can be used for displays such as infrared or surface temperatures.

Computer monitors. LCDs have been, and will likely remain, the standard for laptop computers. They have been used for monitors since the notebook computer was introduced. Because of the low power consumption and thinness of the monitor, their use is likely to continue.

Television sets. The workings of LCDs in televisions have been outlined in the foregoing section. As will be discussed further in the next section, LCDs have had a marked impact on television displays, both in the quality of displays in the home and in

industry overall.

There are several developments that could affect LCD technology in the near future. One example is the development of LEDs for use as backlighting for LCDs. By using LEDs rather than a fluorescent bulb, as LCD technology now uses, LCDs can manifest greater contrast in different areas of the screen. Other areas of LCD development include photoalignment and supertwisted nematic (STN) LCDs.

Grooves are made in glass used for LCDs, but this has raised some concern about possible electric charges, reducing the picture quality. Additionally, photoalignment—a focus on the materials used to align the liquid crystals in the display—should ultimately allow for liquid display screens that are flexible or curved, rather than rigid (as are glass panels).

STN LCDs are modified versions of TN LCDs. Rather than twisting the crystals between the layers

---

FASCINATING FACTS ABOUT LIQUID CRYSTAL TECHNOLOGY

- Although it was the dominant display technology in televisions for more than sixty years, cathode ray tubes were discovered close to twenty years after liquid crystals.

- Partly for his work with liquid crystals, Pierre-Gilles de Gennes was awarded the Nobel Prize in Physics in 1991.

- The world's largest LCDs are in Cowboys Stadium in Arlington, Texas. Each screen of the four-sided display is 72 feet high and 160 feet wide. The screens are backlit by a total of 10.5 million LEDs, and altogether, the display weighs 1.2 million pounds.

- The first LCD television, the black-and-white Casio TV-10, came to market in 1983.

- Introduced in 1982, the Epson HX-20, generally considered the world's first standard-sized laptop, utilized an LCD.

- The discovery of liquid crystals was made by Fried-rich Reinitzer while working with the compound cholesteryl benzoate. In the time since its discovery, thousands of other organic compounds have been found to have a liquid crystal state.

- In addition to use in displays, liquid crystals have some application in polymers. Because they can be oriented, they can produce stronger fibers. As a result, they are used in the production of Kevlar.

a total of 90 degrees, STN LCDs rotate the crystals by 270 degrees within the display. This greater level of twisting allows for a much greater degree of change in the levels of brightness in a display. At the same time, it presents a challenge because the response time for the screen is significantly slower.

IMPACT ON INDUSTRY

The introduction of LCDs allowed for the replacement of CRTs, which had been the standard in the industry. LCDs require less energy than CRTs, and because they do not need the space for the cathode ray, the physical sets are lighter and take up less space for a similar-sized screen. In larger displays, LCDs compete in the consumer market with plasma televisions, which utilize small compartments of gas for the light on their screens. There are still a number of reasons a consumer might choose plasma over LCD, or vice versa.

Because of their polarized light, LCDs do not have the problems with glare that plasma displays may have. However, because LCDs have polarized films on the screen, LCD images vary greatly when they are not viewed straight on. While plasma televisions may have a slightly faster response rate than LCD televisions, LCD televisions consume less energy. While there is some competition between LCDs and plasma screens in the large television market, plasmas are made only in larger sizes, so smaller television sets and other applications such as computer monitors continue to use LCDs.

Political change. With the proliferation of high-definition television sets, whether in plasma or LCD form, came a corresponding need for higher quality signals for broadcast. Cable and satellite television providers have offered packages of channels that are distributed at higher quality.

Additionally, the presence of these televisions in homes was partly responsible for the industry's decision to move to a different broadcast format for television and to free up parts of the broadcast spectrum. Starting in 2005 and 2006 (depending on the size of the set), television sets were required to have digital receivers. As of June 12, 2009, all television signals were broadcast at a high definition (HD) frequency.

CAREERS AND COURSEWORK

Much of liquid crystal technology is oriented toward the production of display screens, but the complexity of the subject leaves a number of career path options. Master's degrees and doctorates are available in the area of liquid crystal research. Programs typically involve interdisciplinary study in chemistry and physics, and in other potentially relevant areas. Liquid crystal technology builds off basic knowledge of physics, chemistry, and organic chemistry.

Some research in the area of liquid crystals focuses on the material of the crystals themselves and in the development of crystals that improve upon current LCD displays. A background in chemical analysis and optics is important. Design of the products themselves involves knowledge about the design of circuits and backlighting.

There are a number of areas for prospective research and product development. These include the design of LCDs themselves, design of the manufacturing process, and the process of creating the molecules used in the displays.

SOCIAL CONTEXT AND FUTURE PROSPECTS

Some of the concerns and problems with LCDs are being confronted by society as a whole. One concern is the high energy consumption of fluorescent lamps used by LCDs. In contrast, LED lights, which use less energy, are being used more and more in LCDs. There is concern, however, about the environmental hazards LEDs may create when they are disposed of in landfills. Another possibility is the use of carbon nanotubes, which would provide LCD backlighting but would use even less energy than LEDs.

Durability concerns may also come to play a role. The grooves in the glass necessary for high definition LCDs also lead to physical wear and tear on the product. Refining the technology further may produce more durable sets while also alleviating some of the concerns about electronics disposal. Future work on LCDs also will involve altering components to overcome picture quality and durability concerns. Given the prominence of the products that utilize liquid crystals, the technology is likely to be important for development for the foreseeable future.

—Joseph I. Brownstein, MS

FURTHER READING

Chandrasekar, Sivaramakrishna. *Liquid Crystals.* 2d ed. New York: Cambridge University Press, 1992. Originally written in 1977, this work is considered one of the classic textbooks in the field and provides early history and an overview of work in liquid crystals.

Chigrinov, Vladimir G., Vladimir M. Kozenkov, and Hoi-Sing Kwok. *Photoalignment of Liquid Crystalline Materials*. Chichester, England: John Wiley & Sons, 2008. This book covers some areas of development in improving screens for LCD devices.

Collings, Peter J., and Michael Hird. *Introduction to Liquid Crystals*. New York: Taylor & Francis, 1997. As an introduction to the field, this book goes through the basics of liquid crystals and then some of the applications, using less technical language than many other texts on the subject.

Delepierre, Gabriel, et al. "Green Backlighting for TV Liquid Crystal Display Using Carbon Nanotubes." *Journal of Applied Physics* 108, no. 4 (September 2010). This article examines the possibility of using carbon nanotubes to backlight LCDs, potentially reducing both production and energy costs.

WEB SITES

Kent State University, Liquid Crystal Institute
http://www.lcinet.kent.edu

Nobel Prize Foundation
.http://www.nobelprize.org

LITHOGRAPHY

FIELDS OF STUDY

Printing; photolithography; photography; physics; chemistry; mathematics; calculus; optics; mechanical engineering; material science; graphic design; electromagnetics; microfabrication; semiconductor manufacturing; laser imaging.

SUMMARY

Lithography is an ink-based printing process that was first used in Europe at the end of the eighteenth century. Unlike an older printing press, in which individual pieces of raised type were pressed down onto sheets of paper, lithography uses a flat plate to transfer an image to a sheet of paper. Nearly all books, newspapers, and magazines being published are printed using lithography, as are posters and packing materials. A specialized subfield of lithography known as photolithography is also used in the making of semiconductors for computers. Career opportunities in lithography are growing in specialized areas but overall are neither increasing or decreasing because of the rise of electronic publishing and marketing.

KEY TERMS AND CONCEPTS

- **emulsion:** Mixture of two chemicals; in lithography, often used on plate surfaces.
- **hydrophilic:** Chemical property on a plate's sur-

face that attracts and holds a water-based ink; the opposite is hydrophobic.
- **image:** Words, pictures, or both on a printing plate.
- **imagesetter:** Device that transfers an image from a computer directly to a plate without the use of photographic negatives.
- **offset:** Transfer of an image from a plate to a secondary surface, often a rubber mat, that reverses it before final printing.
- **photolithography:** Process that uses high-precision equipment and light-sensitive chemicals to make products such as semiconductors.
- **photomask:** Flat surface into which holes have been cut to allow light to pass through; used in photolithography.
- **Plate:** Printing surface, made of metal or stone, on which areas have been chemically treated to attract or repel ink.

DEFINITION AND BASIC PRINCIPLES

Lithography is the process of making an image on a flat stone or metal plate and using ink to print the image onto another surface. Areas of the plate are etched or treated chemically in order to attract or repel ink. The ink is then transferred, directly or indirectly, to the surface where the final image appears.

Unlike a process such as letterpress, where raised letters or blocks of type are coated with ink and pressed against a surface such as paper, lithographic printing yields a result that is smooth to the touch.

Lithography differs from photocopying in that plates must be created and ink applied before prints can be made. Photocopying uses a process known as xerography, in which a tube-shaped drum charged with light-sensitive material picks up an image directly from a source. Laser printing is another application of xerography and is not the same as lithography.

Photolithography is a process that imitates traditional lithography in several ways but is not identical. Its high level of precision–a photolithographic image can be accurate down to the level of a micrometer or smaller–is useful in applications such as the manufacturing of computer components.

BACKGROUND AND HISTORY

Lithography was invented in 1798 by Alois Senefelder, a German playwright. Senefelder, who was looking for a way to publish his plays cheaply, discovered that printing plates could be made by writing on a flat stone block with grease pencil and etching away the stone surface around the writing. Eventually Senefelder developed a process by which ink adhered only to the parts of a flat surface not covered by grease. He later expanded the process to include multiple ink colors and predicted that lithography would one day be advanced enough to reproduce works of fine art.

German and French printers in the early 1800's made additional innovations. A patent was issued in 1837 to artist Godefroy Engelmann in France for a process he called chromalithography, in which colors were layered to create book illustrations. Interest in lithography and color printing also spread to North America, where printers in Boston invented new technologies that made the mass production of lithographic prints both high quality and economical. The process quickly spread from books to greeting cards, personal and business cards, posters, advertisements, and packaging labels. Lithography is still the leading process by which mass-produced reading material and packaging are printed.

HOW IT WORKS

Lithography in the context of printing follows a different set of steps than photolithography as used to make microprocessors.

Offset lithography. While there are many ways to print on paper or packaging using lithographic techniques, most items involve a process known as offset lithography. The term "offset" refers to the fact that the printing plate does not touch the paper or item itself.

In offset lithography, a plate is first created with the image to be printed. The plate may be made of metal, paper, or a composite such as polyester or Mylar. Lithographic printing plates were flat at one time, but modern printing presses use plates shaped like cylinders, with the image on the outside. To transfer the image to the plate, the surface of the plate is roughened slightly and covered with a light-sensitive chemical emulsion. A sheet of photo film with a reverse, or negative, of the image is laid over the emulsion. When an ultraviolet light is shone on the negative, the light filters through the image only in the areas where the negative is translucent. The result is a positive image–essentially, a negative of the negative–left on the printing plate.

The plate is treated again with a series of chemicals that make the darker areas of the image more likely to pick up ink, which is oil based. The lighter areas of the image are made to be hydrophilic, or water loving. Because oil and water do not mix, water blocks ink from being absorbed by these areas. A water-based mixture called fountain solution is applied to the surface of the plate and is picked up by the hydrophilic areas of the image. Rollers then coat the plate with ink, which adheres only to the hydrophobic (water-fearing) areas that will appear darker on the final image. Once the plate is inked, the press rolls it against a rubber-covered cylinder known as a blanket. The ink from the plate is transferred to the blanket in the form of a negative image. Excess water from the ink as well as fountain solution is removed in the process. The blanket is rolled against the sheet of paper or other item that will receive the final image. Finally, the paper carrying the newly inked image passes through an oven, followed by a set of water-chilled metal rollers, to set the image and prevent the ink from smudging.

Photolithography. Like lithography, the process of photolithography depends on the making of a plate coated with a light-sensitive substance. The plate is known as the substrate, while the light-sensitive chemical is known as the photoresist. Instead of a photo negative with the image to be printed, a photomask is used to shield the photoresist from light in some areas and expose it in others.

The similarities to traditional lithography end

methods of printing that came earlier. Over time, lithography came to be associated with lower-cost editions of books and other printed matter intended to be short-lived, such as newspapers, magazines, and catalogs. Lithography has also evolved as a method of artistic printmaking that can produce works of great beauty and high value. On the photolithography side, the technology has kept pace with the needs of generations of computers.

Web-fed offset printing. Large numbers of copies of a printed work–in the range of 50,000 copies and up–require printing processes that can run quickly and efficiently. Web-fed offset printing takes its name from the way in which paper is fed into the press. A web press uses a roll of paper, known as a web, which is printed and later cut into individual sheets. The largest web presses stand nearly three stories tall, print images on both sides of a sheet at once, and can print at a rate of 20,000 copies per hour. Major newspapers and magazines as well as best-selling books with high print runs are printed on web presses. One of the disadvantages of using a web press is that post-print options, such as folding and binding, are limited. Page sizes are highly standardized and cannot be changed easily to meet the needs of an individual print run. Image quality also is not as high as that offered by other types of lithographic presses.

Sheet-fed offset printing. As its name suggests, sheet-fed offset printing uses a paper supply of individually cut sheets rather than a paper roll. Each press has a mechanism that feeds paper sheets into the machine, one at a time. This process is less efficient than web-fed printing and can lead to a higher rate of mechanical problems, such as damage to the rubber blanket when more than one sheet is fed into the press in error. However, sheet-fed printing allows for a greater degree of customization for each printing job. The size and type of the paper can be changed, as can the area of the page on which each image is to be printed. A paper of heavier, higher-quality grade may be used in a sheet-fed printer. A wider range of post-print options are also available. Sheet-fed print runs can be bound using a number of different methods, including lamination and glue. These features make it more suitable for products such as sales brochures, corporate annual reports, coffee-table books, and posters.

Lithography in art. When it first appeared in the United States in the mid-1800's, lithography was

Maurits (M.C.) Escher was considered a master of lithography. Photo taken around 23 Nov. 1971 by Hans Peters

here, however. In photolithography, the substrate—rather than a sheet of paper or packaging material—is the final product. Once the image is transferred through the photo mask onto the photoresist, the substrate is treated with a series of chemicals that engraves the image into the surface. In lithographic printing, the image is never engraved directly onto the plate. Unlike printing plates and blankets, which are cylindrical, substrates are always flat. The result is a thin sheet of silicon, glass, quartz, or a composite etched precisely enough to be used as a microprocessing component.

APPLICATIONS AND PRODUCTS

Lithography as a printing technology has developed in multiple, almost opposing, directions throughout its history. Because lithographic plates can be used to make large numbers of impressions, the development of lithography allowed for printing of images and type on a mass scale that was commercially viable, a major change from the letterpress and intaglio associated with high-quality printing, particularly reproductions of works of art. The later introduction of technologies such as photogravure printing eventually made lithographic illustrations in mass-produced printed matter obsolete. At the same time, a number of artists on both sides of the Atlantic Ocean were making advances in lithographic printing as an art form. Henri Toulouse-Lautrec depended on lithography to achieve the bold lines and fields of color in his iconic posters for the Moulin Rouge and other French cabarets in the late 1800's. Another surge of interest in lithography came in the 1920's with works from painters Wassily Kandinsky, Georges Braque, and Pablo Picasso. In some cases, such as Toulouse-Lautrec's posters, these works were originally commercial in nature and intended to be reproduced in large print runs. Artists who experimented with lithography in the twentieth century were more likely to be drawn to the medium for its visual characteristics and possibilities for expression, not for its ability to generate copies. Paris was a major center of lithographic art until World War II, at which point many artists relocated to New York. A revival of the technique emerged in the 1950's with new prints from artists such as Sam Francis, Jasper Johns, and Robert Rauschenberg. Lithography is taught in many fine-arts schools. Some artists prefer to work directly with the stone or metal printing plates, while others draw or paint images and rely on third parties to transfer the work from the page to the plate.

Semiconductor manufacturing. Photolithography has been used to manufacture semiconductors and microprocessing components for about fifty years. When it was first developed, photolithography depended on the use of photomasks that came into direct contact with the photoresist. This contact often damaged the photomasks and made the manufacturing process costly. Next, a system was developed in which photomasks were suspended a few microns above the photoresist without touching it. This strategy reduced damage, but also lowered the precision with which a photomask could project an image. Since the 1970's manufacturing plants have used a system known as projection printing, in which an image is reflected through an ultrahigh-precision lens onto a photoresist. This technology has allowed manufacturers to fit increasingly higher numbers of integrated circuits onto a single microchip. In 1965, Gordon Moore, a technology executive who would go on to cofound Intel, predicted that the number of transistors that could be placed on a microchip would double about every two years. The prediction has been so accurate that the principle is now known as Moore's law.

IMPACT ON INDUSTRY

Commercial Printing. The market for commercial lithographic printing is a mature one and is not expected to see significant growth in the next several years. Unlike many other areas of technology, lithography in printing does not receive government research funding or have programs at academic institutions devoted to its study.

The commercial printing industry in North America is divided into tiers by company size and niche. As a market, lithographic printing on a large scale is led by RR Donnelley followed by Quad/Graphics. These firms and their competitors dominate market segments such as books, magazines, directories, catalogs, and direct-mail marketing pieces. Beyond corporations such as these, however, the commercial printing industry is made up primarily of small businesses with local clientele. According to the U.S. Bureau of Labor Statistics, seven out of ten companies offering lithographic printing services have fewer than ten employees. Taken as a whole, commercial printers earned about $100 billion in revenue in 2009. The roughly 1,600 daily newspapers in the United States make up another major market segment in lithographic printing. Because most newspapers own and operate their own printing facilities, they are seen as belonging to a related but separate industry, and their revenues are not included in most printing-industry estimates.

Most of the innovations in large-scale lithographic printing that have occurred since the 1990's involve the use of digital technology. Rather than using film and photo negatives to create reverse images, computers allow typesetters and printers to transfer an image directly onto the surface of a printing plate. However, many developments in commercial printing methods largely involve technologies that do not rely on lithography. When this trend is taken into account, along with the migration of many books and news sources from print formats to electronic ones, the future of commercial lithographic printing seems very limited.

Photolithography. Prospects for photolithography

and semiconductor manufacturing are much brighter. The semiconductor industry reported sales worldwide of $298.3 billion in 2010, reflecting a 32 percent increase over 2009. Much of the growth was due to a rise in microchip purchases by customers in the Asia Pacific region, which makes up slightly more than half of the global market by volume. Microchip buyers in this context are not consumers, but rather companies that manufacture computers, mobile phones, and other types of hardware. The leading semiconductor manufacturers also reflect the global nature of this industry. Intel tops the list by volume. Its competitors include Analog Devices, Texas Instruments, and Micron Technology in the United States; STMicroelectronics and NXP Semiconductors in Europe; and, Samsung, Toshiba, NEC Electronics, and Taiwan Semiconductor Manufacturing in Asia.

For many years, industry sources have predicted that photolithography would be replaced by other technologies because of an increasing need for precision in the making of microchips. Innovations such as excimer laser technology have allowed lithography to become so precise that features smaller than a single wavelength of light can be printed accurately. The vast expense of microtechnology on this scale prevents many nonprofit institutions such as universities from devoting significant resources to its study. Instead, advances are most likely to come directly from manufacturing companies themselves, which reinvest about 15 percent of sales into research and development each year. Research funding is also supplied by government sources such as the National Science Foundation, the National Institute of Standards and Technology, the U.S. Department of Energy, and the U.S. Department of Defense.

CAREERS AND COURSE WORK

The course work required for a career in lithography varies widely with the nature of the product and the stage in the printing process.

In traditional lithography, one major professional area is media printing. Books, magazines, and newspapers must be designed and laid out page by page before lithographic plates can be created. Many of the professionals who hold these jobs have earned bachelor's or master's degrees in academic areas such as art, graphic design, industrial and product design, and journalism. A background of this type could include course work in typography, color theory, digital imaging, or consumer marketing. Students seeking opportunities in media design also pursue internships with publishers and other companies in their fields of interest.

The mechanical process of lithography has become highly automated. Fewer employees are needed in printing plants than before. Most lithographic press operators receive their training on the job and through apprenticeships. Formal education is offered through postsecondary programs at community colleges, vocational and technical schools, and some universities. Students take courses in mechanical engineering and in the maintenance and repair of heavy equipment. Additional course work may include mathematics, chemistry, physics, and color theory.

Lithography as an artistic printing technique is taught in many college and university art departments. While it is considered too specialized by most institutions for a degree, artists may choose to use lithographic printing to create visual works on paper and other materials.

Photolithography is a highly specialized area of technological manufacturing. Its course work and career track are notably different from those in traditional lithography. Professionals working in photolithography have undergraduate or graduate degrees in fields such as engineering, physics, mathematics, and chemistry. An extensive knowledge of micro-technology and the properties of light-sensitive materials is needed. Because most of the world's semiconductor manufacturing takes place outside North America, careers in photolithography can involve frequent travel to areas such as Asia.

SOCIAL CONTEXT AND FUTURE PROSPECTS

As a broad category, lithographic printing offers very limited job growth. Consumers are increasingly concerned about the environmental impact of paper use in catalogs and other sources of bulk mail. In an effort to respond to these concerns, many companies have reduced their use of paper-based marketing campaigns. This change has lowered the demand for commercial offset lithography. A similar trend has affected the printing of checks and invoices, which are being replaced by electronic systems and online banking.

The growth of electronic media, from the Internet to handheld e-book readers, has also lowered the need for lithography in the publishing industry.

Newspapers are reducing the circulation and length of their paper editions and shifting their publishing efforts to Web sites and news feeds. While the demand for paper-based books is not likely to disappear in the near future, sales of new books in electronic formats are growing at a more rapid pace than their print counterparts. In the print segment, new technology is boosting the use of print-on-demand systems for books, which use digital printing techniques rather than lithography.

The prospects for growth in photolithography are more optimistic. Photolithography continues to be one of the most effective and precise ways to make semiconductors. Until it is replaced by a new technology, the field is expected to keep growing with new demand for smaller, faster computers.

—Julia A. Rosenthal, MS

FURTHER READING

Devon, Marjorie. *Tamarind Techniques for Fine Art Lithography*. New York: Abrams, 2009. A hands-on manual of techniques for artists seeking to produce lithographic prints, written by the director of the Tamarind Institute of Lithography at the University of New Mexico in Albuquerque. Landis, Stefan, ed. *Lithography*. Hoboken, N.J.: Wiley-I STE, 2010. A new detailed textbook on lithography as applied to the design and manufacturing of microtechnology.

Meggs, Philip B., and Alston W. Purvis. *Meggs' History of Graphic Design*. 4th ed. Hoboken, N.J.: John Wiley & Sons, 2006. A broad history of graphic design and printing processes throughout the world, including the role of lithography.

Senefelder, Alois. *Senefelder on Lithography: The Classic 1819 Treatise*. Mineola, N.Y.: Dover Publications, 2005. A reproduction of an essay published nearly two centuries ago by the founder of lithography.

Suzuki, Kazuaki, and Bruce W. Smith, eds. *Microlithography: Science and Technology*. 2d ed. Boca Raton, Fla.: CRC Press, 2007. A series of essays by several contributors on the processes behind microlithography, one of the manufacturing techniques used to make semiconductors.

Wilson, Daniel G. *Lithography Primer*. 3d ed. Pittsburgh: GATF Press, 2005. An illustrated overview of each step in the lithographic printing process. Includes chapters on topics such as plate imaging, inks, and papers.

WEB SITES

American Institute of Graphic Arts (AIGA)
http://www.aiga.org

National Association of Litho Clubs
http://www.graphicarts.org

Tamarind Institute
http://tamarind.unm.edu/index.html

See also: Calculus; Mechanical Engineering; Optics; Photography.

M

MAGNETIC RESONANCE IMAGING

FIELDS OF STUDY

Radiology; diagnostics; radiofrequency; pathology; physiology; radiation physics; instrumentation; anatomy; microbiology; imaging; angiography; nuclear physics.

SUMMARY

Magnetic resonance imaging (MRI) is a noninvasive form of diagnostic radiography that produces images of slices or planes from tissues and organs inside the body. An MRI scan is painless and does not expose the patient to radiation, as an X ray does. The images produced are detailed and can be used to detect tiny changes of structures within the body, which are extremely valuable clues to physicians in the diagnosis and treatment of their patients. A strong magnetic field is created around the patient, causing the protons of hydrogen atoms in body tissues to absorb and release energy. This energy, when exposed to a radiofrequency, produces a faint signal that is detected by the receiver portion of the MRI scanner, which transforms it into an image.

KEY TERMS AND CONCEPTS

- **artifact:** Feature in a diagnostic image, usually a complication of the imaging process, that results in an inaccurate representation of the tissue being studied.
- **axial slice:** Horizontal imaging plane that corresponds with right to left and front to back.
- **claustrophobia:** Abnormal and persistent fear of closed spaces, of being closed in or being shut in.
- **computed tomography (CTt) scan:** Image of structures within the body created by a computer from multiple X-ray images.
- **contrast agent:** Dye used to provide contrast, for example, between blood vessels and other tissue.
- **functional magnetic resonance imaging (fMRI):** Use of MRI to study physiological processes rather than just anatomy.
- **gradient coils:** Coils of wire used to generate the magnetic field gradients that are used in MRI.
- **magnetic resonance angiogram (MRA):** Noninvasive complement to MRI to observe anatomy of blood vessels of certain size in the head and neck.
- **sagittal slice:** Vertical imaging plane that corresponds with front-to-back and top-to-bottom.
- **scan:** Single, continuous collection of images.
- **session:** Time that a single subject is in the magnetic resonance scanner; can be two hours for fMRI.
- **Tesla:** Unit of magnetic field strength; named for Nikola Tesla who discovered the rotating magnetic field in 1882.

DEFINITION AND BASIC PRINCIPLES

Magnetic resonance imaging (MRI), sometimes called magnetic resonance tomography, is a noninvasive medical imaging method used to visualize the internal structures and some functions of the body. MRI provides much greater detail and contrast between the different tissues in the body than are available from X rays or computed tomography (CT) without using ionizing radiation. MRI uses a powerful magnetic field to align the nuclear magnetization of protons of hydrogen atoms in the body. A radio frequency alters the alignment of the protons, creating a signal that is detectable by the scanner. The signals are processed through a mathematical algorithm to produce a series of cross-sectional images of the desired area. Image resolution and accuracy can be further refined through the use of contrast agents.

The detailed images produced are extremely valuable in detection and diagnosing of medical conditions and disease. The need for exploratory surgery has been greatly reduced, and surgical procedures and treatments can be more accurately directed by the ability to visualize structures and changes within the body.

BACKGROUND AND HISTORY

Magnetic resonance imaging is a relatively new scientific discovery, and its application to human diagnostics was first published in 1977. Two American scientists, Felix Bloch at Stanford University and Edward Mills Purcell from Harvard University, were both independently successful with their nuclear magnetic resonance (NMR) experiments in 1946. Their work was based on the Larmor relationship, named for Irish physicist Joseph Larmor, which stated that the strength of the magnetic field matched the radiofrequency. Bloch and Purcell found that when certain nuclei were in the presence of a magnetic field, they absorbed energy in the radiofrequency range of the electromagnetic spectrum and emitted this energy when the nuclei returned to their original state. They termed their discovery nuclear magnetic resonance: "nuclear" because only the nuclei of certain atoms reacted, "magnetic" because a magnetic field was required, and "resonance" because of the direct frequency dependence of the magnetic and radiofrequency fields. Bloch and Purcell were awarded the Nobel Prize in Physics in 1952.

NMR technology was used for the next few decades as a spectroscopy method to determine the composition of chemical compounds. In the late 1960's and early 1970's, Raymond Damadian, a State University of New York physician, found that when NMR techniques were applied to tumor samples, the results were distinguishable from normal tissue. His results were published in the journal Science in 1971. Damadian filed patents in 1972 and 1978 for a NMR system large enough to accommodate a human being that would emit a signal if tumor tissue was detected but did not produce an image. Paul Lauterbur, a physicist from State University of New York, devised technology that could run the signals produced by NMR through a computed back projection algorithm, which produced an image. Peter Mansfield, a British physicist from the University of Nottingham, further refined the mathematical analysis, improving the image. He also discovered echo-planar imaging, which is a fast imaging protocol for MRI and the basis for functional MRI. Lauterbur and Mansfield shared the 2003 Nobel Prize in Physiology or Medicine. Some controversy still exists regarding Damadian's exclusion from this honor.

HOW IT WORKS

The human body is made up of about 70 percent water. Water molecules are made up of two hydrogen atoms and one oxygen atom. When exposed to a powerful magnetic field, some of the protons in the nuclei of the hydrogen atoms align with the direction of the field. When a radio frequency transmitter is added, creating an electromagnetic field, a resonance frequency provides the energy required to flip the alignment of the affected protons. Once the field is turned off, the protons return to their original state. The difference in energy between the two states is called a photon, which is a frequency signal detected by the scanner. The photon frequency is determined by the strength of the magnetic field. The detected signals are run through a computerized algorithm to deliver an image. The contrast of the image is produced by differences in proton density and magnetic resonance relaxation time, referred to as T1 or T2.

An MRI scanning machine is a tube surrounded by a giant circular magnet. The patient is placed on an examination table that is inserted through the tube space. Some individuals experience claustrophobia when lying in the closed space of the scanning tube and may be given a mild sedative to reduce anxiety. Children are often sedated or receive anesthesia for an MRI. Patients are required to remain very still during the scan, which normally takes between thirty to ninety minutes to complete. During the scan, patients are

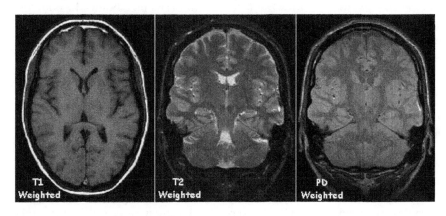

Examples of T1 weighted, T2 weighted and PD weighted MRI scans

usually provided with a hand buzzer or communication device so that they may interact with technicians. The magnetic field is created by passing electric current through a series of gradient coils. The strong magnetic fields are normally safe for patients, with the exception of people with metal implants such as pacemakers, surgical clips or plates, or cochlear implants, making them ineligible for MRI. During the scanning procedure, patients will hear a loud humming, beeping, or knocking noise, which can reach up to 120 decibels. (Patients are often provided with ear protection.) The noise is caused by the interaction of the gradient magnetic fields with the static magnetic field. The gradient coils are subject to a twisting force each time they are switched on and off, and this creates a loud mechanical vibration in the cylinder supporting the coils and surrounding mountings.

To enhance the images, contrast agents can be injected intravenously or directly into a joint. MRI is being used to visualize all parts of the body by producing a series of two-dimensional images that appears as cross sections or slices. These slices can also be reconstructed to create three-dimensional views of the entire body or specific parts.

APPLICATIONS AND PRODUCTS

Research into the applications of magnetic resonance imaging technology beyond basic image generation has been progressing at a tremendous rate. Although the basic images are immensely valuable to physicians and scientists, the application of the scientific principles in the development of specialized scans is reaching far beyond original expectations and benefiting health care delivery and patient care.

Functional MRI (fMRI) is based on the changes in blood flow to the parts of the brain that accompany neural activity, and it provides visualization of these changes. This has been critical in detecting the brain areas involved in specific tasks, processes, or emotions. fMRI does not detect absolute activity of areas of the brain but it detects differences in activity. During the scan, the patient is asked to perform tasks or is presented with stimuli to trigger thoughts or emotions. The detection of the brain areas that are used is based on the blood oxygenation level dependent (BOLD) effect, which creates a variation signal, linked with the concentration of oxy-/deoxy-hemoglobine in each area. These scans are performed every two to three seconds over a period of minutes at a low resolution

and do not often require additional contrast media to be used.

Diffusion MRI can measure the diffusion of water molecules in biological tissues. This is incredibly useful in detecting the movement of molecules in neural fiber, which can enable brain mapping, illustrating connectivity of different regions in the brain, and examination of areas of the brain affected by neural degeneration and demyelination, as in multiple sclerosis. Diffusion MRI, when applied to diffusion-weighted imaging, can detect swelling in brain cells within ten minutes of the onset of ischemic stroke symptoms, allowing physicians to direct reperfusion therapy to specific regions in the brain. Previously, computed tomography would take up to four hours to detect similar findings, delaying cerebral perfusion therapy to salvageable areas.

Interventional magnetic resonance imaging is used to guide medical practitioners during minimally invasive procedures that do not involve any potentially magnetic instruments. A subset of this is intraoperative MRI, which is used during surgical procedures; however, most often images are taken during a break from the procedure in order to track progress and success and further guide ongoing surgery.

Magnetic resonance angiography (MRA) and venography (MRV) provide visualization of arteries and veins. The images produced can help physicians evaluate potential health problems such as narrowing of the vessels or vessel walls at risk of rupture as in an aneurysm. The most common arteries and veins examined are the major vessels in the head, neck, abdomen, kidneys, and legs.

Magnetic resonance spectroscopy (MRS) measures the levels of different metabolites in body tissues, usually in the evaluation of nervous system disorders. Concentrations of metabolites such as N-acetyl aspartate, choline, creatine, and lactate in brain tissue can be examined. Information on levels of metabolites is useful in determining and diagnosing specific metabolic disorders such as Canavan's disease, creatine deficiency, and untreated bacterial brain abscess. MRS has also been useful in the differentiation of high-grade from low-grade brain tumors.

Precise treatment of diseased or cancerous tissue within the body is a tremendous advance in health care delivery. Radiation therapy simulation uses MRI technology to locate tumors within the body and determine their exact location, size, shape, and orientation. The patient is carefully marked with points corresponding to this information, and precise radiation therapy can be delivered to the tumor mass. This drastically reduces excess radiation therapy and limits damage to healthy tissues surrounding the tumor. Similarly, magnetic resonance guided focused ultrasound (MRgFUS) allows ultrasound beams to achieve more precise and complete treatment and the ablation of diseased tissues is guided and controlled by magnetic resonance thermal imaging.

IMPACT ON INDUSTRY

In 1983, the Food and Drug Administration approved MRI scanners for sale in the United States. Magnetic resonance imaging is experiencing rapid growth on the global market and is expected to sustain this trend in the future as more clinics and health care centers use the imaging technique. The aging of the American population has increased the demand for efficient, effective, and noninvasive diagnosis, especially for neurological and cardiovascular diseases. Although a basic MRI system can cost more than $1 million, and the cost of construction of the suite to accommodate it can exceed $500,000, installation of these imaging systems in medical institutions still promises a positive return on investment for the health care providers. Some companies have begun offering refurbished MRI systems at reduced cost. MRI technology is an asset in challenging economic times because it provides advanced and quick diagnoses, leading to faster patient care and greater patient turnover, leading to greater revenues for the institution. In the United States, health care insurers and the federal government provide very good reimbursement for the scan itself and a professional fee for a review of the resulting images by a radiologist.

Commercial development and industrial growth depends on advancements in research, which is being funded and conducted around the world. Some projects that are moving from experimental to commercial and driving the industry are integrated systems combining MRI with another modality, more portable MRI systems, improved contrast agents, advanced image-processing techniques, and magnetic coil technology.

Positron emission tomography (PET) is used in the diagnosis of cancer, but because it lacks anatomical detail, it is used in combination with CT scans. MRI scans provide superior contrast to CT for determining tumor structure and integrating PET and MRI into a single modality would be highly desirable and is in development.

Research is being done to improve the enhancement of the contrast agents used with MRI. Specifically, research involves targeted contrast agents that will bind only to desired tissue at the molecular and cellular level. Visualizing changes at a cellular level would make it possible to detect and diagnose disease sooner and provide opportunities for more focused treatment.

A growing niche market consists of MRI systems that allow patients to remain upright during the scan and open systems that can accommodate patients who previously were unable to have a scan because of their size or their claustrophobia.

Portable and handheld MRI technology has entered the international market. These small devices promise to deliver quality images at low cost, without the need for a special room to accommodate the device.

CAREERS AND COURSE WORK

The most common career choice in the field of magnetic resonance imaging is the MRI technician or technologist, individuals who operate the MRI system to effectively produce the desired images for diagnostic purposes while adhering to radiation safety measures and government regulations. Researchers and government agencies are exploring potential occupational hazards to personnel because of prolonged and frequent exposure to magnetic fields. Technicians first explain the procedure to patients. Then, they ensure that patients do not have any metal present on their person or in their body and position them correctly on the examination table. Some technicians also administer intravenous sedation or contrast media to the patients. During the scan, the technologist observes the patient as well as the display of the area being scanned and makes any needed adjustment to density or contrast to improve picture quality. When the scan is complete, the technologist will evaluate the images to ensure that they are satisfactory for diagnostic purposes. The MRI training program may result in an associate's degree or certificate; some programs require prior completion of radiology or sonography programs and core competencies in writing, math, anatomy, physiology, and psychology. Once admitted to an accredited program, students

FASCINATING FACTS ABOUT MAGNETIC RESONANCE IMAGING

- Magnetic resonance imaging (MRI) is based on nuclear magnetic resonance techniques, but it was named "magnetic" rather than "nuclear" resonance imaging because of the negative connotations associated with the word "nuclear" in the 1970's.
- An overwhelming majority of American physicians identified computed tomography (CT) and MRI as the most important medical innovations for improving patient care during the 1990's.
- In 1977, Raymond Damadian and colleagues built their first magnetic resonance scanner and named it "Indomitable." It is housed in the Smithsonian.
- MRI scanners were primarily developed for use in medicine but are also used to study fossils and historical artifacts.
- Functional MRI can create a video of blood flow in the brain and has been used to study monks serving under the Dalai Llama and the control they exert over mental processes through meditation.
- As of 2009, there were 7,950 magnetic resonance imaging systems in the United States, compared to only 266 in Canada.
- Tattoos received before 1990 may contain small amounts of metal in the ink, which may interfere with or be painful during MRI.
- While using functional magnetic resonance imaging (fMRI) to research children with attention deficit disorder, Pennsylvania psychiatrist Daniel Langleben discovered that deception activates regions in the prefrontal cortex, showing that fMRI could be used as a lie detector.

receive training that includes patient care, magnetic resonance physics, and anatomy and physiology. The American Registry of Magnetic Resonance Imaging

Magnetic Resonance Imaging Technologists requires that students complete one thousand hours of clinical training. To satisfy this clinical training requirement, students are assigned to a specific hospital or are rotated through different hospitals. Becoming an MRI technician can take two to three years.

Diagnostic radiologists are physicians who have specialized in obtaining and interpreting medical images such as those produced by MRI. Becoming a radiologist in the United States requires the completion of four years of college or university, four years of medical school, and four to five years of additional specialized training.

MRI physicists are specialized scientists with a diverse background covering nuclear magnetic resonance (NMR) physics, biophysics, and medical physics, in combination with basic medical sciences, including human anatomy, physiology, and pathology. They also have a good understanding of engineering issues involving advanced hardware, such as large superconducting magnets, high-power radio frequencies, fast digital data processing, and remote sensing and control. Industrial MRI physicists often work in research and development for biotechnology companies or they implement new applications and provide support for equipment already installed in health care centers. Academic MRI physicists work in a university laboratory or in cooperation with a medical center involved in clinical research and training. Academic research may involve basic science in MRI spectroscopy, functional imaging, contrast media, or echo-planar imaging.

SOCIAL CONTEXT AND FUTURE PROSPECTS

Magnetic resonance imaging provides physicians with the ability to see detailed images of the inside of their patients to more easily diagnose and guide treatment. It provides researchers with valuable insight into the metabolism and physiology of the body. Still, there are drawbacks that make this scientific advance unavailable to some patients because of their economic circumstances or body shape. For patients who do not have health care insurance, the price of an MRI scan, which can range from $700 to $2500 depending on body part and type of examination, may be beyond what they can afford. Also, MRI systems have weight and circumference restrictions that make many people unsuitable candidates, limiting the quality of acute care for them. Typically, the weight limit is 350 to 500 pounds for the examination table, and the size of the patients is additionally limited by the diameter of the magnetic tube. As people tend to be larger in size, biotechnology companies are working on scanning systems, such as the upright or open concept scanner, that can accommodate larger people.

—April D. Ingram

FURTHER READING

Blamire, A. M. "The Technology of MRI—The Next Ten Years?" *British Journal of Radiology* 81 (2008): 601-617. Looks at the clinical status and future of MRI.

Filler, Aaron. "The History, Development and Impact of Computed Imaging in Neurological Diagnosis and Neurosurgery: CT, MRI, and DTI." *The Internet Journal of Neurosurgery* 7, no. 1 (2010). Provides a good history of the development of diagnostic imaging techniques.

Haacke, Mark, et al. *Magnetic Resonance Imaging: Physical Principles and Sequence Design.* New York: John Wiley & Sons, 1999. Explains the key fundamental and operational principles of MRI from a physics and mathematical viewpoint.

Simon, Merrill, and James Mattson. *The Pioneers of NMR and Magnetic Resonance in Medicine: The Story of MRI.* Ramat Gan, Israel: Bar-Ilan University Press, 1996. Describes the history of MRI, from its development of scientific principles to application in health care.

Weishaupt, Dominik, Vitor Koechli, and Borut Marincek. *How Does MRI Work? An Introduction to the Physics and Function of Magnetic Resonance Imaging.* New York: Springer, 2006. A good resource for those who wish to familiarize themselves with the workings of magnetic resonance imaging and have some of the challenging concepts explained. It uses conceptual rather than mathematical methods to clarify the physics of MRI.

WEB SITES

Clinical Magnetic Resonance Industry
http://www.cmrs.com

International Society for Magnetic Resonance in Medicine
http://www.ismrm.org

MedlinePlus Magnetic Resonance Imaging
http://www.nlm.nih.gov/medlineplus/mriscans.html

MAGNETIC STORAGE

FIELDS OF STUDY

Computer engineering; computer networking systems; electrical engineering; information technology; materials engineering.

SUMMARY

Magnetic storage is a durable and non-volatile way of recording analogue, digital, and alphanumerical data. In most applications, an electrical current is used to generate a variable magnetic field over a specially prepared tape or disk that imprints the tape or disk with patterns that, when "read" by an electromagnetic drive "head," duplicates the wavelengths of the original signal. Magnetic storage has been a particularly enduring technology, as the original conceptual designs were published well over a century ago.

KEY TERMS AND CONCEPTS

- **biasing:** Process that pre-magnetizes a magnetic medium to reproduce the magnetic flux from the recording head more exactly.
- **bit:** Storage required of one binary digit.
- **byte:** Basic unit, composed of eight bits, of measuring storage capacity and the basis of larger units.
- **ferromagnetic media:** Iron-based media sometimes used as a source of tape, hard, or flexible disks.
- **magnetic anisotrophy:** Tendency, in certain substances, to hold onto a magnetically induced pattern even without the presence of electrical current.
- **magnetization:** Property of an object determining whether it can be affected by magnetism.
- **random access memory (RAM):** Property of certain kinds of storage media where all elements of a sequence may be accessed in the same length of time; also called direct access.
- **sequential access memory:** Property of certain kinds of storage media where remote elements require a longer access time than elements located in the immediate vicinity.

DEFINITION AND BASIC PRINCIPLES

Magnetic storage is a term describing one method in which recorded information is stored for later access. A magnetized medium can be one or a combination of several different substances: iron wire, steel bands, strips of paper or cotton string coated with powdered iron filings, cellulose or polyester tape coated with iron oxide or chromium oxide particles, or aluminum or ceramic disks coated with multiple layers of nonmetallic alloys overlaid with a thin layer of a magnetic (typically a ferrite) alloy. The varying magnetic structures are encoded with alphanumerical data and become a temporary or permanent nonvolatile repository of that data. Typical uses of magnetic storage media range from magnetic recording tape and hard and floppy computer disks to the striping material on the backs of credit, debit, and identification cards as well as certain kinds of bank checks.

BACKGROUND AND HISTORY

American engineer Oberlin Smith's 1878 trip to Thomas Edison's laboratory in Menlo Park, New Jersey, was the source for Smith's earliest prototypes of a form of magnetic storage. Disappointed by the poor recording quality of Edison's wax cylinder phonograph, Smith imagined a different method for recording and replaying sound. In the early 1820's, electrical pioneers such as Hans Ørsted had demonstrated basic electromagnetic principles: Electrical current, when run through a iron wire, could generate a magnetic field, and electrically charged wires affected each other magnetically. Smith toyed with the idea, but did not file a patent—possibly because he never found the time to construct a complete, working model. On September 8, 1888, he finally published a description of his conceptual design, involving a cotton cord woven with iron filings passing through a coil of electrically charged wire, in *Electrical World* magazine. The concept in the article, "Some Possible Forms of Phonograph," though theoretically possible, was never tested.

The first actual magnetic audio recording was Danish inventor Valdemar Poulsen's telegraphone, developed in 1896 and demonstrated at the Exposition Universelle in Paris in 1900. The telegraphone was composed of a cylinder, cut with grooves along its surface, wrapped in steel wire. The electromagnetic head, as it passed over the tightly wrapped iron wire, operated both in recording sound and in playing back the recorded audio. Poulsen, trying to reduce distortion in his recordings, had also made early attempts at biasing (increasing the fidelity of a recording by including a DC current in his phonograph model) but, like Oberlin Smith's earlier model, his recorders, based on wire, steel tape, and steel disks, could not easily be heard and lacked a method of amplification.

Austrian inventor Fritz Pfleumer was the originator of magnetic tape recording. Since Pfleumer was accustomed to working with paper (his business was cigarette-paper manufacturing), he created the original magnetic tape by gluing pulverized iron particles (ferrous oxide) onto strips of paper that could be wound into rolls. Pfleumer also constructed a tape recorder to use his tape. On January 31, 1928, Pfleumer received German patent DE 500900 for his sound record carrier (lautschriftträger), unaware that an American inventor, Joseph O'Neill, had filed a patent—the first—for a device that magnetically recorded sound in December, 1927.

HOW IT WORKS

The theory underlying magnetic storage and magnetic recording is simple: An electrical or magnetic current imprints patterns on the magnetic-storage medium. Magnetic tape, magnetic hard and floppy disks, and other forms of magnetic media operate in a very similar way: An electric current is generated and applied to a demagnetized surface to vary the substratum and form a pattern based on variations in the electrical current. The biggest differences between the three dominant types of magnetic-storage media (tape, rigid or hard disks, and flexible or floppy disks) are the varying speeds at which stored data can be recovered.

Magnetic tape. Magnetic tape used to be employed extensively for archival computer data storage as well as analogue sound or video recording. The ferrous- or chromium-impregnated plastic tape, initially demagnetized, passes at a constant rate over a recording head, which generates a weak magnetic field proportional to the audio or video impulses being recorded and selectively magnetizes the surface of the tape. Although fairly durable, given the correct storage conditions, magnetic tape has the significant disadvantage of being consecutively ordered—the recovery of stored information depends on how quickly the spooling mechanism within the recorder can operate. Sometimes the demand for high-density, cheap data storage outweighs the slower rate of data access. Large computer systems commonly archive information on magnetic tape cassettes or cartridges. Despite the archaic form, advances in tape density allow magnetic tape cassettes to store up to five terabytes (TB) of data in uncompressed formats.

For audio or video applications, sequential retrieval of information (watching a movie or listening to a piece of music) is the most common method, so a delay in locating a particular part is regarded with greater tolerance. Analogue tape was an industry standard for recording music, film, and television until the advent of optical storage, which uses a laser to encode data streams into a recordable media disk and is less affected by temperature and humidity.

Magnetic disks. Two other types of recordable magnetic media are the hard and floppy diskettes—both of which involve the imprinting of data onto a circular disk or platter. The ease and speed of access to recorded information encouraged the development of a new magnetic storage media form for the computer industry. The initial push to develop a nonlinear system resulted, in 1956, with the unveiling of IBM's 350 Disk Storage Unit—an early example of what is currently known as a hard drive. Circular, ferrous-impregnated aluminum disks were designed to spin at a high rate of speed and were written upon or read by magnetic heads moving radially over the disk's surface.

Hard and floppy disks differ only in the range of components available within a standard unit. Hard disks, composed of a spindle of disks and a magnetic read-write apparatus, are typically located inside a metal case. Floppy disks, on the other hand, were packaged as a single or dual-density magnetic disk (separate from the read-write apparatus that encodes them) inside a plastic cover. Floppy disks, because they do not contain recording hardware, were intended to be more portable (and less fragile) than hard disks—a trait that made them extremely popular for home-computer users.

Another variant of the disk-based magnetic storage technology is magneto-optical recording. Like an optical drive, magneto-optical recording operates by burning encoded information with a laser and accesses the stored information through optical means. Unlike an optical-storage medium, a magneto-optical drive directs its laser at the layer of magnetic material.

In 1992, Sony released the MiniDisc, an unsuccessful magneto-optical storage medium.

APPLICATIONS AND PRODUCTS

Applications for magnetic storage range from industrial or institutional uses to private-sector applications, but the technology underlying each of these formats is functionally the same. The technology that created so many different inventions based on electrical current and magnetic imprinting also had a big impact on children's toys. The Magna Doodle, a toy developed in 1974, demonstrates a simple application of the concept behind magnetic storage that can shed light on how the more complex applications of the technology also work. In this toy, a dense, opaque fluid encapsulates fine iron filings between two thin sheets of plastic. The upper layer of plastic is transparent and thin enough that the weak magnetic current generated by a small magnet encased in a cylinder of plastic (a magnetic pen) can make the iron filings float to the surface of the opaque fluid and form a visible dark line. Any images produced by the pen are, like analogue audio signals encoded into magnetic tape, nonvolatile and remain visible until manually erased by a strip of magnets passing over the plastic and drawing the filings back under the opaque fluid.

Magnetic tape drives. It is this basic principle of nonvolatile storage that underlies the usage of the three basic types of magnetic storage media—magnetic tape, hard disks, and floppy disks. All three have been used for a wide variety of purposes. Magnetic tape, whether in the form of steel bands, paper, or any one of a number of plastic formulations, was the original magnetic media and was extensively used by the technologically developed nations to capture audio and, eventually, video signals. It remains the medium of choice for archival mainframe data storage because large computer systems intended for mass data archival require a system of data storage that is both capable of recording vast amounts of information in a minimum of space (high-density) and is extremely inexpensive—two qualities inherent to magnetic tape.

Early versions of home computers also had magnetic tape drives as a secondary method of data storage. In the 1970's, IBM offered their own version of a magnetic cassette tape recorder (compatible with its desktop computer) that used the widely available cassette tape. By 1985, however, hard disks and floppy disks had dominated the market for computer systems designed to access smaller amounts of data frequently and quickly, and cassette tapes became obsolete for home-computer data storage.

Hard disk drives. In 1955, IBM's 350 Disk Storage Unit, one of the computer industry's earliest hard drives, had only a five megabyte (MB) capacity despite its massive size (it contained a spindle of fifty twenty-four-inch disks in a casing the size of a large refrigerator). However, the 350 was just the first of a long series of hard drives with ever-increasing storage capacity. Between the years of 1950 and 2010, the average area density of a hard drive has doubled every few years, starting from about three megabytes to the current high-end availability of three terabytes. Higher-capacity drives are in development, as computer companies such as Microsoft are redefining the basic unit of storage capacity on a hard drive from 512 bytes (IBM's standard unit, established during the 1980's) to a far larger 4 kilobytes (KB). The size of the typical hard drive made it difficult to transport and caused the development, in 1971, of another, similar form of magnetic media—the floppy disk.

Early hard drives such as IBM's 350 were huge (88 cubic feet) and prohibitively expensive (costing about $15,000 per megabyte of data capacity). Given that commercial sales of IBM computers to nongovernmental customers were increasing rapidly, IBM wanted some way to be able to deliver software updates to clients cheaply and efficiently. Consequently, engineers conceived of a way of separating a hard disk's components into two units—the recording mechanism (the drive) and the recording medium (the floppy disk).

Floppy disk drives. The floppy disk itself, even in its initial eight-inch diameter, was a fraction of the weight and size needed for a contemporary hard disk. Because of the rapidly increasing storage needs of the most popular computer programs, smaller disk size and higher disk density became

the end goal of the major producers of magnetic media—Memorex, Shugart, and Mitsumi, among others. As with the hard drive, floppy disk size and storage capacity respectively decreased and increased over time until the 3.5-inch floppy disk became the industry standard. Similarly, the physical dimensions of hard drives also shrunk (from 88 cubic feet to 2.5 cubic inches), allowing the introduction of portable external hard drives into the market. Floppy disks were made functionally obsolete when another small, cheap, portable recording device came on the market—the thumb or flash drive. Sony, the last manufacturer of floppy disks, announced that, as of March, 2011, it would stop producing floppy disks.

Magnetic striping. Magnetic storage, apart from tape, hard and floppy disk, is also widely used for the frequent transmission of small amounts of exclusively personal data—namely, the strip of magnetic tape located on the back of credit, debit, and identification cards as well as the ferrous-impregnated inks that are used to print numbers along the bottom of paper checks. Since durability over time is a key factor, encasing the magnetic stripe into a durable, plastic card has become the industry standard for banks and other lending institutions.

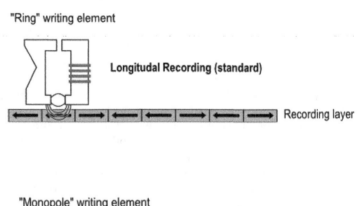

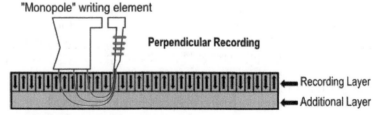

Longitudinal recording and perpendicular recording, two types of writing heads on a hard disk.

IMPACT ON INDUSTRY

At the base of any industry is the need to store information. Written languages, ideographic art, and even musical notation were developed for the purpose of passing knowledge on to others. At the most basic, any individual, company, or government division that uses a computer system relies on some form of magnetic storage to encode and store data.

Government use. Few government entities in even moderately technologically developed countries can operate without using some form of computer system and the requisite magnetic storage of either magnetic tape, internal or external hard drive, or both.

In 1942, the Armour Institute of Technology sold literally thousands of wire sound recorders to the American military, since the Army intelligence service was experimenting with technology that intercepted foreign radio transmissions and wished to record those transmissions for the purpose of decoding them.

In June, 1949, IBM envisioned a new kind of magnetic storage device that could act as a data repository for another new invention—the computer. On May 21, 1952, the IBM 726 Tape Unit with the IBM 701 Defense Calculator was unveiled. On September 13, 1956, a IBM announced their creation of the 305 RAMAC (Random Access Memory Accounting System) computer. These devices (and other, similar inventions) allowed both the government and the private sector to start phasing out punch-card-based computer storage systems. In the following decades, governmental agencies were able to switch from the massive boxes of punch cards to the more efficient database system based on external magnetic tape cassettes and internal magnetic disks. In the United States, for example, the Internal Revenue Service, which records and processes financial data on every working individual in the country identified by a Social Security Number, is one of many governmental agencies that must store massive amounts of sensitive information. As a result, the federal government purchases roughly a million reels every year.

Business use. Besides the magnetic media used in computers, magnetic storage media in the form of magnetic tape has had a dramatic impact on the industrial and business sectors of the developed countries of the world. Miniature cassette recorders, introduced in 1964 by the Philips Corporation, became a common sight in college classrooms, executive offices, and journalists' hands for recording lectures, memos, or interviews because of their small size and adjustable tape speed. These recorders suggested the importance of knowing exactly what another individual has said for reasons of accuracy and truth in reporting.

Entertainment use. The invention of magnetic storage allowed for a more precise system of audio and video recording. Dramatic shifts in popular entertainment and the rise of the major music labels such as Victor or RCA occurred precisely because there were two alternative methods of audio recording that, ultimately, introduced music into vastly different venues and allowed individuals to take an interest in music that was as lucrative to the music company as it was pleasurable to the consumer. Phonograph recordings made on wax, lacquered wood, or vinyl brought music into a family's living room and millions of teenagers' bedrooms, while electromagnetic recordings on tape allowed music to be played as background music in vehicles and workplaces and to be carried from place to place in portable devices. Competition between the rival manufacturers of records (such as Columbia and EMI) and tapes (such as BASF and TDK) was fierce at times, even inspiring copyright-infringement campaigns by such organizations as the British Phonographic Industry (BPI) against the manufacturers and users of blank cassette tapes in the 1980's. Regardless, the portability and shock resistance of magnetic tape-driven music players ensured that magnetic storage was unlikely to be made obsolete by either phonograph technology or music-industry politics.

Consumer use. Magnetic-storage media probably would not have the impact it did on Western society if it had not been so enthusiastically embraced by the willingness of the private individual to purchase and use the technology. For example, although magnetic media gained a reputation for illicit use because of music- and film-industry complaints about some consumers' misuse of blank magnetic media to record illegal (bootleg) copies of phonograph recordings, television programs, and movies broadcast over cable or satellite systems, magnetic-storage media has also given consumers the ability to use blank media as a vehicle for the recording of their own audio and video creations. The technology was robust enough to make a durable recording but flexible enough to allow repeated erasures and rerecordings on a single tape.

CAREERS AND COURSE WORK

Individuals interested in further developing the technology of magnetic media should study computer engineering, electrical engineering, or materials engineering depending on whether they wish to pursue increasing storage capacity of the existing electromagnetic technology or finding alternative methods of using electromagnetic theory on the problems of data storage. Some universities are particularly sensitive to the related nature of computer-component design and electrical-system design and allow for a dual degree in both disciplines. Supplemental course work in information technology would also be helpful, as an understanding of how networks transmit data will more precisely define where future storage needs may be anticipated and remedied.

SOCIAL CONTEXT AND FUTURE PROSPECTS

The current social trend, both locally and globally, is to seek to collect and store vast amounts of data that, nevertheless, must be readily accessible. An example of this is in meteorology, which has been attempting over the past few decades to formulate complex computer models to predict weather trends based on previous trends observed and recorded globally during the past century. Weather and temperature records are increasingly detailed, so computer storage for the ideal weather-predicting computer will need to be able to keep up with the storage requirements of a weather-system archive and meet reasonable deadlines for the access and evaluation of a century's worth of atmospheric data.

Other storage developments will probably center on making more biometric data immediately available in identification-card striping. With surveillance, rather than incarceration, of petty criminals being considered as a method by which state and federal government might cut expenses, an improvement in the design and implementation of identification cards might be one magnetic-storage trend for the future. Another archival-related goal might be the increasing need for quick comparisons of DNA evidence in various sectors of law enforcement.

—*Julia M. Meyers, MA, PhD*

FURTHER READING

Bertram, H. Neal. *Theory of Magnetic Recording.* Cambridge, England: Cambridge University

---

FASCINATING FACTS ABOUT MAGNETIC STORAGE

- One of the earliest magnetic recording devices, Valdemar Poulsen's telegraphone, was not only intended for the recording of telephone calls but was actually constructed out of telephone components.
- In 2011, the highest-capacity desktop hard drive was the three-terabyte Hitachi Deskstar 7K1000. It contains the highest number of hard disk plat-ters—five—and rotates them at 7,200 rpm.
- In 2011, the highest-capacity floppy disk was 3M's 3.5-inch LS-240 SuperDisk, which could store up to 240 megabytes of data.
- The oldest form of magnetic storage media, magnetic tape, still has the highest capacity of all of the current forms of magnetic data storage, at 5 terabytes per cassette.
- Forrest Parry, under contract to the United States government, invented the first security card in 1960—a plastic card with a strip of magnetic tape affixed to the back. "Magstripes" (typically three tracks per stripe) are universally used to record identifying data for a variety of security and financial purposes: credit cards, banking identification cards, gift cards, and government benefit cards.
- Twenty-one states in the United States and six provinces in Canada use driver licenses that have magnetic striping to carry identifying data about the individual user depicted on the card. Some passports also have a magnetic stripe for easier handling by immigration officials.

Press,1994. A complete and thorough handbook onmagnetic recording theory; goes well beyond thetypical introductory-style discussion of magneticand strives for a more advanced discussion of magnetic recording and playback theory.

Hadjipanayis, George C, ed. *Magnetic Storage Systems-Beyond 2000.* Dordrecht, The Netherlands: KluwerAcademic, 2001. This book is the collected papers presented at the "Magnetic Storage Systems Beyond 2000" Conference held by the NATO Advanced Study Institute (ASI) in Rhodes, Greece. Although the papers are technical discussions of the limitations of magnetic technology (such as magnetic heads, particulate media, and systems still in development), the speculative emphasis often describes new areas of possible development. Each paper has a tutorial, which provides an introduction to the ideas discussed.

Mee, C. Denis, and Eric D. Daniel. *Magnetic Recording Technology*. 2d ed. New York: McGraw-Hill, 1996. An older handbook, but one with particularly clear description and analysis of many of the emergent trends in the last few decades in the field of data recording and storage.

National Research Council of the National Academies. *Innovation in Information Technology*. Washington, D.C.: National Academies Press, 2003. A good handbook for the study of storage media and other issues in the field of information technology.

Prince, Betty. *Emerging Memories: Technologies and Trends*. Norwell, Mass.: Kluwer Academic, 2002. A good primer for understanding the development of memory in computing systems. Has application both to data storage and volatile memory forms.

Wang, Shan X., and Alexander Markovich Taratorin. *Magnetic Information Storage Technology*. San Diego: Academic Press, 1999. Provides the basic principles of magnetic storage and digital information and describes the technological need for data recording and the resulting push for increased capacity, faster access rates, and greater durability of media.

WEB SITES

IEEE Magnetics Society
http://www.ieeemagnetics.org

Information Technology Association of America
http://www.itaa.org

International Disk Drive Equipment and Materials Association
http://www.idema.org

See also: Computer Engineering; Computer Science; Electrical Engineering.

MECHANICAL ENGINEERING

FIELDS OF STUDY

Acoustics; algebra; applied mathematics; calculus; chemistry; control theory; drafting; dynamics; economics; electronics; fluid dynamics; graphics; heat transfer; kinematics; materials science; mechanics; optics; physics; product design; robotics; statics; structural analysis; system design; thermodynamics.

SUMMARY

Mechanical engineering is the field of technology that deals with engines, machines, tools, and other mechanical devices and systems. This broad field of innovation, design, and production deals with machines that generate and use power, such as electric generators, motors, internal combustion engines, and turbines for power plants, as well as heating, ventilation, air-conditioning, and refrigeration systems. In many universities, mechanical engineering is integrated with nuclear, materials, aerospace, and biomedical engineering. The tools used by scientists, engineers, and technicians in other disciplines are usually designed by mechanical engineers. Robotics, microelectromechanical systems, and the development of nanotechnology and bioengineering technology constitute a major part of modern research in mechanical engineering.

KEY TERMS AND CONCEPTS

- **computer-aided design (CAD):** Using computer software to design objects and systems and develop and check the graphical representation of a design.
- **computer-aided manufacturing (CAM):** Use of computer software and software-guided machines to perform manufacturing operations starting with computer-aided design.
- **dynamics:** Application of the laws of physics to determine the acceleration and velocity of bodies and systems.
- **engineering economics:** Science focused on determining the best course of action in designing and manufacturing a given system to reduce uncertainty and maximize the return on the investment.
- **fluid mechanics:** Science describing the behavior of fluids.

- **heat transfer:** Science relating the flow of thermal energy to temperature differences and material properties.
- **kinematics:** Science of determining the relationship between the movement of different elements of a system, such as a machine.
- **machine design:** Science of designing the most suitable individual elements and their integration into machines, relating the stresses and loads that the machine must handle, its lifetime, the properties of the materials used, and the cost of manufacturing and using the machine.
- **manufacturing:** Generation of copies of a product, based on good design, in sufficient numbers to satisfy demand at the best return on investment.
- **materials science:** Study of the detailed structure and properties of various materials used in engineering.
- **mechanical system:** Grouping of elements that interact according to mechanical principles.
- **metrology:** Science of measurement dealing with the design, calibration, sensitivity, accuracy, and precision of measuring instruments.
- **robotics:** Science of designing machines that can replace human beings in the execution of specific tasks, such as physical activities and decision making.
- **statics:** Science of how forces are transmitted to, from, and within a structure, determining its stability.
- **strength of materials:** Science of determining the deflection of objects of different shapes and sizes under various loading conditions.
- **thermodynamics:** Science dealing with the relationships between energy, work, and the properties of matter. Thermodynamics defines the best performance that can be achieved with power conversion, generation, and heat transfer systems.

DEFINITION AND BASIC PRINCIPLES

Mechanical engineering is the field dealing with the development and detailed design of systems to perform desired tasks. Developed from the discipline of designing the engines, power generators, tools, and mechanisms needed for mass manufacturing, it has grown into the broadest field of engineering, encompassing or touching most of the disciplines of science and engineering. Mechanical engineers take the laws of nature and apply them using rigorous mathematical principles to design mechanisms. The process of design implies innovation, implementation, and optimization to develop the most suitable solution to the specified problem, given its constraints and requirements. The field also includes studies of the various factors affecting the design and use of the mechanisms being considered.

At the root of mechanical engineering are the laws of physics and thermodynamics. Sir Isaac Newton's laws of motion and gravitation, the three laws of thermodynamics, and the laws of electromagnetism are fundamental to much of mechanical design.

Starting with the Industrial Revolution in the nineteenth century and going through the 1970's, mechanical engineering was generally focused on designing large machines and systems and automating production lines. Ever-stronger materials and larger structures were sought. In the 1990's and first part of the twenty-first century, mechanical engineering saw rapid expansion into the world of ever-smaller machines, first in the field of micro and then nano materials, probes and machines, down to manipulating individual atoms. In this regime, short-range forces assume a completely different relationship to mass. This led to a new science integrating electromagnetics and quantum physics with the laws of motion and thermodynamics. Mechanical engineering also expanded to include the field of system design, developing tools to reduce the uncertainties in designing increasingly more complex systems composed of larger numbers of interacting elements.

BACKGROUND AND HISTORY

The engineering of tools and machines has been associated with systematic processes since humans first learned to select sticks or stones to swing and throw. The associations with mathematics, scientific prediction, and optimization are clear from the many contraptions that humans developed to help them get work done. In the third century BCE, for example, the mathematician Archimedes of Syracuse was associated with the construction of catapults to hurl projectiles at invading armies, who must themselves have had some engineering skills, as they eventually invaded his city and murdered him. Tools and weapons designed in the Middle Ages, from Asia to Europe and Africa, show amazing sophistication. In the thirteenth century, Mesopotamian engineer Al-Jazari invented the camshaft and the cam-slider mechanism and used them in water clocks and water-raising machines. In Italy, Leonardo da Vinci designed many devices, from portable bridges to water-powered engines.

The invention of the steam engine at the start of the Industrial Revolution is credited with the scientific development of the field that is now called mechanical engineering. In 1847, the Institution of Mechanical Engineers was founded in Birmingham, England. In North America, the American Society of Civil Engineers was founded in 1852, followed by the American Society of Mechanical Engineers in 1880. Most developments came through hard trial and error. However, the parallel efforts to develop retrospective and introspective summaries of these trials resulted in a growing body of scientific knowledge to guide further development.

Nevertheless, until the late nineteenth century, engineering was considered to be a second-rate profession and was segregated from the "pure" sciences. Innovations were published through societies such as England's Royal Society only if the author was introduced and accepted by its prominent members, who were usually from rich landed nobility. Publications came from deep intellectual thinking by amateurs who supposedly did it for the pleasure and amusement; actual hands-on work and details were left to paid professionals, who were deemed to be of a lower class. Even in America, engineering schools were called trade schools and were separate from the universities that catered to those desiring liberal arts educations focused on the classics and languages from the Eurocentric point of view.

Rigorous logical thinking based on the experience of hands-on applications, which characterizes mechanical engineering, started gaining currency with the rise of a culture that elevated the dignity of labor in North America. It gained a major boost with the urgency brought about by several wars. From the time of the American Civil War to World War I, weapons such as firearms, tanks, and armored ships saw significant advancements and were joined by airplanes and motorized vehicles that functioned as ambulances. During these conflicts, the individual heroism that had marked earlier wars was eclipsed by the technological superiority and scientific organization delivered by mechanical engineers.

Concomitantly, principles of mass production were applied intensively and generated immense wealth in Europe and America. Great universities were established by people who rose from the working classes and made money through technological enterprises. The Great Depression collapsed the established manufacturing entities and forced a sharp rise in innovation as a means of survival. New engineering products developed rapidly, showing the value of mechanical engineering. World War II and the subsequent Cold War integrated science and engineering inseparably. The space race of the 1960's through the 1980's brought large government investments in both military and civilian aerospace engineering projects. These spun off commercial revolutions in computers, computer networks, materials science, and robotics. Engineering disciplines and knowledge exploded worldwide, and as of 2011 there is little superficial difference between engineering curricula in most countries of the world.

The advent of the Internet accelerated and completed this leveling of the knowledge field, setting up sharper impetus for innovation based on science and engineering. Competition in manufacturing advanced the field of robotics, so that cars made by robots in automated plants achieve superior quality more consistently than those built by skilled master craftsmen. Manufacturing based on robotics can respond more quickly to changing specifications and demand than human workers can.

Beginning in the 1990's, micro machines began to take on growing significance. Integrated micro-electromechanical systems were developed using the techniques used in computer production. One by one, technology products once considered highly glamorous and hard to obtain—from calculators to smart phones—have been turned into mass-produced commodities available to most at an affordable cost. Other products—from personal computers and cameras to cars, rifles, music and television systems, and even jet airliners—are also heading for commoditization as a result of the integration of mechanical engineering with computers, robotics, and micro electromechanics.

How it Works

The most common idea of a mechanical engineer is one who designs machines that serve new and useful functions in an innovative manner. Often these machines appear to be incredibly complex inside or extremely simple outside. The process of accomplishing these miraculous designs is systematic, and good mechanical engineers make it look easy.

System design. At the top level, system design starts with a rigorous analysis of the needs to be satisfied, the market for a product that satisfies those needs, the time available to do the design and manufacturing, and the resources that must be devoted. This step also includes an in-depth study of what has been done before. This leads to "requirements definition," where the actual requirements of the design are carefully specified. Experienced designers believe that this step already determines more than 80 percent of the eventual cost of the product.

Next comes an initial estimate of the eventual system characteristics, performed using simple, commonsense logic, applying the laws of nature and observations of human behavior. This step uses results from benchmarking what has been achieved before and extrapolating some technologies to the time when they must be used in the manufacturing of the design. Once these rudimentary concept parameters and their relationships are established, various analyses of more detailed implications become possible. A performance estimation then identifies basic limits and determines whether the design "closes," meeting all the needs and constraints specified at the beginning. Iterations on this process develop the best design. Innovations may be totally radical, which is relatively rare, or incremental in individual steps or aspects of the design based on new information, or on linking developments in different fields. In either case, extensive analysis is required before an innovation is built into a design. The design is then analyzed for ease and cost of manufacture. The "tooling," or specific setups and machines required for mass manufacture, are considered.

A cost evaluation includes the costs of maintenance through the life cycle of the product. The entire process is then iterated on to minimize this cost. The design is then passed on to build prototypes, thereby gaining more experience on the manufacturing techniques needed. The prototypes are tested extensively to see if they meet the performance required and predicted by the design.

When these improvements are completed and the manufacturing line is set up, the product goes into mass manufacture. The engineers must stay engaged in the actual performance of the product through its delivery to the end user, the customer, and in learning from the customer's experience in order to design improvements to the product as quickly as possible. In modern concurrent engineering practice, designers attempt to achieve as much as possible of the manufacturing process design and economic optimization during the actual product design cycle in order to shorten the time to reach market and the cost of the design cycle. The successful implementation of these processes requires both technical knowledge and experience on the part of the mechanical engineers. These come from individual rigorous fields of knowledge, some of which are listed below.

Engineering mechanics. The field of engineering mechanics integrates knowledge of statics, dynamics, elasticity, and strength of materials. These fields rigorously link mathematics, the laws of motion and gravitation, and material property relationships to derive general relations and analysis methods. Fundamental to all of engineering, these subfields are typically covered at the beginning of any course of study.

In statics, the concept of equilibrium from Newton's first law of motion is used to develop free-body diagrams showing various forces and reactions. These establish the conditions necessary for a structure to remain stable and describe relations between the loads in various elements.

In dynamics, Newton's second law of motion is used to obtain relations for the velocity and acceleration vectors for isolated bodies and systems of bodies and to develop the notions of angular momentum and moment of inertia.

Strength of materials is a general subject that derives relationships between material properties and loads using the concepts of elasticity and plasticity and the deflections of bodies under various types of loading. These analyses help the engineer predict the yield strength and the breaking strength of various structures if the material properties are known. Metals were the preferred choice of material for engineering for many decades, and methods to analyze structures made of them were highly refined, exploiting the isotropy of metal properties. Modern mechanical engineering requires materials the properties of which are much less uniform or exotic in other ways.

Graphics and kinematics. Engineers and architects use graphics to communicate their designs precisely and unambiguously. Initially, learning to draw on paper was a major part of learning engineering skills. As of 2011, students learn the principles of graphics using computer-aided design (CAD) software and computer graphics concepts. The drawing files can also be transferred quickly into machines that fabricate a part in computer-aided manufacturing (CAM). Rapid prototyping methods such as stereo lithography construct an object from digital data generated by computer graphics.

The other use of graphics is to visualize and perfect a mechanism. Kinematics develops a systematic method to calculate the motions of elements, including their dependence on the motion of other elements. This field is crucial to developing, for instance, gears, cams, pistons, levers, manipulator arms, and robots. Machines that achieve very complex motions are designed using the field of kinematics.

Robotics and control. The study of robotics starts with the complex equations that describe how the different parts satisfy the equations of motion with multiple degrees of freedom. Methods of solving large sets of algebraic equations quickly are critical in robotics. Robots are distinguished from mere manipulator arms by their ability to make decisions based on the input, rather than depend on a telepresence operator for commands. For instance, telepresence is adequate to operate a machine on the surface of the Moon, which is only a few seconds of round-trip signal travel time from Earth using electromagnetic signals. However, the round-trip time for a signal to Mars is several minutes, so a rover operating there cannot wait for commands from Earth regarding how to negotiate around an obstacle. A fully robotic rover is needed that can make decisions based on what its sensors tell it, just as a human present on the scene might do.

Entire manufacturing plants are operated using robotics and telepresence supervision. Complex maneuvers such as the rendezvous between two spacecraft, one of which may be spinning out of control, have been achieved in orbits in space, where the dynamics are difficult for a human to visualize. Flight control systems for aircraft have been implemented using robotics, including algorithms to land the aircraft safely and more precisely than human pilots can. These systems are developed using mathematical methods for solving differential equations rapidly, along with software to adjust parameters based on feedback.

Materials. The science of materials has advanced rapidly since the late twentieth century. Wood was once a material of choice for many engineering products, including bridges, aircraft wings, propellers, and train carriages. The fibrous nature of wood required considerable expertise from those choosing how to cut and lay sections of wood; being a natural product, its properties varied considerably from one specimen to another. Metals became much more convenient to use in design and fabrication because energy to melt and shape metals cheaply became available. Various alloys were developed to tailor machinery for strength, flexibility, elasticity, corrosion resistance, and other desirable characteristics. Detailed tables of properties for these alloys were included in mechanical engineering handbooks.

Materials used to manufacture mass-produced items have migrated to molded plastics made of hydrocarbons derived from petroleum. The molds are shaped using such techniques as rapid prototyping and computer-generated data files from design software. Composite materials are tailored with fiber bundles arrayed along directions where high-tensile strength is needed and much less strength along directions where high loads are not likely, thus achieving large savings in mass and weight.

Fluid mechanics. The science of fluid mechanics is important to any machine or system that either contains or must move through water, air, or other gases or liquids (fluids). Fluid mechanics employs the laws of physics to derive conservation equations for specific packets of fluid (the Lagrangian approach) or for the flow through specified control volumes (Eulerian approach). These equations describe the physical laws of conservation of mass, momentum, and energy, relating forces and work to changes in flow properties. The properties of specific fluids are related through the thermal and caloric equations expressing their thermodynamic states. The speed of propagation of small disturbances, known as the speed of sound, is related to the dependence of pressure on density and hence on temperature. Various nondimensional groupings of flow and fluid properties—such as the Reynolds number, Mach number, and Froude number—are used to classify flow behavior. Increasingly, for many problems involving fluid flow through or around solid objects, calculations starting from the conservation equations are able to predict the loads and flow behavior reliably using the methods of computational fluid mechanics (CFD). However, the detailed prediction of turbulent flows remains beyond reach and is approximated through various turbulence models. Fluid-mechanic drag and the movements due to flow-induced pressure remain very difficult to calculate to the accuracy needed to improve vehicle designs.

Methods for measuring the properties of fluids and flows in their different states are important tools for mechanical engineers. Typically, measurements and experimental data are used at the design stage, well before the computational predictions become reliable for refined versions of the product.

Thermodynamics. Thermodynamics is the science behind converting heat to work and estimating the best theoretical performance that a system can achieve under given constraints. The three basic laws of temperature are the zeroth law, which defines temperature and thermal equilibrium; the first law, which describes the exchange between heat, work, and internal energy; and the second law, which defines the concept of entropy. Although these laws were empirically derived and have no closed-form proof, they give results identical to those that come from the law of conservation of energy and to notions of entropy derived from statistical mechanics of elementary particles traced to quantum theory. No one has yet been able to demonstrate a true perpetual-motion machine, and it does not appear likely that anyone will. From the first law, various heat-engine cycles have been invented to obtain better performance suited to various constraints. Engineers working on power-generating engines, propulsion systems, heating systems, and air-conditioning and refrigeration systems try to select and optimize thermodynamic cycles and then use a figure of merit—a means of evaluating the performance of a device or system against the best theoretical performance that could be achieved—as a measure of the effectiveness of their design.

Heat transfer. Heat can be transferred through conduction, convection, or radiation, and all three modes are used in heat exchangers and insulators. Cooling towers for nuclear plants, heat exchangers for nuclear reactors, automobile and home air-conditioners, and the radiators for the International Space Station are all designed from basic principles of these modes of heat transfer. Some space vehicles are designed with heat shields that are ablative. The Thermos flask (which uses an evacuated space

between two silvered glass walls) and windows with double and triple panes with coatings are examples of widely used products designed specifically to control heat transfer.

Machine design. Machine design is at the core of mechanical engineering, bringing together the various disciplines of graphics, solid and fluid mechanics, heat transfer, kinematics, and system design in an organized approach to designing devices to perform specific functions. This field teaches engineers how to translate the requirements for a machine into a design. It includes procedures for choosing materials and processes, determining loads and deflections, failure theories, finite element analysis, and the basics of how to use various machine elements such as shafts, keys, couplings, bearings, fasteners, gears, clutches, and brakes.

Metrology. The science of metrology concerns measuring systems. Engineers deal with improving the accuracy, precision, linearity, sensitivity, signal-to-noise ratio, and frequency response of measuring systems. The precision with which dimensions are measured has a huge impact on the quality of engineering products. Large systems such as airliners are assembled from components built on different continents. For these to fit together at final assembly, each component must be manufactured to exacting tolerances, yet requiring too much accuracy sharply increases the cost of production. Metrology helps in specifying the tolerances required and ensuring that products are made to such tolerances.

Acoustics and vibrations. These fields are similar in much of their terminology and analysis methods. They deal with wavelike motions in matter, their effects, and their control. Vibrations are rarely desirable, and their minimization is a goal of engineers in perfecting systems. Acoustics is important not only because minimizing noise is usually important, but also because engineers must be able to build machines to generate specific sounds, and because the audio signature is an important tool in diagnosing system status and behavior.

Production engineering. Production engineering deals with improving the planning and implementation of the production process, designing efficient and precise tools to produce goods, laying out efficient assembly sequences and facilities, and setting up the flow of materials and supplies into the production line, and the control of quality and throughput rate. Production engineering is key to implementing the manufacturing step that translates engineering designs into competitive products.

APPLICATIONS AND PRODUCTS

Conventional applications. Mechanical engineering is applied to the design, manufacture, and testing of almost every product used by humans and to the machines that help humans build those products. The products most commonly associated with mechanical engineering include all vehicles such as railway trains, buses, ships, cars, airplanes and spacecraft, cranes, engines, and electric or hydraulic motors of all kinds, heating, ventilation and air-conditioning systems, the machine tools used in mass manufacture, robots, agricultural tools, and the machinery in power plants. Several other fields of engineering such as aerospace, materials, nuclear, industrial, systems, naval architecture, computer, and biomedical developed and spun off at the interfaces of mechanical engineering with specialized applications. Although these fields have developed specialized theory and knowledge bases of their own, mechanical engineering continues to find application in the design and manufacture of their products.

An oblique view of a four-cylinder inline crankshaft with pistons

Innovations in materials. Carbon nanotubes have been heralded as a future super-material with strength hundreds of times that of steel for the same mass. As of the first decade of the twenty-first century, the longest strands of carbon nanotubes developed are still on the order of a few centimeters. This is a very impressive length-to-diameter ratio. Composite materials incorporating carbon already find wide use in various applications where high temperatures must be encountered. Metal matrix composites find use in primary structures even for commercial aircraft. Several "smart structures" have been developed, where sensors and actuators are incorporated into a material that has special properties to respond to stress and strain. These enable structures that will twist in a desired direction when bent or become stiffer or more flexible as desired, depending on electrical signals sent through the material. Materials

capable of handling very low (cryogenic) temperatures are at the leading edge of research applications. Magnetic materials with highly organized structure have been developed, promising permanent magnets with many times the attraction of natural magnets.

Sustainable systems. One very important growth area in mechanical engineering is in designing replacements for existing heating, ventilation, and air-conditioning systems, as well as power generators, that use environmentally benign materials and yet achieve high thermodynamic efficiencies, minimizing heat emission into the atmosphere. This effort demands a great deal of innovation and is at the leading edge of research, both in new ways of generating power and in reducing the need for power.

IMPACT ON INDUSTRY

Having developed as a discipline to formalize knowledge on the design of machines for industry, mechanical engineering is at the core of most industries. The formal knowledge and skills imparted by schools of mechanical engineering have revolutionized human industry, bringing about a huge improvement in quality and effectiveness. The disciplined practice

---

FASCINATING FACTS ABOUT MECHANICAL ENGINEERING

- Robotic surgery enables surgeons to conduct very precise operations by eliminating the problems of hand vibrations and by using smaller steps than a human can take.
- The REpower 5M wind turbine in Germany, rated at 5 megawatts, is 120 meters high and has 61.5meter radius blades, more than one and one-half times as long as each wing of an Airbus A380 jetliner.
- The General Electric H System integrates a gas turbine, steam turbine, generator, and heat-recovery steam generator to achieve 60 percent efficiency.
- The nanomotor built in 2003 by Alex Zettl, a physics professor at the University of California, Berkeley, and his research group is 500 nanometers in diameter with a carbon nanotube shaft 5 to 10 nanometers in diameter.
- The crawler-transporter used to move space shuttles on to the launch pad weighs 2,721 metric tons and has eight tracks, two on each corner. Its platform stays level when the crawler moves up a five-degree incline.
- Solar thermo-acoustic cooker-refrigerators use the heat from the Sun to drive acoustic waves in a tube, which convects heat and creates a low temperature on one side.
- Much of the world's telecommunications are carried by undersea fiber-optic cables that connect all continents except Antarctica. The first telegraph cable across the English Channel was laid in 1850.
- In 1650, Thomas Savery invented a steam engine to pump water out of coal mines.

---

of mechanical engineering is responsible for taking innumerable innovations to market success. In the seventeenth through twentieth centuries, rampant industrialization destroyed many long-lasting community skills and occupations, replacing them with mass manufacturing concentrated and collocated with water resources, power sources, and transportation hubs. This has led to many problems as rural communities atrophied and their young people migrated to the unfamiliar and crowded environment of cities in search of well-paying jobs.

The effects on the environment and climate have also been severe. Heightened global concerns about the environment and climate change and new technological innovation may find mechanical engineers

again at the head of a new revolution. This may start a drive to decentralize energy resources and production functions, permitting small communities and enterprises to flourish again.

CAREERS AND COURSE WORK
Mechanical engineers work in nearly every industry, in an innumerable variety of functions. The curriculum in engineering school accordingly focuses on giving the student a firm foundation in the basic knowledge that enables problem solving and continued learning through life. The core curriculum starts with basic mathematics, science, graphics and an introduction to design and goes on to engineering mechanics and the core subjects and specialized electives. In modern engineering schools, students have the opportunity to work on individual research and design projects that are invaluable in providing the student with perspective and integrating their problem-solving skills.

After obtaining a bachelor's degree, the mechanical engineer has a broad range of choices for a career. Traditional occupations include designing systems for energy, heating, ventilation, air-conditioning, pressure vessels and piping, automobiles, and railway equipment. Newer options include the design of bioengineering production systems, microelectromechanical systems, optical instrumentation, telecommunications equipment, and software. Many mechanical engineers also go on to management positions.

SOCIAL CONTEXT AND FUTURE PROSPECTS
Mechanical engineering attracts large numbers of students and offers a broad array of career opportunities. Students in mechanical engineering schools have the opportunity to range across numerous disciplines and create their own specialties. With nano machines and biologically inspired self-assembling robots becoming realities, mechanical engineering has transformed from a field that generally focused on big industry to one that also emphasizes tiny and efficient machines. Energy-related studies are likely to become a major thrust of mechanical engineering curricula. It is possible that the future will unfold a post-industrial age where the mass-manufacture paradigm of the Industrial Revolution that forced the overcrowding of cities and caused extensive damage to the environment is replaced by a widely distributed industrial economy that enables small communities to be self-reliant for essential services and yet

be useful contributors to the global economy. This will create innumerable opportunities for innovation and design.

—*Narayanan M. Komerath, PhD*

FURTHER READING

Avallone, Eugene A., Theodore Baumeister III, and Ali M. Sadegh. *Marks' Standard Handbook for Mechanical Engineers.* 11th ed. New York: McGraw-Hill, 2006. Authoritative reference for solving mechanical engineering problems. Discusses pressure sensors and measurement techniques and their applications in various parts of mechanical engineering.

Calvert, Monte A. *The Mechanical Engineer in America, 1830-1910: Professional Cultures in Conflict.* Baltimore, Md.: Johns Hopkins University Press, 1967. Discusses the life of the mechanical engineer in nineteenth-century America. The author describes the conflict between the shop culture originating in the procedures of the machine shop and the school culture of the engineering colleges that imparted formal education.

Freitas, Robert A., Jr., and Ralph C. Merkle. *Kinematic Self-Replicating Machines.* Georgetown, Tex.: Landes Bioscience, 2004. A review of the theoretical and experimental literature on the subject of self-replicating machines. Discusses the prospects for laboratory demonstrations of such machines.

Hill, Philip G., and Carl R. Peterson. *Mechanics and Thermodynamics of Propulsion.* 2d ed. Reading, Mass.: Addison-Wesley, 1992. This textbook on propulsion covers the basic science and engineering of jet and rocket engines and their components. Also gives excellent sets of problems with answers.

Lienhard, John H., IV, and John H. Lienhard V. *A Heat Transfer Textbook.* 4th ed. Mineola, N.Y.: Dover Publications, 2011. An excellent undergraduate text on the subject.

Liepmann, H. W., and A. Roshko. *Elements of Gas Dynamics.* Reprint. Mineola, N.Y.: Dover Publications, 2001. Classic textbook on the discipline of gas dynamics as applied to high-speed flow phenomena. Contains several photographs of shocks, expansions, and boundary layer phenomena.

Pelesko, John A. *Self Assembly: The Science of Things That Put Themselves Together.* Boca Raton, Fla.: Chapman and Hall/CRC, 2007. Discusses natural self-assembling systems such as crystals and soap films and goes on to discuss viruses and self-assembly of DNA cubes and electronic circuits. Excellent introduction to a field of growing importance.

Shames, Irving H. *Engineering Mechanics: Statics and Dynamics.* 4th ed. Upper Saddle River, N.J.: Prentice Hall, 1997. A classic textbook that integrates both statics and dynamics and uses a vector approach to dynamics. Used by undergraduates and professionals all over the world since the 1970's in its various editions. Extensive work examples.

Shigley, Joseph E., Charles R. Mischke, and Richard G. Budynas. *Mechanical Engineering Design.* 7th ed. New York: McGraw-Hill, 2004. Classic undergraduate textbook showing students how to apply mathematics, physics, thermal sciences, and computer-based analysis to solve problems in mechanical engineering. Includes sections on quality control, and the computer programming sections provide an insight into the logic used with the high-level languages of the 1980's.

Siciliano, Bruno, et al. *Robotics: Modelling, Planning and Control.* London: Springer-Verlag, 2010. Rigorous textbook on the theory of manipulators and wheeled robots, based on kinematics, dynamics, motion control, and interaction with the environment. Useful for industry practitioners as well as graduate students.

WEB SITES

American Society of Heating, Refrigerating and Air-Conditioning Engineers
http://www.ashrae.org

American Society of Mechanical Engineers
http://www.asme.org

National Society of Professional Engineers
http://www.nspe.org/index.html

Society of Automotive Engineers International
http://www.sae.org

Society of Manufacturing Engineers
http://www.sme.org

See also: Acoustics; Algebra; Applied Mathematics; Calculus; Engineering.

METALLURGY

FIELDS OF STUDY

Materials science; materials engineering; mechanical engineering; physical engineering; mining; chemical engineering; electrical engineering; environmental engineering.

SUMMARY

Starting as an art and a craft thousands of years ago, metallurgy has evolved into a science concerned with processing and converting metals into usable forms. The conversion of rocky ores into finished metal products involves a variety of activities. After the ores have been mined and the metals extracted from them, the metals need to be refined into purer forms and fashioned into usable shapes such as rolls, slabs, ingots, or tubing.

Another part of metallurgy is developing new types of alloys and adapting existing materials to new uses. The atomic and molecular structure of materials are manipulated in controlled manufacturing environments to create materials with desirable mechanical, electrical, magnetic, chemical, and heat-transfer properties that meet specific performance requirements.

KEY TERMS AND CONCEPTS

- **Bessemer process:** Steelmaking process in which air is blown through molten pig iron contained in a furnace so that impurities can be removed by oxidation.
- **blast furnace:** Smelting furnace for the production of pig iron in which hot air is blown upward into the furnace as iron ore, coke, and limestone are supplied through the top, producing chemical reactions; the molten iron and slag are collected at the bottom.
- **carburizing:** Process of adding carbon to the surface by exposing a metal to a carbon-rich atmosphere under high temperatures, allowing carbon to diffuse into the surface, making the surface more wear resistant.
- **ductility:** Characteristic of metal that enables it to be easily molded or shaped, without fracturing.

- **extrusion:** Process in which a softened metal is forced through a shaped metal piece or die, creating an unbroken ribbon of product.
- **flux:** Substance added to molten metals to eliminate impurities or encourage fusing.
- **forging:** Process of shaping metal by heating it in a forge, then beating or hammering it.
- **galvanizing:** Process of coating steel with zinc to prevent rust.
- **metal alloy:** Homogeneous mixture of two or
- **more metals. · ore:** Mineral from which metal is extracted.
- **plastic deformation:** Permanent distortion of a metal under the action of applied stresses.
- **recrystallization:** Process by which deformed grains in a metal or alloy are replaced by undeformed grains; reduces the strength and hardness of a material.
- **recrystallization temperature:** Approximate minimum temperature at which a cold-worked metal becomes completely recrystallized within a specified time.
- **sintering:** Process of turning a metal powder into a solid by pressure and heating it to a temperature below its boiling point.
- **slag:** By-product consisting of impurities produced during the refining of ore or melting of metal.

DEFINITION AND BASIC PRINCIPLES

Metallurgy is the science of extracting metals and intermetallic compounds from their ores and working and applying them based on their physical, chemical, and atomic properties. It is divided into two main areas: extractive metallurgy and physical metallurgy.

Extractive metallurgy. Extractive metallurgy, also known as process or chemical metallurgy, deals with mineral dressing, the converting of metal compounds to more treatable forms and refining them. Mineral dressing involves separating valuable minerals of an ore from other raw materials. The ore is crushed to below a certain size and ground to powder. The mineral and waste rock are separated, using a method based on the mineral's properties. After that, water is removed from the metallic concentrate or compound. Because metallic compounds are often complex mixtures (carbonates, sulfides, and oxides), they need to be converted to other forms for easier processing and refining. Carbonates are converted to oxides; sulfides to oxides, sulfates, and chlorides; and oxides to sulfates, and chlorides. Depending on the type of metallic compound, either pyrometallurgy or hydrometallurgy is used for conversion. Both processes involve oxidation and reduction reactions. In oxidation, the metallic element is combined with oxygen, and in reduction, a reducing agent is used to remove the oxygen from the metallic element. The difference between pyrometallurgy and hydrometallurgy, as their names imply, is that the former uses heat while the latter uses chemicals. These two processes also include refining the metallic element in the final stage of extractive metallurgy when heat and chemicals are used. Electrometallurgy refers to the use of the electrolytic process for refining metal elements, precipitating dissolved metal values, and recovering them in solid form.

Physical metallurgy. Physical metallurgy deals with making metal products based on knowledge of the crystal structures and properties (chemical, electrical, magnetic, and mechanical) of metals. Metals are mixed together to make alloys. Heat is used to harden metals, and their surfaces can be protected with metallic coating. Through a process called powder metallurgy, metals are turned into powders, compressed, and heat-treated to produce a desired product. Metals can be formed into their final shapes by such operations as casting, forging, or plastic deformation. Metallography is a subfield of metallurgy that studies the microstructure of metals and alloys by various methods, especially by light and electron microscopes.

BACKGROUND AND HISTORY

Metallurgy came into being in the Middle East around 5000 BCE with the extraction of copper from its ore. The discovery of the first alloy, bronze, the result of melting copper and tin ores together, initiated the Bronze Age (4000-3000 BCE). Melting iron ore with charcoal to obtain iron marked the beginning of the Iron Age in Anatolia (2000-1000 BCE). Gold, silver, and lead were separated from lead-bearing silver in Greece about 500 BCE. Mercury was produced from cinnabar around 100 BCE, and it was later used to recover and refine various metals. Around 30 BCE, brass, the second alloy, was made from copper and zinc in Egypt, and another alloy, steel, was produced

in India.

From the sixth to the nineteenth centuries, metallurgy focused on the development and improvement of the processes involved in obtaining iron, making steel, and extracting aluminum and magnesium from their ores. The blast furnace was developed in the eighth century and spread throughout Europe. During the sixteenth century, the first two books on metallurgy were written by Vannoccio Biringuccio, an Italian metalworker, and by Georgius Agricola, a German metallurgist.

Modern metallurgy began during the eighteenth century. Abraham Darby, an English engineer, developed a new furnace fueled by coke. Another English engineer, Sir Henry Bessemer developed a steelmaking process in 1856. Great Britain became the greatest iron producer in the world, and Spain and France also produced large amounts of iron. About 1886, American chemist Charles Martin Hall and a French metallurgist Paul-Louis-Toussaint Héroult independently developed a way to extract aluminum from its ore, which became known as the Hall-Héroult process. Aluminum soon became an important metal in manufactured goods.

Metallurgy did not emerge as a modern science with two branches, extractive and physical, until the twentieth century. The development and improvement of metallurgy were made possible by the application of knowledge of the chemical and physical principles of minerals.

HOW IT WORKS

Crushing and grinding of ores. In the first step of mineral dressing, two kinds of mechanized crushers are used to reduce ores. Jaw crushes reduce ores to less than 150 millimeters (mm) and cone crushers to less than 10-15 mm. Different kinds of grinding mills are used to reduce crushed ores to powder: cylinder mills filled with grinding bodies (stones or metal balls), autogenous mills (coarse crushed ores grinding themselves), semiautogenous mills using some grinding bodies, and roll crushers, which combine crushing and grinding.

Separating valuable and waste minerals. The process used in the next step of mineral dressing depends on the properties of the minerals. Magnetic separation is used for strongly magnetic minerals such as iron ore and iron-bearing ore. Gold, tin, and tungsten ores require gravity separation. A process called

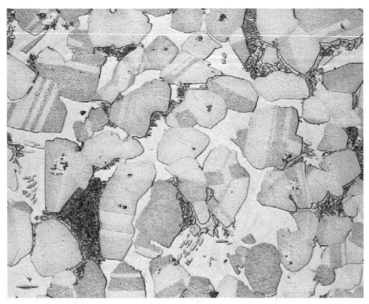

Metallography allows the metallurgist to study the microstructure of metals.

flotation separation is widely used for hydrophilic (water-attracting) intergrown ores containing copper, lead, and zinc. Electrostatic separation work best with particles of different electric charges such as mineral sands bearing zircon, rutile, and monazite.

Pyrometallurgy. Pyrometallurgy is a method of converting metallic compounds to different forms for easier processing and refining by using oxidation and reduction reactions.

The first conversion process, roasting, has two main types: One type changes sulfide compounds to oxides, and the other reduces an oxide to a metal. Other types of roasts convert sulfides to sulfates or change oxides to chlorides. These processes are carried out in different kinds of steel roasters.

The second conversion process, smelting, separates a metallic compound into two parts: an impure molten metal and a molten slag. The two types of smelting are reduction and matte, and the processes are done in many kinds of blast furnaces. Coke is used for fuel and limestone as a flux for making slag. Reduction smelting converts an oxide feed material to a metal and an oxide slag. Matte smelting converts a sulfide feed material to a mixture of nickel, copper, cobalt, and iron sulfides as well as an iron and silicon oxide slag.

Refining, a process of removing any impurities left after roasting or smelting, also can be done in a blast furnace. Iron, copper, and lead can be refined in oxidation reaction that removes impurities as an oxide slag or an oxide gas. Fire refining can separate copper from its impurities of zinc, tin, iron, lead, arsenic, and antimony. Similarly, lead can be separated from such impurities as tin, antimony, and arsenic, and zinc from impurities of cadmium and lead.

Hydrometallurgy. Another method of converting metallic compounds to different forms is hydrometallurgy. It uses several types of leach solvents: ammonium hydroxide for sulfides and carbonates; sulfuric acid, sodium carbonate, or sodium hydroxide for oxides; and sulfuric acid or water for sulfates. The dissolved metal values are then recovered from the leaching solution in solid form. Although numerous recovery processes exist, they usually involve electrolysis. By a process called precipitation, gold that has been dissolved in sodium cyanide and placed in contact with zinc is separated from the solution and gathers on zinc. In another process called electrolytic deposition, or electrowinning, an electric current is passed through the leach solution with dissolved metals, causing metal ions to deposit at the cathode. Copper, zinc, nickel, and cobalt can be obtained this way.

Electrometallurgy. Electrolysis can be used to refine metallic elements as well as to recover them after hydrometallurgical treatment. Copper, nickel, lead, gold, and silver can be refined this way. In this method, for example, impure copper is used as the anode. When the electric current passes through the solution, atoms of pure gold travel to the cathode, acquire electrons, and become neutral copper atoms. Electrolysis is also the process for recovering copper, aluminum, and magnesium in hydrometallurgy.

Alloys. Alloys are made by mixing pure metals together to obtain a substance with increased strength, increased corrosion resistance, lower cost, lower melting points, or desirable magnetic, thermal, or electrical properties. They are usually made by melting the base metals and adding alloying agents. Stainless steel, a mixture of steel, nickel, and chromium, is stronger and more chemically resistant than the base metals from which it was formed.

Powder metallurgy. Powder metallurgy is the process of reducing metals and nonmetals to powder and shaping the powder into a solid form under great heat and pressure. Metal powders are usually produced by atomization of streams of molten metal with a spinning disk or with a jet of water, air, or inert gas. After the powders are cold pressed for initial adhesion, they are heated to temperatures about 80 percent below the melting point of the major component. Friction between powders and pressing dies is reduced by adding lubricants, and porosity in the final product is eliminated by applying pressure.

Metal forming. Metals are usually cast into ingots in iron molds. Casting is also carried out in molds made of sand, plaster of Paris, or glue. Permanent casting uses pressure or centrifugal action. Plastic deformation is performed on metals to change their properties and dimensions. If done below the recrystalization temperature, the process is called cold working; above this temperature but below the melting or burning point, it is called hot working. Techniques involved include rolling, pressing, extrusion, stamping, forging, and drawing. Surface treatments of metals include protective coating and hardening. In metallic coating, zinc and other metals such as chromium, cadmium, lead, and silver are often used. Surface hardening of metals is usually done with heat in a gas rich in carbon or in ammonia and hydrogen.

APPLICATIONS AND PRODUCTS

The most important applications of metallurgy involve common metals and alloys and powder metallurgy technology.

Copper. Copper is ductile and malleable, and it resists corrosion and conducts heat and electricity. Copper and its alloy brass (copper plus zinc) are used to make coins, household fixtures (doorknobs, bolts), and decorative-art objects (statues, sculptures, imitation-gold jewelry). It is also used in transportation vehicles and has many electrical applications (transformers, motors, generators, wiring harnesses). Its alloy bronze (copper plus tin) is used in plumbing and heating applications (water pipes, cooking utensils). Aluminum-bronze is used to make tools and parts for aircraft and automobiles. Manganese-bronze is used to make household fixtures and ship propellers.

Iron. Iron is ductile, malleable, and one of the three magnetic elements (the others are cobalt and nickel). Cast iron is resistant to corrosion and used to make manhole covers and engine blocks for gasoline and diesel engines. Wrought iron is used to make cooking utensils and outdoor household items such as fencing and furniture. Most iron is used to make steel. Steel is used as a structural material in the construction of large, heavy projects (bridges, ships, buildings) and automobile parts (body frames, radial-ply tires). When chromium and nickel are combined with steel, steel becomes stainless, and it is used to make flatware and surgical tools. Steel combined with cobalt is used to make jet engines and gas turbines.

Gold. Applications of gold are based on such properties as its electrical and thermal conductivity, ductility, malleability, resistance to corrosion, and infrared reflectivity. Gold serves as a medium of exchange and money. Its decorative applications include jewelry, golf leaf on the surfaces of buildings, and flourishes on ceramics or glassware. More practical applications include components for electronic devices (telephones, computers), parts for space vehicles, and dental fillings, crowns, and bridges.

Silver. Silver is ductile and very malleable, conducts heat, and has the highest electrical conductivity of all metals. It is used to make cutlery, jewelry, coins, long-life batteries, photographical films, and electronic components (circuits, contacts), and in dentistry. Its alloy, sterling silver (silver plus copper) is also used to make jewelry and tableware. German silver (silver plus nickel) is another alloy used for silverware.

Platinum. This ductile and malleable material is one of the densest and heaviest metals. It is resistant to corrosion and conducts electricity well. It is used to make jewelry, electronic components (hard disk drive coatings, fiber-optics cables), and spark plug components. It is important in making the glass for liquid crystal displays (LCDs), in the petrol industry as an additive and refining catalyst, in medicine (anticancer drugs, implants), and in dentistry. Its alloys (platinum plus cobalt or metals in the platinum groups) are mostly used to make jewelry.

Mercury. Sometimes called quicksilver, mercury is the only common metal that is liquid at ordinary temperatures. It is a fair conductor of electricity and of high density. It is used in barometers and thermometers, to recover gold from its ore, and to manufacture chlorine and sodium hydroxide. Its vapor is used in street lights, fluorescent lamps, and advertizing signs. Mercury compounds have various uses, such as insecticides, rat poisons, disinfectants, paint pigments, and detonators. Mercury easily is alloyed with silver, gold, and cadmium.

Lead. Lead is malleable, ductile, resistant to corrosion, and of high density. Its softness is compensated for by alloying it with such metals as calcium, antimony, tin, and arsenic. Lead is a component in lead-acid batteries, television and computer screens, ammunition, cables, solders, and water drains, and is used as a coloring element in ceramic glazes.

Magnesium. Magnesium is the lightest structural metal, with low density (two-thirds that of aluminum), superior corrosion performance, and good mechanical properties. Because it is more expensive than aluminum, its applications are somewhat limited.

Magnesium and its alloys are used in the bicycle industry, racing car industry (gearbox casings, engine parts), and aerospace industry (engines, gearbox casings, generator housings, wheels).

Manganese. Manganese is a hard but very brittle, paramagnetic metal. Mostly it is used in steel alloys to increase strength, hardness, and abrasion resistance. It can be combined with aluminum and antimony to form ferromagnetic compounds. It is used to give glass an amethyst color, in fertilizers, and in water purification.

Cobalt. Cobalt has a high melting point and retains its strength at high temperatures. It is used as a pigment for glass, ceramics, and paints. When alloyed with chromium and tungsten, it is used to make high-speed cutting tools. It is also alloyed to make magnets, jet engines, and gas turbine engines.

Tungsten. Tungsten has the highest melting point and the lowest thermal expansion of all metals, high electrical conductivity, and excellent corrosion resistance. It is used to make lightbulb filaments, electric contacts, and heating elements; as an additive for strengthening steel; and in the production of tungsten carbide. Tungsten carbide is used to make dies and punches, machine tools, abrasive products, and mining equipment.

Chromium. Chromium is a hard but brittle metal of good corrosion resistance. It is mostly alloyed with other metals, especially steel, to make final products harder and more resistant to corrosion. It is also used in electroplating, leather tanning, and refractory

brick making, and as glass pigments.

Cadmium. Cadmium is resistant to corrosion, malleable, ductile, and of high electrical and thermal conductivity. It is mostly used in rechargeable nickel-cadmium batteries. It is also used to make electronic components and pigments for plastics, glasses, ceramics, enamels, and artists' colors.

Nickel. Nickel is a hard, malleable, and ductile metal that is highly resistant to corrosion. Like chromium, it is used to make stainless steel. Alloyed with copper, it is used for ship propellers and chemical industry plumbing. Other uses include rechargeable batteries, coinage, foundry products, plating, burglar-proof vaults, armor plates, and crucibles.

Aluminum. Aluminum has a density about one-third that of steel, high resistance to corrosion, and excellent electrical and thermal conductivity. Moreover, this nontoxic metal reflects light and heat well. This versatile metal can be used to replace other materials depending on the application. It is widely used in such areas as food packaging and protection (foils, beverage cans), transportation (vehicles, trains, aircraft), marine applications (ships, support structures for oil and gas rigs), and buildings and architecture (roofing, gutters, architectural hardware). Other applications of aluminum include sporting goods, furniture, road signs, ladders, machined components, and lithographic printing plates.

Special alloys. Fusible alloys are mixtures of cadmium, bismuth, lead, tin, antimony, and indium. They are used in automatic sprinklers and in forming and stretching dies and punches. Superalloys are developed for aerospace and nuclear applications: columbium for reactors, tantalum for rocket nozzles and heat exchangers in nuclear reactors, and a nickel-based alloy for jet and rocket engines and electric heating furnaces. The alloy of tin and niobium has superconductivity. It is used in constructing superconductive magnets that generate high field strengths without consuming much power.

Powder metallurgy applications and products. Powder metallurgy was developed in the late 1920's primarily to make tungsten-carbide cutting tools and self-lubricating electric motor bearings. The technique was then applied in the automobile industry, where it is used to make precision-finished machine parts, permanent metal filters, bearing materials, and self-lubricating bearings. It is useful in fabricating products that are difficult to make by other methods, such as tungsten-carbide cutting tools, super magnets of aluminum-nickel alloy, jet and missile applications of metals and ceramics, and wrought powder metallurgy tool steel.

IMPACT ON INDUSTRY

Metallurgy is used in many industries, and the challenges associated with it vary by industry. All industries, for example, seek ways to achieve fuel savings, and where metallurgy is concerned, they want high metallic yields and high-quality products.

Aluminum production requires large amounts of electricity for the electrolytic smelting process. Electricity accounts for 25 percent of the cost of producing aluminum. The techniques of ladle metallurgy (degassing, desulfurization) require high temperatures for preheating, also resulting in high energy costs. Hydrogen and oxides can cause porosity in solidified aluminum, which decreases the mechanical properties of the final product. High oxygen levels ranging from 25 percent in converters to 90 percent in flash-smelting furnaces are used for primary smelting, resulting in low productivity as well as high operating costs.

New technologies in metallurgy not only solve operational problems but also try to offer such benefits as high productivity, availability and reliability of power, safety, and environmental acceptance. For example, Alstom, a company in transport and energy infrastructure, provides air-quality control systems for the aluminum industry. To abate emissions from pot lines, anodes, and green anodes, Alstom systems take the gas from the pots and absorb hydrofluorides on alumina. Additional sulfur is removed from the gas before it is released to the atmosphere. Alstom has also developed an energy recovery system to decrease pot amperage in smelters, thereby ensuring high availability and reliability of power.

Air Liquide provides solutions to the problems caused by the need to preheat the ladle by using oxy-fuel burners, based on the evolution in refractory materials. The necessary calories needed to keep a metal at the suitable pouring temperature during the metal transfer into the ladle are stocked in the refractory. Air Liquide also provides solution to the porosity problem in aluminum. Its technique involves injecting hydrogen through porous plugs. The same company also developed an innovative scheme to lower the use of oxygen in smelting furnaces. It

---

FASCINATING FACTS ABOUT METALLURGY

- Iron is the most abundant element, making up
- 34.6 percent of Earth, and the most used of all metals.
- The magnetic property of steel allows recyclers to reclaim millions of tons of iron and steel from garbage.
- Meteors were the source of much of the wrought iron used in early human history.
- Because gold is so malleable, 1 gram of gold can be hammered into a sheet 1 square meter in size. It can also be made so thin that it appears transparent.
- Nitric acid can be used to determine if gold is present in ore. This "acid test," proving an ore's value, has come to mean a decisive test proving an item's worth or quality.
- Half of the world's gold is held by the Republic of South Africa. The second gold-producing nation, in terms of volume, is the United States.
- A lead pencil might more accurately be called a graphite pencil, as it contains a shaft of graphite, not lead.
- An alloy of equal parts of silver and aluminum is as hard as bronze.
- The amount of platinum mined each year is 133 tons, less than one-tenth of the 1,782 tons of gold mined each year.
- Uranium is very dense. A one-gallon container filled with uranium would weigh about 150 pounds, while that same container filled with milk would weigh around 8 pounds.
- To produce about 40 million kilowatt hours of electricity, it would take 16,000 tons of coal or 1 ton of natural uranium.
- In Japan, all government-subsidized dental alloys must have a palladium content of at least 20 percent.

---

includes a process that allows the direct production of oxygen at 95 percent purity under medium pressure without an oxygen compressor and a balancing system for liquid and gaseous oxygen for handling fast-changing regimes and optimizing working conditions at low operating costs.

Problems in metallurgy are also solved by simply replacing the old technologies. For example, operating problems in the copper and nickel converter include tuyere blockage and refractory erosion at the tuyere line. Using an oxygen injector to replace conventional tuyeres not only permits enrichment levels of up to 60 percent of oxygen and increases converter productivity but also reduces volumes of toxic off-gases.

The metallurgy industry, like many other industries, must constantly deal with new laws and regulations and expends considerable time and money to meet environmental requirements. The industry must adapt its extraction processes and invest in emission-reducing and energy-saving technologies. For example, the steel industry has replaced all its open-hearth furnaces. The aluminum industry has used prebaked instead of Soderberg electrodes in the electrolytic cells. The copper industry has replaced the reverberatory furnace with the flash-smelting furnace. The common reason behind these changes was to reduce toxic off-gases as much as possible. Stacks have become much taller to dispose of sulfur dioxide.

A process based on the combination of biology and hydrometallurgy—also called bioleaching, biohydrometallurgy, microbial leaching, and biomining—has gained some interest from those who want to replace some of the traditional metallurgical processes. Although conventional metallurgy uses smelting of ores at high temperatures, bioleaching involves dissolving metals from ores using microorganisms. Copper, for example, can be leached by the activity of the bacterium *Acidithiobacillus ferrooxidans*. Canada used this process to extract uranium in 1970, and South Africa experimented it with gold during the 1980's. Several countries have used bioleaching for a number of metals (copper, silver, zinc, nickel). The process is helpful in the recovery of low-grade ores, which cannot be economically processed with chemical methods; however, it creates a problem because it results in excessive disposal and handling of a mixture of ferrous hydroxide and gypsum.

CAREERS AND COURSE WORK
High school students who wish to work as metallurgical technicians must take at least two years of mathematics and two years of science, including a physical science. Shop courses of any kind are also helpful. Positions in the metallurgical industry are typically in the areas of production, quality control, and research and development, which share many concerns and often require similar skills from prospective metallurgical technicians. Two years of study in metallurgy or materials science at a community college or technical college is therefore strongly recommended. Metallurgical technicians occupy a middle ground between engineers and skilled trade workers.

Representative entry-level jobs include metallurgical laboratory technicians, metallographers, metallurgical observers, metallurgical research technicians, and metallurgical sales technicians. Students who are interested in these kinds of jobs should have an interest in science and average mathematical ability. Prospective technicians must be willing to participate in a wide variety of work and must be able to communicate well. Companies employing metallurgical technicians can be found in a wide variety of industries. Working environments vary depending on the area of activities.

A number of colleges and universities offer four-year programs in metallurgy. If students wish to become metallurgical engineers, they will need a bachelor's degree in materials science or metallurgical engineering. The first two years of college focus on subjects such as chemistry, physics, mathematics, and introductory engineering. In the following years, courses will focus on metallurgy and related engineering areas. Students who wish to become metallurgical engineers should be interested in nature and enjoy problem solving. They also need to have good communication skills. There are basically three areas in which metallurgical engineers work: extractive, physical, and mechanical metallurgy. Their work environment varies depending on their area of specialty. Companies employing metallurgical engineers include metal-producing and processing companies, aircraft companies, machinery and electrical equipment manufacturers, the federal government, engineering consulting firms, research institutes, and universities.

SOCIAL CONTEXT AND FUTURE PROSPECTS
Metallurgy faces the challenges of reducing the effect of its processes on the air and land, making more efficient use of energy, and increasing the amount of recycling. To these ends, industries are increasingly using clean technologies and developing methods of oxygen combustion that drastically reduce emissions of carbon dioxide and other pollutants. For example, zinc, copper, and nickel are being recovered from their ores through a technique in which pressure leaching is performed in an acid medium, followed by electrolysis in a conventional sulfuric acid medium. The technique produces no dust, no slag, and no sulfur dioxide and therefore is environmentally acceptable. It has been applied for zinc sulfide

concentrates in 1980's, for nickel sulfides in Canada, and for copper sulfide concentrates in the United States. In addition, between 1994 and 2003, the steel industry reduced its releases of chemicals to the air and water by 69 percent. However, it still releases manganese, chromium, and lead to the air, and efforts are concentrating on this problem.

Becoming more energy efficient has long been a goal of the metallurgy industry, especially as it benefits the bottom line. A wide range of approaches has been employed, including lowering generation costs (for example, by generating energy rather than purchasing it), capturing energy (such as gases) produced during various metallurgy processes, making use of energy-efficient equipment and techniques, and monitoring the production process. The steel industry in North America has reduced its consumption of energy by 60 percent since World War II. One way that the industry reduced energy consumption was by using scrap steel instead of natural resources to produce steel.

Metallurgy companies have made efforts to increase recycling. Each year, more steel is recycled in the United States than paper, aluminum, plastic, and glass combined. In 2008, more than 75 million tons of steel were either recycled or exported for recycling. All new steel produced in the United States in 2008 contained at least 25 percent steel scrap on average.

Despite these environmental challenges, metallurgy is an important area and will continue to grow and develop. In addition to developing ways to lessen metallurgy's impact on the environment, engineers are likely to develop improved processes for extraction of ores and processing materials into products. New techniques are likely to develop in response to the need to impart metals with additional qualities and to conserve natural resources by reusing metals and finding uses for by-products. Waste disposal and reduction also are likely to remain areas of research.

—Anh Tran, PhD

FURTHER READING

Abbaschian, Reza, Lara Abbaschian, and Robert E.Reed-Hill. *Physical Metallurgical Principles*. Stamford, Conn.: Cengage Learning, 2009. A comprehensive introduction to physical metallurgy for engineering students.

Boljanovic, Vukota. *Metal Shaping Processes: Casting and Molding, Particulate Processing, Deformation Processes, and Metal Removal*. New York: Industrial Press, 2010. Describes the fundamentals of how metal is shaped into products.

Bouvard, Didier. *Powder Metallurgy*. London: ISTE, 2008. Looks at the thermo-mechanical processes used to turn powdered metal into metal parts and also describes applications.

Brandt, Daniel A., and J. C. Warner. *Metallurgy Fundamentals*. 5th ed. Tinley Park, Ill.: Goodheart-Willcox, 2009. Examines metallurgy, focusing on iron and steel but also covering nonferrous metals.

Pease, Leander F., and William G. West. *Fundamentals of Powder Metallurgy*. Princeton, N.J.: Metal Powder Industries Federation, 2002. A primer on powder metallurgy.

Popov, K. I., Stojan D. Djokić, and Branimir N. Grgur. *Fundamental Aspects of Electrometallurgy*. New York: Plenum, 2002. Examines the theory and mechanisms of electrometallurgy.

Vignes, Alain. *Handbook of Extractive Metallurgy*. Hoboken, N.J.: John Wiley & Sons, 2010. Examines how metals are transformed from ore into liquids ready for pouring.

WEB SITES

American Institute of Mining, Metallurgical, and Petroleum Engineers
http://www.aimeny.org

The Minerals, Metals, and Materials Society
http://www.tms.org

Mining and Metallurgical Society of America
http://www.mmsa.net

See also: Electrometallurgy; Environmental Engineering.

METEOROLOGY

FIELDS OF STUDY

Climatology; hydrology; atmospheric physics; atmospheric chemistry, oceanography.

SUMMARY

Interdisciplinary study of physical phenomena occurring at various levels of the Earth's atmosphere. Practical applications of meteorological findings all relate in some way to understanding longer-term weather conditions. On the whole, however, meteorological weather forecasts are—in contrast to the research goals of climatology—mainly short term in nature. They concentrate on contributing factors, including temperature, humidity, atmospheric pressure, and winds.

KEY TERMS AND CONCEPTS

- **cyclone:** Air mass closed in by spiraling (circular) winds that can become a moderate or violent storm, depending on factors of humidity, temperature, and the force, changing direction, and altitude of the winds.
- **dewpoint:** Temperature at which air becomes saturated and produces dew.
- **front:** Interface between air masses with different temperatures or densities.
- **isotherms:** Graphically recorded lines connecting all points in a given region (large or more limited) having exactly the same temperature.
- **jet stream:** Narrow but very fast air current flowing around the globe at altitudes between 23,000 and 50,000 feet.
- **ozone:** Variant form of oxygen made up of three atoms; forms a layer in the upper atmosphere that helps absorb ultraviolet rays from the Sun.
- **relative humidity:** Percentage value indicating how much water is in an air sample in relation to how much it can hold, or its saturation point, at a given temperature and pressure.
- **saturation point:** Point where the water vapor in the air is at its maximum for a given temperature and pressure; point where condensation occurs.
- **stratosphere:** Atmospheric layer above the tropo-

sphere; temperatures in this layer increase as the altitude becomes greater.
- **temperature inversion:** Phenomenon that is the opposite of normal atmospheric conditions. When air close to ground level remains colder and denser than air at higher levels, the warmer air can form a sort of cover, trapping the colder air, with resultant increases in ground-level fog and smog.
- **troposphere:** Lowest layer of the atmosphere; contains 80 percent of the total molecular mass of the atmosphere.

DEFINITION AND BASIC PRINCIPLES
Meteorology is the study of the Earth's atmosphere, particularly changes in atmospheric conditions. The three main factors affecting change in the atmosphere are humidity, temperature, and barometric pressure. Dynamic short- term interaction among these three atmospheric factors produces the various phenomena associated with changeable weather. Low barometric pressure conditions are generally associated with greater capacity for the atmosphere to absorb water vapor (resulting in various cloud formations), whereas high pressure prevents absorption of humidity.

Meteorological calculation of relative humidity, for example, reveals how much more moisture can be absorbed by the atmosphere at specific temperature levels before reaching the saturation point. Changes in temperature (either up or down) will affect this dynamic process. Rainfall occurs when colder air pushes warmer, moisture-laden air upward into higher altitudes, where the warmer air mass begins to cool. Cooler air cannot hold as much water as warmer air, so the relative humidity of the warmer air mass changes, resulting in condensation and precipitation. This phenomenon is closely associated with the presence of surface winds.

BACKGROUND AND HISTORY
Meteorology, like several other applied sciences that stem from observations of natural phenomena, has a long history. The term "meteorology" comes from the Greek word for "high in the sky." Aristotle's work on meteorology maintained that the Sun attracted two masses of air from the Earth's surface, one humid

and moist (which returned as rain) and the other hot and dry (the source of wind currents). His student, Theophrastus, described distinct atmospheric signs associated with eighty types of rain, forty-five types of wind, and fifty storms.

During the Renaissance, Europeans developed instruments that could refine these ancient Greek theories. The Italian scientist Galileo, for example, used a closed glass container with a system of gauges that showed how air expands and contracts at different temperatures (the principle of the thermometer). The French philosopher Blaise Pascal developed what became the barometer, a device to measure surrounding levels of atmospheric pressure.

Although many important small-scale experiments would be carried out in the eighteenth and nineteenth centuries, a major breakthrough occurred in the first quarter of the twentieth century when the Swede Vilhelm Bjerknes and his son Jacob Bjerknes developed the theory of atmospheric fronts, involving large-scale interactions between cold and warmer air masses close to the Earth's surface. In the 1920's, a Japanese meteorologist first identified what came to be known as jet streams, or fast-moving air currents at altitudes of 23,000 and 50,000 feet.

The turning point for modern meteorology, however, came in April, 1960, when the United States launched TIROS 1, the first in a series of meteorological satellites. This revolutionary tool enabled scientists to study atmospheric phenomena such as radiation flux and balance that were known but had not been measured with high levels of accuracy.

How it Works

Meteorologists employ a variety of basic tools and methods to obtain the data needed to put together a comprehensive picture of local or regional atmospheric conditions and changes. At the most basic level, meteorologists direct their attention to three essential factors affecting the atmosphere: temperature, air pressure, and humidity.

Drawing on empirical data, meteorologists not only analyze the effects of temperature, pressure, and humidity in the area of the atmosphere they are studying but also apply their findings to ever-widening areas of the globe. From their analyses, they are able to predict the weather—for example, the direction and strength of the wind and the nature and the probable intensity of storms heading toward the

area, even if they are still thousands of miles away.

Wind strength and direction. Wind, like rain and snowstorms, is a common weather phenomenon, but the meteorological explanations for wind are rather complicated. All winds, whether local or global, are the product of various patterns of atmospheric pressure. The most common, or horizontal, winds arise when a low pressure area draws air from a higher pressure zone.

To build a complete picture of the likely strengths and directions of winds, however, meteorologists must gather much more than simple barometric data. They must consider, for example, the dynamics of the Coriolis effect (the influence of the Earth's rotation on moving air). Except in the specific latitude of the equator, the Coriolis effect, which is greater near the poles and less near the equator, makes winds curve in a circular pattern. Normal curving from the Coriolis effect can be altered by another force, centrifugal acceleration, which is the result of the movement of air around high and low pressure areas (in opposite directions for high and low pressure areas). Tornadoes and hurricanes (giant cyclones) occur when centrifugal acceleration reaches very high levels.

When a near balance exists between the Coriolis effect and centrifugal acceleration, the resultant (still somewhat curved) wind pattern is called cyclostophic. If no frictional drag (deceleration associated with physical obstacles at lower elevations) exists, something close to a straight wind pattern occurs, especially at altitudes of about two-thirds of a mile and greater. This straight wind, called a geostrophic wind, is characterized by a balance between the pressure gradient and the Coriolis effect.

In some parts of the Northern Hemisphere, massive pressure gradient changes produced by differences in the temperature of the land and the ocean create monsoon winds (typically reversed from season to season), which in the summer are followed by storms and heavy rains. The best-known example of a monsoon occurs in India, where the rising heat of summer creates a subcontinent-wide thermal low pressure zone, which attracts the moisture-laden air from the Indian Ocean into cyclonic wind patterns and much needed, but sometimes catastrophic, heavy rainfall.

Atmospheric absorption and transfer of heat. Meteorologists worldwide use various methods to determine how much heat from the Sun (solar radiation) actually reaches the Earth's surface. Heat values for solar radiation are calculated in relation to a universal reference, the solar constant. The solar constant (1.37 kilowatts of energy per square meter) represents the density of radiation at Earth's mean distance from the Sun and at the point just before the Sun's heat (shortwave infrared waves) enters the Earth's atmosphere. Not all this heat actually reaches Earth's surface. The actual amount is determined by various factors, including latitudinal location, the degree of cloud cover, and the presence in the atmosphere of trace gases that can absorb radiation, such as argon, ozone, sulfur dioxide, and carbon dioxide.

At the same time, data must be gathered to calculate the amount of heat leaving the Earth's surface, mainly in the form of (longwave) infrared rays. Meteorologists attempt to calculate, first on a global scale and then for specific geographic locations, ecologically appropriate energy budgets.

Cyclones. Cyclones include a number of forms of severe weather, the most violent of which is known as a tornado. Cyclones are characterized by circular or turning patterns of air centered on a zone of low atmospheric pressure. The direction of cyclone rotation is counterclockwise in the Northern Hemisphere but clockwise in the Southern Hemisphere. Frontal cyclones, the most common type, usually develop in association with low pressure troughs that form along the polar front, a front that separates arctic and polar air masses from tropical air masses. Typical cyclogenesis occurs when moisture-laden air above the center of a relatively warm low pressure area begins to rotate under the influence of converging and or diverging winds at the surface or at higher levels. Simply stated, the dynamic forces operating within a cyclone can pull broad weather fronts toward them, causing increasingly strong winds and precipitation. Anticylones occur under opposite conditions, originating in areas of high pressure where air masses are pushed down from upper areas of the atmosphere.

APPLICATIONS AND PRODUCTS

Meteorology has both practical and professional, scientific applications. Everyday weather reports are probably the most common application of meteorology. Weather reports and forecasts are available through traditional sources such as newspapers, television, and radio broadcasts and can be obtained by calling the National Oceanic and Atmospheric Administration's National Weather Service, using a smartphone application, or checking one of the many Web sites devoted to weather. Real-time weather reports, including radar, and hourly, daily, and ten-day forecasts are available for the United States and other parts of the world.

Knowledge of existing and forecasted weather conditions is invaluable to companies that provide public transportation, such as airlines. Knowledge of weather conditions and forecasts is essential to ensuring the safety of passengers on airplanes, trains, buses, boats, and ferries. Motorists, whether traveling for pleasure or commuting, need to plan for weather conditions. People who participate in outdoor recreational activities or sports, such as hiking, biking, camping, hang gliding, fishing, and boating, depend on accurate forecasts to ensure that they are adequately prepared for the conditions and to avoid getting into dangerous situations. Those participating in outdoor activities and sports often purchase various meteorological instruments—such as lightning detectors, weather alert radios, and digital weather stations—designed to keep them informed and aware.

Weather equipment. Meteorology involves the measurement of temperature, air pressure, and humidity, and many companies produce equipment for this purpose, ranging from portable products for outdoor use, to products for home use, to products for industry use, to weather balloons and satellites. Some companies, such as Columbia Weather Systems in Oregon, which provides weather station systems and monitoring, focus on the professional level, and others, including Oregon Scientific, Davis Instruments, and La Crosse Technology, concentrate on the many hobbyists who enjoy monitoring the weather. Equipment ranges from simple mechanical rain gauges to digital temperature sensors to complete home weather stations. Some companies offer software that provides specific weather information in a timely manner to companies or individuals who need it.

Scientific applications. The World Data Center in Asheville, North Carolina, distributes meteorological data that it has gathered and processed. Its facilities are open on a limited basis to visiting research scientists sponsored by recognized parent organizations or international programs. A variety of meteorological data are readily available on an exchange basis to counterparts of the center in other countries.

One organization participating in the information exchange is the World Climate Programme, an international program headquartered in Switzerland devoted to understanding the climate system and using its knowledge to help countries that are dealing with climate change and variability.

IMPACT ON INDUSTRY

A wide range of industries and commercial concerns depend on meteorological data to carry out their operations. These include commercial airlines, the fishing industry, and agricultural businesses.

Aviation. Perhaps the most obvious industry that relies on the results of meteorology is the airline industry. No flight, whether private, commercial, or military, is undertaken without a clear idea of the predicted weather conditions from point of departure to point of arrival. Changing conditions and unexpected pockets of air turbulence are a constant concern of pilots and their crews, who are trained in methods of analyzing meteorological data while in flight.

Every day, the National Weather Service issues nearly 4,000 weather forecasts for aviation. Meteorologists at the Aviation Weather Center in Kansas City, Missouri; the Alaska Aviation Weather Center in Anchorage; and at twenty-one Federal Aviation Administration Air Route Traffic Control Centers across the nation use images from satellites as well as real-time information from observation units at major airports and from Doppler radar to create reliable forecasts.

Fishing industry. Rapidly changing and violent weather on the seas and oceans can result in the loss of fishing vessels, their crews, and their catch, or it can cause severe damage to boats and ships, all of which hurt the fishing industry. In addition, when boats get caught in dangerous storms, rescuers must attempt to reach survivors, often endangering themselves. Therefore, accurate, up-to-the-minute forecasts are essential for the fishing industry, which also uses weather reports to help determine the areas where the fish are most likely to be found.

Agriculture. In agriculture, short meteorological predictions are of less importance than longer-term predictions, particularly for agricultural sectors that depend on rainfall rather than irrigation to grow their crops and that rely on predictable windows without rainfall to harvest and dry them. If the possibility of a prolonged drought menaces crops, both large- and small-scale agriculturalists turn to meteorologists to learn if changing weather patterns may bring them relief. However, short-term predictions are helpful, in that they help farmers deal with problems that could create catastrophic losses, such as too much rain in a short period, which could create flooding; extreme heat or cold, which, for example, can ruin the citrus crop; or strong winds and hail, which can flatten corn.

CAREERS AND COURSE WORK

Those seeking a career in meteorology can consider a wide range of possibilities. The American Meteorological Association requires at least twenty semester credits in the sciences, which include geophysics, earth science, physics, and chemistry, as well as computer science and mathematics. Specialized courses beyond such basic science courses (such as atmospheric dynamics, physical meteorology, synoptic meteorology, and hydrology) represent

General Circulation of the Earth's Atmosphere: The westerlies and trade winds are part of the Earth's atmospheric circulation slightly over half the required courses. A number of governmental agencies and private commercial businesses employ full- or part-time meteorologists to provide technical information needed to carry out their operations.

Probably the most familiar job of a meteorologist is to prepare weather reports for broadcast on television or to be published in newspapers. The task of a weather forecaster for a local television station is more complex than it may seem judging from the briefness of the televised forecast. Local predictions are created from data gathered from diverse sources, often sources that gather information for an entire region or the whole country, including pulse-doppler radar operators, who receive special training to enter the profession.

Government agencies. Opportunities for employment in agencies that gather meteorological data are to be found within the National Oceanic and Atmospheric Administration (NOAA), part of the U.S. Department of Commerce. The NOAA contains a number of organizations that deal with meteorology, including the National Environmental Satellite, Data, and Information Service, a specialized satellite technology branch that provides global environmental data and assessments of the environment. Another NOAA organization is the National Weather Service, which provides forecasts, maps, and information on water and air quality. The Office of Oceanic and Atmospheric Research contains the Climate Program Office, which studies climate variability and predictability.

The National Centers for Environmental Prediction, part of the National Weather Service, oversees operations by several key specialized service organizations run by professional meteorologists. The most important of these are the Hydrometeorological Prediction Center in Washington, D.C., the Tropical Prediction Center (includes the National Hurricane Center in Miami, Florida), and the Storm Prediction Center in Norman, Oklahoma, which maintains a constant tornado alert system for vulnerable geographical regions. Other operations include the Ocean Prediction Center, the Aviation Weather Center, the Climate Prediction Center, Environmental Modeling Center, and the Space Weather Prediction Center.

Private commercial operations. A number of private companies are involved in the development and production of instruments used for meteorological

data gathering. The instruments range from simple devices to sophisticated electronic equipment, complete with software. The development of new weather satellites requires instruments that can be used by and can make the most of the gathered information.

SOCIAL CONTEXT AND FUTURE PROSPECTS
The worldwide need for accurate daily and

FASCINATING FACTS ABOUT METEOROLOGY

- The 53rd Weather Reconnaissance Squadron, known as the Hurricane Hunters of the Air Force Reserve, has been flying into tropical storms and hurricanes since 1944. The squadron's planes fly into the storms and send data directly to the National Hurricane Center by satellite.
- Geology departments at some colleges and universities, such as Ball State University in Muncie, Indiana, have led storm chasing groups, which train students in basic storm knowledge and lead them in pursuit of tornadoes and other violent storms.
- The popularity of storm chasing, featured in a Discovery channel television series, has led to the creation of storm chasing commercial tours. One group warns potential customers that while storm chasing, the group may not be able to stop for dinner.
- Without the continual pull of gravity toward the center of the Earth, the atmosphere would gradually disperse into space. Because air is made up of molecules in a gaseous rather than solid material state, it clings to the Earth.
- The chemical content of sedimentary layers on ocean floors and gases trapped in Arctic ice reveal data concerning the composition of the Earth's atmosphere in earlier geological times.
- The amount of oxygen contained in the air in higher mountain environments is markedly less than at sea level or mid-range altitudes. People coming to the mountains from lower elevations can experience serious shortness of breath and even altitude sickness, which is characterized by headaches, loss of appetite, fatigue, and nausea.
- At times, the atmosphere cannot keep up with the Earth it surrounds. This is particularly true at higher levels of the atmosphere, especially near the equator. At the same time, some parts of the atmosphere travel faster than others because of strong jet stream winds.

short-term meteorological forecasts, delivered in various formats, will result in continuing development of more accurate instruments and better predictive models, as well as improved and additional methods of packaging and delivering the information to users. Accurate prediction of where extreme weather, such as hurricanes and tornadoes, will occur and how intense the storms will be has the potential to save many lives and possibly minimize economic damage, and meteorologists will continue to conduct research in this area. Data gathered by meteorologists also are gaining importance in analyzing global issues such as air pollution and changes in the ozone layer. The ability to gather information by satellites allowed meteorology to make significant advances, and future research is likely to focus on increasing satellites' data-sensing ability and the speed and quality of data transmission as well as on software to interpret and analyze the information.

Satellite technology. Development of satellite technology continues to revolutionize the science of meteorology. Various types of polar-orbiting satellites have been devised since the early 1960's. The task of the satellite bus—the computer-equipped part of the satellite without sensitive recording instruments—is to transmit data gathered by an increasingly sophisticated variety of devices designed to collect vital data.

One such scanning device is the advanced very-high-resolution radiometer (AVHRR) used to measure heat radiation rising from localized areas on the Earth's surface. An AVHRR, very much like a telescope, projects a beam that is split by a set of mirrors, lenses, and filters that distribute the work of data recording to several different sensor devices, or channels. To obtain an accurate reading of radiation rising from a given target, the AVHRR must calibrate data received from these different sensors. AVHRR technology produces images that depict the horizontal structure of the atmosphere, and another radiation-recording instrument, the high-resolution infrared radiation sounder (HIRS), produces soundings based on the vertical structure of the atmosphere.

In zones of widespread cloud cover, the microwave sounding unit (MSU) may be used. The area scanned by an MSU is about a thousand times greater than that scanned by a device using infrared wavelengths. Other specialized sounding devices used by meteorological satellites include stratospheric sounding units and solar backscatter ultraviolet radiometers (SBUV), which measure patterns of reflection of solar radiation coming back from the Earth's surface.

The functions of SBUVs in particular became more and more critical as concern about changes in the composition of the ozone layer emerged in the last decades of the twentieth century. Ozone-depleting substances such as chlorofluorocarbons (used for air-conditioning systems and as a propellant for aerosol sprays) have had a negative effect on the protective ozone layer. Ozone depletion allows increased levels of ultraviolet radiation to reach the Earth, raising the incidence of skin cancer and damaging sensitive crops. Beyond the obvious utility of using satellites to predict global weather, these devices play a major role in helping meteorologists monitor the all-important radiation budget (the balance between incoming energy from the Sun and outgoing thermal and reflected energy from the Earth) that ultimately determines the effectiveness of the atmosphere in sustaining life on Earth.

Air pollution. Although the presence of pollutants in the air has been recognized as an undesirable phenomenon since the onset of the Industrial Revolution, by the end of the twentieth century, the question of air quality began to take on new and alarming dimensions. As countries industrialized, the combustion of fossil fuels to power industry and later automobiles released ever-increasing amounts of solar-radiation-absorbing greenhouse gases into the atmosphere. The most alarming effects stem from carbon dioxide, but meteorologists are also concerned about other serious gaseous pollutants and a wide variety of chemical particles suspended in gases (aerosols). Even if industries do not pollute by burning fossil fuels, many of them release sulfur dioxide, asbestos, and silica in quantities large enough to seriously damage air quality. Natural phenomena, including massive volcanic eruptions, also can produce atmospheric chemical imbalances that challenge analysis by meteorologists.

—Byron D. Cannon, MA, PhD

FURTHER READING
Budyko, M. I., A. B. Ronov, and A. L. Yanshin. *History of the Earth's Atmosphere.* Berlin: Springer-Verlag, 1987. Deals with various ways scientists can determine the probable composition of the atmosphere in earlier geological ages.

David, Laurie, and Gordon Cambria. *The Down-to-Earth Guide to Global Warming*. New York: Orchard Books, 2007. A nontechnical discussion of practical factors, many daily and seemingly inconsequential, that contribute to rising levels of carbon dioxide in the atmosphere.

Fry, Juliane L, et al. *The Encyclopedia of Weather and Climate Change: A Complete Visual Guide*. Berkeley: University of California Press, 2010. Covers all aspects of weather, including what it is and how it is monitored, as well as the history of climate change. Many photographs, diagrams, and illustrations.

Grenci, Lee M., Jon M. Nese, and David M. Babb. 5th ed. *A World of Weather: Fundamentals of Meteorology*. Dubuque, Iowa: Kendall/Hunt, 2010. Examines the study of meteorology and how it is monitored.

Hewitt, C.N., and Andrea Jackson, eds. *Handbook of Atmospheric Science*. Oxford, England: Blackwell, 2003. Examines research approaches to and data on the chemical content of the atmosphere.

Kidder, Stanley Q., and Thomas H. Vonder Haar. *Satellite Meteorology*. 1995. Reprint. San Diego: Academic Press, 2008. Covers ways in which satellite instruments can monitor meteorological phenomena at several levels of the atmosphere.

Williams, Jack, and the American Meteorological Society.
The AMS Weather Book: The Ultimate Guide to America's Weather. Chicago: University of Chicago Press, 2009. Examines common weather patterns in the United States and discusses the science of meteorology.

WEB SITES

Air Force Weather Observer
http://www.afweather.af.mil/index.asp

American Meteorological Society
http://www.ametsoc.org

International Association of Meteorology and Atmospheric Sciences
http://www.iamas.org

National Oceanic and Atmospheric Administration-National Weather Service
http://www.weather.gov

World Meteorological Organization
http://www.wmo.int/pages/index_en.html

See also: Aeronautics and Aviation; Air-Quality Monitoring; Atmospheric Sciences; Barometry; Climate Engineering; Climate Modeling; Climatology.

MINERALOGY

FIELDS OF STUDY

Geology; chemistry; geochemistry; physics; petrology; experimental petrology; environmental geology; forensic mineralogy; medical mineralogy; gemology; economic geology; geochronology; descriptive mineralogy; crystallography; crystal chemistry; mineral classification; geologic occurrence; optical mineralogy; mining; chemical engineering.

SUMMARY

Mineralogy is the study of the chemical composition and physical property of minerals, the arrangement of atoms in the minerals, and the use of the minerals. Minerals are naturally occurring elements or compounds. The composition and arrangement of the atoms that make up minerals is reflected in their physical characteristics. For example, gold is a naturally occurring mineral containing one element that has a definite density of 19.3 grams per milliliter and a yellow color and is chemically inactive. Sometimes mineral resources are broadened to refer to oil, natural gas, and coal, although those materials are not technically minerals.

KEY TERMS AND CONCEPTS

- **clay mineral:** Any of a group of tiny silicate minerals (less than 2 micrometers in size) that form by varied degrees of weathering of other silicate minerals.
- **cleavage:** Breaking of a crystallized mineral along a plane, leaving a smooth rather than an irregular

surface.

- **hardness:** Resistance of a mineral to being scratched by another material.
- **hydrothermal deposit:** Mineral deposit precipitated from a hot water or gas solution.
- **igneous rock:** Rock formed from molten rock material.
- **ion:** Atom or group of atoms with a positive or negative charge.
- **lava:** Molten rock material at the Earth's surface.
- **magma:** Molten rock material and suspended mineral crystals below the Earth's surface.
- **major elements:** Elements that make up the bulk of the chemical composition of a mineral or rock.
- **petrology:** Study of the origin, composition, structure, and properties of rocks and the processes that formed the rocks.
- **sedimentary rock:** Rock formed from particles such as sand by the weathering of other rocks at the surface or by precipitation from water.
- **silicate mineral:** Mineral that contains silicon bonded with oxygen to form silicate groups bonded with positive ions; silicates make up 90 percent of the Earth's crust.
- **trace element:** Element present only in tiny amounts (in quantities of parts per million or less) in a mineral or rock.

DEFINITION AND BASIC PRINCIPLES

Minerals are solid elements or compounds that have a definite but often not fixed chemical composition and a definite arrangement of atoms or ions. For instance, the mineral olivine is magnesium iron silicate, with the formula of $(Mg, Fe)_2 SiO_4$, meaning that it has oxygen (O) and silicon (Si) atoms in a ratio of 4:1 and a total of two ions of magnesium (Mg) and iron (Fe) in any ratio. The magnesium and iron component can vary from 100 percent iron with no magnesium to 100 percent magnesium with no iron, and all variations in between. Other minerals, however, such as gold, have nearly 100 percent gold atoms. Minerals are usually formed by inorganic means, but some organisms can form minerals such as calcite (calcium carbonate, with the formula $CaCO_3$).

Minerals make up most of the rocks in the earth, so they are studied in many fields. In geochemistry, for example, scientists determine the chemical composition of the minerals and rocks to derive hypotheses about how various rocks may have formed. Geochemists might study what rocks melt to form magmas or lavas of a certain composition. Environmental geologists attempt to solve problems regarding minerals that pollute the environment. For instance, environmental geologists might try to minimize the effects of the mineral pyrite (iron sulfide, with the formula FeS_2) in natural bodies of water. Pyrite, which is found in coal, dissolves in water to form sulphuric acid and high-iron water, which can kill some organisms. Forensic mineralogists may determine the origin of minerals left at a crime scene. Economic geologists discover and determine the distribution of minerals that can be mined, such as lead minerals (galena), salt, gypsum (for wall board), or granite (for kitchen countertops). Geophysicists may study the minerals below the Earth's surface that cannot be directly sampled. They may, for example, study how the seismic waves given off by earthquakes pass through the ground to estimate the kinds of rocks present.

BACKGROUND AND HISTORY

Archaeological evidence suggests that humans have used minerals in a number of ways for tens of thousands of years. For instance, the rich possessed jewels, red and black minerals were used in cave drawings in France, and minerals such as gold were used for barter. Metals were apparently extracted from ores for many years, but the methods of extraction were conveyed from person to person without being written down. In 1556, Georgius Agricola's *De re metallica* (English translation, 1912) described many of these mineral processing methods. From the late seventeenth century into the nineteenth century, many people studied the minerals that occur in definite crystal forms such as cubes, often measuring the angles between faces.

In the nineteenth century, the fields of chemistry and physics developed rapidly. Jöns Jacob Berzelius developed a chemical classification system for minerals. Another important development was the polarizing microscope, which was used to study the optical properties of minerals to aid in their identification.

During the late nineteenth century, scientists had theorized that the external crystal forms of minerals reflected the ordered internal arrangement of their atoms. In the early twentieth century, this theory was confirmed through the use of X rays. Also, it became possible to chemically analyze minerals and rocks so that chemical mineral classifications could be further developed. Finally, in the 1960's, the use of many instruments such as the electron microprobe allowed geologists to determine variations in chemical composition of minerals across small portions of the minerals so that models for the formation of minerals could be further developed.

HOW IT WORKS

Mineral and rock identification. A geologist may tentatively identify the minerals in a rock using characteristic such as crystal form, hardness, color, cleavage, luster (metallic or non-metallic), magnetic properties, and mineral association. The rock granite, for instance, is composed of quartz (often colorless, harder than other minerals, with rounded crystals, no cleavage, and nonmetallic luster) and feldspars (often tan, softer than quartz, with well-developed crystals, two good nearly right-angle cleavages, and nonmetallic luster), with lesser amounts of black minerals such as biotite (softer than quartz, with flat crystals, one excellent cleavage direction, and shiny nonmetallic luster).

The geologist then slices the rock into a section about 0.03 millimeters thick (most minerals are transparent). The section is examined under a polarizing microscope to confirm the presence of the tentatively identified minerals and perhaps to find other minerals that could not be detected by eye because they were too small or present in very low quantities. The geologist can determine other relationships among the minerals, such as the sequence of crystallization of the minerals within an igneous rock.

Identification using instruments. The minerals in a rock can be analyzed a variety of other ways, depending on the goals of a given study. For instance, X-ray diffraction may be used to identify some minerals. One of the most useful applications of X-ray diffraction is to identify tiny minerals such as clay minerals that are hard to identify using a microscope. The wavelengths of the X rays are similar to the spacing between atoms in the clay minerals, so when X rays of a single wavelength are passed through a mineral, they are diffracted from the minerals at angles that are characteristic of a particular mineral. Thus, the mixture of clay minerals in a rock or soil may be identified.

Mineral magnetite (lodestone), from Tortola, British Virgin Islands

The electron microprobe has enabled the analysis of tiny portions of minerals so that changes in composition across the mineral may be determined. The instrument accelerates masses of electrons into a mineral that releases X rays with energies that are characteristic of a given element so that the elements present can be identified. The amount of energy given off is proportional to the amount of the element in the sample; therefore, the concentration of the element in the mineral can be determined when the results are compared with a standard of known concentration. The electron microprobe can also be used to analyze other materials such as alloys and ceramics.

The scanning electron microscope uses an electron beam that is scanned over a small portion of tiny minerals and essentially takes a photograph of the mineral grains in the sample. Some electron microscopes are set up to determine qualitatively what elements are present in the sample. This information is often enough to identify the mineral.

Other analytical techniques. Many instruments and techniques are used to analyze the major elements, trace elements, and the isotopic composition of minerals and rocks. Commonly used methods are X-ray fluorescence (XRF), inductively coupled plasma mass spectrometer (ICP-MS), and thermal ionization mass spectrometry. X-ray fluorescence is used to analyze bulk samples of minerals, rocks, and ceramics for major elements and many trace elements. A powdered sample or a glass of the sample is compressed and is bombarded by X rays so that an energy spectrum that is distinctive for each element is emitted. The amount of radiation given off by the sample is compared with a standard of known concentration to determine how much of each element is present in the sample.

The inductively coupled plasma mass spectrometer is used to analyze many elements in concentrations as low as parts per trillion by passing vaporized samples into high-temperature plasma so that all elements have positive charges. A mass spectrometer sorts out the ions by their differing sizes and charges in a magnetic field, which permits a determination of the elemental concentrations when the results are compared with a standard of known concentration. Up to seventy-five elements can be rapidly analyzed in a sample at precisions of 2 percent or better.

Thermal ionization mass spectrometers can be used to analyze the isotopic ratios of higher mass elements such as rubidium, strontium, uranium, lead, samarium, and neodymium, which may be used to interpret the geologic age of a rock. The mineral or rock is placed on a heated filament so that the isotopes are ejected into a magnetic field in which the ions are deflected by varied amounts depending on the mass and charge of the isotope. The data may then be used to calculate the amount of a certain isotope in the sample and eventually the isotopic age of the sample.

Other specialized instruments are also available. For instance, gemologists use specialized instruments to study and cut gemstones.

APPLICATIONS AND PRODUCTS

Abundant metals and uses. The most abundant metals are iron, aluminum, magnesium, silicon, and titanium. Iron, which is mostly obtained from several minerals composed of iron and oxygen (hematite and magnetite), accounts for 95 percent by weight of the metals used in the United States. Much of the hematite and magnetite is obtained from large sedimentary rock deposits called banded-iron formations that are up to 700 meters thick and extend for up to thousands of square kilometers. The banded-iron formations formed 1.8 billion to 2.6 billion years ago. They are abundant, for instance, around the Lake Superior region in northern Minnesota, northern Wisconsin, and northwestern Michigan. The banded-iron formations have produced billions of tons of iron ore deposits. The ores are made into pellets and are mixed with limestone and coke to burn at 1,600 degrees Celsius in a blast furnace. The iron produced in this process is molten and can be mixed with small amounts of scarce metals (called ferroalloy metals) to produce steel with useful properties. For instance, the addition of chromium gives steel strength at high temperatures, and it prevents corrosion. The addition of niobium produces strength, and the addition of copper increases the resistance of the steel to corrosion.

Much of the aluminum occurs in clay minerals in which the aluminum cannot be economically removed from the other elements. Thus, most of the aluminum used comes from bauxite, which is a mixture of several aluminum-rich minerals such as gibbsite ($Al(OH)_3$) and boehmite ($AlOOH$). Bauxite forms in the tropics to subtropics by intense chemical weathering from other aluminum-rich minerals such as the clay minerals. The production of aluminum from bauxite is very expensive because the ore is dissolved in molten material at 950 degrees Celsius and the aluminum is concentrated by an electric current. Electricity represents about 20 percent of the total cost of producing aluminum. Recycling of aluminum is economical because the energy expended in making a new can from an old aluminum can is about 5 percent of the energy required to make a new aluminum can from bauxite.

Scarce metals and uses. At least thirty important trace metals occur in the Earth's crust at concentrations of less than 0.1 percent. Therefore, these elements are trace elements in most minerals and rocks until special geologic conditions concentrate them enough so that they become significant portions of some minerals. For instance, gold is present in the Earth's crust at only 0.0000002 percent by weight. Gold often occurs as gold uncombined with other elements and precipitated from hydrothermal solutions as veins. Gold may also combine with the element tellurium to form several kinds of gold-tellurium minerals. Also gold does not react very well chemically during weathering, so it may concentrate in certain streams to form placer deposits. Gold has been used for thousands of years in jewelry, dental work, and coins. In the past, much world exploration has been motivated by the drive to find gold.

Gold is a precious metal. Other precious metals are silver and the platinum group elements (for example, platinum, palladium, and rhodium). Silver, like gold, has been used for thousands of years in jewelry and coins. In modern times, silver also finds uses in batteries and in photographic film and papers because some silver compounds are sensitive to light. Silver occurs in nature as silver sulfide, and it substitutes for copper in copper minerals so that much of the silver produced is a side product from copper mining. The platinum metals occur together as native elements or combined with sulfur and tellurium. They concentrate in some dark-colored igneous rocks or as placers. The platinum group metals are useful as catalysts in chemical reactions, so they are, for example, used in catalytic convertors in vehicles.

Another group of scarce metals are base metals. The base metals are of relatively low economic value compared with the precious metals. The base metals include copper, lead, zinc, tin, mercury, and cadmium. Copper minerals are native copper and various copper minerals combined with sulfate, carbonate, and oxygen. Copper minerals occur in some igneous rocks, including some hydrothermal deposits, mostly as dilute copper-sulfur minerals. These minerals may be concentrated during weathering, often forming copper minerals combined with oxygen and carbonate. Copper conducts electricity very well, and much of it is used in electric appliances and wires. Lead and zinc tend to occur together in hydrothermal deposits in combination with sulfur. Lead is used in automobile batteries, ammunition, and solder. Lead is very harmful to organisms, which restricts its potential use. Zinc is used as a coating on steel to prevent it from rusting, in brass, and in paint.

Fertilizer and minerals for industry. Fertilizers contain minerals that have nitrogen, phosphorus, potassium, and a few other elements necessary for the growth of plants. In Peru, deposits of guano (seabird excrement), which is rich in these elements, have been mined and used as fertilizer. Saltpeter (potassium nitrate) has also been mined in Peru. A calcium phosphate mineral, apatite, occurs in some sedimentary rocks, so some of these have been mined for the phosphorus.

Some minerals are a source for sodium and chlorine. Halite, commonly known as rock salt, is used as a flavoring and to melt ice on roads. Halite is produced in some sedimentary rocks from the slow evaporation of water in closed basins over long periods of time, a process that is occurring in the Great Salt Lake in Utah.

A variety of rocks or sediment—including granite, limestone, marble, sands, and gravels—are used for the exteriors of buildings, construction materials, or for countertops. Sands and gravels may be mixed with cement to make concrete.

IMPACT ON INDUSTRY

The materials of the Earth interact, which means that all societies depend on mineral resources. For example, crops grow in soil containing clay minerals formed from the weathering of rocks. Fertilizers composed of nitrogen, phosphorus, and potassium minerals must be added to the soil to grow enough food to support the growing population of the world. Machinery such as tractors, trains, ships, and trucks made of steel and other materials obtained from processing metal ores must be used to fertilize fields and to move food to markets. Electricity, much of which is obtained by burning coal, must be used to process food, whether in factories or in homes. Most of these mineral resources cannot be renewed quickly, so many are being rapidly used up.

Industry and business. The largest industries directly related to mineralogy are public or private corporations engaged in the exploration, production, and development of minerals. As of 2009, the United States was the world's third largest producer of copper, after Chile and Peru, and of gold, after China and Australia. In 2009, net exports of mineral raw materials and old scrap materials was $10 billion. In 2009, Freeport-McMoRan Copper and Gold's Morenci open-pit mine in Arizona produced about 460,000 pounds of copper (about 40 percent of its total North American production), Barrick Gold Corporation's Goldstrike Complex in Nevada produced 1.36 million ounces of gold, and Teck's Red Dog Mine in Alaska produced 79 percent of the total U.S. production of zinc.

Government regulation. Although most governments do not directly control the mineral resources in their countries, they often regulate mining to ensure the workers' safety and support mining through subsidies and funding. In the United States, government agencies regulating mining include the Mine Safety and Health Administration of the U.S.

Department of Labor and the mining division of the National Institute for Occupational Safety and Health. The U.S. Geological Survey maintains statistics and information on the worldwide supply of and demand for minerals.

Academic research. Engineering departments in colleges and universities engage in mineralogy research such as improving the safety and efficiency of mining operations and searching for better, cleaner ways to process ores. For instance, some engineers

---

FASCINATING FACTS ABOUT MINERALOGY

- In 1822, Friedrich Mohs developed a hardness scale for minerals, using ten minerals. From hardest to softest, they are diamond, corundum, topaz, quartz, potassium feldspar, apatite, fluorite, calcite, gypsum, and talc.
- Diamonds are the hardest natural material on Earth and are the most transparent material known. Ultra-thin diamond scalpels retain their sharp edge for long periods and can transmit laser light to cauterize wounds.
- Blood diamonds, also known as conflict diamonds, are produced or traded by rebel forces opposed to the recognized government in order to fund their activities. Most of the blood diamonds come from Sierra Leone, Angola, the Democratic Republic of Congo, Liberia, and the Ivory Coast.
- The mineral kaolin is used to create a hard clay used in rubber tires for lawn equipment, garden hoses, rubber floor mats and tiles, automotive hoses and belts, shoe soles, and wire and cable.
- In China, ownership of Imperial jade, a fine-grained jade with a rich, uniform green color, was once limited to nobility. Jade is believed to have powers to ward off evil and is associated with longevity.
- The average automobile contains about 0.9 mile of copper wiring. The total amount of copper used in an automobile ranges from 44 to 99 pounds.
- The manufacture of jewelry accounts for 78 percent of the gold used each year. However, because gold conducts electricity and does not corrode, it is used as a conductor in sophisticated electronic devices such as cell phones, calculators, television sets, personal digital assistants, and Global Positioning Systems.

---

analyze ways to break up ores and separate the ore minerals or develop ways to use more of the metals in the ores.

CAREERS AND COURSE WORK

Anyone interested in pursuing a career in the minerals industry should study geology, chemistry, physics, biology, chemical engineering, and mining engineering. A bachelor's degree in one of these subjects is the minimum requirement for working in oil exploration or as a mining technician or water-quality technician. A master's degree or a Ph.D. is required for more challenging and responsible jobs in geochemistry, geophysics, environmental geology, geochronology, forensic mineralogy, and academics. Geochemists should have a background in geology and chemistry. Geophysicists use physics, mathematics, and geology to remotely tell what kinds of rocks are below the surface of the earth. Environmental geologists should have backgrounds in geology, chemistry, and biology. Geochronologists use natural radioactive isotopes to give estimates of the time when some kinds of rocks formed. They should have a background in geology, physics, and chemistry. Forensic mineralogists should have a good background in mineralogy and criminology. Economic geologists should major in geology, with a concentration in courses concerned with ore mineralization. Those interested in an academic career should have a Ph.D. in geology, mining engineering, or chemical engineering. They will be expected to teach and do research in their subject area.

SOCIAL CONTEXT AND FUTURE PROSPECTS

Mineral resources are not evenly distributed throughout the world. In the nineteenth century, the United States did not import many mineral resources, but it has increasingly become an importer of mineral resources. The need to import is driven by many factors, including an increase in the minerals used. U.S. production of final products, including cars and houses, using mineral materials accounted for about 13 percent of the 2009 gross domestic product. Other factors include the depletion of some U.S. sources of minerals and the increased use of minerals that are not naturally available in the United States. In 2009, foreign sources supplied more than 50 percent of thirty-eight mineral commodities consumed in the United States. Nineteen of those minerals were 100 percent foreign sourced. The United States must import much of the manganese, bauxite, platinum group minerals, tantalum, tungsten, cobalt, and petroleum that it uses. In contrast, the United States still has abundant resources of gold, copper, lead, iron, and salt. Japan, however, has few mineral resources, so it must import most of them.

This need to import and export mineral resources has forced countries to cooperate with one another to achieve their needs. The mineral resources traded in the largest quantities, in descending order, are iron-steel, gemstones, copper, coal, and aluminum. The total annual world trade in mineral resources is approaching $700 billion and is likely to rapidly increase as countries such as China begin to use and produce more mineral resources.

As the supply of minerals decreases worldwide and environmental and safety concerns raised by mining gain in importance, it is likely that mineralogy research may turn to improving extraction and processing methods and looking for ways to recycle materials, capture minerals remaining in wastes, and restore mining areas after the minerals have been depleted.

—*Robert L. Cullers, MS, PhD*

FURTHER READING

Craig, James R., David J. Vaughan, and Brian Skinner. *Earth Resources and the Environment.* Upper Saddle River, N.J.: Prentice Hall, 2010. Gives an overview of mineral resources in the Earth and describes environmental problems regarding fossil fuels, metals, fertilizers, and industrial minerals. Appendix, glossary, index, and many illustrations.

Guastoni, Alessandro, and Roberto Appiani. *Minerals.* Buffalo, N.Y.: Firefly Books, 2005. This guide describes the common minerals, gives their chemical formulas, and describes where they are found. Contains illustrations.

Kearny, Philip, ed. *The Encyclopedia of the Solid Earth Sciences.* London: Blackwell Science, 1994. Contains an alphabetical list of defined geologic terms, minerals, and geologic concepts.

Klein, Cornelis, and Barbara Dutrow. *Manual of Mineral Science.* Hoboken, N.J.: John Wiley & Sons, 2008. Describes how to identify minerals using physical properties and modern analytical techniques.

Pellant, Chris. *Smithsonian Handbook: Rocks and Minerals.* London: Dorling Kindersley Books, 2002. This identification guide for minerals and rocks uses many pictures combined with text.

Sinding-Larsen, Richard, and Friedrich-Wilhelm Wellmer, eds. *Non-renewable Resource Issues: Geoscientific and Societal Challenges.* London: Springer, 2010. Looks at minerals as nonrenewable resources and discusses the challenges that societies face.

WEB SITES

Geology and Earth Sciences
http://geology.com

Mineralogical Society of America
http://www.minsocam.org

National Institute for Occupational Safety and Health Mining
http://www.cdc.gov/niosh/mining   U.S. Department of Labor

Mine Safety and Health Administration
http://www.msha.gov

U.S. Geological Survey Minerals Information
http://minerals.usgs.gov/minerals

See also: Electronic Materials Production.

MIRRORS AND LENSES

FIELDS OF STUDY

Physics; astronomy; mathematics; geometry; chemistry; mechanical engineering; optical engineering; aerospace engineering; environmental engineering; medical engineering; computer science; electronics; photography.

SUMMARY

Mirrors and lenses are tools used to manipulate the direction of light and images, mirrors by using flat or curved glass in the reversion, diversion, or formation of images, and lenses by using polished material, usually glass, in the refraction of light. Scientific applications of mirrors and lenses cover a broad spectrum, ranging from photography, astronomy, and medicine to electronics, transportation, and energy conservation. Without mirrors and lenses, it would be impossible to preserve memories in a photograph, view faraway galaxies through a telescope, diagnose diseases through a microscope, or create energy using solar panels. Miniaturized mirrors are also implemented in scanners used in numerous electronic devices, such as copying machines, bar-code readers, compact disc players, and video recorders.

KEY TERMS AND CONCEPTS

- **achromatic lens:** Lens that refracts light without separating its components into the various colors of the spectrum.
- **aperture:** Opening through which light travels in a camera or optical device, determining the amount of light delivered to the lens, and affecting the quality of an image.
- **catadioptric imaging:** Imaging system that uses a combination of mirrors and lenses to create an optical system.
- **catoptric imaging:** Imaging system that uses only a combination of mirrors to create an optical system.
- **diffraction:** Spreading out of light by passing through an aperture or around an object that bends the light.
- **diopter:** Unit of measurement that determines the

optical power of a lens by calibrating the amount of light refracted by the lens, in correlation with its focal length.
- **Fresnel lens:** Flat lens that reduces spherical aberration and increases light intensity due to its numerous concentric circles, often creating a prism effect.
- **negative lens:** Concave lens with thick edges and a thinner center, used most often in eyeglasses for correcting nearsightedness.
- **positive lens:** Convex lens with thin edges and a thicker center, used most often in eyeglasses for correcting farsightedness.
- **refraction:** Bending of light waves, used by a lens to magnify, reduce, or focus images.
- **spherical aberration:** Distortion or blurriness of image caused by the geometrical formation of a spherical lens or mirror.

DEFINITION AND BASIC PRINCIPLES

Mirrors and lenses are instruments used to reflect or refract light. Mirrors use a smooth, polished surface to revert or direct an image. Lenses use a piece of smooth, transparent material, usually glass or plastic, to converge or diverge an image. Because light beams that strike dark or mottled surfaces are absorbed, mirrors must be highly polished in order to reflect light effectively. Likewise, since rough surfaces diffuse light rays in many different directions, mirror surfaces must be exceedingly smooth to reflect light in one direction. Light that hits a mirrored surface at a particular angle will always bounce off that mirrored surface at an exactly corresponding angle, thereby allowing mirrors to be used to direct images in an extremely precise manner.

Images are reflected according to one of three main types of mirrors: plane, convex, and concave. Plane mirrors are flat surfaces and reflect a full-size upright image directly back to the viewer but left and right are inverted. Convex mirrors are curved slightly outward and reflect back a slightly smaller upright image, but in a wider angled view, and left and right are inverted. Concave mirrors are curved slightly inward and reflect back an image that may be larger or smaller, depending on the distance from the object, and the image may be right-side up or upside down.

BACKGROUND AND HISTORY

The ancient Egyptians used mirrors to reflect light into the dark tombs of the pyramids so that workmen could see. The Assyrians used the first known lens, the Nimrud lens, which was made of rock crystal, 3,000 years ago in their work, either as a magnifying glass or as a fire starter. Chinese artisans in the Han dynasty created concave sacred mirrors designed for igniting sacrificial fires, and Incan warriors wore a bracelet on their wrists containing a small mirror for the focusing of light to start fires.

It was not until the end of the thirteenth century, however, when an unknown Italian invented spectacles, that lenses subsequently became used in eyeglasses. Near the end of the sixteenth century, two spectacle makers inadvertently discovered that certain spectacle lenses placed inside a tube made objects that were nearby appear greatly enlarged. Dutch spectacle maker Zacharias Janssen and his son, Hans Janssen, invented the microscope in 1590. In 1608, Zacharias Janssen, in collaboration with Dutch spectacle makers Hans Lippershey and Jacob Metius, invented the first refractory telescope using lenses. However, it was Galileo Galilei who, in 1609, took the first rudimentary refracting telescope invented by the Dutch, drastically improved on its design, and went on to revolutionize history by using his new telescopic invention to observe that the Earth and other planets revolve around the Sun. In 1668, Sir Isaac Newton invented the first reflecting telescope, which, using mirrors, conveyed a vastly superior image than the refractory telescope, since it greatly reduced distortion and spherical aberrations often conveyed by the refractory telescope's lenses.

Almost two hundred years later, in 1839, with the invention of photography by Louis Daguerre, the convex lens became used for the first time in cameras. In 1893, Thomas Edison patented his motion-picture camera, the kinetoscope, which used lenses, and by the 1960's, mirrors were being used in satellites and to reflect lasers.

HOW IT WORKS

The law of reflection states that the angle of incidence is equal to the angle of reflection, meaning that if light strikes a smooth mirrored surface at a 45-degree angle, it will also bounce off the mirror at a corresponding 45-degree angle. Whenever light hits an object, it may be reflected, or it may be absorbed, which

is what happens when light hits a dark surface. Light may also hit a rough shiny surface, in which case the light will be reflected in many different directions, or diffused. Light may also simply pass through an object altogether if the material is transparent.

When light travels from one medium to another, such as air to water, the speed of the light slows down and refracts, or bends. Lenses take advantage of the fact that light bends when changing mediums by manipulating the light to serve a variety of purposes. The angle that light is refracted by a lens depends on the lens's shape, specifically its curvature. A glass or plastic lens can be polished or ground so that it gathers light toward its edges and directs it toward the center of the lens, in which case the lens is concave, and the light rays will be diverged. Conversely, a lens which is convex, because it is thicker in the middle and has thinner edges, will cause light rays to converge.

Both concave and convex lenses rely on the light rays' focal point either to magnify, reduce, or focus an image. The focal point is simply the precise point where light rays come together in a pinpoint and focus an image. The distance from the center of the lens to the exact point where light rays focus is called the lens's focal length. Light that passes through a convex lens is refracted so that the rays join and focus out in front of the lens, creating a convergence. Light rays passing through a concave lens are refracted outward and create a divergence because the focal point for the light rays appears to be originating from behind the lens. Aberrations such as blurriness or color distortion may sometimes occur if lenses are made using glass with impurities or air bubbles or if the lens is not ground or polished properly to make it precisely curved and smooth.

APPLICATIONS AND PRODUCTS

Electronics. Lenses and mirrors are used extensively in consumer electronics, primarily as part of systems to read and write optical media such as CDs and DVDs. These optical disks encode information in microscopic grooves, which are read by a laser in much the same manner as a turntable's needle reads a record. Good-quality lenses are essential to focus the laser beam onto the disk and to capture the reflection from the disk's surface.

Cameras also make use of both mirrors and lenses. In single-lens reflex (SLR) cameras, the film or image sensor is protected from light by a flip mirror.

Telescopes

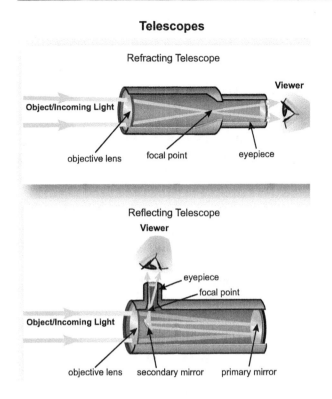

When a photograph is taken, the flip mirror rotates, directing light onto the film or digital sensor to expose the image to a size that will fit onto the film. Modern camera lenses are actually composed of multiple simple lenses, which helps to improve the overall image quality.

Science. Astronomy has long been the driving force behind advances in optics, and many varieties of lenses and mirrors were developed specifically to enable astronomical observation. Large modern telescopes are limited not by the optical quality of their lenses and mirrors but by distortion caused by turbulence in the atmosphere. To address this, state-of-the-art telescopes use adaptive optical systems, which can change the shape of the telescope's primary mirror in response to fluctuations in the atmosphere, which are measured with a powerful laser.

Lens and mirror systems are also indispensable as tools in a larger research apparatus. Microscopes, high-speed cameras, and other digital equipment are commonplace in biological, chemical, and physical science laboratories, and many experimental setups include custom-made imaging equipment all based on combinations of mirrors and lenses.

Retail. Some department stores have installed full-size digital touch-screen mirrors in their stores, revolutionizing the retail experience. The mirrors enable consumers to model clothes, makeup, and jewelry digitally, without actually having to try on the products. Because an extraordinary amount of time and energy is saved by the interactive retail mirrors, customers have more flexibility to shop creatively by sampling a broader range of merchandise. The interactive digital mirrors also offer those shopping for friends or relatives the added convenience of being able to gauge sizes correctly, greatly reducing the number of returns. The increased ease and efficiency of touch-screen digital mirrors has generated more customers and sales for both retailers.

IMPACT ON INDUSTRY

Military. There are numerous military applications for lens- and mirror-based systems, and these have had an enormous influence on industry. Relatively simple applications include binoculars, range-finding systems, and weapon scopes. More complex optics are employed in guidance systems for aircraft and missiles, which often pair telescopic systems with image-recognition software. Reconnaissance satellites are also an important use for mirror and lens systems, and the general concept of a spy satellite is similar to a large space-based telescope, with multiple lenses and mirrors serving to magnify an image before projection onto an imaging sensor. Cutting-edge corporations such as Raytheon, Elbit Systems of America, and Christie were recently awarded multimillion-dollar defense contracts for military-based research involving lens and mirror applications in radar, lasers, and projectors.

Medicine. Lens technology has, of course, revolutionized the practice of optometry, and contact lenses in particular have become a multibillion-dollar industry worldwide. Because eyeglasses are essentially a single-lens optical that corrects for abnormalities in the shape of the eye, the basic geometry of the eyeglass lens has not changed for many years. However, major progress has been made in lens coatings and materials, which improve the aesthetics, durability, weight, and cost of eyeglasses. Additionally, modern manufacturing techniques have allowed greater lens diversity with more complex designs, creating progressive lenses that have a continuously variable amount of focus correction across the lens wearer's field of vision (as opposed to traditional bifocals, which provide only two levels of correction). Bausch + Lomb is at the forefront of engineering progressive lenses and adjustable-focus lenses, which allow the lens wearer to manually adjust the focus correction depending on the distance of the object being viewed.

Government and university research. Undoubtedly, the greatest ongoing government research projects involving lenses and mirrors are undertaken by the National Aeronautics and Space Administration (NASA). The Hubble Space Telescope, launched by NASA in 1990, marks the culmination of decades of scientific research and the apex of mirror engineering and design. NASA's impact on the mirror and lens industry at large is paramount, since the majority of lens and mirror innovations have been a direct result of lens and mirror engineering for telescopes largely developed by NASA scientists. Although the Hubble Space Telescope contains the most advanced and sophisticated mirrors ever created, Hubble's successor, the James Webb Space Telescope, promises to surpass even Hubble by being able to see farther into space by monitoring infrared radiation.

Composite Mirror Applications, in Tucson, Arizona, has worked closely with NASA to build mirrors used in space, most notably building a mirror used by space shuttle Endeavor on its final mission to search for dark matter and anti-matter in the universe. Steward Observatory at the University of Arizona has also worked in close conjunction with Composite Mirror Applications in the development of special optical projects with applications for the telescopic industry.

CAREERS AND COURSE WORK

Courses in physics, astronomy, advanced mathematics, geometry, optical engineering, mechanical engineering, aerospace engineering, computer science, and chemistry provide background knowledge and training needed to pursue a future working with various applications utilizing mirrors and lenses. A bachelor's degree in any of the above fields would assist in the basic applications of mirrors and lenses, but careers involving research or development of lenses in the corporate domain or academia would almost certainly necessitate a graduate degree in one of the above disciplines—ideally a doctorate. Although employment at any university as an educator and researcher is one potential career path,

---

FASCINATING FACTS ABOUT MIRRORS AND LENSES

- In the sixteenth century, mirror makers were held prisoner on the Venetian island of Murano, and they were sentenced to death if they left the island, for fear their secret mirror-making techniques would be revealed to the outside world.

- In 2010, French artist Michel de Broin created the world's largest disco mirror ball. Measuring almost twenty-five feet across and containing more than 1,000 mirrors, the mirror ball was suspended above Paris using a skyscraper crane.

- Other than humans and a handful of other higher primates, such as chimpanzees, orangutans, and bonobos, scientists studying self-awareness have discovered only four species that are able to recognize their own reflection in a mirror: bottlenose dolphins, killer whales, Asian elephants, and European magpies.

- In 1911, at the inaugural race of the Indianapolis 500, winner Ray Harroun used the first known automobile rearview mirror in his race car, the Marmon Wasp.

- The mirrors in the Hubble Space Telescope were ground to within 1/800,000 of an inch of a perfect curve and were the most highly polished mirrors on Earth.

- Roman emperor Nero, who was nearsighted, watched gladiator games at the Roman Colosseum through a large handheld emerald lens centuries before the invention of spectacles.

---

with an emphasis in physics and astronomy, working at an observatory is also a career possibility. Having an additional concentration of course work in aerospace engineering is also outstanding preparation for employment with NASA and government space projects.

Additional career opportunities working with lenses abound in the medical field, especially as an optical engineer. Either a master's or doctorate in physics or optics is a prerequisite for a career as an optical engineer; however, an associate's degree is all that is necessary to become trained to work as a contact lens technician. To work as an optometrist, one must graduate from a college of optometry after completing a bachelor's degree, and to work as an ophthalmologist, one must obtain a medical degree.

Additional career opportunities in the "green" industry using mirrors in solar panels requires a background in environmental engineering, and work in the photographic industry requires extensive knowledge of photography as well as lenses.

SOCIAL CONTEXT AND FUTURE PROSPECTS

Although the increased necessity of using mirrors in green technology (such as solar panels) has become universally recognized, mirrors may conceivably play a key role in another environmental technology in the future. Scientists researching the long-term impact of global warming, increasingly alarmed by the rapid escalation of carbon dioxide buildup in the Earth's atmosphere, have begun seriously investigating the potential for mirrors to help lower the temperature of the Earth's atmosphere. Scientists are studying the possibility of launching a series of satellites to orbit above the Earth, each one containing large mirrors, which could be controlled, like giant window blinds, to reflect back out into space as much of the Sun's rays as desired, thereby regulating the temperature of the Earth's atmosphere like a thermostat.

Another futuristic trend foresees the transformation of mirrors in the home, programming bathroom mirrors, for example, to monitor an individual's health by analyzing a person's physical appearance. Vital signs, such as heart rate, blood pressure, and weight can be measured by the mirror and sent to a computer. Other mirrors may project a person's possible future appearance if certain negative lifestyle habits, such as smoking, overeating, excessive drinking, and lack of exercise, are practiced. Cameras placed throughout the home may record an individual's personal habits and then relay the information to an interactive home mirror linked to a computer. The mirror, in turn, may then portray possible physical results of a negative lifestyle to someone, traits such as obesity, discolored teeth, wrinkled skin, or receding hairline, turning mirrors into modern-day oracles.

—*Mary E. Markland, MA*

FURTHER READING

Andersen, Geoff. *The Telescope: Its History, Technology, and Future.* Princeton, N.J.: Princeton University Press, 2007. A comprehensive survey of telescopes since their invention more than four hundred years ago, plus provides step-by-step instructions describing how to build a homemade telescope and observatory.

Burnett, D. Graham. *Descartes and the Hyperbolic Quest: Lens Making Machines and Their Significance in the Seventeenth Century.* Philadelphia: American Philosophical Society, 2005. Fascinating account of Descartes's lifelong dream of building a machine to manufacture an aspheric hyperbolic lens to end spherical aberration in lenses. Contains extensive drawings of Descartes's designs.

Conant, Robert Alan. *Micromachined Mirrors.* Norwell, Mass.: Kluwer Academic, 2003. Using profuse illustrations, examines how the miniaturization of mirrors has transformed modern technology, especially their use in scanners and in the electronics industry.

Kingslake, Rudolph. *A History of the Photographic Lens.* San Diego: Academic Press, 1989. Documents the development of photographic lenses from the beginning of photography in 1839 to modern telescopic lenses, providing ample pictures and sketches illustrating each lens type.

Pendergrast, Mark. *Mirror Mirror: A History of the Human Love Affair with Reflection.* New York: Basic Books, 2003. An exhaustive examination of mirrors and their influence on science, astronomy, history, religion, art, literature, psychology, and advertising, with an abundance of biographical material, supported by photographs, about scientists and mirror innovators.

Zimmerman, Robert. The Universe in a Mirror: The Saga of the Hubble Space Telescope and the Visionaries Who Built It. Princeton, N.J.: Princeton University Press, 2008. The behind-the-scenes story of the building of the Hubble Space Telescope, from its first conception in the 1940's, to its launch, repair, and triumph in the 1990's. Contains spectacular photographs of outer space sent back to Earth by the Hubble.

WEB SITES

American Astronomic Society
http://aas.org

The International Society for Optics and Photonics
http://spie.org

National Aeronautics and Space Administration The James Webb Space Telescope
http://www.jwst.nasa.gov

See also: Computer Science; Mechanical Engineering; Optics; Photography.

N

NANOTECHNOLOGY

FIELDS OF STUDY

Agriculture; artificial intelligence; bioinformatics; biomedical nanotechnology; business; chemical engineering; chemistry; computational nanotechnology; electronics; engineering; environmental studies; mathematics; mechanical engineering; medicine; molecular biology; microelectromechanical systems; molecular scale manufacturing; nanobiotechnology; nanoelectronics; nanofabrication; molecular nanoscience; molecular nanotechnology; nanomedicines; pharmacy; physics; toxicology.

KEY TERMS AND CONCEPTS

- **bottom-up nanofabrication:** The creation of nanoparticles that will combine to create nanostructures that meet the requirements of the application.
- **buckyballs (Fullerenes):** Molecules made up of carbon atoms that make a soccer ball shape of hexagons and pentagons on the surface. Named for the creator of the geodesic dome, Buckminster Fuller.
- **carbon nanotubes:** Molecules made up of carbon atoms arranged in hexagonal patterns on the surface of cylinders.
- **mechanosynthesis:** Placing atoms or molecules in specific locations to build structures with covalent bonds.
- **molecular electronics:** Electronics that depend upon or use the molecular organization of space.
- **Moore's law:** Transistor density on integrated circuits doubles about every two years. American chemist Gordon Moore attributed this effect to shrinking chip size.
- **nano:** Prefix in the metric system that means "10^{-9}" or "1 billionth." The word comes from the Greek *nanos*, meaning "dwarf." The symbol is *n*.
- **nanoelectronics:** Electronic devices that include components that are less than 100 nanometers (nm) in size.
- **nanoionics:** Materials and devices that depend upon ion transport and chemical changes at the nanoscale.
- **nanolithography:** Printing nanoscale patterns on a surface.
- **nanometer:** 10^{-9} or 1 billionth of a meter. The symbol is *nm*.
- **nanorobotics:** Theoretical construction and use of nano-scaled robots.
- **nanosensors:** Sensors that use nanoscale materials to detect biological or chemical molecules.
- **scanning probe microscopy:** The use of a fine probe to scan a surface and create an image at the nanoscale.
- **scanning tunneling microscopy (STM):** Device that creates images of molecules and atoms on conductive surfaces.
- **self-assembly:** Related to bottom-up nanofabrication. A technique that causes functionalized nanoparticles to chemically bond with or repel other atoms or molecules to assemble in specified patterns without external manipulation.
- **top-down nanofabrication:** Manufacturing technique that removes portions of larger materials to create nano-sized materials.

SUMMARY

Nanotechnology is dedicated to the study and manipulation of structures at the extremely small nano level. The technology focuses on how particles of a substance at a nanoscale behave differently than particles at a larger scale. Nanotechnology explores how those differences can benefit applications in a variety of fields. In medicine, nanomaterials can be used to deliver drugs to targeted areas of the body needing treatment. Environmental scientists can use nanoparticles to target and eliminate pollutants in the water and air. Microprocessors and consumer products also benefit from increased use of nanotechnology, as components and associated products become exponentially smaller.

DEFINITION AND BASIC PRINCIPLES

Nanotechnology is the science that deals with the study and manipulation of structures at the nano level. At the nano level, things are measured in nanometers (nm), or one billionth of a meter (10^{-9}). Nanoparticles can be produced using various techniques known as top-down nanofabrication, which starts with a larger quantity of material and removes portions to create the nanoscale material. Another method is bottom-up nanofabrication, in which individual atoms or molecules are assembled to create nanoparticles. One area of research involves developing bottom-up self-assembly techniques that would allow nanoparticles to create themselves when the necessary materials are placed in contact with one another.

Nanotechnology is based on the discovery that materials behave differently at the nanoscale, less than 100 nm in size, than they do at slightly larger scales. For instance, gold is classified as an inert material because it neither corrodes nor tarnishes; however, at the nano level, gold will oxidize in carbon monoxide. It will also appear as colors other than the yellow for which it is known.

Nanotechnology is not simply about working with materials such as gold at the nanoscale. It also involves taking advantage of the differences at this scale to create markers and other new structures that are of use in a wide variety of medical and other applications.

BACKGROUND AND HISTORY

In 1931, German scientists Ernst Ruska and Max Knoll built the first transmission electron microscope (TEM). Capable of magnifying objects by a factor of up to one million, the TEM made it possible to see things at the molecular level. The TEM was used to study the proteins that make up the human body. It was also used to study metals. The TEM made it possible to view particles smaller than 200 nm by focusing a beam of electrons to pass through an object, rather than focusing light on an object, as is the case with traditional microscopes.

In 1959 the noted American theoretical physicist Richard Feynman brought nanoscale possibilities to the forefront with his talk "There's Plenty of Room at the Bottom," presented at the California Institute of Technology in 1959. In this talk, he asked the audience to consider what would happen if they could arrange individual atoms, and he included a discussion of the scaling issues that would arise. It is generally agreed that Feynman's reputation and influence brought increased attention to the possible uses of structures at the atomic level.

In the 1970s scientists worked with nanoscale materials to create technology for space colonies. In 1974 Tokyo Science University professor Norio Taniguchi coined the term "nano-technology." As he defined it, nanotechnology would be a manufacturing process for materials built by atoms or molecules.

In the 1980s the invention of the scanning tunneling microscope (STM) led to the discovery of fullerenes, or hollow carbon molecules, in 1986. The carbon nanotube was discovered a few years later. In 1986, K. Eric Drexler's seminal work on nanotechnology, *Engines of Creation*, was published. In this work, Drexler used the term "nanotechnology" to describe a process that is now understood to be molecular nanotechnology. Drexler's book explores the positive and negative consequences of being able to manipulate the structure of matter. Included in his book are ruminations on a time when all the works in the Library of Congress would fit on a sugar cube and when nanoscale robots and scrubbers could clear capillaries or whisk pollutants from the air. Debate continues as to whether Drexler's vision of a world with such nanotechnology is even attainable.

In 2000 the US National Nanotechnology Initiative was founded. Its mandate is to coordinate federal nanotechnology research and development. Great growth in the creation of improved products using nanoparticles has taken place since that time. The creation of smaller and smaller components—which reduces all aspects of manufacture, from the amount of materials needed to the cost of shipping the finished product—is driving the use of nanoscale materials in the manufacturing sector. Furthermore, the ability to target delivery of treatments to areas of the body needing those treatments is spurring research in the medical field.

The true promise of nanotechnology is not yet known, but this multidisciplinary science is widely viewed as one that will alter the landscape of fields from manufacturing to medicine.

HOW IT WORKS

Basic tools. Nanoscale materials can be created for specific purposes, but there exists also natural nanoscale material, like smoke from fire. To create nanoscale material and to be able to work with it requires specialized tools and technology. One essential piece of equipment is an electron microscope. Electron microscopy makes use of electrons, rather than light, to view objects. Because these microscopes have to get the electrons moving, and because they need several thousand volts of electricity, they are often quite large.

One type of electron microscope, the scanning electron microscope (SEM), requires a metallic sample. If the sample is not metallic, it is coated with gold. The SEM can give an accurate image with good resolution at sizes as small as a few nanometers.

For smaller objects or closer viewing, a TEM is more appropriate. With a TEM, the electrons pass through the object. To accomplish this, the sample has to be very thin, and preparing the sample is time consuming. The TEM also has greater power needs than the SEM, so SEM is used in most cases, and the TEM is reserved for times when a resolution of a few tenths of a nanometer is absolutely necessary.

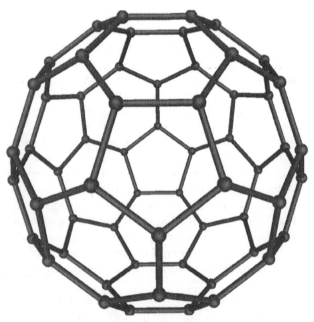

Buckminsterfullerene C60, also known as the buckyball, is a representative member of the carbon structures known as fullerenes. Members of the fullerene family are a major subject of research falling under the nanotechnology umbrella.

The atomic force microscope (AFM) is a third type of electron microscope. Designed to give a clear image of the surface of a sample, this microscope uses a laser to scan across the surface. The result is an image that shows the surface of the object, making visible the object's "peaks and valleys."

Moving the actual atoms around is an important part of creating nanoscale materials for specific purposes. Another type of electron microscope, the scanning tunneling microscope (STM), images the surface of a material in the same way as the AFM. The tip of the probe, which is typically made up of a single atom, can also be used to pass an electrical current to the sample, which lessens the space between the probe and the sample. As the probe moves across the sample, the atoms nearest the charged atom move with it. In this way, individual atoms can be moved to a desired location in a process known as quantum mechanical tunneling.

Molecular assemblers and nanorobots are two other potential tools. The assemblers would use specialized tips to form bonds with materials that would make specific types of materials easier to move. Nanorobots might someday move through a person's blood stream or through the atmosphere, equipped with nanoscale processors and other materials that enable them to perform specific functions.

Bottom-up nanofabrication. Bottom-up nanofabrication is one approach to nanomanufacturing. This process builds a specific nanostructure or material by combining components of atomic and molecular scale. Creating a structure this way is time consuming, so scientists are working to create nanoscale materials that will spontaneously join to assemble a desired structure without physical manipulation.

Top-down nanofabrication. Top-down nanofabrication is a process in which a larger amount of material is used at the start. The desired nanomaterial is created by removing, or carving away, the material that is not needed. This is less time consuming than bottom-up nanofabrication, but it produces considerable waste.

Specialized processes. To facilitate the manufacture of nanoscale materials, a number of specialized processes are used. These include nanoimprint lithography, in which nanoscale features are stamped or printed onto a surface; atomic layer epitaxy, in which a layer that is only one atom thick is deposited on a surface; and dip-pen lithography, in which the tip of an atomic force microscope writes on a surface after being dipped into a chemical.

APPLICATIONS AND PRODUCTS

Smart materials. Smart materials are materials that react in ways appropriate to the stimulus or situation they encounter. Combining smart materials with nanoscale materials would, for example, enable scientists to create drugs that would respond when encountering specific viruses or diseases. They could also be used to signal problems with other systems, such as nuclear power generators or pollution levels.

Sensors. The difference between a smart material and a sensor is that the smart material will generate a response to the situation encountered, while the sensor will generate an alarm or signal that there is something that requires attention. The capacity to incorporate sensors at a nanoscale greatly enhances the ability of engineers and manufacturers to create structures and products with a feedback loop that is not cumbersome. Nanoscale materials can easily be incorporated into the product.

Medical uses. The potential uses of nanoscale materials in the field of medicine are of particular interest to researchers. Theoretically, nanorobots could be programmed to perform functions that would eliminate the possibility of infection at a wound site. They could also speed healing. Smart materials could be designed to dispense medication in appropriate doses when a virus or bacteria is encountered. Sensors could be used to alert physicians to the first stages of malignancy. There is great potential for nanomaterials to meet the needs of aging populations without intrusive surgeries requiring lengthy recovery and rehabilitation.

Energy. Nanomaterials also hold promise for energy applications. With nanostructures, components of heating and cooling systems could be tailored to control temperatures with greater efficiency. This could be accomplished by engineering the materials so that some types of atoms, such as oxygen, can pass through, while others, such as mold or moisture, cannot. With this level of control, living conditions could be designed to meet the specific needs of different categories of residents.

Extending the life of batteries and prolonging their charge has been the subject of decades of research. With nanoparticles, researchers at Rutgers University and Bell Labs have been able to better separate the chemical components of batteries, resulting in longer battery life. With further nanoscale research, it may be possible to alter the internal composition of batteries to achieve even greater performance.

Light-emitting diode (LED) technology uses 90 percent less energy than conventional, non-LED lighting. It also generates less heat than traditional metal-filament light bulbs. Nanomanufacture would make it possible to create a new generation of efficient LED lighting products.

Electronics. Moore's law states that transistor density on integrated circuits doubles about every two years. With the advent of nanotechnology, the rate of miniaturization has the potential to double at a much greater rate. This miniaturization will profoundly affect the computer industry. Computers will become lighter and smaller as nanoparticles are used to increase everything from screen resolution to battery life while reducing the size of essential internal components, such as capacitors.

SOCIAL CONTEXT AND FUTURE PROSPECTS

Whether nanotechnology will ultimately be good or bad for the human race remains to be seen, as it continues to be incorporated into more and more products and processes, both common and highly specialized. There is tremendous potential associated with the ability to manipulate individual atoms and

---

FASCINATING FACTS ABOUT NANOTECHNOLOGY

- Approximately 80,000 nanos equal the width of one strand of human hair.
- Gold is considered an inert material because it does not corrode or tarnish. At the nano level, this is not the case, as gold will oxidize in carbon monoxide.
- The transmission electron microscope (TEM) allowed the first look at nanoparticles in 1931. The TEM was built by German scientists Ernst Ruska and Max Knoll.
- German physicist Gerd Binning and Swiss physicist Heinrich Rohrer, colleagues at IBM Zurich Research Laboratory, invented the scanning tunneling microscope in 1981. The two scientists were awarded the Nobel Prize in Physics in 1986 for their invention.
- Nanoparticles can be enclosed in an outer covering or coat.

molecules, to deliver medications to a disease site, and to build products such as cars that are lighter yet stronger than ever. Much research is devoted to using nanotechnology to improve fields such as pollution mitigation, energy efficiency, and cell and tissue engineering. However, there also exists the persistent worry that humans will lose control of this technology and face what Drexler called a "gray goo" scenario, in which self-replicating nanorobots run out of control and ultimately destroy the world.

Despite fears linked to cutting-edge technology, many experts, including nanotechnology pioneers, consider such doomsday scenarios involving robots to be highly unlikely or even impossible outside of science fiction. More worrisome, many argue, is the potential for nanotechnology to have other unintended negative consequences, including health impacts and ethical challenges. Some studies have shown that the extremely small nature of nanoparticles makes them susceptible to being breathed in or ingested by humans and other animals, potentially causing significant damage. Structures including carbon nanotubes of graphene have been linked to cancer. Furthermore, the range of possible applications for nanotechnology raises various ethical questions about how, when, and by whom such technology can and should be used, including issues of economic inequality and notions of "playing God." These risks, and the potential for other unknown negative impacts, have led to calls for careful regulation and oversight of nanotechnology, as there has been with nuclear technology, genetic engineering, and other powerful technologies.

—*Gina Hagler, MBA*

FURTHER READING

Berlatsky, Noah. *Nanotechnology*. Greenhaven, 2014.

Binns, Chris. *Introduction to Nanoscience and Nanotechnology*. Hoboken: Wiley, 2010. Print.

Biswas, Abhijit, et al. "Advances in Top-Down and Bottom-Up Surface Nanofabrication: Techniques, Applications & Future Prospects." *Advances in Colloid and Interface Science* 170.1–2 (2012): 2–27. Print.

Demetzos, Costas. *Pharmaceutical Nanotechnology: Fundamentals and Practical Applications*. Adis, 2016.

Drexler, K. Eric. *Engines of Creation: The Coming Era of Nanotechnology*. New York: Anchor, 1986. Print.

Drexler, K. Eric. *Radical Abundance: How a Revolution in Nanotechnology Will Change Civilization*. New York: PublicAffairs, 2013. Print.

Khudyakov, Yury E., and Paul Pumpens. *Viral Nanotechnology*. CRC Press, 2016.

Ramsden, Jeremy. *Nanotechnology*. Elsevier, 2016.

Ratner, Daniel, and Mark A. Ratner. *Nanotechnology and Homeland Security: New Weapons for New Wars*. Upper Saddle River: Prentice, 2004. Print.

Ratner, Mark A., and Daniel Ratner. *Nanotechnology: A Gentle Introduction to the Next Big Idea*. Upper Saddle River: Prentice, 2003. Print.

Rogers, Ben, Jesse Adams, and Sumita Pennathur. *Nanotechnology: The Whole Story*. Boca Raton: CRC, 2013. Print.

Rogers, Ben, Sumita Pennathur, and Jesse Adams. *Nanotechnology: Understanding Small Systems*. 2nd ed. Boca Raton: CRC, 2011. Print.

Stine, Keith J. *Carbohydrate Nanotechnology*. Wiley, 2016.

NUCLEAR TECHNOLOGY

FIELDS OF STUDY

Energy; chemistry; physics; medicine; environmental science; government; international politics.

KEY TERMS AND CONCEPTS

- **alpha decay:** Decay of matter emitting an alpha particle composed of two protons and two neutrons, a property of unstable radioactive elements.

- **atom:** Microscopic building block of ordinary matter, composed of negatively charged electrons, positively charged protons, and neutrons, which have no charge.

- **beta decay:** Radioactive decay of matter emitting a beta particle when a proton is converted into a neutron in the atom's nucleus and released as an electron.

- **fission:** Energy-generating process by which the nucleus of a heavy atom is split into two or more lighter atoms by the absorption of a neutron, and in which a dense number of heavy atoms will generate an atom-splitting chain reaction.

- **fusion:** Process by which two light atoms are combined at extraordinarily high temperatures into a single, heavy nucleus, resulting in the release of much greater amounts of energy than produced by nuclear fission.
- **gamma decay:** Release of gamma rays; electromagnetic radiation in the form of energy, not matter.
- **laser:** Device producing light in the form of electromagnetic radiation.
- **plutonium:** Heavy, manmade radioactive element useful in the production of both nuclear energy and nuclear weapons.
- **radioactivity:** Spontaneous emission of radiation from an atom's nucleus.
- **radium:** Highly radioactive element found in uranium ores.
- **uranium:** Radioactive metallic substance, and one of the most abundant elements found in nature, used to produce the fissile isotope uranium-235 (U-235), the principal fuel of nuclear reactors.
- **x ray:** High-frequency electromagnetic radiation emitted when atomic electrons drop to lower energy levels.

SUMMARY

Nuclear technology focuses on the particles composing the atom to produce reactions and radioactive materials that have practical use in such areas as agriculture, industry, medicine, and consumer products, as well as in the generation of electrical power and the construction of nuclear weapons.

DEFINITION AND BASIC PRINCIPLES

One of most significant developments of the twentieth century, nuclear technology focuses on practical applications of the atomic nuclei reactions that result in the Earth from the decay of uranium and from the artificial stimulation of uranium particles, the atomic nuclei of which are normally separate from one another because they contain positive electrical charges that cause them to repel one another.

The scientific principles employed in nuclear technology grow out of the initial research on radium conducted during the early twentieth century by Henri Becquerel and Marie and Pierre Curie and

their daughter, Irène Joliet-Curie. This research involved the alpha, beta, and gamma ray activity of radium. The subsequent research of their successors focused on manipulating the relationship between the proton, neutron, and electron properties of the atom of radioactive elements to produce chain reactions and radioactive isotopes of value in numerous fields. Most of the scientific principles involving the peaceful use of nuclear technology relate to the process of fission, an energy- and neutron-releasing process in which the nucleus of an atom is split into two relatively equal parts. The production of nuclear weapons also builds on these principles but in addition to fission, it involves the fusion of several small nuclei into a single one, the mass of which is less than the sum of the small nuclei used in its creation.

Since the use of atomic weapons in World War II, a second set of principles relating to the technology has also emerged—one designed to govern its application in a manner beneficial to humanity. The most important of these applied principles pertain to the beneficial and responsible use of nuclear technologies in a manner mindful of human and environmental safety, the technology's continued improvement in terms of efficiency and safety, and the securing of research information and nuclear material from acquisition by those who might use them for destructive purposes.

BACKGROUND AND HISTORY

Most accounts of the nuclear age's roots begin with the 1896 work by the French physicist Henri Becquerel, who is credited with discovering radioactivity while exploring uranium's phosphorescence. The research center in the field remained in France for decades thereafter, most notably in the groundbreaking work of Pierre Curie, and that of his wife, Marie, and their daughter Irène Joliot-Curie following Pierre's death in 1906. It is Pierre Curie who is credited with coining the term "radioactivity," and it was his and his wife's work on the properties of decaying uranium that provided the foundation for the research on nuclear fission and nuclear fusion, which eventually led to the creation of the atomic bomb, nuclear electricity-generating power plants, and the radioisotopes so abundantly useful in medicine, industry, and daily life.

By the 1940's, research conducted a decade earlier by British physicist James Chadwick, Italian physicist

Enrico Fermi, German chemist Otto Hahn, and others on the unstable property of the atom had progressed sufficiently for scientists to envision the development of nuclear weapons that could, through an induced chain reaction, release enormous amounts of destructive energy. A wartime race to produce the atomic bomb ensued between Germany and the eventual winners, the German, British, and American scientists who collaborated in the United States' Manhattan Project under the direction of American physicist Robert Oppenheimer. Using reactors constructed in Hanford, Washington, to create weapons-grade uranium, U-238 and U-235, the project's scientists tested the first atomic bomb on July 16, 1945. Shortly thereafter, detonation of atomic bombs on the Japanese cities of Hiroshima (August 6) and Nagasaki (August 9) led to Japan surrendering unconditionally, ending World War II. The resultant peace, however, was short lived. By 1947, the wartime alliance between the United States and the Soviet Union had dissolved into an intense competition for global influence. When the Soviet Union exploded its first atomic weapon in 1949, it began a nuclear arms race between the two countries that both threatened the safety of the world and kept it in check out of the mutual recognition by the two superpowers that given such weaponry, no nuclear war could be "won" in any meaningful sense.

Meanwhile, research in the field of nuclear technology began to focus on the use of nuclear energy to generate electricity. This captured the attention of world leaders concerned with their ability to meet the anticipated postwar demand for electricity in their growing cities. That feat was first accomplished on a test basis near Arco, Idaho, on December 20, 1951. The first nuclear power station went online in the Soviet Union on June 27, 1954, and two years later the first commercial nuclear power station opened in Sellafield, England. The United States joined the world, commercially producing nuclear power in December, 1957, with the opening of the Shippingport Atomic Power Station in Pennsylvania. By then, the United Nations (UN) had already convened in 1955 a conference to explore the peaceful use of nuclear technology, six Western European countries were banding together to form the European Atomic Energy Community (Euratom), a supranational body committed to the cooperative development of nuclear power in Europe. The UN

had created the International Atomic Energy Agency (IAEA) to encourage the peaceful use of nuclear technology on an even wider basis.

In retrospect, it appears that the early zeal in opening nuclear energy plants inadvertently paved the way for the technology's declining appeal by the end of the century. The early emphasis was on constructing and making operational a growing number of power plants, with a resultant neglect of safety issues in the choice of reactor design, the construction of the plants, and the training of the technicians charged with operating them. The United States Atomic Energy Commission (AEC), for example, was charged with both promoting and regulating the commercial use of nuclear energy. It was a dual mandate in which the first charge invariably got the better of the second, most notably demonstrated in 1979's Three Mile Island accident in Middletown, Pennsylvania, when the rush to get the plant online before all tests were performed resulted in a construction-flaw-induced accident compounded by human error that did much to dampen the appeal of nuclear power in the United States. Seven years later, human error merged with the flawed design of the Soviet nuclear power station in Chernobyl to undercut the appeal of nuclear power throughout most of Western Europe.

The military utility of the atom demonstrated during World War II also encouraged its continued pursuit in the military field in the United States, where nuclear reactors were harnessed to propel fleets of nuclear submarines and aircraft carriers that remain a mainstay of the U.S. Navy. Nonetheless, the real diffusion of nuclear technology has occurred in the civilian field, and as a result of scientific and political events separated by nearly a generation. In 1934, Irène Joliot-Curie and her husband, Frédéric Joliot-Curie, discovered that radium-like elements could be created by bombarding materials with neutrons ("induced radioactivity"), a discovery that eventually led to the inexpensive production of radioisotopes.

After World War II ended, though, there was a concerted postwar effort by governments to control tightly all research pertaining to nuclear technology.

It was not until 1953, when President Dwight Eisenhower proposed a broad sharing of information in his "Atoms for Peace" speech at the UN. That was a significant declassification of information after which the fruits of research in the field of nuclear

Delivering Nuclear Energy

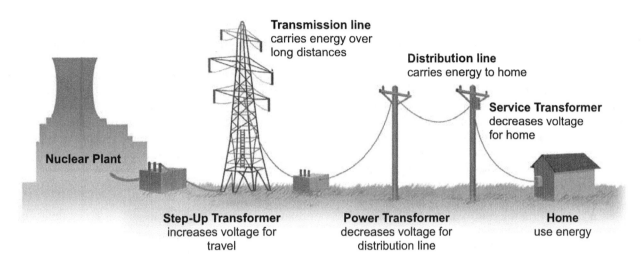

Transmission line carries energy over long distances

Distribution line carries energy to home

Service Transformer decreases voltage for home

Nuclear Plant

Step-Up Transformer increases voltage for travel

Power Transformer decreases voltage for distribution line

Home use energy

technology became widely available and began to be utilized in myriad areas.

How It Works

The means by which nuclear technology is applied in the various arenas surveyed below varies from sector to sector. In general, however, nuclear technology produces its benefits either by altering the activity and/or weight of nuclear particles, or by exposing nonnuclear matter to radiation.

The nuclear power plants that generate electrical power, for example, like those that burn fossil fuel, function by heating water into steam in order to turn the turbines that produce electricity. The fuel consists of uranium oxide (commonly known as yellowcake) processed into solid ceramic pellets and packaged into long vertical tubes that are inserted into reactors to produce a controlled fissile chain reaction. Either pressure or cold water is utilized to control reactor heat and the intensity of the reaction.

By contrast, atomic weapons rely on generating a uranium chain reaction of an intentionally uncontrolled nature for maximum destructive effect, with the principle of fusion (the compressing of the atom into a smaller particle in order to produce energy) being exclusively utilized in the production of the more powerful thermonuclear bombs.

Elsewhere, the industrial use of radioisotopes rests on the fact that radiation loses energy as it passes through substances. Manufacturers have consequently been able to develop gauges to measure the thickness and density of products and, using radioisotopes as imaging devices, to check finished products for flaws and other sources of weakness. For their part, the fossil fuel industries involved both in mining and oil and gas exploration are using radioactive waves that measure density to search for resource deposits beneath the soil and sea. The medical community, the agriculture industry, and the producers of consumer goods that use nuclear technology largely rely on radioisotopes—more specifically, on exposing selected "targets" to radioisotope-containing chemical elements that can either be injected into a patient's body to "photograph" how an organ is functioning or employed to destroy undesirable or harmful elements.

Applications and Products

Although for some the mention of nuclear technology is most likely to conjure up threatening images of mushroom clouds or out-of-control nuclear power plants, nuclear technology has become a daily part of the lives of citizens in much of the developed world.

Nuclear power industry. The nuclear-based power industry that has emerged in the United States,

United Kingdom, France, and more than twenty other countries around the globe encompasses more than 440 nuclear power plants and produces nearly one-fifth of the world's electrical output. Moreover, despite a slowdown in the construction of new plants, the slightly more than 100 U.S. nuclear power stations still generate more electricity than any fuel source and have assisted with the electrical needs of a steadily growing population.

Space exploration. Space exploration has also substantially profited from nuclear technology—in particular, the development of the radioisotope thermoelectric generators (RTGs) that have used plutonium-generated heat to produce electrical power for unmanned spaced travel ever since the launch of Voyager I in 1977.

Medicine. Apart from the growing area of laser optical surgery, the existing applications of nuclear technology in medicine principally involve the use of positron emission tomography (PET) scans, other forms of imagining, and X rays in diagnostics, and radiation in the treatment of cancer.

Industry. The centerpiece of nuclear technology in the industrial field revolves around the diagnostic use of lasers and radioisotopes in order to improve the quality of goods, including the quality of the steel used in the automotive industry and the detection of flaws in jet engines.

Agriculture and the pharmaceutical industry. Nuclear technology is also used in these sectors to test the quality of products. The U.S. Food and Drug Administration (FDA), for example, requires testing of all new drugs, and 80 percent of that testing employs radioisotopes. But radiation is also widely used to treat products, especially in agriculture, where an irradiation process exposes food to gamma rays from a radioisotope of cobalt 60 to eliminate potentially harmful or disease-causing elements. Even livestock products are covered. Like its counterparts in at least ten other countries, the FDA approves the use of irradiation for pork, poultry, and red meat as well as for fruits, vegetables, and spices in order to kill bacteria, insects, and parasites that can lead to such diseases as salmonella and cholera.

Mining, and oil and gas exploration. The process of searching for valuable natural resources has been radically altered in the last generation by the introduction of radiation wave-based exploratory techniques. Nuclear technology is also important to

resource recovery and transportation in these industries. Lateral drilling, for example, relies on radiation wave directives to tap into small oil deposits, and construction and pipeline crews routinely use radiation waves to test the durability of welds and the density of road surfaces.

Consumer products. Virtually every American home contains several consumer products using nuclear technology, from nonstick pans treated with radiation to prolong the life span of their surfaces, to photocopiers and computer disks that use small amounts of radiation to eliminate static, to cosmetics, bandages, contact-lens solutions and hygiene products sterilized with radiation to remove allergens.

CAREERS AND COURSE WORK

Given the breadth of the applications of nuclear technology, career options lie in almost every field, but almost all require a college degree involving specific technical training, especially in fields such as nuclear chemistry, nuclear physics, nuclear engineering, and nuclear medicine. For the more specialized areas, a career in either nuclear-technology research or in development and application tends to require one or more advanced degrees. Given the complex, cutting-edge nature of such work, graduate training is often important for even nuclear-technician positions. Jobs nonetheless remain plentiful in most sectors for those who have that training, both in government (maintaining and operating the Navy's nuclear fleet) and in the private sector. The power industry routinely advertises its need for design engineers, process control engineers, technical consultants, civil, mechanical, and electrical nuclear engineers, and nuclear work planners; the medical community constantly seeks radiologists and other personnel trained in nuclear technology; and nuclear technicians remain in high demand in consumer product manufacturing, mining, and agriculture.

Public administration careers should not be ignored. National and local government entities such as the U.S. Department of Energy, the Nuclear Regulatory Commission, oversight agencies at the state level, and their counterparts in other countries are also career outlets for those combining business administration or public administration training with a knowledge of nuclear technologies.

FASCINATING FACTS ABOUT NUCLEAR TECHNOLOGY

- J. Robert Oppenheimer, who headed the Manhattan Project that produced the atomic bomb, was later denied security clearance when he lobbied for international control of atomic energy and openly opposed the further development of nuclear weapons and the dangers he saw lining that path.

- It is widely believed that it was the Balance of Terror–the likelihood that a hot war between the United States and the Soviet Union would result in absolute mutually assured destruction (the MAD system)–that kept the peace between the two superpowers throughout the Cold War.

- Caught up in the optimism of the 1950's atomic age, in 1958 the Ford Motor Company unveiled the design of its future Nucleon line of atomic cars, which were to be powered by small reactors in their trunks.

- Outdoor malls across the United States often feature clocks with dials lit by radiation emanating from the low level of nuclear material that they contain, but not so low a level that Homeland Security does not fret every time one disappears, lest its radiation be used to create a dirty bomb.

- By the twenty-first century the products of nuclear technology had become so pervasive that most Western homes unknowingly contained them in such common, everyday devices as smoke detectors and DVD players.

SOCIAL CONTEXT AND FUTURE PROSPECTS

The application of nuclear technology is basically progressing on three pathways, the first of which is of serious global concern.

Students of international affairs have long been concerned with the problem of "runaway" nuclear proliferation: the acquisition of nuclear weapons by so many states that others will also feel the need to acquire them, trebling the number of nuclear-armed states in a short time and making an accidental or intentional nuclear war more likely. The presence of stateless terrorist organizations who are willing to engage in extremist activity involving high kill numbers has significantly elevated this concern. Until the 2000's, the pace of proliferation was incremental and those who acquired the weaponry were sometimes loath to publicize the fact. The acquisition of atomic weapons by Pakistan and North Korea, and the apparent pursuit of nuclear arms by Iran, have been highly publicized and may have pushed the world to that "runaway" point in which countries with fast breeder reactors—whose recycled nuclear fuel can be brought to weapons-grade quality—may be encouraged to develop nuclear weapons themselves.

The second track holds considerably more potential for good: a renewed interest in nuclear power to meet the world's growing electrical needs. As a source of electrification, nuclear power fell largely out of fashion during the late twentieth century in much of the world as a result of: the cost of building and maintaining nuclear power plants compared with the cheap cost of imported energy between 1984 and 2003; the antinuclear movement and the public's concern about the construction of nuclear power plants in their backyards; and the appeal of environmentally friendly, renewable green energy sources during the era of rising oil prices that followed the US-led invasion of Iraq in 2003.

That noted, the prospect for employment in the field of nuclear power remains good for three reasons. First, research and development activity has resulted in the application of techniques that have prolonged the life span of existing nuclear power plants well beyond their intended use cycle. Trained personnel are needed at all levels to continue that research and safely operate those nuclear power plants. Second, the green technologies being pursued are unlikely to be able to power the giant electrical grids that are increasingly being demanded by the megacities that are emerging, especially in third world countries. Such technologies will not be able to meet the existing global demand for ever more electrical power, which is increasing at about 1 percent per year in the United States, and at a far higher rate in developing areas. Large-scale nuclear power plants can meet those needs, while also adhering to the environmental standards to which northern and southern hemisphere states have committed themselves.

Finally, there is the broad, umbrella area of civilian societal applications, where a continuing, high demand for nuclear technology in the field of medicine, virtually every area of industry, agriculture, and consumer products can be predicted with a far greater assurance than the future demand for nuclear power as a source of electrification. In fact, so assured is the presumption of a steadily growing demand for nuclear-based products in medicine alone that it is driving much of the interest in constructing new reactor fa-cilities just to produce the materials used in radiation-based therapies.

—Joseph R. Rudolph, Jr., PhD

FURTHER READING

Angelo, Joseph A., Jr. *Nuclear Technology*. Westport: Greenwood, 2004. Print.

Morris, Robert C. *The Environmental Case for Nuclear Power: Economic, Medical and Political Considerations*. St. Paul: Paragon House, 2000. Print.

Shackett, Peter. *Nuclear Medicine Technology: Procedures and Quick Reference*. 2nd ed. Philadelphia: Lippincott, 2008. Print.

Stanculescu, A., ed. *The Role of Nuclear Power and Nuclear Propulsion in the Peaceful Exploration of Space*. Vienna: Intl. Atomic Energy Agency, 2005. Print.

United States Congress, House Committee on Foreign Affairs, Subcommittee on Terrorism, Nonproliferation, and Trade. *Isolating Proliferators, and Sponsors of Terror: The Use of Sanctions and the International Financial System to Change Regime Behavior*. Washington, DC: Government Printing Office, 2007. Print.

Yang, Chi-Jen. *Belief-Based Energy Technology Development in the United States: A Comparative Study of Nuclear Power and Synthetic Fuel Policies*. Amherst: Cambria, 2009. Print.

NUMERICAL ANALYSIS

FIELDS OF STUDY

Chemistry; computer science; mathematics; engineering; physics.

KEY TERMS AND CONCEPTS

- **algorithm:** Finite set of steps that provide a definite solution to a mathematical problem.
- **derivative:** Instantaneous rate of change with respect to time of a dependent variable.
- **differential equation:** Equation including a derivative; if the equation has at least one partial derivate, it is called a partial differential equation.
- **eigenvalue:** Solution of the linear equation $T(x) = x$ for a linear operator T.
- **error analysis:** Attaching a value to the difference between the actual solution of a system and a numerical approximation to the solution.
- **finite difference:** Approximating a derivative, such as y', by a difference, such as y/x.
- **finite difference method:** Replacing derivatives in an equation by their finite difference approximation and solving for one of the resulting variables.
- **partial derivative:** Instantaneous rate of change of a dependent variable of at least two independent variables with respect to one of the independent variables.
- **solution:** Value, or set of values that, when substituted into an equation, or system of equations, satisfies the equation or system.

SUMMARY

Numerical analysis is the study of how to design, implement, and optimize algorithms that provide approximate values to variables in mathematical expressions. Numerical analysis has two broad subareas: first, finding roots of equations, solving systems of linear equations, and finding eigenvalues; and second, finding solutions to ordinary and partial differential equations (PDEs). Much of this field involves using numerical methods (such as finite differences) to solve sets of differential equations. Examples are Brownian motion of polymers in solution, the kinetics of phase transition, the prediction of material microstructures, and the development of novel methods for simulating earthquake mechanics.

DEFINITION AND BASIC PRINCIPLES

Most of the phenomena of science have discrete or continuous models that use a set of mathematical equations to represent the phenomena. Some of the equations have exact solutions as a number or set of numbers, but many do not. Numerical analysis provides algorithms that, when run a finite number of times, produce a number or set of numbers that approximate the actual solution of the equation or set of equations. For example, since ϖ is transcendental, it has no finite decimal representation. Using English mathematician Brook Taylor's series for the arctangent, however, one can easily find an approximation of ϖ to any number of digits. One can also do an error analysis of this approximation by looking at the tail of the series to see how closely the approximation came to the exact solution.

Finding roots of polynomial equations of a single variable is an important part of numerical analysis, as is solving systems of linear equations using Gaussian elimination (named for German mathematician Carl Friedrich Gauss) and finding eigenvalues of matrices using triangulation techniques. Numeric solution of ordinary differential equations (using simple finite difference methods such as Swiss mathematician Leonhard Euler's formula, or more complex methods such as German mathematicians C. Runge and J. W. Kutta's Runge-Kutta algorithm) and partial differential equations (using finite element or grid methods) are the most active areas in numerical analysis.

BACKGROUND AND HISTORY

Numerical analysis existed as a discipline long before the development of computers. By 1800, Lagrange polynomials, named for Italian-born mathematician Joseph-Louis Lagrange, were being used for general approximation, and by 1900 the Gaussian technique for solving systems of equations was in common use. Ordinary differential equations with boundary conditions were being solved using Gauss's method in 1810, using English mathematician John Couch Adams's difference methods in 1890, and using the Runge-Kutta algorithm in 1900. Analytic solutions of partial differential equations (PDEs) were being developed by 1850, finite difference solutions by 1930, and finite element solutions by 1956.

The classic numerical analysis textbook *Introduction to Numerical Analysis* (1956), written by American mathematician Francis Begnaud Hildebrand, had substantial sections on numeric linear algebra and ordinary differential equations, but the algorithms were computed with desktop calculators. In these early days, much time was spent finding multiple representations of a problem in order to get a representation that worked best with desktop calculators. For example, a great deal of effort was spent on deriving Lagrange polynomials to be used for approximating curves. The early computers, such as the Electronic Numerical Integrator And Computer (ENIAC), built by John Mauchly and John Presper Eckert in 1946, were immediately applied to the existing numerical analysis algorithms and stimulated the development

of many new algorithms as well.

Modern computer-based numerical analysis really got started with John von Neumann and Herman Goldstine's 1947 paper "Numerical Inverting of Matrices of High Order," which appeared in the *Bulletin of the American Mathematical Society*. Following that, many new and improved techniques were developed for numerical analysis, including cubic-spline approximation, sparse-matrix packages, and the finite element method for elliptic PDEs with boundary condition.

HOW IT WORKS

Numerical analysis has many fundamental techniques. Below are some of the best known and most useful.

Approximation and error. It is believed that the earliest examples of numerical analysis were developed by the Greeks and others as methods of finding numerical quantities that were approximations of values of variables in simple equations. For example, the length of the hypotenuse of a 45-degree right triangle with side of length 1 is $\sqrt{2}$, which has no exact decimal representation but has rational approximations to any degree of accuracy. The first three elements of the standard sequence of approximations of $\sqrt{2}$ are 1.4, 1.414, and 1.4142. Another fundamental idea is error (a bound of the absolute difference between a value and its approximation); for example, the error in the approximation of $\sqrt{2}$ by 1.4242, to ten digits, is 0.000013563.

There are many examples of approximation in numerical analysis. The Newton-Raphson method, named for English physicist Sir Isaac Newton and English mathematician Joseph Raphson, is an iterative method used to find a real root of a differentiable function and is included in most desktop math packages. A function can also be approximated by a combination of simpler functions, such as representing a periodic function as a Fourier series (named for French mathematician Jean-Baptiste Joseph Fourier) of trigonometric functions, representing a piecewise smooth function as a sum of cubic splines (a special polynomial of degree 3), and representing any function as a Laguerre series (a sum of special polynomials of various degrees, named for French mathematician Edmond Laguerre).

Solution of systems of linear equations. Systems of linear equations were studied shortly after the

Babylonian clay tablet YBC 7289 (c. 1800–1600 BC) with annotations. The approximation of the square root of 2 is four sexagesimal figures, which is about six decimal figures. $1 + 24/60 + 51/60^2 + 10/60^3 = 1.41421296...$

introduction of variables; examples existed in early Babylonia. Solving a system of linear equations involves determining whether a solution exists and then using either a direct or an iterative method to find the solution. The earliest iterative solutions of linear system were developed by Gauss, and newer iterative algorithms are still published. Many of the problems of science are expressed as systems of linear equations, such as balancing chemical equations, and are solved when the linear system is solved.

Finite differences. Many of the equations used in numerical analysis contain ordinary or partial derivatives. One of the most important techniques used in numerical analysis is to replace the derivatives in an equation, or system of equations, with equivalent finite differences of the same order, and then develop an iterative formula from the equation. For example, in the case of a first-order differential equation with an initial value condition, such as $f(x) = F[x, f(x)]$, $y_0 = f(x_0)$, one can replace the derivative by using equivalent differences, $f(x_1) = [f(x_1) - f(x_0)]/(y_1 - x_0)$ and solve the resulting equation for y_1. After getting y_1, one can use the same technique to find approximations for yi given xi for i greater than 2. There are many examples of using finite differences in solving differential equations. Some use forward differences, such as the example above; others use backward

differences, much like antiderivatives; and still others use higher-order differences.

Grid methods. Grid methods provide a popular technique for solving a partial differential equation with n independent variables that satisfies a set of boundary conditions. A grid vector is a vector of the independent variables of a partial differential equation, often formed by adding increments to a boundary point. For example, from (t, x) one could create a grid by adding t to t and x to x systematically. The basic assumption of the grid method is that the partial differential equation is solved when a solution has been found at one or more of the grid points. While grid methods can be very complex, most of them follow a fairly simple pattern. First, one or more initial vectors are generated using the boundary information. For example, if the boundary is a rectangle, one might choose a corner point of the rectangle as an initial value; otherwise one might have to interpolate from the boundary information to select an initial point. Once the initial vectors are selected, one can develop recursive formulas (using techniques like Taylor expansions or finite differences) and from these generate recursive equations over the grid vectors. Adding in the information for the initial values yields sufficient information to solve the recursive equations and thus yields a solution to the partial differential equation. For example, given a partial differential equation and a rectangle the lower-left corner point of which satisfies the boundary condition, one often generates a rectangular grid and set of recursive equations that, when solved, yield a solution of the partial differential equation. Many grid methods support existence and uniqueness proofs for the PDE, as well as error analysis at each grid vector.

APPLICATIONS AND PRODUCTS

The applications of numerical analysis, including the development of new algorithms and packages within the field itself, are numerous. Some broad categories are listed below.

Packages to support numerical analysis. One of the main applications of numerical analysis is developing computer software packages implementing sets of algorithms. The best known and most widely distributed package is LINPACK, a software library for performing numeric linear algebra developed at

---

FASCINATING FACTS ABOUT NUMERICAL ANALYSIS

- In 1923, New Zealand-born astronomer Leslie Comrie taught the first numerical analysis course at Swarthmore College in Pennsylvania.
- In 1970, the first IMSL was released; its subroutines were called from FORTRAN programs.
- American electrical engineer Seymour Cray used an approximation to division, rather than the actual floating point division, for the CDC 6600 to get a faster arithmetic logic unit (ALU).
- The simplex method for optimization of well-formed problems was introduced by American mathematician George Dantzig and demonstrates the value of numerical analysis in business.
- The finite element method was introduced in 1942 by German mathematician Richard Courant.

Stanford University by students of American mathematician George Forsythe. Originally developed in FORTRAN for supercomputers, it is now available in many languages and can be run on large and small computers alike, although it has largely been superseded by LAPACK (Linear Algebra Package). Another software success story is the development of microcomputer numerical analysis packages. In 1970, a few numerical analysis algorithms existed for muMath (the first math package for microcomputers); by the twenty-first century, almost all widely used math packages incorporated many numerical analysis algorithms.

Astronomy, biology, chemistry, geology, and physics. Those in the natural sciences often express phenomena as variables in systems of equations, whether differential equations or PDEs. Sometimes a symbolic solution is all that is necessary, but numeric answers are also sought. Astronomers use numeric integration to estimate the volume of Saturn a million years ago; entomologists can use numeric integration to predict the size of the fire ant population in Texas twenty years from the present. In physics, the solutions of differential equations associated with dynamic light scattering can be used to determine the size of polymers in a solution. In geology, some of the Earth's characteristics, such as fault lines, can

be used as variables in models of the Earth, and scientists have predicted when earthquakes will occur by solving these differential equations.

Medicine. Many of the phenomena of medicine are represented by ordinary or partial differential equations. Some typical applications of numerical analysis to medicine include estimating blood flow for stents of different sizes (using fluid flow equations), doing a study across many physicians of diaphragmatic hernias (using statistical packages), and determining the optimal artificial limb for a patient (solving some differential equations of dynamics).

Engineering. Engineers apply the natural sciences to real-world problems and often make heavy use of numerical analysis. For example, civil and mechanical engineering have structural finite element simulations and are among the biggest users of computer time at universities. In industry, numerical analysis is used to design aerodynamic car doors and high-efficiency air-conditioner compressors. Electrical engineers, at universities and in industry, always build a computer model of their circuits and run simulations before they build the real circuits, and these simulations use numeric linear algebra, numeric solution of ordinary differential equations, and numeric solution of PDEs.

Finite element packages. For some problems in numerical analysis, a useful technique for solving the problem's differential equation is first to convert it to a new problem, the solution of which agrees with that of the original equation. The most popular use of the finite element, to date, has been to solve elliptic partial PDEs. In most versions of finite element packages, the original equations are replaced by a new system of equations that agree at the boundary points. The new system of differential equations is easier to solve "inside" the boundary than is the original system and is proved to be close to the solution of the original system. Much care is taken in selecting the original finite element grid and approximating functions. Examples of finite element abound, including modeling car body parts, calculating the stiffness of a beam that needs to hold a number of different weights and simulating the icing of an airplane wing.

CAREERS AND COURSE WORK

Students who enter one of the careers that use numerical analysis typically major in mathematics or physics. One needs substantial course work in computer science, mathematics, and physics. A bachelor's degree is sufficient for those seeking positions as programmers, although a master's degree is helpful. For a position involving development of new algorithms, one generally needs a master's or doctoral degree. A formal minor in an area that regularly uses numerical analysis, such as biology, is recommended.

Those seeking careers involving numerical analysis take a wide variety of positions. These include programmers, algorithm designers, managers of scientific labs, and computer science instructors.

SOCIAL CONTEXT AND FUTURE PROSPECTS

Advances in numerical algorithms have made a number of advances in science possible, such as improved weather forecasting, and have improved life for everyone. If scientific theories are to live up to their full potential in the future, ways of finding approximations to values of the variables used in these theories are needed. The increased complexity of these scientific models is forcing programmers to design, implement, and test new and more sophisticated numerical analysis algorithms.

—George M. Whitson, III, MS, PhD

FURTHER READING

Burden, Richard L., and J. Douglas Faires. *Numerical Analysis.* 9th ed. Boston: Brooks, 2011. Print

Hildebrand, F. B. *Introduction to Numerical Analysis.* 2nd ed. New York: McGraw, 1974. Print.

Iserles, Arieh. *A First Course in the Numerical Analysis of Differential Equations.* 2nd ed. New York: Cambridge UP, 2009. Print.

Moler, Cleve B. *Numerical Computing with MATLAB.* Philadelphia: Soc. for Industrial and Applied Mathematics, 2008. Print.

Overton, Michael L. *Numerical Computing with IEEE Floating Point Arithmetic.* Philadelphia: Soc. for Industrial and Applied Mathematics, 2001. Print.

Ralston, Anthony, and Philip Rabinowitz. *A First Course in Numerical Analysis.* 2nd ed. Mineola: Dover, 2001. Print.

Strauss, Walter A. *Partial Differential Equations: An Introduction.* 2nd ed. Hoboken: Wiley, 2008. Print.

O

OCEAN AND TIDAL ENERGY TECHNOLOGIES

FIELDS OF STUDY

Engineering; oceanography; fluid mechanics; hydrodynamics; atmospheric physics; energy conversion; computer control systems.

KEY TERMS AND CONCEPTS

- **barrage:** Artificial obstruction in a watercourse, such as a bridge or dam.
- **biomass:** Mass of living organisms, such as kelp, from which energy can be obtained.
- **greenhouse gases:** Atmospheric gases thought to contribute to global warming.
- **ocean temperature energy conversion (OTEC):** Obtains energy from water-temperature differences.
- **reversing tidal power plant:** Power plant that generates power both on rising and falling tides.
- **spring tide:** High tide that is unusually high, occurring twice monthly because of the cycle of the Moon.
- **tidal range:** Vertical distance between the highest and lowest stands of the tide.
- **tidal stream:** Fast-moving ocean or estuary current generated by the tides.
- **tide mill:** Grinding mill for grain that uses tidal power to turn a water wheel.

SUMMARY

Every continent on the planet is surrounded by a cleaner, safer, more efficient energy resource. As conventional energy supplies are depleted, means are being developed, and in some cases are in actual operation, to convert the energy found in waves, tidal currents, ocean and river currents, ocean thermal gradients, and offshore wind into usable electric power for utility-scale grids, independent power producers, and the public sector.

DEFINITION AND BASIC PRINCIPLES

The tides were the earliest source of obtaining power from the ocean. The requirements were simple: a dam to contain a head of water that was brought in by high tide and a water wheel to turn as the water was let out, thus generating the power. The obtaining of power from crashing waves was another dream of oceanographers, and it has finally been realized in a variety of modest projects that generate power in a number of ingenious ways. Another dream of oceanographers has been deriving power from fast-moving currents in estuaries on in the ocean itself. Several so-called in-stream devices have now been developed, but the ultimate goal of obtaining power from Florida's famous Gulf Stream, the fastest-moving ocean current in the world, is still elusive although study is under way. Another interesting project was the famous ocean temperature energy conversion (OTEC) project to obtain power from the temperature difference between warm- and cold-ocean waters. This project had a brief success in Hawaii during the 1980's, was dropped, but is being looked at again by several countries. And a spectacular newcomer for obtaining power from the ocean is the development of offshore wind turbines. These giant windmills, whirling high on stilts, are now enjoying great success in the coastal waters of Europe, although development in the United States as of 2011 has been slow.

BACKGROUND AND HISTORY

Power has been generated from the tides in Europe since at least the Middle Ages, and tide mills were common in Great Britain, France, Ireland, and along the east coast of the United States until the middle of the nineteenth century. As of 2011, the world has four power stations generating electricity from the tides. The largest is located on the Rance River estuary in France and is rated at 240 megawatts. Others are on the Annapolis River in Nova Scotia, Canada (rated at up to twenty megawatts), the Kislaya Guba, Russia, plant (rated at one to two megawatts), and the Xiamen, China, plant (rated at up to three mega-watts). The generation of power from the waves has been a more recent development. Small-scale installations are now generating power in Scotland and at other locations, and projects involving various new methods are under way.

HOW IT WORKS

The model for a successful tidal barrage plant is on the Rance River estuary in Brittany, France. Here the tide range is about forty feet. As the rising tide passes through the circular openings in the barrage, which is located near the mouth of the estuary, rotors spaced at regular intervals in the circular openings generate power as the water level rises. After six hours, the tide will have turned and the flow of water is in the opposite direction. Then the rotors are turned 180 degrees so that the plant can generate power as the tide flows out to sea.

Wave power. Many ingenious devices have been designed to obtain electric power from waves. One type is the point absorber, a bottom-mounted or floating structure that can absorb wave energy coming from all directions. A second type is the terminator, which reflects or absorbs all the wave energy coming at it. Another type is the linear absorber, which is oriented parallel to the direction of the oncoming waves. It is composed of interlocking sections, and the pitching and yawing of these sections, because of the waves, pressurizes a hydraulic fluid that turns a turbine. A fourth device is an oscillating water-column terminator, which is a partially submerged chamber with air trapped above the water's surface. As entering waves push the air column up, the compression of the air will act as a piston to drive a turbine. All of these devices are in the developmental stage, but no single design has been judged the best. Factors that must be considered are the corrosive and occasionally violent marine environment and biofouling, which begins the moment any device is placed in the ocean.

Power from tidal and ocean currents. The dream of obtaining power from the ocean's fast-moving currents may finally be nearing realization. Tidal generators are being tested in several estuaries and ocean channels, using either bottom-mounted rotors or rotors suspended from floating barges. Power generation is nearly continuous, with the rotors turning for both flood and ebb currents. The greatest challenge will be harnessing power from the ocean's famous

Gulf Stream. A group of researchers at Florida Atlantic University in Boca Raton, Florida, plans to mount one or more giant rotors in this current as it flows between Florida and the Bahamas. Water depths approach 2,500 feet there, so attaching the rotors to the seabed will be a challenge. In addition, there will be the usual problems of biofouling, corrosion, and getting the power to shore. The rotors will constitute a potential hazard for passing ships, submarines, and large marine mammals, such as whales, so these concerns will also have to be addressed as well. The amount of power generated might well approach 10,000 megawatts.

Ocean thermal energy conversion (OTEC). An OTEC plant operated successful in Hawaii during the 1980's and is being considered by several European countries, despite its high cost. This plant derives power from the differential between 40-degree-Fahrenheit deep water and 80-degree-Fahrenheit surface water. The 40-degree-Fahrenheit water, drawn from thousands of feet down, is used to condense ammonia, which is then brought in contact with 80-degree-Fahrenheit surface water, causing it to vaporize explosively, driving a turbine. One problem is the disposal of the 40-degree-Fahrenheit water after it has been warmed in the vaporization process. It cannot be returned to the ocean, where it would kill reefs and tropical fish, so the solution in Hawaii was to pipe it through the soil where it fooled cool-weather crops such as strawberries and asparagus into growing in a tropical climate.

Wind power. The latest method for obtaining energy from the ocean is the giant turbines turning in the near-shore, shallow waters of Europe, where more than 1,500 megawatts of power is being generated. These huge turbines function much like windmills on land, except that they are firmly anchored to the seafloor, often in plain sight of coastal residents. Environmental and aesthetic concerns have so far held up the installation of such turbines along the east coast of the United States, but construction has finally been approved for a group of wind turbines off the coast of Massachusetts.

APPLICATIONS AND PRODUCTS

The number of tidal barrage power plants in the world is extremely limited because of the large tidal range required—a minimum of a ten-foot rise between low tide and high tide. Few coasts in the world have that great a rise in a body of water narrow enough to be dammed, and even fewer have it in an area that is sufficiently populated to provide a market for the power generated. In addition to the four tidal power plants already mentioned, an additional plant has been proposed at the Severn Estuary in southwestern England (rated at 1,200 to 4,000 megawatts). One problem faced by all tidal barrage plants is that the power generated is intermittent. The time suitable for power generation shifts steadily as the Moon orbits around the Earth. This means that the supply of power and the demand for power will not always coincide. For several nights customers will have ample power to cook their supper and enjoy evening activities, but the next few nights they will have no power at all.

Power generation from waves. Installations generating a total of about 4 megawatts of power have now been created worldwide, mostly on an experimental or demonstration project basis, but one that has been generating power successfully since 2000 is the Limpet, the world's first commercial, wave power station, on the rockbound coast of the Island of Islay in Scotland. The plant is of the oscillating water column design with a fortress-like exterior fronting the waves just at sea level along the rocky shore. When a breaking wave enters the long, concrete tube, with its opening just below water level, it drives air in and out of a pressure chamber through a specially designed

The world's first commercial-scale and grid-connected tidal stream generator, Sea-Gen, in Strangford Lough. The strong wake shows the power in the tidal current.

air turbine, generating electricity. The Limpet produces 0.5 megawatts of power, which is fed into the island's power grid. The design makes the Limpet easy to build and install, and its low visible profile does not intrude on the coastal landscape or the ocean views.

Power from currents. Besides the proposed giant rotors in the Gulf Stream, which would be an open-ocean device, a number of projects have been designed for obtaining power from tidal currents in estuaries and other constricted passages using bottom- or surface-mounted turbines. These turbines do not require construction of an expensive barrage. The flow of water simply turns the turbine as the tide comes in, and, if the turbine is reversed 180 degrees, it can also generate power as the tide goes out. An experimental array of six turbines was installed in New York City's East River in 2007 and has thus far proved successful. Similar arrays are being installed in Ireland, the United Kingdom, Italy, Korea, and Canada. The turbines strongly resemble torpedoes, with the rotor at one end, and they stand on a stout pedestal firmly attached to a channel bed or they can be suspended from floating barges. Power is transmitted to shore by means of cables.

Wind farms. The major components for a wind turbine are a tower, a rotor with hub and blades, a gear box, and a generator. Offshore systems are larger than those on land because of the greater cost to install and service them. Most are rated at three to five megawatts, compared with one and one-half to three megawatts for those on land. They are usually fixed in water depths of fifteen to seventy-five feet and require a foundation driven deep into the seabed to support the weight of the tower and the rotor. Frequently the turbines are arranged in arrays to reduce maintenance and cabling costs. The power generated is sent to shore through a high-voltage cable buried in the seabed. A service area is always required, with boats and a hoist crane for repairs and maintenance. By the end of 2008, 1,471 megawatts of turbines have been installed worldwide, primarily in the United Kingdom, Germany, Denmark, the Netherlands, and Sweden. The U.S. projects will be located along the North and Central Atlantic coasts, and several are in the permit process. Wind farms in the Great Lakes, the Gulf of Mexico, and eastern Canada will probably come next.

CAREERS AND COURSE WORK

College-level students seeking careers in marine renewable energy are advised to take a general oceanography course in order to familiarize themselves with the characteristics of the environment in which they will be working. In addition, they should supplement this study with basic courses in math and physics so that they can understand the technology involved in the design of the various marine-energy projects. For those students planning to go on to graduate study, engineering courses are highly recommended, especially for those students interested in designing and building ocean energy generators. As of 2011, the jobs available in the United States for students seeking opportunities in the field of marine-renewable energy are still somewhat limited because most of the American marine energy projects operating are just getting under way or are still in developmental stages. The country's first wind farm project was approved for Massachusetts in late 2010, but construction has not yet begun. Additional wind farm projects are planned but the permit process for them is a lengthy one. Several coastal wave energy projects are now operating, as well as a few tidal stream projects in rivers and estuaries with strong tidal flows, but these projects are all in developmental stages. Once they mature into full-time operations, job opportunities in these areas should begin to appear.

SOCIAL CONTEXT AND FUTURE PROSPECTS

For many years the United States has been totally dependent on traditional energy sources, such as oil, natural gas, coal, water power, and atomic energy. Attention is turning to the generation of electricity from marine renewable energy sources. Projects in this field are still in the developmental stages, with tidal stream and wind farm installations showing the most promise. The potential for them to make a significant contribution to U.S. energy needs is great because nearly half the states have access to the oceans along their borders. Although the United States has ample supplies of coal, the burning of coal harms the environment, and U.S. reserves of oil and natural gas are on a downward trend. Attention to marine-energy sources cannot help but increase in the coming years. The recent approval of a wind farm for Massachusetts is an encouraging sign. Wind farms are already making a significant contribution to energy needs in Europe, and coastal conditions along

---

FASCINATING FACTS ABOUT OCEAN AND TIDAL ENERGY TECHNOLOGIES

- In its World Energy Outlook for 2008, the International Energy Agency predicts that if present trends continue, the world's overall energy demands will almost double by 2050.
- As of 2011, only about 7 percent of the U.S. energy supply comes from renewable sources, and most of this is derived from biomass and hydropower. The rest of U.S. energy needs come from oil, coal, natural gas, and nuclear power. Marine renewable energy sources accounted for only about 0.5 percent of U.S. energy needs in 2008.
- Thirty states in the United States border either an ocean or one of the Great Lakes. These states generate and consume 75 percent of the nation's electricity, but the development of renewable marine power resources by these states so far is practically nil.
- The fastest growing renewable energy source in the world as of 2011 is wind energy, with a yearly growth rate of 20 to 30 percent. Germany and the United Kingdom are two of the largest wind-power generators, and Germany's Innogy Nordsee 1 project, being developed in the North Sea, is expected to provide power for nearly 1,000,000 homes.
- Limpet, the world's first commercial wave power station, generates 0.5 megawatt of power–enough to light about 500 homes.

---

the Atlantic and Gulf coasts of the United States are equally favorable. The cost of these projects is large, but they will provide many jobs for the installation and maintenance of the equipment, as well as for the manufacture of the turbines.

—Donald W. Lovejoy, MS, PhD

FURTHER READING

Bedard, Roger, et al. "An Overview of Ocean Renewable Energy Technologies." *Oceanography* 23, no. 2 (June, 2010): 22–31. Highlights the development status of the various marine-energy conversion technologies. Contains many helpful color photographs and explanatory diagrams.

Boon, John D. *Secrets of the Tide: Tide and Tidal Current*

Application and Prediction, Storm Surges and Sea Level Trends. Chichester, England: Horwood, 2004. Describes the so-called tide mills used in many locations along the East Coast of the United States in the eighteenth century for grinding grain into flour or meal.

Carter, R. W. G. *Coastal Environments: An Introduction to the Physical, Ecological, and Cultural Systems of Coastlines.* San Diego: Academic Press, 1988. Includes an outstanding analysis of tidal power, with a description of the mechanics of power generation and a discussion of the environmental and ecological effects.

Segar, Douglas A. *Introduction to Ocean Sciences.* 2d ed. New York: W. W. Norton, 2007. A well-respected textbook with solid foundational information.

Sverdrup, Keith A., and E. Virginia Armbrust. *An Introduction to the World's Oceans.* 10th ed. New York: McGraw-Hill, 2009. Provides information on how energy is now being obtained, or may someday be obtained, from ocean waves, currents, and tides. Many useful diagrams.

Thresher, R., and W. Musial. "Ocean Renewable Energy's Potential Role in Supplying Future Electrical Needs." *Oceanography* 23, no. 2 (June, 2010): 16–21. A summary of the nation's and the world's energy needs and the sources from which these energy needs have been met and will be met in the future.

Ulanski, Stan. *Gulf Stream: Tiny Plankton, Giant Bluefin, and the Amazing Story of the Powerful River in the Atlantic.* Chapel Hill: University of North Carolina Press, 2010. Describes the world's fastest ocean current, summarizing the physical characteristics of location and speed that make this current a prime candidate for energy generation.

OPTICS

FIELDS OF STUDY

Astronomy; geometrical optics; ophthalmology; optometry; photography; physical optics; physics; quantum optics; quantum physics.

KEY TERMS AND CONCEPTS

- **coherence:** Light rays that are in synchronous phase.
- **diffraction:** Change in direction of a light ray when it hits an obstruction.
- **geometric optics:** Use of ray diagrams and associated mathematical equations to describe light behavior.
- **interference:** Interaction between light rays under specific circumstances that leads to an increase or decrease in light intensity.
- **laser:** Light amplification by stimulated emission of radiation; monochromatic, coherent light used in various applications.
- **ophthalmology:** Practice of examining eye health, diagnosing disease, prescribing medical and surgical treatment for eye conditions by a specialized medical doctor.
- **optics:** Study of the behavior of light.
- **optometry:** Practice of examining eye health and prescribing corrective lenses by a doctor of optometry.
- **photoelectric effect:** Release of electrons that occurs when light rays hit a metallic surface.
- **photon:** Particle released because of the photoelectric effect.
- **photonics:** Study of photons as they relate to light energy in applications such as telecommunications and sensing.
- **quantum optics:** Branch of quantum physics focused on the study of the dual wave and particle characteristics of light.
- **refraction:** Bending of light waves that occurs when light traverses from one media to another, such as from air to water.

SUMMARY

Optics is the study of light. It includes the description of light properties that involve refraction, reflection, diffraction, interference, and polarization of electromagnetic waves. Most commonly, the word "light" refers to the visible wavelengths of the electromagnetic spectrum, which is between 400 and 700 nanometers (nm). Lasers use wavelengths that vary from the ultraviolet (100 nm to 400 nm) through the visible spectrum into the infrared spectrum (greater than 700 nm). Optics can be used to understand and study mirrors, optical instruments such as telescopes and microscopes, vision, and lasers used in industry and medicine.

DEFINITION AND BASIC PRINCIPLES

Optics is the area of physics that involves the study of electromagnetic waves in the visible-light spectrum, between 400 and 700 nm. Optics principles also apply to lasers, which are used in industry and medicine. Each laser has a specific wavelength. There are lasers that use wavelengths in the 100–400 nm range, others that use a wavelength in the visible spectrum, and some that use wavelengths in the infrared spectrum (greater than 700 nm).

Light behaves as both a wave and a particle. This duality has resulted in the division of optics into physical optics, which describes the wave properties of light; geometric optics, which uses rays to model light behavior; and quantum optics, which deals with the particle properties of light. Optics uses these theories to describe the behavior of light in the form of refraction, reflection, interference, polarization, and diffraction.

When light and matter interact, photons are absorbed or released. Photons are a specific amount of energy described as the sum of Planck's constant, h (6.626×10^{-34}) ,and the wavelength of the light. The formula to describe the energy of photons is $E = hf$. Photons have a constant speed in a vacuum. The speed of light is $c = 2.998 \times 10^8$. The constant speed of light in a vacuum is an important concept in astronomy. The speed of light is used in the measurement of astronomic distances in the unit of light-years.

BACKGROUND AND HISTORY

Optics dates back to ancient times. The three-thousand-year-old Nimrud lens is crafted from natural crystal, and it may have been used for magnification or to start fires. Early academics such as Euclid in 300 BCE theorized that rays came out of the eyes in order to produce vision. Greek astronomer Ptolemy later described angles in refraction. In the thirteenth century, English philosopher Roger Bacon suggested that the speed of light was constant and that lenses might be used to correct defective vision.

By the seventeenth century, telescopes and microscopes were being developed by scientists such as Hans

Lippershey, Johannes Kepler, and Galileo Galilei. During this time, Dutch astronomer Willebrord Snellius formulated the law of refraction to describe the behavior of light traveling between different media, such as from air to water. This is known as Snell's law, or the Snell-Descartes law, although it was previously described in 984 by Persian physicist Ibn Sahl.

Sir Isaac Newton was one of the most famous scientists to put forward the particle theory of light. Dutch scientist Christiaan Huygens was a contemporary of Newton's who advocated the wave theory of light. This debate between wave theory and particle theory continued into the nineteenth century. French physicist Augustin-Jean Fresnel was influential in the acceptance of the wave theory through his experiments in interference and diffraction.

The wave-particle debate continued into the next century. The wave theory of light described many optical phenomena; however, some findings, such as the emission of electrons when light strikes metal, can be explained only using a particle theory. In the early twentieth century, German physicists Max Planck and Albert Einstein described the energy released when light strikes matter as photons with the development of the formula $E = hv$, which states that the photon energy equals the sum of the wavelength and Planck's constant.

In the early twenty-first century, it is generally accepted that both the wave and the particle theories are correct in describing optical events. For some optical situations, light behaves as a wave, and for others, the particle theory is needed to explain the situation. Quantum physics tries to explain the wave-particle duality, and it is possible that future work will unify the wave and particle theories of light.

HOW IT WORKS

Physical optics. Physical optics is the science of understanding the physical properties of light. Light behaves as both a particle and a wave. According to the wave theory, light waves behave similarly to waves in water. As light moves through the air the electric field increases, decreases, and then reverses direction. Light waves generate an electric field perpendicular to the direction the light is traveling and a magnetic field that is perpendicular both to the direction the light is traveling and to the electric field.

Interference and coherence refer to the

interactions between light rays. Both interference and coherence are often discussed in the context of a single wavelength or a narrow band of wavelengths from a light source. Interference can result either in an increased intensity of light or a reduction of intensity to zero. The optical phenomenon of interference is used in the creation of antireflective films.

Coherence occurs when light is passed through a narrow slit. This produces waves that are in phase with the waves exactly lined up or waves that are out of phase but have a constant relationship with one another. Coherence is an important element to the light emitted by lasers and allows for improved focusing properties necessary to laser applications.

Polarization involves passing light waves through a filter that allows only wavelengths of a certain orientation to pass. For example, polarized sunglasses allow only vertical rays to pass and stop the horizontal rays, such as light reflected from water or pavement. In this way, polarized sunglasses can reduce glare.

Diffraction causes light waves to change direction as light encounters a small opening or obstruction. Diffraction becomes a problem for optical systems of less than 2.5 millimeters (mm) for visible light. Telescopes overcome the diffraction effect by using a larger aperture, however, for very large-diameter telescopes the resolution is then limited due to atmospheric conditions. Space telescopes such as the Hubble are unaffected by these conditions as they are operating in a vacuum.

Scattering occurs when light rays encounter irregularities in their path, such as dust in the air. The increased scattering of blue light due to particles in the air is responsible for the blue color of the sky.

Illumination is the quantitative measurement of light. The watt is the measurement unit of light power. Light can also be measured in terms of its luminance as it encounters the eye. Units of luminance include the lumen, the candela, and the now-obsolete apostilb.

The photoelectric effect that supports the particle theory of light was discovered by German physicist Heinrich Rudolph Hertz in 1887 and later by Albert Einstein. When light waves hit a metallic surface, electrons are emitted. This effect is used in the generation of solar power.

Geometric optics. Geometric optics describes optical behavior in the form of rays. In most ordinary situations, the ray can accurately describe the movement of light as it travels through various media such as glass or air and as it is reflected from a surface such as a mirror. Geometric optics can describe the basics of photography. The simplest way to make an image of an object is to use a pinhole to produce an inverted image. When lenses and mirrors are added to the pinhole a refined image can be produced.

Reflection and refraction are two optical phenomena in which geometric optics applies. Reflections from plane (flat) mirrors, convex mirrors, and concave mirrors can all be described using ray diagrams. A plane mirror creates a virtual image behind the mirror. The image is considered virtual because the light is not coming from the image but only appears to because of the direction of the reflected rays. A convex mirror can create a real image in front of the lens or a virtual image behind the lens depending on where the object is located. If an object is past the focal point of the convex mirror then the image is real and located in front of the mirror. If the object is between the focal point and the convex mirror then the image is virtual and located behind the mirror. A convex lens will create a virtual image. Geometric optics involves ray diagrams that will allow the determination of image size (magnification or minification), location of the image, and if it is real or virtual.

Refraction of light happens when light passes between two different substances such as air and glass or air and water. Snell's law expresses refraction of light as a mathematical formula. One form of Snell's law is: $n_i \sin_i = n_t \sin_t$ where n_i is the refractive index of the incident medium, $_i$ is the angle of incidence, n_t is the refractive index of the refracted medium, and $_t$ is the angle of transmission. This formula, along with its variations, can be used to describe light behavior in nature and in various applications such as manufacturing corrective lenses. Refraction also occurs as light travels from the air into the eye and as it moves through the various structures inside the eye to produce vision.

Magnification or minification can be a product of refraction and reflection. Geometric optics can be applied to both microscopes and telescopes, which use lenses and mirrors for magnification and minification.

Quantum optics. Quantum optics is a division of physics that comes from the application of mathematical models of quantum mechanics to the dual wave and particle nature of light. This area of optics has applications in meteorology, telecommunications, and other industries.

APPLICATIONS AND PRODUCTS

Vision and vision science. There is a vast network of health-care professionals and industries that study and measure vision and vision problems as well as correct vision. Optometrists measure vision and refractive errors in order to prescribe corrective spectacles and contact lenses. Ophthalmologists are medical doctors who specialize in eye health and vision care. Some ophthalmologists specialize in vision-correction surgery, which uses lasers to reduce the need for glasses or contact lenses. In order to perform vision-correction surgeries there are a number of optical instruments, including wave-front mapping analyzers, that may be used.

The industries that support optometry and ophthalmology practices include laser manufacturers, optical diagnostic instruments manufacturers, and lens manufacturers. Lenses are used for diagnosis of vision problems as well as for vision correction.

Development of new lens technology in academic institutions and industry is ongoing, including multifocal lens implants and other vision-correction technologies.

Research. Many areas of research, including astronomy and medicine, use optical instruments and optics theory in the investigation of natural phenomena. In astronomy, distances between planets and galaxies are measured using the characteristics of light traveling through space and expressed as light-years. Meteorological optics is a branch of atmospheric physics that uses optics theory to investigate atmospheric events. Both telescopes and microscopes are optimized using optical principles. Many branches of medical research use optical instruments in the investigations of biological systems.

Medicine. Lasers have become commonplace in medicine, from skin-resurfacing and vision-correction procedures to the use of carbon-dioxide lasers in general surgery.

Industry. As noted above, there is an industry sector that is dedicated to the manufacture and

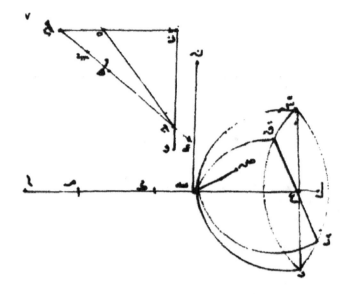

Reproduction of a page of Ibn Sahl's manuscript from c. 984 showing his knowledge of the law of refraction.

development of vision-correction and diagnostic lenses and tools. Optics is an important part of the telecommunications industry, which uses fiber optics to transmit images and information. Photography, from the manufacture of cameras and lenses to their use by photographers, involves applied optics. Lasers are also used for precision manufacturing of a variety of products.

CAREERS AND COURSE WORK

A career in an optics field can be as varied as the applications. An interest in optics might lead to a career in physics, astronomy, meteorology, vision care, or photography. Depending on the specific position desired, the training may range from a high school diploma and on the job training to a university degree and postgraduate work.

FASCINATING FACTS ABOUT OPTICS

- English surgeon Sir Harold Ridley decided on polymethyl-methacrylate (PMMA) as suitable material for intraocular lens implants after observing Royal Air Force pilots with pieces of the PMMA airplane canopy in their eyes after accidents. He noticed that this material was not rejected in the eye, and it was used for subsequent decades to implant lenses after cataract surgery.

- The first images received from the Hubble Space Telescope in 1990 were blurry because of spherical aberration caused by a flaw the size of one-fiftieth of a sheet of paper in the focusing mirror. NASA scientists designed a series of small mirrors that were installed by a team of astronauts in 1993 to overcome this flaw. The subsequent Hubble images were free from the aberration and had the excellent resolution expected from a space-based telescope.

- The stereo images produced by the Mars Pathfinder's cameras functioned similarly to stereo vision produced by binocular vision in humans. Two sets of cameras produced individual images that were fused used prisms. The successor to the Pathfinder is NASA's Opportunity, which is still sending images from Mars.

- Geckos' eyes have 350 times more sensitivity to color in dim light than human eyes.

- Newer-generation excimer laser systems used for vision-correction surgery have the capacity to measure and correct higher-order optical aberrations of the human eye. Iris recognition with a rotational adjustment is also available on some lasers.

- The different colors of the northern lights are created when solar energy in the form of solar flares enter the Earth's magnetic sphere and collide with atmospheric gases. These collisions cause the gases to emit light. Collisions with oxygen will tend to cause a red color, while nitrogen or helium will produce blue or green colors.

Optics involves a combination of math and physics. An understanding of human eye anatomy is also essential for a career in vision care. For all of optics-related fields it is important to have a strong background in high school mathematics. For occupations in allied health care such as opticians or ophthalmic technicians, a high school diploma and technical training is required post high school. Photographers may pursue formal training through a university or art school or might develop skills through experience or an apprenticeship.

Many careers in physics, astronomy, and meteorology require at least a bachelor's degree and most require a master's or doctoral degree. University course work in these fields includes mathematics and physics. To become an optometrist a bachelor's degree plus a doctor of optometry degree is required. An ophthalmologist will need a bachelor's degree, a degree in medicine, and residency training.

SOCIAL CONTEXT AND FUTURE PROSPECTS

The advancements in optics theory and application have changed the fabric of life in industrialized countries, from the way people communicate to how the universe is understood. It is almost impossible to imagine what future advances will occur in optics, since the last fifty years has brought profound changes in the fields of photography, medicine, astronomy, manufacturing, and a number of other fields.

As wireless technology advances it seems possible that this technology may replace some of the millions of miles of fiber-optic telecommunications cables that currently exist. Because of their reliability, fiber optics will continue to be used for the foreseeable future. Existing lasers will continue to be to optimized, and most likely new lasers will be developed.

Refinements in optical systems will aid in research in a variety of fields. For example, oceanographers already apply optics theory to the study of low-light organisms and to the development of techniques for conducting research in low light. Improved optical systems will likely have a positive impact on this and other research.

Quantum computers using photonic circuits are a possible future development in the field of optics. A quantum computer that takes advantage of the photoelectric effect may be able to increase the capacity of computation over conventional computers. Optics and photonics may also be applied to chemical sensing, imaging through adverse atmospheric conditions, and solid-state lighting.

Some scientists have commented that the wave and particle theories of light are perhaps a temporary solution to the true understanding of light behavior. The area of quantum optics is dedicated to furthering the understanding of this duality of light. It is possible that in the future a more unified theory will lead to applications of optics and the use of light energy in ways that have not yet been imagined.

—*Ellen E. Anderson Penno, MS, MD, FRCSC, Dip. ABO*

FURTHER READING

Azar, Dimitri T. *Clinical Optics, 2014–2015*. San Francisco: Amer. Acad. of Ophthalmology, 2014. Print. Basic and Clinical Science Course 3.

Meschede, Dieter. *Optics, Light and Lasers: The Practical Approach to Modern Aspects of Photonics and Laser Physics*. 2nd ed. Weinheim: Wiley, 2007. Print.

Pedrotti, Frank L., Leno M. Pedrotti, and Leno S. Pedrotti. 3d ed. *Introduction to Optics*. 3rd ed. Upper Saddle River: Prentice, 2007. Print.

Siciliano, Antonio. *Optics: Problems and Solutions*. Singapore: World Scientific, 2006. Print.

Tipler, Paul A., and Gene Mosca. *Physics for Scientists and Engineers*. 6th ed. New York: Freeman, 2008. Print.

Wolfe, William J. *Optics Made Clear: The Nature of Light and How We Use It*. Bellingham: SPIE, 2007. Print.

P

PARALLEL COMPUTING

FIELDS OF STUDY

Computer engineering; computer programming; computer science; engineering; mathematics; networking; physics; information systems; information technology.

KEY TERMS AND CONCEPTS

- **computer:** Set of nodes functioning as a unit that uses a network or high-speed bus for communication.
- **flop:** Floating point operation; generally a central processing unit addition, subtraction, multiplication, or division.
- **grid computing:** Distributing a single computer job over multiple independent computers.
- **MIMD:** Stands for multiple instruction multiple data. Computer with multiple processors each of which executes an instruction stream using a local data set.
- **node:** Set of processors functioning as a unit that is tightly coupled by a switch or high-speed network.
- **processor:** Unit of a computer that performs arithmetic functions and exercises control of the instruction stream.
- **SIMD:** Stands for single instruction multiple data. Computer with multiple processors each of which executes the same instructions on their local data set.
- **symmetric multiprocessing:** Use of shared memory for processing by two or more independent processors of a node of a computer.
- **vector processor:** Central processing unit that executes an array of data, processing the data sequentially by decomposing the arithmetic units into components that execute multiple numbers in parallel.

SUMMARY

Parallel computing involves the execution of two or more instruction streams at the same time. Parallel computing takes place at the processor level when multiple threads are used in a multi-core processor or when a pipelined processor computes a stream of numbers, at the node level when multiple processors are used in a single node, and at the computer level when multiple nodes are used in a single computer. There are many memory models used in parallel computing, including shared cache memory for threads, shared main memory for symmetric multiprocessing (SMP) and distributed memory for grid computing. Parallel computing is often done on supercomputers that are capable of achieving processing speeds of more than 10^6 Gflop/s.

DEFINITION AND BASIC PRINCIPLES

Several types of parallel computing architectures have been used over the years, including pipelined processors, specialized SIMD machines and general MIMD computers. Most parallel computing is done on hybrid supercomputers that combine features from several basic architectures. The best way to define parallel processing in detail is to explain how a program executes on a typical hybrid parallel computer.

To perform parallel computing one has to write a program that executes in parallel; it executes more than one instruction simultaneously. At the highest level this is accomplished by decomposing the program into parts that execute on separate computers and exchanging data over a network. For grid computing, the control program distributes separate programs to individual computers, each using its own data, and collects the result of the computation. For in-house computing, the control program distributes separate programs to each node (the node is actually a full computer) of the same parallel computer and, again, collects the results. Some of the control programs manage parallelization themselves, while

523

others let the operating system automatically manage parallelization.

The program on each computer of a grid, or node of a parallel computer, is itself a parallel program, using the processors of the grid computer, or node, for parallel processing over a high-speed network or bus. As with the main control program, parallelization at the node can be under programmer or operating system control. The main difference between control program parallelization at the top level and the node level is that data can be moved more quickly around one node than between nodes.

The finest level of parallelization in parallel programming comes at the processor level (for both grid and in-house parallel programming). If the processor uses pipelining, then streams of numbers are processed in parallel, using components of the arithmetic units, while if the processor is multi-core, the processor program decomposes into pieces that run on the different threads.

BACKGROUND AND HISTORY

In 1958, computer scientists John Cocke and Daniel Slotnick of IBM described one of the first uses of parallel computing in a memo about numerical analysis. A number of early computer systems supported parallel computing, including the IBM MVS series (1964–1995), which used threadlike tasks; the GE Multics system (1969), a symmetric multiprocessor; the ILLIAC IV (1964–1985), the most famous array processor; and the Control Data Corporation (CDC) 7600 (1971–1983), a supercomputer that used several types of parallelism.

American computer engineer Seymour Cray left CDC and founded Cray Research in 1972. The first Cray 1 was delivered in 1977 and marked the beginning of Cray's dominance of the supercomputer industry in the 1980s. Cray used pipelined processing as a way to increase the flops of a single processor, a technique also used in many of the Reduced Instruction Set Computing (RISC), such as the i860 used in the Intel Hypercube. Cray developed several MIMD computers, connected by a high-speed bus, but these were not as successful as the MIMD Intel Hypercube series (1984–2005) and the Thinking Machines Corporation SIMD Connection Machine (1986–1994). In 2004, multi-core processors were introduced as the latest way to do parallel computing, running a different thread on each core, and by

2011, many supercomputers were based on multi-core processors.

Many companies and parallel processing architectures have come and gone since 1958, but the most popular parallel computer in the twenty-first century consists of multiple nodes connected by a high-speed bus or network, where each node contains many processors connected by shared memory or a high-speed bus or network, and each processor is either pipelined or multi-core.

HOW IT WORKS

The history of parallel processing shows that during its short lifetime (1958 through the present), this technology has taken some very sharp turns resulting in several distinct technologies, including early supercomputers that built super central processing units (CPUs), SIMD supercomputers that used many processors, and modern hybrid parallel computers that distribute processing at several levels. What makes this more interesting is that all of these technologies remain active, so a full explanation of parallel computing must include several rather different technologies.

Pipelined and super processors. The early computers had a single CPU, so it was natural to improve the CPU to provide an increase in speed. The earliest computers had CPUs that provided control and did integer arithmetic. One of the first improvements to the CPU was to add floating point arithmetic to the CPU (or an attached coprocessor). In the 1970s, several people, most notably Seymour Cray, developed pipelined processors that could process arrays of floating point numbers in the CPU by having CPU processor components, such as part of the floating point multiplier, operate on numbers in parallel with the other CPU components. The Cray 1, X-MP and Y-MP were the leaders in supercomputers for the next few years. Cray, and others, considered using gallium arsenide rather than silicon for the next speed improvement for the CPU, but this technology never worked. A number of companies attempted to increase the clock speed (the number of instructions per second executed by the CPU) using other techniques such as pipelining, and this worked reasonably well until the 2000's.

A number of companies were able to build a CPU chip that supported pipelining, with Intel's i860 being one of the first. While chip density increased

IBM's Blue Gene/P massively parallel supercomputer.

as predicted by Moore's law (density doubles about every two years), signal speed on the chip limited the size of pipelined CPU that could be put on a chip. In 2005, Intel introduced its first multi-core processor that had multiple CPUs on a chip. Applications software was then developed that decomposed a program into components that could execute a different thread on each processor, thus achieving a new type of single CPU parallelism.

Another idea has developed for doing parallel processing at the processor level. Some supercomputers are being built with multiple graphics processing units (GPUs) on a circuit board, and some have been built with a mix of CPUs and GPUs on a board. A variety of techniques is being used to combine these processors into nodes and computers, but they are similar to the existing hybrid parallel computers.

Data transfer. Increasing processor speed is an important part of supercomputing, but increasing the speed of data transfers between the various components of a supercomputer is just as important. A processor is housed on a board that also contains local memory and a network or bus connection module. In some cases, a board houses several processors that communicate via a bus, shared memory, or a board-level network. Multiple boards are combined to create a node. In the node, processors exchange data via a (backplane) bus or network. Nodes generally exchange data using a network, and if multiple computers are involved in a parallel computing program, the computers exchange data via

a transmission-control protocol/Internet protocol (TCP/IP) network.

Flynn's taxonomy. In 1972, American computer scientist Michael J. Flynn described a classification of parallel computers that has proved useful. He divided the instruction types as single instruction (SI) and multiple instruction (MI); and divided the data types as single data (SD) and multiple data (MD). This led to four computer types: SISD, SIMD, MISD and MIMD. SISD is an ordinary computer, and MISD can be viewed as a pipelined processor. The other classifications described architectures that were extremely popular in the 1980's and 1990's and are still in use. SIMD computers are generally applied to special problems, such as numeric solution of partial differential equations, and are capable of very high performance on these problems. Examples of SIMD computers include the ILLIAC IV of the University of Illinois (1974), the Connection Machine (1980s), and several supercomputers from China. MIMD computers are the most general type of supercomputer, and the new hybrid parallel processors can be seen as a generalization of these computers. While there have been many successful MIMD computers, the Intel Hypercube (1985) popularized the architecture, and some are still in use.

Software. Most supercomputers use some form of Linux as their operating system, and the most popular languages for developing applications are FORTRAN and C++. Support for parallel processing at the operating-system level is provided by operating-system directives, which are special commands to the operating system to do something, such as use the maximum number of threads within a code unit. At the program level, one can use blocking/unblocking message passing, access a thread library, or use a section of shared memory with the OpenMP API (application program interface).

APPLICATIONS AND PRODUCTS

Weather Prediction and Climate Change. One of the first successful uses of parallel computing was in predicting weather. Information, like temperature, humidity, and rainfall, has been collected and used to predict the weather for more than 500 years. Many early computers were used to process weather data, and as the first supercomputers were deployed in the 1970s, some of them were used to provide faster and more accurate weather forecasts. In 1904, Norwegian physicist and meteorologist Vilhelm Bjerknes proposed a differential-equation model for weather forecasting that included seven variables, including temperature, rainfall, and humidity. Many have added to this initial model since its introduction, producing complex weather models with many variables that are ideal for supercomputers, and this has led to many government agencies involved in weather prediction using supercomputers. The European Centre for Medium-Range Weather Forecasts (ECMWF) was using a CDC 6600 by 1976; the National Center for Atmospheric Research (NCAR) used an early model by Cray; and in 2009, the National Oceanic and Atmospheric Administration (NOAA) announced the purchase of two IBM 575s to run complex weather models, which improved forecasting of hurricanes and tornadoes. Supercomputer modeling of the weather has also been used to determine previous events in weather, such as the worldwide temperature of the Earth during the age of the dinosaurs, as well as to predict future phenomena such as global warming.

Efficient wind turbines. Mathematical models describing the characteristics of airflow over a surface, such as an airplane wing, using partial differential equations (Claude-Louis Navier, George Gabriel Stokes, Leonhard Euler, and others) have existed for more than a hundred years. The solution of these equations for a variable, such as the lift applied to an airplane wing for a given airflow, has used computers since their invention, and as soon as supercomputers appeared in the 1970s, they were used to solve these types of problems. There also has been great interest in using wind turbines to generate electricity. Researchers are interested in designing the best blade to generate thrust without any loss due to vortices, rather than developing a wing with the maximum lift. The EOLOS Wind Energy Research Consortium at the University of Minnesota used the Minnesota Supercomputer Institute's Itasca supercomputer to perform simulations of airflow over a turbine (three blades and their containing device) and as a result of these simulations was able to develop more efficient turbines.

Multispectral image analysis. Satellite images consist of large amounts of data. For example, Landsat

7 images consists of seven tables of data, where each entry in a table represents a different magnetic wavelength (blue, green, red, or thermal-infrared) for a 30-meter-square pixel of the Earth's surface. A popular use of Landsat data is to determine what a particular set of pixels represents, such as a submerged submarine. One approach used to classify Landsat pixels is to build a backpropagation neural network and train it to recognize pixels (determining the difference between water over the submarine and water that is not). A number of neural networks has been implemented on supercomputers over the years to identify multispectral images.

Biological modeling. Many applications of supercomputers involve the modeling of biological processes on a supercomputer. Both continuous modeling, involving the solution of differential equations, and discrete modeling, finding selected values of a large set of values, has made use of supercomputers., Dr. Dan Siegal-Gaskins of the Ohio State University built a continuous model of cress cell growth, consisting of seven differential equations and twelve unknown factors, to study why some cress cells divide into trichomes, a cell that assists in growth, as opposed to an ordinary cress cell. The model was run on the Ohio Supercomputer Center's IBM 1350 Cluster, which has 9,500 core CPUs and a peak computational capability of 75 teraflops. After running a large number of models, and comparing the results of the models to the literature, Siegal-Gaskins decided that three proteins were actively involved in determining whether a cell divided into a trichome or ordinary cells. Many examples of discrete modeling in biology using supercomputers are also available. For example, many DNA and protein-recognition programs can only be run on supercomputers because of their computational requirements.

Astronomy. There are many applications of supercomputers to astronomy, including using a supercomputer to simulate events in the future, or past, to test astronomical theories. The University of Minnesota Supercomputer Center simulated what a supernova explosion originating on the edge of a giant interstellar molecular gas cloud would look like 650 years after the explosion.

CAREERS AND COURSE WORK

A major in computer science, engineering, mathematics, or physics is most often the course selected to prepare for a career in parallel computing. It is advisable for those going into parallel processing to have a strong minor in an application field such as biology, medicine, or meteorology. One needs substantial course work in mathematics and computer science, especially scientific programming, for a career in parallel computing. There are a number of positions available for those interested in extending operating systems and building packages to be used by application programmers, and for these positions, one generally needs a master's or doctoral degree. Most universities teach courses in computer science and mathematics. Those parts of parallel computing related to the construction of devices are generally taught in computer science, electrical engineering, or computer engineering, while those involved in developing systems and application software are usually taught in computer science and mathematics. Taking a number of courses in programming and mathematics is advisable for anyone seeking a career in parallel computing.

Those seeking careers in parallel computing take a wide variety of positions. A few go to work for companies that develop hardware such as Intel, AMD, and Cray. Others go to work for companies that build system software such as Cray or IBM, or develop parallel computing applications for a wide range of organizations.

SOCIAL CONTEXT AND FUTURE PROSPECTS

Parallel processing can be used to solve some of societies' most difficult problems, such as determining when, and where, a hurricane is going to hit, thus improving people's standard of living. An interesting phenomenon is the rapid development of supercomputers for parallel computing in Europe and Asia. While this will result in more competition for the United States' supercomputer companies, it should also result in a wider use in the world of supercomputers, and improve the worldwide standard of living.

Supercomputers have always provided technology for tomorrow's computers, and technology developed for today's supercomputers will provide faster processors, system buses, and memory for future home and office computers. Looking at the growth of the supercomputer industry in the beginning of the twenty-first century and the state of computer technology, there is good reason to believe that the supercomputer industry will experience just as much growth in the 2010s and beyond. If quantum computers become a reality in the future, that will only increase the power and variety of future supercomputers.

—*George M. Whitson, III, MS, PhD*

FURTHER READING

Culler, David, Jaswinder Pal Singh, and Anoot Gupta. *Parallel Computer Architecture: A Hardware/Software Approach.* San Francisco:Kaufmann, 1999. Print.

Hwang, Kai, and Doug Degroot, eds. *Parallel Processing for Supercomputers and Artificial Intelligence.* New York: McGraw, 1989. Print.

Kirk, David, and Wen-mei W. Hwu. *Programming Massively Parallel Processors: A Hands-On Approach.* Burlington: Kaufmann, 2010. Print.

Rauber, Thomas, and Gudula Rünger. *Parallel Programming: For Multicore and Cluster Systems.* New York: Springer, 2010. Print.

Varoglu, Sevin, and Stephen Jenks. "Architectural support for thread communications in multi-core processors." *Parallel Computing* 37.1 (2011): 26–41. Print.

PATTERN RECOGNITION (SCIENCE)

FIELDS OF STUDY

Statistics; mathematics; physics; chemistry; engineering; computer science; artificial intelligence; machine learning; psychology; marketing and advertising; economics; forensics; physiology; complexity science; biology.

KEY TERMS AND CONCEPTS

- **Algorithm:** Rule or set of rules that describe the steps taken to solve a certain problem.
- **Artificial intelligence:** Field of computer science and engineering seeking to design computers that have the capability to use creative problem-solving strategies.
- **Character recognition:** Programs and devices used to allow machines and computers to recognize and interpret data symbolized with numbers, letters, and other symbols.
- **Computer-aided diagnosis (CAD):** Field of research concerned with using computer analysis to more accurately diagnose diseases.
- **Machine learning:** Field of mathematics and computer science concerned with building machines and computers capable of learning and developing systems of computing and analysis.
- **Machine vision:** Field concerned with developing systems that allow computers and other machines to recognize and interpret visual cues.
- **Neural network:** Patterns of neural connections and signals occurring within a brain or neurological system.
- **Pattern:** Series of distinguishable parts that has a recognizable association, which in turn forms a relationship among the parts.

SUMMARY

Pattern recognition is a branch of science concerned with identifying patterns within any type of data, from mathematical models to visual and auditory information. Applied pattern recognition aims to create machines capable of independently identifying patterns and using patterns to perform tasks or make decisions. The field involves cooperation between statistical analysis, mechanical and electrical engineering, and applied mathematics. Pattern recognition technology emerged in the 1970's from work in advanced theoretical mathematics. The development of the first pattern recognition systems for computers occurred in the 1980's and early 1990's. Pattern recognition technology is found both in complex analyses of systems such as economic markets and physiology and in everyday electronics such as personal computers.

DEFINITION AND BASIC PRINCIPLES

Pattern recognition is a field of science concerned with creating machines and programs used to categorize objects or bits of data into various classes. A pattern is broadly defined as a set of objects or parts that have a relationship. Recognizing patterns therefore requires the ability to distinguish individual parts, identify the relationship between the parts, and remember the pattern for future applications. Pattern recognition is closely linked to machine learning and artificial intelligence, which are fields concerned with creating machines capable of learning new information and using it to make decisions. Pattern recognition is one of the tools used by learning machines to solve problems.

One example of a pattern recognition machine is a computer capable of facial recognition. The computer first evaluates the face using a visual sensor and then divides the face into parts, including the eyes, lips, and nose. Next, the system assesses the relationship between individual parts, such as the distance between the eyes, and notes features such as the length of the lips. Once the machine has evaluated an individual's face, it can store the data in memory and later compare the information against other facial scans. Facial recognition machines are most often used to confirm identity in security applications.

BACKGROUND AND HISTORY

The earliest work on pattern recognition came from theoretical statistics. Early research concentrated on creating algorithms that would later be used to control pattern recognition machines. Pattern recognition became a distinct branch of mathematics and statistics in the early 1970's. Engineer Jean-Claude

Simon, one of the pioneers of the field, began publishing papers on optical pattern recognition in the early 1970's. His work was followed by a number of researchers, both in the United States and abroad. The first international conference on pattern recognition was held in 1974, followed by the creation of the International Association for Pattern Recognition in 1978.

In 1985, the United States spent $80 million on the development of visual and speech recognition systems. During the 1990's, pattern recognition systems became common in household electronics and were also used for industrial, military, and economic applications. By the twenty-first century, pattern recognition had become a robust field with applications ranging from consumer electronics to neuroscience. The field continues to evolve along with developments in artificial intelligence and computer engineering.

HOW IT WORKS

Pattern recognition systems are based on algorithms, or sets of equations that govern the way a machine performs certain tasks. There are two branches of pattern recognition research: developing algorithms for pattern recognition programs and designing machines that use pattern recognition to perform a function.

Obtaining data. The first step in pattern recognition is to obtain data from the environment. Data can be provided by an operator or obtained through a variety of sensory systems. Some machines use optical sensors to evaluate visual data, and others use chemical receptors to detect molecules or auditory sensors to evaluate sound waves.

Most pattern recognition computers focus on evaluating data according to a few simple rules. For example, a visual computer may be programmed to recognize only red objects and to ignore objects of any other color, or it may be programmed to look only at objects larger than a target length.

Translating data. Once a machine has intercepted data, the information must be translated into a digital format so that it can be manipulated by the computer's processor. Any type of data can be encoded as digital information, from spatial relationships and geometric patterns to musical notes.

A character recognition program is programmed to recognize symbols according to their spatial geometry. In other words, such a program can distinguish the letter *A* from the letter *F* based on the unique organization of lines and spaces. As the machine identifies characters, these characters are encoded as digital signals. A user can then manipulate the digital signals to create new patterns, using a computer interface such as a keyboard.

A character recognition system that is familiar to many computer users is the spelling assistant found on most word-processing programs. The spelling assistant recognizes patterns of letters as words and can therefore compare each word against a preprogrammed list of words. If a word is not recognized, the program looks for a word that is similar to the one typed by the user and suggests a replacement.

Memory and repeated patterns. In addition to recognizing patterns, machines must also be able to use patterns in problem solving. For example, the spelling assistant program allows the computer to compare patterns programmed into its memory against input given by a user. Word processors can also learn to identify new words, which become part of the machine's permanent memory.

Advanced pattern recognition machines must be able to learn without direct input from a user or engineer. Certain learning robots, for instance, are programmed with the capability to change their own programming according to experience. With repeated exposure to similar patterns, the machines can become faster and more accurate.

APPLICATIONS AND PRODUCTS

Computer-aided diagnosis. Pattern recognition technology is used by hospitals around the world in the development of computer-aided diagnosis (CAD) systems. CAD is a field of research concerned with using computer analysis to more accurately diagnose disease. CAD research is usually conducted by radiologists and also involves participation from computer and electrical engineers. CAD systems can be used to evaluate the results taken from a variety of imaging techniques, including radiography, computed tomography (CT), magnetic resonance imaging (MRI), and ultrasound. Radiologists can use CAD systems to evaluate disorders affecting any body system, including the pulmonary, cardiac, neurologic, and gastrointestinal systems.

At the University of Chicago Medical Center, the radiology department has obtained more than seventy patents for CAD systems and related technology. Among other projects, specialists have been working on systems to use CAD to evaluate potential tumors.

Speech recognition technology. Some pattern recognition systems allow machines to recognize and respond to speech patterns. Speech recognition computers function by recording and analyzing sound waves; individual speech patterns are unique, and the computer can use a recorded speech pattern for comparison. Speech recognition technology can be used to create security systems in which an individual's speech pattern is used as a passkey to gain access to private information. Speech recognition programs are also used to create dictation machines that translate speech into written documents.

Neural networks. Many animals, including humans, use pattern recognition to navigate their environments. By examining the way that the brain behaves when confronted with pattern recognition problems, engineers are attempting to design artificial neural networks, which are machines that emulate the behavior of the brain.

Biological neural networks have a complex, nonlinear structure, which makes them unpredictable and adaptable to new problems. Artificial neural networks are designed to mimic this nonlinear function by using sets of algorithms organized into artificial neurons that imitate biological neurons. The artificial network is designed to be adaptive, so that repeated exposure to similar problems creates strong connections among the artificial neurons—similar to memory in the human brain.

Although artificial neural networks have just begun to emerge, they could potentially be used for any pattern recognition application from economic analysis to fingerprint identification. Many engineers have come to believe that artificial neural networks are the future of pattern recognition technology and will eventually replace the linear algorithms that have been used most frequently.

Military applications. Pattern recognition is one of the most powerful tools in the development of military technology, including both surveillance and offensive equipment. For example, the Tomahawk cruise missile, sometimes called a smart bomb, is an application of pattern recognition used for offensive military applications. The missile uses a digital scene area matching correlation (DSAMC) system to guide the missile toward a specific target identified by the pilot. The missile is equipped with sensors, an onboard computer, and flight fins that can be used to adjust its trajectory. After the missile is fired, the DSAMC adjusts the missile's flight pattern by matching images from its visual sensors with the target image.

CAREERS AND COURSE WORK

The most direct route to achieve a career in pattern recognition would be to receive advanced training in electrical engineering. Professional statisticians, mathematicians, neurobiologists, and physicists also participate in pattern research. Alternatively, some medical professionals work with pattern recognition, most notably radiologists who participate in CAD research and development.

Texts and training materials in pattern recognition generally require a strong background in mathematics and statistics. Basic knowledge of statistical analysis is a prerequisite for the most basic college courses in statistical engineering. Those hoping to work at the forefront of pattern recognition research will also need experience with machine learning, artificial intelligence, artificial neural network design, and related areas.

SOCIAL CONTEXT AND FUTURE PROSPECTS

Pattern recognition technology has become familiar to many consumers. Voice-activated telephones, fingerprint security for personal computers, and character analysis in word-processing software are just a few of the many applications that affect daily life. Advances in medicine, military technology, and economic analysis are further examples of how pattern recognition has come to shape the development of society.

Projects that may represent the future of pattern recognition technology include the development of autonomous robots, space exploration, and evaluating complex dynamics. In the field of robotics, pattern recognition is being used in an attempt to create robots with the ability to locate objects in their environment. The National Aeronautics and Space Administration has begun research to create probes capable of using pattern recognition to find objects or sites of interest on alien landscapes. Combined with research on artificial neural networks, automated systems may soon be capable of making complex decisions based on the recognition of patterns.

—Micah L. Issitt, MA

FURTHER READING

Brighton, Henry, and Howard Selina. Introducing Artificial Intelligence. Edited by Richard Appignanesi. 2003. Reprint. Cambridge, England: Totem Books, 2007. Introduces many aspects of artificial intelligence, including machine learning, geometric algorithms, and pattern recognition, and provides examples of how specialists in artificial intelligence and pattern recognition create applications from their research.

Frenay, Robert. Pulse: The Coming Age of Systems and Machines Inspired by Living Things. 2006. Reprint. Lincoln: University of Nebraska Press, 2008. An introduction to bioengineering, artificial intelligence, and other organism-inspired machines. Provides an introduction to the philosophy and politics of artificial intelligence research.

Marques de Sá, J. P. Pattern Recognition: Concepts, Methods and Applications. New York: Springer Books, 2001. A nonspecialist treatment of issues surrounding pattern recognition. Looks at many of the principles underlying the discipline as well as discussions of applications. Although full understanding requires knowledge of advanced statistics and mathematics, the book can be understood by students with a basic science background.

McCorduck, Pamela. Machines Who Think: A Personal Inquiry into the History and Prospects of Artificial Intelligence. 2d ed. Natick, Mass.: A. K. Peters, 2004. An introduction to the history and development of artificial intelligence written for the general reader. Contains information about pattern recognition systems and applications.

Singer, Peter Warren. Wired for War: The Robotics Revolution and Conflict in the Twenty-first Century. New York: Penguin Press, 2009. An exploration of the use of intelligent machines in warfare and defense. Contains information on a variety of military applications for pattern recognition systems.

Theodoridis, Sergio, and Konstantinos Koutroumbas. Pattern Recognition. 4th ed. Boston: Academic Press, 2009. A complex introduction to the field, with information about pattern recognition algorithms, neural networks, logical systems, and statistical analysis.

PETROLEUM EXTRACTION AND PROCESSING

FIELDS OF STUDY

Engineering; petroleum engineering; chemical engineering; mechanical engineering; environmental engineering; control engineering; geosciences; geology; geophysics; hydrogeology; mining; seismology; chemistry; physics; mathematics; business management.

KEY TERMS AND CONCEPTS

- **barrel:** Standard unit of measuring the volume of oil as well as oil and gas products; equals 42 gallons or 159 liters.
- **crude oil:** Liquid part of petroleum, a mix of different hydrocarbon chains.
- **distillation:** Physical separation by heating of mixed components with different volatilities.
- **downstream:** Processing of crude oil and natural gas, transporting and marketing the processed products to consumers.
- **exploration:** Searching for geological formations likely to hold petroleum reservoirs and accessing them.
- **feedstock:** Chemical compound to be processed into a higher value chemical in a petrochemical plant.
- **field:** Area covered by an underground petroleum reservoir.
- **fraction:** End product of refining petroleum.
- **natural gas:** Volatile part of petroleum.
- **refining:** Processing crude oil through distillation, cracking of fractions, reforming, and blending.
- **reservoirs:** Large underground deposits of fossilized hydrocarbons.
- **rig:** Mechanism to drill an oil or gas well.
- **upstream:** Exploring and producing petroleum.

SUMMARY

Petroleum extraction and processing is the human exploitation of fossil fuels consisting of hydrocarbons in the form of crude oil and natural gas. The primary use of processed petroleum products is as powerful fuels, particularly for transportation in the form of gasoline, diesel fuel, or jet fuel, and for heating purposes, which account for 84 percent of petroleum use. The remainder of processed petroleum products is used in the petrochemical industry as fuel additives and to create applications such as plastics, specialty chemicals, solvents, fertilizer, pesticides, and pharmaceuticals. Worldwide, about 37 billion barrels of crude oil and natural gas are extracted each year.

DEFINITION AND BASIC PRINCIPLES

Petroleum extraction and processing encompass the activities through which crude oil and natural gas are taken from their natural reservoirs below the surface of the earth on land and sea and treated so they can be used as fuels and materials for the petrochemical industry. Reservoirs can be detected by applying the results of petroleum geology, the science describing under what circumstances oil and gas reservoirs were created during ancient times and how they can be found.

Once a likely reservoir is identified, exploratory drilling through the ground begins, either on land or on sea. When a well strikes a reservoir, the size of its area is estimated. Depending on whether crude oil or natural gas is dominant, the reservoir is called an oil or gas field. Many wells are drilled to extract oil and gas from fields that can be quite vast, extending for hundreds or thousands of square kilometers (or miles) under the surface. As of 2010, there were about 40,000 producing oil and gas fields in the world.

Usually, natural pressure forces gas and oil to the surface. In mature wells, however, pressure has to be added by technical means. After gas and oil are extracted, both are sent by pipelines for further processing.

Processing natural gas separates almost all other components from its key ingredient, methane. Processing crude oil occurs in a refinery. Because the different components of crude oil vaporize at different temperatures, they can be separated by distillation. To obtain the most desirable end products, the components are further treated through chemical processes called cracking and reforming.

BACKGROUND AND HISTORY

Modern petroleum use was based on demand for kerosene for lighting. Polish pharmacist Ignacy

Łukasiewicz discovered how to distill kerosene from petroleum in 1853. He dug the first modern oil well in 1846 and the world's first refinery in 1856.

Retired railway conductor Edwin Drake dug the first oil well in the United States at Titusville, Pennsylvania, on August 28, 1859. His drilling method, using a pipe to cover a drill suspended from a rig and leading the pipe into the earth to prevent collapse of the borehole, was widely copied as he did not patent it. It is still the basic drilling mechanism. Horizontal rather than vertical drilling, which allows easier access to the reservoir of an oil field, revolutionized crude oil extraction in the 1990's.

The appearance and spread of the internal combustion engine by the late nineteenth century led to crude oil being refined to produce gasoline. The quest to extract more gasoline led to the application of thermal cracking in 1913, and by 1975, many subsequent new cracking techniques had culminated in residual hydrocracking. Beginning in 1916, processes were invented to decrease the unwanted sulfur content of crude oil. Since 1939, visbreaking has been used to make smoother flowing products, and isomerization and alkylation have increased gasoline yield and quality since 1940. Modern refineries try to make these processes as efficient and clean as possible.

HOW IT WORKS

Exploration. Almost all exploration of new oil and gas fields is done by geophysical means such as seismic reflection, or by other seismic, gravity, or magnetic surveys. These surveys use the different densities of subterranean rock formations to identify source rock formations likely to hold hydrocarbons. Based on this information, three-dimensional computer models are generated and analyzed. Once petroleum engineers and geologists have decided on a promising prospect, exploratory drilling begins.

Extraction. Oil and gas deposits are accessed by drilling a well from a rig, either on land or in the sea. Rigs are adapted to their particular environment but share some common elements. Suspended from a metal tower called a derrick, the drilling bit hangs from its drilling string over the well bore, the borehole where the well is dug. As the drilling bit goes deeper, the borehole is fitted with a pipe to prevent its collapse. The pipe is placed in a larger steel casing, leaving space between the pipe and casing. Drilling mud descends through the pipe to cool the drilling

bit as it bores into the rock and percolates back to the surface between the pipe and casing. On the sea, the rig is placed on an offshore oil platform. Modern drills can dig thousands of meters (or feet) deep.

Drilling is successful when the well strikes petroleum deposits below their seal rock. Natural pressure ejects gas and oil through the pipe. A wellhead is placed on the top of the well, and natural gas is separated from crude oil in a separator unit. Both products are collected and transported by intermediary pipelines connecting the many wells of a single field.

As long as natural pressure ejects petroleum, primary recovery continues. Once this pressure decreases, water or gas is pumped into the reservoir to create artificial pressure during secondary recovery. When maintaining artificial pressure is no longer economical, the well is considered exhausted.

Processing. Natural gas has to be cleaned in gas processing plants typically built close to the point of extraction. It is separated from its water and natural gas condensate (also called natural gasoline), and the latter is sent to a refinery. Then it is stripped of its acid gases (carbon dioxide, hydrogen sulfide, and other gases with sulfur content), which are desulfurized and often burned as tail gases. To gain as much pure methane, its main component, as is possible, natural gas undergoes a further series of processes to remove contaminants (such as mercury) and nitrogen. Natural liquid gases (NLGs) are removed for use as feedstock in petrochemical plants. Then, consumer-grade natural gas is ready for sale.

Crude oil is processed at a refinery. Refineries need not be close to oil fields as crude oil can be transported by pipeline or ocean tanker. A refinery, which is a highly integrated chemical plant, seeks to create the most desirable products out of crude oil. These products are hydrocarbon chains with few rather than many carbon atoms (called light instead of heavy).

First, crude oil is desalted, then it is separated into its various components, called fractions, by atmospheric distillation. As crude oil is heated, its different fractions vaporize at different temperatures and are removed separately. All fractions are processed further. The lightest fraction, distilled gas, is subjected to sulfur extraction. The heaviest fraction, called atmospheric bottom, undergoes vacuum distillation. Most distilled fractions undergo hydrotreatment, which is infusion of hydrogen to remove sulfur.

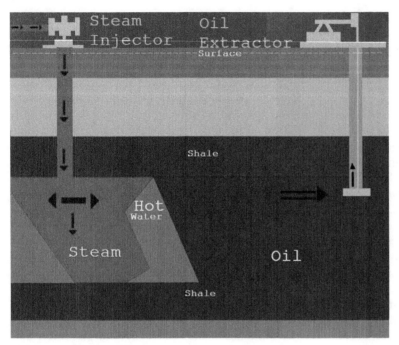

Steam is injected into many oil fields where the oil is thicker and heavier than normal crude oil

At the heart of a refinery are its fluid catalytic cracker and hydrocracker, which break down heavy (long) hydrocarbon chains into lighter (shorter) ones. A catalytic reformer is used to create higher octane reformate, a more valuable naphtha distillate. Isomerization and alkylation are two chemical processes designed to boost the distillate's octane rating. Refineries send their products to end users or the petrochemical industry.

APPLICATIONS AND PRODUCTS

Gasoline. The most common and valuable petroleum product is gasoline. Typically, about 46 percent of crude oil is processed into gasoline of various qualities. As fuel for the internal combustion engine, gasoline is essential for any industrial society. It is used by most cars and many light trucks. Aviation gasoline fuels the piston engines of planes.

Refineries seek to optimize their gasoline output and strive to create blends that minimize damaging spontaneous combustion in engines (known as knocking). Lead was used until the 1970's, but its use has been prohibited in most industrialized nations because of its harmful effects. In its stead, chemists have developed a variety of alternative additives.

Diesel and fuel oils. The second most important petroleum products, which account for about 26 percent of crude oil, are diesel and other fuel oils. These fuel oils have longer hydrocarbon chains than gasoline and thus boil at higher temperatures. Diesel is used as fuel for trucks, buses, ships, locomotives, and automobiles. In Europe, diesel engines are very popular for automobiles as they are more fuel efficient. Environmentalists have expressed concern about burning these fuels because the process creates the pollutant sulfur dioxide.

Heavy fuel oils account for another 4 percent of petroleum products processed from crude oil. Among them, bunker oil is the heaviest. It is used to fuel large ship engines.

Natural gas. Natural gas accounts for about 35 percent of the value of all petroleum products. It is used for generation of electric power, domestic and industrial heating, and increasingly transportation by bus or automobile. Europe has a dense networks of pipelines bringing natural gas, primarily from Russia, to industrial and domestic consumers.

Natural gas can be liquefied (as liquefied natural gas, LNG) for transport by gas tankers across oceans. This is the preferred method to bring natural gas from gas fields in the Middle East and North Africa to Europe. After reaching the destination, LNG is gasified and pumped into a pipeline at the port terminal.

Jet fuel. Jet fuel powers gas-turbine engines of airplanes (jets). Its production amounts to about 9 percent of processed crude oil. Commercial jet fuel is produced according to international standards. It is either Jet A or Jet A-1, with Jet A freezing at -40 degrees Celsius (-40 degrees Fahrenheit) and A-1 at -47 degrees Celsius (-52.6 degrees Fahrenheit). Jet B is a lighter, more flammable jet fuel for cold regions. Military jet fuel is produced according to individual national standards.

Liquefied petroleum gas. The result of distilling the remaining gas from crude oil at a refinery, liquefied petroleum gas (LPG) is widely used for domestic heating and cooking in Asia, South America, and

Eastern Europe, as well as a global fuel alternative for automobiles. LPG provides about 3 percent of worldwide fuel demand.

Petrochemicals. Petrochemicals are often by-products of gasoline production. They are grouped as olefins or aromatics. The value of petrochemicals, which can account for up to 8 percent of petroleum products made of crude oil, has been rising consistently as new chemical processes have found additional applications.

After they leave a refinery, most petrochemicals are used as feedstock for petrochemical plants. Among olefins, ethylene and propylene are building blocks for plastics and specialty chemicals. The olefin butadiene is processed to eventually form synthetic rubber. Among aromatics, benzene is used in the production of synthetic dyes, detergents, and such innovative products as polycarbonates that form light hard plastic shells in many electronic items such as mobile phones. Toluene and xylene are used for solvents or as building blocks for other chemicals, creating, for example, polyester fibers.

Petrochemicals are also used to create adhesives, food ingredients, and pharmaceuticals. They have become a source of much innovative chemical research.

Lubricating oils. Although their volume accounts for only 1 percent of petroleum products leaving a refinery, lubricating oils and greases are important for the machines and engines of industrialized nations. Refineries create a great variety of lubricating oils through blending with different additives.

Asphalt. The heaviest product from crude oil, asphalt makes up about 3 percent of refinery products. Concrete asphalt for durable roads and runways accounts for about 85 percent of North American asphalt use; asphalt roof shingles are the next most common use.

CAREERS AND COURSE WORK
Students interested in the petroleum industry should take science and mathematics courses, particularly chemistry and physics, in high school. In college, the aim should be a bachelor of science. The choice of science to study depends on whether the student wishes to pursue a career in exploration and production or processing and petrochemicals.

For the former, the geosciences, particularly geology, geophysics, mining, hydrology, and hydrogeology, as well as chemistry, are useful. Computer literacy and ability to work with specific applications in geology, particularly three-dimensional sensing and modeling, is important. Oil companies will welcome successful graduates who hold a bachelor of science in geology, hydrogeology and engineering, chemical engineering, geology and mining, or even physics. A master's or engineering degree in these fields will provide additional opportunities. For example, a master's of science in petroleum engineering or in environmental engineering geology can lead to an advanced entrance position as reservoir geologist. One of the most important criteria for top positions, however, is years of practical work experience. Because increasing responsibilities lead toward management, it would not be uncommon for a senior geologist to hold a master of business administration degree in addition to science degrees.

For a career in petroleum processing at a refinery or petrochemical plant, a bachelor of science in chemistry or a related field is a good foundation. A master's or doctoral degree in chemistry or chemical engineering, as well as in mechanical, environmental, or control engineering, will serve for an advanced position.

The oil and gas industry also employs many laboratory scientists and technicians. Corporate research and development departments are typically headed by scientists with doctorates in chemistry and related sciences.

Because of the cyclical nature of the oil and gas industry, there are periods in which skilled employees such as drilling engineers are in high demand, but these workers also experience layoffs during downturns. Work can be at remote locations, and global mobility is often expected.

SOCIAL CONTEXT AND FUTURE PROSPECTS
The late-nineteenth-century discovery that crude oil and natural gas could be turned into powerful fuels for industrialized society meant that those nations where oil and gas were found possessed a very valuable natural resource. However, use and distribution of wealth from petroleum was a huge social challenge.

The industrialized nations' dependence on petroleum products has significantly influenced global politics since the early twentieth century. Since the late 1980's, there has been increasing concern that the massive burning of hydrocarbon fuel causes global warming because of the accumulation of greenhouse gases in the atmosphere. As of 2010, no global consensus had been reached on how to address and mitigate this issue.

Oil spills, whether from wells or when crude oil tankers run aground, can be major environmental disasters. On April 20, 2010, an explosion destroyed the Deepwater Horizon oil drilling platform in the Gulf of Mexico. The ensuing spill of gas and crude oil from a well on the ocean floor, about 1,500 meters (about 5,000 feet) below the water, became the world's biggest spill, releasing between 35,000 to 60,000 barrels per day and proving very hard to contain.

There is ongoing scientific debate over when the Earth's petroleum resources will be depleted. In 1956, American geoscientist Marion King Hubbert published his peak oil theory, in which he tried to calculate when the volume of petroleum extracted would exceed the volume of remaining petroleum reserves. A concerted effort involving new exploration and extraction technologies seeks to discover even more remote reserves, particularly deep under the sea, and to develop oil shale and oil sands more efficiently and as economically as possible. Scientists have also predicted that future alternative fuels, including those derived from hydrogen or renewable resources, will one day take the role of petroleum.

—*R. C. Lutz, MA, PhD*

FURTHER READING

Burdick, Donald. *Petrochemicals in Nontechnical Language*. 3d ed. Tulsa, Okla.: Penn Well, 2010. An accessible introduction to the field that relates serious science concepts in an informative fashion. Figures and tables.

Duffield, John. *Over a Barrel: The Costs of U.S. Foreign Oil Dependence*. Stanford, Calif.: Stanford University Press, 2008. Addresses economic and foreign policy in the United States as well as military responses to the problem. Concludes with a proposal to lower the costs of dependence. Tables, figures, maps.

Gary, James, et al. *Petroleum Refining: Technology and Economics*. 5th ed. New York: CRC Press, 2007. A well-written textbook that provides a good study of refining. The last two chapters cover economic issues. Five appendixes, index, photographs.

FASCINATING FACTS ABOUT PETROLEUM EXTRACTION AND PROCESSING

- The term "petroleum" comes from Latin *petra* for rock and *oleum* for oil, and literally means rock oil. In China, since 1008 CE, crude oil has been called *shiyou*, which means rock (*shi*) oil (*you*).
- The earliest known extraction of petroleum was before 2000 BCE from oil pits near Babylon, in modern Iraq, for use as asphalt for city walls.
- By 347 CE, the Chinese had drilled the first oil well and laid bamboo pipelines to salt springs. Crude oil was burned to evaporate brine and gain salt.
- Petroleum reservoirs were formed when ancient zooplankton and algae died and sank to ocean floors. There, the Earth's heat and pressure transformed them into oil and gas.
- The United States rose to power in part because of its unique combination of industrial assets and rich petroleum resources, being the world's leading petroleum producer until the mid-1970's.
- In 2010, the United States was still the third largest producer of petroleum, behind Saudi Arabia and Russia, producing 8.5 billion barrels per day.
- The relatively thinly populated oil-producing countries of the Middle East saw a massive jump in per capital income after the Organization of Petroleum Exporting Countries hiked prices in 1973.
- Power plants that use natural gas in the combined cycle mode, using both a gas and a steam turbine, are among the most fuel-efficient and least polluting means of electricity generation.
- Liquefied petroleum gas (LPG) is a primary domestic heating fuel in Asia. Trucks providing LPG canisters to households are a common sight even in highly industrialized Japan.
- The petroleum industry is cost intensive. Digging a deep offshore oil well costs more than $100 million, and a new world-class refinery costs up to $4 billion.

Hyne, Norman. *Nontechnical Guide to Petroleum Geology, Exploration, Drilling, and Production*. 2d ed. Tulsa, Okla.: Penn Well, 2001. A comprehensive description of all activities that take place before petroleum is processed. Tables, figures, index.

Raymond, Martin, and William Leffler. *Oil and Gas*

Production in Nontechnical Language. Tulsa, Okla.: Penn Well, 2005. An accessible introduction that covers all major aspects. Lighthearted but very informative style.

Yeomans, Matthew. *Oil: Anatomy of an Industry.* New York: The New Press, 2004. A critical look at the oil industry, particularly in the United States. Provides a good historical overview and addresses issues of America's dependency on oil, world conflicts caused by oil, and the question of alternatives such as hydrogen fuel.

PHOTOGRAPHY SCIENCE

FIELDS OF STUDY

Chemistry; physics; optics; art and architecture; visual media; visual design; environmental science; history and aesthetics of images; journalism; computer science; computer graphics.

KEY TERMS AND CONCEPTS

- **charge-coupled device (CCD):** Type of sensor that converts electromagnetic radiation into electric charges that are processed to form a digital image.
- **color temperature:** Color that appears when a hypothetical body, called a blackbody, is heated to a specific temperature.
- **complementary metal-oxide semiconductor (CMOS):** Type of sensor that converts electromagnetic radiation into electric charges that are processed to form a digital image.
- **dynamic range:** Range of light contrast in a scene.
- **electromagnetic radiation:** Wave energy emitted by any substance that possesses heat. The temperature of the substance determines the wavelength of emission; for visible light, the wavelengths range from about 0.4 to 0.7 micrometers (μm).
- **high dynamic range (HDR):** Combination of two or more images of the same scene but with different exposures to form a single image that more faithfully represents the scene's dynamic range.
- **pixel:** Picture element; the smallest unit that contains light-sensitive information on a digital imaging sensor or display screen.
- **silver halide:** Light-sensitive compound used in photographic film and paper.
- **single lens reflex (SLR):** Mechanical mirror system in a film or digital camera that reflects the light entering the lens so that the view through the camera's eyepiece matches the scene recorded on the film or digital sensor.

SUMMARY

Photography is the process of forming images on surfaces that are sensitive to electromagnetic radiation. These surfaces are usually silver halide-based film or an electronic photosensor, and the radiation recorded is usually visible light, infrared radiation, ultraviolet radiation, or X rays. Photography is an important technological tool for examining and documenting the natural and human-built world. For more than 150 years, photography has added vital knowledge to the physical sciences, environmental sciences, biological sciences, medical sciences, forensic sciences, materials sciences, and engineering sciences.

DEFINITION AND BASIC PRINCIPLES

Photography is the process of forming images on surfaces that are sensitive to electromagnetic radiation. Usually the surfaces are either a silver halide-based film or an electronic (digital) photosensor that has been designed to be sensitive to visible light. Depending on the application, the film or electronic sensor may be designed to form images in response to infrared radiation, ultraviolet radiation, or X rays.

BACKGROUND AND HISTORY

The word "photography" derives from the Greek words *photos* and *graphos*, which together mean "light drawing." Experiments in the precursor technologies of photography date back to ancient times, including the pinhole camera and later the camera obscura, a device used by artists to assist in drawing. Thomas Wedgwood allegedly made photographs as far back as the 1790s, but had no way to stabilize the image to make them permanent. The first permanent photographic images were made by French inventor Nicéphore Niépce using photosensitive bitumen of Judea, including the first image of a natural scene, taken in 1826. In 1833 Hercules Florence, working independently in Brazil, produced prints of drawings during experiments with light-sensitive silver salts (his pioneering work would remain largely unknown until the 1970s). Meanwhile Niépce and Louis Daguerre in France and William Henry Fox Talbot in England continued to improve methods of capturing and preserving images. It was Talbot's work that led preeminent scientist John Herschel to coin the word "photography" in 1839 to describe Talbot's process for fixing an image. Because of Herschel's broad connections to the scientific community, he is traditionally credited with introducing the words "photography" and "photograph."

The invention of photography was announced to the world on January 7, 1839, at a meeting of the French Academy of Sciences in Paris, France. François Arago, a physicist and secretary of the academy, announced that Daguerre had employed the action of light to permanently fix an image with the aid of a camera obscura. Daguerre's image was made on a highly polished, light-sensitized silver plate, which produced a one-of-a-kind image called a daguerreotype.

Several months after Arago's announcement of the daguerreotype, Talbot announced his process for fixing an image by the action of light. In contrast to Daguerre's process, which relied on a light-sensitized silver plate, Talbot's process relied on light-sensitized paper. Talbot's process, which he called photogenic drawing, produced a stable paper negative from which multiple prints could be made. Talbot went on to improve his process to produce images called calotypes, after the Greek word *kalos*, meaning beautiful. Although the daguerreotype was sharper and produced greater detail than the calotype, the paper-based calotype is more similar to the modern photograph.

HOW IT WORKS

The camera is a device that basically consists of three components: a light-tight box containing a light-sensitive surface (such as film or a digital sensor), a lens for focusing the light onto the surface, and a means for controlling the exposure, such as a mechanical or electronic shutter. With a film-based camera, image processing takes place outside the camera, often in a darkroom or image-processing laboratory. With digital photography, image capture, processing, and storage all take place within the camera.

Film-based cameras are designed in a variety of formats, although the three main formats are 35 millimeter (mm), medium format, and large format view. The type of camera format plays an important role in image quality; typically, larger format cameras produce higher quality images. Because each format differs sharply in portability, the choice of format depends largely on the application. Traditional photographic techniques are based on chemistry. In most film-based cameras, the film is coated with silver halide, a light-sensitive compound that is suspended in a gelatin. The film characteristics, such as grain, light sensitivity, and overall image quality, depend on the size, shape, and distribution of the silver-halide crystals. The sensitivity of the film to light is denoted by its International Organization for Standardization (ISO) number. Film characterized as slow has a low ISO number, such as ISO 50, whereas film characterized as fast has a high ISO number, such as ISO 400. The film grain becomes increasingly apparent as the ISO number increases. Correspondingly, the image quality tends to decrease as the ISO number increases.

Digital cameras are designed with a light-sensitive electronic sensor that records image information. The sensor is made up of an array of photosensors that convert light to electric signals. The signals are proportional to the intensity of the light and are assigned numbers, which the image processor uses to create individual picture elements called pixels. The pixels contain information on brightness (luminance) and color (chrominance). The digital information is then processed by the camera's computer chips to produce the image that is stored in the camera's memory.

Digital cameras generally use two types of digital sensors: the charge-coupled device (CCD) and the complementary metal-oxide semiconductor (CMOS). The CCD and CMOS sensors differ primarily in how they convert charge to voltage. In the CMOS sensor, the conversion takes place at each photosensor, whereas in the CCD sensor, the conversion takes place at a common output amplifier. The advantages of the CCD sensor over the CMOS are that it has a higher dynamic range and very low noise levels. The advantages of the CMOS sensor over the CCD are that it is less costly to produce and requires less power.

Digital cameras are designed with image sensors of varying sizes. The size of the sensor and the number of pixels contained on the sensor play an important role in image quality. Rapid technological advances in sensor design have resulted in dramatic improvements in the dynamic range recorded by the sensors and the reduction of digital noise in long exposure, low-light conditions.

Digital photography has several advantages over film photography. For example, image composition and exposure can be reviewed immediately; if necessary, images can be deleted to free up file space on the camera's memory card, and many more photographs can be stored compared to a roll of film. There is no need for film or chemical processing in a darkroom. Color temperature can be changed with different lighting situations to ensure color fidelity on the image. Images can be transferred almost instantly and wirelessly to remote locations or uploaded onto the Internet. Digital images are easily edited using image processing software. The flexibility of digital images, along with the increasing image quality and

Lens and mounting of a large-format camera

decreasing price of equipment through the 2000s, led to the increasing popularity of digital photography in virtually all fields of photography. However, film photography remains in use in some applications, especially as an aesthetic choice in artistic photography, though mostly as a niche market.

APPLICATIONS AND PRODUCTS

Aerial photography. First accomplished with balloons in the nineteenth century and with aircraft and satellites in the twentieth century, aerial photography has enabled scientists to acquire information from platforms above the Earth's surface. That information is used to characterize and track changes in land use, soil erosion, agricultural development, water resources, vegetation distribution, animal and human populations, and ecosystems. It is also used to detect water pollution, monitor oil spills, assess habitats, and provide the basis for geologic mapping. Because aerial photographs can record wavelengths of electromagnetic radiation that are invisible to the human eye, such as thermal infrared radiation, plant canopy temperatures can be measured and displayed on an aerial photograph that characterize the plant's stress due to environmental conditions. Aerial photography is a form of remote sensing.

By applying photogrammetric methods, whereby spatial relationships on an aerial photograph are related to spatial relationships on Earth's surface, analysts can relate distances on the photograph to distances on the ground. Object heights and terrain elevations can also be obtained by comparing photographs made from two different vantage points, each with a different line of sight.

Additional information can be gleaned from aerial photographs by examining tonal changes and shadow distributions within the photograph. Tonal changes are related to surface texture, which can be used to distinguish between vegetation type, soil type, and other surface features. Because the shapes of shadows change with time of day and are unique to particular objects, such as bridges, trees, and buildings, the shadows can be used to aid in the identification of the objects.

Environmental Photography. As the environment undergoes changes because of human activities and natural forces such as floods and earthquakes, documenting those changes has become increasingly important. Photography has played, and will continue to play, an important role in that documentation process. For example, scenes photographed in the nineteenth century by such pre-eminent photographers as William Henry Jackson, Timothy O'Sullivan, and Carleton Watkins as part of government-sponsored surveys of the American West were used to acquire scientific data about the geology and geomorphology of the land. In the twenty-first century, those same scenes are being photographed from the same vantage points and under similar lighting conditions to document the environmental changes that have occurred since the mid-nineteenth century. Environmental photographs made from land-based and elevated platforms, such as airplanes and satellites, provide valuable visual information for monitoring present-day and future environmental changes involving ecological systems, snowpacks, forests, deserts, soils, water sources, and the human-built landscape. Such information aids in habitat restoration, land-use planning, and environmental policy making.

Medical photography. Photography has been an integral part of medical science since the mid-nineteenth century. In 1840, one year after the invention of photography was announced to the world, Alfred Donné, a Paris-based physician, used a microscope-daguerreotype to photograph bone and dental tissue. In 1865, in a presentation to the Royal Society, British physician Hugh Welch Diamond advocated the use of photography to document mental patients for later analysis. In the late nineteenth century, Frederick Glendening, a London-based pioneer in medical photography, used clinical photographs of the human body to assist in the diagnosis of disease. At around the same time, Thomas R. French and George Brainerd collaborated to produce the first photographs of the larynx, while W. T. Jackman and J. D. Webster are believed to have made the first photographs of the human retina. The first X-ray photographs were made by Wilhelm Conrad Röntgen in 1895. Since then, X-ray photographs have become a routine tool for medical diagnostics.

Medical photography, also known as biomedical photography, has become a highly specialized field that requires precision and accuracy to be an effective diagnostic tool. Because medical photography is deeply rooted in digital technology, involving, for example, X rays, magnetic resonance imaging (MRI), and positron emission tomography (PET), practitioners must be well versed in digital imaging techniques. Additional training in the biological sciences, medical sciences, photogrammetry, and photomicrography may be required depending on the area of specialization.

Scientific photography. In addition to aerial photography, environmental photography, and medical photography, there are many other specialties that fall under scientific photography. Kirilian photography involves using electrophotography to form a contact print, for example, by applying high-voltage but low current to an object on a photosensitive surface. Some uses are primarily for record keeping or documentation, including archaeological photography (images of past human remains and artifacts taken on-site at excavations or in laboratories), forensic photography (documentary images that can be used in the legal process or admitted as evidence in court), and time-lapse photography (images taken at specified time intervals to show movement or change over time. Botanical photography involves taking images of plants, paying particular attention to their botanical and morphological adaptations, to aid in the identification and study of environmental stresses, diseases, and their interactions with other plants and animals. In biological photography, which sometimes uses optical and electron microscopy, images are taken of plant and animal specimens from micro to macro scales and in land and aquatic environments. In underwater photography, waterproof cameras or traditional cameras in unmanned or manned submersibles are used to photograph underwater scenes. Astrophotography uses both Earth-based and satellite platforms to reproduce images of celestial objects.

CAREERS AND COURSE WORK

Photography is a broad field with many specialty areas and applications. The best preparation for a career in photography is to earn a bachelor of fine arts degree or a bachelor of science degree with elective courses in fields related to specific photographic specializations. Beyond a core of liberal arts courses, general course work should include classes in the materials and processes of photography, general art classes, the history of art and architecture, studio photography, studio drawing, two-dimensional design, and computer science. Although a master's degree is not required for most careers in photography, the additional graduate-level training may be beneficial for some areas of specialization, such as fine arts photography, architectural photography, and biomedical photography.

For photographic specialties in the physical, medical, or engineering sciences, additional course work is recommended in chemistry, physics, mathematics, applied computer science, data analysis, imaging systems, and technical writing.

For a photography career specializing in the environmental sciences, additional course work in meteorology, climate science, soil sciences, ecology, and hydrology is recommended. For a photography career in the biological sciences, perhaps photographing wildlife, plants, or insects, additional courses in wildlife management, biology, entomology, and related fields would be beneficial. A career in forensic photography, which focuses on documentation of accident and crime scenes for law enforcement and disaster scenes for the insurance industry, may require additional training in underwater photography, the principles of photogrammetry, and criminal justice.

Because academia, industry, and government will continue to have an increasing demand for accurate visual information and visual communication skills, career opportunities in photography should continue to grow for those individuals who are best prepared academically.

SOCIAL CONTEXT AND FUTURE PROSPECTS

Since Daguerre "imprisoned" light in the mid-nineteenth century, photography has had a profound influence on art and science. In the hands of the artist, the camera has heightened awareness of the aesthetic qualities of space and light while revealing hidden truths about culture and society. From centuries-old experiments in optics and chemistry to the digital revolution, the camera has relied on science for its development while also serving as an essential scientific tool for probing and documenting the natural and human-built world.

In its relatively short life, photography has evolved rapidly and profoundly. During the nineteenth century, the slow exposure times associated with the daguerreotype and calotype processes were greatly improved by the newer wet plate collodian process. Film was eventually introduced, shortening the exposure times even more. The shorter exposure times combined with improved lens optics enabled scientists of the nineteenth century to study phenomena that were too quick for the unaided eye to see. Photographs of galloping horses, humans in motion, birds in flight, lightning, and distant galaxies were among the phenomena studied through photography.

At the beginning of the twenty-first century, digital photography largely supplanted film photography. Digital photography continues on a trajectory of rapid growth and development. It has enabled nearly

---

FASCINATING FACTS ABOUT PHOTOGRAPHY

- The mammoth camera built in 1900 for the Chicago and Alton Railway to photograph trains—along with the holder for the 8-by-4.5-foot photographic plates—weighed 1,400 pounds and is believed to be the largest camera ever built.

- In 2008, about 100 million digital cameras were sold, 100 billion photographic images were made, and more than 1 trillion digital images were posted on the World Wide Web.

- The first image taken by Wilhelm Conrad Röntgen, who discovered X rays in 1895, was of his wife's hand. It showed her opaque flesh, bones, and rings.

- In 1890, Austrian physiologist Sigmund Exner gazed through a microscope and photographed through the amputated eye of a firefly. The photomicrograph showed a dreamlike image of a barn framed by a window. Exner's groundbreaking photograph provided proof for a then-controversial theory that the multiple images produced by the firefly's compound eye are resolved into a single upright image.

- In 1872, Leland Stanford, former governor of California, president of the Central Pacific Railroad, and founder of Stanford University, commissioned English photographer Eadweard Muybridge to solve a debate regarding whether all four hooves of a galloping horse were ever in the air at the same time. The momentary airborne state of the horse, not visible to the human eye, was revealed by Muybridge's pioneering high-speed photography.

- Photographic manipulation was commonplace in the nineteenth century, when images were combined in the darkroom to overcome the inability of the photographic emulsions to resolve detail in both the sky and the landscape.

immediate sharing of visual information, which has greatly aided in the monitoring of the environment, the diagnostics of diseases and health problems, the documentation of crime scenes by law enforcement, and the dissemination of images by the news media. In the coming years, digital cameras will continue to diminish in size and price and improve in resolution and noise reduction. The ease of digital image making, especially as digital cameras become smaller and less conspicuous, has also raised issues concerning personal privacy and image making in public venues.

—*Terrence R. Nathan, PhD*

FURTHER READING

Easby, Rebecca Jeffrey. "Early Photography: Niépce, Talbot, and Muybridge." *Khan Academy*. Khan Academy, 2015. Web. 26 Feb. 2015.

Frankel, Felice. *Envisioning Science: The Design and Craft of the Science Image*. Cambridge: MIT P, 2004. Print.

"History of Photography." *WGBH*. PBS Online, WGBH, 2000. Web. 26 Feb. 2015.

Hunter, Fil, Steven Biver, and Paul Fuqua. *Light, Science and Magic: An Introduction to Photographic Lighting*. 3d ed. Burlington: Focal, 2007. Print.

Keller, Corey, ed. *Brought to Light: Photography and the Invisible, 1840–1900*. New Haven: San Francisco Museum of Modern Art, Yale U P, 2008. Print.

Mante, Harald. *The Photograph: Composition and Color Design*. Santa Barbara: Rocky Nook and Verlag Photographie, 2008. Print.

Peat, F. Davis. "Photography and Science: Conspirators." *Photography's Multiple Roles: Art, Document, Market, Science*, ed. by Terry A. Neff. Chicago: Museum of Contemporary Photography, 1998. Print.

Rosenblum, Naomi. *A World History of Photography*. New York: Abbeville, 2007. Print.

Thomas, Ann. *Beauty of Another Order: Photography in Science*. New Haven: Yale U P, 1997. Print.

PHOTONICS

FIELDS OF STUDY

Optics; quantum physics; fiber optics; holography; information processing; electro-optics; laser optics; solar cells.

KEY TERMS AND CONCEPTS

- **diffraction grating:** Series of tiny slits that allow light to pass through.
- **holography:** Image construction of an object using interference effects that produce an apparent three-dimensional image.
- **laser:** Light amplification and stimulated emission of radiation; atoms with electrons in excited states that all transition to lower energy states by emitting the same frequency photons.
- **optical fiber:** Thin, flexible, transparent material that serves as a wave guide for transmitting light.
- **optical waveguide:** Structure with a high index of refraction that allows light to pass through without side transmission losses.
- **photoelectric effect:** Conversion of light energy into electric energy.
- **photons:** Localized, massless bundles traveling at the speed of light with energy proportional to their frequency.
- **q-bit:** Basic unit cell in a quantum computer.

SUMMARY

Photonics is a rapidly emerging field that uses the quantum interpretation that light has both wave and particle aspects that generate, detect, and modify it. Photonics covers the full range of the electromagnetic spectrum, but most applications are in the visible and infrared. Photonic systems are replacing electricity in the transmission, reception, and amplification of telecommunication information. Photonic applications include lasers, photovoltaic solar cells, sensors, detectors, and quantum computers.

DEFINITION AND BASIC PRINCIPLES

Photonics is the application of the scientific idea that electromagnetic radiation in all its forms, from radio waves to cosmic rays, exhibits both wave and particle behavior. However, these different behaviors cannot be observed simultaneously. Which is observed depends on the physical arrangement at the time of detection. Since this radical idea was presented, science has come to accept this dualism, and photonics is the practical use of this wave-particle duality in instruments and measurement methodologies. Light traveling from source to destination follows the rules of wave motion, but at its emission and reception points it behaves as a particle. Particles emit light energy in localized bundles called photons, and photons transfer their energy when they interact with other particles.

Photonics uses this property to develop instruments sensitive to the interaction of photons with particles in the transmission, detection, and modulation of light. Two properties of light enable it to replace electricity in applications involving information technology and power transmission. Light travels at the fastest speed possible in nature, and through the use of optical waveguides in fiber-optic material, there is almost no loss in the signal.

Background and History

In 1905, Albert Einstein expanded German physicist Max Planck's idea of quantized energy units to explain the photoelectric effect. It was known that certain metals had the ability to produce an electric current (photocurrent) when radiated by light energy. The classical understanding posits that if there were sufficient intensity of the light, it would provide enough energy to free bound electrons in the metal's atoms and produce electric current. What Einstein realized that instead of a wave of diffuse intensity interacting with the metal's electrons, a local bundle of light carrying one quanta of energy could release the electron if its energy exceeded the energy holding the electron in the atom. This bundle of light was named a photon, and the reaction was viewed as a particle-particle effect instead of a wave-particle interaction.

Soon afterward, other phenomena such as Compton scattering, x-ray production, and pair creation and annihilation were interpreted successfully using a photon model of light. Light still retains its wavelike properties as it travels through space. It assumes its photon or particle-like behavior only when it interacts with matter in a detector or at a target.

How It Works

Generation and emission. An atom's electrons are placed in excited energy states, and as they drop to

lower states, photons are emitted. In lasers, an amplification of light is achieved through stimulated emission in which electrons are excited to a particular higher-energy state, and then they all emit the same photon when they drop in energy. This produces a coherent source of light at a particular frequency.

Transmission. Transmission is the process of sending and receiving a signal from one point to another. In photonics, the signal is sent over an optical-fiber waveguide to ensure the integrity of the signal. A transmitted signal may be altered by digitization or modulation in coding for security or error control.

Modulation.

Modulation is the varying of any time-input signal (carrier wave) by an accompanying signal (modulating wave) to produce some information that can be processed. In photonics, the two types of modulations are digital and analog. In digital modulation of a laser diode, the output signal is zero when the input current (bias) is at the minimum (threshold) frequency. When the input frequency is greater than threshold, a constant positive value is produced by the output. In analog modulation, the output signal varies in step with the input frequency.

Signal Processing.

Signal processing represents the operations performed on input waveforms that provide amplification, coding, and information. The inputs are either analog or digital representations of time-varying quantities. In photonics, the modulation of a light signal determines the type of processing performed.

Switching. Fiber-optical switches are useful in redirecting the optical signal in an optical network. A two-position switch reroutes a signal to one of two output channels. Factors determining the efficiency of a two-position switch are speed, reproducibility, and cross talk. Speeds of a few milliseconds are possible with electromechanical switches. Reproducibility provides the same intensity in the signal every time a switch is made. Cross talk measures how uncoupled one output channel is from the other in a multichannel optical system.

Amplification. There exists an array of optical amplifiers to increase the signal. Optical communications use fiber optic and semiconductor amplifiers. For research, there are Raman and quantum-dot amplifiers.

Photodetection and sensing. Photodetectors are devices that take light radiation and directly convert it to electrical signals varying the electric current or voltage to replicate the changes in the input light source. In one type, electrons are emitted from the surface of a metal using the photoelectric effect. Photodiodes and photomultipliers operate under this effect. Another type is made of junctions of semiconductors. Electrons or electron holes (positive current) are emitted on the device's absorption of radiant energy. The p-n junction photodiode, the PIN photodiode, and the avalanche photodiode work under this property. Most fiber-optic communication systems employ a PIN or an avalanche photodiode. The effectiveness of a photodetector is measured by the ratio of the output electric current (I) over the input optical power (P).

APPLICATIONS AND PRODUCTS

Fiber optics. This includes all the various technologies that use transparent materials to transmit light waves. A traditional fiber-optic cable consists of a bundle of glass threads, each of which has the capability of transmitting messages in light waves that have been modulated in some fashion. The advantages to transmitting electrons through conducting wire include their travel at the speed of light, less signal loss due to optical waveguides, and greater bandwidth. The data can be transmitted digitally, and the fiber-optic materials are typically lighter weight than metal cable lines. Fiber optics is used most heavily for local-area networks in data and telecommunications.

The heavy reliance on fiber-optic technology makes necessary the continual development of more efficient optical-fiber materials with ever-increasing switching speeds and more bandwidth to accommodate users' increasing video demands.

Quantum optics. The peculiar properties of quantum systems using photons make them candidates for quantum computing devices. The basic q-bit state has the ability through the property of superposition to be more than one value simultaneously. This can lead to properties of a computing system that can perform certain tasks such as code breaking faster and more efficiently than existing binary computers. Using photons that can be polarized into two states, an optical computer can be designed to take advantages of these light quanta. Whether such systems will ever have the stability to serve as computational

devices and have the speed and low power consumption of the electronic computers commonly being used remains an active research question.

Telecommunications. Optical telecommunication devices send coded information from one location to another through optical fibers. The astonishing growth of the Internet and the ever-increasing demands for more information delivered faster with more efficiency have spurred the development of optical transmission networks. There are optical networks laid underneath the Earth's vast oceans as well as extensive ground-based systems connecting continental communication systems.

Holography. Holography is used optically to store, retrieve, and process information. Its ability to project three-dimensional (3-D) images has allowed for such videos to be more accessible for public viewing. The use of holograms in data storage inside crystals or photopolymers has increased the amount of memory that can be encoded in these structures. Holographic devices are used as security scanners in assessing contents of packages, in determining the authenticity of art, and for examining material structures.

Micro-optics. Microphotonics uses the properties of certain materials to reduce light to microscopic size so that it can be used in optical networking applications. Light waves are confined to move in materials because of total internal reflection using wave guides. The materials have a high index of refraction decreasing the critical angle. This enhances the total reflection capabilities. A photonic crystal has several reflections inside the material. Optical waveguides, optical microcavities, and waveguide gratings represent different materials and geometries.

Biophotonics. Biophotonics encompasses all the various interactions of light with biological systems. It refers especially to the effect of photons (quanta of light) on cells, tissues, and organisms. These interactions include emission, detection, absorption, reflection, modification, and creation of radiation from living tissue and materials produced from biological organisms. Areas of application include medicine, agriculture, and environmental science.

Medicine. Medical uses for photonic technologies include laser surgery, vision correction, and endoscopic examinations.

Laser surgery uses laser light to remove diseased tissue or to treat bleeding blood vessels. Lasers are also extensively used in correcting problems in human vision. Laser-assisted in situ keratomileusis (LASIK) is a technique that uses a microkeratome laser to cut flaps in the cornea and remove excess tissue to correct myopia (near-sightedness). An alternative procedure is photorefractive keratectomy (PRK), which uses an excimer laser to reshape the corneal surface. Other optical uses of lasers include the removal of cataracts and the reduction of excess ocular pressure in the treatment of glaucoma.

Using a fiber-optic flexible tube and a suitable light source, a physician can obtain visual images of internal organs without more invasive surgery or high-energy x-rays.

Military. Photonic devices have found use in military operations in terms of sensors, particularly infrared. Through the use of light-emitting diodes (LEDs) and lasers, photonics technologies are being developed for the infantry soldier on the battlefield and the field officer in the command center. This technology is also utilized in diverse areas such as navigation, search and rescue, mine laying, and detection. Applications range from an optical scope that enables soldiers to see around obstacles during night operations using a flexible fiber-optic tube to weapons such as low-, medium-, and high-power lasers in the millimeter (microwave) wavelength region.

CAREERS AND COURSE WORK

Photonics is a multidisciplinary field. It has roots in physics through classical optics and quantum theory. The explosion in the applications has been driven by the use of engineering to develop instruments and devices using the particular properties of photons for transmitting, sensing, and detecting.

Career paths include optical engineering, illumination engineering, and optoelectronics. There are more than one hundred universities in the United States offering degree programs or conducting research in photonics. There are also a number of community colleges that offer associate's degrees for careers as laser technicians. The basic undergraduate major would be physics with some emphasis in optics. A typical master's or doctoral program would concentrate on physics and quantum optics or optical engineering with research work in lasers or photonics.

The number of industries using photonics technology is growing. Photonics is prevalent in telecommunications, medicine, industrial manufacturing, energy, lighting, remote sensing, security, and defense. Job titles include research physicist, optical engineer, light-show director, laser-manufacturing technician, industrial laser technician, medical laser technician, and fiber-optic packaging and manufacturing engineer.

SOCIAL CONTEXT AND FUTURE PROSPECTS

The photonics industry is an important component in the ever-growing use of handheld devices for voice, video, and data. The job growth in photonics is anticipated to be in the design and manufacture of display screens for television sets, computer monitors, mobile phones, handheld video-game systems, personal digital assistants, navigation systems, electronic-book readers, and electronic tablets such as the iPad. These systems have traditionally used semiconductor light sources such as light-emitting and superluminescent diodes (LEDs and SLDs), fluorescent lamps,

---

FASCINATING FACTS ABOUT PHOTONICS

- In 1990, only about 10 percent of all telephone calls in the United States were carried by optical fibers. In 2010, more than 90 percent employed optical cables.

- Continued advances in photovoltaic conversion of solar energy to electricity may enable solar energy to produce about 50 percent of the world's electricity by 2050.

- Nanoplasmonics is an emerging field that studies the effect of light on the edges of metal surfaces. Applications include reduction of tumors, solar energy conversion, and detection of potential explosive reactions.

- The term "photonic engineer" is derived from the word "photon," which refers to a quanta or particle of light.

- Most photonic engineers work for large telecommunications firms and optical-fiber manufacturers. Optical physicists work for research institutions and universities.

- The demand for better visual displays for mobile (handheld) communication devices fuels the growth in the photonic industry. It is anticipated that by 2015 there will be more than 7 billion such mobile devices, which will be equivalent to the world's population.

- By 2015, two-thirds of the mobile device business will be for video transmission, thus increasing the need for better video displays.

and cathode ray tubes (CRTs). Plasma display panels (PDPs) and liquid crystal displays (LCDs) are in great demand. Green photonics develops organic light-emitting diodes (OLEDs) and light-emitting polymers (LEPs).

The design and development of media such as glass or plastic fibers for transmission is another career path in photonics. There is a need for engineers to develop new photonic crystals, photonic crystal fibers, and metal surfaces (nanoplasmonics).

There is also demand for better photodetectors that range from very fast photodiodes (PDs) for communications to charge-coupled devices (CCDs) for digital cameras to solar cells that are used to collect solar energy.

—Joseph Di Rienzi, PhD

FURTHER READING

Cvijetic, Milorad. *Optical Transmission: Systems Engineering.* Norwood: Artech, 2004. Print.

Hecht, Jeff. *Beam: The Race to Make the Laser.* New York: Oxford UP, 2005. Print.

Longdell, Jevon. "Quantum Information: Entanglement on Ice." *Nature* 469.7331 (2011): 475–76. Print.

Menzel, Ralf. *Photonics: Linear and Nonlinear Interactions of Laser Light and Matter.* 2nd ed. Berlin: Springer, 2007. Print.

Rogers, Alan. *Understanding Optical Fiber Communications.* Norwood: Artech, 2001. Print.

PLANETOLOGY AND ASTROGEOLOGY

FIELDS OF STUDY

Astronomy; astrophysics; astrobiology; artificial intelligence; aeronautics; astronautics; biology; evolutionary biology; microbiology; biochemistry; chemistry; isotope geochemistry; computer science; computer programming; electronics; electrical engineering; geology; glaciology; metallurgy; geomorphology; meteorology; meteoritics; mineralogy; petrology; planetary geology; vulcanology; mathematics; mechanical engineering; field robotics; space robotics; space science.

KEY TERMS AND CONCEPTS

- **asteroid:** Small solid body composed of either ice or rock that is between 30 meters and 600 kilometers in diameter and is the parent body for most meteorites.
- **asteroid belt:** Concentration of hundreds of thousands of asteroids positioned roughly between the orbits of Mars and Jupiter.
- **astrobiology:** Branch of biology that deals with the possible life-forms that may exist on other planets or in extreme environments.
- **cosmochemistry:** Branch of chemistry that deals with the chemical composition of extraterrestrial materials and the composition of stars.
- **dwarf planet:** Planet such as Pluto or Eris that is relatively small, is mainly composed of exotic ices, and orbits at the extreme limit of the solar system.
- **extremophile life-form:** Microbial life-form that exists under extreme conditions of acidity, temperature, or salinity.
- **jovian planet:** Planet that has a very large mass, large size with a low density, and is composed mainly of hydrogen and helium gas, much like Jupiter.
- **kuiper belt:** Region of the solar system beyond Neptune where a large number of comets and dwarf planets exist.
- **meteorite:** Piece of rock or metal that originated in the asteroid belt and has survived its fiery plunge through Earth's atmosphere and reached the surface intact.
- **near-earth object:** Comet or asteroid that follows orbits that cross the path of the Earth and is a candidate for a possible impact event.
- **runaway greenhouse effect:** Effect produced when a planet builds up an intense surface temperature like that on Venus because its atmosphere tends to trap heat rather than allow it to escape to space.
- **terrestrial planet:** Relatively small, cold, solid planet composed of iron and silicate minerals with surface features similar to those of Earth.

SUMMARY

Planetology and astrogeology are separate branches of science that examine the physical and chemical characteristics of the planets and minor bodies in the solar system. The principal difference between these two scientific disciplines is that planetology is inclusive of all planetary bodies, while astrogeology concentrates on those worlds that are basically similar to the Earth. Scientists within the field of planetology can study a variety of topics that include planetary atmospheres, interiors, orbital characteristics, the potential for life, and all aspects of planet formation and evolution. In comparison, astrogeology essentially concentrates on the various surface features and geological processes of the Earth as seen on other worlds.

DEFINITION AND BASIC PRINCIPLES

The terms "planetology" and "astrogeology" are used to describe the scientific disciplines that study the planets in the solar system and those objects, believed to be planets, that orbit other distant stars. "Planetology" is a more general term that includes the study of all planets in every respect.

"Astrogeology" is actually somewhat of a misnomer; during the space program in the 1960s, the term was used to describe geological situations on other planets. If taken in a literal sense, astrogeology refers to the geology of the stars, which is not the case. Geology is the science that deals with the history of the Earth and life as recorded in rocks. The prefix "astro" indicates a place of origin beyond the Earth and is used in conjunction with many other sciences such as astrobiology, astrophysics, and astronautics.

The unmanned spacecraft missions of the 1960s to Mercury, Venus, the Moon, and Mars literally created the field of astrogeology. Traditional geologists could compare their understanding of Earth processes to what they were seeing on these other planets. Geologists wondered how the craters had formed on the Moon, but during the 1960s, it could not be determined with any certainty. The scientific community was equally divided between whether the craters were the result of impacts or volcanic in origin. A definitive answer had to wait for the astronauts of the Apollo program, who brought back rock samples. For the first time, scientists had a verifiable piece of the Moon to compare with Earth rocks. Since the Apollo program, spacecraft have landed on Venus, Mars, Phobos (one of Mars's moons), Titan (one of Saturn's moons), and asteroids and comets, but none of these has brought back any materials. Spacecraft

have also made intentional impacts into the atmosphere of Jupiter, but these missions did not make a landing, per se.

BACKGROUND AND HISTORY

The planets in the solar system have attracted the attention of humans since before recorded history. One of the first observations was the recognition that stars remain in fixed positions while planets move in the sky. With the development of writing and numbers, early astronomers were able to accurately calculate and predict planetary motion. This was the limit of planetary science until the invention of the telescope in 1607.

The telescope transformed the planets from bright little points of light into actual worlds with definable surface features and cloud formations. In 1610, Galileo's telescopic observations of Jupiter revealed a "miniature solar system" of revolving moons, and his discovery of the phases of Venus helped support the heliocentric model of the solar system (the plants revolving around the sun). Subsequent improvements in telescope design and quality led to the discovery of Saturn's rings and numerous moons, as well as the planets Uranus and Neptune. Further discoveries were limited only by technology.

The next major breakthrough came in the 1950s with the development of rocket technology and spacecraft design. For the first time, scientists were able to extend their observations beyond Earth's atmosphere and send instruments to the Moon and to various planets. By 1989, all the gas giant planets had been visited by spacecraft, leaving only Pluto as an unexplored world until 2015, when the *New Horizons* probe completed a flyby of the dwarf planet. In addition to gathering an enormous amount of data on each of these planets, the spacecraft also examined all of their major moons. In fact, many of these moons turned out to be more interesting than their planets.

HOW IT WORKS

The scientific disciplines of planetology and astrogeology attempt to answer fundamental questions concerning the origin and evolution of the planets in the solar system. To observe the planets from Earth requires the cooperation of several different sciences and technologies. Modern planetology includes such disciplines as astronomy, astrobiology, astrogeology,

Planetary geologist and NASA astronaut Harrison "Jack" Schmitt collecting lunar samples during the Apollo 17 mission in early-December 1972

astrophysics, and cosmochemistry, coupled with various technologies such as computer science, electronic engineering, and mechanical engineering. Their approach to problem solving involves a combination of direct observation and data collection with various laboratory experiments and computer simulation models.

Remote sensing. Before the electronic age, planetary studies were limited by the vision of astronomers and the optics of their telescopes. In virtually all aspects of astronomy-related science, researchers must depend on indirect observations through various electronic instruments used for remote sensing, particularly if they are studying distant stars and galaxies. Astronomers must understand how the basic principles of light and other forms of electromagnetic energy affect what they are seeing. Planetary scientists, by employing a combination of Earth-based telescopic images and direct spacecraft observations, can actually see events happening on these planets in real time.

Interference from the Earth's atmosphere has always been a problem for optical astronomers, especially when attempting to view surface details on the terrestrial planets such as Mercury or Mars. Planets with dense atmospheres such as Venus also present a problem for astronomers because their thick cloud layers prevent direct surface observations. To overcome this difficulty, in the 1950s astronomers developed a technique using radar imaging to reveal surface features. Radar signals can easily penetrate clouds and are reflected back by items they hit. By timing the rate of return for these signals, astronomers assembled computer-generated maps indicating the high and low elevations. Although the quality of these early surface radar images was quite poor, it gave astronomers an idea of the nature of the geology of Venus. Later, orbiting spacecraft provided much higher quality images, which were used to construct a complete geological surface map of Venus. A similar technique employing laser technology has been used to map the elevations of geological features on Mars with pinpoint accuracy.

Direct observation. The geologist primarily depends on fieldwork to construct geological maps and determine the location of mineral resources. However, extraterrestrial fieldwork has been limited to the Moon. All twelve of the Apollo astronauts who went to the Moon were trained in geology, but only one, Harrison Schmitt, was a professional geologist. Supplied with detailed lunar maps and reports of surface materials, the Apollo astronauts were able to successfully land at six locations and collect more than 400 kilograms of rock. Scientists continue to examine these materials and make exciting new discoveries with technologies that did not exist at the time of the Moon landings.

Before the Apollo Moon landings, geologists did have the opportunity to study extraterrestrial materials in the form of meteorites. By studying the chemical and mineralogical composition of meteorites and comparing them with Earth rocks, scientists were able to confirm their extraterrestrial origin. Meteorites proved to be older than the Earth and are believed to have originated in the asteroid belt between Mars and Jupiter. They represent some of the oldest solid material in the solar system and are the building blocks of the terrestrial planets.

APPLICATIONS AND PRODUCTS
Global resource management. One of the major

benefits derived from the study of the other planets in the solar system is the ability to turn that technology around and study the Earth. Observing the Earth from space offers scientists the opportunity to view the Earth as a single entity rather than as a collection of apparently unrelated components. The technique of multispectral imaging has been used to map the mineral composition of the Moon's surface and to search for evidence of water. Similar technology has also been employed in mineral exploration on Earth. Vast regions of the Earth's surface, including parts of Siberia and central Australia, remain virtually unexplored and are believed to contain great mineral wealth. The use of remote-sensing satellites has enabled geologists to assess a site's potential without actually setting foot on the ground. Although determining a site's true value still requires fieldwork, remote-sensing data obtained from space certainly can make the work more efficient and reduce expenses.

Food production management is another area that can directly benefit from planetary monitoring technology. As of 2016, the world's population was more than 7.3 billion and was expected to reach 9 billion by 2050. Space-age technology, by providing information about global conditions, can help farmers increase food production capabilities and manage resources to meet these demands. Fisherman can employ satellite data to help track schools of fish to increase the efficiency and productivity of their efforts. They can also use this information to monitor their fishing grounds and preserve them.

Planetary monitoring can help address another concern, the availability of an abundant supply of drinkable water, by keeping track of global water resources. The problem of maintaining an adequate supply of water affects the world's population in that water is also used for many other purposes, including irrigating crops.

Meteorology. Meteorologists use part of the data that planetologists have derived from their studies of planets with dense atmospheres. Satellite technology originally designed to study the atmosphere of another planet has been adapted to monitor the Earth's dynamic weather. Studying Venus, with its extremely dense atmosphere and its runaway greenhouse effect, provides Earth scientists with working models to use when trying to understand the effects of greenhouse gases in the Earth's atmosphere. Observing the various weather systems in the atmospheres of the

Jovian planets also helps meteorologists understand wind and weather patterns on the Earth. Jupiter's Great Red Spot is essentially a 350-year-old hurricane, and Neptune exhibits the highest velocity winds of any planet in the solar system. Meteorologists have benefited from studying the dust devils seen blowing across the surface of Mars. These small dust storms closely resemble tornadoes on Earth. Periodically, Mars also experiences global dust storms that lift huge quantities of dust high into the atmosphere and block out most of its surface features for months on end.

Climate change. The effects of climate change—whether the shrinking of the polar ice caps or the expansion of the deserts—can be seen clearly from space. The deforestation of the Brazilian rain forest as well as the amount of sediment that a river carries into the ocean each year can be precisely measured from satellite observations. Oceanographers can use the data from satellite observations in their studies of ocean currents and the effects of pollution on surface water and coastlines. City planners can use satellite technology to monitor urban sprawl and help develop better methods of waste management and disposal. Climatologists can study the localized weather patterns that develop over major cities and how they affect the smaller communities down wind. In the United States, these localized weather patterns are most apparent in the Great Lakes region and on the eastern seaboard. Similarly, the inadequacy of environmental controls in emerging industrial countries such as India and China is apparent when viewed from space. The pollution generated affects not only populations in India and China but also those in neighboring areas.

Analytical instrumentation. Before the Apollo Moon landings, meteorites were the only extraterrestrial materials available for astrogeologists to study. Usually the classification of a meteorite requires a certain amount of destructive analysis. In many cases, the most interesting and rare meteorites are available only in very small quantities, thereby limiting the amount of material available for analysis. Similarly, only small amounts of the rocks recovered from the Moon were available for analysis. The National Aeronautics and Space Administration (NASA) deliberately preserved a large quantity of lunar material for future scientists to study with instruments not yet invented. They realized that another trip to the

FASCINATING FACTS ABOUT PLANETOLOGY AND ASTROGEOLOGY

- Saturn's moon Enceladus has active geysers that spew giant plumes of water hundreds of kilometers out into space. Scientists have been able to detect the presence of various organic molecules in these plumes.

- The twelfth man to set foot on the Moon in December, 1972, was a scientist. The first eleven were military pilots, but Harrison H. Schmitt was a professional geologist who later learned to fly aircraft.

- On October 6, 2008, astronomer Richard Kowalski discovered that a small asteroid was headed for an impact with Earth. Later calculations predicted that it would hit within the next thirteen hours somewhere in the Sudan. It did, and by December, more than forty-seven meteorite fragments had been collected.

- In the search for extraterrestrial life, Mars always seemed the most promising place to look. However, strong evidence of warm oceans of water beneath the icy surface of Jupiter's moon Europa make it the more likely location for life.

- Scientists were certain that planets had been hit by comets and other objects, but they never expected to see it happen. In July, 1994, more than twenty large fragments of comet Shoemaker-Levy 9 plunged into the atmosphere of Jupiter, leaving behind huge dark scars that marked the points of impact.

- Clyde William Tombaugh, the astronomer who discovered Pluto in 1930, died in 1997. A small portion of his ashes was later placed onboard the New Horizons spacecraft, which is bound for Pluto.

- The Cassini spacecraft successfully landed a small probe on the surface of Saturn's moon Titan in 2005. The probe discovered features that resemble lakes, rivers, and shorelines on Earth, but on Titan, they were formed by liquid methane instead of water.

- The Moon has long been thought of as a waterless world. Studies have provided evidence that water ice may be present at the bottom of deep craters near the Moon's south pole that are never exposed to sunlight.

- The Earth is not the only planet with meteorites on its surface. The rovers *Spirit* and *Opportunity* photographed meteorites sitting among the rocks they encountered as they traversed the surface of Mars.

Moon might not occur for many years.

To cope with the limited availability, astrogeologists had to develop techniques to gain the maximum amount of data from the least amount of material. They needed the help of technicians to invent the electronic equipment and to develop the procedures needed to analyze the material. Radiation became an important component in modern analytical technology. Geologists often employ X-ray diffraction and X-ray fluorescence to identify the minerals in a rock. By using a mass spectrometer, the half-life decay rates of certain radioactive isotopes can be measured to obtain the age of Moon rocks or to determine the cosmic-ray exposure age of meteorites. Other instruments such as the electron microprobe, an instrument that can analyze particles as small as a few microns in size, are the primary tools of the scientists who study extraterrestrial materials.

The scanning electron microscope is another valuable tool that gives the scientist the ability to magnify the object that they are studying to an incredibly high power. This instrumentation is especially useful to astrobiologists who are attempting to prove that fossils of microbial life-forms are present in certain types of meteorites. If their theory is correct, then these extremophile life-forms could have originated somewhere else in the solar system and have been brought to Earth by either comets or meteorites very early in Earth's history.

CAREERS AND COURSE WORK

Students who are interested in making either planetology or astrogeology their career must first complete an undergraduate degree program in one of the fundamental sciences, which include biology, chemistry, geology, or physics. Adding computer science, mathematics, or electrical or mechanical engineering as a double major would increase job prospects. Most students pursue graduate work in a specialty. A master's degree is usually the minimum requirement for most technicians, while a doctorate is more appropriate for a senior scientist position. There are many ways to become employed in space science research, and students should possess a variety of technical skills to compliment their academic training. The industrial job market for highly skilled scientists and technicians in space science is quite unpredictable, but positions are likely to be available for the best applicants.

University teaching positions present an opportunity for scientists with doctorates to find employment and still pursue their own individual research interests. Major universities expect their professors to conduct independent research and encourage collaborative efforts with government agencies or major museums. Postdoctorate positions are usually available to recent graduates so that they can work with and learn from senior scientists in their field. Federal grants are available to scientists to support their research, although such grants can be difficult to obtain. Governmental agencies such as NASA, the National Oceanographic and Atmospheric Administration (NOAA), National Science Foundation (NSF), and the US Geological Survey (USGS) all employ earth scientists and geologists in various positions. The USGS actually has an astrogeology branch in Flagstaff, Arizona, where maps are created from the data returned from many of the planetary missions.

SOCIAL CONTEXT AND FUTURE PROSPECTS

Historians have often stated that the twentieth century will most likely be remembered for its two world wars and for the realization of space travel. The wars and the weapons race are probably responsible for humankind's venturing into space. However, in addition to powerful rockets and brave astronauts, sending people into space required major advancements in technology to create the necessary hardware. Technology originally developed in connection with the space program can be found in almost every modern technological necessity, such as cell phones, personal computers, and medical diagnostic instrumentation.

Although the primary motivation for sending a man to the Moon was political, it momentarily opened the eyes of the world to something greater than nationalism. The images of Neil Armstrong and Buzz Aldrin on the Moon on July 21, 1969, show what humankind is capable of achieving. The subsequent lunar landings drew little public interest, and humankind has remained fixed in Earth orbit since 1972. Although human exploration of the solar system appears to be on hold for the foreseeable future, many more robotic missions are planned. Planetary probes have provided unimaginable visions of worlds. Perhaps it will take a major discovery for humankind to once again become fascinated with space exploration.

—Paul P. Sipiera, PhD

FURTHER READING

Chyba, Christopher. "The New Search for Life in the Universe." *Astronomy* 38.5 (2010): 34–39. Print.

Geotz, Walter. "Phoenix on Mars." *American Scientist* 98.1 (2010): 40–47. Print.

Greeley, Ronald, and Raymond Batson. *The Compact NASA Atlas of the Solar System.* New York: Cambridge UP, 2001. Print.

Publications and Graphics Department-NASA. *Spinoff: Fifty Years of NASA-Derived Technologies, 1958–2008.* Washington, DC: NASA Center for Aero-Space Information, 2008. Print.

Schmitt, Harrison H. *Return to the Moon: Exploration, Enterprise, and Energy in the Human Settlement of Space.* New York: Copernicus, 2006. Print.

Sparrow, Giles. *The Planets: A Journey Through the Solar System.* London: Quercus, 2006. Print.

Talcott, Richard. "How We'll Explore Pluto." *Astronomy* 38.7 (2010): 24–29. Print.

POLYMER SCIENCE

FIELDS OF STUDY

Chemistry; physics; chemical engineering; materials science; organic chemistry; inorganic chemistry; physical chemistry; analytical chemistry; differential equations; computer science; structures and reactions of macromolecular compounds; kinetics and mechanisms of polymer synthesis; plastics engineering; synthetic rubber engineering; biopolymers; molecular biology; biochemistry; biophysics.

KEY TERMS AND CONCEPTS

- **addition polymerization:** Chemical chain-reaction process in which monomers bond to each other in the step-by-step growth of a polymer, without generating any by-products.
- **biopolymer:** Naturally occurring polymer.

- **copolymer:** Polymer made from more than one kind of monomer, resulting in a macromolecule with more than one monomeric unit (or "mer") in the backbone.
- **elastomer:** Substance composed of chemically (or physically) cross-linked linear polymers possessing the elastic properties of natural rubber.
- **free radical:** Short-lived and highly reactive molecular fragment with one or more unpaired electrons.
- **inorganic polymer:** Polymer made from inorganic monomers, such as silicates in nature and artificially made cements and lubricants.
- **monomer (monomeric unit):** Largest atomic array (with possible projecting groups) constituting the basic building block of a polymer.
- **nucleic acid:** One of several nitrogenous compounds found in living cells and viruses, which, in their polymeric form as polynucleotides, contain the following principal parts: a base (either a purine or pyrimidine), a sugar, and a phosphoric-acid group.
- **plastic:** Organic polymer, generally synthetic, often combined with additives such as colorants and curatives, capable of being molded or cast under heat and pressure into various shapes.
- **polycondensation:** Chemical process in which nonidentical monomers react to form a polymer, with the concomitant formation of leaving molecules.
- **polymerization:** Chemical reaction, usually catalyzed, in which a monomer or mixture of monomers form a chain-like macromolecule (or polymer).
- **protein:** Complex polymer composed of chains of amino acids (groups that contain carbon, hydrogen, oxygen, nitrogen, and, sometimes, sulfur).

SUMMARY

Polymer science is a specialized field concerned with the structures, reactions, and applications of polymers. Polymer scientists generate basic knowledge that often leads to various industrial products such as plastics, synthetic fibers, elastomers, stabilizers, colorants, resins, adhesives, coatings, and many others. A mastery of this field is also essential for understanding the structures and functions of polymers found in living things, such as proteins and deoxyribonucleic acid (DNA).

DEFINITION AND BASIC PRINCIPLES

Polymers are very large and often complex molecules constructed, either by nature or by humans, through the repetitive yoking of much smaller and simpler units. This results in linear chains in some cases and in branched or interconnected chains in others. The polymer can be built up of repetitions of a single monomer (homopolymer) or of different monomers (heteropolymer). The degree of polymerization is determined by the number of repeat units in a chain. No sharp boundary line exists between large molecules and the macromolecules characterized as polymers. Industrial polymers generally have molecular weights between ten thousand and one million, but some biopolymers extend into the billions.

Chemists usually synthesize polymers by condensation (or step-reaction polymerization) or addition (also known as chain-reaction polymerization). A good example of chain polymerization is the free-radical mechanism in which free radicals are created (initiation), facilitating the addition of monomers (propagation), and ending when two free radicals react with each other (termination). A general example of step-reaction polymerization is the reaction of two or more polyfunctional molecules to produce a larger grouping, with the elimination of a small molecule such as water, and the consequent repetition of the process until termination.

Besides free radicals, chemists have studied polymerizations utilizing charged atoms or groups of atoms (anions and cations). Physicists have been concerned with the thermal, electrical, and optical properties of polymers. Industrial scientists and engineers have devoted their efforts to creating such new polymers as plastics, elastomers, and synthetic fibers. These traditional applications have been expanded to include such advanced technologies as biotechnology, photonics, polymeric drugs, and dental plastics. Other scientists have found uses for polymers in such new fields as photochemistry and paleogenetics.

BACKGROUND AND HISTORY

Nature created the first polymers and, through chemical evolution, such complex and important macromolecules as proteins, DNA, and polysaccharides. These were pivotal in the development of increasingly multifaceted life-forms, including Homo sapiens, who, as this species evolved, made better and better use of such polymeric materials as pitch, woolen and linen fabrics, and leather. Pre-Columbian Native Americans used natural rubber, or *cachucha*, to waterproof fabrics, as did Scottish chemist Charles Macintosh in nineteenth century Britain.

The Swedish chemist Jöns Jakob Berzelius coined the term "polymer" in 1833, though his meaning was far from a modern chemist's understanding. Some scholars argue that the French natural historian Henri Braconnot was the first polymer scientist since, in investigating resins and other plant products, he created polymeric derivatives not found in nature. In 1836, the Swiss chemist Christian Friedrich Schönbein reacted natural cellulose with nitric and sulfuric acid to generate semisynthetic polymers. In 1843 in the United States, hardware merchant Charles Goodyear accidentally discovered "vulcanization" by heating natural rubber and sulfur, forming a new product that retained its beneficial properties in cold and hot weather. Vulcanized rubber won prizes at the London and Paris Expositions in 1850, helping to launch the first commercially successful product of polymer scientific research.

In the early twentieth century, the Belgian-American chemist Leo Baekeland made the first totally synthetic polymer when he reacted phenol and formaldehyde to create a plastic that was marketed under the name Bakelite. The nature of this and other synthetic polymers was not understood until the 1920's and 1930's, when the German chemist Hermann Staudinger proved that these plastics (and other polymeric materials) were extremely long molecules built up from a sequential catenation of basic units, later called monomers. This enhanced understanding led the American chemist Wallace Hume Carothers to develop a synthetic rubber, neoprene, that had numerous applications, and nylon, a synthetic substitute for silk. Synthetic polymers found wide use in World War II, and in the postwar period the Austrian-American chemist Herman Francis Mark founded the Polymer Research Institute at Brooklyn Polytechnic, the first such facility in the United States. It helped foster an explosive growth in polymer science and a flourishing commercial polymer industry in the second half of the twentieth century.

HOW IT WORKS

After more than a century of development, scientists and engineers have discovered numerous techniques for making polymers, including a way to make them using ultrasound. Sometimes these techniques depend on whether the polymer to be synthesized is inorganic or organic, fibrous or solid, plastic or elastomeric, crystalline or amorphous. How various polymers function depends on a variety of properties, such as melting point, electrical conductivity, solubility, and interaction with light. Some polymers are fabricated to serve as coatings, adhesives, fibers, or thermoplastics. Scientists have also created specialized polymers to function as ion-exchange resins, piezoelectrical devices, and anaerobic adhesives. Certain new fields have required the creation of specialized polymers like heat-resistant plastics for the aerospace industry.

Condensation polymerization. Linking monomers into polymers requires the basic molecular building blocks to have reaction sites. Carothers recognized that most polymerizations fall into two broad categories, condensation and addition. In condensation, which many scientists prefer to call step, step-growth, or stepwise polymerization, the polymeric chain grows from monomers with two or more reactive groups that interact (or condense) intermolecularly, accompanied by the elimination of small molecules, often water. For example, the formation of a polyester begins with a bifunctional monomer, containing a hydroxyl group (OH, oxygen bonded to hydrogen) and a carboxylic acid group (COOH, carbon bonded to an oxygen and an OH group). When a pair of such monomers reacts, water is eliminated and a dimer formed. This dimer can now react with another monomer to form a trimer, and so on. The chain length increases steadily during the polymerization, necessitating long reaction times to get "high polymers" (those with large molecular weights).

Addition polymerization. Many chemists prefer to call Carothers's addition polymerization chain, chain-growth, or chain-wise polymerization. In this process, the polymer is formed without the loss of molecules, and the chain grows by adding monomers repeatedly, one at a time. This means that monomer concentrations decline steadily throughout the polymerization, and high polymers appear quickly. Addition polymers are often derived from unsaturated monomers (those with a double bond), and in the polymerization process the monomer's double bond is rearranged in forming single bonds with other molecules. Many of these polymerizations also require the use of catalysts and solvents, both of which have to be carefully chosen to maximize yields. Important examples of polymers produced by this mechanism are polyurethane and polyethylene.

APPLICATIONS AND PRODUCTS

Since the start of the twentieth century, the discoveries of polymer scientists have led to the formation of hundreds of thousands of companies worldwide that manufacture thousands of products. In the United States, about 10,000 companies are manufacturing plastic and rubber products. These and other products exhibit phenomenal variety, from acrylics to zeolites. Chemists in academia, industry, and governmental agencies have discovered many applications for traditional and new polymers, particularly in such modern fields as aerospace, biomedicine, and computer science.

Elastomers and plastics. From its simple beginnings manufacturing Bakelite and neoprene, the plastic and elastomeric industries have grown rapidly in the quantity and variety of the polymers their scientists and engineers synthesize and market. Some scholars believe that the modern elastomeric industry began with the commercial production of vulcanized rubber by Goodyear in the nineteenth century. Such synthetic rubber polymers as styrene-butadiene, neoprene, polystyrene, polybutadiene, and butyl rubber (a copolymer of butylene and isoprene) began to be made in the first half of the twentieth century, and they found extensive applications in the automotive and other industries in the second half.

Although an early synthetic plastic derived from cellulose was introduced in Europe in the nineteenth century, it was not until the twentieth century that the modern plastics industry was born, with the introduction of Bakelite, which found applications in the manufacture of telephones, phonograph records, and a variety of varnishes and enamels. Thermoplastics, such as polyethylene, polystyrene, and polyester, can be heated and molded, and billions of pounds of them are produced in the United States annually. Polyethylene, a low-weight, flexible material, has many applications, including packaging, electrical insulation, housewares, and toys. Polystyrene has found uses as an electrical insulator and, because of

its clarity, in plastic optical components. Polyethylene terephthalate (PET) is an important polyester, with applications in fibers and plastic bottles. Polyvinyl chloride (PVC) is one of the most massively manufactured synthetic polymers. Its early applications were for raincoats, umbrellas, and shower curtains, but it later found uses in pipe fittings, automotive parts, and shoe soles.

Carothers synthesized a fiber that was stronger than silk, and it became known as nylon and led to a proliferation of other artificial textiles. Polyester fibers, such as PET, have become the world's principal man-made materials for fabrics. Polyesters and nylons have many applications in the garment industry because they exceed natural fibers, including cotton and wool, in such qualities as strength and wrinkle resistance. Less in demand are acrylic fibers, but, because they are stronger than cotton, they have had numerous applications by manufacturers of clothing, blankets, and carpets.

Optoelectronic, aerospace, biomedical, and computer applications. As modern science and technology have expanded and diversified, so, too, have the applications of polymer science. For example, as researchers explored the electrical conductivity of various materials, they discovered polymers that have exhibited commercial potential as components in environmentally friendly battery systems. Transparent polymers have become essential to the fiber optics industry. Other polymers have had an important part in the improvement of solar-energy devices through products such as flexible polymeric film reflectors and photovoltaic encapsulants. Newly developed polymers have properties that make them suitable for optical information storage. The need for heat-resistant polymers led the U.S. Air Force to fund the research and development of several such plastics, and one of them, polybenzimidazole, has achieved commercial success not only in aerospace but also in other industries as well.

Following the discovery of the double-helix structure of DNA in 1953, a multiple of applications followed, starting in biology and expanding into medicine and even to such fields as criminology. Nondegradable synthetic polymers have had multifarious medical applications as heart valves, catheters, prostheses, and contact lenses. Other polymeric materials show promise as blood-compatible linings for cardiovascular prostheses. Biodegradable synthetic polymers have found wide use in capsules that release drugs in carefully controlled ways. Dentists regularly take advantage of polymers for artificial teeth, composite restoratives, and various adhesives. Polymer scientists have also contributed to the acceleration of computer technology since the 1980's by developing electrically conductive polymers, and, in turn, computer science and technology have enabled polymer scientists to optimize and control various polymerization reactions.

CAREERS AND COURSE WORK

Building on a base of advanced courses in chemistry, mathematics, and chemical engineering, undergraduates generally take an introductory course in polymer science. Graduate students in polymer science usually take courses in line with their chosen career goal. For example, students aspiring to positions in the plastics industry would need to take advanced courses in macromolecular synthesis and the chemical engineering of polymer syntheses. Students interested in biotechnology or bioengineering would need to take graduate courses in molecular biology, biomolecular syntheses, and so on.

Many opportunities are available for graduates with degrees in polymer science. The field is expanding, and careers in research can be forged in government agencies, academic institutions, and various industries. Chemical, pharmaceutical, biomedical, cosmetics, plastics, and petroleum companies hire polymer scientists and engineers. Because of concerns raised by the modern environmental movement, many companies are hiring graduates with expertise in biodegradable polymers. The rapid development of the computer industry has led to a need for graduates with an understanding of electrically charged polymeric systems. In sum, traditional and new careers are accessible to polymer scientists and engineers both in the United States and many foreign countries.

SOCIAL CONTEXT AND FUTURE PROSPECTS

Barring a total global economic collapse or a cataclysmic environmental or nuclear-war disaster, the trend of expansion in polymer science and engineering, well-established in the twentieth century, should continue throughout the twenty-first. As polymer scientists created new materials that contributed to twentieth-century advances in such areas

as transportation, communications, clothing, and health, so they are well-positioned to meet the challenges that will dominate the twenty-first century in such areas as energy, communications, and the health of humans and the environment. Many observers have noted the increasing use of plastics in automobiles, and polymer scientists will most likely help to create lightweight-plastic vehicles of the future. The role of polymer science in biotechnology will probably exceed its present influence, with synthesized polymers to monitor and induce gene expression or as components of nanobots to monitor and even improve the health of vital organs in the human body. Environmental scientists have made the makers of

---

FASCINATING FACTS ABOUT POLYMER SCIENCE

- Life would not exist without polymers, since a lack of proteins would mean no enzymes for essential chemical reactions, and a lack of nucleic acids would mean that life forms could not replicate themselves.
- Many chemists initially rejected the idea that gargantuan polymers exist, and one eminent chemist stated that it was as if zoologists "were told that somewhere in Africa an elephant was found that was 1,500 feet long and 300 feet high."
- Lord Alexander Todd, the winner of the 1957 Nobel Prize in Chemistry, once stated that the development of polymerization was "the biggest thing that chemistry has done," since this has had "the biggest effect on everyday life."
- There are more chemists researching and developing synthetic polymers than in all other areas of chemistry combined.
- It takes about $1 billion to discover, develop, and introduce a new polymer into the marketplace.
- In the twenty-first century, employment figures show that more than half of American chemical industrial employment is in synthetic polymers.
- The plastic now known as Silly Putty was called Nutty Putty by the chemical engineer who invented it in the 1940's.
- In the new generation of commercial airplanes, such as the Boeing 787, about half of the construction materials come from polymers, polymer-derived fabrics, and composites.

plastics aware that many of their products end up as persistent polluters of the land and water, thus fostering a search that will likely continue throughout the twenty-first century for biodegradable polymers that will serve both the needs of advanced industrialized societies and the desire for a sustainable world.

—*Robert J. Paradowski, MS, PhD*

FURTHER READING

Carraher, Charles E., Jr. *Giant Molecules: Essential Materials for Everyday Living and Problem Solving.* 2d ed. Hoboken, N.J.: John Wiley & Sons, 2003. Called an "exemplary book" for general readers interested in polymers, *Giant Molecules* facilitates understanding with an apt use of illustrations, figures, and tables along with relevant historical materials.

_____. *Introduction to Polymer Chemistry.* 2d ed. Boca Raton, Fla.: CRC Press, 2010. This text, intended for undergraduates, explains polymer principles in the contexts of actual applications, while including material on such new areas as optical fibers and genomics. Summaries, glossaries, exercises, and further readings at the ends of chapters.

Ebewele, Robert O. *Polymer Science and Technology.* Boca Raton, Fla.: CRC Press, 2000. The author covers polymer fundamentals, the creation of polymers, and polymer applications, with an emphasis on polymer products.

Morawetz, Herbert. *Polymers: The Origins and Growth of a Science.* Mineola, N.Y.: Dover, 2002. This illustrated paperback reprint makes widely available an excellent history of the subject.

Painter, Paul C., and Michael M. Coleman. *Essentials of Polymer Science and Engineering.* Lancaster, Pa.: DEStech Publications, 2009. This extensively and beautifully illustrated book is intended for newcomers to the polymer field. Recommended readings at the ends of chapters.

Scott, Gerald. *Polymers and the Environment.* Cambridge, England: Royal Society of Chemistry, 1999. From an environmental viewpoint, this book introduces the general reader to the benefits and limitations of polymeric materials as compared with traditional materials.

Seymour, Raymond B., ed. *Pioneers in Polymer Science: Chemists and Chemistry.* Boston: Kluwer Academic Publishers, 1989. This survey of the history of polymer science emphasizes the scientists responsible for the innovations. Several of the chapters were written by scientists who were directly involved in these developments.

PROPULSION TECHNOLOGIES

FIELDS OF STUDY

Physics; chemistry; thermodynamics; gas dynamics; aerodynamics; heat transfer; materials; controls.

KEY TERMS AND CONCEPTS

- **cycle efficiency:** Ratio of the useful work imparted by a thermodynamic system to the heat put into the system.
- **delta-v:** Velocity increment, expressed in units of speed. Measure of the energy required for propulsion from one orbit or energy level to another.
- **equivalent exhaust velocity:** Total thrust divided by the propellant mass flow rate.
- **fuel-to-air ratio:** Ratio of the fuel mass burned per unit time to the air mass flow rate through the burner per unit time.
- **geostationary earth orbit (GEO):** Type of geosynchronous orbit, 35,785 kilometers above the equator, in which the orbiting object appears to stay at a fixed point relative to the Earth's surface at all times.
- **low earth orbit (LEO):** Orbit occurring between roughly 100 and 1,500 kilometers above the Earth, with the high point sometimes reaching 2,000 kilometers.
- **mass ratio:** Ratio of the initial mass at launch to the mass left when propellant is expended.
- **overall pressure ratio:** Ratio between the highest and the lowest pressure in a propulsion cycle.
- **propulsive efficiency:** Ratio of kinetic energy imparted to the propulsion system of the vehicle to the net work done by the propulsion system.
- **specific impulse:** Equivalent exhaust speed, divided by the standard value of acceleration due to gravity, to give units of seconds.
- **thermal efficiency:** Ratio of work done by the pro-

pulsion system to heat put into the system.

- **thrust:** The force exerted by the propulsion system on the vehicle.

SUMMARY

The field of propulsion deals with the means by which aircraft, missiles, and spacecraft are propelled toward their destinations. Subjects of development include propellers and rotors driven by internal combustion engines or jet engines, rockets powered by solid- or liquid-fueled engines, spacecraft powered by ion engines, solar sails or nuclear reactors, and matter-antimatter engines. Propulsion system metrics include thrust, power, cycle efficiency, propulsion efficiency, specific impulse, and thrust-specific fuel consumption. Advances in this field have enabled humanity to travel across the world in a few hours, visit space and the Moon, and send probes to distant planets.

DEFINITION AND BASIC PRINCIPLES

Propulsion is the science of making vehicles move. The propulsion system of a flight vehicle provides the force to accelerate the vehicle and to balance the other forces opposing the motion of the vehicle. Most modern propulsion systems add energy to a working fluid to change its momentum and thus develop force, called thrust, along the desired direction. A few systems use electromagnetic fields or radiation pressure to develop the force needed to accelerate the vehicle itself. The working fluid is usually a gas, and the process can be described by a thermodynamic heat engine cycle involving three basic steps: First, do work on the fluid to increase its pressure; second, add heat or other forms of energy at the highest possible pressure; and third, allow the fluid to expand, converting its potential energy directly to useful work, or to kinetic energy in an exhaust.

In the internal combustion engine, a high-energy fuel is placed in a small closed area and ignited by compression. This produces expanding gas, which drives a piston and a rotating shaft. The rotating shaft drives a transmission whose gears transfer the work to wheels, rotors, or propellers. Rocket and jet engines operate on the Brayton thermodynamic cycle. In this cycle, the gas mixture is compressed adiabatically (no heat added or lost during compression). Heat is added externally or by chemical reaction to the fluid, ideally at constant pressure. The expanding gases are exhausted, with a turbine extracting some work. The gas then expands out through a nozzle.

BACKGROUND AND HISTORY

Solid-fueled rockets developed in China in the thirteenth century achieved the first successful continuous propulsion of heavier-than-air flying machines. In 1903, Orville and Wilbur Wright used a spinning propeller driven by an internal combustion engine to accelerate air and develop the reaction force that propelled the first human-carrying heavier-than-air powered flight.

As propeller speeds approached the speed of sound in World War II, designers switched to the gas turbine or jet engine to achieve higher thrust and speeds. German Wernher von Braun developed the V2 rocket, originally known as the A4 for space travel, but in 1944, it began to be used as a long-range ballistic missile to attack France and England. The V2 traveled faster than the speed of sound, reached heights of 83 to 93 kilometers, and had a range of more than 320 kilometers. The Soviet Union's 43-ton Sputnik rocket, powered by a LOX/RP2 engine generating 3.89 million Newtons of thrust, placed a 500-kilogram satellite in low Earth orbit on October 4, 1957.

The United States' three-stage, 111-meter-high Saturn V rocket weighed more than 2,280 tons and developed more than 33.36 million Newtons at launch. It could place more than 129,300 kilograms into a low-Earth orbit and 48,500 kilograms into lunar orbit, thus enabling the first human visit to the Moon in July, 1969. The reusable space shuttle weighs 2,030 tons at launch, generates 34.75 million Newtons of thrust, and can place 24,400 kilograms into a low-Earth orbit. In January, 2006, the New Horizons spacecraft reached 57,600 kilometers per hour as it escaped from Earth's gravity. Meanwhile, air-breathing engines have grown in size and become more fuel efficient, propelling aircraft from hovering through supersonic speeds.

HOW IT WORKS

Rocket. The rocket is conceptually the simplest of all propulsion systems. All propellants are carried on board, gases are generated with high pressure, heat is added or released in a chamber, and the gases are exhausted through a nozzle. The momentum of the working fluid is increased, and the rate of increase of this momentum produces a force. The reaction to this force acts on the vehicle through the mounting structure of the rocket engine and propels it.

Jet propulsion. Although rockets certainly produce jets of gas, the term "jet engine" typically denotes an engine in which the working fluid is mostly atmospheric air, so that the only propellant carried on the vehicle is the fuel used to release heat. Typically, the mass of fuel used is only about 2 to 4 percent of the mass of air that is accelerated by the vehicle. Types of jet engines include the ramjet, the turbojet, the turbofan, and the turboshaft.

Propulsion system metrics. The thrust of a propulsion system is the force generated along the desired direction. Thrust for systems that exhaust a gas can come from two sources. Momentum thrust comes from the acceleration of the working fluid through the system. It is equal to the difference between the momentum per second of the exhaust and intake flows. Thrust can also be generated from the product of the area of the jet exhaust nozzle cross section and the difference between the static pressure at the nozzle exit and the outside pressure. This pressure thrust is absent for most aircraft in which the exhaust is not supersonic, but it is inevitable when operating in the vacuum of space. The total thrust is the sum of momentum thrust and pressure thrust. Dividing the total thrust by the exhaust mass flow rate of propellant gives the equivalent exhaust speed. All else being equal, designers prefer the highest specific impulse, though it must be noted that there is an optimum specific impulse for each mission. LOX-LH2 rocket engines achieve specific impulse of more than 450 seconds, whereas most solid rocket motors cannot achieve 300 seconds. Ion engines exceed 1,000 seconds. Air-breathing engines achieve very high values of specific impulse because most of the working fluid does not have to be carried on-board.

The higher the specific impulse, the lower the mass ratio needed for a given mission. To lower the mass ratio, space missions are built up in several stages. As each stage exhausts its propellant, the propellant tank and its engines are discarded. When all the propellant is gone, only the payload remains. The relation connecting the mass ratio, the delta-v, and specific impulse, along with the effects of gravity and drag, is called the rocket equation.

Propulsion systems, especially for military applications, operate at the edge of their stable operation envelope. For instance, if the reaction rate in a solid propellant rocket grows with pressure at a greater than linear rate, the pressure will keep rising until the rocket blows up. A jet engine compressor will stall, and flames may shoot out the front if the blades go past the stalling angle of attack. Diagnosing and solving the problems of instability in these powerful systems has been a constant concern of developers since the first rocket exploded.

APPLICATIONS AND PRODUCTS

Many kinds of propulsion systems have been developed or proposed. The simplest rocket is a cold gas thruster, in which gas stored in tanks at high pressure is exhausted through a nozzle, accelerating (increasing momentum) in the process. All other types of rocket engines add heat or energy in some other form in a combustion (or thrust) chamber before exhausting the gas through a nozzle.

Solid-fueled rockets are simple and reliable, and can be stored for a long time, but once ignited, their thrust is difficult to control. An ignition source decomposes the propellant at its surface into gases whose reaction releases heat and creates high pressure in the thrust chamber. The surface recession rate is thus a measure of propellant gas generation. The thrust variation with time is built into the rocket grain geometry. The burning area exposed to the hot gases in the combustion chamber changes in a preset way with time. Solid rockets are used as boosters for space launch and for storable missiles that must be launched quickly on demand.

Liquid-fueled rockets typically use pumps to inject propellants into the combustion chamber, where the propellants vaporize, and a chemical reaction releases heat. Typical applications are the main engines of space launchers and engines used in space, where the highest specific impulse is needed.

Hybrid rockets use a solid propellant grain with a liquid propellant injected into the chamber to vary the thrust as desired. Electric resistojets use heat generated by currents flowing through resistances. Though simple, their specific impulse and thrust-to-weight ratio are too low for wide use. Ion rocket engines use electric fields or, in some cases, heat to ionize a gas and a magnetic field to accelerate the ions through the nozzle. These are preferred for long-duration space missions in which only a small level of thrust is needed but for an extended duration because the electric energy comes from solar photovoltaic panels. Nuclear-thermal rockets generate heat from nuclear fission and may be coupled with ion propulsion. Proposed matter-antimatter propulsion systems use the annihilation of antimatter to release heat, with extremely high specific impulse.

Pulsed detonation engines are being developed for some applications. A detonation is a supersonic shock wave generated by intense heat release. These engines use a cyclic process in which the propellants come into contact and detonate several times a second. Nuclear-detonation engines were once proposed, in which the vehicle would be accelerated by shock waves generated by nuclear explosions in space to reach extremely high velocities. However, international law prohibits nuclear explosions in space.

Ramjets and turbomachines. Ramjet engines are used at supersonic speeds and beyond, where the deceleration of the incoming flow is enough to generate very high pressures, adequate for an efficient heat engine. When the heat addition is done without slowing the fluid below the speed of sound, the engine is called a scramjet, or supersonic combustion ramjet. Ramjets cannot start by themselves from rest. Turbojets add a turbine to extract work from the flow leaving the combustor and drive a compressor to increase the pressure ratio. A power turbine may be used downstream of the main turbine. In a turbofan engine, the power turbine drives a fan that works on a larger mass flow rate of air bypassing the combustor. In a turboprop, the power is taken to a gearbox to reduce revolutions per minute, powering a propeller. In a turboshaft engine, the power is transferred through a transmission as in the case of a helicopter rotor, tank, ship, or electric generator. Many applications combine these concepts, such as a propfan, a turboramjet, or a rocket-ramjet that starts off as a solid-fueled rocket and becomes a ramjet when propellant consumption opens enough space to ingest air.

Gravity assist. A spacecraft can be accelerated by sending it close enough to another heavenly body (such as a planet) to be strongly affected by its gravity field. This swing-by maneuver sends the vehicle into a more energetic

orbit with a new direction, enabling surprisingly small mass ratios for deep space missions.

Tethers. Orbital momentum can be exchanged using a tether between two spacecraft. This principle has been proposed to efficiently transfer payloads from Earth orbit to lunar or Martian orbits and even to exchange payloads with the lunar surface. An extreme version is a stationary tether linking a point on Earth's equator to a craft in geostationary Earth orbit, the tether running far beyond to a countermass. The electrostatic tether concept uses variations in the electric potential with orbital height to induce a current in a tether strung from a spacecraft. An electrodynamic tether uses the force that is exerted on a current-carrying tether by the magnetic field of the planet to propel the tether and the craft attached to it.

Solar and plasma sails. Solar sails use the radiation pressure from sunlight bounced off or absorbed by thin, large sails to propel a craft. Typically, this works best in the inner solar system where radiation is more intense. Other versions of propulsion sails, in which lasers focus radiation on sails that are far away from the Sun, have been proposed. In mini magnetospheric plasma propulsion (M2P2), a cloud of plasma (ionized gas) emitted into the field of a magnetic solenoid creates an electromagnetic bubble around 30 kilometers in diameter, which interacts with the solar

Armadillo Aerospace's quad rocket vehicle showing visible banding (shock diamonds) in the exhaust plume from its propulsion system

---

FASCINATING FACTS ABOUT PROPULSION TECHNOLOGIES

- Thirty-two launches of the Saturn V rocket system were conducted in the 1960's and early 1970's. All succeeded.

- The Galileo mission to Jupiter and beyond obtained nearly 5 of the required 9 kilometers per second delta-v from one flyby of Venus and two flybys of Earth. This is only slightly higher than the delta-v required to reach lunar orbit.

- The hydrogen airships of the 1930's carried 110 people across the Atlantic in three days, at a speed of around 135 kilometers per hour. The Airbus A380, the largest modern airliner, carries 500 to 800 people at 900 kilometers per hour.

- The turbopump of the space shuttle's main engines uses liquid hydrogen at -250 degrees Celsius, but at the end of combustion, the temperature climbs to more than 3,300 degrees Celsius. The turbine of the turbopump is driven by combustion gases and is connected to the impeller, where liquid hydrogen comes in and is pressurized.

- In the linear aerospike nozzle used in the X-33 experimental vehicle, the contoured nozzle surface is on the inside, while the outer flow boundary adjusts itself, thus avoiding the cost of a large variable geometry nozzle.

- In space, to change the plane or direction of an orbit, a spacecraft must add a velocity component sideways, enough to make the resultant velocity vector point along the new direction. This is best done at the apogee of an elliptical orbit, where the orbital speed is lowest.

- A spacecraft can harvest electricity from the potential variation in the Earth's electromagnetic field by extending an electrostatic tether down from its orbit. A current-carrying electrodynamic tether between two spacecraft can be used to gain propulsive thrust because of the Faraday force exerted by the magnetic field on the tether.

- Like all high-performance systems, propulsion systems often push the edge of stable operation. Some descriptive terms include the "pogo" instability of rockets, the "screech" instability of jet engines, the supersonic inlet "buzz," compressor "surge," and "sloshing" in satellite fuel tanks.

---

wind of charged particles that travels at 300 to 800 kilometers per second. The result is a force perpendicular to the solar wind and the (controllable) magnetic field, similar to aerodynamic lift. This system has been proposed to conduct fast missions to the outer reaches of the solar system and back.

CAREERS AND COURSE WORK

Propulsion technology spans aerospace, mechanical, electrical, nuclear, chemical, and materials science engineering. Aircraft, space launcher, and spacecraft manufacturers and the defense industry are major customers of propulsion systems. Workplaces in this industry are distributed over many regions in the United States and near many major airports and National Aeronautics and Space Administration centers. The large airlines operate engine testing facilities. Propulsion-related work outside the United States, France, Britain, and Germany is usually in companies run by or closely related to the government. Because propulsion technologies are closely related to weapon-system development, many products and projects come under the International Traffic in Arms Regulations.

Students aspiring to become rocket scientists or jet engine developers should take courses in physics, chemistry, mathematics, thermodynamics and heat transfer, gas dynamics and aerodynamics, combustion, and aerospace propulsion.

Machinery operating at thousands to hundreds of thousands of revolutions per minute requires extreme precision, accuracy, and material perfection. Manufacturing jobs in this field include specialist machinists and electronics experts. Because propulsion systems are limited by the pressure and temperature limits of structures that must also have minimal weight, the work usually involves advanced materials and manufacturing techniques. Instrumentation and diagnostic techniques for propulsion systems are constantly pushing the boundaries of technology and offer exciting opportunities using optical and acoustic techniques.

SOCIAL CONTEXT AND FUTURE PROSPECTS

Propulsion systems have enabled humanity to advance beyond the speed of ships, trains, balloons, and gliders to travel across the oceans safely, quickly, and comfortably and to venture beyond Earth's atmosphere. The result has been a radical transformation of global society since the early 1900's. People travel overseas regularly, and on any given day, city centers on every continent host conventions with thousands of visitors from all over the world. Jet engine reliability has become so established that jetliners with only two engines routinely fly across the Atlantic and Pacific oceans.

Propulsion technologies are just beginning to grow in their capabilities. As of 2010, specific impulse values were at best a couple of thousand seconds; however, concepts using radiation pressure, nuclear propulsion, and matter-antimatter promise values ranging into hundreds of thousands of seconds. Air-breathing propulsion systems promise specific impulse values of greater than 2,000 seconds, enabling single-stage trips by reusable craft to space and back. As electric propulsion systems with high specific impulse come down in system weight because of the use of specially tailored magnetic materials and superconductors, travel to the outer planets may become quite routine. Spacecraft with solar or magnetospheric sails, or tethers, may make travel and cargo transactions to the Moon and inner planets routine as well. These technologies are at the core of human aspirations to travel far beyond their home planet.

—*Narayanan M. Komerath, PhD*

FURTHER READING

Faeth, G. M. *Centennial of Powered Flight: A Retrospective of Aerospace Research.* Reston, Va.: American Institute of Aeronautics and Astronautics, 2003. Traces the history of aerospace and describes important milestones.

Henry, Gary N., Wiley J. Larson, and Ronald W. Humble. *Space Propulsion Analysis and Design.* New York: McGraw-Hill 1995. Combines short essays by experts on individual topics with extensive data and analytical methods for propulsion system design.

Norton, Bill. *STOL Progenitors: The Technology Path to a Large STOL Aircraft and the C-17A.* Reston, Va.: American Institute of Aeronautics and Astronautics, 2002. Case study on a short takeoff and landing aircraft development program. Describes the various steps in a modern context, considering cost, technology, and military requirements.

Peebles, C. *Road to Mach 10: Lessons Learned from the X-43A Flight Research Program.* Reston, Va.: American Institute of Aeronautics and Astronautics, 2008. Case study of a supersonic combustion demonstrator flight program authored by a historian.

Shepherd, D. *Aerospace Propulsion.* New York: Elsevier, 1972. A prescient, simple, and lucid book that has inspired generations of aerospace scientists and engineers.

R

RADIO

FIELDS OF STUDY

Broadcasting; communications; electronics; Global Positioning System; microwave technology; radar technology; radio technology; radio astronomy.

KEY TERMS AND CONCEPTS

- **amplification:** Process of increasing the strength of an electronic transmission or a sound wave.
- **amplitude:** Refers to the height of a radio wave.
- **antenna:** Device that either converts an electric current into an electromagnetic radiation (transmitter) or converts electromagnetic radiation into an electric current (receiver).
- **microphone:** Device that converts sound into an electrical signal.
- **modulation:** Process of varying one or more properties of an electromagnetic wave; three parameters can be altered via modulation: amplitude (height), phase (timing), and frequency (pitch). Two common forms of radio modulation are amplitude modulation (AM) and frequency modulation (FM).
- **radio frequency:** Oscillation of a radio wave in the range of 3 kilohertz (3,000 cycles per second) to 300 gigahertz (300,000,000,000 cycles per second).
- **radio wave:** Electromagnetic radiation, which travels at the speed of light; radio waves are of a longer wavelength than infrared light.

SUMMARY

Radio is a technology that involves the use of electromagnetic waves to transmit and receive electric impulses. Since its inception as a method of wirelessly transmitting Morse code, radio communications technology has had a tremendous impact on society. Although television has supplanted radio to a significant extent for public broadcasting, radio continues to play a significant role in this arena. Radio broadcasts may be delivered via technologies such as satellites, cable networks, and the Internet. Although the term "radio" commonly brings to mind listening to news broadcasts, music, and other audio signals, electromagnetic-wave transmission encompasses other fields, such as radar, radio astronomy, radio control, and microwave technology.

DEFINITION AND BASIC PRINCIPLES

In contrast to sound waves, which require a medium such as air or water for propagation, electromagnetic waves can travel through a vacuum. In a vacuum such as outer space they travel at the speed of light (299,800 kilometers per second). In space, electromagnetic waves conform to the inverse-square law: the power density of an electromagnetic wave is proportional to the inverse of the square of the distance from a point source. Thus, all radio waves weaken as they travel a distance. When traveling through air the intensity of the waves is weakened. At some point, depending on the strength of the signal, the electromagnetic wave will no longer be discernible. Interference can also weaken or destroy a radio signal. Other radio transmitters and accidental radiators (such as automobile ignition systems) produce interference and static. Frequency modulation (FM) radio signals are much more resistant to interference and static than amplitude modualtion (AM) radio signal. Radio waves travel in a straight line; therefore, the curvature of the earth limits their range. However, radio waves can be reflected by the ionosphere, which extends their range.

The following steps occur in radio transmission: A transmitter modulates (converts) sound to a specific radio frequency; an antenna broadcasts the electromagnetic wave; the wave is received by an antenna; and a receiver tuned to the radio frequency demodulates the electromagnetic energy back into sound. Radio waves range from 3 kilohertz (kHz) to 300 gigahertz (GHz) and are categorized as: very low frequency (VLF; 3 to 30 kHz); low frequency (LF; 30 to

300 kHz); medium frequency (MF; 300 to 3,000 kHz); high frequency (HF; 3 to 30 megahertz [MHz]); very high frequency (VHF; 30 to 300 MHz); ultra high frequency (UHF; 300 to 3,000 MHz); super high frequency (SHF; 3 to 30 GHz); and extremely high frequency (EHF; 30 to 300 GHz). Each frequency range has unique characteristics and unique applications for which they are best suited.

Background and History

Electromagnetic waves were discovered in 1877 by the German physicist Heinrich Hertz, whose name is used to describe radio frequencies in cycles per second. Eight years later, American inventor Thomas Alva Edison obtained a patent for wireless telegraphy by discontinuous (intermittent) wave. A far superior system was developed in 1894 by the Italian inventor Guglielmo Marconi. Initially, he transmitted telegraph signals over a short distance on land. Subsequently, an improved system was capable of transmitting signals across the Atlantic Ocean. At the start of the twentieth century, Canadian inventor Reginald Aubrey Fessenden began experimenting with voice transmission via discontinuous waves while employed by the United States Weather Bureau, and he eventually pioneered radio broadcasting. In 1902, he switched to continuous waves and successfully transmitted voice and music. In 1906, history was made when Fessenden transmitted voice and music from Massachusetts that was heard as far away as the West Indies. The same year, Lee de Forest developed the Audion tube (later known as the triode vacuum tube), which became a key component of radios.

In 1920, the first radio news program was broadcast by station 8MK in Detroit. Also in 1920, the station WRUC in New York began broadcasting a series of Thursday night concerts, the initial range of which was 100 miles. However, it was soon expanded to 1,000 miles. WRUC also began sports broadcasts in the same year. In the early 1930's, frequency modulation (FM) and single sideband shortwave radio were invented by amateur radio operators. In 1954, the companies Texas Instruments and Regency embarked on a joint venture to launch the Regency TR-1, the first portable transistor radio, which was powered by a 22.5-volt battery.

Radio also played a role in other fields. Radio astronomy began in the 1930s and expanded greatly after World War II. The broadcast of television over radio waves, first demonstrated in the 1920s, became increasingly popular in the 1950s. The cellular radio telephone was invented in 1947 by Bell Laboratories, although it was not until the late 1990s and early 2000s that cell phones became a major cultural influence. Radio technology played a significant role in the space race, and satellites themselves increased the broadcasting of signals. In the later twentieth and early twenty-first centuries, digital technology shaped the radio industry.

How It Works

Basic example of a radio receiver. The humble crystal set, which first appeared at the close of the nineteenth century, is a radio receiver in its simplest form. It consists of an antenna, a tuned circuit, a crystal detector, and earphones. The tuned circuit consists of a tuning coil (a sequentially wound coil that can be tapped at any point) connected to a capacitor. This pair of components allows tuning of the receiver to a specific frequency, known as the resonant frequency. Only signals at the resonant frequency pass through the tuned circuit; other frequencies are blocked. The crystal is a semiconductor, which extracts the audio signal from the radio frequency carrier wave. This is accomplished by allowing current to pass in just one direction, blocking half the oscillations of the radio wave. This rectifies, or changes, the wave into a pulsing direct current, which varies with the audio signal. The earphones then convert the direct current into sound. The sound power is solely derived from the radio station that originated it. The electrical power and circuitry of more complex receivers serve to amplify this extremely weak signal to one that can power loudspeakers.

Amplitude modulation (AM). With AM radio, the amplitude (height) of the transmitted signal is made proportional to the sound amplitude captured (transduced) by the microphone. The transmitted frequency remains constant. AM transmission is degraded by static and interference because sources of electromagnetic transmission such as lightning and automobile ignitions, which are at the same frequency, add their amplitudes to that of the transmitted signal. AM radio stations in the United States and Canada are limited to 50 kilowatts (kW). Early twentieth century, U.S. stations had powers up to 500 kW, some of which could be heard worldwide. Conventional AM transmission involves the use of a carrier signal, which is an inefficient use of power. The carrier can be removed (suppressed) from the AM signal. This produces a reduced-carrier transmission, which is termed a double-sideband suppressed-carrier (DSB-SC) signal. A sideband refers to one side of the mirror-image radio signal. DSB-SC has three times more power efficiency than an unsuppressed AM signal. Radio receivers for DSB-SC signals must reinsert the carrier. Another AM refinement is single-sideband modulation in which both one sideband and the carrier are stripped out. This modification doubles the effective power of the signal.

Frequency modulation (FM). With FM, the variation in amplitude from the microphone causes fluctuations in the transmitter frequency. FM broadcasts have a higher fidelity than AM broadcasts and are resistant to static and interference. FM requires a wider bandwidth to operate and is transmitted in the VHF (30 to 300 MHz) range. These high frequencies travel in a straight line, and the reception range is generally limited to about 50 to 100 miles. FM signals broadcast from a satellite back to Earth do not have this distance limitation. An FM transmission can contain a subcarrier in which secondary signals are transmitted in a piggyback together with the main program. The subcarrier allows stereo broadcasts to be transmitted. The subcarrier can also transmit other information such as station identification and the title of the current song being played.

Digital radio. Digital audio transmission consists of converting the analogue audio signal into a digital code of zeros and ones in a process known as digitizing. This technology allows for an increase in the number of radio programs in a given frequency range, improved fidelity, and reduction in fading (signal loss). Satellite radio is a digitized signal and can cover a distance in excess of 22,000 miles.

APPLICATIONS AND PRODUCTS

Radio applications are widespread and include commercial radio, amateur radio, marine radio, aviation radio, radar, navigation, and microwave.

Commercial radio. Radio broadcasts are available throughout the globe. They offer a variety of products, such as news, music, and political opinion. Many broadcasts are free of charge and available to receivers in range of the station. The station derives its revenue from advertising. An hour of broadcasting time typically contains ten to twenty minutes

of advertising. Cable television and Internet services rebroadcast these stations either at no charge or with a fee. Cable television services and satellite radio broadcast a variety of commercial-free programs. In these cases, the subscription fee covers the cost of the broadcast.

Amateur radio. Amateur radio (also known as ham radio) is the licensed use of designated radio bands for noncommercial exchange of messages, private recreation, emergency communication, and experimentation. Ham operators maintain and operate their own equipment. Through the years, they have made numerous contributions to radio technology, including FM and single sideband. They also perform a number of public services at no cost. For example, during the Vietnam War, the Military Auxiliary Radio System (MARS) allowed military personnel to call friends and relatives in the United States. A broadcast emanating from Vietnam was received by a US-based amateur operator who patched the communication into the phone lines. MARS is still active; however, many of its services have been supplanted by the Internet and e-mail.

Shortwave radio. Shortwave radio, which was first theorized in 1919, has many applications for long-range communication. The term "shortwave" refers to the wavelength of the frequency spectrum in which it operates: high frequency (3,000 to 30,000 kHz). The high-frequency wavelength is shorter than the ones first used for radio communications: medium frequency and low frequency. Shortwave broadcasts are readily transmitted over distances of several thousand kilometers, allowing intercontinental communication and ship-to-shore communication. Low-cost shortwave radios are available worldwide and facilitate the transfer of information to individuals where other forms of media are controlled for political reasons. A major disadvantage of shortwave radio is that it is subject to significant interference problems, such as atmospheric disturbances, electrical interference, and overcrowding of wave bands. The Internet and satellite radio have impacted shortwave radio, but it is still useful in areas where those services are unavailable or too expensive.

Marine radio. All large ships and most small oceangoing vessels are equipped with a marine radio. Its purposes included calling for aid, communicating with other vessels, and communicating with shore-based facilities, such as harbors, marinas, bridges, and locks. Marine radio operates in VHF, from 156 to 174 MHz. Channels from 0 to 88 are assigned to specific frequencies. Channel 16 (156.8 MHz) is designated as the international calling and distress channel.

Aircraft band. The aircraft band (or air band) operates in the VHF range. Different sections of the band are used for commercial and general aviation aircraft, air traffic control, radio-navigational aids, and telemetry (remote transmission of flight information). VHF omnidirectional range (VOR) or an instrument landing system operates in the aircraft band. A VOR is a ground-based station that transmits a magnetic bearing of the aircraft from the station. Unmanned aircraft (drones) are navigated by a radio system in which a pilot sits at a console on the ground and directs the flight.

Radio telescope. A radio telescope is a large, directional, parabolic (dish) antenna that collects data from space probes and Earth-orbiting satellites. The first purpose-built example was created in 1937 by Grote Reber, an engineer and amateur astronomer. Many astronomical objects emit electromagnetic radiation in the radio frequency range, therefore radio telescopes can image astronomical objects such as galaxies. Some researchers are using radio telescopes to search for intelligent life forms in the universe.

Navigation. Radio navigation encompasses a number of applications including radio direction finding (RDF), radar, loran, and Global Positioning System (GPS).

An RDF system homes in on a radio transmission, which can be a specialized antenna that directs aircraft or commercial transmitters that have a known location.

Radar is an acronym for radio detecting and ranging. The device consists of a transmitter and receiver. The transmitter emits radio waves, which are deflected from a fixed or moving object. The receiver, which can be a dish or an antenna, receives the wave. Radar circuitry then displays an image of the object in real time. The screen displays the distance of the object from the radar. If the object is moving, consecutive readings can calculate the speed and direction of the object. If the object is airborne, and the radio is so equipped, the altitude is displayed. Radar is invaluable in foggy weather when visibility can be severely reduced.

Loran is an acronym for long-range navigation. The system relies on land-based low-frequency radio transmitters. The device calculates a ship's position by the time difference between the receipt of signals from two radio transmitters. The device can display a line of position, which can be plotted on a nautical chart. Most loran convert the data into longitude and latitude. Since GPS became available, the use of loran has markedly declined.

FASCINATING FACTS ABOUT RADIO

- Radio pioneer Guglielmo Marconi initially transmitted radio signals over water. He was unable to transmit over land because prevailing laws in Europe gave government postal services a monopoly on message delivery.

- Another radio pioneer, Lee de Forest, who developed the Audion tube (triode vacuum tube) was indicted for using the mails to defraud because he was promoting "a worthless device." He was subsequently acquitted and his invention continued to be a key component of radios until vacuum tubes were supplanted by transistors.

- In this high-tech age, many deem Morse code to be an antiquated, inferior form of message transmission. However, on May 20, 2005, the *Tonight Show with Jay Leno* show featured two teams: a pair of amateur radio operators (hams) who were proficient at Morse code and another duo who were skilled at text messaging. The texters were out-messaged handily by the hams, who transmitted and received at a rate of twenty-nine words per minute.

- Radio gained acceptance more rapidly in the United States among amateur radio operators than the general public. In 1913, 322 amateurs were licensed; however, by 1917 there were 13,581, primarily boys and young men. Many older individuals considered radio to be a fad. They reasoned that listening to dots and dashes or occasional experimental broadcast of music or speech over earphones was a worthless endeavor.

- The microwave oven was a by-product of radar research. Raytheon engineer Percy Spencer was testing a new vacuum tube called a magnetron and found that a candy bar in his pocket had melted. Deducing that electromagnetic energy was responsible, he placed some popcorn kernels near the tube. The kernels sputtered and popped. His next trial was with an egg, which exploded in short order.

GPS is a space-based global navigation satellite system that provides accurate location and time information at any location on Earth where there is an unobstructed line of sight to four or more GPS satellites. GPS can function under any weather condition and anywhere on the planet. The technology depends on triangulation, just as a land-based system such as loran employs. GPS is composed of three segments: the space segment; the control segment; and the user segment. The US Air Force operates and maintains both the space and control segments. The space segment is made up of satellites, which are in medium-space orbit. The satellites broadcast signals from space, and a GPS receiver (user segment) uses these signals to calculate a three-dimensional location (latitude, longitude, and altitude). The signal also transmits the current time, accurate within nanoseconds.

CAREERS AND COURSE WORK

Many technical and nontechnical careers are available in radio. The technical fields require a minimum of a bachelor's degree in engineering or other scientific field. Many also require a postgraduate degree such as a master's or doctorate. Course work should include mathematics, engineering, computer science, and robotics. Positions are available in both the government and private sector. The ability to be a team player is often of value for these positions because ongoing research is often a collaborative effort. Less technical positions such as operation or maintenance of broadcasting equipment can be achieved with less training, which can be obtained at a technical college or trade school. A bachelor's degree in communications or related fields can qualify one for employment in radio or television broadcasting.

SOCIAL CONTEXT AND FUTURE PROSPECTS

In view of the continuous advances in radio technology, further advances are extremely likely. Radio and related technologies are an integral component of everyday life in most societies. Radio has political implications, particularly in nations governed by tyrants and dictators, where radio may be used as a propaganda tool and use of shortwave radio and the Internet is strongly discouraged. The Voice of America (VOA), which began broadcasting in 1942, is an international multimedia broadcasting service funded by the US government. The VOA broadcasts about 1,500 hours of news, information, educational, and cultural programming each week to an estimated worldwide audience of 123 million people. In some repressed nations, the broadcast is jammed, which involves broadcasting noise at the same frequency to prevent reception. Even in the United States, political movements exist that wish to ban certain types of radio broadcasts (such as talk radio), which are deemed too conservative or too liberal.

Many associate the term microwave with a useful appliance to heat a frozen meal, but microwave is in the radio wavelength (0.3 GHz to 300 GHz) and is also used in communication. The microwave oven illustrates the fact that electromagnetic waves are a form of energy capable of heating—and damaging—tissue. For example, standing near or touching a powerful radio transmitter can result in severe burns. Microwave ovens are shielded to prevent exposure; other devices are not. The heating effect of an electromagnetic wave varies with its power and the frequency. The heating effect is measured by its specific absorption rate (SAR) in watts per kilogram. Many national governments as well as the Institute of Electrical and Electronics Engineers (IEEE) have established safety limits for exposure to various frequencies of electromagnetic energy based on their SAR.

—*Robin L. Wulffson, MD, FACOG*

FURTHER READING

Hallas, Joel. *Basic Radio: Understanding the Key Building Blocks.* Newington: American Radio Relay League, 2005. Print.

Kaempfer, Rick, and John Swanson. *The Radio Producer's Handbook.* New York: Allworth, 2004. Print.

Keith, Michael C. *The Radio Station: Broadcast, Satellite and Internet.* 8th ed. Burlington: Focal, 2010. The standard guide to radio as a medium. Discusses the various departments in all kinds of radio stations in detail. Print.

Rudel, Anthony. *Hello, Everybody: The Dawn of American Radio.* Orlando: Houghton Mifflin Harcourt, 2008. Print.

Silver, H. Ward. *ARRL Ham Radio License Manual: All You Need to Become an Amateur Radio Operator.* Newington: American Radio Relay League, 2010. Print.

RADIO ASTRONOMY

FIELDS OF STUDY

Astronomy; physics; electronics; mathematics; electrical engineering; computer science; radio technology.

SUMMARY

Radio astronomy is the branch of astronomy associated with studying the heavens using radio frequency electromagnetic radiation, generally those frequencies below 300 gigahertz. Visual light composes only a tiny portion of the electromagnetic spectrum. Many astrophysical systems emit more strongly in radio waves than in visual light, and other systems emit only radio waves. Radio emissions also carry different information about the radiating system than do visual light waves, so without using radio astronomy, astronomers are unable to fully study the universe. Radio astronomers use radio telescopes, antennas dedicated to radio astronomy, sometimes used individually and sometimes grouped in arrays. The techniques of radio astronomy have been adapted to other fields, including low-intensity radio communication and image analysis.

KEY TERMS AND CONCEPTS

- **antenna temperature:** Theoretical temperature at which a blackbody would emit thermal radiation equivalent to the electrical noise of the antenna.
- **aperture synthesis:** Technique to combine the signals of multiple smaller radio telescopes to simulate the characteristics of a much larger instrument.
- **array:** Set of radio telescopes arranged to work together to yield more information than the individual telescopes working separately.
- **beamwidth:** Angular size measuring the resolution of an antenna.
- **brightness temperature:** Theoretical temperature at which a blackbody would radiate thermal energy at the rate detected by a radio telescope.
- **intensity:** Energy per unit time per unit area.
- **interferometry:** Technique of combining signals between multiple radio telescopes to increase the angular resolution of the telescopes.
- **jansky:** Unit of radio power flux used in radio astronomy; equal to $10-26$ watts per square meter per hertz.
- **radio window:** Range of electromagnetic radiation at which the atmosphere is mostly transparent, extending roughly from 5 megahertz to 300 gigahertz.
- **21-centimeter radiation:** Wavelength of the characteristic emission of neutral hydrogen atoms undergoing a spin-flip transition.

DEFINITION AND BASIC PRINCIPLES

Radio astronomy is the study of the heavens using radio frequency radiation. Physical processes often produce electromagnetic radiation. Although visual light is a form of electromagnetic radiation, it is only a very small part of the electromagnetic spectrum. Many physical processes produce radio radiation in addition to or sometimes instead of visual light. Thus, to fully study the universe, astronomers must observe more than just visual light.

Radio astronomy uses techniques for detection of radio signals from space similar to those used by radio communication systems. The radio telescope is basically the antenna that focuses the radio flux onto a detector. Electronic circuitry amplifies the detected signal. As with radio communication, the signal from multiple antennas, or multiple radio telescopes, can be combined electronically to yield more information than that which could be detected by one radio telescope alone. The similarity between the two technologies has permitted many advances in radio astronomy to be used in the field of commercial communication and vice versa.

Visual telescopes can be fitted with cameras to take images. That is not possible, however, with radio telescopes, which measure only the intensity of radio flux coming from whatever direction the radio telescope is pointing. Thus, images must be constructed from multiple measurements, and several techniques have been developed for creating images using different types of radio telescopes. This is far more time-consuming than simply taking a photograph and often involves powerful computer software. This software can be adapted to build images of other systems from a series of measurements, such as with medical imaging.

BACKGROUND AND HISTORY

In 1930, Karl G. Jansky, an engineer with Bell Telephone Laboratories, began work to isolate sources of interference with long distance radio telephone communications. By 1932, Jansky had determined that one source of interference, a steady hiss of static, originated external to Earth. The following year, he published a paper showing that the extraterrestrial radio interference originated in the Milky Way, with the most intense interference coming from the general direction of the center of the galaxy. After hearing about Jansky's discovery, Chicago radio engineer and amateur radio operator Grote Reber decided to build his own radio telescope in his backyard in 1937. Reber's antenna was the first radio antenna constructed solely to monitor signals from space for scientific purposes and thus is often regarded as the first true radio telescope.

World War II brought an end to many basic scientific investigations, including radio astronomy. However, radar operators during the war monitored interference with radar systems caused by solar activity. After the war, military surplus radar technology became readily available to radio astronomers, and there was a surge in radio astronomy activity. Radio telescopes, however, must be extremely large, and thus extremely expensive, to compete with optical telescopes in resolution and sensitivity. The cutting edge of radio astronomy was too expensive for individual institutions to pay for without government aid. The establishment of the Jodrell Bank Experimental Station, with its giant radio telescope, in the United Kingdom in 1945 helped spur the United States and other nations to build their own radio astronomy facilities, giving rise to America's National Radio Astronomy Observatory in 1956, with its 300-foot-diameter radio telescope, and the 1,000-foot-diameter Arecibo Observatory radio dish in Puerto Rico in 1963.

HOW IT WORKS

The heart of a radio telescope is a radiometer, a device that measures the strength of radio signals gathered by the telescope. The output of the radiometer is measured and recorded. The output of a radio telescope is essentially analogue, so older systems used a chart recorder to record the signal strength. More modern systems have used magnetic tape or other similar media. The radio telescope itself is essentially an antenna that concentrates and amplifies the incoming signals for the radiometer. Though the dish shape similar to a satellite dish is most commonly portrayed in photographs of radio telescopes, any radio antenna can be used as a radio telescope. The most effective antenna design depends on the wavelength and type of observation. The advantage of the dish shape is that the dish acts as a reflector to focus radio radiation striking the entire dish area toward the receiver that is located at the focus of the dish. The radiometer connects to the receiver, and often the two parts are a unit subassembly. The receiver is designed to measure the intensity of radio radiation across a particular radio frequency band. Unlike with many communication radio receivers, changing the frequency band is more complicated with a radio telescope than just pressing a button or turning a dial. Often, the entire receiver subassembly must be replaced. Other shapes, such a Yagi antenna radio telescope (a long boom with crosspieces, a example is a typical very-high-frequency, or VHF, external television antenna), are more easily constructed and handled than a dish design but are typically constructed to operate best at only one set of radio frequencies. A dish system is more versatile.

Image construction. Radio telescopes measure intensity only. The radio intensity is governed by the amount of radio power concentrated by the antenna onto the receiver. The most useful antennas tend to be directional, so that the antenna preferentially concentrates radio energy from a particular direction in the sky, determined by the orientation of the antenna. Thus, radio telescopes are often mounted in such a way that they can be readily pointed in different directions as needed. However, there is no such thing as a radio camera or a radio imaging system for the radio telescope. Therefore, radio astronomers must take many measurements from slightly different orientations of the radio telescope to construct an image. This image is constructed using sophisticated computer software. The data analysis technique required to produce an image depends on the type of radio telescope system, and much of the work of a radio astronomer is done on the computer analyzing the data collected by the radio telescopes.

The Very Large Array, a radio interferometer in New Mexico, USA

Aperture synthesis. One of the problems of radio telescopes is that radio waves are quite long compared with other forms of electromagnetic radiation. The longer the wavelength, the larger the instrument must be to achieve comparable resolution, which is the ability to distinguish between nearby objects. The higher the resolution, the higher quality the images that can be produced and the more precisely the source of radio signals can be determined. The resolution of the radio telescope is the beamwidth of its antenna and is determined by the frequency band being used and the shape of the antenna. For a single dish radio telescope to have a resolution comparable with the human eye, it would need to be nearly a mile across, which is not feasible.

However, because of the wave nature of electromagnetic radiation, the principle of interferometry can be used to achieve greater resolution than would be possible with a single radio telescope. The signals from two radio telescopes can be combined electronically. Radio waves have to travel different paths to get to the two antennas, and thus the signals are slightly out of phase, depending on the location of the source relative to the antennas. If they are out of phase by one-half wavelength, then the signals cancel. The farther apart the two antennas, the greater the effect, and thus the more precisely the position of the source can be determined. This effectively allows the two radio telescopes to have a resolution along one direction equal to that of a radio telescope the diameter of the distance between the two instruments. By extending this technique to an array of many radio telescopes and combining the signals in the proper manner, radio astronomers can effectively simulate the resolution of a radio telescope equal in size to the size of the array in a technique called aperture synthesis. The National Radio Astronomy Observatory's Very Large Array (VLA) near Socorro, New Mexico, is an example of a radio telescope array that uses aperture synthesis to produce high-resolution images.

APPLICATIONS AND PRODUCTS
Active galaxies. Some early radio survey objects were identified with galaxies, giving rise to the term "radio galaxies" to describe these objects. In the 1950's, however, a few radio sources were identified with what appeared to be stars. Analysis of these radio stars showed that they were not stars at all. They became known as quasi-stellar objects, or quasars. Later research showed that quasars and radio galaxies, along with other active galaxies, are powered by supermassive black holes in their centers. Radio astronomy continues to play an important role in understanding galactic formation and dynamics.

Interstellar medium. The galaxy is filled with a very thin hydrogen gas. Much of this gas is in the form of individual neutral hydrogen atoms and does not shine with any visual light. However, when an electron flips its spin to a lower energy state in neutral hydrogen, it can emit radio waves having a wavelength of 21 centimeters. These radio emissions can be studied using radio telescopes. Observations of the 21-centimeter radiation have yielded a great deal of information about the density of material between stars and have been used to map the Milky Way and other galaxies. Studies of the 21-centimeter radiation in other galaxies are important in determining the rotational curves of those galaxies, an important tool in the study of dark matter in the universe. Rotational velocity measurements in other galaxies are also an important tool in calibrating the distance measurements needed to measure the cosmological expansion of the universe.

Pulsars. Jocelyn Bell and Antony Hewish discovered pulsating radio sources in 1967. These objects, named pulsars, turned out to be the first observational evidence of neutron stars, collapsed remnants of massive stars remaining after supernova explosions. Radio astronomy continues to play an important role in the study of pulsars, and knowledge of neutron star formation and properties is important in understanding how massive stars die and heavy elements form and are distributed throughout the universe.

Search for Extraterrestrial Intelligence. By the twentieth century, scientists were beginning to seriously consider the possibility that intelligent life may have evolved elsewhere in the universe. In the 1950's, some scientists suggested that perhaps such life could be detected by its radio transmissions. The first search for radio signals from intelligent life beyond Earth was conducted by Frank Drake in 1960. Since then, radio telescopes have routinely been used to search for extraterrestrial intelligence, but not all searches have been through dedicated radio telescopes. Several instrument packages piggyback on other scientific observations, looking for coherently modulate signals while the radio telescope is otherwise engaged in scientific research.

Radar observations. Radar observations within the solar system fall under radio astronomy. The Arecibo Observatory has a powerful radar transmitter that has been used to refine the orbits of a number of near-Earth asteroids. Furthermore, radar reflections have been used to map the surface of Venus and the topography of several nearby asteroids, tasks that are either impossible or extremely difficult to do with visual astronomy.

IMPACT ON INDUSTRY

Jansky's discovery of radio signals coming from outer space opened the doorway for an entirely new way to study the heavens. With the large radio telescopes built in the latter half of the twentieth century, radio astronomy has become a major tool of astronomical research. Some astrophysical processes produce radio emissions rather than visual light, and radio telescopes are able to peer through interstellar media that block other wavelengths of electromagnetic radiation. Thus, radio telescopes are able to observe things that are not visible in other ways. Comprehensive studies of any celestial object have come to include observations in multiple wavelengths: visual, infrared, ultraviolet, and radio.

Government and university research. Radio astronomy's biggest users are astronomers conducting research. The majority of research astronomers work either as faculty members at universities or researchers at government-funded laboratories. Radio observatories run or funded by the government operate on a shared-user basis. Astronomers submit proposals to use the facilities, and a review board judges which proposals are most qualified to warrant use of the facilities. Those researchers are then awarded time on the radio telescopes. Sometimes the researchers go the observatory to collect data, but other times data are collected by technicians at the observatory according to instructions from the researchers.

Most of the funding to operate large radio astronomy facilities comes from government agencies. The largest source of funding for radio astronomy in the United States is the National Science Foundation. Some smaller facilities associated with individual institutions were constructed with grant money, primarily from the foundation, but were operated out of state or local university funds. Budget shortfalls early in the twenty-first century, however, put pressure on many institutions to close their radio telescope facilities.

Some private organizations, such as the Planetary Society and the SETI Institute have secured private funds through donations or grants to fund their own radio astronomy studies. Generally, these organizations work with other facilities to fund operations of an existing radio telescope otherwise slated for decommissioning or to build new facilities, such as the Allen Telescope Array built by the SETI Institute in cooperation with University of California, Berkeley, at the university's Hat Creek Radio Observatory.

Industry and business. There is very little economic incentive for private industry to conduct radio astronomy research for scientific gain. However, a small specialized group of companies works with the large government-funded observatories to construct new radio telescopes or to refurbish older ones. The physical structure of such large instruments requires mechanical engineering expertise to properly construct them, and several companies have found an economic niche supplying this expertise.

The technology developed for radio astronomy is virtually the same as that used for commercial communication receivers. Therefore, many of the advances in radio astronomy have commercial applications in the form of smaller, more sensitive, and less expensive satellite communication equipment. Many satellite communication companies keep abreast of the developments in radio astronomy technology to adapt this technology to produce better communication systems.

CAREERS AND COURSE WORK

Amateur radio astronomy is possible with simply a familiarity with electronics and astronomy; however, a career in radio astronomy generally requires more. Radio astronomy is a branch of astronomy, which is associated with physics and related to various engineering fields. Therefore, radio astronomers need courses in advanced mathematics, physics, computer programming, electronics, and astronomy. Chemistry and engineering courses are also useful. A radio astronomer should have a graduate degree in physics or astronomy, typically a doctorate, to do research. However, the field is quite broad, and most astronomers do not specialize in radio astronomy. Large radio telescopes, such as those that professional astronomers use, require mechanical engineers to design the structures and electrical engineers to design the receivers. Computer programmers are required to design the software to operate the telescope and analyze the data. These tasks often do not require a doctorate in the field but do require familiarity with astronomy. Because many astronomers who use radio telescopes are not themselves radio astronomers, technicians typically operate the telescopes. Technician positions generally require a bachelor's or master's degree in physics, astronomy, or electrical engineering.

Many astronomers work in academic settings as college professors. These positions always require advanced degrees. Astronomers often use many resources to study the heavens, including radio astronomy. However, technicians work directly with radio telescopes, either in construction or operation of the instruments. Therefore, technicians typically are employed directly by the radio telescope operators. A few private research laboratories or universities operate radio telescopes; however, most

---

FASCINATING FACTS ABOUT RADIO ASTRONOMY

- After hearing about the discovery of pulsars, a tabloid dubbed the objects "little green men," or LGMs, speculating that they were beacons placed in space by aliens.
- The giant radio telescopes at Arecibo have been featured in several motion pictures, including the James Bond film *Golden Eye* (1995), and several television programs, including an episode of *The X-Files* (1993–2002).
- The jansky, a unit of radio power flux used in radio astronomy, was named for Karl G. Jansky, who is first credited with discovering extraterrestrial radio signals.
- The largest fully steerable radio telescope dish, the Green Bank Telescope, has a diameter about as large as a football field.
- When Grote Reber built his radio telescope in Chicago, it was on the eve of World War II, and it was rumored that he was working on an antiaircraft death ray for the U.S. Army.
- Simple radio telescopes can be built very inexpensively by amateurs using commercially available parts without modification. An old analogue television set and an ultra-high-frequency (UHF) antenna can be used to detect the Crab Nebula, and amateur radio equipment can be used to detect Jupiter.

---

are operated by government laboratories or government-funded laboratories. There are comparatively few jobs available for radio telescope technicians, but there are also even fewer people who specifically choose this career path. Most radio telescope technicians start off as engineers or students in related fields who migrate to the field of radio astronomy.

SOCIAL CONTEXT AND FUTURE PROSPECTS

Radio astronomy continues to be an important tool in understanding the universe. However, the sensitive radio receivers of the radio telescopes are increasingly subject to interference from commercial communication systems. Additional technologies are being developed to deal with this interference, but the interference will most likely continue to be a problem and become worse at established radio observatories. Plans to build new radio observatories in even more remote locations will probably forestall the problem of interference but will not end it for ground-based radio astronomy.

The National Aeronautics and Space Administration (NASA) and several other space agencies have studied the feasibility of space-based radio telescopes. Several radio astronomy satellites have been launched into Earth orbit, where interference from terrestrial interference is much less. These instruments have been quite successful, and even more capable radio astronomy satellites are likely to be launched throughout the twenty-first century. Furthermore, several feasibility studies have been done on the prospect of building a radio astronomy observatory on the far side of the Moon, where the radio telescopes would be completely shielded from terrestrial interference. As yet, there are no plans for such a facility, but it is likely to eventually be built, probably at some point within the century. Naturally, such developments will bring new challenges and new opportunities to the field of radio astronomy.

—Raymond D. Benge, Jr., MS

FURTHER READING

Burke, Bernard F., and Francis Graham-Smith. *An Introduction to Radio Astronomy.* 3d ed. New York: Cambridge University Press, 2009. A very thorough but also quite technical overview of the theory of radio astronomy and applications to astronomical observations.

Carr, Joseph J. *Radio Science Observing.* 2 vols. Indianapolis, Ind.: Prompt Publications, 1999. Contains radio astronomy ideas, circuits, and projects suitable for amateur radio operators or advanced high school or college students.

Kellermann, Kenneth I. "Radio Astronomy in the Twenty-first Century." *Sky and Telescope* 93, no. 2 (February, 1997): 26–34. An excellent overview of the development of radio astronomy through the twentieth century with a look forward to prospects for developments through the twenty-first century.

Lonc, William. *Radio Astronomy Projects.* Louisville, Ky.: Radio-Sky Publishing, 1996. Detailed instructions on the construction of various radio telescope systems and information on projects using those systems for the amateur astronomer with a good understanding of electronics. Suitable for engineering students.

Malphrus, Benjamin K. *The History of Radio Astronomy and the National Radio Astronomy Observatory: Evolution Toward Big Science.* Malabar, Fla.: Krieger, 1996. A chronicle of radio astronomy history from the early days to the development of large national radio astronomy laboratories.

Shostak, Seth. "Listening for a Whisper." *Astronomy* 32, no. 9 (September, 2004): 34–39. A very brief overview of how radio telescopes are used in the search for extraterrestrial intelligence.

Sullivan, Woodruff T., III. *Cosmic Noise: A History of Early Radio Astronomy.* New York: Cambridge University Press, 2009. A very thorough historical record of the first decades of radio astronomy, the people and instruments involved, and the discoveries made.

Verschuur, Gerrit L. *The Invisible Universe: The Story of Radio Astronomy.* 2d ed. New York: Springer, 2007. An excellent introduction and overview of the history of radio astronomy and discoveries made by radio astronomers.

S

SCANNING PROBE MICROSCOPY

FIELDS OF STUDY

Mathematics; physics; chemistry; biology; applied mathematics; surface science; material science; tribology.

KEY TERMS AND CONCEPTS

- **nanoAmpere:** The typical unit of measurement of the tunneling current between probe tip and surface. One nanoAmpere equals 10^{-9} Ampere.
- **piezo-electric element:** A component used to control the vertical movement of the scanning probe tip to maintain a constant force or current.
- **probe deconvolution:** Processing of image data to correct the resolution of data measured for steep gradients parallel to the probe tip.
- **quantum dot:** A small group of atoms that share an electron charge as though they were a single atom.
- **Van Der Waals force:** A weak force acting between atoms that varies from weakly attractive to moderately attractive to intensely repulsive as the distance between the two atoms decreases.

SUMMARY

Scanning probe microscopy is a methodology that allows direct observation of structures and properties at the atomic and molecular scales. The techniques are applicable to a wide variety of purposes, providing information that is otherwise inaccessible. Scanning probe microscopy is particularly appropriate to nanotechnology, permitting the direct construction of nanoscale objects in an atom-by-atom manner.

DEFINITION AND BASIC PRINCIPLES

Scanning probe microscopy is a methodology that interfaces the macroscale of human observation (1–10^{-2} meters) to the atomic scale of the physical world, providing direct observation of features in the range of 100 micrometers to 10 picometers (10^{-4}–10^{-11} m).

Though the upper range of resolution for scanning probe microscopy is well within the range of other methods of microscopy, it is the lower range that is of most interest, because it is at this range that direct observation of atomic and molecular scale properties is possible.

The basic principles of scanning probe microscopy are founded in the quantum mechanical properties of atoms. The methods use the measurement of electronic properties (current, voltage, or "atomic force") as the means of observing the nature of surfaces and surface phenomena.

Quantum mechanics describes and defines the behavior of electrons in atoms. One of the rules of quantum behavior is that electrons are constrained to specific locales known as orbitals within the structure of an atom and are not allowed to exist at the boundaries of those locales. An observable property called quantum mechanical tunneling occurs, however, which permits electrons to move from one locale to another across the orbital boundaries. In scanning probe microscopy the miniscule electrical current due to quantum mechanical tunneling between the atoms at a surface and the atoms at the tip of an atomic-scale probe is measured as a function of their relative positions. This provides a corresponding atomic scale map of the surface structure.

BACKGROUND AND HISTORY

The scanning tunneling microscope (STM) was invented in 1982 by German physicist Gerd Binnig and Swiss physicist Heinrich Rohrer. Their device could be conceived of as a sort of quantum mechanical phonograph, in which an exceedingly sharp metallic needle scans a surface in a manner similar to the way a phonograph needle scans the groove of a vinyl phonograph record. The needle would ideally taper down to a single atom at the point, enabling atom-to-atom interaction at the surface. Sensitive digital-electronic measurement devices would measure the electronic tunneling current between the tip and the surface. The devices would then relay that data directly to a computer that would then correlate the values according to the relative dimensions and spatial relationships of the probe tip and the surface. The result would be displayed as an image having resolution of atomic scale features.

In 1986, Binnig, along with American electrical engineer Calvin Quate and Swiss physicist Christoph Gerber, introduced the scanning force microscope (SFM), which is also known as the atomic force microscope, or AFM, a variation on the STM that maintained a constant force between the scanning tip and the surface. This allowed any surface to be scanned, whereas the STM could only be used with electrically conductive surfaces. More recently, scanning near-field optical microscopy (SNOM) was developed, using measurement of short-range components of electromagnetic fields (a very small light source) between tip and surface to produce the equivalent of a photographic representation of the surface features.

How It Works

To appreciate the operation of scanning probe microscopy, it is necessary to understand the scale on which it operates. The unaided human eye can discern detail as small as approximately 0.1 millimeter. Optical microscopes can extend this to a resolution of about 0.0001 meter. Scanning electron microscopes can typically produce images with a resolution as fine as 10 micrometers (0.00001 meter). This is the range at which scanning probe microscopes only begin to work, and they typically provide information with a resolution of as little as 10 picometers (0.00000000001 meter).

At this scale, the operation is in the realm of quantum mechanical physics rather than classical physics, and the effects associated with that scale are very different from those that occur on a larger scale. The most important difference lies in what is meant by the word "surface."

Quantum mechanical physics. On scales that are significantly larger than atomic and molecular diameters, a "surface" is solid matter, analogous to a smooth tabletop. At the atomic scale of quantum mechanics, however, there is no such thing as a hard surface in that sense.

Quantum mechanics describes the structure of atoms as a very small, dense nucleus of massive protons and neutrons, surrounded by a cloud of electrons that is 100,000 times greater in diameter than the nucleus. The electron cloud is therefore very diffuse. The electrons in a neutral atom are equal in number to the protons contained in the nucleus, and are confined to specific three-dimensional regions, called orbitals, around the nucleus. and are allowed to have only very specific energies according to the orbitals they occupy. At this scale of operation, a scanning probe microscope measures the electromagnetic interaction of the electron clouds in the atoms of the probe tip and atoms of the surface being scanned.

Scanning tunneling microscopes (STMs). The STM operates by moving the atoms-wide point of the scanning tip across a metallic, and therefore electrically conducting, surface at a distance of less than one nanometer (10^{-9} meter). The device measures the magnitude of the "tunneling current" that arises between the probe tip and the surface atoms. This current is exceedingly small, and its measurement requires extremely sensitive digital sampling and amplification electronics, and computers to process the measured data according to the relative geometries of the tip and the surface. As the probe scans, the angle between the tip and the atomic surface changes, as does the distance between them. The tunneling current measurement thus has three-dimensional vector-field properties, and because the probe tip maintains a constant orientation, variations in the tunneling current are presumably caused by the three-dimensional shape of the atoms being scanned.

Atomic force microscopes (AFMs). The AFM operates in essentially the same manner as the STM, except that its function is to maintain a constant measured electrical force between the probe tip and the atomic surface being scanned. In this function, the probe tip follows the shape of the atomic surfaces directly, rather than measuring a property difference that changes according to the shape of the surface. Several different modes of operation are available within this context, such as constant contact, non-contact, intermittent contact, lateral force, magnetic force, and thermal scanning. Each mode provides a different type of information about the surface atoms.

Scanning near-field optical microscopes (SNFOMs). The SNFOMs use an extremely small-point light source as the probe, rather than a physical tip. Measurement of the effect that the surface has on the light provides the image data. The technique can employ a broad range of wavelengths to investigate different properties of the atoms being scanned.

APPLICATIONS AND PRODUCTS

The field of scanning probe microscopy is a high technology research practice. Its direct applications are limited to the analytical study of surface phenomena and structures. The nature of the techniques of scanning probe microscopy makes it the method of choice for examination and study of surfaces that are not amenable to any other means of close examination, especially those of certain biological materials. Scanning probe microscopy, with its ability to probe and manipulate single atoms and so to form molecule-sized structures, is invaluable in chemistry and in the development of nanotechnology.

Surface chemistry. Chemical reactions take place at the level of the outermost electronic orbitals of atoms, or at the electronic surface of the atoms involved. In catalyst-mediated reactions, the interaction among chemical species takes place on the surface of the catalyst material, where the atoms of the reactants have interacted electronically with the atoms of the catalytic material. This lowers the energy barriers that must be overcome for the reaction to occur, with the result that the desired reaction is facilitated.

Scanning probe microscopy has a spectroscopy mode that allows the direct measurement of electron energies in single atoms and of single atomic bonds between atoms in a molecule. The methodology provides a better understanding of the chemistry that takes place at surfaces, in turn contributing to the development of new and improved chemical processes. Perhaps the most important aspect of this is the enhanced knowledge of the mechanisms of surface phenomena such as oxidation and chemical corrosion.

Electronic materials and integrated circuits. How small can a functional transistor be? Technology enables the construction of several million transistors on the surface of the small silicon chip that is the central processing unit (CPU) of a computer. Scanning probe microscopy enables the physical study of such structures in minute detail. It also enables the construction and study of transistor structures that are orders of magnitude smaller. The research in this field examines possible ways to construct integrated circuits and computers that exceed existing capabilities of production.

The available methods of production of integrated circuits have essentially reached the physical limit of their capabilities to construct viable semiconductor transistors. Atomic force microscopy is being used to investigate the construction of transistor-like structures based on quantum dots rather than on semiconductor junctions.

Another area of research is the construction of molecule-sized transistors made from graphene or carbon nanotubes and other materials. These technologies, when fully developed, will completely change the nature of computing by enabling the construction of a quantum computer, a device that could carry out in seconds calculations that existing computers would require possibly billions of years to complete.

Data storage would also be revolutionized by these innovations. Research indicates that data storage will soon reach capacities measured in terabits per square centimeter. The ultimate binary data storage density would have each bit stored in the space of one atom, a density that can be envisioned only with scanning probe microscope technology.

Nanostructures and nanodesign. Nanotechnology works at the nanometer scale of 10^{-9} meters. To appreciate this scale, imagine the length of one millimeter divided into one million segments, each of which would be one nanometer. The concept of nanotechnology is to produce physical machines constructed to that scale. Because scanning probe microscopy can manipulate single atoms, it can be used to construct nanoscale, and even picoscale, physical mechanisms. The latter are essentially individual molecules whose physical structures imitate those of much larger devices and mechanisms, such as gears.

In September, 2011, researchers at Tufts University reported the successful use of low-temperature scanning tunneling microscopy to construct a working electric motor consisting of a single molecule. As can be imagined, this is a complex field of research, because quantum effects play a significant role in the interoperability of such small devices.

One application of scanning probe microscopy that is of immediate importance is the study of friction and abrasion at the atomic level, which is where those processes take place. Atomic force microscopy can be used to literally scratch the surface of a material, providing detailed information about how friction and abrasion actually work and about what

---

FASCINATING FACTS ABOUT SCANNING PROBE MICROSCOPY

- The scanning tunneling microscope was invented in 1982 by Gerd Binnig and Heinrich Rohrer, for which they were awarded a Nobel Prize in Physics in 1986.
- Scanning probe microscopes can achieve resolution of structural surface details as small as 10 picometers.
- Scanning probe microscopes can move and assemble individual atoms to construct molecule-sized structures.
- Using a low-temperature scanning tunneling microscope, chemists at Tufts University successfully assembled and operated an electric motor consisting of a single molecule.
- Scanning probe microscopy can be used to examine biological molecules and materials that cannot be studied by any other method.
- The spectroscopy mode of scanning probe microscopes allows examination of the energy of single electrons in an atom and direct measurement of the energy of a bond between two atoms in a molecule.
- The tip of a scanning probe has a typical radius as small as 3 nanometers and can be as fine as a single atom.
- The probe tip of an atomic force microscope is attached to a cantilevered leaf spring. The up-and-down movements of the spring as the probe tip moves across the atoms of a surface are measured electronically.

---

might be done to lessen or prevent those effects.

Biological studies. Scanning electron microscopy (SEM) has been the workhorse of biological research since its invention, providing detailed images of extremely small biological structures. The technology has some practical limits, however, because of the principles on which it functions. Many biological materials that are of interest cannot be studied in detail using SEM, but are amenable to study using scanning probe microscopy. The methods are useful in measuring the forces that exist among functional groups in biological and organic chemical structures.

SOCIAL CONTEXT AND FUTURE PROSPECTS

Scanning probe microscopy is a field that will have very little direct social context because of the extremely small scale of its subject matter. However, the secondary effects of the knowledge and technology derived from research and development in this field could have a large social impact, primarily because of the economic benefits from control of friction and from new technologies for integrated circuits and magnetic memory media. Any real predictions for the future prospects of the field of scanning probe microscopy are entirely conjectural.

—Richard M. Renneboog, MSc

FURTHER READING

Bhushan, Bharat, Harald Fuchs, and Masahiko Tomitori, eds. *Applied Scanning Probe Methods VIII: Scanning Probe Microscopy Techniques (NanoScience and Technology)*. Berlin: Springer, 2008. Provides the most up-to-date information on this rapidly evolving technology and its applications.

Howland, Rebecca, and Lisa Benatar. "A Practical Guide to Scanning Probe Microscopy." ThermoMicroscopes. March 2000. Web. Accessed September, 2011. This introductory guide to scanning probe microscopy describes several techniques and operating modes of the devices, as well as their structure and principles of operation, and discusses the occurrence of image artifacts in their use.

Meyer, Ernst, Hans Josef Hug, and Roland Bennewitz. *Scanning Probe Microscopy: The Lab on a Tip*. Berlin: Springer, 2004. This book provides an excellent overview of scanning probe microscopy before delving into more detailed discussions of the various techniques.

Mongillo, John. *Nanotechnology 110*. Westport, Conn.: Greenwood, 2007. Demonstrates the essential relationship between and value of scanning probe microscopy to nanotechnology.

SPACE ENVIRONMENTS FOR HUMANS

FIELDS OF STUDY

Aeronautics; astrobiology; space engineering; life sciences; astronomy; computer science; environmental science; nanotechnology; robotics and embedded systems.

KEY TERMS AND CONCEPTS

- **extraplanetary resources:** Resources that are available in off-Earth environments; these could include natural resources used for powering spacecraft and settlements.
- **extravehicular activity (EVA):** Activity that occurs outside the protective environments of spacecraft, such as a spacewalk.
- **life-support system:** System developed to allow human and other biological life-forms to live in space environments.
- **space engineering:** Science of engineering as applied to off-Earth environments. By their nature, these environments have different physical properties and therefore require knowledge specifically related to space science and physics.
- **space habitat:** Semipermanent or permanent human living environment on off-Earth planetary bodies (such as satellites or planets) or aboard orbiting spacecraft.
- **space radiation:** Cosmic and solar energy or matter that could have detrimental effects on space travelers, as well as their spacecraft, equipment, and environments.
- **space tourism:** Use of space environments for leisure activities such as short- and long-term trips in orbit around Earth, as well as vacations aboard spacecraft or in human settlements on planetary bodies.

SUMMARY

Space environments for humans concerns the potential development of habitats, spacecraft, and techniques that will allow humans to create and live in sustainable environments in space, such as space stations, or on planets and satellites, such as the Moon or Mars. Advances in space science, engineering, and technology since the mid-twentieth century resulted in plausible scenarios for human colonization of space, as well as means for harvesting and using extraplanetary resources for purposes of human settlement. Although no definitive off-Earth scenario that addresses how humans will live in space yet exists, it does appear possible that human environments in space will remain a subject of concern for scientists, engineers, space agencies, and technologists into the future.

DEFINITION AND BASIC PRINCIPLES

Although human settlement in space has not become a scientific reality, three major space initiatives allowed scientists and technologists to realize the first stages of creating long-term human living environments in space. The first was a series of National Aeronautics and Space Administration (NASA) programs—Mercury, Gemini, and Apollo—that culminated in a successful lunar landing in 1969 and subsequent return trips to the Moon. The second and third developments were the successful deployment of NASA's space shuttle program and several Earth-orbiting space stations—including, most notably, the Russian Mir and the International Space Station (ISS). The data collected from each of these projects broadened understanding of the possibilities of long-term survival of humans in various space environments and opened up possibilities for continued ventures in space.

Areas of continuing research related to human environments in space include the effects of solar and other types of radiation on spacefaring humans and their spacecraft, the availability of extraplanetary resources that could sustain human life in space and power the spacecraft, the creation of materials to protect humans in space or on other planets, realistic colonization prospects for humans on other planets or satellites, funding and management of space projects related to human environments in space, and the determination of whether living in space is a realistic option for human beings.

As scientists, space agencies, and technologists grapple with such questions, scientific research continues. In large part, this is the result of scientific curiosity, and the knowledge gained from living and working in space adds to understanding of how people interact with their home planet. However, because of environmental issues such as global warming and possible energy shortages, there are also major concerns about the continued sustainability of Earth and its resources for human habitation.

BACKGROUND AND HISTORY

A long-established subject for science-fiction and fantasy writers, the dream of living and traveling widely in space and on other planets precedes the scientific reality of being able to do so. Much of what has been discovered about the needs of humans in space environments was learned during the space race in the 1950's and 1960's, when the Soviet Union and the United States competed to see who would be the first nation to put a man in space and return him safely to Earth. Also adding to knowledge of the effects of spaceflight on humans were the subsequent missions in NASA's space shuttle program and the many space stations launched since the 1970's. By this time, scientists had recognized that space habitats for humans could become viable options for space programs.

In the years before space travel, however, many scientific visionaries considered the special needs of environments for humans in space. Russian scientist Konstantin Tsiolkovsky, for example, was a pioneer in theorizing about spaceflight. Among Tsiolkovsky's numerous influential ideas were the space elevator (a contraption anchored to Earth that would allow travel high into Earth's orbit without the use of rockets) and the notion of colonizing space as an answer to the potential dangers faced by humans on Earth, such as catastrophic impacts from solar bodies. Similar concerns are still used as justification by some parties in insisting that space colonization is an important next step for humankind.

In the twenty-first century, the idea of building space environments for humans is still alive and, more than ever before, has taken on an international tone. Numerous nations, including the United States, Russia, China, Japan, and India, have been considering flights to the Moon and Mars, and some have made plans for the creation of permanent or semipermanent lunar and Martian bases, as well as long-term journeys within the solar system.

HOW IT WORKS

Because of the harsh conditions awaiting humans in off-Earth environments, special consideration must be given to the protective clothing worn by spacefarers, as well as the protective materials that will be used to build their stationary and floating habitats. Just as important are the design of habitats in which humans are to live and work, as well as the resources that will power their habitats and vehicles and possibly even feed entire colonies. Likewise, spacecraft engineering is vital for transportation to bases or colonies or for craft used as living quarters.

For space environments for humans to become feasible, further research and development must take place in three main areas: life-support systems and shielding, types of space-bound habitats, and resources to sustain habitat populations. Given the vast number of possibilities within these scenarios, there are many considerations that will take precedence over others as humankind progresses toward a more developed future in space. For example, the operating conditions, power resources needs, and amenities aboard a habitat built for long-term, near-Earth orbit will be very different from those for a round-trip to the Moon or Mars. Despite these differences, there are many other areas in which research and development could uncover certain universally applicable materials and techniques.

Space habitats. Since it was first understood that it is possible to live in an off-Earth environment, many designs have been proposed to accommodate humans in diverse space environments. The only living and working environments ever used by humans in space have been the capsules used to transport astronauts to and from the Moon and space shuttles and space stations launched into orbit from Earth. In the case of the lunar capsules, each was designed for a short voyage of less than a month's duration. Similarly, space shuttles, while offering a wider range of movement than capsules, as well as more storage capacity, were not designed for human habitation out of Earth's atmosphere. Space stations, however, because of their size and the ability to add modules (modular compartments, designed for many purposes, whether personal, scientific, or operational), offer a glimpse into what orbiting habitats will probably be like during the initial stages if spacefaring becomes more common. They are also durable and can be built to the discreet specifications of the sponsor of the vessel.

Several types of habitats may be needed as

housing, storage, and work areas for humans in space. NASA and other space agencies have been considering the establishment of base camps on the Moon and perhaps even Mars. One plan uses torus-shaped designs that can be inflated and connected together in a manner similar to inflated inner tubes connected to each other by portals. The tori would be constructed of highly durable protective fabric that is light enough to transport from Earth to the Moon. Other possible designs include the use of bunkers dug into the surface of the Moon; the advantage of this design is that, instead of being exposed to the extremely cold air on the Moon, the bunkers could be temperature-regulated and would have a natural barrier—the surface regolith (soil)—against cosmic rays. Other proposed habitats include designs for mobile vehicles that can be joined together for periods of time to form base camps and separated for movement when appropriate.

Life-support systems and shielding. The physiological effects of the weather and atmospherics awaiting astronauts in space and on extraplanetary bodies are different from those found on Earth. For example, long-term exposure to solar radiation can cause major health problems for astronauts. In addition, continuous living and working in zero-gravity environments can lead to cardiovascular problems, loss of bone density, muscle atrophy, and many other issues that can lessen the productivity of astronauts and even cause life-threatening conditions. A further consideration is protection from meteoroids that can damage machinery and habitats. With these things in mind, designs for special protective suits and gear are being created that will lower exposure to solar particles and energy. The protective suits will also encourage efficient oxygen circulation and regulate temperature and air pressure, flexibility and mobility, and weight and comfort.

Life-sustaining space resources. Because of the very high cost of supplying life-sustaining resources such as oxygen, water, food, and power from Earth, it is very likely that vessels and habitats will need to become self-reliant. Gathering, storing, and manufacturing these resources once initial launch supplies are depleted remains a significant issue. Although orbiting space vessels will remain reliant on terrestrial sources, lunar (and possibly even Martian) colonies may be able to extract some of these things from the space resources available at hand. Scientists think

that both water and oxygen can be processed from lunar regolith, giving those settlements a nearly endless source of these resources. Power generation could easily be solved by using solar units, but only during the lunar day, which lasts for fourteen days. During the other fourteen days, lunar settlers would probably have to use a different source of power, such as battery cells and possibly even power resources beamed in from satellites.

APPLICATIONS AND PRODUCTS

Human-inhabited environments in space have many potential uses, and like many of humankind's earliest journeys on Earth, they will be expensive and potentially hazardous. They may also be extremely rewarding and alter the course of human endeavor for centuries to come. However, as of the early twenty-first century, space programs were directed toward exploration rather than settlement.

Lunar and Martian bases. The space shuttles and manned space station traversing the sky are used for scientific research. Lunar or Martian bases would most likely be used in the same way but could offer much more data on a wider variety of subjects. They would be constructed in environments that allow study of such topics as the capabilities of natural resource development in space, the long-term effects of non-Earth atmospheres and conditions on humans (also plants and animals), the social and psychological effects of spacefaring, and the technological capabilities of the human race. All of these are very broad subjects, but they are vastly important to an understanding of the requirements of continuing to voyage into space.

Mining and harvesting of space resources. Human settlements in space could help solve specific resource needs faced by humans on Earth, such as environmental and power needs. One proposed endeavor would be to use lunar regolith as a source of energy that could be beamed back to Earth through microwaves. Other possibilities include using the helium-3 found in the lunar soil as a fusion-power source to fuel rockets. If proper transport was arranged, helium-3 could also be used for Earth's power needs. Perhaps the most important use for resources found in space would be to sustain human settlements and help them become self-sufficient.

Space tourism. Since 2001, a handful of wealthy, private individuals have traveled into space as tourists,

FASCINATING FACTS ABOUT SPACE ENVIRONMENTS FOR HUMANS

- In zero-gravity environments, human height increases by a few centimeters. This is because of the effects of reduced spinal curvature and the expansion of intervertebral discs in the back. The downside of this is lower back pain for astronauts.

- Lunar dust can pose pesky problems for astronauts working on the Moon. At sunrise and sunset, lunar dust tends to levitate and stick to surfaces. If lunar dust seeps into machinery, it can cause system failures, making this an important engineering concern for future lunar base builders.

- Being a space tourist is extremely expensive. Space visitor Guy Laliberté paid about $35 million to travel aboard a Russian shuttle to the International Space Station. It is estimated that, someday, such trips will be less expensive and will accommodate more tourists.

- If large-scale space habitats for humans are ever built, it is likely that they will be based at least somewhat on designs such as the O'Neill cylinder and the Bernal sphere (similar to ships featured in the 1968 film *2001: A Space Odyssey* and 1994-1998 television series *Babylon 5*, respectively). The designs for these habitats promise living space for thousands of people and enough resources to have an agricultural industry in space.

- Becoming an astronaut is not easy. Aside from the professional and academic credentials needed for the job, prospective astronauts must also have perfect eyesight, stand between 62 and 75 inches high, have excellent blood pressure, and be able to tread water for ten minutes while wearing a flight suit.

all of them aboard Russian spacecraft. Willing to pay tens of millions of dollars, these so-called space tourists are the first clients of a potentially lucrative—although extremely dangerous—international space tourism industry that may someday offer trips to low-Earth-orbit hotels, space stations, and lunar or Martian bases. The cost of such trips is likely to decrease over time until, for example, short-term trips into orbit cost tens of thousands of dollars. Before space tourism becomes common, it is likely that this industry will be heavily regulated.

Living and traveling in space. Each new space venture helps determine whether people can live in orbit or on other planetary bodies. Once the feasibility is determined, it will open the door for many other considerations that have remained merely speculation. Questions and areas for research include whether it will be possible to create environments that can grow into self-sustaining human settlements, perhaps with farms and industry as a source of income; space law; whether human settlements need governance in the same way colonies once did; and whether space colonies might seek independence from or become hostile to governments on Earth. Though these questions seem to be the stuff of science fiction, they could become real issues affecting people on Earth.

CAREERS AND COURSE WORK
Many possible career paths are available in the area of space environments for humans. For those interested in creating and designing these habitats, focus should be given to areas of space biology, space engineering, computer science, robotics and embedded systems, nanotechnology, and aeronautics, because space environments will draw on the knowledge of workers from all these areas. To become involved in the use of these environments, some of these same areas of focus will be needed, but the quickest path to doing so is to become an astronaut. To be a NASA astronaut, one needs to meet basic science and math requirements as well as to hold a bachelor's degree or higher in an area of biological or life science, engineering, or math. Professional experience in the military and aviation could be of use to students seeking a career in piloting spacecraft.

The industry that has grown up around the needs related to space environments for humans, although relatively small, is a highly technical field. There are many very specialized areas of focus for technicians, project managers, and other workers, many of whom will have to acquire advanced degrees in their fields of study to be considered for even entry-level jobs. Some careers, such as robotics and embedded systems engineers, require degrees in computer science, while others, such as environment designers, need a degree in engineering or physics.

SOCIAL CONTEXT AND FUTURE PROSPECTS
Humans may someday use environments in space as regularly as they use terrestrial spaces. Space

environments may be used as private living spaces for entire families and public spaces for commerce. Of course, these are merely dreams, but by identifying what humans need to survive in space—whether it is clothing, habitats, or means of transportation—and by building these things, humans will move one step closer to making those dreams a reality.

The first phases of learning about the needs of humankind in space have come and gone. The days of the first space race between the United States and the Soviet Union taught space scientists and astronauts the fundamentals of taking humans into space and returning them safely to Earth. However, research is allowing people to envision far more than short-term journeys. For example, the space suits designed for NASA's Constellation program are made for a variety of purposes and time frames. Similarly, the types of habitats that have been envisioned appear to be small-scale versions of what could someday become cities in space (either floating or stationary). In short, scientists no longer worry about whether humans can survive in space environments but instead about which designs will be used first and for what purposes.

However, humankind is not yet living in space. Unlike science-fiction stories in which entire civilizations are packed into starships and shipped out across the galaxy and beyond, plausible plans for living in space are limited to the concerns of the small groups of astronauts who will be involved in the initial test runs of these plans. Growth in the space science and exploration industry is most likely to occur as a result not only of national and international interests but also of corporate concerns and the desires of people to make the next small steps toward human-friendly space environments.

—*Craig Belanger, MST*

FURTHER READING

Harris, Philip. *Space Enterprise: Living and Working Offworld in the Twenty-First Century.* New York: Springer, 2009. Print.

Howe, A. Scott, and Brent Sherwood, eds. *Out of This World: The New Field of Space Architecture.* Reston: American Inst. of Aeronautics and Astronautics, 2009. Print.

Seedhouse, Erik. *Tourists in Space: A Practical Guide.* New York: Springer, 2008. Print.

Thirsk, Robert, et al. "The Space Flight Environment: The International Space Station and Beyond." *Canadian Medical Association Journal* 180.12 (June, 2009): 1216–1220. Print.

Zubrin, Robert. *Entering Space: Creating a Spacefaring Civilization.* New York: Putnam, 2000. Print.

SPACE SCIENCE

FIELDS OF STUDY

Astronomy; physics; chemistry; geology; meteorology; space weather; computer science; communication technology; radio communications; electronics; aerospace engineering; electrical engineering; mechanical engineering; chemical engineering; biomedical sciences; medicine; environmental engineering; astrobiology; robotics; remote sensing.

KEY TERMS AND CONCEPTS

- **astrodynamics:** Study of orbits and trajectories.
- **astronautics:** Engineering field associated with spacecraft engineering, especially manned spacecraft.
- **booster:** Rocket used to lift off from the surface of a planetary body.
- **cosmic ray:** Ultra-high-energy particle radiation from space.
- **exobiology:** Study of life, or the theoretical study of conditions for life, beyond Earth.
- **geomagnetic storms:** Disruptions in Earth's magnetic field caused by interactions with the Sun and particles emitted by the Sun.
- **geosynchronous orbit:** Orbit around Earth at a radius of 42,164 kilometers that has a period equal to the rotational period of the Earth.
- **low Earth orbit (LEO):** Orbit at an altitude of only a few hundred kilometers above the Earth's surface.
- **micrometeoroid:** Tiny particle, often the size of a grain of sand, that moves through space at a very high speed.

- **near-Earth object (NEO):** Asteroid or comet whose orbit comes near or crosses the orbit of the Earth.
- **satellite:** Natural or artificial object that orbits a planet.
- **space exploitation:** Use of space for monetary, commercial, or intellectual gain.
- **space weather:** Solar-terrestrial interactions, typically within Earth's magnetosphere, that can give rise to geomagnetic activity, radio blackouts, or radiation storms.

SUMMARY

Space science is an all-encompassing term describing many different, often multidisciplinary fields focused on the study of anything beyond the Earth's atmosphere. At first, space science was limited to observational astronomy; however, developments in rocket technology have allowed the field to expand to experimental planetary studies and to astronautics. Astronomy has advanced to the point where several subfields, including astrophysics and cosmology, have become important fields of study in their own right. Applications of space technology have led to the exploitation of near space in the areas of communication, navigation, and space tourism. Space science is used to describe all these areas of study and technologies.

DEFINITION AND BASIC PRINCIPLES

Space science is not a single field of study. Rather, the term "space science" applies to any field that studies whatever is external to the Earth. These fields include planetary studies, stellar and galactic astronomy, and solar astronomy. Originally, astronomy involved simply looking at the skies, either with the naked eye or through telescopes, and recording what was observed. It has since become a branch of physics, with mathematical analysis and application of physical laws in an attempt to understand astrophysical processes. Additionally, robotic spacecraft have allowed planetary scientists to collect data on other planets that are useful to geologists and meteorologists. Discoveries of phenomena unknown on Earth brought many disciplines into the field of space science. Furthermore, theoretical investigations, such as models of potential planetary systems or the search for life in space, also are part of space science. The science of studying phenomena beyond

Earth employs nearly every aspect, field, and subfield of science.

Space science also encompasses the technology developed to further the aims of space science, such as rocketry and astronautics. Additionally, it includes the exploitation of space, with such technologies as satellites for communication and navigation, weather, and military and intelligence surveillance. It also includes possible future technologies, such as mining asteroids and establishing space settlements and colonies.

BACKGROUND AND HISTORY

Throughout most of history, space was inaccessible to humans. The only way to study extraterrestrial objects was through observation. Astronomy began with these simple observations. In the seventeenth century, people began to use telescopes to view the sky, and physics and chemistry began to be important in interpreting their observations. In the nineteenth century, scientists discovered the existence of cosmic rays and learned the nature of meteorites, but the study of space remained largely limited to observations from the surface of the Earth.

In 1903, however, Russian scientist Konstantin Tsiolkovsky proposed that rockets could allow people to engage in space travel. The development of rocketry over the next half century eventually led to the creation of a rocket powerful enough to reach the edge of the Earth's atmosphere, allowing the edge of space to be studied directly. On October 4, 1957, the Soviet Union launched the first human-made Earth-orbiting satellite, *Sputnik 1.* The following month, the Soviets launched a larger satellite, *Sputnik 2,* with a dog aboard to study the effect of space travel on a life-form. The United States launched *Explorer 1* on January 31, 1958. The data gathered from *Explorer 1* helped scientists discover the Van Allen radiation belts surrounding Earth.

With the use of rockets, space exploitation became possible, leading to a flurry of interplanetary space missions by robotic spacecraft in the 1960s and 1970s, followed by much larger and more capable robotic spacecraft. In the 1960s, the first communication and weather satellites paved the way for the communication, navigation, and Earth-monitoring satellites that have become mainstays of modern society. Manned exploration of space began with simple capsules on top of converted missiles, evolved to more sophisticated Apollo craft that carried astronauts to the Moon, and then to reusable spacecraft such as the space shuttle. Manned space stations evolved from the simple single-mission systems launched by the Soviet Union in the 1970s to the International Space Station (ISS), constructed during the first decade of the twenty-first century. Even telescopes took to space, with a number of astronomical satellites being launched, including the Hubble Space Telescope in 1990.

HOW IT WORKS

Astronomy, astrophysics, and cosmology. Astronomy is an observational science. Observations are made by a variety of instruments, including optical telescopes, radio telescopes, infrared and ultraviolet telescopes, and gamma-ray and x-ray telescopes. Instead of simply taking photographs to study, modern astronomers measure spectra, intensities, and many other properties to understand the objects of their study. Some of the instruments they use are located at ground-based observatories with large telescopes, and others are located in orbit around Earth in space-based observatories, the most famous of which is the Hubble Space Telescope.

Astrophysics mathematically studies the physical processes of objects and develops theories based on physical laws. Cosmology involves the study of the universe and space itself. Both cosmology and astrophysics involve considerable computer modeling and mathematical analysis of astronomical observations.

Planetary sciences. Until the 1960s, astronomers were limited to making observations of planets from Earth. However, the advent of interplanetary robotic spacecraft allowed scientists to study other planets in much the same way that they study Earth. Robotic spacecraft employ a battery of instruments, including cameras, spectrometers, neutron sensors, and magnetometers, to study planets from orbit. Spacecraft have landed on the moon, Venus, Mars, and Saturn's moon Titan to study their surfaces.

Astrobiology/exobiology. Since about the end of the sixteenth century, when astronomers were beginning to guess the nature of the planets, scientists have speculated on whether life could exist on other planets. Robotic spacecraft have searched for life on Mars, and rocks returned to Earth by the Apollo missions to the Moon have been studied for life. The basic essentials for life have been found in space,

but as of the early twenty-first century, no definitive evidence of life has been found on other worlds. Biologists, working with astronomers, have studied the nature of life and probed its possible origins and the conditions necessary for life. This study was originally called astrobiology, although the term "exobiology" has slowly gained in popularity. Exobiology is largely theoretical.

Rocketry. Rockets are key to space study. Rockets are entirely self-contained and do not need to take in air to operate. They may be either liquid or solid fueled. Rockets operate on the principle of conservation of momentum. High-speed gases exiting the rocket carry momentum, so the rocket must have momentum in the other direction to conserve momentum. The rate of change of momentum is force and is the thrust of the rocket.

Astronautics. The space environment is harsh and hostile, not only to life but also to human-made devices. Thus, great care and redundancy must go into any spacecraft design. Aerospace engineers must design spacecraft that can operate in extremes of hot and cold, microgravity, and the vacuum of space, as well as under intense radiation exposure and despite repeated micrometeoroid impacts. These conditions are all difficult to reproduce on Earth. For manned spacecraft or space stations, the additional requirement is that a habitable environment must exist within the spacecraft. The spacecraft must be shielded from radiation as much as possible to protect its occupants, and systems must be reliable enough to keep the human occupants safe for the duration of the mission. This requires skill at mechanical and environmental engineering, as well as with advanced electronics systems.

Space weather. Earth's magnetic field extends a long way above the surface of the planet. The region of the solar system in which Earth's magnetic field dominates is called Earth's magnetosphere. Particles trapped in the magnetosphere create regions of intense particulate radiation surrounding the planet. These regions were discovered in 1958 using instruments aboard the American *Explorer* spacecraft by physicist James Van Allen and are known as the Van Allen radiation belts.

Earth's magnetosphere is not static; it is constantly shifting and adjusting to the interplanetary space environment, which is constantly being affected by the sun. The sun continually emits particles that travel

Apollo 11 crewmember Buzz Aldrin on the Moon, 1969

outward through the solar system. This stream of particles, called the solar wind, varies in intensity, density, and speed based on solar activity. Earth's magnetosphere adjusts accordingly.

Occasionally, very large bursts of energy, called solar flares, cause portions of the sun's corona to detach and move across the solar system as a large bubble of plasma. When such a coronal mass ejection affects Earth's magnetosphere, a significant shift in the magnetic field occurs over a short period of time. Often solar flares are associated with significant elevations in solar cosmic rays, events called solar radiation storms. Passengers in aircraft and spacecraft can receive significant levels of radiation exposure during such radiation storms.

Variations in Earth's magnetic field are called geomagnetic storms. Large geomagnetic storms resulting from large, rapid shifts in Earth's magnetic field can induce damaging electric currents in pipelines, power lines, and telephone lines. Geomagnetic storms are often associated with radio communication interference and even blackouts. Collectively, the study of the variations of Earth's magnetosphere and the solar-terrestrial interactions is called space weather.

APPLICATIONS AND PRODUCTS

Much of space science involves basic scientific research. Astronomers and planetary scientists seek to learn about stars, planets, and other celestial bodies, in particular how they came to be the way that they are and why they have the properties that they have. Although many aspects of these studies have applications beyond academic studies, the goal of the scientist is to understand the system and to let other scientists or engineers apply the knowledge and technology to other purposes.

Weather and remote sensing. Being at high altitude permits an observer to get a bigger and broader picture than is available from the ground. Early researchers used balloons and then aircraft to better study the Earth. The advantages of observations from orbit were obvious to the very first space engineers in both the United States and the Soviet Union. Therefore, the first weather satellite, TIROS 1, was launched on April 1, 1960, less than three years after the first artificial satellite was launched into orbit around the Earth. Early weather satellites used film that was returned to Earth in canisters. Soon, engineers perfected the technology that enabled images to be transmitted electronically. In the twenty-first century, real-time or nearly real-time weather images have become essential tools for meteorologists. Modern weather satellites do not just take images but also contain instruments to make numerous measurements.

In addition to providing information on weather, satellites can monitor aspects of Earth itself such as land use, ocean currents, and atmospheric pollution. Collectively, this technology is called remote sensing. Originally, government agencies operated these remote-sensing orbital satellites; however, the demand for such data by researchers and private industry spurred several private companies to deploy their own remote-sensing satellites. Dozens of such satellites continually monitor the Earth from space.

Space weather. Space weather storms have disabled satellites, navigation, and ground-based communication and power distribution grids. Protective measures are sometimes possible given ample warning. The military, recognizing the dangers associated with communication blackouts during times of national tension or during armed conflicts, was the first to start monitoring the sun for signs of solar flares that might trigger geomagnetic storms and communication blackouts. Both the United States and the Soviet Union established global networks of solar observatories for this purpose.

As society has become increasingly dependent on electronic technology that could be interrupted by such events, other government agencies have supplemented the military's observations of the sun. By 2000, the Space Weather Prediction Center, operated by the National Oceanographic and Atmospheric Administration (NOAA), had become one of the top clearinghouses for solar observation data and predictions of space weather events. Most of the center's data are available over the Internet. These data are important for satellite operators, electric utilities, airlines, and many other industries. The Space Weather Prediction Center issues space weather watches and warnings in much the same way that the National Weather Service, another NOAA division, issues thunderstorm and tornado watches and warnings. Many other technologically advanced nations have similar space weather centers.

Communication. When the first satellites were launched, microwave relay towers were being constructed to carry voice communications without the need for wires. These towers could be placed only so far apart because the curvature of the Earth would interfere with transmissions. Satellites, however, can act as ultra-high towers. The very first test of satellite communications occurred with the launch of the Project SCORE satellite in 1958. SCORE was not capable of real-time communication; it recorded an incoming signal and transmitted it later. Soon, however, technology was developed to permit satellites to relay signals in real time. By 1963, the first geosynchronous satellite, the Syncom 3, was launched. A geosynchronous satellite remains over a single position on the ground, making it easy to aim antennas and eliminating the need for expensive and complicated satellite-tracking systems. By 2010, several hundred geosynchronous communication satellites had been launched.

Although geosynchronous satellites have many advantages over low Earth orbit (LEO) satellites that track across the sky in a few minutes, LEO satellites have the advantage of being able to use less powerful signals. A number of LEO communication satellites exist, notably in the form of satellite telephones, systems similar to cell phones but using satellites instead of cell phone towers. Satellite phones are useful in remote regions where cell phone service

is not available. The military often relies on satellite phones for battlefield operations.

NAVIGATION.
Celestial navigation, used for millennia, requires a skilled navigator who is able to actually see the sky. Beginning in the 1930s, networks of radio transmitters were set up to aid in navigation. These systems become more advanced but were still limited to areas that had transmitters, typically the land areas of technologically advanced nations, and were unavailable at sea. Satellites overcame this problem in much the same way that they did for communication relay towers. The first satellite navigation satellite, Transit 1B, was launched in 1960.

Satellite navigation culminated with the US Department of Defense's Global Positioning System (GPS). GPS technology, originally deployed as NAVSTAR strictly for military use in 1978, was eventually made available in civilian form in 1983 on the order of President Ronald Reagan. In 1993, GPS was made available to anyone free of charge, and it has been fully operational since 1995. Free use of the GPS signals for navigation spurred a number of private companies to sell commercial handheld GPS navigation systems. Software to display position on a map and to give audible turn-by-turn driving directions using maps and GPS position data made handheld GPS navigation systems very popular by the 2000s. Additional enhancements to GPS navigation are planned by both the Department of Defense and private companies.

Space tourism. Government space agencies have long been the only avenue for space exploration and manned space travel. However, there has been a gradual privatization of space technology, and many of the functions previously done only by government agencies are now carried out by private companies under government contract. A few private space tourism companies have formed to provide recreational space travel to wealthy individuals. Several people have purchased rides in the Russian Soyuz spacecraft to the former Mir space station and the International Space Station through the American company Space Adventures. Although these private companies are working with the space agencies, other companies are developing their own spacecraft. Some of these companies aim to sell their services to the International Space Station, but others, such as

Virgin Galactic, intend to provide rides into space for those willing and able to pay for a ticket. Other companies are planning to place private space stations into orbit as tourist destinations.

Military applications. Even before the first satellite was launched, military leaders recognized the importance of space. Military reconnaissance satellites (often called spy satellites) were among the first satellites launched after the developmental flights. In the 1960s, many feared that orbital weapons platforms capable of dropping nuclear bombs anywhere on Earth would be deployed. This fear was never realized. Although long-range missiles lift their warheads into suborbital trajectories, they are not space-based weapons systems. The United States and the Soviet Union both began developing antisatellite weapon systems to counter the military advantage of reconnaissance satellites. The first test of such a system was conducted in 1963 by the Soviet Union. In 1985, the United States tested its first antisatellite weapon, a missile designed to destroy low Earth orbit satellites. In 2007, China conducted an antisatellite weapon test, and Israel, India, and Russia are reportedly working on advancing such technology.

CAREERS AND COURSE WORK

The field of space science is really a great many different fields, with no single career path. Careers range from machinists building parts for rockets to assembly workers building rockets to engineers designing rockets and spacecraft. The field includes scientists studying the heavens, planetary scientists studying planets, and robotics engineers designing rovers to land on other worlds. It also includes businesses specializing in communications and remote sensing and the engineers who design the satellites to perform those functions. Some space businesses need employees with business degrees and experience rather than science backgrounds. Even astronauts do not all have the same qualifications. Some astronauts are pilots with extensive high-performance aircraft experience, and others are scientists with multiple graduate degrees. The growth of space science to include virtually every discipline means that there is no single career path.

However, space science is a very technical field, and therefore, a typical space science career generally requires many courses in science and mathematics. Physics and chemistry are needed for almost all space science careers. For applied space science work, a background in either electrical or mechanical engineering is probably necessary. For basic research, a background in physics or astrophysics is most common, but planetary scientists can have backgrounds in geology, meteorology, or any similar field. Exobiology would, of course, require a biology background. Almost all space science jobs that result in advancement require advanced degrees beyond the bachelor's degree.

SOCIAL CONTEXT AND FUTURE PROSPECTS

Space science has evolved from simply the science of astronomy to real, practical applications. Planetary science helps scientists understand how planets function, yielding valuable insights about Earth as a planet. This knowledge gains in importance as issues such as global warming and ozone depletion in the stratosphere are more widely examined. Studies of stars help scientists understand how the Sun works, which is important in the realm of space weather.

FASCINATING FACTS ABOUT SPACE SCIENCE

- In September, 1859, one of the most powerful geomagnetic storms in history knocked out telegraph systems around the world.
- The word *sputnik* means "fellow traveler."
- Rockets do not need to push against anything to fly through space.
- Objects on the Moon weigh about one-sixth of what they do on Earth.
- The first communication satellite, Project SCORE, was launched December 18, 1958, and the first message broadcast was a holiday peace message from President Dwight Eisenhower.
- The United States' first space station, Skylab, was constructed from the third stage of a Saturn V rocket.
- Much of the universe is a vacuum, but other parts of space contain clouds of interstellar dust and tiny particles (the solar wind), many isolated particles and hydrogen atoms (which sometimes form nebulae), planetoids, asteroids, radiation bursting forth in solar flares, and beams of light, heat, and X rays radiating through space.

The exploitation of space provides useful services for the inhabitants of Earth. Modern society relies very heavily on rapid communication and the transmittal of a vast amount of data across far distances on a regular basis. Satellite communication facilitates this need, and many common everyday activities, such as credit or debit card purchases or ATM withdrawals, use this technology. GPS navigation has become very common and probably will become even more common, as more and more automobiles come equipped with GPS systems as a standard feature. Accurate weather and storm forecasts would be impossible without weather satellite data.

Space travel used to be fanciful speculation, then evolved into expensive government endeavors. However, in the early twenty-first century, several private companies were on the verge of providing space tourism at prices within the reach of many wealthy individuals. If the trend continues, space travel may eventually become almost as common and inexpensive as air travel.

—*Raymond D. Benge, Jr., MS*

FURTHER READING

Angelo, Joseph A., Jr. *Space Technology.* Westport: Greenwood, 2003. Print.

Berinstein, Paula. *Making Space Happen: Private Space Ventures and the Visionaries behind Them.* Medford: Plexus, 2002. Print.

Carlowicz, Michael J., and Ramon E. Lopez. *Storms from the Sun: The Emerging Science of Space Weather.* Washington: Henry, 2002. Print.

Comins, Neil F. *The Hazards of Space Travel: A Tourist's Guide.* New York: Villard, 2007. Print.

Freedman, Roger A., Robert M. Geller, and William J. Kaufmann III. *Universe.* 10th ed. New York: Freeman, 2015. Print.

Harra, Louise K., and Keith O. Mason, eds. *Space Science.* London: Imperial Coll. P, 2004. Print.

Matloff, Gregory L. *Deep Space Probes: To the Outer Solar System and Beyond.* New York: Springer, 2005. Print.

Swinerd, Graham. *How Spacecraft Fly: Spaceflight without Formulae.* New York: Copernicus, 2008. Print.

SPECTROSCOPY

FIELDS OF STUDY

Analytical chemistry; physical chemistry; organic chemistry; biochemistry, molecular biology; immunology; cell biology; physics; astronomy; inorganic chemistry; geology; forensic science; materials science and engineering; chemical engineering; biomedical engineering; toxicology; agricultural engineering; electrochemistry; polymer chemistry; military science.

KEY TERMS AND CONCEPTS

- **absorption:** Process that occurs when an atom absorbs radiation.
- **analyte:** Component of a system that is being analyzed.
- **Beer's law:** Law that states that absorbance of a solution is proportional to its concentration.
- **electromagnetic spectrum:** Entire range of wavelengths of light.
- **emission:** Process that occurs when an atom or molecule emits radiation.
- **excitation:** Process that occurs when matter absorbs radiation and is promoted to a higher energy state.
- **photon:** Massless, finite unit of electromagnetic energy that behaves as both a particle and a wave.
- **scattering:** Process that occurs when light does not bounce off a molecule at the same wavelength at which it started but instead undergoes a change in wavelength.
- **synchrotron:** Particle accelerator that produces ultra-intense light.
- **wavelength:** Distance a photon travels in the time that it takes for it to complete one period of its wave motion.

SUMMARY

Spectroscopy is the study of the interactions of electromagnetic radiation, or light, with matter in order to gain information about the atoms or bonds present within the system. There are many different types of spectroscopic techniques; however, most of the techniques are based on the absorption or emission of photons from the material being studied. The applications of spectroscopy span a variety of disciplines and can allow scientists to, among countless other things, determine the elemental composition of a nearby dwarf star, the chemical identity of an unknown white powder sample, whether a transfected gene has been expressed, or the types of individual bonds within a molecule.

DEFINITION AND BASIC PRINCIPLES

Spectroscopy is the study of how light interacts with matter. It allows scientists in a broad array of fields to study the composition of both extremely large and extremely small systems. Each spectroscopic technique is unique; however, most of the widely used techniques are based on one of three phenomena: the absorption of light by matter, the emission of light by matter, or the scattering of light by matter. A photon can behave as both a particle and a wave. For most spectroscopic techniques, the wave nature of the photon is the most critical because the wavelength of light being emitted, absorbed, or scattered is where the information about the sample is contained.

Absorption spectroscopy involves the absorption of photons by matter and can give information about the types of atoms or bonds in a molecule. Typically, a given material will absorb specific wavelengths of light and will reflect or transmit all the other wavelengths. Emission spectroscopy involves the emission of photons from a sample on excitation. Scattering deals with light that is inelastically scattered from a sample, meaning that the wavelength of light bouncing off the sample is not the same as the wavelength of light that was shined on the sample.

Related disciplines include electron spectroscopy and mass spectroscopy. Electron spectroscopic techniques generally involve either irradiating a sample with light and causing the emission of photoelectrons or bombarding a sample with electrons and causing the emission of X rays. Either the photoelectrons or X rays can be detected to gain chemical information about a sample. Mass spectroscopy is unique in that various methods can be employed to liberate ions from a sample, the masses of which are then detected to gain a chemical fingerprint of the molecules that are present.

BACKGROUND AND HISTORY

In 1666, Sir Isaac Newton became the first person to discover that ambient light could be separated into a continuous band of varying colors using a prism. This discovery was the beginning of all spectroscopic research. In 1803, Thomas Young performed his famous slit experiment demonstrating the wave nature of light. William Hyde Wollaston observed the first absorption bands in solar radiation in 1802; however, it was Joseph von Fraunhofer who assigned these lines (now known as Fraunhofer lines) significance in 1817, thus paving the way for the fields of spectroscopy and astrophysics. In 1848, French physicist Jean Bernard Leon Focault determined that the locations of the lines were characteristic of the elements that were present. Lord Rayleigh (John William Strutt) investigated the elastic scattering of light in 1871, and Albert Einstein followed in his footsteps in 1910. However, a wavelength shift caused by inelastic scattering of light was not observed until 1928, a discovery for which Sir Chandrasekhara Venkata Raman won the Nobel Prize in Physics. In 1960, Theodore Harold Maiman built the first successful optical laser, which would later prove essential for modern spectroscopic instruments.

HOW IT WORKS

Absorption techniques. Fourier transform infrared spectroscopy (FTIR) is a vibrational spectroscopy technique that has the potential to elucidate what types of bonds are present within a sample. In transmittance mode, the sample is placed between an infrared light source and a detector and is irradiated with infrared light. Attenuated total reflectance (ATR) is a closely related technique that has minimal sample preparation. A sample is placed in contact with a crystal, such as diamond or germanium, and an evanescent infrared wave is sent through the crystal. Most bonds will absorb a specific wavelength of infrared light that causes them to vibrate or bend, and these wavelengths of light are absorbed in these processes and do not reach the detector. Polar bonds, such as carbonyls and ethers, are the bonds most easily detected using FTIR. The output of both FTIR and ATR is a spectrum with bands showing which wavelengths of light were absorbed by the sample. These wavelengths can then be interpreted to determine what types of bonds or functional groups were present in the sample.

Ultraviolet visible (UV-Vis) spectroscopy is similar to FTIR in that it involves shining light of a known wavelength range, in this case visible and ultraviolet, at a sample and determining which wavelengths are absorbed by the sample. UV-Vis does not provide as much specific information about bond character as FTIR; however, it is a very useful and fast technique for quantifying the concentration of a known analyte in solution according to Beer's law.

Atomic absorption spectroscopy (AAS) is a quantitative elemental analysis technique that relies on the reproducible absorption of specific wavelengths of light by a given element on excitation. A sample in solution is atomized by drawing the solution into a heat source and then irradiated with a specific wavelength of light selected to excite a particular element. The amount of light absorbed by the sample at that wavelength is used to calculate the concentration of the analyte in solution. AAS is very sensitive and is often used for detection of trace elements. Several types of heat sources, including an open flame, a graphite furnace, or an acetylene torch, can be employed to atomize the sample.

Astronomical spectroscopy is a method that astronomers use to determine the elemental composition of far-away celestial bodies that cannot be sampled in a laboratory. These bodies emit electromagnetic radiation across the entire electromagnetic spectrum; however, each element absorbs several characteristic wavelengths of light. So, by measuring which wavelengths of light are not being emitted by the very large sample in question, a fingerprint of all the elements present can be obtained.

Emission techniques. Atomic emission spectroscopy (AES) is a technique that provides similar information to the information obtained using AAS but by essentially opposite means. In AES, a sample is atomized and excited by the same heat source, often a flame or a plasma, which is a highly ionized and energized gas. If a plasma is used, the technique is referred to as inductively coupled plasma AES (ICP-AES). The excited analyte will emit light at wavelengths that are characteristic of the elements that are present, so by detecting the emitted wavelengths of light and quantifying them, the elements in the sample can be identified and quantified. Multiple elements can be detected at once in AES.

Fluorescence spectroscopy is a family of techniques that uses higher energy radiation, such as ultraviolet light or X rays, to excite the atoms of a sample. The excited state is not stable, so atoms must release lower energy photons to relax back to a more energetically favored state. Depending on the type of incident radiation, information about the sample can be obtained to varying degrees of specificity. For some analyses, it is enough to induce fluorescence so that the emitted photons can be imaged, such as is the case in many biological applications. In other methods, such as X-ray fluorescence (XRF), specific elemental and chemical information can be obtained.

Scattering techniques. Raman spectroscopy is a nondestructive vibrational spectroscopic technique based on inelastic light scattering. In Raman spectroscopy, a monochromatic laser is shined on the surface of a sample. Most of the light is elastically scattered,

A huge diffraction grating at the heart of the ultra-precise ESPRESSO spectrograph.

meaning that the light bounces off the sample at the same wavelength at which it entered; however, a very small fraction of the light is inelastically scattered, meaning that the light bounces off the sample at a different wavelength than that at which it entered because of interactions with the bonds in the sample. These interactions cause the bonds to vibrate and bend, similar to what happens in FTIR. The magnitude of the wavelength shift is characteristic of the bonds that caused it, and the resulting spectra are quite similar to FTIR spectra. However, whereas FTIR is most sensitive to polar bonds, Raman spectroscopy is most sensitive to polarizable bonds, such as carbon-carbon double bonds and aromatic rings.

Other spectroscopic methods. There are many other spectroscopic techniques that are in use across industry and academia; however, most are based on the same principles as the more common types. Examples of other techniques include circular dichroism, dynamic light scattering, and spectroscopic ellipsometry. One widely used spectroscopic method that is not based on conventional absorption, emission, or scattering methods is nuclear magnetic resonance (NMR). In organic chemistry, NMR is used to study the organization of bonds in a molecule and can give information about the number and location of hydrogens, carbons, or other elements being studied in a sample.

APPLICATIONS AND PRODUCTS
Spectroscopy is widely used in industrial and government-funded scientific endeavors for a variety of purposes, including pharmaceutical analysis, failure analysis, materials science, reverse engineering, and toxicology. Vibrational spectroscopy is widely used to aid in the identification of unknown materials. For example, an analytical chemist may use FTIR to identify foreign material found on a manufacturing line or in a finished product, materials used in a competitor's product, or unknown material recovered from a crime scene. Searchable libraries of vibrational spectra can be consulted in conjunction with spectral interpretation to determine the identity of the material in question. There are many commercially available libraries, but libraries can also be constructed from in-house samples and standard materials.

Raman spectroscopy is a nondestructive, surface-sensitive technique and therefore is often used to identify pigments in historical texts, art, and textiles

to learn more about the people who created them and to verify their authenticity. Glass is largely transparent to Raman spectroscopy, making this technique ideal for identifying unknown materials found in glass containers without opening the container, thereby reducing the risks associated with exposure to an unknown material. This technique is especially useful in the identification of unknown flammable liquids. Ahura manufactures a portable Raman spectrometer for use in the field to allow emergency workers to evaluate potentially hazardous materials, such as chemical warfare agents or explosives. Techniques such as FTIR and Raman are used in airport screenings, as they are fast and reliable techniques for quickly identifying potentially harmful compounds.

Spectroscopic methods can also be used to quantify known materials. UV-Vis spectrometers can be used alone or as detectors in a chromatographic system in the pharmaceutical industry to quantify concentrations of drugs, biologics, and excipients in solution, as well as to verify the identity, purity, and stability of such compounds. Toxicologists use methods based on spectroscopy to quantify drugs in biological tissues and fluids.

Production facilities will often place spectroscopic instrumentation in line to continuously monitor products. For example, spectroscopic ellipsometry, which has the potential to measure thin films on the order of nanometers, can be placed in a production setting to monitor the thickness of vapor-deposited thin films.

NMR is typically used by synthetic chemists to determine or verify the structure of the compounds they synthesized, and it is often used in polymer science, drug chemistry, and materials science. In the medical community, NMR is used to image various types of soft tissue in situ and is better known as magnetic resonance imaging (MRI). Hospitals also use spectroscopy to measure blood counts, as well as coupling spectroscopy with immunoassays to screen for drugs or other compounds in urine.

Various elemental analysis techniques, such as AAS or AES, are often used along with Raman spectroscopy to analyze the composition of geological samples and meteorites. By understanding what elements and compounds are present in these samples, geologists can gain a better understanding of their formation and hypothesize about the processes that would

FASCINATING FACTS ABOUT SPECTROSCOPY

- Consumer products incorporating spectroscopy are becoming more prevalent, and one gadget Web site markets a keychain that detects and indicates the levels of ultraviolet light at a given moment.
- Environmental chemists often employ spectroscopy to check for levels of pollutants in water and air samples.
- Ocean Optics manufactures a line of handheld spectrometers that are ideal for performing analyses on the go.
- Any object or being that radiates heat, such as an animal, star, bonfire, or car engine, emits a spectrum of light.
- The wavelength range over the entire electromagnetic spectrum is 1,000 meters (longwaves) to 0.1 angstroms (gamma rays). The visible range of the spectrum lies between about 400 and 700 nanometers in wavelength.
- Sir Chandrasekhara Venkata Raman, the winner of the Nobel Prize in Physics in 1930, was so confident that his discovery of Raman scattering would win him the prize that he booked tickets to Stockholm even before the award winners were announced.
- Biologists amplify DNA in the presence of fluorescent markers, then use capillary electrophoresis with a fluorescence detector to determine the nucleic acid sequence of the DNA strand.
- Thomas Young's famous double slit experiment proved that light behaves as both a particle and a wave.

create such a specimen. For example, researchers at Université de Lyon in France used Raman spectroscopy to identify two new types of ultrahard diamonds, one predicted and one unexpected, in the Havaro meteorite, which indicated that the parent asteroid had undergone an extremely violent collision.

Biologists use fluorescence spectroscopy to determine whether gene expression has occurred. By labeling the gene they wish to introduce into an organism with green fluorescent protein (GFP), they can visualize everywhere that the gene of interest was expressed. GFP can be visualized in cells, tissues, and organisms by exciting the GFP with blue light, which causes the emission of green light. Real-time Polymerase chain reaction (PCR), a method for making many copies of DNA, often uses fluorescent probes that bind to the new double-stranded DNA.

By using fluorescence spectroscopy to quantify the amount of fluorescence, biologists can monitor how much DNA they have synthesized.

Several spectroscopic techniques, such as Raman, FTIR, and NMR, can be used to chemically map out surfaces and volumes of materials. By taking individual spectra at various points across a sample, the intensity of a given signal or combination of signals can be tracked and mapped on an intensity scale. The resolution of these techniques, that is, the smallest features that can be resolved, depends on the wavelength of the light that is used. Thus, Raman microscopy typically has much better resolution than FTIR because of the shorter wavelengths of light used in Raman spectroscopy. Chemical mapping can reveal details that would not otherwise be seen in a sample, such as where a specific protein may be expressed in a type of tissue or cell.

CAREERS AND COURSE WORK

Students interested in spectroscopy have a wide variety of career paths from which to choose, and their final career goal should determine their course work. Completion of a bachelor of science degree in one of the core sciences such as chemistry or physics serves as preparation for graduate school, which, although not required in every case, greatly improves one's prospects of obtaining a career in spectroscopic research. A bachelor's degree qualifies an individual for a position at the level of a laboratory technician; such an employee is usually trained on the operation of a spectroscopic instrument and performs mainly routine tasks. A master's degree or doctorate and appropriate research experience qualify the holder for a position involving advanced use and maintenance of spectroscopic instruments; design of spectroscopy experiments; spectroscopic data analysis, interpretation, and communication; or possibly even the design of new instruments.

The careers that involve spectroscopy are as diverse as the spectroscopic methods themselves. Chemists, physicists, astronomers, materials scientists, chemical engineers, biologists, and geologists can all be spectroscopists. Careers involving spectroscopy can be found in industries such as pharmaceutical companies, in colleges and universities, and with the government, perhaps at a national laboratory or at a Federal Bureau of Investigation crime laboratory. Despite the diversity of these options, it is imperative that applicants possess a strong foundation in basic science, including chemistry and physics, as well as in mathematics. Course work in analytical and physical chemistry is highly pertinent to most careers involving spectroscopy.

Social Context and Future Prospects

Spectroscopic techniques are powerful methods for identifying and sometimes quantifying components of a system, and applications of these techniques are likely to diversify further in the future. Handheld spectrometers probably will become commonplace in field forensic and military work because they allow personnel to identify the compounds with which they are dealing and to know what safety precautions to take. Spectrometers will also most likely find their way into everyday-use products and continue to play a significant role in space missions, such as looking for the presence of water or amino acids on distant planets.

An area of spectroscopy of interest is research using synchrotron light. Synchrotrons such as the Canadian Light Source and the European Synchrotron Radiation Facility use particle accelerators to produce ultra-high-intensity light of many wavelengths, which can then be diverted from the particle accelerator down a beam line and into a spectroscopy facility using a set of mirrors, gratings, attenuators, and lenses. Synchrotron light can penetrate deeper into materials and space, allowing information to be gained from farther away than before.

The resolution of spectroscopic chemical maps had been limited by physics, with maximum resolution being obtained using synchrotron light and operating right at the limit of diffraction. However, several commercial techniques are able to image below the diffraction limit of light. One example is the nanoIR by Anasys Instruments, which uses an atomic-force microscope tip to gain infrared information about very small areas of a sample. Typically, ATR resolution is on the order of three to ten microns, but with nanoIR, resolution can be obtained at the submicron level. A great deal of research at universities, government-funded labs, and at spectroscopy companies has been focusing on achieving better resolution so that smaller and smaller features can be accurately probed. In 2014, scientists Stefan Hell, Eric Betzig, and William Moerner were recognized for their individual contributions to the development of super-resolved fluorescence microscopy, receiving the Nobel Prize in Chemistry. Their microscopy and spectroscopy techniques allow scientists to study living cells in real time at the molecular level, providing the opportunity to more accurately investigate the proteins involved in diseases such as Alzheimer's and Huntington's.

—Lisa LaGoo, MS

Further Reading

Chang, Kenneth. "Nobel Laureates Pushed Limits of Microscopes." *New York Times.* New York Times, 8 Oct. 2014. Web. 26 Feb. 2015.

Harris, Daniel C. *Quantitative Chemical Analysis.* New York: Freeman, 2003. Print.

Hashemi, Ray H., William G. Bradley, and Christopher J. Lisanti. *MRI: The Basics.* 3rd ed. Philadelphia: Lippincott, 2010. Print.

Robertson, William C. *Light.* Arlington: Natl. Science Teachers Assn., 2003. Print.

Robinson, Keith. *Spectroscopy: The Key to the Stars—Reading the Lines in Stellar Spectra.* New York: Springer, 2007. Print.

SURFACE AND INTERFACE SCIENCE

FIELDS OF STUDY

Materials science and engineering; chemistry; physics; inorganic chemistry; forensic science; electrical engineering; semiconductors; biomedical engineering; chemical engineering; toxicology; biology.

KEY TERMS AND CONCEPTS

- **adhesion:** Attraction process that brings two or more materials together in direct contact.
- **corrosion:** Oxidation of a metal surface; for example, rust on iron-containing metals.
- **functionalization:** Attaching new functional groups to a surface to impart new properties to the

surface.

- **interface:** Region in a system where two or more materials meet.
- **plasma:** Highly activated and ionized gas.
- **surface:** Uppermost 0.5 to 3 nanometers of a sample or material.
- **surface chemistry:** Molecular and atomic species and bonds present at the surface of a material.
- **topography:** Peaks and valleys of a surface and their relationship to each other in three-dimensional space.

SUMMARY

Surface and interface science examines the properties of materials and products at the surface and at the locations where two or more materials meet. The discipline typically focuses on surface analysis, surface modification, and interface science. Surface modification changes the properties of a surface, such as the hydrophobicity, lubricity, chemistry, or topography, and surface analysis tells what those properties are for the material being studied. Interface science can include studying interactions between materials and how well the materials adhere to one another, as well as how to achieve desired interactions. The interface between a material and a solution or air can also be studied.

Definition and Basic Principles

Surface and interface science encompasses several disciplines, including surface analysis, interface analysis, and surface modification.

Surface analysis is the study of the chemistry, crystal structure, and morphology of surfaces, using a combination of various surface sensitive analytical methods. Surface sensitivity can mean different things for different applications; however, a good working definition would be the uppermost 0.5 to 3 nanometers (nm) of a surface, or two to ten layers of atoms. Because surfaces are often modified with films in the tens to hundreds of nanometers range, the uppermost 100 nm can be considered the surface for some applications. Any surface thicker than 100 nm is typically considered bulk material.

Interface analysis is the study of the morphology and chemistry of internal interfaces, or regions where two or more materials meet. A good understanding of the interface between two materials will

provide scientists and engineers with a wealth of information, including how well these materials are bonded, whether a certain amount of force is likely to separate the materials, whether any new species have been formed, or whether the two materials are diffusing into each other.

Surface modification is the process by which scientists and engineers change the chemistry, physical properties, or topography of a surface to improve the function of the material, such as changing the coefficient of friction, the lubricity, the hydrophobicity, the roughness, or bondability of the material.

Surface analysis and surface modification are both quite sensitive to contamination, so much care must be taken that the surfaces of interest do not come in contact with materials that could potentially transfer to the surfaces. However, even if every precaution is taken to prevent transfer, contamination can occur from simple exposure to air, which is why many surface modifications and analyses are conducted under vacuum. A vacuum is also essential for many of the surface techniques to prevent interactions between air molecules and the species being analyzed.

BACKGROUND AND HISTORY

Surface science is a relatively new area of science. The first X-ray photoelectron spectrum was collected in 1907 by P. D. Innes, and the first vacuum photoelectron spectrum was taken in 1930. American physicists Clinton Davisson and Lester Germer pioneered low energy electron diffraction (LEED) in 1927 and observed diffraction patterns on crystalline nickel. Ultra-high vacuum techniques for surface analysis and modification (specifically welding) were invented in the 1950's. In 1968, a surface science division of the American Vacuum Society was proposed, the same year that Auger electron spectroscopy began to be used. German scientist Alfred Benninghoven introduced time-of-flight secondary ion mass spectrometry (TOF-SIMS) in the 1980's, about the same time that scanning tunneling microscopy began to be used.

HOW IT WORKS

Surface modification. Surface modification can be accomplished in many ways, including methods that modify the chemistry of a surface, the topography of a surface, or both. Films of material can be grown on a substrate in a vacuum or can be deposited using an ultrasonic spray-coating, spin-coating, or dip-coating process. Anodization is a process whereby an oxide layer is deposited on a metal using oxidation and reduction reactions. It has the potential to grow oxide films in a very controlled manner. Passivation is a process that changes the oxide layer on a metal to make it more resistant to oxidation and is typically carried out in an acid solution. Surface topography can be modified using laser or acid etching, which roughens the surface, or by using electropolishing, which uses electricity to smooth conductive surfaces.

Plasma treatment is a method of modifying the chemistry and often the topography of a surface. It uses a highly ionized, activated gas to react with the molecules of a surface. The plasma gas can vary from an inert gas, such as argon or helium, which would be expected to cause the species at the surface to react with one another, or a polymeric monomer, which could polymerize on the surface and create a thin plasma-treated layer. Plasmas can also be employed to clean surfaces before modification. Chemical vapor deposition is a technique in which the sample is exposed to a vapor that reacts with the surface to modify it.

Surface analysis techniques. There are many surface analytical techniques. The most common techniques are complementary to one another, and several are often used in combination to gain all the necessary information about a surface.

Scanning electron microscopy (SEM) is a technique that uses a focused electron beam that scans across the surface of a sample and allows for the construction of a very detailed map of surface topography. The surface sensitivity of this technique varies based on the sample and the instrument parameters, which can make the instrument range from simply surface sensitive to able to gain information up to microns within a sample.

Electron spectroscopy for chemical analysis (ESCA), also known as X-ray photoelectron spectroscopy (XPS), uses X rays to generate electrons from the surface of a sample. These electrons can then be detected and traced back to whatever elements are present at the surface. XPS gives quantitative elemental information about the uppermost 5 to 10 nm of a sample surface and can also give some information about the types of bonds present. Auger electron spectroscopy (AES) is a surface-sensitive technique that uses an electron beam to gain surface elemental information similar to that obtained by XPS. AES is less quantitative than XPS; however, higher resolution mapping is possible with AES than with XPS.

Time-of-flight secondary ion mass spectrometry (TOF-SIMS) is an extremely surface-sensitive analytical technique that uses a beam of liquid metal ions to probe the top one to three molecular layers of a sample, causing the emission of ions from the sample to be detected in a time-of-flight mass spectrometer, which gives detailed information about the specific molecules and atoms at a surface.

White light interferometry, sometimes called profilometry, uses white light interference patterns to construct detailed maps of surface topography very quickly. It can detect minute surface features in the nanometer range. Spectroscopic ellipsometry is often used to measure the thickness of surface films, such as oxides or plasma treatments on materials. Atomic force microscopy (AFM) and scanning tunneling microscopy (STM) both employ a very sharp tip called a cantilever that rasters across a sample and yields high-resolution images of surface topography. AFM is typically employed for nonconducting samples, and STM is useful for conducting samples.

Interface science. The interface of two materials can be probed using several techniques, including some of the techniques used for surface analysis. Depth profiling is a method by which information is gained from within the sample. If the interface in question is near the surface, several techniques can be used. Both TOF-SIMS and XPS use beams of ions or carbon 60 (Buckministerfullerene) to sputter material away from the surface to obtain information about interfacial layers and boundaries. An advantage to these techniques is that they are performed under vacuum; therefore, the interface is not exposed to ambient conditions before analysis. Confocal Raman microscopy has the potential to probe interfaces up to 100 microns below the surface of a sample by focusing the laser on the interface and using a pinhole to eliminate out-of-focus data.

APPLICATIONS AND PRODUCTS

Surface modification. Surface modification has a very wide range of applications. Surfaces are modified to impart desirable properties and often change properties such as lubricity, surface energy, the coefficient of friction, or the functional groups present on a surface. Self-assembled monolayers can be deposited, and crystals can be grown in a highly controlled manner. Sensors and catalysts can also be fabricated using surface modification. Surface modification is applied across a wide variety of industries, including the automobile, aerospace, medical device, textile, chemical, steel, and electrical industries.

A major application for surface modification is making materials more corrosion and wear resistant. Steel is used in everything from skyscrapers to motorcycles, so increasing the stability of this crucial material is a priority not only for many people in the steel industry but also for industries that use the steel to fabricate other products.

In the medical device community, the surface of a material is extremely important because it is what is in contact with the body. For example, the components of implants may be plasma treated to make their surfaces more biocompatible, which reduces cell adhesion and the formation of fibrous tissue around the implant. Implantable metal devices are often passivated to make the device resistant to corrosion when subjected to the aqueous environment inside the body.

In the automobile industry, surfaces of parts that must be bonded to other parts to provide structural integrity to the vehicle are sometimes plasma treated to promote better adhesion and reliability. The surfaces of exhaust systems are modified with a catalyst that greatly reduces toxic emissions.

Surface analysis. Surface analysis can be employed in many situations, but it is particularly well suited for the analysis of contamination or surface damage. In many industries, contamination at the nanometer scale can spell disaster for a process. For example, semiconductor materials have very predictable conductive behavior, which is essential for designing microelectronics that work properly. The addition of contaminants to the system will change the behavior of the materials and can cause failures, so it is essential that surface analysis be employed in the development of a product and sometimes during manufacturing stages to ensure that the materials are clean and reliable.

Surface analysis is critical in the study of nanomaterials, as these materials and structures are often so minute that the majority of the mass of a sample is considered surface material. Less sensitive techniques would not be able to effectively study these extremely small materials.

Interface science. The interface and interactions between two solid materials are very interesting to materials scientists; however, interface science also includes the study of the interactions between materials and a gas, such as air, or a liquid, such as water. Proteins such as insulin are often sold in solution form. Because proteins are unstable under most conditions, it is important for drug companies to understand the interactions between proteins and the surfaces of the containers in which they are stored.

Adhesion specialists often need to determine whether the two materials they are trying to bond are well bonded. Mechanical tests, such as peel tests or scratch tests, are often employed to determine the integrity of the bond in combination with a chemical analysis that has the ability to look for new bond types between the layers.

CAREERS AND COURSE WORK

A wide variety of careers is available in surface science, and the desired career will determine the proper course work. To work in surface science requires a bachelor of science degree in a core science such as chemistry or physics, and having a graduate degree greatly improves a student's chances of finding work in the field. A bachelor's degree is adequate qualification for becoming a laboratory technician, which generally involves being trained on the operation of an instrument and performing mainly routine tasks. A master's or doctoral degree and appropriate research experience are the necessary qualifications for a position that involves the advanced use and maintenance of surface analysis instruments, experiment design, and data analysis, interpretation, and communication, as well as possibly the development of new instruments.

Careers involving surface science can be found in industry (such as at an automobile company), in academia, and in government laboratories. Surface and interface science is critical in materials science and engineering and is likely to play a major role in the development of the field. Regardless of the specific career choice, a strong foundation in basic science, including chemistry and physics, as well as in mathematics, is required.

SOCIAL CONTEXT AND FUTURE PROSPECTS

Surface science is a growing field that will continue to play a major role in the development of materials science and engineering, especially in light of the focus on nanomaterials and the potential of surface science to make a significant contribution to that field. In many industries, the focus is on making devices smaller and smaller, which naturally leads to the surface becoming a more significant part of the device because of the increase in the surface area to volume ratio.

New applications are being discovered for surface modification every day. Plasma treatment is becoming more common in many industries, such as the automobile and medical device industries, because it can make it easier to bond one material to another. The process increases quality and is relatively fast, which suits production in the fast-paced cultures of these industries.

Additionally, as governing agencies such as the Food and Drug Administration step up their scrutiny of medical devices and drug-device combination products with respect to biocompatibility and stability, surface analysis will probably play an increasing role in characterizing the surfaces that come into direct contact with a patient.

—*Lisa LaGoo, MS*

FURTHER READING

D'Agostino, Riccardo, et al., eds. *Plasma Processes and Polymers*. Weinheim, Germany: Wiley, 2005. Brings together papers from a conference on plasma treatments for polymers.

Hudson, John B. *Surface Science: An Introduction*. New York: John Wiley & Sons, 1998. Contains information on surface analysis, interface interactions, and crystal growth.

Kim, Hyun-Ha. "Nonthermal Plasma Processing for Air-Pollution Control." *Plasma Processes and Polymers* 1, no. 2, (September, 2004): 91–110. Discusses the history and future of using plasma treatment to reduce air pollution.

Kolasinski, Kurt W. *Surface Science: Foundations of Catalysis and Nanoscience*. 2d ed. Hoboken, N.J.: John Wiley & Sons, 2009. A detailed, more technical resource for those interested in cutting-edge applications of surface science.

Rivière, John C., and S. Myhra, eds. *Handbook of Surface and Interface Analysis: Methods for Problem-Solving*. 2d ed. Boca Raton, Fla.: CRC Press, 2009. Discusses common surface analysis methods and problems; chapter 7 deals with synchrotron-based techniques.

Vickerman, John C., ed. *Surface Analysis: The Principal Techniques*. 2d ed. Chichester, England: John Wiley & Sons, 2009. Describes the techniques involved in surface analysis. The introduction contains a basic overview of surface analysis that is easy to read and informative.

FASCINATING FACTS ABOUT SURFACE AND INTERFACE SCIENCE

- Researchers at IBM inadvertently modified the tip of an atomic force microscope (AFM) with a carbon monoxide molecule, which ended up allowing them to image the structure of interlocking benzene molecules with more clarity than ever before.
- Ion scattering spectroscopy (ISS) has the potential to determine the structure of the uppermost layer of atoms of a sample.
- Textiles are sometimes plasma-treated to make the fabric resistant to wear, bacteria, fungi, and shrinkage.
- Varying the thickness of the oxide layer on titanium parts changes the color of the part.
- A single monolayer of contamination can obscure the results of surface analysis techniques.
- One of the most common surface contaminants is silicone oil.
- Attenuated total reflectance (ATR) and Raman spectroscopy are not true surface techniques, but both can be modified to give surface chemistry information about the specific types of functional groups present.
- Surface enhanced Raman spectroscopy (SERS) involves adsorbing molecules in a metal surface or depositing metal nanoparticles on a surface. The amount of Raman signal obtained can be enhanced by 10_{15}, making what is normally a bulk technique extremely surface sensitive.

T

TELEPHONE TECHNOLOGY AND NETWORKS

FIELDS OF STUDY

Electrical engineering; information technology; computer programming; computer science.

KEY TERMS AND CONCEPTS

- **analogue signal:** Continuously variable signal that can vary in frequency of amplitude in response to physical phenomenon.
- **bit:** Either of the digits 1 or 0 in a binary numeration system; also known as binary digit.
- **broadband:** Circuit permitting high-speed data to be transmitted across a network.
- **cellular radio network:** Network of base stations that form large pockets (cells) of radio signal coverage for mobile phones.
- **circuit switching:** Method of establishing a total end-to-end circuit between two parties.
- **digital signal:** Signal transmitted in binary code (1's and 0's) indicating pulses of varying levels of magnetic energy.
- **fiber optics:** Technology of sending pulses of light down very slender, very pure strands of glass.
- **hierarchy of switching systems:** Pyramid-like structure of switching centers used to connect two parties in the public switched telephone network.
- **Internet:** Vast network of communications channels used for data transmission.
- **1G, 2G, 3G, and 4G:** Four generations of technology in the cell phone field; "G" stands for "generation."
- **packet switching:** Method for transmitting data by means of small bundles of 1's and 0's that make up a portion of a data message.

SUMMARY

A country without a solid telecommunications infrastructure faces insurmountable obstacles. Telephones and their networks are an essential part of telecommunications. The ability to connect individuals and government agencies by telephone is necessary to allow nations and economies to flourish. Continuing advances in technology have advanced telephones from large wooden boxes fixed to walls to small handheld portable devices that can do much more than simply enable people to talk at a distance.

DEFINITION AND BASIC PRINCIPLES

Any communications between two parties—be they people or machines—requires an originator and a receiver, as well as a network connecting the two. The simplest network tying these two parties together is a simple wire (in electrical terms) or a pair of wires. However, as the number of parties increases to three, three hundred, or three thousand, the required network becomes impossibly large. In the legacy telephone network, a hierarchy of switching systems has been developed to allow communications from any user to any other user.

As technology has advanced, other means of communication have been developed. In many cases, the pair of wires connecting the transmitting party and the receiving party has been replaced with radio transmission systems (such as satellite communications, cell phones, microwave systems), coaxial cables (a physical structure that permits high-speed transmission), or fiber optics (which converts electrical signals to light).

Just as the technology used for telecommunications has advanced, so has the structure of the telephone network. The Internet is rapidly replacing the circuit-switched network (the public switched telephone network) that has been in use for more than a century for transmission of not only data but also sound.

BACKGROUND AND HISTORY

The telephone industry began on February 14, 1876, when Alexander Graham Bell filed his first patent for a telephone. One month later, on March 10, 1876, he and his assistant Thomas Augustus Watson produced a model that worked, and the immortal words "Mr. Watson, come here, I want you," were spoken over the device.

In the years that followed, numerous technological advances were made, and the telephone became a practical device. Telephone companies sprang up by the thousands, frequently two or more firms in the same city. Connections among these companies were the exception rather than the rule, and competition became ruthless.

By the middle of the twentieth century, the industry had reached some degree of stability. The Bell System, the largest of the country's telephone companies, had managed to capture about 80 percent of the telephone service in the country.

The Bell System survived a number of government-initiated lawsuits over the years, but in 1982, it agreed to split itself up into two major entities, the telephone operating companies and the parent company, AT&T, which brought with it the long-distance

arm, manufacturing facility, and research and development units.

Once again, the industry became highly competitive, with new companies appearing and disappearing each year.

The new technologies embraced by the many startups caused the industry to change dramatically. Advances in radio technology permitted satellite communications, microwave transmissions, and cellular radio. The development of fiber optics allowed data to be transmitted almost at the speed of light and increased the capacity of the country's network. Advances in electronics changed the way information was transmitted. Digital signals replaced analogue. In the 1960s packet switching between a network of computer networks, or internet, changed things dramatically.

How It Works

The modern telephone has four major components: the transmitter (which converts sound waves into electrical signals), the receiver (which converts received electrical signals into audio signals), an alerting device (generally called a ringer, although the alerting signal is usually some sort of tone), and a signaling device (either a rotary dial or a set of push-buttons).

The ringer rides the telephone line whenever the telephone is on-hook and is removed from the line when the telephone is in use. The ringing signal for most telephones is 90 volts, alternating current.

A major part of the circuitry of the telephone deals with sidetone. Because the telephone usually operates on a two-wire circuit, sounds spoken into a telephone tend to be heard not only at the distant end of the circuit but also in the local receiver. The circuitry used to combat this reduces the volume of the conversation that is fed back to the local receiver. It is interesting to note that this fed-back conversation is not electronically eliminated entirely, lest the user think the telephone is not operating.

A call is initiated by removing the handset from the cradle. This action removes the ringer from the circuit and connects the telephone's electronic apparatus. A circuit is established through this apparatus, which is detected by the central office. Equipment at the central office sends a dial

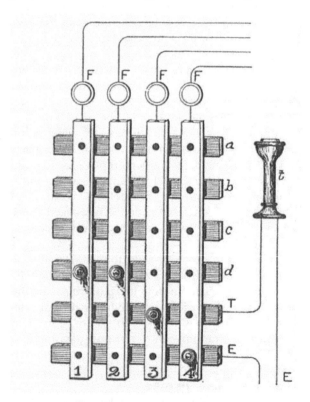

1903 manual crosspoint switch

tone to the subscriber. Hearing this dial tone, the subscriber dials a number—either with a true dial, which interrupts the two-wire circuit the indicated number of times, or a keypad, which transmits two tones to the central office.

The transmission medium. The modern transmission medium is a far cry from the two-wire connections of yesteryear. Copper still predominates, but coaxial cable, fiber optics, and radio communications are increasingly used.

Coaxial cable has a greater capacity than does twisted-pair copper and is still the prime transmission medium for cable television. The cable is, in fact, coaxial. A single strand of copper is surrounded by insulation, which is encased in metallic sheathing. This configuration keeps desired signals in and undesirable signals out.

Fiber optics has become the transmission medium of choice. A hair-thin strand of ultrapure glass carries pulses of light at gigabit rates. By using electronic techniques, thousands of conversations can be multiplexed onto a single strand of fiber.

Several forms of radio communications—satellites, cellular radio, and Wi-Fi—are commonly in use. Satellite communications can be used for voice, television, or data. When used for voice, satellites often result in an undesirable confusion because the time required to bounce a signal off a satellite (parked at an altitude of 22,300 miles, in geostationary orbit) is 0.125 second each way. Therefore, there is a built-in 0.25-second delay in any conversation.

A more talked-about use of radio is the wireless communication provided by cellular radio. The wireless segment of a cell phone conversation is remarkably short; simply the two- or three-mile segment from the cell phone to the cell phone tower. At this point, the signal travels down the tower through copper or fiber optics and goes to the local telephone central office.

A third use of radio transmissions is in local area networks. A radio transmission connects a device attached to a telephone line and a computer. Frequently, this is identified as Wi-Fi. Although the technology is different, a cordless telephone operates in much the same way.

Transmission path selection. Some sort of a selection device—for instance, a telephone dial—is required to select the destination of whatever message is being sent. The method of transmitting this

address to the network and the network's ability to establish a connection are equally important.

Commonly used transmission devices are telephones, which increasingly are cell phones, smartphones, or handheld computers. The dial is usually a set of pushbuttons (the keypad). The system converts the analogue signals of the unit to a digital signal (a series of ones and zeros).

Since its establishment, the hierarchy of switching systems has served its function well. A transmission path is established from a local switching office to a higher class office and then (if required) to a still higher class office. Then the path is extended downward to finally reach the called party. Thus, a total circuit is established, and the overall process is logically called circuit switching.

This circuit switching network is being replaced with a packet switching network. Instead of establishing and holding a total circuit for the duration of a call (a wasteful process), a message (voice or data) is digitized and then broken into small chunks (called packets). Each packet of this digitized data is preceded with supervisory material that indicates the importance of the packet, its destination, and other information. Then the switch (which of course bears a remarkable resemblance to a computer) establishes a connection to a distant switch and transmits its packet. Each packet, therefore, can travel a different path to the ultimate destination. It is left for the final switch to reassemble the packets in the appropriate sequence.

APPLICATIONS AND PRODUCTS

The technologies used in the first hundred years of the telephone industry were primarily mechanical and electromechanical. Massive amounts of equipment were necessary, a great deal of space was required for central offices, and the speed of transmission was limited. The advent of electronics—primarily the transistor—changed all this, and new technologies continue to be developed. Some of the technologies that have made major differences in most people's lives are the transistor, the integrated circuit, and central office switching.

The transistor. In 1947, three scientists from Bell Telephone Laboratories, Walter Brattain, John Bardeen, and William Shockley, invented the transistor, a solid-state device that would replace the vacuum tube. The tiny device was a solid and did not employ a vacuum or special gas. It consumed very little power and could perform amplification or switching just as the vacuum tube could. Furthermore, its life was almost unlimited. Over the years, it also became very inexpensive.

Clearly, the transistor could be used to replace the vacuum tubes used in amplifiers in the telephone network. Also, a rethinking of the switching function in the telephone central offices made possible a new device, the digital switch, and, in later years, the Internet and packet switching.

Although the invention of the transistor was important (it has been called the invention of the century), it by no means ended technological advances in electronics. Transistor technology advanced, and devices were made both smaller and faster, and the result was called the integrated circuit. These advances have not stopped, and there is no evidence that they will. In 1965, Gordon E. Moore, director of engineering at Fairchild Semiconductor, analyzed the advances being made in transistor technology and determined that the capability of the solid-state device would most likely double each eighteen months. This doubling of capacity is called Moore's law.

Central office switching. In the conventional telephone switching system, many telephone lines enter the central office. The switch, based on input instructions (the dialed number), connects one of those lines to another, thus establishing a circuit, a process called circuit switching. Such a circuit is maintained between the two parties (calling and called) for the duration of the telephone call.

Digital switching takes a different approach. The analogue voice signal is sampled very rapidly and converted into a series of ones and zeros. These ones and zeros trigger electronics within the switch, and the identical signal is transmitted from the central office to the next office. Such a switch is more efficient than the legacy analogue switch.

Certainly the transistor, the integrated circuit, and the technology of digital switching were of immense value to the telecommunications industry. Of equal importance, however, was the change that took place in the use of the network itself.

Packet switching. Circuit switching is inherently inefficient because the dedicated circuit is idle about half the time. A better method, although orders of magnitude more complicated, is packet switching. With packet switching, little bundles, or packets,

of ones and zeros are transmitted from computer to computer across the country. Supervisory information preceding each packet of data identifies its priority, its destination, and other pertinent information.

Each packet of information follows its own path to the destination. Because of traffic on the network and the number of stops along the way, a packet sent earlier than another packet may arrive later. If the packets are chunks of digitized voice, the order matters, and it is left to the computer at the receiving point to put the packets in the proper order before delivering them to the recipient.

Two points are extremely important in such a process. The first has to do with the speed of transmission through the network and the time that each packet takes to pass through a node (this is called latency). In actual use, latency has been reduced to about one millisecond. The second point demonstrates the improvement in network efficiency. Each time a packet passes through a node, headed for a transmission link to the next node, the previous link is released. In telephone terms, the computer hangs up. So except for the supervisory information attached to each packet, the efficiency of the system approaches 100 percent.

Wireless communications. One technology, wireless communications, has become supremely important for telephones. The process of operating telephones over radio is not new, nor is it complicated. Radio transmitting stations, whether public or private, simply mount a very tall antenna on the highest piece of land available and apply enough power to reach all potential users (such as the fire department, police department, radio stations, and taxicab companies). As long as these users stay within range of the antenna, all is well. As many channels as allowed are provided with the system (this number is easily determined with radio and television stations). For a telephone system, however, this poses problems. Just how many channels are required? More important, how many channels are available? All too frequently, too few channels are available to satisfy an entire city. The result would be a lengthy wait for dial tone.

What is the solution? A number of organizations and companies, in the United States and abroad, thought they had an answer. Instead of one large antenna and one set of frequency channels, they would provide a plurality of frequency channels, each assigned to a subset of the geographic community heretofore covered. These geographic subsets would be called cells, and the antenna covering each would be reduced in strength to the point where it would not impinge on nearby cells. This would make it possible to reuse those particular frequency channels in cells somewhat removed from the original cell.

A major problem with this approach, of course, was one of mobility. Would a taxicab driver in one cell, talking on one frequency, after moving to a different cell, have to redial? Certainly, this was not acceptable. By using very sophisticated electronics, the cell phone in use could transfer the call to the cell site antenna it was approaching and drop it from the cell it was leaving. This handoff capability was one of the most important characteristics of the proposed system.

FASCINATING FACTS ABOUT TELEPHONE TECHNOLOGY AND NETWORKS

- The telephone was invented by people from several countries. In the United States, papers from Alexander Graham Bell and Elisha Gray were filed on the same day in 1876. Bell received the patent, and this patent has withstood the test of time.
- Undertaker Almon Brown Strowger developed the first automatic switching device (patented in 1891), allowing a telephone user to dial directly, because he thought that telephone operators were sending potential customers to his competitors.
- A group of phreakers (telephone system "hackers") found that a whistle from a box of Cap'n Crunch cereal would trigger the electronics in a telephone central office and allow free access to the long-distance network. John T. Draper, self-labeled "Captain Crunch," was arrested for toll fraud after an article was published in *Esquire* in 1971.
- In 1973, Motorola researcher Morton Cooper made the first analogue mobile phone call, using a prototype.
- In 1981, 150 users in Baltimore, Maryland, and Washington, D.C., began a two-year trial of a mobile phone network, with seven towers.
- In 1984, customers lined up to purchase the first mobile phone, Motorola's DynaTAC 8000X, which weighed 2 pounds, was 13 inches long, and cost $3,995.
- In May, 2001, NTT's DoCoMo launched a trial of the third generation (3G) mobile phone network in the Tokyo area. Five months later, it followed with a commercial network.

The cell phone has become very popular. About two-thirds of the people in the world have access to a cell phone. In the United States, more and more people with cell phones no longer have a conventional land line. The number of subscribers to conventional phone service is dropping by about 10 percent per year.

Spectrum. This expansion of wireless communications is making it more difficult to find sufficient electromagnetic spectrum to support it. In the United States, toward the end of the twentieth century, the Federal Communications Commission began auctioning off huge chunks of spectrum. They were to be used "to provide a wide array of innovative wireless services and technologies, including voice, data, video and other wireless broadband services." The value of such spectrum was not lost on the telecommunications service providers. In one such auction, begun August 9, 2006, there were 161 rounds of bidding. The Federal Government took in more than $13 billion from this auction.

1G, 2G, 3G, and 4G. Wireless networks are placed into generations known as 1G, 2G, 3G, and 4G. 1G describes the original, analogue cell phone systems. From 2G on, the systems were digital. 3G improved on 2G, and 4G, which began to appear in 2010, is even more powerful.

CAREERS AND COURSE WORK
The services provided by telephone companies and their suppliers no longer amount to what is called POTS (plain old telephone service) but rather are very sophisticated. Anyone who chooses to work in telecommunications, at least on the technological side, must be very conversant with electrical engineering, computer engineering, and information technology.

New terms for new concepts and advancements are constantly being generated. For example, a webmaster is a person who structures, maintains, and expands a Web site. Also, many companies have created a chief information officer (CIO) position. These positions are evidence of the expanding telephone industry.

SOCIAL CONTEXT AND FUTURE PROSPECTS
Moore's law predicts continuing development in computer technology, which is likely to translate to technical advances in the field of telecommunications.

In terms of telecommunications, each site on Earth is the same distance from any other site. The cost of transmitting information—voice, data, or video—is extremely low. Communication among people, especially the young, is very frequently through electronic means.

All of this has changed the shape of many companies and institutions. One large computer manufacturer does not have a home office. Dozens of major corporations have offices far removed from their manufacturing facilities. Major universities are offering classes, lesson plans, and other student resources online. Some universities in Florida are set up to allow students to "attend" class even during emergencies, such as hurricanes, when the campus is closed.

Of course, the reliance on online communications does come with some danger. Supposedly secure information frequently turns out to be not secure, and important information (such as credit card numbers) ends up in the hands of unscrupulous people. Systems to avoid problems of this nature are being developed every day but so are means to bypass these new security measures. However, the conveniences outweigh the dangers for most people, who will continue to use a mixture of landlines, cell phones, smart phones, laptops, and desktops to communicate.

—*Robert E. Stoffels, MBA*

FURTHER READING

Bigelow, Stephen J., Joseph J. Carr, and Steve Winder. *Understanding Telephone Electronics.* 4th ed. Boston: Newnes, 2001.

Brooks, John. *Telephone: The First Hundred Years.* New York: Harper & Row, 1976.

Crystal, David. *Txtng: The Gr8 Db8.* New York: Oxford University Press, 2008.

Mercer, David. *The Telephone: The Life Story of a Technology.* Westport, Conn.: Greenwood Press, 2006.

Murphy, John. *The Telephone: Wiring America.* New York: Chelsea House, 2009.

Poole, Ian. *Cellular Communications Explained: From Basics to 3G.* Oxford, England: Newnes, 2006.

TELESCOPY

FIELDS OF STUDY

Electronics; electrical engineering; systems engineering; aerospace engineering; mechanical engineering; physics; optical physics; atomic physics; particle physics; astrophysics; astronomy; astrometry; cosmology; mathematics; quantitative analysis; statistics; image analysis; interferometry; communications; electromagnetism; spectroscopy; robotics; quantum electrodynamics; meteorology; environmental science; earth science; computer science; photography.

KEY TERMS AND CONCEPTS

- **adaptive optics:** Technique of increasing the resolving power of a telescope by correcting for errors caused by atmospheric effects such as turbulence.
- **angular resolution:** For any given telescope, the smallest angle between two closely situated objects that can be clearly distinguished as being separate. Measured in arcseconds and also known as a telescope's diffraction limit.

- **aperture:** Size of the opening in a telescope that collects light or other electromagnetic radiation.
- **arcsecond:** 1/3600 of a degree.
- **catadioptric Telescope:** Optical telescope that makes use of both reflective and refractive properties.
- **interference:** Process in which multiple electromagnetic waves are superimposed and combine to form a single wave. Depending on whether the waves have the same or different amplitudes, they can either reinforce or cancel each other.
- **interferometry:** Process in which the signals from multiple telescopes are combined to form a single image with a higher resolving power than that of any of the individual telescopes.
- **optical telescope:** Telescope that refracts or reflects light to create a clear, bright magnified image of a distant object.
- **radio telescope:** Telescope that bends or reflects radio waves to create a clear, bright magnified image of a distant object.
- **spectrograph:** Instrument designed to record or photograph various properties of electromagnetic

radiation; often used as an auxiliary device on astronomical telescopes.

- **very long baseline interferometry:** Technique in which astronomical observations are carried out by separate radio telescopes thousands of miles apart, then combined to form a single image.

SUMMARY

Telescopy is the science behind the creation and use of telescopes, devices that clearly render objects that are too dim or distant to be seen by the naked eye. The word "telescope" has its origins in two ancient Greek words meaning "to watch" and "from afar." By enabling people to study and accurately pinpoint stars such as Polaris and Sirius, planets such as Jupiter and Mars, and other structures in the sky, telescopy radically transformed humankind's ability to map locations and navigate from place to place. The use of telescopes in astronomy and cosmology has also profoundly deepened knowledge of the complexities of the universe and its origins.

DEFINITION AND BASIC PRINCIPLES
Telescopy is the field concerned with the development, improvement, and practical application of telescopes. A telescope is any device that enables the viewing or photographing of objects that are either too dim or too far away to be seen without aid. Although they are built to operate in many different ways, all telescopes make use of information gathered from various parts of the electromagnetic spectrum. The electromagnetic spectrum is made up of various forms of radiation—energy that comes from a particular source and travels through space or some other material in the form of a wave. Various types of radiation have different wavelengths and frequencies. The higher the frequency of a wave of radiation and the shorter its wavelength, the higher its energy as it travels.

The most common kinds of telescopes, known as refractors and reflectors, use systems of mirrors and lenses to gather and focus visible light. Both types of telescopes fall under the umbrella of optical telescopy. Radio telescopes gather information not from light but from radio waves, which have the longest wavelengths on the electromagnetic spectrum. X-ray telescopes and gamma-ray telescopes use the kinds of radiation with the shortest wavelengths. Specialized

microscopes detect other types of electromagnetic radiation, including ultraviolet light and infrared light, and are used for specific purposes. For example, infrared telescopes are similar to reflecting optical telescopes in construction, but they are designed to collect radiation that is invisible to the naked eye. One reason these telescopes are so useful is that infrared radiation is able to travel through thick clouds of dust and gas in a way that visible light cannot. Thus, infrared telescopes allow scientists to gain insight into the phenomena taking place within hidden regions of space. With these tools and others, modern telescopy is able to detect every region of the electromagnetic spectrum. Researchers use telescopy to create clear images of stars, planets, galaxies, and other celestial objects.

No matter how it is constructed, a telescope is designed to serve three basic functions. First, it should effectively collect large amounts of electromagnetic radiation and focus, or concentrate, that radiation. This makes objects that would otherwise appear very dim seem much brighter and easier to see. Second, it should resolve, or clearly distinguish between, the small details of an image. This makes objects that would otherwise appear blurry seem sharp and focused. Third, it should magnify the image it creates, so that objects at a distance appear larger. Although many people think of magnification as being the primary purpose of a telescope, it is in fact the least important function—if an image is not bright or clearly resolved, no matter how much it is magnified, it will not be useful.

BACKGROUND AND HISTORY
Although it is difficult to trace the invention of the telescope to a single individual, the first person to have tried to patent a basic telescope (in 1608) was a German-Dutch lensmaker named Hans Lippershey, who used a combination of two lenses separated by a tube to magnify objects by about three or four times. At about the same time, Italian mathematician Galileo built a very similar instrument using a combination of a concave (inwardly curving) lens and a convex (outwardly curving) lens. Galileo promptly showed his version of the telescope to the chief magistrate of Venice and became famous for using it to conduct astronomical observations showing, among other things, that the Earth revolved around the Sun, rather than the other way around. As a result of these

well-known activities, Galileo is often wrongly credited as the inventor of the telescope. While Galileo did not invent the telescope, he may have been the first to call the instrument a telescope.

The devices made by Lippershey and Galileo were both refractor telescopes, which rely primarily on lenses that gather and focus light. In the second half of the seventeenth century, English physicist Isaac Newton was among the early pioneers of the reflecting telescope, which relies primarily on curved mirrors that bend light.

Over the next three hundred years, optical telescopes underwent vast technological improvements. For example, achromatic lenses were invented to compensate for the errors in color caused by older lenses that failed to treat all the colors of visible light in the same way. Radio telescopes were first developed in the twentieth century, based on the discovery that faraway celestial bodies were constantly emitting faint amounts of radiation in the form of radio waves. This new form of telescopy was soon applied to both military radar operations and astronomical research.

Other technological advances have affected the field of telescopy. The development of photography enabled astronomers to create permanent still images of the celestial bodies they were observing and to use light-sensitive plates to gather, over long periods of time, even more light than could be collected by lenses or mirrors. Similarly, the invention of increasingly sensitive electronic devices for capturing light, such as the charge-coupled device (CCD), revolutionized modern telescopy. In addition, advances in computer technology allow astronomers and the military to constantly monitor selected portions of the sky using computers that alert human overseers if anything unusual is detected.

How It Works

Optical telescopes. Optical telescopes are designed to gather and focus light that radiates from distant objects. The two main types of optical telescopes are reflectors and refractors. Each type is based on a different principle derived from the physics of light. A third type, a catadioptric telescope, is a hybrid of reflectors and refractors.

Refraction is a phenomenon by which light is bent as it travels from a medium of a certain density to a medium of a different density (such as moving from air into glass). Basic refracting telescopes use a combination of two lenses to refract light. As light rays from a distant object, such as a star, approach a telescope, they travel in nearly perfectly parallel lines. The first lens, known as the primary, bends or refracts these parallel rays of light so that they converge on a single point. This creates an intermediary image of the object that is both bright and in focus. The purpose of the second lens, known as the secondary, is to take that bright, focused image and magnify it by spreading the light rays once more, enabling them to form a larger image on the retina of the eye. In the course of this process, the rays of light cross (light from the top of the object is bent downward and light from the bottom of the object is bent upward), so the image is upside down. Many refracting telescopes use another pair of lenses to render the image right side up.

Reflection telescopes are based on the principle that if light waves meet a surface that will not absorb them, they are redirected away from the surface at the same angle at which they were originally traveling. The angles at which light meets and are deflected from a surface are called the angles of incidence and reflection. Basic reflecting telescopes use a combination of two mirrors rather than lenses to reflect light. The mirrors are usually coated with a thin film of a shiny metal such as aluminum, which makes them more reflective. As light enters the tube of a simple reflective telescope, it is reflected off the primary mirror and travels back in the direction from which it came to form a bright, focused image just as in a refraction telescope. A secondary mirror in a reflection telescope functions similarly to the secondary lens in a refraction telescope, creating a magnified image focused comfortably on the retina.

Some optical telescopes use a combination of reflecting and refracting techniques; these are known as catadioptric telescopes. One of the most common types of catadioptric telescopes is the Schmidt-Cassegrain telescope, which takes its name from two scientists whose work informed its design. This type of telescope contains a deeply curved concave primary mirror at the back of the tube, which reflects light toward a convex secondary mirror at the front of the tube. Schmidt-Cassegrain telescopes also contain a corrective lens that helps counteract the optical aberrations caused by the mirrors, such as making points of light look like disks.

Radio telescopes. Rather than manipulating light,

radio telescopes collect and focus radio waves—the same kinds of electromagnetic radiation that are used to transmit radio, television, and cell phone signals. The reason radio telescopes are useful for purposes such as astronomical observation is that faraway celestial objects, including stars and quasars (incredibly bright, high-energy bodies that resemble stars), are constantly emitting radio waves. There are many different kinds of radio telescopes, but each is made up of the same fundamental parts. The first is a radio antenna, which often looks like a huge, curved television satellite dish. The greater the surface area of the antenna, the more sensitive it can be to the relatively weak radio waves being transmitted from cosmic sources and the fainter and more distant the objects it can detect. The second basic part of a radio telescope is a radiometer, also known as an amplifier. This instrument is placed at the central focusing point of the antenna, and its purpose is to receive and amplify the signal produced by the antenna, convert it to a lower frequency, and transmit it via cable to an output device that charts or displays the information collected by the telescope.

Many radio telescopes use a technique known as interferometry to increase their angular resolving power, which is relatively weak compared with that of optical telescopes. (The reason for the relative weakness is that the angular resolving power of a telescope is defined by the wavelength of the radiation it measures divided by the telescope's diameter. Radio waves, with much longer wavelengths than those of visible light, require telescopes with very large diameters to achieve the same angular resolution as optical telescopes.) An interferometer is a device that takes advantage of the interference phenomenon to electronically combine the signals from multiple telescopes and create a single image. For instance, the National Radio Astronomy Observatory's Very Large Array (VLA), a prominent radio astronomy observatory in New Mexico, has nearly thirty radio antennae each measuring 75 feet in diameter. By spacing the antennae far apart, the observatory has created an array that functions like a single telescope with a diameter as wide as the distance between the first and the last antenna. When the signals from the antennae are combined through interferometry, the array is able to resolve details in

Here is the primary mirror assembly of an infrared space telescope under construction. This is a segmented mirror and its coated with gold to reflect infrared light

the sky at a much greater power.

Spectrographs. Spectrographs are important auxiliary instruments that are often attached to optical telescopes. Their primary function is to split up the light collected by a telescope and separate it into its individual wavelengths, thereby creating a spectrum. Spectrographs can be extremely complicated, but their basic construction involves an entrance slit, two lenses, a prism, and a charge-coupled device (CCD). The entrance slit is designed to reduce the interference of any background light not coming from the particular star being observed. The first lens is designed to direct the rays of light coming from the star into the prism, which then breaks up the light into its different wavelengths. After the light exits the prism as a spectrum, it is directed by the second lens onto the CCD, which produces a readout of how much light of each wavelength is coming from the star. This information can then be used to analyze various important characteristics of the object under observation. For example, spectrographs can help astronomers learn the chemical composition of a star as well as its temperature and rotation speed.

APPLICATIONS AND PRODUCTS

Astronomical observations. Perhaps the most important scientific application of telescopy is its use in facilitating astronomical observations. Telescopes are the fundamental tools used by astronomers and astrophysicists to further their understanding of space, celestial objects, and the universe as a whole. Measurements produced with the aid of telescopes, for example, revealed the shape of the Milky Way galaxy and the location of Earth within it. Decades of careful observations through the Hooker optical telescope at the Mount Wilson Observatory in Pasadena, California, enabled astronomer Edwin Hubble to prove not only that the Galaxy is just one among many such systems in the universe but also that the universe itself is expanding as these galaxies move farther apart. Without the help of telescopes, no human eye would ever have laid sight on such astonishing phenomena as the icy rings that surround Saturn, the gigantic high-pressure storm on Jupiter known as the Great Red Spot, the craggy craters on the far side of the Moon, or the brilliant azure of the atmosphere around Neptune caused by the reflection of blue light by methane gas.

The highest angular resolution achievable by ground-based telescopes is limited by the fact that radiation of some wavelengths does not travel well through the Earth's atmosphere but is absorbed by water vapor and carbon dioxide as it travels. Ground-based telescopes are also affected by atmospheric turbulence—small, irregularly moving air currents—which can cause blurry images. However, because space telescopes orbit the Earth at a high altitude, they are not affected by these problems. Therefore, some astronomical observations can be carried out only by telescopes located above the atmosphere of the Earth.

The Hubble Space Telescope, which was launched into orbit by the National Aeronautics and Space Administration (NASA) in 1990, is the most advanced space-based telescope system ever conceived. It is a large optical visible-ultraviolet reflecting telescope (its primary mirror is nearly 8 feet in diameter) that travels around the Earth several times a day, collecting images of star systems, planets, comets, galaxies, and other celestial bodies. The Hubble Space Telescope is also equipped with a wide-field planetary camera that can record images of space at resolutions several times higher than any telescope based on Earth. In addition, the telescope has a faint-object camera designed to detect extremely dim celestial objects, a faint-object spectrograph that collects information about the chemical composition of these objects, and a high-resolution spectrograph that gathers ultraviolet light from very distant objects.

Although telescopes are not generally used for navigation purposes, one of their most important early contributions was in helping sailors and explorers pinpoint their exact locations on the seas by finding the position of known stars or planets in the sky. Astronomers still rely on telescopes to pinpoint the positions of celestial bodies. In doing so, they are able to create detailed and systematic surveys, or maps, of the sky. For example, since 2000, the Sloan Digital Sky Survey has been using a large reflecting telescope with charge-coupled devices to pinpoint the location of distant galaxies and quasars.

Military surveillance. For centuries, telescopes have provided army and navy surveillance teams with an invaluable tool by enabling military personnel to detect the movements of hostile forces from a distance. Initially, only ground-based telescopes were used, but in the late eighteenth century, it became possible to greatly increase the visual range of telescopes by placing them on board hot-air balloons. In World War I, European military forces used refractor telescopes mounted on airplanes to perform aerial surveillance, and in World War II, surveillance airplanes were equipped with sophisticated telescopes with powerful lenses and cameras that produced high-resolution images of military bases and enemy territories far below. Modern aerial surveillance techniques typically involve telescopes with extraordinarily good angular resolutions mounted on unmanned aircraft systems (remotely piloted aircraft) such as the United States Air Force's Global Hawk. In addition, ground-based telescopes are still used by

FASCINATING FACTS ABOUT TELESCOPY

- Between 1990 and 2009, the Hubble Space Telescope traveled around the Earth more than 100,000 times, covering a total distance of nearly 3 billion miles.

- One story about the invention of the telescope claims that German-Dutch lensmaker Hans Lippershey was inspired to develop an early version of a refractor telescope by watching two children playing with lenses in his shop.

- Scientists are working on making a miniature telescope that can be implanted behind the iris of the eye. It projects magnified images onto the retina, helping patients with the type of blindness known as macular degeneration see once again.

- Satellites equipped with telescopes that capture images of the surface of the Earth are used by governments to spy on other nations, as well as to observe the effects of fires, flooding, and deforestation.

- The first planet to have been discovered with the aid of the telescope was the large blue-green orb known as Uranus.

- It is possible to build a very simple refractor telescope using two magnifying glasses and a cardboard tube.

- Since light travels at a constant, finite speed, the further away an object is, the longer it takes for light from that object to reach the Earth. To look at a very distant star or galaxy is to see it as it was thousands, millions, or even billions of years ago. In that sense, telescopes are a little like time machines.

- In the late nineteenth century, many telescopes contained spider webs. The fine, strong silk produced by spiders made the perfect material for crosshairs, a pair of crossed lines in the optical viewfinder that help the viewer position and focus a telescope.

- Some telescopes exist just to help other telescopes do their job. Astronomical transit instruments are small refracting telescopes that precisely determine the locations of specific stars and planets, enabling larger telescopes to focus directly on them.

countries around the world to keep an eye on objects in the sky, such as enemy aircraft, missiles and other weapons, and satellite surveillance equipment belonging to other nations. For example, there are two optical surveillance sites in the United States, one in Hawaii and one in New Mexico, equipped with the latest in adaptive optics telescopes.

Increasingly, military surveillance is also conducted in space, with telescopes mounted on satellites. Satellite telescopes can move at vastly greater speeds than airplanes and have the advantage of being able to navigate to any region above the desired surveillance target without having to contend with national airspace boundaries. Although detailed information about the tools used by military surveillance units is closely guarded, it is generally thought that most satellite-based telescopes travel in low-Earth-orbit altitudes, about 62 to 310 miles above sea level. They probably conduct observations by collecting electromagnetic radiation with short wavelengths, such as infrared light and green light and are likely to be about 20 to 26 feet in diameter. Based on these parameters, experts have calculated that military satellite telescopes are most likely capable of distinguishing details that are less than an inch apart on the Earth—enough resolving power to read a newspaper headline.

Environmental applications. Governments and other organizations also use satellite-based telescopes in surveillance applications. One of the most common uses is to monitor changes in the environment. NASA's Earth-observing system satellite *Terra* carries multiple sophisticated telescopes onboard that are used to detect such phenomena as volcanic activity, emerging forest fires, and floods. *Terra* also provides scientists with images so that they may track the effects of climate change on the Earth's surface; for example, scientists track the melting of the ice sheets in the Arctic over time. Brazil, a country whose rain forests have been reduced by centuries of cattle ranching and other agricultural activity, uses satellite-based telescopes to closely monitor the extent of deforestation and to evaluate how well its efforts to preserve the rain forests are working.

Communications. One application of telescopy that is still largely in the research-and-development stage is its potential use in laser communications between deep space and the Earth. A laser is a device that uses excited atoms or molecules to emit a powerful beam of electromagnetic radiation in a single wavelength (called monochromatic light). The light produced by lasers is intense and directed, meaning that the light rays do not spread out very quickly. This makes lasers a useful tool for transmitting a secure information-carrying signal directly to a receiver in a specific location. NASA, for example, has been looking into using laser signals to transmit data (including photographs, radar images, and analyses of space dust) collected by space probes such as the *Cassini*, which is orbiting Saturn. However, collection of the laser signal on Earth would require a huge telescope.

Recreational applications. Telescopes are far from just a tool for scientists, military personnel, or government officials. Durable, lightweight general-purpose optical telescopes known as spotting scopes are used frequently in everyday life in a wide variety of recreational applications. These instruments have greater magnification and resolving powers than binoculars. Naturalists, for instance, use spotting scopes to identify plumage markings and observe the behavior of bird species at a distance, without alerting the birds as to the presence of humans. Airplane and train spotters use them to distinguish fine details in faraway vehicles. Long-range game hunters use spotting scopes to view their prey as they take aim, and sharpshooters use them to check on the position of their targets.

CAREERS AND COURSE WORK

Preparation for a career in any of the major fields related to telescopy, namely astronomy, astrophysics, meteorology, or military surveillance, should begin with a complete course of high school mathematics, up to and including precalculus. In addition, chemistry and physics are important subjects to cover in high school. Outside the academic environment, astronomy clubs or observatories are excellent places to gain practical experience using telescopes and to learn about the details of their operation. At the undergraduate level, a student interested in a career involving telescopy should work toward a bachelor of science degree with a concentration in a field such as physics, astronomy, mathematics, or computer science. No matter what major is chosen, additional course work in optics, electromagnetism, thermodynamics, mechanics, atomic physics, cosmology, statistics, and calculus provides essential background knowledge for further study—an important consideration, because almost all research positions in astronomy or related sciences, as well as any job involving the development of telescope technology itself, require the completion of a graduate degree, preferably a doctorate.

The typical career path for a student interested

in telescopy involves pursuing work as an astronomer either in an academic or a government-based observatory such as the National Radio Astronomy Observatory or the Mauna Kea Observatories at the University of Hawaii. Most astronomers who work at universities teach in addition to conducting research. Because jobs for practicing astronomers can be relatively scarce, students may benefit from taking one or more short-term positions such as a paid internship or postdoctoral fellowship to gain experience and contacts in the field before seeking a more permanent appointment.

Another career option is to approach telescopy from the point of view of engineering rather than scientific research. Electrical engineers and other technicians are essential members of the teams at astronomical observatories. These jobs, which involve repairing, upgrading, testing, and maximizing the efficiency of high-powered telescopes, generally do not require graduate degrees. A bachelor's degree in engineering or electronics and a strong background in mathematics and physics are sufficient qualifications to pursue this kind of telescopy career.

SOCIAL CONTEXT AND FUTURE PROSPECTS

Over the four hundred years of its existence, the telescope has enabled humankind to transcend not just visual limitations but also mental ones. The telescope has been the impetus for a flood of astonishingly deep revelations (and further questions) about the origin of matter itself, the place of the Earth within the universe, and the future of the universe.

For the entire length of recorded human history, humans have constructed stories and mythologies about how the world came into being. Telescopes have provided a way to approach that issue from a scientific point of view. With the help of ever larger telescopes such as the Giant Magellan Telescope in Chile (scheduled for completion in 2018), physicists hope to be able to see what the universe looked like just a few hundred million years after the big bang and thereby to gain an understanding of how the very first stars, planets, and galaxies were formed. For millennia, humankind believed that Earth held a central place in the universe. Telescopes have turned that worldview on its head by showing that, in fact, there may be billions of Earth-like planets in the Milky Way galaxy alone and countless more across the entire universe. Although scientists once believed that the universe was unchanging, they have come to know—because of telescopy—that it is dynamic and expanding.

In an age in which the devastating effects of environmental pollution and the growing impact of climate change dominate the headlines, telescopy—by giving humans a holistic view from afar of the beautiful, vulnerable planet they inhabit—has a particularly important role to play in inspiring those who live on Earth to preserve and protect it for future generations.

FURTHER READING

Andersen, Geoff. *The Telescope: Its History, Technology,* *and Future.* Princeton: Princeton UP, 2007. Print.

Burke, Bernard F., and Sir Francis Graham-Smith. *An Introduction to Radio Astronomy.* 3rd ed. New York: Cambridge UP, 2010. Print.

Koupelis, Theo. *In Quest of the Universe.* 6th ed. Sudbury: Jones, 2010. Print.

Pugh, Philip. *The Science and Art of Using Telescopes.* New York: Springer, 2009. Print.

Schilling, Govert, and Lars Lindberg Christensen. *Eyes on the Skies: Four Hundred Years of Telescopic Discovery.* Chichester: Wiley, 2009. Print.

Zirker, Jack B. *An Acre of Glass: A History and Forecast of the Telescope.* Baltimore: Johns Hopkins UP, 2006. Print.

TEMPERATURE MEASUREMENT

FIELDS OF STUDY

Chemistry; electricity; electronics; heat transfer; optics; physics; thermodynamics.

KEY TERMS AND CONCEPTS

- **absolute zero:** Lowest imaginable temperature, a state at which all motion ceases even at the subatomic level.
- **blackbody temperature:** Equilibrium temperature corresponding to the spectral distribution of radiation from a perfectly absorbing body.
- **kinetic temperature:** Temperature inferred from the amount and distribution of kinetic energy of random motion of the centers of mass of molecules.
- **kinetic theory:** Analytical framework relating gas properties to the statistical effects of individual molecules moving in random directions and colliding with each other.
- **Seebeck effect:** Generation of an electromotive force between two junctions of wires made of certain types of metals, when the junctions are at different temperatures.
- **Stefan-Boltzmann constant:** Constant of proportionality between the energy radiated per unit time per unit area of a blackbody and the fourth power of absolute temperature. The value of the constant in International System of Units is 5.6704×10^{-8} W/m$^2 \cdot$K^4.

- **thermal equilibrium:** State at which the net rate of heat transfer in any direction is zero.
- **vibrational temperature:** Value of temperature inferred from the amount of energy in the vibrational levels of a gas. This is equal to the kinetic temperature at equilibrium.

SUMMARY

Temperature is a measure of energy. Objects are said to be in thermal equilibrium when they have the same temperature. Accurate temperature measurements are essential to determine the rate of chemical reactions, the weather, and the health of a living creature, to cite a few examples. Measurements are said to be made in absolute units when they are referred to absolute zero or in relative units when referred to some other standard state. Methods of measuring temperature include those using mechanical probes, those using chemical paints, and nonintrusive optical and sonic methods. The temperature range of interest depends on the application. Temperatures in the Earth's atmosphere range from around 200 to 320 Kelvin (K), while that on Mercury varies from around 70 K on the dark side to more than 670 K on the side facing the Sun. The Sun's surface temperature is nearly 6,000 K.

DEFINITION AND BASIC PRINCIPLES

Temperature measurement, or thermometry, is the process of determining the quantitative value of the degree of heat, or the amount of sensible thermal energy contained in matter. The temperature of matter is a manifestation of the motion of atoms and molecules, either relative to each other or because of motions within themselves. Absolute zero is a temperature that creates a state in which molecules come to complete rest.

Temperature can be measured from heat transfer by conduction, convection, or radiation. Household thermometers use either the expansion of metals or other substances or the increase in resistance with temperature. Thermocouples measure the electromotive force generated by temperature difference. Pyrometers measure infrared radiation from a heat source. Spectroscopic thermometry compares the spectrum of radiation against a blackbody spectrum. Temperature-sensitive paints and liquid crystals change intensity of radiation in certain wavelengths with temperature.

Temperature is measured in degrees on either an absolute or a relative scale, with the value of a degree differing from one system of units to another. Two major systems of units in use are degrees Fahrenheit (F) and degrees Celsius (C). The absolute scales of temperature using these units are called the Rankine and the Kelvin scales respectively, both of which refer to the absolute zero of temperature. A change of 5 Kelvin or 5 degrees Celsius corresponds to a change of 9 degrees Rankine or 9 degrees Fahrenheit. Since the freezing point of water at normal pressures is set at 32 degrees Fahrenheit and 0 degrees Celsius, temperatures in degrees Celsius are converted to degrees Fahrenheit by multiplying by 1.8 and adding 32.

BACKGROUND AND HISTORY

Danish astronomer Ole Christensen Rømer used the expansion of red wine as the temperature indicator in a thermometer he created in the seventeenth century. He set the zero as the temperature of a salt-ice mixture, 7.5 as the freezing point of water, and 60 as its boiling point. German physicist Daniel Gabriel Fahrenheit invented a mercury thermometer and the Fahrenheit scale, where 0 and 100 were roughly the coldest and hottest temperatures encountered in the European winter and summer. Swedish astronomer Anders Celsius invented the inverted centigrade scale, which was converted to the current Celsius scale by Swedish botanist Carl Linnaeus in the nineteenth century so that 0 represents water's

freezing point and 100 its boiling point. British physicist William Thomson, also known as Lord Kelvin, devised the absolute scale in which absolute zero (0 K) corresponds to −273.15 C. Thus, 273.16 K is the triple point of water, defined as the temperature where all three states—ice, liquid water, and water vapor—can coexist. Scottish engineer William John Macquorn Rankine devised an absolute scale measured in units of degrees F, with absolute zero being -459.67 F.

HOW IT WORKS

Probe thermometers. Volume expansion thermometers use the expansion of liquids with rising temperature through a narrow tube. The expansion coefficient, defined as the increase in volume per unit volume per unit rise in temperature, is 0.00018 per Kelvin for mercury and 0.00109 per Kelvin for ethyl alcohol colored with dye. Calculating temperature from the actual random thermal motion velocity of every molecule, or the energy contained in a vibrational excitation of every molecule, is impractical. So temperature is measured indirectly in most applications. Different metals expand to different extents when their temperature rises. This difference is used to measure the bending of two strips of metal attached to one another in outdoor thermometers. Thermocouples use the Seebeck or thermoelectric effect discovered by German physicist Thomas Johann Seebeck, in which a voltage difference is produced between two junctions between wires of different metals, when the two junctions are at different temperatures. Some metal combinations produce a voltage that is linear with temperature, and these are used to produce thermocouple sensors. Resistance temperature detectors (RTDs) use the temperature sensitivity of the resistance of certain materials for measurement by including them in an electrical "bridge" circuit of interconnected resistances, such as a Wheatstone bridge, and measuring the voltage drop in a part of the circuit when the resistances are equalized. Probes using carbon resistors are the most stable at low temperatures.

Radiation thermometers. Infrared thermometers use the different rates of emission in the infrared part of the electromagnetic spectrum from objects, compared to the emissivity of a perfectly absorbing blackbody. The device is calibrated by comparing the radiation coming from the object against that from a reference object of known emissivity at a known temperature. The dependence on emissivity is eliminated using the two-color ratio infrared thermometer, which compares the emission at two different parts of the spectrum against the blackbody spectrum.

Ultrasonic thermometer. Ultrasonic temperature measurement considers the change in the frequency of sound of a given wavelength traveling through a medium. The frequency increases as the speed of sound increases proportional to the square root of temperature. This technique is difficult to implement with good spatial resolution since the changes in temperature over the entire path of the beam affect the frequency. However, it is suitable for techniques based on mapping multidimensional temperature fields.

Relation between temperature and internal energy. Nonintrusive laser-based methods generally use the fundamental properties of atoms and molecules. Temperature is a measure of the kinetic energy of the random motion of molecules and is hence proportional to the square of the speed of their random motion. Some energy also goes into the rotation of a molecule about axes passing through its center of mass and into the potential and kinetic energy of vibration about its center of mass. Other energy goes into the excitation of electrons to different energy levels. Quantum theory holds that each of these modes of energy storage occurs only in discrete steps called energy levels. Absorption and emission of energy, which may occur during or after collision with another atom, molecule, subatomic particle, or photon of energy, occurs with a transition by the molecule or atom from one energy level to another. The quantum of energy released or absorbed in such a transition is equal to the difference between the values of energy of the two levels involved in the transition. If the energy is released as radiation, the energy of the transition can be gauged from the frequency of the photon, using Planck's constant.

APPLICATIONS AND PRODUCTS

Temperature measurement is used in a multitude of applications. Volume expansion thermometers work in the range from about 250 to 475 K, but each thermometer is usually designed for a much narrower range for specific purposes. Examples include the measurement of human body temperature, the atmospheric temperature, and the temperature in ovens used for cooking.

Mercury thermometers were used widely to measure room temperature as well as the temperature of the human body. Human body temperature is usually 98.6 to 99 degrees Fahrenheit, and variations of a few degrees either way may indicate illness or hypothermia. Alcohol thermometers are used in weather sensing and in homes. The atmospheric temperature varies between roughly 200 K in the cold air above the polar regions and more than 330 K above the deserts. With regard to food preparation, temperatures vary from the 256 K of a refrigerator freezer to the 515 K of an oven.

Since the 1990's, expansion thermometers for the lower temperature ranges have mostly been replaced with solid-state devices generally using RTDs with liquid crystal digital (LCD) displays. Platinum film RTDs offer fast and stable response with a constant temperature coefficient of resistance (0.024 ohms per K), whereas coil sensors offer higher sensitivity. Pt100 (platinum-coiled) RTDs have a standard 100 ohms resistance at 273 K and a 0.385 ohms per degree Celsius sensitivity. RTDs are generally limited to temperatures well below 600 K. Thermocouples are made of various metal alloy pairs. Type K (chromel-alumel) thermocouples provide a sensitivity of 41 microvolts per Kelvin in the range −200 C to 1350 C. Type E (chromel-constantan) generates 68 microvolts per Kelvin and is suited to cryogenic temperatures. Types B, R, and S are platinum or platinum-rhodium thermocouples offering around 10 microvolts per Kelvin with high stability and temperature ranges up to 2,030 Kelvin. In using thermocouples, the chemical reaction or catalytic effect of the thermocouple with the environment is important. Some metals, such as tungsten, may get oxidized, while others, such as platinum, catalyze reactions containing hydrogen.

Light waves scattered from different parts of an object interfere with each other. In the case of Rayleigh scattering, where the object such as a molecule is much smaller than the wavelength, the scattering intensity is independent of the direction in which it is scattered. It is much lower in intensity than the Mie scattering that occurs when the particle size is comparable to wavelength. Rayleigh scattering measurement can work only where the Mie scattering from relatively huge dust particles does not drown it out. Rayleigh scattering thermometry uses the fact that in a flow where the static pressure is mostly constant, such as a subsonic jet flame, the amount of light scattered by molecules from an incident laser beam is proportional to the density and inversely proportional to temperature if the composition of the gas does not vary. This simple technique is used to obtain high frequency response; however, there is often a substantial error in assuming that the

A typical Celsius thermometer measures a winter day temperature of –17°C

gas composition is constant. Rayleigh scattering occurs at the same wavelength as the incident radiation, but the frequency band of the scattered spectrum is broadened by the Doppler shift of molecules because of their speed relative to the observer. Where the Doppler shift is due to the random motion, such as in the stagnant or slow-moving air of the atmosphere, the Doppler broadening gives the temperature. This fact is used in high spectral resolution lidar (HSRL) measurements of atmospheric conditions used from ground-based weather stations as well as from satellites.

More sophisticated nonlinear laser-based diagnostic techniques are used to measure temperature in several research applications. The technique of Raman spectroscopy uses the phenomenon of Raman scattering that occurs when molecules change from one state of vibration to another after being excited into a higher vibrational level by laser energy. A variant of this technique is called coherent anti-Stokes Raman spectroscopy (CARS), where the signature of the emission from molecules in given narrow spectral bands is compared against databases of the emission signatures of various gases to determine the composition and temperature of a given gas mixture at high temperatures. This technique is used in developing burners for jet engines and rocket engines. In laser-induced fluorescence, a strong laser pulse excites the molecules in a given small "interrogation volume." Within a few microseconds, the excited molecules release energy as photons at a different frequency than that of the laser pulse energy. A highly sensitive camera relates the spectral distribution to the temperature.

Infrared thermometers are used where nonintrusive sensors are needed or where other electromagnetic fields might interfere with thermocouples. Infrared thermometry is used to capture the change in temperature due to the change in skin friction between laminar and turbulent regions of the flow in the boundary layer over the skin of the space shuttle to determine if there are regions where the flow has separated because of missing tiles or protuberances. Since the space shuttle returned to flight after the *Columbia* disaster in 2003, such diagnostics are performed using ground-based telescope cameras or from chase planes during the upper portions of the liftoff and during orbital passes in order to alert mission controllers if repairs are needed in orbit.

FASCINATING FACTS ABOUT TEMPERATURE MEASUREMENT

- The color of stars depends on the shape of their emission spectra and the temperature near their surface. Blue stars, which emit more high-energy radiation, are hotter than white stars, which are hotter than yellow or red ones.
- Glowworms emit visible radiation through a phenomenon called phosphorescence, which occurs with very efficient conversion from thermal energy to emitted radiation, keeping the temperature low.
- The reaction rate in a flammable vapor rises exponentially with temperature per the Arrhenius rate expression. A 3 percent change in temperature can lead to a doubling of reaction rate, leading to a sharp increase in temperature and heat release, causing a flame or explosion.
- When very high frequency sound passes through a liquid, tiny bubbles are formed by the expansion regions of the sound pressure wave, which then collapse suddenly when the compression regions arrive. This collapse generates supersonic shock waves where the local temperature rises to very high levels, accompanied by the emission of visible light. This is called sonoluminescence. Some researchers believe that sonoluminescence can be used to generate conditions suitable for nuclear fusion reactions.
- Researchers have reached very close to the absolute zero of temperature by forcing molecules to come to a stop using the pressure exerted by laser beams.

Thermal equilibrium occurs very rapidly when molecules are allowed to collide with each other for some time. At equilibrium, the energy tied up in each of the various modes of energy storage has a specific value. In this case, the temperature measured from the instantaneous value of vibrational energy in the gas will give the same answer as that measured from the rotational or translational energy. However, in some situations, where the pressure and temperature change extremely rapidly or energy is added to a specific mode of storage, the temperature measured from the value of one type of energy storage may be very different from that measured from the translational energy. Examples are strong shocks in supersonic flows and rapid expansion through the nozzle of a gas dynamic laser. A common example is that of a fluorescent light bulb, where energy is added to the

electronic excitation of molecules without much exciting of the translational, vibrational, or rotational levels. The gas glows as if it were at a temperature of 10,000 K (the electronic temperature is 10,000 K) whereas the translational temperature may be only 400 K.

Temperature-sensitive paints of various kinds are used to capture the temperature distribution over a surface using either the changes in color (wavelength of emitted radiation) or the intensity at a particular wavelength. Typically these use fluorescent materials. Devices range from simple stick-on tapes to expensive paints that are used on models in high-speed wind tunnel tests.

CAREERS AND COURSE WORK

A wide variety of physical phenomena can be used to measure temperature. The basic principles come from college-level physics, including optics, and from chemistry. Heat transfer is important, as are material science and electronics, with digital signal processing and image processing used to analyze data and control theory used to turn the measurements into feedback systems to control the temperature of a process or to compensate for the imperfections of an inexpensive measuring system and obtain results as good as those from a much more complicated system. Although few careers can be imagined where temperature measurement is the primary job, several careers involve skills in temperature measurement. Medical doctors, nurses, mothers, and cooks use temperature measurement, as do heating, ventilation, and air-conditioning technicians, nuclear and chemical plant technicians, weather forecasters, astronomers, combustion engineers, and researchers. Preparation for temperature measurement includes courses in physics, chemistry, electrical engineering, and heat transfer.

SOCIAL CONTEXT AND FUTURE PROSPECTS

Temperature measurement will continue to be an area that demands curiosity and scientific thinking across many disciplines. Probes, gauges, and nonintrusive instruments, as well as paints, are all still in heavy use, and each technique appears to be suitable to some particular problem. This diversity appears to be increasing rather than decreasing, so that the field of temperature measurement may be expected to expand and broaden. As energy technologies move away from fossil fuel combustion, where temperatures rarely exceed 2,200 K, into systems based on hydrogen combustion and nuclear fission and fusion, techniques that can measure much higher temperatures may be expected to become much more common.

—*Narayanan M. Komerath, PhD*

FURTHER READING

Baker, Dean H., E. A. Ryder, and N. H. Baker. *Temperature Measurement in Engineering, Vol. 1.* New York: John Wiley & Sons, 1963. Essential reading and an excellent reference, guiding the user through the various issues.

Benedict, Robert P. *Fundamentals of Temperature, Pressure, and Flow Measurements.* 3d ed. New York: John Wiley & Sons, 1984. Used all over the world in chemical and mechanical engineering courses, this book is a valuable resource for engineers and technicians preparing for professional certification. Contains practical information on instrument interfacing as well as measurement techniques.

Chang, Hasok. *Inventing Temperature: Measurement and Scientific Progress.* New York: Oxford University Press, 2007. Discusses the history of temperature measurement and how researchers managed to establish the reliability and accuracy of the various methods.

Childs, P. R. N., J. R. Greenwood, and C. A. Long. "Review of Temperature Measurement." *Review of Scientific Instruments* 71, no. 8 (2000): 2959–2978. Succinct discussion of the various techniques used in temperature measurement in various media. Discusses measurement criteria and calibration techniques and provides a guide to select techniques for specific applications.

Michalski, L., et al. *Temperature Measurement.* 2d ed. New York: John Wiley & Sons, 2001. Covers basic temperature-measurement techniques at a level suitable for high school students. A section on fuzzy-logic techniques used in thermostats is a novel addition.

Richmond, J. C. "Relation of Emittance to Other Optical Properties." *Journal of Research of the National Bureau of Standards* 67C, no. 3 (1963): 217–226. Regarded as a pioneering piece of work in developing optical measurement techniques for temperature based on the emittance of various materials and media.

TERRESTRIAL MAGNETISM

FIELDS OF STUDY

Geophysics; Electromagnetism

KEY TERMS AND CONCEPTS

- **center of mass:** the mass weighted center of an object or system of objects.
- **current:** the rate at which an electric charge, usually in the form of electrons, moves through a wire or other conductive material.
- **electromagnetic field:** a physical field consisting of a combined electric field (generated by stationary electric charges) and magnetic field (generated by moving electric charges) that affects the behavior of charged objects in its vicinity.
- **friction:** the force created by resistance to relative motion between solid surfaces.
- **paleomagnetism:** the magnetic signature remaining in rocks that indicates the strength and direction of Earth's magnetic field in the past.
- **sea-floor spreading:** the gradual movement of sea-floor bedrock away from an undersea rift where magma continuously wells up from the mantle and solidifies due to tectonic activity in Earth's crust.

SUMMARY

Terrestrial magnetism has been known and used for navigation purposes for thousands of years, though it was not understood in any scientific sense until 1600, when William Gilbert determined that Earth is in fact a huge magnet. Sea-floor spreading and continental drift demonstrated the dynamic nature of Earth's interior, and subsequent research indicates that 90 percent of Earth's magnetism is generated by the outer core due to the induced electrical current caused by the rotation of the solid inner core of the planet. Terrestrial magnetism reverses its polarity about every 500,000 years, possibly as the result of the Earth colliding with large meteors.

HOW IT WORKS

Earth's magnetic field. People have understood for hundreds of years that this planet possesses a magnetic field, although it was not appreciated as such for much of that time. A naturally occurring form of iron ore called 'lodestone' was used to make a guidance device, or compass, during the Han Dynasty in China as much as 2,300 years ago. Similar devices may have been used by the Olmecs, in Central America, as much as a thousand years earlier, although this is not yet conclusively proven. European navigators were using lodestone, and later, dry magnetic compasses more than fifteen hundred years later. Yet, in all that time, the principle that made a compass function remained unrealized. It wasn't until 1600 that William Gilbert (1540 – 1603) systematically investigated the phenomenon of magnetism, and determined that Earth is itself a source of magnetism. Still, a working model of the cause of magnetism was not determined until the development of the model of atomic structure based on quantum mechanics early in the 20th century. Although Gauss and Maxwell had been able to describe the behavior of an electromagnetic field more than a century earlier, that work did not posit a viable cause for magnetism. In the intervening years, it came to be understood that magnetic fields are caused by the movement of electrons, as an electric current, through a conductive material. Earth's magnetic field must therefore be generated by the same cause somehow, within Earth's interior, and although the intensity of the magnetic field is weak, it is nevertheless global in scale. It was also believed to be constant over time, with only minor variations due to the local composition of Earth's crustal matter. In the mid-1960s, this concept was proven false.

Continental drift and magnetic pole reversals. In 1915, Alfred Wegener (18890 – 1930) published a theory to explain the observation that the coastlines of the continents can be fitted together like the pieces of a planet-sized jigsaw puzzle. According to that theory, the continents, particularly Africa and South America, were at one time joined together in a single 'supercontinent' and through the phenomena of sea-floor spreading and continental drift have moved away from each other to their present locations. The theory was not accepted at the time, partly because Wegener was a meteorologist and not a geologist, but primarily because the scientific establishment maintained a strong adherence to the 'steady state' principle. Earth was deemed to be in an unchanging condition, rather than as a dynamic structure. Wegener's theory did not gain universal acceptance until the mid-1960s, as a direct result of mapping the magnetic signature, or paleomagnetism, of sea-floor bedrock along the Pacific coast of California. The mapping revealed a pattern of bands of reversed magnetic polarities that could be matched perfectly to the pattern obtained on the opposite side of a seismic faultline. Similar research in the Atlantic Ocean revealed the existence of the Mid-Atlantic Ridge, about midway between Africa and the American continents. The mechanism of sea-floor spreading was observed directly at that location. Taken together, continental drift and the polarity reversals in paleomagnetism demonstrate that the interior of Earth is a dynamic system, rather than a steady-state system. This dynamism is therefore responsible for the generation of Earth's magnetic field. Given that magnetism is generated by the movement of electrons through a conductive medium, the internal structure of Earth must therefore function as a dynamo.

Earth's interior as a dynamo. The interior structure of Earth cannot be examined directly for several reasons. Firstly, the crustal layer is tens of kilometers thick (1 kilometer = 0.625 mile). Secondly, the material lying below the crust is too hot for any material to withstand. But most importantly, the interior is dynamic and in constant motion, however slow. Essentially all information about the interior of Earth is obtained from analysis of seismic waves. This has indicated that the interior of Earth has a layered structure below the crust. There are basically three layers below the crust: the mantle, the outer core, and the inner core. Sea-floor bedrock is essentially the solidified surface material of the mantle. The friction caused by rotation of the core about the center of mass of the planet is sufficient to keep the mantle layer in a fluid state. The core itself consists mostly of iron, with some nickel and a few minor elements. The inner and outer cores are separated by a 'transition layer' that is several hundred kilometers in thickness. The outer core is in a fluid state, and is believed to produce more than 90% of the terrestrial magnetism. The inner core is believed to be a single enormous crystal of solid iron, at a temperature well above its melting point but kept in a solid state by pressure. Seismic analysis indicates that the inner core rotates at a slightly faster rate than the outer core, and so generates the electrical current in the outer core that produces the terrestrial magnetism. The cause and mechanism of magnetic polarity reversals are still unknown, though it is known that they occur approximately every half-million years.

WHY TERRESTRIAL MAGNETISM IS IMPORTANT

Earth is under a constant state of bombardment from charged particles emitted by the Sun, and essentially the entire universe. These 'cosmic rays' are capable of doing great harm to living systems as they are able to pass through matter and damage the DNA molecules that define living organisms on Earth. The atmosphere is incapable of absorbing them and preventing them from reaching the surface. But because they are electrically charged, they interact with magnetic fields such as the terrestrial magnetism. As they encounter Earth's magnetic field, the charged particles become trapped by the interaction and do not reach the surface of the planet. Individual charged particles enter into a spiral motion about the particular particular magnetic 'lines of force' that they have matched with, and move along those lines until their kinetic energy has dissipated. They then are able to enter the atmosphere as 'cosmic dust'. It is estimated that hundreds of tonnes of cosmic dust add to the mass of Earth each year. While losing their energy, however, the charged particles emit energy as light, the Northern and Southern Lights (the Aurora Borealis and the Aurora Australis).

—*Richard M. Renneboog MSc*

FURTHER READING

Merrill, Ronald T., McElhinny, Michael W. and McFadeen, Phillip L., The Magnetic Field of the Earth. Paleomagnetism, the Core, and the Deep Mantle San Diego, CA: Academic Press, 1998

FASCINATING FACTS ABOUT TERRESTRIAL MAGNETISM

- The study of past magnetic field of the Earth is known as paleomagnetism.
- Humans have used compasses for direction finding since the 11th century AD and for navigation since the 12th century.
- The geomagnetic field changes on time scales from milliseconds to millions of years. Shorter time scales mostly arise from currents in the ionosphere (ionospheric dynamo region) and magnetosphere, and some changes can be traced to geomagnetic storms or daily variations in currents. Changes over time scales of a year or more mostly reflect changes in the Earth's interior, particularly the iron-rich core.
- Studies of lava flows on Steens Mountain, Oregon, indicate that the magnetic field could have shifted at a rate of up to 6 degrees per day at some time in Earth's history, which significantly challenges the popular understanding of how the Earth's magnetic field works.
- The Earth's magnetic field is believed to be generated by electric currents in the conductive material of its core, created by convection currents due to heat escaping from the core. However the process is complex, and computer models that reproduce some of its features have only been developed in the last few decades.

Tauxe, Lisa, Essentials of Paleomagnetism Berkeley, CA: University of California Press, 2010.

Cox, A. and Hrt, Robert Brian, Plate Tectonics. How It Works Palo Alto, CA: Blackwell Scientific Publications, 2008.

Merrill, Ronald T., Our Magnetic Earth. The Science of Geomagnetism Chicago, IL: University of Chicago Press, 2010.

Lowrie, William A Student's Guide to Geophysical Equations New York, NY: Cambridge University Press, 2011.

Rikitake, Tsuneji Electromagnetism and the Earth's Interior New York, NY: Elsevier Publishing, 1966.

TIME MEASUREMENT

FIELDS OF STUDY

Horology; metrology; mathematics; physics; particle physics; classical mechanics; quantum mechanics; geophysics; geology; nuclear physics; astronomy; cosmology; navigation; psychology; physiology; neurobiology; telecommunications; computer science; mechanical engineering; physical engineering.

KEY TERMS AND CONCEPTS

- **accuracy:** Degree to which the pulse or signal measured by a particular clock conforms to a reference pulse or signal, such as coordinated universal time. The opposite of accuracy is drift.
- **atomic clock:** Clock that either contains an internal atomic oscillator or uses an external atomic oscillator as a reference.
- **atomic oscillator:** Oscillator that uses changes in the energy level of atoms to determine the intervals between the signals it produces.
- **coordinated universal time (UTC):** Twenty-four-hour international reference time scale, based on the atomic definition of a second that is the standard timekeeping reference for the world; measures the hour, minute, and second at the prime meridian (0 degrees longitude) located near Greenwich, England; and is sometimes called Greenwich Mean Time, in casual, nonscientific usage.
- **frequency standard:** Oscillator whose signals are used by other devices as a reference for measuring time intervals.
- **oscillator:** Device that generates a constant, periodic signal that varies in magnitude around a central point.
- **quality factor (Q):** Number of vibrations an oscillator can make before it requires an additional burst of energy to continue working.
- **resonance:** Vibration or oscillation of an atom as it emits electromagnetic radiation. Any atom of a given element will resonate at precisely the same frequency, or rate.
- **second:** Time required for a cesium-133 atom to oscillate 9,192,631,770 times; can be further subdivided into units such as nanoseconds (one billionth of a second), picoseconds (one trillionth of a second), and femtoseconds (one quadrillionth of a second).
- **stability:** Degree to which a particular clock can produce a repeating pulse or signal at exactly the same frequency over a given period of time.

SUMMARY

Time measurement is the science and practice of counting the repetitions of recurring phenomena and subdividing the intervals between each repetition into smaller units that are capable of being measured by a variety of devices, including mechanical and atomic clocks. Time measurement is an important part of everyday life. It enables people to schedule activities, measure distances and speed, and navigate from place to place. The accurate measurement of immensely small units of time is essential to a huge number of fields of basic and applied science, including physics, computing, and medicine.

DEFINITION AND BASIC PRINCIPLES

The science of time measurement involves devising methods to count the number of iterations of phenomena that repeat themselves at regular intervals. For example, the rotation of Earth on its axis, the shifts in the phases of the moon, and the changing of the seasons are all familiar units of time based on observable changes in material objects.

Time itself is a slippery concept that has no simple scientific definition, though physicists consider it one of the four fundamental dimensions of the universe, along with length, width, and depth. Although scientists have not managed to satisfactorily define time, they have developed extraordinarily accurate ways of quantifying its passage, so that individual events can be referred to on a consistent scale. The goal of time measurement is to supply information about time of day, or the instant at which an event takes place; time interval, or the duration between two events; and frequency, or the rate at which a repeated event takes place.

One of the basic principles of time measurement is the notion that time is not a physical constant but rather can shrink or expand in response to other forces. According to Albert Einstein's general theory of relativity, strong gravitational fields can stretch the interval between the signals given by a clock. This phenomenon can affect both observations of the simultaneity of events and measurements of the duration of events. Clocks located in space, such as the ones on the International Space Station or on satellites orbiting the Earth, must be adjusted to correct for the effects of relativity. The special theory of relativity shows that two observers moving in relation to one another will observe duration and simultaneity differently; the differences between their observations are negligible, however, except when the speeds involved approach the speed of light.

BACKGROUND AND HISTORY

Attempts to mark the passage of days, months, and seasons using the movements of the sun, moon, and stars long predate attempts to measure shorter periods of time. Calendars, in other words, are older than clocks. Early clocks such as sundials and obelisks made use of the changing length of shadows over the course of the day to mark intervals that roughly corresponded to modern hours. Other primitive means of keeping track of the passage of time include water clocks, which were designed to drip at a relatively constant rate; hourglasses, which used the same principle but incorporated sand instead of water; and clocks that used the burning of candles or incense to measure time.

In the fourteenth century, mechanical (machine-powered) clocks began to appear. Some relied on a device known as an escapement, which controlled an unwinding spring that rotated a series of gears that, in turn, caused the hands of the clock to tick forward steadily. Others used a pendulum (a weight on a string), the back-and-forth motion of which served as a natural oscillator. Over the next several hundred years, inventors and engineers labored to reduce inaccuracies in timekeeping by compensating for such factors as friction, changes in temperature, and interference from other moving parts within the clocks.

The twentieth century saw two major advances in time measurement: the development of quartz clocks, which used electric circuits to generate constant electrical vibrations in quartz crystals; and the invention of atomic clocks, which take advantage of the natural resonance frequency of atoms to create an immensely stable oscillator. The atomic clock

has become the standard tool for modern scientific timekeeping.

HOW IT WORKS

Physical time scales span a dazzlingly wide range. The smallest scale that physicists are able to work with mathematically is Planck time, a single unit of which (about 10^{-43} second) is defined as the length of time it takes for a photon traveling at the speed of light to cross a distance of one Planck length (about 10^{-33} centimeter). In contrast, the cosmological time scale, on which events such as the beginning of the universe and the formation of stars and planets is marked, consists of periods as vast as tens of billions of years. Only a portion of the time scales that exist within this range can be accurately measured using the technology described in this essay.

Household clocks. The majority of clocks, watches, and small electronic circuits used in everyday life are built on oscillators that use quartz crystals to generate a consistent pulse. Quartz crystals, whether natural or synthetic, exhibit a property known as piezo-electricity, which means that the crystals expand and contract—that is, vibrate—when they receive an electric force. The combination of a quartz crystal with a battery that applies and then reverses an electric voltage produces a regular oscillation, the frequency of which depends on the size of the crystal and the form into which it is cut. A quartz oscillator found in an ordinary household clock will probably have a quality factor (Q) of about 10^4 to 10^6 and be accurate to about a few seconds per month—perfectly adequate for everyday use.

Scientific clocks. Scientific and industrial purposes demand atomic clocks with far more accuracy and stability than that of household clocks or watches. Atomic clocks make use of the fact that all atoms are capable of existing at a number of different discrete (noncontinuous) levels of energy. As an atom jumps back and forth between a higher and a lower energy level, it resonates at a particular frequency, and this frequency is exactly the same for every atom of a given element. For example, cesium-133 atoms (one of the two types of atoms most commonly used in atomic clocks; rubidium is the other) resonate between two particular energy levels at 9,192,631,770 cycles per second. In fact, the time it takes for this number of oscillations of a cesium-133 atom to take place serves as the definition of a second in the

International System of Units (SI). This stable vibration can be thought of as paralleling the swinging back and forth of the pendulum in an old-fashioned mechanical clock.

There are many different forms of atomic clocks, but the basic mechanics involved are relatively consistent. A laser beam is shone into a cloud of atoms, tossing them high into the air. The frequency of light at which the laser shines is adjusted until the vast majority of the atoms experience a change in energy state. This process tunes the laser's frequency to match the resonance frequency of the atoms themselves, and it can then be used to mark the passage of time with great accuracy and reliability. Most atomic clocks also make use of a technique known as laser cooling, in which the movement of atoms is slowed by dropping the temperature within the clock to something very close to absolute zero. This lengthens the period of time during which the atoms can be properly observed.

Standards in time measurement. Within the field of time measurement, standards refer to devices or signals that serve as benchmarks for particular measurements, such as time intervals or frequencies. Standards allow other clocks to be precisely adjusted so that they all keep the same time and can be

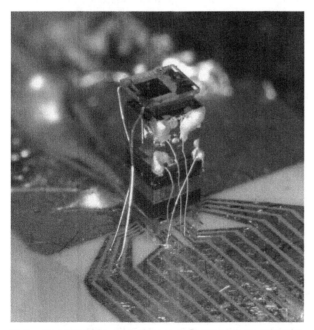

Chip-scale atomic clocks, such as this one unveiled in 2004, are expected to greatly improve GPS location.

recalibrated according to the same measure if they should happen to gain or lose time. For example, the National Institute of Standards and Technology's (NIST) cesium fountain atomic clock NIST-F1, located in Boulder, Colorado, was introduced in 2000 as the standard atomic cesium clock on which all other clocks in the United States were to be calibrated. NIST-F1 is one of hundreds of highly accurate atomic clocks around the world that together define the interval between each second in UTC, the official global time of day. NIST-F1 will gain or lose a single second only once every hundred million years or so. In 2014, NIST introduced the NIST-F2, a more advanced cesium fountain atomic clock that operates alongside NIST-F1 and is approximately three times more accurate, gaining or losing a second once every three hundred million years or so.

APPLICATIONS AND PRODUCTS

Navigation. The ability to measure time accurately and to precisely synchronize more than one clock is essential for many forms of navigation, including the US-based Global Positioning System (GPS, or NAVSTAR) used in many cars, boats, airplanes, missile-guidance systems, cell phones, watches, computer clocks, network clocks, and other devices. To determine a location, a GPS receiver calculates the time it takes for signals from four separate satellites to reach it, multiplies that time by the speed of a radio wave, and uses this figure to determine the exact distance between the receiver and each satellite. Then the receiver creates four imaginary spheres, each of which has a radius that is the same length as the distance between the receiver and one of the satellites. The latitude, longitude, and altitude at the point where the spheres cut across each other is the latitude, longitude, and altitude of the device. This process is known as triangulation. The United States is not the only country with a global navigation satellite system; Russia, China, and the European Union are among those working on developing similar applications.

Even the tiniest inaccuracies in the measurement of the time interval between when the signal is sent and when it is received can cause substantial errors—differences of several feet or more—in the triangulation calculation that determines the user's physical location. As a result, GPS receivers rely heavily on standard atomic frequency references. Each satellite has a small handful of local atomic clocks onboard (multiple clocks are used to provide a system of redundancies, a way to ensure reliability and accuracy). Each clock is calibrated to a master atomic clock measuring UTC and managed by the US Naval Observatory. In general, each clock on a GPS receiver, no matter where on Earth it is located, will be no more than 500 nanoseconds to 1 millisecond out of sync with UTC. Time measurement not only facilitates the movement of cars, boats, and airplanes on the earth and sea and in the sky but also is fundamental to interplanetary space travel. Every National Aeronautics and Space Administration (NASA) spacecraft is equipped with standards-referenced atomic clocks that enable accurate and reliable navigation.

Telecommunications. Every telecommunications system in use—including standard telephone systems, wireless communication networks, radio and television broadcasting systems, cable networks, satellite telephones, and the Internet—relies on the ability to synchronize exactly the signals sent between transmitters and receivers. For example, wireless telephones transmit information to each other in the form of small chunks of data called packets. The signal between the transmitter and receiver must be perfectly synchronized to be able to identify where each packet begins and ends. In addition, the transmitter and receiver must process data at the same speed so as to prevent lags or data loss. Finally, for multiple telephones to send data using the same frequency channel, each telephone is assigned a particular time slot. Individual transmitters send their signals in series, never interfering with one another's data even though they share the same medium of transmission. (Radios broadcast their signals using exactly the same system, which is known as time division multiple access, or TDMA.) Because cell phones move from location to location as their users travel, they also require a precise time-measurement system, usually pegged to the atomic clocks on GPS satellites. These systems help make minute changes to the timing of the cell phone signals to compensate for the fact that the distance between the telephone and the base station is changing.

Electronics and computers. Every computer contains a small, built-in quartz-based oscillator that serves as an internal marker of time intervals for the machine. The central processing unit (CPU) uses this clock to determine the intervals at which its

microprocessor is directed to complete instructions, as well as for purposes such as scheduling automatic processes and time-stamping events. It is also important for computers that are sending and receiving information over a network or the Internet to be highly synchronized with each other for data to be transmitted accurately. Since the piezoelectric qualities of quartz crystals change with temperature, however, computer clocks tend to drift as the machinery inside them heats up with use. For this reason, most computer networks are equipped with a Network Time Protocol (NTP) server that uses an atomic frequency standard such as the signal produced by NIST-F1 to synchronize the internal clocks of the computers to a more accurate and stable time signal.

CAREERS AND COURSE WORK

A student contemplating a career as an academic or industry-based researcher in time measurement or metrology (the study of measurement itself) or planning to enter a related field such as astronomy, geophysics, or engineering should begin by pursuing a rigorous course of study in science, technology, and mathematics, preferably leading to a bachelor of science degree in physics. Of prime importance within the field of physics are subjects such as atomic structure, special and general relativity, electromagnetics, cosmology, the physics of the solar system, and quantum mechanics. In mathematics, geometry and trigonometry are especially instructive for the science of time measurement.

An interest in time measurement might lead down a number of career paths. For instance, one might enter the field of horology and become a designer of timekeeping apparatuses such as clocks, watches, timers, and marine chronometers. One might become a staff researcher investigating the physics of time at a government or university laboratory or observatory such as the National Institute of Standards and Technology, the US Naval Observatory, or the NMi Van Swinden Laboratorium in the Netherlands. Alternatively, one might pursue a job developing, maintaining, or repairing time measurement instrumentation for scientific or industrial purposes, in which case, course work in mechanical and electrical engineering would also be required.

SOCIAL CONTEXT AND FUTURE PROSPECTS

Scientists are working on a new generation of optical

FASCINATING FACTS ABOUT TIME MEASUREMENT

- As the Moon slowly spirals away from the Earth, its gravitational pull on the planet lessens, making each day last a tiny fraction longer and, over time, causing each year to contain fewer days.

- In 1761, a horologist named John Harrison won $20,000 from the British government for building a chronometer that remained accurate and stable even on the turbulent waters of the sea, thus enabling sailors to calculate longitude to within one-third of a degree.

- The body's internal molecular timekeeper, known as the circadian clock, regulates cycles of sleep and wakefulness. When it is out of sync, it can cause disorders such as jet lag and depression.

- Conventional time-of-day measurements are based on the changing position of the Sun in the sky, but shifts in the positions of stars can also be used to define a kind of time known as sidereal time. One sidereal day lasts a few minutes less than one solar day.

- Highly accurate time-of-day information, based on atomic frequency standards, is available through National Institute of Science and Technology radio broadcasts twenty-four hours a day.

- Station WWVB in Colorado broadcasts on 60 kilohertz; stations WWV and WWVH are located in Colorado and Hawaii, respectively, and broadcast on five frequencies from 2.5 to 20 megahertz.

- A leap second is added to UTC every so often (usually less than once per year) to bring it into sync with the more stable astronomical time, which is based not on atomic clocks but on the rate at which the Earth rotates.

Time-measurement research has the potential to revolutionize a host of fields. Such devices could allow military navigators to bypass jammed GPS signals, enable e-mails to be encrypted at a far safer and more complex level than ever before, or be used in inexpensive portable medical imaging devices to scan patients' hearts, brains, and other organs at high resolution even outside hospitals.

—*M. Lee, MA*

FURTHER READING

Audoin, Claude, and Bernard Guinot. *The Measurement of Time: Time, Frequency, and the Atomic Clock.* Trans. Stephen Lyle. New York: Cambridge UP, 2001. Print.

Dunlap, Jay C., Jennifer J. Loros, and Patricia J. DeCoursey, eds. *Chronobiology: Biological Timekeeping.* Sunderland: Sinauer, 2004. Print.

Eidson, John C. *Measurement, Control, and Communication Using IEEE 1588.* London: Springer, 2006. Print.

Jones, Tony. *Splitting the Second: The Story of Atomic Time.* Philadelphia: IOP, 2000. Print.

Kaplan, Elliot D., and Christopher J. Hegarty, eds. *Understanding GPS: Principles and Applications.* 2nd ed. Boston: Artech, 2006. Print.

Lombardi, Michael A. "Radio Controlled Wristwatches." *Horological Journal* 148.5 (2006): 187–92. Print.

Lombardi, Michael A., Thomas P. Heavner, and Steven R. Jefferts. "NIST Primary Frequency Standards and the Realization of the SI Second." *Measure* 2.4 (2007): 74–89. Print.

Moskvitch, Katia. "Atomic Clocks to Become Even More Accurate." *LiveScience.* Purch, 14 June 2013. Web. 25 Feb. 2015.

Ost, Laura. "A New Era for Atomic Clocks." *NIST Physical Measurement Laboratory.* NIST, 4 Feb. 2014. Web. 25 Feb. 2015.

atomic clocks that use atoms such as strontium-87 and mercury to produce oscillations at optical frequencies about 100,000 times faster than conventional cesium clocks, which oscillate at microwave frequencies between 0.3 and 300 gigahertz. These clocks are still experimental but hold the potential to be about 100 times more accurate than NIST-F1—in other words, only gaining or losing a second once every ten billion years—and capable of measuring time at higher resolutions (subdividing it into smaller units).

TRANSISTOR TECHNOLOGIES

FIELDS OF STUDY

Physics; chemistry; solid-state physics; semiconductor physics; semiconductor devices; microelectronics; materials science; engineering; electronics.

KEY TERMS AND CONCEPTS

- **bipolar transistor:** Transistor that has both electron and positive holes.
- **chip:** Integrated electronic circuit (for example, a microprocessor) containing very large numbers of transistors and other circuit elements interconnected with very thin, very narrow, metallic interconnections on a piece of silicon.
- **field effect transistor:** Transistor with a source, gate, and drain with an electric field from the gate controlling the flow of electrons from the source to drain in a semiconductor; or the flow of holes from source to drain in a different transistor.
- **hole:** A defect in a crystal caused by an electron leaving its position in one of the crystal's bonds; equivalent to a positively charged particle.
- **junction transistor:** Has two p-n junctions electrically producing potential barriers between emitter and base and between base and collector. Abbreviations include NPN and PNP.
- **n-type:** Electric conduction using electrons, which have a negative charge.
- **p-type:** Electric conduction with positive holes.
- **single crystal silicon:** Nearly perfect silicon starting material from which transistors are manufactured.
- **transistor:** Three-terminal amplifier and on/off switch made from a semiconductor, most frequently silicon but sometimes germanium, germanium-silicon, gallium arsenide, or silicon carbide.
- **unipolar transistor:** Uses electrons only or holes only; field effect transistor.

SUMMARY

Transistor technologies require the expertise of an interdisciplinary group that includes physicists, chemists, and engineers. The semiconductor transistor was invented by physicists, and may be the single most important invention of the twentieth century. As an electronic amplifier and on/off electronic switch, the transistor has revolutionized electronics and by so doing has revolutionized many fields of industry including banking, manufacturing, automobiles, aircraft, military systems, space exploration, medical instrumentation, household appliances, and communication of all sorts. Transistors have been so greatly miniaturized in size that semiconductor chips can contain up to one billion transistors on a silicon chip no larger than a thumbnail.

DEFINITION AND BASIC PRINCIPLES

NPN and PNP are symbols for the two most important junction transistors. There are three legs or wires leading from each of these transistors, one each from the three separate regions. NPN means that a p-type region with positive holes separates two n-type regions with electrons carrying the electric current. Thus, there are two junctions. They both have natural potential barriers that prevent electrons in the n region from entering the p region and prevent holes from entering the n region until external voltage is applied. When external voltage is applied the transistor functions. The p region is called the base of the transistor. When negative voltage is applied between the first n region and the base, the first n region becomes an emitter of electrons into the base region. Positive voltage is applied to the other n region, between it and the base. That n region and its junction become the collector of the transistor, because it collects the electrons emitted from the emitter. Its potential barrier becomes even larger and an excellent collector of electrons. An electrical signal applied to the emitter modulates the current from emitter to the collector. The input signal is amplified in the collector circuit, which is a high-resistance circuit.

A common current is flowing from a low-voltage input to a high-voltage output. There is both voltage and power gain. There could be current gain if the input signal source were placed in the base leg and the emitter were grounded. The p region has mobile holes, and electrons can be captured by the positive holes in what is called electron-hole recombination. This is detrimental. Recombination is minimized by creating very thin regions and by using semiconductor material of very high quality. The

PNP transistor is identical to the NPN, save for the changing places of the electrons and holes.

BACKGROUND AND HISTORY

Walter H. Brattain and John Bardeen, physicists at Bell Laboratories, discovered the transistor effect in December, 1947. They had attempted to produce a semiconductor surface field effect amplifier but failed. In the process of studying surfaces to find the cause of failure, they probed the germanium surfaces with a sharp metal point. They discovered that an electrical signal into one point contact produced an amplified signal in the circuit of a second point contact close to the first. The point contact transistor was born. The name "transistor" was chosen by John R. Pierce, a fellow physicist at Bell Labs. Pierce reasoned that because the new device had current input and voltage output, it should be viewed as a transresistance. Since other devices (conductor, resistor, varistor, thermistor) ended in "-or" the name of the new device should end in "-or," hence the name transistor. Brattain's and Bardeen's supervisor was physicist William B. Shockley, who was not included among the inventors. Shockley went on to invent what would become an even better transistor: a p-n junction transistor. The properties of this transistor could be much more readily designed and controlled than those of a surface-point contact device because Shockley's transistor depended on the "inside" bulk properties of the semiconductor, which were much more easily controlled than the surface device. Some point-contact transistors were manufactured by the early industry, but Shockley's junction transistor dominated the growing industry into the 1960's. As a result, all three, Bardeen, Brattain, and Shockley, received the Nobel Prize in Physics in 1956 for research in semiconductors and the discovery of the transistor effect.

HOW IT WORKS

Producing junction transistors. Junction transistors can be produced during the original crystal growing process for the silicon (or germanium) crystal by adding known n-type and p-type impurities to the molten semiconductor as the solid crystal is slowly pulled from the melt.

An improved method for producing junction transistors uses metal-alloying techniques. For example, aluminum as a dopant (desired impurity) makes silicon p-type. With a small n-type chip of silicon, a small ball of aluminum can be placed onto each of the two surfaces of the silicon chip and the temperature raised to a level high enough to melt the aluminum but not melt the silicon. However, the molten aluminum does alloy with a small surface region of the silicon, and a PNP transistor is produced.

One of the very best ways for producing NPN and PNP junction transistors is by diffusing dopants into the silicon. Diffusion of impurity atoms into silicon is a slow process and must be done at high temperatures without melting the silicon. Double diffusions and triple diffusions have been found useful. One double diffusion method might start with n-type silicon into which a p-type impurity is diffused (such as boron). This produces one junction. An n-type impurity (such as phosphorus) can be diffused on top of the p-type diffusion.

The n diffusion is made to go less deep than the p diffusion, but it must be of higher concentration to overcome and change part of the p region back into n, thereby producing an NPN transistor.

Another way to produce NPN and PNP junction transistors is by means of ion implantation. In this case, atoms of a dopant are ionized and then shot into the surface region of the silicon with very high energy driving the impurity into the silicon to whatever depth is desired in the design. Because this process is very energetic it damages the crystal structure of the silicon. The silicon must then be annealed at some higher temperature in order to remove the damage but at the same time leave the impurity in place. All p-n junction transistors are called bipolar transistors because both n-type and p-type regions are used in each transistor.

Surface field effect transistors. Surface field effect transistors are called unipolar transistors because in any particular transistor only n-type regions and electrons are used, or only p-type regions and holes are used. An external electric field is applied to the surface of the silicon and modulates the electrical conductivity. In place of an emitter-base-collector of the junction transistors, the field effect transistor has a source, gate, and drain. The gate is a thin metal region separated from the silicon by a thin layer of insulating dielectric silicon dioxide. The gate applies the electric field. The source and drain are both n-type if the current flowing through the device is electrons. The source and drain are both p-type if the current

What Bardeen, Brattain, and Shockley invented in 1947 was the first point-contact transistor and were jointly awarded the 1956 Nobel Prize in Physics "for their researches on semiconductors and their discovery of the transistor effect"

and consumer products. Transistors are used in all modern electronics. Many different forms of the junction transistor were created at many companies during the first fifteen years of the industry. Engineers at Fairchild Semiconductor learned how to produce reliable surface field effect transistors, and the entire industry moved in that direction in the late 1960's and continues to do so. The industry had also started working with the semiconductor silicon, a more desirable choice than germanium. Many companies used both germanium and silicon producing both diodes and transistors. National Semiconductor was formed in 1959 as the first company to use only silicon and to produce only transistors and its own early integrated circuits. Integrated circuits were originally invented by Robert Noyce at Fairchild, and later Intel, and by Jack Kilby at Texas Instruments in the late 1950's. The integrated circuit (the so-called chip) sealed the future success of the electronics revolution. Intel has become a major force in the semiconductor industry.

flowing is holes. Surface field effect transistors have become the dominant type of transistor used in integrated circuits, which can contain up to one billion transistors plus resistors, capacitors, and the very thinnest of deposited connection wires made from aluminum, copper, or gold. The field effect transistors are simpler to produce than junction transistors and have many favorable electrical characteristics. The names of various field effect transistors go by the abbreviations MOS (metal-oxide semiconductor), PMOS (p-type metal-oxide semiconductor), NMOS (n-type metal-oxide semiconductor), CMOS (complementary metal-oxide semiconductor—uses both p-type unipolar and n-type unipolar).

APPLICATIONS AND PRODUCTS

One of the great advantages of the transistor, either in the form of a single transistor or many transistors on a chip, is that it is generic. The transistor or chip can be used in any electronic product requiring amplifiers and on-off switches with small size, ruggedness, low power loss, desirable frequency characteristics, and high reliability. The transistor can be made larger for higher power usage, smaller for high-frequency usage, extremely high reliability for military and space usage, and inexpensive for household

It would not be an exaggeration to say that every manufacturing company, every service organization, most homes in the United States, all advanced defense and military equipment, all aircraft, all research, all modern automobiles, and all modern communications use electronic equipment of some kind. In each of thousands upon thousands of applications, the electronic circuitry is different, but the generic transistor in single units or in massive numbers of transistors on single chips are common to all. These transistors and transistor chips are "hidden" in small packages, which in turn are hidden in the larger containers that house the particular electronic equipment. Intel has tried to change the hidden perspective by having computer manufacturers place a label displaying "Intel inside" on the outside covers of personal computers (PCs). Perhaps that helped PC users to realize that what was going on inside the personal computer was not magic but the result of one of the greatest inventions of the twentieth century—the transistor.

Here are a few examples of where transistors and

transistor-packed chips are used.

Aircraft: The entire control instrumentation and "fly by wire" flight systems.

Automobiles: Engine controls and accessory controls

Business and personal communication: Communications equipment, cell phones, all of the electronics of the Internet, smart phones.

Financial institutions: Large computers.

Household appliances: Microwave oven controls, washing machines, timer controls, televisions.

Manufacturing: Automatic controls of many products.

Medical instrumentation: All types of body scanners.

Military systems: Intercontinental ballistic missiles' computers, inertial guidance systems, and telemetering systems and unmanned drone aircraft.

Personal computing: Personal computers: desktops, laptops, notepads.

Research and meteorology: Supercomputers.

Space exploration: Everything for both manned and unmanned flights.

Software is needed to tie all of the electronics together and to tell the electronics what to do, and the software will be different from one application to another. The software and the electronics hardware are married. The transistors are operating at lightning speed as on/off switches and as amplifiers of smaller electric signals, but switches must be told what to do. All forms of computers have central processing units (semiconductor chips), semiconductor memories, and analogue-to-digital and digital-to-analogue semiconductor chips for interfacing with the real world.

CAREERS AND COURSE WORK

The semiconductor transistor chip industry probably has the largest percentage of employees with doctoral degrees in all of industry. Highly educated and trained people are also needed in the related industries that supply equipment and materials to the semiconductor industry. A semiconductor fabrication facility (fab plant) is so advanced with environmental requirements and manufacturing equipment that final capital costs can total several billion dollars for just one facility. If nanoscience and nanotechnology are successful in developing mass production processes for nanotransistors on chips, the requirements for extremely clean facilities and for very advanced

manufacturing equipment will be huge.

The undergraduate will find many colleges and universities with courses in solid-state physics, microelectronics, and materials science and engineering. These courses are necessary for those expecting to enter the modern field of semiconductors. Following

---

FASCINATING FACTS ABOUT TRANSISTOR TECHNOLOGIES

- The January 2011 issue of *IEEE Spectrum* listed its choices of the top eleven technologies of the decade. All of these eleven applications, except for light-emitting diodes (LEDs), require the use of advanced semiconductor transistors, most in the form of chips with huge numbers of very small transistors. LEDs, however, are produced from semiconductors (of a different kind than silicon or germanium).

- Flexible AC transmission allows more efficient use of huge quantities of electric power to different destinations and allows better incorporation of electricity from renewable—and unreliable—resources such as solar and wind.

- Smart phones contain an abundance of chips and are able to perform an abundance of functions—in addition to talking to another person, from composing e-mail and watching videos to reading the morning paper.

- The telephones and videophones that make up Voice-over Internet Protocol (VoIP) owe their success to chips.

- LEDs use chips of two special semiconductors that give off light and are the future of lighting.

- Multi-core central processing units (CPUs) use several transistor chips in a single package to give more powerful performance.

- Semiconductor chips factor heavily into the success of cloud computing, which allows many computers to use the power of servers located elsewhere.

- The Mars Rovers, Spirit and Opportunity, contain semiconductor chips that receive commands from Earth's Jet Propulsion Lab in Pasadena, California.

- The 2009 Nobel Prize in Physics went to the inventors of the semiconductor charge-coupled device (CCD). The CCD is used in digital cameras and is responsible for recording the image to be photographed.

- Class-D Audio, the very best audio for car stereos, television sets, and personal computers, uses the most advanced transistor chips.

that preparation, it is important to gain experience in one of the nation's many nanoscience and nanotechnology centers, each of which is usually part of a university. University of California, Berkeley, Los Angeles, and Santa Barbara; The Johns Hopkins University; State University of New York at Albany; Purdue University; Carnegie Mellon University; Rice University; Illinois Institute of Technology; Cornell University; Harvard University; University of Massachusetts at Amherst; Columbia University; and the University of Pennsylvania are a few of the many schools that have nanoscience centers.

SOCIAL CONTEXT AND FUTURE PROSPECTS

It would not be an overstatement to claim that virtually every manufacturing company, every service organization, and most homes in the United States use electronic equipment containing transistor-loaded semiconductor chips. Research in nanoscience and nanotechnology will continue to create a culture and a society even more heavily dependent on the availability of advanced electronics.

—Edward N. Clarke, MS, PhD

FURTHER READING

Bondyopadhyay, Probir K., Pallab K. Chatterjee, and Utpal K. Chakrabarti, eds. *Proceedings of the IEEE: Special Issue on the Fiftieth Anniversary of the Transistor* 86, no. 1 (January, 1998). Comprehensive collection of papers concerning the invention of the transistor; Moore's law; patents, letters, and notes by William Shockley, John Bardeen, and Walter Brattain; and life in the early days of Silicon Valley.

Callister, William D., Jr., and David G. Rethwisch. *Materials Science and Engineering: An Introduction.* 8th ed. Hoboken, N.J.: John Wiley & Sons, 2010. A popular book for undergraduate engineering students, this text covers many aspects of materials, and has well written sections on semiconductors and semiconductor devices.

Muller, Richard S., and Theodore I. Kamins, with Mansun Chan. *Device Electronics for Integrated Circuits.* 3d ed. New York: John Wiley & Sons, 2003. Very clear description of materials, manufacture, and many types of transistors.

Reid, T. R. *The Chip: How Two Americans Invented the Microchip and Launched a Revolution.* Rev. ed. New York: Simon & Schuster, 2001. The very readable story about Jack Kilby of Texas Instruments and Robert Noyce of Fairchild Semiconductor and later Intel, who independently came up with different approaches to the invention of the microchip.

Suplee, Curt. *Physics in the Twentieth Century.* Edited by Judy R. Franz and John S. Rigden. New York: Harry N. Abrams, in association with the American Physical Society and the American Institute of Physics, 1999. A well-written and illustrated book prepared on the occasion of the centennial of the American Physical Society; the story of the transistor is found in Chapter 4.

Sze, S. M., and Kwok K. Ng. *Physics of Semiconductor Devices.* 3d ed. Hoboken, N.J.: John Wiley & Sons, 2007. Written by physicists at Bell Labs, this textbook is often used in college courses.

TRANSURANIC ELEMENTS

FIELDS OF STUDY

Atomic physics; nuclear physics; particle physics

SUMMARY

Stable atoms have a specific number of protons and neutrons in their nuclei. Variations from this in atoms that have more or fewer neutrons result in instability and those atoms tend to break down through a radioactive decay process. All elements above atomic number 92, uranium, are unstable and radioactive, and do not occur naturally. These elements require very high energies for their synthesis. Currently 26 transuranic elements have been identified, and research to find even heavier elements continues, with the possibility of a range of superheavy elements being found that are stable.

KEY TERMS AND CONCEPTS

- **element:** a form of matter consisting only of atoms of the same atomic number
- **half-life:** the length of time required for one-half

of a given amuount of material tp decompose or be consumed through a continuous decay process.

- **isotope:** an atom of a specific element that contains the usual number of protons in its nucleus but a different number of neutrons.
- **radioactive decay:** the loss of particles from the nucleus of an unstable atom in the form of ionizing radiation.
- **standard model:** a generally accepted unified framework of particle physics that explains electromagnetism, the weak interaction, and the strong interaction as products of interactions between different types of elementary particles.
- **man-made element:** an element or isotope that does not occur naturally but is synthesized in high-energy patricle accelerators by bombarding other elements with streams of nuclear particles.

DEFINITIONS AND BASIC PRINCIPLES

In the standard model, atoms consist of a very small, dense nucleus composed of protons and neutrons that is surrounded by a very large, and very diffuse,

FASCINATING FACTS ABOUT TRANSURANIC ELEMENTS

- Any atoms of the transuranic elements, if they ever were present at the Earth's formation, have long since decayed.
- Essentially all the transuranium elements have been discovered at four laboratories: Lawrence Berkeley National Laboratory in the United States (elements 93–101, 106, and joint credit for 102–105), the Joint Institute for Nuclear Research in Russia (elements 114–118, and joint credit for 102–105), the GSI Helmholtz Centre for Heavy Ion Research in Germany (elements 107–112), and RIKEN in Japan (element 113).
- As of 2008, weapons-grade plutonium cost around $4,000/gram, and californium cost $60,000,000/gram.
- Einsteinium is the heaviest transuranic element that has ever been produced in macroscopic quantities.
- The nuclear fusion of californium-249 and carbon-12 creates rutherfordium-261. These elements are created in quantities on the atomic scale and no method of mass creation has been found.

cloud of electrons. The identity of any particular atom as an element is defined by the number of protons that are contained within its nucleus. The number of neutrons present in a particular nucleus determines the mass properties of that atom and determine its identity as an isotope of the corresponding element. Stable isotopes are characterized by having only the appropriate number of neutrons, and atoms that vary from this by having either more or fewer neutrons than this optimum number are generally unstable and break down spontaneously by some form of radioactive decay process. All such decay processes are characterized by a half-life for the particular isotope, which can be as short as just a few milliseconds to several millions of years. It is a very curious fact that 99% of the known matter in the universe is simple hydrogen. The Sun, for example, is comprised essentially of just hydrogen, yet it accounts for 98% of the matter in the Solar System. Only 92 elements are known to exist naturally in the universe, and all have been formed from hydrogen through nuclear fusion processes, In fusion, two separate nuclei combine to form one larger nucleus. The principal reaction occurring in the Sun, for example, is the fusion of two hydrogen atoms into one helium atom (the process is somewhat more complex than this, as two more hydrogen atoms must be condensed into neutrons in order to complete the helium nucleus). In contrast, fission processes involve the splitting of a nucleus into two smaller nuclei. This is the fundamental process of radioactivity. Both fusion and fission are typically accompanied by the release of large amounts of energy

TRANSURANIC ELEMENTS

The heaviest naturally-occurring stable element is uranium, element number 92 on the periodic table. All known elements having a greater number of protons in the nucleus are unstable, and radioactive. They are termed 'transuranic elements' because they are positioned beyond uranium in the periodic table and accordingly have atomic numbers beyond that of uranium. Other proper terms for these elements are the 'actinides' and 'lanthanides' for their chemical similarities to the elements actinium and lanthanum. Transuranic elements are normally known as man-made elements, although their synthesis in stellar processes elsewhere is not precluded. Being unstable, however, their natural synthesis would normally be followed by a decay process that ultimately produces smaller stable atoms. The first transuranic elements were discovered through research geared to the discovery of materials suitable for use in nuclear weapons. Uranium used as the fuel of nuclear reactors is known to undergo a complex series of nuclear reactions that result in the formation of a number of transuranic elements from atomic numbers 93 to 98. Tests of the residues from the detonation of the first hydrogen bomb – a fusion-based device as opposed to the fission-based atomic bomb – revealed the presence of two additional transuranic elements, numbers 99 and 100. As of 2016, the number of known elements has increased and currently is officially at 118, bringing the number of known transuranic elements to twenty-six. Research continues in an effort to synthesize even heavier elements.

THE ISLAND OF STABILITY

There are three levels of chemistry possible in the standard model. Normal chemistry, the reactions between atoms and molecules, takes place at the level of the outermost electrons. Nuclear chemistry takes place when nuclei undergo fission or fusion processes. The third level is sub-nuclear chemistry, and involves the interactions of subnuclear particles such as quarks to produce the subatomic particles that make up atoms. This is the field that gives rise to the transuranic elements beyond element 100. These elements are produced by high-energy collisions between nuclei that have been accelerated to speeds that approach the speed of light. In some cases, the element produced has been identified by the traces left by no more than four or five atoms. The study of the elements produced by such collisions, in regard to the number of protons and neutrons in the atom and its relative stability, suggests that a range of superheavy elements exists beyond what is currently known, in which the atoms are particularly stable. As yet, no such elements have been synthesized, but the search continues.

—*Richard M. Renneboog MSc*

FURTHER READING

Krebs, Robert E. *The History and Use of Our Earth's Chemical Elements. A Reference Guide* 2nd ed., Westport, CT: Greenwood Press, 2006

Morss, Lester R., Edelstein, Norman M. and Fuger, Jean, eds. *The Chemistry of the Actinide and Transactinide Elements, Vol. 1*, 3rd ed. New York, NY: Springer, 2008.

Angelo, Joseph A. *Nuclear Technology* Westport, CT: Greenwood Press, 2004.

Seaborg, Glenn T. and Loveland, Walter D. *The Elements Beyond Uranium* Hoboken, NJ: John Wiley & Sons, 1990.

Vértes, Attila, Nagy, Sándor, Klencsár, Zoltán, Lovas, Rezs G. and Rösch, Frank, eds. *Handbook of Nuclear Chemistry* 2nd ed. New York, NY: Springer, 2011.

WEIGHT AND MASS MEASUREMENT

FIELDS OF STUDY

Physics; mathematics; chemistry; materials science; transportation; economics; aeronautics; logistics; electronics

KEY TERMS AND CONCEPTS

- **austenitic steel:** A form of stainless steel containing a high degree of austenite, noted for its resistance to corrosion.
- **drift:** A slow, steady movement of a scale reading from its initial determination, caused by equilibration of the scale components.
- **force restoration transducer:** A device that uses electronic feedback to adjust the current in an electromagnet until it balances the force put against it by a mass, permitting a highly accurate determination of weight.
- **hysteresis:** Variation of readings in a scale according to whether the weight being measured is increasing or decreasing.

SUMMARY

Mass is an intrinsic property of matter and remains constant, regardless of the force of gravity, whereas an object's weight varies according to variation in the force of gravity acting upon it. Both mass and weight are relative quantities, rather than absolute. They are fundamental to quantifying human economic and social transactions. The development of standardized systems of weights and measures has been necessary for advancements in science, technology, and commerce. The recognized standard of weights and measures in use today is the metric system.

DEFINITION AND BASIC PRINCIPLES

It is not possible to state the absolute mass or weight of any material object. At the most fundamental level, one can only state that a hydrogen atom has the mass of a single proton and a single electron; it is impossible to know the absolute masses of those subatomic particles. This basic relationship, however, provides the means whereby other atoms can be ascribed their corresponding masses and the effects of accumulated mass become observable.

Weight is described as the product of an object's mass and the gravitational force acting on that mass. The hydrogen atom that on Earth has the mass of one proton plus one electron must by definition have the same mass wherever it is located. The gravitational force experienced by that atom will be different in other places, and at each location it will have a somewhat different weight while maintaining the same relative mass to other atoms. The same logic applies to larger quantities of matter because of this atomic relationship.

When the gravitational force experienced by different masses is common to their weights, it is expedient, though not absolutely correct, to use the terms "mass" and "weight" interchangeably in common usage. To quantify these properties in a meaningful and useful way, it also is necessary to define some standard quantity of each to use as a reference to which other masses and weights can be compared. Devices calibrated to correspond to the standard value can be used to measure other material quantities. International standardization of weights and measures works to assist trade fairness and facilitates the comprehension of scientific, technological, and theoretical work in disparate locations.

BACKGROUND AND HISTORY

In prehistoric times, measured amounts of materials most likely consisted of handfuls and other amounts, with little or no consideration beyond the equivalency of perceived value. There is an intrinsic conflict in this sort of measurement, as weight versus volume. At some point it would be realized that a container of large seeds had a sensibly different weight than the same container filled with small seeds, arrowheads, fish, or whatever might be traded at the time. The concept of fair exchange on an equivalent weight or equivalent volume basis, or the development of some means of relating weight and volume, would be the beginning of standardization of weights and measures.

History does not record the beginning of standardization, although ancient records reveal that certain standard measurements were used in ancient times, such as cubits for distance and talents and carats for weights. Each measurement related to some basic definition, which was often merely the distance between two body parts and varied from person to person, place to place. Through the sixteenth, seventeenth, and eighteenth centuries, the Imperial or British system of measurement was the accepted standard around the world; it continues to be used today.

In 1790, a commission of the French Academy of Science developed a standardized system of weights and measures that defined unitary weights, volumes, and distances on quantities that were deemed unchangeable and that related to each other in some basic manner. This metric system has since developed into the Système international d'unités (International System of Units; abbreviated SI) and is the universal standard recognized and used around the world.

How It Works

All weight and mass measurements are relative and cannot be known as absolute values. The definition of standard weights and measures provides a reference framework in which units can be treated as though they are absolute measurements.

For the measurement of mass, the metric standard unit is the kilogram (kg), corresponding to 1,000 grams (g). The gram, in turn, is defined as the mass of 1 cubic centimeter (cm^3) of pure water at its temperature of maximum density. Under those conditions, therefore, 1 kg of water occupies a volume of 1,000 cm^3, called 1 cubic decimeter (dm^3) or 1 liter (l). This is a readily understandable definition, but in order to have an unchanging standard to which other weights may be readily compared a solid object is required.

The international prototype standard kilogram is housed at the International Bureau of Weights and Measures in Sèvres, France. It is a small cylinder exactly 39 millimeters (mm) in diameter and length, and the alloy is of 90 percent platinum and 10 percent iridium. To protect the constant value of its mass and weight, it is stored under inert gas in a specially constructed triple bell jar to protect it from exposure that could affect the mass of the object. A minimum of sixty countries around the world hold registered precise copies of the prototype standard kilogram, similarly protected. These have traditionally been taken by special secure courier to France, through a series of special precautionary procedures, to be compared with the true prototype kilogram for calibration. Following this procedure the copies are returned to the appropriate national government, where they then serve as the legal standard of measurement until their next calibration.

Given the security measures in place for air transportation and international entry, the method has associated with it a very high degree of fallibility. For all but the most legally sensitive of applications, less stringent standards are normally used. It is sufficient for essentially all purposes that a

A balance-type weighing scale is unaffected by the strength of gravity.

less precise comparison be used. This permits devices used to determine mass and weight to be designed with preset comparisons and to be arbitrarily adjustable to accepted weight standards. This is typically achieved through the use of standard weight sets that—though neither as stable nor as stringently controlled as the prototype standard—are nevertheless uniformly consistent and conform to the tolerances stated by governing regulations. Typically, these are standardized stainless steel objects or specific electromagnetic conditions known to exert a specific force.

APPLICATIONS AND PRODUCTS

Because mass is an intrinsic property of matter inextricably tied to weight, its proper measurement is essential in many different fields. The applications and products that have been developed to accommodate the various fields all use the same basic principle of mass comparison, as indeed they must.

Scales and balances. The measurement of mass or weight (these terms, though often used interchangeably, are technically different) is carried out using scales and balances. A balance functions by directly comparing the mass to be determined with masses that are known. The two are deemed equal when a properly constructed balancing scale indicates equality.

The simplest device, called a two-pan beam balance, consists of a crossbeam from which two pans are suspended to hold the masses to be compared. The crossbeam is balanced on a knife-edge fulcrum and attached to a device that indicates deflection from the neutrality point. The unknown mass is placed on one of the pans and known masses are placed on the other pan until the scale balances at the neutrality point again. At that point, the unknown mass on one pan is equal to the sum of the known masses on the other pan.

Variations of the pan balance have been constructed and used in mass capacities ranging from grams to several tonnes (1 tonne [t], or metric ton = 1,000 kg). The smallest of these have been those used primarily in scientific research that involves small quantities of matter, while the largest variations have been used for weighing large quantities of goods for transport.

With the application of electronics, beam balances are replaced by scale balances (commonly just called scales) that do not directly compare masses. Instead, a scale uses the beam balance principle to quantify an unknown mass by comparing it with a calibrated scale against a known force. The amount of deflection of the scale applied by the unknown mass against the known force determines the quantity of the unknown mass. This tends to simplify the procedure by eliminating the requirement to maintain a set of known comparative masses (a set of standard weights) that are always subject to physical damage, chemical attack, and loss. However, it also requires that the functional force and electromagnetic feedback systems be routinely tested and calibrated.

Science and engineering. Mass and weight are central concepts of science and engineering, especially the disciplines of chemistry and physics. In chemistry, mass is central to all chemical reactions and processes. All masses are relative, beginning with the mass of the hydrogen atom. This provides the basis for assigning a corresponding mass to every other type of atom that can be identified.

Because of this relative relationship, it is possible to relate the number of atoms or molecules in a sample to a weight that can be measured. The quantity of hydrogen atoms required to make up a bulk mass equal to 1 g must be exactly the same as the number of atoms of carbon (assigned atomic mass = 12) required to make up a bulk mass equal to 12 g, and so on. This equivalency is called the gram-equivalent weight, and the number of atoms or molecules required to make up a gram-equivalent weight is termed 1 mole. Every properly balanced chemical equation specifies the number of moles of each different material in the process, corresponding exactly to the number of atoms or molecules of each that is involved in each single instance of the reaction. Without this concept, the science of modern chemistry would not exist. It is worth noting that this concept would not have developed had medieval alchemists not applied the measurement of material weights in their studies.

Similarly, physics has developed to its present state through the concepts of mass and weight and the application of them as a means to measure forces. In any practical sense a force can only be quantified by measuring its effect relative to a mass that has itself been quantified. In the basic principles of Newtonian mechanics, a force is equal to the product of mass and acceleration ($F = ma$). Newtonian mechanics, however, is not sufficient to describe the small structure of atoms and molecules. This is the realm of quantum mechanics, of which Newtonian mechanics is a subset.

Mass at the atomic and molecular level is especially relevant in the analytical procedure called mass spectrometry. In this methodology, the masses of molecular fragments carrying a positive electrical charge are determined by the magnitude of the interaction of that charge with an applied electromagnetic field. Another procedure, called isotopic labeling, is used to measure the effect of atomic mass on vibrational frequency of specific bonds in spectroscopic analyses and on the rates of reactions in chemical kinetics. These methods, which depend on highly specialized techniques of measuring mass, provide a great deal of essential information about molecular structures and behaviors.

SOCIAL CONTEXT AND FUTURE PROSPECTS

The world is a material one, and the concepts of mass and weight are inseparable from matter. All aspects of society are impacted by mass and weight in some manner. The single greatest economic expenditure of commerce and society is the energy required for the movement and transportation of mass. This is closely related to other physical aspects, such as tribology, the study of friction and lubrication, and applied physics, the study of the properties and utilization of condensed matter. Tribological effects, all of which are intimately associated with the movement of mass and the corresponding weight, are estimated to carry an immediate economic cost of about 4 percent of a nation's gross national product, with one-third of all energy expenditures made to overcome friction. If the frictional effects involved in transporting mass could be reduced or eliminated, the economic and social benefits would be immense. Even the movement of heating materials, such as natural gas, exacts an economic cost because of that material's mass and weight.

The green movement, which includes energy-efficient building methods, works toward minimizing the economic impact of mass movement, but at the same time it engenders its own costs in that area. Trade and commerce will always demand mass and weight measurement practices and practitioners to maintain stability, productivity, efficiency, and fairness.

—*Richard M. Renneboog, MSc*

FASCINATING FACTS ABOUT WEIGHT AND MASS MEASUREMENT

- The weight of an object can vary by as much as 0.8 percent according to its location on the planet, but its mass remains absolutely constant.
- Weighing an object determines the force of gravity acting on that object, rather than the mass of the object.
- Mass and weight are used interchangeably within the same gravitational field, though they technically have different meanings.
- Mass is an intrinsic property of matter and remains constant. Its weight is not, and it varies according to the force of gravity acting on that matter.
- The vibrational effect of substituting a single atom of deuterium for a single atom of hydrogen in a molecule is the most readily detected isotope effect because the deuterium atom has twice the mass of a hydrogen atom.
- Beam-balance scales used to weigh large trucks work on the same principle as the smallest pan balances.
- The international standard prototype kilogram is a cylinder made of 90 percent platinum and 10 percent iridium. The cylinder is precisely 39 mm high and 39 mm in diameter.
- Copies of the standard kilogram are kept in more than sixty countries, with each copy housed in a specially designed triple bell jar to prevent changes in its mass from oxidation or the loss of even a single atom.
- The masses of electrically charged pieces of molecules and atoms are directly measured by mass spectrometry.

FURTHER READING

Davidson, S., M. Perkin, and M. Buckley. *The Measurement of Mass and Weight.* Measurement Good Practice Guide 71. Teddington, England: National Physical Laboratory, 2004. Describes the best practices for weighing technique in the context of scientific and technological application.

_____. *Specifications, Tolerances, and Other Technical Requirements for Weighing and Measuring Devices.* http://ts.nist.gov/WeightsAndMeasures/Publications/upload/HB44-10-Complete.pdf. Accessed October, 2011. The official statement of regulations governing the function and operating requirements of weighing and measuring devices in the United States.

Appendices

TIMELINE

The Time Line below lists milestones in the history of applied science: major inventions and their approximate dates of emergence, along with key events in the history of science. The developments appear in boldface, followed by the name or names of the person(s) responsible in parentheses. A brief description of the milestone follows.

2,500,000 BCE	Stone tools	Stone tools, used by Homo habilis and perhaps other hominids, first appear in the Lower Paleolithic age (Old Stone Age).
400,000 BCE	Controlled use of fire	The earliest controlled use of fire by humans may have been about this time.
200,000 BCE	Stone tools using the prepared-core technique	Stone tools made by chipping away flakes from the stones from which they were made appear in the Middle Paleolithic age.
100,000-50,000 BCE	Widespread use of fire by humans	Fire is used for heat, light, food preparation, and driving off nocturnal predators. It is later used to fire pottery and smelt metals.
100,000-50,000 BCE	Language	At some point, language became abstract, enabling the speaker to discuss intangible concepts such as the future.
16,000 BCE	Earliest pottery	The earliest pottery was fired by putting it in a bonfire. Later it was placed in a trench kiln. The earliest ceramic is a female figure from about 29,000 to 25,000 BCE, fired in a bonfire.
10,000 BCE	Domesticated dogs	Dogs seem to have been domesticated first in East Asia.
10,000 BCE	Agriculture	Agriculture allows people to produce more food than is needed by their families, freeing humans from the need to lead nomadic lives and giving them free time to develop astronomy, art, philosophy, and other pursuits.
10,000 BCE	Archery	Archery allows human hunters to strike a target from a distance while remaining relatively safe.
10,000 BCE	Domesticated sheep	Sheep seem to have been domesticated first in Southwest Asia.
9000 BCE	Domesticated pigs	Pigs seem to have been domesticated first in the Near East and in China.
8000 BCE	Domesticated cows	Cows seem to have been domesticated first in India, the Middle East, and sub-Saharan Africa.

7500 BCE	Mud bricks	Mud-brick buildings appear in desert regions, offering durable shelter. The citadel in Bam, Iran, the largest mud-brick building in the world, was built before 500 BCE and was largely destroyed by an earthquake in 2003.
7500 BCE	Domesticated cats	Cats seem to have been domesticated first in the Near East.
6000 BCE	Domesticated chickens	Chickens seem to have been domesticated first in India and Southeast Asia.
6000 BCE	Scratch plow	The earliest plow, a stick held upright by a frame and pulled through the topsoil by oxen, is in use.
6000 BCE	Electrum	The substance is a natural blend of gold and silver and is pale yellow in color like amber. The name "electrum" comes from the Greek word for amber.
6000 BCE	Gold	Gold is discovered—possibly the first metal to be recognized as such.
6000-4000 BCE	Potter's wheel	The potter's wheel is developed, allowing for the relatively rapid formation of radially symmetric items, such as pots and plates, from clay.
5000 BCE	Wheel	The chariot wheel and the wagon wheel evolve—possibly from the potter's wheel. One of humankind's oldest and most important inventions, the wheel leads to the invention of the axle and a bearing surface.
4200 BCE	Copper	Egyptians mine and smelt copper.
4000 BCE	Moldboard plow	The moldboard plow cut a furrow and simultaneously lifted the soil and turned it over, bringing new nutrients to the surface.
4000 BCE	Domesticated horses	Horses seem to have been domesticated first on the Eurasian steppes.
4000 BCE	Silver	Silver can be found as a metal in nature, but this is rare. It is harder than gold but softer than copper.
4000 BCE	Domesticated honeybees	The keeping of bee hives for honey arises in many different regions.
4000 BCE	Glue	Ancient Egyptian burial sites contain clay pots that have been glued together with tree sap.

3500 BCE	Lead	Lead is first extruded from galena (lead sulfide), which can be made to release its lead simply by placing it in a hot campfire.
c. 3100 BCE	Numerals	Numerals appeared in Sumerian, Proto-Elamite, and Egyptian hieroglyphics.
3000 BCE	Bronze	Bronze, an alloy of copper and tin, is developed. Harder than copper and stronger than wrought iron, it resists corrosion better than iron.
3000 BCE	Cuneiform	The method of writing now known as cuneiform began as pictographs but evolved into more abstract patterns of wedge-shaped (cuneiform) marks, usually impressed into wet clay. This system of marks made complex civilization possible, since it allowed record keeping to develop.
3000 BCE	Fired bricks	Humans begin to fire bricks, creating more durable building materials that (because of their regular size and shape) are easier to lay than stones.
3000 BCE	Pewter	The alloy pewter is developed. It is 85 to 99 percent tin, with the remainder being copper, antimony, and lead; copper and antimony make the pewter harder. Pewter's low melting point, around 200 degrees Celsius, makes it a valuable material for crafting vessels that hold hot substances.
2700 BCE	Plumbing	Earthenware pipes sealed together with asphalt first appear in the Indus Valley civilization. Greeks, Romans, and others provided cities with fresh water and a way to carry off sewage.
2650 BCE	Horse-drawn chariot (Huangdi)	Huangdi—a legendary patriarch of China—is possibly a combination of many men. He is said to have invented—in addition to the chariot—military armor, ceramics, boats, and crop rotation.
2600 BCE	Inclined plane	Inclined planes are simple machines and were used in building Egypt's pyramids. Pushing an object up a ramp requires less force than lifting it directly, although the use of a ramp requires that the load be pushed a longer distance.
c. 2575-c. 2465 BCE	Pyramids	Pyramids of Giza are built in Egypt.
1750 BCE	Tin	Tin is alloyed with copper to form bronze.
1730 BCE	Glass beads	Red-brown glass beads found in South Asia are the oldest known human-formed glass objects.

1600 BCE	Mercury	Mercury can easily be released from its ore (such as cinnabar) by simply heating it.
1500 BCE	Iron	Iron, stronger and more plentiful than bronze, is first worked in West Asia, probably by the Hittites. It could hold a sharper edge, but it had to be smelted at higher temperatures, making it more difficult to produce than bronze.
1500 BCE	Zinc	Zinc is alloyed with copper to form brass, but it will not be recognized as a separate metal until 1746.
1000 BCE	Concrete	The ancient Romans build arches, vaults, and walls out of concrete.
1000 BCE	Crossbow	The crossbow seems to come from ancient China. Crossbows can be made to be much more powerful than a normal bow.
1000 BCE	Iron Age	Iron Age begins. Iron is used for making tools and weapons
700 BCE	Magnifying glass	An Egyptian hieroglyph seems to show a magnifying glass.
350 BCE	Compass	Ancient Chinese used lodestones and later magnetized needles mostly to harmonize their environments with the principles of feng shui. Not until the eleventh century are these devices used primarily for navigation.
350-100 BCE	Scientific method (Aristotle)	Aristotle develops the first useful set of rules attempting to explain how scientists practice science.
300 BCE	Screw	Described by Archimedes, the screw is a simple machine that appears to be a ramp wound around a shaft. It converts a smaller turning force to a larger vertical force, as in a screw jack.
300 BCE	Lever	Described by Archimedes, the lever is a simple machine that allows one to deliver a larger force to a load than the force with which one pushes on the lever.
300 BCE	Pulley	Described by Archimedes, the pulley is a simple machine that allows one to change the direction of the force delivered to the load.
221-206 BCE	Compass	The magnetic compass is invented in China using lodestones, a mineral containing iron oxide.
215 BCE	Archimedes' principle (Archimedes of Syracuse)	Archimedes describes his law of displacement: A floating body displaces an amount of fluid the weight of which is equal to the weight of the body.

200 BCE	Astrolabe	A set of engraved disks and indicators becomes known as the astrolabe. When aligned with the stars, the astrolabe can be used to determine the rising and setting times of the Sun and certain stars, establish compass directions, and determine local latitude.
40 CE	Ptolemy's geocentric system (Ptolemy)	A world system with the Earth in the center, and the Moon, Venus, Mercury, Sun, Mars, Jupiter, Saturn, and fixed stars surrounding it. The geocentric Ptolemaic system would remain the most widely accepted cosmology for the next fifteen hundred years.
90 CE	Aeolipile (Hero of Alexandria)	The aeolipile—a steam engine that escaping steam causes to rotate like a lawn sprinkler—is developed.
105 CE	Paper and papermaking (Cai Lun)	Although papyrus paper already existed, Cai Lun creates paper from a mixture of fibrous materials softened into a wet pulp that is spread flat and dried. The material is strong and can be cheaply mass-produced.
250 CE	Force pump (Ctesibius of Alexandria)	Ctesibius develops a device that shoots a jet of water, like a fire extinguisher.
815 CE	Algebra (al-Khwrizm)	al-Khwrizm develops the mathematics that solves problems by using letters for unknowns (variables) and expressing their relationships with equations.
877 CE	Maneuverable glider (Abbas ibn Firnas)	A ten-minute controlled glider flight is first achieved.
9th century	Gunpowder	Gunpowder is invented in China.
1034	Movable type	Movable type made of baked clay is invented in China.
1170	Water-raising machines (al-Jazari)	In addition to developing machines that can transport water to higher levels, al-Jazari invents water clocks and automatons.
1260	Scientific method (Roger Bacon)	Bacon develops rules for explaining how scientists practice science that emphasize empiricism and experimentation over accepted authority.
1284	Eyeglasses for presbyopia (Salvino d'Armate)	D'Armate is credited with making the first wearable eyeglasses in Italy with convex lenses. These spectacles assist those with farsightedness, such as the elderly.

1439	Printing press (Johann Gutenberg)	Gutenberg combined a press, oil-based ink, and movable type made from an alloy of lead, zinc, and antimony to create a revolution in printing, allowing mass-produced publications that could be made relatively cheaply and disseminated to people other than the wealthy.
1450	Eyeglasses for the nearsighted (Nicholas of Cusa)	Correcting nearsightedness requires diverging lenses, which are more difficult to make than convex lenses.
1485	Dream of flight (Leonardo da Vinci)	On paper, Leonardo designed a parachute, great wings flapped by levers, and also a person-carrying machine with wings to be flapped by the person. Although these flying devices were never successfully realized, the designs introduced the modern quest for aeronautical engineering.
1543	Copernican (heliocentric) universe	Copernicus publishes *De revolutionibus* (*On the Revolutions of the Heavenly Spheres*), in which he refutes geocentric Ptolemaic cosmology and proposes that the Sun, not Earth, lies at the center of the then-known universe (the solar system).
1569	Mercator projection (Gerardus Mercator)	The Mercator projection maps the Earth's surface onto a series of north/south cylinders.
1594	Logarithms (John Napier)	Napier's logarithms allow the simplification of complex multiplication and division problems.
1595	Parachute (Faust Veranzio)	Veranzio publishes a book describing sixty new machines, one of which is a design for a parachute that might have worked.
1596	Flush toilet (Sir John Harington)	Harington's invention is a great boon to those previously assigned to empty the chamber pots.
1604	Compound microscope (Zacharias Janssen)	Janssen, a lens crafter, experiments with lenses, leading to both the microscope and the telescope.
1607	Air and clinical thermometers (Santorio Santorio)	Santorio develops a small glass bulb that can be placed in a person's mouth, with a long, thin neck that is placed in a beaker of water. The water rises or falls as the person's temperature changes.
1608	Refracting telescope (Hans Lippershey)	Lippershey is one of several who can lay claim on developing the early telescope.
1609	Improved telescope (Galileo Galilei)	Galileo grinds and polishes his own lenses to make a superior telescope. Galileo will come to be known as the father of modern science.

1622	Slide rule (William Oughtred)	English mathematician and Anglican minister Oughtred invents the slide rule.
1629	Steam turbine (Giovanni Branca)	Branca publishes a design for a steam turbine, but it requires machining that is too advanced to be built in his day.
1642	Mechanical calculator (Blaise Pascal)	Eighteen-year-old Pascal invents the first mechanical calculator, which helps his father, a tax collector, count taxes.
1644	Barometer (Evangelista Torricelli)	Torricelli develops a mercury-filled barometer, in which the height of the mercury in the tube is a measure of atmospheric pressure.
1650	Vacuum pump (Otto von Guericke)	After demonstrating the existence of a vacuum, von Guericke explores its properties with other experiments.
1651	Hydraulic press (Blaise Pascal)	Pascal determines that hydraulics can multiply force. For example, a 50-pound force applied to the hydraulic press might exert 500 pounds of force on an object in the press.
1656	Pendulum clock (Christiaan Huygens)	Huygens discovers that, for small oscillations, a pendulum's period is independent of the size of the pendulum's swing, so it can be used to regulate the speed of a clock.
1662	Demography (John Graunt)	Englishman Graunt develops the first system of demography and publishes *Natural and Political Observations Mentioned in the Following Index and Made Upon the Bills of Mortality*, which laid the groundwork for census taking.
1663	Gregorian telescope (James Gregory)	The Gregorian telescope produces upright images and therefore becomes useful as a terrestrial telescope.
1666	The calculus (Sir Isaac Newton)	Newton (and independently Gottfried Wilhelm Leibniz) develop the calculus in order to calculate the gravitational effect of all of the particles of the Earth on another object such as a person.
1670	Spiral spring balance watch (Robert Hooke)	Hooke is also credited as the author of the principle that describes the general behavior of springs, known as Hooke's law.
1672	Leibniz's calculator (Gottfried Wilhelm Leibniz)	Leibniz develops a calculator that can add, subtract, multiply, and divide, as well as the binary system of numbers used by computers today.
1674	Improvements to the simple microscope (Antoni van Leeuwenhoek)	Leeuwenhoek, a lens grinder, applies his lenses to the simple microscope and uses his microscope to observe tiny protozoa in pond water.

1681	Canal du Midi opens	The 150-mile Canal du Midi links Toulouse, France, with the Mediterranean Sea.
1698	Savery pump (Thomas Savery)	Savery's pump was impractical to build, but it served as a prototype for Thomas Newcomen's steam engine.
1699	Eddystone Lighthouse (Henry Winstanley)	English merchant Winstanley designs the first lighthouse in England, located in the English Channel fourteen miles off the Plymouth coast. Winstanley is moved to create the light-house after two of his ships are wrecked on the Eddystone rocks.
1700	Piano (Bartolomeo Cristofori)	Cristofori, a harpsichord maker, constructs an instrument with keys that can be used to control the force with which hammers strike the instrument's strings, producing sound that ranges from piano (soft) to forte (loud)—hence the name "pianoforte," later shortened to "piano."
1701	Tull seed drill (Jethro Tull)	Before the seed drill, seeds were still broadcast by hand.
1709	Iron ore smelting with coke (Abraham Darby)	Darby develops a method of smelting iron ore by using coke, rather than charcoal, which at the time was becoming scarce. Coke is made by heating coal and driving off the volatiles (which can be captured and used).
1712	Atmospheric steam engine (Thomas Newcomen)	Newcomen's engine is developed to pump water out of coal mines.
1714	Mercury thermometer, Fahrenheit temperature scale (Daniel Gabriel Fahrenheit)	Fahrenheit uses mercury in a glass thermometer to measure temperature over the entire range for liquid water.
1718	Silk preparation	John Lombe, owner of the Derby Silk Mill in England, pat-ents the machinery that prepared raw silk for the loom.
1729	Flying shuttle (John Kay)	On a loom, the shuttle carries the horizontal thread (weft or woof) and weaves it between the vertical threads (warp). Kay develops a shuttle that is named "flying" because it is so much faster than previous shuttles.
1738	Flute Player and Digesting Duck automatons (Jacques de Vaucanson)	De Vaucanson builds cunning, self-operating devices, or automatons (robots) to charm viewers.
1740	Steelmaking	Benjamin Huntsman invents the crucible process of making steel.
1742	Celsius scale (Anders Celsius)	Celsius creates a new scale for his thermometer.

1745-1746	Leiden jar (Pieter van Musschenbroek and Ewald Georg von Kleist)	Von Kleist (1745) and Musschenbroek (1746) independently develop the Leiden jar, an early type of capacitor used for storing electric charge.
1746	Clinical trials prove that citrus fruit cures scurvy (James Lind)	Others had suggested citrus fruit as a cure for scurvy, but Lind gives scientific proof. It still will be another fifty years before preventive doses of foods containing vitamin C are routinely provided for British sailors.
1752	Franklin stove (Benjamin Franklin)	Franklin develops a stove that allows more heat to radiate into a room than go up the chimney.
1752	Lightning rod (Benjamin Franklin)	Franklin devises a iron-rod apparatus to attach to houses and other structures in order to ground them, preventing damage during lightning storms.
1756	Wooden striking clock (Benjamin Banneker)	Banneker's all-wood striking clock operates for the next fifty years. Banneker also prints a series of successful scientific almanacs during 1790's.
1757	Nautical sextant (John Campbell)	When used with celestial tables, Campbell's sextant allows ships to navigate to within sight of their destinations.
1762	Marine chronometers (John Harrison)	An accurate chronometer was necessary to determine a ship's position at sea, solving the pressing quest for longitude.
1764	Spinning jenny (James Hargreaves)	Hargreaves develops a machine for spinning several threads at a time, transforming the textile industry and laying a foundation for the Industrial Revolution.
1765	Improved steam engine (James Watt)	A steam condenser separate from the working pistons make Watt's engine significantly more efficient than Newcomen's engine of 1712.
1767	Spinning machine (Sir Richard Arkwright)	Arkwright develops a device to spin fibers quickly into consistent, uniform thread.
1767	Dividing engine (Jesse Ramsden)	Ramsden develops a machine that automatically and accurately marks calibrated scales.
1770	Steam dray (Nicolas-Joseph Cugnot)	Cugnot builds his three-wheeled fardier à vapeur to move artillery; the prototype pulls 2.5 metric tons at 2 kilometers per hour.
1772	Soda water (Joseph Priestley)	Priestley creates the first soda water, water charged with carbon dioxide gas. The following year he develops an apparatus for collecting gases by mercury displacement that would otherwise dissolve in water.

1775	Boring machine (John Wilkinson)	Wilkinson builds the first modern boring machine used for boring holes into cannon, which made cannon manufacture safer. It was later adapted to bore cylinders in steam engines.
1776	Bushnell's submarine (David Bushnell)	Bushnell builds the first attack submarine; used unsuccessfully against British ships in the Revolutionary War, it nevertheless advances submarine technology.
1779	Cast-iron bridge	Abraham Darby III and John Wilkinson build the first cast-iron bridge in England.
1779	Spinning mule (Samuel Crompton)	Crompton devises the spinning mule, which allows the textile industry to manufacture high-quality thread on a large scale.
1781	Uranus discovered (Sir William Herschel)	Herschel observes what he first believes to be a comet; further observation establishes it as a planet eighteen times farther from the Sun than the Earth is.
1782	Hot-air balloon (Étienne-Jacques and Joseph-Michel Montgolfier)	Shaped like an onion dome and carrying people aloft, the Montgolfiers' hot-air balloon fulfills the fantasy of human flight.
1782	Oil lamp (Aimé Argand)	Argand's oil lamp revolutionizes lighthouse illumination.
1783	Parachutes (Louis-Sébastien Lenormand)	Lenormand jumps from an observatory tower using his parachute and lands safely.
1783	Wrought iron (Henry Cort)	Cort converts crude iron into tough malleable wrought iron.
1784	Improved steam engine (William Murdock)	In an age when much focus was on steam technology, Murdock works to improve steam pumps that remove water from mines. He will go on to invent coal-gas lighting in 1794.
1784	Bifocals (Benjamin Franklin)	Tired of changing his spectacles to see things at close range as opposed to objects farther away, Franklin designs eyeglasses that incorporate both myopia-correcting and presbyopia-correcting lenses.
1784	Power loom (Edmund Cartwright)	Cartwright's power loom forms a major advance in the Industrial Revolution.
1785	Automated flour mill (Oliver Evans)	Evans's flour mill lays the foundation for continuous production lines. In 1801, he will also invent a high-pressure steam engine.
1790	Steamboat (John Fitch)	Fitch not only invents the steamboat but also proves its practicality by running a steamboat service along the Delaware River.

1792	Great clock (Thomas Jefferson)	Jefferson's clock, visible and audible both within Monticello and outside, across his plantation, is designed to maintain efficiency. He also invented an improved portable copying press (1785) and will go on to invent an improved ox plow (1794).
1792	Coal gas (William Murdock)	Murdock develops methods for manufacturing, storing, and purifying coal gas and using it for lighting.
1793	Cotton gin (Eli Whitney)	Whitney's engine to separate cotton seed from the fiber transformed the American South, both bolstering the institution of slavery and growing the "cotton is king" economy of the Southern states. Five years later, Whitney develops an assembly line for muskets using interchangeable parts.
1793	Semaphore (Claude Chappe)	Chappe invents the semaphore.
1796	Smallpox vaccination (Edward Jenner)	Jenner's vaccine will save millions from death, culminating in the eradication of smallpox in 1979.
1796	Rumford stove (Benjamin Thompson)	The Rumford stove—a large, institutional stove—uses several small fires to heat the stove top uniformly.
1796	Hydraulic press (Joseph Bramah)	Bramah builds a practical hydraulic press that operates by a high-pressure plunger pump.
1796	Lithography (Aloys Senefelder)	Senefelder invents lithography and a process for color lithography in 1826.
1799	Voltaic pile/electric battery (Alessandro Volta)	Volta creates a pile—a stack of alternating copper and zinc disks separated by brine-soaked felt—that supplies a continuous current and sets the stage for the modern electric battery.
1800	Iron printing press (Charles Stanhope)	Stanhope invents the first printing press made of iron.
1801	Pattern-weaving loom (Joseph M. Jacquard)	Jacquard invents a loom for pattern weaving.
1804	Monoplane glider (George Cayley)	Cayley develops a heavier-than-air fixed-wing glider that inaugurates the modern field of aeronautics. Later models carry a man and lead directly to the Wright brothers' airplane.
1804	Amphibious vehicle (Oliver Evans)	Evans builds the first amphibious vehicle, which is used in Philadelphia to dredge and clean the city's dockyards.

1805	Electroplating (Luigi Brugnatelli)	Brugnatelli develops the method of electroplating by connecting something to be plated to one pole of a battery (voltaic pile) and a bit of the plating metal to the other pole of the battery, placing both in a suitable solution.
1805	Morphine (Friedrich Setürner)	Setürner, a German pharmacist, isolates morphine from opium, but it is not widely used for another ten years.
1806	Steam locomotive (Richard Trevithick)	After James Watt's patent for the steam engine expires in 1800, Trevithick develops a working steam locomotive. By 1806 he has developed his improved steam engine, named the Cornish engine, which sees worldwide dissemination.
1807	Internal combustion engine (François Isaac de Rivaz)	De Rivaz builds the first vehicle powered by an internal combustion engine.
1807	Paddle-wheel steamer (Robert Fulton)	Fulton's steamboat becomes far more commercially successful than those of his competitors.
1808	Law of combining volumes for gases (Joseph-Louis Gay-Lussac)	Gay-Lussac discovers that, when gaseous elements combine to make a compound, the volumes involved are always simple whole-number ratios.
1810	Preserving food in sealed glass bottles (Nicolas Appert)	Appert answers Napoleon's call to preserve food in a way that allows his soldiers to carry it with them: He processes food in sealed, air-tight glass bottles.
1810	Preserving food in tin cans (Peter Durand)	Durand follows Nicolas Appert in preserving food for the French army, but he uses tin-coated steel cans in place of breakable bottles.
1815	Miner's safety lamp (Sir Humphry Davy)	Davy devises a miner's safety lamp in which the flame is surrounded by wire gauze to cool combustion gases so that the mine's methane-air mixture will not be ignited.
1816	Macadamization (John Loudon McAdam)	McAdam designs a method of paving roads with crushed stone bound with gravel on a base of large stones. The roadway is slightly convex, to shed water.
1816	Kaleidoscope (Sir David Brewster)	The name for Brewster's kaleidoscope comes from the Greek words *kalos* (beautiful), *eidos* (form), and *scopos* (watcher). "Kaleidoscope," therefore, literally means "beautiful form watcher."
1816	Stirling engine (Robert Stirling)	The Stirling engine proves to be an efficient engine that uses hot air as a working fluid.

1818	First photographic images (Joseph Nicéphore Niépce)	Niépce creates the first lasting photographic images.
1819	Stethoscope (René-Théophile-Hyacinthe Laënnec)	Laënnec invents the stethoscope to avoid the impropriety of placing his ear to the chest of a female heart patient.
1820	Dry "scouring" (Thomas L. Jennings)	Jennings discovered that turpentine would remove most stains from clothes without the wear associated with washing them in hot water. His method becomes the basis for modern dry cleaning.
1821	Diffraction grating (Joseph von Fraunhofer)	Von Fraunhofer's diffraction grating separates incident light by color into a rainbow pattern. The various discrete patterns reveal the structure of specific atomic nuclei, making it possible to identify the chemical compositions of various substances.
1821	Braille alphabet (Louis Braille)	Braille develops a tactile alphabet—a system of raised dots on a surface—that allows the blind to read by touch.
1821	Electromagnetic rotation (Michael Faraday)	Faraday publishes his work on electromagnetic rotation, which is the principle behind the electric motor.
1822	Difference engine (Charles Babbage)	Babbage's "engine" was a programmable mechanical device used to calculate the value of a polynomial—a precursor to today's modern computers.
1823	Waterproof fabric is used in raincoats (Charles Macintosh)	Macintosh patents a waterproof fabric consisting of soluble rubber between two pieces of cloth. Raincoats made of the fabric are still often called mackintoshes (macs), especially in England.
1824	Astigmatism-correcting lenses (George Biddell Airy)	Airy develops cylindrical lenses that correct astigmatism. An astronomer, Airy will go on to design a method of correcting compasses used in ship navigation and the altazimuth telescope. He becomes England's astronomer royal in 1835.
1825	Electromagnet (William Sturgeon)	Sturgeon builds a U-shaped, soft iron bar with a coil of varnished copper wire wrapped around it. When a voltaic current is passed through wire, the bar becomes magnetic—the world's first electromagnet.
1825	Bivalve vaginal speculum (Marie Anne Victoire Boivin)	Boivin develops the tool now widely used by gynecologists in the examination of the vagina and cervix.
1825	"Steam waggon" (John Stevens)	Stevens builds the first steam locomotive to be manufactured in the United States.

1826	Color lithography (Aloys Senefelder)	Senefelder invents color lithography.
1827	Matches (John Walker)	Walker coats the ends of sticks with a mixture of antimony sulfide, potassium chlorate, gum, and starch to produce "strike anywhere" matches.
1827	Water turbine (Benoît Fourneyron)	Fourneyron builds the first water turbine; it has six horse-power. His larger, more efficient turbines powered many factories during the Industrial Revolution.
1828	Combine harvester (Samuel Lane)	Patent is granted to Lane for the combine harvester, which combines cutting and threshing.
1829	Rocket steam locomotive (George Stephenson)	Stephenson builds the world's first railway line to use a steam locomotive.
1829	Boiler (Marc Seguin)	Seguin improves the steam engine with a multiple fire-tube boiler.
1829	Polarizing microscope (William Nicol)	Nicol invents the polarizing microscope, an important forensic tool.
1830	Steam locomotive (Peter Cooper)	Cooper's four-wheel locomotive with a vertical steam boiler, the *Tom Thumb*, demonstrates the possibilities of steam locomotives and brings Cooper national fame. His other inventions and good management enable Cooper to become a leading industrialist and philanthropist.
1830	Lawn mower (Edwin B. Budding)	Budding, an English engineer, invents the lawn mower.
1830	Paraffin (Karl von Reichenbach)	Von Reichenbach, a German chemist, discovers paraffin.
1830	Creosote (Karl von Reichenbach)	Von Reichenbach distills creosote from beachwood tar. It is used as an insecticide, germicide, and disinfectant.
1831	Alternating current (AC) generator (Michael Faraday)	Faraday constructs the world's first electric generator.
1831	Mechanical reaper (Cyrus Hall McCormick)	McCormick's reaper can harvest a field five times faster than earlier methods.
1831	Staple Bend Tunnel	The first railroad tunnel in the United States is built in Mineral Point, Pennsylvania.

1832	Electromagnetic induction (Joseph Henry)	Henry discovers that changing magnetic fields induce voltages in nearby conductors.
1832	Codeine (Pierre-Jean Robiquet)	French chemist Robiquet isolates codeine from opium. Because of the small amount found in nature, most codeine is synthesized from morphine.
1834	Hansom cab (Joseph Aloysius Hansom)	English architect Hansom builds the carriage bearing his name.
1835	Colt revolver (Samuel Colt)	The Colt revolver becomes known as "one of the greatest advances of self-defense in all of human history."
1835	Photography (Joseph Nicéphore Niépce)	Niépce codevelops photography with Louis-Jacques-Mandé Daguerre.
1836	Daniell cell (John Frederic Daniell)	Daniell invents the electric battery bearing his name, which is much improved over the voltaic pile.
1836	Acetylene (Edmund Davy)	Davy creates acetylene by heating potassium carbonate to high temperatures and letting it react to water.
1837	Electric telegraph (William Fothergill Cooke and Charles Wheatstone)	Wheatstone and Cooke devise a system that uses five pointing needles to indicate alphabetic letters.
1837	Steam hammer (James Hall Nasmyth)	Nasmyth develops the steam hammer, which he will use to build a pile driver in 1843.
1837	Steel plow (John Deere)	Previously, plows were made of cast iron and required frequent cleaning. Deere's machine is effective in reducing the amount of clogging farmers experienced when plowing the rich prairie soil.
1837	Threshing machine (Hiram A. and John A. Pitts)	The Pitts, brothers, develop the first efficient threshing machine.
1838	Fuel cell (Christian Friedrich Schönbein)	Schönbein's fuel cell might use hydrogen and oxygen and allow them to react, producing water and electricity. There are no moving parts, but the reactants must be continuously supplied.
1838	Propelling steam vessel (John Ericsson)	Swedish engineer Ericsson invents the double screw propeller for ships allowing them to move much faster than those relying on sails.
1839	Nitric acid battery (Sir William Robert Grove)	The Grove cell delivered twice the voltage of its more expensive rival, the Daniell cell.

1839	Daguerreotype (Jacques Daguerre)	Improving on the discoveries of Joseph Nicéphore Niépce, Daguerre develops the first practical photographic process, the Daguerreotype.
1839	Vulcanized rubber (Charles Goodyear)	Adding sulfur and lead monoxide to rubber, Goodyear processes the batch at a high temperature. The process, later called vulcanization, yields a stable material that does not melt in hot weather or crack in cold.
1840	Electrical telegraph (Samuel F. B. Morse)	Others had already built telegraph systems, but Morse's system was superior and soon replaced all others.
1841	Improved electric clock (Alexander Bain)	With John Barwise, Bain develops an electric clock with a pendulum driven by electric impulses to regulate the clock's accuracy.
1841	First negatives in photography (William Henry Fox Talbot)	Talbot, an English polymath, invents the calotype process, which produces the first photographic negative.
1842	Commercial fertilizer (John B. Lawes)	Lawes develops superphosphate, the first commercial fertilizer.
1843	Rotary printing press (Richard March Hoe)	Patented in 1847, the steam-powered rotary press is far faster than the flatbed press.
1843	Multiple-effect vacuum evaporator (Norbert Rillieux)	Rillieux develops an efficient method for refining sugar using a stack of several pans of sugar syrup in a vacuum chamber, which allows boiling at a lower temperature.
1845	Suspension bridges (John Augustus Roebling)	A manufacturer of wire cable, Roebling wins a competition for an aqueduct over the Allegheny River and goes on to design other aqueducts and suspension bridges, culminating in the Brooklyn Bridge, which his son, Washington Augustus Roebling, completes in 1883.
1845	Sewing machine (Elias Howe)	Howe develops a machine that can stitch straight, strong seams faster than those sewn by hand.
1846	Neptune discovered (John Galle)	German astronomer Galle observes a new planet, based on irregularities in the orbit of Uranus calculated the previous year by England's John Couch Adams and France's Urbain Le Verrier.
1847	Nitroglycerin (Ascanio Sobrero)	Italian chemist Sobrero creates nitroglycerin.

1847	Telegraphy applications (Werner Siemens)	Siemens refines a telegraph in which a needle points to the alphabetic letter being sent.
1849	Laryngoscope (Manuel P. R. Garcia)	Spanish singer and voice teacher, Garcia, known as the father of laryngology, devises the first laryngoscope.
1851	Foucault's pendulum (Léon Foucault)	Foucault's pendulum proves that Earth rotates.
1851	Sewing machine (Isaac Merritt Singer)	Singer improves the sewing machine and successfully markets it to women for home use.
1851	Ophthalmoscope (Hermann von Helmholtz)	Helmholtz invents a device that can be used to examine the retina and the vitreous humor. In 1855, he will invent an ophthalmometer, an instrument that measures the curvature of the eye's lens.
1854	Kerosene (Abraham Gesner)	Canadian geologist Gesner distills kerosene from petroleum.
1852	Hypodermic needle (Charles G. Pravaz)	French surgeon Pravaz devises the hypodermic syringe.
1855	Bunsen burner (Robert Wilhelm Bunsen)	Bunsen—along with Peter Desaga, an instrument maker, and Henry Roscoe, a student—develops a high-temperature laboratory burner, which he and Gustav Kirchhoff use to develop the spectroscope (1859).
1855	Bessemer process (Sir Henry Bessemer)	Bessemer creates a converter that leads to a process for inexpensively mass-producing steel.
1856	Synthetic dye (William H. Perkin)	British chemist Perkin produces the first synthetic dye. The color is mauve, which triggers a mauve fashion revolution.
1857	Safety elevator (Elisha Graves Otis)	Otis's safety elevator automatically stops if the supporting cable breaks.
1858	Internal combustion engine (étienne Lenoir)	Lenoir's engine, along with his invention of the spark plug, sets the stage for the modern automobile.
1858	Transatlantic cable (Lord Kelvin)	Kelvin helps design and install the under-ocean cables for telegraphy between North America and Europe, serving as a chief motivating force in getting the cable completed.
1859	Signal flares (Martha J. Coston)	Coston's brilliant and long-lasting white, red, and green flares will be adopted by the navies of several nations.
1859	Lead-acid battery (Gaston Planté)	French physicist Planté invents the lead-acid battery, which led to the invention of the first electric, rechargeable battery.

1860	Refrigerant (Ferdinand Carré)	French inventor Carré introduces a refrigerator that uses ammonia as a refrigerant.
1860	Electric incandescent lamp (Joseph Wilson Swan)	Swan produces and patents an incandescent electric bulb; in 1880, two years after Edison's light bulb, Swan will produce a more practical bulb.
1860	Web rotary printing press (William Bullock)	Bullock's press has an automatic paper feeder, can print on both sides of the paper, cut the paper into sheets, and fold them.
1860	Henry rifle (Tyler Henry)	American gunsmith Henry designs the Henry rifle, a repeating rifle, the year before the Civil War begins.
1860	First mail service	Pony Express opens overland mail service. The service eventually expands to include more than 100 stations, 80 riders, and more than 400 horses.
1861	Machine gun (Richard Gatling)	Gatling develops the first machine gun, called the Gatling gun. It has six barrels that rotate into place as the operator turns a hand crank; the shells were automatically chambered and fired.
1861	First color photograph	Thomas Sutton develops the first color photo based on Scottish physicist James Clerk Maxwell's three-color process.
1861-1862	USS *Monitor* (John Ericsson)	Ericsson develops the first practical ironclad ship, which will be use during the Civil War. He goes on to develop a torpedo boat that can fire a cannon from an underwater port.
1862	Pasteurization (Louis Pasteur)	Pasteur's germ theory of disease leads him to develop a method of applying heat to milk products in order to kill harmful bacteria. He goes on to develop vaccines for rabies, anthrax, and chicken cholera (1867-1885).
1863	Subway	The first subway opens in London; it uses steam locomotives. It does not go electric until 1890.
1865	*Pioneer* (Pullman) sleeping car (George Mortimer Pullman	Pullman began working on sleeping cars in 1858, but the *Pioneer* is a luxury car with an innovative folding upper birth to allow the passenger to sleep while traveling.
1866	Self-propelled torpedo (Robert Whitehead)	English engineer Whitehead develops the modern torpedo.
1866	Transatlantic telegraph cable	The first successful transatlantic telegraph cable is laid; it spans 1,686 nautical miles.

1867	Dynamite (Alfred Nobel)	Nobel mixes clay with nitroglycerin in a one-to-three ratio to create dynamite (Nobel's Safety Powder), an explosive the ignition of which can be controlled using Nobel's own blasting cap. He goes on to patent more than three hundred other inventions and devotes part of the fortune he gained from dynamite to establish and fund the Nobel Prizes.
1867	Baby formula (Henri Nestlé)	Nestlé combines cow's milk with wheat flour and sugar to produce a substitute for infants whose mothers cannot breast-feed.
1867	Steam velocipede motorcycle (Sylvester Roper)	Roper spent his lifetime making steam engines lighter and more powerful in order to make his motorized bicycles faster. His velocipede eventually reaches 60 miles per hour.
1867	Flat-bottom paper bag machine (Margaret E. Knight)	Knight designs a machine that can manufacture flat-bottom paper bags, which can stand open for easy loading.
1867	Dry-cell battery (Georges Leclanché)	French engineer Lelanche invents the dry-cell battery.
1868	Typewriter (Christopher Latham Sholes)	American printer Sholes produces the first commercially successful typewriter.
1869	Periodic table of elements (Dmitry Ivanovich Mendeleyev)	The periodic table, which links chemical properties to atomic structure, will prove to be one of the great achievements of the human race.
1869	Air brakes for trains (George Westinghouse)	In 1867, Westinghouse developed a signaling system for trains. The air brake makes it easier and safer to stop large, heavy, high-speed trains.
1869	Transcontinental railroad	The United States transcontinental railroad is completed.
1869	Celluloid (John Wesley Hyatt)	American inventor Hyatt produces celluloid, the first commercially successful plastic, by mixing solid pyroxylin and camphor.
1869	Suez Canal opens	The canal, 101 miles long, took a decade to build and connects the Red Sea with the eastern Mediterranean Sea.
1871	Fireman's respirator (John Tyndall)	The respirator grows from Tyndall's studies of air pollution.
1871	Commercial generator (Zénobe T. Gramme)	Belgian electrical engineer Gramme builds the Gramme machine, the first practical commercial generator for producing alternating current.

1872	Blue jeans (Levi Strauss)	Miners tore their pockets when they stuffed too many ore samples in them. Strauss makes pants using heavy-duty material with riveted pocket corners so they will not tear out.
1872	Burbank russet potato (Luther Burbank)	Burbank breeds all types of plants, using natural selection and grafting techniques to achieve new varieties. His Burbank potato, developed from a rare russet potato seed pod, grows better than other varieties.
1872	Automatic lubricator (Elijah McCoy)	McCoy uses steam pressure to force oil to lubricate the pistons of steam engines.
1872	Vaseline (Robert A. Chesebrough)	Chesebrough, an American chemist, patents his process for making petroleum jelly and calls it Vaseline.
1873	QWERTY keyboard (Christopher Latham Sholes)	After patenting the first practical typewriter, Sholes develops the QWERTY keyboard, designed to slow the fastest typists, who otherwise jammed the keys. The basic QWERTY design remains the standard on most computer keyboards.
1874	Barbed wire (Joseph Farwell Glidden)	An American farmer, Glidden invents and patents barbed wire. Barbed-wire fences make farming and ranching of the Great Plains practical. Without effective fences, animals wandered off and crops were destroyed. At the time of his death in 1906, Glidden is one of the richest men in the country.
1874	Medical nuclear magnetic resonance imaging (Raymond Damadian)	Damadian and others develop magnetic resonance imaging (MRI) for use in medicine.
1876	Four-stroke internal combustion engine (Nikolaus August Otto)	In order to deliver more horsepower, Otto's engine compresses the air-fuel mixture. His previous engines operated near atmospheric pressure.
1876	Ammonia-compressor refrigeration machine (Carl von Linde)	Breweries need refrigeration so they can brew year-round. Linde refines his ammonia-cycle refrigerator to make this possible.
1876	Telephone (Elisha Gray)	Gray files for a patent for the telephone the same day that Alexander Graham Bell does so. While the case is not clear-cut, and Gray fought with Bell for years over the patent rights, Bell is generally credited with the telephone's invention.
1877	Phonograph (Thomas Alva Edison)	Edison invents the phonograph—an unexpected outcome of his telephone research.

1878	First practical lightbulb (Thomas Alva Edison)	Twenty-two people have invented lightbulbs before Edison and Joseph Swan, but they are impractical. Edison's is the first to be commercially viable. Eventually, Swan's company merges with Edison's.
1878	Loose-contact carbon microphone (David Edward Hughes)	Hughes's carbon microphone advances telephone technology. In 1879, he will invent the induction balance, which will be used in metal detectors.
1878	Color photography (Frederic Eugene Ives)	American inventor Ives develops the halftone process for printing photographs.
1879	Saccharin (Ira Remsen)	Remsen synthesizes a compound that is up to three hundred times sweeter than sugar; he also establishes the important *American Chemical Journal*, serving as its editor until 1915.
1880	Milne seismograph (John Milne)	Milne invents the first modern seismograph for measuring earth tremors. He will come to be called the father of modern seismology.
1881	Improved incandescent lightbulb (Lewis Howard Latimer)	Latimer develops an improved way to manufacture and to attach carbon filaments in lightbulbs.
1881	Sphygmomanometer (Karl Samuel Ritter von Basch)	Von Basch invents the first blood pressure gauge.
1882	Induction motor (Nikola Tesla)	Tesla's theories and inventions make alternating current (AC) practical.
1882	Two-cycle gasoline engine (Gottlieb Daimler)	Daimler builds a small, high-speed two-cycle gasoline engine. He will also build a successful motorcycle in 1885 and (with Wilhelm Maybach) an automobile in 1889.
1883	Solar cell (Charles Fritts)	American scientist Fritts designs the first solar cell.
1883	Shoe-lasting machine (Jan Ernst Matzeliger)	The machine sews the upper part of the shoe to the sole and reduces the cost of shoes by 50 percent.
1884	Fountain pen (Lewis Waterman)	The commonly told story is that Waterman was selling insurance and lost a large contract when his pen leaked all over it, prompting him to invent the leak-proof fountain pen.
1884	Vector calculus (Oliver Heaviside)	Heaviside develops vector calculus to represent James Clerk Maxwell's electromagnetic theory with only four equations instead of the usual twenty.

1884	Roll film (George Eastman)	Roll film will replace heavy plates, making photography both more accessible and more convenient. In 1888, Eastman and William Hall invent the Kodak camera. These developments open photography to the masses.
1884	Roll film (George Eastman)	Roll film will replace heavy plates, making photography both more accessible and more convenient. In 1888, Eastman and William Hall invent the Kodak camera. These developments open photography to the masses.
1884	Steam turbine (Charles Parsons)	Designed for ships, Parsons's steam turbine is smaller, more efficient, and more durable than the steam engines in use.
1884	Census tabulating machine (Herman Hollerith)	Hollerith's machine uses punch cards to tabulate 1890 census data. He goes on to found the company that later becomes International Business Machines (IBM).
1885	Machine gun (Hiram Stevens Maxim)	Maxim patents a machine gun that can fire up to six hundred bullets per minute.
1885	Bicycle (John Kemp Starley)	English inventor Starley is responsible for producing the first modern bicycle, called the Rover.
1885	First gasoline-powered automobile (Carl Benz)	Benz not only manufactures the first gas-powered car but also is first to mass-produce automobiles.
1885	Incandescent gas mantle (Carl Auer von Welsbach)	The Austrian scientist invents the incandescent gas mantle.
1886	Dictaphone (Charles Sumner Tainter)	Tainter, an American engineer who frequently worked with Alexander Graham Bell, designs the Dictaphone.
1886	Dishwasher (Josephine Garis Cochran)	Like modern washers, Cochran's dishwasher cleans dishes with sprays of hot, soapy water and then air-dries them.
1886	Electric transformer (William Stanley)	Stanley, working at Westinghouse, builds the first practical electric transformer.
1886	Gramophone (Emile Berliner)	A major contribution to the music recording industry, Berliner's gramophone uses flat record discs for recording sound. Berliner goes on to produce a helicopter prototype (1906-1923).
1886	Linotype machine (Ottmar Mergenthaler)	Pressing keys on the machine's keyboard releases letter molds that drop into the current line. The lines are assembled into a page and then filled with molten lead.

1886	Electric-traction system (Frank J. Sprague)	Sprague's motor can propel a tram up a steep hill without its slipping.
1886	Hall-Héroult electrolytic process (Charles Martin Hall and Paul Héroult)	The industrial production of aluminum from bauxite ore made aluminum widely available. Prior to the electrolytic process, aluminum was a precious metal with a value about equal to that of silver.
1886	Coca-Cola (John Stith Pemberton)	Developed as pain reliever less addictive than available opiates, the original Coca-Cola contains cocaine from cola leaves and caffeine from kola nuts. It achieves greater success as a beverage marketed where alcohol is prohibited.
1886	Yellow pages (Reuben H. Donnelly)	Yellow paper was used in 1883 when the printer ran out of white paper. Donnelly now purposely uses yellow paper for business listings.
1886	Fluorine (Henri Moissan)	French chemist Moissan isolates fluorine and is awarded the Nobel Prize in Chemistry in 1906. Compounds of fluorine are used in toothpaste and in public water supplies to help prevent tooth decay.
1887	Radio transmitter and receiver (Heinrich Hertz)	Hertz will use these devices to discover radio waves and confirm that they are electromagnetic waves that travel at the speed of light; he also discovers the photoelectric effect.
1887	Distortionless transmission lines (Oliver Heaviside)	Heaviside recommends that induction coils be added to telephone and telegraph lines to correct for distortion.
1887	Olds horseless carriage (Ransom Eli Olds)	Olds develops a three-wheel horseless carriage using a steam engine powered by a gasoline burner.
1887	Synchronous multiplex railway telegraph (Granville T. Woods)	Woods patents a variation of the induction telegraph that allows messages to be sent between moving trains and between trains and railway stations. He will eventually obtain sixty patents on electrical and electromechanical devices, most of them related to railroads and communications.
1888	Cordite (Sir James Dewar)	Dewar, with Sir Frederick Abel, invents cordite, a smokeless gunpowder that is widely adopted for munitions.
1888	Pneumatic rubber tire (John Boyd Dunlop)	Dunlop's pneumatic tires revolutionize the ride for cyclists and motorists.
1888	Kodak camera	George Eastman, founder of Eastman Kodak, introduces the first Kodak camera.

1889	Electric drill (Arthur James Arnot)	Arnot's drill is used to cut holes in rock and coal.
1889	Bromine extraction (Herbert Henry Dow)	Dow's method for extracting bromine from brine enables bromine to be widely used in medicines and in photography.
1889	Rayon (Louis-Marie-Hilaire Bernigaud de Chardonnet)	Bernigaud de Chardonnet, a French chemist, invents rayon, the first artificial fiber, as an alternative to silk.
1889	Celluloid film	George Eastman replaces paper film with celluloid.
1890	Improved carbon electric arc (Hertha Marks Ayrton)	The carbon arc produces an intense light that is used in streetlights.
1890	Pneumatic (air) hammer (Charles B. King)	A worker with a pneumatic hammer can break up a concrete slab many times faster than can a worker armed with only a sledgehammer.
1890	Smokeless gunpowder (Hudson Maxim)	Maxim (perhaps with brother Hiram) develops a version of smokeless gunpowder that is adopted for modern firearms; he goes on to develop a smokeless cannon powder that will be used during World War I.
1890	Rubber gloves in the operating room	American surgeon William Stewart Halsted introduces the use of sterile rubber gloves in the operating room.
1891	Rubber automobile tires (André and édouard Michelin)	The Michelin brothers manufacture air-inflated tires for bicycles and later automobiles, which leads to a successful ad campaign, featuring the Michelin Man (Bibendum).
1891	Carborundum (Edward Goodrich Acheson)	Attempting to create artificial diamonds, Acheson instead synthesizes silicon carbide, the second hardest substance known. He will develop an improved graphite-making process in 1896.
1892	Kinetoscope (Thomas Alva Edison)	Edison completes Kinetoscope; the first demonstration is held a year later.
1892	Calculator (William Seward Burroughs)	Burroughs builds the first practical key-operated calculator; it prints entries and results.
1892	Dewar flask (Sir James Dewar)	Dewar invents the vacuum bottle, a vacuum-jacketed vessel for storing and maintaining the temperature of hot or cold liquids.
1892	Artificial silk (Charles F. Cross and Edward J. Bevan)	British chemists Cross and Bevan create viscose artificial silk (cellulose acetate).

1893	Color photography plate (Gabriel Jonas Lippmann)	Also known as the Lippmann plate for its inventor, the color photography plate uses interference patterns, rather than various colored dyes, to reproduce authentic color.
1893	Alternating current calculations (Charles Proteus Steinmetz)	Steinmetz's calculations make it possible for engineers to determine alternating current reliably, without depending on trial and error, when designing a new motor.
1894	Cereal flakes (John Harvey Kellogg)	Kellogg, a health reformer who advocates a diet of fruit, nuts, and whole grains, invents flaked breakfast cereal with the help of his brother, Will Keith Kellogg. In 1906 Kellogg established a company in Battle Creek, Michigan, to manufacture his breakfast cereal.
1894	Automatic loom (James Henry Northrop)	Northrop builds the first automatic loom.
1895	Streamline Aerocycle bicycle (Ignaz Schwinn)	Through hard work and dedication, Schwinn develops a bicycle that eventually makes his name synonymous with best of bicycles.
1895	Victrola phonographs (Eldridge R. Johnson)	Johnson develops a spring-driven motor for phonographs that provides the constant record speed necessary for good sound reproduction.
1895	Cinématographe (Auguste and Louis Lumière)	The Lumière brothers' combined motion-picture camera, printer, and projector helps establish the movie business. Using a very fine-grained silver-halide gelatin emulsion, they cut photographic exposure time down to about one minute.
1895	Antenna	Aleksandr Stepanovich Popov demonstrated radio reception with a coherer, which he also used as a lightning detector.
1896	Wireless telegraph system (Guglielmo Marconi)	Marconi is the first to send wireless signals across the Atlantic Ocean, inaugurating a new era of telecommunications.
1896	Aerodromes (Samuel Pierpont Langley)	Langley's "Aerodrome number 6," using a small gasoline engine, makes an unmanned flight of forty-eight hundred feet.
1896	Four-wheel horseless carriage (Ransom Eli Olds)	Oldsmobile patents Olds's internal combustion engine and applies it to his four-wheel horseless carriage, naming it the "automobile."
1896	High-frequency generator and transformer (Elihu Thomson)	Thomson produces an electric air drill, which advances welding to improve the construction of new appliances and vehicles. He will also invent other electrical devices, including an improved X-ray tube.

1896	X-ray tube (Wilhelm Conrad Röntgen)	After discovering X radiation, Röntgen mails an X-ray image of a hand wearing a ring and paving the way for the medical use of X-ray imaging—one of the most important discoveries ever made for medical science.
1896	Better sphygmomanometer (Scipione Riva-Rocci)	Italian pediatrician Riva-Rocci develops the most successful and easy-to-use blood-pressure gauge.
1897	Modern submarine (John Philip Holland)	Holland's submarine is the first to use a gasoline engine on the surface and an electric engine when submerged.
1897	Oscilloscope (Karl Ferdinand Braun)	The oscilloscope is an invaluable device used to measure and display electronic waveforms.
1897	Jenny coupler (Andrew Jackson Beard)	Beard's automatic coupler connects the cars in a train without risking human life. The introduction of automatic couplings reduces coupling-related injuries by a factor of five.
1897	Escalator (Charles Seeberger)	Before Seeberger built the escalator in its now-familiar form, it was a novelty ride at the Coney Island amusement park.
1897	Automobile components (Alexander Winton)	The Winton Motor Carriage Company is incorporated, and Winton begins manufacturing automobiles. His popular "reliability runs" helps advertise automobiles to the American market. He will produce the first American diesel engine in 1913.
1897	Diesel engine (Rudolf Diesel)	Diesel's internal combustion engine rivals the efficiency of the steam engine.
1897	Electron discovered (J. J. Thomson)	Thomson uses an evacuated tube with a high voltage across electrodes sealed in the ends. Invisible particles (later named electrons) stream from one of the electrodes, and Thomson establishes the particles' properties.
1898	Flashlight (Conrad Hubert)	Hubert combines three parts—a battery, a lightbulb, and a metal tube—to produce a flashlight.
1898	Mercury vapor lamp (Peter Cooper Hewitt)	Hewitt's mercury vapor lamp proves to be more efficient than incandescent lamps.
1899	Alpha particle discovered (Ernest Rutherford)	Rutherford detects the emission of helium 4 nuclei (alpha particles) in the natural radiation from uranium.
1900	Aspirin	Aspirin is patented by Bayer and sold as a powder. In 1915 it is sold in tablets.

1900	Dirigibles (Ferdinand von Zeppelin)	Von Zeppelin flies his airship three years before the Wright brothers' airplane.
1900	Gamma ray discovered (Paul Villard)	Villard discovers gamma rays in the natural radiation from uranium. They resemble very high-energy X rays.
1900	Brownie camera:	George Eastman introduces the Kodak Brownie camera. It is sold for $1 and the film it uses costs 15 cents. The Brownie made photography an accessible hobby to almost everyone.
1901	Acousticon hearing aid (Miller Reese Hutchison)	Hutchison invents a battery-powered hearing aid in the hopes of helping a mute friend speak.
1901	Vacuum cleaner (H. Cecil Booth)	Booth patents his vacuum cleaner, a machine that sucks in and traps dirt. Previous devices, less effective, had attempted to blow the dirt away.
1901	String galvanometer (electrocardiograph) (Willem Einthoven)	Einthoven's device passes tiny currents from the heart through a silver-coated silicon fiber, causing the fiber to move. Recordings of this movement can show the heart's condition.
1901	Silicone (Frederick Stanley Kipping)	English chemist Kipping studies the organic compounds of silicon and coins the term "silicone."
1902	Airplane engine (Charles E. Taylor)	Taylor begins building engines for the Wright brothers' airplanes.
1902	Lionel electric toy trains (Joshua Lionel Cowen)	Cowen publishes the first Lionel toy train catalog. Lionel miniature trains and train sets become favorite toys for many years and are prized by collectors to this day.
1902	Air conditioner (Willis Carrier)	Whole-house air-conditioning becomes possible.
1903	Windshield wipers (Mary Anderson)	At first, the driver operated the wiper with a lever from inside the car.
1903	Wright Flyer (Wilbur and Orville Wright)	The Wright Flyer is the first heavier-than-air machine to solve the problems of lift, propulsion, and steering for controlled flight.
1903	Safety razor with disposable blade (King Camp Gillette)	Gillette's razor used a disposable and relatively cheap blade, so there was no need to sharpen it.

1903	Space-traveling projectiles (Konstantin Tsiolkovsky)	Tsiolkovsky publishes "The Exploration of Cosmic Space by Means of Reaction-Propelled Apparatus," in which he includes an equation for calculating escape velocity (the speed required to propel an object beyond Earth's field of gravity). He is also recognized for the concept of rocket propulsion and for the wind tunnel.
1903	Ultramicroscope (Richard Zsigmondy)	Zsigmondy builds the ultramicroscope to study colloids, mixtures in which particles of a substance are dispersed throughout another substance.
1903	Spinthariscope (Sir William Crookes)	Crookes invents a device that sparkles when it detects radiation. He also develops and experiments with the vacuum tube, allowing later physicists to identify alpha and beta particles and X rays in the radiation from uranium.
1903	Crayola crayons (Edwin Binney)	With his cousin C. Harold Smith, Binney invents dustless chalk and crayons marketed under the trade name Crayolas.
1903	Motorcycle	Harley-Davidson produces the first motorcycle, built to be a racer.
1903	Electric iron	Earl Richardson introduces the lightweight electric iron.
1904	Glass bottle machine	American inventor Michael Joseph Owens designs a machine that produces glass bottles automatically.
1905	Novocaine (Alfred Einkorn)	While researching a safe local anesthetic to use on soldiers, German chemist Einkorn develops novocaine, which becomes a popular dental anesthetic.
1905	Special relativity (Albert Einstein)	At the age of twenty-six, Einstein uses the constancy of the speed of light to explain motion, time, and space beyond Newtonian principles. During the same year, he publishes papers describing the photoelectric effect and Brownian motion.
1905	Intelligence testing	French psychologist Alfred Binet devises the first of a series of tests to measure an individual's innate ability to think and reason.
1906	Hair-care products (Madam C. J. Walker)	Walker trains a successful sales force to go door-to-door and sell directly to women. Her saleswomen, beautifully dressed and coiffed, are instructed to pamper their clients.
1906	Broadcast radio (Reginald Aubrey Fessenden)	In broadcast radio, sound wave forms are added to a carrier wave and then broadcast. The carrier wave is subtracted at the receiver leaving only the sound.

1906	Klaxon horn (Miller Reese Hutchison)	Hutchison files a patent application for the electric automobile horn.
1906	Chromatography (Mikhail Semenovich Tswett)	Tswett, a Russian botanist, invents chromatography.
1906	Freeze-drying (Jacques Arsène d'Arsonval and George Bordas)	D'Arsonval and Bordas invent freeze-drying, but the practice is not commercially developed until after World War II.
1907	Sun valve (Nils Gustaf Dalén)	Dalén's device uses sunlight to activate a lighthouse beacon. His other inventions make automated acetylene beacons in lighthouses possible.
1907	Mantoux tuberculin skin test (Charles Mantoux)	French physician Mantoux develops a skin-reaction test to diagnose tuberculosis. He builds on the work of Robert Koch and Clemens von Pirquet.
1908	Helium liquefaction (Heike Kamerlingh Onnes)	Kamerlingh Onnes produces liquid helium at a temperature of about 4 kelvins. He will also discover superconductivity in several materials cooled to liquid helium temperature.
1908	"Tin Lizzie" (Model T) automobile (Henry Ford)	Ford's development of an affordable automobile, manufactured using his assembly-line production methods, revolutionize the U.S. car industry.
1908	Electrostatic precipitator (Frederick Gardner Cottrell)	The electrostatic precipitator is invaluable for cleaning stack emissions.
1908	Geiger-Müller tube (Hans Geiger)	Geiger invents a device, popularly called the Geiger counter, that is a reliable, portable radiation detector. Later his student Walther Müller helps improve the instrument.
1908	Vacuum cleaner (James Murray Spangler)	Spangler receives a patent on his electric sweeper, and his Electric Suction Sweeper Company eventually becomes the Hoover Company, the largest such company in the world.
1908	Cellophane (Jacques Edwin Brandenberger)	Brandenberger builds a machine to mass-produce cellophane, which he has earlier synthesized while unsuccessfully attempting to develop a stain-resistant cloth.
1908	Water treatment	Chlorine is used to purify water for the first time in the United States, in New Jersey, helping to reduce waterborne illnesses such as cholera, typhoid, and dysentery.

1908	Audion (Lee De Forest)	De Forest invents a vacuum tube used in sound amplification. In 1922, he will develop talking motion pictures, in which the sound track is imprinted on the film with the pictures, instead of on a record to be played with the film, leading to exact synchronization of sound and image.
1909	Synthetic fertilizers (Fritz Haber)	Haber also invents the Haber process to synthesize ammonia on a small scale.
1909	Maxim silencer (Hiram Percy Maxim)	The silencer reduces the noise from firing the Maxim machine gun.
1909	pH scale	Danish chemist Søren Sørensen introduces the pH scale as a standard measure of alkalinity and acidity.
1910	Chlorinator (Carl Rogers Darnall)	Major Darnall builds a machine to add liquid chlorine to water to purify it for his troops. His method is still widely used today.
1910	Bakelite (Leo Hendrik Baekeland)	Bakelite is the first tough, durable plastic.
1910	Neon lighting (Georges Claude)	Brightly glowing neon tubes revolutionize advertising displays.
1910	Syphilis treatment	German physician Paul Ehrlich and Japanese physician Hata Sahachirò discover the effective treatment of arsphenamine (named Salvarsan by Ehrlich) for syphilis.
1911	Colt .45 automatic pistol (John Moses Browning)	Commonly called the Colt Model 1911, an improved version of the Colt Model 1900, the Colt .45 is the first autoloading pistol produced in America. Among Browning's other inventions are the Winchester 94 lever-action rifle and the gas-operated Colt-Browning machine gun.
1911	Gyrocompass (Elmer Ambrose Sperry)	Sperry receives a patent for a nonmagnetic compass that indicates true north.
1911	Atomic nucleus identified (Ernest Rutherford)	Rutherford discovered the nucleus by bombarding a thin gold foil with alpha particles. Some were deflected through large angles showing that something small and hard was present.
1911	Ductile tungsten (William David Coolidge)	Coolidge also invented the Coolidge tube, an improved X-ray producing tube.
1911	Ochoaplane (Victor Leaton Ochoa)	In addition to inventing this plane with collapsible wings, Ochoa also developed an electricity-generating windmill.

1911	Automobile electric ignition system (Charles F. Kettering)	Kettering invents the first electric ignition system for cars.
1912	Automatic traffic signal system (Garrett Augustus Morgan)	Morgan also invents a safety hood that served as a rudimentary gas mask.
1913	Gyrostabilizer (Elmer Ambrose Sperry)	Sperry develops the gyrostabilizer, a device to control the roll, pitch, and yaw of a moving ship. He will go on to invent the flying bomb, which is guided by a gyrostabilizer and by radio control.
1913	Erector set (Alfred C. Gilbert)	Erector sets provide hands-on engineering experience for countless children.
1913	Zipper (Gideon Sundback)	While others had made zipper-like devices but had never successfully marketed them, Sundback designs a zipper in approximately its present form. He also invents a machine to make zippers.
1913	Improved electric lightbulb (Irving Langmuir)	Langmuir fills his lightbulb with a low-pressure inert gas to retard evaporation from the tungsten filament.
1913	Industrialization of the Haber process (Carl Bosch)	Bosch scales up Haber's process for making ammonia to an industrial capacity. The process comes to be known as the Haber-Bosch process.
1913	Bergius process (Friedrich Bergius)	Bergius develops high-pressure, high-temperature process to produce liquid fuel from coal.
1913	Electric dishwasher	The Walker brothers of Philadelphia produce the first electric dishwasher.
1913	Stainless steel (Harry Brearley)	Brearley invents stainless steel.
1913	Thermal cracking (William Burton and Robert Humphreys)	Standard Oil chemical engineers Burton and Humphreys discover thermal cracking, a method of oil refining that significantly increases gasoline yields.
1914	Backless brassiere (Caresse Crosby)	The design of a new women's undergarment leads to the expansion of the U.S. brassiere industry. Caresse was originally a marketing name that Mary Phelps Jacob eventually adopted as her own.
1915	Panama Canal opens	The passageway between the Atlantic and Pacific oceans creates a boon for the shipping industry.

1915	General relativity (Albert Einstein)	Einstein refines his 1905 theory of relativity (now called special relativity) to describe the theory that states that uniform accelerations are almost indistinguishable from gravity. Einstein's theory provides the basis for physicists' best understanding of gravity and of the framework of the universe.
1915	Jenny (Glenn H. Curtiss)	The Jenny becomes a widely used World War I biplane, and Curtis becomes a general manufacturer of airplanes and airplane engines.
1915	Pyrex	Corning's brand name for glassware is introduced.
1915	Warfare	Depth-charge bombs are first used by the Allies against German submarines
1916	By-products of sweet potatoes and peanuts (George Washington Carver)	Carver publishes his famous bulletin on 105 ways to prepare peanuts.
1919	Proton discovered (Ernest Rutherford)	After bombarding nitrogen gas with alpha particles (helium 4 nuclei), Rutherford observes that positive particles with a single charge are knocked loose. They are protons.
1919	Toaster (Charles Strite)	Strite invents the first pop-up toaster.
1920	Microelectrode (Ida H. Hyde)	Hyde's electrode is small enough to pierce a single cell. Chemicals can also be very accurately deposited by the microprobe.
1921	Ready-made bandages	Johnson & Johnson puts Band-Aids on the market.
1921	Antiknock solution (Thomas Midgley, Jr.)	While working at a General Motors subsidiary, American mechanical engineer Midgley develops an antiknock solution for gasoline.
1921	Insulin	University of Toronto researchers Frederick Banting, J. J. R. Macleod, and Charles Best first extract insulin from a dog, and the first diabetic patient is treated with purified insulin the following year. Banting and Macleod win the 1923 Nobel Prize in Physiology or Medicine for their discovery of insulin.
1923	Improved telephone speaker (Georg von Békésy)	Békésy's studies of the human ear lead to an improved telephone earpiece. He will also construct a working model of the inner ear.
1923	Quick freezing (Clarence Birdseye)	Birdseye's quick-freezing process preserves food's flavor and texture better than previously used processes.

1924	Coincidence method of particle detection (Walther Bothe)	Bothe's method proves invaluable in the use of gamma rays to discover nuclear energy levels.
1924	Ultracentrifuge (Theodor Svedberg)	Svedberg's ultracentrifuge can separate isotopes, such as uranium 235 from uranium 238, from each other—a critical step in building the simplest kind of atomic bomb.
1924	EEG	German scientist Hans Berger records the first human electroencephalogram (EEG), which shows electrical patterns in the brain.
1925	Leica I camera	Leitz introduces the first 35-millimeter Leica camera at the Leipzig Spring Fair.
1925	First U.S. television broadcast	Charles Francis Jenkins transmits the silhouette image of a toy windmill.
1926	Automatic power loom (Sakichi Toyoda)	Toyoda's loom helps Japan catch up with the western Industrial Revolution.
1926	Liquid-fueled rocket (Robert H. Goddard)	A solid-fueled rocket is either on or off, but a liquid-fueled rocket can be throttled up or back and can be shut off before all the fuel is expended.
1927	Aerosol can (Erik Rotheim)	Norwegian engineer Rotheim patents the aerosol can and valve.
1927	Adiabatic demagnetization (William Francis Giauque)	Adiabatic demagnetization is part of a refrigeration cycle that, when used enough times, can chill a small sample to within a fraction of a kelvin above absolute zero.
1927	All-electronic television (Philo T. Farnsworth)	Farnsworth transmits the first all-electronic television image using his newly developed camera vacuum tube, known as the image dissector. Previous systems combined electronics with mechanical scanners.
1927	First flight across the Atlantic	Charles Lindbergh flies the Spirit of St. Louis across the Atlantic. He is the first to make a solo, nonstop flight across the ocean.
1927	Iron lung (Philip Drinker)	Drinker, a Harvard medical researcher, assisted by Louis Agassiz Shaw, devises the first modern practical respirator using an iron box and two vacuum cleaners. Drinker calls the device the iron lung.

1927	Garbage disposal (John W. Hammes)	American architect Hammes develops the first garbage disposal to make cleaning up the kitchen easier for his wife. It is nicknamed the "electric pig" when it first goes on the market.
1927	Adjustable-temperature iron	The Silex Company begins to sell the first iron with an adjustable temperature control.
1927	Analogue computer (Vannevar Bush)	Bush builds the first analogue computer. He is also the first person to describe the idea of hypertext.
1928	Sliced bread (Otto F. Rohweddeer)	Bread that came presliced was advertised as "the greatest forward step in the baking industry since bread was wrapped." Today the phrase "the greatest thing since sliced bread" is used to describe any innovation that has a broad, positive impact on daily life.
1928	First television programs	First regularly scheduled television programs in the United States air. They are produced out of a small, experimental station in Wheaton, Maryland.
1928	Link Trainer (Edwin Albert Link)	Link's flight simulator created realistic conditions in which to train pilots without the expense or risk of an actual air flight. Link also developed a submersible decompression chamber.
1928	New punch card	IBM introduces a new punch card that has rectangular holes and eight columns.
1928	Radio network	NBC establishes the first coast-to-coast radio network in the United States.
1928	Pap smear (George N. Panpanicolaou)	Greek cytopathologist Panpanicolaou patents the pap smear, a test that helps detect uterine cancer.
1928	Portable offshore drilling (Louis Giliasso)	Giliasso creates an efficient portable method of offshore drilling by mounting a derrick and drilling outfit onto a submersible barge.
1929	Iconoscope (Vladimir Zworykin)	Zworykin claims that he, not Philo T. Farnsworth, should be credited with the invention of television.
1929	Strobe light (Harold E. Edgerton)	Edgerton's strobe is used as a flash bulb. He pioneers the development of high-speed photography.
1929	Dymaxion products (R. Buckminster Fuller)	Fuller's "Dymaxion" products feature an energy-efficient house using prefabricated, easily shipped parts.

1929	Van de Graaff generator (Robert Jemison van de Graaff)	Van de Graaff invents the Van de Graaff generator, which accumulates electric charge on a moving belt and deposits it in a hollow glass sphere at the top.
1929-1936	Cyclotron (Ernest Orlando Lawrence and M. Stanley Livingston)	Lawrence and Livingston are studying particle accelerators and develop the cyclotron, which consists of a vacuum tank between the poles of a large magnet. Alternating electric fields inside the tank can accelerate charged particles to high speeds. The cyclotron is used to probe the atomic nucleus or to make new isotopes of an element, including those used in medicine.
1930	Schmidt telescope (Bernhard Voldemar Schmidt)	Schmidt's telescope uses a spherical main mirror and a correcting lens at the front of the scope. It can photograph large fields with little distortion.
1930	Pluto discovered (Clyde Tombaugh)	Tombaugh observes a body one-fifth the mass of Earth's moon. Pluto comes to be regarded as the ninth planet of the solar system, but in 2006 it is reclassified as one of the largest-known Kuiper Belt objects, a dwarf planet.
1930	Freon refrigeration and air-conditioning (Charles F. Kettering)	After inventing an electric starter in 1912 and the Kettering Aerial Torpedo in 1918 (the world's first cruise missile), Kettering and Thomas Midgley, Jr., use Freon gas in their cooling technology. (Freon will later be banned because of the effects of chlorofluorocarbons on Earth's ozone layer.)
1930	Synthetic rubber (Wallace Hume Carothers)	Carothers synthesizes rubber and goes on to develop nylon in 1935. His work professionalizes polymer chemistry as a scientific field.
1930	Scotch tape (Richard G. Drew)	After inventing masking tape, Drew invents the first waterproof, see-through, pressure-sensitive tape that also acted as a barrier to moisture.
1930	Military and commercial aircraft (Andrei Nikolayevich Tupolev)	Tupolev emerges as one of the world's leading designers of military and civilian aircraft. His aircraft set nearly eighty world records.
1930's	Washing machine (John W. Chamberlain)	Chamberlain invents a washing machine that enables clothes to be washed, rinsed, and have the water extracted from them in a single operation.
1931	Electric razor (Jacob Schick)	Schick introduces his first electric razor, which allows dry shaving. It has a magazine of blades held in the handle.
1931	Radio astronomy (Karl G. Jansky)	One of the founders of the field of radio astronomy, Janksy detects radio static coming from the Milky Way's center.

1932	Positron discovered (Carl D. Anderson)	Anderson discovers the positron, a positive electron and an element of antimatter.
1932	Neoprene (Julius Nieuwland)	The first synthetic rubber is marketed.
1932	Neutron discovered (James Chadwick)	Chadwick detects the neutron, an atomic particle with no charge and a mass only slightly greater than that of a proton. Except for hydrogen 1, the atomic nuclei of all elements consist of neutrons and protons.
1932	Phillips-head screw (Henry M. Phillips)	The Phillips-head screw has an X-shaped slot in the head and can withstand the torque of a machine-driven screwdriver, which is greater than the torque that can be withstood by the conventional screw.
1932	Duplicating device for typewriters (Beulah Louise Henry)	Henry's invention uses three sheets of paper and three ribbons to produce copies of a document as it is typewritten. Henry also develops children's toys—for example, a doll the eye color of which can be changed.
1932	Cockroft-Walton accelerator (John Douglas Cockcroft and Ernest Thomas Sinton Walton)	The Cockroft-Walton accelerator is used to fling charged particles at atomic nuclei in order to investigate their properties.
1932	Richter scale (Charles Francis Richter)	Richter develops a scale to describe the magnitude of earthquakes; it is still used today.
1932	Neutron (Sir James Chadwick)	Chadwick proves the existence of neutrons; he is awarded the 1935 Nobel Prize in Physics for his work.
1933	Nuclear chain reaction (Leo Szilard)	Szilard conceives the idea of a nuclear chain reaction. He becomes a key figure in the Manhattan Project, which eventually builds the atomic bomb.
1933	Magnetic tape recorder (Semi Joseph Begun)	Begun builds the first tape recorder, a dictating machine using wire for magnetic recording. He also develops the first steel tape recorder for mobile radio broadcasting and leads research into telecommunications and underwater acoustics.
1933	Electron microscope (Ernst Ruska)	Ruska makes use of the wavelengths of electrons—shorter than those of visible light—to build a microscope that can image details at the subatomic level.
1933	Recording	Alan Dower Blumlein's patent for stereophonic recording is granted.

1933	Polyethylene (Eric Fawcett and Reginald Gibson)	Fawcett and Gibson of Imperial Chemical Industries in London accidentally discover polyethylene. Hula hoops and Tupperware are just two of the products made with the substance.
1933	Modern airliner	Boeing 247 becomes the first modern airliner.
1933	Solo flight	Wiley Post makes the first around-the-world solo flight.
1934	First bathysphere dive	Charles William Beebe and Otis Barton make the first deep-sea dive in the Beebe-designed bathysphere off the Bermuda coast.
1934	Langmuir-Blodgett films (Katharine Burr Blodgett)	A thin Langmuir-Blodgett film deposited on glass can make it nearly nonreflective.
1934	Passenger train	The Burlington Zephyr, America's first diesel-powered streamlined passenger train, is revealed at the World's Fair in Chicago.
1935	Frequency modulation (Edwin H. Armstrong)	Armstrong exploits the fact that, since there are no natural sources of frequency modulation (FM), FM broadcasts are static-free.
1935	Diatometer (Ruth Patrick)	Patrick's diatometer is a device placed in the water to collect diatoms and allow them to grow. The number of diatoms is sensitive to water pollution.
1935	Kodachrome color film (Leopold Mannes and Leopold Godowsky, Jr.)	Mannes and Godowsky invent Kodachrome, a color film that is easy to use and produces vibrant colors. (With the digital revolution of the late twentieth century, production of Kodachrome is finally retired in 2009.)
1935	Physostigmine and cortisone (Percy Lavon Julian)	Julian synthesizes physostigmine, used to treat glaucoma, and cortisone, used for arthritis. He will hold more than 130 patents and will become the first African American chemist inducted into the National Academy of Sciences.
1935	Mobile refrigeration (Frederick McKinley Jones)	Mobile refrigeration enables the shipping of heat-sensitive products and compounds, from blood to frozen food.
1935	Radar-based air defense system (Sir Robert Alexander Watson-Watt)	Watson-Watt's technical developments and his efforts as an administrator will be so important to the development of radar that he will be called the "father of radar."

1935	Fallingwater (Frank Lloyd Wright)	Wright designs and builds a showcase house blending its form with its surroundings. One of the greatest architects of the twentieth century, he will produce many architectural innovations in structure, materials, and design.
1936	Field-emission microscope (Erwin Wilhelm Müller)	Müller completes his dissertation, "The Dependence of Field Electron Emission on Work Function," and goes on to develop the field-emission microscope, which can resolve surface features as small as 2 nanometers.
1936	Pentothal (Ernest Volwiler and Donalee Tabern)	Pentothal is a fast-acting intravenous anesthetic.
1937	Muon discovered (Seth Neddermeyer)	Neddermeyer, working with Carl Anderson, J. C. Street, and E. C. Stevenson discover the muon (a particle similar to a heavy electron) while examining cosmic-ray tracks in a cloud chamber.
1937	Concepts of digital circuits and information theory (Claude Elwood Shannon)	Shannon's most important contributions were electronic switching and using information theory to discover the basic requirements for data transmission.
1937	X-ray crystallography (Dorothy Crowfoot Hodgkin)	Hodgkin uses X-ray crystallography to reveal the structure of molecules. She goes on to win the 1964 Nobel Prize in Chemistry.
1937	Model K computer (George Stibitz)	The model K, an early electronic computer, employs Boolean logic.
1937	Artificial sweetener	American chemist Michael Sveda invents cyclamates, which is used as a noncaloric artificial sweetener until it is banned by the U.S. government in 1970 because of possible carcinogenic effects.
1937	First pressurized airplane cabin	The first pressurized airplane cabin is achieved in the United States with Lockheed's XC-35.
1937	Antihistamines (Daniel Bovet)	Swiss-born Italian pharmacologist Bovet discovers antihistamines. He is awarded the 1957 Nobel Prize in Physiology or Medicine for his work.
1938	Teflon (Roy J. Plunkett)	Plunkett accidentally synthesizes polytetrafluoroethylene (PTFE), now commonly known as Teflon, while researching chlorofluorocarbon refrigerants.

1938	Electron microscope (James Hillier and Albert Prebus)	Adapting the work of German physicists, Hillier and Prebus develop a prototype of the electron microscope; and in 1940 Hillier produces the first commercial electron microscope available in the United States.
1938	Xerography (Chester F. Carlson)	Xerography uses electrostatic charges to attract toner particles to make an image on plain paper. A hot wire then fuses the toner in place.
1938	Walkie-talkie (Alfred J. Gross)	Gross's portable, two-way radio allows the user to move around while sending messages without remaining tied to a bulky transmitter. Gross invents a pager in 1949 and a radio tuner in 1950 that automatically follows the drift in carrier frequency due to movement of a sender or receiver.
1937-1938	Analogue computer (George Philbrick)	Philbrick builds the Automatic Control Analyzer, which is an electronic analogue computer.
1939	Helicopter (Igor Sikorsky)	Sikorsky, formerly the chief construction engineer and test pilot for the first four-engine aircraft, tests his helicopter, the Vought-Sikorsky 300, which after improvements will emerge as the world's first working helicopter.
1939	Jet engine (Hans Joachim Pabst von Ohain)	The first jet-powered aircraft flies in 1939, while the first jet fighter will fly in 1941.
1939	Atanasoff-Berry Computer (John Vincent Atanasoff and Clifford Berry)	The ABC, the world's first electronic digital computer, uses binary numbers and electronic switching, but it is not programmable.
1939	DDT (Paul Hermann Müller)	Müller discovers the insect-repelling properties of DDT. He is awarded the 1948 Nobel Prize in Physiology or Medicine.
1940's	Solar technology (Maria Telkes)	Telkes develops the solar oven and solar stills to produce drinking water from ocean water.
1940	Cavity magnetron (Henry Boot and John Randall)	Boot and Randall develop the cavity magnetron, which advances radar technology.
1940	Penicillin	Sir Howard Walter Florey and Ernst Boris Chain isolate and purify penicillin. They are awarded, with Sir Alexander Fleming, the 1945 Nobel Prize in Physiology or Medicine.
1940	Blood bank (Charles Richard Drew)	Drew establishes blood banks for World War II soldiers.

1940	Color television (Peter Carl Goldmark)	Goldmark produces a system for transmitting and receiving color-television images using synchronized rotating filter wheels on the camera and on the receiver set.
1940	Paintball gun (Charles and Evan Nelson)	The gun and paint capsules, invented to mark hard-to-reach trees in the forest, are eventually used for the game of paintball (1981), in which people shoot each other with paint.
1940	Audio oscillator (William Redington Hewlett)	Hewlett invents the audio oscillator, a device that creates one frequency (pure tone) at a time. It is the first successful product of his Hewlett-Packard Company.
1940	Antibiotics (Selman Abraham Waksman)	Waksman, through study of soil organisms, finds sources for the world's first antibiotics, including streptomycin and actinomycin.
1940	Plutonium (Glenn Theodore Seaborg)	Seaborg synthesizes one of the first transuranium elements, plutonium. He becomes one of the leading figures on the Manhattan Project, which will build the atomic bomb. While he and others urged the demonstration of the bomb as a deterrent, rather than its use on the Japanese civilian population, the latter course was taken.
1940	Thompson submachine gun (John T. Thompson)	Thompson works with Theodore Eickhoff and Oscar Payne to invent the American version of the submachine gun.
1940	Automatic auto transmission	General Motors offers the first modern automatic automobile transmission.
1941	Jet engine (Sir Frank Whittle)	Whittle develops the jet engine independent of Hans Joachim Pabst von Ohain in Germany. After World War II, they meet and become good friends.
1941	Solid-body electric guitar (Les Paul)	Paul's guitar lays the foundation for rock music. He also develops multitrack recording in 1948.
1941	Z3 programmable computer (Konrad Zuse)	Zuse and his colleagues complete the first general-purpose, programmable computer, the Z3, in December. In 1950, Zuse will sell a Z4 computer—the only working computer in Europe.
1941	Velcro (Georges de Mestral)	Burrs sticking to his dog's fur give de Mestral the idea for Velcro, which he perfects in 1948.
1941	Dicoumarol	The anticoagulant drug dicoumarol is identified and synthesized.

1941	RDAs	The first Recommended Dietary Allowances (RDAs), nutritional guidelines, are accepted.
1942	Superglue (Harry Coover and Fred Joyner)	After developing superglue (cyanoacrylate), Coover rejects it as too sticky for a 1942 project. Coover and Joyner rediscover superglue in 1951, when Coover recognizes it as a marketable product.
1942	Aqua-Lung (Jacques-Yves Cousteau and émile Gagnon)	The Aqua-Lung delivers air at ambient pressure and vents used air to the surroundings.
1942	Controlled nuclear chain reaction (Enrico Fermi)	In 1926 Fermi helped develop Fermi-Dirac statistics, which describe the quantum behavior of groups of electrons, protons, or neutrons. He now produces the first sustained nuclear chain reaction.
1942	Synthetic vitamins (Max Tishler)	After synthesizing several vitamins during the 1930's, Tishler and his team develop the antibiotic sulfaquinoxaline to treat coccidiosis. He also develops fermentation processes to produce streptomycin and penicillin.
1942	Bazooka	The United States military first uses the bazooka during the North African campaign in World War II.
1943	Meteorology	Radar is first used to detect storms.
1944	Electromechanical computer (Howard Aiken and Grace Hopper)	The Mark series of computers is built, designed by Aiken and Hopper. The U.S. Navy uses it to calculate trajectories for projectiles.
1944	Colossus	Colossus, the world's first vacuum-tube programmable logic calculator, is built in Britain for the purpose of breaking Nazi codes.
1944	Phased array radar antennas (Luis W. Alvarez)	Alvarez's phased array sweeps a beam across the sky by turning hundreds of small antennas on and off and not by moving a radar dish.
1944	V-2 rocket (Wernher von Braun)	Working for the German government during World War II, von Braun and other rocket scientists develop the V-2 rocket, the first long-range military missile and first suborbital missile. Arrested for making anti-Nazi comments, he later emigrates to the United States, where he leads the team that produces the Jupiter-C missile and launches vehicles such as the Saturn V, which help make the U.S. space program possible.

1944	Quinine	Robert B. Woodward and William von Eggers Doering synthesize quinine, which is used as an antimalarial.
1945	Automatic Computing Engine (Alan Mathison Turing)	While the Automatic Computing Engine (ACE) was never fully built, it was one of the first stored-program computers.
1945	Atomic bomb (J. Robert Oppenheimer)	Oppenheimer, the scientific leader of the Manhattan Project, heads the team that builds the atomic bomb. On the side of military use of the bomb to end World War II quickly, Oppenheimer saw this come to pass on August 6, 1945, when the bomb was dropped over Hiroshima, Japan, killing and maiming 150,000 people; a similar number of casualties ensued in Nagasaki on August 9, when the second bomb was dropped. Japan surrendered on August 14.
1945	Dialysis machine (Willem Johan Kolff)	Kolff designs the first artificial kidney, a machine that cleans the blood of patients in renal failure, and refuses to patent it. He will construct the artificial lung in 1955.
1945	Radioimmunoassay (RIA) (Rosalyn Yalow)	RIA required only a drop of blood (rather than the tens of milliliters previously required) to find trace amounts of substances.
1945	Electronic Sackbut (Hugh Le Caine)	Le Caine builds the first music synthesizer, joined by the Special Purpose Tape Recorder in 1954, which could simultaneously change the playback speed of several recording tracks.
1945	ENIAC computer (John William Mauchly and John Presper Eckert)	The Electronic Numerical Integrator and Computer, ENIAC, is the first general-purpose, programmable, electronic computer. (The Z3, developed independently by Konrad Zuse from 1939 to 1941 in Nazi Germany, did not fully exploit electronic components.) Built to calculate artillery firing tables, ENIAC is used in calculations for the hydrogen bomb.
1945	Microwave oven (Percy L. Spencer)	The microwave oven grew out of the microwave generator, the magnetron tube, becoming more affordable.
1946	Tupperware (Earl S. Tupper)	Tupper exploits plastics technology to develop a line of plastic containers that he markets at home parties starting in 1948.
1946	Carbon-14 dating (Willard F. Libby)	Libby uses the half-life of carbon 14 to develop a reliable means of dating ancient remains. Radiocarbon dating has proven to be invaluable to archaeologists.

1946	Magnetic tape recording (Marvin Camras)	Camras develops a magnetic tape recording process that will be adapted for use in electronic media, including music and motion-picture sound recording, audio and videocassettes, floppy disks, and credit card magnetic strips. For many years his method is the primary way to record and store sound, video, and digital data.
1946-1947	Audiometer (Georg von Békésy)	Békésy invents a pure-tone audiometer that patients themselves can control to measure the sensitivity of their own hearing.
1946	Radioisotopes for cancer treatment	The first nuclear-reactor-produced radioisotopes for civilian use are sent from the U.S. Army's Oak Ridge facility in Tennessee to Brainard Cancer Hospital in St. Louis.
1947	Transistor (John Bardeen, Walter H. Brattain, and William Shockley)	Hoping to build a solid-state amplifier, the team of Bardeen, Brattain, and Shockley discover the transistor, which replaces the vacuum tube in electronics. Bardeen is later part of the group that develops theory of superconductivity.
1947	Platforming (Vladimir Haensel)	American chemical engineer Haensel invents platforming, a process that uses a platinum catalyst to produce cleaner-burning high-octane fuels.
1947	Tubeless tire	B.F. Goodrich announces development of the tubeless tire.
1948	Holography (Dennis Gabor)	Gabor publishes his initial results working with holograms in Nature. Holograms became much more spectacular after the invention of the laser.
1948	Long-playing record (LP) (Peter Carl Goldmark)	Goldmark demonstrates the LP playing the cello with CBS musicians. The musical South Pacific is recorded in LP format and boosts sales, making the LP the dominant form of recorded sound for the next four decades.
1948	Gamma-ray pinhole camera (Roscoe Koontz)	Working to make nuclear reactors safer, Koontz invents the gamma-ray pinhole camera. The pinhole should act like a lens and form an image of the gamma source.
1948	Instant photography (Edwin Herbert Land)	Land develops the simple process to make sheets of polarizing material. He perfects the Polaroid camera in 1972.
1948	Synthetic penicillin (John C. Sheehan)	Sheehan develops the first total synthesis of penicillin, making this important antibiotic widely available.
1949	First peacetime nuclear reactor	Construction on the Brookhaven Graphite Research Reactor at Brookhaven Laboratory on Long Island, New York, is completed.

1949	Magnetic core memory (Jay Wright Forrester)	Core memory is used from the early 1950's to the early 1970's.
1950's	Fortran (John Warner Backus)	Backus develops the computer language Fortran, which is an acronym for "formula translation." Fortran allows direct entry of commands into computers with Englishlike words and algebraic symbols.
1950	Planotron (Pyotr Leonidovich Kapitsa)	Kapitsa invents a magnetron tube for generating microwaves. He becomes a corecipient of the Nobel Prize for Physics in 1978 for discovering superfluidity in liquid helium.
1950	Purinethol (Gertrude Belle Elion)	Elion develops the first effective treatment for childhood leukemia, 6-mercaptopurine (Purinethol). Elion later discovers azathioprine (Imuran), an immunosuppressive agent used for organ transplants.
1950	Artificial pacemaker (John Alexander Hopps)	Hopps develops a device to regulate the beating of the heart to treat patients with erratic heartbeats. By 1957, the device is small enough to be implanted.
1950	Contact lenses (George Butterfield)	Oregon optometrist Butterfield develops a lens that is molded to fit the contours of the cornea.
1951	Fiber-optic endoscope (fibroscope) (Harold Hopkins)	Hopkins fastened together a flexible bundle of optical fibers that could convey an image. One end of the bundle could be inserted into a patient's throat, and the physician could inspect the esophagus.
1951	The Pill (Carl Djerassi)	The birth-control pill, which becomes the world's most popular and is possibly most widely used contraceptive, revolutionizes not only medicine but also gender relations and women's status in society. Its prolonged use is later revealed to have health consequences.
1951	Field-emission microscope (Erwin Wilhelm Müller)	Müller develops the field-ion microscope, followed by an atom-probe field-ion microscope in 1963, which can detect individual atoms.
1951	Maser (Charles Hard Townes)	The maser (microwave amplification by stimulated emission of radiation) is a "laser" for microwaves. Discovered later, the "laser" patterned its name the acronym "maser."
1951	Artificial heart valve (Charles Hufnagel)	Hufnagel develops an artificial heart valve and performs the first heart-valve implantation surgery in a human patient the following year.

1951	UNIVAC (John Mauchly and John Presper Eckert)	Mauchly and Eckert invent the Universal Automatic Computer (UNIVAC). UNIVAC is competitor of IBM's products.
1952	Bubble chamber (Donald A. Glaser)	In a bubble chamber, bubbles form along paths taken by sub-atomic particles as they interact, and the bubble trails allow scientists to deduce what happened.
1952	Photovoltaic cell (Gerald Pearson)	The photovoltaic cell converts sunlight into electricity.
1952	Improved electrical resistor (Otis Boykin)	Boykin's resistor had improved precision, and its high-frequency characteristics were better than those of previous resistors.
1952	Language compiler (Grace Murray Hopper)	Hopper invents the compiler, an intermediate program that translates English-language instructions into computer language, followed in 1959 by Common Business Oriented Language (COBOL), the first computer programming language to translate commands used by programmers into the machine language the computer understands.
1952	Amniocentesis (Douglas Bevis)	British physician Bevis develops amniocentesis.
1952	Gamma camera (Hal Anger)	Nuclear medicine pioneer Anger creates the first prototype for the gamma camera. This leads to the inventions of other medical imaging devices, which detect and diagnose disease.
1953	Medical ultrasonography (Inge Edler and Carl H. Hertz)	Edler and Hertz adapt an ultrasound probe used in materials testing in a shipyard for use on a patient. Their technology makes possible echograms of the heart and brain.
1953	Inertial navigation systems (Charles Stark Draper)	Draper's inertial navigation system (INS) is designed to determine the current position of a ship or plane based on the initial location and acceleration.
1953	Heart-lung machine (John H. Gibbon, Jr.)	American surgeon Gibbon conducts the first successful heart surgery using a heart-lung machine that he constructed with the help of his wife, Mary.
1953	First frozen meals	Swanson develops individual prepackaged frozen meals. The first-ever meal consists of turkey, cornbread stuffing, peas, and sweet potatoes.
1954	Geodesic dome ® (Buckminster Fuller)	After developing the geodesic dome, Fuller patents the structure, an energy-efficient house using prefabricated, easily shipped parts.

1954	Atomic absorption spectroscopy (Sir Alan Walsh)	Atomic absorption spectroscopy is used to identify and quantify the presence of elements in a sample.
1954	Synthetic diamond (H. Tracy Hall)	Hall synthesizes diamonds using a high-pressure, high-temperature belt apparatus that can generate 120,000 atmospheres of pressure and sustain a temperature of 1,800 degrees Celsius in a working volume of about 0.1 cubic centimeter.
1954	Machine vision (Jerome H. Lemelson)	Machine vision allows a computer to move and measure products and to inspect them for quality control.
1954	Hydrogen bomb (Edward Teller)	The first hydrogen bomb, designed by Teller, is tested at the Bikini Atoll in the Pacific Ocean.
1954	Silicon solar cells (Calvin Fuller)	Silicon solar cells have proven to be among the most efficient and least expensive solar cells.
1954	First successful kidney transplant (Joseph Edward Murray)	American surgeon Murray performs the first successful kidney transplant, inserting one of Ronald Herrick's kidneys into his twin brother, Richard. Murray shares the 1990 Nobel Prize for Physiology or Medicine with E. Donnall Thomas, who developed bone marrow transplantation.
1954	Transistor radio	The first transistor radio is introduced by Texas Instruments.
1954	IBM 650	The IBM 650 computer becomes available. It is considered by IBM to be its first business computer, and it is the first computer installed at Columbia University in New York.
1954	First nuclear submarine	The United States launches the first nuclear-powered submarine, the USS *Nautilus*.
1955	Color television's RGB system (Ernst Alexanderson)	The RGB system uses three image tubes to scan scenes through colored filters and three electron guns in the picture tube to reconstruct scenes.
1955	Floppy disk and floppy disk drive (Alan Shugart)	Working at the San Jose, California, offices of International Business Machines (IBM), Shugart develops the disk drive, followed by floppy disks to provide a relatively fast way to store programs and data permanently.
1955	Hovercraft (Sir Christopher Cockerell)	Cockerell files a patent for his hovercraft, an amphibious vehicle. He earlier invented several important electronic devices, including a radio direction finder for bombers in World War II.

1955	Pulse transfer controlling device (An Wang)	The device allows magnetic core memory to be written or read without mechanical motion and is therefore very rapid.
1955	Polio vaccine (Jonas Salk)	Salk's polio vaccine, which uses the killed virus, saves lives and improves the quality of life for millions afflicted by polio.
1956	Fiber optics (Narinder S. Kapany)	Kapany, known as the father of fiber optics, coins the term "fiber optics." In high school, he was told by a teacher that light moves only in a straight line; he wanted to prove the teacher wrong and wound up inventing fiber optics.
1956	Scotchgard (Patsy O'Connell Sherman)	Sherman develops a stain repellent for fabrics that is trade-marked as Scotchgard.
1956	Ovonic switch (Stanford Ovshinsky)	Ovshinsky invents a solid-state, thin film switch meant to mimic the actions of neurons.
1956	Videotape recorder (Charles P. Ginsburg)	The video recorder allows programs to be shown later, to provide instant replays in sports, and to make a permanent record of a program.
1956	Liquid Paper (Bette Nesmith Graham)	Graham markets her "Mistake Out" fluid for concealing typographical errors.
1956	Dipstick blood sugar test (Helen M. Free)	Free and her husband Alfred coinvent a self-administered urinalysis test that allows diabetics to monitor their sugar levels and to adjust their medications accordingly.
1956	350 RAMAC	IBM produces the first computer disk storage system, the 350 RAMAC, which retrieves data from any of fifty spinning disks.
1957	Wankel rotary engine (Felix Wankel)	Having fewer moving parts, the Wankel rotary engine ought to be sturdier and perhaps more efficient than the common reciprocating engine.
1957	Laser (Gordon Gould, Charles Hard Townes, Arthur L. Schawlow, Theodore Harold Maiman)	Having conducted research on using light to excite thallium atoms, Gould tries to get funds and approval to build the first laser, but he fails. Townes (inventor of the maser) and Schawlow of Bell Laboratories will first describe the laser, and Maiman will first succeed in building a small optical maser. Gould coins the term "laser," which stands for light amplification by stimulated emission of radiation.
1957	Intercontinental ballistic missile (ICBM)	The Soviet Union develops the ICBM.
1957	First satellite	The Soviet Union launches Sputnik, the first man-made satellite.

1958	CorningWare	CorningWare cookware is introduced. It is based on S. Donald Stookey's 1953 discovery that a heat-treatment process can transform glass into fine-grained ceramics.
1958	Integrated circuit (Robert Norton Noyce and Jack St. Clair Kilby)	The microchip, independently discovered by Noyce and Kilby, proves to be the breakthrough that allows the miniaturization of electronic circuits and paves the way for the digital revolution.
1958	Ultrasound	Ultrasound becomes the most common method for examining a fetus.
1958	Planar process (Jean Hoerni)	Hoerni develops the first planar process, which improves the integrated circuit.
1960's	Lithography	Optical lithography, a process that places intricate patterns onto silicon chips, is used in semiconductor manufacturing.
1960	Measles vaccine (John F. Enders)	Enders, an American physician, develops the first measles vaccine. It is tested the following year and is hailed a success.
1960	Echo satellite (John R. Pierce)	The first passive-relay telecommunications satellite, Echo, reflected signals. The signals, received from one point on Earth, "bounce" off the spherical satellite and are reflected back down to another, far distant, point on Earth.
1960	Automatic letter-sorting machine (Jacob Rabinow)	Rabinow's machine greatly increased the speed and efficiency of mail delivery in the United States. He also invented an optical character recognition (OCR) scanner.
1960	Ruby laser (Theodore Harold Maiman)	Maiman produces a ruby laser, the world's first visible light laser.
1960	Helium-neon gas laser (Ali Javan)	Javan produces the world's second visible light laser.
1960	Chardack-Greatbatch pacemaker (Wilson Greatbatch and William Chardack)	Greatbatch and Chardack create the first implantable pacemaker.
1960	Radionuclide generator	Powell Richards and Walter Tucker and their colleagues at Brookhaven Laboratory in New York invent a short half-life radionuclide generator for use in nuclear medicine diagnostic imaging procedures.

1961	Audio-animatronics (Walt Disney)	Disney established WED, a research and development unit that developed the inventions he needed for his various enterprises. WED produced the audio-animatronic robotic figures that populated Disneyland, the 1964-1965 New York World's Fair, films, and other attractions. Audio-animatronics enabled robotic characters to speak or sing as well as move.
1961	Ruby laser	The ruby laser is first used medically by Charles Campbell and Charles Koester to excise a patient's retinal tumor.
1961	First person in space	Soviet astronaut Yuri Gagarin becomes the first person in space when he orbits the Earth on April 12.
1962	Soft contact lenses (Otto Wichterle)	Wichterle's soft contacts can be worn longer with less discomfort than can hard contact lenses.
1962	Continuously operating ruby laser (Willard S. Boyle and Don Nelson)	The invention relies on an arc lamp shining continuously (rather than the flash lamp used by Theodore Maiman in 1960).
1962	Light-emitting diode (Nick Holonyak, Jr.)	Holonyak makes the first visible-spectrum diode laser, which produces red laser light but also stops lasing yet remains a useful light source. Holonyak has invented the red light-emitting diode (LED), the first operating alloy device—the "ultimate lamp."
1962	Telstar satellite (John R. Pierce)	The first satellite to rebroadcast signals goes into operation, revolutionizing telecommunications.
1962	Quasar 3C 273 (Maarten Schmidt)	Schmidt shows that this quasar is very distant and hence very bright. Further research shows quasars to be young galaxies with active, supermassive black holes at their centers.
1962	First audiocassette	The Philips company of the Netherlands releases the audiocassette tape.
1962	Artificial hip (Sir John Charnley)	British surgeon Charnley invents the low-friction artificial hip and develops the surgical techniques for emplacing it.
1963	Learjet (Bill Lear)	The Learjet, a small eight-passenger jet with a top speed of 560 miles (900 kilometers) per hour, can shuttle VIPs to meetings and other engagements.
1963	Self-cleaning oven	General Electric introduces the self-cleaning electric oven.

1963	Artificial heart (Paul Winchell)	Winchell receives a patent (later donated to the University of Utah's Institute for Biomedical Engineering) for an artificial heart that purportedly became the model for the successful Jarvick-7.
1963	6600 computer (Seymour Cray)	The 6600 was the first of a long line of Cray supercomputers.
1963	Carbon fiber (Leslie Philips)	British engineer Philips develops carbon fiber, which is much stronger than steel.
1964	Three-dimensional holography (Emmett Leith)	Leith and Juris Upatnieks present the first three-dimensional hologram at the Optical Society of America conference. The hologram must be viewed with a reference laser. The hologram of an object can then be viewed from different angles, as if the object were really present.
1964	Moog synthesizer (Robert Moog)	The Moog synthesizer uses electronics to create and combine musical sounds.
1964	Cosmic background radiation (Arno Penzias and Robert Wilson)	Penzias and Wilson detect the cosmic background radiation, which corresponds to that which would be radiated by a body at 2.725 kelvins. It is thought to be greatly redshifted primordial fireball radiation left over from the big bang.
1964	BASIC programming language (John Kemeny and Thomas Kurtz)	Kemeny and Kurtz develop the BASIC computer programming language. BASIC is an acronym for Beginner's All-purpose Symbolic Instruction Code.
1965	Minicomputer (Ken Olsen)	Perhaps the first true minicomputer, the PDP-8 is released by Digital Equipment Corporation. Founder Olsen makes computers affordable for small businesses.
1965	Aspartame (James M. Schlatter)	Schlatter discovers aspartame, an artificial sweetener, while trying to come up with an antiulcer medication.
1965	First space walk	Soviet astronaut Aleksei Leonov is the first person to walk in space.
1966	Gamma-electric cell (Henry Thomas Sampson)	Sampson works with George H. Miley to produce the gamma-electric cell, which converts the energy of gamma rays into electrical energy.
1966	Handheld calculator (Jack St. Clair Kilby)	While working for Texas Instruments, Kilby does for the adding machine what the transistor had done for the radio, inventing a handheld calculator that retails at $150 and becomes an instant commercial success.

1966	First unmanned moon landing	Soviet spacecraft Luna 9 lands on the moon.
1967	Electrogasdynamic method and apparatus (Meredith C. Gourdine)	Gourdine develops electrogasdynamics, which involves the production of electricity from the conversion of kinetic energy in a moving, ionized gas.
1967	Pulsars (Jocelyn Bell and Antony Hewish)	Pulsars, rapidly rotating neutron stars, are discovered.
1968	Practical liquid crystal displays (James Fergason)	Fergason develops an liquid crystal display (LCD) screen that has good visual contrast, is durable, and uses little electricity.
1968	Lasers in medicine	Francis L'Esperance begins using the argon-ion laser to treat patients with diabetic retinopathy.
1968	Computer mouse (Douglas Engelbart)	Engelbart presents the computer mouse, which he had been working on since 1964.
1968	Apollo 7	Astronauts on Apollo 7, the first piloted Apollo mission, take photographs and transmit them to the American public on television.
1968	Interface message processors	Bolt Beranek and Newman Incorporated win a Defense Advanced Research Projects Agency (DARPA) contract to develop the packet switches called interface message processors (IMPs).
1969	Rubella vaccine	The rubella vaccine is available.
1969	First person walks on the moon	Neil Armstrong, a member of the U.S. Apollo 11 spacecraft, is the first person to walk on the moon.
1969	Boeing 747	The Boeing 747 makes its first flight, piloted by Jack Waddell.
1969	Concorde	The Concorde makes its first flight, piloted by André Turcat.
1969	Charge-coupled device (Willard S. Boyle and George E. Smith)	Boyle and Smith develop the charge-coupled device, the basis for digital imaging.
1969	ARPANET launches	The Advanced Research Projects Agency starts ARPANET, which is the precursor to the Internet. UCLA and Stanford University are the first institutions to become networked.
1970's	Digital seismology	Digital seismology is used in oil exploration and increases accuracy in finding underground pools.

1970's	Mud pulse telemetry	Mud pulse telemetry becomes an oil-industry standard; pressure pulses are relayed through drilling mud to convey the location of the drill bit.
1970	Optical fiber (Robert Maurer and others)	Maurer, joined by Donald Keck, Peter Schultz, and Frank Zimar, produces an optical fiber that can be used for communication.
1970	Compact disc (James Russell)	The compact disc (CD) revolutionizes the way digital media is stored.
1970	UNIX (Dennis Ritchie and Kenneth Thompson)	Bell Laboratories employees Ritchie and Thompson complete the UNIX operating system, which becomes popular among scientists.
1970	Network Control Protocol	The Network Working Group deploys the initial ARPANET host-to-host protocol, called the Network Control Protocol (NCP), establishing connections, break connections, switch connections, and control flow over the ARPANET.
1971	Computerized axial tomography (Godfrey Newbold Hounsfield)	In London, doctors performed the first CAT scan of a living patient and detected a brain tumor. In a CAT (or CT) scan, X rays are taken of a body like slices in a loaf of bread. A computer then assembles these slices into a detail-laden three-dimensional image.
1971	First videocassette recorder	Sony begins selling the first videocassette recorder (VCR) to the public.
1971	Microprocessor (Ted Hoff)	The computer's central processing unit (CPU) is reduced to the size of a postage stamp.
1971	Electronic switching system for telecommunications (Erna Schneider Hoover)	Hoover's system prioritizes telephone calls and fixes an efficient order to answer them.
1971	Intel microprocessors	Intel builds the world's first microprocessor chip.
1971	Touch screen (Sam Hurst)	Hurst's touch screen can detect if it has been touched and where it was touched.
1972	First recombinant DNA organism (Stanley Norman Cohen, Paul Berg, and Herbert Boyer)	The methods to combine and transplant genes are discovered when this team successfully clones and expresses the human insulin gene in the Escherichia coli.
1972	Far-Ultraviolet Camera (George R. Carruthers)	The Carruthers-designed camera is used on the Apollo 16 mission.

1972	Cell encapsulation (Taylor Gunjin Wang)	Wang develops ways to encapsulate beneficial cells and introduce them into a body without triggering the immune system.
1972	Pioneer 10	The U.S. probe Pioneer 10 is launched to get information about the outer solar system.
1972	Networking goes public	ARPANET system designer Robert Kahn organizes the first public demonstration of the new network technology at the International Conference on Computer Communications in Washington, D.C.
1972	Pong video game (Nolan K. Bushnell and Ted Dabney)	Bushnell and Dabney register the name of their new computer company, Atari, and issue Pong shortly thereafter, marking the rise of the video game industry.
1973	Automatic computerized transverse axial (ACTA) whole-body CT scanner (Robert Steven Ledley)	The first whole-body CT scanner is operational. Ledley goes on to spend much of his career promoting the use of electronics and computers in biomedical research.
1973	Packet network interconnection protocols TCP/IP (Vinton Gray Cerf and Robert Kahn)	Cerf and Kahn develop transmission control protocol/ Internet protocol (TCP/IP), protocols that enable computers to communicate with one another.
1973	Automated teller machine (Don Wetzel)	Wetzel receives a patent for his ATM. To make it a success, he shows banks how to generate a group of clients who would use the ATM.
1973	Food processor	The Cuisinart food processor is introduced in the United States.
1973	Air bags in automobiles	The Oldsmobile Tornado is the first American car sold equipped with air bags.
1973	Space photography	Astronauts aboard Skylab, the first U.S. space station, take high-resolution photographs of Earth using photographic remote-sensing systems. The astronauts also take photographs with handheld cameras.
1974	Kevlar (Stephanie Kwolek)	Kwolek receives a patent for the fiber Kevlar. Bullet-resistant Kevlar vests go on sale only one year later.
1975	Ethernet (Robert Metcalfe and David Boggs)	Metcalfe and Boggs invent the Ethernet, a system of software, protocols, and hardware allowing instantaneous communication between computer terminals in a local area.

1975	Semiconductor laser	Scientists working at Diode Labs develop the first commercial semiconductor laser that will operate continuously at room temperature.
1976	First laser printer	IBM's 3800 Printing System is the first laser printer. The ink jet is invented in the same year, but it is not prevalent in homes until 1988.
1976	Apple computer (Steve Jobs)	Jobs cofounds Apple Computer with Steve Wozniak.
1976	Jarvik-7 artificial heart (Robert Jarvik)	The Jarvik-7 allows a calf to live 268 days with the artificial heart. Jarvik combined ideas from several other workers to produce the Jarvik-7.
1976	Apple II (Steve Wozniak)	Wozniak develops the Apple II, the best-selling personal computer of the 1970's and early 1980's.
1976	First Mars probes	The National Aeronautics and Space Administration (NASA) launches Viking 1 and Viking 2, which land on obtain images of Mars.
1976	Kurzweil Reading Machine (Ray Kurzweil)	Kurzweil develops an optical character reader (OCR) able to read most fonts.
1976	Microsoft Corporation (Bill Gates)	Gates, along with Paul Allen, found Microsoft, a software company. Gates will remain head of Microsoft for twenty-five years.
1976	Conductive polymers	Hideki Shirakawa, Alan G. MacDiarmid, and Alan J. Heeger discover conductive polymers. They are awarded the 2000 Nobel Prize in Chemistry.
1977	Global Positioning System (GPS) (Ivan A. Getting)	The first GPS satellite is launched, designed to support a navigational system that uses satellites to pinpoint the location of a radio receiver on Earth's surface.
1977	Fiber-optic telephone cable	The first fiber-optic telephone cables are tested.
1977	Echo-planar imaging (Peter Mansfield)	British physicist Mansfield first develops the echo-planar imaging (EPI).
1977	Gossamer Condor (Paul MacCready)	MacCready designs the Gossamer Albatross, which enables human-powered flight.

1978	Smart gels (Toyoichi Tanaka)	Tanaka discovers and works with "smart gels," polymer gels that can expand a thousandfold, change color, or contract when stimulated by minor changes in temperature, magnetism, light, or electricity. This capacity makes them useful in a broad range of applications.
1978	Charon discovered (James Christy)	Charon is discovered as an apparent bulge on a fuzzy picture of Pluto. Its mass is about 12 percent that of Pluto.
1978	First cochlear implant surgery	Graeme Clark performs the first cochlear implant surgery in Australia.
1978	The first test-tube baby	Louise Brown is born in England.
1978	First MRI	The first magnetic resonance image (MRI) of the human head is taken in England.
1979	First laptop (William Moggridge)	Moggridge, of Grid Systems in England, designs the first laptop computer.
1979	First commercially successful application	The VisiCalc spreadsheet for Apple II, designed by Daniel Bricklin and Bob Frankston, helps drive sales of the personal computer and becomes its first successful business application.
1979	USENET (Tom Truscott, Jim Ellis and Steve Belovin)	Truscott, Ellis, and Belovin create USENET, a "poor man's ARPANET," to share information via e-mail and message boards between Duke University and the University of North Carolina, using dial-up telephone lines.
1979	In-line roller skates (Scott Olson and Brennan Olson)	After finding some antique in-line skates, the Olson brothers begin experimenting with modern materials, creating Rollerblades.
1980's	Controlled drug delivery (Robert S. Langer)	Langer develops the foundation of controlled drug delivery technology used in cancer treatment.
1980	Alkaline battery (Lewis Urry)	Eveready markets alkaline batteries under the trade name Energizer. Urry's alkaline battery lasts longer than its predecessor, the carbon-zinc battery.
1980	Interferon (Charles Weissmann)	Weissmann produces the first genetically engineered human interferon, which is used in cancer treatment.
1980	TCP/IP	The U.S. Department of Defense adopts the TCP/IP suite as a standard.

1981	Ablative photodecomposition (Rangaswamy Srinivasan)	Srinivasan's research on ablative photodecomposition leads to multiple applications, including laser-assisted in situ keratomileusis (LASIK) surgery, which shapes the cornea to correct vision problems.
1981	Scanning tunneling microscope (Heinrich Rohrer and Gerd Binnig)	The scanning tunneling microscope shows surfaces at the atomic level.
1981	Improvements in laser spectroscopy (Arthur L. Schawlow and Nicolaas Bloembergen)	Schawlow shares the Nobel Prize in Physics with Nicolaas Bloembergen for their work on laser spectroscopy. While most of Schawlow's inventions involved lasers, he also did research in superconductivity and nuclear resonance.
1981	First IBM personal computer	The first IBM PC, the IBM 5100, goes on the market with a $1,565 price tag.
1982	Compact discs appear	Compact discs are now sold and will start replacing vinyl records.
1982	First artificial heart	Seattle dentist Barney Clark receives the first permanent artificial heart, and he survives for 112 days.
1983	Cell phone (Martin Cooper)	The first mobile (wireless) phone, the DynaTAC 8000X, receives approval by the Federal Communications Commission (FCC), heralding an age of wireless communication.
1983	Internet	ARPANET, and networks attached to it, adopt the TCP/IP networking protocol. All networks that use the protocol are known as the Internet.
1983	Cyclosporine	Immunosuppressant cyclosporine is approved for use in transplant operations in the United States.
1983	Polymerase chain reaction (Kary B. Mullis)	While driving to his cottage in Mendocino, California, Mullis develops the idea for the polymerase chain reaction (PCR). PCR will be used to amplify a DNA segment many times, leading to a revolution in recombinant DNA technology and a 1993 Nobel Prize in Chemistry for Mullis.
1984	Domain name service is created	Paul Mockapetris and Craig Partridge develop domain name service, which links unique Internet protocol (IP) numerical addresses to names with suffixes such as .mil, .com, .org, and .edu.
1984	Mac is released	Apple introduces the Macintosh, a low-cost, plug-and-play personal computer with a user-friendly graphic interface.

1984	CD-ROM	Philips and Sony introduce the CD-ROM (compact disc read-only memory), which has the capacity to store data of more than 450 floppy disks.
1984	Surgery in utero	William A. Clewall performs the first successful surgery on a fetus.
1984	Cloning	Danish veterinarian Steen M. Willadsen clones a lamb from a developing sheep embryo cell.
1984	AIDS blood test (Robert Charles Gallo)	Gallo and his colleagues identify the virus HTLV-3/LAV (later renamed human immunodeficiency virus, or HIV) as the cause of acquired immunodeficiency syndrome, or AIDS. Gallo creates a blood test that can identify antibodies specific to HIV. This blood test is essential to keeping the supply in blood banks pure.
1984	Imaging X-ray spectrometer (George Edward Alcorn)	Alcorn patents his device, which makes images of the source using X rays of specific energies, similar to making images with a specific wavelength (color) of light. It is used in acquiring data on the composition of distant planets and stars.
1984	DNA profiling (Alec Jeffreys)	Noticing similarities and differences in DNA samples from his lab technician's family, Jeffreys discovers the principles that lead to DNA profiling, which has become an essential tool in forensics and the prosecution of criminal cases.
1985	Windows operating system (Bill Gates)	The first version of Windows is released.
1985	Implantable cardioverter defibrillator	The U.S. Food and Drug Administration (FDA) approves Polish physician Michel Mirowski's implantable cardioverter defibrillator (ICD), which monitors and corrects abnormal heart rhythms.
1985	Industry Standard Architecture (ISA) bus (Mark Dean and Dennis Moeller)	Dean and Moeller design the standard way of organizing the central part of a computer and its peripherals, the ISA bus, which is patented in this year.
1985	Atomic force microscope	Calvin Quate, Christoph Gerber, and Gerd Binnig invent the atomic force microscope, which becomes one of the foremost tools for imaging, measuring, and manipulating matter at the nano scale.
1986	Mir	The Soviet Union launches the Mir space station, the first permanent space station.

1986	Burt Rutan's Voyager	Dick Rutan (Burt's brother) and Jeana Yeager make the first around-the-world, nonstop flight without refueling in the Burt Rutan-designed Voyager. The Voyager is the first aircraft to accomplish this feat.
1986	High-temperature superconductor (J. Georg Bednorz and Karl Alexander Müller)	Bednorz and Müller show that a ceramic compound of lanthanum, barium, copper, and oxygen becomes superconducting at 35 kelvins, a new high- temperature record.
1987	Azidothymidine	The FDA approves azidothymidine (AZT), a potent antiviral, for AIDS patients.
1987	Echo-planar imaging	Echo-planar imaging is used to perform real-time movie imaging of a single cardiac cycle.
1987	Parkinson's treatment	French neurosurgeon Alim-Louis Benabid implants a deep-brain electrical-stimulation system into a patient with advanced Parkinson's disease.
1987	First corneal laser surgery	New York ophthalmologist Steven Trokel performs the first laser surgery on a human cornea. He had refined his technique on a cow's eye. Trokel was granted a patent for the Excimer laser to be used for vision correction.
1987	UUNET and PSINet	Rick Adams forms UUNET and Bill Schrader forms PSINet to provide commercial Internet access.
1988	Transatlantic fiber-optic cable	The first transatlantic fiber-optic cable is installed, linking North America and France.
1988	Laserphaco probe (Patricia Bath)	Bath's probe is used to break up and remove cataracts.
1989	Method for tracking oil flow underground using a supercomputer (Philip Emeagwali)	Emeagwali receives the Gordon Bell Prize, considered the Nobel Prize for computing, for his method, which demonstrates the possibilities of computer networking.
1989	World Wide Web (Tim Berners-Lee and Robert Cailau)	Berners-Lee finds a way to join the idea of hypertext and the young Internet, leading to the Web, coinvented with Cailau.
1989	First dial-up access	The World debuts as the first provider of dial-up Internet access for consumers.
1990's	Environmentally friendly appliances	Water-saving and energy-conserving washing machines and dryers are introduced.

1990	Hubble Space Telescope	The Hubble Space Telescope is launched and changes the way scientists look at the universe.
1990	Human Genome Project begins	The U.S. Department of Energy and the National Institutes of Health coordinate the Human Genome Project with the goal of identifying all 30,000 genes in human DNA and determining the sequences of the three billion chemical base pairs that make up human DNA.
1990	BRCA1 gene discovered (Mary-Claire King)	King finds the cancer-associated gene on chromosome 17. She demonstrates that humans and chimpanzees are 99 percent genetically identical.
1991	Nakao Snare (Naomi L. Nakao)	The Snare is a device that captures polyps that have been cut from the walls of the intestine, solving the problem of "lost polyp syndrome."
1991	America Online (AOL)	Quantum Computer Services changes its name to America Online; Steve Case is named president. AOL offers e-mail, electronic bulletin boards, news, and other information.
1991	Carbon nanotubes (Sumio Iijima)	Although carbon nanotubes have been seen before, Iijima's 1991 paper establishes some basic properties and prompts other scientists' interest in studying them.
1991	The first hot-air balloon crosses the Pacific (Richard Branson and Per Lindstrad)	Branson and Lindstrad, who teamed up in 1987 to cross the Atlantic, make the 6,700-mile flight in 47 hours and break the world distance record.
1992	Newton	Apple introduces Newton, one of the first handheld computers, or personal digital assistants, which has a liquid crystal display operated with a stylus.
1993	Mosaic (Marc Andreessen)	Andreessen launches Mosaic, followed by Netscape Navigator in 1995—the first Internet browsers. Both Mosaic and Netscape allow novices to browse the World Wide Web.
1993	Flexible tailored elastic airfoil section (Sheila Widnall)	Widnall applies for a patent for this device, which addresses the problem of being able to measure fluctuations in pressure under unsteady conditions. She serves as secretary of the Air Force (the first woman to lead a branch of the military) and also serves on the board investigating the space shuttle Columbia accident of 2003.
1993	Light-emitting diode (LED) blue and UV (Shuji Nakamura)	Nakamura's blue LED makes white LED light possible (a combination of red, blue, and green).

1994	Genetically modified (GM) food	The Flavr Savr tomato, the first GM food, is approved by the FDA.
1994	Channel Tunnel	Channel Tunnel, or Chunnel, opens, connecting France and Britain by a railway constructed beneath the English Channel.
1995	51 Pegasi (Michel Mayor and Didier Queloz)	Mayor and Queloz detect a planet orbiting another normal star, the first extrasolar planet (exoplanet) to be found. As of June, 2009, 353 exoplanets were known.
1995	Saquinavir	The FDA approves Saquinavir for the treatment of AIDS. It is the first protease inhibitor, which reduces the ability of the AIDS virus to spread to new cells.
1995	iBot (Dean Kamen)	Kamen invents iBOT, a super wheelchair that climbs stairs and helps its passenger to stand.
1995	Global Positioning System (Ivan A. Getting)	The GPS becomes fully operational.
1995	Illusion transmitter (Valerie L. Thomas)	A concave mirror can produce a real image that appears to be three-dimensional. Thomas's system uses a concave mirror at the camera and another one at the television receiver.
1996	LASIK	The first computerized excimer laser (LASIK), designed to correct the refractive error myopia, is approved for use in the United States.
1996	First sheep is cloned	Scottish scientist Ian Wilmut clones the first mammal, a Finn Dorset ewe named Dolly, from differentiated adult mammary cells.
1997	Robotic vacuum	Swedish appliance company Electrolux is the first to create a prototype of a robotic vacuum cleaner.
1998	PageRank (Larry Page)	The cofounder of Google with Sergey Brin, Page devises PageRank, the count of Web pages linked to a given page and a measure how valuable people find that page.
1998	UV Waterworks (Ashok Gadgil)	The device uses UV from a mercury lamp to kill waterborne pathogens.

1998	Napster	College dropout Shawn Fanning creates Napster, an extremely popular peer-to-peer file-sharing platform that allowed users to download music for free. In 2001 the free site was shut down because it encouraged illegal sharing of copyrighted properties. The site then became available by paid subscription.
1999	Palm VII	The Palm VII organizer is on the market. It is a handheld computer with 2 megabytes of RAM and a port for a wireless phone.
1999	BlackBerry (Research in Motion of Canada)	A wireless handheld device that began as a two-way pager, the BlackBerry is also a cell phone that supports Web browsing, e-mail, text messaging, and faxing—it is the first smart phone.
2000	Hoover-Diana production platform	A joint venture by Exxon and British Petroleum (BP), the Hoover-Diana production platform goes into operation in the Gulf of Mexico. Within six months it is producing 20,000 barrels of oil a day.
2000	Clone of a clone	Japanese scientists clone a bull from a cloned bull.
2000	Minerva	The Library of Congress initiates a prototype system called Minerva (Mapping the Internet Electronic Resources Virtual Archives) to collect and preserve open-access Web resources.
2000	Supercomputer	The ASCI White supercomputer at the Lawrence Livermore National Laboratory in California is operational. It can hold six times the information stored in the 29 million books in the Library of Congress.
2001	XM Radio	XM Radio initiates the first U.S. digital satellite radio service in Dallas-Ft. Worth and San Diego.
2001	Human cloning	Scientists at Advanced Cell Technology in Massachusetts clone human embryos for the first time.
2001	iPod (Tony Fadell)	Fadell introduces the iPod, a portable hard drive-based MP3 player with an Internet-based electronic music catalog, for Apple.
2001	Segway PT (Dean Kamen)	Kamen introduces his personal transport device, a self-balancing, electric-powered pedestrian scooter.
2003	First digital books	Lofti Belkhir introduces the Kirtas BookScan 1200, the first automatic, page-turning scanner for the conversion of bound volumes to digital files.

2003	Aqwon (Josef Zeitler)	The hydrogen-powered scooter Aqwon can reach 30 miles (50 kilometers) per hour. Its combustion product is water.
2003	Human Genome Project is completed	After thirteen years, the 25,000 genes of the human genome are identified and the sequences of the 3 million chemical base pairs that make up human DNA are determined.
2004	Stem cell bank	The world's first embryonic stem cell bank opens in England.
2004	SpaceShipOne and SpaceShipTwo (Burt Rutan)	Rutan receives the U.S. Department of Transportation's first license issued for suborbital flight for SpaceShipOne, which shortly thereafter reaches an altitude of 328,491 feet. Rutan's rockets are the first privately funded manned rockets to reach space (higher than 100 kilometers above Earth's surface).
2004	Columbia supercomputer	The NASA supercomputer Columbia, built by Silicon Graphics and Intel, achieves sustained performance of 42.7 trillion calculations per second and is named the fastest supercomputer in the world. It is named for those who lost their lives in the explosion of the space shuttle Columbia in 2003. Because technology evolves so quickly, the Columbia will not be the fastest for very long.
2005	Blue Gene/L supercomputer	The National Nuclear Security Administration's BlueGene/L supercomputer, built by IBM, performs at 280.6 trillion operations per second and is now the world's fastest supercomputer.
2005	Eris (Mike Brown)	Working with C. A. Trujillo and D. L. Rabinowitz, Brown discovers Eris, the largest known dwarf planet and a Kuiper Belt object. It is 27 percent more massive than Pluto, another large Kuiper Belt object.
2005	Nix and Hydra discovered (Pluto companion team)	The Hubble research team—composed of Hal Weaver, S. Alan Stern, Max Mutchler, Andrew Steffl, Marc Buie, William Merline, John Spencer, Eliot Young, and Leslie Young—finds these small moons of Pluto.
2006	Digital versus film	Digital cameras have almost wholly replaced film cameras. *The New York Times* reports that 92 percent of cameras sold are digital.
2007	First terabyte drive	Hitachi Global Storage Technologies announces that it has created the first one-terabyte (TB) hard disk drive.

2007	iPhone (Apple)	Apple introduces its smart phone, a combined cell phone, portable media player (equal to a video iPod), camera phone, Internet client (supporting e-mail and Web browsing), and text messaging device, to an enthusiastic market.
2008	Roadrunner	The Roadrunner supercomputer, built by IBM and Los Alamos National Laboratory, can process more than 1.026 quadrillion calculations per second. It works more than twice as fast as the Blue Gene/L supercomputer and is housed at Los Alamos in New Mexico.
2008	Mammoth Genome Project	Scientists sequence woolly mammoth genome, the first of an extinct animal.
2008	Columbus lands	The space shuttle Atlantis delivers the Columbus science laboratory to the International Space Station. The twenty-three-foot long laboratory is able to conduct experiments both inside and outside the space station.
2008	Retail DNA test (Anne Wojcicki)	Wojcicki (wife of Google founder Sergey Brin) offers an affordable DNA saliva test, 23andMe, to determine one's genetic markers for ninety traits. The product heralds what *Time* magazine dubs a "personal-genomics revolution."
2009	Large Hadron Collider	The Large Hadron Collider (LHC) becomes the world's highest energy particle accelerator.
2009	Hubble Space Telescope repairs (NASA)	STS-125 astronauts conducted five space walks from the space shuttle Atlantis to upgrade the Hubble Space Telescope, extending its life to at least 2014.
2009	AIDS vaccine	Scientists in Thailand create a vaccine that seems to reduce the risk of contracting the AIDS virus by more than 31 percent.
2010	Jaguar supercomputer	The Oak Ridge National Laboratory in Tennessee is home to Jaguar, the world's fastest supercomputer, the peak speed of which is 2.33 quadrillion floating point operations per second.
2010	Superbugs	NDM-1 superbug decoded
2010	Mass of stars	Scientists solve mystery of mass in variable stars Grzegorz Pietrzyski, of the Universidad de Concepcion in Chile

2010	Antimatter	Researchers at CERN's Geneva labs have recently managed to trap a sizeable amount of antihydrogen using a magnetic trap in a vacuum for a tenth of a second.
2010	Synthetic life	Creation of first self-replicating synthetic life
2010	Antihydrogen	Researchers at CERN's Geneva labs have recently managed to trap a sizeable amount of antihydrogen using a magnetic trap in a vacuum for a tenth of a second.
2011	Solar system	Kepler space telescope discovers a solar system of six planets orbiting the star Kepler-11
2011	Artificial intelligence	In IBM's Watson supercomputer defeats two humans on the Jeopardy! quiz show.
2011	Space exploration	The MESSENGER becomes the first spacecraft to orbit the planet Mercury.
2011	Stem cell research	Scientists in Japan grow working retinas from mouse stem cells.
2011	Genetics	A modified anti-malaria gene is successfully introduced to a population of mosquitoes.
2012	Spaceflight	California's SpaceX became the first private company to successfully fly a spaceship to the International Space Station.
2012	Space exploration	Scientists discovered a very hot "carbon-rich" exoplanet, perfect for turning carbon into diamonds.
2012	Higgs Boson	Scientists at Europe's CERN announced they'd found a particle that could be the famous "Higgs boson" particle, believed to be responsible for all mass in the universe
2012	Curiosity Rover	NASA successfully landed the $2.5 billion Mars Curiosity Rover on the red planet.
2013	Cancer treatment	Scientists test a new cancer treatment that uses sickle cells to kill off tumours by starving them of their blood supply.
2013	Bionics	The United States Food and Drug Administration approves the first functional commercial bionic eye, the Argus II, for the treatment of blindness. The device uses a combination of ocular implants and camera-equipped eyeglasses to restore vision to people blinded by retinitis pigmentosa.

2013	Dark matter	NASA reports that the Alpha Magnetic Spectrometer has detected possible signs of the elusive phenomenon known as dark matter.
2013	Optics	Researchers induce "twisted light" beams in optical fibers, allowing for extremely high-bandwidth data transfer.
2014	Climate change	New computer models show that climate change is more sensitive to the effects of cloud formation than previously thought.
2014	Space exploration	The Very Large Telescope discovers the largest known yellow star, HR 5171A, which is 1,300 times the diameter of our Sun, with a companion star that orbits so close, the two stars are almost merged. [
2014	Graphene	Samsung has developed a new method of growing large area, single crystal wafer scale graphene, a major development that will accelerate the commercialization of this material.
2014	Climate change	A carbon dioxide "sponge" that could help absorb man-made emissions from power plants has been announced by the American Chemical Society. [
2014	Internet of Things	Engineers at Stanford University have created ant-sized radios-on-a-chip, powered by incoming electromagnetic waves, that could be used for the Internet of Things.
2015	Neuroscience	As part of the Open Worm Project, scientists have mapped the brain of a roundworm (C. Elegans), created software to mimic its nervous system and uploaded it to a lego robot, which seeks food and avoids obstacles.
2015	Nanoparticles	Researchers used biodegradable nanoparticles to kill brain cancer cells in animals and lengthen their survival.
2015	Batteries	Tesla Motors reveals a new large-scale battery technology for homes and businesses, which will provide a means of storing energy from localized renewables and a reliable backup system during power outages. [
2015	Artificial intelligence	Google demonstrates a new AI chatbot that is able to "remember facts, understand contexts and perform common sense reasoning, all with fewer hand-crafted rules."

2015	Medicine	An Ebola vaccine developed by the Public Health Agency of Canada is found to be 100% successful in an initial trial.
2016	Transistor technology	A new method to produce transistors is presented, based on nanocrystal 'inks'. This allows them to be produced on flexible surfaces, possibly with 3D printers.
2016	Nanotechnology	Scientists at Rice University characterize how single-molecule "nanocars" move in open air, which they claim will help the kinetics of molecular machines in ambient conditions over time.
2016	Gravitational waves	Scientists announce detecting a second gravitational wave event (GW151226) resulting from the collision of black holes.
2016	Lasers	A new "vortex" laser that travels in a corkscrew pattern is shown to carry 10 times or more the information of conventional lasers, potentially offering a way to extend Moore's Law.
2016	Algorithms	Researchers at University College London devise a software algorithm able to scan and replicate almost anyone's handwriting.
2016	Autophagy	2016 Nobel Prize in Physiology or Medicine is awarded to Yoshinori Ohsumi of Japan for discoveries about autophagy

Charles W. Rogers, Southwestern Oklahoma State University, Department of Physics; updated by the editors of Salem Press

BIOGRAPHICAL DICTIONARY OF SCIENTISTS

Alvarez, Luis W. (1911-1988): A physicist and inventor born in San Francisco, Alvarez was associated with the University of California, Berkeley, for many years. He explored cosmic rays, fusion, and other aspects of nuclear reaction. He invented time-of-flight techniques and conducted research into nuclear magnetic resonance for which he was awarded the 1968 Nobel Prize in Physics. He contributed to radar research and particle accelerators, worked on the Manhattan Project, developed the ground-controlled approach for landing airplanes, and proposed the theory that dinosaurs were rendered extinct by a massive meteor impacting Earth.

Archimedes (c. 287-c. 212 BCE): A Greek born at Syracuse, Sicily, Archimedes is considered a genius of antiquity, with interests in astronomy, physics, engineering, and mathematics. He is credited with the discovery of fluid displacement (Archimedes' principle) and a number of mathematical advancements. He also developed numerous inventions, including the Archimedes screw to lift water for irrigation (still in use), the block-and-tackle pulley system, a practical odometer, a planetarium using differential gearing, and several weapons of war. He was killed during the Roman siege of Syracuse.

Babbage, Charles (1791-1871): An English-born mathematician and mechanical engineer, Babbage designed several machines that were precursors to the modern computer. He developed a difference engine to carry out polynomial functions and calculate astronomical tables mechanically(which was not completed) as well as an analytical engine using punched cards, sequential control, branching and looping, all of which contributed to computer science. He also made advancements in cryptography, devised the cowcatcher to clear obstacles from railway locomotives, and invented an ophthalmoscope.

Bacon, Sir Francis (1561-1626): A philosopher, statesman, author, and scientist born in England, Bacon was a precocious youth who at the age of thirteen began attending Trinity College, Cambridge. Later a member of Parliament, a lawyer, and attorney general, he rejected Aristotelian logic and advocated for inductive reasoning—collecting data, interpreting information, and carrying out experiments—in his major work, *Novum Organum* (*New Instrument*), published in 1620, which greatly influenced science

from the seventeenth century onward. A victim of his own research, he experimented with snow as a way to preserve meat, caught a cold that became bronchitis, and died.

Baird, John Logie (1888-1946): A Scottish electrical engineer and inventor, Baird successfully transmitted black-and-white (in 1925) and color (in 1928) moving television images, and the BBC used his transmitters to broadcast television from 1929 to 1937. He had more than 175 patents for such far-ranging and forward-thinking concepts as big-screen and stereo TV sets, pay television, fiber optics, radar, video recording, and thermal socks. Plagued with ill health and a chronic lack of financial backing, Baird was unable to develop his innovative ideas, which others later perfected and profited from.

Bardeen, John (1908-1991): A Wisconsin-born electrical engineer and physicist, Bardeen worked for Gulf Oil, researching magnetism and gravity, and later studied mathematics and physics at Princeton University, where he earned a doctoral degree. While working at Bell Laboratories after World War II he, Walter Brattain (1902-1987), and William Shockley (1910-1989) invented the transistor, for which they shared the 1956 Nobel Prize in Physics. In 1972, Bardeen shared a second Nobel Prize in Physics for a jointly developed theory of superconductivity; he is the only person to win the same award twice.

Barnard, Christiaan (1922-2001): A heart-transplant pioneer born in South Africa, Barnard was a cardiac surgeon and university professor. He performed the first successful human heart transplant in 1967, extending a patient's life by eighteen days, and subsequent transplants—using innovative operational techniques he devised—allowed new heart recipients to survive for more than twenty years. He was one of the first surgeons to employ living tissues and organs from other species to prolong human life and was a contributor to the effective design of artificial heart valves.

Bates, Henry Walter (1825-1892): A self-taught naturalist and explorer born in England, Bates accompanied anthropologist-biologist Alfred Russel Wallace (1823-1913) on a scientific expedition to South America between 1848 and 1852, which he described in his 1864 work, *The Naturalist on the River Amazons*. He collected thousands of plant and animal species, most of them unknown to science, and was the first to study the survival phenomenon of insect mimicry. For nearly thirty years he was secretary of the Royal Geographical Society and also served as president of the Entomological Society of London.

Becquerel, Antoine-Henri (1852-1908): A French physicist and engineer born into a family boasting several generations of scientists, Becquerel taught applied physics at the National Museum of Natural History and at the Polytechnic University, both in Paris, and also served as primary engineer overseeing French bridges and highways. He served as president of the French Academy of Sciences and received numerous awards for his work investigating polarization of light, magnetism, and the properties of radioactivity, including the 1903 Nobel Prize in Physics, which he shared with Pierre and Marie Curie.

Bell, Alexander Graham (1847-1922): A Scottish engineer and inventor whose mother and wife were deaf, Bell researched hearing and speech throughout his life. He began inventing practical solutions to problems as a child. His experiments with acoustics led to his creation of the harmonic telegraph, which eventually resulted in the first practical telephone in 1876 and spawned Bell Telephone Company. Bell became a naturalized American citizen and also invented prototypes of flying vehicles, hydrofoils, air conditioners, metal detectors, and magnetic sound and video recording devices.

Benz, Carl (1844-1929): A German engineer and designer born illegitimately as Karl Vaillant, Benz designed bridges before setting up his own foundry and mechanical workshop. In 1888, he invented, built and patented a gas-powered, engine-driven, three-wheeled horseless carriage named the Benz Motorwagen, which was the first automobile available for purchase. In 1895, he built the first trucks and buses and introduced many technical innovations still found in modern automobiles. The Benz Company merged with Daimler in the 1920's and introduced the famous Mercedes-Benz in 1926.

Berzelius, Jons Jakob (1779-1848): A physician and chemist born in Sweden, Berzelius was secretary of the Royal Swedish Academy of Sciences for thirty years. He is credited with discovering the law of constant proportions for inorganic substances and was the first to distinguish organic from inorganic compounds. He developed a system of chemical symbols and a table of relative atomic weights that are still

in use. In addition to coining such chemical terms as "protein, "catalysis," "polymer," and "isomer," he identified the elements cerium, selenium, silicon, and thorium.

Bessemer, Henry (1813-1898): The English engineer and inventor is chiefly known for development of the Bessemer process, which eliminated impurities from molten pig iron and lowered costs in the production of steel. Holder of more than one hundred patents, Bessemer also invented items to improve the manufacture of glass, sugar, military ordnance, and postage stamps, and built a test model of a gimballed, hydraulic-controlled steamship to eliminate seasickness. His steel-industry creations led to the development of the modern continuous casting process of metals.

Birdseye, Clarence (1886-1956): The naturalist, inventor, and entrepreneur was born in Brooklyn, New York. He began experimenting in the early 1920's with flash-freezing fish. Using a patented process, he was eventually successful in freezing meats, poultry, vegetables, and fruits, and in so doing changed consumers' eating habits. Birdseye sold his process to the company that later became General Foods Corporation, for whom he continued to work in developing frozen-food technology. His surname—split in two for easy recognition—became a major brand name that is still familiar.

Bohr, Niels (1885-1962): A Danish theoretical physicist, Bohr introduced the concept of atomic structure, in which electrons orbit the nucleus of an atom, and laid the foundations of quantum theory, for which he was awarded the 1922 Nobel Prize in Physics. He later identified U-235, an isotope of uranium that produces slow fission. During World War II, after escaping from Nazi-occupied Denmark, he worked as consultant to the Manhattan Project. Following the war, he returned to Denmark and became a staunch advocate for the nondestructive uses of atomic energy.

Bosch, Carl (1874-1940): The German-born chemist, metallurgist, and engineer devised a high-pressure chemical technique (the Haber-Bosch process) to fix nitrogen, used in mass-producing ammonia for fertilizers, explosives, and synthetic fuels. He was awarded (along with Friedrich Bergius, 1884-1949) the 1931 Nobel Prize in Chemistry for his work. He was a founder and chairman of the board of IG Farben, for a time the largest chemical company in the world, but was ousted in the late 1930's for criticizing the Nazis.

Brahe, Tycho (1546-1601): A nobleman born of Danish heritage in what is modern-day Sweden, Brahe became interested in astronomy while studying at the University of Copenhagen. He made improvements to the primitive observational instruments of the day but never had access to the telescope. Nonetheless, he was able to study the positions of stars and planets accurately and produced useful catalogs of celestial bodies, particularly for the planet Mars, which helped Johannes Kepler (1571-1630) to formulate the laws of planetary motion. Craters on the Moon and on Mars are named in Brahe's memory.

Brunel, Isambard Kingdom (1806-1859): A British-born civil engineer and inventor, Brunel designed and built tunnels, bridges, and docks—many still in use—often devising ingenious solutions to problems in the process. He is best remembered for developing the SS *Great Britain*, the largest and most modern ship of its time and the first ocean-going iron ship driven by a propeller. Brunel was also a railroad pioneer, serving as chief engineer for Great Western Railway, for which he specified a broad-gauge track to allow higher speeds, improved freight capacity, and greater passenger comfort.

Burbank, Luther (1849-1926): Despite having only an elementary-school education, the Massachusetts-born botanist and horticulturist was a pioneer in the field of agricultural science. Working from a greenhouse and experimental fields in Santa Rosa, California, Burbank developed more than 800 varieties of plants, including new strains of flowers, peaches, plums, nectarines, cherries, peaches, berries, nuts, and vegetables, as well as new crossbred products such as the plumcot. One of his most useful creations, the Russet Burbank, became the potato of choice in food processing, particularly for French fries.

Calvin, Melvin (1911-1997): A Minnesota-born chemist of Russian heritage, Calvin taught molecular biology for nearly fifty years at the University of California, Berkeley, where he founded and directed the Laboratory of Chemical Biodynamics (later the Structural Biology Division) and served as associate director of the Lawrence Berkeley National Laboratory. He and his research team traced the path of carbon-14 through plants during photosynthesis, greatly enhancing understanding of how sunlight stimulates chlorophyll to create organic compounds. He was awarded the 1961 Nobel Prize in Chemistry for his work.

Carnot, Sadi (1796-1832): A French physicist and military engineer, Carnot was an army officer before becoming a scientific researcher, specializing in the theory of heat as produced by the steam engine. His *Reflections on the Motive Power of Fire* focused on the relationship between heat and mechanical energy and provided the foundation for the second law of thermodynamics. His work greatly influenced scientists such as James Prescott Joule (1818-1889), William Thomson (Lord Kelvin, 1824-1907), and Rudolf Diesel (1858-1913) and made possible more practically and efficiently designed engines later in the nineteenth century. Carnot's career was cut short by his death from cholera.

Carson, Rachel (1907-1964): A marine biologist and author born in Springdale, Pennsylvania, Carson worked for the U.S. Bureau of Fisheries before turning full-time to writing about nature. Her popular and highly influential articles, radio scripts and books, including *The Sea Around Us, The Edge of the Sea, Under the Sea-Wind,* and *Silent Spring* enlightened the public about the wonders of nature and the dangers of pesticides such as DDT, which was eventually banned in the United States. Carson is credited with spurring the modern environmental movement.

Celsius, Anders (1701-1774): A Swedish astronomer, Celsius studied the aurora borealis and was the first to link the phenomena to the Earth's magnetic field. He also participated in several expeditions designed to measure the size and shape of the Earth. Founder of the Uppsala Astronomical Observatory, he explored star magnitude, observed eclipses, and compiled star catalogs. He is perhaps best known for the Celsius international temperature scale, which accounts for atmospheric pressure in measuring the boiling and freezing points of water.

Clausewitz, Carl von (1780-1831): As a Prussian-born soldier and military scientist, Clausewitz participated in numerous campaigns, beginning in the early 1790's, and fought in the Napoleonic Wars. After his appointment in 1818 to major general, he taught at the Prussian military academy and helped reform the state army. His principal written work, *On War*, unfinished at the time of his death from cholera, is still considered relevant and continues to influence military thinking via its practical approach to command policies, instruction for soldiers, and methods of planning for strategists.

Colt, Samuel (1814-1862): An inventor born in Hartford, Connecticut, Colt designed a workable multi-shot pistol while working in his father's textile factory. In the mid-1830's he patented a revolver and set up an assembly line to produce machine-made weapons featuring interchangeable parts. The perfected product, the Colt Peacemaker, was used in the Seminole and Mexican-American wars and became popular during America's western expansion, and Colt became a millionaire. Colt's Manufacturing Company continues to produce a wide variety of firearms for civilian, military, and law-enforcement purposes.

Copernicus, Nicolaus (1473-1543): The Polish mathematician, physician, statesman, artist, linguist, and astronomer is credited with beginning the scientific revolution. His major work, published the year of his death, *De revolutionibus orbium coelestium* (*On the Revolutions of the Heavenly Spheres*), was the first to propose a heliocentric model of the solar system. The book inspired further research by Tycho Brahe (1546-1601), Galileo Galilei (1564-1642), and Johannes Kepler (1571-1630) and stimulated the birth of modern astronomy.

Cori, Gerty Radnitz (1896-1957): A biochemist born in Prague (now the Czech Republic), Cori came to the United States in 1922 and became a naturalized American citizen in 1928. She worked with her husband Carl at what is now Roswell Park Cancer Institute in Buffalo, New York, researching carbohydrate metabolism and discovered how glycogen is broken down into lactic acid to be stored as energy, a process now called the Cori cycle. She was awarded the 1947 Nobel Prize in Physiology or Medicine, the first American woman so honored.

Cousteau, Jacques (1910-1997): A French oceanographer, explorer, filmmaker, ecologist, and author, Cousteau began underwater diving in the 1930's, and it became a lifelong obsession. He coinvented the Aqua-Lung in the 1940's—the precursor to modern scuba gear—and began making nature films during the same decade. He founded the French Oceanographic Campaigns in 1950 and aboard his ship *Calypso* explored and researched the world's oceans for forty years. In the 1970's he created the Cousteau Society, which remains a strong ecological advocacy organization.

Crick, Francis (1916-2004): An English molecular biologist and physicist, Crick designed magnetic and acoustic mines during World War II. He was later part of a biological research team at the Cavendish Laboratory. Focusing on the X-ray crystallography of proteins, he identified the structure of deoxyribonucleic acid (DNA) as a double helix, a discovery that greatly advanced the study of genetics. He and his colleagues, American James D. Watson (b. 1928) and New Zealander Maurice Wilkins (1916-2004), shared the 1962 Nobel Prize in Physiology or Medicine for their groundbreaking work.

Curie, Marie Sklodowska (1867-1934) and Pierre Curie (1859-1906): Polish-born chemist-physicist Marie was the first woman to teach at the University of Paris. She married French physicist-chemist Pierre Curie in 1895, and the couple collaborated on research into radioactivity, discovering the elements polonium and radium. She and her husband shared the 1903 Nobel Prize in Physics for their work; she was the first woman so honored. After her husband died, she continued her research and received the 1911 Nobel Prize in Chemistry, the first person to receive the award in two different disciplines. She founded the Radium (later Curie) Institute.

Daimler, Gottlieb (1834-1900): The German-born mechanical engineer, designer, inventor, and industrial magnate was an early developer of the gasoline-powered internal combustion engine and the automobile. He and fellow industrial designer Wilhelm Maybach (1846-1929) began a partnership in the 1880's to build small, high-speed engines incorporating numerous devices they patented—flywheels, carburetors, and cylinders—still found in modern engines. After creating the first· motorcycle they founded Daimler Motors and began selling automobiles in the early 1890's. Their Phoenix model won history's first auto race.

Darwin, Charles Robert (1809-1882): An English naturalist and geologist, Darwin participated in the five-year-long worldwide surveying expedition of the HMS *Beagle* during the 1830's, observing and collecting specimens of animals, plants, and minerals. The voyage inspired numerous written works, particularly *On Natural Selection, On the Origin of the Species,* and *The Descent of Man,* which collectively supported his theory that all species have evolved from common ancestors. Though modern science has virtually unanimously accepted Darwin's findings, his theory of evolution remains a controversial topic among various political, cultural, and religious groups.

Davy, Sir Humphry (1778-1829): A chemist, teacher, and inventor born in England, Davy began conducting scientific experiments as a child. As a teen he worked as a surgeon's apprentice and became addicted to nitrous oxide. He later researched galvanism and electrolysis, discovered the elements sodium, chlorine, and potassium, and contributed to the discovery of iodine. He invented the Davy safety lamp for use in coal mines, was a founder of the Zoological Society of London, and served as president of the Royal Society.

Diesel, Rudolf (1858-1913): A French-born mechanical engineer and inventor of German heritage, Diesel designed an innovative refrigeration system for an ice plant in Paris and improved the efficiency of steam engines. His self-named, patented diesel engine introduced the concept of fuel injection. The efficient diesel engine later became commonplace in trucks, locomotives, ships, and submarines and, after redesign to reduce weight, in modern automobiles. Diesel disappeared while on a ship. His body was later discovered floating in the sea, but it is still unknown whether he fell overboard, committed suicide, or was murdered.

Edison, Thomas Alva (1847-1931): A scientist, inventor and entrepreneur born in Milan, Ohio, Edison worked out of New Jersey in the fields of electricity and communication and profoundly influenced the world. Credited with more than 1,000 patents, he is best known for creating the first practical incandescent light bulb, which has illuminated the lives of humans since 1879. Other inventions include the stock ticker, a telephone transmitter, electricity meters, the mimeograph, an efficient storage battery, and the phonograph and the kinetoscope, which he combined to produce the first talking moving picture in 1913.

Einstein, Albert (1879-1955): A German-born theoretical physicist and author of Jewish heritage, Einstein came to the United States before World War II. Regarded as a genius, and one of the world's most recognized scientists, he was awarded the 1921 Nobel Prize in Physics. He developed general and special theories of relativity, particle and quantum theories, and proposed ideas that continue to influence numerous fields of study, including energy, nuclear power, heat, light, electronics, celestial mechanics,

astronomy, and cosmology. Late in life, he was offered the presidency of Israel but declined the honor.

Euclid (c. 330-c. 270 BCE): A Greek mathematician who taught in Alexandria, Egypt, Euclid is considered the father of plane and solid geometry. He is remembered principally for his major extant work, *The Elements*—a treatise containing definitions, postulates, geometric proofs, number theories, discussions of prime numbers, arithmetic theorems, algebra, and algorithms—which has served as the basis for the teaching of mathematics for two thousand years. He also explored astronomy, mechanics, gravity, moving bodies, and music and was one of the first scientists to write about optics and perspective.

Everest, Sir George (1790-1866): A geographer born in Wales, Everest participated for 25 years in the Great Trigonometrical Survey of the Indian subcontinent—which surveyed an area encompassing millions of square miles while locating, measuring, and naming the Himalayan Mountains—and served as superintendent of the project from 1823 to 1843. Later knighted, he served as vice president of the Royal Geographical Society. The world's tallest peak, Mount Everest in Nepal (known locally as Chomolungma), was named in his honor.

Faraday, Michael (1791-1867): A self-educated British chemist and physicist, Faraday served an apprenticeship with chemist Sir Humphry Davy (1778-1829), during which time he experimented with liquefied gases, alloys, and optical glasses. He invented a prototype of what became the Bunsen burner, discovered benzene, and performed experiments that led to his discovery of electromagnetic induction. He built the first electric dynamo, the precursor to the power generator, researched the relationship between magnetism and light, and made numerous other contributions to the studies of electromagnetism and electrochemistry.

Farnsworth, Philo (1906-1971): An inventor born in Utah, Farnsworth became interested in electronics and mechanics as a child. He experimented with television during the 1920's, and late in the decade he demonstrated an electronic, nonmechanical scanning system for image transmissions. During the early 1930's, he worked for Philco but left to carry out his own research. In addition to significant contributions to television, Farnsworth held more than 300 patents and devised a milk-sterilizing process, developed fog lights, an infrared telescope, a prototype of an air traffic control system, and a fusion reaction tube.

Fermi, Enrico (1901-1954): An Italian-born experimental and theoretical physicist and teacher who became an American citizen in 1944, Fermi studied mechanics and was instrumental in the advancement of thermodynamics and quantum, nuclear, and particle physics. Awarded the 1938 Nobel Prize in Physics for his research into radioactivity, he was a member of the team that developed the first nuclear reactor in Chicago in the early 1940's and served as a consultant to the Manhattan Project, which produced the first atomic bomb. He died of cancer from sustained exposure to radioactivity.

Feynman, Richard (1918-1988): A physicist, author, and teacher born in New York, Feynman participated in the Manhattan Project and made numerous contributions to a diverse field of specialized scientific disciplines including quantum mechanics, supercooling, genetics, and nanotechnology. He shared the 1965 Nobel Prize in Physics—with Julian Schwinger (1918-1994) and Sin-Itiro Tomonaga (1906-1979)—for work in quantum electrodynamics, particularly for his lucid explanation of the behavior of subatomic particles. He was a popular and influential professor for many years at California Institute of Technology.

Fleming, Alexander (1881-1955): A Scottish-born biologist, Fleming served in the Royal Army Medical Corps during World War I and witnessed the deaths of many wounded soldiers from infection. He was a professor of bacteriology at a teaching hospital, and he specialized in immunology and chemotherapy research. In 1928 he discovered an antibacterial mold, which over the next decade was purified and mass-produced as the drug penicillin, which played a large part in suppressing infections during World War II. A major contributor to the development of antibiotics, he shared the 1945 Nobel Prize in Physiology or Medicine.

Forrester, Jay Wright (1918-2016): An engineer, teacher, and computer scientist born in Nebraska, Forrester built a wind-powered electrical system while in his teens. Associated with the Massachusetts Institute of Technology as a researcher and professor for many years, he developed servomechanisms for military use, designed aircraft flight simulators, and air defense systems. He founded the field of system dynamics to produce computer-generated

mathematical models for such tasks as determining water flow, fluid turbulence, and a variety of mechanical movements.

Fourier, Joseph (1768-1830): A French-born physicist, mathematician, and teacher Fourier accompanied Napoleon's expedition to Egypt, where he served as secretary of the Egyptian Institute and was a major contributor to *Description of Egypt,* a massive work describing the scientific findings that resulted from the French military campaign. After returning to France, Fourier explored numerous scientific fields but is best known for his extensive research on the conductive properties of heat and for his theories of equations, which influenced later physicists and mathematicians.

Franklin, Benjamin (1706-1790): A Boston-born author, statesman, scientist, and inventor, Franklin worked as a printer in his youth and from 1733 to 1758 published *Poor Richard's Almanack.* A key figure during the American Revolution and a founding father of the United States, he established America's first lending library and Pennsylvania's first fire department, served as first U.S. postmaster general and as minister to France and Sweden. He experimented with electricity and is credited with inventing the lightning rod, bifocal glasses, an odometer, the Franklin stove, and a musical instrument made of glass.

Freud, Sigmund (1856-1939): An Austrian of Jewish heritage, Freud studied neurology before specializing in psychopathology and conducted extensive research into hypnosis and dream analysis to treat hysteria. Considered the father of psychoanalysis and a powerful influence on the field, he originated such psychological concepts as repression, psychosomatic illness, the unconscious mind, and the division of the human psyche into the id, ego, and superego. He fled from the Nazis and went to London. Riddled with cancer from years of cigar smoking, he took morphine to relieve his suffering and hasten his death.

Frisch, Karl von (1886-1982): The Vienna-born son of a surgeon-university professor, von Frisch initially studied medicine before switching to zoology and comparative anatomy. Working as a teacher and researcher out of Munich, Rostock, Breslau, and Graz universities, he focused his research on the European honeybee. He made many discoveries about the insect's sense of smell, optical perception, flight patterns, and methods of communication that have since proved invaluable in the fields of apiology and botany. He was awarded the 1973 Nobel Prize in Physiology of Medicine in recognition of his pioneering work.

Fuller, R. Buckminster (1895-1983): The Massachusetts-born architect, philosopher, engineer, author, and inventor developed systems for lightweight, weatherproof, and fireproof housing while in his twenties. Teaching at Black Mountain College in the late 1940's, Fuller perfected the geodesic dome, built of aluminum tubing and plastic skin, and afterward developed numerous designs for inventions aimed at providing practical and affordable shelter and transportation. He coined the term "synergy" and advocated exploiting renewable sources of energy such as solar power and wind-generated electricity. He was awarded the Presidential Medal of Freedom in 1983.

Galen (129-c. 199): An ancient Roman surgeon, scientist, and philosopher, Galen traveled and studied widely before serving as physician to Roman emperors Marcus Aurelius (121-180), Commodus (161-192), Septimus Severus (146-211), and Caracalla (188-217). In the course of his education he explored human and animal anatomy, became an advocate of proper diet and hygiene, and advanced the practice of surgery by treating the wounds of gladiators and ministering to plague victims. His medical discoveries and healing methods, detailed in numerous written works, influenced medicine for more than 1,500 years.

Galilei, Galileo (1564-1642): A physicist, astronomer, mathematician, and philosopher born in Pisa, Italy, Galileo is known as the father of astronomy and the father of modern science. A keen astronomical observer, he made significant improvements to the telescope, through which he studied the phases of Venus, sunspots, and the Milky Way, and he discovered Jupiter's four largest moons. He risked excommunication and death championing the heretical Copernican heliocentric view of the solar system. He also invented a military compass and a practical thermometer and experimented with pendulums and falling bodies.

Galton, Francis (1822-1911): An anthropologist, geographer, meteorologist, and inventor born in England, Galton was a child prodigy. He traveled widely, exploring the Middle East and Africa, and wrote about his expeditions. Fascinated by numbers, he devised the first practical weather maps for use in newspapers. He also contributed to the science of statistics, studied heredity—coining the term "eugenics" and the phrase "nature or nurture"—and was an early advocate of using fingerprints in criminology. Galton is responsible for inventing a high-frequency whistle used in training dogs and cats.

Gates, Bill (b. 1955): An entrepreneur, philanthropist, and author born in Seattle, Gates became interested in computers as a teenager. He left Harvard in 1975 to cofound and to serve as chairman (until 2006) of Microsoft, which developed software for IBM and other systems before launching its own system in 1985. The result, Microsoft Windows, became the dominant software product in the worldwide personal computer market. Profits from his enterprise made Gates one of the world's richest people, and he has used his vast wealth to assist a wide variety of charitable causes.

Goddard, Robert H. (1892-1945): A physicist, engineer, teacher, and inventor born in Massachusetts, Goddard became interested in science as a child and experimented with kites, balloons, and rockets. He received the first of more than 200 patents in 1914 for multistage and liquid-fuel rockets, and during the 1920's he conducted successful test flights using liquid fuel. Goddard experimented with solid fuels and ion thrusters and is credited with developing tail fins, gyroscopic guidance systems, and many other basics of rocketry that greatly influenced the designs of rocket scientists who came after him.

Grandin, Temple (b. 1947): An animal scientist born in Massachusetts, Grandin was diagnosed with autism as a child. As an adult she earned advanced degrees before receiving a doctorate from the University of Illinois in 1989. A professor at Colorado State University, an author, and an autism advocate, she has made numerous humane improvements to the design of livestock-handling facilities that have been incorporated into meat-processing plants worldwide to reduce or eliminate animal stress, pain, and fear.

Haber, Fritz (1868-1934): A chemist and teacher of Jewish heritage (later a convert to Christianity) born in Germany, Haber developed the Haber process to produce ammonia used in fertilizers, animal feed, and explosives, for which he was awarded the 1918 Nobel Prize in Chemistry. At Berlin's Kaiser Wilhelm

Institute (later the Haber Institute) between 1911 and 1933, he developed chlorine gas used in World War I, experimented with the extraction of gold from seawater, and oversaw production of Zyklon B, the cyanide-based pesticide that was employed at extermination camps during World War II.

Halley, Edmond (1656-1742): An English astronomer, mathematician, physicist, and meteorologist, Halley wrote about sunspots and the solar system while a student at Oxford. In the 1670's, he cataloged the stars of the Southern Hemisphere and charted winds and monsoons. Inventor of an early diving bell and a liquid-damped magnetic compass, a colleague of Sir Isaac Newton (1642-1727), and leader of the first English scientific expedition, Halley is best remembered for predicting the regular return of the comet that bears his name.

Heisenberg, Werner (1901-1976): A theoretical physicist and teacher born in Germany, Heisenberg conducted research in quantum mechanics with Niels Bohr (1885-1962) at the University of Copenhagen. There he developed the uncertainty principle, which proves it is impossible to determine the position and momentum of subatomic particles at the same time. Awarded the 1932 Nobel Prize in Physics for his work, he also contributed research on positrons, cosmic rays, spectral frequencies, matrix mechanics, nuclear fission, superconductivity, and plasma physics to the continuing study of atomic theory.

Herschel, William (1738-1822) and Caroline Herschel (1750-1848): A German-born astronomer, composer, and telescope maker who moved to England in his teens, William spent the early part of his career as a musician, playing cello, oboe, harpsichord, and organ and wrote numerous symphonies and concerti. In the 1770's he began building his own large reflecting telescopes and with his diminutive (4 feet, 3 inches) but devoted sister Caroline spent countless hours observing the sky while cataloging nebulae and binary stars. The Herschels are credited with discovering two of Saturn's moons, the planet Uranus and two of its moons, and coining the word "asteroid."

Hersey, Mayo D. (1886-1978): A mechanical engineer born in Rhode Island, Hersey was a preeminent expert on tribology, the study of the relationship between interacting solid surfaces in motion, the adverse effects of wear, and the ameliorating effects of lubrication. He worked as a physicist at the National

Institute of Standards and Technology (1910-1920) and the U.S. Bureau of Mines (1922-1926) and taught at the Massachusetts Institute of Technology (1910-1922). He was a consultant to the Manhattan Project and won numerous awards for his contributions to lubrication science.

Hippocrates (c. 460-c. 377 BCE): An ancient Greek physician born on the island of Kos, Hippocrates was the first of his time to separate the art of healing from philosophy and magical ritual. Called the father of Western medicine, he originated the belief that diseases were not the result of superstition but of natural causes, such as environment and diet. Though his concept that illness was the result of an imbalance in the body's fluids (called humors) was later discredited, he pioneered such common modern clinical practices as observation and documentation of patient care. He originated the Hippocratic Oath, which for many centuries served as the guiding principle governing the behavior of doctors.

Hooke, Robert (1635-1703): A brilliant, multitalented British experimental scientist with interests in physics, astronomy, chemistry, biology, geology, paleontology, mechanics, and architecture, Hooke was instrumental as chief surveyor in rebuilding the city of London following the Great Fire of 1666. Among many accomplishments in diverse fields he is credited with inventing the compound microscope—via which he discovered the cells of plants and formulated a theory of fossilization—devised a balance spring to improve the accuracy of timepieces, and either created or refined such instruments as the barometer, anemometer, and hygrometer.

Howlett, Freeman S. (1900-1970): A horticulturist born in New York, Howlett was associated with the Ohio State University as teacher, administrator, and researcher for more than forty-five years and was considered an expert on the history of horticulture. His investigations focused on plant hormones, embryology, fruit setting, reproductive physiology, and foliation for a variety of crops, including fruits, vegetables, and nuts. He created five new varieties of apples popular among consumers. A horticulture and food science building at Ohio State is named in his honor.

Hubble, Edwin Powell (1889-1953): An astronomer born in Missouri, Hubble was associated with the Mount Wilson Observatory in California for more than thirty years. Using what was then the world's largest telescope, he was the first to discover galaxies beyond the Milky Way, which greatly expanded science's concept of the universe. He studied red shifts in formulating Hubble's law, which confirmed the big bang or expanding universe theory. The American space telescope launched in 1990 was named for him, and he was honored in 2008 with a commemorative postage stamp.

Huygens, Christiaan (1629-1695): Born in the Netherlands, Huygens was an astronomer, physicist, mathematician, and prolific author. He made early telescopic observations of Saturn and its moons and was the first to suggest that light is made up of waves. He discovered centrifugal force, proposed a formula for centripetal force, and developed laws governing the collision of celestial bodies. An inveterate inventor, he patented the pendulum clock and the pocket watch and designed an early internal combustion engine. He is also considered a pioneer of science fiction for writing about the possibility of extraterrestrial life.

Jacquard, Joseph Marie (1752-1834): A French inventor, Jacquard created a series of mechanical looms in the early nineteenth century. His experiments culminated in the Jacquard loom attachment, which could be programmed, via punch cards, to weave silk in various patterns, colors, and textures automatically. The labor-saving device became highly popular in the silk-weaving industry, and its inventor received royalties on each unit sold and became wealthy in the process. The loom inspired scientists to incorporate the concept of punch cards for computer information storage.

Jenner, Edward (1749-1823): A surgeon and anatomist born in England, Jenner experimented with cowpox inoculations in an attempt to prevent smallpox, a virulent infectious disease of ancient origin with a high rate of mortality that killed millions of people. In the early nineteenth century, Jenner successfully developed a method of vaccination that provided immunity from smallpox and late in life became personal physician to King George IV. The smallpox vaccination was made compulsory in England and elsewhere, and the disease was declared eradicated worldwide in 1979.

Jobs, Steven (b. 1955-2011): An inventor and entrepreneur of Syrian and American heritage born in San Francisco, Jobs worked at Hewlett-Packard as a teenager and was later employed at Atari designing circuit boards. In 1976, he and coworker Steve Wozniak (b. 1950) and others founded Apple, which designed, built, and sold a popular and highly successful line of personal computers. A multibillionaire and holder of more than 200 patents, Jobs continued to make innovations in interfacing, speakers, keyboards, power adaptation, and myriad other components related to modern computer science until his death in late 2011

Kepler, Johannes (1571-1630): A German mathematician, author and astronomer, he became interested in the cosmos after witnessing the Great Comet of 1577. He worked for Tycho Brahe (1546-1601) for a time and after Brahe's death became imperial mathematician in Prague. A major contributor to the scientific revolution, Kepler studied optics, observed many celestial phenomena, provided the foundation for Sir Issac Newton's theory of gravitation, and developed a set of laws governing planetary motion around the Sun—including the discovery that the orbits of planets are elliptical—that were confirmed by later astronomers.

Krebs, Sir Hans Adolf (1900-1981): A biochemist and physician born the son of a Jewish surgeon in Germany, Krebs was a clinician and researcher before moving to England after the rise of the Nazis. As a professor at Cambridge University, he explored metabolism, discovering the urea and citric acid cycles—biochemical reactions that promote understanding of organ functions in the body and explain the cellular production of energy—for which he shared the 1953 Nobel Prize in Medicine or Physiology. He was also knighted in 1958 for his work.

Lawrence, Ernest O. (1901-1958): A South Dakota-born physicist and teacher, Lawrence researched the photoelectric effect of electrons at Yale. In 1928, he became a professor at the University of California, Berkeley, where he invented the cyclotron particle accelerator, for which he was awarded the 1939 Nobel Prize in Physics. During World War II, he was involved in the Manhattan Project. Lawrence popularized science and was a staunch advocate for government funding of significant scientific projects. After his death, laboratories at the University of California and the chemical element lawrencium were named in his honor.

Leakey, Louis B. (1903-1972) and Mary Nicol Leakey (1913-1996): Louis, born in Kenya, was an archaeologist, paleontologist, and naturalist who married London-born anthropologist and archaeologist Mary Nicol. Together and often with their sons, Jonathan, Richard, and Philip, they excavated at Olduvai Gorge in East Africa, where they unearthed the tools and fossils of ancient hominids. Their discoveries of the remains of Proconsul africanus, Australopithecus boisei, Homo habilis, Homo erectus, and other large-brained, bipedal primates effectively proved Darwin's theory of evolution and extended human history by several million years.

Leonardo da Vinci (1452-1519): An Italian genius considered the epitome of the Renaissance man, da Vinci was a superb artist, architect, engineer, mathematician, geologist, musician, mapmaker, inventor, and writer. Creator of such famous paintings as the *Mona Lisa* and *The Last Supper,* he is credited with imagining the helicopter, solar power, and the calculator centuries before their invention. His far-ranging mind explored such subjects as anatomy, optics, vegetarianism, and hydraulics, and his journals, written in mirror-image script, are filled with drawings, ideas, and scientific observations that are still closely studied.

Linnaeus, Carolus (1707-1778): A Swedish botanist, zoologist, physician, and teacher, Linnaeus began studying plants as a child. As an adult, he embarked on expeditions throughout Europe observing and collecting specimens of plants and animals and wrote numerous works about his findings. He devised the binomial nomenclature system of classification for living and fossil organisms—called taxonomy—still used in modern science, which provides concise Latin names of genus and species for each example. Linnaeus also cofounded the Royal Swedish Academy of Science.

Lippershey, Hans (c. 1570-c. 1619): A master lens grinder and spectacle maker born in Germany who later became a citizen of the Netherlands, Lippershey is credited with designing the first practical refracting telescope (which he called "perspective glass"). After fruitlessly attempting to patent the device, he built several prototypes for sale to the Dutch government, which distributed information about the telescope across Europe. Other scientists, such as Galileo, soon duplicated and improved upon Lippershey's invention, which became a primary instrument in the science of astronomy.

Lumière, Auguste (1862-1954) and Louis Lumière (1864-1948): The French-born brothers worked at their father's photographic business and devised the dry-plate process for still photographs. From the early 1890's, they patented several techniques—including perforations to guide film through a camera and a color photography process—that greatly advanced the development of moving pictures. From 1895 to 1896, they publicly screened a series of short films to enthusiastic audiences in Asia, Europe, and North and South America, demonstrating the commercial potential of the new medium and launching what would become the multibillion-dollar film industry.

Maathai, Wangari Muta (1940-2011): An environmental and political activist of Kikuyu heritage born in Kenya, Maathai studied biology in the United States before becoming a research assistant and anatomy teacher at the University of Nairobi, where she was the first East African woman to earn a PhD She founded the Green Belt Movement, an organization that plants trees, supports environmental conservation, and advocates for women's rights. A former member of the Kenyan Parliament and former Minister of Environment, she was awarded the 2004 Nobel Peace Prize for her work and is the first African woman to receive the award.

McAdam, John Loudon (1756-1836): A Scottish engineer, McAdam became a surveyor in Great Britain and specialized in road building. He devised an effective method—called "macadam" after its inventor—of creating long-lasting roads using gravel on a foundation of larger stones, with a camber to drain away rainwater, which was adopted around the world. He also introduced hot tar as a binding agent (dubbed "tarmac," an abbreviation of tarmacadam) to produce smoother road surfaces. Modern road builders still use many of the techniques he innovated.

Mantell, Gideon (1790-1852): A British surgeon, geologist, and paleontologist, Mantell began collecting fossil specimens from quarries as a child. As an adult, he was a practicing physician and pursued geology in his spare time. He discovered fossils that were eventually identified as belonging to the Iguanodon and Hylaeosaurus—which he named Megalosaurus and Pelorosaurus—and he became a recognized authority on dinosaurs. His major works were *The Fossils of South Downs: Or, Illustrations of the Geology of Sussex* (1822) and *Notice on the Iguanodon: A Newly Discovered Fossil Reptile* (1825).

Marconi, Guglielmo (1874-1937): An Italian-born electrical engineer and inventor, Marconi experimented with electricity and electromagnetic radiation. He developed a system for transmitting telegraphic messages without the use of connecting wires and by the early twentieth century was sending transmissions across the Atlantic Ocean. His devices eventually evolved into radio, and the transmitter at his factory in England was the first in 1920 to broadcast entertainment to the United Kingdom; he shared the 1909 Nobel Prize in Physics with German physicist Ferdinand Braun (1850-1918).

Maxwell, James Clerk (1831-1879): A Scottish-born mathematician, theoretical physicist, and teacher, Maxwell had an insatiable curiosity from an early age and as a teenager began presenting papers to the Royal Society of Edinburgh. He experimented with color, examined hydrostatics and optics, and wrote about Saturn's rings. His most significant work, however, was performed in the field of electromagnetism, in which he showed that electricity, magnetism, and light are all results of the electromagnetic field, a concept that profoundly affected modern physics.

Mendel, Gregor Johann (1822-1884): Born in Silesia (now part of the Czech Republic), Mendel became interested in plants as a child. In the 1840's he entered an Augustinian monastery, where he studied astronomy, meteorology, apiology, and botany. Called the father of modern genetics, he is best known for his experiments in hybridizing pea plants, which evolved into what later were called Mendel's laws of inheritance. Though his work exerted little influence during his lifetime, his concepts were rediscovered early in the twentieth century and have since proven invaluable to the study of heredity.

Mendeleyev, Dmitri Ivanovich (1834-1907): A Russian chemist, teacher, and inventor, Mendeleyev studied the properties of liquids and the spectroscope before becoming a professor in Saint Petersburg and later serving as director of weights and measures. He created a periodic table of the sixty-three elements then known arranged by atomic mass and the similarity of properties (a revised form of which is still employed in modern science) and used the table to correctly predict the characteristics of elements and isotopes not yet found. Element 101, mendelevium, discovered in 1955, was named in his honor.

Meng Tian (259-210 BCE): A general serving under Qin Shi Huang, first emperor of the Qin Dynasty (221-207 BCE), Meng Tian led an army of 100,000 to drive warlike nomadic tribes north out of China. Descended from architects, he oversaw building of the Great Wall to prevent invasions, cleverly incorporating topographical features and natural barriers into the defensive barricade, which he extended for more than 2,000 miles along the Yellow River. After a coup following Emperor Qin's death, Meng Tian was forced to commit suicide. The Qin Dynasty fell just three years later.

Montgolfier, Joseph Michel (1740-1810) and Jacques-Etienne Montgolfier (1745-1799): Born in France to a prosperous paper manufacturer, the Mongolfier brothers designed and built a hot-air balloon, and in 1783 Jacques-Etienne piloted the first manned ascent in a lighter-than-air craft. The French Academy of Science honored the brothers for their exploits, which inspired further developments in ballooning. The Montgolfier brothers subsequently wrote books on aeronautics and continued experimenting. Joseph is credited with designing a calorimeter and a hydraulic ram, and Jacques-Etienne invented a method for the manufacture of vellum.

Morse, Samuel F. B. (1791-1872): An artist and inventor born in Massachusetts, Morse painted portraits and taught art at the City University of New York before experimenting with electricity. In the mid-1830's, he designed the components of a practical telegraph—a sender, receiver, and a code to translate signals into numbers and words—and in 1844 sent the first message via wire. Within a decade, the telegraph had spread across America and subsequently around the world. The invention would inspire such later advancements in communication as radio, the Teletype, and the fax machine.

Nernst, Walther (1864-1941): A German physical chemist, physicist, and inventor, Nernst discovered the Third Law of Thermodynamics—defining the chemical reactions affecting matter as temperatures drop toward absolute zero—for which he was awarded the 1920 Nobel Prize in Chemistry. He also invented an electric lamp, and developed an electric piano and a device using rare-earth filaments that significantly advanced infrared spectroscopy. He made numerous contributions to the specialized fields of electrochemistry, solid-state chemistry, and photochemistry.

Newton, Sir Isaac (1642-1727): The English physicist, mathematician, astronomer, and philosopher is considered one of the most gifted and scientifically influential individuals of all time. He developed theories of color and light from studying prisms, was instrumental in creating differential and integral calculus, and formulated still-valid laws of celestial motion and gravitation. He was knighted in 1705, the first British scientist so honored. From 1699 until his death he served as master of the Royal Mint and during his tenure devised anticounterfeiting measures and moved England from the silver to the gold standard.

Nobel, Alfred (1833-1896): A Swedish chemist and chemical engineer, Nobel invented dynamite while studying how to manufacture and use nitroglycerin safely. In the course of building a manufacturing empire based on the production of cannons and other armaments, he experimented with combinations of explosive components, also producing gelignite and a form of smokeless powder, which led to the development of rocket propellants. Late in his life, he earmarked the bulk of his vast estate for the establishment of the Nobel Prizes, annual monetary awards given in recognition of outstanding achievements in science, literature, and peace.

Oppenheimer, J. Robert (1904-1967): A brilliant theoretical physicist, researcher, and teacher born to German immigrants in New York City, Oppenheimer was the scientific director of the Manhattan Project, which developed the atomic bombs dropped on Japan during World War II. Following the war, he was primary adviser to the U.S. Atomic Energy Commission and director of the Institute for Advanced Study in Princeton, New Jersey. He contributed widely to the study of electrons and positrons, neutron stars, relativity, gravitation, black holes, quantum mechanics, and cosmic rays.

Owen, Richard (1804-1892): An English biologist, taxonomist, anti-Darwinist, and comparative anatomist, Owen founded and directed the natural history department at the British Museum. He originated the concept of homology, a similarity of structures in different species that have the same function, such as the human hand, the wing of a bat, and the paw of an animal. He also cataloged many living and fossil specimens, contributed numerous discoveries to zoology, and coined the term "dinosaur." Owen advanced the theory that giant flightless birds once inhabited New Zealand long before their remains were found there.

Paré, Ambroise (c. 1510-1590): A French royal surgeon, Paré revolutionized battlefield medicine, developing techniques and instruments for the treatment of gunshot wounds and for performing amputations.

He greatly advanced knowledge of human anatomy by studying the effects of violent death on internal organs. He pioneered the lifesaving practices of vascular ligating and herniotomies, designed prosthetics to replace amputated limbs, and was the first to create realistic artificial eyes from such substances as glass, porcelain, silver, and gold.

Pasteur, Louis (1822-1895): A chemist, microbiologist, and teacher born in France, Pasteur focused on researching the causes of diseases and methods for preventing them after three of his children died from typhoid. He proposed a germ theory, demonstrating that microorganisms affect foodstuffs. This ultimately led to his invention of pasteurization—a method of killing bacteria in milk, which was later applied to other substances. A pioneer in immunology, he also developed vaccines to combat anthrax, rabies, and puerperal fever.

Pauli, Wolfgang (1900-1958): An Austrian theoretical physicist of Jewish heritage who converted to Catholicism, Pauli earned a PhD at the age of twenty-one. While lecturing at the Niels Bohr Institute for Theoretical Physics, he researched relativity and quantum physics. He discovered a new law governing the behavior of atomic particles and the characteristics of matter, called the Pauli exclusion principle, for which he was awarded the 1945 Nobel Prize in Physics. During World War II, he moved to the United States and became an American citizen but later relocated to Zurich.

Pauling, Linus (1901-1994): Born in Portland, Oregon, Pauling earned advanced degrees in chemical engineering, physical chemistry, and mathematical physics. A Guggenheim Fellow, he studied quantum mechanics in Munich, Copenhagen, and Zurich before teaching at the California Institute of Technology. He specialized in theoretical chemistry and molecular biology and greatly advanced understanding of the nature of chemical bonds. A political activist who warned of the dangers of nuclear weapons, he became one of a handful of scientists to receive Nobel Prizes in two fields: the 1954 prize in chemistry and the 1982 peace prize.

Pavlov, Ivan (1849-1936): A Russian physiologist and psychologist, Pavlov began investigating the digestive system, which led to experiments with the effects of behavior on the nervous system and the body's automatic functions. He used animals in researching conditioned reflex actions to a variety of visual, tactile, and sound stimuli—including bells, whistles, and electric shocks—to discover the relationship between salivation and digestion and was able to make dogs drool in anticipation of receiving food. He was awarded the 1904 Nobel Prize in Physiology or Medicine for his work.

Planck, Max (1858-1947): A German theoretical physicist credited with founding quantum theory—which affects all matter in the universe—Planck earned a doctoral degree at the age of twenty-one before becoming a professor at the universities of Kiel and Berlin. He explored electromagnetic radiation, quantum mechanics, thermodynamics, blackbodies, and entropy. He formulated the Planck constant, which describes the proportions between the energy and frequency of a photon and provides understanding of atomic structure. He was awarded the 1918 Nobel Prize in Physics for his discoveries.

Ptolemy (c. 100-c. 178): A mathematician, astronomer, and geographer of Greek heritage who worked in Roman-ruled Alexandria, Egypt, Ptolemy wrote several treatises that influenced science for centuries afterward. His *Almagest*, written in about 150, contains star catalogs, constellation lists, Sun and Moon eclipse data, and planetary tables. Ptolemy's eight-volume *Geographia* (*Geography*) followed and incorporates all known information about the geography of the Earth at the time and helped introduce the concept of latitudes and longitudes. His work on astrology influenced Islamic and medieval Latin worlds, and his writings on music theory and optics pioneered study in those fields.

Pythagoras (c. 580-c. 500 BCE.): An ancient Greek philosopher and mathematician from Samos, Pythagoras traveled widely seeking wisdom and established a religious-scientific ascetic community in Italy around 530 BCE. He had interests in music, astronomy, medicine, and mathematics, and though none of his writings survived, he is credited with the discovery of the Pythagorean theorem governing right triangles (the square of the hypotenuse is equal to the sum of the squares of the other two sides). His life and philosophy exerted considerable influence on Plato (c. 427-347 BCE) and through Plato greatly affected Western thought.

Reiss, Archibald Rodolphe (1875-1929): A chemist, photographer, teacher, and natural scientist born in Germany, Reiss founded the world's first school of forensic science at the University of Lausanne, Switzerland, in 1909. He published numerous works that greatly influenced the new discipline, including *La photographie judiciaire* (*Forensic photography*, 1903) and *Manuel de police scientifique. I Vols et homicides* (*Handbook of Forensic Science: Thefts and Homicides*, 1911). During World War I he investigated alleged atrocities in Serbia and lived there for the rest of his life. The institute he founded more than a century ago has become a major school offering numerous courses in various forensic sciences, criminology, and criminal law.

Röntgen, Wilhelm Conrad (1845-1923): A German physicist, Röntgen studied mechanical engineering before teaching physics at the universities of Strassburg, Giessen, Würzburg, and Munich. He experimented with fluorescence and electrostatic charges. In the process of his work he discovered X rays—and also discovered that lead could effectively block the rays—meanwhile laying the foundations of what would become radiology: the medical specialty that uses radioactive imaging to diagnose disease. He was awarded the first Nobel Prize in Physics in 1901. Element 111, roentgenium, was named in his honor in 2004.

Rutherford, Ernest (1871-1937): A chemist and physicist born in New Zealand, Rutherford studied at the University of Cambridge before teaching physics at McGill University in Montreal and at the University of Manchester. He made some of the most significant discoveries in the field of atomic science, including the relative penetrating power of alpha, beta, and gamma rays, the transmutation of elements via radioactivity, and the concept of radioactive half-life. His work, for which he received the 1908 Nobel Prize in Chemistry, was instrumental in the development of nuclear energy and carbon dating.

Sabin, Albert Bruce (1906-1993): A microbiologist born of Jewish heritage as Albert Saperstein in Russia, Sabin later became an American citizen and changed his name. Trained in internal medicine, he conducted research into infectious diseases and assisted in the development of a vaccine to combat encephalitis. His major contribution to medicine was an effective oral polio vaccine, which was administered in mass immunizations during the 1950's and 1960's and eventually led to the eradication of the disease worldwide. Among other honors, he received the Presidential Medal of Freedom in 1986.

Sachs, Julius von (1832-1897): A German botanist,

writer, and teacher, Sachs made great strides in the investigation of plant physiology, morphology, heliotropism, and germination while professor of botany at the University of Würzburg. In addition to numerous written works on photosynthesis, water absorption, and chloroplasts that significantly advanced the science of botany, he also invented a number of devices useful to research, including an auxanometer to measure growth rates, and the clinostat, a device that rotates plants to compensate for the effects of gravitation on botanical growth.

Sakharov, Andrei (1921-1989): A Russian nuclear physicist and human rights activist, Sakharov researched cosmic rays, particle physics, and cosmology. He was a major contributor to the development of the hydrogen bomb but later campaigned against nuclear proliferation and for the peaceful use of nuclear power. He received the 1975 Nobel Peace Prize, and though he received several international honors in recognition of his humanitarian efforts, he spent most of the last decade of his life in exile within the Soviet Union. A human rights center and a scientific prize are named in his honor.

Scheele, Carl Wilhelm (1742-1786): A chemist born in a Swedish-controlled area of Germany, Scheele became a pharmacist at an early age. Though he discovered oxygen through experimentation, he did not publish his findings immediately, and the discovery was credited to Antoine-Laurent Lavoisier (1743-1794) and Joseph Priestly (1733-1804), though science later gave the Scheele recognition he deserved. Scheele also discovered the elements barium, manganese, and tungsten, identified such chemical compounds as citric acid, glycerol, and hydrogen cyanide, experimented with heavy metals, and devised a method of producing phosphorus in quantity for the manufacture of matches.

Shockley, William (1910-1989): A physicist and inventor born to American parents in England, Shockley was raised in California. After earning a doctoral degree, he conducted solid-state physics research at Bell Laboratories. During World War II, he researched radar and anti-submarine devices. Following the war, he was part of the team that invented the first practical solid-state transistor, for which he shared the 1956 Nobel Prize in Physics with John Bardeen (1908-1991) and Walter Brattain (1902-1987). He later set up a semiconductor business that was a precursor to Silicon Valley. His major

work, *Electrons and Holes in Semiconductors* (1950), greatly influenced many scientists.

Sikorsky, Igor (1889-1972): A Ukrainian engineer and test pilot who immigrated to the United States and became a naturalized American citizen, Sikorsky was a groundbreaking designer of both airplanes and helicopters. Inspired as a child by the drawings of Leonardo da Vinci (1452-1519), he created and flew the first multi-engine fixed-wing aircraft and the first airliner in the 1910's. He built the first flying boats in the 1930's—the famous Pan Am Clippers—and in 1939 designed the first practical helicopter, which introduced the system of rotors still used in modern helicopters.

Spilsbury, Sir Bernard Henry (1877-1947): The first British forensic pathologist, Spilsbury began performing postmortems in 1905. He investigated cause of death in many spectacular homicide cases—including those of Dr. Crippen and the Brighton trunk murders—that resulted in convictions and enhanced the science of forensics. He was a consultant to Operation Mincemeat, a successful World War II ruse (dramatized in the 1956 film *The Man Who Never Was*) involving the corpse of an alleged Allied courier, which deceived the Axis powers about the invasion of Sicily. Spilsbury was found dead in his laboratory—a victim of suicide.

Stephenson, George (1781-1848): A British mechanical and civil engineer, Stephenson invented a safety lamp for coal mines that provided illumination without the risk of explosions from firedamp. He designed a steam-powered locomotive for hauling coal, which evolved into the first public railway line in the mid-1820's, running on his specified track width of 4 feet, 8.5 inches. This measurement became the worldwide standard railroad gauge. He worked on numerous rail lines, in the process making many innovations in the design and construction of locomotives, tracks, viaducts, and bridges that greatly advanced railroad transport.

Teller, Edward (1908-2003): An outspoken theoretical physicist born in Hungary, Teller came to the United States in the 1930's and taught at George Washington University while researching quantum, molecular, and nuclear physics. A naturalized American citizen, he was a member of the atomic-bomb-building Manhattan Project. A strong supporter for nuclear energy development and testing for both wartime and peacetime purposes, he cofounded and directed Lawrence Livermore National Laboratory and founded the department of applied science at the University of California, Davis.

Tesla, Nikola (1856-1943): Born in modern-day Croatia, the brilliant if eccentric Tesla came to the United States in 1884 to work for Thomas Edison's company and later for Edison's rival George Westinghouse (1846-1914). In 1891, Tesla became a naturalized American citizen. A physicist, mechanical and electrical engineer, and an inventor specializing in electromagnetism, he created fluorescent lighting, pioneered wireless communication, built an alternating- current induction motor, and developed the Tesla coil, variations of which have provided the basis for many modern electrical and electronic devices.

Vavilov, Nikolai Ivanovich (1887-1943): A Russian botanist and plant geneticist, Vavilov served for two decades as director of the Institute of Agricultural Sciences (now the N. I. Vavilov Research Institute of Plant Industry) in Leningrad (now Saint Petersburg). During his tenure, he collected seeds from around the world, establishing the world's largest seed bank—with more than 200,000 samples—and conducted extensive research on genetically improving grain, cereal, and other food crops to produce greater yields to better feed the world. Arrested during World War II for disagreeing with Soviet methods of agronomy, he died of complications from starvation and malnutrition.

Vesalius, Andreas (1514-1564): A physician and anatomist born as Andries van Wesel in the Habsburg Netherlands (now Belgium), Vesalius taught surgery and anatomy at the universities of Padua, Bologna, and Pisa in Italy. Dissatisfied at the inaccuracies in the standard texts of the day—based solely on the 1,400-year-old work of ancient physician Galen, since Rome had long discouraged performing autopsies— he dissected a human corpse in the presence of artists from Titian's studio. This resulted in the seven-volume illustrated work, *De humani corporis fabrica libri septem* (*On the Fabric of the Human Body*, 1543), which served as the foundation for modern anatomy.

Vitruvius (c. 80-c. 15 BCE): A Roman architect and engineer, Vitruvius served in many campaigns under Julius Caesar (100-44 BCE), for whom he designed and built mechanical military weapons, such as the ballista (a projectile launcher) and siege machines. His major written work, *De architectura* (*On Architecture,*

c. 27 BCE.), set the standard for building structures solidly, usefully, and attractively. The book covers the construction of machines—including cranes, pulleys, sundials, and water clocks. It discusses construction materials and describes ancient Roman building innovations that greatly influenced later architects, particularly during the Renaissance.

Watt, James (1736-1819): A Scottish mechanical and civil engineer, Watt designed a steam engine to pump water out of mines. Refinements of his engine were used in grinding, milling, and weaving, and further improvements—including gauges, throttles, gears, and governors—enhanced the engine's efficiency and safety, making it the prime mover of the Industrial Revolution and the power source of choice for early trains and ships. Watt also devised an early copying machine and discovered a method for producing chlorine for bleaching. The unit of electrical power is named for him.

Wegener, Alfred (1880-1930): A German meteorologist, climatologist, and geophysicist, Wegener was one of the first to employ weather balloons. He was first to advance the theory of continental drift, proposing that the Earth's continents were once a single mass that he called Pangaea; his ideas, however, were not accepted until long after his death. From 1912, he worked in remote areas of Greenland examining polar airflows and drilling into the ice to study past weather patterns. He died in Greenland during his last ill-fated expedition.

Westinghouse, George (1846-1914): Born in Central Bridge, New York, Westinghouse was an engineer, inventor, entrepreneur, and a rival of Thomas Edison (1847-1931). He built a rotary steam engine while still a teenager and in his youth patented several devices—including a fail-safe compressed-air braking system—to improve railway safety. He developed an alternating-current power distribution network that proved superior to Edison's direct-current scheme, invented a power meter still in use, built several successful hydroelectric generating plants, and devised shock absorbers for automobiles.

Whittle, Sir Frank (1907-1996): Born the son of an engineer in England, Whittle joined the Royal Air Force as an aircraft mechanic and advanced to flying officer and test pilot before eventually rising to group captain. While in the Royal Air Force, he began designing aircraft engines that used turbines rather than pistons. In the mid-1930's, he formed a partnership, Power Jets, which produced the first effective turbojet design before the company was nationalized. He later developed a self-powered drill for Shell Oil and wrote a text on gas turbine engines.

Wiener, Norbert (1894-1964): A mathematician born in Missouri, Wiener was a child prodigy. He began college at the age of eleven, earned a bachelor's degree in math at fourteen, and a doctorate in philosophy from Harvard at the age of eighteen. During World War I, he researched ballistics at Aberdeen Proving Ground and afterward spent his career teaching mathematics at Massachusetts Institute of Technology. A pioneer of communication theory, he is credited with the development of theories of cybernetics, robotics, automation, and computer systems, and his work greatly influenced later scientists.

Woodward, John (1665-1728): An English naturalist, physician, paleontologist, and geologist, Woodward was an early collector of fossils, which served as the basis for his *Classification of English Minerals and Fossils* (1729), a work that influenced geology for many years. He also conducted pioneering research into the science of hydroponics. His collection of specimens formed the foundation of Cambridge University's Sedgwick Museum, and his estate was sold to provide a post in natural history, now the Woodwardian Chair of Geology at Cambridge.

Wright, Orville (1871-1948) and Wilbur Wright (1867-1912): The Wright brothers were American aviation pioneers who began experimenting with flight in their teens. In the early 1890's they opened a bicycle sales and repair shop, which financed their research into manned gliders. They soon progressed to designing powered aircraft. They eventually invented and built the first practical fixed-wing aircraft and piloted the world's first sustained powered flight—a distance of more than 850 feet over nearly a minute—in 1903 at Kitty Hawk, North Carolina. The Wright Company later became part of Curtiss-Wright Corporation, a modern high-tech aerospace component manufacturer.

Zeppelin, Ferdinand von (1838-1917): Born in Germany, Zeppelin served in the Prussian army and made a balloon flight while serving as a military observer in the American Civil War. After returning to Europe, he designed and constructed airships and devised a transportation system using lighter-than-air craft. He created a rigid, streamlined, engine-powered dirigible in 1900 and was instrumental in the creation of duralumin, which later led to lightweight all-metal airframes. By 1908, he was providing commercial air service to passengers and mail, which had an enviable record for safety until the *Hindenburg* disaster in 1937.

Zworykin, Vladimir (1889-1982): A Russian who emigrated to the United States after World War I, Zworykin worked at the Westinghouse laboratories in Pittsburgh. An engineer and inventor who patented a cathode ray tube television transmitting and receiving system in 1923, he later worked in development for the Radio Corporation of America (RCA) in New Jersey, where his inventions were perfected in time to be used to telecast the 1936 Olympic Games in Berlin. He also contributed to the development of the electron microscope.

GENERAL BIBLIOGRAPHY

"History of Photography." *WGBH*. PBS Online, WGBH, 2000. Web. 26 Feb. 2015.

Aaboe, Asger. *Episodes from the Early History of Astronomy.* New York: Springer-Verlag, 2001.

Abbate, Janet. *Inventing the Internet.* Cambridge, Mass.: MIT Press, 2000.

Abell, George O., David Morrison, and Sidney C. Wolff. *Exploration of the Universe.* 5th ed. Philadelphia: Saunders College Publishing, 1987.

Achilladelis, Basil, and Mary Ellen Bowden. *Structures of Life.* Philadelphia: The Center, 1989.

Ackerknecht, Erwin H. *A Short History of Medicine.* Rev. ed. Baltimore: The Johns Hopkins University Press, 1982.

Aczel, Amir D. *Fermat's Last Theorem: Unlocking the Secret of an Ancient Mathematical Problem.* Reprint. New York: Four Walls Eight Windows, 1996.

Adler, Robert E. *Science Firsts: From the Creation of Science to the Science of Creation.* Hoboken, N.J.: John Wiley & Sons, 2002.

Alberts, Bruce, et al. *Molecular Biology of the Cell.* 2d ed. New York: Garland, 1989.

Alcamo, I. Edward. *AIDS: The Biological Basis.* 3d ed. Boston: Jones and Bartlett, 2003.

Aldersey-Willliams, Hugh. *The Most Beautiful Molecule: An Adventure in Chemistry.* London: Aurum Press, 1995.

Alexander, Arthur F. O'Donel. *The Planet Saturn: A History of Observation, Theory, and Discovery.* 1962. Reprint. New York: Dover, 1980.

Alioto, Anthony M. *A History of Western Science.* 2d ed. Upper Saddle River, N.J.: Prentice Hall, 1993.

Allen, Oliver E., and the editors of Time-Life Books. *Atmosphere.* Alexandria, Va.: Time-Life Books, 1983.

Ames, W. F., and C. Rogers, eds. *Nonlinear Equations in the Applied Sciences.* San Diego: Academic Press, 1992.

Andersen, Geoff. *The Telescope: Its History, Technology, and Future.* Princeton: Princeton UP, 2007. Print.

Andriesse, Cornelis D. *Christian Huygens.* Paris: Albin Michel, 2000.

Angelo, Joseph A., Jr. *Nuclear Technology.* Westport: Greenwood, 2004. Print.

Angelo, Joseph A., Jr. *Space Technology.* Westport: Greenwood, 2003. Print.

Angier, Natalie. *Natural Obsessions: Striving to Unlock the Deepest Secrets of the Cancer Cell.* Boston: Mariner Books/Houghton Mifflin, 1999.

Annaratone, Donnatello. *Transient Heat Transfer.* New York: Springer, 2011.

Anstey, Peter R. *The Philosophy of Robert Boyle.* London: Routledge, 2000.

Anton, Sebastian. *A Dictionary of the History of Science.* Pearl River, N.Y.: Parthenon Publishing, 2001.

Archimedes. *The Works of Archimedes.* Translated by Sir Thomas Heath. 1897. Reprint. New York: Dover, 2002.

Arms, Karen, and Pamela S. Camp. *Biology: A Journey into Life.* 3d ed. Philadelphia: Saunders College Publishing, 1987.

Armstrong, Neil, Michael Collins, and Edwin E. Aldrin. *First on the Moon.* New York: Williams Konecky Associates, 2002.

Arrizabalaga, Jon, John Henderson, and Roger French. *The Great Pox: The French Disease in Renaissance Europe.* New Haven, Conn.: Yale University Press, 1997.

Arsuaga, Juan Luis. *The Neanderthal's Necklace: In Search of the First Thinkers.* Translated by Andy Klatt. New York: Four Walls Eight Windows, 2002.

Artmann, Benno. *Euclid: The Creation of Mathematics.* New York: Springer- Verlag, 1999.

Asimov, Isaac. *Exploring the Earth and the Cosmos.* New York: Crown, 1982.

Asimov, Isaac. *Jupiter, the Largest Planet.* New York: Ace, 1980.

Asimov, Isaac. *The History of Physics.* New York: Walker, 1984.

Aspray, William. *John von Neumann and the Origins of Modern Computing.* Boston: MIT Press, 1990.

Astronomical Society of the Pacific. *The Discovery of Pulsars.* San Francisco: Author, 1989.

Audesirk, Gerald J., and Teresa E. Audesirk. *Biology: Life on Earth.* 2d ed. New York: Macmillan, 1989.

Audoin, Claude, and Bernard Guinot. *The Measurement of Time: Time, Frequency, and the Atomic Clock.* Trans. Stephen Lyle. New York: Cambridge UP, 2001. Print.

Aughton, Peter. *Newton's Apple: Isaac Newton and the English Scientific Revolution.* London: Weidenfeld & Nicolson, 2003.

Aujoulat, Norbert. *Lascaux: Movement, Space, and Time.* New York: Harry N. Abrams, 2005.

Aveni, Anthony F., ed. *Skywatchers*. Rev. ed. Austin: University of Texas Press, 2001.

Azar, Dimitri T. *Clinical Optics, 2014–2015*. San Francisco: Amer. Acad. of Ophthalmology, 2014. Print. Basic and Clinical Science Course 3.

Baggott, Jim. *Perfect Symmetry: The Accidental Discovery of Buckminsterfullerene*. New York: Oxford University Press, 1994.

Baine, Celeste. *Is There an Engineer Inside You? A Comprehensive Guide to Career Decisions in Engineering*. 2d ed. Belmont, Calif.: Professional Publications, 2004.

Baker, Dean H., E. A. Ryder, and N. H. Baker. *Temperature Measurement in Engineering, Vol. 1*. New York: John Wiley & Sons, 1963. Essential reading and an excellent reference, guiding the user through the various issues.

Baker, John. *The Cell Theory: A Restatement, History and Critique*. New York: Garland, 1988.

Baldwin, Joyce. *To Heal the Heart of a Child: Helen Taussig, M.D.* New York: Walker, 1992.

Barbieri, Cesare, et al., eds. *The Three Galileos: The Man, the Spacecraft, the Telescope: Proceedings of the Conference Held in Padova, Italy on January 7-10, 1997*. Boston: Kluwer Academic, 1997.

Barkan, Diana Kormos. *Walther Nernst and the Transition to Modern Physical Science*. New York: Cambridge University Press, 1999.

Barrett, Peter. *Science and Theology Since Copernicus: The Search for Understanding*. Reprint. Dorset, England: T&T Clark, 2003.

Bartusiak, Marcia. *Thursday's Universe*. New York: Times Books, 1986.

Basta, Nicholas. *Opportunities in Engineering Careers*. New York: McGraw-Hill, 2003.

Bates, Charles C., and John F. Fuller. *America's Weather Warriors, 1814-1985*. College Station: Texas A&M Press, 1986.

Bazin, Hervé. *The Eradication of Smallpox: Edward Jenner and the First and Only Eradication of a Human Infectious Disease*. Translated by Andrew Morgan and Glenise Morgan. San Diego: Academic Press, 2000.

Beatty, J. Kelly, and Andrew Chaikin, eds. *The New Solar System*. 3d rev. ed. New York: Cambridge University Press, 1990.

Becker, Wayne, Lewis Kleinsmith, and Jeff Hardin. *The World of the Cell*. New York: Pearson/Benjamin Cummings, 2006.

Bedard, Roger, et al. "An Overview of Ocean Renewable Energy Technologies." *Oceanography* 23, no. 2 (June, 2010): 22–31. Highlights the development status of the various marine-energy conversion technologies. Contains many helpful color photographs and explanatory diagrams.

Benedict, Robert P. *Fundamentals of Temperature, Pressure, and Flow Measurements*. 3d ed. New York: John Wiley & Sons, 1984. Used all over the world in chemical and mechanical engineering courses, this book is a valuable resource for engineers and technicians preparing for professional certification. Contains practical information on instrument interfacing as well as measurement techniques.

Berinstein, Paula. *Making Space Happen: Private Space Ventures and the Visionaries behind Them*. Medford: Plexus, 2002. Print.

Berlatsky, Noah. *Nanotechnology*. Greenhaven, 2014.

Berlinski, David. *A Tour of the Calculus*. New York: Vintage Books, 1997.

Bernstein, Jeremy. *Three Degrees Above Zero: Bell Labs in the Information Age*. New York: Charles Scribner's Sons, 1984.

Bernstein, Peter L. *Against the Gods: The Remarkable Story of Risk*. New York: John Wiley & Sons, 1996.

Bertolotti, M. *Masers and Lasers: An Historical Approach*. Bristol, England: Adam Hilger, 1983.

Bhushan, Bharat, Harald Fuchs, and Masahiko Tomitori, eds. *Applied Scanning Probe Methods VIII: Scanning Probe Microscopy Techniques (NanoScience and Technology)*. Berlin: Springer, 2008. Provides the most up-to-date information on this rapidly evolving technology and its applications.

Bickel, Lennard. *Florey: The Man Who Made Penicillin*. Carlton South, Victoria, Australia: Melbourne University Press, 1995.

Bigelow, Stephen J., Joseph J. Carr, and Steve Winder. *Understanding Telephone Electronics*. 4th ed. Boston: Newnes, 2001.

Binns, Chris. *Introduction to Nanoscience and Nanotechnology*. Hoboken: Wiley, 2010. Print.

Biswas, Abhijit, et al. "Advances in Top-Down and Bottom-Up Surface Nanofabrication: Techniques, Applications & Future Prospects." *Advances in Colloid and Interface Science* 170.1–2 (2012): 2–27. Print.

Bizony, Piers. *Island in the Sky: Building the International Space Station*. London: Aurum Press Limited,

1996.

Blackwell, Richard J. *Galileo, Bellarmine, and the Bible.* London: University of Notre Dame Press, 1991.

Bliss, Michael. *The Discovery of Insulin.* Chicago: University of Chicago Press, 1987.

Blumenberg, Hans. *The Genesis of the Copernican World.* Translated by Robert M. Wallace. Cambridge, Mass.: MIT Press, 1987.

Blunt, Wilfrid. *Linnaeus: The Compleat Naturalist.* Princeton, N.J.: Princeton University Press, 2001.

Bodanis, David. *Electric Universe: The Shocking True Story of Electricity.* New York: Crown Publishers, 2005.

Bohm, David. *Causality and Chance in Modern Physics.* London: Routledge & Kegan Paul, 1984.

Bohren, Craig F. *Clouds in a Glass of Beer: Simple Experiments in Atmospheric Physics.* New York: John Wiley & Sons.

Boljanovic, Vukota. *Applied Mathematics and Physical Formulas: A Pocket Reference Guide for Students, Mechanical Engineers, Electrical Engineers, Manufacturing Engineers, Maintenance Technicians, Toolmakers, and Machinists.* New York: Industrial Press, 2007.

Bolt, Bruce A. *Inside the Earth: Evidence from Earthquakes.* New York: W. H. Freeman, 1982.

Bond, Peter. *The Continuing Story of the International Space Station.* Chichester, England: Springer-Praxis, 2002.

Bondyopadhyay, Probir K., Pallab K. Chatterjee, and Utpal K. Chakrabarti, eds. *Proceedings of the IEEE: Special Issue on the Fiftieth Anniversary of the Transistor* 86, no. 1 (January, 1998). Comprehensive collection of papers concerning the invention of the transistor; Moore's law; patents, letters, and notes by William Shockley, John Bardeen, and Walter Brattain; and life in the early days of Silicon Valley.

Boon, John D. *Secrets of the Tide: Tide and Tidal Current Application and Prediction, Storm Surges and Sea Level Trends.* Chichester, England: Horwood, 2004. Describes the so-called tide mills used in many locations along the East Coast of the United States in the eighteenth century for grinding grain into flour or meal.

Boorstin, Daniel J. *The Discoverers.* New York: Random House, 1983.

Bottazzini, Umberto. *The Higher Calculus: A History of Real and Complex Analysis from Euler to Weierstrass.* New York: Springer-Verlag, 1986.

Bourbaki, Nicolas. *Elements of the History of Mathematics.* Translated by John Meldrum. New York: Springer, 1994.

Bowler, Peter J. *Charles Darwin: The Man and His Influence.* Cambridge, England: Cambridge University Press, 1996.

Bowler, Peter J. *Evolution: The History of an Idea.* Rev. ed. Berkeley: University of California Press, 1989.

Bowler, Peter J. *The Mendelian Revolution: The Emergence of Hereditarian Concepts in Modern Science and Society.* Baltimore: The Johns Hopkins University Press, 1989.

Boyer, Carl B. *A History of Mathematics.* 2d ed., revised by Uta C. Merzbach. New York: John Wiley & Sons, 1991.

Bracewell, Ronald N. *The Fourier Transform and Its Applications.* 3d rev. ed. New York: McGraw-Hill, 1987.

Brachman, Arnold. *A Delicate Arrangement: The Strange Case of Charles Darwin and Alfred Russel Wallace.* New York: Times Books, 1980.

Bredeson, Carmen. *John Glenn Returns to Orbit: Life on the Space Shuttle.* Berkeley Heights, N.J.: Enslow, 2000.

Brighton, Henry, and Howard Selina. *Introducing Artificial Intelligence.* Edited by Richard Appignanesi. 2003. Reprint. Cambridge, England: Totem Books, 2007. Introduces many aspects of artificial intelligence, including machine learning, geometric algorithms, and pattern recognition, and provides examples of how specialists in artificial intelligence and pattern recognition create applications from their research.

Brock, Thomas, ed. *Milestones in Microbiology, 1546-1940.* Washington, D.C.: American Society for Microbiology, 1999.

Brock, William H. *The Chemical Tree: A History of Chemistry.* New York: W. W. Norton, 2000.

Brooks, John. *Telephone: The First Hundred Years.* New York: Harper & Row, 1976.

Brooks, Paul. *The House of Life: Rachel Carson at Work.* 2d ed. Boston: Houghton Mifflin, 1989.

Browne, Janet. *Charles Darwin: The Power of Place.* New York: Knopf, 2002.

Brush, Stephen G. *Cautious Revolutionaries: Maxwell, Planck, Hubble.* College Park, Md.: American Association of Physics Teachers, 2002.

Brush, Stephen G., and Nancy S. Hall. *Kinetic Theory of Gases: An Anthology of Classic Papers With Historical Commentary.* London: Imperial College Press,

2003.

Bryant, Stephen. *The Story of the Internet.* London: Pearson Education, 2000.

Buffon, Georges-Louis Leclerc. *Natural History: General and Particular.* Translated by William Smellie. Avon, England: Thoemmes Press, 2001.

Burden, Richard L., and J. Douglas Faires. *Numerical Analysis.* 9th ed. Boston: Brooks, 2011. Print

Burdick, Donald. *Petrochemicals in Nontechnical Language.* 3d ed. Tulsa, Okla.: Penn Well, 2010. An accessible introduction to the field that relates serious science concepts in an informative fashion. Figures and tables.

Burger, Edward B., and Michael Starbird. *Coincidences, Chaos, and All That Math Jazz: Making Light of Weighty Ideas.* New York: W. W. Norton, 2005.

Burke, Bernard F., and Francis Graham-Smith. *An Introduction to Radio Astronomy.* 3d ed. New York: Cambridge University Press, 2009. A very thorough but also quite technical overview of the theory of radio astronomy and applications to astronomical observations.

Burke, Terry, et al., eds. *DNA Fingerprinting: Approaches and Applications.* Boston: Birkhauser, 2001.

Byrne, Patrick H. *Analysis and Science in Aristotle.* Albany: State University of New York Press, 1997.

Calder, William M., III, and David A. Traill, eds. *Myth, Scandal, and History: The Heinrich Schliemann Controversy.* Detroit: Wayne State University Press, 1986.

Calinger, Ronald. *A Contextual History of Mathematics.* Upper Saddle River, N.J.: Prentice Hall, 1999.

Callister, William D., Jr., and David G. Rethwisch. *Materials Science and Engineering: An Introduction.* 8th ed. Hoboken, N.J.: John Wiley & Sons, 2010. A popular book for undergraduate engineering students, this text covers many aspects of materials, and has well written sections on semiconductors and semiconductor devices.

Canning, Thomas N. *Galileo Probe Parachute Test Program: Wake Properties of the Galileo.* Washington, D.C.: National Aeronautics and Space Administration, Scientific and Technical Information Division, 1988.

Cantor, Geoffrey. *Michael Faraday: Sandemanian and Scientist: A Study of Science and Religion in the Nineteenth Century.* New York: St. Martin's Press, 1991.

Carlisle, Rodney. *Inventions and Discoveries: All the Milestones in Ingenuity—from the Discovery of Fire to the Invention of the Microwave.* Hoboken, N.J.: John Wiley & Sons, 2004.

Carlowicz, Michael J., and Ramon E. Lopez. *Storms from the Sun: The Emerging Science of Space Weather.* Washington: Henry, 2002. Print.

Carlson, Elof Axel. *Mendel's Legacy: The Origin of Classical Genetics.* Woodbury, N.Y.: Cold Spring Harbor Laboratory Press, 2004.

Carola, Robert, John P. Harley, and Charles R. Noback. *Human Anatomy and Physiology.* New York: McGraw-Hill, 1990.

Carpenter, B. S., and R. W. Doran, eds. *A. M. Turing's ACE Report of 1946 and Other Papers.* Cambridge, Mass.: MIT Press, 1986.

Carpenter, Kenneth J. *The History of Scurvy and Vitamin C.* Cambridge England: Cambridge University Press, 1986.

Carr, Joseph J. *Radio Science Observing.* 2 vols. Indianapolis, Ind.: Prompt Publications, 1999. Contains radio astronomy ideas, circuits, and projects suitable for amateur radio operators or advanced high school or college students.

Carraher, Charles E., Jr. *Giant Molecules: Essential Materials for Everyday Living and Problem Solving.* 2d ed. Hoboken, N.J.: John Wiley & Sons, 2003. Called an "exemplary book" for general readers interested in polymers, *Giant Molecules* facilitates understanding with an apt use of illustrations, figures, and tables along with relevant historical materials.

_____.*Introduction to Polymer Chemistry.* 2d ed. Boca Raton, Fla.: CRC Press, 2010. This text, intended for undergraduates, explains polymer principles in the contexts of actual applications, while including material on such new areas as optical fibers and genomics. Summaries, glossaries, exercises, and further readings at the ends of chapters.

Carrigan, Richard A., and W. Peter Trower, eds. *Particle Physics in the Cosmos.* New York: W. H. Freeman, 1989.

_____.*Particles and Forces: At the Heart of the Matter.* New York: W. H. Freeman, 1990.

Carter, R. W. G. *Coastal Environments: An Introduction to the Physical, Ecological, and Cultural Systems of Coastlines.* San Diego: Academic Press, 1988. Includes an outstanding analysis of tidal power, with a description of the mechanics of power generation and a discussion of the environmental and ecological effects.

Cassanelli, Roberto, et al. *Houses and Monuments of*

Pompeii: The Works of Fausto and Felice Niccolini. Los Angeles: J. Paul Getty Museum, 2002.

Caton, Jerald A. *A Review of Investigations Using the Second Law of Thermodynamics to Study Internal-Combustion Engines.* London: Society of Automotive Engineers, 2000.

Chaikin, Andrew. *A Man on the Moon: The Voyages of the Apollo Astronauts.* New York: Penguin Group, 1998.

Chaisson, Eric J. *The Hubble Wars.* New York: HarperCollins, 1994.

Chaisson, Eric J., and Steve McMillan. *Astronomy Today.* 5th ed. Upper Saddle River, N.J.: Pearson Prentice Hall, 2004.

Chandrasekhar, Subrahmanyan. *Eddington: The Most Distinguished Astrophysicist of His Time.* Cambridge, England: Cambridge University Press, 1983.

Chang, Hasok. *Inventing Temperature: Measurement and Scientific Progress.* New York: Oxford University Press, 2007. Discusses the history of temperature measurement and how researchers managed to establish the reliability and accuracy of the various methods.

Chang, Kenneth. "Nobel Laureates Pushed Limits of Microscopes." *New York Times.* New York Times, 8 Oct. 2014. Web. 26 Feb. 2015.

Chang, Laura, ed. *Scientists at Work: Profiles of Today's Groundbreaking Scientists from "Science Times."* New York: McGraw-Hill, 2000.

Chant, Christopher. *Space Shuttle.* New York: Exeter Books, 1984.

Chapman, Allan. *Astronomical Instruments and Their Users: Tycho Brahe to William Lassell.* Brookfield, Vt.: Variorum, 1996.

Chase, Allan. *Magic Shots.* New York: William Morrow, 1982.

Check, William A. *AIDS.* New York: Chelsea House, 1988.

Cheng, K. S., and G. V. Romero. *Cosmic Gamma-Ray Sources.* New York: Springer-Verlag, 2004.

Childs, P. R. N., J. R. Greenwood, and C. A. Long. "Review of Temperature Measurement." *Review of Scientific Instruments* 71, no. 8 (2000): 2959–2978. Succinct discussion of the various techniques used in temperature measurement in various media. Discusses measurement criteria and calibration techniques and provides a guide to select techniques for specific applications.

Christianson, John Robert. *On Tycho's Island: Tycho Brahe and His Assistants, 1570-1601.* New York:

Cambridge University Press, 2000.

Chung, Deborah D. L. *Applied Materials Science: Applications of Engineering Materials in Structural, Electronics, Thermal and Other Industries.* Boca Raton, Fla.: CRC Press, 2001.

Chyba, Christopher. "The New Search for Life in the Universe." *Astronomy* 38.5 (2010): 34–39. Print.

Clark, Ronald W. *The Life of Ernst Chain: Penicillin and Beyond.* New York: St. Martin's Press, 1985.

_____.*The Survival of Charles Darwin: A Biography of a Man and an Idea.* New York: Random House, 1984.

Cline, Barbara Lovett. *Men Who Made a New Physics.* Chicago: University of Chicago Press, 1987.

Clos, Lynne. *Field Adventures in Paleontology.* Boulder, Colo.: Fossil News, 2003.

Clugston, M. J., ed. *The New Penguin Dictionary of Science.* 2d ed. New York: Penguin Books, 2004.

Coffey, Patrick. *Cathedrals of Science: The Personalities and Rivalries That Made Modern Chemistry.* New York: Oxford University Press, 2008.

Cohen, I. Bernard, and George E. Smith, eds. *The Cambridge Companion to Newton.* New York: Cambridge University Press, 2002.

_____.*Benjamin Franklin's Science.* Cambridge, Mass.: Harvard University Press, 1990.

_____.*The Newtonian Revolution.* New York: Cambridge University Press, 1980.

Cole, K. C. *The Universe and the Teacup: The Mathematics of Truth and Beauty.* Fort Washington, Pa.: Harvest Books, 1999.

Cole, Michael D. *Galileo Spacecraft: Mission to Jupiter: Countdown to Space.* New York: Enslow, 1999.

Collin, S. M. H. *Dictionary of Science and Technology.* London: Bloomsbury Publishing, 2003.

Comins, Neil F. *The Hazards of Space Travel: A Tourist's Guide.* New York: Villard, 2007. Print.

Connor, James A. *Kepler's Witch: An Astronomer's Discovery of Cosmic Order Amid Religious War, Political Intrigue, and the Heresy Trial of His Mother.* San Francisco: HarperSanFrancisco, 2004.

Conrad, Lawrence, et al., eds. *The Western Medical Tradition, 800 B.C. to A.D. 1800.* New York: Cambridge University Press, 1995.

Cook, Alan. *Edmond Halley: Charting the Heavens and the Seas.* New York: Oxford University Press, 1998.

Cooke, Donald A. *The Life and Death of Stars.* New York: Crown, 1985.

Cooper, Geoffrey M. *Oncogenes.* 2d ed. Boston: Jones

and Bartlett, 1995.

Cooper, Henry S. F., Jr. *Imaging Saturn: The Voyager Flights to Saturn.* New York: H. Holt, 1985.

Corsi, Pietro. *The Age of Lamarck: Evolutionary Theories in France, 1790-1830.* Berkeley: University of California Press, 1988.

Coulthard, Malcolm, and Alison Johnson, eds. *The Routledge Handbook of Forensic Linguistics.* New York: Routledge, 2010.

Craven, B. O. *The Lebesgue Measure and Integral.* Boston: Pitman Press, 1981.

Crawford, Deborah. *King's Astronomer William Herschel.* New York: Julian Messner, 2000.

Crease, Robert P., and Charles C. Mann. *The Second Creation: Makers of the Revolution in Twentieth Century Physics.* New York: Macmillan, 1985.

Crewdson, John. *Science Fictions: A Scientific Mystery, A Massive Cover-Up, and the Dark Legacy of Robert Gallo.* Boston: Little, Brown, 2002.

Crick, Francis. *What Mad Pursuit: A Personal View of Scientific Discovery.* New York: Basic Books, 1988.

Crump, Thomas. *A Brief History of Science as Seen Through the Development of Scientific Instruments.* New York: Carroll & Graf, 2001.

Crystal, David. *Txtng: The Gr8 Db8.* New York: Oxford University Press, 2008.

Culler, David, Jaswinder Pal Singh, and Anoot Gupta. *Parallel Computer Architecture: A Hardware/Software Approach.* San Francisco:Kaufmann, 1999. Print.

Cunningham, Andrew. *The Anatomical Renaissance: The Resurrection of the Anatomical Projects of the Ancients.* Brookfield, Vt.: Ashgate, 1997.

Cutler, Alan. *The Seashell on the Mountaintop: A Story of Science, Sainthood, and the Humble Genius Who Discovered a New History of the Earth.* New York: Dutton/Penguin, 2003.

Cvijetic, Milorad. *Optical Transmission: Systems Engineering.* Norwood: Artech, 2004. Print.

D'Agostino, Riccardo, et al., eds. *Plasma Processes and Polymers.* Weinheim, Germany: Wiley, 2005. Brings together papers from a conference on plasma treatments for polymers.

Dalrymple, G. Brent. *The Age of the Earth.* Stanford, Calif.: Stanford University Press, 1991.

Darrigol, Oliver. *Electrodynamics from Ampère to Einstein.* Oxford, England: Oxford University Press, 2000.

Dash, Joan. *The Longitude Prize.* New York: Farrar, Straus and Giroux, 2000.

Daston, Lorraine. *Classical Probability in the Enlightenment.* Princeton, N.J.: Princeton University Press, 1988.

Davidson, S., M. Perkin, and M. Buckley. *Specifications, Tolerances, and Other Technical Requirements for Weighing and Measuring Devices.*

_____.*The Measurement of Mass and Weight.* Measurement Good Practice Guide 71. Teddington, England: National Physical Laboratory, 2004. Describes the best practices for weighing technique in the context of scientific and technological application.

Davies, John K. *Astronomy from Space: The Design and Operation of Orbiting Observatories.* New York: John Wiley & Sons, 1997.

Davies, Paul. *The Edge of Infinity: Where the Universe Came from and How It Will End.* New York: Simon & Schuster, 1981.

Davis, Joel. *Flyby: The Interplanetary Odyssey of Voyager 2.* New York: Atheneum, 1987.

Davis, Martin. *Engines of Logic: Mathematicians and the Origin of the Computer.* New York: W. W. Norton, 2000.

Davis, Morton D. *Game Theory: A Nontechnical Introduction.* New York: Dover, 1997.

Davis, William Morris. *Elementary Meteorology.* Boston: Ginn, 1894.

Dawkins, Richard. *River Out of Eden: A Darwinian View of Life.* New York: Basic Books, 1995.

_____.*The Ancestor's Tale: A Pilgrimage to the Dawn of Evolution.* New York: Houghton Mifflin, 2004.

Day, Michael H. *Guide to Fossil Man.* 4th ed. Chicago: University of Chicago Press, 1986.

Day, William. *Genesis on Planet Earth.* 2d ed. New Haven, Conn.: Yale University Press, 1984.

De Jonge, Christopher J., and Christopher L. R. Barratt, eds. *Assisted Reproductive Technologies: Current Accomplishments and New Horizons.* New York: Cambridge University Press, 2002.

Dean, Dennis R. *James Hutton and the History of Geology.* Ithaca, N.Y.: Cornell University Press, 1992.

Debré, Patrice. *Louis Pasteur.* Translated by Elborg Forster. Baltimore: The Johns Hopkins University Press, 1998.

DeJauregui, Ruth. *100 Medical Milestones That Shaped World History.* San Mateo, Calif.: Bluewood Books, 1998.

Delaporte, François. *The History of Yellow Fever: An Essay on the Birth of Tropical Medicine.* Cambridge,

Mass.: MIT Press, 1991.

Demetzos, Costas. *Pharmaceutical Nanotechnology: Fundamentals and Practical Applications.* Adis, 2016.

Dennett, Daniel C. *Darwin's Dangerous Idea: Evolution and the Meanings of Life.* New York: Simon & Schuster, 1995.

Dennis, Carina, and Richard Gallagher. *The Human Genome.* London: Palgrave Macmillan, 2002.

DeVorkin, David H. *Race to the Stratosphere: Manned Scientific Ballooning in America.* New York: Springer-Verlag, 1989.

Dewdney, A. K. *The Turing Omnibus.* Rockville, Md.: Computer Science Press, 1989.

Diamond, Jared. *The Third Chimpanzee: The Evolution and Future of the Human Animal.* New York: Harper-Collins, 1992.

DiCanzio, Albert. *Galileo: His Science and His Significance for the Future of Man.* Portsmouth, N.H.: ADASI, 1996.

Dijksterhuis, Eduard Jan. *Archimedes.* Translated by C. Dikshoorn, with a new bibliographic essay by Wilbur R. Knorr. Princeton, N.J.: Princeton University Press, 1987.

Dijksterhuis, Fokko Jan. *Lenses and Waves: Christiaan Huygens and the Mathematical Science of Optics in the Seventeenth Century.* Dordrecht, the Netherlands: Kluwer Academic, 2004.

Dimmock, N. J., A. J. Easton, and K. N. Leppard. *Introduction to Modern Virology.* 5th ed. Malden, Mass.: Blackwell Science, 2001.

Dore, Mohammed, Sukhamoy Chakravarty, and Richard Goodwin, eds. *John Von Neumann and Modern Economics.* New York: Oxford University Press, 1989.

Drake, Stillman. *Galileo: A Very Short Introduction.* New York: Oxford University Press, 2001.

_____. *Galileo: Pioneer Scientist.* Toronto: University of Toronto Press, 1990.

Drexler, K. Eric. *Engines of Creation: The Coming Era of Nanotechnology.* New York: Anchor, 1986. Print.

_____. *Radical Abundance: How a Revolution in Nanotechnology Will Change Civilization.* New York: PublicAffairs, 2013. Print.

Dreyer, John Louis Emil, ed. *The Scientific Papers of Sir William Herschel.* Dorset, England: Thoemmes Continuum, 2003.

Duck, Ian. *One Hundred Years of Planck's Quantum.* River Edge, N.J.: World Scientific, 2000.

Dudgeon, Dan E., and Russell M. Mersereau.

Multidimensional Digital Signal Processing. Englewood Cliffs, N.J.: Prentice Hall, 1984.

Duffield, John. *Over a Barrel: The Costs of U.S. Foreign Oil Dependence.* Stanford, Calif.: Stanford University Press, 2008. Addresses economic and foreign policy in the United States as well as military responses to the problem. Concludes with a proposal to lower the costs of dependence. Tables, figures, maps.

Dunham, William. *Euler: The Master of Us All.* Washington, D.C.: Mathematical Association of America, 1999.

_____. *Journey Through Genius.* New York: John Wiley & Sons, 1990.

_____. *The Calculus Gallery: Masterpieces from Newton to Lebesgue.* Princeton, N.J.: Princeton University Press, 2005.

Dunlap, Jay C., Jennifer J. Loros, and Patricia J. DeCoursey, eds. *Chronobiology: Biological Timekeeping.* Sunderland: Sinauer, 2004. Print.

Durham, Frank, and Robert D. Purrington. *Frame of the Universe.* New York: Cambridge University Press, 1983.

Easby, Rebecca Jeffrey. "Early Photography: Niépce, Talbot, and Muybridge." *Khan Academy.* Khan Academy, 2015. Web. 26 Feb. 2015.

Easton, Thomas A. *Careers in Science.* 4th ed. Chicago: VGM Career Books, 2004.

Ebewele, Robert O. *Polymer Science and Technology.* Boca Raton, Fla.: CRC Press, 2000. The author covers polymer fundamentals, the creation of polymers, and polymer applications, with an emphasis on polymer products.

Edelson, Edward. *Gregor Mendel: And the Roots of Genetics.* New York: Oxford University Press, 2001.

Edey, Maitland A., and Donald C. Johanson. *Blueprints: Solving the Mystery of Evolution.* Boston: Little, Brown, 1989.

Edwards, Robert G., and Patrick Steptoe. *A Matter of Life.* New York: William Morrow, 1980.

Ehrenfest, Paul, and Tatiana Ehrenfest. *The Conceptual Foundations of the Statistical Approach in Mechanics.* Mineola, N.Y.: Dover, 2002.

Ehrlich, Melanie, ed. *DNA Alterations in Cancer: Genetic and Epigenetic Changes.* Natick, Mass.: Eaton, 2000.

Eidson, John C. *Measurement, Control, and Communication Using IEEE 1588.* London: Springer, 2006. Print.

Eisen, Herman N. *Immunology: An Introduction to Molecular and Cellular Principles of the Immune Responses.* 2d ed. Philadelphia: J. B. Lippincott, 1980.

Espejo, Roman, ed. *Biomedical Ethics: Opposing Viewpoints.* San Diego: Greenhaven Press, 2003.

Evans, James. *The History and Practice of Ancient Astronomy.* New York: Oxford University Press, 1998.

Fabian, A. C., K. A. Pounds, and R. D. Blandford. *Frontiers of X-Ray Astronomy.* London: Cambridge University Press, 2004.

Faeth, G. M. *Centennial of Powered Flight: A Retrospective of Aerospace Research.* Reston, Va.: American Institute of Aeronautics and Astronautics, 2003. Traces the history of aerospace and describes important milestones.

Fara, Patricia. *An Entertainment for Angels: Electricity in the Enlightenment.* New York: Columbia University Press, 2002.

_____. *Newton: The Making of a Genius.* New York: Columbia University Press, 2002.

_____. *Sex, Botany, and the Empire: The Story of Carl Linnaeus and Joseph Banks.* New York: Columbia University Press, 2003.

Farber, Paul Lawrence. *Finding Order in Nature: The Naturalist Tradition from Linnaeus to E. O. Wilson.* Baltimore: The Johns Hopkins University Press, 2000.

Fauvel, John, and Jeremy Grey, eds. *The History of Mathematics: A Reader.* 1987. Reprint. Washington, D.C.: The Mathematical Association of America, 1997.

Feferman, S., J. W. Dawson, and S. C. Kleene, eds. *Kurt Gödel: Collected Works.* 2 vols. New York: Oxford University Press, 1986-1990.

Feldman, David. *How Does Aspirin Find a Headache?* New York: HarperCollins, 2005.

Ferejohn, Michael. *The Origins of Aristotelian Science.* New Haven, Conn.: Yale University Press, 1991.

Ferguson, Kitty. *The Nobleman and His Housedog: Tycho Brahe and Johannes Kepler—The Strange Partnership That Revolutionized Science.* London: Headline, 2002.

Ferris, T. *Coming of Age in the Milky Way.* New York: Doubleday, 1989.

Ferris, Timothy. *Galaxies.* New York: Harrison House, 1987.

Field, George, and Donald Goldsmith. *The Space Telescope.* Chicago: Contemporary Books, 1989.

Field, J. V. *The Invention of Infinity: Mathematics and Art in the Renaissance.* New York: Oxford University Press, 1997.

Fincher, Jack. *The Brain: Mystery of Matter and Mind.* Washington, D.C.: U.S. News Books, 1981.

Finlayson, Clive. *Neanderthals and Modern Humans: An Ecological and Evolutionary Perspective.* New York: Cambridge University Press, 2004.

Finocchiaro, Maurice A., ed. *The Galileo Affair: A Documentary History.* Berkeley: University of California Press, 1989.

Fischer, Daniel, and Hilmar W. Duerbeck. *Hubble Revisited: New Images from the Discovery Machine.* New York: Copernicus Books, 1998.

Fischer, Daniel. *Mission Jupiter: The Spectacular Journey of the Galileo Spacecraft.* New York: Copernicus Books, 2001.

Fisher, Richard B. *Edward Jenner, 1741-1823.* London: Andre Deutsch, 1991.

Flowers, Lawrence O., ed. *Science Careers: Personal Accounts from the Experts.* Lanham, Md.: Scarecrow Press, 2003.

Ford, Brian J. *Single Lens: The Story of the Simple Microscope.* New York: Harper & Row, 1985.

_____. *The Leeuwenhoek Legacy.* London: Farrand, 1991.

Fournier, Marian. *The Fabric of Life: Microscopy in the Seventeenth Century.* Baltimore: The Johns Hopkins University Press, 1996.

Fowler, A. C. *Mathematical Models in the Applied Sciences.* New York: Cambridge University Press, 1997.

Foyer, Christine H. *Photosynthesis.* New York: Wiley-Interscience, 1984.

Frängsmyr, Tore, ed. *Linnaeus: The Man and His Work.* Canton, Mass.: Science History Publications, 1994.

Frankel, Felice. *Envisioning Science: The Design and Craft of the Science Image.* Cambridge: MIT P, 2004. Print.

Franklin, Benjamin. *Autobiography of Benjamin Franklin.* New York: Buccaneer Books, 1984.

Freedman, Roger A., Robert M. Geller, and William J. Kaufmann III. *Universe.* 10th ed. New York: Freeman, 2015. Print.

Frenay, Robert. *Pulse: The Coming Age of Systems and Machines Inspired by Living Things.* 2006. Reprint. Lincoln: University of Nebraska Press, 2008. An introduction to bioengineering, artificial intelligence, and other organism-inspired machines. Provides an introduction to the philosophy and politics of artificial intelligence research.

French, A. P., and P. J. Kennedy, eds. *Niels Bohr: A Centenary Volume.* Cambridge, Mass.: Harvard University Press, 1985.

French, Roger. *William Harvey's Natural Philosophy.* New York: Cambridge University Press, 1994.

Fridell, Ron. *DNA Fingerprinting: The Ultimate Identity.* New York: Scholastic, 2001.

Friedlander, Michael W. *Cosmic Rays.* Cambridge, Mass.: Harvard University Press, 1989.

Friedman, Meyer, and Gerald W. Friedland. *Medicine's Ten Greatest Discoveries.* New Haven, Conn.: Yale University Press, 2000.

Friedman, Robert Marc. *Appropriating the Weather: Vilhelm Bjerknes and the Construction of a Modern Meteorology.* Ithaca, N.Y.: Cornell University Press, 1989.

Friedrich, Wilhelm. *Vitamins.* New York: Walter de Gruyter, 1988.

Friedrichs, Günter, and Adam Schaff. *Microelectronics and Society: For Better or for Worse, a Report to the Club of Rome.* New York: Pergamon Press, 1982.

Frist, William. *Transplant.* New York: Atlantic Monthly Press, 1989.

Fuchs, Thomas. *The Mechanization of the Heart: Harvey and Descartes.* Rochester, N.Y.: University of Rochester Press, 2001.

Gallo, Robert C. *Virus Hunting: AIDS, Cancer, and the Human Retrovirus: A Story of Scientific Discovery.* New York: Basic Books, 1993.

Galston, Arthur W. *Life Processes of Plants.* New York: Scientific American Library, 1994.

Gamow, George. *The New World of Mr. Tompkins.* Cambridge, England: Cambridge University Press, 1999.

Gani, Joseph M., ed. *The Craft of Probabilistic Modeling.* New York: Springer-Verlag, 1986.

García-Ballester, Luis. *Galen and Galenism: Theory and Medical Practice from Antiquity to the European Renaissance.* Burlington, Vt.: Ashgate, 2002.

Gardner, Eldon J., and D. Peter Snustad. *Principles of Genetics.* 7th ed. New York: John Wiley & Sons, 1984.

Gardner, Robert, and Eric Kemer. *Science Projects About Temperature and Heat.* Berkeley Heights, N.J.: Enslow Publishers, 1994.

Garner, Geraldine. *Careers in Engineering.* 3d ed. New York: McGraw-Hill, 2009.

Gartner, Carol B. *Rachel Carson.* New York: Frederick Ungar, 1983.

Gary, James, et al. *Petroleum Refining: Technology and Economics.* 5th ed. New York: CRC Press, 2007. A well-written textbook that provides a good study of refining. The last two chapters cover economic issues. Five appendixes, index, photographs.

Gasser, James, ed. *A Boole Anthology: Recent and Classical Studies in the Logic of George Boole.* Dordrecht, the Netherlands: Kluwer, 2000.

Gay, Peter. *Freud: A Life for Our Time.* New York: W. W. Norton, 1988.

_____. *The Enlightenment: The Science of Freedom.* New York: W. W. Norton, 1996.

Gazzaniga, Michael S. *The Social Brain: Discovering the Networks of the Mind.* New York: Basic Books, 1985.

Geison, Gerald. *The Private Science of Louis Pasteur.* Princeton, N.J.: Princeton University Press, 1995.

Gell-Mann, Murray. *The Quark and the Jaguar: Adventures in the Simple and the Complex.* New York: W. H. Freeman, 1994.

Georgotas, Anastasios, and Robert Cancro, eds. *Depression and Mania.* New York: Elsevier, 1988.

Geotz, Walter. "Phoenix on Mars." *American Scientist* 98.1 (2010): 40–47. Print.

Gerock, Robert. *Mathematical Physics.* Chicago: University of Chicago Press, 1985.

Gest, Howard. *Microbes: An Invisible Universe.* Washington, D.C.: ASM Press, 2003.

Gesteland, Raymond F., Thomas R. Cech, and John F. Atkins, eds. *The RNA World: The Nature of Modern RNA Suggests a Prebiotic RNA.* 2d ed. Cold Spring Harbor, N.Y.: Cold Spring Harbor Laboratory Press, 1999.

Gigerenzer, Gerd, et al. *The Empire of Chance: How Probability Theory Changed Science and Everyday Life.* New York: Cambridge University Press, 1989.

Gilder, Joshua, and Anne-Lee Gilder. *Heavenly Intrigue: Johannes Kepler, Tycho Brahe, and the Murder Behind One of History's Greatest Scientific Discoveries.* New York: Doubleday, 2004.

Gillispie, Charles Coulston, Robert Fox, and Ivor Grattan-Guinness. *Pierre-Simon Laplace, 1749-1827: A Life in Exact Science.* Princeton, N.J.: Princeton University Press, 2000.

Gingerich, Owen. *The Book Nobody Read: Chasing the Revolutions of Nicolaus Copernicus.* New York: Walker, 2004.

_____. *The Eye of Heaven: Ptolemy, Copernicus, Kepler.* New York: Springer-Verlag, 1993.

Glashow, Sheldon, with Ben Bova. *Interactions: A Journey Through the Mind of a Particle Physicist and the Matter of This World.* New York: Warner Books, 1988.

Glass, Billy. *Introduction to Planetary Geology.* New York: Cambridge University Press, 1982.

Gleick, James. *Chaos: Making a New Science.* New York: Penguin Books, 1987.

_____.*Isaac Newton.* New York: Pantheon Books, 2003.

Glen, William. *The Road to Jaramillo: Critical Years of the Revolution in Earth Science.* Stanford, Calif.: Stanford University Press, 1982.

Glickman, Todd S., ed. *Glossary of Meteorology.* 2d ed. Boston: American Meteorological Society, 2000.

Goddard, Jolyon, ed. *National Geographic Concise History of Science and Invention: An Illustrated Time Line.* Washington, D.C.: National Geographic, 2010.

Goding, James W. *Monoclonal Antibodies: Principles and Practice.* New York: Academic Press, 1986.

Godwin, Robert, ed. *Mars: The NASA Mission Reports.* Burlington, Ont.: Apogee Books, 2000.

_____.*Mars: The NASA Mission Reports.* Vol. 2. Burlington, Ont.: Apogee Books, 2004.

_____.*Space Shuttle STS Flights 1-5: The NASA Mission Reports.* Burlington, Ont.: Apogee Books, 2001.

Goetsch, David L. *Building a Winning Career in Engineering: 20 Strategies for Success After College.* Upper Saddle River, N.J.: Pearson/Prentice Hall, 2007.

Gohlke, Mary, with Max Jennings. *I'll Take Tomorrow.* New York: M. Evans, 1985.

Gold, Rebecca. *Steve Wozniak: A Wizard Called Woz.* Minneapolis: Lerner, 1994.

Goldsmith, Donald. *Nemesis: The Death Star and Other Theories of Mass Extinction.* New York: Berkley Publishing Group, 1985.

Goldsmith, Maurice, Alan Mackay, and James Woudhuysen, eds. *Einstein: The First Hundred Years.* Elmsford, N.Y.: Pergamon Press, 1980.

Golinski, Jan. *Science as Public Culture: Chemistry and Enlightenment in Britain, 1760-1820.* Cambridge, England: Cambridge University Press, 1992.

Golthelf, Allan, and James G. Lennox, eds. *Philosophical Issues in Aristotle's Biology.* Cambridge, England: Cambridge University Press, 1987.

Gooding, David, and Frank A. J. L. James, eds. *Faraday Rediscovered: Essays on the Life and Work of Michael Faraday, 1791-1867.* New York: Macmillan, 1985.

Gordin, Michael D. *A Well-Ordered Thing: Dmitrii Mendeleev and the Shadow of the Periodic Table.* New York: Basic Books, 2004.

Gornick, Vivian. *Women in Science: Then and Now.* New York: Feminist Press at the City University of New York, 2009.

Gould, James L., and Carol Grant Gould. *The Honey Bee.* New York: Scientific American Library, 1988.

Gould, Stephen Jay. *Time's Arrow, Time's Cycle: Myth and Metaphor in the Discovery of Geological Time.* Cambridge, Mass.: Harvard University Press, 1987.

_____.*Wonderful Life: The Burgess Shale and the Nature of History.* New York: W. W. Norton, 1989.

Govindjee, J. T. Beatty, H. Gest, and J.F. Allen, eds. *Discoveries in Photosynthesis.* Berlin: Springer, 2005.

Gow, Mary. *Tycho Brahe: Astronomer.* Berkley Heights, N.J.: Enslow, 2002.

Graham, Loren R. *Science, Philosophy, and Human Behavior in the Soviet Union.* New York: Columbia University Press, 1987.

Grattan-Guinness, Ivor. *The Norton History of the Mathematical Sciences.* New York: W. W. Norton, 1999.

Gray, Robert M., and Lee D. Davisson. *Random Processes: A Mathematical Approach for Engineers.* Englewood Cliffs, N.J.: Prentice-Hall, 1986.

Greeley, Ronald, and Raymond Batson. *The Compact NASA Atlas of the Solar System.* New York: Cambridge UP, 2001. Print.

Greene, Mott T. *Geology in the Nineteenth Century: Changing Views of a Changing World.* Ithaca, N.Y.: Cornell University Press, 1982.

Gregory, Andrew. *Harvey's Heart: The Discovery of Blood Circulation.* London: Totem Books, 2001.

Gribbin, John, ed. *The Breathing Planet.* New York: Basil Blackwell, 1986.

Gribbin, John. *Deep Simplicity: Bringing Order to Chaos and Complexity.* New York: Random House, 2005.

_____.*Future Weather and the Greenhouse Effect.* New York: Delacorte Press/Eleanor Friede, 1982.

_____.*In Search of Schrödinger's Cat: Quantum Physics and Reality.* New York: Bantam Books, 1984.

_____.*In Search of the Big Bang.* New York: Bantam Books, 1986.

_____.*The Hole in the Sky: Man's Threat to the Ozone Layer.* New York: Bantam Books, 1988.

_____.*The Omega Point: The Search for the Missing Mass and the Ultimate Fate of the Universe.* New York: Bantam Books, 1988.

_____.*The Scientists: A History of Science Told Through the Lives of Its Greatest Inventors.* New York: Random House, 2002.

Gutkind, Lee. *Many Sleepless Nights: The World of Organ*

Transplantation. New York: W. W. Norton, 1988.

Hackett, Edward J., et al., eds. *The Handbook of Science and Technology Studies.* 3d ed. Cambridge, Mass.: MIT Press, 2008.

Hald, Anders. *A History of Mathematical Statistics from 1750 to 1950.* New York: John Wiley & Sons, 1998.

_____. *A History of Probability and Statistics and Their Applications Before 1750.* New York: John Wiley & Sons, 1990.

_____. *The Scientific Revolution, 1500-1750.* 3d ed. New York: Longman, 1983.

Hallas, Joel. *Basic Radio: Understanding the Key Building Blocks.* Newington: American Radio Relay League, 2005. Print.

Halliday, David, and Robert Resnick. *Fundamentals of Physics: Extended Version.* New York: John Wiley & Sons, 1988.

Halliday, David, Robert Resnick, and Jearl Walker. *Fundamentals of Physics.* 7th ed. New York: John Wiley & Sons, 2004.

Hankins, Thomas L. *Science and the Enlightenment.* Reprint. New York: Cambridge University Press, 1991.

Hanlon, Michael, and Arthur C. Clarke. *The Worlds of Galileo: The Inside Story of NASA's Mission to Jupiter.* New York: St. Martin's Press, 2001.

Hanson, Earl D. *Understanding Evolution.* New York: Oxford University Press, 1981.

Hargittai, István. *Martians of Science: Five Physicists Who Changed the Twentieth Century.* New York: Oxford University Press, 2006.

Harland, David M. *Jupiter Odyssey: The Story of NASA's Galileo Mission.* London: Springer-Praxis, 2000.

_____. *Mission to Saturn: Cassini and the Huygens Probe.* London: Springer-Praxis, 2002.

_____. *The Space Shuttle: Roles, Missions, and Accomplishments.* New York: John Wiley & Sons, 1998.

Harland, David M., and John E. Catchpole. *Creating the International Space Station.* London: Springer-Verlag, 2002.

Harra, Louise K., and Keith O. Mason, eds. *Space Science.* London: Imperial Coll. P, 2004. Print.

Harrington, J. W. *Dance of the Continents:* New York: V. P. Tarher, 1983.

Harrington, Philip S. *The Space Shuttle: A Photographic History.* San Francisco: Brown Trout, 2003.

Harris, Daniel C. *Quantitative Chemical Analysis.* New York: Freeman, 2003. Print.

Harris, Henry. *The Birth of the Cell.* New Haven, Conn.: Yale University Press, 1999.

Harris, Philip. *Space Enterprise: Living and Working Offworld in the Twenty-First Century.* New York: Springer, 2009. Print.

Harrison, Edward R. *Cosmology: The Science of the Universe.* Cambridge England: Cambridge University Press, 1981.

Hart, Michael H. *The 100: A Ranking of the Most Influential Persons in History.* New York: Galahad Books, 1982.

Hart-Davis, Adam. *Chain Reactions: Pioneers of British Science and Technology and the Stories That Link Them.* London: National Portrait Gallery, 2000.

Hartmann, William K. *Moons and Planets.* 5th ed. Belmont, Calif.: Brooks-Cole Publishing, 2005.

_____. *The Cosmic Voyage: Through Time and Space.* Belmont, Calif: Wadsworth, 1990.

Hartwell, L. H., et al. *Genetics: From Genes to Genomes.* 2d ed. New York: McGraw-Hill, 2004.

Harvey, William. *The Circulation of the Blood and Other Writings.* New York: Everyman's Library, 1990.

Hashemi, Ray H., William G. Bradley, and Christopher J. Lisanti. *MRI: The Basics.* 3rd ed. Philadelphia: Lippincott, 2010. Print.

Haskell, G., and Michael Rycroft. *International Space Station: The Next Space Marketplace.* Boston: Kluwer Academic, 2000.

Hathaway, N. *The Friendly Guide to the Universe.* New York: Penguin Books, 1994.

Havil, Julian. *Gamma: Exploring Euler's Constant.* Princeton, N.J.: Princeton University Press, 2003.

Hawking, Stephen W. *A Brief History of Time.* New York: Bantam Books, 1988.

Haycock, David. *William Stukeley: Science, Religion, and Archeology in Eighteenth-Century England.* Woodbridge, England: Boydell Press, 2002.

Hazen, Robert. *The Breakthrough: The Race for the Superconductor.* New York: Summit Books, 1988.

Headrick, Daniel R. *Technology: A World History.* New York: Oxford University Press, 2009.

Heath, Sir Thomas L. *A History of Greek Mathematics: From Thales to Euclid.* 1921. Reprint. New York: Dover Publications, 1981.

Hecht, Jeff. *Beam: The Race to Make the Laser.* New York: Oxford UP, 2005. Print.

Heilbron, J. L. *Electricity in the Seventeenth and Eighteenth Centuries: A Study in Early Modern Physics.* Mineola, N.Y.: Dover Publications, 1999.

_____. *Elements of Early Modern Physics.* Berkeley:

University of California Press, 1982.

_____.*Geometry Civilized: History, Culture, and Technique.* Oxford, England: Clarendon Press, 1998.

_____.*The Dilemmas of an Upright Man: Max Planck As a Spokesman for German Science.* Berkeley: University of California Press, 1986.

Heilbron, J. L., and Robert W. Seidel. *Lawrence and His Laboratory: A History of the Lawrence Berkeley Laboratory.* Berkeley: University of California Press, 1989.

Heisenberg, Elisabeth. *Inner Exile: Recollections of a Life with Werner Heisenberg.* Translated by S. Cappelari and C. Morris. Boston: Birkhäuser, 1984.

Hellegouarch, Yves. *Invitation to the Mathematics of Fermat-Wiles.* San Diego: Academic Press, 2001.

Henig, Robin Marantz. *The Monk in the Garden: The Lost and Found Genius of Gregor Mendel, the Father of Genetics.* New York: Mariner Books, 2001.

Henry, Gary N., Wiley J. Larson, and Ronald W. Humble. *Space Propulsion Analysis and Design.* New York: McGraw-Hill 1995. Combines short essays by experts on individual topics with extensive data and analytical methods for propulsion system design.

Henry, Helen L., and Anthony W. Norman, eds. *Encyclopedia of Hormones.* 3 vols. San Diego: Academic Press, 2003.

Henry, John. *Moving Heaven and Earth: Copernicus and the Solar System.* Cambridge, England: Icon, 2001.

Herrmann, Bernd, and Susanne Hummel, eds. *Ancient DNA: Recovery and Analysis of Genetic Material from Paleographic, Archaeological, Museum, Medical, and Forensic Specimens.* New York: Springer-Verlag, 1994.

Hershel, Sir John Frederic William, and Pierre-Simon Laplace. *Essays in Astronomy.* University Press of the Pacific, 2002.

Hildebrand, F. B. *Introduction to Numerical Analysis.* 2nd ed. New York: McGraw, 1974. Print.

Hillar, Marian, and Claire S. Allen. *Michael Servetus: Intellectual Giant, Humanist, and Martyr.* New York: University Press of America, 2002.

Hobson, J. Allan. *The Dreaming Brain.* New York: Basic Books, 1988.

Hodge, Paul. *Galaxies.* Cambridge, Mass.: Harvard University Press, 1986.

Hodges, Andrew. *Alan Turing: The Enigma.* 1983. Reprint. New York: Walker, 2000.

Hofmann, James R., David Knight, and Sally Gregory

Kohlstedt, eds. *André-Marie Ampère: Enlightenment and Electrodynamics.* Cambridge, England: Cambridge University Press, 1996.

Holland, Suzanne, Karen Lebacqz, and Laurie Zoloth, eds. *The Human Embryonic Stem Cell Debate: Science, Ethics, and Public Policy.* Cambridge, Mass.: MIT Press, 2001.

Holmes, Frederic Lawrence. *Antoine Lavoisier, the Next Crucial Year: Or, the Sources of His Quantitative Method in Chemistry.* Princeton, N.J.: Princeton University Press, 1998.

Horne, James. *Why We Sleep.* New York: Oxford University Press, 1988.

Hoskin, Michael A. *The Herschel Partnership: As Viewed by Caroline.* Cambridge, England: Science History, 2003.

_____.*William Herschel and the Construction of the Heavens.* New York: Norton; 1964.

Howe, A. Scott, and Brent Sherwood, eds. *Out of This World: The New Field of Space Architecture.* Reston: American Inst. of Aeronautics and Astronautics, 2009. Print.

Howland, Rebecca, and Lisa Benatar. "A Practical Guide to Scanning Probe Microscopy." ThermoMicroscopes. March 2000. Web. Accessed September, 2011. This introductory guide to scanning probe microscopy describes several techniques and operating modes of the devices, as well as their structure and principles of operation, and discusses the occurrence of image artifacts in their use.

Howse, Derek. *Greenwich Time and the Discovery of Longitude.* New York: Oxford University Press, 1980.

Hoyt, William G. *Planet X and Pluto.* Tucson: University of Arizona Press, 1980.

Hsü, Kenneth J. *The Great Dying.* San Diego: Harcourt Brace Jovanovich, 1986.

http://ts.nist.gov/WeightsAndMeasures/Publications/upload/HB44-10-Complete.pdf. Accessed October, 2011. The official statement of regulations governing the function and operating requirements of weighing and measuring devices in the United States.

Hudson, John B. *Surface Science: An Introduction.* New York: John Wiley & Sons, 1998. Contains information on surface analysis, interface interactions, and crystal growth.

Huerta, Robert D. *Giants of Delft, Johannes Vermeer and the Natural Philosophers: The Parallel Search for Knowledge During the Age of Discovery.* Lewisburg, Pa.:

Bucknell University Press, 2003.

Hummel, Susanne. *Fingerprinting the Past: Research on Highly Degraded DNA and Its Applications.* New York: Springer-Verlag, 2002.

Hunter, Fil, Steven Biver, and Paul Fuqua. *Light, Science and Magic: An Introduction to Photographic Lighting.* 3d ed. Burlington: Focal, 2007. Print.

Hunter, Michael, ed. *Robert Boyle Reconsidered.* New York: Cambridge University Press, 1994.

Hwang, Kai, and Doug Degroot, eds. *Parallel Processing for Supercomputers and Artificial Intelligence.* New York: McGraw, 1989. Print.

Hyne, Norman. *Nontechnical Guide to Petroleum Geology, Exploration, Drilling, and Production.* 2d ed. Tulsa, Okla.: Penn Well, 2001. A comprehensive description of all activities that take place before petroleum is processed. Tables, figures, index.

Hynes, H. Patricia. *The Recurring Silent Spring.* New York: Pergamon Press, 1989.

Ihde, Aaron J. *The Development of Modern Chemistry.* New York: Dover, 1984.

Irwin, Patrick G. J. *Giant Planets of Our Solar System: Atmospheres, Composition, and Structure.* London: Springer-Praxis, 2003.

Isaacson, Walter. *Benjamin Franklin: An American Life.* New York: Simon & Schuster, 2003.

Iserles, Arieh. *A First Course in the Numerical Analysis of Differential Equations.* 2nd ed. New York: Cambridge UP, 2009. Print.

Jackson, Myles. *Spectrum of Belief: Joseph Fraunhofer and the Craft of Precision Optics.* Cambridge, Mass.: MIT Press, 2000.

Jacobsen, Theodor S. *Planetary Systems from the Ancient Greeks to Kepler.* Seattle: University of Washington Press, 1999.

Jacquette, Dale. *On Boole.* Belmont, Calif.: Wadsworth, 2002.

Jaffe, Bernard. *Crucibles: The Story of Chemistry.* New York: Dover, 1998.

James, Ioan. *Remarkable Mathematicians: From Euler to Von Neumann.* Cambridge, England: Cambridge University Press, 2002.

Janowsky, David S., Dominick Addario, and S. Craig Risch. *Psychopharmacology Case Studies.* 2d ed. New York: Guilford Press, 1987.

Jeffreys, Diarmuid. *Aspirin: The Remarkable Story of a Wonder Drug.* London: Bloomsbury Publishing, 2004.

Jenkins, Dennis R. *Space Shuttle: The History of the National Space Transportation System: The First 100 Missions.* Stillwater, Minn.: Voyageur Press, 2001.

Johanson, Donald C., and James Shreeve. *Lucy's Child: The Discovery of a Human Ancestor.* New York: William Morrow, 1989.

Johanson, Donald C., and Maitland A. Edey. *Lucy: The Beginnings of Humankind.* New York: Simon & Schuster, 1981.

Johanson, Donald, and B. Edgar. *From Lucy to Language.* New York: Simon and Schuster, 1996.

Johnson, George. *Strange Beauty: Murray Gell-Mann and the Revolution in Twentieth-Century Physics.* New York: Alfred A. Knopf, 1999.

Jones, Henry Bence. *Life and Letters of Faraday.* 2 vols. London: Longmans, Green and Co., 1870.

Jones, Meredith L., ed. *Hydrothermal Vents of the Eastern Pacific: An Overview.* Vienna, Va.: INFAX, 1985.

Jones, Sheilla. *The Quantum Ten: A Story of Passion, Tragedy, and Science.* Toronto: Thomas Allen, 2008.

Jones, Tony. *Splitting the Second: The Story of Atomic Time.* Philadelphia: IOP, 2000. Print.

Jones, W. H. S., trans. *Hippocrates.* 4 vols. 1923-1931. Reprint. New York: Putnam, 1995.

Jordan, Paul. *Neanderthal: Neanderthal Man and the Story of Human Origins.* Gloucestershire, England: Sutton, 2001.

Joseph, George Gheverghese. *The Crest of the Peacock: The Non-European Roots of Mathematics.* London: Tauris, 1991.

Jungnickel, Christa, and Russell McCormmach. *Cavendish: The Experimental Life.* Lewisburg, Pa.: Bucknell University Press, 1999.

Kaempfer, Rick, and John Swanson. *The Radio Producer's Handbook.* New York: Allworth, 2004. Print.

Kaplan, Elliot D., and Christopher J. Hegarty, eds. *Understanding GPS: Principles and Applications.* 2nd ed. Boston: Artech, 2006. Print.

Kaplan, Robert. *The Nothing That Is: A Natural History of Zero.* New York: Oxford University Press, 2000.

Kargon, Robert H. *The Rise of Robert Millikan: Portrait of a Life in American Science.* Ithaca, N.Y.: Cornell University Press, 1982.

Katz, Jonathan. *The Biggest Bang: The Mystery of Gamma-Ray Bursts.* London: Oxford University Press, 2002.

Keith, Michael C. *The Radio Station: Broadcast, Satellite and Internet.* 8th ed. Burlington: Focal, 2010. The standard guide to radio as a medium. Discusses the various departments in all kinds of radio stations in detail. Print.

Keller, Corey, ed. *Brought to Light: Photography and the Invisible, 1840–1900*. New Haven: San Francisco Museum of Modern Art, Yale U P, 2008. Print.

Kellermann, Kenneth I. "Radio Astronomy in the Twenty-first Century." *Sky and Telescope* 93, no. 2 (February, 1997): 26–34. An excellent overview of the development of radio astronomy through the twentieth century with a look forward to prospects for developments through the twenty-first century.

Kellogg, William W., and Robert Schware. *Climate Change and Society: Consequences of Increasing Atmospheric Carbon Dioxide*. Boulder, Colo.: Westview Press, 1981.

Kelly, Thomas J. *Moon Lander: How We Developed the Apollo Lunar Module*. Washington, D.C.: Smithsonian Books, 2001.

Kemper, John D., and Billy R. Sanders. *Engineers and Their Profession*. 5th ed. New York: Oxford University Press, 2001.

Kepler, Johannes. *New Astronomy*. Translated by William H. Donahue. New York: Cambridge University Press, 1992.

Kermit, Hans. *Niels Stensen: The Scientist Who Was Beatified*. Translated by Michael Drake. Herefordshire, England: Gracewing 2003.

Kerns, Thomas A. *Jenner on Trial: An Ethical Examination of Vaccine Research in the Age of Smallpox and the Age of AIDS*. Lanham, Md.: University Press of America, 1997.

Kerrod, Robin. *Hubble: The Mirror on the Universe*. Richmond Hill, Ont.: Firefly Books, 2003.

_____. *Space Shuttle*. New York: Gallery Books, 1984.

Kevles, Bettyann. *Naked to the Bones: Medical Imaging in the Twentieth Century*. Reading, Mass.: Addison Wesley, 1998.

Khudyakov, Yury E., and Paul Pumpens. *Viral Nanotechnology*. CRC Press, 2016.

Kiessling, Ann, and Scott C. Anderson. *Human Embryonic Stem Cells: An Introduction to the Science and Therapeutic Potential*. Boston: Jones and Bartlett, 2003.

Kim, Hyun-Ha. "Nonthermal Plasma Processing for Air-Pollution Control." *Plasma Processes and Polymers* 1, no. 2, (September, 2004): 91–110. Discusses the history and future of using plasma treatment to reduce air pollution.

King, Helen. *Greek and Roman Medicine*. London: Bristol Classical, 2001.

_____. *Hippocrates' Woman: Reading the Female Body in Ancient Greece*. New York: Routledge, 1998.

Kirk, David, and Wen-mei W. Hwu. *Programming Massively Parallel Processors: A Hands-On Approach*. Burlington: Kaufmann, 2010. Print.

Kirkham, M. B. *Principles of Soil and Plant Water Relations*. St. Louis: Elsevier, 2005.

Kline, Morris. *Mathematical Thought from Ancient to Modern Times*. New York: Oxford University Press, 1990.

Klotzko, Arlene Judith, ed. *The Cloning Sourcebook*. New York: Oxford University Press, 2001.

Knipe, David, Peter Howley, and Diane Griffin. *Field's Virology*. 2 vols. New York: Lippincott Williams and Wilkens, 2001.

Knowles, Richard V. *Genetics, Society, and Decisions*. Columbus, Ohio: Charles E. Merrill, 1985.

Koestler, Arthur. *The Sleepwalkers*. New York: Penguin Books, 1989.

Kolasinski, Kurt W. *Surface Science: Foundations of Catalysis and Nanoscience*. 2d ed. Hoboken, N.J.: John Wiley & Sons, 2009. A detailed, more technical resource for those interested in cutting-edge applications of surface science.

Kolata, Gina Bari. *Clone: The Road to Dolly, and the Path Ahead*. New York: William Morrow, 1998.

Komszik, Louis. *Applied Calculus of Variations for Engineers*. Boca Raton, Fla.: CRC Press, 2009.

Koupelis, Theo. *In Quest of the Universe*. 6th ed. Sudbury: Jones, 2010. Print.

Kramer, Barbara. *Neil Armstrong: The First Man on the Moon*. Springfield, N.J.: Enslow, 1997.

Krane, Kenneth S. *Modern Physics*. New York: John Wiley & Sons, 1983.

La Thangue, Nicholas B., and Lasantha R. Bandara, eds. *Targets for Cancer Chemotherapy: Transcription Factors and Other Nuclear Proteins*. Totowa, N.J.: Humana Press, 2002.

Lagerkvist, Ulf. *Pioneers of Microbiology and the Nobel Prize*. River Edge, N.J.: World Scientific Publishing, 2003.

Lamarck, Jean-Baptiste. *Lamarck's Open Mind: The Lectures*. Gold Beach, Ore.: High Sierra Books, 2004.

_____. *Zoological Philosophy: An Exposition with Regard to the Natural History of Animals*. Translated by Hugh Elliot with introductory essay by David L. Hull and Richard W. Burckhardt, Jr. Chicago: University of Chicago Press, 1984.

Landes, Davis S. *Revolution in Time: Clocks and the Making of the Modern World*. Rev. ed. Cambridge, Mass.: Belknap Press, 2000.

Langone, John. *Superconductivity: The New Alchemy.* Chicago: Contemporary Books, 1989.

Lappé, Marc. *Broken Code: The Exploitation of DNA.* San Francisco: Sierra Club Books, 1984.

_____. *Germs That Won't Die.* Garden City, N.Y.: Doubleday, 1982.

Laudan, R. *From Mineralogy to Geology: The Foundations of a Science, 1650-1830.* Chicago: University of Chicago Press, 1987.

Lauritzen, Paul, ed. *Cloning and the Future of Human Embryo Research.* New York: Oxford University Press, 2001.

Le Grand, Homer E. *Drifting Continents and Shifting Theories.* New York: Cambridge University Press, 1988.

Leakey, Mary D. *Disclosing the Past.* New York: Doubleday, 1984.

Levy, David H. *Clyde Tombaugh: Discoverer of Planet Pluto.* Tucson: University of Arizona Press, 1991.

Lewin, Benjamin. *Genes III.* 3d ed. New York: John Wiley & Sons, 1987.

_____. *Genes IV.* New York: Oxford University Press, 1990.

Lewin, Roger. *Bones of Contention: Controversies in the Search for Human Origins.* New York: Simon & Schuster, 1987.

Lewis, Richard S. *The Voyages of Columbia: The First True Spaceship.* New York: Columbia University Press, 1984.

Lindberg, David C. *The Beginnings of Western Science.* Chicago: University of Chicago Press, 1992.

Lindley, David. *Degrees Kelvin: A Tale of Genius, Invention, and Tragedy.* Washington, D.C.: Joseph Henry Press, 2004.

Linzmayer, Owen W. *Apple Confidential: The Real Story of Apple Computer, Inc.* San Francisco: No Starch Press, 1999.

Lloyd, G. E. R., and Nathan Sivin. *The Way and the Word: Science and Medicine in Early China and Greece.* New Haven, Conn.: Yale University Press, 2002.

Logan, J. David. *Applied Mathematics.* 3d ed. Hoboken, N.J.: John Wiley & Sons, 2006.

Logsdon, John M. *Together in Orbit: The Origins of International Participation in the Space Station.* Washington, D.C.: National Aeronautics and Space Administration, 1998.

Lombardi, Michael A. "Radio Controlled Wristwatches." *Horological Journal* 148.5 (2006): 187–92. Print.

Lombardi, Michael A., Thomas P. Heavner, and Steven R. Jefferts. "NIST Primary Frequency Standards and the Realization of the SI Second." *Measure* 2.4 (2007): 74–89. Print.

Lonc, William. *Radio Astronomy Projects.* Louisville, Ky.: Radio-Sky Publishing, 1996. Detailed instructions on the construction of various radio telescope systems and information on projects using those systems for the amateur astronomer with a good understanding of electronics. Suitable for engineering students.

Longdell, Jevon. "Quantum Information: Entanglement on Ice." *Nature* 469.7331 (2011): 475–76. Print.

Longrigg, James. *Greek Medicine: From the Heroic to the Hellenistic Age.* New York: Routledge, 1998.

_____. *Greek Rational Medicine: Philosophy and Medicine from Alcmaeon to the Alexandrians.* New York: Routledge, 1993.

Lorenz, Edward. *The Essence of Chaos.* Reprint. St. Louis: University of Washington Press, 1996.

Lorenz, Ralph, and Jacqueline Mitton. *Lifting Titan's Veil: Exploring the Giant Moon of Saturn.* London: Cambridge University Press, 2002.

Loudon, Irvine. *The Tragedy of Childbed Fever.* New York: Oxford University Press, 2000.

Luck, Steve, ed. *International Encyclopedia of Science and Technology.* New York: Oxford University Press, 1999.

Lutgens, Frederick K., and Edward J. Tarbuck. *The Atmosphere: An Introduction to Meteorology.* 2d ed. Englewood Cliffs, N.J.: Prentice-Hall, 1982.

Lyell, Charles. *Elements of Geology.* London: John Murray, 1838.

_____. *Principles of Geology, Being an Attempt to Explain the Former Changes of the Earth's Surface by Reference to Causes Now in Operation.* 3 vols. London: John Murray, 1830-1833.

_____. *The Geological Evidences of the Antiquity of Man with Remarks on Theories of the Origin of Species by Variation.* London: John Murray, 1863.

Lynch, William T. *Solomon's Child: Method in the Early Royal Society of London.* Stanford, Calif.: Stanford University Press, 2000.

Ma, Pearl, and Donald Armstrong, eds. *AIDS and Infections of Homosexual Men.* Stoneham, Mass: Butterworths, 1989.

MacDonald, Allan H., ed. *Quantum Hall Effect: A Perspective.* Boston: Kluwer Academic Publishers,

1989.

MacHale, Desmond. *George Boole: His Life and Work.* Dublin: Boole Press, 1985.

Machamer, Peter, ed. *The Cambridge Companion to Galileo.* New York: Cambridge University Press, 1998.

Mactavish, Douglas. *Joseph Lister.* New York: Franklin Watts, 1992.

Mader, Sylvia S. *Biology.* 3d ed. Dubuque, Iowa: Win. C. Brown, 1990.

Magner, Lois. *A History of Medicine.* New York: Marcel Dekker, 1992.

Mahoney, Michael Sean. *The Mathematical Career of Pierre de Fermat, 1601-1665.* 2d rev. ed. Princeton, N.J.: Princeton University Press, 1994.

Malphrus, Benjamin K. *The History of Radio Astronomy and the National Radio Astronomy Observatory: Evolution Toward Big Science.* Malabar, Fla.: Krieger, 1996. A chronicle of radio astronomy history from the early days to the development of large national radio astronomy laboratories.

Mammana, Dennis L., and Donald W. McCarthy, Jr. *Other Suns, Other Worlds? The Search for Extrasolar Planetary Systems.* New York: St. Martin's Press, 1996.

Mandelbrot, B. B. *Fractals and Multifractals: Noise, Turbulence, and Galaxies.* New York: Springer-Verlag, 1990.

Mann, Charles, and Mark Plummer. *The Aspirin Wars: Money, Medicine and 100 Years of Rampant Competition.* New York: Knopf, 1991.

Mante, Harald. *The Photograph: Composition and Color Design.* Santa Barbara: Rocky Nook and Verlag Photographie, 2008. Print.

Marco, Gino J., Robert M. Hollingworth, and William Durham, eds. *Silent Spring Revisited.* Washington, D.C.: American Chemical Society, 1987.

Margolis, Howard. *It Started with Copernicus.* New York: McGraw-Hill, 2002.

Marques de Sá, J. P. *Pattern Recognition: Concepts, Methods and Applications.* New York: Springer Books, 2001. A nonspecialist treatment of issues surrounding pattern recognition. Looks at many of the principles underlying the discipline as well as discussions of applications. Although full understanding requires knowledge of advanced statistics and mathematics, the book can be understood by students with a basic science background.

Marshak, Daniel R., Richard L. Gardner, and David Gottlieb, eds. *Stem Cell Biology.* Woodbury, N.Y.: Cold Spring Harbor Laboratory Press, 2002.

Martzloff, Jean-Claude. *History of Chinese Mathematics.* Translated by Stephen S. Wilson. Berlin: Springer, 1987.

Massey, Harrie Stewart Wilson. *The Middle Atmosphere as Observed by Balloons, Rockets, and Satellites.* London: Royal Society, 1980.

Masson, Jeffrey M. *The Assault on Truth: Freud's Suppression of the Seduction Theory.* New York: Farrar, Straus and Giroux, 1984.

Mateles, Richard I. *Penicillin: A Paradigm for Biotechnology.* Chicago: Canadida Corporation, 1998.

Matloff, Gregory L. *Deep Space Probes: To the Outer Solar System and Beyond.* New York: Springer, 2005. Print.

Mayo, Jonathan L. *Superconductivity: The Threshold of a New Technology.* Blue Ridge Summit, Pa.: TAB Books, 1988.

McCarthy, Shawn P. *Engineer Your Way to Success: America's Top Engineers Share Their Personal Advice on What They Look for in Hiring and Promoting.* Alexandria, Va.: National Society of Professional Engineers, 2002.

McCay, Mary A. *Rachel Carson.* New York: Twayne, 1993.

McCorduck, Pamela. *Machines Who Think: A Personal Inquiry into the History and Prospects of Artificial Intelligence.* 2d ed. Natick, Mass.: A. K. Peters, 2004. An introduction to the history and development of artificial intelligence written for the general reader. Contains information about pattern recognition systems and applications.

McDonnell, John James. *The Concept of an Atom from Democritus to John Dalton.* Lewiston, N.Y.: Edwin Mellen Press, 1991.

McEliece, Robert. *Finite Fields for Computer Scientists and Engineers.* Boston: Kluwer Academic, 1987.

McGraw-Hill Concise Encyclopedia of Science and Technology. 6th ed. New York: McGraw-Hill, 2009.

McGraw-Hill Dictionary of Scientific and Technical Terms. 6th ed. New York: McGraw-Hill, 2002.

McGrayne, Sharon Bertsch. *Nobel Prize Women in Science: Their Lives, Struggles and Momentous Discoveries.* 2d ed. Washington, D.C.: Joseph Henry Press, 1998.

McIntyre, Donald B., and Alan McKirdy. *James Hutton: The Founder of Modern Geology.* Edinburgh: Stationery Office, 1997.

McLester, John, and Peter St. Pierre. *Applied Biomechanics: Concepts and Connections.* Belmont, Calif.:

Thomson Wadsworth, 2008.

McMullen, Emerson Thomas. *William Harvey and the Use of Purpose in the Scientific Revolution: Cosmos by Chance or Universe by Design?* Lanhan, Md.: University Press of America, 1998.

McQuarrie, Donald A. *Quantum Chemistry.* Mill Valley, Calif: University Science Books, 1983.

Menard, H. W. *The Ocean of Truth: A Personal History of Global Tectonics.* Princeton, N.J.: Princeton University Press, 1986.

Menzel, Donald H., and Jay M. Pasachoff. *Stars and Planets.* Boston: Houghton Mifflin Company, 1983.

Menzel, Ralf. *Photonics: Linear and Nonlinear Interactions of Laser Light and Matter.* 2nd ed. Berlin: Springer, 2007. Print.

Mercer, David. *The Telephone: The Life Story of a Technology.* Westport, Conn.: Greenwood Press, 2006.

Merrell, David J. *Ecological Genetics.* Minneapolis: University of Minnesota Press, 1981.

Meschede, Dieter. *Optics, Light and Lasers: The Practical Approach to Modern Aspects of Photonics and Laser Physics.* 2nd ed. Weinheim: Wiley, 2007. Print.

Mettler, Lawrence E., Thomas G. Gregg, and Henry E. Schaffer. *Population Genetics and Evolution.* 2d ed. Englewood Cliffs, N.J.: Prentice-Hall, 1988.

Meyer, Ernst, Hans Josef Hug, and Roland Bennewitz. *Scanning Probe Microscopy: The Lab on a Tip.* Berlin: Springer, 2004. This book provides an excellent overview of scanning probe microscopy before delving into more detailed discussions of the various techniques.

Meyers, Robert A., ed. *Encyclopedia of Physical Science and Technology.* 18 vols. San Diego: Academic Press, 2005.

Meyerson, Daniel. *The Linguist and the Emperor: Napoleon and Champollion's Quest to Decipher the Rosetta Stone.* New York: Ballantine Books, 2004.

Michalski, L., et al. *Temperature Measurement.* 2d ed. New York: John Wiley & Sons, 2001. Covers basic temperature-measurement techniques at a level suitable for high school students. A section on fuzzy-logic techniques used in thermostats is a novel addition.

Middleton, W. E. Knowles. *A History of the Thermometer and Its Use in Meteorology.* Ann Arbor, Mich.: UMI Books on Demand, 1996.

Miller, Ron. *Extrasolar Planets.* Brookfield, Conn.: Twenty-First Century Books, 2002.

Miller, Stanley L. *From the Primitive Atmosphere to the Prebiotic Soup to the Pre-RNA World.* Washington, D.C.: National Aeronautics and Space Administration, 1996.

Mishkin, Andrew. *Sojourner: An Insider's View of the Mars Pathfinder Mission.* New York: Berkeley Books, 2003.

Mlodinow, Leonard. *Euclid's Window: A History of Geometry from Parallel Lines to Hyperspace.* New York: Touchstone, 2002.

Moler, Cleve B. *Numerical Computing with MATLAB.* Philadelphia: Soc. for Industrial and Applied Mathematics, 2008. Print.

Mongillo, John. *Nanotechnology 110.* Westport, Conn.: Greenwood, 2007. Demonstrates the essential relationship between and value of scanning probe microscopy to nanotechnology.

Monmonier, Mark. *Air Apparent: How Meteorologists Learned to Map, Predict, and Dramatize Weather.* Chicago: University of Chicago Press, 2000.

Moore, Keith L. *The Developing Human.* Philadelphia: W. B. Saunders, 1988.

Moore, Patrick. *Eyes on the University: The Story of the Telescope.* New York: Springer-Verlag, 1997.

_____.*Patrick Moore's History of Astronomy.* 6th rev. ed. London: Macdonald, 1983.

Morawetz, Herbert. *Polymers: The Origins and Growth of a Science.* Mineola, N.Y.: Dover, 2002. This illustrated paperback reprint makes widely available an excellent history of the subject.

Morell, V. *Ancestral Passions: The Leakey Family and the Quest for Humankind's Beginnings.* New York: Simon & Schuster, 1995.

Morgan, Kathryn A. *Myth and Philosophy from the Presocratics to Plato.* New York: Cambridge University Press, 2000.

Moritz, Michael. *The Little Kingdom: The Private Story of Apple Computer.* New York: Morrow, 1984.

Morris, Peter, ed. *Making the Modern World: Milestones of Science and Technology.* 2d ed. Chicago: KWS Publishers, 2011.

Morris, Richard. *The Last Sorcerers: The Path from Alchemy to the Periodic Table.* Washington, D.C.: Joseph Henry Press, 2003.

Morris, Robert C. *The Environmental Case for Nuclear Power: Economic, Medical and Political Considerations.* St. Paul: Paragon House, 2000. Print.

Morrison, David, and Jane Samz. *Voyage to Jupiter.* NASA SP-439. Washington, D.C.: Government Printing Office, 1980.

Morrison, David, and Tobias Owen. *The Planetary System.* 3d ed. San Francisco: Addison Wesley, 2003.

Morrison, David. *Voyages to Saturn.* NASA SP-451. Washington, D.C.: National Aeronautics and Space Administration, 1982.

Moskvitch, Katia. "Atomic Clocks to Become Even More Accurate." *LiveScience.* Purch, 14 June 2013. Web. 25 Feb. 2015.

Moss, Ralph W. *Free Radical: Albert Szent-Györgyi and the Battle over Vitamin C.* New York: Paragon House, 1988.

Muirden, James. *The Amateur Astronomer's Handbook.* 3d ed. New York: Harper & Row, 1987.

Muller, Richard S., and Theodore I. Kamins, with Mansun Chan. *Device Electronics for Integrated Circuits.* 3d ed. New York: John Wiley & Sons, 2003. Very clear description of materials, manufacture, and many types of transistors.

Mullis, Kary. *Dancing Naked in the Mind Field.* New York: Pantheon Books, 1998.

Mulvihill, John J. *Catalog of Human Cancer Genes: McKusick's Mendelian Inheritance in Man for Clinical and Research Oncologists.* Foreword by Victor A. McKusick. Baltimore: The Johns Hopkins University Press, 1999.

Murphy, John. *The Telephone: Wiring America.* New York: Chelsea House, 2009.

Nahin, Paul J. *Oliver Heaviside: Sage in Solitude.* New York: IEEE Press, 1987.

Ne'eman, Yuval, and Yoram Kirsh. *The Particle Hunters.* New York: Cambridge University Press, 1986.

Netz, Reviel. *The Shaping of Deduction in Greek Mathematics: A Study in Cognitive History.* New York: Cambridge University Press, 2003.

Neu, Jerome, ed. *The Cambridge Companion to Freud.* New York: Cambridge University Press, 1991.

North, John. *The Norton History of Astronomy and Cosmology.* New York: W. W. Norton, 1995.

Norton, Bill. *STOL Progenitors: The Technology Path to a Large STOL Aircraft and the C-17A.* Reston, Va.: American Institute of Aeronautics and Astronautics, 2002. Case study on a short takeoff and landing aircraft development program. Describes the various steps in a modern context, considering cost, technology, and military requirements.

Nutton, Vivian, ed. *The Unknown Galen.* London: Institute of Classical Studies, University of London, 2002.

Nye, Mary Jo. *Before Big Science: The Pursuit of Modern Chemistry and Physics, 1800-1940.* New York: Twayne, 1996.

Nye, Robert D. *Three Psychologies: Perspectives from Freud, Skinner, and Rogers.* Pacific Grove, Calif.: Brooks-Cole, 1992.

Oakes, Elizabeth H. *Encyclopedia of World Scientists.* Rev ed. New York: Facts on File, 2007.

Olson, James S., and Robert L. Shadle, ed. *Encyclopedia of the Industrial Revolution in America.* Westport, Conn.: Greenwood Press, 2002.

Olson, Steve. *Mapping Human History: Genes, Races and Our Common Origins.* New York: Houghton Mifflin, 2002.

Ost, Laura. "A New Era for Atomic Clocks." *NIST Physical Measurement Laboratory.* NIST, 4 Feb. 2014. Web. 25 Feb. 2015.

Overton, Michael L. *Numerical Computing with IEEE Floating Point Arithmetic.* Philadelphia: Soc. for Industrial and Applied Mathematics, 2001. Print.

Ozima, Minoru. *The Earth: Its Birth and Growth.* Translated by Judy Wakabayashi. Cambridge, England: Cambridge University Press, 1981.

Pagels, Heinz R. *Perfect Symmetry: The Search for the Beginning of Time.* New York: Simon & Schuster, 1985.

_____. *The Cosmic Code.* New York: Simon & Schuster, 1982.

_____. *The Cosmic Code: Quantum Physics As the Law of Nature.* New York: Bantam Books, 1984.

Pai, Anna C. *Foundations for Genetics: A Science for Society.* 2d ed. New York: McGraw-Hill, 1984.

Painter, Paul C., and Michael M. Coleman. *Essentials of Polymer Science and Engineering.* Lancaster, Pa.: DEStech Publications, 2009. This extensively and beautifully illustrated book is intended for newcomers to the polymer field. Recommended readings at the ends of chapters.

Pais, Abraham. *The Genius of Science: A Portrait Gallery of Twentieth-Century Physicists.* New York: Oxford University Press, 2000.

Palmer, Douglas. *Neanderthal.* London: Channel 4 Books, 2000.

Parker, Barry R. *The Vindication of the Big Bang: Breakthroughs and Barriers.* New York: Plenum Press, 1993.

Parslow, Christopher Charles. *Rediscovering Antiquity: Karl Weber and the Excavation of Herculaneum, Pompeii, and Stabiae.* New York: Cambridge University Press, 1998.

Parson, Ann B. *The Proteus Effect: Stem Cells and Their*

Promise. Washington, D.C.: National Academies Press, 2004.

Peat, F. Davis. "Photography and Science: Conspirators." *Photography's Multiple Roles: Art, Document, Market, Science,* ed. by Terry A. Neff. Chicago: Museum of Contemporary Photography, 1998. Print.

Pedrotti, Frank L., Leno M. Pedrotti, and Leno S. Pedrotti. 3d ed. *Introduction to Optics.* 3rd ed. Upper Saddle River: Prentice, 2007. Print.

Pedrotti, L., and F. Pedrotti. *Optics and Vision.* Upper Saddle River, N.J.: Prentice Hall, 1998.

Peebles, C. *Road to Mach 10: Lessons Learned from the X-43A Flight Research Program.* Reston, Va.: American Institute of Aeronautics and Astronautics, 2008. Case study of a supersonic combustion demonstrator flight program authored by a historian.

Peitgen, Heinz-Otto, and Dietmar Saupe, eds. *The Science of Fractal Images.* New York: Springer-Verlag, 1988.

Peltonen, Markku, ed. *The Cambridge Companion to Bacon.* New York: Cambridge University Press, 1996.

Penrose, Roger. *The Emperor's New Mind: Concerning Computers, Minds, and the Laws of Physics.* New York: Oxford University Press, 1989.

Persaud, T. V. N. *A History of Anatomy: The Post-Vesalian Era.* Springfield, Ill.: Charles C. Thomas, 1997.

Peterson, Carolyn C., and John C. Brant. *Hubble Vision: Astronomy with the Hubble Space Telescope.* London: Cambridge University Press, 1995.

_____.*Hubble Vision: Further Adventures with the Hubble Space Telescope.* 2d ed. New York: Cambridge University Press, 1998.

Pfeiffer, John E. *The Emergence of Humankind.* 4th ed. New York: Harper & Row, 1985.

Piggott, Stuart. *William Stukeley: An Eighteenth-Century Antiquary.* New York: Thames and Hudson, 1985.

Pike, J. Wesley, Francis H. Glorieux, David Feldman. *Vitamin D.* 2d ed. Academic Press, 2004.

Plionis, Manolis, ed. *Multiwavelength Cosmology.* New York: Springer, 2004.

Plotkin, Stanley A., and Edward A. Mortimer. *Vaccines.* 2d ed. Philadelphia: W. B. Saunders, 1994.

Polter, Paul. *Hippocrates.* Cambridge, Mass.: Harvard University Press, 1995.

Poole, Ian. *Cellular Communications Explained: From Basics to 3G.* Oxford, England: Newnes, 2006.

Popper, Karl R. *The World of Parmenides: Essays on the Presocratic Enlightenment.* Edited by Arne F. Petersen

and Jørgen Mejer. New York: Routledge, 1998.

Porter, Roy, ed. *Eighteenth Century Science.* Vol. 4 in *The Cambridge History of Science.* New York: Cambridge University Press, 2003.

Porter, Roy. *The Greatest Benefit to Mankind: A Medical History of Humanity, from Antiquity to the Present.* New York: W. W. Norton, 1997.

Poundstone, William. *Prisoner's Dilemma.* New York: Doubleday, 1992.

Poynter, Margaret, and Arthur L. Lane. *Voyager: The Story of a Space Mission.* New York: Macmillan, 1981.

Principe, Lawrence. *The Aspiring Adept: Robert Boyle and His Alchemical Quest.* Princeton, N.J.: Princeton University Press, 1998.

Prochnow, Dave. *Superconductivity: Experimenting in a New Technology.* Blue Ridge Summit, Pa.: TAB Books, 1989.

Publications and Graphics Department-NASA. *Spinoff: Fifty Years of NASA-Derived Technologies, 1958–2008.* Washington, DC: NASA Center for Aero-Space Information, 2008. Print.

Pugh, Philip. *The Science and Art of Using Telescopes.* New York: Springer, 2009. Print.

Pullman, Bernard. *The Atom in the History of Human Thought.* New York: Oxford University Press, 1998.

Pycior, Helena M. *Symbols, Impossible Numbers, and Geometric Entanglements: British Algebra Through the Commentaries on Newton's Universal Arithmetick.* New York: Cambridge University Press, 1997.

Ralston, Anthony, and Philip Rabinowitz. *A First Course in Numerical Analysis.* 2nd ed. Mineola: Dover, 2001. Print.

Ramsden, Jeremy. *Nanotechnology.* Elsevier, 2016.

Rao, Mahendra S., ed. *Stem Cells and CNS Development.* Totowa, N.J.: Humana Press, 2001.

Ratner, Daniel, and Mark A. Ratner. *Nanotechnology and Homeland Security: New Weapons for New Wars.* Upper Saddle River: Prentice, 2004. Print.

Ratner, Mark A., and Daniel Ratner. *Nanotechnology: A Gentle Introduction to the Next Big Idea.* Upper Saddle River: Prentice, 2003. Print.

Rauber, Thomas, and Gudula Rünger. *Parallel Programming: For Multicore and Cluster Systems.* New York: Springer, 2010. Print.

Raup, David M. *The Nemesis Affair: A Story of the Death of Dinosaurs and the Ways of Science.* New York: W. W. Norton, 1986.

Raven, Peter H., and George B. Johnson. *Biology.* 2d ed. St. Louis: Times-Mirror/Mosby, 1989.

Raven, Peter H., Ray F. Evert, and Susan E. Eichhorn. *Biology of Plants*. 6th ed. New York: W. H. Freeman, 1999.

Raymond, Martin, and William Leffler. *Oil and Gas Production in Nontechnical Language*. Tulsa, Okla.: Penn Well, 2005. An accessible introduction that covers all major aspects. Lighthearted but very informative style.

Reader, John. *Missing Links: The Hunt for Earliest Man*. Boston: Little, Brown, 1981.

Reichhardt, Tony. *Proving the Space Transportation System: The Orbital Flight Test Program*. NASA NF-137-83. Washington, D.C.: Government Printing Office, 1983.

Reid, T. R. *The Chip: How Two Americans Invented the Microchip and Launched a Revolution*. Rev. ed. New York: Simon & Schuster, 2001. The very readable story about Jack Kilby of Texas Instruments and Robert Noyce of Fairchild Semiconductor and later Intel, who independently came up with different approaches to the invention of the microchip.

Remick, Pat, and Frank Cook. *21 Things Every Future Engineer Should Know: A Practical Guide for Students and Parents*. Chicago: Kaplan AEC Education, 2007.

Repcheck, Jack. *The Man Who Found Time: James Hutton and the Discovery of the Earth's Antiquity*. Reading, Mass.: Perseus Books, 2003.

Rescher, Nicholas. *On Leibniz*. Pittsburgh, Pa.: University of Pittsburgh Press, 2003.

Reston, James. *Galileo: A Life*. New York: HarperCollins, 1994.

Rhodes, Richard. *The Making of the Atomic Bomb*. New York: Simon & Schuster, 1986.

Richmond, J. C. "Relation of Emittance to Other Optical Properties." *Journal of Research of the National Bureau of Standards* 67C, no. 3 (1963): 217–226. Regarded as a pioneering piece of work in developing optical measurement techniques for temperature based on the emittance of various materials and media.

Rigutti, Mario. *A Hundred Billion Stars*. Translated by Mirella Giacconi. Cambridge, Mass.: MIT Press, 1984.

Ring, Merrill. *Beginning with the Presocratics*. 2d ed. New York: McGraw-Hill, 1999.

Riordan, Michael. *The Hunting of the Quark*. New York: Simon & Schuster, 1987.

Rivière, John C., and S. Myhra, eds. *Handbook of Surface and Interface Analysis: Methods for Problem-Solving*. 2d ed. Boca Raton, Fla.: CRC Press, 2009. Discusses common surface analysis methods and problems; chapter 7 deals with synchrotron-based techniques.

Roan, Sharon. *Ozone Crisis: The Fifteen-Year Evolution of a Sudden Global Emergency*. New York: John Wiley & Sons, 1989.

Robertson, William C. *Light*. Arlington: Natl. Science Teachers Assn., 2003. Print.

Robinson, Daniel N. *An Intellectual History of Psychology*. 3d ed. Madison: University of Wisconsin Press, 1995.

Robinson, Keith. *Spectroscopy: The Key to the Stars—Reading the Lines in Stellar Spectra*. New York: Springer, 2007. Print.

Rogers, Alan. *Understanding Optical Fiber Communications*. Norwood: Artech, 2001. Print.

Rogers, Ben, Jesse Adams, and Sumita Pennathur. *Nanotechnology: The Whole Story*. Boca Raton: CRC, 2013. Print.

Rogers, Ben, Sumita Pennathur, and Jesse Adams. *Nanotechnology: Understanding Small Systems*. 2nd ed. Boca Raton: CRC, 2011. Print.

Rogers, J. H. *The Giant Planet Jupiter*. New York: Cambridge University Press, 1995.

Rose, Frank. *West of Eden: The End of Innocence at Apple Computer*. New York: Viking, 1989.

Rosenblum, Naomi. *A World History of Photography*. New York: Abbeville, 2007. Print.

Rosenthal-Schneider, Ilse. *Reality and Scientific Truth: Discussions with Einstein, von Laue, and Planck*. Detroit: Wayne State University Press, 1980.

Rossi, Paoli. *The Birth of Modern Science*. Translated by Cynthia De Nardi Ipsen. Oxford, England: Blackwell, 2001.

Rowan-Robinson, Michael. *Cosmology*. London: Oxford University Press, 2003.

Rowland, Wade. *Galileo's Mistake: A New Look at the Epic Confrontation Between Galileo and the Church*. New York: Arcade, 2003.

Rudel, Anthony. *Hello, Everybody: The Dawn of American Radio*. Orlando: Houghton Mifflin Harcourt, 2008. Print.

Rudin, Norah, and Keith Inman. *An Introduction to Forensic DNA Analysis*. Boca Raton, Fla.: CRC Press, 2002.

Rudwick, M. J. S. *The Meaning of Fossils: Episodes in the*

History of Paleontology. Chicago: University of Chicago Press, 1985.

Rudwick, Martin, J. S. *The Great Devonian Controversy: The Shaping of Scientific Knowledge Among Gentlemanly Specialists.* Chicago: University of Chicago Press, 1985.

Ruestow, Edward Grant. *The Microscope in the Dutch Republic: The Shaping of Discovery.* New York: Cambridge University Press, 1996.

Ruse, Michael. *The Darwinian Revolution: Science Red in Tooth and Claw.* 2d ed. Chicago: University of Chicago Press, 1999.

Ruspoli, Mario. *Cave of Lascaux.* New York: Harry N. Abrams, 1987.

Sagan, Carl, and Ann Druyan. *Comet.* New York: Random House, 1985.

Sagan, Carl. *Cosmos.* New York: Random House, 1980.

Sandler, Stanley I. *Chemical and Engineering Thermodynamics.* New York: John Wiley & Sons, 1998.

Sang, James H. *Genetics and Development.* London: Longman, 1984.

Sargent, Frederick. *Hippocratic Heritage: A History of Ideas About Weather and Human Health.* New York: Pergamon Press, 1982.

Sargent, Rose-Mary. *The Diffident Naturalist: Robert Boyle and the Philosophy of Experiment.* Chicago: University of Chicago Press, 1995.

Sauer, Mark V. *Principles of Oocyte and Embryo Donation.* New York: Springer, 1998.

Schaaf, Fred. *Comet of the Century: From Halley to Hale-Bopp.* New York: Springer-Verlag, 1997.

Schatzkin, Paul. *The Boy Who Invented Television: A Story of Inspiration, Persistence, and Quiet Passion.* Silver Spring, Md.: TeamCon Books, 2002.

Schiffer, Michael Brian. *Draw the Lightning Down: Benjamin Franklin and Electrical Technology in the Age of the Enlightenment.* Berkeley: University of California Press, 2003.

Schilling, Govert, and Lars Lindberg Christensen. *Eyes on the Skies: Four Hundred Years of Telescopic Discovery.* Chichester: Wiley, 2009. Print.

Schlagel, Richard H. *From Myth to Modern Mind: A Study of the Origins and Growth of Scientific Thought.* New York: Peter Lang Publishing, 1996.

Schlegel, Eric M. *The Restless Universe: Understanding X-Ray Astronomy in the Age of Chandra and Newton.* London: Oxford University Press, 2002.

Schliemann, Heinrich. *Troy and Its Remains: A Narrative of Researches and Discoveries Made on the Site of Ilium and in the Trojan Plain.* London: J. Murray, 1875.

Schmitt, Harrison H. *Return to the Moon: Exploration, Enterprise, and Energy in the Human Settlement of Space.* New York: Copernicus, 2006. Print.

Schneider, Stephen H. *Global Warming: Are We Entering the Greenhouse Century?* San Francisco: Sierra Club Books, 1989.

Schofield, Robert E. *The Enlightened Joseph Priestley: A Study of His Life and Work from 1773 to 1804.* University Park: Pennsylvania State University Pres, 2004.

Scholz, Christopher, and Benoit B. Mandelbrot. *Fractals in Geophysics.* Boston: Kirkäuser, 1989.

Schonfelder, V. *The Universe in Gamma Rays.* New York: Springer-Verlag, 2001.

Schopf, J. William, ed. *The Earth's Earliest Biosphere.* Princeton, N.J.: Princeton University Press, 1983.

Schorn, Ronald A. *Planetary Astronomy: From Ancient Times to the Third Millennium.* College Station: Texas A&M University Press, 1999.

Schwinger, Julian. *Einstein's Legacy.* New York: W. H. Freeman, 1986.

Scott, Gerald. *Polymers and the Environment.* Cambridge, England: Royal Society of Chemistry, 1999. From an environmental viewpoint, this book introduces the general reader to the benefits and limitations of polymeric materials as compared with traditional materials.

Sears, M., and D. Merriman, eds. *Oceanography: The Past.* New York: Springer-Verlag, 1980.

Seavey, Nina Gilden, Jane S. Smith, and Paul Wagner. *A Paralyzing Fear: The Triumph over Polio in America.* New York: TV Books, 1998.

Seedhouse, Erik. *Tourists in Space: A Practical Guide.* New York: Springer, 2008. Print.

Segalowitz, Sid J. *Two Sides of the Brain: Brain Lateralization Explored.* Englewood Cliffs, N.J.: Prentice-Hall, 1983.

Segar, Douglas A. *Introduction to Ocean Sciences.* 2d ed. New York: W. W. Norton, 2007. A well-respected textbook with solid foundational information.

Segrè, Emilio. *From X-Rays to Quarks.* San Francisco: W. H. Freeman, 1980.

Sekido, Yataro, and Harry Elliot. *Early History of Cosmic Ray Studies: Personal Reminiscences with Old Photographs.* Boston: D. Reidel, 1985.

Seymour, Raymond B., ed. *Pioneers in Polymer Science: Chemists and Chemistry.* Boston: Kluwer Academic Publishers, 1989. This survey of the history of polymer science emphasizes the scientists responsible for the innovations. Several of the chapters were written by scientists who were directly involved in these developments.

Sfendoni-Mentzou, Demetra, et al., eds. *Aristotle and Contemporary Science*. 2 vols. New York: P. Lang, 2000-2001.

Shackett, Peter. *Nuclear Medicine Technology: Procedures and Quick Reference*. 2nd ed. Philadelphia: Lippincott, 2008. Print.

Shank, Michael H. *The Scientific Enterprise in Antiquity and the Middle Ages*. Chicago: University of Chicago Press, 2000.

Sharratt, Michael. *Galileo: Decisive Innovator*. Cambridge, Mass.: Blackwell, 1994.

Shectman, Jonathan. *Groundbreaking Scientific Experiments, Investigations, and Discoveries of the Eighteenth Century*. Westport, Conn.: Greenwood Press, 2003.

Sheehan, John. *The Enchanted Ring: The Untold Story of Penicillin*. Cambridge, Mass.: MIT Press, 1982.

Shepherd, D. *Aerospace Propulsion*. New York: Elsevier, 1972. A prescient, simple, and lucid book that has inspired generations of aerospace scientists and engineers.

Shilts, Randy. *And the Band Played On: Politics, People, and the AIDS Epidemic*. New York: St. Martin's Press, 1987.

Shostak, Seth. "Listening for a Whisper." *Astronomy* 32, no. 9 (September, 2004): 34–39. A very brief overview of how radio telescopes are used in the search for extraterrestrial intelligence.

Siciliano, Antonio. *Optics: Problems and Solutions*. Singapore: World Scientific, 2006. Print.

Silk, Joseph. *The Big Bang*. Rev. ed. New York: W. H. Freeman, 1989.

Silver, H. Ward. *ARRL Ham Radio License Manual: All You Need to Become an Amateur Radio Operator*. Newington: American Radio Relay League, 2010. Print.

Silverstein, Arthur M. *A History of Immunology*. San Diego: Academic Press, 1989.

Simmons, John. *The Scientific Hundred: A Ranking of the Most Influential Scientists, Past and Present*. Secaucus, N.J.: Carol, 1996.

Simon, Randy, and Andrew Smith. *Superconductors: Conquering Technology's New Frontier*. New York: Plenum Press, 1988.

Simpson, A. D. C., ed. *Joseph Black, 1728-1799: A Commemorative Symposium*. Edinburgh: Royal Scottish Museum, 1982.

Singer, Peter Warren. *Wired for War: The Robotics Revolution and Conflict in the Twenty-first Century*. New York: Penguin Press, 2009. An exploration of the use of intelligent machines in warfare and defense. Contains information on a variety of military applications for pattern recognition systems.

Singh, Simon. *Fermat's Enigma: The Epic Quest to Solve the World's Greatest Mathematical Problem*. New York: Anchor, 1998.

Slayton, Donald K., with Michael Cassutt. *Deke! U.S. Manned Space: From Mercury to the Shuttle*. New York: Forge, 1995.

Smith, A. Mark. *Ptolemy and the Foundations of Ancient Mathematical Optics*. Philadelphia: American Philosophical Society, 1999.

Smith, G. C. *The Boole-De Morgan Correspondence, 1842-1864*. New York: Oxford University Press, 1982.

Smith, Jane S. *Patenting the Sun: Polio and the Salk Vaccine*. New York: Anchor/Doubleday, 1991.

Smith, Robert W. *The Space Telescope: A Study of NASA, Science, Technology and Politics*. New York: Cambridge University Press, 1989.

Smyth, Albert Leslie. *John Dalton, 1766-1844*. Aldershot, England: Ashgate, 1998.

Snider, Alvin. *Origin and Authority in Seventeenth-Century England: Bacon, Milton, Butler*. Toronto: University of Toronto Press, 1994.

Sobel, Dava. *Galileo's Daughter: A Historical Memoir of Science, Faith, and Love*. New York: Penguin Books, 2000.

_____. *Longitude: The True Story of a Lone Genius Who Solved the Greatest Scientific Problem of His Time*. New York: Penguin Books, 1995.

Spangenburg, Ray, and Diane Kit Moser. *Modern Science: 1896-1945*. Rev ed. New York: Facts on File, 2004.

Sparrow, Giles. *The Planets: A Journey Through the Solar System*. London: Quercus, 2006. Print.

Spilker, Linda J., ed. *Passage to a Ringed World: The Cassini-Huygens Mission to Saturn and Titan*. Washington, D.C.: National Aeronautics and Space Administration, 1997.

Stanculescu, A., ed. *The Role of Nuclear Power and Nuclear Propulsion in the Peaceful Exploration of Space*. Vienna: Intl. Atomic Energy Agency, 2005. Print.

Stanley, H. Eugue, and Nicole Ostrowsky, eds. *On Growth and Form: Fractal and Non-Fractal Patterns in Physics*. Dordrecht, the Netherlands: Martinus

Nijhoff, 1986.

Starr, Cecie, and Ralph Taggart. *Biology*. 5th ed. Belmont, Calif.: Wadsworth, 1989.

Stefik, Mark J., and Vinton Cerf. *Internet Dreams: Archetypes, Myths, and Metaphors*. Cambridge, Mass.: MIT Press, 1997.

Stein, Sherman. *Archimedes: What Did He Do Besides Cry Eureka?* Washington, D.C.: Mathematical Association of America, 1999.

Steiner, Robert F., and Seymour Pomerantz. *The Chemistry of Living Systems*. New York: D. Van Nostrand, 1981.

Stewart, Ian, and David Tall. *Algebraic Number Theory and Fermat's Last Theorem*. 3d ed. Natick, Mass.: AK Peters, 2002.

Stigler, Stephen M. *The History of Statistics*. Cambridge, Mass.: Harvard University Press, 1986.

Stine, Gerald. *AIDS 2005 Update*. New York: Benjamin Cummings, 2005.

Stine, Keith J. *Carbohydrate Nanotechnology*. Wiley, 2016.

Strathern, Paul. *Mendeleyev's Dream: The Quest for the Elements*. New York: Berkeley Books, 2000.

Strauss, Walter A. *Partial Differential Equations: An Introduction*. 2nd ed. Hoboken: Wiley, 2008. Print.

Streissguth, Thomas. *John Glenn*. Minneapolis, Minn.: Lerner, 1999.

Strick, James. *Sparks of Life: Darwinism and the Victorian Debates over Spontaneous Generation*. Cambridge, Mass.: Harvard University Press, 2000.

Strogatz, Steven H. *Nonlinear Dynamics and Chaos: With Applications to Physics, Biology, Chemistry and Engineering*. Reading, Mass.: Perseus, 2001.

Struik, Dirk J. *The Land of Stevin and Huygens: A Sketch of Science and Technology in the Dutch Republic During the Golden Century*. Boston: Kluwer, 1981.

Stryer, Lubert. *Biochemistry*. 2d ed. San Francisco: W. H. Freeman, 1981.

Stukeley, William. *The Commentarys, Diary, & Commonplace Book & Selected Letters of William Stukeley*. London: Doppler Press, 1980.

Sturtevant, A. H. *A History of Genetics*. 1965. Reprint. Woodbury, N.Y.: Cold Spring Harbor Laboratory Press, 2001.

Sullivan, Woodruff T., ed. *Classics in Radio Astronomy*. Boston: D. Reidel, 1982.

_____. *The Early Years of Radio Astronomy. Reflections Fifty Years After Jansky's Discovery*. New York: Cambridge University Press, 1984.

_____. *Cosmic Noise: A History of Early Radio Astronomy*. New York: Cambridge University Press, 2009. A very thorough historical record of the first decades of radio astronomy, the people and instruments involved, and the discoveries made.

Sulston, John, and Georgina Ferry. *The Common Thread: A Story of Science, Politics, Ethics, and the Human Genome*. Washington, D.C.: Joseph Henry Press, 2002.

Suplee, Curt. *Physics in the Twentieth Century*. Edited by Judy R. Franz and John S. Rigden. New York: Harry N. Abrams, in association with the American Physical Society and the American Institute of Physics, 1999. A well-written and illustrated book prepared on the occasion of the centennial of the American Physical Society; the story of the transistor is found in Chapter 4.

Sutton, Christine. *The Particle Connection*. New York: Simon & Schuster, 1984.

Suzuki, David T., and Peter Knudtson. *Genethics*. Cambridge, Mass.: Harvard University Press, 1989.

Sverdrup, Keith A., and E. Virginia Armbrust. *An Introduction to the World's Oceans*. 10th ed. New York: McGraw-Hill, 2009. Provides information on how energy is now being obtained, or may someday be obtained, from ocean waves, currents, and tides. Many useful diagrams.

Swanson, Carl P., Timothy Merz, and William J. Young. *Cytogenetics: The Chromosome in Division, Inheritance, and Evolution*. 2d ed. Englewood Cliffs, N.J.: Prentice-Hall, 1980.

Swetz, Frank, et al., eds. *Learn from the Masters*. Washington, D.C.: Mathematical Association of America, 1995.

Swinerd, Graham. *How Spacecraft Fly: Spaceflight without Formulae*. New York: Copernicus, 2008. Print.

Sze, S. M., and Kwok K. Ng. *Physics of Semiconductor Devices*. 3d ed. Hoboken, N.J.: John Wiley & Sons, 2007. Written by physicists at Bell Labs, this textbook is often used in college courses.

Talcott, Richard. "How We'll Explore Pluto." *Astronomy* 38.7 (2010): 24–29. Print.

Tanford, Charles. *Franklin Stilled the Waves*. Durham, N.C.: Duke University Press, 1989.

Tarbuck, Edward J., and Frederick K. Lutgens. *The Earth: An Introduction to Physical Geology*. Columbus, Ohio: Charles E. Merrill, 1984.

Tattersall, Ian. *The Last Neanderthal: The Rise, Success,*

and Mysterious Extinction of Our Closest Human Relatives. New York: Macmillan, 1995.

Taub, Liba Chaia. *Ptolemy's Universe: The Natural Philosophical and Ethical Foundations of Ptolemy's Astronomy.* Chicago: Open Court, 1993.

Tauber, Alfred I. *Metchnikoff and the Origins of Immunology: From Metaphor to Theory.* New York: Oxford University Press, 1991.

Taubes, Gary. *Nobel Dreams: Power, Deceit and the Ultimate Experiment.* New York: Random House, 1986.

Taylor, Michael E. *Partial Differential Equations I: Basic Theory.* 2d ed. New York: Springer, 2011.

Taylor, Peter Lane. *Science at the Extreme: Scientists on the Cutting Edge of Discovery.* New York: McGraw-Hill, 2001.

The Rand McNally New Concise Atlas of the Universe. New York: Rand McNally, 1989.

Theodoridis, Sergio, and Konstantinos Koutroumbas. *Pattern Recognition.* 4th ed. Boston: Academic Press, 2009. A complex introduction to the field, with information about pattern recognition algorithms, neural networks, logical systems, and statistical analysis.

Thirsk, Robert, et al. "The Space Flight Environment: The International Space Station and Beyond." *Canadian Medical Association Journal* 180.12 (June, 2009): 1216–1220. Print.

Thomas, Ann. *Beauty of Another Order: Photography in Science.* New Haven: Yale U P, 1997. Print.

Thomas, John M. *Michael Faraday and the Royal Institution: The Genius of Man and Place.* New York: A. Hilger, 1991.

Thompson, A. R., James M. Moran, and George W. Swenson, Jr. *Interferometry and Synthesis in Radio Astronomy.* New York: John Wiley & Sons, 1986.

Thompson, D'Arcy Wentworth. *On Growth and Form.* Mineola, N.Y.: Dover, 1992.

Thoren, Victor E., with John R. Christianson. *The Lord of Uraniborg: A Biography of Tycho Brahe.* New York: Cambridge University Press, 1990.

Thresher, R., and W. Musial. "Ocean Renewable Energy's Potential Role in Supplying Future Electrical Needs." *Oceanography* 23, no. 2 (June, 2010): 16–21. A summary of the nation's and the world's energy needs and the sources from which these energy needs have been met and will be met in the future.

Thrower, Norman J. W., ed. *Standing on the Shoulders of Giants: A Longer View of Newton and Halley.*

Berkeley: University of California Press, 1990.

Thurman, Harold V. *Introductory Oceanography.* 4th ed. Westerville, Ohio: Charles E. Merrill, 1985.

Tietjen, Jill S., et al. *Keys to Engineering Success.* Upper Saddle River, N.J.: Prentice-Hall, 2001.

Tillery, Bill W., Eldon D. Enger, and Frederick C. Ross. *Integrated Science.* New York: McGraw-Hill, 2001.

Tiner, John Hudson. *Louis Pasteur: Founder of Modern Medicine.* Milford, Mich.: Mott Media, 1990.

Tipler, Paul A., and Gene Mosca. *Physics for Scientists and Engineers.* 6th ed. New York: Freeman, 2008. Print.

Todhunter, Isaac. *A History of the Mathematical Theory of Probability: From the Time of Pascal to that of Laplace.* Sterling, Va.: Thoemmes Press, 2001.

Tombaugh, Clyde W., and Patrick Moore. *Out of Darkness: The Planet Pluto.* Harrisburg, Pa.: Stackpole Books, 1980.

Toulmin, Stephen, and June Goodfield. *The Fabric of the Heavens: The Development of Astronomy and Dynamics.* Chicago: University of Chicago Press, 1999.

Townes, Charles H. *How the Laser Happened: Adventures of a Scientist.* New York: Oxford University Press, 1999.

Traill, David A. *Schliemann of Troy: Treasure and Deceit.* London: J. Murray, 1995.

Trefil, James S. *From Atoms to Quarks: An Introduction to the Strange World of Particle Physics.* New York: Charles Scribner's Sons, 1980.

_____. *Space, Time, Infinity: The Smithsonian Views the Universe.* New York: Pantheon Books, 1985.

Trefil, James S. *The Dark Side of the Universe. Searching for the Outer Limits of the Cosmos.* New York: Charles Scribner's Sons, 1988.

_____. *The Unexpected Vista.* New York: Charles Scribner's Sons, 1983.

Trefil, James, and Robert M. Hazen. *The Sciences: An Integrated Approach.* New York: John Wiley & Sons, 2003.

Trefil, James, ed. *The Encyclopedia of Science and Technology.* New York: Routledge, 2001.

Trento, Joseph J. *Prescription for Disaster: From the Glory of Apollo to the Betrayal of the Shuttle.* New York: Crown, 1987.

Trinkhaus, Eric, ed. *The Emergence of Modern Humans: Biocultural Adaptations in the Later Pleistocene.* Cambridge, England: Cambridge University Press, 1989.

Trounson, Alan O., and David K. Gardner, eds. *Handbook of In Vitro Fertilization*. 2d ed. Boca Raton, Fla.: CRC Press, 1999.

Tucker, Tom. *Bolt of Fire: Benjamin Franklin and His Electrical Kite Hoax*. New York: Public Affairs Press, 2003.

Tucker, Wallace H., and Karen Tucker. *Revealing the Universe: The Making of the Chandra X-Ray Observatory*. Cambridge, Mass.: Harvard University Press, 2001.

Tunbridge, Paul. *Lord Kelvin: His Influence on Electrical Measurements and Units*. London, U.K.: P. Peregrinus, 1992.

Tuplin, C. J., and T. E. Rihll, eds. *Science and Mathematics in Ancient Greek Culture*. New York: Oxford University Press, 2002.

Turnill, Reginald. *The Moonlandings: An Eyewitness Account*. New York: Cambridge University Press, 2003.

Ulanski, Stan. *Gulf Stream: Tiny Plankton, Giant Bluefin, and the Amazing Story of the Powerful River in the Atlantic*. Chapel Hill: University of North Carolina Press, 2010. Describes the world's fastest ocean current, summarizing the physical characteristics of location and speed that make this current a prime candidate for energy generation.

United States Congress, House Committee on Foreign Affairs, Subcommittee on Terrorism, Nonproliferation, and Trade. *Isolating Proliferators, and Sponsors of Terror: The Use of Sanctions and the International Financial System to Change Regime Behavior*. Washington, DC: Government Printing Office, 2007. Print.

United States Office of the Assistant Secretary for Nuclear Energy. *The First Reactor*. Springfield, Va.: National Technical Information Service, 1982.

University of Chicago Press. *Science and Technology Encyclopedia*. Chicago: Author, 2000.

Van Allen, James A. *Origins of Magnetospheric Physics*. Expanded ed. 1983. Reprint. Washington, D.C.: Smithsonian Institution Press, 2004.

Van Dulken, Stephen. *Inventing the Nineteenth Century: One Hundred Inventions That Shaped the Victorian Age*. New York: New York University Press, 2001.

Van Heijenoort, Jean. *From Frege to Gödel: A Source Book in Mathematical Logic, 1879-1931*. Cambridge, Mass.: Harvard University Press, 2002.

Varoglu, Sevin, and Stephen Jenks. "Architectural support for thread communications in multi-core processors." *Parallel Computing* 37.1 (2011): 26–41. Print.

Verschuur, Gerrit L. *Hidden Attraction: The History and Mystery of Magnetism*. New York: Oxford University Press, 1993.

_____. *The Invisible Universe: The Story of Radio Astronomy*. 2d ed. New York: Springer, 2007. An excellent introduction and overview of the history of radio astronomy and discoveries made by radio astronomers.

Vickerman, John C., ed. *Surface Analysis: The Principal Techniques*. 2d ed. Chichester, England: John Wiley & Sons, 2009. Describes the techniques involved in surface analysis. The introduction contains a basic overview of surface analysis that is easy to read and informative.

Villard, Ray, and Lynette R. Cook. *Infinite Worlds: An Illustrated Voyage to Planets Beyond Our Sun*. Foreword by Geoffrey W. Marcy and afterword by Frank Drake. Berkeley: University of California Press, 2005.

Viney, Wayne. *A History of Psychology: Ideas and Context*. Boston: Allyn & Bacon, 1993.

Vogt, Gregory L. *John Glenn's Return to Space*. Brookfield, Conn.: Millbrook Press, 2000.

Von Bencke, Matthew J. *The Politics of Space: A History of U.S.-Soviet/Russian Competition and Cooperation in Space*. Boulder, Colo.: Westview Press, 1996.

Wagener, Leon. *One Giant Leap: Neil Armstrong's Stellar American Journey*. New York: Forge Books, 2004.

Wakefield, Robin, ed. *The First Philosophers: The Presocratics and the Sophists*. New York: Oxford University Press, 2000.

Waldman, G. *Introduction to Light*. Englewood Cliffs, N.J.: Prentice Hall, 1983.

Walker, James S. *Physics*. 2d ed. Upper Saddle River, N.J.: Pearson Prentice Hall, 2004.

Wallace, Robert A., Jack L. King, and Gerald P. Sanders. *Biosphere: The Realm of Life*. 2d ed. Glenview, Ill.: Scott, Foresman, 1988.

Waller, John. *Einstein's Luck: The Truth Behind Some of the Greatest Scientific Discoveries*. New York: Oxford University Press, 2002.

_____. *Fabulous Science: Fact and Fiction in the History of Science Discovery*. Oxford, England: Oxford University Press, 2004.

Walt, Martin. *Introduction to Geomagnetically Trapped Radiation*. New York: Cambridge University Press, 1994.

Wambaugh, Joseph. *The Blooding*. New York: Bantam Books, 1989.

Wang, Hao. *Reflections on Kurt Gödel*. Cambridge, Mass.: MIT Press, 1985.

Watson, James D. *The Double Helix: A Personal Account of the Discovery of the Structure of DNA*. Reprint. New York: W. W. Horton, 1980.

Watson, James D., and John Tooze. *The DNA Story*. San Francisco: W. H. Freeman, 1981.

Watson, James D., et al. *Molecular Biology of the Gene*. 4th ed. Menlo Park, Calif.: Benjamin/Cummings, 1987.

Weber, Robert L. *Pioneers of Science: Nobel Prize Winners in Physics*. 2d ed. Philadelphia: A. Hilger, 1988.

Weedman, Daniel W. *Quasar Astrophysics*. Cambridge, England: Cambridge University Press, 1986.

Wells, Spencer. *The Journey of Man: A Genetic Odyssey*. Princeton, N.J.: Princeton University Press, 2002.

Westfall, Richard S. *Never at Rest: A Biography of Isaac Newton*. New York: Cambridge University Press, 1980.

Wheeler, J. Craig. *Cosmic Catastrophe: Supernovae and Gamma-Ray Bursts*. London: Cambridge University Press, 2000.

Whiting, Jim, and Marylou Morano Kjelle. *John Dalton and the Atomic Theory*. Hockessin, Del.: Mitchell Lane, 2004.

Whitney, Charles. *Francis Bacon and Modernity*. New Haven, Conn.: Yale University Press, 1986.

Whyte, A. J. *The Planet Pluto*. New York: Pergamon Press, 1980.

Wilford, John Noble. *The Mapmakers*. New York: Alfred A. Knopf, 1981.

_____. *The Riddle of the Dinosaur*. New York: Alfred A. Knopf, 1986.

Wilkie, Tom, and Mark Rosselli. *Visions of Heaven: The Mysteries of the Universe Revealed by the Hubble Space Telescope*. London: Hodder & Stoughton, 1999.

Will, Clifford M. *Was Einstein Right?* New York: Basic Books, 1986.

Williams, F. Mary, and Carolyn J. Emerson. *Becoming Leaders: A Practical Handbook for Women in Engineering, Science, and Technology*. Reston, Va.: American Society of Civil Engineers, 2008.

Williams, Garnett P. *Chaos Theory Tamed*. Washington, D.C.: National Academies Press, 1997.

Williams, James Thaxter. *The History of Weather*. Commack, N.Y.: Nova Science, 1999.

Williams, Trevor I. *Howard Florey: Penicillin and After*.

London: Oxford University Press, 1984.

Wilmut, Ian, Keith Campbell, and Colin Tudge. *The Second Creation: The Age of Biological Control by the Scientists That Cloned Dolly*. London: Headline, 2000.

Wilson, Andrew. *Space Shuttle Story*. New York: Crescent Books, 1986.

Wilson, Colin. *Starseekers*. Garden City, N.Y.: Doubleday, 1980.

Wilson, David B. *Kelvin and Stokes: A Comparative Study in Victorian Physics*. Bristol, England: Adam Hilger, 1987.

Wilson, Jean D. *Wilson's Textbook of Endocrinology*. 10th ed. New York: Elsevier, 2003.

Windley, Brian F. *The Evolving Continents*. 2d ed. New York: John Wiley & Sons.

Wojcik, Jan W. *Robert Boyle and the Limits of Reason*. New York: Cambridge University Press, 1997.

Wolf, Fred Alan. *Taking the Quantum Leap*. San Francisco: Harper & Row, 1981.

Wolfe, William J. *Optics Made Clear: The Nature of Light and How We Use It*. Bellingham: SPIE, 2007. Print.

Wollinsky, Art. *The History of the Internet and the World Wide Web*. Berkeley Heights, N.J.: Enslow, 1999.

Wolpoff, M. *Paleoanthropology*. 2d ed. Boston: McGraw-Hill, 1999.

Wood, Michael. *In Search of the Trojan War*. Berkeley: University of California Press, 1988.

Wormald, B. H. G. *Francis Bacon: History, Politics, and Science, 1561-1626*. New York: Cambridge University Press, 1993.

Yang, Chi-Jen. *Belief-Based Energy Technology Development in the United States: A Comparative Study of Nuclear Power and Synthetic Fuel Policies*. Amherst: Cambria, 2009. Print.

Yen, W. M., Marc D. Levenson, and Arthur L. Schawlow. *Lasers, Spectroscopy, and New Ideas: A Tribute to Arthur L. Schawlow*. New York: Springer-Verlag, 1987.

Yeomans, Matthew. *Oil: Anatomy of an Industry*. New York: The New Press, 2004. A critical look at the oil industry, particularly in the United States. Provides a good historical overview and addresses issues of America's dependency on oil, world conflicts caused by oil, and the question of alternatives such as hydrogen fuel.

Yoder, Joella G. *Unrolling Time: Huygens and the Mathematization of Nature*. New York: Cambridge University Press, 2004.

Yolton, John W. ed. *Philosophy, Religion, and Science in*

the Seventeenth and Eighteenth Centuries. Rochester, N.Y.: University of Rochester Press, 1990.

Zeilik, Michael. *Astronomy: The Evolving Universe.* 4th ed. New York: Harper & Row, 1985.

Zirker, Jack B. *An Acre of Glass: A History and Forecast of the Telescope.* Baltimore: Johns Hopkins UP, 2006. Print.

Zubrin, Robert. *Entering Space: Creating a Spacefaring Civilization.* New York: Putnam, 2000. Print.

INDEX

A

Absolute zero 133, 134, 227, 228, 609, 610, 613, 619, 667, 711

Absorption techniques 588

Abundant metals and uses 489

Abutment 159

Achromatic lens 493

Acoustics and vibrations 468

Activated sludge process 352

Active galaxies 569

Active matrix liquid crystal display 441

Addition polymerization 551, 553

Adenosine triphosphate 127

Advanced ceramics 145, 146, 147, 148, 150

Aegis 33, 34, 35, 36, 37

Aerial photography 539

Aerodynamic lift 10, 11, 560

Aerodynamics 10, 11, 13, 15, 112, 138, 368, 369, 370, 556, 560

Aeroelasticity 13

Aeronautics and aviation v, 10, 14

Aerosol 17, 58, 60, 143, 175, 347, 485, 667

Aerospace industry applications 231

Aerospace structures 12

Agricultural and industrial ecology 349

Aircraft band 565

Aircraft testing 88

Airfoil 10, 13, 370, 693

Air masses 61, 178, 180, 480, 481, 482

Air pollutants 17, 18, 60, 62, 64, 102

Air pollution 17, 18, 19, 20, 21, 22, 58, 62, 64, 96, 175, 180, 334, 352, 353, 354, 485, 596, 653, 730

Air-pollution control 596, 730

Air quality index (AQI) 17

Air-quality monitoring 21

Airspeed indicator 78, 79, 80

Albedo 166

Algebra of propositions (logic or proposition calculus) 28

Algebra of sets 24, 28

Algorithm 8, 38, 53, 54, 105, 106, 107, 214, 233, 235, 236, 237, 238, 452, 453, 508, 509, 512

Algorithmic science 334, 335, 336

Alloys 5, 29, 121, 146, 250, 296, 308, 311, 314, 337, 458, 467, 472, 473, 475, 476, 477, 488, 705

Amateur radio 563, 564, 565, 568, 571, 572, 720

Ampere 280, 301, 573

Amplification 67, 68, 70, 317, 428, 429, 458, 517, 542, 543, 574, 600, 664, 678, 681

Analogue signal processing 66, 74

Analytical distillation 260

Analytical instrumentation 319

Ancillary devices 281

Aneroid barometer 85, 86, 87

Angular motion 110, 116

Anisotropic 441

Anode 252, 253, 254, 293, 294, 295, 296, 306, 307, 308, 309, 373, 374, 375, 376, 434, 435, 474

Anthrosphere 345, 351

Antiballistic missile defense systems v, 31, 34, 35

Antiderivative 140, 141, 142

Aperiodic 151, 153

Aperture 493, 519, 569, 602

Aperture synthesis 567, 569

Aphygmomanometers for blood pressure 88

Application programming interface (API) 199, 203

Application software 410, 527

Applied physics v, 9, 17, 45, 48, 52, 78, 132, 251, 305, 316, 327, 344, 422, 446

Aqueduct 159, 406, 650

Arch bridge 135, 136

Arc length 142

Aromatic compound 284

Artificial horizon 78, 79, 80

Artificial intelligence 44, 52, 53, 54, 56, 57, 58, 214, 321, 329, 342, 410, 413, 498, 528, 529, 531, 546, 719, 724, 732

Asperity 264

Asphalt 260, 267, 535, 536, 637

Assistive device manufacturers 70

Astrobiology/exobiology 583

Astronautics 16, 377, 561, 581, 583, 724, 728, 734, 735

Astronomical observations 572, 583, 603, 606, 720

Atmospheric absorption and transfer of heat 481

Atmospheric sciences v, 17, 22, 58, 62, 65, 91, 171, 178, 184, 486

Atmospheric water 58, 180

Atomic force microscopes (AFMs) 574

Attenuation 7, 8, 358

Attractor 151, 152

Audiologist 66, 67, 70

Audiology and hearing aids v, 66, 69, 99
Auditory nerve 66, 67, 68, 69, 112
Auger electron 323
Auger electron spectroscopy (AES) 324, 594
Automated processes and servomechanisms v, 72, 77, 322
Automatic theorem proving 52
Automotive applications 148, 231
Automotive glass 387
Aviation 10, 14, 37, 52, 82, 83, 84, 311, 483, 484, 486, 534
Avionics and aircraft instrumentation v, 17, 78, 81, 196
Avionics and navigation 14
Axial slice 452
Azeotropic extractive distillation 260

B

Backfill 159
Baghouse 352
Ballistic missile 31, 32, 35, 232, 557, 681
Balmer series 323
Bandwidth 6, 192, 200, 358, 359, 421, 544, 564
Bar 85, 243, 253, 424, 428, 431, 432, 440, 493, 565, 647
Barcinogen 345
Barometry v, 66, 85, 86, 87, 184, 486
Baseband 191, 193
Basic analogue measuring devices 281
Basic digital measuring devices 281
basic input-output system (BIOS) 197, 198, 209
Batch and continuous distillation 259
Batching and mixing 146
Batteries and fuel cells 287, 291, 293
Beam bridge 135, 137
Beam splitter 423, 424
Beneficiation 146
Bernoulli's theorem 406, 408
Bessemer process 472, 651, 702
Bhlorofluorocarbon (CFC) 345
Bias frame 155
Biasing 311, 418, 419, 420, 457, 458
Bicycles and scooters 289
Binary system 214, 215, 641
Biochar 169, 171
Biochemical engineering 99, 334
Biochemical oxygen demand (BOD) 352, 354
Biocompatible material 91
Biodiesel 99, 100, 101, 102, 103, 104, 400

Biofluidics 110
Biofuels and synthetic fuels v, 99, 102, 132, 171, 404
Biogas 99, 100, 101, 102, 103, 104, 331
Bioinformatics 28, 91, 105, 214, 335, 336, 498
Biological modeling 109
Biologics 121, 124, 590
Biomagnification 345, 348
Biomass 62, 99, 100, 102, 103, 104, 169, 174, 180, 331, 333, 348, 513, 516
Biomaterials 121, 125, 126, 138
Biomathematics v, 99, 105, 106, 107, 108, 109, 110
Biomechanical engineering v, 110, 115, 334
Biomechanical modeling 118
Biomedical applications 94
Biomolecular structures 129
Bionanotechnology 94, 121
Bionics and biomedical engineering v, 52, 99, 110, 115, 121, 123, 126, 278
Biophotonics 544
Biophysics 105, 127, 128, 130, 131, 132, 456, 551
Bioreactor 92
Bioremediation 92
Bipolar transistors and MOSFETs 419
Bits and bytes 247
Blast furnace 472, 473, 474, 489
Blooming 155, 156
Boolean algebra 23, 24, 26, 27, 29, 246, 247, 248, 319
Boolean searches 26
Boost phase 31
Boundary conditions 38, 509, 510
Boundary layer 10, 16, 369, 370, 406, 471, 612
Bourdon tubes for household barometry 88
Bridge design and barodynamics v, 135, 138, 166, 271
Broadcast television and radio 275
Butterfly effect 43, 151
Bypass ratio 10
Byte 211, 247, 457

C

Cable television and satellite radio 194
Cadmium 20, 286, 288, 295, 297, 307, 345, 348, 474, 475, 476, 490
Calculators and digital watches 444
C and C++ 210, 211, 220
Cantilever bridge 135
Capacitance 86, 87, 89, 271, 278, 279, 280, 300, 318
Carbon dioxide removal (CDR) 168
Carbon monoxide (CO) 312
Carburizing 472

Carnot efficiency 373

CAS software 236

Catadioptric imaging 493

Catalytic cracking 221, 222, 223, 224, 225, 226, 227, 260

Catch basin 352

Cathode rays 317, 318

Catoptric imaging 493

CD spectroscopy 128, 129

Cellular phones 82, 185, 191

Cellulose fibers 364

Central office switching 599

Ceramic dispersions 243

Channeling 186

Channel rays 317, 318

Chaotic Systems v, 150, 154

Charge-coupled devices v, 155, 158, 159

Chromium 287, 302, 309, 458, 474, 475, 476, 479, 489

Ciphertext 234, 235, 237

Circuit boards 121, 158, 256, 309, 319, 331, 709

Circuit design and connectivity 199

Circuit protection 255

Circular dichroism (CD) 127

Cisjunction 23

Civil engineering tools 161

Claustrophobia 452, 453, 455

Clay mineral 486

Cleartext 234, 235

Cleavage 486, 488

Climate engineering v, 66, 99, 166, 171, 178, 184, 486

Climate modeling v, 66, 171, 172, 176, 177, 184, 486

Climate modification 167, 168, 172

Climate system 172, 173, 174, 177, 483

Climatology 172, 177, 182, 338, 480

Climograph 178

Clock speed 246, 248, 249, 524

Clone 55, 121, 695

Cloning 123, 125, 691, 730, 731

Cloning and stem cells 125

Closed heat exchanger 394, 397

Cloud computing 200, 272, 410, 415, 416, 625

CNC (Computer Numeric Control) 72

Coaching 120

Cobalt 146, 286, 309, 474, 475, 492, 506

Cochlea 1, 4, 8, 66, 67, 68, 69, 70, 93, 94

Cochlear implants 66, 68, 69, 70, 330, 454

Codes and design criteria 162, 354

Cofferdam 135, 136, 160

Cogeneration 373, 374, 375

Coking 221, 224, 225, 227, 376

Collaboration 7, 95, 201, 332, 344, 350, 393, 494

Color temperature 434

Combustor 559

Communication barriers 186

Communications satellite technology v, 191, 194

Compiler 209, 210, 211, 212, 216, 679

Complement of a set 23

Complex system 96, 98, 151, 153, 431

Component integration and programming 411

Computational mathematics 25, 40

Computational science 29, 30, 335, 336, 338

Computer-aided design (CAD) 27, 203, 413, 463, 467

Computer-aided diagnosis 528, 529

Computer-aided manufacturing (CAM) 463, 467

Computer algebra systems 27

Computer animation 29

Computer circuits 26

Computer graphics 15, 203, 204, 206, 207, 208, 214, 383, 385, 467, 537

Computer hardware 26, 41, 197, 204, 209, 212, 410, 411, 413, 414

Computer integrated manufacturing 413

Computer languages, compilers, and tools v, 31, 58, 203, 208, 213, 221

Computer models 4, 172, 174, 462, 533, 616

Computer monitors 275, 440, 444, 445, 545

Computer organization and architecture 215

Computer security management 414

Computer vision 52

Condensation polymerization 553

Condensed matter 45, 46, 632

Condensers 259, 296, 395, 396, 398, 399

Conductive hearing loss 67

Conic section 379

Constructive interference 423

Consumer electronics 201, 212, 275, 494, 529

Consumer optics 388

Continental drift and magnetic pole reversals 615

Continuity 24, 140, 161, 369, 370, 390, 391, 405, 407

Continuous emission monitoring 17

Continuum 301, 334, 369, 380

Contrast agent 452

Control loop 72

Conventional applications 469

Converting electricity into motion 287

Cordage 366, 367

Core sampling 266

Coriolis effect 58, 59, 481
Corrosion control and dielectric materials 296
Counterflow 394
Cqual sets 23
Cquivalent sets 23
Cracking v, 221, 222, 224, 225, 226, 227, 263
Criteria air pollutant 17
Critical micelle concentration (CMC) 240, 242
Crude oil 96, 221, 222, 223, 224, 225, 226, 244, 262,
     532, 533, 534, 535, 536
Cryocooler 228
Cryogenic processing 228, 230, 231, 232
Cryogenic refrigeration 229
Cryogenics v, 227, 228, 229, 231, 233
Cryogenic tempering 228
Cryopreservation 228, 233
Cryosphere 172, 173
Cryptology and cryptography v, 233, 237
Cryptosystem 234, 237, 238
Curie temperature xi, 300
Cycle efficiency 557
Cyclone 480, 482
Czochralski method 311

D

Dark frame 155
Data analyst or data miner 337
Database management 410, 413, 415
Data transfer 359
Deaerator (direct-contact heater) 394
Deafness 8, 66, 69
Decibel (db) 1
Definite integral 140, 142
Delta V 10
Denier 363, 366, 367
Dentistry 117, 231, 475
Denuded zone 311
Deoxyribonucleic acid (DNA) 297, 552, 704
Design and development 319, 331, 336, 341, 546
Design storm 160
Destructive distillation 145, 258, 260, 261, 262
Destructive interference 1, 7, 423
Detergents 224, 226, 240, 241, 242, 243, 244, 245,
     348, 535
Dewpoint 480
dichloro-diphenyl-trichloroethane (DDT) 345, 356
Diesel and fuel oils 534
Differential calculus 44, 141, 142
Differential equation 40, 41, 105, 508, 510, 511

Diffraction 488, 493, 517, 518, 519, 550, 589, 592,
     593, 602, 647
Digested sludge 352
Digital communications 249, 420
Digital electronics 418, 420
Digital logic 73, 246, 247, 248, 249, 250, 251, 273
Digital radio 564
Digital signal processing 14, 66, 74, 87, 271, 272, 613
Digital signal processing (DSP) 272
Digital signatures 233, 236, 237, 238
Dimensional measurements 424
Diode technology v, 252, 257, 438, 439
Diopter 493
Direct current (DC) 252, 272, 280
Direct methanol fuel cells (DMFCs) 374
Discotic liquid crystal 441
Distilland 258
Distillate 223, 224, 225, 258, 259, 260, 534
Distillation 100, 145, 221, 222, 223, 224, 225, 258,
     259, 260, 261, 262, 263, 532, 533
Distilled water and desalination 261
Doppler effect 1, 4
Doppler radar 63, 88, 182, 483
Downlinking 192
Drivetrain 273, 284, 400
Driving range 399, 401
Drug discovery 130
Drying 147
Ductility 472, 475
Duration 73, 89, 160, 278, 339, 340, 341, 430, 558,
     578, 583, 599, 600, 617, 618
Dynamical system 151

E

Earth-Moon differences 61
Earthquake engineering v, 139, 264, 266, 268, 269,
     270, 271, 343, 344
Earth's available water 180
Earth's global energy balance 61, 179
Earth's magnetic field xi, 581, 703
EchoStar and high-speed internet access 195
Ecological footprint 171, 345
Efficient wind turbines 526
Effluent 352, 354
Elastomers and Plastics 554
Electrical gauges 88
Electrical measurement v, 278, 282, 283, 284, 322
Electrical power generation 302
Electric automobile technology v, 284, 291, 378, 404

Electricity generation 536
Electric trains and buses 275
Electrocatalysis 373, 378
Electrochemical cell 284, 293, 294, 373, 399
Electrochemistry 278, 291, 292, 293, 294, 295, 296, 297, 298, 299, 300, 310, 345, 373, 377, 587, 705, 711
Electrode 306, 373
Electrodeposition, electroplating, and electrorefining 296
Electrodes 8, 95, 123, 224, 232, 253, 295, 296, 300, 307, 316, 422, 478, 660
Electroforming applications 309
Electroluminescence 435
Electrolysis 102, 293, 294, 295, 296, 306, 307, 308, 309, 310, 474, 478, 705
Electrolyte 242, 286, 293, 294, 295, 306, 307, 308, 309, 373, 374, 375
Electrolyte pocesses 295
Electromagnetic pulse (EMP) 31
Electromagnetic waves xi, 129, 301, 517, 518, 562, 566, 602, 657
Electromagnetism xi, 133, 273, 278, 300, 301, 302, 303, 304, 305, 391, 392, 464, 602, 608, 627, 705, 711, 715
Electromagnet technologies v, 233, 300, 305, 322, 327
Electrometallurgy v, 284, 296, 300, 306, 307, 309, 310, 311, 327, 473, 474, 479
Electronic games 56, 444
Electronic materials production v, 257, 311, 315, 316, 322, 422, 492
Electronic medical record 412
Electron microscopy 128, 327, 540, 576, 594
Electron spectroscopy analysis v, 323, 326
Electron spin resonance (ESR) 324
Electron-volt 323
Electrophoresis 130, 292, 293, 591
Electroplating applications 309
Electrorefining applications 309
Electrostatic precipitator 352, 663
Electrowinning of aluminum 306
Electrowinning of other metals 308
Emergent properties 152
Emission techniques 589
Emulsion 246, 446, 447, 659
Endoscope 358, 678
End-user support 412
Engineering economics 165, 463

Engineering mathematics v, 31, 45, 145, 251, 334, 337, 338, 385
Engineering mechanics 159, 160, 162, 165, 351, 466, 470
Engineering seismology 264, 339, 340, 341, 342, 343, 344
Engineering seismology societies 343
Engineering tools 354
Enhanced oil recovery (EOR) 243
Environmental applications 607
Environmental chemistry v, 246, 345, 349, 350, 351
Environmental photography 539
Epicenter 264, 265, 266, 339, 340, 342
Epi reactor 311
Epitaxial enhancement 313
Equipment and machinery 124
Equivalent exhaust velocity 556
Ergonomics 115, 331
Essential oils 258
Ethanol 99, 100, 101, 103, 104, 105, 129, 260, 286, 288, 402
Etymology 184
Euclidean 379, 380, 384, 385
evaporative cooling 228, 229
Exoatmospheric 31, 34, 36
Experimentation xi, 39, 89, 106, 123, 305, 564, 639, 714
Extractive metallurgy 472, 473
Extreme applications of barometry 89
Extrusion 472, 475

F
Fabrics and textiles 365
Fan-out 417, 420
Faraday's law 293, 295
Feasibility study 328
Feedwater heaters 397
Ferromagnetic media 457
Fiberglass 365, 387, 388
Fiber-optic cable 200, 388, 389
Fiber-optic communications 362
Fiberscope 358, 359, 360
Fiber technologies v, 363, 367
Fibonacci series 379
Financial analyst 29
Financial impact 175
Finite differences 508, 510
Finite element packages 511
Firing or sintering 147

First-order predicate calculus 52, 54, 55, 57, 321
Fischer-tropsch process 99
Fishing 176, 177, 387, 482, 483, 549
Fixed service satellites 194
Flat field 155
Flexicoking 224
Flight dynamics and controls 12
Flight instruments 78, 79, 80, 83, 84
Flip-flop 199, 417
Floppy disk drives 460
Fluid catalytic cracker (FCC) 221
Fluid catalytic cracking 224
Fluid dynamics 38, 40, 41, 88, 90, 172, 221, 368, 369, 370, 371, 372, 384, 463
Fluid kinematics 404
Flux xi, 89, 177, 272, 457, 472, 474, 481, 567, 571
Flynn's taxonomy 525
Focal depth 339
Food and beverage applications 230
Force balance in flight 11
Forensics 28, 45, 117, 214, 321, 528, 691, 714
Forging 472, 473, 475
Forklifts 289
FORTRAN and COBOL 210, 211
Forward bias 252, 254, 255, 434
Fossil fuel 22, 65, 167, 168, 170, 184, 262, 291, 375, 399, 505, 614
Fouling 395
Fractional distillation 258, 259, 260
Fractionator 221, 223, 224
Freeboard 160
Frequency modulation (FM) 562, 563, 671
Fresnel lens 493
Froude number 372, 404, 407, 467
Fuel cell v, 289, 292, 373, 376, 377, 378
Fuel cell technologies v, 373, 376
Fuel-to-air ratio 556
Functional magnetic resonance imaging (fMRI) 452, 456
Function of detergency 241
Fuselage 10, 80, 365

G
Galvanizing 472
Game development 205
Game programming and animation 383
Game theory 45, 722
Gasification 99, 101, 375
Gate 135, 136, 162

Gearbox 400, 476, 559
General circulation model 172
Generation and emission 543
Genetics 66, 91, 105, 106, 108, 109, 127, 128, 129, 131, 132, 337, 704, 706, 711
Geoengineering 92, 97, 98, 166, 167, 168, 169, 170, 171, 338
Geographic information management 413
Geometric isomers 379
Geometric optics 517, 518, 519
Geostationary orbits 193
Geostationary satellites 193
Gettering 311, 316
Girder 136
GIS and GPS 382
Glass and glassmaking v, 385, 388
Glazier 386, 389
Glazing 145, 146
Global positioning system (GPS) 14, 79, 212, 331, 565, 585, 688
Global resource management 548
Golden ratio 43, 379, 384, 385
Gradient 54, 140, 452, 454, 481
Gradient coils 452, 454
Graphical user interface (GUI) 203
Graphics and kinematics 466
Graphics pipeline 204
Gravitational wave detection 425
Gravity assist 14, 16, 559
Gravity assist maneuver 14
Gravity flow 404, 407
Greenhouse effect xii, 59, 61, 167, 178, 179, 180, 181, 345, 346, 549
Greenhouse gases xi, 44, 58, 65, 100, 101, 167, 168, 172, 175, 177, 179, 183, 309, 348, 375, 485, 513, 536, 549
Green machining 147
Grid methods 509, 510
Ground-based avionics 80
Ground motion 267, 269, 339, 341, 342

H
Haptics 184
Hard disk drives 304, 305
Hardness 46, 261, 306, 309, 354, 386, 387, 472, 476, 487, 488, 491
Hash function algorithms 236
Hazardous waste 92, 260, 262, 345, 355
Hazardous-waste management 350, 351, 352, 356

Hearing aids v, 66, 69, 99
Heat conduction 228, 229
Heat-exchanger technologies v, 394, 398
Heat generation 102
Heat transfer 221, 233, 338, 371, 394, 395, 396, 397,
    398, 463, 464, 468, 556, 560, 609, 610, 613
Heavy metal 345
Hemoglobin 293, 337
Hertz (Hz) 1
Heterodyne detection 423, 424
Hit-to-kill 32, 33, 34
Hohmann Transfer 12
Holography 542, 684
Homodyne detection 423, 424
horizontal situation indicator (HSI) 78, 80
Household clocks 618
Hybrid vehicle 285, 292, 399, 400, 401, 402, 403
Hybrid vehicle technologies 292
Hydraulic calculations 395
Hydraulic engineering v, 78, 166, 404, 408, 409
Hydraulic machinery 404
Hydraulic structures 407
Hydraulic transients in pipes 407
Hydrocracking 221, 222, 223, 224, 227, 533
Hydrodynamics 371, 372, 406
Hydrogen fuel cell vehicles 375
Hydrology 353
Hydrometallurgy 473, 474, 478
Hydrophilic 129, 240, 242, 347, 446, 447, 474
Hydrophilic-lipophilic Balance (HLB) 240
Hydrosphere 348
Hydrostatic pressure and force 142
Hydrostatics 404, 406, 711
Hydrothermal deposit 487
Hyperlinks 192
Hypocenter 264, 266, 267, 339, 340

I
Icon 728
Ideal fluid flow 404
Ideal fluids 369
Identification using instruments 488
Igneous rock 487, 488
Image construction 542
Imagesetter 446
Imaging systems 124, 157, 159, 455, 456, 541
Incompressible flows 369, 370
Incompressible fluid 369, 404, 405
Indefinite integral 140

Indirect applications 281
Inductance 272, 278, 279, 280, 300, 318
Induction 279, 281, 285, 300, 649, 655, 657, 705, 715
Industrial optics 387, 388
Inertial navigation systems (INS) 81
Information assurance 410
Information management 214, 410, 411, 413
Information security, privacy, and assurance 412
Information technology 410, 411, 412, 414, 415
Information technology companies 415
Information transmittal 359
Infrared absorbers 172
Infrared radiation 172, 314, 485, 496, 537, 539, 603,
    610
Infrasound 1
Innovations in materials 469
Insertion 70, 175, 199, 223, 231
Instability 152
Instrument flight 78, 79, 84
Instrument landing system (ILS) 78
Integral calculus 141, 142, 712
Integrated-circuit design v, 251, 284, 316, 322, 417,
    421
Integrated circuit (IC) 272, 319, 417
Intelligent agent 52, 56
Intelligent tutor systems 55
Intercontinental telephone service 194
Intercoolers 397
Interface science vi, 316, 422, 502, 592, 596, 718
Interleaf 246, 249
Internal combustion engine (ICE) 399, 400
Internet access and Cisco 195
Internet service provider (ISP) 192, 197, 217
Inter-network service architecture, interfaces, and
    inter-network interfaces 200
Interpersonal communication 184, 187
Interpreter 209, 212, 213
Interstellar medium 569
Invert 160, 325, 417
Inviscid fluid 369
Ionization potential 323
Ionosphere xi, 192, 562, 616
Iridium satellite phones 194
Iron fertilization of the oceans 169
Isotherms 179, 480

J
Jet fuel 221, 225, 226, 273, 332, 532, 534
Jet propulsion 16, 192, 558, 625

Jet stream 480, 484
Joule-Thomson effect 228, 229, 231

K

Karnaugh map 246, 248
Kelvin temperature scale (k) 228
Kepler's laws 143
Keystone 135, 136, 137
Kill vehicle 32, 34, 35
Kiln 145, 635
Kilobit 192
Kinematics 110, 115, 116, 240, 404, 406, 409, 463, 464, 466, 467, 468, 471
Kinesics 184
Kinesiology 110, 115, 116
Kinetics 110, 116
Knee voltage 252
Knowledge representation 52, 54

L

Laminar flow 40, 133, 134, 404
Laparoscopic surgery 428
Laser 3-D scanners 430
Laser ablation 429
Laser cutting 429
Laser diodes 255, 256, 257
Laser guidance 428, 431, 432
Laser interferometry v, 393, 423, 426
Laser lighting displays 429, 431
Laser printing 429, 447
Laser technologies v, 428, 432
Lava 367, 487, 616
Layers of the atmosphere 59
LC oscillation 300
Left-hand rule 279
Life-sustaining space resources 579
Lift to drag ratio 10
Light-emitting diode (LED) 252, 359, 683
Limit 4, 87, 98, 119, 140, 141, 142, 174, 220, 259, 260, 316, 342, 365, 456, 546, 547, 575, 592, 602
Linguistics 7, 184, 187, 190
Liquefied petroleum gas 224, 534
Liquid crystal technology v, 251, 316, 422, 441
Lithium-ion battery 285, 376, 399
Lithography 425, 446, 447, 448, 449, 450, 451, 467, 500, 645, 648, 682
Logical topology 197
Logic programming 52, 54
Love waves and Rayleigh waves 266

Low earth orbits 193
Lubricating Oils 535
Lunar and Martian bases 578
Lyotropic liquid crystal 441

M

Machine operational control 302
Magma 487, 614
Magnesium 241, 250, 293, 299, 304, 306, 308, 309, 354, 473, 474, 487, 489
Magnetic anisotrophy 457
Magnetic damping 300
Magnetic disks 459
Magnetic levitation 45, 48
Magnetic media 302
Magnetic resonance angiogram (MRA) 452
Magnetic resonance imaging (MRI) 94, 452, 456
Magnetic Storage vi, 305, 457, 462
Magnetic striping 460
Magnetic tape 176, 430, 458, 459, 460, 461, 462, 568, 677
Magnetic tape drives 459
Magnetization xi, 452, 457
Mainframe 197, 199, 209, 210, 459
Manganese 286, 287, 353, 354, 479, 492, 714
Marine cloud whitening 169
Marine radio 564
Markov chain 38
Mass communication 184, 188, 191
Mass dampers 267, 268
Material handling 300
Matrix 13, 38, 54, 105, 155, 338, 365, 375, 429, 439, 440, 441, 443, 469, 509, 708
Measures of standards and calibration 424
Measuring temperature 91, 609
Mechanical system 284
Mechanobiology 110, 111
Medical imaging 94, 124, 126
Medical photography 540
Medical technology 9, 50, 153
Melting and crystallization 312
Membrane structure and transport 129
Memory and repeated patterns 529
Metabolism 101, 109, 111, 456, 704, 709
Metal forming 475
Metallurgy 221, 292, 306, 311, 323, 326, 327, 472, 473, 475, 476, 477, 478, 479, 546
Metrology 278, 316, 383, 423, 425, 427, 464, 468, 617, 620

Micelle formation and critical micelle concentration (CMC) 242
Microclimate 172
Microelectronics 272, 313, 595, 622, 625
Micro-optics 544
Microscale distillation 260
Microsoft.NET 211
Military surveillance 606
Mineralogy vi, 486, 491, 731
Mining and harvesting of space resources 579
Mining and mineral processing 243
Mirrors and lenses vi, 390, 493
Missiles 11, 31, 32, 33, 34, 35, 36, 37, 78, 81, 83, 331, 360, 428, 431, 495, 557, 558, 582, 586, 607, 625
Mixed hearing loss 67
Mobile computing and the cloud 413
Modeling a biological function 105
Modern communications 359
Modulation 543, 562, 563, 564, 671
Molecular distillation 259
Molecular genetics 127
Molecular hydrogen 99, 100, 102
Molinya orbits 193
Molten carbonate fuel cells (MCFCs) 375
Monochromatic 423, 443, 517, 589, 607
Multispectral image analysis 526
Musical acoustics 2, 3, 4, 6
Mutagen 345

N
Nanobiology 129, 130
Nanomaterials 138, 292, 293, 298, 299, 427, 498, 501, 595, 596
Nanoprobe 45
Nanoscience 283, 327, 502, 596, 718, 730
Nanostructures and nanodesign 575
Natural gas 5, 99, 100, 102, 167, 179, 225, 260, 288, 400, 405, 486, 516, 532, 533, 534, 535, 536, 632
Natural language processing 52, 53
Negation 23, 25, 26
Negative lens 493
Nematic state 441
Networking and web systems 411
Network management 413, 415
Neural network 52, 53, 54, 526, 531
Newtonian fluids 369
Newtonian mechanics 46, 51, 173, 393, 631, 632
Newton's laws of motion 45, 117, 464
Next Generation air transportation system (NextGen) 82
NMR spectroscopy 128
Nonpoint source 352
Nonvisual uses for LEDs 352
NOTs, ANDs, ORs, and other gates 420
Nozzles 145, 148, 476
Nuclear explosion sensors 89
Nuclear magnetic resonance (NMR) 41, 45, 128, 453, 456, 590
Nuclear Power 269, 508, 733, 738, 742
Nuclear technology 149, 502, 503, 504, 505, 506, 507
Numerical Analysis vi, 31, 45, 145, 339, 508, 509, 511, 512, 720, 728, 729, 735

O
Oblique shock 10
Occluded front 178
Ocean acidification 167, 168
Ocean and tidal energy technologies vi, 513, 516
Ocean thermal energy conversion (OTEC) 514
Offset 22, 168, 255, 446, 447, 448, 450
Ohm's law 279, 280, 282
Oncology 105
Open-channel flow 404, 406
Open heat exchanger 394
Operating system 197, 198, 199, 200, 201, 202, 209, 210, 213, 215, 216, 220, 415, 416, 524, 525, 686, 691
Operational amplifier (op-amp) 317
Ophthalmology 425, 522, 718
Optical metrology 425
Optical telescopes 568, 583, 603, 604, 605, 608
Optimization 29, 40, 54, 140, 142, 195, 197, 334, 337, 464, 466, 511
Orbital mirrors and space sunshades 169
Orbital missions 12
Organisms and agriculture 107
Ototoxicity 66, 69
Overall pressure ratio 556
Oxidation pond 352

P
Packaging 308, 315, 332, 348, 386, 387, 388, 389, 437, 447, 448, 476, 485, 545, 554
Packet switching 597, 598, 599, 600
Paralinguistics 184
Parallel computing vi, 58, 221, 523, 527, 741
Parallel flow (co-current) 394
Parallel system hybrid 399

Paramagnetism 323

Particle-in-a-box model 38, 40

Pascal 85, 86, 210, 481, 641, 740

Pattern recognition (science) vi, 528

Pellucidity 386

Percolation test 160

Personal computers and televisions 443

Personal vehicles 288

Pesticide 345, 347, 348, 349, 356, 383, 708

Petrochemical 148, 208, 221, 222, 224, 225, 226, 227, 258, 259, 260, 532, 533, 534, 535

Petroleum extraction and processing 532

Petrology 486, 487, 546

pH 95, 167, 293, 295, 353, 664

Phase difference 279

Phonetics 4, 185, 187

Photochemical etching 419

Photochemical smog 352

Photodetection and sensing 544

Photodiodes 255, 544

Photoelectric effect 155, 156, 315, 517, 519, 521, 543, 544, 657, 662, 709

Photoelectron emission 323

Photoelectron spectroscopy 324

Photography science vi, 537

Photolithography 446, 447, 448, 449, 450, 451

Photomask 446, 447, 449

Photon 323, 428, 434, 435, 453, 517, 518, 543, 545, 587, 588, 611, 618, 713

Photonics vi, 257, 316, 439, 497, 522, 542, 543, 545, 546, 733

Physical acoustics 1, 7

Physical metallurgy 472, 479

Physical optics 517, 518

Physiological and psychological acoustics 3, 6

Phytoplankton 167, 169

Pi 137, 142, 379, 380, 384

Piezoelectric material 85

Pixel 88, 155, 156, 314, 440, 441, 443, 526

Planes and surfaces 381

Planetary sciences 583

Planetology and astrogeology vi, 546, 550

Plastic deformation 472, 473

Plate heat exchanger 394

Platinum 286, 287, 295, 296, 424, 475, 477, 490, 492, 611, 630, 632, 677

Pm 10 and Pm 2.5 17

P-n junction 252, 254, 255, 256, 257, 434, 435, 436, 439, 544, 623

Polarization 285, 287, 347, 424, 426, 440, 517, 518, 702

Polarized light 127, 128, 445

Polishing 313

Pollution prevention 349

Polyalphabetic cipher 234

Polygon 379, 383

Polyhedron 379

Polymer science vi, 551, 555, 556, 723, 734, 737

Polysilicon (metallurgical grade silicon) 311

Pontoon bridge 135, 137

Population dynamics and epidemiology 41

Portability 209, 213, 461, 538

Positive lens 493

Pot residue 258

Pounds per square inch absolute (Psia) 85

Pounds per square inch Gauge (PsiG) 85

Powder metallurgy 473, 475, 477, 479

Power dissipation 252, 434

Power from currents 515

Power from tidal and ocean currents 514

Power generation from waves 515

Power grid 272, 273, 275, 277, 515

Power supplies 274

Prandtl-Meyer expansion 10

Predictability 38, 113, 174, 484

Pressure flow 404, 407

Pressure-sensitive paint 85

Prime number 234, 236

Probability 38, 143, 722, 725, 727, 740

Probe thermometers 610

Production engineering 468, 469

Professional organizations 219

Programmable logic controller 248

Propfan 559

Proprietary information 234

Propulsion system metrics 557

Propulsion technologies 560

Propulsive efficiency 556

Prosthesis 70, 94, 121, 123, 126

Prosthetics 45, 94, 115, 121, 431, 712

Protocol 175, 348, 376, 620, 625, 686

Proton 373, 666

Proton exchange membrane fuel cell 285, 374, 376

Proxemics 185

Proximity sensor 72, 73

Public key algorithms 235, 236

P wave 264, 266

Pyramids 43, 160, 328, 494, 637

Pyrometallurgy 306, 309, 473
Pythagorean 379, 380, 382, 384, 713

Q

Qualitative movement 116
Quantitative movement 116
Quantum efficiency 155
Quantum mechanical physics 574
Quantum optics 517, 518, 521, 545

R

Radar altimeter 78
Radar observations 570
Radiant efficiency 434
Radiation budget 61, 485
Radiation noise 155
Radiation thermometers 610
Radiative cooling 172
Radiative forcing 167
Radio astronomy 562, 567, 568, 570, 571, 572, 605,
    669, 720, 730, 732, 739, 741
Radio telescope 565
Rain gauges 62, 181, 182, 482
Ram 78, 79, 80, 711
Ramjets and turbomachines 559
Rammed earth 264, 267
Random access memory (RAM) 215, 457
Randomness 151, 153
Rapid sand filter 160
Raster graphics 203
Raw materials 145, 146, 222, 224, 226, 244, 262, 367,
    472, 490
Rayleigh wave 264
Readout noise 155
Real fluid flow 404
Reasoning ix, 16, 23, 24, 25, 26, 52, 53, 54, 55, 56, 57,
    77, 321, 701
Recombinant DNA 93, 122, 124, 686, 690
Recreational applications 608
Recrystallization temperature 472
Rectifiers 252, 254, 256
Redox ractions 295
Reflective roofs 169
Reflux ratio 258
Refraction xi, 424, 493, 517, 518, 519, 520, 542, 544,
    604
Regenerative medicine 92
Regenerator 222, 223, 224
Regional haze 17

Reinforced concrete 135
Relation between temperature and internal energy
    611
Relative humidity 480
Remote-sensing techniques 181
Rendering 203, 204, 205, 207, 326
Replacements for colored incandescent lights 436
Residue 221, 222, 224, 225, 258, 260, 313
Responses to climate change 63
Retrofitting old structures 266
Reverse breakdown voltage 252, 434
Reynolds number 370, 404, 467
Richter scale 265, 339, 342, 343, 670
Riemann sum 140
Ring lasers and gyroscopes 425
Robotics and control 467
Rocketry 79, 370, 582, 707
Roving 363

S

Sagittal slice 452
Sampling 8, 17, 174, 217, 249, 266, 274, 281, 317,
    320, 328, 361, 431, 495, 574
Sanitary landfill 352, 355
Satellite-based systems 80
Satellite images 157, 526
Satellite technology 485, 549
Satellite trucks and portable satellites 194
Saturation point 480
Scales and balances 631
Scanning near-field optical microscopes (SNFOMs)
    574
Scanning probe microscopy 316, 498, 573, 574, 575,
    576, 728, 733
Scanning tunneling microscopes (STMs) 574
Scarce metals and uses 489
Scattering techniques 589
Scientific clocks 618
Scientific photography 540
Scrubbers and artificial trees 169
Search for extraterrestrial intelligence 570
Sedimentary rock 487, 489
Seismic design software 341
Semantics 55, 185
Semiconductor laser 358, 359, 429, 688
Semiconductor manufacturing 155, 158, 252, 256,
    257, 434, 437, 438, 446, 450, 682
Semiotics 184, 185, 190
Semipermanent low-pressure centers 178

Sensorineural hearing loss 67
Separating valuable and waste minerals 473
Sequential access memory 457
Series system hybrid 399
Servomechanisms v, 72, 74, 75, 77, 249, 251, 322
Set intersection 23
Set theory 23, 24, 28, 44
Set union 23
Seven principles 117
Sharp-crested weir 160
Shear strain 369
Sheet-fed offset printing 448
Shell 53, 54, 55, 56, 58, 133, 134, 253, 294, 301, 324, 394, 395, 397, 398, 399, 435
Shell and tube heat exchanger 394
Shifting charges 156
Shortwave radio 563, 564, 566
Shottky diodes 255
Signal processing 7, 8, 14, 29, 36, 66, 74, 85, 87, 88, 155, 248, 249, 271, 272, 613
Silica 307, 359, 386, 485
Silicate mineral 487
Silicon manufacture 418
Simple distillation 258
Simple LCDs 442
Sintering 146, 147, 472
Sinusoidal 6, 300, 301, 417
Skin-cleansing bars 243
Slag 347, 472, 474, 478
Small power generation for the portable electronic devices 376
Smart materials 501
Smart pressure transmitters for automatic control systems 89
Smectic state 441
Software integration 410
Soil mechanics 159, 160, 161, 162, 165, 270, 351, 352, 353, 356
Solar and plasma sails 559
solar radiation management (SRM) 167, 168
Solar technology 314, 673
Solid oxide fuel cells (SOFCs) 375
Solid-state diodes 252, 253, 254, 435
Solid-waste management 351, 355, 356
Solution of systems of linear equations 509
Sound amplifiers 274
Sound spectrum 1
Source parameters 339, 340, 341
Space environments for humans vi, 577, 580

Space exploration 83, 315, 531, 551, 585, 622
Space science 379, 546, 550, 577, 581, 582, 584, 586
Space tourism 577, 580, 582, 585, 587
Space weather 581, 582, 584, 585, 586
Sparse system 38
Spatial coherence 423
Special alloys 476
Specialized processes 500
Specific impulse 557, 558, 559, 561
Spectrogram 1, 4
Spectrographs 48, 605
Spectrometry 45, 48, 323, 488, 593, 594, 632
Speech acoustics 2, 4, 6
Speech recognition technology 530
Spherical aberration 493, 497, 521
Stall 10, 558
Stand-alone computers 199
Static Stability 12
Static system 151
Stationary power plants and hybrid power systems 375
Statistical analysis and actuarial science 42
statistical process control 42, 44, 383
Steady flow 404, 408
Steam cracking 221, 224
Steam distillation 260
Steam generators 14, 394, 395, 397, 398, 399
Strange attractors 152
Stratosphere 11, 60, 65, 81, 167, 169, 171, 183, 297, 480, 586
Stratospheric sulfur aerosols 169
Streamline 369, 370
Strength calculations 395
Strength of materials 10, 159, 160, 161, 162, 165, 464, 466
Stress drop 339, 340, 341
String theory 38
Structural engineering 135, 159, 160, 161, 162, 165, 265, 269, 270, 339, 343
Structural fibers 365
structured query language (SQL) 200, 209
Structure of air-monitoring programs 19
Subcritical flow 404, 407
Sublimation 178, 258
Superconducting device 228
Superconducting magnet 228, 229
Superconductivity applications 230
Supercritical flow 404, 407
Supplier side 244

Surface analysis techniques 594
Surface and interface science 593, 595
Surface area and volume 142
Surface chemistry 245, 326, 593, 596
Surface field effect transistors 623, 624
Surface modification 593, 595, 596
Surface tension 240, 242, 243, 246, 307, 386
Surfactant 240, 241, 242, 243, 244, 246
Suspension bridge 135, 137, 162
S wave 266
Switching 24, 26, 249, 279, 287, 289, 318, 417, 419, 437, 544, 597, 598, 599, 600, 601, 672, 673, 686, 706
Symmetric key algorithms 236
Synfuel 99, 100
Syntax 185
Synthesis gas 99
Synthetic biology 129
Synthetic polymers 348, 363, 364, 553, 554, 555
System design 462, 463, 464, 466, 468, 561, 728
Systems biology 92, 129, 132, 150
Systems integration 410

T
Takeoff gross weight 10
Taylor series 140
Telephone technology and networks vi, 191, 363, 417, 597, 601
Telescopy vi, 602, 603, 607
Television sets 69, 319, 439, 442, 443, 444, 445, 491, 545, 625
Temperature inversion 19, 480
Temperature measurement vi, 66, 184, 609, 613, 614, 718, 721, 733
Temperature monitors 444
Tempered glass 386, 387
Temporal coherence 423, 428
Tension member 160
Terminal phase 32
Terrestrial magnetism vi, 614, 616
Tesla 51, 272, 276, 278, 288, 290, 318, 452, 655, 715
Test equipment 274
Tethers 559
Theorem 52, 53, 54, 55, 142, 379, 380, 382, 384, 406, 408, 713
Theorem provers 55
Theoretical computer 214, 414
Theoretical plate 258
Thermal calculations 395

Thermal cracking 221, 222, 223, 224, 225, 533, 665
Thermal rfficiency 556
Thermal power dissipation 434
Thermionic diode 252, 253
Thermotropic liquid crystals 441
Thin film transistor (TFT) 441
Third-generation hybrid 399
Three-dimensional (3-D) model 203
Thrust 11, 12, 137, 159, 470, 526, 556, 557, 558, 560, 583
thrust-specific fuel consumption 557
Time measurement vi, 617, 621
Tools, equipment, and instrument applications 231
Top-down nanofabrication 498, 499
Toroid 300
Torque 112, 116, 118, 284, 285, 291, 670
Torr 324
Torricelli barometer 85
Torus 379, 579
Trace element 487
Tracking systems 231, 585
Traditional ceramics 146, 147
Traffic control 77, 78, 79, 82, 83, 200, 565, 706
transducer 1, 5, 629
Transgenic organism 92
Transistor technologies vi, 251, 316, 322, 422, 622, 625
Transmission medium 599
Transmission path selection 599
Transmitter 63, 69, 93, 182, 358, 359, 453, 562, 564, 565, 566, 570, 598, 620, 657, 673, 694, 705, 711
Transponder 79, 192, 193
Transuranic elements vi, 626
Trickling filter 352
Troposphere 17, 59, 60, 61, 167, 178, 180, 480
Truss bridge 135, 137
Tube sheet 394, 398
Tungsten 146, 365, 473, 476, 477, 492, 611, 664, 665, 714
Turbojet 558, 716
Turboprop 559
Turboshaft 558, 559
Turbulent flow 133, 134, 404
Turn and bank indicator 78, 79, 80
Twisted-nematic effect (TNE) 441
Twisted-nematic liquid crystal display (TN LCD) 441

U
Ultimate bearing capacity 160

Ultrasonics 2, 5, 6, 8
Ultrasonic thermometer 611
Ultrasound 1, 3, 5, 8, 94, 275, 455, 529, 553, 679
Underwater sound 4, 6
Uniform flow 10, 404, 409
Units of measurement 280
Universal set 23, 25, 28
Unsteady flow 405, 409
Uplinking 192, 193
Urban heat island 59
Utility vehicles and trucks 289

V

Vacuum distillation 225, 259, 533
Vaporization 242, 258, 514
Vapor pressure 64, 258, 259, 260
Varied flow 405, 407, 409
Vector graphics 203, 204
Vehicular emissions 19
Vibrational measurements 425
Video games 203, 205, 206, 208, 213, 383, 430
Visbreaking 224, 225, 227, 533
Viscosity 40, 222, 224, 228, 243, 307, 369, 370, 386, 395, 404, 406
Vision and vision science 520
Vitreous 386, 651
Volatile organic compounds (VOCs) 17

W

Wafers 312, 313, 314, 418, 421
Warhead 32, 33, 34, 35, 49
Warp clock 363
Wastewater treatment 165, 243, 351, 352, 353, 354, 356

Water distribution systems 353, 404, 407, 408
Water pollution 347
Water resources engineering 163, 404
Water treatment 160, 351, 352, 356
Wavelet minimization 38
Wave power 409, 515, 516
Wave propagation path 339, 340
Weather and remote sensing 584
Weather equipment 482
Weather forecasting 64, 88, 89, 91, 193, 195, 206, 512, 526
Weather prediction and climate change 525
Web-fed offset printing 448
Web site development 414
Web systems 410, 411, 415
Weight and mass measurement vi, vii, 629, 632
Weld cell 72, 73, 74
Why terrestrial magnetism is important 616
Wind farms 515, 516
Wind power 514
Wind strength and direction 481
Wind tunnel 10, 13, 36, 89, 370, 613, 662
Wing 10, 15, 79, 80, 151, 370, 372, 408, 470, 511, 526, 645, 712, 714, 716
Wireless communications 49, 50, 600, 601

X

X-braces and pneumatic dampers 268
X-ray crystallography 128, 129, 672, 704

Z

Zener diodes 254, 255
Zener voltage 252, 317